ASTRONOMY

```
QB
43.2      Zeilik, Michael.
.Z43          Astronomy, the
                cosmic perspective
1983
```

**LIBRARY
FLORIDA KEYS COMMUNITY COLLEGE**
5901 W. JUNIOR COLLEGE ROAD
KEY WEST, FL 33040

ASTRONOMY
THE COSMIC PERSPECTIVE

QB
43.2
.Z43
1983

MICHAEL ZEILIK • JOHN GAUSTAD
THE UNIVERSITY OF NEW MEXICO   SWARTHMORE COLLEGE

1817

HARPER & ROW, PUBLISHERS, New York
Cambridge, Philadelphia, San Francisco,
London, Mexico City, São Paulo, Sydney

Sponsoring Editor: Malvina Wasserman
Project Editor: David Nickol
Designer: T. R. Funderburk/R. Bull
Production Manager: Jeanie Berke
Compositor: York Graphic Services, Inc.
Printer and binder: R. R. Donnelley & Sons Company
Art Studio: J&R Art Services, Inc.
Cover: R. Bull

ASTRONOMY: The Cosmic Perspective

Copyright © 1983 by Michael Zeilik and John Gaustad

All rights reserved. Printed in the United States of America. No part of this book may be used or reproduced in any manner whatsoever without written permission, except in the case of brief quotations embodied in critical articles and reviews. For information address Harper & Row, Publishers, Inc., 10 East 53d Street, New York, NY 10022.

Library of Congress Cataloging in Publication Data

Zeilik, Michael.
  Astronomy, the cosmic perspective.

  Includes index.
  1. Astronomy. 2. Cosmology. 3. Evolution.
I. Gaustad, John E. II. Title.
QB43.2.Z43   1983   520   82-23254
ISBN 0-06-047387-8

CONTENTS

Preface

PART ONE CHANGING CONCEPTIONS OF THE UNIVERSE

CHAPTER 1 Observing the Sky: From Chaos to Cosmos — 3

Learning Objectives — 3
- 1.1 The Visible Sky — 4
- 1.2 The Motions of the Sun — 8
- 1.3 The Motions of the Moon — 12
- 1.4 The Motions of the Planets — 13
  - FOCUS 1.1 Precession of the Equinoxes — 15
- 1.5 Eclipses — 19
- 1.6 Observing the Sky — 24
- 1.7 Prehistoric Astronomy: The Anasazi Achievement — 25

Summary — 28
Key Words — 29
Review Questions — 29
Problems — 30
Beyond This Book . . . — 30

CHAPTER 2 The Birth of Cosmological Models — 31

Learning Objectives — 31
- 2.1 Scientific Models — 32
- 2.2 Babylonian Skywatching: The Seeds of a Science — 34
  - FOCUS 2.1 Astrology — 36
- 2.3 Greek Models of the Cosmos — 36
- 2.4 Claudius Ptolemy: A Complete Geocentric Model — 47
  - FOCUS 2.2 Eclipses and the Distance to the Moon — 48

Summary — 56
Key Words — 56
Review Questions — 56
Problems — 57
Beyond This Book . . . — 58

CHAPTER 3 The New World Order: A Sun-Centered Model — 59

Learning Objectives — 59
- 3.1 Copernicus the Conservative — 60
- 3.2 The Heliocentric Model — 63

3.3	Tycho Brahe: Supporter of the Geocentric View	74
3.4	Johannes Kepler and the Cosmic Harmonies	78
3.5	Scientific Models and Cosmological Revolutions	86

Summary 87
Key Words 88
Review Questions 88
Problems 89
Beyond This Book . . . 89

CHAPTER 4 The Clockwork Universe: A Physical Cosmos 91

Learning Objectives 91
4.1	Galileo: Advocate of the Heliocentric Model	92
4.2	Newton: A Physical Model of the Cosmos	103
	FOCUS 4.1 Centripetal Acceleration and Force	112
4.3	Cosmic Consequences of Universal Laws	117

Summary 123
Key Words 124
Review Questions 124
Problems 125
Beyond This Book . . . 126

CHAPTER 5 Einstein and the Evolving Universe 127

Learning Objectives 127
5.1	Natural Motion Reexamined	129
5.2	The Rise of Relativity	131
5.3	The Geometry of Spacetime	136
5.4	Geometry and the Universe	143
5.5	Relativity and the Cosmos	150

Summary 156
Key Words 157
Review Questions 157
Problems 158
Beyond This Book . . . 158

PART TWO STARS—THE FOCUS OF COSMIC EVOLUTION

CHAPTER 6 Our Sun: Home Star 163

Learning Objectives 163
6.1	A Solar Physical Checkup	164
	FOCUS 6.1 Energy and Work	168
6.2	Sunlight and Spectroscopy	171
6.3	Spectra and Atoms	177
6.4	Messages from Sunlight	188
6.5	The Quiet Sun	196
6.6	Energy from the Solar Interior	200

6.7	The Active Sun	208
	FOCUS 6.2 Magnetic Fields and Forces	210

Summary 217
Key Words 219
Review Questions 220
Problems 221
Beyond This Book . . . 221

CHAPTER 7 Gathering Light from Space 223

Learning Objectives 223
7.1	Observations and Models	225
7.2	Visible Astronomy: Optical Telescopes	226
7.3	Invisible Astronomy	233
	FOCUS 7.1 Understanding Contour Maps	238

Summary 245
Key Words 245
Review Questions 245
Problems 246
Beyond This Book . . . 246

CHAPTER 8 The Stars: Other Suns 247

Learning Objectives 247
8.1	Some Messages of Starlight	248
8.2	Stellar Distances	255
8.3	Stellar Colors, Temperatures, and Sizes	261
8.4	Spectral Classification of Stars	267
8.5	The Hertzsprung-Russell Diagram	273
8.6	Weighing and Sizing Stars: Binary Systems	279

Summary 294
Key Words 294
Review Questions 294
Problems 296
Beyond This Book . . . 296

CHAPTER 9 Starbirth and the Interstellar Medium 297

Learning Objectives 297
9.1	The Interstellar Medium: Gas	298
9.2	The Interstellar Medium: Dust	318
9.3	Starbirth: Theoretical Ideas	329
	FOCUS 9.1 Hydrostatic Equilibrium	334
9.4	Starbirth: Observational Clues	340

Summary 350
Key Words 350
Review Questions 351
Problems 351
Beyond This Book . . . 352

CHAPTER 10 The Lives of Stars — 353

Learning Objectives — 353
10.1 Stellar Evolution and the Hertzsprung-Russell Diagram — 354
10.2 Stellar Anatomy — 357
10.3 Star Models — 359
    FOCUS 10.1 The Equations of Stellar Structure — 360
10.4 Energy Generation in Stars — 363
    FOCUS 10.2 Physical Basis of the Mass-Luminosity Law — 364
10.5 Theoretical Evolution of a Solar-Mass Star — 367
10.6 Theoretical Evolution of a 5-Solar-Mass Star — 374
10.7 Theoretical Evolution of Very Massive Stars — 377
10.8 Chemical Composition and Evolution — 379
10.9 Observational Evidence for Stellar Evolution — 382
10.10 The Synthesis of Elements in Stars — 394
Summary — 396
Key Words — 396
Review Questions — 397
Problems — 398
Beyond This Book . . . — 398

CHAPTER 11 Stardeath — 399

Learning Objectives — 399
11.1 White Dwarf Stars: Common Corpses — 400
11.2 Neutron Stars: Compact Remains of Massive Stars — 407
11.3 Novas: Mild Stellar Explosions — 409
    FOCUS 11.1 Distances to Novas — 411
11.4 Supernovas: Cataclysmic Explosions — 416
11.5 Pulsars: Neutron Stars in Rotation — 427
11.6 The Manufacture of Heavy Elements — 441
Summary — 444
Key Words — 444
Review Questions — 444
Problems — 445
Beyond This Book . . . — 446

CHAPTER 12 Black Holes — 447

Learning Objectives — 447
12.1 Black Holes: Warps in Spacetime — 448
    FOCUS 12.1 Tidal Gravitational Forces — 452
12.2 Observing Black Holes — 458
Summary — 470
Key Words — 470
Review Questions — 471
Problems — 471
Beyond This Book . . . — 472

PART THREE THE EVOLUTION OF PLANETS

CHAPTER 13 The Planets: Past and Present — 475

Learning Objectives — 475
13.1 Stardeath and the Sun's Birth — 476
13.2 The Formation of the Planets: A Preview — 482
13.3 The Planets Today — 483
13.4 The New Solar System: A Preview — 490
Summary — 497
Key Words — 497
Review Questions — 497
Problems — 498
Beyond This Book . . . — 498

CHAPTER 14 The Earth: Home Planet — 499

Learning Objectives — 499
14.1 The Mass and Density of the Solid Earth — 500
14.2 The Interior of the Earth — 502
14.3 The Age of the Earth — 503
FOCUS 14.1 Sounding Out the Earth's Interior — 504
14.4 The Magnetic Field of the Earth — 506
14.5 The Blanket of the Atmosphere — 508
14.6 The Atmosphere and Incoming Radiation — 514
14.7 The Restless Earth: Evolution of the Crust — 517
14.8 Evolution of the Atmosphere and Oceans — 520
14.9 An Overview of the Evolution of the Earth — 524
Summary — 527
Key Words — 527
Review Questions — 528
Problems — 529
Beyond This Book . . . — 529

CHAPTER 15 The Moon and Mercury: Dead Worlds — 531

Learning Objectives — 531
15.1 Orbital and Physical Characteristics of the Moon — 532
15.2 The Surface Environment of the Moon — 540
15.3 The Surface of the Moon: Pre-Apollo — 543
15.4 The Moon Close Up: Apollo — 547
15.5 The Origin of the Moon — 554
15.6 Cratering and the Evolution of the Terrestrial Planets — 557
15.7 Orbital and Physical Characteristics of Mercury — 557
15.8 The Surface Environment of Mercury — 561
15.9 The Surface of Mercury Close Up — 562
15.10 The Evolution of the Moon and Mercury Compared — 565
15.11 The Mystery of Mercury's Magnetic Field — 566

Summary 566
Key Words 567
Review Questions 567
Problems 567
Beyond This Book . . . 568

CHAPTER 16 Venus and Mars: Evolved Worlds 569

Learning Objectives 569
16.1 Orbital and Physical Characteristics of Venus 570
16.2 The Atmosphere of Venus 573
16.3 The Surface of Venus 576
16.4 The Evolution of Venus 582
16.5 General Characteristics of Mars 583
16.6 The Martian Atmosphere and Surface Temperature 586
16.7 The Martian Surface Viewed from the Earth 588
16.8 The Invasion of Mars by the Earth 591
16.9 The Moons of Mars 597
16.10 Venus, Earth, and Mars: An Evolutionary Comparison of Kindred Planets 599
Summary 600
Key Words 601
Review Questions 601
Problems 601
Beyond This Book . . . 602

CHAPTER 17 The Jovian Planets: Primitive Worlds 603

Learning Objectives 603
17.1 Jupiter: Lord of the Heavens 604
17.2 The Many Moons of Jupiter 614
17.3 Saturn: Jewel of the Solar System 621
17.4 Uranus: The First New World 635
17.5 Neptune: Guardian of the Deep 639
17.6 Pluto: Guardian of the Dark 641
    FOCUS 17.1 Is Pluto an Escaped Moon of Neptune? 644
Summary 646
Key Words 646
Review Questions 646
Problems 647
Beyond This Book . . . 648

CHAPTER 18 Solar System Debris: Clues to the Past 649

Learning Objectives 649
18.1 Asteroids: The Minor Planets 651
    FOCUS 18.1 The Titius-Bode "Law" 652
18.2 Comets: Snowballs in Space 658
18.3 Meteors and Meteorites 665
18.4 Interplanetary Gas and Dust 673

Summary	674
Key Words	675
Review Questions	675
Problems	675
Beyond This Book . . .	676

CHAPTER 19 The Origin and Evolution of the Solar System — 677

Learning Objectives	677
19.1 Pieces and Puzzles	678
19.2 Early Ideas of the Origin of the Solar System	681
19.3 An Overview of Nebular Models	684
19.4 Formation of the Planets	685
Summary	694
Key Words	695
Review Questions	695
Problems	696
Beyond This Book . . .	696

PART FOUR GALAXIES—ISLANDS OF STARS

CHAPTER 20 The Milky Way: Home Galaxy — 699

Learning Objectives	699
20.1 A Tour of the Milky Way	700
20.2 Herschel Maps the Milky Way	704
20.3 Stars Traveling Through Space	705
FOCUS 20.1 Proper Motion, Transverse Velocity, and Distance	708
20.4 The Discovery of the Galaxy	713
20.5 Stellar Populations and the Structure of the Galaxy	721
20.6 The Disk of the Galaxy	723
20.7 The Nucleus of the Galaxy	726
20.8 The Halo of the Galaxy	734
Summary	736
Key Words	737
Review Questions	737
Problems	737
Beyond This Book . . .	738

CHAPTER 21 The Evolution of the Galaxy — 739

Learning Objectives	739
21.1 An Overview of the Galaxy's Structure	740
21.2 Galactic Rotation: Stars in Motion	743
21.3 Galactic Structure from Optical Observations	750
21.4 Exploring Galactic Structure by Radio Astronomy	751
21.5 The Evolution of the Spiral Structure of the Galaxy	754
FOCUS 21.1 Galactic Structure from 21-cm Observations	756
21.6 A Possible History of Our Galaxy	759

Summary	761
Key Words	762
Review Questions	762
Problems	763
Beyond This Book . . .	763

CHAPTER 22 Beyond the Milky Way: Galaxies — 765

Learning Objectives	765
22.1 The Resolution of the Shapley-Curtis Debate	766
22.2 The Galaxian Zoo	770
22.3 Surveying the Universe of Galaxies	775
22.4 General Characteristics of Galaxies	783
22.5 Clusters of Galaxies	790
22.6 Intergalactic Matter	800
Summary	802
Key Words	803
Review Questions	803
Problems	803
Beyond This Book . . .	804

CHAPTER 23 Cosmic Violence: Active Galaxies and Quasars — 805

Learning Objectives	805
23.1 Violence in the Nucleus of the Galaxy	806
23.2 Active Galaxies	808
23.3 Quasars: Mysterious Energy Emitters	823
Summary	842
Key Words	843
Review Questions	843
Problems	843
Beyond This Book . . .	843

CHAPTER 24 The Origin and Evolution of the Universe — 845

Learning Objectives	845
24.1 Fundamental Assumptions of Cosmology	847
24.2 A Brief Review of Cosmology	850
24.3 Contemporary Cosmological Models	851
24.4 The Primeval Fireball	858
24.5 The End of Time	868
24.6 From Big Bang to Galaxies	870
24.7 Speculations on the Evolution of Quasars and Active Galaxies	876
24.8 Cosmic Neutrinos with Mass: Masters of the Universe?	877
Summary	878
Key Words	879
Review Questions	879
Problems	879
Beyond This Book . . .	880

PART FIVE COSMIC SPECULATION

CHAPTER 25 Life in the Universe — 883

Learning Objectives — 883
25.1 The Nature of Life on Earth — 884
25.2 Biochemistry Simplified (a Lot!) — 885
25.3 Ideas About the Genesis of Life on the Earth — 887
25.4 The Spark of Life — 894
25.5 Amino Acids from Space — 898
25.6 From Molecule to Organism — 900
25.7 The Solar System as an Abode of Life — 901
25.8 The Galaxy as an Abode of Life — 906
25.9 Neighboring Solar Systems? — 911
Summary — 914
Key Words — 915
Review Questions — 915
Problems — 916
Beyond This Book . . . — 916

CHAPTER 26 First SETI, Then CETI — 917

Learning Objectives — 917
26.1 The Central Issue: How Many Technical Civilizations? — 918
    FOCUS 26.1 Greetings to the Cosmos — 920
26.2 How to SETI — 924
26.3 SETI Strategies — 928
26.4 SETI Results to Date — 933
26.5 Where Are They? — 934
26.6 UFOs: Evidence of ETI? — 935
26.7 CETI — 937
Summary — 939
Key Words — 940
Review Questions — 940
Problems — 940
Beyond This Book . . . — 940

CHAPTER 27 Cosmic Futures — 941

Learning Objectives — 941
27.1 The Future of the Universe — 942
27.2 The Future of the Milky Way Galaxy — 946
27.3 The Future of the Earth and Sun — 947
27.4 The Future of Humankind — 948
Epilogue — 953
Key Words — 953
Review Questions — 953
Problems — 953
Beyond This Book . . . — 954

Appendix A Units	A-1
Appendix B Planetary Data	A-3
Appendix C Physical Constants, Astronomical Data	A-6
Appendix D Stars Within 13 Light Years	A-7
Appendix E The Twenty Brightest Stars	A-8
Appendix F Celestial Coordinate Systems	A-9
Appendix G Periodic Table of the Elements	A-12
Glossary	G-1
Index	I-1

PREFACE

Astronomers practice the ultimate archaeology. As they look out into space, they look back into time and see the cosmos as it was. This awesome journey into the past reveals the history of the cosmos conveyed by light. Modern astronomy tries to read this record of time to fathom the evolution of the universe and all that it contains. This cosmic search has had remarkable success. In the time that we have been astronomers, our understanding has opened to vistas that were unimagined and unexpected. The purpose of this book is to share that cosmic perspective with you, so that you may acquire at least a glimmer of the connected beauty of the astronomical universe.

INTENDED AUDIENCE

Our years of teaching experience have been distilled into this text. We have designed it for use in a two-semester, introductory astronomy course for students who want a comprehensive view of the sweep of modern astronomy. We expect that you will be science-oriented students, though not necessarily science majors. You may be engineering or math majors, or students in the humanities who have some science and math background. Mostly we expect that you will have a deep curiosity about astronomy and a desire to learn about the cosmos.

This book uses some math—the basics of algebra, trigonometry, and geometry. (We also expect that you are adept at reading and drawing graphs.) An important aspect of the process of science is that it makes predictions—not vague ones, but very specific ones in quantitative ways. The rigor and logic of mathematical development ensures the correctness of these predictions, which are then confirmed (or rejected) by observations. So mathematics lies at the basis of how astronomers understand the cosmos. We have assumed that you are familiar with the basics and are willing to work through material in a mathematical fashion.

OVERALL STRUCTURE

To achieve our goal, we have selected material that relates to one central theme: cosmic evolution—how the universe and its contents have changed with time. We have taken great care to present the material in a way that you—with a little effort—can understand. If you expend that effort you will gain the insight of a modern cosmic perspective.

To interweave the material around the central theme, we have divided the book into five parts. Part One investigates how ideas about the cosmos have developed in the western intellectual tradition. It starts with the basics—observing the sky—and ends with the most modern of ideas—Einstein's vision of the cosmos. This part introduces the central conceptual process of science: the scientific model. Part Two turns to the key figures in the play of cosmic evolution—the stars. It examines how we know their physical properties (starting out with the sun as the nearest example of a typical star), and how we infer their births, lives, and deaths. It shows how the touch of the stars works deep in space and time. Part Three focuses on local space—the planets (and other bodies) of the solar system. Here we first probe the familar—the earth—the most evolved planet. Then we compare the other planets to the earth in both physical properties and possible histories. This trail back into time runs into the question of the formation of the solar system, an event that seems naturally tied up with the birth of the sun. Part Four leaps out into a cosmic vista populated by galaxies—enormous realms of billions of stars. Again, we start with the familar: the Milky Way Galaxy, the one in which our solar system resides. We compare others to it and find that some galaxies partake in cosmic violence far beyond the energy output of our Galaxy. We push out into the actual tapestry of the universe—the reaches of galaxies gathered in clusters, which in turn are also joined in superclusters. Here our cosmic vista ends—at the beginning of the universe in the greatest explosion of all. Part Five ends the book with some speculation about life elsewhere in the universe and the means of communication with it. This part deals with unconfirmed and unfinished ideas—an appropiate ending, since our ideas about the universe are always changing and are never final in their formulation.

AIDS FOR THE STUDENT

You will probably run into some material that you find difficult as you delve into this book. That's to be expected, given the range of the subject. Yet, the beauty of the astronomical perspective is this: All the seemingly separate entities of the cosmos are interrelated. To help you see these cosmic connections, we have made a concerted effort to write a book to aid your understanding. We have selected fundamental astronomical and physical concepts for use throughout our explanations of how we infer the properties of astronomical objects. Basic physical ideas are presented and developed in an intuitive way. We introduce physical ideas first, and then key a mathematical treatment that serves to reinforce the basic argument.

To emphasize the most basic material, we begin each chapter with a set of learning objectives—approximately ten statements of the most basic material in the chapter. Of course, the chapters contain more material than outlined in the learning objectives—material related but not as critical. Read the learning objectives before you read the chapter for the first time so you know what the key ideas are. After reading the chapter, go back and review the learning objectives, perhaps checking off the material in the chapter that pertains to each. Next, try the Review Questions at the end of the chapter; note that these do not require any math and are keyed to the learning objectives. Finally, do the Problems, which require some kind of basic computation or mathematical argument.

SPECIAL NOTES

Two comments, one about names and one on units. As you read, you'll come across the names of many people—some familiar, such as Aristotle and Newton, and some not as well known, such as Hubble and Sandage. We did not include these names to be memorized by rote. Some are people who have strongly affected the western intellectual heritage; others are astronomers of the past who invented seminal ideas. Still others are astronomers working today at the frontiers of the field whose ideas may or may not pan out. We give their names mainly to remind you that astronomy is done by living people. We make no claim to be complete; credit for our modern view of the cosmos belongs to many more people than we have mentioned.

As far as units of measurable quantities are concerned, astronomers are a perverse lot. We make frequent use of units that have no physical basis, such as parsecs and magnitudes. And astronomers have not yet consistently adopted the international system of units. We have used the international system except when we believe that a unit from another system—such as atmospheres for atmospheric pressure—will be easier to grasp physically. We also tend to use light travel time, a physically fundamental unit, rather than parsecs to measure distances and sizes.

ACKNOWLEDGMENTS

Many people have helped us directly and indirectly in this project. We thank: James DePresto, Bruce Dietrich, Georgeanne Caughlan, Paul Hodge, Owen Gingerich, Dave Latham, Marc Price, Jack Burns, Brian Flannery, Bruce Elmegreen, William Buscombe, Larry Cram, Lawrence Frederick, Thomas Jones, Kenneth Janes, Robert Mutel, Robert Dukes, Joe Veverka, Krystyne Jaworowska, David King, Bruce Partridge, Richard Sears, Robert Robbins, Dave Wilkinson, Charlie Lada, Tom Balonek, Paul Heckert, Gary Henson, Frank Shu, Hyron Spinrad, Leo Blitz, Joseph Silk, Stuart Vogel, Donald Terndrup, Howard Penner, Sammie Lee, Alan Peterson, and Ko Hummel.

We also thank our students, graduate and undergraduate, who, over the years of our teaching, have always directed us to new insights to the cosmos and so have expanded our astronomical perspective.

MICHAEL ZEILIK
Department of Physics and Astronomy, The University of New Mexico
Albuquerque, New Mexico 87131

JOHN GAUSTAD
Department of Astronomy, Swarthmore College
Swarthmore, Pennsylvania 19081

ASTRONOMY

PART ONE
CHANGING CONCEPTIONS OF THE UNIVERSE

OBSERVING THE SKY:
From Chaos to Cosmos

1

LEARNING OBJECTIVES
After studying this chapter, you should be able to:
1. Describe the motions of the sun and the moon, as seen from the earth, relative to the stars of the zodiac.
2. Describe the motions of the planets, as seen from the earth, relative to the stars of the zodiac, with special attention to retrograde motion.
3. Describe the daily motions of the sun, stars, moon, and planets relative to the horizon.
4. Describe the seasonal motion of the sun relative to the horizon at noon and at sunrise/sunset.
5. Tell what astronomical events or cycles set the following time intervals: day, month, and year.
6. Describe the astronomical conditions necessary for the occurrence of a total solar and total lunar eclipse.
7. Describe the concept of angular diameter, and relate angular diameter to the true diameter and distance of an object.
8. Describe the concept of angular speed, and relate angular speed to the actual speed and distance of an object.
9. Argue, on the basis of naked-eye observations of planetary motion and clearly stated assumptions, an order of the planets from the earth.
10. Identify at least one astronomical achievement of a prehistoric culture.

CENTRAL QUESTION:
Which astronomical objects can you see
without a telescope, and how do their
positions change with time?

Do the heavens have an order? Can we make sense of the objects and events in the sky? Early skywatchers felt overwhelmed by what they saw. For some, their initial wonder gave way to a desire to find some order in the apparent chaos of the sky.

This quest drove them to concepts of space and time naturally derived from astronomy. Space was the arching heavens, cushioned by the invisible air and are far removed from the earth. Time was marked on the cosmic clock of the cycles of celestial motions. By watching the heavens for a long time they began to recognize patterns of stars in the heavens and basic cycles of the movement of these patterns. These discoveries of the early astronomers brought a sense of order to the heavens, a structure in space and a sequence in time. What at first seemed to be chaos acquired order and led to discoveries of how the world is put together.

This chapter deals with observations you can make without optical equipment—the same observations made by early astronomers. From such observations you can sense the regular cycles of motions in the heavens. Long-term observations can establish the periods of celestial cycles with amazing accuracy. The recognition of these rhythms marked a crucial step in the development of astronomy and early concepts of the cosmos. This chapter will not attempt to explain those motions. (That explanation will come in Chapters 2, 3, and 4.) Here we present only the facts observed by the ancients, facts which must be explained by any model of the cosmos. In the next chapters we will consider models designed to explain these facts, and how these models have changed through history.

1.1 THE VISIBLE SKY

Have you ever looked carefully at the night sky from a location far away from a city? Perhaps you have, from a mesa in New Mexico or the High Sierra in California. You

1.1 THE VISIBLE SKY

probably were amazed by the sparkle of stars set against the deep velvet of the sky. At first glance, you can't count the stars or find any order to them. You have no way to judge their distances, except to say that they are far away.

Constellations

If you study the stars for a while, their apparent chaos falls into patterns, designs imposed by the human mind (Fig. 1.1). Ancient astronomers perceived stellar patterns and gave them names; such patterns are called *constellations*. Early constellations marked a convenient group of stars, with ill-defined boundaries, that outlined a mythological or realistic figure (Fig. 1.2). The oldest known constellations originated about 3000 B.C., somewhere in the Tigris-Euphrates valley of Mesopotamia. Here people enchanted by the heavens dreamed up stories about the celestial figures they saw. (The endpapers of this book show the figures of some modern constellations.)

During the development of Babylonian astronomy the constellations came into use as

FIGURE 1.1 A time exposure of the stars in the constellation Orion and others nearby. The three bright stars making a diagonal line pointing to the top right form Orion's belt. The very bright star down right of center is Rigel. This photo shows many faint stars invisible to the eye. See the constellation endpapers for winter to find Orion. [*Courtesy Mt. Wilson Observatory, Carnegie Institute of Washington.*]

FIGURE 1.2 The mythological figure of Orion, the Hunter, from the star atlas *Uranometria*, published in 1603 by Johann Bayer. Albrecht Dürer made the engraving.

signposts in the sky to mark the positions of the sun, moon, and planets. The Egyptians took over some constellations from the Babylonians, the Greeks borrowed some from the Egyptians, and we indirectly received these celestial heirlooms from the Romans.

If you observe the stars night after night, you will see that the constellations rise and set, but do not change their shape. In fact if you were patient enough to watch for your whole life, you would notice no change. The stars appear to maintain fixed positions relative to one another.

Angular Measurement

Suppose you want to make a map of a constellation. You need to measure how far apart the stars appear in the sky. How can you do this? You need a standard sighting device. The extent of your hand held at arm's length will do for a crude measurement, but a large protractor or surveyor's instrument is better. You can then measure the angle between the line of sight to one star and the line to another. This angle is the *angular separation* or *angular distance* between two stars.

Angular measurement involves determining the sizes of angles. Angles are commonly (but not always) measured in the *sexagesimal system* (developed by the Babylonians). The sexagesimal system is based on the number 60 rather than on the number 10 of our decimal system. In this system a circle contains 360 degrees, each degree contains 60 minutes, and each minute contains 60 seconds. The sexagesimal system also survives in measurements of time (60 seconds equal 1 minute, 60 minutes equal 1 hour). To distinguish angular minutes and seconds from units of time, we refer to *minutes of arc* (or *arcminutes*) and *seconds of arc* (or *arcseconds*).

At arm's length your fist covers about 10° of sky and each fingertip about 1° (Fig. 1.3). This crude, but handy angular measuring instrument can mark out the angular separation of the pointer stars in the Big Dipper (about 5°, or half a fist; Fig. 1.4).

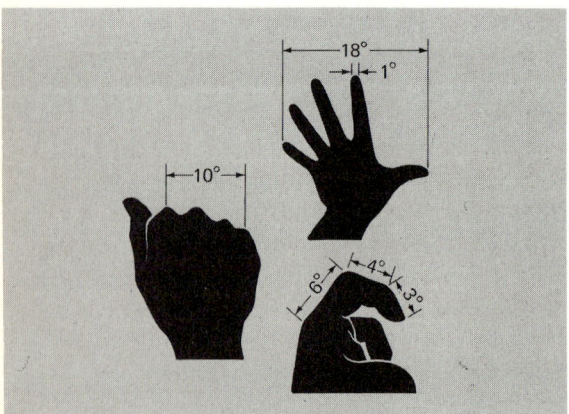

FIGURE 1.3 Angular measurements made with a hand extended at arm's length. The angles are approximate and typical for an average adult.

FIGURE 1.4 Measuring angles on the sky with a hand at arm's length. The angle between the two pointer stars in the Big Dipper is about 5°, or half a fist.

1.1 THE VISIBLE SKY

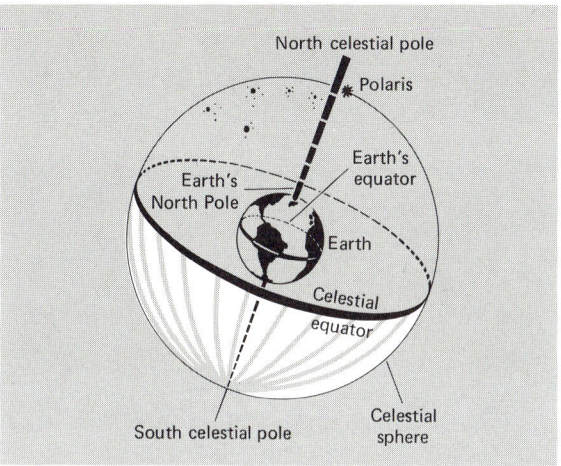

FIGURE 1.5 A simplified geocentric model of the earth and sky. The round earth rests in the center of a larger sphere of stars called the *celestial sphere*. The celestial poles are points on the celestial sphere directly above the earth's poles, and the celestial equator is a circle on the celestial sphere right above the earth's equator.

FIGURE 1.6 The motion of the stars relative to the horizon. This photo is a time exposure of a few hours; it shows the *east-to-west* motion of the stars. [*Courtesy Lick Observatory.*]

The Celestial Sphere

It's not hard to see the sky as a dome to which the stars are attached. Imagine another half to this apparent dome, the half below your horizon. This is the *celestial sphere* (Fig. 1.5), a concept first devised by the Greeks. The celestial sphere appears to be centered on the earth, and the stars seem to be fixed to it. This picture of the cosmos, typical of older ideas, is *geocentric*, or centered on the earth.

If you watch the stars for a few hours, you will notice that they move relative to your horizon. *The stars rise in the east, travel slowly in arcs against the celestial sphere, and set in the west* (Fig. 1.6). If you are in the northern hemisphere and observe the stars to the north (or in the southern hemisphere and face south), you will find that some stars never dive below the horizon. Instead, they trace complete circles above it (Fig. 1.7a); we call them *circumpolar stars*. As you watch the circumpolar stars swing around, they draw concentric circles like the rings of a bull's-eye (Fig. 1.7b). The center of these rings marks the *celestial pole*, the point about which the stars appear to pivot. (A moderately bright star called *Polaris* now lies close to the north celestial pole; no bright star now falls close to the position of the south celestial pole.) A circle drawn around the sky halfway between the poles is the *celestial equator*. It crosses the horizon due east and due west.

The constellations shift their positions relative to the sun with the change of seasons. For example, from mid-northern latitudes, on a certain day in winter, you can see Orion due south at midnight. It is then just opposite the sun (which is due south a half day later, at noon). Look due south every night at midnight. Orion creeps slowly to the west toward the sun. In summer you can't see Orion at all because it's close to the sun, well below the horizon at midnight. In winter, a year later, Orion again lies due south at midnight. Every constellation takes one year to return to its initial place in the sky relative to the sun.

(a) (b)

FIGURE 1.7 The motion of the stars around the north celestial pole. (a) Time exposure showing apparent motion of the stars *counterclockwise (east to west)* about the north pole star, Polaris. The small arc traced by Polaris, near the center, indicates that it is not exactly at the north celestial pole. [*Courtesy Lick Observatory.*] (b) Amateur astronomers under the night sky. This time exposure is shorter than that in (a). It shows shorter star trails, an airplane's lights (near horizon), and the movement of astronomers' flashlights. [*Courtesy T. Gondola.*]

(If you look at the endpaper charts, you will see that different constellations are visible at night at different times of the year.)

In summary, the stars, over a human lifetime, do not move noticeably with respect to each other. They do move, daily, from east to west relative to the horizon. And they also move, annually, westward with respect to the sun.

1.2 THE MOTIONS OF THE SUN

The sun dominates all objects visible in the sky and establishes the most fundamental time cycle of our world: day and night. Every day the sun rises above the eastern horizon, traces an arched path across the sky, and falls below the western horizon. Midway between sunrise and sunset the sun ascends to its highest point relative to the horizon. This daily event defines *noon*, a fundamental reference in the measurement of time. The interval from one noon to the next sets the length of the *solar day*.

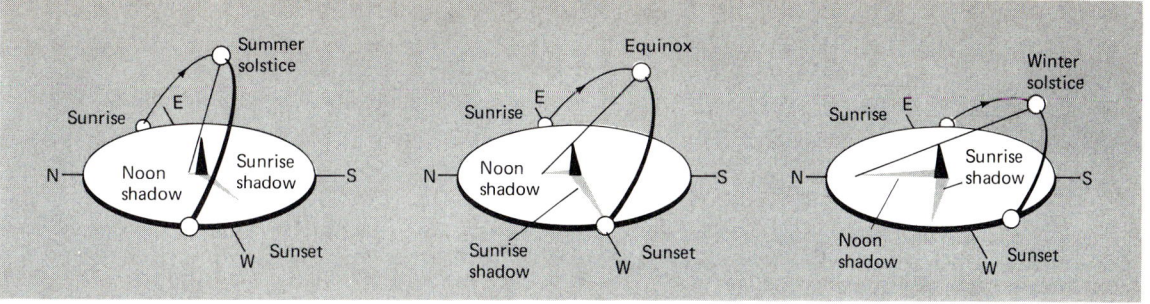

FIGURE 1.8 Gnomon shadows and the position of the sun in the sky. The higher the sun relative to the horizon, the shorter the shadow. The shortest noon shadow of the year occurs on the day of the summer solstice (left), the longest on the winter solstice (right). At the spring and fall equinoxes (middle), the length of the noon shadow falls between that for winter and summer.

Motion Relative to the Horizon

If you place a stick vertically in a flat place on the ground, you have constructed an instrument to study the sun's daily and seasonal motion in the sky relative to your horizon. This device is called a *gnomon* and was first developed by the Babylonians. Look at the shadow cast by the gnomon. The tip of the gnomon's shadow marks the end of a line that connects the shadow's tip, the top of the stick, and the sun (Fig. 1.8). The shadow points in the direction *opposite* to the position of the sun in the sky. The shadow's length is related to the angle of the sun above the horizon. When the sun hangs low in the sky, the shadow is long; when it is high, the shadow is short. At noon the shadow has its shortest length for that day. It then points north in the northern latitudes, south in southern latitudes, and north or south, depending on the time of year, near the equator.

Motion Relative to the Stars

The sun also moves with respect to the stars. This motion is hard to observe, for you can't see the stars during the day. Try this: Pick out a bright constellation visible above the western horizon after sunset (Fig. 1.9). Look again at the same time about a week later; the constellation will be closer to the horizon and also to the sun. So, relative to the stars, the sun appears to move to the *east*. (This is the same relative motion as that described in the previous section—the stars moving westward with respect to the sun—but here we are looking at it from a different point of view.) In one year the sun returns to the same position relative to the stars; that's 360° in a year or about 1° in a day *eastward*. This rate of change of the sun's position in the sky (1°/day) is its *angular speed*.

Warning: Don't confuse this slow, annual motion with the much faster daily rising and setting of the sun. The sun moves *eastward* with respect to the *stars*, once around in a *year*. At the same time it and the stars move *westward* with respect to the *horizon*, once around in a *day*.

If you recorded the sun's position with respect to the stars for a year and drew an imaginary line through these points, you would trace out a complete circle on the sky; it is called the *ecliptic* (Fig. 1.10). The ecliptic, a great circle on the celestial sphere, is tilted with respect to the celestial equator. Half of it lies north and half south of the equator.

Because the sun's motion along the ecliptic takes it north and south of the celestial

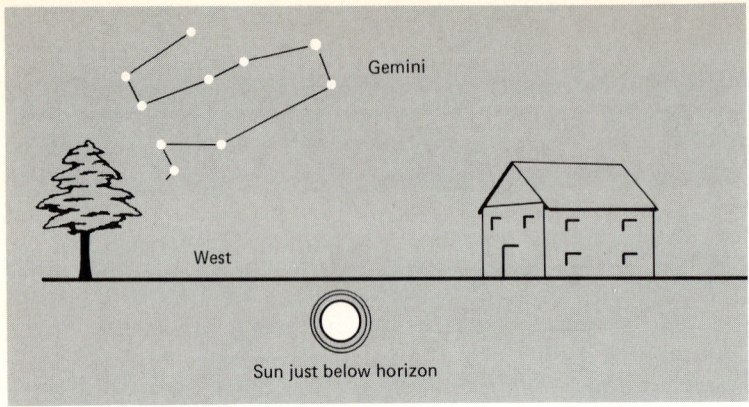
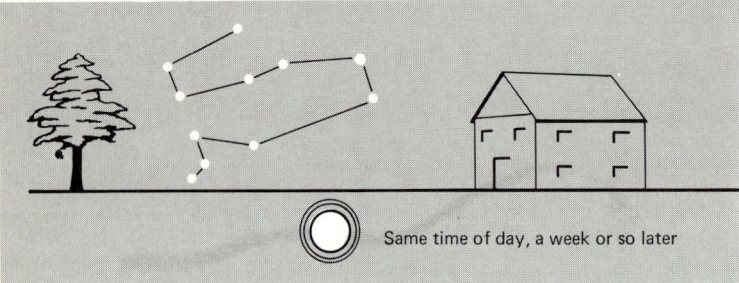

FIGURE 1.9 The motion of the sun relative to the stars. Find a constellation (such as Gemini, shown here) near the western horizon at sunset (top). About a week later look for the same stars just after sunset (bottom). They will appear *closer* to the western horizon. Eventually, they will set before the sun goes down. So the stars seem to move to the *west* with respect to the sun. Or you can think of the sun as moving *east* with respect to the stars.

equator, the height of the sun in the sky *at noon* varies from summer to winter. For example, in Albuquerque, New Mexico, at noon on March 21 or September 21, a gnomon gives an angular height of the sun of 55°; on December 21 at noon, a height of only 31.5°; at noon on June 21, 78.5°.

How do we get these angles from the gnomon? We measure the length of the shadow (*l*) and the height of the gnomon (*h*), and use the trigonometric formula for right triangles:

$$\tan \alpha = \frac{h}{l}$$

If we know h and l, we can calculate α.

The shadow at noon is shortest on the *summer solstice*, the day of the year with the greatest number of daylight hours. The sun is then at its most northern point on the ecliptic and the noon sun at its highest point in the sky for the year. The noon shadow stretches longest on the *winter solstice*, the day of the year with the fewest daylight hours. The sun has dropped to its lowest point in the sky for the year and is farthest south on the ecliptic. At the first day of spring and autumn the gnomon casts a shadow with a length between its summer minimum and winter maximum. At those times, called the *equinoxes* (vernal equinox on March 21, autumnal equinox on September 21), the sun is at the point where the ecliptic crosses the equator, and the days and nights have equal lengths. This cycle of the gnomon's noon shadow defines the year of seasons, a second fundamental unit of time (the first being the day). (*Note:* The terms solstice and equinox refer both to the time of year and a position on the ecliptic.)

The constellations through which you see the sun move define the *zodiac* (Fig. 1.10), the most ancient and coherent set of constellations (Table 1.1). The zodiac probably developed from a desire to mark the sun's position with respect to the stars. The sun's

1.2 THE MOTIONS OF THE SUN

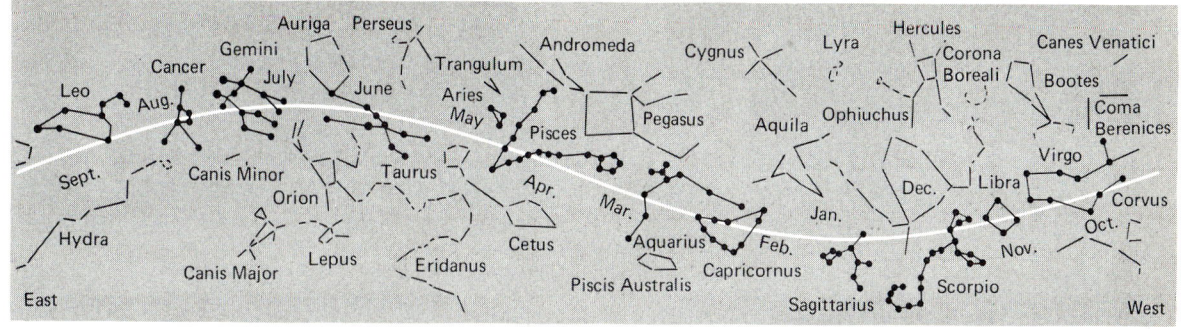

FIGURE 1.10 The ecliptic and zodiacal stars. With no atmosphere, you could see the sun's changing position with respect to background stars. The sun's path is called the *ecliptic*. It is a great circle on the celestial sphere, but in this mercator projection, it appears as a curved line. This diagram indicates the sun's position by date along the ecliptic and shows the major constellations. The zodiacal stars are indicated by darker lines. Note that the sun moves west to east among the stars.

position along the ecliptic is labeled with reference to these constellations. For instance, to say that the sun is "in the sign of Taurus" specifies the sun's approximate position along the ecliptic.

Taurus the Bull, called by the Babylonians the "Bull in Front," is probably the oldest zodiacal constellation. In Babylonian times the sun rose on the day of the vernal equinox with the stars of Taurus—the sun was "in Taurus." This astronomical event signaled the renewal of spring.

Today Taurus no longer marks the vernal equinox. The positions of the equinoxes and solstices shift over time. To say that the sun is "in the *sign* of Taurus" indicates that it is

TABLE 1.1 The Zodiacal Constellations

Constellation	Astronomical Symbol	Mesopotamian-Euphratean Identity
Aries the Ram	♈	Ram, Messenger
Taurus the Bull	♉	Bull in Front, Bull of Heaven
Gemini the Twins	♊	Great Twins
Cancer the Crab	♋	Workman of the River Bed
Leo the Lion	♌	Lion
Virgo the Virgin	♍	Proclaimer of the Rain
Libra the Balance	♎	Life-Maker of Heaven
Scorpius the Scorpion	♏	Scorpion of Heaven
Sagittarius the Archer	♐	Star of the Bow
Capricornus the Goat	♑	Goat-Fish
Aquarius the Water Bearer	♒	Urn
Pisces the Fish	♓	Cord-Place Joining the Fish

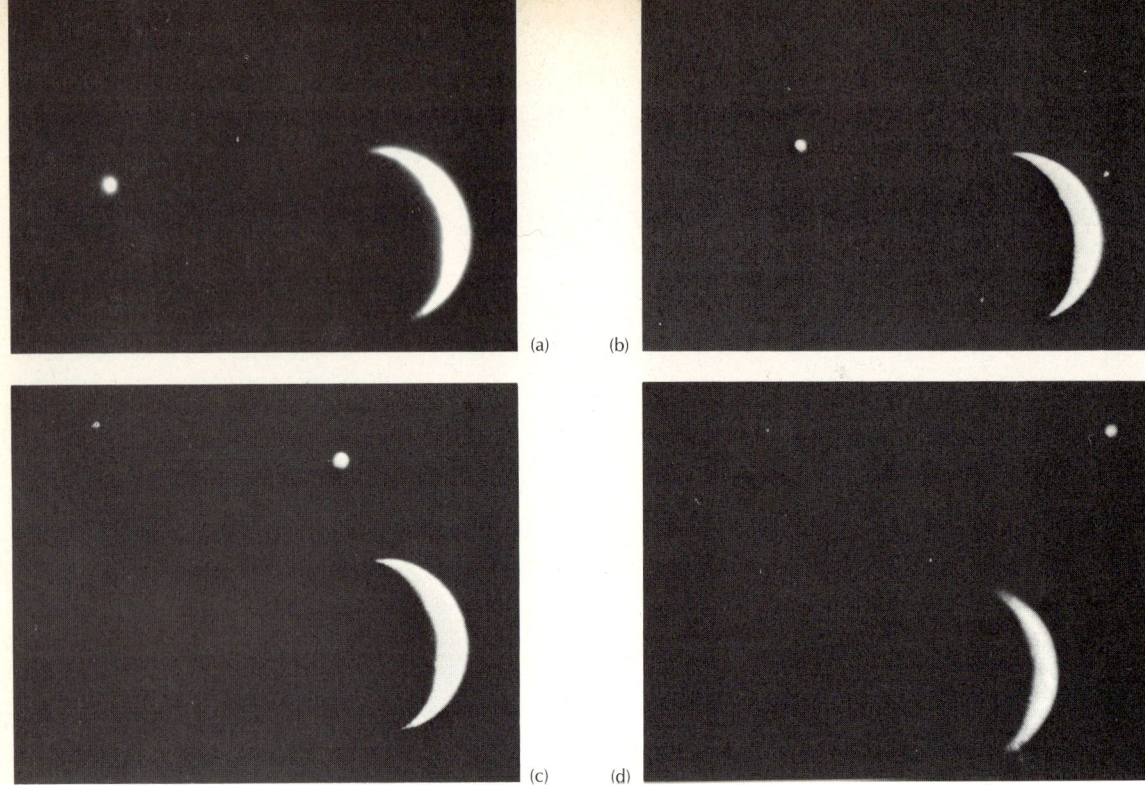

FIGURE 1.11 The motion of the moon with respect to the planet Venus. The time span over all four photos is 2½ hours. North is at the top and east is at the left. [*Courtesy D. Hoff.*]

on a certain section of the ecliptic, but it is *not* superimposed on the *constellation* Taurus as it would have been in 500 B.C. The position of the sun has moved since Babylonian times so that now Pisces marks the vernal equinox (see Focus 1.1).

Warning: The stars that form the constellations are not necessarily close to each other in space; they appear close together because they're all more or less in the same direction along our line of sight. So the statement "the sun is in Pisces" doesn't mean that the sun is surrounded by that clump of stars, but that it's *in the same direction in the sky* as the constellation. Knowing that the sun, the moon, or a planet is "in" a particular constellation tells you where to look for it in the sky, but that's about all.

To sum up, as seen from the earth, the sun displays (1) daily motion from east to west with respect to the horizon and (2) annual motion from west to east with respect to the stars through the zodiac (including, because of the tilt of the ecliptic, the seasonal motion from north to south of the equator).

1.3 THE MOTIONS OF THE MOON

If you watch the moon throughout a warm summer's night, you can spot two of its celestial motions. First, like the sun and stars, the moon daily rises in the east and sets in the west. Second, the moon journeys *eastward* against the backdrop of zodiacal stars. Wait until the moon appears just east of a bright star or planet (Fig. 1.11). Observe the moon again a few hours later. You will notice that the moon has moved farther east of the

FIGURE 1.12 A close-up of the full moon. [*Courtesy Lick Observatory.*]

FIGURE 1.13 A crescent moon, a few days after new moon. [*Courtesy Lick Observatory.*]

star or planet, at an angular speed of about $\frac{1}{2}°$ per hour. At this rate the moon completes a circuit of the zodiac in about 27 days. (*Note*: The moon's path does not lie exactly along the ecliptic, but it does fall within 5° of it and so stays within the zodiac.)

You have probably noticed the moon's changing appearance in the sky, its *phases*, which follow a regular sequence. When the moon rises at sunset, it glares as a full moon (Fig. 1.12). About 14.5 days after the full moon, the moon cannot be seen as it rises; it in the new moon phase. (The term *new moon* derives from legends in which the moon died at the end of its cycle of phases and was reborn. People did not recognize the moon as a reflector of sunlight until the fourth century B.C.) After this vanishing act, the moon reappears, low in the west at sunset, as a crescent moon (Fig. 1.13). Next comes first-quarter phase, when the moon is half illuminated, followed by gibbous phase, when it is more than half illuminated, but not yet full. The phases follow in order: new, crescent, first quarter, gibbous, full, gibbous, last quarter, crescent, new. A complete cycle of phases, say from one full moon to the next, takes about 29.5 days. It defines a third fundamental unit of time: the month of phases.

The different phases of the moon correspond to specific alignments of the sun and the moon in the sky. At new moon the sun and the moon are close together. At first quarter the moon lies 90° *east* of the sun (Fig. 1.14a); at full moon it is 180° from the sun (Fig. 1.14b); and at last quarter it is 90° *west* of the sun (Fig. 1.14c). (*Note*: The term *quarter* refers to the position of the moon in the sky, one-quarter of a full circle from the sun, not to the degree of illumination of the moon; the moon at quarter phase looks half full.)

To summarize, the moon moves westward daily with respect to the horizon and also eastward, 360° in about one month, with respect to the stars.

1.4 THE MOTIONS OF THE PLANETS

If you observe the sky often enough, you can quickly pick out objects that don't belong to the familiar stellar patterns. Five of these move in more or less regular ways among the stars of the zodiac; these are the *planets*. You can see them easily without optical aid:

Mercury, Venus, Mars, Jupiter, and Saturn. These planets were known to the oldest civilizations that practiced astronomy or astrology. (Uranus, Neptune, and Pluto were discovered much later with telescopes.)

Retrograde Motion

The planets display a peculiar celestial motion that sets them apart from all other objects in the sky. Suppose you observe Mars every night for several months around the time when Mars appears brightest in the sky. First, you will notice that Mars moves through the zodiac on or close to the ecliptic. This motion is at first *eastward* with respect to the stars (Fig. 1.15). At some point Mars falters in its eastward motion and stops. Then for about three months Mars travels *westward*, opposite to its normal direction. In the middle of this westward motion Mars appears brightest. Then Mars again slows down,

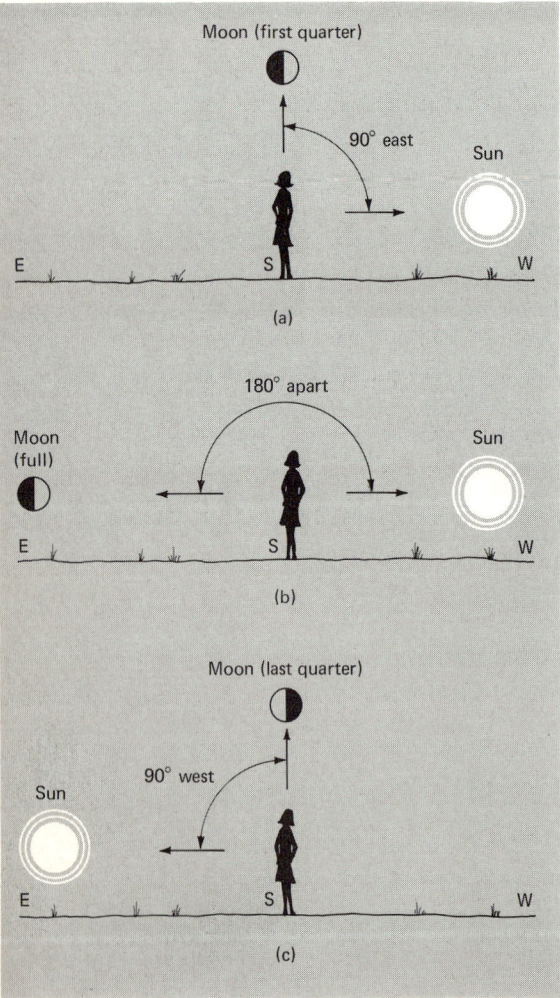

FIGURE 1.14 The orientation of the sun and moon in the sky for different phases of the moon. At first quarter (a) the moon is 90° *east* of the sun (due south as the sun sets). At full (b) it is 180° away rising as the sun sets. At last quarter (c) the moon is 90° *west* of the sun (due south as the sun rises). The view here is facing south.

Note on focus sections: The material in the focus sections is set off from the text so that it does not interrupt the flow of main ideas in the chapter. These sections contain more advanced development of astronomical concepts and mathematical detail or occasional sidelights. A few of them are essential to understanding later material, but most are not. Your instructor may assign specific focus sections for you to read. Otherwise, check each focus as it is referred to in the text and read it carefully if it interests you or contains material new to you.

FOCUS 1.1 PRECESSION OF THE EQUINOXES

In Sumerian times the sun appeared in Taurus at the vernal equinox. Today you see the sun in Pisces at the beginning of spring. In the passage of 5000 years the position of the vernal equinox in the zodiac has moved to the west out of Taurus, through Aries, and into Pisces. Knowing this you can estimate how long it would take the vernal equinox to circuit the zodiac. If in 5000 years it has moved through two constellations, or one-sixth of the zodiac, then it would take six times as long, or about 30,000 years, to cover the entire zodiac. This slow, westward drift of the equinoxes is called the *precession of the equinoxes*.

The precession of the equinoxes changes the zodiacal location of the sun at the equinoxes and solstices (Fig. F.1). Precession results in another, less obvious but important, effect: The celestial poles move in the sky, so the pole star changes (Fig. F.2). The north celestial pole now lies near the star Polaris at the end of the handle of the Little Dipper. About 3000 years ago the north celestial pole was near the star Thuban in Draco. Approximately 12,000 years from now precession will have carried the north celestial pole near the bright star Vega in Lyra.

Precession is hard to observe because it takes place so slowly. However, if a culture kept astronomical records for a few centuries, its astronomers could notice the shift of the equinoxes and solstices in the zodiac. Giorgio de Santillana and Hertha von Dechend argue in *Hamlet's Mill* that a few ancient cultures may have been aware of precession. If so, it demonstrates a remarkable astronomical achievement and the serious attention paid to celestial motions by early skywatchers.

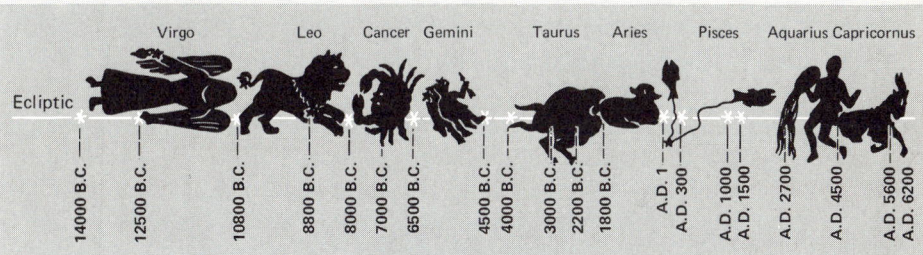

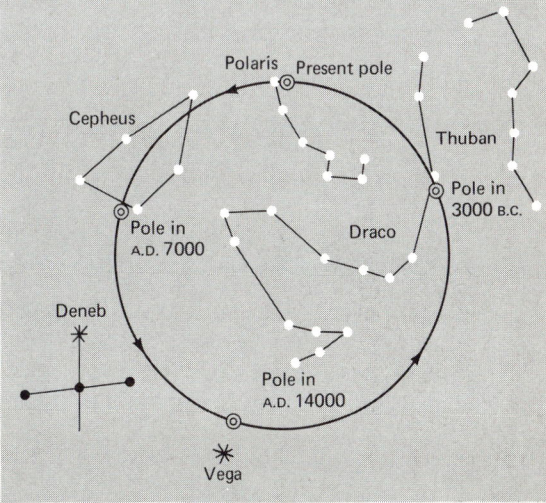

FIGURE F.1 Precession and the change of position of the vernal equinox with respect to the zodiacal stars.

FIGURE F.2 Precession and the change of position of the north celestial pole with respect to the stars. The pole's motion completes a circle in the sky in about 26,000 years. About 12,000 years from now, the pole will be near the bright star Vega.

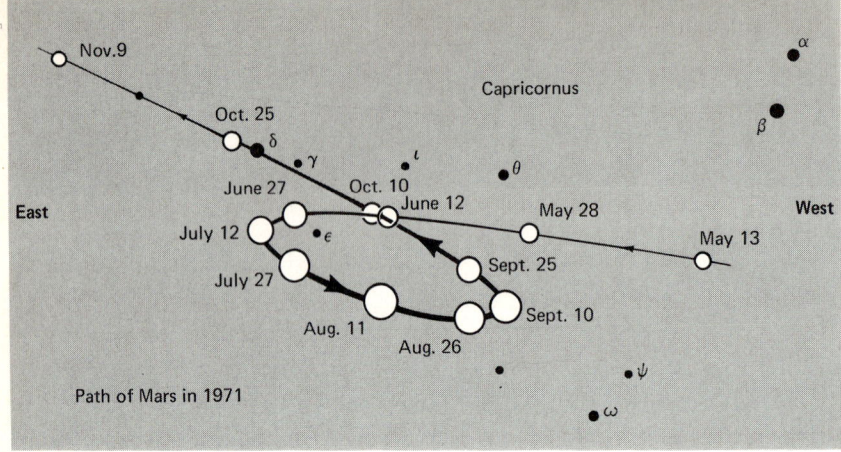

FIGURE 1.15 The retrograde motion of Mars, relative to the stars, during its August 1971 opposition. The stars of the constellation Capricornus are labeled by Greek letters. Note that from July 12 until September 10, Mars moved from east to west; this was the time of its retrograde motion. The size of the circles indicates how the brightness of Mars varied; the larger the circle, the brighter the planet appeared.

stops its westward motion, and resumes its normal eastward course. The planet's backward swing to the *west* is called *retrograde motion*.

All planets loop along or near the ecliptic in retrograde motion, but generally not at the same time or for the same duration. For instance, Mars takes about 83 days to go through its retrograde motion, but Saturn takes 139 days (Table 1.2). In 1980 Mars was in the middle of a retrograde loop on February 25, Saturn on March 14.

Ancient astronomers were startled by the planets' retrograde motion. In general, the planets travel eastward along or near the ecliptic, within the zodiac, as do the sun and moon, but the moon and sun never exhibit such westward motion with respect to the stars.

In addition to these motions, the planets share with everything else in the sky the daily motion from east to west with respect to the horizon. So the planets display three motions in the sky: east to west daily with respect to the horizon; generally eastward in the zodiac; and occasionally westward in retrograde loops, always on or near the ecliptic.

One more point about retrograde motion. The alignment of the sun and planet at the time of retrograde motion, as seen from the earth, separates the visible planets into two groups. Mercury and Venus make up one group and Mars, Jupiter, and Saturn the other.

TABLE 1.2 Fundamental Observations of the Visible Planets

Planet	Typical Duration of Retrograde Motion	Period Around Ecliptic
Mercury	34 days	1 year
Venus	43 days	1 year
Mars	83 days	2 years
Jupiter	118 days	12 years
Saturn	139 days	30 years

Note: The duration of retrograde motion (from start of westward displacement to renewal of eastward displacement) may vary a little from the above values from retrograde to retrograde.

1.4 THE MOTIONS OF THE PLANETS

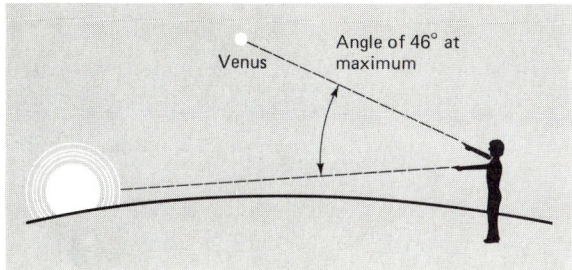

FIGURE 1.16 Measuring the greatest elongation of Venus at sunset. At this time the angle between Venus and the sun is about 46°. The same observation for Mercury gives a maximum angle of about 23°.

Mercury and Venus never stray very far along the ecliptic from the sun. Because they stick close to the sun, they are visible only as morning or evening "stars" (but keep in mind that they are *not* actually stars), that is, they are visible in the eastern sky just before sunrise (if they are west of the sun) or in the western sky just after sunset (if they lie east of the sun).

The angular separation of a planet from the sun is called the *elongation* of the planet. (You can measure this angle crudely with your fist.) When a planet is at its closest point to the sun, it is in an alignment called *conjunction*. At conjunction, the elongation of the planet is 0°. When Mercury or Venus reaches its greatest angular separation from the sun, it is at *maximum elongation* (Fig. 1.16). For Mercury the average maximum elongation is 23° (about two fists), and for Venus it is about 46° (about four and a half fists). Mercury and Venus begin their retrograde motions a little while after they have swung farthest east of the sun as evening stars. They then move westward, pass the sun, and reappear as morning stars west of the sun at dawn (Fig. 1.17).

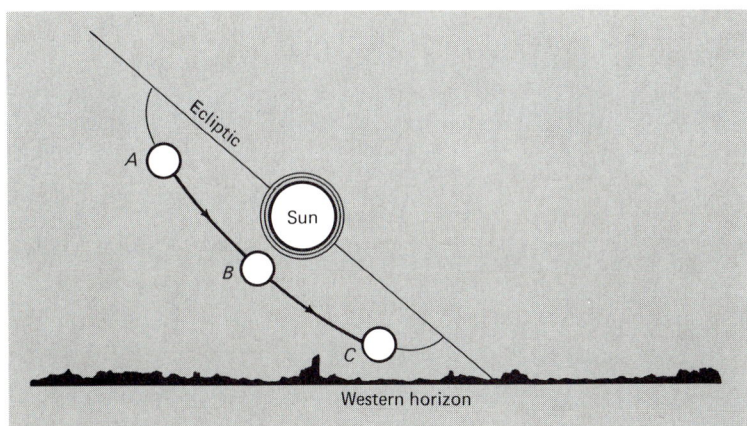

FIGURE 1.17 Retrograde motion of Venus. The upper part of the diagram shows how Venus moves (from A to C, east to west) with respect to the sun and ecliptic during retrograde motion. Venus goes through retrograde motion when it switches from being an evening "star" (A) to a morning "star" (C), that is, from being on the east side of the sun to the west side.

The second group of planets, Mars, Jupiter, and Saturn, freely move anywhere along the ecliptic with respect to the sun. Unlike Mercury and Venus, they are not restricted to a region near the sun. The planets of this group retrograde when they stand in *opposition* to the sun, which means they are opposite the sun in the sky as seen from the earth. At opposition, the elongation of the planet is 180°. A planet at opposition rises just as the sun sets. If you extend one arm to point at the rising planet and the other at the setting sun, your arms will be 180° apart (they make a straight line). When at opposition, each planet is in the middle of its retrograde loop and shines its brightest.

Mercury and Venus can *never* be in opposition to the sun; they retrograde near conjunction after their greatest eastern elongation. Mars, Jupiter, and Saturn can be in opposition or conjunction, but they retrograde at opposition, not conjunction. In later chapters, we'll see how different models of the solar system explain these differences.

Note: The alignments of conjunction and opposition are not restricted to the planets and sun. When any two celestial objects are seen close together in the sky (within a few degrees), they are in conjunction. When any two are roughly 180° apart in the sky, they are in opposition.

Relative Distances of the Planets from the Earth

The differing amounts of time that each planet spends in retrograde motion and that each takes to circle the zodiac provide a clue to the relative distances of the planets from the earth.

Suppose one night you see the lights of two airplanes moving across the sky. Using your watch and fist, you can time how long it takes each to cover a certain angular distance in a given amount of time. For example, one plane may move at 1° per second, the other at 2° per second. Their rates of angular motion are their angular speeds.

But how fast is each plane really traveling? And which plane is closer? To answer the first question requires that you know the distance to the planes. When you see an object move in the sky—an airplane or the sun—you typically do not know how fast it is actually moving (say, in kilometers per hour). All you can tell directly is its angular speed. You cannot determine its actual speed unless you know its distance. (That's why so many reports of the "fantastic" speeds of UFO's are not believable; usually the observer does not know the distance to the UFO.)

Although angular speeds give you actual speed only when you know distances, a comparison of angular speeds can allow you to estimate relative distances, if it is at all reasonable to assume that the actual speeds involved are roughly the same. Assuming that both planes fly at the same speed (that's reasonable), then the one with the greater angular speed is the closer of the two (Fig. 1.18). The faster plane is at half the distance of the farther one, for it has twice the angular speed. If one plane had three times the angular speed of the other, the slower is at three times the distance of the faster.

If we assumed that the planets move at the same real speed, then the planet which appears from the earth to move slowest would be the most distant, and the one that appears to be the swiftest would be the nearest. In other words a planet with a greater angular speed would be closer to the earth than one with a smaller angular speed. And, roughly speaking, if one planet moves with half the angular speed of another, it would be about twice as far away.

Greek astronomers during the third century B.C. applied this argument and established the order of Mercury, Venus, the sun, Mars, Jupiter, and Saturn. Because the moon moves more swiftly than any other heavenly body, the Greeks placed the moon between the earth and Mercury. The ordering of the planets determined in this way is almost correct, even though we now know that the assumption of equal speed is wrong. Mars *is* closer to the earth than Jupiter or Saturn, but Mercury is farther from the earth than Venus, not closer.

This wondrous whirl of celestial bodies seemed to center on the earth, so ancient people naturally pictured the cosmos as geocentric. But retrograde motion posed a sticky problem. The explanation of this contrary motion was difficult in a geocentric cosmos and its true cause eluded astronomers for centuries. The evolution of our ideas about planetary motions plays a main theme in the next two chapters.

1.5 ECLIPSES

Eclipses, especially of the sun (Fig. 1.19) have always awed people. According to ancient historians, in 585 B.C. the armies of the Medes and the Lydians were locked in bitter combat. Suddenly the sun disappeared in an eclipse. This omen struck people so strongly that the armies ceased their fight and established a peace.

A less dramatic but still powerful impact occurred during the March 1970 solar eclipse. In New Haven, Connecticut, the eclipse appeared almost total. In the courtyard of a Yale college a group of black students had set up drums and were beating out deep, long rhythms. The combination of the drumbeats and dimming sun aroused an excitement and fear in the crowd, even in one (MZ) who was to become a professional astronomer.

Solar eclipses occur when the moon, as seen from the earth, passes in front of the sun (Fig. 1.20). Although the moon is much smaller than the sun, it's closer by just the right amount so that the angular diameters of the sun and moon in the sky are almost the

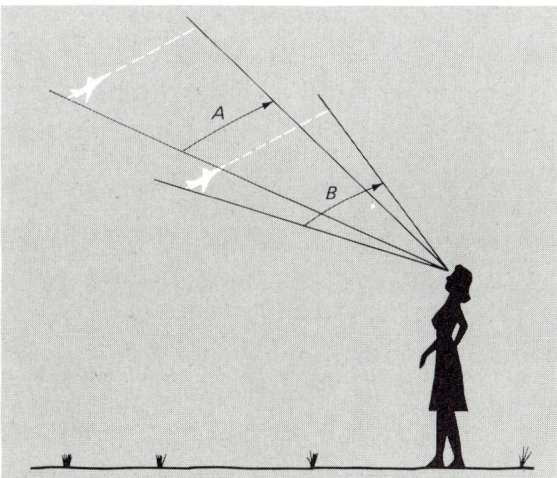

FIGURE 1.18 Judging relative distances from angular speeds. In the same time plane *A* covers a smaller angle than plane *B*. So plane *B* must be the closer, if both planes fly at the same speed.

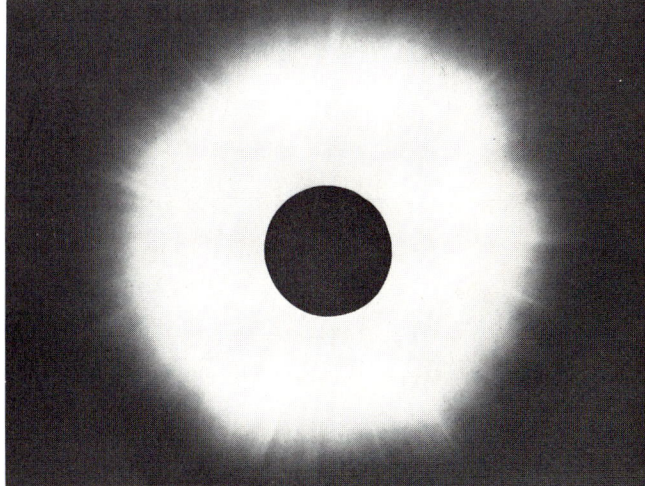

FIGURE 1.19 A total eclipse of the sun. As viewed from the earth, the moon covers the sun's visible disk, so just the sun's outer atmosphere is visible. [*Courtesy Yerkes Observatory.*]

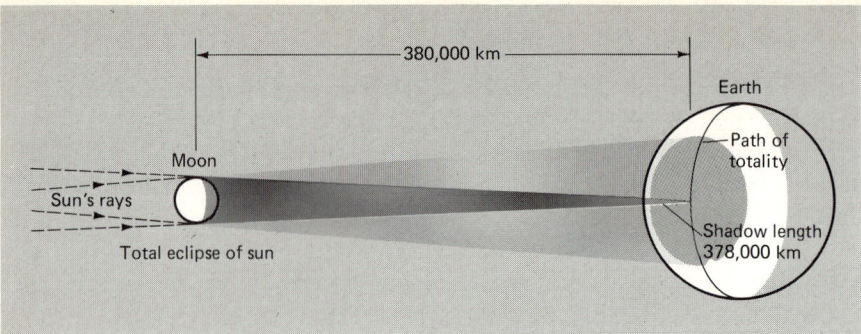

FIGURE 1.20 Alignment of the moon, sun, and earth for a total solar eclipse. The moon must be new and on or very close to the ecliptic. The length of the moon's central shadow is usually long enough to hit the earth; a total eclipse is seen in the path where the central shadow moves along the earth.

same—about $\frac{1}{2}°$ (one-half a fingertip). So the moon can just cover the sun's disk when it passes directly between the sun and earth, as it may do at new moon.

Why don't we have an eclipse every month? The moon's path in the sky relative to the stars does *not* coincide with the ecliptic, but is tilted at an angle of about 5° (Fig. 1.21). Only at or near the two points where the ecliptic and the moon's path intersect will the sun and moon pass close enough for a solar eclipse to occur. The points of intersection are called the *nodes*. Remember, the sun and moon each appear only $\frac{1}{2}°$ in angular diameter, so they must be within $\frac{1}{2}°$ of each other for the moon to block out at least part of the sun's disk.

The moon is at a distance such that its shadow just barely reaches the earth. When a total solar eclipse does take place, the moon's shadow on the earth is at most 300 km wide. Only people in the narrow band where the shadow sweeps across the earth can see a total solar eclipse. People just outside this band see only a partial eclipse.

An eclipse of the moon (or a *lunar eclipse*) occurs when the moon passes directly through the earth's shadow, so that the sun's illumination is cut off from the moon (Fig. 1.22). A total eclipse of the moon can take place only when the moon is full, that is, when it is on the opposite side of the earth from the sun, where the shadow is (Fig. 1.23). Again, the moon must also be close to the ecliptic; otherwise it will miss the earth's shadow.

Eclipses happen less frequently than the other celestial events described in this chapter. Their spectacular nature sets them apart and has motivated people to study them. The Greeks, for example, found that eclipses were predictable from the motions of the sun and moon in the sky. They also noted that solar eclipses demonstrate that the moon must be closer to the earth than the sun is and that therefore the sun must be larger than the moon.

If you've been puzzled about why the sun's path relative to the stars is called the

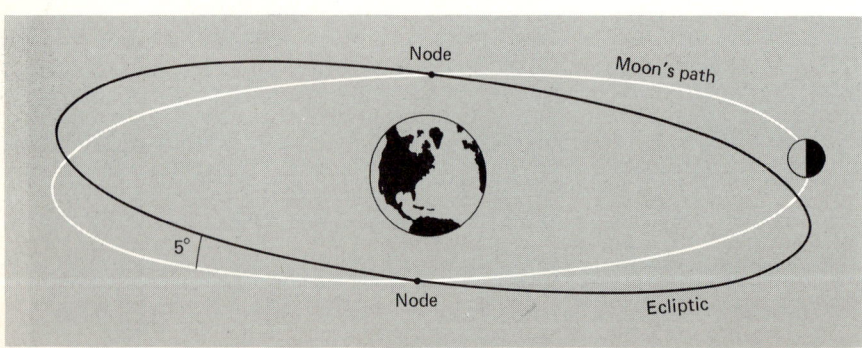

FIGURE 1.21 The position of the moon's path in the sky relative to the ecliptic. Note the moon crosses the ecliptic at two points; these are called *nodes*.

FIGURE 1.22 Time sequence of some stages of a total lunar eclipse (July 1982). From left to right, these individual exposures (made on the same negative) show the full moon emerging from the earth's shadow. [Courtesy Brian Walski.]

ecliptic, it should now become clear: Only when the moon lies on or close to the *ecliptic* can *eclipses* occur.

We mentioned briefly above that the sun and moon have the same angular diameters, because, even though the moon is smaller, it is closer than the sun. Let's examine more closely the relation between angular diameter, true diameter, and distance.

If you take a coin and move it farther away, it appears to get smaller; the angular diameter of an object decreases as its distance increases. Similarly, if you examine two different coins at the same distance, the one with the larger true diameter also has the larger angular diameter; the angular diameter of an object increases as its true diameter increases. These two common observations can be expressed in a simple equation. Let *s*

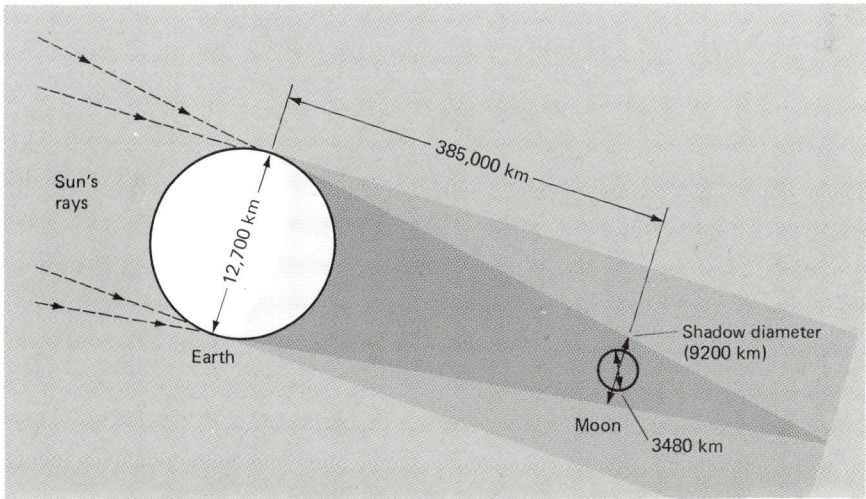

FIGURE 1.23 Alignment of the sun, moon, and earth for a total lunar eclipse. The moon must be full and on or close to the ecliptic. The total lunar eclipse is visible to everyone on the night side of the earth.

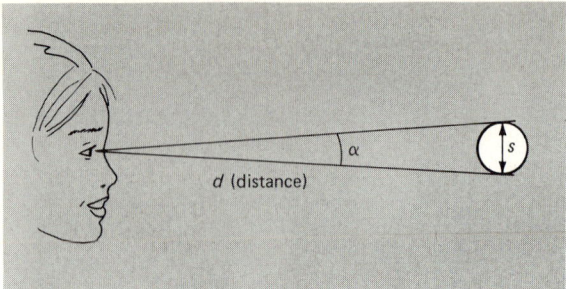

FIGURE 1.24 Angular size (α), distance (d), and true size (s) of an object.

be the true diameter of some object at a distance d; it has an angular diameter α (Fig. 1.24). The true diameter, distance, and angle are related by the equation

$$\tan \alpha = \frac{s}{d}$$

where tan is the trigonometric function *tangent*. Using this equation, you can calculate, for example, the true diameter of the moon. The moon's distance is about 384,000 km, and its angular diameter is roughly $\frac{1}{2}°$. Since $\alpha = \frac{1}{2}°$ and $d = 384{,}000$ km,

$$\tan (\tfrac{1}{2}°) = \frac{s}{384{,}000 \text{ km}}$$

$$0.0087 = \frac{384{,}000 \text{ km}}{s}$$

$$s = 3400 \text{ km}$$

Because the sun and the moon have the same angular diameter and the sun is farther away, it must be larger than the moon (see Problem 3).

Skinny Triangles

Many astronomical problems involve the so-called skinny triangle—a triangle with an angle (and side) very much smaller than the other sides and angles. With the small angles typical in astronomy (usually less than 1°) some simple approximations can be used for the common trigonometric functions, particularly the sine and tangent. To use these approximations, we must express the angles in *radian measure*.

Radians (rad) are a dimensionless measure of angles. They are related to degree measure by the fact that a circle (360°) contains 2π radians. This relationship arises from the definition of a radian: the angle at the center of a circle subtended by an arc along the circumference equal to the circle's radius. For example, suppose the radius of a circle is 1.0 (in some arbitrary system of units); this fits into the circumference 2π times. So the angle subtended by this radius is $2\pi/360° = 57.3°$ (rounded off to three significant figures), that is, 1 radian = 57.3°. Conversely, $1° = 1.75 \times 10^{-2}$ radians, $1' = 2.92 \times 10^{-4}$ radians, and $1'' = 4.86 \times 10^{-6}$ radians.

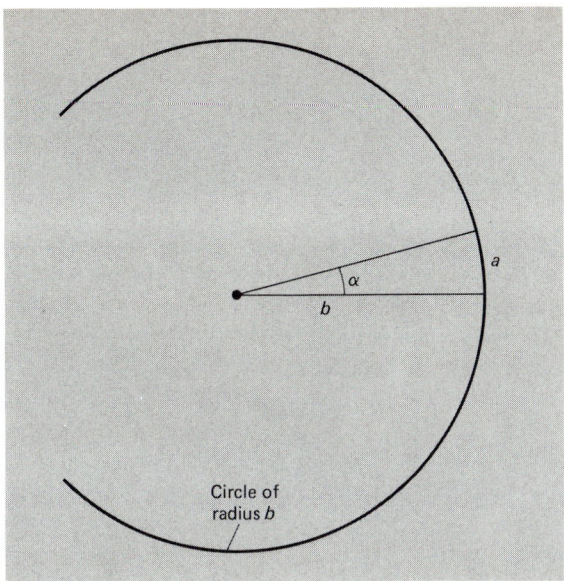

FIGURE 1.25 A skinny triangle as part of a circle.

If an angle, say α, is the small angle of a skinny triangle (Fig. 1.25), the side opposite this angle, a, to a good approximation is equal to the arc subtended by the angle on a circle with radius b, the length of the other side of the triangle. Then α in radians is simply the ratio of a and b or $\alpha = a/b$. This ratio is also the definition of the tangent of α, so

$$\tan \alpha = \frac{a}{b} \approx \alpha$$

As long as the side a is small compared to side b (which is another way of saying that angle α is small), the angle in radians is equal to the tangent of the angle. Similarly, the definition of $\sin \alpha$ is a/c. However, side c and side b are almost equal, so that

$$\sin \alpha = \frac{a}{c} \approx \frac{a}{b} = \tan \alpha \approx \alpha$$

The other common trigonometric function, the cosine, becomes even simpler, for since sides c and b are almost equal,

$$\cos \alpha = \frac{b}{c} \approx 1$$

So, if α is small, the sine and tangent are both equal to the angle expressed in radians, and the cosine is approximately equal to 1. These are simple relationships for the most often used trigonometric functions, valid for small angles.

At what angles do these approximations break down? That depends on the accuracy needed. A table of trig functions shows that at $10° = 1.745 \times 10^{-1}$ rad, $\tan 10° = 1.763 \times 10^{-1}$, $\sin 10° = 1.736 \times 10^{-1}$, and $\cos 10° = 0.9848$. The tangent and the angle differ by 1 percent, the sin and the angle differ by 0.5 percent, and the cosine differs from 1 by 1.5 percent. At 20°, the sine and the angle differ by 2 percent. So for a 1 percent accuracy the skinny triangle method can approximate angles as large as 10° or so.

1.6 OBSERVING THE SKY

It is hard to visualize the motions of the sun, moon, and planets in the sky simply from reading about them. The best way to understand these phenomena is to observe them for yourself. Try to detect the motions listed in Table 1.3 (in the summary at the end of the chapter). The daily motions are easy; the slower motions take more time, but noticeable changes can be detected in a few weeks. Observe the changing locations of the sunset or sunrise. You might also want to make your own gnomon (or sundial) and keep track of the shadow at noon for a few months.

If you are in the northern hemisphere, try to locate Polaris (Fig. 1.26). A line drawn between the stars in the end of the bowl of the Big Dipper points toward Polaris. Once you have found it, use your fist to measure the height of the star above your horizon. This angle is equal to your latitude. Draw a picture of your horizon, Polaris, and the stars of the Big Dipper. Wait a few hours and draw another picture. How have the stars of the Big Dipper moved? Has Polaris moved by any noticeable amount?

Observe the phases of the moon from night to night. Also try to find the moon in the daytime. Where is the moon relative to the sun at different phases? What time does it rise or set at each phase? When is it highest in the sky?

You might also want to find some of the brighter planets, such as Venus, Mars, Jupiter, and Saturn. Check your local newspaper or astronomy magazines, such as *Sky and Telescope* or *Astronomy*, to find out what planets are visible in the sky. Look for these planets when the moon is also visible. If you draw an imaginary line through the planets and the moon, you can get a rough idea of where the ecliptic is relative to the stars of the zodiac.

Draw a rough picture of the positions of the planets with respect to the stars. Wait about a week and make another sketch. What changes do you notice in the planets' positions?

Be careful to describe directions in the sky properly. If you look up at the sky while

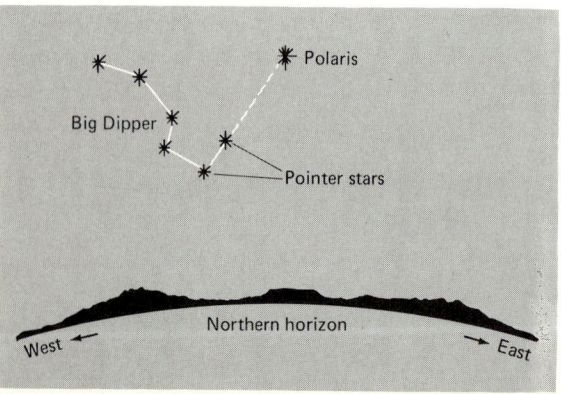

FIGURE 1.26 Pointer stars in the Big Dipper used to find Polaris.

facing south, as you would do to observe the moon and planets (in the northern hemisphere), north is up, south is down, east is to your left, and west is to your right. Because you are looking up, not down, the order of the compass directions is opposite to that on a terrestrial map: N, E, S, and W run counterclockwise.

Try to learn some constellations. Star charts are printed inside the covers of this book. To use them pick the chart for the month closest to the current date. Go outside and hold the chart so that the direction you are facing is at the bottom. The center of the chart represents the zenith (the point directly overhead), so as you look in different directions, rotate the chart around to keep the appropriate direction at the bottom. Other charts can be found in *Sky and Telescope* and *Astronomy* magazines.

1.7 PREHISTORIC ASTRONOMY: THE ANASAZI ACHIEVEMENT

How much astronomy did ancient peoples know? Certainly, they had dark skies unlit by local lights, and the celestial cycles presented one of the few entertainments available at night. Yet, to define the subtle rhythms of the heavens, to see the long cycles, takes years of observations and the means to record them. Prehistory leaves no written records. How did these people inform themselves about celestial motions?

They probably did so by word of mouth in the form of symbolic stories. Such communication leaves no records, yet fossils of oral traditions do exist. The Polynesian islanders knew how to navigate from Tahiti to Hawaii by stories of the sky. In the pueblos of New Mexico sun watchers still spin their seasonal tales. We are now uncovering many archeological sites that seem to have had astronomical purposes. One of the authors (MZ) visited one such site: Chaco Canyon in New Mexico. Here is his report.

Snow greeted me on my first night in Chaco Canyon: a late November storm that blew in a gray fury out of the San Juan mountains of Colorado to engulf the mesas of northwestern New Mexico. The next day I huddled in the ruins of Pueblo Bonito (Fig. 1.27a), the largest of the dwellings in the canyon. At its height in A.D. 850–1150, Chaco contained perhaps 10,000 inhabitants; 2,000 of these in Pueblo Bonito alone. As I toured the fascinating ruins, I came upon an upper-story room with a corner window high upon the south wall (Fig. 1.27b). How strange! Why would the Indians have put in such an unusual design feature?

I didn't think about that window again until I came across an article by Jonathan Reyman, who argued that careful solar observations must have formed an important aspect of the farming strategy in prehistoric Chaco. As careful attention to the seasons was critical to successful crops, the former inhabitants of Chaco may have introduced astronomical alignments in their buildings. Reyman pointed out two rooms in Pueblo Bonito that had high, exterior corner windows—well placed for watching the sun rise on the winter solstice [Fig. 1.27c]. (At the Zuni pueblo today, the winter solstice marks the start of the seasonal year.) Reyman set up cameras to photograph the winter solstice sunrise from both rooms and discovered that the sun rose in the centers of both, parallel to the jambs, as the lower part of the sun just appeared to touch the horizon. So from these rooms, early sun watchers in New Mexico could have spied the winter solstice and then announced the event to all.

FIGURE 1.27 The southeastern corner of the ruins of Pueblo Bonito in Chaco Canyon just after dawn on the winter solstice, 1981. (a) Corner windows (arrows) in the second story align with the winter solstice sunrise. (b) Sunlight from the window on the left in (a) hitting the corner of the interior room where a student photographs the view out the window. (c) Sunlight streaming through the window on right in (a), about half an hour after dawn on the winter solstice. [*Photos by M. Zeilik.*]

Ray Williamson, who has investigated Chaco and other southwestern sites for solstice markers, has noted that Pueblo Bonito has other solstice alignments. Rob Roosen and I have noticed that the main axes of the largest kiva in Pueblo Bonito is aligned to a true north-south line (toward Polaris). The Anasazi ("Old Ones" in Navajo) appear to have had a firm grasp of astronomical knowledge and practice.

A new discovery has recently come to light on a new aspect of Anasazi astronomy: a probable *noon* marker of the summer solstice. At the southern end of Chaco Canyon erupts the sandstone mass of Fajada Butte (Fig. 1.28). In June of 1977 Anna Sofaer, an American artist interested in Indian rock art (petroglyphs), climbed past the rattlesnakes to the top of the butte. There she found three large rock slabs (each about 3 m high) that allowed a dagger-shaped beam of sunlight to fall on a spiral petroglyph on the rock wall behind the slabs (Fig. 1.29).

The time was near noon (and it was also near the summer solstice). The dagger landed near the spiral's center. Suspecting this to be a solstice marker, Sofaer informed Volker

1.7 PREHISTORIC ASTRONOMY: THE ANASAZI ACHIEVEMENT

FIGURE 1.28 Fajada Butte in Chaco Canyon, viewed from the northeast. The noon solar marker is at the top (arrow) near the center. [*Photo by M. Zeilik.*]

FIGURE 1.29 Three rock slabs, each about 3 m long, against the rock wall at the top of Fajada Butte. [*Courtesy W. Wampler.*]

Zinser, an architect, and Rolf Sinclair, on the staff of the National Science Foundation. They hiked up the butte on the solstice and discovered that at noon the light dagger slashed through the center of the spiral (Fig. 1.30). The flat rocks and spiral make a precise marker of noon at the summer solstice. By the clever manipulation of sunlight, the Anasazi showed their careful knowledge of the sun's motion in the sky.

We will return to Chaco Canyon in Chapter 11, where we show evidence of a possible astronomical record of the dramatic explosion of a star.

FIGURE 1.30 A close-up of the noon sunlight dagger, on the day after the summer solstice, on the spiral petroglyph. [*Courtesy W. Wampler.*]

SUMMARY

Simple observational astronomy leads to ideas of space and time. The heavens appear far away from the earth, in a distinct, eternal realm of their own. The motions of the sun and moon in regular cycles set the fundamental time periods of day, month, and year. The stars appear fixed and unchanging; they form distinct patterns called constellations. The backdrop to the constellations is the celestial sphere, the apparent dome of the sky.

The most important constellations are the twelve through which the sun journeys each year; these make up the zodiac. With reference to these stars, the sun annually traces a complete circle in the sky. This path is called the ecliptic. The moon and visible planets hug the ecliptic as they travel eastward with respect to the stars. However, the planets regularly slow down, halt their eastward motion, and move westward for a period of time. They eventually stop their westward motion and move eastward again. This contrary, westward motion of the planets is called retrograde motion. It is unique to the planets and so sets them apart from other celestial objects. (See Table 1.3 for a summary of visible celestial motions.)

TABLE 1.3 A Summary of Major Celestial Motions Visible Without Optical Aid

Object	Daily Motion	Long-Term Motion
Sun	E to W in about 12 h from sunrise to sunset. Day length varies from season to season.	W to E along ecliptic 1° per day. Height of sun in sky at noon is maximum in summer, minimum in winter.
Moon	E to W in about 12 h 25 min from moonrise to moonset. Moonrise is about 50 min later each day.	W to E within 5° of ecliptic. It takes 27.3 days to travel 360° relative to the stars. Phases repeat in cycles of 29.5 days.
Planets	E to W in about 12 h from rising to setting.	W to E within 7° of ecliptic. Time around ecliptic varies, shortest for Mercury and longest for Saturn. Retrograde motion from E to W at a time specific to each planet.
Stars	E to W in about 12 h from starrise to starset. Starrise is about 4 min earlier each day. Circumpolar constellations never set; their motion centers on the celestial pole.	In fixed positions with respect to each other. Relative to the sun, a constellation returns to the same position in 1 yr. Position of the celestial pole changes slowly, returning to its initial position in about 26,000 yr (Focus 1.1).

In any picture of the cosmos the retrograde motions of the planets emerge as the most difficult motions to explain. The question of how the planets move baffled astronomers for centuries and will be dealt with as a central theme in the next three chapters.

Key Words

celestial sphere	noon	elongation
constellation	solar day	retrograde motion
circumpolar stars	solstice	nodes
celestial pole	equinox	maximum elongation
Polaris	skinny triangle	conjunction
celestial equator	phases of moon	opposition
angular separation	eclipse	geocentric
angular diameter	zodiac	gnomon
angular speed	ecliptic	radian measure
	precession	

Review Questions

1. a. You go outside one night about 9:00 P.M., and face south. The moon is off to your right and near the horizon. Is it rising or setting?
   b. The next night you go out again at 9:00 P.M. Where is the moon? Is it higher, lower, or not up at all? Did it move east or west?
   (Objective 1)
2. Which of the following are possible uses of a gnomon? (Mark all correct answers.)
   a. To indicate that a year has gone by
   b. To indicate that another day has slipped away
   c. To predict the moon's rising time
   d. To indicate south
   e. To indicate north
   f. To indicate when the summer solstice occurs
   (Objectives 3–5)
3. What *two* reasons did the ancient Greeks have for believing (correctly) that the moon was closer to the earth that the sun? (Objective 9)
4. Indicate whether the following statements about astronomical events are always true (A), sometimes true (S), or never true (N).
   a. Mercury is near the ecliptic during retrograde motion.
   b. Venus is in opposition to the sun at full moon.
   c. About a month after an eclipse of the sun, the moon is new.
   d. Jupiter is just rising as the sun sets.
   e. Circumpolar stars rise in the east and set in the west.
   (Objectives 2, 3, and 6)
5. Tell how you would find the ecliptic and the zodiac in the sky. (Objectives 1 and 2)
6. What celestial bodies never exhibit retrograde motion? (Objective 1)
7. Into what two groups can the planets be divided on the basis of their retrograde motion? (Objective 2)

8. When Mars is at opposition, at about what time will the planet rise? Set? (Objective 2)
9. Why don't we have an eclipse of the moon every full moon? (Objective 5)
10. Where is Saturn in the sky with respect to the sun, as seen from the earth, when it goes through retrograde motion? Venus? (Objective 2)

Problems

1. What is the angular speed (in degrees per day) of the sun's eastward motion relative to the stars? The moon's eastward motion? (Objectives 1 and 7)
2. What is the angular size of the sun? The moon? What do solar eclipses tell you about the relative distances of the sun and moon? (Objectives 5 and 6)
3. The distance of the sun is about 150 million km. Use the skinny triangle method to calculate the sun's true diameter. (Objective 7)
4. A stick 10 cm high casts a shadow 3.5 cm long at noon. What is the sun's height above the horizon? (Objective 4)

BEYOND THIS BOOK . . .

* *Astronomy* and *Sky and Telescope* magazines contain monthly star maps and planetary locations to help you observe.
* *Hamlet's Mill* (Gambit, Boston, 1969) by G. de Santillana and H. von Dechend has an intriguing analysis of the astronomy of preliterate people incorporated in oral myths.
* In *The Roots of Civilization* (McGraw-Hill, New York, 1972), A. Marshack argues that the need to keep track of time, especially the cycle of the month, led to the development of symbolic notation and language.
* For a return to Stonehenge and a look at other possible ancient astronomical observatories, read *Beyond Stonehenge* (Harper & Row, New York, 1973) by G. Hawkins. An excellent summary about prehistoric sites in Great Britain and Europe is in *Sun, Moon, and Standing Stones* (Oxford University Press, New York, 1978) by J. E. Wood.
* For articles on Chaco Canyon, see "Solstice-Watchers of Chaco" by K. Frazier, *Science News*, vol. 114 (August 26, 1978), p. 148; "Astronomy, Architecture and Adaptation at Pueblo Bonito" by J. E. Reyman, *Science*, vol. 193 (September 10, 1976), p. 957; and "The Anasazi Sun Dagger" by K. Frazler, *Science 80*, vol. 1 (November–December, 1980), p. 56. A general reference is *Native American Astronomy* (University of Texas Press, Austin, 1977), edited by A. Aveni.
* A good book to help with finding objects in the sky is *Whitney's Starfinder* (Knopf, New York, 1974).

THE BIRTH OF COSMOLOGICAL MODELS

2

LEARNING OBJECTIVES
After studying this chapter, you should be able to:
1. Describe the essential characteristics of a scientific model and evaluate astronomical models in light of these ideas.
2. List two important astronomical achievements of the Babylonians.
3. Illustrate the distinction between a mythical and a mechanical cosmological model with specific examples from ancient astronomy.
4. Use at least one specific case to show how geometrical and aesthetic concepts influenced Greek ideas about the cosmos.
5. Describe the Aristotelian cosmological model in terms of its physical reasoning and explanations of astronomical observations.
6. Use at least two examples to show how simple models and astronomical observations lead to new information about the cosmos.
7. State the assumptions and physical ideas behind Ptolemy's model of the cosmos.
8. Sketch the Ptolemaic model for the inner planets (Mercury and Venus), the sun, and the outer planets (Mars, Jupiter, and Saturn), and show how the epicycle and deferent explained retrograde motion.
9. Describe how each geometrical part of the Ptolemaic model explained a specific astronomical observation.
10. Evaluate the assets of the Ptolemaic model which led to its wide acceptance.
11. Explain how the lack of an observed annual stellar parallax supported a geocentric model.

CENTRAL QUESTION:
What is a scientific model, and how did early cosmological models cope with astronomical observations?

In the previous chapter we described what can be observed in the sky without optical aid, specifically the motions of the sun, moon, and planets. In this and the next three chapters we will discuss the historical development of models to explain these facts.

Most ancient cultures viewed the cosmos as finite and geocentric. Some boundary, usually a shell of stars, closed off the universe with the earth enthroned in the center. Usually ancient cosmologies paid little attention to the details of celestial motions, even if the cycles were carefully observed, as was done by the Babylonians.

The Greeks were the first to attempt to take their cosmological ideas beyond the skeleton of a finite, geocentric cosmos. Grappling with the problem of planetary motions, Greek philosophers fleshed out the bare cosmological structure with mechanical devices to account for the celestial cycle. These schemes marked the first earnest models: mental constructions that exhibited features like those observed in nature. Such systematic efforts to explain the natural world culminated in the geocentric model of Claudius Ptolemy. His cosmological model marks the first careful effort to represent accurately the observed celestial motions. So well did Ptolemy succeed that his system was not seriously challenged for over fourteen centuries.

This chapter examines Babylonian and Greek cosmologies (1) to see how closely early cosmological ideas related to actual astronomical observation; (2) to see how these cultures tried to make sense of what they saw in the sky; (3) to contrast cosmic ideas grounded in myths to those grounded in aesthetic, physical, and geometrical ideas—those which are scientific models; (4) to investigate the birth of the first comprehensive model of the universe, one that eventually led to the models used today.

2.1 SCIENTIFIC MODELS

Chapter 1 described naked-eye astronomical observations you can make from the earth. As you read the text or actually studied the sky, you may have had the urge to find a grand design in these observations, a model for the operation of the heavens, even though we tried to present the observations in a way that was not tied down to any explanation of heavenly motions.

Most people share this natural drive to make sense of what they see, and astronomical observations drove people to create models of the cosmos—conceptual plans that attempt to explain what is seen in nature. A model attempts to come to grips with a seemingly chaotic world by casting it in familiar terms. It strives to make sense of what is observed. The earliest models tended to be mythical. Like all natural things, the concept of a scientific model sprung from these older visions and evolved as we discovered more about our world.

A scientific model is a mental picture based on geometrical ideas, physical concepts, and aesthetic notions, which attempts to explain what is seen in nature. What goes into constructing a scientific model (Fig. 2.1)? First, our sense impressions about the world provide the raw information that sparks our curiosity. Then we try to make sense of this input by using what seem to be intuitive notions: geometry, physics, and aesthetics. Ideas from geometry establish a visual framework for the model. Physical ideas attempt to cope with the motions and interactions of the various parts of the model. And aesthetic ideas—gut judgments of what is appealing—select the simplest, most pleasing models from the flock a fertile mind manufactures. Then comes the crucial test: How well do the features of the model correspond to what can be seen in nature? If this correspondence is good (within the errors of observation), we have a growing confidence that the model is appropriate. If not, we change various features of the model to get a better fit.

A scientific model has two key functions: (1) It *explains* what is seen, and (2) it *predicts* what can be seen. The explanations must be anchored in basic physical ideas. The predictions must relate directly to observations, and they must do so with sufficient accuracy to be convincing. Explanation and prediction are common to all good scientific models. (One, however, may overshadow the other.)

The power of prediction prompts the drive to confirm a model by finding out how well the predictions fit observations. The art of confirmation can make or break a model, depending on how well observations and the model's predictions agree.

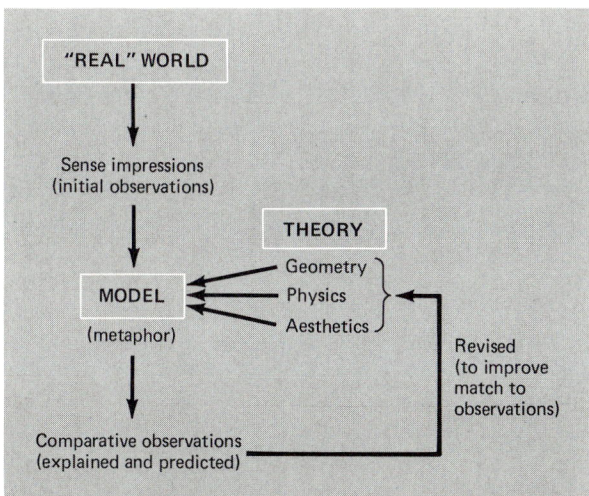

FIGURE 2.1 A schematic flowchart of the process of scientific model making.

2.2 BABYLONIAN SKYWATCHING: THE SEEDS OF A SCIENCE

Washed by time and rivers, Mesopotamian civilizations came to flood and ebb in a succession of cultures, each enriched by its predecessors. The oldest was that of the Sumerians, a people who used clay tablets, bearing notations in the cuneiform alphabet, to preserve written records. When Hammurabi of Babylon (1725–1686 B.C.) rose to power, the fabled Babylonians adopted much of the culture, mythology, and technology of the Sumerians.

Babylonian Astronomy

About 1600 B.C. the Babylonians compiled the first star catalogs and began making crude records of planetary motions. By 800 B.C. the Babylonian astronomers were able to fix planetary locations with respect to the stars. They kept records of planetary positions that compared a planet's position with that of recognized constellations. Their early observations included the motions of Venus, Jupiter, and Mars. From the ninth century B.C. the Babylonians kept continuous astronomical records on clay tablets.

For what reasons did the Babylonians become careful observers? In part, the development of astronomy relied on state support as far back as the time of Hammurabi. Astronomical information was needed for both the calendar and the practice of astrology (Focus 2.1). These problems stimulated the development of arithmetical techniques that the Babylonians used to predict planetary positions. In addition, observations preserved on cuneiform tablets enabled the Babylonians to find the daily, monthly, and annual cycles—the main themes of celestial motions. (These are listed in Table 1.3.)

Such permanent records indicate that the Babylonian astronomers also knew of the variations in the celestial cycles. For example, the angular size of a planet's retrograde loop and the duration of its retrograde motion vary from one time of opposition to the next (Fig. 2.2). The Babylonian astronomers had lists of these cycles, with each major cycle represented by a table of consecutive numbers. The departures from the average cycle were also tabulated. A Babylonian astronomer could sum the set of numbers of all major and minor cycles to predict, for example, the next time of retrograde motion.

So Babylonian astronomers could predict future planetary motions from their tables of past cycles. This procedure did not require an explanation of the cycles, merely a knowledge of their existence over a long period of time.

Babylonian Cosmology

In Babylon, the priests were the astronomers. They occupied the holy ziggurats, towers that also served as observatories. The existence of the priest-astronomer fostered the continuity of astronomical knowledge, but it also divorced Babylonian cosmology from astronomy. In the cosmic picture the gods created, ordered, and controlled the world, and these divine functions were considered far beyond human comprehension. Consequently, the greatest store of observed knowledge was held by men who believed that it was not explicable except as religious myth.

2.2 BABYLONIAN SKYWATCHING: THE SEEDS OF A SCIENCE

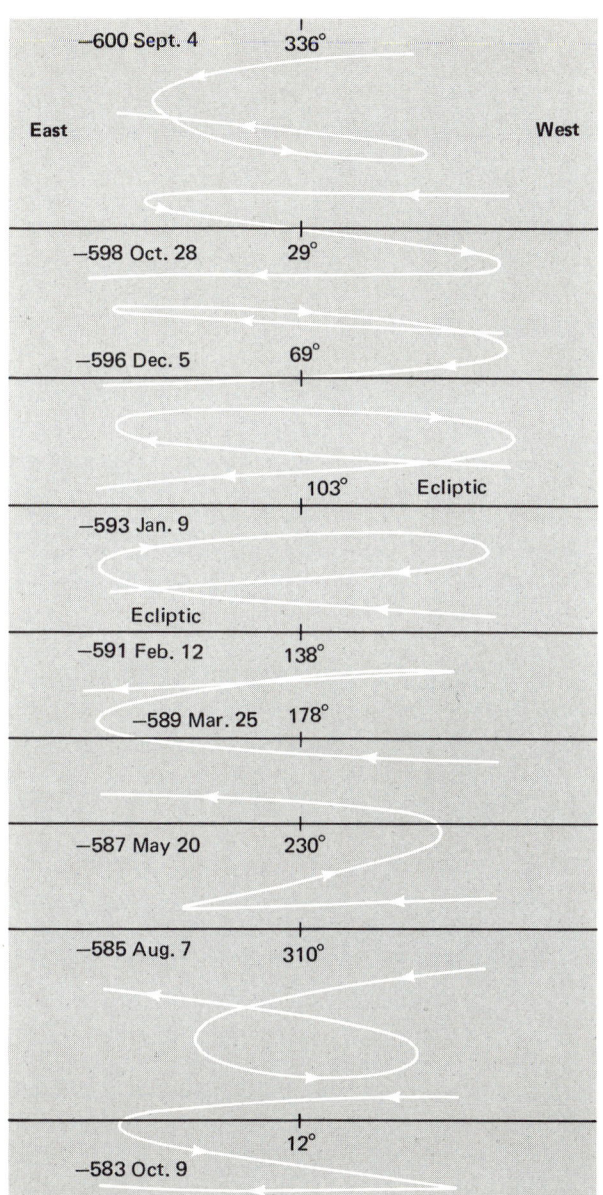

FIGURE 2.2 Retrograde motions of Mars in Babylonian times. This figure shows Mars at different positions along the ecliptic (indicated in degrees) for nine different times of retrograde motion. Note how the shape and size (and so the duration) of the retrograde loops vary. The motions shown here are relative to the stars; east is to the left, west to the right. [*Courtesy O. Gingerich.*]

The fantastic Babylonian tale of genesis, the *Enuma Elish* (literally, "when above"), carefully pictures the creation of space and the ordering of time by the god Marduk. The details of astronomical motions pale before the grand scheme. The actual predictions were purely arithmetical; no geometrical framework provided any skeletal support to the natural appearances. The Babylonian astronomers knew well the periods of the planets, sun, and moon through the zodiac and the occurrence of retrograde motion. They did not, however, attempt to account for the causes of these motions beyond a religious, mythical explanation. They were able to predict, but not to explain, in the modern sense

FOCUS 2.1 ASTROLOGY

Some ancient cultures believed that the planets were actually gods and that these gods had powers over people and events on earth. Such beliefs crystallized in the practice of astrology, which sought a reasonable connection between celestial and terrestrial events.

Both astronomy and astrology start with the recognition of cycles in celestial events. The early Babylonians practiced astrology to understand how the planets influenced the fate of rulers and nations. Later the Greeks gave birth to the idea that each person, not just the mighty, fell under the influence of the stars, especially their positions at the time of his or her birth. This belief and practice is called *natal astrology*.

Ptolemy's clever scheme for predicting planetary positions was devised in part for astrological purposes. His writing, the *Tetrabiblios*, synthesized the astrological knowledge of the day and established the traditional horoscope interpretations that most western schools of astrology rely upon. (The *horoscope* charts the planetary and zodiacal positions for the time and place of an individual's birth and sets the keystone of natal astrology; Fig. F.3.)

The list of astronomers who practiced astrology—whether for money, prestige, or personal curiosity—includes Ptolemy, Tycho Brahe, and Johannes Kepler. For Kepler, who divined the structure of the cosmos in the harmony of the spheres, astrology struck an inner chord to which he responded with mixed feelings of belief and skepticism. He thoroughly spurned customary rules of the astrology of his day, which derived from the work of Ptolemy. Yet Kepler did advise the use of astrology for political purposes. Although financial gain strongly motivated Kepler, personal beliefs also compelled him to cast over 1800 horoscopes. He wrote:

> That the heaven does something in people one sees clearly enough; but what it does specifically remains hidden.

The scientific revolution in Europe in the seventeenth and eighteenth centuries demolished most scientists' belief in astrology. Almost all modern astronomers view astrology as at best a pseudoscience and at worst an outright fraud, if they bother to consider it at all. In the context of modern scientific attitudes, astrology cannot stand up as a science because astrologers must depend on unknown forces, exerted by the planets in geocentric lineups, to influence the destiny of human beings. Since the time of Copernicus (Section 3.1) the earth has been removed from its former geocentric glory. Astronomers suspect geocentric claims because the universe, as they see it today, has no unique central place for the earth.

---

of a scientific explanation. Their ideas lacked the notion of physical cause—a concept central to modern science. Instead, religious myths secured the structure of their world.

2.3 GREEK MODELS OF THE COSMOS

The Greeks took number and geometry seriously, but they paid only passing heed to careful observational work such as the Babylonians did. Yet the Greek philosophers accomplished what the Babylonian astronomers probably never dreamed of: They invented

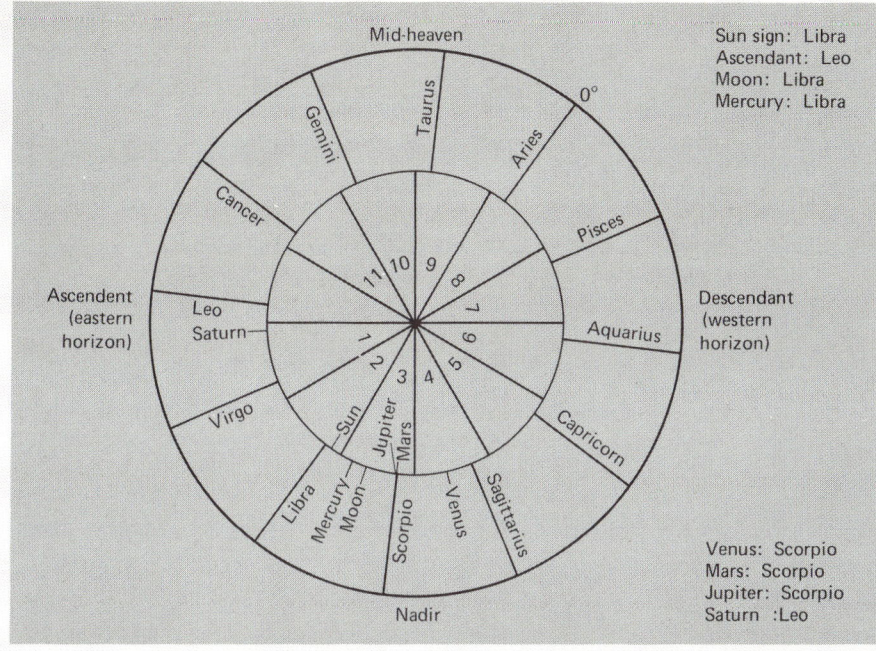

FIGURE F.3 A simplified horoscope for M. Zeilik, showing the stars and planets relative to the horizon at his birth.

Does astrology have a scientific basis? At present we must say no. No astrological school we're familiar with has demonstrated a link-by-link chain of physical influence between heavenly bodies and the earth. There are dedicated astrologers who are searching for such links, however.

Don't take a denial of a scientific basis to modern astrology as a rejection of the idea that heavenly bodies have influences on the earth. They do. The most obvious are the natural cycles of day, month, and year to which all living creatures respond. In addition, the moon produces tides, and the sun appears to affect our climate in direct and subtle ways that we are just beginning to understand.

Despite today's rejection of astrology by astronomers, the historical fact remains that astrology motivated the early development of astronomy. (It also provided the bread-and-butter employment for people like Kepler!) And certainly, the posture of astrology—to place us in an ordered universe, in a cosmos rather than a chaos—is akin to the cosmic perspective of modern astronomy.

a geometrical, physical model of the universe. The Babylonians made sense of the world through myths. The Greeks had myths, too, but their philosophies drove them to develop models to understand reality. Let's look at Greek astronomy and cosmology to see how the idea of a scientific model was born.

The Music of the Spheres

Numbers reveal the nature of things. This theme reverberates throughout western thought. The Pythagoreans, for example, found harmony in numbers, especially 1 and

10. They believed that these sacred numbers generated all other numbers, which made up the *kosmos*, the "good array." The word we use today to describe an ordered universe—cosmos—derives directly from the Greek *kosmos* and conveys a sense of harmony and symmetry. These aesthetic notions are based on a faith of order in nature.

The Pythagorean picture of the cosmos (Fig. 2.3) was the first to incorporate some aspects of a scientific model. According to the Pythagoreans, the earth had a spherical shape. This figure was supported not only from arguments of symmetry, but also from the shape of the earth's shadow on the moon during lunar eclipses (look back to Fig. 1.22). The stars were fixed to a spherical shell, enclosing the cosmos, and smaller spheres carried the planets around. The stellar sphere, carrying the other spheres with it, rotated daily east to west, and the planetary spheres within rotated west to east with the period of each planet's circuit time of the zodiac (Table 1.3). The earth was not placed in the center of the cosmos, but moved about a central fire once every 24 hours. (This fire was *not* the sun.)

Note that the Pythagorean model contains geometrical and aesthetic elements and a crude correspondence to observations. But, it lacks physical ideas, and it does not cope with matching observations in careful detail. For example, the model fails to account for retrograde motion.

In their model the Pythagoreans relied strongly on the aesthetic notions of harmony

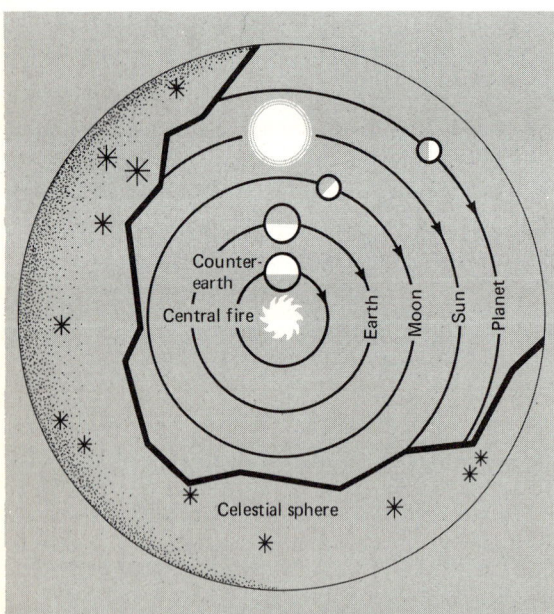

FIGURE 2.3 An artist's representation of the Pythagorean model of the cosmos. From symmetry, the Pythagoreans argued that the shapes of the celestial bodies and the entire cosmos must be spherical. Note that the earth is *not* in the center; rather, it revolves around a central fire (not the sun) once in 24 hours. [*Adapted from a figure by R. E. Ridley.*]

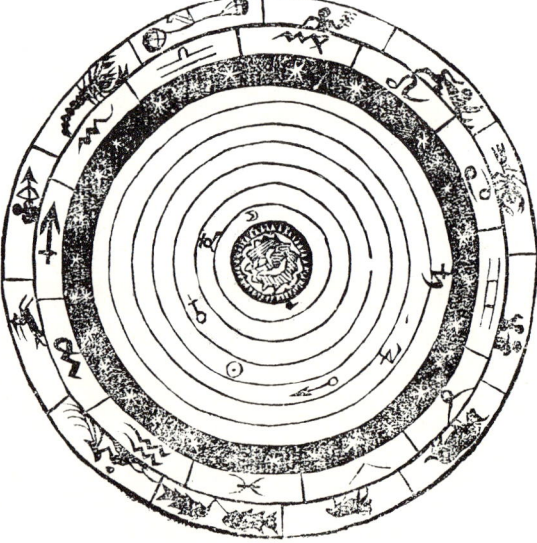

FIGURE 2.4 A geocentric model of the cosmos in the tradition of Aristotle. This is a medieval picture of Aristotle's system of geocentric spheres. The earth lies in the center; above it are the natural realms of water, air, and fire below the sphere of the moon. Beyond the moon are the heavenly spheres to which the planets and the stars are attached. Only one sphere for each planet is shown here; the actual scheme used a number of spheres to account for all the motions of each planet. [*Courtesy Yerkes Observatory.*]

and symmetry. These principles recur in cosmology—the study of the universe—even today.

The ideas of the philosopher Plato (428–348 B.C.) derived from Pythagorean concepts. His questions concerning astronomy painted an influential cosmological picture. Plato saw the perfection of the universe in the form of a three-dimensional sphere. In keeping with this intuitive harmony, all motions of the heavenly bodies must be composed of *uniform, circular motions*. Plato succinctly described the problem of the planets: The goal of an astronomical model was to "save the appearances," to devise a model that explained natural phenomena such as the motions of the planets. This goal preoccupied astronomers for centuries.

A Mechanical Model

Eudoxus (ca. 370 B.C.), a friend and student of Plato, devised the first geometrical system based on uniform, circular motion to account for the celestial cycles. He synthesized Greek astronomical knowledge into a scheme that attempted to save the appearances, as Plato had demanded. He saw that just a few heavenly spheres could not account for the complicated behavior of the planets, especially their retrograde motions. His ingenious model required 27 spheres centered on the earth.

Although Eudoxus's scheme did not reproduce the actual retrograde patterns well, his model stands as the first serious effort to construct a geometrical cosmos that exhibited some imitation of actual planetary motions. Even more important, Eudoxus's spheres directly influenced the astronomical ideas of Aristotle (384–322 B.C.), the most famous of Plato's pupils.

Aristotle added 28 more spheres to Eudoxus's basic system. Even then, his model (Fig. 2.4) described poorly the intricate celestial ballet. Although Aristotle was unhappy with his model in this respect, he was pleased because the astronomy fitted comfortably into his aesthetic and physical ideas about the universe. Let's see what these were.

In *De caelo (About the Heavens)* Aristotle anchored his model in physical principles about motion. (This use of physics made his work more scientific in the modern sense of the word.) His cosmos had two distinct realms—the heavens and the earth—in which *different* physical laws of motion applied. Below the sphere of the moon was the corruptible region of the universe; above was the incorruptible heaven made of the *quintessence*, an immutable, transparent substance that formed the heavenly spheres. Aristotle proposed the idea that all corruptible bodies were made of the four basic elements: earth, air, fire, and water. Each of these elements had its own natural motion toward its natural place in the universe: earth to the center, fire to the greatest heights, air below fire, and water between earth and air.

In the incorruptible realm of the heavenly spheres the natural motion was to rotate. *No forces* were needed to keep the planets moving in their paths. In the terrestrial realm the nature of motion was quite different. Except for the natural downward motion of earth and upward motion of water, air, and fire, motion required the application of a force to persist. For example, in order to keep a cart moving, a person must push it. When the pushing ceases, the cart rolls a bit, but eventually stops because no force is being applied. Aristotle called such motion *forced motion* in contrast to the *natural motion* of the four elements. Aristotle's idea that forces were needed for terrestrial motions, but not for celestial motions, was the crux of his physical reasoning.

From this foundation Aristotle reasoned that the earth must be stationary and in the center of the universe. First, the natural motion of earthly material is to fall toward the center of the universe. This explains the central location of the earth. Second, if the earth moved, bodies thrown upward would not fall back to their points of departure. Yet "heavy objects, if thrown forcibly upwards in a straight line, come back to their starting places," so the earth must be stationary. (Note that these arguments spring from Aristotle's basic understanding of natural motion. He coupled physical ideas to aesthetic ones to arrive at a cosmological conclusion.)

A model gains acceptance when it leads to the explanation of observations. In this respect Aristotle's finite, geocentric universe was a successful model. Two important observations which the model explained were the lack of an annual parallax of stars and the spherical shape of the earth. Aristotle noted that if the earth moved around the sun, the stars (which are a finite distance away) must display an annual shift in position, called *parallax*. No one observed this change, so Aristotle concluded that the earth did not move around the sun. (You will see in Chapter 8 that there *is* a parallax, that the earth does move around the sun, but this parallax is too small to be detected with the naked eye.) Aristotle's model explained the facts as they were known at the time.

Let's look at this parallax concept in a bit more detail. *Stellar parallax* is an apparent shift in the positions of stars. In a heliocentric model it arises from the earth's revolution around the sun, so it is usually called *heliocentric parallax*. The details of heliocentric parallax differ depending on whether the stars in space are confined to a thin shell (as in the Greek picture) or are spread more or less throughout space (as in modern ideas).

Consider what happens if the stars are stuck in a thin shell that closes off the universe, as in the heliocentric picture of Aristarchus (Fig. 2.5). Pick out two stars close together on

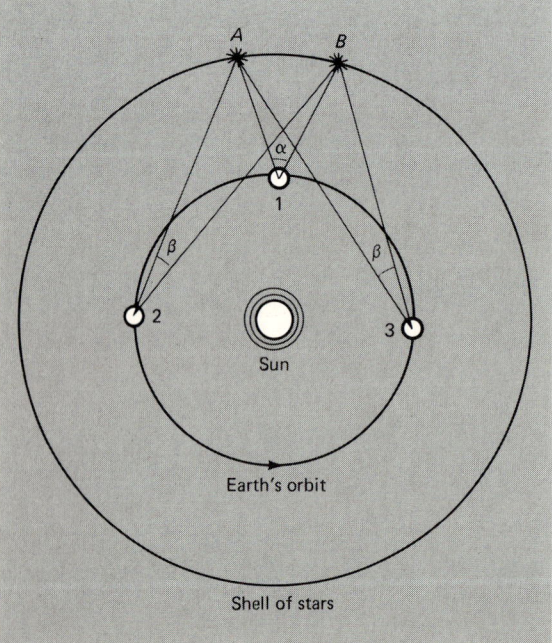

FIGURE 2.5 Stellar parallax in a finite, heliocentric model. A and B are two stars fixed in the stellar sphere. The earth moves from 2 to 3 in 6 months.

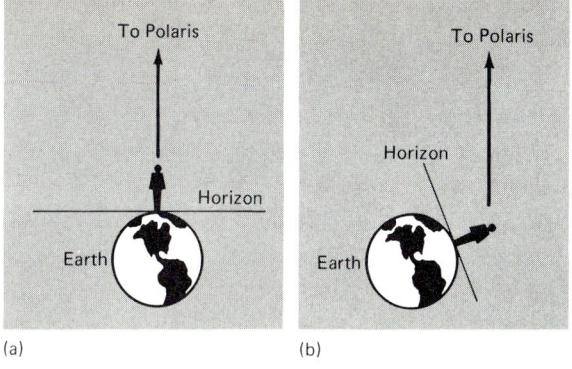
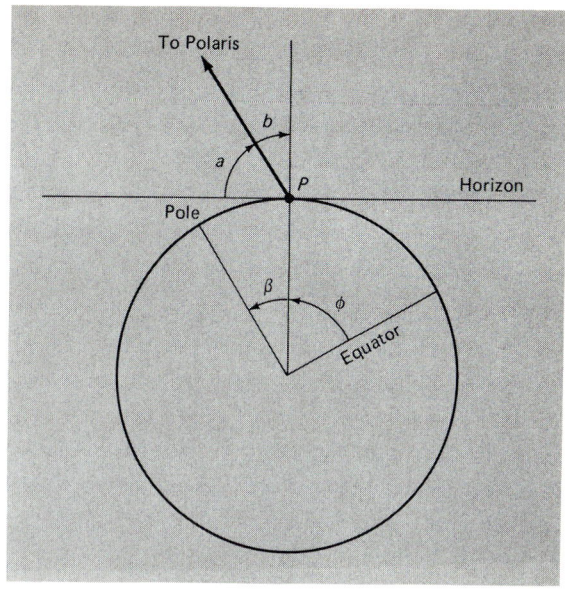

FIGURE 2.6 The relationship between latitude on the earth and the angular height of Polaris above the northern horizon. (a) At the North Pole Polaris is overhead. (b) At latitudes south of the North Pole Polaris is closer to the northern horizon.

FIGURE 2.7 The basic geometry for latitude and height of Polaris above the northern horizon.

the celestial sphere (A and B in Fig. 2.5). Observe them at midnight (position 1 in Fig. 2.5); they will appear some angular distance, α, apart. Just after sunset three months later, observe the stars again (position 2 in Fig. 2.5). Their angular separation is now β. They will appear closer together (angle β is less than angle α), partially because you're now seeing them at an angle, rather than face on. If you had observed them three months before your first observation (position 3 in Fig. 2.5), their angular separation would have again been β.

Now imagine that you observe these stars in a six-month cycle (from positions 3, 2, and 1). You will see the stars close together (β), then farther apart (α), and then close together again (β). This cyclical shift (from β to α to β) is the heliocentric parallax. Note that the greater the distance to the stars, compared with the earth-sun distance, the smaller the observed parallax.

Following Pythagorean ideas, Aristotle argued that the earth must be spherical in shape. He noted that at every total eclipse of the moon, the earth's shadow on the moon had a circular edge. Only a spherical earth could produce such a shape.

He also had heard from travelers that the angular height of the north pole star, Polaris, above the northern horizon varies as you travel north or south. If you travel due north, Polaris climbs higher relative to the northern horizon (Fig. 2.6). If you move south, it sinks closer to the northern horizon. Aristotle concluded that the curvature of a spherical earth causes such effects.

If you stand at the north pole, Polaris is directly over your head—an angular height of 90° above the horizon. As you go south, Polaris moves closer to the northern horizon until at the equator it sits just on the horizon, due north. More generally, at any point on the earth the angular height of Polaris equals the latitude. In Fig. 2.7, P is a point on the surface of the earth, the angle ϕ between a line from the center of the earth to P and a line to the equator is the latitude, and the angle a is the altitude of the celestial pole. The angle $a + b$ is 90°, because the horizon is perpendicular to a line from the center of the earth. The angle $\phi + \beta$ is 90° because the pole is at latitude 90°. Therefore,

$$a + b = \phi + \beta$$

But the line to the celestial pole is parallel to the earth's axis, so $\beta = b$ (because corresponding angles of parallel lines are equal). Therefore,

$$a = \phi$$

Surveying the Earth

The actual measurement of the earth's size derived from the belief that the earth was, in fact, a sphere. Around 200 B.C. the Greek astronomer and geographer Eratosthenes, believing the earth to be round, measured its circumference and derived a diameter of about 13,400 km. Eratosthenes worked in the great library in Alexandria. While on a vacation trip in Syene, Egypt, he noticed that on June 21 (the summer solstice) sunlight fell directly down a well at noon. This indicated that the sun was directly overhead. At noon on the same date in another year, he observed that in Alexandria (located directly north of Syene) the shadow of a pillar indicated that the sun was about 7° south of the point directly overhead (Fig. 2.8). Because the circumference of the earth encompasses 360°, he concluded that the distance from Syene to Alexandria must be roughly 7/360, or about 1/50, of the earth's circumference (Fig. 2.9).

To determine the length of the circumference, Eratosthenes needed to know the distance from Alexandria to Syene. He may have known that, as Herodotus recorded, the trip by camel took about 50 days. Because the average camel traveled about 100 stadia a day, it covered 5000 stadia for the entire journey. (The *stadium* was an ancient unit of length, about 1/6 km, although its exact length was different in different parts of the ancient world.) Eratosthenes calculated the earth's circumference as $50 \times 5000 = 250,000$ stadia. If we take the length of a stadium to be 1/6 km, then the Alexandria-Syene distance as determined by Eratosthenes was about 1000 km, and the circumference of the earth about 42,000 km, surprisingly close to the modern value of 40,030 km. (This numerical

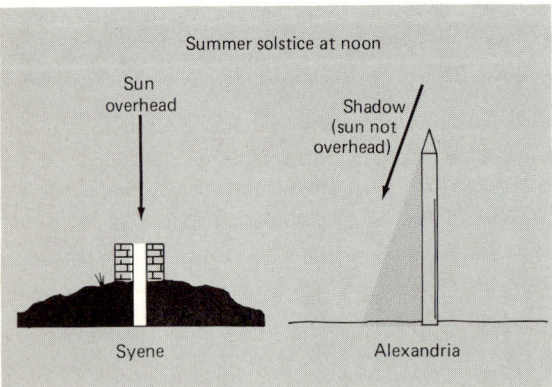

FIGURE 2.8 Eratosthenes' solar observations at noon at Alexandria and Syene (Aswan). At Syene on the summer solstice the noon sun's light came down from directly overhead. In another year he saw that in Alexandria on the summer solstice the noon sun's light came down at an angle of 7° from the vertical.

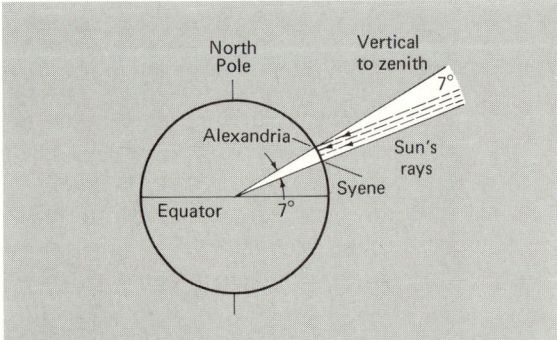

FIGURE 2.9 The geometrical basis for calculating the earth's circumference. If the sun is so far away from the earth that its incoming rays are parallel, only a curved earth could simply account for the difference in the noon angle at the two locations.

correspondence may be mere chance, because the precise length of the stadium is not known.) Knowing the circumference, we can easily compute the earth's radius by dividing it by 2π; the result is about 6700 km.

Note in this instance how the adoption of a certain model, that of a round earth, resulted in an observation that not only confirmed the model, but also provided new information about the earth.

A Contrary View: The First Heliocentric Model

After the time of Alexander the Great, the scientific tradition, carried by the Greeks through the known world, centered upon the library at Alexandria. Eratosthenes, who directed the library, was one of the astronomers in this new Hellenistic tradition.

Another Hellenistic astronomer, Aristarchus, who lived during the third century B.C., went so far as to propose a *heliocentric* (sun-centered) rather than geocentric model for the cosmos (Fig. 2.10). This heliocentric model explained the apparent daily motion of the stars by the rotation of the earth on its axis once a day; the sphere of the stars did not move. Since we stand on the earth and do not feel it moving, we think that the heavens rather than the earth rotate. Aristarchus argued that if the earth rotated from west to east, it would appear to us that the heavens moved east to west. In addition, Aristarchus believed that the earth also revolved around the sun in one year. The sun's motion along the ecliptic was simply a reflection of the actual motion of the earth. In these ideas Aristarchus ran far ahead of his time.

Unfortunately, the major writings of Aristarchus have been lost, and we know his ideas only from comments made by other people and from one surviving work. In this piece Aristarchus worked out the earth-sun distance relative to the earth-moon distance, and from this result he inferred that the sun was a body much larger than the earth (see below). But we have no direct record that Aristarchus worked out the details of planetary motions with his model.

The heliocentric scheme was attacked during Greek times from two bases. First, it contradicted both common sense and Aristotle's physics in stating that the earth moved. Second, it required an annual parallax of stars that was not, in fact, observed. Because of these philosophical and observational objections and because of the influence of Aristotle's ideas, Aristarchus's model was ignored.

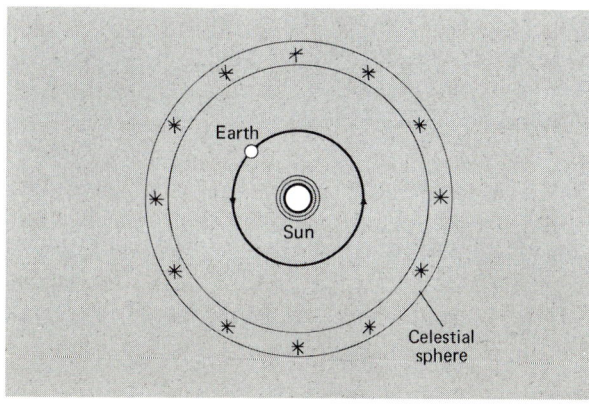

FIGURE 2.10 A simplified sketch of the heliocentric model of Aristarchus. The earth rotating on its axis explained the daily motion of the sky; its yearly revolution around the sun explained the sun's annual motion along the ecliptic.

Surveying the Cosmos

About 250 B.C. Aristarchus devised an elegant way to work out the distances to the sun with the distance to the moon as the unit of measure. From solar eclipses (Section 1.5) Aristarchus knew that the moon was closer to the earth than the sun. But how much closer was it?

The astronomy of Greek times pictured the moon as a dark sphere illuminated by the sun. Every few weeks, the moon appears as half a disk in the sky (at first quarter and last quarter phases). Aristarchus recognized that at these times the angle at the moon between the directions to the earth and sun was 90° (Fig. 2.11). He saw that if he could measure the angle between the sun and the moon as seen from the earth, he could use simple geometry to determine the earth-sun distance in units of the earth-moon distance.

Let's do this calculation. Aristarchus measured an angle of 87° between the sun and moon. The angle at the moon is 90°; so the angle at the sun must be 3°, since the sum of the angles of a triangle is 180°. Then the earth-moon distance (EM) divided by the earth-sun distance (ES) is the sine of 3°, or

$$\sin 3° = \frac{EM}{ES}$$

$$0.052 = \frac{EM}{ES}$$

$$ES = 19\ EM$$

For these figures, the sun lies 19 times more distant from the earth than does the moon.

What are the drawbacks to this method? First, the moon-sun angle measured from the earth is so little different from 90° that it is difficult to measure accurately. (Aristarchus's measurement was too small; the actual value is 89.25°.) Lines drawn from the sun to the earth and to the moon are almost exactly parallel (Fig. 2.12), so a very small error in

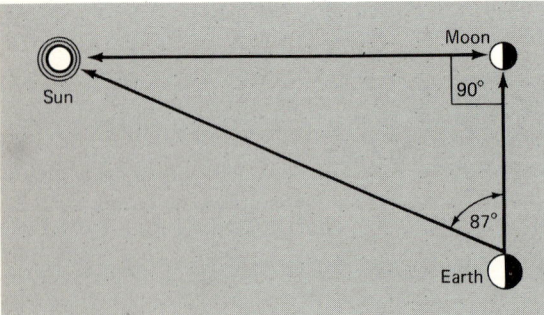

FIGURE 2.11 Geometry of the sun, moon, and earth if the earth-sun distance is not much larger than the earth-moon one.

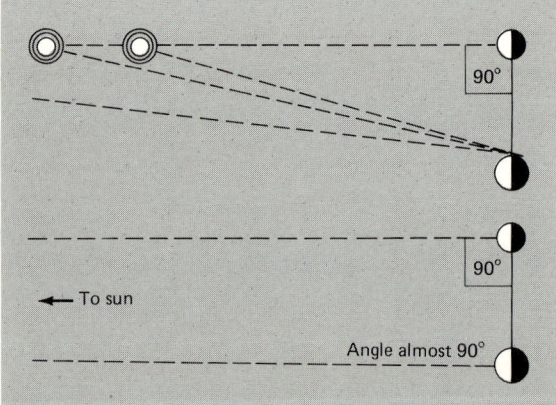

FIGURE 2.12 Effect of the sun's distance on the moon-earth-sun angle for a quarter phase of the moon.

measuring the angle results in an enormous error in the distance. Second, it's very hard to determine the time when the moon is exactly half illuminated. Third, since the sun is a bright object in the sky, it's hard to observe directly. So, although Aristarchus's method is fine on paper, it doesn't work out well in practice.

From his observations Aristarchus concluded that the sun was 18 to 20 times as distant as the moon. Therefore, he thought the sun must be 18 to 20 times the diameter of the moon, since they both have the same angular diameter in the sky.

We know now that the sun is actually about 400 times more distant than the moon. But you shouldn't scoff at Aristarchus's results; it's his method that counts, along with the underlying belief in a simple geometrical model that pictured the distance as capable of being measured. As far as we know, Aristarchus kicked off the surveying of the cosmos beyond the earth, an activity carried on by astronomers today.

Expanding the Mechanical, Geocentric Model

Another distinguished Hellenistic astronomer, Hipparchus, lived and worked at Rhodes from 160 to 127 B.C. His most useful accomplishment was the organization of the observations available from Babylonia. Hipparchus codified the ancient records because he wisely recognized that to devise a better cosmological system required a careful accumulation of observations; these observations made up the raw materials from which a model could be molded. Although Hipparchus gathered relevant information, he did not assimilate it completely enough to develop a comprehensive model of planetary motion. He did, however, use geometrical devices called *eccentrics*, *deferents*, and *epicycles*, which enabled him to explain planetary motions with some accuracy. Each device accounted for some aspect of planetary motion.

Following the older ideas of Plato, Hipparchus demanded that the planets move at a constant speed along circular paths. Yet he knew from observations that their motion in the sky is not uniform, but varies regularly. Hipparchus explained the variation in a planet's speed along the ecliptic by the *eccentric* (Fig. 2.13a). The earth was displaced from the center of a planet's motion. Then as the planet moved, it appeared to go faster when closer to the earth and slower when farther away (Fig. 2.13b). However, the planet's motion remained uniform when viewed from the center of its circular path.

The eccentric did not provide an explanation of retrograde motion. To cope with this observed fact, Hipparchus used the epicycle and the deferent (Fig. 2.14). The *deferent* was a large circle centered more or less on the earth. (The deferent could be an eccentric; if so, then the earth was a bit off-center.) Around the circumference of the deferent moved the center of a smaller circle called the *epicycle*. The planet moved on the epicycle, so its motion was a combination of circular motion about the epicycle and the epicycle's circular motion about the deferent. If the epicycle and deferent turn in the same direction, the combination seen from the deferent's center imitates retrograde motion. As the planet moves on the part of the epicycle interior to the circumference of the deferent, it moves opposite the deferent's motion. Hipparchus had the deferent represent the planet's general motion west to east along the ecliptic. The reverse swing on the epicycle then represented the east-to-west retrograde motion (Fig. 2.15).

Taking up a suggestion of Aristarchus, Hipparchus measured the earth-moon distance from observations during a total lunar eclipse (Focus 2.2). To do this, he had to assume

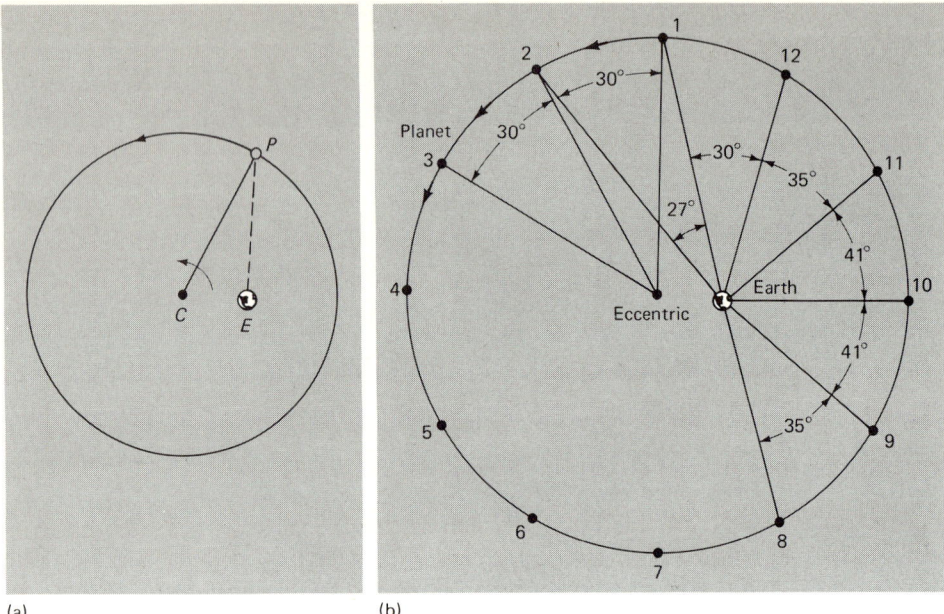

FIGURE 2.13 The eccentric. (a) A planet (P) revolves with uniform circular motion about the center (C) of its path. The earth (E) is displaced from the circle's center, so the planet's motion as seen from the earth is not uniform. (b) The effect of the eccentric for modeling a planet's motion. As seen from the circle's center, the planet moves through 30° angles in equal amounts of time. But as seen from the earth, the planet covers different angles in the same times.

that the earth and moon were round and that the moon orbited the earth. This step at surveying the cosmos was crucial in establishing the size of the universe—a task continued by modern astronomers. Note the similarity here to Eratosthenes; both men accepted geometrical models of the universe and, because of these models, made fruitful observations. Without the models they had no reason to make such observations. In this sense scientific models preset what observations are made and what is expected from them.

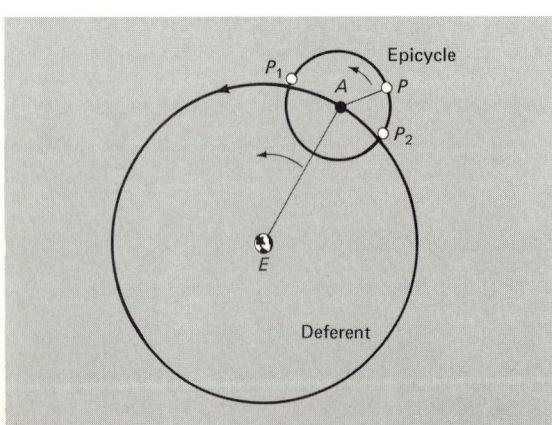

FIGURE 2.14 The epicycle and deferent for retrograde motion. A planet (P) is attached to a small circle (*epicycle*) whose center rides on a larger circle (*deferent*). The earth (E) lies in the center of the deferent. The radius of the epicycle turns in the same direction as the radius of the deferent. So when the planet moves on the inside of the deferent (from P_1 to P_2), it moves *opposite* its normal motion with respect to the stars; this is its retrograde motion.

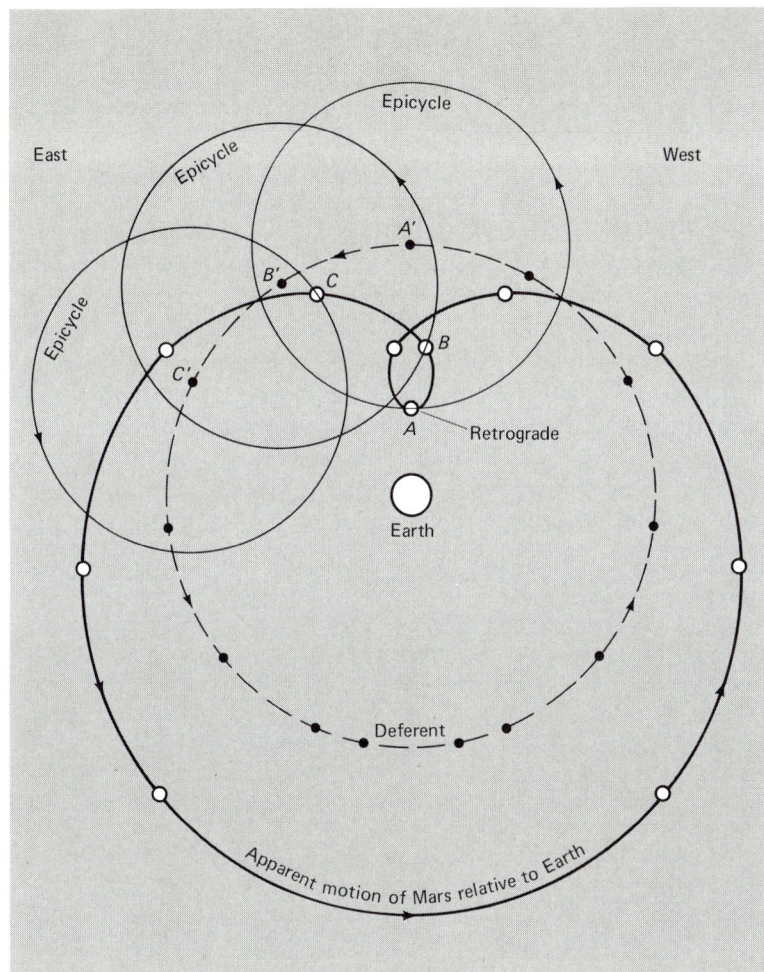

FIGURE 2.15 The motion of Mars, as seen from the earth, produced by an epicycle and deferent. Three positions of the epicycle are shown. Mars is at A, B, and C when the center of the epicycle is at A', B', and C'. Note that Mars is closest to the earth in the middle of its retrograde loop, and so it appears at its brightest then. The period of the deferent is the average time it takes Mars to circuit the ecliptic; the period of the epicycle is the time between retrograde motions.

2.4 CLAUDIUS PTOLEMY: A COMPLETE GEOCENTRIC MODEL

Two and a half centuries after Hipparchus, Claudius Ptolemy (Fig. 2.16) worked in Alexandria and so had access to the records in the famous library. Ptolemy molded the monumental records and ideas of the astronomy then known into a comprehensive model that would endure for centuries.

We know little of Ptolemy's personal life; in fact, we do not even know the dates of his birth and death. From his observations, however, we know that he worked around A.D. 125. His most noted astronomical work was the *Almagest*. The original Greek title is translated as the *Mathematical Composition of Claudius Ptolemy*, but because the impact of this book on western science came through the Arabic text, it is usually referred to as the *Almagest*, a transliteration of Arabic words meaning "the greatest."

FOCUS 2.2 ECLIPSES AND THE DISTANCE TO THE MOON

Hipparchus refined a method first developed by Aristarchus to determine the distance to the moon and the moon's diameter in terms of the earth's. From direct observations Hipparchus knew that the sun and moon have about the same angular diameter—$\frac{1}{2}°$. He also knew that the longest total lunar eclipses lasted about 2.5 hours. Now, because the moon moves about $\frac{1}{2}°$ per hour relative to the stars, Hipparchus inferred that the earth's shadow had a diameter of about 1.25° at the moon's distance from the earth. (Note that he believed that the moon revolved around the earth.) Finally, Hipparchus needed to use Aristarchus's estimate that the sun was about 19 times more distant than the moon.

With this information Hipparchus constructed a scale diagram of the earth, moon, and sun (Fig. F.4) with the proper angles drawn in. From it he found that the moon's distance from the earth was about 30 times the earth's diameter, and that the moon was about $\frac{1}{3}$ the diameter of the earth. Both figures are close to modern values. From this result he inferred that the sun was about 7 times the earth's diameter. This result was wrong. The correct number is 109 times, because the sun is much farther away than he believed.

These relative diameters and distances can be converted to standard measures if the earth's diameter is known, and it was, by the method of Eratosthenes (Section 2.3). So

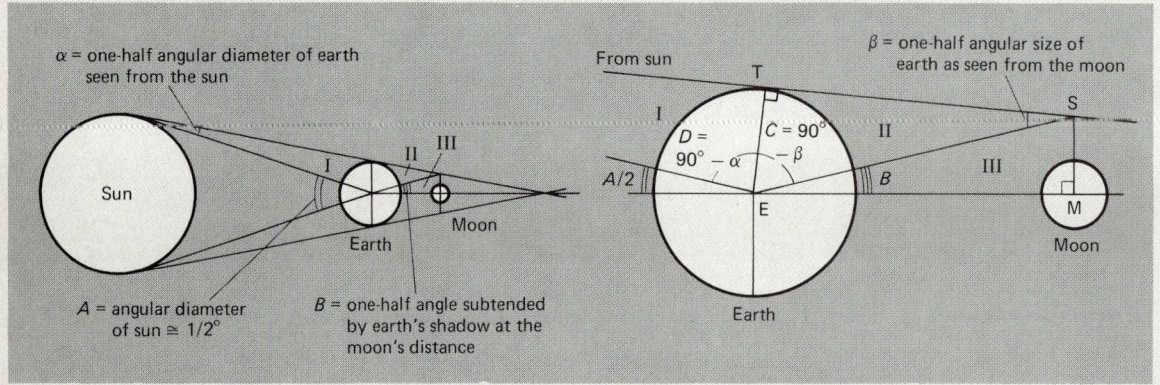

FIGURE F.4 Geometry for finding the earth-moon distance from observations of a total lunar eclipse.

Many of the technical devices that Ptolemy used did not originate with him, but he was the first person to design a complete system to describe and predict planetary motions. The *Almagest* was the first professional astronomy textbook. Most important, his model worked; Ptolemy's many circles whirled the planets around the ecliptic with an error of usually not more than 5° and generally less. The careful application of geometry explained the celestial motions.

A Mechanical, Geocentric Model

In the opening pages of the *Almagest*, Ptolemy pays his respects to Aristotle and then defines the scope of the problem:

from a simple model of the moon orbiting the earth and naked-eye observations, Greek astronomers figured out the diameter of and distance to the moon.

You can duplicate Hipparchus's results by drawing your own scale model. Or you can work it out as follows. In Fig. F.4a are three triangles (I, II, and III) that meet at the earth's center (Fig. F.4b). Note that the angles $A/2 + D + C + B$ around the earth's center amount to a 180° total. Now $A/2$ is $(1/2°) \times (1/2)$, or $1/4°$. Angle D, since I is a right triangle, equals $90° - \alpha$; but if the sun is far away, α is very small, and we can call it zero. Angle C, since II is a right triangle, is $90° - \beta$. And B, from eclipse observations is $(1/2) \times (5/4°)$, or $5/8°$. Then

$$A/2 + D + C + B = 180°$$
$$1/4° + 90° + (90° - \beta) + 5/8° = 180°$$
$$\beta = 7/8°$$

Now the tangent of β is ET/TS, where TS is almost the same length as EM. So

$$\tan 7/8° = \frac{ET}{TS} \approx \frac{ET}{EM}$$

or

$$0.0153 \approx \frac{ET}{EM}$$

$$EM \approx 65\ ET$$

earth-moon distance $\approx$ 33 earth diameters

The result from the Greeks' observations and calculations was an earth-moon distance of about 30 earth diameters, essentially the modern value!

> . . . we wish to find the evident and certain appearances from observations of the ancients and our own, and apply the consequences of these conceptions by means of geometrical demonstrations.

Note that Ptolemy is following Plato's dictum to "save the appearances."

Ptolemy starts with the assumptions that the earth is spherical, in the center of the heavens, has no motions, and is much smaller than the sphere containing the stars. In addition, his system requires that *uniform motion around the center of circles* be the only proper motion for the celestial spheres carrying the planets. (In practice, Ptolemy ends up violating this precept.)

How well does the system work? The first major problem arises from his last assump-

FIGURE 2.16 Claudius Ptolemy (with the goddess Astronomy). " . . . We shall only report what was rigorously proved by the ancients. . . ." [*Courtesy O. Gingerich.*]

tion. If the planetary motion is in fact uniform along circles, how do you explain the observed variation in the motions in the sky? Not only do the planets have retrograde motion, but these retrograde motions vary. For example, Fig. 2.2 shows nine retrograde motions of Mars, all of which have different angular sizes, shapes, and durations. Also, the planets do not move uniformly along the ecliptic, but their rates of motion vary throughout the year. To provide geometrical explanations of all these variations Ptolemy used a combination of eccentrics, epicycles, deferents, and his new invention, *equants*.

Like Hipparchus, Ptolemy used the deferent and the epicycle to account for retrograde motion. The periods for the epicycle and deferent are set from observations. The average time it takes the planet to return to the same place in the zodiac (recall Table 1.2) is the period of the deferent. Mars, for example, has a deferent period of almost 2 years. The period of the epicycles is just the average time between retrograde motions. For example, at the midpoint of its retrograde motion, Mars is at opposition. The next midpoint of retrograde motion takes place at the same relative alignment of the sun, earth, and planet—the next opposition. The epicycle's period equals the time from opposition to opposition. For Mars this time is 780 days.

Ptolemy also noted that the average angular speed of a planet is faster in one region of the ecliptic and slower in the region opposite. He offset the center of the planetary deferents from the center of the universe (where the earth was located) to account for this variation in angular speed. This geometrical setup was the eccentric.

To these older geometrical schemes Ptolemy added his new device, the *equant*. Start with an eccentric. Imagine a point in the circle placed the same distance away from the center as the eccentric point, but located on the opposite side from it (Fig. 2.17a). Ptolemy demanded that the planet move on the circle in such a way that an imaginary

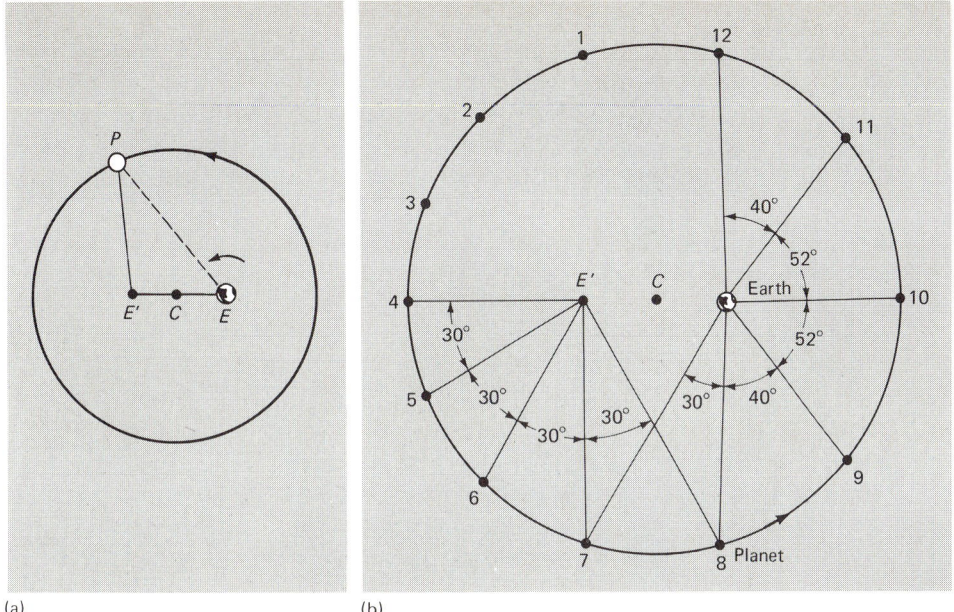

FIGURE 2.17 Ptolemy's equant. (a) Here C is the center of the circle and the earth is at the eccentric point (E). E' is the equant point, an imaginary point on the opposite side of C from E. A planet P moves on the circle. (b) If P were viewed from E', it would appear to cover equal angles (30°) in equal times, but as seen from C or the earth, it moves through different angles in equal times.

observer at this point (E' in Fig. 2.17a) sees uniform motion. This point is called the *equant point*. Note that the planet no longer moves along the circle with uniform speed (Fig. 2.17b). The equant is a nonphysical, totally geometrical device.

We are not completely sure why Ptolemy introduced the equant, but it helped him to calculate more accurate positions. Ptolemy's equant was a significant break with past astronomical traditions because the planetary motion was *no longer uniform about the center of a circle*.

Note that Ptolemy violated the precept of uniform motion about the center of a circle to be able to construct a model consistent with his aesthetic ideals about motion along circles and one that predicted planetary positions reasonably well.

How did the system look when it was all fitted together? (A simplified diagram is shown in Fig. 2.18.) First consider the inner planets. Recall from Section 1.4 that Mercury never strays more than about 23° from the sun and Venus never more than about 46°. Ptolemy could account for these observations by requiring that the size of the epicycle for Venus be larger than that for Mercury and by insisting that the centers of the epicycles always lie on the line connecting the earth and the sun.

For other planets the plan is somewhat different. Recall that Mars, Jupiter, and Saturn can be found far away from the sun. Therefore the centers of the epicycles can be anywhere around the deferent. But to ensure that the planets go through retrograde motion when, and only when, they are at opposition, the model requires that the radii of the epicycles must all line up with the radius of the earth-sun deferent.

Although the Ptolemaic model (Fig. 2.19) lost acceptance some centuries ago, you should not condemn it as obviously wrong. This model remained in use for some 1,400 years, and not solely because of the purported stagnation of intellectual life during the

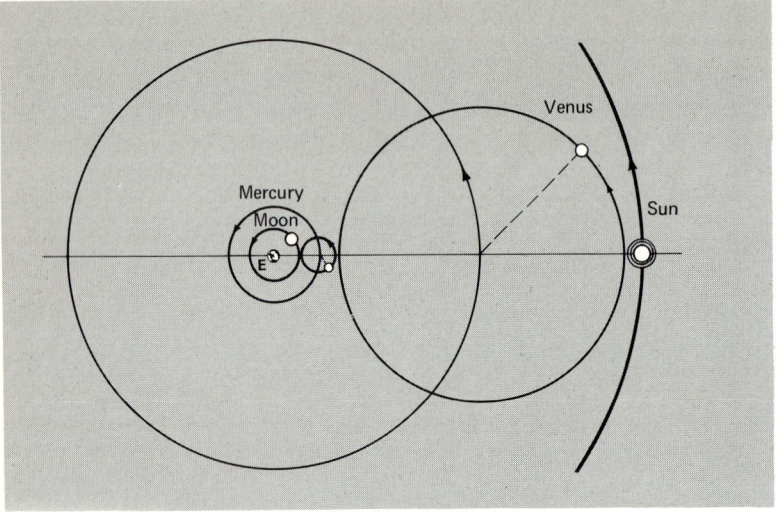

FIGURE 2.18a The Ptolemaic model for Mercury and Venus. The centers of the epicycles of these planets are fixed along a line with the sun; this explains the fact that they are never very far from the sun. The size of each epicycle accounts for the measured size of the maximum elongation angles.

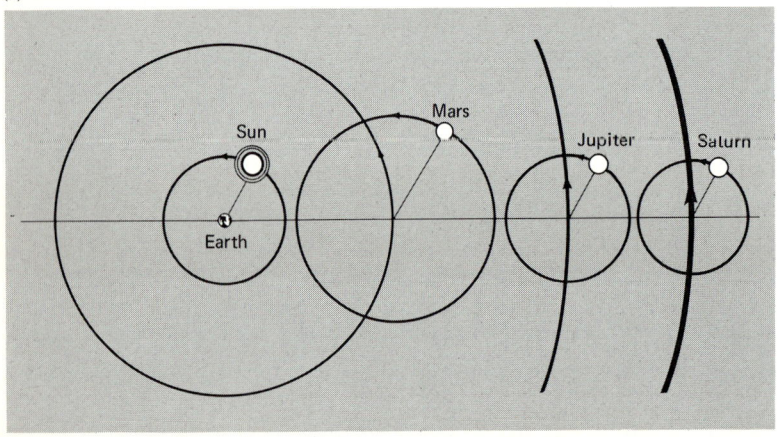

FIGURE 2.18b The Ptolemaic model for Mars, Jupiter, and Saturn. To account for the observed sizes of the retrograde loops, the epicycles of these planets decrease in size, so that Mars has the largest and Saturn the smallest. The radii of these epicycles must align with the earth-sun radius for the retrogrades to occur at opposition. The arrangement shown here emphasizes that point; in general, these planets can lie anywhere on their deferents.

Middle Ages. The Ptolemaic system incorporated useful technical methods for calculating planetary positions with some accuracy. Ptolemy's model also agreed in most details with the philosophical doctrines of the Greeks and had a commonsense geocentric appeal. Later, it fitted in with St. Thomas Aquinas's synthesis of Christian belief and Aristotelian physics. It survived because no one else advanced as comprehensive a system to compete effectively with it.

Describing Observations with the Ptolemaic Model

It may help you to comprehend the model of Ptolemy if you try to connect it with real observations of the sky. Suppose you go out at night and observe Mars rising just as the sun sets. Jupiter is 45° up in the eastern sky. You also observe Venus in the western sky, 46° east of the sun (greatest eastern elongation). The moon is at first quarter phase. The scene is like that shown in Fig. 2.20a.

2.4 CLAUDIUS PTOLEMY: A COMPLETE GEOCENTRIC MODEL

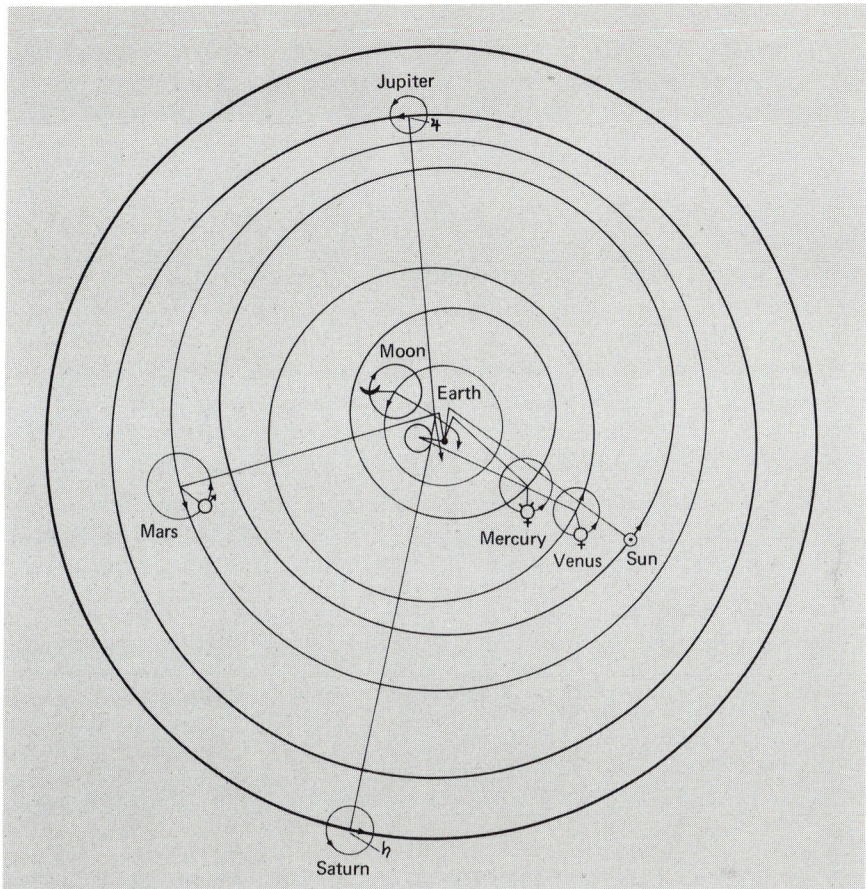

FIGURE 2.19 A view of the complete Ptolemaic model. Note that the centers of the deferents are offset from the earth; these have been made into eccentrics. This figure does *not* show the proper relative sizes of the epicycles or their correct alignments; these are given in Figure 2.18. [*Adapted from a diagram by W. Stahlman.*]

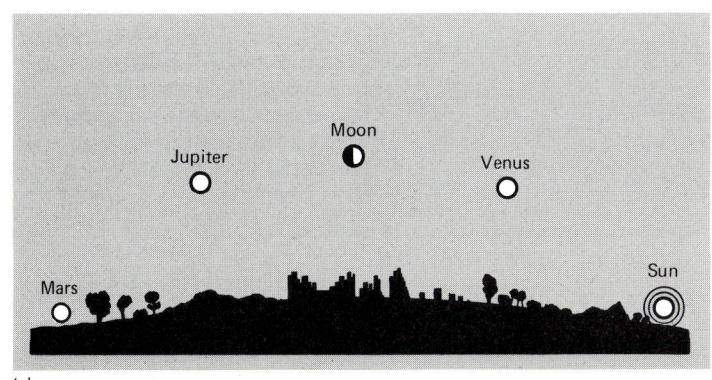
(a)

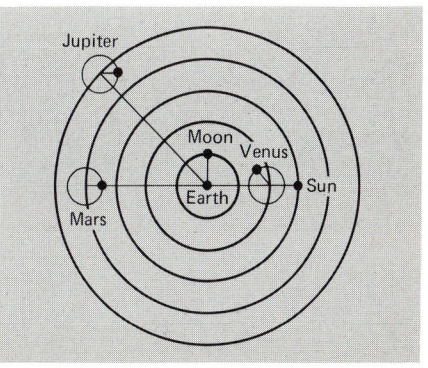
(b)

FIGURE 2.20 (a) A night scene. (b) Explanation of these observations by a Ptolemaic model arrangement.

To describe this situation in Ptolemy's model, let's draw a picture. Put the earth at the center and draw five circles around it to represent the deferents of the moon, Venus, the sun, Mars, and Jupiter. Imagine that you are looking at the earth from above the north pole. Let's put the sun off to the right, because (if you face south) that's where it is setting. Next put in the center of Venus' epicycle (which must always be on the earth-sun line). Now put Venus on its epicycle at the point where a line from the earth is tangent to the epicycle, because this orientation gives the greatest angle of Venus away from the sun (greatest elongation). As Mars is just rising at sunset, it must be opposite to the sun (at opposition). Because the radius of its epicycle must be parallel to the earth-sun line, the center of its epicycle must be on the same direct line as that from the earth to Mars. The moon is observed overhead, halfway around the sky between the sun and Mars, and so must be on its deferent 90° around from the earth-sun line. What about Jupiter? Draw a line 45° clockwise from the earth-Mars line. Now place the center of Jupiter's epicycle on its deferent such that the radius to Jupiter is parallel to the earth-sun line. You should now have something that looks like Fig. 2.20b.

The Size of the Cosmos

Ptolemy's cosmos was finite, but how large was it? For the sun and moon he gives the distances in terms of earth radii. Both were determined from the apparent angular sizes of the bodies and the lunar distance established by Hipparchus. To go farther out he assumed that no space is wasted between the heavenly spheres. Then with the earth-moon distance set at 60 earth radii, the distances out to Saturn can be laid out if all the spheres nest tightly together. The sphere of the fixed stars was then 20,000 earth radii from the earth. It was a small universe, only about the actual presently known distance of the earth from the sun! But it marked a reasonable attempt to establish the scale of the cosmos.

Observations and Scientific Models

What makes one model more scientific than another? A paramount consideration is correspondence with observations. If a particular model does not predict observations to a satisfactory degree of accuracy, it must be replaced by a better one.

But ability to predict is not enough. A second criterion is that the model be based on physical principles, simple statements about the nature of reality and how it works. There must be "reasons" behind the model. Aristotle's model of the universe was based on his physics (for example, his principle that the natural motion of earthly material was toward the center of the universe). So we consider it more "scientific" than the cosmology of the Babylonians, even though the Babylonians, with their tables, could predict planetary positions quite well.

Note the use of the word *simple* above. The fewer basic statements of physics in a model, the better it is, provided it also adequately predicts the observations. Ptolemy strove to use the simplest principles in his model (uniform motion about the center of circles), but was forced by the need for better correspondence with observations to adopt a more complex scheme (the equant). This conflict, between the drive for simplicity and the need for accurate prediction, arises time and again in the development of scientific thought. Watch for it in future chapters.

2.4 CLAUDIUS PTOLEMY: A COMPLETE GEOCENTRIC MODEL

TABLE 2.1 A Summary of Ptolemy's Model

Observation	*Explanation*
Motion of entire sky E to W in about 24 h	Daily motion E to W of the sphere of stars, carrying all other spheres with it
Sun	
Motion yearly W to E along ecliptic	Motion of sun's sphere W to E in a year
Nonuniform rate along ecliptic	Eccentric (earth displaced from center of sun's circle)
Moon	
Monthly motion W to E compared to stars	W to E motion of the moon's sphere in a month
Planets	
General motion W to E through zodiac	Motion of deferent W to E. Period is set by observation of period of planet to go around ecliptic.
Retrograde motion	Motion of epicycle in same direction as deferent. Period is the time between retrograde motions.
Variations in speed through zodiac, in retrograde motions	Eccentrics, equants
Mercury, Venus	
Average greatest elongations of 23° and 46°	Size of deferents set by those angles
Appear fixed to sun	Centers of deferents on earth-sun line
Mars, Jupiter, Saturn	
Retrograde at opposition, when brightest	Radii of epicycles aligned with earth-sun radius

Notes: (1) Motions of heavenly spheres were natural motions and did not require a force. (2) Use of equants *violated* the precept of uniform, circular motion for all heavenly bodies. (3) Error of predictions was usually 5° or less, occasionally larger.

By the criterion of correspondence to observations, just how good was Ptolemy's model of the cosmos? The error of his predictions was 5° or less. This amounts to a rather large error. Recall (Fig. 1.4) that the distance between the two pointer stars in the bowl of the Big Dipper is about 5°. A careful observer certainly could have noticed such a discrepancy. Apparently Ptolemy's predictions were considered satisfactorily accurate, even though the errors were easily observable.

Did Ptolemy think his system was *real*? Did he believe the heavens actually were filled with spheres of quintessence? We're not sure. In modern usage a model *represents* reality, but it isn't reality. A model acts like a metaphor in a poem. It has features that resemble those characteristics of the world it tries to explain. We now look on the Ptolemaic system as a model, not reality. Whether or not Ptolemy believed the planets actually moved on epicycles and deferents, his system did explain and predict, with some accuracy, the motion of the planets. A remarkable achievement for a person without a telescope!

SUMMARY

This scan of astronomy from Babylonian to Greek times has picked out one important aspect of cosmological ideas: the evolution of a mechanical model of the universe. The invention of a model was lacking in Babylonian astronomy, even though the Babylonians were excellent observers and developed arithmetical methods to predict planetary positions. The cosmos was explained in terms of myths, such as the *Enuma Elish*.

Cosmological model making began in Greek times. The early cosmological model of the Pythagoreans included geometrical and aesthetic elements (spheres and the harmony of the spheres), but lacked physical ideas. In these attempts by Plato and Eudoxus to "save the appearances," mechanical devices, usually with a simple geometry, were used to explain the actual observed motions of the planets. Aristotle injected physics into his cosmological model, but knew that the correspondence of his model with some observations, especially retrograde motion, just wasn't terribly good.

Ptolemy put together the first comprehensive cosmological model, one that incorporated all three key elements: geometry from the Greeks, aesthetics of his own, and physics from Aristotle. More important, his model conformed to observations within sufficient accuracy for his day (Table 2.1). The *Almagest* provided a conceptual framework for and a practical approach to the geocentric cosmos. So Ptolemy's model worked and was complete. That's why it survived for so long.

Key Words

scientific model	earth, air, fire, water	eccentric
cosmos	quintessence	epicycle
forced motion	heliocentric	deferent
natural motion	geocentric	equant
	parallax	

Review Questions

1. Which of the following are valid criticisms of the heliocentric theory? (Mark all correct answers.)
    a. Two observations at midnight six months apart should not enable you to see all the stars in the sky.
    b. A star's brightness should depend on the time of year.
    c. The angle between any two stars should vary over the course of the year.
    d. If the earth were moving around the sun, a falling object would get left behind.
    (Objective 11)
2. How do we currently evaluate Aristotelian and Babylonian astronomy with respect to the characteristics of a scientific model? (Mark all correct answers.)
    a. We consider Aristotelian astronomy scientific because it is based on physical laws and a physical system. We consider Babylonian astronomy unscientific because it is not.
    b. Aristotelian astronomy is unscientific because Aristotle's physical principles are incorrect. Babylonian astronomy is more scientific because their observations were accurate and they had a system of cosmology which explained them.

PROBLEMS

c. Aristotelian astronomy is scientific because the scheme, though incorrect, accurately predicts planetary positions. Babylonian astronomy is unscientific because its predictions, though good at the time, were much less accurate than Aristotle's.

d. We consider Aristotelian astronomy unscientific because it does not accurately predict the positions of the planets. The essence of a scientific system is its ability to predict, and the Aristotelian system gave predictions inaccurate by as much as two moon diameters, an easily measurable error. We consider Babylonian astronomy unscientific, too, for the same reason.
(Objective 1)

3. Indicate whether the following statements are an assumption underlying the Ptolemaic theory (A), a facet of Ptolemaic theory included to improve agreement with observations (O), or false, in the Ptolemaic view (F). Think carefully about the (great) difference between an initial assumption of a model and later tinkering to make that model agree better with observation.
   a. The earth is at the center of the universe.
   b. The retrograde motions of the planets result from motions along their epicycles.
   c. All heavenly motions must be along perfect circles.
   d. This motion may, however, be nonuniform as seen from the center of the circle.
   e. Even so, the motions must be uniform as seen from somewhere!
   f. Planets are made of earth, too, with the lighter elements (water, air, and fire) on top. The sun is mostly fire.
   (Objectives 7 and 8)

4. What was an important observational achievement of Babylonian astronomers? (Objective 2)

5. Contrast Babylonian and Greek astronomy in terms of their observational achievements. (Objectives 2, 3, and 4)

6. How did the model of Aristotle explain:
   a. Apparent lack of motion of the earth
   b. Daily motion of the stars
   c. Annual motion of the sun
   (Objective 5)

7. How did Aristotle argue against a heliocentric system? (Objectives 4 and 5)

8. In the Ptolemaic system how do observations establish the period of the epicycle? The deferent? (Objectives 7 and 8)

9. Why and how did Ptolemy treat Mercury and Venus differently from Mars, Jupiter, and Saturn? (Objective 8)

10. How did Ptolemy violate his own precepts? (Objectives 7, 8, and 9)

Problems

1. Suppose, in Aristarchus's method to find the earth-sun distance, he had observed a sun-moon angle of 88° from the earth. Using this observation, how far would the sun be? How large would the sun be, relative to the moon? (Objective 6)

2. Now suppose the measured angle is 87°. How much different are the results from those above? What do you conclude about the method? (Objective 6)

3. Suppose that the shell of stars had a radius of 10 earth-sun distances in a heliocentric model such as that of Aristarchus. How large an angular change would you expect to see for two stars which are 1° apart, as seen from the sun? (Objective 11)
4. An unknown Greek tried to duplicate Eratosthenes's observations and measured a noon sun angle of 8° in Alexandria. What would this person calculate as the earth's circumference and radius? (Objectives 4 and 6)
5. The sun does *not* move at a uniform angular speed along the ecliptic. Over the half year centered on the summer, it covers 180° in 186 days; over the winter half, 180° in 179 days. Using an eccentric (Fig. 2.13) to model this motion, calculate how far away from the circle's center the earth must be. (Objectives 8 and 9)

BEYOND THIS BOOK . . .

* A classic work on the astronomical achievements of the Babylonians is *The Exact Sciences in Antiquity* (Dover, New York, 1969) by O. Neugebauer.
* Another classic with a wider view is *A History of Science: Ancient Science Through the Golden Age of Greece* (Harvard University Press, Cambridge, Mass., 1952) by G. Sarton.
* You can find an English translation of the *Almagest* in volume 16 of Great Books of the Western World (Encyclopaedia Britannica, Chicago, 1952).
* A good collection of Greek philosophical ideas is in *The Origins of Scientific Thought* (Mentor Books, New American Library, New York, 1961) by G. de Santillana.
* For a controversial look at Ptolemy, read *The Crime of Claudius Ptolemy* (Johns Hopkins University Press, Baltimore, 1977) by R. Newton.
* For a case against astrology, read *Astrology Disproved* (Prometheus Books, Buffalo, N.Y., 1977) by L. E. Jerome.

THE NEW WORLD ORDER: A Sun-Centered Model

3

LEARNING OBJECTIVES
After studying this chapter, you should be able to:

1. List the arguments and assumptions that Copernicus used to support his model and refute the Ptolemaic one.
2. Explain, with a simple diagram, retrograde motion in the Copernican model.
3. Demonstrate how, in Copernicus's model, you can determine the planets' relative distances from the sun from observations and simple geometry.
4. Derive and make use of the relation between synodic and sidereal periods in the Copernican model.
5. Evaluate the assets and weaknesses of the Copernican model compared to the Ptolemaic one.
6. Present one example of the influence of Copernican ideas on the work of Kepler.
7. Argue, using at least two examples, how Kepler rather than Copernicus deserves the title of the "first astrophysicist."
8. Describe and use important geometrical properties of ellipses.
9. State Kepler's three laws of planetary motion and apply them to appropriate astronomical situations.
10. Compare and contrast the Keplerian model to the Copernican one.

CENTRAL QUESTION:
How did the sun-centered model explain the motions of the planets?

Ptolemy's model superseded all cosmological competitors, basically because the essential Ptolemaic plan worked quite well for predicting planetary positions. Practicing astronomers had no reason to be unhappy with it.

Although most medieval astronomers were satisfied with the Ptolemaic model, one of them was not. He was a quiet Polish monk, Nicolaus Copernicus (1473–1543). Copernicus was offended aesthetically by Ptolemy's use of the equant, for it violated the ancient precept that the motions of the planets must be uniform along circles. In the year of his death Copernicus's great work was published, *De revolutionibus orbium coelestium (On the Revolution of the Heavenly Spheres)*, in which the earth was shaken from its static place at the center of the universe into an orbit around the sun and became another planet in the cosmos. The model of Copernicus was *heliocentric*.

Copernicus's view did not immediately wrench the minds of astronomers away from their geocentric notions. (In fact, his model was not more accurate in predicting planetary positions than the Ptolemaic one.) Some, like the Danish observer Tycho Brahe, considered the Copernican claims, but finally clung to a modified geocentric model. Others, like Johannes Kepler, found an essential harmony in Copernicus's model. Kepler's efforts to provide an explanation for that harmony led to the introduction of a new concept into models of the cosmos: a physical force in the heavens to drive the motions of the planets.

3.1 COPERNICUS THE CONSERVATIVE

In the sixteenth and seventeenth centuries in Europe a new model of the cosmos was forged. Nicolaus Copernicus (Fig. 3.1) marks the center of this revolution. We highlight here some events in Copernicus's life that may have influenced him in developing a new model of the universe.

FIGURE 3.1 Nicolaus Copernicus. "In the center rests the sun." [*Sixteenth century woodcut by T. Stimmer.*]

Following the traditional liberal arts curriculum at the Collegium Maius in Crakow, Copernicus studied astronomy with more interest than the average student. His copy of the *Alfonsine Tables* still exists, with many notes in the margins. (These tables gave predictions of planetary positions based on the Ptolemaic system.) Copernicus made observations from which he noted that some Ptolemaic predictions of planetary conjunctions were fairly inaccurate, especially those of Jupiter and Saturn, one of which occurred in 1504. He found that the conjunction of 1504 took place at a time 10 days different from the predicted time. These inaccuracies may have driven Copernicus to reconsider the Ptolemaic model and eventually to devise a heliocentric one.

How did the sun come to play a central role in his model? Copernicus may have first begun to consider the importance of the sun when he was at Crakow and read the work of a Neoplatonic scholar, who published a new edition of Plato. In Italy he became familiar with works of Aristotle, Pythagoras, and Plato. An offshoot of Plato's philosophy, Neoplatonism, asserted that the sun is the source of the godhead and of all knowledge. Based on Plato's writings, the Neoplatonists singled out the sun as a body quite different from the planets and perhaps encouraged Copernicus to consider a heliocentric model. Copernicus later wrote:

> In the middle of all sits the sun enthroned. . . . He is rightly called the Lamp, the Mind, the Ruler of the universe . . . the sun sits upon a royal throne ruling his children the planets which circle round him.

About 1529 Copernicus wrote a general description of his heliocentric cosmology, which he circulated in manuscript copy to some of his friends. In his new model the sun replaced the earth as the center of the cosmos, and the earth not only revolved around the sun but also rotated daily on its axis (Fig. 3.2). (These ideas were not completely new; they were actually first outlined by Aristarchus, Section 2.3.) Copernicus also asserted that all the celestial motions must be composed of uniform, circular motions. Viewing the equant as a violation of this principle, Copernicus attacked the Ptolemaic model as "not sufficiently pleasing to the mind." His new system would reinstate the rule of uniform motion to its proper unadulterated status and result in a more pleasing model.

Just what compelled Copernicus to offer a new cosmological and computational system still puzzles us. The basic Ptolemaic model generally worked well enough to predict

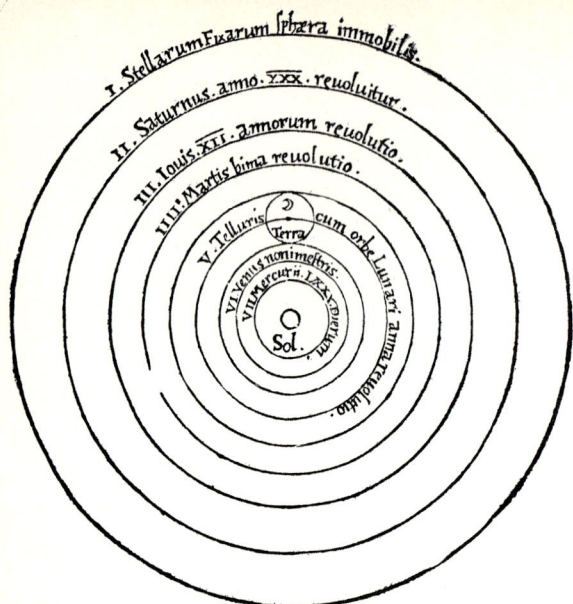

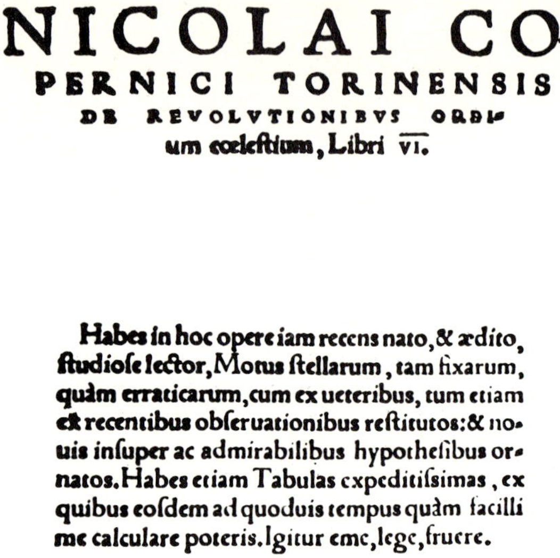

NICOLAI CO-
PERNICI TORINENSIS
DE REVOLVTIONIBVS ORBI-
um coelestium, Libri VI.

Habes in hoc opere iam recens nato, & ædito, studiose lector, Motus stellarum, tam fixarum, quàm erraticarum, cum ex ueteribus, tum etiam ex recentibus obseruationibus restitutos: & no-uis insuper ac admirabilibus hypothesibus or-natos. Habes etiam Tabulas expeditissimas, ex quibus eosdem ad quoduis tempus quàm facilli me calculare poteris. Igitur eme, lege, fruere.

FIGURE 3.2 A simplified diagram of the Copernican model. This figure emphasizes one departure of the Copernican from the Ptolemaic model: the sun is placed in the center of the cosmos with the planets, including the earth, revolving around it. Note that, like Ptolemy, Copernicus placed the sphere of stars just beyond the sphere of Saturn. [*Courtesy Yerkes Observatory.*]

FIGURE 3.3 Title page of *De Revolutionibus*. The words *Orbium Coelestium* were added by Osiander to emphasize that it was the celestial spheres that were moving and so to play down the motions of the earth. [*Courtesy O. Gingerich.*]

the planetary positions. Contrary to some stories, the Ptolemaic model did *not* require the continual addition of circles upon circles to match the observations and so did not grow to a monstrosity. Copernicus would take 20 years to work out the details of his new system, and even then his predictions came out no better than those based on the Ptolemaic system.

Copernicus asserted that his model surpassed the Ptolemaic from an aesthetic standpoint. He said:

> So we find underlying this ordination an admirable commensurability in the Universe, and a clear harmonious bond between the motion and magnitude of the spheres such as can be discovered in no other way.

Copernicus focused on the notions of harmony and commensurability. What were these new, aesthetically striking aspects of his model?

The Plan of *De revolutionibus*

Long after Copernicus had privately distributed his ideas among his friends, his work became known to two astronomers at the University of Wittenberg, Georg Joachim Rheticus and Erasmus Reinhold. Fascinated by Copernicus's new model, Rheticus visited the aging astronomer, who showed him the manuscript copy for *De revolutionibus*. Awed by this work, Rheticus begged Copernicus to have it published for all to read. (Not until Copernicus's lifetime had printed texts, rather than manuscript copies, become common.) Copernicus agreed, and after much trouble with the printer, the book was published around April 1543 (Fig. 3.3). Meanwhile, Copernicus, who was then 70, suffered a stroke and was confined to his bed. Although the completed book was delivered to him before he died in June, he probably did not read it in finished form.

Copernicus would have been amazed to find that the final version of his life's work had been given a new title. The work was probably originally titled *De revolutionibus (On the Revolutions)*, but the published title sported two additional words: *De revolutionibus orbium coelestium (On the Revolutions of the Heavenly Spheres)*. He would have been even more surprised to find a new anonymous preface, which stated that the work contained a new hypothesis, of a heliocentric universe, proposed for the computation of planetary positions. It was not, however, to be taken as a statement of reality, because the astronomer "cannot by any line of reasoning reach the true causes of these movements. . . . let no one expect anything in the way of certainty from astronomy, since astronomy can offer us nothing certain."

For some years people took this preface as a statement by Copernicus that his model had no basis in reality. However, Johannes Kepler (whom you will meet later in this chapter) discovered that the preface had been written by Andrew Osiander, a Nuremberg clergyman, who oversaw completion of the book's publication. Osiander had needed the disclaimer to elicit Protestant approval for the publication of a book by a Catholic. He may have changed the title in order to emphasize the motions of the heavens rather than those of the earth.

De revolutionibus imitated the *Almagest* not only in style and outline, but also in basic intention: to explain the planetary motions using only combinations of uniform, circular motions, the precept of the Greeks (Section 2.3). Copernicus thought that although Ptolemy's theory was consistent with observations, it clashed with the Greek ideas about heavenly motions because it required equants. Copernicus was offended by the equant, because it did not allow for uniform motion of a planet around the center of a circle (the planet's deferent). In a basically conservative mood he wished to devise a system that was more faithful to the uniform, circular motion of the Greeks. While eliminating the offensive equant, Copernicus discovered a natural explanation for retrograde motion and a new fundamental harmony for the celestial spheres, which related the planets' distances to their periods of revolution around the sun.

3.2 THE HELIOCENTRIC MODEL

In the introduction to *De revolutionibus* Copernicus, like Ptolemy, lays out his assumptions before looking at the details. He requires that the planets move in circular paths around the sun at uniform speeds and that the closer a planet's orbit is to the sun, the greater its velocity. Except for using the sun as the center of motion, these points matched ideas of Ptolemy. For the rest, however, Copernicus treads a different ground. Here's a short summary of the fundamental ideas for his heliocentric picture.

1. All the heavenly spheres revolve around the sun, and the sun is therefore the center of the universe. (In actual detail, Copernicus does not place the sun exactly at the center of the planetary paths.)

2. The distance from the earth to the sphere of fixed stars is much greater than the distance from the earth to the sun.
3. The apparent daily motion of heavenly bodies results from the earth's motion about its axis.
4. The apparent annual motion of the sun results from the revolution of the earth around the sun.
5. The planets' retrograde motions occur because of the motion of the earth relative to the planets.

Let's expand upon these points to emphasize the differences between the Copernican and Ptolemaic models.

Take the first point, that the cosmos is heliocentric rather than geocentric. In Section 2.4 we described how Ptolemy would explain the following observational situation (Fig. 2.20a): Mars rising at sunset, Venus at greatest eastern elongation, the moon at first quarter phase, and Jupiter 45° above the eastern horizon. How would Copernicus explain the same observations? Get a piece of paper, and put the sun at the center. Draw four circles around it (if you want a good scale model, draw them with radii in the ratio of 0.7, 1.0, 1.5, and 5.2). Place the earth on its path (the second circle). Draw a line from the earth tangent to the first circle and place Venus at the point of tangency. Mars, since it is at opposition, must be placed on the third circle exactly on an extension of the earth-sun line. Draw a line, making a 45° angle with the sun-earth-Mars line, outward from the earth to the fifth circle and place Jupiter at the intersection. Finally, draw a small circle around the earth and place the moon on it 90° counterclockwise from the earth-sun line (Fig. 3.4). Compare this to the drawing of the same observations in the Ptolemaic model, Fig. 2.20b. (Both figures have left out some of the geometrical complications.)

In point 2 Copernicus tries to get around a problem inherent in a heliocentric model, the stellar parallax, the lack of which undermined the model of Aristarchus (Section 2.3).

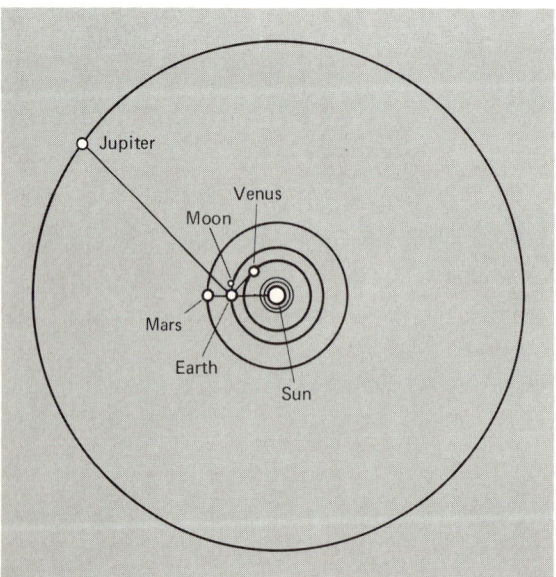

FIGURE 3.4 Explanation of the nighttime observations in Fig. 2.20a by a Copernican model arrangement.

3.2 THE HELIOCENTRIC MODEL

Point 2 resolves this issue. If the stars are far away compared with the earth-sun distance then the parallax angle is too small to be detected by naked-eye observations. (Later work with telescopes showed that Copernicus's intuition about the vast distances to stars was correct.)

In point 3 Copernicus asserts that the daily rising-setting motions of all celestial objects arise from the earth's rotation rather than from the rotation of the celestial spheres. This idea has aesthetic appeal, for it replaces the rotations of many celestial spheres with the rotation of one sphere—the earth. But how can the earth rotate and objects on its surface not fly off? This objection was voiced against Copernicus's model, and he had no good answer for it. (Now we explain it by gravity.)

Point 4 explains how in a heliocentric system the sun still appears to move annually around the earth relative to the stars. Suppose looking out from the earth you see the sun in Leo (Fig. 3.5). (Note that from the sun you would see the earth in Aquarius.) As the earth revolves counterclockwise, the background stars change; after one month the sun appears in Virgo. It seems to have moved, relative to the stars, counterclockwise. Actually, the earth has moved, and it's our line of sight from the earth to the sun to the stars that has changed.

Point 5 has an important consequence for the heliocentric model, for it makes retrograde motion a natural result of the planets' revolutions. This eliminates the need for the epicycle to explain retrograde motion. (Copernicus retained small epicycles to explain

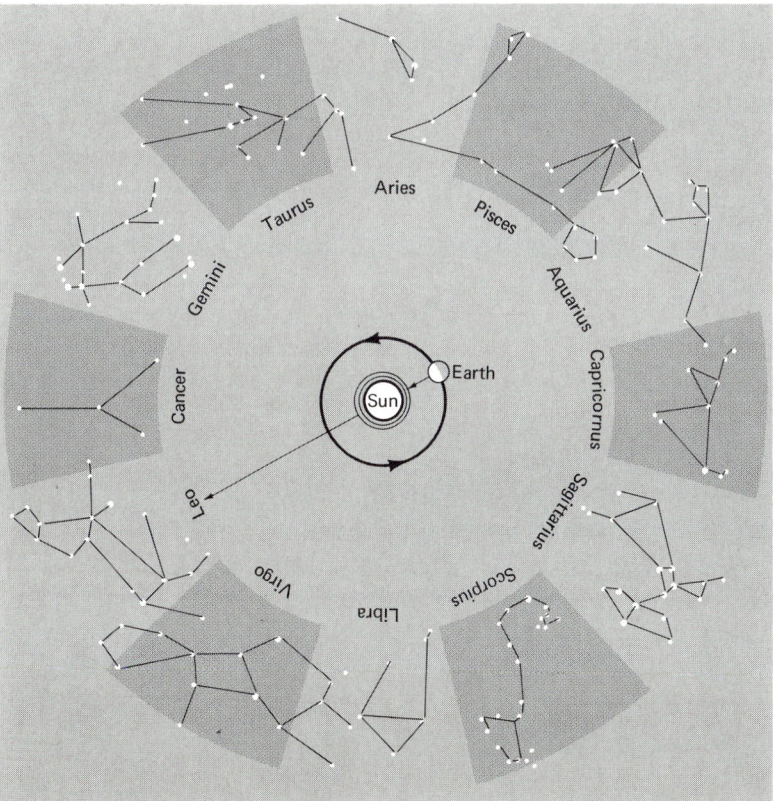

FIGURE 3.5 The sun's apparent motion through the zodiac in a heliocentric model. As the earth travels around the sun (counterclockwise; west to east), the line of sight to the sun and toward the background stars moves in the same direction (west to east). For example, start with the sun in Leo. A month later the earth will have moved eastwardly enough so that Virgo lies behind the sun. The sun seems to move eastwardly through the zodiacal stars at the rate of about one constellation a month.

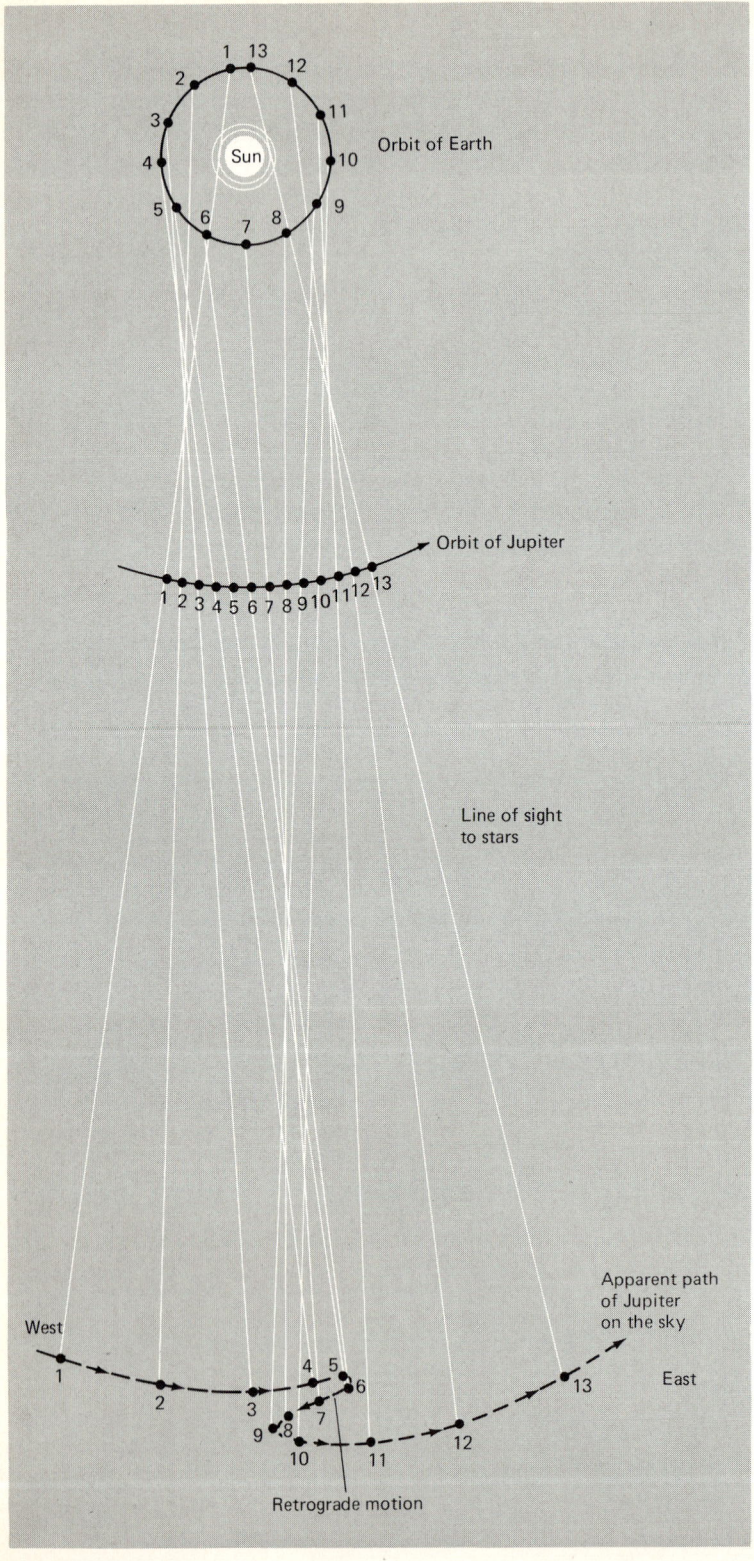

FIGURE 3.6 Retrograde motion of Jupiter in the Copernican model. As the earth comes around the same side of the sun as Jupiter, it is moving faster along its path and so passes Jupiter (point 7). During this passing interval Jupiter appears to move backward (to the west) with respect to the stars (points 5–9). Note that Jupiter appears in the middle of its retrograde motion just as the earth passes it, so Jupiter is in opposition to the sun as viewed from the earth. Also, it is closest to the earth, and so appears its brightest in the sky. The same basic diagram applies to Mars and Saturn; they undergo retrograde when the earth passes them. The illusion of retrograde motion results from the passing situation.

3.2 THE HELIOCENTRIC MODEL

seen from the interior one). The synodic period (let's call it S) is the time until they are aligned again. What is the relationship of this time to the sidereal periods of the inner and outer planets (let's call them P_i and P_o)?

From the starting position the inner planet, moving at $360/P_i$ degrees per day, travels faster than the outer one. The outer planet, moving at the slower rate of $360/P_o$ degrees per day, lags behind the inner planet more and more each day. For example, at the end of one day the inner planet has gained an angle of $(360/P_i - 360/P_o)$ degrees on the outer one. Since in one synodic period (S) the inner planet gains $360°$ on the outer one, this one day's gain is equal to $360/S$ degrees. In algebraic terms

$$\frac{360°}{S} = \frac{360°}{P_i} - \frac{360°}{P_o}$$

or, by dividing by $360°$,

$$\frac{1}{S} = \frac{1}{P_i} - \frac{1}{P_o}$$

This relation holds as long as we use the same base unit for S, P_i, and P_o. Let's use years. Suppose the earth is the inner planet. Then, since the earth's sidereal period is one year, $P_i = 1$ and

$$\frac{1}{S} = 1 - \frac{1}{P_o}$$

where P_o and S, the sidereal and synodic periods of this outer planet, are in years.

Now imagine the case of the earth as the outer planet. Then $P_o = 1$ is the earth's sidereal period, and

$$\frac{1}{S} = \frac{1}{P_i} - 1$$

where all the quantities are in years. Let's do this one out completely for Venus. The synodic period of Venus is 585 days, or $585/365 = 1.60$ years. Using the equation just above,

$$\frac{1}{P_{Venus}} = 1 + \frac{1}{1.60}$$

$$\frac{1}{P_{Venus}} = 1.62$$

$$P_{Venus} = 0.62 \text{ year}$$

Making this transformation from synodic to sidereal periods, Copernicus found that the planetary order, from Mercury to Saturn, fell into a *natural* sequence, going from the shortest sidereal period (Mercury) to the longest (Saturn). He saw a harmony in this

TABLE 3.1 Copernicus's Relative Distances of the Planets

Planet	Copernicus's Value (AU)*	Modern Value (AU)
Mercury	0.38	0.387
Venus	0.72	0.723
Earth	1.00	1.00
Mars	1.52	1.52
Jupiter	5.22	5.20
Saturn	9.17	9.54

*AU = Astronomical Unit; equivalent to the average earth-sun distance.

sequence, which later scientists such as Kepler and Galileo considered an essential elegance of the heliocentric model.

Relative Distances in the Copernican Model

Copernicus went one important step further. He not only calculated the order of the planets, but he also calculated their actual relative distances from the sun. Copernicus had no way to fix his system's distance scale (in units like kilometers or miles) from observations, for all he could do was observe angular speed and direction. But he could establish from observation the scale of distances *relative to* the earth-sun distance (Table 3.1). Copernicus was extremely pleased by this result; here he found the commensurability in the motions of the planets and their distances from the sun. Let's see how this was done.

The method Copernicus used differs for the planets interior to the earth and those exterior. For Mercury and Venus, the interior planets, the method is a direct geometrical one and rests on the observed maximum elongation angle of the planet from the sun. This angle averages 23° for Mercury and 46° for Venus. Take Venus as a specific case.

In a heliocentric model maximum elongation for Venus occurs when the line of sight to Venus is a line tangent to its orbit (Fig. 3.9); the radius of Venus's orbit is then at right angles to the line of sight at the circumference of the orbit. The angle of maximum elongation (α) is then one angle of a right triangle, with the radius of Venus's orbit being

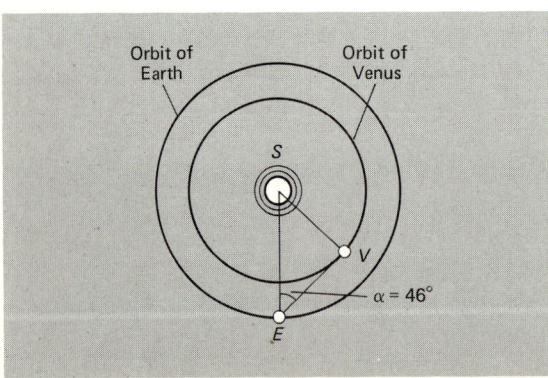

FIGURE 3.9 The distance of Venus from the sun. As seen from the earth at its time of greatest elongation, Venus makes an angle (α) of about 46° from the sun. The line of sight from the earth then touches tangent to the orbit of Venus. So the radius of Venus's orbit drawn to this point makes a right triangle with one side the sun-Venus distance (SV) and the other the earth-sun distance (ES).

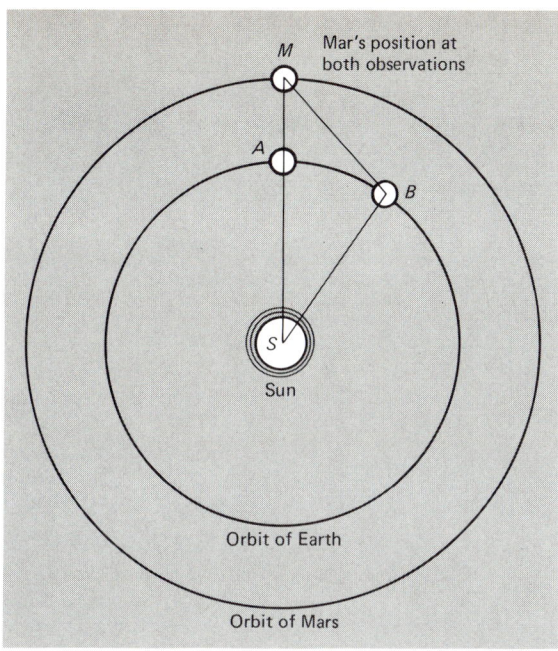

FIGURE 3.10 The distance of Mars from the sun. Start with the earth at A and Mars at M at opposition. Wait one sidereal period of Mars. Then Mars returns to the same place (M) in its orbit (as seen from the sun), but the earth has gone around almost twice (to B). The line of sight from the earth to Mars crosses the Martian orbit at the same spot that the earlier one did (from A). Call the earth-sun distance 1 unit. Then the distance from the sun to Mars is known to this scale because all the angles are known.

one side (side SV), and the radius of the earth's orbit being another side (side SE). (Make sure you notice where the 90° angle is in this triangle. It is at Venus, not at the sun.) The ratio of side SV to side SE is the sine of angle α, so

$$\sin \alpha = \frac{SV}{SE}$$

The angle α is known from observation to be 46°; the sine of 46° is approximately 0.72. So the ratio of Venus's distance from the sun (SV) to the earth's distance from the sun (SE) is about 0.72. Let us define the average earth-sun distance as 1 *Astronomical Unit* (AU). Then Venus must be 0.72 AU from the sun. Similarly, you can find the size of Mercury's orbit, taking 23° as the maximum elongation angle, to be 0.39 AU. Note that it takes only one observation for each of the inner planets to establish their distances from the sun (relative to the earth-sun distance) in the heliocentric model.

The method of finding the distances for the outer planets is less direct. Let's use Mars to illustrate the method. Start with Mars at opposition (earth at position A in Fig. 3.10). Then Mars must be somewhere along the line SA extended away from the sun. Wait until one sidereal period of Mars (about 1.9 years) has passed, so that Mars returns to the same position in its orbit. (Remember that in the Copernican system you can derive the sidereal period from the observed synodic period.) The earth has gone around less than twice (position B in Fig. 3.10). Now draw a line at the proper angle with respect to the line SB to indicate the direction of Mars as seen from B. Where the two lines intersect must be the position of Mars (M in Fig. 3.10). If you have drawn an accurate scale diagram, taking the earth-sun distance as 1 AU, you can measure the sun-Mars distance (SM) in AU's from the diagram. Or you can solve for the side SM using trigonometry, because you know all the angles in the triangle MSB. A similar approach applies to the other outer planets. The distances found by these methods (Table 3.1) are fixed by the planetary periods and the geometry of the heliocentric model.

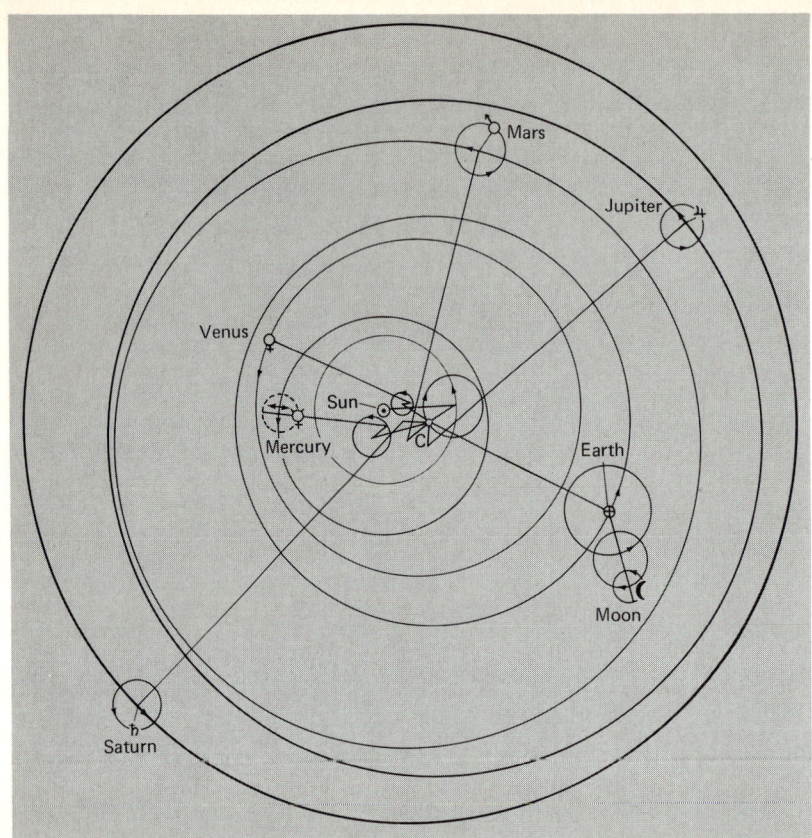

FIGURE 3.11 Details of the Copernican model. Observations forced Copernicus to add eccentrics and small epicycles (called epicyclets) to his model to account for irregularities in the planets' motions. The overall result resembled the complete Ptolemaic model. Compare to Figure 2.19. [*Adapted from a diagram by W. Stahlman.*]

Predictions of Planetary Motions

Copernicus eliminated the epicycles as an explanation of retrograde motion and at the same time removed equants from the system, but he still needed small epicycles to explain the variable speeds of the planets. In practice the heliocentric model predicted planetary positions no better than that of Ptolemy. Judged by a simple count of circles, the heliocentric model was not even simpler than the Ptolemaic one. Although Copernicus eliminated five circles that had been required for retrograde motion, he had to introduce more small ones to replace the equants. The actual Copernican model came out somewhat more complicated than the original model of Ptolemy (Fig. 3.11). Nonetheless, the Copernican model displayed a fundamental, compelling unity which was lacking in the Ptolemaic one.

Copernicus also sidestepped two important issues. First, his system violated Aristotelian physics (for example, in stating that earthlike material, the earth itself, did *not* fall to the center of the universe, but revolved around it). Second, the earth's revolution required a heliocentric stellar parallax (Section 2.3) which was *not*, in fact, observed. The Copernican cosmos was spherical and finite, with the stars in a shell at some fixed distance from the sun. Copernicus explained away the failure to observe stellar parallax by assuming that the sphere of fixed stars was so far off compared with the earth-sun distance that stellar parallax was undetectable. As for physical ideas, Copernicus did not refute Aristotle's views in detail. A new physics still needed development. Copernicus had provided only a new geometrical model, not a new physical one.

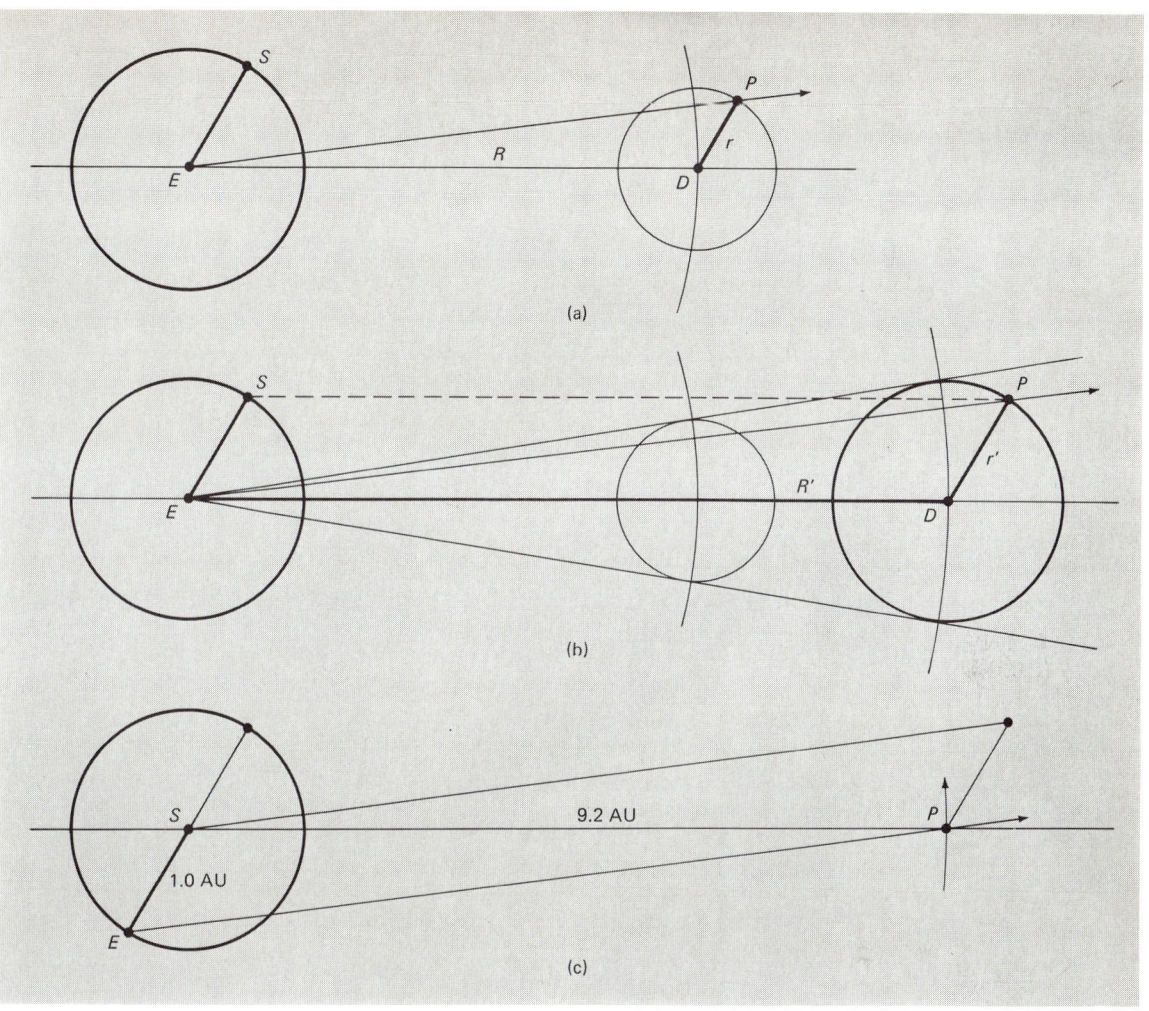

FIGURE 3.12 The geometrical equivalence of the Copernican and Ptolemaic models. Starting with the Ptolemaic scheme (a) for a planet (P) farther from the sun than the earth (E), geometrical transformations (b) preserve all angles and result in a Copernican scheme (c).

The Copernican model did *not* predict planetary positions any better than an updated Ptolemaic system. Because Copernicus maintained the assumption of circular, uniform motion, his heliocentric model was still similar to the Ptolemaic geocentric one. (Compare Figure 2.19 with Figure 3.11.) Except that a different point was regarded as the center, *the models were geometrically the same.* Here's a demonstration. Choose an outer planet (Fig. 3.12a). Note in the Ptolemaic model that the earth-sun line (ES) is parallel to the radius of the planet's epicycle (r). Imagine expanding the planet's epicycle until it is the same size as the earth-sun distance (from r to r'). At the same time expand the planet's deferent (from R to R') so that the angular size of the epicycle remains the same as viewed from the earth (Fig. 3.12b). The epicycle and deferent have been enlarged to retain the same proportions, and so all *angular* relationships have been preserved. Then push the planet (along r') until it lies in the center of its enlarged epicycle, and at the same time push the sun to the earth's position and the earth out to the sun's circle (Fig. 3.12c).

Again, all angular relationships are kept the same, since all motions have been along parallel lines. What do you see? The Copernican setup with the earth-sun and planet-sun distances in their correct relative proportions. This example should convince you that the Ptolemaic and Copernican models are more geometrical than physical in their construction.

Warning: Do not allow present knowledge of gravity to distort your understanding of the Copernican system. The Copernican model did not include a force between the sun and the planets. Copernicus probably believed in the reality of the heavenly spheres of quintessence and certainly held that the planetary motions resulted from the natural, uniform, and circular motions of the celestial spheres. As in the Ptolemaic model, they did not require a force.

The Impact of the Heliocentric Cosmos

Although Copernicus's claims clashed with those of the scholastic scientists and Christian theologians, some astronomers in the middle of the sixteenth century took his model seriously and tried to test the new setup of the planets. They hoped to improve upon the *Alfonsine Tables*, which were known to contain some inaccurate predictions of planetary conjunctions.

Foremost among those willing to try the new techniques was Erasmus Reinhold. Reinhold quickly calculated planetary position tables based on the Copernican model. These were called the *Prutenic Tables*, and these were adopted by almanac makers. However, Reinhold did not swing to wholehearted adoption of the heliocentric system. He was not a wholly convinced Copernican; he saw the system as a geometrical model rather than as reality.

If astronomers did not rush to accept the sun-centered cosmology, why did Copernicus bother to develop the idea at all? Certainly he was strongly motivated by aesthetic reasons; he saw a new harmony in the "design of the universe and fixed symmetry of its parts" that was "pleasing to the mind." The "fixed symmetry" referred to the fact that in the Copernican model the spacing and order of the planets are fixed by observation; in the Ptolemaic model each planet's circles could be treated independently of the others.

3.3 TYCHO BRAHE: SUPPORTER OF THE GEOCENTRIC VIEW

Tycho Brahe (1546–1601) rejected the heliocentric system on both physical and observational grounds. He proposed an alternative cosmological model that was a compromise of the classical geocentric universe and the Copernican one. His work reveals the attitudes of a typical professional astronomer at the time just after the publication of *De revolutionibus*; it emphasizes the fact that the reality of the Copernican cosmology escaped the astronomers of the day, who saw the heliocentric model as merely a geometrical device.

Tycho Brahe (Fig. 3.13) was born in Denmark into a family of noble standing. While Tycho was a child, his uncle, who had no children and wanted an heir, stole him from his parents, supported the boy, and decided that his adopted son should become a lawyer, a suitable profession for a Danish nobleman. Young Tycho was sent off with a private tutor

FIGURE 3.13 A portrait of Tycho Brahe. It shows the part of his nose that was cut off in a fencing duel and replaced with a silver piece.

to study law in Germany, but he secretly worked on astronomy, spent his allowance on astronomy books, and crept out at night to make observations.

When Tycho returned to Copenhagen, he lectured on astrology, a subject in which he had a deep interest. His fame as an astronomer reached the Danish court. In November 1572 a new star burst into view in the constellation Cassiopeia (Fig. 3.14). Such an apparently new star was called a *nova* from the Latin *stella nova*, "new star." (Today we know that a nova is a star normally too faint to be conspicuous that suddenly increases in brightness so that it becomes visible to the naked eye. We also know that this new star, now called Tycho's star, marked a very violent outburst we call a *supernova*.) The nova was so bright that it could be seen in the daytime.

Tycho used this rare event in a remarkable way. He collected observations of the nova from all over Europe and showed that the new star occupied the same position relative to the stars in the constellation as seen from all observation points. He concluded that since the nova did not have an observable shift in position, it must be a great distance from the earth—certainly beyond the moon. This observation directly contradicted the Aristotelian model, which required that all changes in the sky take place within the orbit of the moon. Tycho's positioning of the new star beyond the moon collided with the traditional doctrine that the heavens must be immutable, since they are perfect.

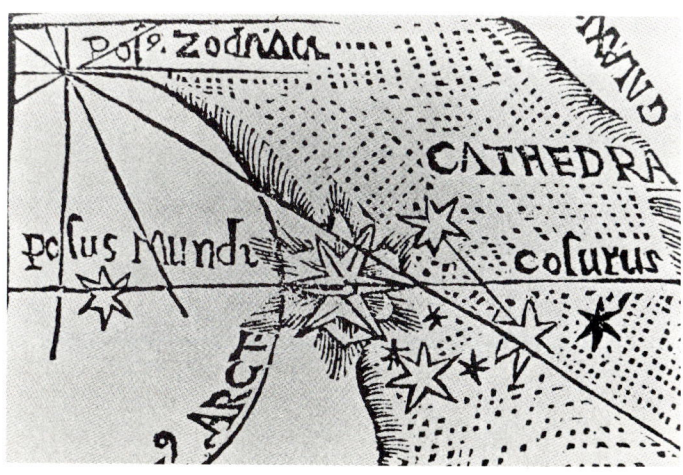

FIGURE 3.14 The location of Tycho's "new star" of 1572 (large star near center) in the constellation of Cassiopeia. [*From Stella Peregrinae by Cornelius Gamma, 1573.*]

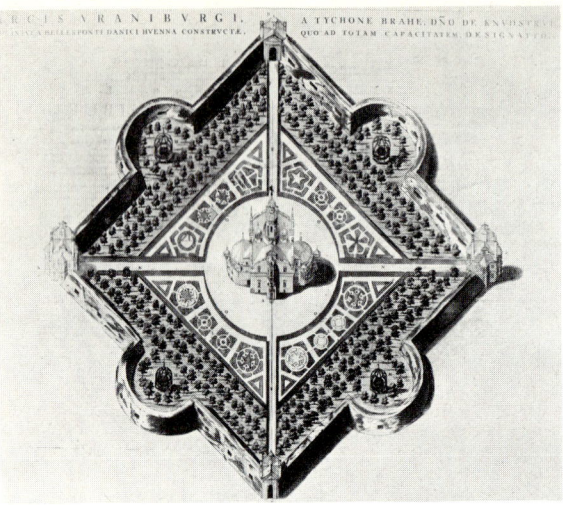

FIGURE 3.15 Tycho's observatory grounds on the island of Hveen. [*Courtesy Yerkes Observatory.*]

Tycho's work on the nova of 1572 catapulted him to fame and persuaded King Frederick II of Denmark to give him his own observatory on an island site over which Tycho would rule as feudal master (Fig. 3.15). The king gave Tycho the use of the island Hveen across the water from Hamlet's castle at Elsinore. With royal funds, estimated by him to be more than a ton of gold, Tycho constructed on Hveen the first modern research observatory: Uraniborg, Castle of the Heavens (Fig. 3.16). Here he worked in grand style. Not only did he have the finest observing equipment (much of which he designed himself) and a bevy of assistants, but he also had his own paper mill to provide the paper for his own printing press, which published his observations (along with some of his inspired poetry). An extensive wine cellar provided for the entertainment of visiting dignitaries, while a separate building in the garden housed a prison for people who annoyed him.

FIGURE 3.16 A view of Uraniborg, Tycho's research observatory.

3.3 TYCHO BRAHE: SUPPORTER OF THE GEOCENTRIC VIEW

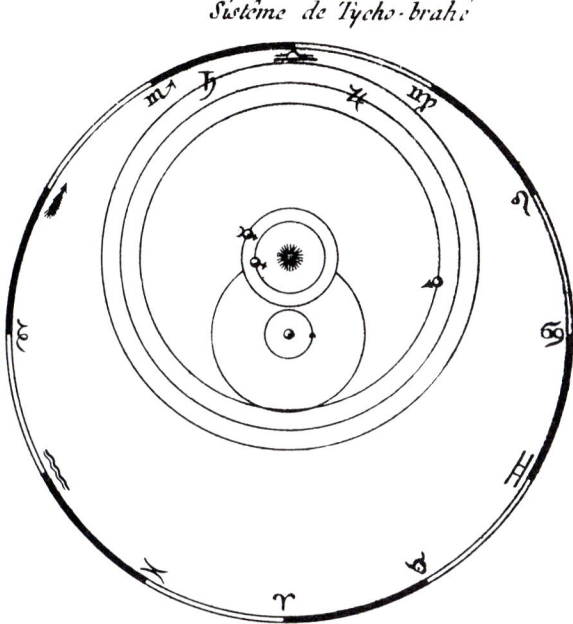

FIGURE 3.17 A simplified version of Tycho's model for the cosmos. It is a geocentric modification of the Copernican one. The sun moves around the earth, and all the planets circle around the sun. [*Courtesy Yerkes Observatory.*]

Convinced that the classical geocentric model was untenable, Tycho devised his own, which retained the earth's central position. Although Tycho viewed the Copernican system simply as a hypothetical, geometric model to describe planetary motions, he took very seriously Copernicus's dictum that all celestial motions must be uniform and circular. Tycho could not accept the revolution of the earth around the sun, for he thought that "the earth, that hulking, lazy body," was "unfit for a motion as quick as that of the ethereal torches." His model was geocentric, but geometrically identical (Section 3.2) to the Copernican system—all that changed was the center of motion (Fig. 3.17). In Tycho's model the moon and sun revolve around the earth, and all the planets revolve around the sun.

Tycho also had an observational objection to a heliocentric model. With his precision instruments he tried to measure an annual heliocentric parallax (Section 2.3), but failed to detect it. Tycho's equipment could pick up an angular change of as little as 1'. The lack of a detectable parallax of at least 1' fortified Tycho's belief that the Copernican system was invalid.

Tycho's main contribution to astronomy was not his planetary system, but rather his extensive collection of observations of planetary positions. He made a careful effort to observe planetary positions to the limit of naked-eye ability (about 1 minute of arc), not only at the times of importance, such as during retrograde motion, but also at intermediate times. For the first time in astronomical history a continuous record (1576–1597) was compiled of precise planetary positions.

When Frederick II died, Tycho fell from royal favor because of his despotic ways. He therefore moved with his servants and his equipment to Prague, to work for Emperor Rudolph of Bohemia. Here he took on the young Johannes Kepler as an assistant to do theoretical work. When Tycho died in 1601, an era in astronomy ended. A new one began when Kepler created a new astronomy based on Tycho's observations.

FIGURE 3.18 Johannes Kepler. "Astronomy has two ends, to save the appearances and to contemplate the true form of the ediface of the world."

3.4 JOHANNES KEPLER AND THE COSMIC HARMONIES

Despite its heroic shifting of the earth and the sun, the Copernican system had a conservative intention: to develop a model of planetary motions more in line with the classical ideas of uniform, circular motion than the Ptolemaic model. Copernicus accomplished this goal, but his system was as technically complicated as Ptolemy's. Also, the Copernican model lacked a basis in physics, which the Ptolemaic system did have. The heliocentric model of this time was more geometrical than physical.

Johannes Kepler (1571–1630) forged the new ideas for planetary motion that form the foundations of modern concepts about the nature of the planetary orbits (Fig. 3.18). Kepler made the Copernican model a truly heliocentric one, with the sun cast in a central, *physical* role that is absent from the geometrical plan of *De revolutionibus*. Kepler was the first to understand that the sun determines the planetary orbits by some physical force. Not only did he inject essential physics into the heliocentric model, but he also improved its usefulness for planetary predictions.

The Harmonies of the Spheres

Kepler was born on December 27, 1571, in the small town of Weil der Stadt (near modern Stuttgart, in southwestern Germany). Kepler worked out the time and date of his conception, 4:37 A.M. on May 16, 1571, because he considered it an important date in the casting of his own horoscope. (Remember that in Kepler's day astronomy and astrology were considered one discipline. Much of a professional astronomer's duty at the time concerned astrological matters.)

In 1589 Kepler entered the University of Tübingen, where his superior mental ability was formally recognized by the school senate. After achieving his B.A. and M.A. degrees, Kepler enrolled in the three-year theological program with the intention of becoming a Lutheran clergyman. In his last year as a theology student he was selected to replace the teacher of mathematics in the Protestant high school in Graz, Austria. Since Kepler was a

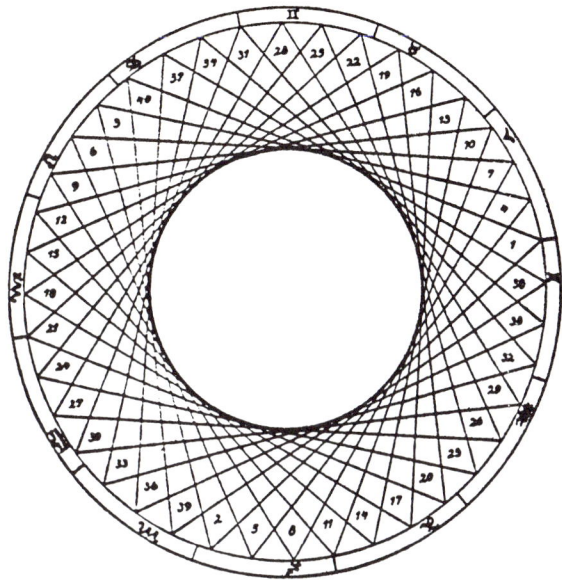

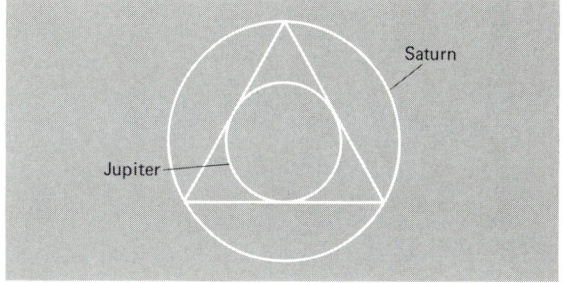

FIGURE 3.19 Kepler's drawing of the cycle of conjunctions of Jupiter and Saturn. These conjunctions take place about every 20 years, moving in position about one-third of the way around the zodiac. Because they do not occur at exactly 120° intervals, a set of three is followed by three that are shifted a little in angle. The vertices of these triangles drawn in sets of three conjunctions move around the zodiac. A circle is formed where the lines of the triangles cross. The ratio of this circle to the larger one formed by the vertices of the triangles is almost the same as that of the distances of Jupiter and Saturn from the sun.

FIGURE 3.20 The orbits of Jupiter and Saturn drawn within (Jupiter) and around (Saturn) a triangle.

scholarship student, he was at the bidding of the duke, who in consultation with the Tübingen faculty decided that young Kepler was the suitable man for the job.

Kepler quickly made his mark in Graz. Part of his job was the preparation of a small almanac that included meteorological, political, and astrological predictions. To Kepler's surprise, three of his major predictions were fulfilled. His fame as an astrologer stemmed from these first lucky successes.

His fame as an astronomer also began accidentally. One day while teaching his class, Kepler was struck by the most compelling insight of his life. He had drawn a series of triangles within two circles while explaining to his students the conjunctions of Jupiter and Saturn (Fig. 3.19). As he inscribed the triangles, he realized that the ratio of the radii of the outer and inner circles, which is 2 to 1 (Fig. 3.20), almost equaled the ratio of the distances of Jupiter and Saturn from the sun, which is 1.83 to 1. He found this result exciting, for Saturn was the first planet in from the sphere of stars and the triangle is the first (simplest, because it has only three sides) plane figure. He stumbled upon an idea that directed much of his later astronomical thinking: a geometrical design of the solar system.

Kepler believed in the Copernican system. While at Tübingen he had been taught

FIGURE 3.21 The five regular solids. They are the only possible solid figures made with faces of regular polygons.

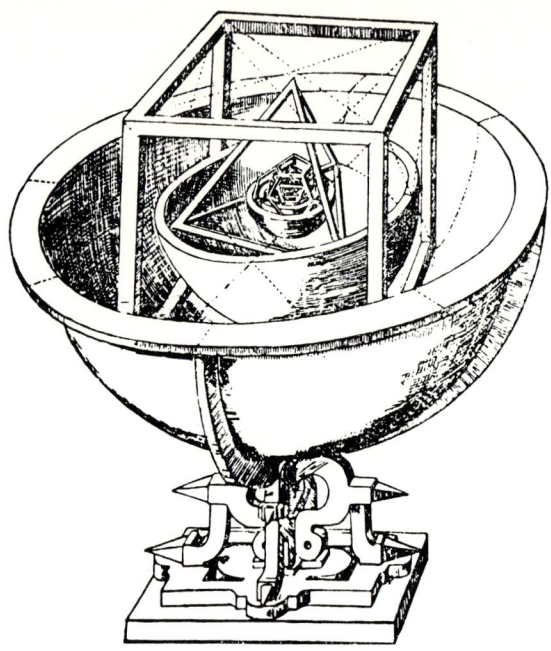

FIGURE 3.22 Kepler's model for the spacing of the planets. During Kepler's life six planets were known, so five spaces existed between them. Kepler knew that only five regular solids exist (Figure 3.21). By arranging these solids between the spheres holding the planets, Kepler thought he explained the "ediface of the world." [*Courtesy O. Gingerich.*]

astronomy by Michael Maestlin, one of the first people to read *De revolutionibus* carefully. Although Maestlin did not actively promote the Copernican cosmology, he did refer to Copernicus and used some of the tables from *De revolutionibus*. Kepler apparently was excited by the harmonies he found in the heliocentric system, which he quickly adopted in his thinking.

One of these harmonies appears in the spacing of the planets. Copernicus had noted this relationship but had not offered an explanation. Kepler thought he could. For three days he worked without solving the problem. Then he realized that the system was actually three-dimensional rather than two-dimensional and so required solid rather than plane figures. From classical geometry Kepler knew that there are only five possible regular solids—solid figures with all faces having the same kind of regular polygons (Fig. 3.21). The known planets numbered six, separated by five spaces. Kepler thought the five solids might be used to establish the planetary spacings and worked out the details of nesting the solids and spheres (Fig. 3.22). The finished scheme accounted for the distances of the planets with an accuracy within 10 percent of the values accepted at the time. Kepler's intuitive insight resulted in a geometrical, heliocentric structure for the solar system. His scheme was wrong, the agreement with observation a mere coincidence, but this geometrical model energized much of his study of planetary motions.

3.4 JOHANNES KEPLER AND THE COSMIC HARMONIES

Excitedly, Kepler wrote to his former teacher, Maestlin, telling him of both the spacing system and another idea: that the sun in the middle of the universe must have some power or force to propel the planets around in their orbits. This view seemed reasonable to him because Mercury, being closer to the sun where the force would be stronger, moves faster in its orbit than Venus, and so on. The model was firmly implanted in his mind and motivated his later work.

In 1594 he published the result of his work in *Mysterium cosmographicum (The Cosmic Mystery)*, which announced the details of his geometrical system and asserted his adherence to the Copernican system—the first such printed treatise by any professional astronomer after Copernicus. His intuition that the sun physically directs the planetary motions both justified the Copernican model and set the stage for the next development of astronomy in terms of physical laws. Here we see that Kepler believed the heliocentric model to be *real*, not just a calculational device.

The Battle with Mars

In his foreword to *Mysterium cosmographicum*, Maestlin expressed his appreciation of the new cosmos and his hope that it might soon be tested by comparison with actual observations. Kepler felt somewhat insulted by this remark, for he thought that his blueprint explained the observations adequately. However, he did think that he needed better values for the eccentricities that determined the thickness of the spherical shells. To obtain them Kepler turned to the foremost observer of the day: Tycho Brahe.

Tycho had read Kepler's *Mysterium cosmographicum* and, although he did not like its mystical approach to astronomy, recognized it as the work of a genius. When he received Kepler's request for the orbital eccentricities, Tycho did not honor it, for he felt reluctant to send his hard-earned observations to a believer in the Copernican system. However, he did invite Kepler to discuss the matter personally. By then Tycho had left Denmark for Prague, and Kepler traveled there to meet him.

Kepler and Tycho met on February 4, 1600, an encounter that was fateful to Kepler and to the history of modern science. During the next 10 months the personalities of these two men clashed so strongly that they accomplished little astronomical work.

One night while at a party with a baron, Tycho drank too much. Because it was considered impolite to take leave during the affair, Tycho was not able to relieve himself and suffered a prostate infection, which resulted in his death. In the few days before he died, Tycho urged Kepler to justify Tycho's cosmological system from Tycho's observations. Upon Tycho's death Kepler was promoted to Tycho's position (but with only one-third the nobleman's salary). Kepler was allowed more access to the unpublished observational records, but Tycho's heirs believed them to have some financial value and so reserved censorship rights until the emperor actually paid for the observing books. After some bargaining with the heirs, Kepler attacked the problem of the orbit of Mars.

Kepler worked for four years on the motions of Mars. Starting with a heliocentric scheme that used a circular orbit with an equant and an eccentric, Kepler fitted Tycho's observations within 2′ along the ecliptic. However, the predictions of the planet's position above or below the ecliptic were wide of the mark. This discrepancy struck Kepler as important, for he realized that the planetary orbits must be viewed in three dimensions. He then moved the center of the earth's orbit and found that he could correctly predict the positions above and below the ecliptic, but not along it, where he was off by 8′. Kepler

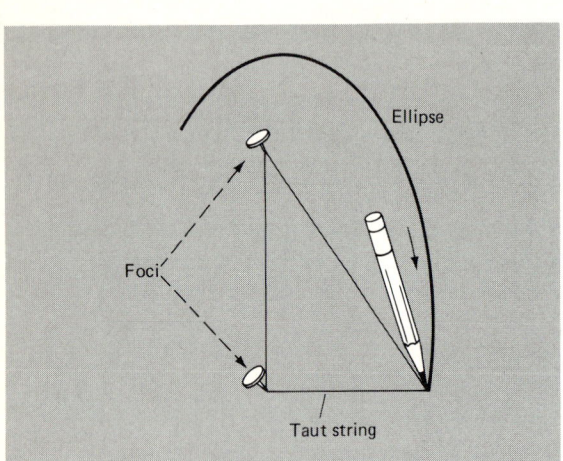

FIGURE 3.23 Drawing an ellipse with a string loop and tacks.

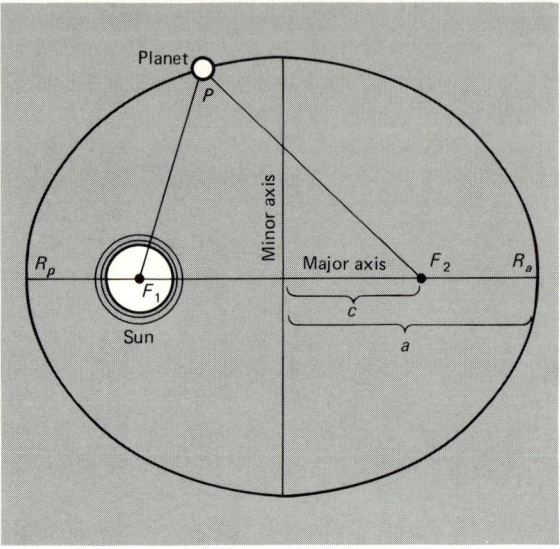

FIGURE 3.24 An ellipse. The sum of the distances of any point on the ellipse from the two foci (F_1 and F_2) is a constant.

was dissatisfied with the result, for he knew that Tycho's observations were often as accurate as 1'. He had nevertheless discovered that the plane of the earth's orbit (and eventually of the other planetary orbits as well) passed through the sun. The discovery marked a critical prelude to Kepler's recognition of his other laws of planetary motion, for it gave the sun a special status in the solar system.

Kepler sought a single physical explanation for planetary motion driven by the sun. What force did the sun use to keep the planets in their orbits? Influenced by William Gilbert's work *De magnete*, Kepler envisioned the sun as a source of magnetic force. Using the observations of Tycho, Kepler returned to the Martian orbit and applied his magnetic-force analysis to find that it worked *if* the true orbit of Mars was *elliptical* rather than circular. He wrote, "With reasoning derived from physical principles agreeing with experience, there is no figure left for the orbit of the planet except a perfect ellipse." (We now know that it is gravity, rather than magnetism, which controls the planets, but this force also requires elliptical orbits.) By this process Kepler found the first of his famous three laws of planetary motion, which rested on a physical explanation of the accurate observations of Tycho. Further analysis led to statements about the speed of the planets in their orbits.

Properties of Ellipses

To grasp Kepler's discoveries, you need to understand some properties of ellipses. To draw an ellipse take a loop of string; fasten it to a board in two places with tacks so that the part of the loop between the tacks forms a straight line; pull the loop taut on the board with a pencil (Fig. 3.23). Keeping the string taut with the pencil, draw a curve around the tacks. That curve is an ellipse (Fig. 3.24). The two tacks mark the two *foci* of the ellipse.

Each point on the ellipse has the property that the sum of its distances from the two foci (F_1 to P and F_2 to P in Fig. 3.24) is the same. (That's why the taut string construction works.)

The line through the foci to both sides of the ellipse (R_a to R_p) is called the *major axis*. Half this length is the *semimajor axis*, usually designated a. So the major axis has length $2a$. Note this equals the length $R_p F_1 + R_p F_2$ or ($R_a F_1 + R_a F_2$), so

$$F_1 P + F_2 P = 2a$$

The distance from the center of the ellipse to a focus is designated c; so the distance between the two foci is $2c$.

How an ellipse differs from a circle (how "squashed" it looks) is determined by the ellipse's *eccentricity*, e. It is defined as

$$e = \frac{c}{a}$$

Imagine you took the two tacks at the foci and moved them closer together. The ellipse would become more circular; c, and thus e, becomes smaller. When the two tacks are exactly together the ellipse becomes a circle; its eccentricity then is zero. As the two tacks are moved farther apart, the eccentricity increases until the tacks sit at opposite ends of the major axis. The distance c is then equal to a and the eccentricity is 1.

If the sun is imagined at one focus (F_1) of a planet's orbit, then R_p, the closest point to the sun, is the *perihelion* point in the orbit. The perihelion distance $R_p F_1$ is $a - c = a - ae = a(1 - e)$. R_a, the farthest point from the sun, is the *aphelion* point. The aphelion distance $R_a F_1$ is $a + c = a + ae = a(1 + e)$. Note that the sum of the perihelion and aphelion distances equals the major axis:

$$R_p F_1 + R_a F_1 = a(1 - e) + a(1 + e) = 2a$$

or, the average of this greatest and smallest distance equals the semimajor axis, a.

Kepler's Laws of Planetary Motion

Kepler's fame today rests primarily on his three laws of planetary motion, yet these laws were but small fragments in his wider search for harmonies in the physics of celestial motions. In modern terminology Kepler's laws are:

1. **Law of Ellipses (1609).** The orbit of each planet is an ellipse, with the sun located at one focus (Fig. 3.25). (The other focus is located in space and not centered on any body.) Note the distance from the sun to the planet varies as the planet moves along its elliptical orbit.
2. **Law of Equal Areas (1609).** The line drawn from the planet to the sun sweeps out equal areas in equal times. Law 2 notes that the orbital velocities are nonuniform, but vary in a regular fashion. The farther a planet is from the sun, the more slowly it moves in its orbit (Fig. 3.26).

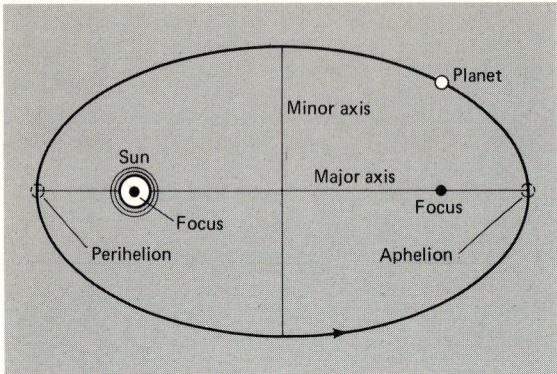

FIGURE 3.25 Kepler's first law. The shape of the planetary orbits is elliptical (greatly exaggerated here) with the sun located at one focus of the ellipse.

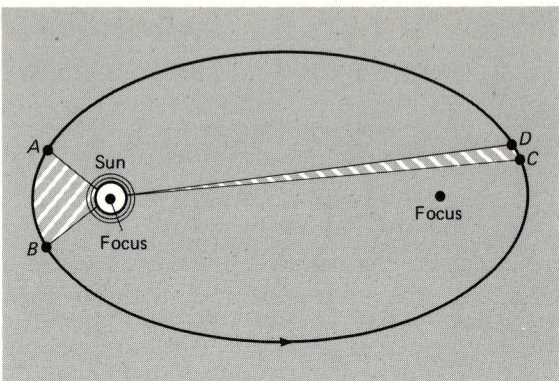

FIGURE 3.26 Kepler's second law. Consider two equal periods of time: one when the planet is closest to the sun (AB), and one when it is farther away (CD). At AB the planet-sun distance is shortest, and the planet moves fastest in its orbit. At CD the planet moves more slowly because it is farther from the sun. In both cases the areas (shaded areas AB to the sun and CD to the sun) covered by the line drawn from the planet to the sun are equal.

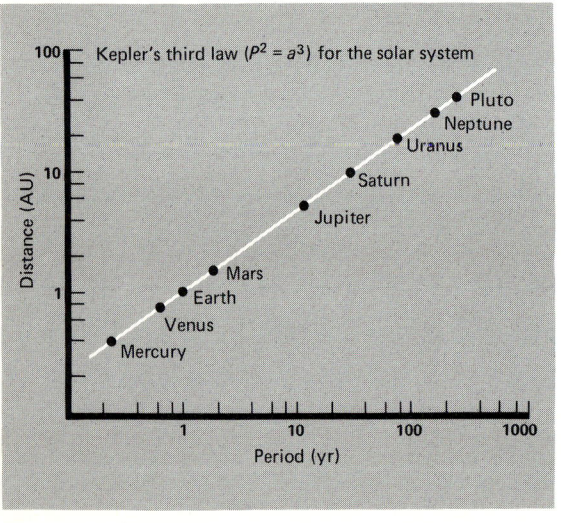

FIGURE 3.27 Kepler's third law. A planet's orbital period (once around the sun) is related to its average distance from the sun. This diagram shows for each (now known) planet a plot of its orbital period (in years) squared to its distance (in AU) cubed.

3. Harmonic Law (1618). The square of the period of a planet is directly proportional to the cube of the planet's average distance from the sun, that is, to the cube of the semimajor axis of the elliptical orbit (Fig. 3.27). Law 3 also points out that planets move more slowly, in a predictable fashion, the greater their distance from the sun, a fact that implies that any sun-planet force decreases with distance.

The dates given above are those of publication, not of discovery.

Kepler discovered the second law before the first, when he was grappling with the orbit of Mars and Tycho's observations. Even before finding the second law, Kepler had arrived at the crucial conclusion that the planes of all the planetary orbits pass through the sun. This discovery set the stage for the first law.

3.4 JOHANNES KEPLER AND THE COSMIC HARMONIES

Algebraically, the third law may be written

$$P^2 = ka^3$$

If P is the period of the planet's revolution about the sun in years and a is the average distance from the sun in AU's, then the constant k is equal to 1. An example: Mars has an average distance of 1.5 AU from the sun. How long does it take Mars to orbit the sun once? Solution:

$$P^2 = (1.5)^3 = 3.4$$
$$P = (3.4)^{1/2} = 1.8 \text{ yr}$$

Note that the third law applies to *any* body (even a spacecraft) in orbit around the sun and subject only to its gravitational force.

For systems of bodies other than the sun and planets, the constant k has a different value that depends on the units chosen for a and P, and also on the mass of the central body. (Chapter 4 discusses this dependence.)

What do these laws mean? They describe all the essential features of planetary motion. The first law points out that the shape of the planetary orbits is elliptical, so a planet's distance from the sun varies. The second law notes that as a planet's distance varies, so does its orbital speed. The closer a planet is to the sun, the faster it goes. You get the sense that the sun pulls a planet toward it, the planet whips around the sun, and then the sun slows it down as it moves away again. The third law says that the farther a planet's orbit is from the sun, the slower its orbital motion will be. Both the second and third law imply that the force between the sun and planets is weaker at greater distances. Hidden in the second and third law is an exact description of how the sun-planet force works, how it weakens with distance, but Kepler was not able to figure it out. Not long after Kepler, Isaac Newton solved this final puzzle about the ancient problem of the motion of the planets.

The New Astronomy

As he announced in his *Astronomia nova (The New Astronomy)* in 1609, Kepler had, in his opinion, developed an astronomy based not only on geometry, but also on physical causes. With the discovery of the ellipse as the shape of the planetary orbits, he satisfied both the observations of Tycho Brahe and his own conviction that the dance of the planets could be conducted by some physical cause. From this approach Kepler calculated the *Rudolphine Tables* (1627), which supplanted both the *Prutenic Tables* based on the Copernican system and the venerable *Alfonsine Tables* calculated from the Ptolemaic system. The *Rudolphine Tables* completely revised the previously shoddy science of theoretical astronomy where a few degrees of error were considered quite accurate. Kepler made predictions far more accurate than any earlier ones based on the Copernican model (Fig. 3.28) or on the Ptolemaic model.

Kepler broke the ancient spell cast by the idea of perfect circles and uniform motion that had mesmerized astronomers for over 2,000 years. In his search for the physical design underlying the harmonies of the heavens, Kepler found three laws which accu-

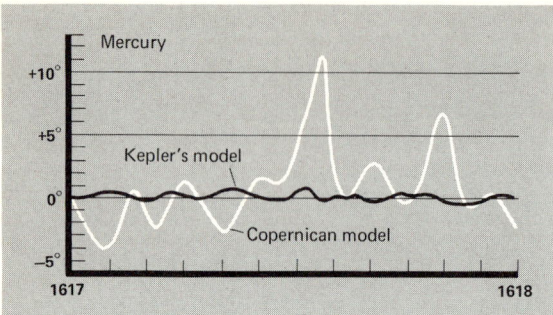

FIGURE 3.28 The accuracy of predictions of planetary positions. This diagram shows the difference between the predicted positions and the actual positions of Mercury in Kepler's model (heliocentric, elliptical orbits, nonuniform speeds) and the Copernican one (heliocentric, circular orbits, uniform speeds). In the Copernican model the error is as large as 10°. For Kepler's scheme the maximum error is 10 times smaller. [*Courtesy O. Gingerich and B. Welther.*]

rately describe planetary motions. He rewove the fabric of the heavens into a pattern that proclaimed the end of an ancient era and the birth of a modern science.

3.5 SCIENTIFIC MODELS AND COSMOLOGICAL REVOLUTIONS

What drives the evolution of cosmological models? Recall that any model tries to explain and predict observations in the simplest way possible. Copernicus may have been motivated to propose a new model because of discrepancies in predictions by the Ptolemaic model. Yet, in the end, his model did no better with predictions than did Ptolemy's. Copernicus did feel that his explanations of celestial motions was more pleasing than Ptolemy's. In particular, he was able to get rid of the equant and provide a natural explanation of retrograde motion with a heliocentric scheme.

To what degree was the Copernican model more pleasing? Today we find the heliocentric scheme quite natural, to a large extent because the original model has been replaced by ones in which there is a physical justification for the heliocentricity. But in Copernicus's time, a moving earth seemed clumsy and unphysical. What is simple and elegant in one age is viewed differently by another. Aesthetics also evolves.

Note that although agreement with observations is a key requirement for a useful model, the correspondence is by itself insufficient for prolonged success. Kepler's geometrical design, the five regular solids nested within six spheres, gave predictions accurate to a few percent. But we now reject this model, despite its accuracy of predictions, because it lacks a physical basis. (Discovery of other planets would have invalidated it anyway, for there are five and only five regular solids.)

This analysis brings up another point: the changing attitude toward the accuracy of predictions as a model develops. When is a prediction "good enough"? It is not just a matter of predicting to the accuracy that can be observed. The Ptolemaic model gave predictions that were several degrees off—an error which is easily observable with the naked eye. Yet that was considered satisfactory for many centuries. Modern astrophysicists, groping for an understanding of quasars, for example, would consider their models a success if they came within a factor of 10 of predicting the observed properties of quasars. Yet Kepler rejected several of his models because they gave predictions that were off by 8′ of arc from Tycho's observations, which he knew were accurate to about 1′. And we suppose the quasar theorists of future generations will not be satisfied until they can explain all the properties of quasars to an accuracy of a few percent or better.

Tycho's failure to observe a parallax for the stars was a much tougher test of the heliocentric model than anything Ptolemy might have argued. The fact is that the stars are much farther away than anyone in earlier centuries could possibly have imagined. The parallax of the nearest of them is only about 1 sec of arc, 60 times smaller than anything Tycho could have observed.

The crucial element of all scientific models is change. They must evolve as we learn more about the world. Sometimes the change is no more than tinkering with a small part of a model. At other times, as when the Copernican model took over from the Ptolemaic one, a real revolution occurs. The impulse behind these cosmological revolutions seems to come from aesthetics that are finally confirmed by observations.

SUMMARY

This chapter covered a crucial episode in the evolution of astronomy. Copernicus, in a return to traditional Greek ideas, attacked the Ptolemaic model and replaced it with a heliocentric one. Copernicus apparently presented his new cosmology on aesthetic grounds, for his system was "pleasing to the mind," because it placed the sun in its appropriate central position in the universe, described celestial motions as uniform without equants, and established a natural order of planets from the sun based on their increasing sidereal periods.

But the Copernican model had no direct observational support, for it required a heliocentric parallax that was not observed and it predicted planetary positions with no more accuracy than the Ptolemaic system. In addition, the motions of the earth violated the essentials of the physics of the day. Note that from naked-eye observations alone (those described in Chapter 1), you cannot decide between Copernicus's heliocentric and Ptolemy's geocentric model.

Tycho Brahe's observational legacy formed the basis of Kepler's transformation of the Copernican model into a *physical* cosmology. Copernicus's system lacked a physical basis and predicted planetary positions no better than Ptolemy's model; so in terms of scientific model-making it had two critical weak points. Kepler shored up both. His elliptical orbits led to more accurate positions, and his goal of physical causes invested the sun with an active role in keeping the planets in their places. Although Kepler's use of magnetic forces later proved unworkable, his demand for physical explanations resulted in a cosmological model closely tied to observations. This demand arose because Kepler believed the heliocentric model to be real. Kepler was driven to find a force whirling the planets around in part because of his discovery that the orbits were, in fact, elliptical. Such shapes ruled out a simple model of nested material spheres.

The overt connection of a model with reality as defined by observation lies at the heart of modern scientific explanation. Kepler stands out as the first to require such an interlocking, the first to pull together astronomy and cosmology with the rope of physical laws (Table 3.2).

TABLE 3.2 A Comparison of the Ptolemaic (P), Copernican (C), and Keplerian (K) Systems

Observed "Fact"	Explanation
Motion of entire heavens daily from E to W	P: Motion of all heavenly spheres E to W C: Reflection of rotation of earth from W to E K: Same
Annual motion of sun W to E through zodiac	P: Rotation of sun's sphere W to E in a year C: Reflection of annual revolution of the earth about the sun K: Same
Nonuniform motion of sun through zodiac	P: Orbit of sun eccentric, with uniform speed C: Same K: Elliptical orbit of the earth, with nonuniform speed
Retrograde motions of the planets	P: Epicycles and deferents C: Relative motions of planets, including earth, around sun K: Same
Variations in retrograde motions	P: Equant, eccentrics C: Small epicycles, eccentrics K: Nonuniform orbital motion and elliptical shape of orbits
Distances of planets	P: Arbitrary as long as angular relationships are correct C: Relative distances set by observations K: Distances related to periods
"Cause" of planetary motions	P: Natural motion of celestial spheres, no force C: Same K: Magnetic force from sun
Accuracy of predictions	P: Error typically 5° or less, sometimes 10° C: Same K: Error generally about 10', sometimes as large as 1°

Key Words

conjunction
sidereal period
synodic period
Astronomical Unit (AU)
ellipse
Kepler's laws of planetary motion
perihelion
aphelion
semimajor axis
eccentricity

Review Questions

1. The Copernican system was superior to the Ptolemaic one in that it (mark all correct answers):
   a. Required fewer circles.
   b. Used no epicycles.
   c. Eliminated the unaesthetic equant.
   d. Explained retrograde motion in a simpler way.
   e. Not only established the order of the planets but gave their relative distances.
   f. Predicted planetary predictions with far greater accuracy.
   g. Was not considered to be only a calculational device, but was thought to describe the absolute reality of the cosmos.
   (Objective 5)
2. Which of the following are found in Kepler's model and *not* in the Copernican one?
   a. Elliptical orbits
   b. Nonuniform motion
   c. Force between the sun and the planets
   d. Prediction of heliocentric stellar parallax
   (Objective 10)
3. What was Copernicus's *primary* objection to the Ptolemaic system? (Objective 1)
4. How did Copernicus account for retrograde motion in his system? (Objective 2)
5. If you were on Mars, under what astronomical circumstances would you see Jupiter retrograde? The earth? (Objective 2)
6. What is the difference between a planet's *synodic* and *sidereal* periods? How can sidereal periods be found from synodic periods? (Objective 4)
7. In *your* opinion, what was the major advantage, if any, of the Copernican system over the Ptolemaic one? (Objectives 1 and 5)
8. What keeps the planets moving in the Copernican system? (Objectives 6 and 7)
9. Give at least two differences between the models of Copernicus and Kepler. (Objective 7)
10. What advantages did Kepler's system have over that of Copernicus? (Objective 7)

Problems

1. If the average maximum elongation of Mercury from the sun is 23°, what is the sun-Mercury distance in Astronomical Units? (Objective 3)
2. Take the sun-Mercury distance you calculated in the question above and use one of Kepler's laws to find Mercury's orbital period. (Objective 9)
3. Mercury's synodic period is 116 days. What is its sidereal period? Compare with your answer to Question 2. (Objective 4)
4. If heliocentric parallax is less than 1', how far away (at least) must the stars lie from the sun? (Objective 5)
5. The semimajor axis of Mars's orbit is 1.52 AU and its eccentricity is 0.093. The eccentricity of the earth's orbit is 0.017. How far is the sun from the centers of each orbit? What are the perihelion and aphelion distances of the earth and Mars? What is their maximum and minimum separation at opposition? (Objective 8)

BEYOND THIS BOOK . . .

* For an excellent analysis of the work of Copernicus and the framework of Western thought in which he developed, read *The Copernican Revolution* (Vintage Books, New York, 1959) by T. Kuhn.
* For a modern variety of views about Copernicus and the impact of his work, try *The Nature of Scientific Discovery* (Smithsonian Institution Press, Washington, D.C., 1975), edited by O. Gingerich.
* An excellent biography of Tycho Brahe is *Tycho Brahe: A Picture of Scientific Life and Work in the Sixteenth Century* (Dover, New York, 1963) by J. L. E. Dreyer.
* A fine biography of Kepler is in *Kepler* (Collier Books, New York, 1962) by M. Casper, translated by C. D. Hellman. A very technical look at Kepler's work can be found in *An Account of the Astronomical Discoveries of Kepler* (University of Wisconsin Press, Madison, Wisc., 1963) by R. Small. The *Dictionary of Scientific Biography* has a very informative article on Kepler by O. Gingerich.
* You can try to tackle a controversial view of Copernicus, Tycho Brahe, Kepler, and Galileo in *The Sleepwalkers* (Grosset & Dunlap, New York, 1963) by A. Koestler; in particular, read part III, "The Timid Canon," and part IV, "The Watershed."
* Volume 16 of *Great Books of the Western World* (Encyclopaedia Britannica, Chicago, 1952) contains a translation of Copernicus's *De revolutionibus*. It also contains excerpts from two of Kepler's books.

THE CLOCKWORK UNIVERSE: A Physical Cosmos

4

LEARNING OBJECTIVES
After studying this chapter, you should be able to:
1. Describe Galileo's important telescopic discoveries and their impact on the controversy between the Copernican and Ptolemaic systems.
2. Indicate Galileo's purpose in developing a new science of terrestrial motions.
3. Describe the difference between accelerated and unaccelerated motion.
4. Use simple equations to describe the motions of bodies for the case of constant velocity and uniform acceleration.
5. Contrast Galileo's astronomy and cosmology to that of Copernicus and Kepler.
6. Cite Newton's three laws of motion, describe each in simple terms using specific examples, and apply them to astronomical situations.
7. Contrast Newton's concept of natural motion to that of Aristotle.
8. Describe Newton's law of gravitation.
9. Make use of Newton's laws of motion and gravitation in algebraic form.
10. Outline how the moon-apple test confirms Newton's law of gravitation.
11. Define the concept of center of mass, and show how the center-of-mass relation follows from Newton's laws.
12. Use the concept of centripetal force to show that Kepler's third law leads to a law of gravity in which the force varies as the inverse square of the distance.
13. Contrast Newton's cosmology to that of Copernicus, and use Newton's physical ideas to support a Copernican model.

CENTRAL QUESTION:
How did Newton's laws of motion and gravitation unify the universe physically?

What makes the world go 'round? For the planets, according to Isaac Newton, it's inertia and gravitation. Newton achieved the goal that had eluded Kepler: a mechanical model of the cosmos controlled by a single physical force. Newton saw the force of gravitation as the prime mover in the clockwork universe.

Newton arrived at his great achievement after important groundwork had been laid by Copernicus, Kepler, and Galileo. Copernicus made a crucial move in placing the sun, rather than the earth, at the center of the universe. To do so, Copernicus had to violate the physics of his day and offered no good substitute. Kepler, motivated by mystical insights, sought to establish a clockwork cosmos driven by a single force. In his choice of a magnetic force he failed. But his demand for a force from the sun to drive the planets in their orbits directly influenced Newton's vision. Galileo recognized that the Copernican model needed a new physics to back it up. He searched for new physical laws by studying the motion of falling bodies near the earth. Galileo did describe the effects of gravity, but he never extended these ideas from the earth to the heavens.

Newton shattered the old separation of the earth and heavens. He linked gravity—terrestrial physics—to the orbital motion of the planets—celestial physics. The links he forged were his laws of motion and gravitation. With Newton's ideas the universe took on a new appearance. No more the closed, finite universe of Ptolemy and Copernicus. In Newton's vision the universe grew to an infinite expanse, connected by a single force: *gravitation*.

4.1 GALILEO: ADVOCATE OF THE HELIOCENTRIC MODEL

The Italian scientist Galileo Galilei (1565–1642) desired to establish celestial physics on a firm experimental and mathematical basis. In contrast to Kepler, Galileo (Fig. 4.1) concerned himself almost exclusively with terrestrial motions, especially those of falling bodies. He felt that to establish the laws of terrestrial motions would do more to cement the structure of the Copernican cosmos than any observations, including those made with a

FIGURE 4.1 Galileo Galilei. "By denying scientific principles, one may maintain any paradox." [*Courtesy Yerkes Observatory.*]

telescope. He failed to achieve his goal, but his discoveries guided the later work of Newton.

The Magical Telescope

Galileo used his telescope to bolster the Copernican model. He did *not* invent optical lenses nor their use in a telescope. Rather, he promoted his astronomical ideas by utilizing the novelty and shock value of telescopic observations.

By Galileo's time glass lenses had been known for about 300 years. Their date and place of origin are not clear, but they were used by eyeglass makers to correct defects of vision. The opticians of the day had no physical understanding of how their glasses functioned and had adopted a purely experimental approach to their construction.

In 1609 a messenger returning to Venice brought Galileo the news that a Dutchman had constructed a spyglass that made distant objects appear to be nearby. Although Galileo had little experience with optics, he set to work immediately in his workshop to duplicate the instrument (Fig. 4.2). Sparing neither labor nor expense, he ultimately succeeded in constructing an optical device that made objects appear 30 times closer than when viewed with the naked eye (Chapter 7). He put this marvelous tube to astronomical use. Within a few weeks in 1609 and 1610 he made a series of astronomical discoveries that began a new era in astronomy. Although Galileo was not the first person to build a telescope, he first recognized that the telescope increased our power to perceive reality in the heavens.

Galileo first observed the moon. He saw that the moon's surface was not smooth and spherical, as Aristotelian ideas required for perfect heavenly bodies, but rough, with chains of mountains and valleys and many craters (Fig. 4.3). He determined the height of a lunar mountain from the length of its shadow. The height of about 6 km was the astonishing (and correct) result.

Galileo next peered at the stars. His instrument's power fragmented the faint band of the Milky Way into innumerable stars, many more than could be seen individually with

FIGURE 4.2 (a) The lens from Galileo's largest telescope. It was accidently broken by him and mounted in this ivory frame in 1677. (b) Two telescopes made by Galileo. The upper one magnifies 14 times, the lower 20 times. [*Courtesy M. L. Righini Bionelli, Florence Museum of the History of Science.*]

the unaided eye (Fig. 4.4). This observation refuted the Aristotelian idea that the sky contained only a certain number of stars and that this known number could not change.

Then Galileo hit on what he considered to be his most important discovery: four new "planets" that no one had seen before. What he found were not actually new planets, but rather the four brightest moons of Jupiter. (At least 16 moons are now known; see Chapter

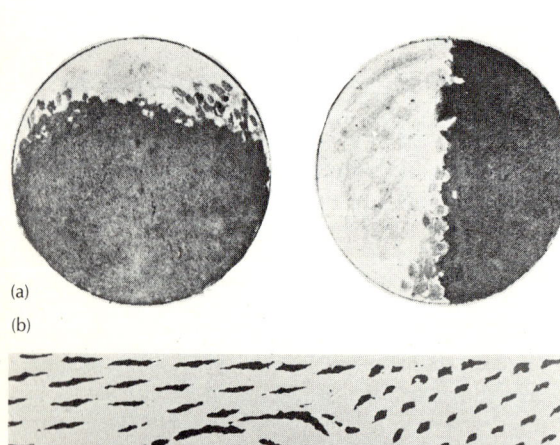

FIGURE 4.3 (a) Galileo's drawings of the moon as seen through his telescope. Note the craters and mountains. (b) A detailed drawing of a lunar crater done by Galileo in 1611.

FIGURE 4.4 Galileo's drawing of the stars in the region of the Pleiades. Here he saw many more stars than visible to the unaided eye. [*Courtesy O. Gingerich.*]

17.) To his amazement and joy, Galileo found that these four bodies revolved around Jupiter (Fig. 4.5). Here he found another argument against the scholastics, for Jupiter and its satellites resembled a miniature solar system. This fact required a second center of revolution in the cosmos, a notion that contradicted the Aristotelian doctrine that only the center of the universe could be the center of the revolution and that that center was the earth.

Galileo described his novel observations in *Siderius nuncius (The Starry Messenger)*, which was published on March 12, 1610. His telescopic discoveries, especially that of Jupiter's moons, grabbed the public imagination. The book sold as fast as it could be printed; it brought fame to Galileo and a demand for the telescopes produced in his private workshop. However, he was still cautious in print; although he used his observations to demolish the Aristotelian cosmos, he did not openly advocate the Copernican scheme or support it with his observations.

Galileo also discovered with his telescope that Venus exhibits phases just like the moon. Venus is illuminated by the sun, and we can see only the part which is lit up (Fig. 4.6). So, when we look at Venus from different angles, we will see it as new, crescent, gibbous, or full—the whole array of phases, just like for the moon. The difference is that full phase is hard to observe for it occurs when Venus is on the other side of the sun as viewed from the earth. The boxes in Fig. 4.6 show how Venus would look through a telescope when it is at the positions marked.

The observation of these phases for Venus gives strong support to the Copernican model. In the Ptolemaic model Venus would also show phases, but they would be different. For example, there would never be a full phase, because in Ptolemy's scheme Venus is never on the other side of the sun. Refer to Figure 2.18 and see if you can decide how Venus would look through a telescope at various points along its epicycle. Contrast these phases with those predicted by the Copernican model.

Warning: The observations of the phases of Venus do favor the Copernican model over the Ptolemaic one, but they do *not* in general show that a heliocentric model is better than a geocentric one. For example, Tycho's geocentric model (Fig 3.17) produces the same phases for Venus as the heliocentric model of Copernicus.

Critics quickly scorned Galileo's work by suggesting that he was seeing atmospheric phenomena or false images produced by flaws in his telescope's lenses, arguments that were reasonable enough in the seventeenth century. The art of lens making was not far advanced, so many lenses did indeed generate extra, so-called ghost images. (The optics in Galileo's telescopes were far inferior to those in a cheap modern telescope.) His opponents said that although Galileo may have honestly reported what he had seen, his observations had no direct connection with reality.

Galileo, however, was convinced by the regularity of the telescopic phenomena that he was viewing reality and not illusion. In 1611 Kepler, who was well regarded in scientific circles, backed up Galileo with observations of his own and also developed a theory of optics to support the validity of telescopic observations.

Goaded by the opposition to what he considered indisputable facts, Galileo continued to scan the skies for new marvels. By projecting the image of the sun onto a piece of paper, he observed sunspots. As defects on the supposedly perfect sun, the sunspots dealt another blow to the tenets of Aristotelian cosmology. In 1613 Galileo published his results in *The Letters on Sunspots* and made his first direct, printed declaration of his belief in the Copernican system.

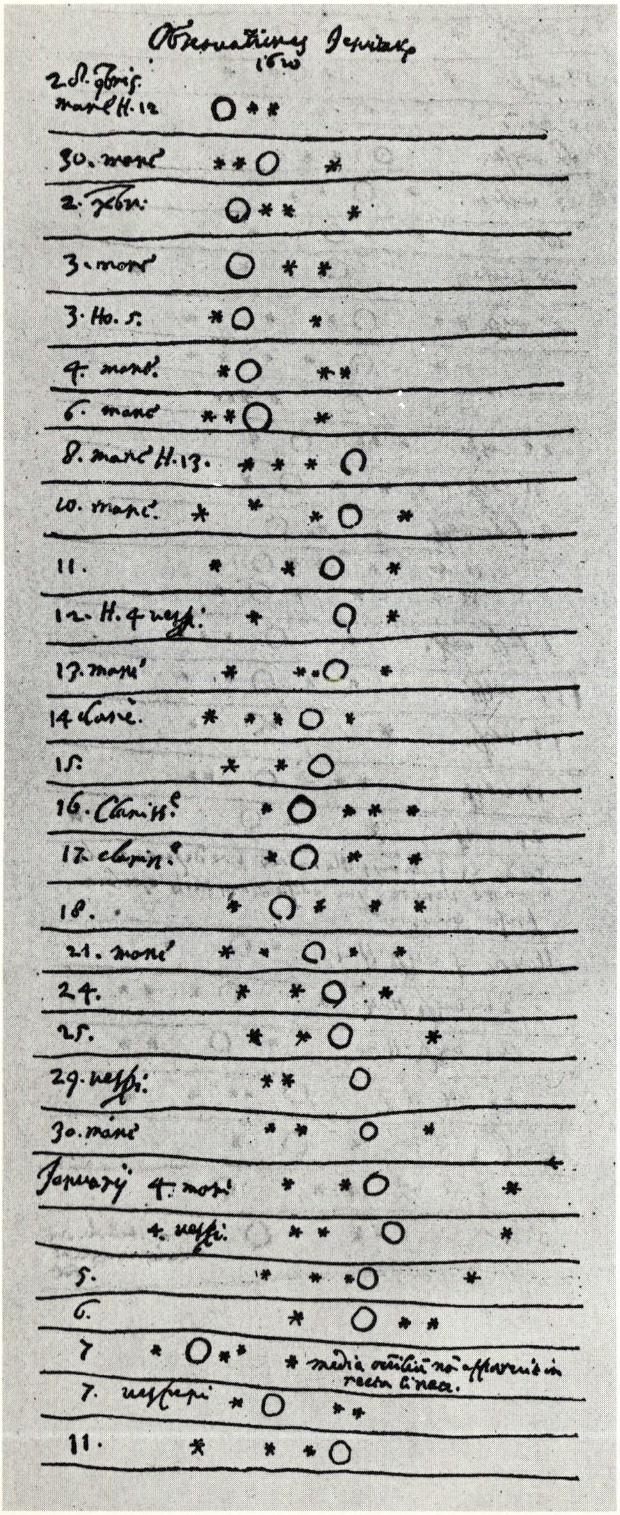

FIGURE 4.5 Some of Galileo's observations of the moons of Jupiter. The large circle represents Jupiter; the "stars" are the moons. Their changing positions next to the planet indicated that they were revolving around Jupiter and were not just background stars. [*Courtesy Yerkes Observatory.*]

FIGURE 4.6 The phases of Venus in the Copernican model. The view through a telescope at each position is shown in the boxes.

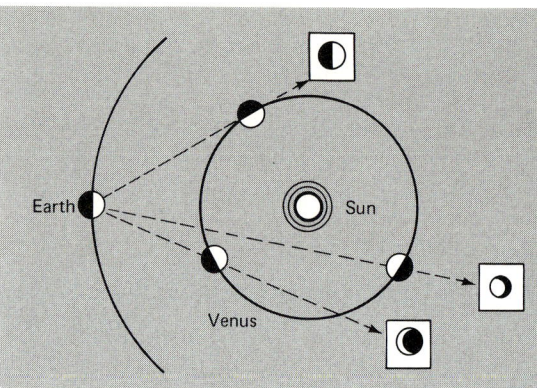

4.1 GALILEO: ADVOCATE OF THE HELIOCENTRIC MODEL

The New Science of Mechanics

Despite his observational triumphs Galileo still felt that the Copernican system lacked the anchor of a physical understanding of motion such as Aristotle had provided for his own system. Kepler had made an important step by his mathematical description of *celestial* motions. Although Galileo ignored most of Kepler's work, he took the next step by devising a mathematical description of *terrestrial* motions, particularly those of bodies falling under the influence of gravity. Galileo aimed to justify the Copernican scheme by a systematic, experimental search for physical ideas of motion. He unraveled the details of motion produced by applied forces; this study is called *mechanics*. To found the new science of mechanics Galileo needed an important tool: a description of the motion of falling bodies, that is, the motion of masses under the influence of gravity. But before we explain Galileo's contribution, let's describe motion more precisely.

Acceleration, Velocity, and Speed

When you step on a car's accelerator, it does just that—accelerates. If you start out at rest, the car goes faster and faster, as you can tell from the speedometer. Or, if you are cruising on the highway, you overtake a car by increasing your speed. In these instances your velocity changes; that's the meaning of acceleration.

But what is velocity? It's not simply speed, which tells just how fast you're going. Velocity involves *direction* as well as speed. For example, if you drive from Albuquerque to Santa Fe, a distance of about 100 km, in an hour, your speed is about 100 km/h. When you drive back at the same speed, your velocity is different because you're heading in a *different direction*.

Imagine now that you're riding on the outside horse of a merry-go-round which turns at a constant speed. Are you accelerating? Yes, because your velocity is constantly changing *direction* as you turn around the circle of the merry-go-round, even though your speed does not change.

Of course, you are also accelerating if *both* your speed and direction change.

So speed and velocity have different technical meanings. Speed is the average rate of travel measured in units of distance per unit of time, for example, miles per hour or meters per second. Velocity is speed with direction added. (Quantities which involve direction as an essential part of their meaning are called *vectors*.) Velocity is also measured in units of distance per units of time, with direction implied. Acceleration is the rate at which velocity (not speed) *changes*. It is measured in units of velocity per unit of time, for example, meters per second per second or miles per hour per second. It also is a vector, for the velocity changes in a certain direction.

Uniform Acceleration

Galileo's work on mechanics focused on finding the relationship between distance and time for the case of uniform acceleration. Let's look at this concept in modern terms.

Take the simple case of an object moving with a constant velocity, which means it has zero acceleration (Fig. 4.7a). If its velocity is 10 m/sec, at the end of 1 sec the object has

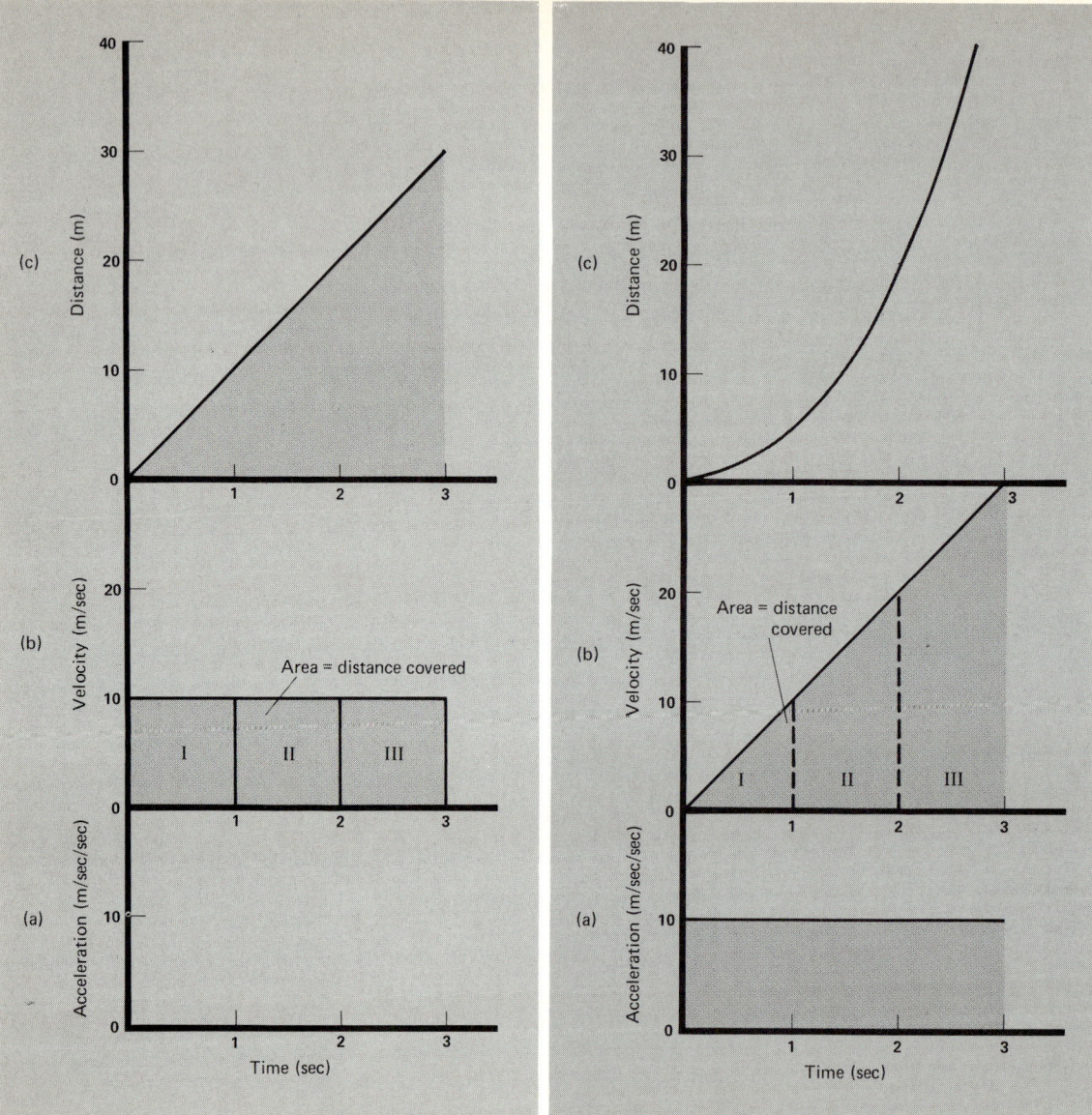

FIGURE 4.7 Motion at a constant velocity of 10 m/sec.

FIGURE 4.8 Motion at a constant acceleration of 10 m/sec/sec.

gone 10 m, after 2 sec 20 m, and so on. We can state this fact algebraically with the equation

$$d = vt \quad \text{(constant velocity)}$$

where d is the distance traveled, v is the velocity, and t is the time. Note that the distance traveled (Fig. 4.7c) is equal to the *area* under the line in the *velocity-time plot* (Fig. 4.7b), since each square (I, II, III) has an area of vt.

Now consider the case of uniform (or constant) acceleration. If the acceleration is 10 m/sec/sec (Fig. 4.8a), at the end of 1 sec the object has a velocity of 10 m/sec, after

2 sec 20 m/sec, and so on (Fig. 4.8b). Algebraically, the relationship between velocity and time is

$$v = at$$

What will the *distance-time plot* be for this case? In the case of uniform velocity the area under the line in the velocity-time plot (Fig. 4.7b) turned out to be the distance traveled. The same approach works here. In Figure 4.8b area I is a triangle with an area of $\frac{1}{2}$ (base) × (height), or $\frac{1}{2}$ vt, which equals $\frac{1}{2}$ × (10 m/sec) × (1 sec), or 5 m. Now note that areas I and II together also form a triangle, which has an area of $\frac{1}{2}$ × (20 m/sec) × (2 sec), or 20 m. That's how far the object's gone after 2 sec of travel. We've plotted these points (and a few more) to make a distance-time graph in Figure 4.8c. Notice that the points do not fall on a straight line, but make a curve (a parabola) in which the distance changes as the *square* of the time. So we infer from this plot that for the case of uniform acceleration the distance traveled is proportional to the time squared. The constant of proportionality is the acceleration divided by 2. Algebraically, it is stated as

$$d = \tfrac{1}{2} at^2 \quad \text{(uniform acceleration)}$$

where d is the distance, a the acceleration, and t the time.

Another simple way to obtain this result is to multiply the *average velocity* (v_{avg}) by the time. At the start the velocity is 0; after time t it is at. So the average velocity is $(0 + at)/2$, and the distance is

$$d = v_{avg} t = \frac{(0 + at)}{2} t = \tfrac{1}{2} at^2$$

Natural Motion Revisited

One of the nagging problems of Aristotelian physics was that of *inertia*, the tendency of a body forced into motion to retain that motion or of a body at rest to remain at rest. Aristotle had divided motions into two categories: natural and forced (Section 2.3). The fall of a rock to the earth was attributed to the natural tendency of earthly material to seek the central point of the cosmos, and it required no force. But the throwing of a rock, a motion at right angles to the natural downward motion, required a continuously acting force to keep up the unnatural, forced motion. However, as you know, this statement is not strictly true, for after you throw an object, the motion will continue for a while even after the force is removed.

Galileo inverted Aristotle's ideas of natural and forced motion. He concluded that the downward motion of objects resulted from the attractive force of gravity, so falling objects underwent *forced* motion. In addition, he viewed the horizontal motion of objects flying through the air as due to their inertia. This inertial motion, he argued, was a *natural* one and would continue if no forces, such as air resistance, acted.

To take this important step, Galileo had to arrive at a description of inertia. He explained his ideas as follows (Fig. 4.9): If a perfectly smooth ball were placed on a hard, flat surface which sloped, the ball would roll down the slope forever, if the surface were

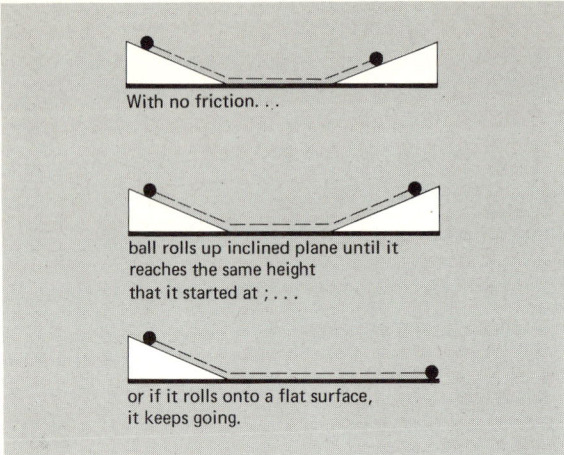

FIGURE 4.9 Galileo's concept of inertia (natural motion). The trick is to ignore friction in this thought experiment of rolling a ball down an inclined plane.

infinitely long. The greater the slope, the faster the ball would roll. By the same argument, a ball traveling up a slope would eventually stop. If the surface had no slope and the ball were placed on the level with no horizontal velocity, the ball would remain at rest. If a ball on a level surface were pushed, it would continue to move straight ahead at a constant velocity forever, since it is not slowed down by an ascent or speeded up by a descent.

With this concept of inertia Galileo dealt with the problem of falling bodies. As we have said, he did not view this motion as natural (as had Aristotle), but as motion due to a force. From a combination of experiments and intuition Galileo concluded that such motion takes place at a *constant* acceleration, with the object's velocity changing at a constant rate as it falls. This means that falling objects cover distances proportional to the squares of the times of fall. For example, after 1 sec a falling body drops 4.9 m, after 2 sec 19.6 m, after 3 sec 44.1 m, and so on. This observation implies that the acceleration is uniform and equal to 9.8 m/sec/sec. That's the acceleration due to gravity at the earth's surface; it is usually denoted by the symbol g.

So for objects falling near the earth and starting out at rest, the relation between the distance fallen (d), acceleration of gravity (g), and time of fall (t) is

$$d = \tfrac{1}{2} g t^2$$

For instance, since g is 9.8 m/sec/sec, at the end of 5 sec an object falls

$$d = \tfrac{1}{2}(9.8)(5)^2$$
$$= 122 \text{ m}$$

Galileo also found that *all* falling masses (not subject to air resistance) have the *same* acceleration. When dropped, they reach the same velocity after the same time and also fall the same distance in the same time. The conclusion that all masses fall with the same acceleration near the earth directly contradicted the scholastic teaching that a more massive body falls faster than a less massive one and so moves a greater distance in a given time.

4.1 GALILEO: ADVOCATE OF THE HELIOCENTRIC MODEL

Legend has it that Galileo climbed to the top of the Leaning Tower of Pisa and dropped two different objects, which, to the astonishment of the skeptical professors gathered around the tower's base, hit the ground at the same time. As far as historians can determine, Galileo did not actually attempt this experiment, although a friend of his may have. It would have been a risky demonstration, for, more likely than not, the masses would not have struck the ground simultaneously because of air resistance. (The rumored Pisa experiment may not be the only one that Galileo did not do, even though he reported the results in his writings. Some experiments were really mental exercises, but Galileo intuitively reached the correct results more often than not.) Try the experiment yourself. Use objects of different masses but the same basic shape—a couple of coins will do—so that air friction will not confound the results. Apollo astronauts did this experiment on the moon (which has no atmosphere) with a feather and a hammer. Both hit the surface at the same time.

The key points, found by Galileo, are that the motion of falling bodies near the earth is one of *uniform* acceleration and that all objects fall with the *same* acceleration.

The Crime of Galileo

Galileo had publicly declared himself a Copernican in 1613. His intellectual vehemence and enthusiasm irritated many of his opponents, some of whom enjoyed substantial power in the Church. In 1616 an official of the Inquisition apparently warned Galileo to stop teaching the Copernican theory as truth rather than as a hypothesis (as Osiander had done in his preface to *De revolutionibus*). Galileo's position was held to be contrary to Holy Scripture. At the same time *De revolutionibus* was placed on the Index of Forbidden Books until "corrections" had been made. (Only about 8 percent of the copies were actually changed.)

These two events, whose implications Galileo failed to grasp, initiated a complex chain of events that historians have yet to unravel completely. Galileo, a faithful and obedient Catholic, avoided teaching the Copernican theory as true. At the same time he entered into arguments about scientific truth versus revealed truth. His argumentative spirit promoted his downfall.

In 1623 Cardinal Barberini, a friend of Galileo's and a patron of the sciences, was elected Pope Urban VIII. Galileo thought he saw his chance and had a long audience with the new Pope, during which Urban discussed the decree forbidding the heliocentric teachings. Feeling that he had the support of the Pope and also of the Jesuit astronomers (many of whom adhered to the Copernican system), Galileo wrote the *Dialogue on the Two Chief World Systems* (Fig. 4.10). With the approval of the Catholic censors in Florence, the book was published in 1632, after a few minor corrections had been made. The book supposedly relates an objective debate about the relative merits of the Ptolemaic and Copernican systems, with the judgment being rendered in favor of the traditional view. In reality, the text is a thinly veiled polemic favoring the Copernican cosmos.

Immediately Galileo found himself in trouble, for his opponents swiftly countered his claims with theological arguments. The Pope himself was incensed because his favorite arguments had been put in the mouth of Simplicio, the supporter of the geocentric viewpoint in the *Dialogue*. (Galileo did not draw Simplicio kindly.) Some copies of the book were seized before they left the printers, and the Inquisition summoned Galileo to Rome and forced him to publicly recant his scientific beliefs. His friends were afraid to

FIGURE 4.10 Frontispiece to Galileo's *Dialogue*. Pictured, from left to right, are Aristotle, Ptolemy, and Copernicus.

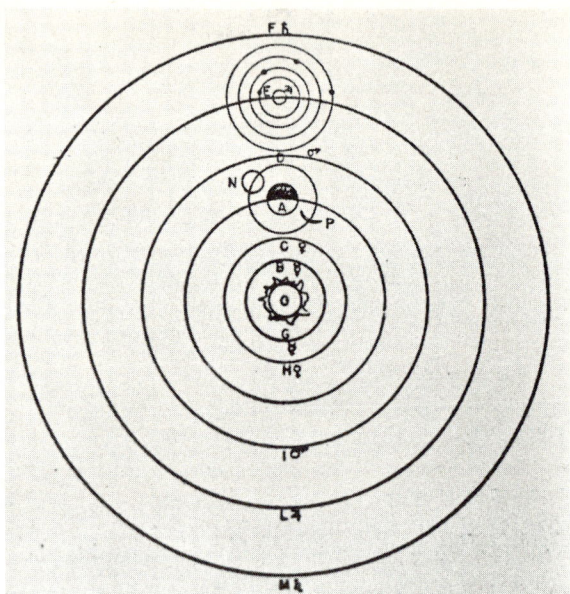

FIGURE 4.11 A drawing of Galileo's heliocentric model. Note that four moons orbit Jupiter. Also, the scheme lacks a sphere of stars just beyond Saturn; Galileo left open the possibility that the stars extend throughout space. All the orbits are circular, despite the discovery by Kepler that they are actually elliptical.

come to his support. As punishment he was placed under perpetual house arrest and denied the Church sacraments. The *Dialogue* remained forbidden to Catholics until 1835, when the works of Galileo, Copernicus, and Kepler were finally removed from the Index. Recently, the Roman Catholic Church decided to reexamine the case against Galileo.

Galileo's trial by the Inquisition, although motivated as much by internal Church politics and Galileo's abrasive personality as by conviction that the Copernican system was heretical, frightened intellectuals in regions where the Church exerted power. His confrontation with the Church surprised Galileo, who thought he had divorced theology from science. Despite his public difficulties, Galileo's views on motion became a pivot on which modern physics turned.

Galileo's Cosmology

In his *Dialogue* Galileo draws a model of the planetary system. It is essentially a Copernican scheme with no evidence of the ideas of Kepler, such as elliptical orbits. In addition, Galileo makes no attempt to apply his terrestrial mechanics to the motions of the planets.

He presents a simplified Copernican model, lacking eccentrics and small epicycles (Fig. 4.11). Here is evidence that Galileo paid little attention to the use of the Copernican system in predicting the planetary positions. What struck Galileo was the Copernican cosmology—the order of the universe—rather than the detailed astronomy of planetary motions. He apparently felt, with the same conviction as Copernicus, that the harmony of the planetary order was established from observation. Of course, Galileo relied heavily on his telescopic observations, along with the general arguments of Copernicus (Section 3.1), to support his beliefs.

Galileo placed the fixed stars in a thick shell far beyond the planets. This stellar shell remained from Greek ideas. However, his spokesman in the *Dialogue* keeps open the possibility that the universe is not spherical and finite, but open and infinite, with the stars sprinkled "through the immense abyss of the universe." Here we find hints of the idea of the cosmos as an infinite universe rather than a closed space, a view that would be supported by Newton.

4.2 NEWTON: A PHYSICAL MODEL OF THE COSMOS

Despite his interest in and insight into terrestrial motions, Galileo did not worry about the details of celestial motions. He never bothered to apply his new science of mechanics and understanding of gravity near the earth to the problem of predicting planetary positions. By this omission he failed in his quest for a physical foundation for the Copernican model.

Sir Isaac Newton (1642–1727) emerged as the genius destined to fuse the terrestrial and celestial realms and so to end the long-standing separation initiated by the Greeks (Fig. 4.12). The publication in 1687 of Newton's *Principia*, containing his new physics of motion and the concept of gravitation, resulted in a truly physical view of the universe.

The Prodigious Young Newton

In the small English village of Woolsthorpe, on Christmas Day 1642, Newton's widowed mother gave birth to a sickly child. The fragile baby was so small at birth that it is said he could have fit into a quart mug. At the age of 12, Newton enrolled at King's School, a 7-mile walk from his home. Because of the distance, he frequently lodged with an apothecary who encouraged the lad's experimentation. As a child, however, Newton did not display an outstanding genius. Absentmindedness was one of his more obvious traits; one day he walked home holding a bridle from which the horse had escaped, but he did not notice this fact until he arrived at his house.

At 18 Newton enrolled in Trinity College, Cambridge University. He first intended to study mathematics as applied to astrology, but a meeting with Professor Isaac Barrow, who sensed Newton's abilities, encouraged him to study physics. In 1665 the bubonic plague overwhelmed England, and the University shut down. Newton returned to his home and mother at Woolsthorpe and, in quiet isolation, made discoveries in mathematics, optics, and the science of mechanics. As he wrote, "In those days I was in the prime of my invention, and minded mathematics and philosophy more than at any other time since."

This fertile period in Woolsthorpe generated the legend of the falling apple. As is common in a creative flash of genius, Newton, whether he in fact saw an apple fall or not, linked two seemingly unrelated phenomena: the fall of an apple and the orbit of the moon

FIGURE 4.12 Sir Isaac Newton. "I have laid down the principles of philosophy; principles not philosophical but mathematical. . . ." [*Courtesy Yerkes Observatory.*]

FIGURE 4.13 Edmund Halley. He inspired Newton to put together the *Principia*. Later, he used Newton's laws to compute the orbit of the comet that bears his name. [*Courtesy Yerkes Observatory.*]

(see below). He recalled to his friend and biographer William Stukeley that he was puzzled by the fall of objects, such as apples, and wondered about the nature of the force that attracts masses, such as the moon, to the earth's center.

On his return to Trinity Newton showed his work to Barrow, who soon resigned his position so that Newton could be elected to it. Because of his interest in optics, Newton came to invent the *reflecting telescope* (Chapter 7) which uses a mirror as the primary light gatherer. His design was communicated to the Royal Society of London; after he constructed a small reflector for them, he was elected a Fellow of the Society. His election was not a completely happy one, for his work on light, which was also communicated to the Society, brought bitter controversy. Newton resolved never to publish his ideas again.

The Magnificent *Principia*

Newton broke his vow of nonpublication about 10 years later, when Edmund Halley (1656–1742) requested his advice on the problem of elliptical orbits described by Kepler's laws. Halley (Fig. 4.13) queried Newton on the nature of the force between the sun and planets required to produce such orbits and was surprised to hear that Newton had solved the problem in exact detail. Absentminded as ever, Newton had misplaced the solution and could not find it at the moment; he promised Halley that he would send it along later.

Recognizing the importance of Newton's discovery, Halley cajoled his introverted friend to publish the studies and promised to oversee and finance their publication. New-

4.2 NEWTON: A PHYSICAL MODEL OF THE COSMOS

ton, stimulated intellectually, labored for two years to complete his *Philosophiae naturalis principia mathematica (The Mathematical Principles of Natural Philosophy)*. Published in 1687, the *Principia* presented the solution to the problem of planetary motions.

At the beginning of the *Principia*, Newton states his intention:

> For the whole burden of philosophy seems to consist of this—from the phenomena of motions to investigate the forces of nature and then from these forces to demonstrate all other phenomena.

After stating his purposes, Newton defines mass, velocity, and acceleration. He then expounds his three "Axioms, or Laws of Motion." In modern terms, these famous laws are:

1. *The Inertia Law.* A body at rest or in motion at a constant velocity along a straight line remains in that state of rest or motion unless acted upon by a net outside force (Fig. 4.14).
2. *The Force Law.* The change in a body's velocity (its acceleration) due to an applied net force is in the same direction as the force and proportional to it, but is inversely proportional to the body's mass (Fig. 4.15).
3. *The Reaction Law.* For every applied force, a force of equal size, but opposite direction arises (Fig. 4.16).

Newton's first law takes a logical step beyond Galileo's concept of inertia by postulating that constant, uniform motion is the *natural* state of a moving mass anywhere in the universe. The first law gives you a way to judge whether or not a net force is acting on an object: You are told to look for an acceleration, a change in an object's speed, or in the direction of its motion, or in both speed and direction.

The second law extends the concept from recognition of a force to the recognition of its consequences. A force (which you can think of simply as a push or a pull) accelerates an object; it slows it down, speeds it up, or changes the direction of its motion. The direction of the acceleration is in the same direction as the applied force. Furthermore, the amount of acceleration depends directly on the size of the force and inversely on the mass of the accelerated object. Note that because forces involve direction, they, like accelerations and velocities, are vectors.

For example, suppose you are floating in space next to a bunch of small objects. You push one. It accelerates and moves away, and it travels in the direction in which you pushed it. You can measure its acceleration, its change in velocity. Push the same object with twice as much force, and it accelerates to twice the velocity. Now push another mass with the same force. You measure its acceleration and find that it is half as much. You applied the same force to both objects, so the second must have twice the mass of the first. Newton's second law, as well as describing the consequences of the application of forces to masses, provides a means of *measuring* mass.

In algebraic form Newton's second law is

$$F = ma$$

where F is the net applied force, m the object's mass, and a the acceleration resulting from the force.

4 THE CLOCKWORK UNIVERSE: A PHYSICAL COSMOS

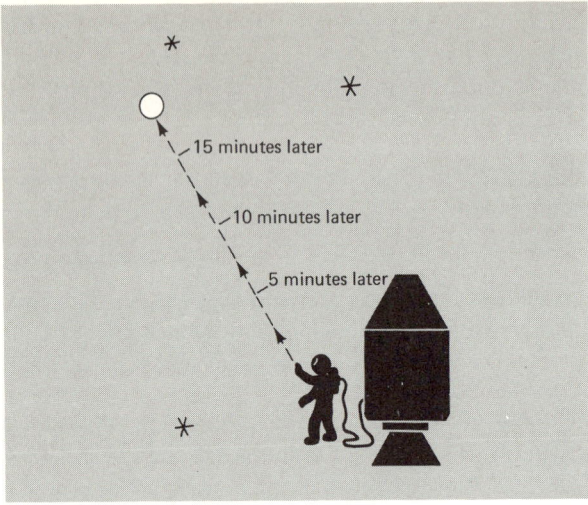

FIGURE 4.14 Newton's first law. An object thrown by an astronaut moves at a constant speed along a straight line (no friction slows it down).

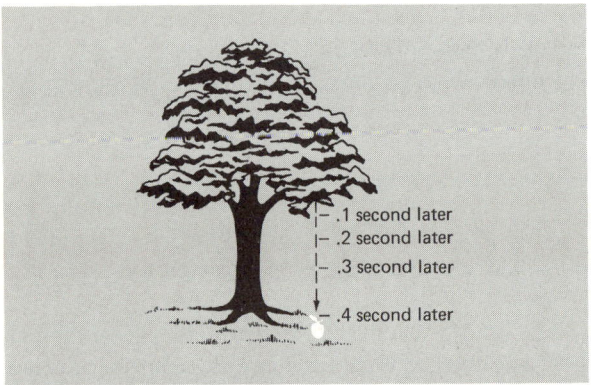

FIGURE 4.15 Newton's second law. A net force applied at an object changes its speed, direction of motion, or both in the direction of the applied force. Here, gravity pulls on an apple and accelerates it to the ground. As the apple accelerates, it falls faster so that it covers a greater distance in the same time (0.1 sec).

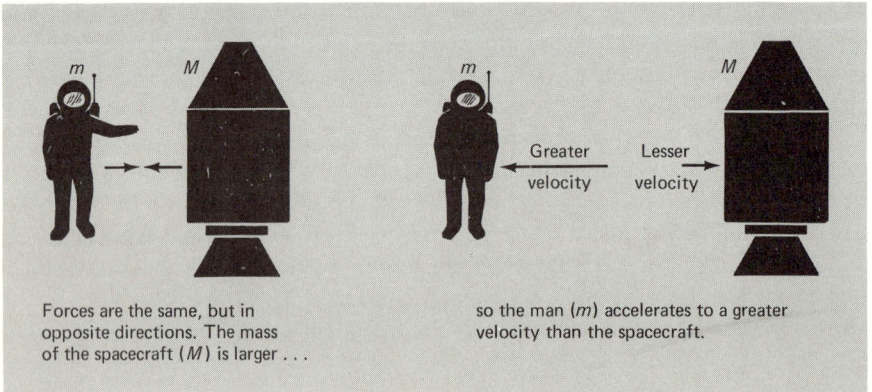

FIGURE 4.16 Newton's third law. Imagine you are in space next to a spacecraft that has a larger mass than you. Push it; it pushes back on you with a force equal to the force you apply to it, but in the opposite direction. So you and the spacecraft move apart. You end up moving with a greater velocity than the spacecraft because you have less mass accelerated by the same force.

The third law recognizes that forces are interactions and must act in simultaneous pairs. If you were at rest in space and pushed against a massive spacecraft, the ship would react to your applied force with an equal, but oppositely directed force, pushing you away, as described by Newton's third law. The forces applied to you and the spacecraft are the same (though in opposite directions), but the resulting accelerations are different. According to Newton's second law, the acceleration is greater for you than for the spacecraft, because your mass is less. As a result, you would move away from the spacecraft quickly, while it would hardly budge. Also, you and the spacecraft would be moving in *opposite* directions.

After developing these ideas Newton attacked the problem of planetary motion by devising the law of gravitation. Two questions needed to be answered. In what *direction* does the force of gravity act, and what is the *amount* of the force? The first question involves a recognition of the general nature of the force, and the second involves a recognition of the physical properties that determine the force's strength. To answer these questions, Newton stood on the shoulders of Kepler. Newton's procedure combined his laws of motion with Kepler's planetary laws to arrive at a law of universal gravitation.

Newton demonstrated that the type of force that causes the elliptical orbits of Kepler's first law must be a *central force*, one directed to the center of the motion. Also, he showed that planets moving in orbits under the influence of a central force follow Kepler's second law, the law of areas. Finally, Newton showed from the geometrical properties of ellipses that the force may be described by an inverse square law and then derived Kepler's third law. In this manner he ensured that his procedure fell in line with planetary laws as they were known in his time.

How do the moon and the famous apple enter into this scheme? Newton recognized that gravity causes the apple's fall. He wondered if it extended beyond the neighborhood of the earth's surface (Fig. 4.17). Although they knew about gravity, earlier scientists had

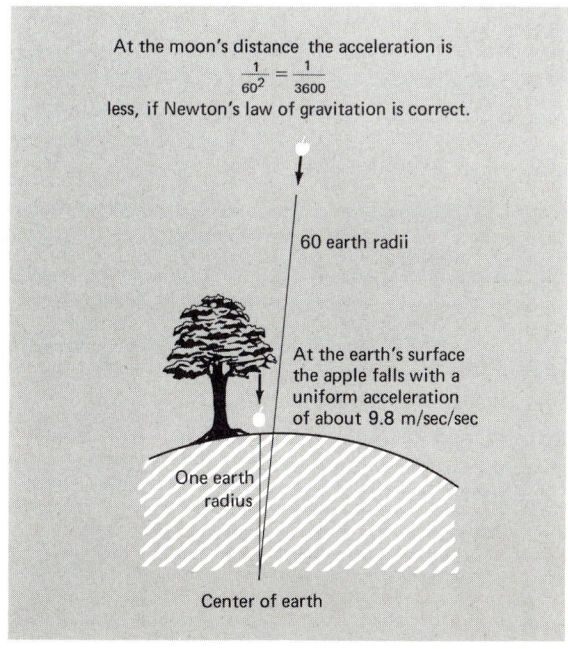

FIGURE 4.17 The moon and the apple. The apple is 1 earth radius from the earth's center and falls with an acceleration of 9.8 m/sec/sec. If an inverse square law for gravity is correct, then the moon accelerates at a rate 1/3600 less than the apple.

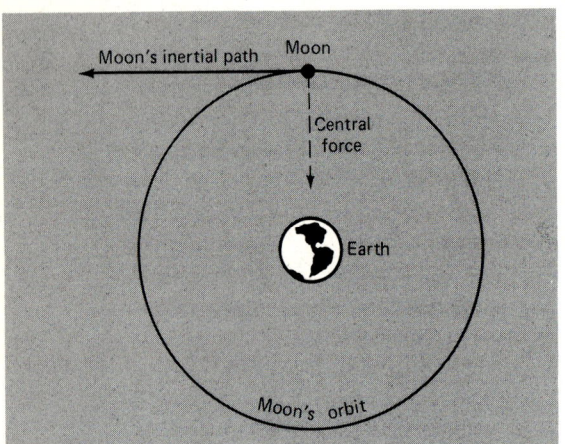
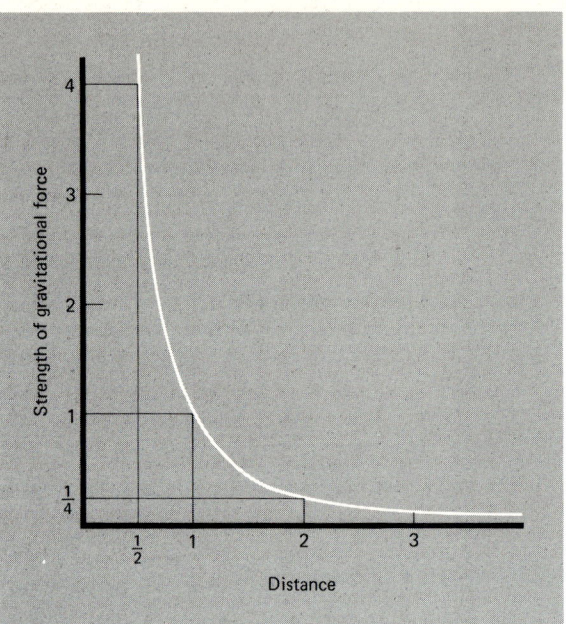

FIGURE 4.18 Central force and the moon's orbit. Since the moon moves neither at a constant speed nor along a straight line, according to Newton's first law, a force must be acting on it. This force pulls the moon toward the earth in a direction toward the center of the moon's orbit.

FIGURE 4.19 An inverse square law. This graph shows how the strength of the gravitational force changes with the distance between masses. In the graph at a separation of 1 unit of distance, the force has 1 unit of strength. When the masses are twice as far apart, a distance of 2, the force has only $\frac{1}{4}$ the original strength.

made no attempt to allow gravity to work far from the earth. Newton's creative leap came despite traditional opinions that motions on earth and in the heavens were distinctly different. Might it not be, he thought, that the earth's gravity, pulling on the moon, also keeps the moon in its orbit? For simplicity, assume that the moon's orbit is circular. Now, the direction of the moon's orbital motion changes constantly; the moon stays on a curved path rather than moving along a straight line away from the earth (Fig. 4.18). Newton's first law tells us that there must be acting on the moon a force that, according to the second law, results in an acceleration toward the earth's center. Such a centrally directed acceleration is called a *centripetal acceleration*. (*Centripetal* means "directed toward the center.") But what force causes this centripetal acceleration, and just how can it be described?

Newton generalized from the apple to the moon: "I began to think of gravity extending to the orb of the moon." He went even further, making the creative leap toward a new insight that *every body in the universe attracts every other body with a gravitational force*. This statement became the first *universal* physical law.

In modern algebraic form Newton's law of gravitation is

$$F = \frac{Gm_1m_2}{R^2}$$

where F is the gravitational attraction between two bodies of mass m_1 and m_2, which are

4.2 NEWTON: A PHYSICAL MODEL OF THE COSMOS

separated by a distance R. The symbol G is a constant, a number whose value is assumed not to vary with time and location in the universe. The value for G in the mks (meter-kilogram-second) system is 6.67×10^{-11}. (See Appendix A for units.)

What does Newton's law of gravitation mean? First, all masses in the universe attract all other masses. (This force can only attract; it does not repel.) Second, if you consider just two masses for a moment, the amount of the gravitational force depends directly on the amount of material *each* mass has. If you double the mass of one and keep the distance between the two the same, the force doubles. (Note that the kind of material that makes up the masses does not matter.) Third, masses at greater separations have a smaller gravitational force than those closer together, and this drop-off of force with distance happens as the inverse square of the distance (Fig. 4.19). Consider, for example, two masses 1 m apart. A certain amount of gravitational force attracts one to the other. If the masses are moved to 2 m apart, the force is less by the square of 2, or 4.

Newton, the Apple, and the Moon

Newton used the moon and apple to test the validity of the law of gravitation. He knew that the earth's gravity at its surface (1 earth radius from the center) caused the apple to fall. The earth-moon distance is about 60 earth radii, so if an inverse square law correctly describes gravitational forces, the acceleration of the moon toward the earth must be $1/(60^2)$ or 1/3600 as much as the acceleration of the apple. Newton then compared his predicted centripetal acceleration with the centripetal acceleration derived from observations of the moon's orbit. As he put it, the predicted and the observed accelerations were "pretty nearly" the same. Newton concluded that the cause of the moon's centripetal acceleration is the same as that of the apple's—the earth's gravity. This force, extended out to the distance of the moon, kept the moon in its orbit.

Let's look at this critical point in detail. The acceleration due to gravity at the earth's surface is 9.80 m/sec/sec. For the moon, then, since it is 60 earth radii away, the predicted acceleration is (9.80 m/sec/sec)/3600, or about 2.72×10^{-3} m/sec/sec.

How does this *predicted* value compare with the *actual* rate of the moon's fall? To answer this question, Newton had to find a way to calculate the amount of the moon's centripetal acceleration. Let's first try to get a feel for the quantities that determine it. We all experience centripetal acceleration when we turn a corner in a car. It's clear that the faster you are going, the greater the acceleration in turning a corner; if you try turning while going too fast, the friction of the tires on the road cannot produce a strong enough force to make the car change direction and you go into a skid. Now consider the size of the circle. If you are on the freeway making a long gentle curve, even at 90 km/h, very little force is needed to change the car's direction. But if you try to turn into your driveway at 90 km/h, that is, on a circle with a much smaller radius, the force required to produce that much acceleration would be too great for the tires to handle, and you would probably roll over. So a higher speed or a smaller radius of the circle both mean a larger acceleration.

In modern algebraic terms Newton showed that for a circular orbit the centripetal acceleration has a value

$$a = \frac{v^2}{R}$$

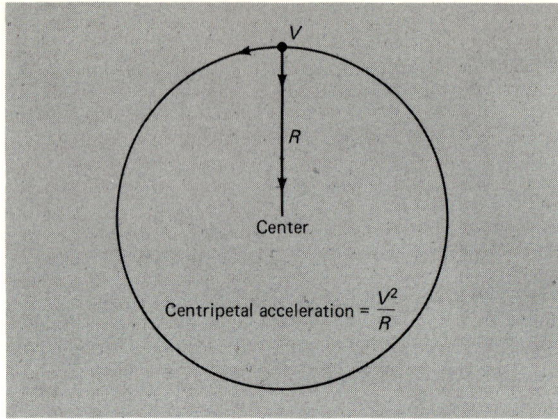

FIGURE 4.20 Centripetal acceleration and centripetal force are directed toward the *center* of circular motion.

where a is the centripetal acceleration, R the radius of the orbit, and v the orbital velocity (Fig. 4.20). (See Focus 4.1.) For the moon R equals 3.84×10^8 m. Its velocity is the distance it travels in one orbit divided by the period for one orbit, or

$$v = \frac{2\pi R}{P}$$

The moon's period is 27.3 days. Since 1 day contains $24 \times 60 \times 60$, or 8.64×10^4 sec, the moon's orbital velocity is

$$v = \frac{2\pi (3.84 \times 10^8 \text{ m})}{(27.3 \text{ days})(8.64 \times 10^4 \text{ sec/day})}$$

$$= 1.023 \times 10^3 \text{ m/sec}$$

Then the moon's orbital acceleration is

$$a = \frac{(1.023 \times 10^3 \text{ m/sec})^2}{3.84 \times 10^4 \text{ m}}$$

$$= 2.72 \times 10^{-3} \text{ m/sec/sec}$$

The result is as predicted.

Newton did not have the modern values used above for the period and size of the moon's orbit and the acceleration due to gravity. He chose a value of the moon's distance that made his results compare closely. Newton did feel assured that his approach was correct, even though he fudged the figures slightly.

Note that Newton did not need to know the mass of the apple or of the moon because their accelerations do not depend on their masses. Galileo (Section 4.1) had reached this same conclusion from his experiments. Newton made crucial use of this result in the moon-apple test.

Let's look at Galileo's law of free-fall from Newton's standpoint. Imagine you're going to repeat Galileo's rumored experiment at the Leaning Tower of Pisa by dropping a

4.2 NEWTON: A PHYSICAL MODEL OF THE COSMOS

cannonball and a tennis ball from its uppermost story. The earth exerts a much greater gravitational force on the cannonball than on the tennis ball because of the cannonball's much greater mass. However, when the two are dropped, they fall side by side and land at the same time (if you neglect air resistance). So the "effect" of gravity has been in some way the same on both objects; more precisely, the *acceleration* of each is the same, because they speed up at the same rate. Although the forces are different, the accelerations turn out to be the same, because gravitational forces are proportional to the masses of the falling objects, but accelerations are inversely proportional to masses (Newton's force law), so the mass cancels out.

Mathematically, you can see this by combining Newton's law of gravitation

$$F = \frac{GMm}{R^2}$$

(where M is the mass of the earth and m is the mass of the falling object) with the force law

$$F = ma$$

to obtain

$$ma = \frac{GMm}{R^2}$$

or, dividing by m,

$$a = \frac{GM}{R^2} (= g)$$

See how the mass of the falling object cancels out? This quantity is the acceleration due to gravity, usually denoted as g. (Look at the expression for g. It suggests a method for obtaining the mass of the earth. If G is known, from a laboratory measurement, and if R is measured, as Eratosthenes did, for example, then simply by measuring g we can calculate M, the mass of the earth!)

Don't confuse mass and weight! Most people are familiar enough with space travel to know that if you are far enough away from any large mass and if your spaceship is not accelerating, you will be *weightless*. What does that mean (besides the fact that you'll float around)? Simply, if you placed a scale beneath you, it would read zero. Now, when you stand on a scale on the earth, what are you measuring? In Newton's terms you are reading the amount of *gravitational force* exerted on you by the earth's mass. So *weight is a force*.

Using Newton's second law, $F = ma$, at the earth's surface, where $a = g$, we have

$$W = mg$$

where W is an object's weight and m is its mass.

Mass relates to the inertial properties of matter; it is *not* a force. But forces can tell you how much mass you have. For example, in order to get around in a spaceship, even if you are weightless, you have to put out an effort. Suppose you want to go from one side of the

FOCUS 4.1 CENTRIPETAL ACCELERATION AND FORCE

Because his system was based on forces and accelerations, Newton first had to find the acceleration of a body moving in a circle, its *centripetal acceleration*. To see which way this force (and acceleration) is directed, imagine a ball attached to a string and swung in a circle over your head. By Newton's first law, the ball would move off in a straight line tangent to the circle if it were not held in by the string. The string is applying a centripetal force directed toward the center. The rate of change of the velocity toward the center is the centripetal acceleration.

Let's calculate the amount of the acceleration. Assume an object undergoes uniform circular motion. If r represents its position, then its velocity v is the rate of change with time of its position: $v = \Delta r/\Delta t$. (The symbol Δ means "change in.") In Fig. F.5a the position (r) at equally spaced time intervals (Δt) and the changes in the positions (Δr) are

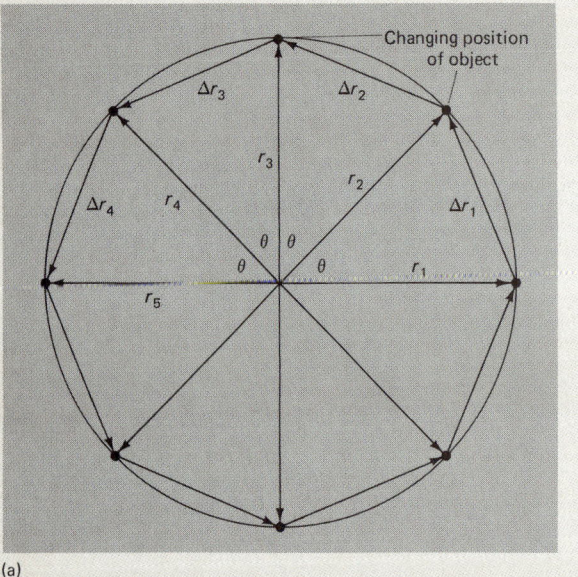
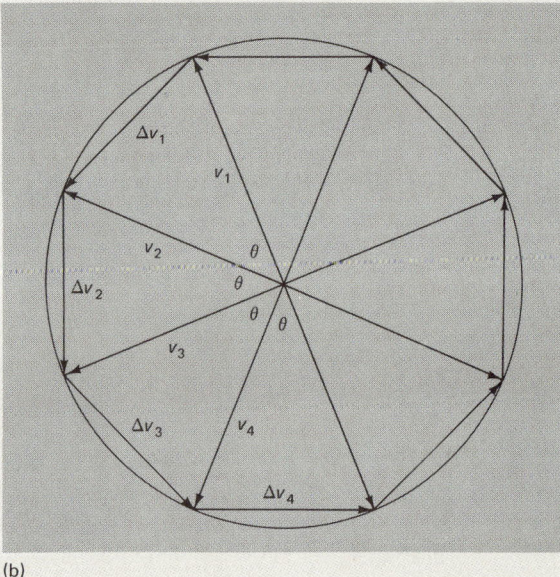

(a) (b)

FIGURE F.5 Changing position of an object in circular motion (a) and the changing direction of its velocity (b).

ship to the other. The easiest way is to push yourself off a wall with a small amount of force. You'll then drift to the other side and in order to stop yourself you'll need to push against the opposite wall with the same amount of force. If you measure the amount of force and your acceleration, you can use Newton's second law to calculate your mass. No matter where you are—on the earth, or the moon, or in space—your mass is the same.

Not so for your weight. Imagine you did go to the moon, with your trusty bathroom scale. Your weight there would be about one-sixth that on the earth. Why? Look again at Newton's law of gravitation. The moon is smaller than the earth, about one-fourth the earth's diameter. Since you are 4 times closer to the center of the moon, you might expect the gravitational force to be 16 times greater. But, since the moon has only about 1/81 the earth's mass, the gravitational force at its surface is 5 times less than on the earth.

Let's compare the moon's gravitational force on you to that of the earth's.

shown. If we make the time interval smaller and smaller, each of the triangles becomes "skinny." Using the skinny triangle approximation, $\Delta r = r\theta$, where θ is given in radians. To obtain θ, note that its size just equals the angular speed of the object's position as it goes around (ω) multiplied by the time interval (Δt). Since the object completes a circuit of 2π radians in the period of the motion T, the angular speed is $\omega = 2\pi/T$. So the change in the position is $\Delta r = r\omega\Delta t = r(2\pi/T)\Delta t$. Dividing both sides by Δt, we find that the value of the velocity (the speed) is $\Delta r/\Delta t = v = 2\pi r/T$. This is just as we expected; the speed is just equal to the total distance traveled (the circumference) divided by the total time (the period).

To calculate the centripetal acceleration, we must determine the rate of change of the velocity with time: $a_c = \Delta v/\Delta t$. Figure F.5b shows the changes in the velocity as the object goes around the circle. Note that the direction of v is the same as Δr; only its amount is different ($v = 2\pi r/T$). Using the same skinny triangle argument as before, the amount of the velocity change is $\Delta v = v\theta$ (if Δt is small). Again $\theta = \omega\Delta t$, and $\Delta v = v(2\pi/T)\Delta t$. So $\Delta v/\Delta t = v(2\pi/T)$. But this is just the amount of the acceleration, a_c. Multiplying and dividing by r,

$$a_c = \frac{v(2\pi r/T)}{r}$$

or, since $v = 2\pi r/T$,

$$a_c = \frac{v^2}{r}$$

Using Newton's second law ($F = ma$), the centripetal force required to keep a mass in uniform circular motion is

$$F_c = m\frac{v^2}{r}$$

$$F_{moon} = \frac{GmM_m}{R_m^2} \quad \text{and} \quad F_{earth} = \frac{GmM_e}{R_e^2}$$

$$\frac{F_{moon}}{F_{earth}} = \frac{GmM_m/R_m^2}{GmM_e/R_e^2} = \left(\frac{M_m}{M_e}\right)\left(\frac{R_e}{R_m}\right)^2$$

$$= \left(\frac{7.4 \times 10^{22} \text{ kg}}{6.0 \times 10^{24} \text{ kg}}\right)\left(\frac{6400 \text{ km}}{1700 \text{ km}}\right)^2$$

$$= 0.17$$

where m is your mass, G Newton's gravitational constant, M_m the moon's mass, R_m the moon's radius, M_e the earth's mass, and R_e the earth's radius.

Newton's Version of Kepler's Third Law

Let's see how the concept of centripetal acceleration combined with Newton's laws results in Kepler's third law. To do so, we need to introduce the concept of *center of mass*.

If you have played on a seesaw, you have experienced this concept. The center of mass is the balance point in a system of joined masses. If you and a friend with the same mass sit on the ends of a seesaw, in order to balance it the pivot must be placed at the seesaw's center. You will both be the same distance from it (Fig. 4.21). But if your friend has *twice* your mass, the balance point will be closer to him or her, so that your distance to the pivot will be twice that of your friend's. Let m_1 be your mass, m_2 your friend's mass, r_1 your distance from the center of mass, and r_2 your friend's distance. Then the algebraic relationship for center of mass is

$$m_1 r_1 = m_2 r_2$$

The same concept applies to any two masses joined by gravity (see next section).

Consider now a planet orbiting the sun. It actually orbits around the center of mass of the sun-planet system, and so does the sun. Let r_{planet} be the distance of the planet from the center of mass, and r_{sun} the sun's distance, so the total distance between the sun and planet, R, is $r_{planet} + r_{sun}$ (Fig. 4.22). The gravitational force between them is

$$F = \frac{GM_{sun}m_{planet}}{R^2}$$

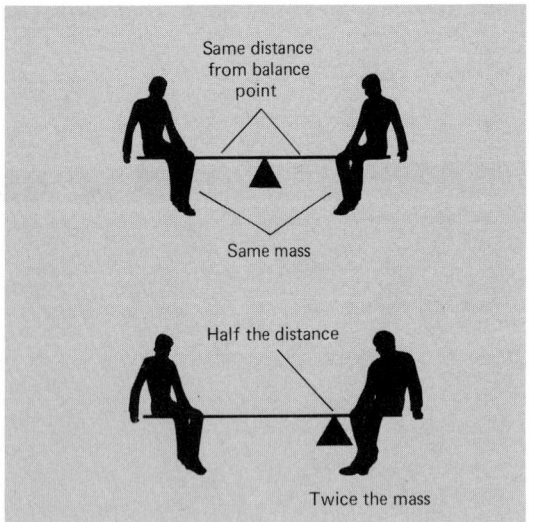

FIGURE 4.21 A seesaw and its balance point—the center of mass. If the two people on the seesaw have the same mass, they are located at the same distance from the pivot. But if one is more massive, he is closer to the balance point than the other.

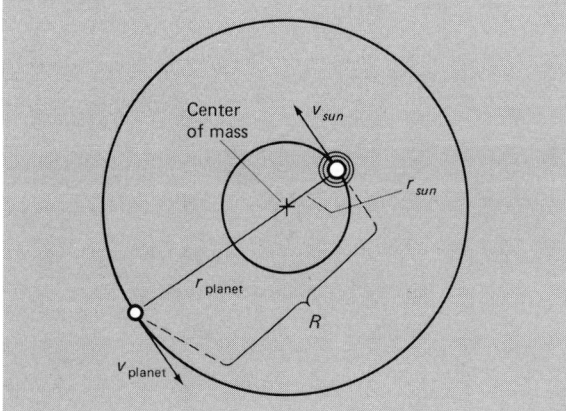

FIGURE 4.22 Motions of the sun and a planet about their center of mass.

4.2 NEWTON: A PHYSICAL MODEL OF THE COSMOS

According to Newton's third law, this force is the same for the planet acting on the sun as for the sun acting on the planet. As a result, both bodies accelerate about the center of mass.

To simplify matters, assume the orbital paths are circular. Then the centripetal acceleration of the planet about the center of mass is

$$a_{planet} = \frac{v_{planet}^2}{r_{planet}}$$

For a circular orbit, $v = 2\pi r/P$, where r is the radius and P the orbital period; so

$$a_{planet} = \frac{(2\pi r_{planet}/P)^2}{r_{planet}}$$

$$= \frac{4\pi^2}{P^2} r_{planet}$$

For the sun,

$$a_{sun} = \frac{v_{sun}^2}{r_{sun}}$$

$$= \frac{4\pi^2}{P^2} r_{sun}$$

(Note that P is the same for the sun and the planet.)

Now apply Newton's second law to the planet:

$$F = m_{planet} a_{planet}$$

$$\frac{GM_{sun} m_{planet}}{R^2} = m_{planet} \frac{4\pi^2}{P^2} r_{planet}$$

$$\frac{GM_{sun}}{R^2} = \frac{4\pi^2}{P^2} r_{planet}$$

and to the sun:

$$F = M_{sun} a_{sun}$$

$$\frac{GM_{sun} m_{planet}}{R^2} = M_{sun} \frac{4\pi^2}{P^2} r_{sun}$$

$$\frac{GM_{planet}}{R^2} = \frac{4\pi^2}{P^2} r_{sun}$$

Add these two equations to get

$$\frac{G(M_{sun} + m_{planet})}{R^2} = \frac{4\pi^2}{P^2}(r_{planet} + r_{sun})$$

But $r_{planet} + r_{sun} = R$, so

$$\frac{G(M_{sun} + m_{planet})}{R^2} = \frac{4\pi^2 R}{P^2}$$

or

$$P^2 = \frac{4\pi^2 R^3}{G(M_{sun} + m_{planet})}$$

This equation is Kepler's third law,

$$P^2 = kR^3$$

with $k = 4\pi^2/G(M_{sun} + m_{planet})$. Note that the constant relates to the masses of the *two* bodies involved. So a $1/R^2$ force law for gravity and Newton's laws of motion result in Kepler's third law.

Confident of his description of gravitation, Newton launched a massive attack on astronomical problems in the third part of the *Principia* (see next section). One doubt nagged at the back of his mind: Although he had described gravitational force, what was its cause? His own description required that masses interact over large distances. This action-at-a-distance notion ran contrary to mechanical views of nature, which explained phenomena in terms of the local collisions of objects. Newton ultimately sidestepped the question of the fundamental nature of gravitational forces, but his avoidance of the issue was at least consistent with the rest of his philosophy:

> I have not yet disclosed the cause of gravity, nor have I undertaken to explain it, since I could not understand it from phenomena.

In other words, he wished to base his scientific knowledge on direct observation or experiment (the phenomena), rather than on a philosophical argument. If he could not do so, as in the case of gravity, he would make no attempt to offer an explanation. In spite of this evasion, he saw that the cosmos was linked together by eternal, invisible chains.

Warning: Newton's laws of motion and his law of gravitation appear quite simple. Indeed they are, and that is their beauty. But be careful that you learn precisely what they do and do not say. For example, is the gravitational force of the moon on the earth bigger or smaller than the gravitational force of the earth on the moon? The correct answer is they are *exactly* the same, for the law of gravity says that the force between objects is proportional to the product of *both* masses.

As another example, what is the reaction force, equal but of opposite direction, corresponding to the force of gravity pulling you downward as you sit in your chair? You might think that it is the force of the chair pushing up on you and keeping you from falling to the floor, but the correct answer is that it is the force which you exert on the earth. After all, if you took the chair away, you would fall to the floor, but there still must be a reaction force to the force of gravity pulling you down, which is the force you exert on the earth, pulling it slightly upward toward you. (Think: Why do we say the earth only moves *slightly*, even though the forces are the same?)

4.3 COSMIC CONSEQUENCES OF UNIVERSAL LAWS

Newton's central discovery that the gravitational force of the earth causes the centripetal acceleration of the moon resulted in a new understanding of planetary motion. Newton had found the physical interaction between the sun and planets first sought by Kepler. The sun's gravity locks the planets in their elliptical orbits. He derived a new form of Kepler's third law that included the gravitational constant G and the masses of the interacting bodies in the constant k. Related to basic physical laws, the revised third law became a potent tool in determining the masses of the planets (if the planet has at least one satellite) and of the sun—quantities never known before! Newton answered in detail the ancient question of how the planets moved, and he answered it precisely. His predictions of planetary positions were far more accurate than previous ones. Newton's ideas provided the physical support sorely needed for the Copernican model.

The Earth's Rotation

One objection to the Copernican model was that objects not tied down to the earth should fly off because of its rotation. Newtonian physics explained that objects on the earth had inertia. If thrown upward, they did not lose their inertial motion, but continued to move with the ground. So they landed at their starting points. A rotating earth did not leave behind unattached objects.

One demonstration of the earth's rotation was exhibited in Paris in 1851. A French physicist, Jean B. L. Foucault (1819–1868), suspended a cannonball on a long steel wire in the Pantheon in Paris. Foucault pulled this sphere over to one side and, after all the cable vibrations had ceased, released the sphere to swing freely in a long arc. As the pendulum swung back and forth, a pointer on the lower part of the sphere traced the plane of the swing on a sand table under the sphere. Slowly, the plane of the pendulum's swing shifted clockwise and completed a circle in about 32 hours.

Why did this happen? The answer relates to Newton's first and second laws of motion. In the absence of outside force, the plane of swing of a pendulum, which is free to move in any direction, tends to retain its original direction in space with respect to the stars. If the pendulum were suspended above the north pole, the earth would rotate counterclockwise underneath the pendulum with respect to the stars in 23 hours, 56 minutes. During this time the pendulum's direction of swing does not change because the only force acting on it—the force of gravity—pulls the pendulum bob toward the earth's center. So the pendulum's inertia constrains it to swing in the same plane.

The table beneath the pendulum, which serves to mark the direction of swing, turns with the earth. From our point of view, fixed to the earth and moving with its rotation, the

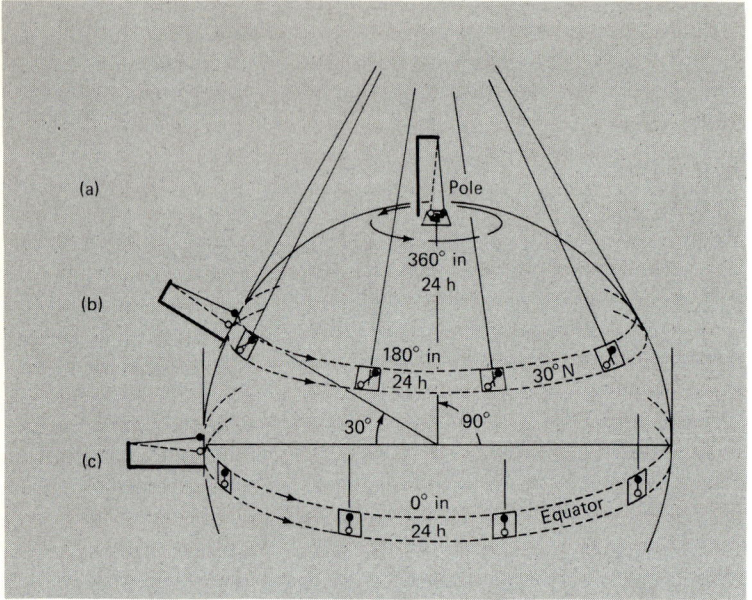

FIGURE 4.23 The motion of a pendulum at the earth's north pole (a), 30° latitude (b), and the equator (c). At the pole the plane of the pendulum's swings covers 360° in 24 h; at 30° latitude it covers 180° in 24 h; at the equator it shows no motion. [*Adapted from a figure by A. Strahler*, The Earth Sciences, *2nd ed., Harper & Row, New York, 1971.*]

plane of swing appears to turn clockwise. Actually, the earth's counterclockwise rotation creates the pendulum's apparent motion. As you move from the pole toward the equator, the period for one apparent revolution of the pendulum becomes progressively longer (Fig. 4.23). The reason is that the direction of gravity is at an angle to the axis of rotation. At the equator the plane of swing does not move with respect to the ground, that is, the period of rotation is infinite.

Note that this analysis of the pendulum's motion and its implications about the earth's rotation depend on Newton's laws. The key to the reasoning was the conclusion that no forces acted on the pendulum to change the direction of swing.

The Earth's Revolution and the Sun's Mass

You can use the period and radius of the earth's orbit to find the sun's mass. The earth-sun distance, a, is 1.50×10^{11} m. The earth's period, P, is 365.25 days, or 3.16×10^7 sec. Because the mass of the sun is much larger than that of the earth, we can approximate $M_S + m_E$ by M_S. Then Newton's revision of Kepler's third law gives

$$M_S = \frac{4\pi^2}{G} \frac{a^3}{P^2}$$

$$= \frac{39.5}{6.67 \times 10^{-11}} \frac{(1.50 \times 10^{11})^3}{(3.16 \times 10^7)^2}$$

$$= 1.99 \times 10^{30} \text{ kg}$$

Note that you need to know the value of G to do this calculation.

4.3 COSMIC CONSEQUENCES OF UNIVERSAL LAWS

If you don't know G, you can still use the third law to find *relative* masses. Use the earth again, but this time you also need the moon. The moon's distance is 3.84×10^8 m, or roughly 2.6×10^{-3} AU, and the moon's period is 27.3 days or 7.5×10^{-2} yr. For the earth and moon

$$m_E + m_M = \frac{4\pi^2}{G} \frac{a_{EM}^3}{P_{EM}^2}$$

and for the sun and earth

$$M_S + m_E = \frac{4\pi^2}{G} \frac{a_{ES}^3}{P_{ES}^2}$$

Divide the second equation by the first to get

$$\frac{(M_S + m_E)}{(m_E + m_M)} = \frac{\dfrac{4\pi^2}{G} \dfrac{a_{ES}^3}{P_{ES}^2}}{\dfrac{4\pi^2}{G} \dfrac{a_{EM}^3}{P_{EM}^2}}$$

As before, approximate $M_S + m_E$ by M_S, and $m_E + m_M$ by m_E. Then

$$\frac{M_S}{m_E} = \left\{ \frac{1 \text{ AU}}{2.6 \times 10^{-3} \text{ AU}} \right\}^3 \left\{ \frac{7.5 \times 10^{-2} \text{ yr}}{1 \text{ yr}} \right\}^2$$

$$= (5.7 \times 10^7)(5.6 \times 10^{-3})$$

$$= 3.3 \times 10^5$$

You have found the sun's mass relative to the earth's mass.

The knowledge of these masses clinched the truth of the Copernican model in the framework of Newtonian physics. Newton's laws show that the sun has roughly 3.3×10^5 the mass of the earth. Newton's third law of motion requires equal gravitational forces between the sun and the earth; the force of the sun on the earth equals that of the earth on the sun. However, because the earth's mass is much less than that of the sun, the second law demands that the earth's acceleration be much greater than the sun's:

$$a_{earth} = \frac{F}{m_{earth}}$$

$$a_{sun} = \frac{F}{M_{sun}}$$

$$\frac{a_{earth}}{a_{sun}} = \frac{M_{sun}}{m_{earth}}$$

Now, we showed in the previous section in deriving Kepler's third law, that the accelerations can be written as

$$a_{earth} = \frac{v_{earth}^2}{r_{earth}} = \frac{4\pi^2}{P^2} r_{earth}$$

$$a_{sun} = \frac{v_{sun}^2}{r_{sun}} = \frac{4\pi^2}{P^2} r_{sun}$$

where r_{earth} and r_{sun} are the distances of the earth and sun from their common center of mass. Therefore, combining the above equations,

$$\frac{a_{earth}}{a_{sun}} = \frac{r_{earth}}{r_{sun}} = \frac{M_{sun}}{m_{earth}} = 3.3 \times 10^5$$

We have not only proved the center of mass relation holds for gravitating bodies

$$m_{earth} r_{earth} = M_{sun} r_{sun}$$

but also that the earth is much farther from the center of mass than is the sun. In fact, the center of mass is deep inside the sun. So the application of Newton's laws shows that it makes sense to say, as Copernicus did, that the earth orbits the sun, rather than the other way around.

The Precession of the Earth's Axis

Focus 1.1 described the precession of the equinoxes. You can understand this effect if you picture the earth as spinning like a top (Fig. 4.24a). If you place a spinning top on a

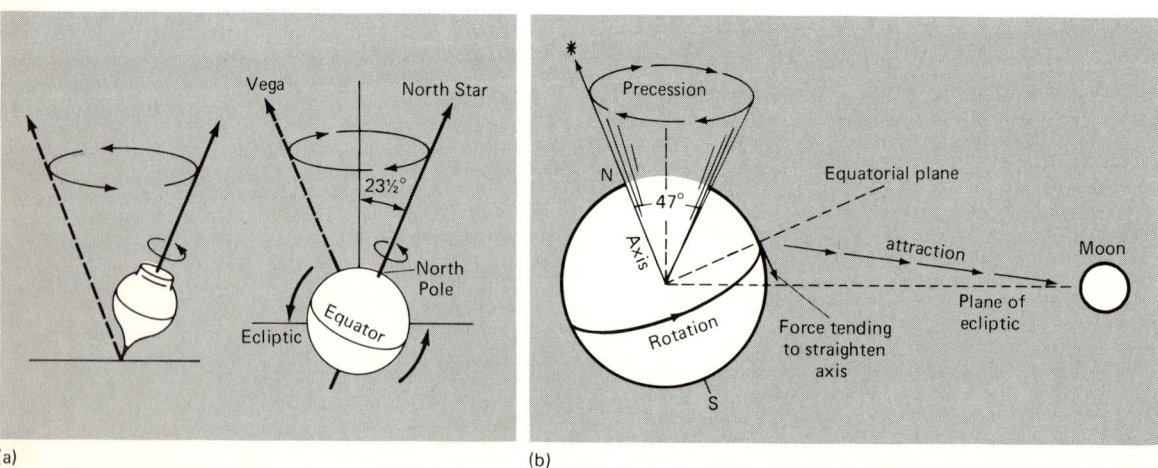

FIGURE 4.24 Precession of the earth's axis. (a) Precession of a top and the earth compared. (b) The moon's gravitational attraction on the earth's equatorial bulge. [*Adapted from a figure by A. Strahler, The Earth Sciences, 2nd ed., Harper & Row, New York, 1971.*]

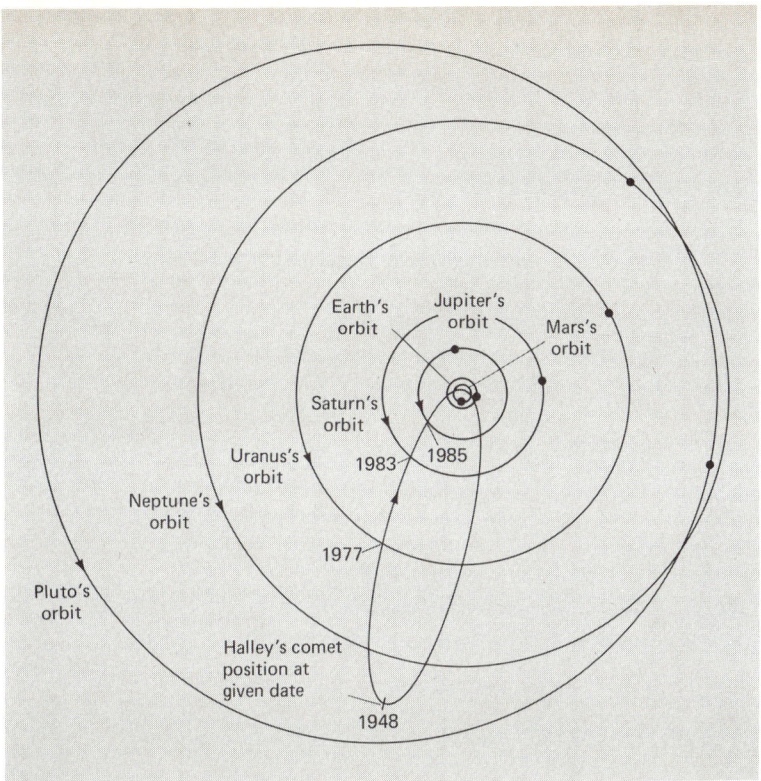

FIGURE 4.25 The orbit of Halley's comet. It will return to perihelion again in 1985–1986.

table, its axis of rotation moves in a circle, an effect called precession. The precessional motion occurs in response to gravity trying to pull the top down.

The earth, because it is spinning, has an equatorial bulge. (Newton actually predicted the amount of the oblateness.) Because the moon's orbit does not lie in the same plane as the earth's equator, but within 5° of the ecliptic (Fig. 4.24b), the moon's gravitational force attracts the earth unevenly; it tends to pull the bulge closest to it more than the part of the bulge on the opposite side of the earth. The net force tries to tip the earth's axis of rotation from its orientation in space. The spinning earth responds by a precession of its axis, which we see as the precession of the equinoxes. (The sun's gravity also takes part, but its effect is much less than the moon's.)

Gravity and Orbits

To add to its achievements, Newton's physics correctly described the orbits of comets. Through antiquity and the Middle Ages astronomers believed that comets were objects confined to the earth's atmosphere. Newton and Halley decisively demonstrated that comets orbit the sun in accordance with the law of gravitation (Fig. 4.25). In fact, Halley correctly predicted the return of the comet that bears his name, but he did not live to see it return. (More on comets in Chapter 18.)

Newton's ideas also led to the discovery of Neptune, in 1846, long after the first publication of the *Principia*. Neptune was the first planet to be found by its gravitational effects on another. Newton's laws predicted its existence *before* it was observed.

The discovery of Neptune rested upon observed irregularities in the orbit of Uranus. Astronomers had noted small discrepancies between the observed positions of Uranus and those predicted from Newton's laws. Such irregularities occur because all planets attract one another. Jupiter's tug, for instance, influences the orbit of Uranus so that it differs from that expected if Uranus and the sun were the only attracting bodies. The undiscovered Neptune revealed itself when Uranus's motion deviated from the path predicted from the effects of the known planets.

Applying Newton's laws to explain the deviations as due to a planet beyond Uranus, the Englishman John C. Adams in 1845 estimated where the unknown planet should be in the sky. He communicated his result to the Astronomer Royal, Sir George Airy, hoping that a search for the new planet would begin immediately. But Airy did not believe the result, partly because he was one of those who thought that the gravitational force deviated from an inverse square law at large distances.

At almost the same time Urbain J. J. Leverrier in Paris made similar calculations and transmitted them to Johann Galle in Berlin, who found the planet on September 23, 1846. Newton's laws triumphed. In the twentieth century discrepancies in Neptune's orbit stimulated the search for other planets. The search resulted in the discovery of Pluto (Chapter 17), even though we now know that Pluto actually has too small a mass to affect Neptune significantly.

An equally dramatic discovery, which extended the validity of Newton's laws beyond the bounds of the solar system, was the observation, in the late eighteenth century, of binary star systems. (A binary star system consists of two stars, held together by their mutual gravity, orbiting each other.) William Herschel (1768–1822) and his sister Caroline Herschel (1750–1848) (Fig. 4.26) observed many pairs of stars over a long time, looking for heliocentric parallax. Unexpectedly, the Herschels observed the orbital motions of some pairs. These observations showed that some pairs of stars in the sky lie close together not simply because they both happen to be in adjacent lines of sight from the earth, but because they are actually linked by gravitational forces. Calculations of the period of binary star systems and of the separation between the stars confirmed that their motions followed Kepler's laws. These laws can be derived from Newton's laws of motion and gravitation. Since binary star systems obey Kepler's laws, they indirectly confirm Newton's laws. (See Chapter 8 for more details on binary star systems.)

(a)

(b)

FIGURE 4.26 (a) William Herschel. (b) Caroline Herschel. [*Courtesy Yerkes Observatory.*]

Triumph after triumph hammered Newtonian ideas into people's minds. In terms of a scientific model Newton's synthesis was complete. Its predictive success was astounding, compared with previous systems. Time after time the numbers, not just the general ideas, came out right. Planetary positions could be predicted precisely. And Newton's ideas were *universal*. No longer were separate laws needed for falling bodies, the moon, or binary stars. One force binds them all—gravity!

Newton's Cosmology

The physical ideas that Newton found in the universe resulted in a new conception of the cosmos. For the Greeks, Ptolemy, and Copernicus, the cosmos appeared finite and bounded. Galileo had left open the possibility that the universe might be infinite, but did not strongly push his case. Newton argued that his laws required that the universe be infinite in extent. He argued that if the universe were finite, gravitation would eventually pull all matter to the center, and only one large mass would exist. However, we see other masses, such as stars. In an infinite universe matter would be pulled into an infinite number of small condensations. Newton believed that this picture was more like the real world, so he concluded that the universe was infinite.

Not all things in heaven and earth sat happily in Newton's infinite universe. The innermost planet, Mercury, posed an annoying problem: the axis of its elliptical orbit rotated in space. This motion could not be completely explained by the attraction of existing planets. Some astronomers thought that the excess rotation was caused by a planet between Mercury and the sun. Although observations of this hypothetical body, called Vulcan, have been reported, they have so far proved to be mistaken. The supposed Vulcan has never been seen, even with the most modern observational techniques. (The problem of Mercury's orbit was later solved by Einstein; see Chapter 5.)

Newton was also sorely disturbed by the mutual forces among the planets, which he thought must eventually lead to the disintegration of the solar system. To avoid this awful event, Newton envisioned the hand of God occasionally descending to reset the clockwork mechanisms of planetary motions, like a conscientious craftsperson making adjustments. The order of the mechanical universe, ordained by the Divine Being, was maintained by His intervention, and the expanses of Newtonian space were benevolently watched by the distant God.

Newton's evaluation of his great achievements was rather modest.

> I do not know what I may appear to the world; but to myself I seem to have been only like a boy, playing on the seashore, and diverting myself in now and then finding a smoother pebble or a prettier shell than ordinary, while the great ocean of truth lay all undiscovered before me.

SUMMARY

This chapter presented the revolution in the development of a *physical* model of the cosmos. This transformation was ignited by the Copernican model, which itself lacked unifying physical ideas. Galileo promoted acceptance of the Copernican model with his telescopic observations. He also searched for physical laws to support the system and found laws for motion under gravity, but he did not apply them to the heavens. Motivated more by the design of the Copernican cosmos than by the details of planetary motion, Galileo ignored the discoveries of Kepler and saw the planets as swinging in circular orbits.

With the publication of Newton's *Principia* the Copernican model finally acquired its physical foundation. But to build this support, Newton (and Galileo) had to challenge Aristotle's ideas about motion, especially his concept of natural motion. To Aristotle a falling rock followed its natural motion, so no force was needed. By developing a clear concept of inertia, Newton turned around Aristotle's ideas about natural motion and saw a falling rock as *forced* motion, with the force being gravity. Coaxed by the apple and the moon, he then connected motions in the heavens and on the earth with the ties of gravitation to produce a unified model of the cosmos. Note how the idea of natural motion lies at the heart of his fundamental physical ideas and also, finally, at the conception of the cosmos that evolved from it.

From observations such as those described in Chapter 1, you cannot tell the difference between a heliocentric and a geocentric model. Physics, however, can help you decide. Newton did not have direct observational support for a heliocentric model, but such a system fell in line with his physical ideas. His comprehensive picture of the world so persuaded people that they accepted its heliocentric framework *before* they had direct observational evidence of its validity.

Newton's universe was infinite, but as carefully crafted as a clock. His laws of motion and gravitation describe what makes the world go 'round. Newton even anticipated the launching of satellites into orbit. These universal laws worked so well that his viewpoint went unchallenged until the twentieth century, when Albert Einstein created a new theory of gravitation.

Key Words

mechanics	centripetal force	acceleration
inertia	centripetal acceleration	binary star
natural motion	mass	vector
laws of motion	density	center of mass
universal law of gravitation	velocity	inverse square law

Review Questions

1. Suppose a contemporary of Galileo had made the following (hypothetical) discoveries with his new telescope. Which of them would favor the Copernican model, which the Ptolemaic model, and which neither model?
    a. A crescent phase for Jupiter
    b. Moons orbiting Venus
    c. A shift in position of some stars over the course of a year
    d. Binary stars in orbit about one another
    e. Absence of a "full" phase for Mercury
    f. Mountains on Mars
      (Objective 1)
2. The proof that gravity holds the solar system together came when Newton compared the effect of gravity on an apple to its effect on the moon. This test showed that (mark all correct answers):

a. The moon and the apple feel forces of identical size.
b. The moon and the apple experience identical accelerations.
c. The acceleration of any object by the earth is inversely proportional to the square of its distance from the earth.
d. The moon is accelerated in the direction of its motion with an acceleration $1/r^2$ times that of the apple.
e. The mass of the moon is much smaller than that of the earth.
(Objective 10)
3. Copernicus thought that the stars were in a shell surrounding the earth. Where did Galileo and Newton think they were and what reasons did each give for his belief? (Objectives 5 and 13)
4. Which of Galileo's telescopic discoveries support the Copernican system and refute the Aristotelian-Ptolemaic one? (Objective 1)
5. What important discovery of Kepler's was ignored by Galileo? (Objective 5)
6. Describe Galileo's concept of inertia. (Objective 2)
7. You have a lead ball and a wooden ball of the same size and shape. You hold them up at the same height and drop them simultaneously. What happens? (Objectives 2, 6, and 7)
8. Suppose you're out in space and push away from you an object that has the same mass as you. What happens? (Objective 6)
9. Describe two ways in which Newton's system differed from that of Copernicus. (Objective 13)
10. Why, according to Newtonian ideas, do objects thrown upward fall back to the same spot, even though the earth rotates beneath them? (Objective 6)
11. How do Newton's laws support the Copernican notion that the earth revolves around the sun, rather than vice versa? (Objective 13)
12. What observation shows that a law of gravitation in which the force varies as the inverse *first* power of distance, $1/R$, cannot be valid? (Objective 10)

Problems

1. Given that the moon's period is 27.3 days, its distance from the earth is 3.84×10^5 km, and $G = 6.67 \times 10^{-11}$ in mks units, determine the mass of the earth. (Objectives 9 and 12)
2. Demonstrate by a physical or mathematical argument that all masses near the earth's surface fall with the same acceleration. (Objectives 4 and 9)
3. Imagine you put a golf ball at the moon's distance from the earth. Compare its centripetal acceleration to that of the moon. Compare the gravitational force of the earth on the golf ball with the force of the earth on the moon. (Objectives 9 and 10)
4. Show that Kepler's third law implies an inverse square force law for gravitation. Hint: Use circular orbits. (Objective 12)
5. Given the masses of the earth and moon ($m_{earth} = 6.0 \times 10^{24}$ kg, $m_{moon} = 7.3 \times 10^{22}$ kg) and the average distance between their centers ($R = 3.8 \times 10^5$ km), find the distance of the center of mass from the center of the earth, and compare that distance to the radius of the earth. (Objective 11)

6. TV satellites are generally put in orbits such that the satellite stays above a certain place on the earth. Calculate how far away from the earth's center such a satellite must orbit. (Objective 9)
7. How much would you weigh on Mercury? (Objective 9)
8. Suppose you jump off a 10-meter diving board. How long does it take you to reach the water? What is your velocity at that time? What is the acceleration of the earth upward toward you (you will need to estimate your mass)? What is the earth's velocity when you hit the water? How far has it moved? (Objectives 6, 9, and 11)

BEYOND THIS BOOK...

* Galileo's conflict with the Church and his vibrant personality make him perhaps the most famous of astronomers in the popular eye. For a fictionalized look into Galileo the man, read *Galileo* (Grove Press, New York, 1966) by B. Brecht (better yet, see a production of the play) or *The Star-Gazer* (Putnam, New York, 1939) by Z. de Harsanyi, translated by P. Tabor.
* For a detective-style analysis of Galileo's problems with religious authority, try *The Crime of Galileo* (University of Chicago Press, Chicago, 1959) by G. de Santillana.
* For a view of Newton's work, nothing beats the *Principia: Mott's Translation Revised* (University of California Press, Berkeley, 1966), translated by F. Cajori.
* You can get to know Galileo by reading *Dialogue Concerning the Two World Systems* (University of California Press, Berkeley, 1967), translated by S. Drake. *Discoveries and Opinions of Galileo*, also by Drake, contains a translation of *The Starry Messenger*.
* For insight into how Galileo actually arrived at the correct analysis of the motion of falling bodies, read "Galileo's Discovery of the Law of Free Fall" by S. Drake in *Scientific American*, May 1973.
* A short biography of Newton is "Isaac Newton" by I. B. Cohen, *Scientific American*, December 1955; see also "Newton's Discovery of Gravity," March 1981.
* For details about how Neptune was found, look at *The Discovery of Neptune* (Harvard University Press, Cambridge, Mass., 1962) by M. Grosser.

EINSTEIN AND THE EVOLVING UNIVERSE

5

LEARNING OBJECTIVES
After studying this chapter, you should be able to:
1. Contrast Aristotle's, Newton's, and Einstein's concepts of natural motion for bodies falling near the earth and for the motions of heavenly bodies.
2. Decribe some basic ideas of the special and general theories of relativity.
3. Write down and make use of Einstein's equation that relates matter and energy, $E = mc^2$.
4. Describe what is meant by the term *spacetime* and sketch simple space-time diagrams.
5. State the principle of the equivalence.
6. Show how the principle of equivalence leads to the local cancellation of gravitational forces.
7. Argue that concepts of natural motion must be coupled to a notion of the geometry of spacetime.
8. Describe the Hubble law for the radial velocities and distances of galaxies, and derive from it the value of the Hubble constant.
9. Interpret the Hubble law as a consequence of the uniform expansion of the universe.
10. Outline how to use the Hubble law to estimate the age of the universe and to determine its geometry—hyperbolic, spherical (closed), or flat.
11. Define *escape velocity* and use it to describe the expansion of the universe and relate the universe's average density to its geometry.

CENTRAL QUESTION:
How did Einstein's ideas about gravitation lead to a dynamic model of the cosmos?

The previous chapters highlighted important changes in people's ideas of the universe. Aristotle's and Newton's schemes differed dramatically in their concepts of natural motion, and these different foundations resulted in distinct pictures of the nature of the universe. Newton's cosmos was infinite, yet was tied together by a universal physical law; Aristotle's cosmos was finite, but had different laws for the terrestrial and celestial realms. However, both models had one aspect in common: The universe was static.

Newton's grand model gained the authority of success after success. His infinite space, established firmly by his laws and intertwined with gravitation, seemed invincible. Nevertheless, Newton's physics had its detractors. Continental scientists chided him for his mysterious gravitational force, which worked instantaneously over astronomical distances like some magical quality.

Albert Einstein (1879–1955) (Fig. 5.1) greatly admired Newton's synthesis. But he wondered if the last word had been said about the clockwork universe. Some observations, such as the amount of the precession of Mercury's orbit, were not explained by Newton's laws. And Newton had failed to discuss the nature of gravitational force, although he did describe its effects. In addition to pondering gravity, Einstein was puzzled by the nature of light. From these two seemingly separate realms, gravity and light, Einstein synthesized a new view of the universe in his theory of relativity. And his theory had one surprising result; in it, *gravity is no longer a force.*

In Einstein's theory space, time, mass, and natural motion take on new and crucial relationships from which a new conception of the cosmos emerges, one in which the universe itself *evolves.* This chapter looks again at the concepts underlying Newtonian physics to find the themes of relativity and the consequences of relativistic ideas for the cosmos.

FIGURE 5.1 Albert Einstein. "What is inconceivable about the universe is that it is at all conceivable." [*Courtesy Yerkes Observatory*.]

FIGURE 5.2 Matching speeds makes relative velocities become zero. Two spacecraft can dock by maneuvering to zero relative velocity.

5.1 NATURAL MOTION REEXAMINED

We live in a Newtonian world. By this statement we mean that most people see the world the way Newton did. If an object falls or moves, we assume that a force acts on it. We are then unconsciously applying Newton's laws. We *know* a force is acting because we see its effect, a change in motion. The force seems real because we can see the acceleration which results from it. As a stone falls, its velocity continually increases; it exhibits a constant acceleration. There must be a force acting on it, because this is *not* the stone's *natural motion*, which, according to Newton's definition, is to be at rest or moving at a constant speed along a straight line (Section 4.2).

Newton's definition of natural motion leads directly to his concept of force, particularly gravitational force. Newton saw constant motion as ordinary, inconsequential, and so needing no explanation. To him, accelerated motion demanded explanation. After all, without accelerated motion apples would not fall from trees, the moon would not orbit the earth, and the earth would not revolve around the sun.

To cope with accelerations, Newton had to deal with forces. Forces result in changes in velocity. This emphasis on change of velocity rather than velocity itself is no accident. All velocities are relative, so you can always find a frame of reference in which any given velocity disappears. For example, when you are stopped in your car and about to enter a highway, other cars go by you at about 100 km/h. When you have accelerated to 100 km/h, too, the velocities of the other cars relative to you are reduced to zero. The original 100 km/h difference has vanished. Relative velocities can always be made zero with respect to some reference frame (Fig. 5.2).

Now imagine that forces were related directly to velocities rather than to accelerations.

Then, by matching velocities, you could always make the effects of a force vanish. No effects, no force! To be able to make forces disappear in such an arbitrary fashion would have been contrary to Newton's philosophy of nature. He believed that forces were *real*, that they could not be made to disappear arbitrarily. With forces connected to accelerations (and not velocities), Newton seemed to have nailed down their absolute existence.

However, Newton had made an assumption which you may not have noticed. Recall in his definition of natural motion that he talks of velocity along a *straight line*. But what is a straight line? When you state the definition as the shortest distance between two points, you probably have an intuitive picture of a line drawn on a flat surface, as Euclid described it, and that is the same geometrical assumption that Newton made. His "straight line" means a Euclidean straight line, for in his time Euclidean geometry was thought to be the only possible geometry.

Can you imagine straight lines that aren't Euclidean? Consider a curved surface. The shortest distance between two points on the earth's surface differs from a straight line on a flat surface. Look at a flat (Mercator) map of the world (Fig. 5.3). Traveling on a line of constant latitude is not the shortest distance between two points of the same latitude. Airplanes, for instance, travel along part of a *great circle*, a circle on the earth's surface whose center is the earth's center; this path is a straight line on a sphere.

To make straight lines on a sphere, stretch a string tightly against a globe holding it at the two places you want to connect (Fig. 5.4). Note that this path curves relative to latitude lines. On a flat, Mercator map such a line does not look straight. But it is! On a *curved* surface.

The geometry of a curved surface is not the same as that of a flat surface, nor is the geometry of a curved space the same as that of a flat space. In his definition of natural motion, Newton assumed that our universe was flat and had the properties described by Euclidean geometry. Einstein challenged his assumption.

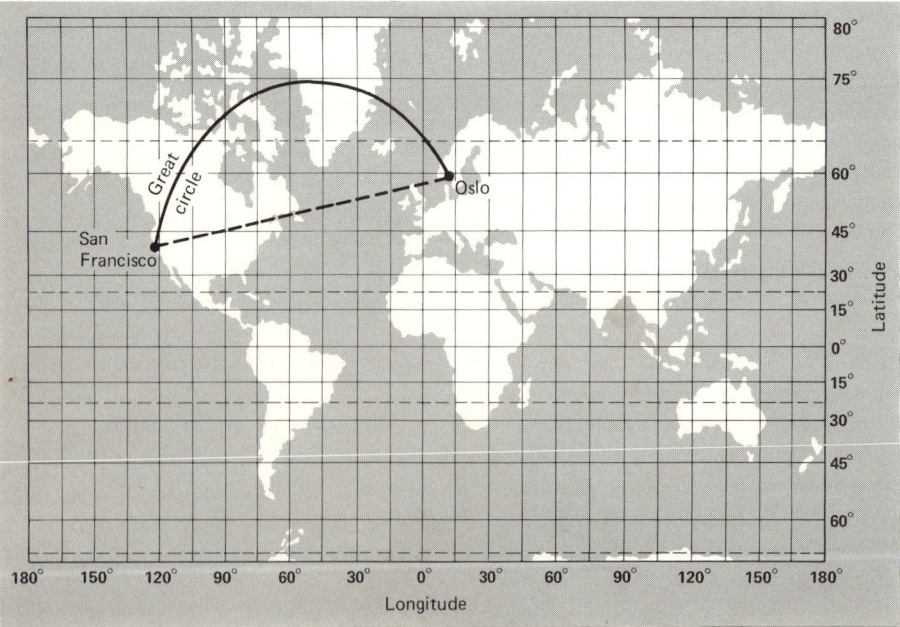

FIGURE 5.3 Straight lines and geometry on the earth. This Mercator map artificially makes the round earth appear flat. So straight lines on such a map are *not* the shortest distances between two points on the earth. If you use a ruler on this map to draw a straight line between San Francisco and Oslo, you do not get the path that is the shortest distance between these cities. That path is a great circle line, which appears curved on this flat map.

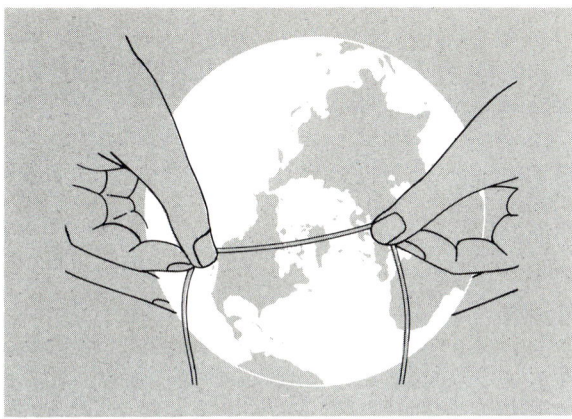

FIGURE 5.4 Straight lines on a sphere. You can check that great circle distances on a globe are the shortest distances (and so straight lines in a spherical geometry) by stretching a string along the globe's surface to connect two points.

5.2 THE RISE OF RELATIVITY

At the beginning of this century Albert Einstein created a new vision of the cosmos. This picture emerged in two stages: the special theory of relativity and the general theory. The special theory dealt with the laws of physics as seen by *unaccelerated* observers; those who experience uniform motion. The general theory goes beyond the special theory to cope with the nature of gravitation and so deals with *accelerated* observers.

Throughout the nineteenth century scientists had puzzled over the nature of light. Einstein was displeased by the theories they had developed. To satisfy his aesthetic qualms he was forced to reexamine Newton's concept of space and time. From this reexamination arose Einstein's special theory of relativity.

The Special Theory of Relativity

Einstein's special theory can be summed up in one sentence: *The laws of physics are the same for all observers experiencing uniform motion*. We won't treat the special theory in any detail, but only describe a few of its major consequences.

Einstein based the special theory on two fundamental postulates. First, the laws of physics hold true for all observers traveling at constant velocities. Second, the speed of light is a universal constant, which, when measured, has the same value to all such observers.

Einstein showed that from these two postulates follows the famous formula relating mass and energy to the speed of light:

$$E = mc^2$$

where E is the energy in joules (J), m is the mass in kilograms, and c is the speed of light in meters per second. This relationship means that you can think of matter as "frozen energy." It also means you can think of all kinds of energy as possessing mass. This mass produces gravity and also shows up as inertia.

Small masses convert to large amounts of energy. For instance, suppose we could completely transform 1 kg to energy.

$$E = mc^2$$
$$= (1 \text{ kg})(3 \times 10^8 \text{ m/sec})^2$$
$$= 9 \times 10^{16} \text{ J}$$

This amount of energy is enough to keep a 100-watt light bulb lit for about 30 *million* years!

Another consequence of the special theory is that no mass can travel faster than light. More accurately, an object cannot be accelerated from a speed less than that of light to a speed greater than that of light. Einstein's theory shows that as you accelerate an object closer and closer to the speed of light, it takes a greater and greater force to accelerate it more; its apparent mass gets larger at higher speeds. To get the object to go at the speed of light would require an *infinite* force, and that's impossible!

You may have heard about hypothetical particles that are believed to travel *only faster* than light. These particles, called tachyons, are a speculation at the moment; they have not been observed. Such particles would not contradict the special theory, because they are *always* traveling faster than light.

Spacetime

The special theory also resulted in the unification of the concepts of space and time. Previously, scientists had looked at these as separate aspects of the universe (similar to the ancient astronomical separation of earth and sky). Newton, for instance, dealt with space and time as totally unrelated. In Einstein's view all events in the universe involve space and time together. He made special note of what seems so obvious: When anything in the world takes place, it happens in *both* space and time. He called such a happening an *event*. You mark an event by noting where it took place (space) and when it took place (time). These events make up points in *spacetime*, a four-dimensional system consisting of three dimensions of space and one of time.

To help you grasp the intimate connection of space and time, consider this question. Are the events you see at any one moment really happening all at the *same* time? Remember, light takes a finite time to reach you to let you see what's going on. So you can never see events happening at the same instant if they are spread out in space. You see more distant objects as they were in the past, compared to nearer ones. When you look at the moon, you see it as it *was* about 1 second ago. When you look at the sun, your view is 8 minutes old. The stars give you a deeper look into the past. Sirius, as you see it now, shines as it did 8 years ago. As you look out into space, you look back in time. You now receive light emitted from different objects at many different times. In this sense, a telescope acts as a time machine that allows us to witness events that occurred in the past.

Spacetime Diagrams

If you're still having trouble picturing the world in spacetime, perhaps a visual tool will enlighten you. A *spacetime diagram* is a two-dimensional map that represents the four-dimensional world.

Consider an ordinary two-dimensional map, such as that in Fig. 5.3. The two direc-

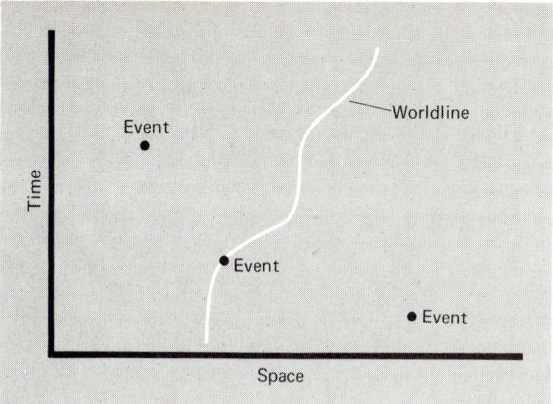 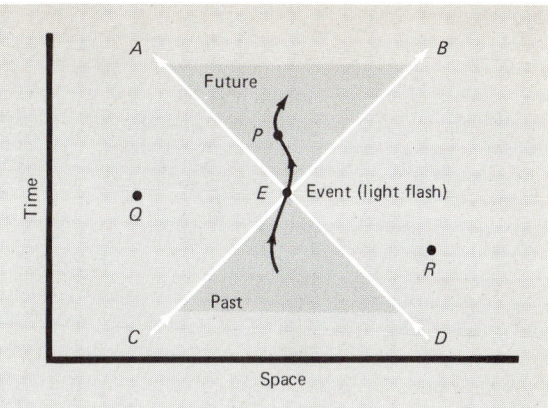

FIGURE 5.5 A spacetime map. Any point on this map is an *event*, which involves a specific place and time. A series of events makes up a *worldline*.

FIGURE 5.6 Light paths in a spacetime map. Light travels on straight lines in spacetime. A light flash at E travels along the paths contained by the lines EA and EB; the region within these lines is the future as seen from E. Likewise, the region within the lines CE and ED is the past as viewed from E.

tions in that map are latitude (north or south of the equator) and longitude (around the equator). Every point on the earth can be identified uniquely by giving its longitude and latitude. Now examine Fig. 5.5, which is a spacetime map. Here the vertical direction is time and the horizontal one is space, for which you can use any one of the three spatial directions. (To be complete, we should put in all three spatial dimensions, but such a map couldn't be drawn on a two-dimensional sheet of paper.) A point on this map is an event, something happening at a given place and given time. A path on this map that connects many different events is a history, a sequence of events, called a *worldline*. Consider, for example, your life throughout the day that has just passed. It was a series of events, a worldline in spacetime.

This spacetime diagram, which was first invented by the Russian mathematician Hermann Minkowski (1864–1909), has a crucial feature in the context of special relativity. Only certain events can affect other events (we say they are *causally connected*). Imagine this event: you send out a flash of light (point E in Fig. 5.6). If both length and time are expressed in the same units (such as meters of length and meters of light travel time, where 1 meter of light travel time equals roughly 3×10^{-9} sec), the paths of the light make 45° angles with the space and time axes. Now look to the future from point E. To go into the future from E outside of the light lines requires you to travel faster than the speed of light, which you cannot do. By the same argument, to get to E from past events outside of the light lines also means traveling faster than light. The event of sending out the light flash is causally connected only to the events within the intersecting light paths (shaded region in Fig. 5.6). In special relativity there are limited regions of spacetime that can affect us (the past) and which we can affect (the future).

The General Theory of Relativity

From 1905 to 1915 Einstein tackled the problem of widening the scope of the special theory to include *accelerated* observers. This investigation led to the *general theory of relativity*, which dealt with gravitation.

Einstein questioned Newton's concepts of mass and inertia. He saw that Newton defined mass by two different operations: in the second law and in the law of gravitation. Suppose you apply a known force to an object and measure its acceleration. Using the force and acceleration, Newton's second law (Section 4.2) gives you the object's mass. Mass determined this way is called *inertial mass*. Now take the same object and weigh it. Weight is a force, the amount of gravitational force with which the earth (in this case) attracts the object. Mass measured this way is called *gravitational mass*.

Newton believed (and so did not make the distinction) that an object's inertial mass and gravitational mass were the same. He knew this from Galileo's experiments with falling bodies (Section 4.1) as well as from his own careful experiments. These results demonstrated that, near the earth, *all masses fall with the same acceleration*.

Section 4.2 showed mathematically why the accelerations of all falling bodies are the same (which Galileo proved experimentally). One of the steps involved canceling the mass from both sides of an equation. But that was a cheat! Those two "masses" are not logically or physically the same. To see this, let's do that calculation again. Take Newton's law of gravitation

$$F = \frac{GMm_g}{R^2}$$

where m_g is the *gravitational mass*, and Newton's force law

$$F = m_i a$$

where m_i is the *inertial mass*, to obtain

$$m_i a = \frac{GMm_g}{R^2}$$

We can obtain the final result

$$a = \frac{GM}{R^2}$$

only if $m_i = m_g$.

Experimentally, this equality of gravitational and inertial mass holds true to a very high degree of accuracy. No difference has ever been detected to the limits of sensitivity with which tests can be done. The best experiment to date, done by V. B. Braginsky and V. I. Panov at Moscow University, found the inertial and gravitational masses of gold and platinum to be the same to within one part in 10^{12}.

Einstein felt that the equality of gravitational and inertial mass was no accident. He saw it as a fundamental fact about the universe and gave it a special place in the general theory as the *principle of equivalence*: You cannot distinguish accelerations due to gravitation from accelerations due to other kinds of forces.

Here is Einstein's own example of the principle of equivalence. Imagine that you are on the earth in a spacecraft with no windows (Fig. 5.7). You drop objects in the space-

5.2 THE RISE OF RELATIVITY

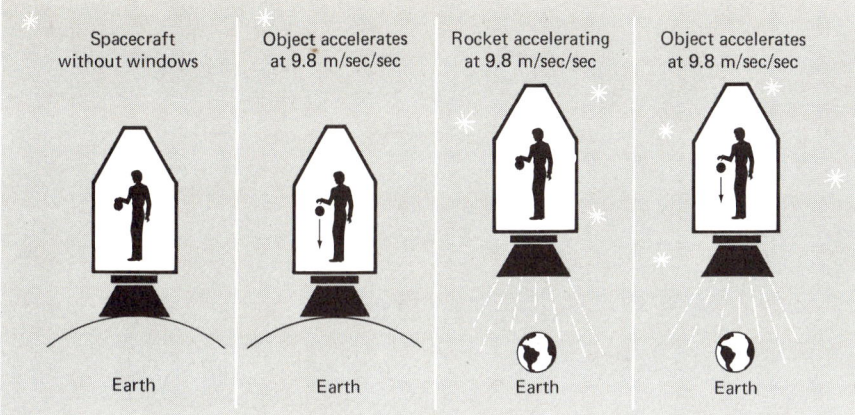

FIGURE 5.7 The principle of equivalence. Imagine you are in a small spacecraft with no windows. Drop a mass and measure its acceleration. If the spacecraft is on the earth, the acceleration will be 9.8 m/sec/sec. If the ship is in space far away from the earth and its engine is accelerating it at 9.8 m/sec/sec, the same experiment will yield the same results. You can't tell from it whether you're on the earth or in an accelerating ship.

craft, measure their accelerations, and find that they all fall with the same acceleration, 9.8 m/sec/sec. Now suppose that without your knowledge you and the spacecraft are instantly transported out in space and constantly accelerated at 9.8 m/sec/sec. As you continue your experiment, there is no change in the acceleration of the falling objects. It remains 9.8 m/sec/sec. Only by being able to look out of a window could you tell that you had left the earth. You cannot, by your experiments, distinguish between the effects due to a gravitational force and those due to the force of a rocket engine.

Let's examine the principle of equivalence from the viewpoint of a spacetime diagram. Imagine you are standing on the ground watching an object fall. Its path in spacetime (Fig. 5.8a) is a curve (in fact, a parabola) from your perspective. Why? Because the object is accelerating. Now imagine that you fall also. Again draw the object's path in spacetime (Fig. 5.8b), and note the difference. In the view of the falling observer the object's path is a straight line rather than a curve. To the falling observer, the object does not accelerate, as both the object and the observer free-fall together.

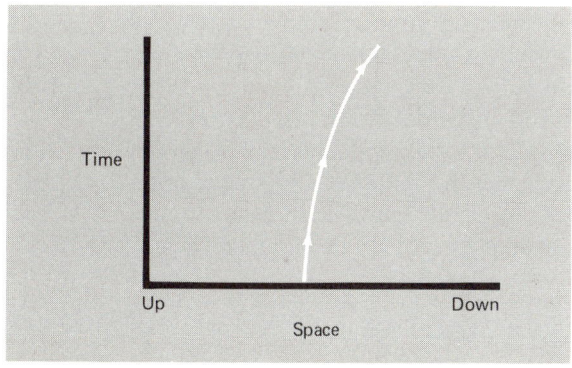
(a)

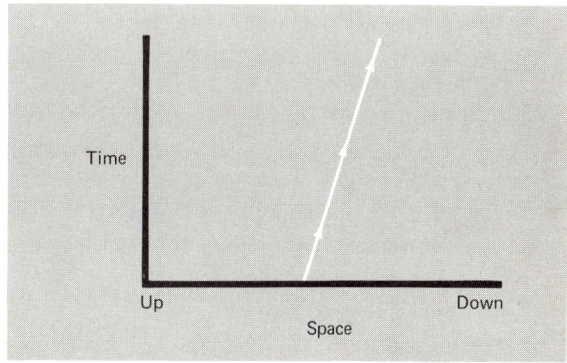
(b)

FIGURE 5.8 (a) Spacetime path of a falling object as seen by a stationary observer on the ground. The path is curved (it's a parabola). (b) Spacetime path of the same object as seen by an observer free-falling. The path is now a straight line.

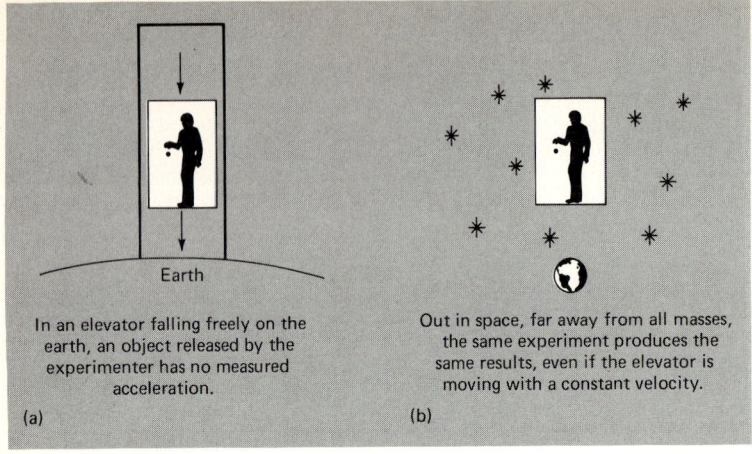

FIGURE 5.9 The local elimination of gravity, using the principle of equivalence.

(a) In an elevator falling freely on the earth, an object released by the experimenter has no measured acceleration.

(b) Out in space, far away from all masses, the same experiment produces the same results, even if the elevator is moving with a constant velocity.

These examples of the principle of equivalence seem simple enough, yet the principle has profound consequences. You now have a way to cancel out gravity locally. Put yourself in an elevator in a tall building. Let the elevator free-fall (Fig. 5.9). You find yourself weightless; gravity has vanished! Transport yourself into space far away from any large masses. Your condition is the same—you are weightless without gravity.

You might object that in this example of a free-fall, gravity will reappear quite dramatically when the elevator hits the ground. But to view this situation correctly, imagine a long tunnel drilled through the earth, so that the elevator never hits the ground. Then the elevator would free-fall back and forth from one side of the earth to the other, completing a cycle in about 88 min. Throughout this swing no accelerations would be felt inside the elevator, even though it's passing through the earth!

Newton considered gravity a force, the same no matter what the situation of the observer. Yet by free-falling, an observer can make this "absolute" force disappear. For Newton, gravity was mysterious, even though he was astute enough to describe its effects. To Einstein the falling of objects was not mysterious at all. They are simply following their *natural motion in spacetime.* In Einstein's view, gravity is *not* a force.

Here's another way of stating this: *When you are weightless, you are following your natural motion in spacetime.* Having weight is, in this sense, "unnatural."

Einstein's view of natural motion matches Aristotle's (Section 2.3) much more closely than Newton's. Drop an object. Aristotle says, "That's natural motion." Newton declares, "That's forced motion due to gravity." Einstein states, "That's natural motion in spacetime with no force." Their views of falling objects applied as well to the motion in the heavens. Aristotle saw it as natural motion, Newton as motion influenced by the force of gravity, and Einstein as natural motion without force. But if the moon's motion around the earth, for example, is its natural motion, why is the orbit curved rather than straight? To answer this question requires a look at geometry and its relationship to physics.

5.3 THE GEOMETRY OF SPACETIME

How can geometry and physics relate? Newton assumed that the geometry of the universe was Euclidean. Einstein makes no such assumption. He puts this question up to *experimental* confirmation. Einstein wrote:

5.3 THE GEOMETRY OF SPACETIME

> Geometry is . . . evidently a natural science; we may in fact regard it as the most ancient branch of physics. Its affirmations rest on induction from experience. . . . whether the . . . geometry of the universe is Euclidean or not has a clear meaning, and its answer can only be furnished by experience. . . . I attach special importance to this view of geometry . . . for without it I should have been *unable to formulate the theory of relativity*.

What does Einstein mean here?

Geometry and Physics

Geometry, developed by the Greeks into mathematics, derived from the practical surveying techniques of the Egyptians. So geometry, as we understand it, was developed from experience. For centuries people held that Euclidean geometry, and *only* Euclidean geometry, applied throughout space to physical measurements. Newton made this assumption too, for no other geometry had been devised by his time.

Because of Euclid's parallel line postulate (essentially, two parallel lines when extended to infinity will remain the same distance apart and will never meet), Euclidean geometry is flat. Some consequences of its flatness in two dimensions are the Pythagorean theorem for right triangles and the statement that the sum of the angles of any triangle equals 180° (Fig. 5.10a).

In 1829 the Russian mathematician Nikolai I. Lobachevski (1793–1856) pointed out that Euclid's parallel line postulate was not the only possible one. Lobachevski proposed a new postulate that allowed parallel lines to diverge but still resulted in a self-consistent geometry. In two dimensions the surface of a saddle-shaped object has properties of his geometry, in which all parallel lines diverge when extended; the geometry is curved. This geometry is sometimes termed *hyperbolic geometry*. When a triangle is drawn on a hyperbolic surface, the sum of its angles is *less than* 180° (Fig. 5.10b). Hyperbolic geometry is infinite, because extended parallel lines never meet. Both hyperbolic and flat geometries are for this reason sometimes called *open geometries*.

In 1854 Georg F. B. Riemann (1826–1866) devised yet another self-consistent geometry having a different parallel postulate. Riemann allowed parallel lines to converge when

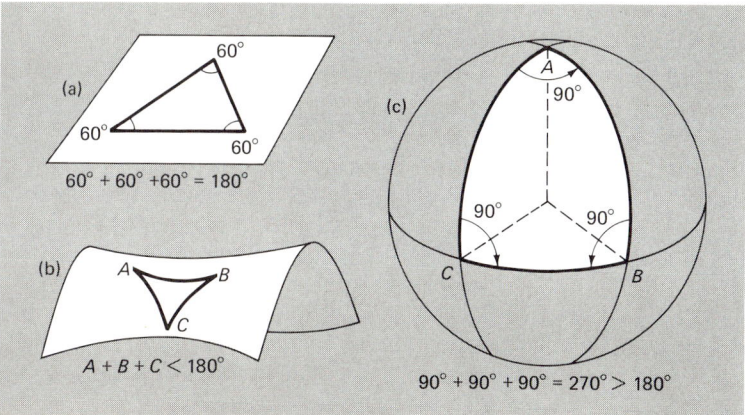

FIGURE 5.10 Properties of different geometries. (a) In a flat geometry the sum of the angles of a triangle always *equals* 180°. (b) In a hyperbolic (open) geometry the sum of the angles is always *less than* 180°. (c) In a spherical geometry the sum of the angles is always *greater than* 180°. For example, the triangle drawn from two points on the earth's equator to the north pole contains 270°.

extended. In two dimensions the surface of a sphere has properties of his geometry (Fig. 5.10c). Consider the idealized earth's surface with its lines of longitude and latitude. Two lines of longitude, both of which are perpendicular to the equator (and hence parallel to each other), intersect at the poles. Here the sum of a triangle's angles is *greater than* 180°. Such a geometry is sometimes called *spherical geometry*, or *closed geometry*.

Both of these non-Euclidean geometries may be characterized by their curvature (or bending), in contrast to the flatness of Euclidean geometry. Hyperbolic geometry has a negative curvature. It bends away from itself and therefore extends infinitely far. Spherical geometry has a positive curvature. It bends in on itself and therefore is finite, but unbounded, that is, it has a definite size, but no edge. For example, you can travel around the earth's surface as many times as you like and in any direction you want without ever discovering a boundary, yet the surface of the earth has a definite area (5×10^{14} m^2). (Keep in mind that we are using two-dimensional examples here because they are easy to visualize, but the physical world exists in the four dimensions of spacetime. That's harder to picture, but the idea is the same.)

So by Einstein's time three general categories of geometry—hyperbolic, spherical, and flat—were available. (Newton had only one.) Which one is the appropriate choice to apply to the physical world? And how does this choice relate to gravity?

Local Geometry and Gravity

Imagine that you live in two dimensions on the earth's surface. You have no concept of a third dimension and no experience of it. You cannot conceive of an "up" that is off the surface. And, you believe that the geometry of the world you live in is flat.

You and a friend do an experiment. You both stand on the equator some distance apart (Fig. 5.11). You both walk away from the equator on paths that are at right angles to the equator. As you walk, being very careful to keep on lines at right angles to the equator, you find that you are moving closer together. Yet by all the precepts of the Euclidean geometry you should be traveling on parallel lines and so stay the same distance apart. What's happening?

You might stick with the belief that your world is flat. Then you could explain your coming together by saying that there is a strange force of attraction bringing you and your friend together. You might even call this force "gravity" and see it as mysterious. Then you would be thinking like Newton.

Or you could say to your friend: "Maybe our assumption about this world's being flat is wrong. Maybe it's actually curved. We are moving on straight lines at right angles to another line. These lines should be parallel. Yet we move closer together. That's not what we should expect if the world is flat. So our experiment is telling us that our world is *not* flat, but curved. Then our moving together has a natural explanation: It's our world informing us it's curved." Now you would be thinking like Einstein. No need for a mysterious force! An experiment has revealed the local geometry of the world in the region where you do the experiment.

Einstein views four-dimensional spacetime in the same way. Whether it is curved or flat and the degree of its curvature are questions to be answered by experiment.

How curved is spacetime, and why? In Einstein's general theory of relativity the distribution of mass (and energy) determines the geometry of spacetime. A massive object

5.3 THE GEOMETRY OF SPACETIME   • 139

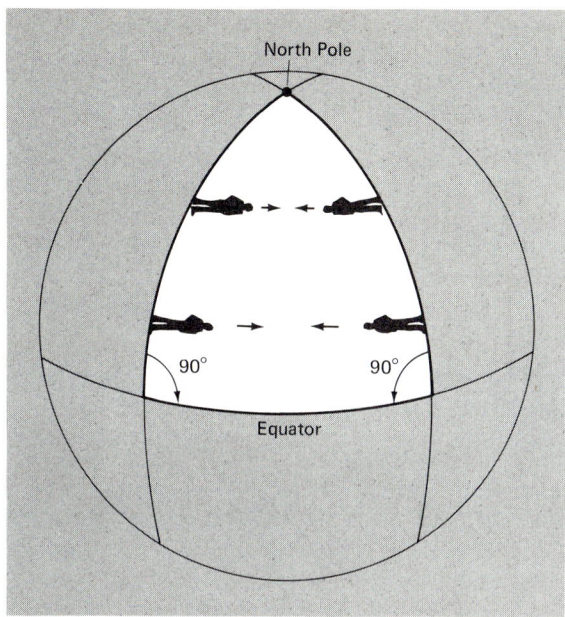

FIGURE 5.11   Following straight lines in a spherical geometry.

produces a curvature of nearby spacetime. And that curvature shows itself by accelerated motion, which Newton would say was caused by gravitational forces.

Let us try to explain this notion of spacetime curvature using falling bodies. Imagine a bunch of small particles in space surrounding the earth (Fig. 5.12a). Let them all fall toward the earth's center, and imagine they are not stopped by the ground. What will a spacetime diagram of their paths look like? The paths converge until all the particles collide (Fig. 5.12b). This set of paths resembles the convergence of longitude lines on the earth's surface to the poles. You know why that occurs; the earth's surface is curved. Likewise, the spacetime diagram of the falling particles tells you that spacetime in and

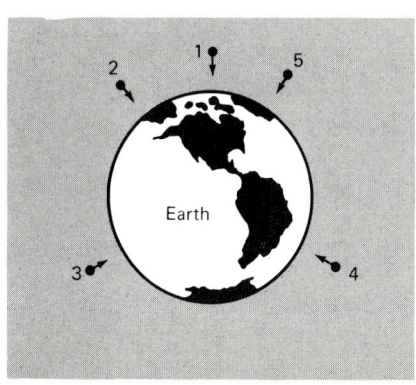
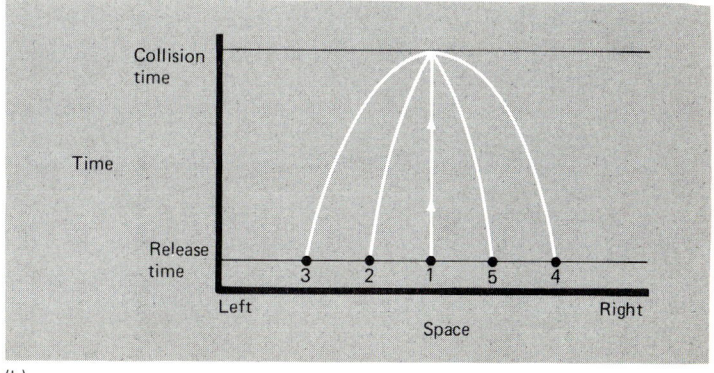

FIGURE 5.12   (a) Masses free-falling toward the earth. (b) A spacetime map of the paths of these falling objects as seen by an observer on the ground.

near the earth is also curved. The paths the particles follow, however, are straight lines in a curved spacetime. In fact, for any one observer at one of the particles, spacetime looks flat in free-fall. (Recall Fig. 5.8.)

Note in the above example that *any* particles made of *any* material of *any* shape and *any* mass fall along the same paths. No matter what particles you use to test spacetime's curvature, you get the same result. This example suggests that the distortion of particle paths is a property of spacetime itself, not of the objects moving in it.

Einstein's theory of general relativity also predicts that light rays will be bent by the curvature of spacetime. To see this effect, go back into our Einstein elevator out in space, which is accelerated by its rocket engine at 9.8 m/sec/sec. Put a laser in the front of the elevator to fire across its width at a target at the same level on the other side (Fig. 5.13a). Will the light beam hit the target? No, it will strike below it (Fig. 5.13b). As the light travels from one side of the elevator to the other, the elevator continues to accelerate "upward." The light path appears curved to an observer inside the elevator. Now, by the principle of equivalence, if you moved the elevator to the earth's surface, you shouldn't be able to tell the difference by an experiment inside the elevator. The light path will also bend and by the same amount as before (Fig. 5.13c). So light paths should be deflected in all regions of curved spacetime.

To test this prediction of general relativity on the earth would be a difficult job because at the speed of light the distances and times involved are so small. The amount of bending in the elevator example above will be extremely small, because light crosses the width of the elevator in an extremely short time. For instance, if the elevator is 3 m wide, the light transit time is a mere 10^{-8} sec, and during this time the elevator moves only 5×10^{-16} m.

Deflection of Light by the Sun

The most warped region of spacetime in the solar system is where the most mass is located, at the sun. Relativity predicts that the sun's mass deflects light rays away from Euclidean, straight-line paths. This prediction can be tested by comparing a picture of the stars around the sun when the sun's light is cut off during a total solar eclipse with a picture of the same stars when the sun is not in this part of the sky. With the sun present the angular separation of two stars close to but on opposite sides of the sun is *greater* than when the sun is not there (Fig. 5.14). This bending is greatest for light that just skirts the edge of the sun. So a solar eclipse is a natural event in which to try to measure the deflection of light from stars.

However, these are hard observations to make, for precise photographs must be made of the sun and stars during eclipse and also of the stars without the sun in the vicinity. Also, total solar eclipses (Section 1.5) don't usually conveniently fall at established observatories, so astronomers must lug their telescopes to fairly distant locations. Errors are hard to estimate and may be as large as 20 percent. General relativity predicts that the deflection should amount to 1.75 arcseconds, and observed results have ranged from 1.43 to 2.7 arcseconds. The scatter in the observed values reflects the difficulty of doing this observation.

Modern technology permits a similar experiment to be done more precisely by radio astronomers. Radio and light are essentially the same, so these deflections should also occur for radio signals from distant objects. Some quasars (Chapter 23) are intense radio

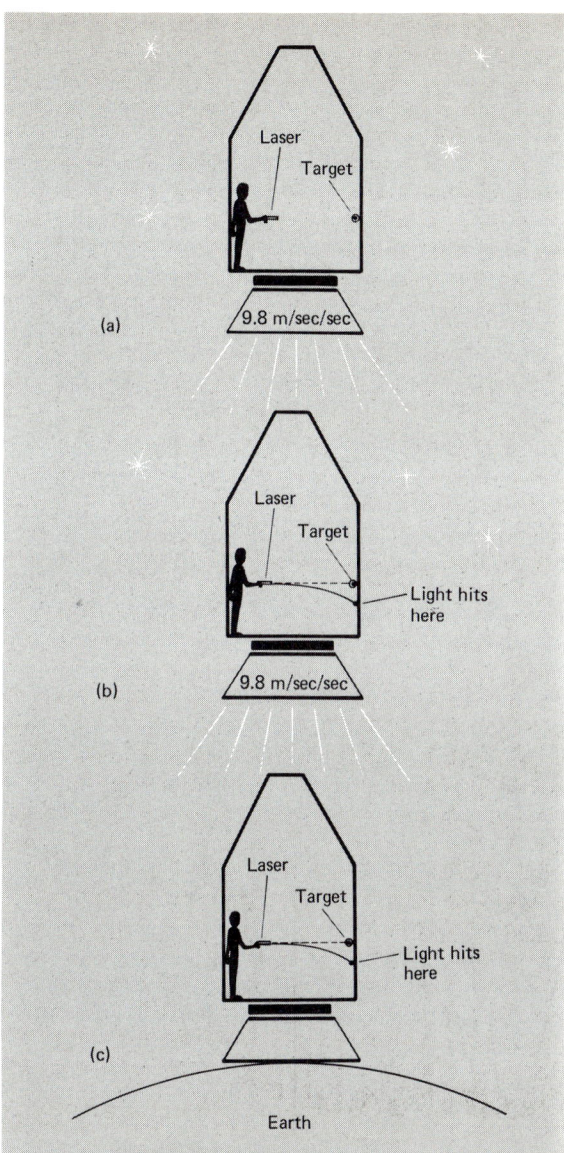

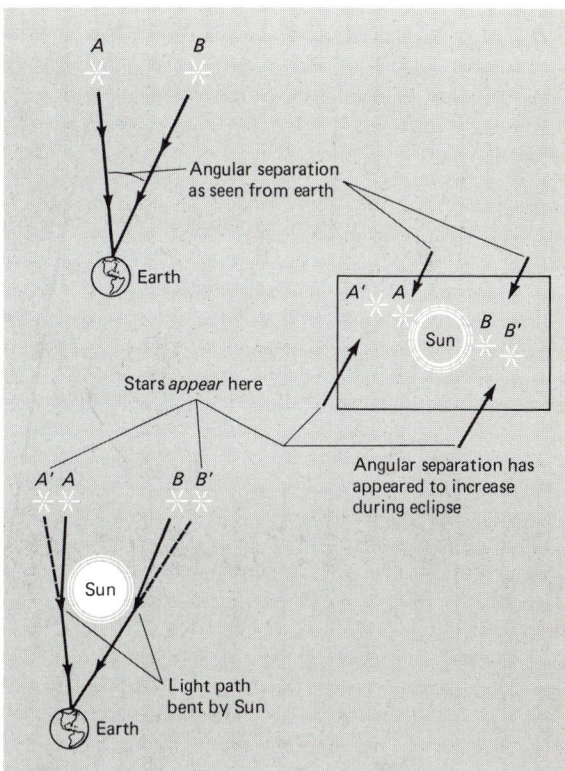

FIGURE 5.13 The bending of light in a gravitational field. In an accelerating spacecraft (a), a laser beam shot at a target on the opposite wall will hit below the target (b). By the principle of equivalence, the same will happen if a spacecraft is at rest on the earth (c).

FIGURE 5.14 The effect of the sun on the apparent angular separation of two stars. Consider two stars (A and B) that have a measured angular separation when the sun is not nearby. During an eclipse, when the sun lies between the two stars, they appear to be *farther apart* due to the bending of the light paths.

sources. Each October the sun eclipses one of these sources, the quasar 3C 279. By monitoring the position of 3C 279 relative to a nearby quasar (3C 273), radio astronomers, using a technique called interferometry (Chapter 7), can measure the angular separation between the two quasars to an accuracy of about 0.1 arcseconds. This experiment has been done a number of times, and the observed change in position of 3C 279 is within 10 percent of the value predicted by general relativity.

Have we been fair to Newton? If he had known that light has an equivalent mass ($E = mc^2$), he could also have calculated, using his law of gravitation, a deflection from the sun's gravity. But the result is only *one-half* Einstein's value. Einstein wins in this experimental test!

What causes the deflection of the starlight? In Einstein's view, light pursues a straight line in spacetime. But the spacetime in which it travels is curved by the sun's mass. You

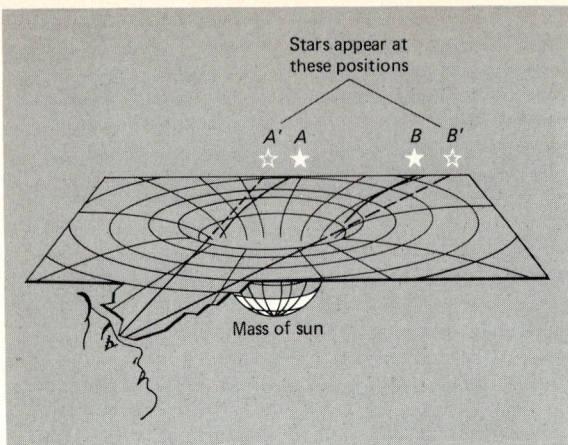

FIGURE 5.15 Einstein's explanation for the change in two stars' angular separation during an eclipse. The mass of the sun curves the spacetime around it; the closer to the sun, the greater the curvature. The light from the stars follows this curved spacetime when the sun lies between them as seen from the earth (A'B'). When the sun is not there, the light paths travel through a flatter region of spacetime, and so the stars appear closer together (AB). [*Adapted from a diagram by C. Misner, K. Thorne, and J. Wheeler.*]

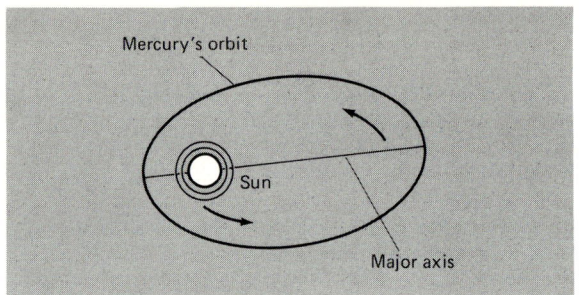

FIGURE 5.16 Precession of Mercury's orbit. The major axis of the orbit rotates in space with respect to the stars.

can picture the sun's mass as creating a warp in the geometry of spacetime (Fig. 5.15). Light crossing the warp appears to us to take a curved path, although it is actually traveling on the shortest distance between two points through the nonflat region of spacetime.

Imagine the situation as analogous to that of a golf ball moving along a poor putting green filled with depressions. In between depressions the surface is flat and the ball moves along a straight line. When it crosses a depression, the ball follows a path that bends compared to the path on the flat surface. The amount of bending depends on the depth of the depression and how fast the ball travels. Similarly, in Einstein's general theory of relativity, mass and energy create local warps in spacetime. The amount of mass-energy determines the amount of warping. As an object moves through a warped region of spacetime, it follows a straight-line, natural-motion path. In warped spacetime this path appears curved. But no force acts on the object. This is essentially how Einstein explained the nature of the gravitational "force" locally.

Einstein considered the orbits of the planets in the same way. All the planets move on natural-motion, straight-line paths in spacetime. We view these paths as elliptical orbits in three dimensions. The paths appear curved to us because spacetime is warped, not flat. The amount of warping decreases as the distance from the sun increases. That's why the earth, for example, follows an orbit more strongly curved than that of Jupiter.

Precession of Mercury's Orbit

The orbits calculated from Einstein's theory do not differ much from those calculated by Newtonian theory, except in the case of Mercury. Astronomers have known for a long time that the major axis of Mercury's elliptical orbit does not remain fixed in space with respect to the stars (Fig. 5.16). The axis rotates around, or *precesses*, in the plane of the orbit. Part of this shifting arises from the gravitational attraction of the other planets on Mercury. But when this effect and others are taken into account, there remains a residual shift of 41 arcseconds a century (which means the orbit turns through an extra 360° in about 3 million years).

What causes this precession? Newtonians attributed it to the undiscovered planet Vulcan orbiting within the orbit of Mercury. But no such planet exists. General relativity predicts a precession because of the strong curvature of spacetime close to the sun. (Precession due to curvature also happens for all the planets, but it is largest for Mercury.) The predicted value for Mercury, *if* the sun is perfectly spherical, is 43 arcseconds per century. So the observed and predicted results agree to within 4 percent.

The hooker here is the *if*. If the sun is nonspherical, the distribution of mass will also cause Mercury's orbit to precess. Because the sun is so bright, it's quite difficult to find out if it is oblate. Early experiments done at Princeton by Robert Dicke and M. Goldenberg indicated a mild solar oblateness. Later work by Henry Hill and colleagues, however, found no measurable oblateness. So it appears that general relativity accurately accounts for the precession of Mercury's orbit.

In summary, a major difference between Newton's and Einstein's models for gravity is that Newton concentrates on *forces*, while Einstein concentrates on *courses*. Newton sees gravity as due to a force acting instantaneously between all matter, a force whose strength depends on the mass of the attracted object. But this force causes an acceleration independent of the mass of the object, since acceleration is force divided by mass, and the inertial and gravitational masses happen to be equal. Einstein focuses on the paths of free-falling objects in spacetime, which don't depend on the mass of the object. The question of equality of gravitational and inertial mass does not arise. The paths that free-falling objects take, their natural motion, are determined by the local geometry of spacetime.

Einstein's model for gravitation would be no more than a fascinating idea if it did not make numerical predictions that could be tested experimentally of effects unknown or unexplained by Newton's model for gravitation. The deflection of light in curved spacetime near the sun and the precession of Mercury's orbit are two such tests.

Warning and notice: Einstein's geometrical ideas about gravity and spacetime will *not* be used, in general, in the rest of this book. For most purposes, Newton's model of gravitation is a perfectly good one, because in most cases the curvature of spacetime is very small. So you do not need to master Einstein's ideas to the same depth and competency as you do Newton's. Einstein's general theory crops up only when we deal with the universe as a whole (the rest of this chapter and Chapter 24) and with black holes (Chapter 12).

5.4 GEOMETRY AND THE UNIVERSE

So far we've described Einstein's picture of gravity in a small region of spacetime, such as near the sun. But his general theory of relativity applies also to the universe. It allows you to picture the geometry of the cosmos as it relates to the dynamics of the whole universe. Keep in mind that three basic geometries are possible: hyperbolic, spherical, and flat. Remember also that Einstein does not state which geometry must apply; he leaves this open to experimental confirmation.

Imagine again living in two dimensions on the surface of a sphere. Start at any point and walk on a straight line away from it. Eventually you'll return to your starting point, because a sphere has so much curvature that it comes back on itself. Suppose the universe has the same geometrical properties (in spacetime) as a sphere's surface. Then if we sent out light signals, they would eventually return to us. Just like the two-dimensional surface, our four-dimensional spacetime, if curved enough, will close back on itself. This also means that it does not have an edge or boundary. Remember that you can go around a sphere's surface many times and never find an edge. We don't know for sure that our universe *is* closed—that depends on whether it contains enough matter and energy to curve it sufficiently. But if it does contain enough it is finite and unbounded, just as the surface of a sphere is finite and unbounded, and light can go all the way around. Using this predicted circuit might be a method of finding out if the universe is indeed closed, but it is obviously impractical. Because the universe is large and the speed of light finite, it would take too long for light to go around. We shall consider another method in Section 5.5.

In 1917 Einstein used general relativity to calculate a model of the universe with a closed geometry. How different from Newton's model, which had to be infinite (Section 4.3)! In a geometrical sense Einstein's initial model was akin to the finite, closed picture of Aristotle (Section 2.3), and it was also static. Our conception of the cosmos seems to have come full circle and started around again. A few years after Einstein proposed his static model, astronomers discovered that the universe is *expanding*. Though Einstein's general relativity is still correct, his particular application of it to a *nonchanging* universe was wrong.

The Expansion of the Universe

The evidence that the universe is expanding comes from the study of the motions and distances of *galaxies*. Galaxies will be discussed in detail in Chapter 22. Here you need only know that galaxies are systems of billions of stars, each more or less like the sun. The galaxy of which our sun is a part we call the Milky Way Galaxy. The galaxies extend throughout the universe, and each galaxy is a marker for a point in the universe. These markers are found to be moving apart, which means the universe is expanding.

In 1912 Vesto M. Slipher (1875–1969) made the first measurement of a galaxy's velocity. (The technique for measuring velocities is discussed in Chapter 8.) He discovered that the nearby Andromeda galaxy is moving toward us at a speed of about 170 km/sec. By 1917 Slipher had measured the velocities of 15 galaxies. Unlike the Andromeda galaxy, all but two were receding from our Galaxy. By 1928 Slipher had measured over 40 galaxies, and the trend was becoming clear: Most galaxies were apparently moving away from the Milky Way Galaxy.

The Hubble Law

At about the same time Edwin P. Hubble (1899–1953) had determined the distances to some galaxies. (The technique for measuring distances of galaxies is discussed in Chapter 22; it is based on the principle that fainter galaxies are generally farther away.) Because the distances to stars and galaxies are so large, new units of measurement are needed. The common ones are the *lightyear* (ly), the distance light travels in a year; the *parsec* (pc), the

5.4 GEOMETRY AND THE UNIVERSE

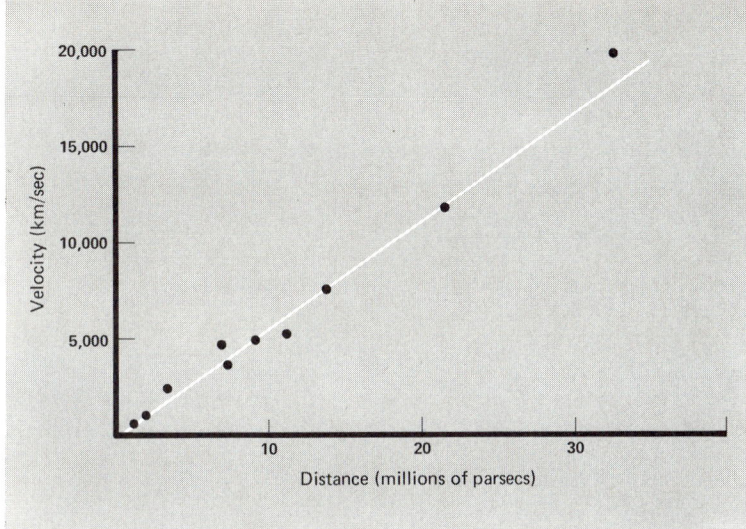

FIGURE 5.17 Hubble and Humason's distance-velocity relationship, now called the Hubble law. Plotted here are the recessional velocities (in km/sec) of galaxies against their distances (in millions of parsecs, or megaparsecs). The straight line indicates the trend of the data points.

distance at which the radius of the earth's orbit would have an angular diameter of 1 second of arc (see Section 8.2); and the *megaparsec* (*Mpc*), 1 million parsecs. The relationships between these units are: 1 ly = 9.5×10^{12} km; 1 pc = 3.26 ly = 3.09×10^{13} km. When Hubble compared his distance measurements with the velocities measured by Slipher, he noted an unexpected direct relationship between velocity and distance. For every million lightyears farther out, the galaxies had 170 km/sec more velocity, or, equivalently, for every megaparsec farther out, the galaxies had 550 km/sec more velocity (Fig. 5.17). (Modern measurements give a much smaller value.) In later collaborative work Hubble and Milton L. Humason (1891–1972), using the new 100-inch telescope, added more data to support the trend. This relationship, now known as the *Hubble law*, states that the distance to a galaxy and its recessional velocity are directly related, as expressed by the equation

$$v = Hd$$

The number connecting the distance and velocity is called the *Hubble constant* and is usually indicated as *H*. It is the slope of the line on a velocity-distance plot such as Fig. 5.17. The steeper the slope, the larger the value of *H*. *H* is the ratio of the velocities and distances of galaxies, or,

$$H = \frac{v}{d}$$

where *v* is the velocity (in km/sec) along the line of sight and *d* the distance (in Mpc). (The part of an object's velocity along the line of sight is called its *radial velocity*; so *v* in the above equations is a galaxy's radial velocity.) Note that we can turn this equation around:

$$d = \frac{v}{H}$$

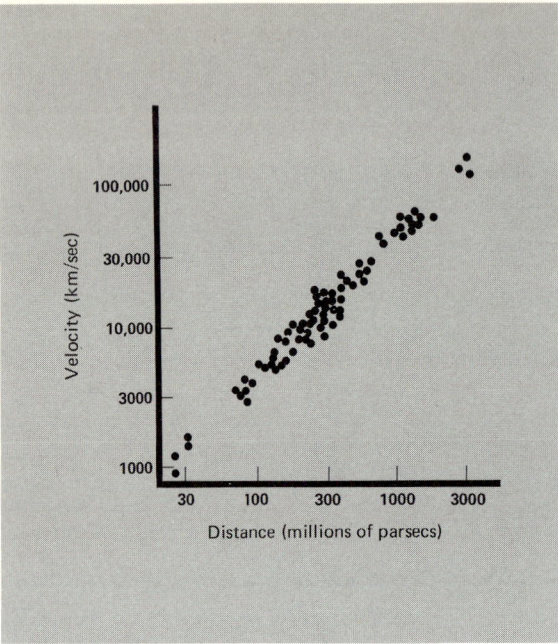

FIGURE 5.18 A contemporary Hubble diagram [*data by A. Sandage*]. Shown here are the recessional velocities plotted against distance. From these data, Sandage concluded in 1972 that the Hubble constant is 55 km/sec/Mpc.

So if we know a galaxy's radial velocity and the Hubble constant, we can estimate its distance.

Hubble and Humason thought that H was 550 km/sec/Mpc. Today astronomers believe that H lies between 50 and 100 km/sec/Mpc (Fig. 5.18). The exact value is disputed. (A review by James Peebles at an astronomical meeting presented a value of 68 km/sec/Mpc as the average value determined from all the different methods of finding H. Chapter 22 deals with the controversy over H.) This book uses 50 km/sec/Mpc. For example, a galaxy at 50 Mpc travels at 50×50, or 2500 km/sec velocity. Using the law in reverse, a galaxy with a measured velocity of 10,000 km/sec is at a distance of 2000 Mpc.

The way we stated Hubble's law—the farther away a galaxy is from us, the faster it is moving away from us—might seem to imply that there is some mysterious force that accelerates galaxies as they move away from our own. (Remember Newton's first two laws.) Astronomers don't much care for mysterious forces if they can come up with a simpler explanation, so let's see if we can find one for the Hubble law. Try turning the law around; it then says things that are moving fast are far away. That actually makes pretty good sense. We think all the galaxies in the universe were formed together and have been moving away from each other for the same amount of time, so the ones that started out moving fastest relative to us will be the farthest away, and that's exactly what Hubble's law says.

The discovery of this trend, that more distant galaxies move faster, caused a radical revision of Einstein's static model for the universe. It appeared that the entire universe was expanding. As Hubble quipped, "The history of astronomy is the history of receding horizons." Einstein later admitted that his 1917 static model was the "biggest mistake of my life." The discovery of the cosmic expansion forced him to revise his original, static model into a dynamic, expanding one.

5.4 GEOMETRY AND THE UNIVERSE

The Meaning of the Hubble Law

Hubble and Humason were careful to term the galaxies' rush away as "apparent." Taken at face value the velocities leave us fixed in the center of the universe. Having been rudely thrust away from the center by Copernicus, astronomers felt somewhat uncomfortable at being repositioned there. Was the Milky Way Galaxy now enthroned as the center of the universe?

A simple argument demonstrates that our Galaxy does not really have a privileged status. From our viewpoint the rest of the universe appears to be retreating from our Galaxy. But if the expansion is *uniform*, then the view from any other galaxy will be the same (Fig. 5.19). Transported to another galaxy with the usual bundle of tools, an astronomer would plot the same Hubble diagram as he or she does from our Galaxy. Another galaxy appears, to those in it, as the "center" of the expansion, so no privileged position actually exists.

As an analogy, imagine a jungle gym with a person located at each intersection of the bars (Fig. 5.20). Suppose the bars expanded at the same constant rate. Each observer would see other observers moving away from him or her. How fast they moved would depend on their distance. One twice as far away as another would move twice as fast. (There are twice as many bars in between, each of which is expanding.) Since the expansion is uniform, all the observers would describe the expansion in the same way, and each would think that he or she was in the center. If everyone plotted a graph of distance versus

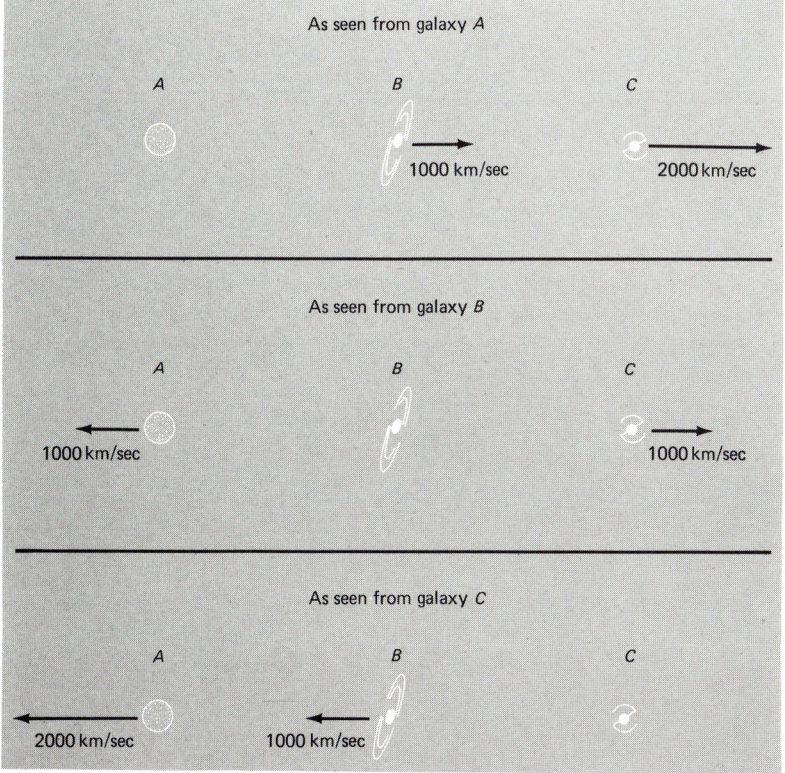

FIGURE 5.19 Uniform expansion in two dimensions. Imagine three galaxies (A, B, and C) equal distances apart and arranged in a line. Picture yourself on galaxy A. If these galaxies expand uniformly, then B appears to be receding at 1000 km/sec and C at 2000 km/sec. Now move to B. From there A appears to be receding at 1000 km/sec and so does C. From galaxy C, B appears to recede at 1000 km/sec and A at 2000 km/sec.

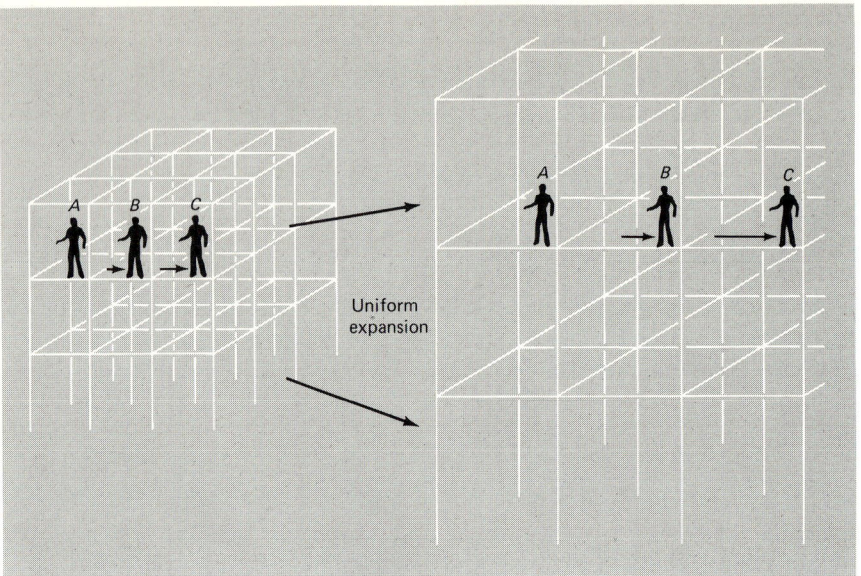

FIGURE 5.20 Uniform expansion in three dimensions. Imagine three people (A, B, and C) lined up at three intersections of a huge jungle gym. If it expands uniformly, then each person at each intersection sees the others recede, and each infers the same distance-velocity relationship.

velocity, all the graphs would look the same, all would resemble the Hubble law for galaxies, and all the values for H obtained this way would all be the same.

Warning: When first confronted by the expansion of the universe, people commonly picture the universe expanding "into empty space," but that image is a misconception. It is *space itself* that is expanding, and therefore it does not expand into "something" or even into "nothing." The phrase "expand into" is meaningless in this context.

How does this expansion look in a spacetime diagram? Let's look at three galaxies (A, B, C), with us placed at galaxy B (Fig. 5.21). From our perspective galaxies A and C move away from us on paths inclined at some angle with respect to ours (but less than 45°). Now imagine these spacetime paths in the past; at some point they cross, that is, the paths originate at the same point! That point is an event that marks the beginning of the expansion, the origin of the universe. Astronomers call this beginning of the universe's expansion the Big Bang.

The Hubble Constant and the Age of the Universe

You can infer from the measured value of H the time since the expansion of the universe began, if you assume that the expansion rate is uniform over time. *Warning*: This assumption does *not* apply to the universe as we know it. For any realistic model velocities must decrease with time; however, the assumption allows a simple calculation of the oldest possible age of the universe.

For an object moving at a constant velocity we know that the distance traveled (d) is equal to the velocity (v) multiplied by the time (t).

5.4 GEOMETRY AND THE UNIVERSE

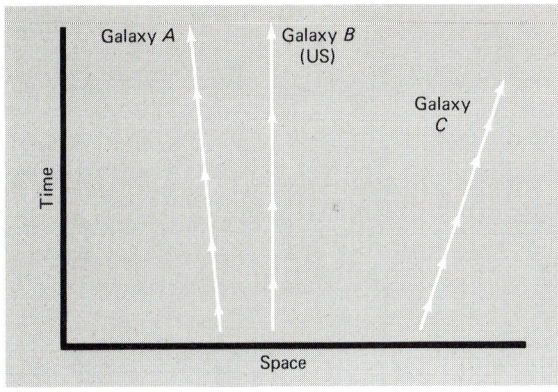

FIGURE 5.21 Spacetime map of an expanding universe. From any galaxy all other galaxies appear to be receding.

In algebraic form, $$d = vt$$

Conversely, the travel time is $$t = \frac{d}{v}$$

The Hubble law is $$v = Hd$$

where v is the velocity of a galaxy, d its distance, and H the Hubble constant. Now compare the Hubble law written as

$$\frac{1}{H} = \frac{d}{v}$$

to the trip formula above. We see that the travel time is simply

$$t = \frac{1}{H}$$

This is the time since the expansion started. Using a value of 50 km/sec/Mpc for H and using conversion factors to get everything in the same units, we obtain the value of $t = 1/H$ in years.

$$t = \frac{1}{H} = \frac{1}{50 \text{ km/sec/Mpc}}$$

$$= \frac{1}{50 \text{ km/sec/Mpc}} \times 10^6 \text{ pc/Mpc} \times (3 \times 10^{13} \text{ km/pc})$$

$$= 6 \times 10^{17} \text{ sec}$$

$$= \frac{6 \times 10^{17} \text{ sec}}{3 \times 10^7 \text{ sec/yr}}$$

$$= 2 \times 10^{10} \text{ yr}$$

So the expansion of the universe started about 20 billion years ago!

We assumed uniform expansion in order to arrive at this figure. However, even if the expansion were a bit faster in the past than it is now, the value of 2×10^{10} yr is still in the right vicinity of the actual time since the expansion began. (In the case of a universe with the critical density, see Section 5.5, the actual age is only 13 billion years.)

5.5 RELATIVITY AND THE COSMOS

With the discovery of the expanding universe, cosmologists began devising models that would account for the expansion. These models had to be consistent with Einstein's general theory of relativity and with the observed Hubble expansion law.

In these models the average cosmological density, the Hubble constant, and the geometry of spacetime (whether flat, hyperbolic, or spherical) are physically and mathematically interrelated. Let us show you how, using the concept of escape velocity.

Escape Velocity

Newton was the first to work on the problems of putting objects into orbit. From his analysis, which was based on his laws of motion and gravitation, arises the concept of *escape velocity*, the minimum velocity an object needs to escape the gravitational bonds of another.

Imagine, as Newton did, a giant cannon placed on the top of a very high mountain and aimed parallel to the ground (Fig. 5.22). When you fire a cannonball, it travels some distance and then hits the ground (A). If you use more powder in the cannon, the ball

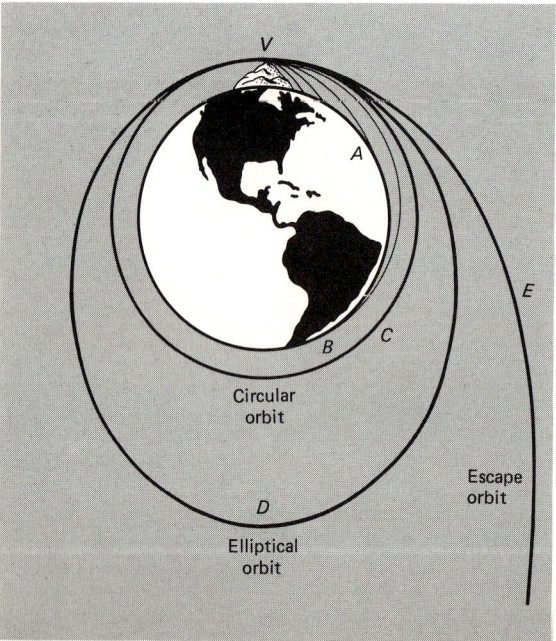

FIGURE 5.22 Launching of earth satellites, according to Newton. Imagine firing cannonballs from a high mountain (V). At low speeds the ball hits the earth (A and B). A certain minimum speed is needed to put the ball into a circular orbit (C). Higher velocities result in elliptical orbits (such as D). At some velocity the ball will escape from the earth (E).

5.5 RELATIVITY AND THE COSMOS

travels farther along the earth before it hits the ground (B). A large enough charge sends the ball completely around the earth in a *circular* orbit, returning to the cannon (C). The inertial motion of the ball just compensates for the falling due to gravity. With a still larger charge, one that creates a starting velocity greater than that needed for a circular orbit, the cannonball travels around the earth in an *elliptical* orbit (D). Increasing the starting velocity makes the orbit more elliptical. Eventually the orbit becomes so elliptical that the semimajor axis is infinitely long (E), and it would take the ball an infinite time to return. So it never returns. The velocity producing such an orbit is called the *escape velocity*. Any velocity larger than the escape velocity produces the same effect: The ball leaves the gravitational grip of the earth, never to return.

What determines the escape velocity from an object? Consider the earth. Its escape velocity is about 11 km/sec. Suppose that you kept the earth at its present size, but increased its mass. Its escape velocity would increase. If you kept its mass the same, but decreased its radius, the escape velocity would again increase. Both an object's mass and its size determine its escape velocity. Greater mass results in a greater escape velocity, greater radius in a smaller one. Note that both greater mass and smaller size mean a higher density. So for two objects of the same mass, the one with the higher density has the higher escape velocity (because it must have a smaller radius). Similarly, for two objects of the same radius, the one with the higher density has the higher escape velocity (because it must have a higher mass).

Algebraically, the escape velocity formula is

$$V_e = \left(\frac{2GM}{R}\right)^{1/2}$$

where V_e is the escape velocity, M the mass of the object you want to escape from, and R its radius. (We've assumed the mass is spherical.) Note that the mass of the escaping object does *not* enter into the calculation of escape velocity. (Surprised? Would Galileo have been surprised? Think about it.)

Let's calculate, as an example, the earth's escape velocity. The mass of the earth is 6.0×10^{24} kg, its radius is 6.4×10^6 m, and $G = 6.7 \times 10^{-11}$ in mks units. So

$$V_e = \left[\frac{2 \times (6.7 \times 10^{-11}) \times (6.0 \times 10^{24})}{6.4 \times 10^6}\right]^{1/2}$$

$$= 1.1 \times 10^4 \text{ m/sec} = 11 \text{ km/sec}$$

Consider throwing a ball off the earth's surface. If the ball's velocity is less than the escape velocity, it slows down and falls back, regardless of how large or small its mass is. When thrown with a velocity greater than the escape velocity, the ball slows down as it leaves the earth, but eventually coasts out toward infinity.

Turn to the universe and the galaxies within it. Any one galaxy in the universe is analogous to the ball. We know that the galaxies were once all "thrown away" from one another, because we now see an expansion. Take a galaxy at a distance R and ourselves at the apparent center of the expansion. Newton showed that there is no net gravitational force on the galaxy produced by any matter farther away than R, and that the net effect of all matter within the distance R is as if this total mass were concentrated at the center (as

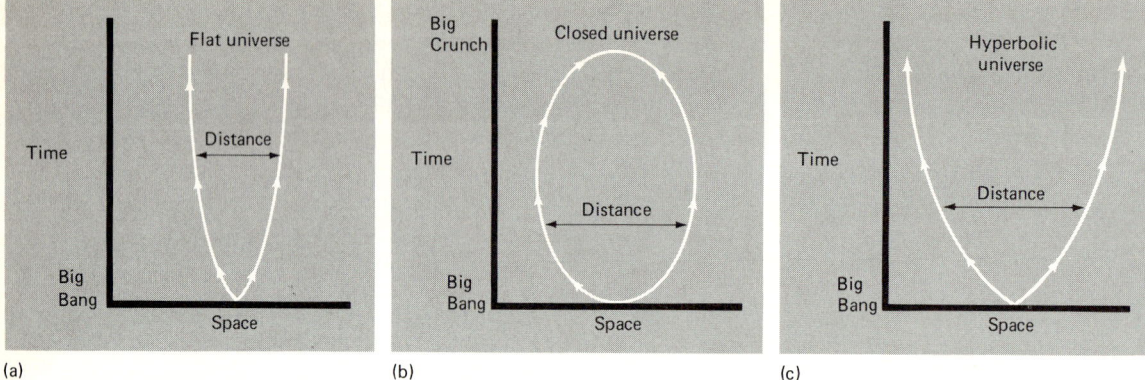

FIGURE 5.23 (a) Spacetime map for two galaxies in a flat (open) universe. (b) Spacetime map for two galaxies in a spherical (closed) universe. (c) Spacetime map for two galaxies in a hyperbolic (open) universe.

long as the matter is distributed uniformly). If there is enough mass within the distance R, the escape velocity will be larger than the expansion velocity. Note that "enough mass within the distance R" means a high enough density. So if the density is large enough, the galaxies will not have escape velocity; the expansion velocities will decrease with time and eventually reverse. If the average density is too low, the galaxies will have more than escape velocity; gravity will never bring the galaxies to a halt, and the expansion will continue indefinitely.

Let's look at the cases of expansion forever and eventual collapse with spacetime diagrams. For simplicity, consider the worldlines of just two galaxies (A and B). Both start close together at the event that's the beginning of the expansion, the Big Bang (Fig. 5.23a). In one case the paths diverge and then eventually run along until parallel. Here, the galaxies just had escape velocity, so after a very long time (infinity!), the expansion stops, and the galaxies stay the same distance apart. Note that the galaxies' paths then run parallel, which is geometrically equivalent to saying that spacetime is flat.

In the case in which the galaxies have less than escape velocity (Fig. 5.23b), they reach a time when they are a maximum distance apart, and then they draw together again until their paths intersect. That event, symmetrical to the Big Bang, marks the end of the universe. Note that the paths resemble those in the spacetime diagram of particles falling to the earth (Fig. 5.12) and in the diagram of longitude lines on the earth. By analogy, in this instance, the geometry of spacetime is spherical and closed.

The third possible geometry is hyperbolic. It works essentially like the flat case, but after an infinite time the galaxies are still moving away from each other (Fig. 5.23c).

So spherical, hyperbolic, and flat geometries correspond physically to the cases of less than, equal to, and greater than escape velocity.

Escape Velocity and Critical Density

If material is distributed in a spherically symmetrical way, you can work out the relationship between escape velocity and density and between the density of the universe and the Hubble constant.

5.5 RELATIVITY AND THE COSMOS

For a sphere, the mass and density are related by

$$M = \frac{4}{3}\pi R^3 \rho$$

where ρ is the average density of matter and energy in the space with radius R. (How to find it? Add up all the masses and divide by the volume, $4\pi R^3/3$.) Substitute this expression for M in the escape velocity equation.

$$V_e = \left(\frac{2GM}{R}\right)^{1/2} = \left[\frac{2G}{R}\left(\frac{4}{3}\pi R^3 \rho\right)\right]^{1/2}$$

$$= \left(\frac{8\pi}{3}G\rho\right)^{1/2} R$$

$$\frac{V_e}{R} = \left(\frac{8\pi}{3}G\rho\right)^{1/2}$$

Apply this result to two galaxies, distance R apart, moving away from each other at just the escape velocity V_e. Then V_e/R equals H!
So

$$H = \left(\frac{8\pi}{3}G\rho\right)^{1/2}$$

or

$$\rho_c = \frac{3H^2}{8\pi G}$$

This is the critical density (ρ_c) which makes the escape velocity just equal to the expansion velocity. It's the density that divides the cases of open and closed universe. It says if we know H, we can calculate the average density needed for the universe to have a flat geometry; this is the critical density. More and the universe is closed; less and it's open.

Using $H = 50$ km/sec/Mpc results in 4.7×10^{-27} kg/m^3 as the critical density. This *calculated* density can then be compared with the *observed* cosmological density (which must include energy as well, since $E = mc^2$). If the observed cosmological density is greater than the critical density, the universe is spherical and closed. If less, the universe is hyperbolic and open. If they are exactly equal, the universe is flat. So if we can observe the average density of matter and energy in the universe, we then have an *experimental* basis for finding out the appropriate geometry of spacetime! Observations of the Hubble constant combined with general relativity provide us with a way to find out if the universe is open or closed.

Making such an observation is not easy, however. First, astronomers can see only to about 10 billion lightyears (3000 Mpc). Second, we can see only those masses that radiate or absorb energy intensely enough to register on the instruments we have devised. Third, we need some method to determine the mass of observed objects. Note that our methods of observation cannot detect "invisible" objects. A density derived from estimating the mass in galaxies, about 4×10^{-28} kg/m^3, neglects the contribution of such nonradiating objects and all other unnoticed masses. We get this value by estimating the masses of all objects we can observe within a large volume of space, then dividing the total mass by the volume to get the density. The observed density amounts to about 12 times less than is needed to close the universe, but it is still tantalizingly near the necessary value. For cosmologists who prefer a closed universe, the close-but-not-quite results have spurred a search for the "missing matter" needed to close the universe. (The matter is not missing at all if the universe has an open geometry.)

Geometry and the Hubble Diagram

The observed Hubble diagram is not the simple straight line implied by the relation $v = Hd$. First of all, even in a uniformly expanding universe the Hubble constant will decrease slowly as the universe grows older. The velocities stay the same, but the distances always get bigger. Therefore $H = v/d$ will always get smaller, and the slope of the velocity-distance relation will decrease with time. If, in addition, the velocities are decreasing due to the deceleration of gravity, H will decrease even more rapidly. Now to this notion add the fact that as we look out into space, we are looking backward in time. We see a galaxy not as it is now, but as it was in the past when the light we are measuring left it. It was closer in the past than it is now, and if the expansion is decelerating, it was moving faster than it is now. So the observed point representing a galaxy on the Hubble diagram will be to the left, and possibly above, the true position on the straight line Hubble law (Fig. 5.24a).

Applying this idea to all galaxies, we expect that the observed Hubble relation will not be a straight line, but a curve that bends up, the amount of bending depending on the rate at which the velocities have changed in the past (Fig. 5.24b). The greater the density of the universe, the more rapidly the velocities change, and the greater the curvature in the Hubble diagram. Here then we have another way of measuring the density of the universe and hence determining its geometry. If the Hubble line curves up enough, the density is high, and the universe is closed and spherical. If it doesn't curve up enough, the universe is open.

The above effects are geometrical ones, but there's another complication arising from looking far back in time. There's an assumption being made here, which may be of

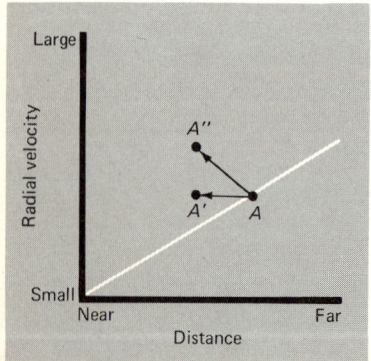

(a)

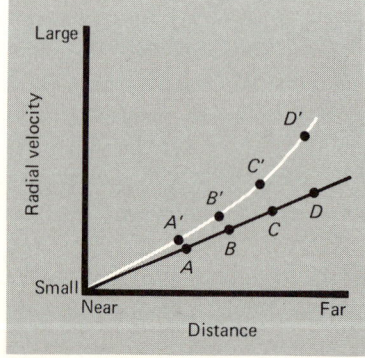

(b)

FIGURE 5.24 (a) A galaxy now at a true distance and velocity represented by point A will be observed at a closer distance (A′), or if the expansion is decelerating, at a closer distance and higher velocity (A″). (b) Galaxies now with velocities and distances represented by the points A, B, C, and D are observed with velocities and distances represented by A′, B′, C′, and D′. So the observed Hubble diagram curves upward compared to a straight line.

5.5 RELATIVITY AND THE COSMOS

questionable validity when dealing with galaxies at very great distances. In order to use the faintness of objects to compare their distances, we have to know that they are giving off the same amount of energy, that is, that they are the same kind of object. With nearby galaxies we have some clues (see Chapter 22). But when we look very far out, we cannot see the details. More important, we know that galaxies are made of stars, and that stars evolve, changing their luminosity with time. So we expect that evolutionary effects will change galactic luminosities. The question is how much and over what period of time? We probably will have to develop a better theoretical understanding of how galaxies evolve before we can be confident in using the Hubble diagram to learn about the geometry of spacetime.

The rate of change of the Hubble constant is related to a number called the *deceleration parameter*. The value for the deceleration parameter, usually written as q, depends directly on the geometry of the universe. In a cosmos that is just closed (actual density just equal to the critical density), $q = \frac{1}{2}$. If q is less than $\frac{1}{2}$ (and greater than zero), the universe is open; if it is greater than $\frac{1}{2}$, the universe is closed. In all evolving cases q must be greater than zero.

What's the future for the universe in these various cases? In all three there is some moment when the expansion begins. In a closed universe the rate of expansion slows down and eventually stops, and then the universe begins to contract. In a hyperbolic universe the rate of expansion does not slow down as rapidly, and it never stops. Even after infinite time the galaxies are still moving apart at a finite velocity. In the borderline case of a flat universe the expansion slows down just enough so that it comes to a stop after infinite time.

Modern cosmology can be considered as the grand search for two numbers: the Hubble constant H and the deceleration parameter q. If we can accurately find out their values, we can then trace out the entire history of the universe.

The Future of the Universe

We seem to have two possible cosmic destinies: In one the expansion grinds on forever, and the universe gradually thins out. In the other the expansion slows, stops, reverses, and the universe collapses. The galaxies rush together into a dense conglomeration with (theoretically) zero radius. The universe would then crush everything into high-energy light (Chapter 24).

Which fate does the Hubble diagram display for us (Fig. 5.25)? Unfortunately, the observations have not yet settled the issue. As far out as galaxies can be observed, the observational errors are as large as the effect we are looking for, so deviations from a straight-line Hubble plot are hard to see. Also, as mentioned above, it's difficult to separate the evolution of galaxies from the evolution of the universe.

Results to date have been a mixed bag, some supporting an open universe, others closed. Jerome Kristian, Allan Sandage, and James Westphal have extended the Hubble diagram out to radial velocities of 75 percent the speed of light (Fig. 5.26). At face value the observations indicate the universe is closed. But if galaxies evolve rapidly with time, the results are consistent with a flat model. These astronomers state that "the case is not yet settled."

But don't worry! Even if the universe is closed, it will be billions of years before it comes crashing together.

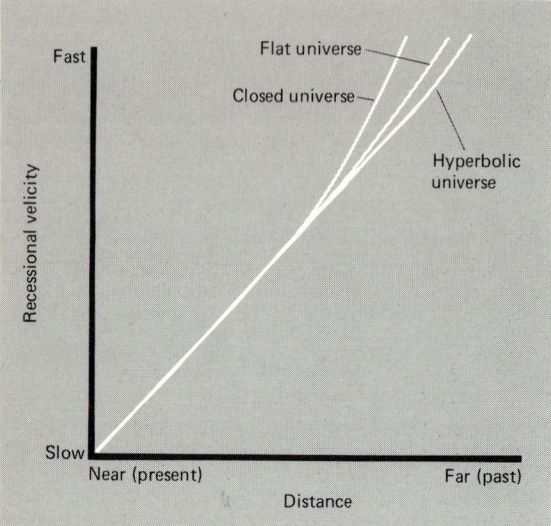

FIGURE 5.25 The geometry of spacetime and the Hubble diagram. If we can look far enough out in space (and so back in time), the Hubble graph can be used to infer the overall geometry of the cosmos. For example, if the universe is closed, the graph will bend sharply up compared to the plot for a flat universe. This upward bend means that the expansion rate (the Hubble constant) was greater in the past than now.

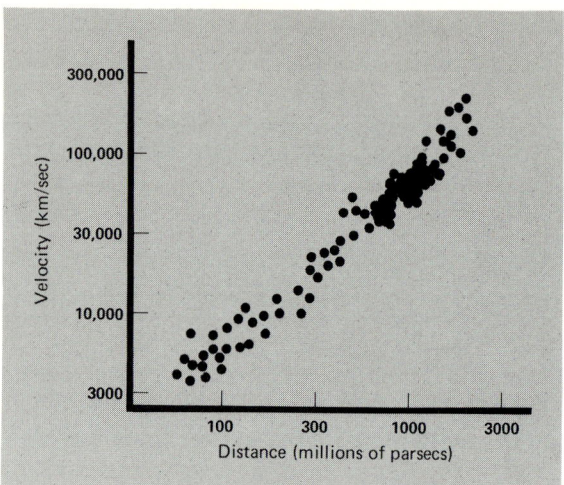

FIGURE 5.26 A Hubble diagram for galaxies with radial velocities as large as 75% the speed of light. The horizontal axis is the galaxy's distance. The vertical axis is the recessional velocity. [*Adapted from J. Kristian, A. Sandage, and J. Westphal*, Astrophysical Journal, *vol. 221, p. 383, copyright 1978 by the American Astronomical Society.*]

SUMMARY

To describe the universe properly requires a consideration of the geometry of spacetime. Mathematics by itself provides a wide range of choices. Only experiments and observations can confirm the geometry suitable for the universe. The choices can be boiled down to three: flat, hyperbolic, or spherical (Table 5.1). Present observations cannot tell which is correct.

Newton assumed that the geometry of Euclid applies to the universe. This assumption of flat geometry implicitly underlies his first law of motion and concept of inertia. Einstein, in contrast, demanded that the geometry of the universe be put to an experimental test. His general theory of relativity allows the cosmos to have any geometrical properties; it is not limited to only one choice.

Einstein arrived at this view from a reconsideration of space, time, and gravitation. In the process he replaced the Newtonian view of gravity as a force with the view of gravity as a manifestation of the local geometry of spacetime. From where does mass get its moving orders? Newton would have answered: from the forces of other masses, acting at a distance. Einstein says instead: from the local geometry of spacetime. (Think again of the bumpy putting green and a ball rolling over it.)

This transformation of the concept of gravitation led to new models of the universe. The most important observation related to these models was the discovery that other galaxies are moving away from us. This recessional motion follows a trend known as the

REVIEW QUESTIONS

TABLE 5.1 Geometry of the Universe and Observations

	Hyperbolic	Spherical	Flat
Extent in space	Infinite	Finite	Infinite
Extent in time	Infinite	Finite	Infinite
Bounded or unbounded	Unbounded	Bounded	Unbounded
Average density (H = 50 km/sec/Mpc)	Less than 5×10^{-27} kg/m^3	More than 5×10^{-27} kg/m^3	Equals 5×10^{-27} kg/m^3
	(Observed average density is about 10^{-28} kg/m^3)		
Hubble plot	Bends slightly upward	Bends strongly upward	Bends upward
Age to date (H = 50 km/sec/Mpc)	More than 13 billion years	Less than 13 billion years	Equals 13 billion years
Future	Expansion forever	Expansion stops, collapse	Expansion forever

Hubble law: The farther away a galaxy is, the faster it is receding. It implies that the universe is expanding. When combined with the general theory of relativity, Hubble's law demands that *the universe itself evolves*.

Contemporary cosmological models have evolved a long way from the conceptions of the Greeks. No more are the earth and sky separated. No more are the heavens static for eternity. We now perceive the universe as an evolving entity, understandable because of universal physical laws.

Key Words

natural motion
reference frame
straight line
great circle
special theory of relativity
spacetime
event
spacetime diagram
worldline

causally connected
principle of equivalence
inertial mass
gravitational mass
general theory of relativity
flat geometry
hyperbolic geometry
spherical geometry
closed geometry

open geometry
lightyear
parsec
megaparsec
Hubble law
Hubble constant
deceleration parameter
escape velocity
critical density

Review Questions

1. Suppose it were discovered that, through some mistake or recalibration of basic data, all of the measured distances of galaxies were half as large as had been thought previously. What would happen to:
   a. The value of the Hubble constant.
   b. The age of the universe.
   c. The velocities observed for all galaxies.
   (Objectives 8 and 10)

2. According to the theory of general relativity (mark all correct answers):
   a. No forces other than gravitational forces are real.
   b. Falling objects follow their natural motion in spacetime.
   c. Gravitational effects can always be canceled by measuring things with respect to a freely falling reference frame.
   d. Euclidian geometry applies to three-dimensional space, but not to four-dimensional spacetime.
   e. Space is more highly curved near massive objects than it is far away from such objects.
   (Objectives 1, 5, 6, and 7)
3. What keeps you on the earth in Newton's view and in Einstein's? (Objective 1)
4. Describe orbiting the earth in Newton's terms. Contrast this to Einstein's description. (Objective 1)
5. Sketch the earth's orbit around the sun from Newton's viewpoint. Now from the point of view of an observer orbiting the earth, sketch a spacetime diagram of the earth's motion. (Objectives 1 and 4)
6. In Einstein's model how is the geometry of the universe related to the average density of the universe? (Objective 11)
7. Suppose you were an astronomer in the Andromeda galaxy with the same tools as an earth-bound astronomer. You measure the distances and radial velocities of other galaxies. What would you see? (Objectives 8 and 9)

Problems
1. Suppose you hear on the radio that astronomers have discovered a galaxy 500 Mpc from us. Use the Hubble law to estimate how fast that galaxy is receding from us. Assume that $H = 50$ km/sec/Mpc. (Objective 8)
2. Again assume that $H = 50$ km/sec/Mpc. At what distance would the recessional velocity equal half the speed of light (150,000 km/sec)? (Objective 8)
3. Suppose that $H = 100$ km/sec/Mpc. What is the value of the critical density? How does it compare to the actual estimated density? What, then, is the geometry of the universe? (Objective 11)
4. What would the value of the Hubble constant have to be so that the observed density of matter in the universe (4×10^{-28} kg/m^3) would be equal to the critical density? (Objective 11)
5. A typical house in the United States uses about a million Joules of energy in a month in the form of electricity. What is the mass equivalent of this amount of energy? (Objective 3)
6. What would be the maximum age of the universe for a Hubble constant of 550 km/sec/Mpc, Hubble's original value? (Objective 10)

BEYOND THIS BOOK . . .

*You are probably excited and tremendously confused by Einstein's theory of relativity. Go slowly, think about it, and the basics will come to you, even though it may take a while. For a little help, try *Relativity* (Crown, New York, 1961) by A. Einstein; *The Meaning of Relativity* (Princeton University Press, Princeton, N. J., 1956) also by Einstein; *Relativity and Common Sense* (Doubleday, Garden City, N. Y., 1964) by H. Bondi; and *Albert Einstein: Creator and Rebel* (Viking, New York, 1972) by B. Hoffman in collaboration with H. Dukas.

BEYOND THIS BOOK . . .

* The best book we've found on the nonscientific side of Einstein is R. Clark's *Einstein: The Life and Times* (World, New York, 1971).
* Having trouble picturing four dimensions? For a satirical analogy in fewer dimensions, read *Flatland* (Dover, New York, 1952) by E. A. Abbott.
* After you've read *Flatland*, try "The Curvature of Space in a Finite Universe" by J. J. Callahan in *Scientific American*, August 1976.
* For a witty exposition of Einstein's ideas with an emphasis on time, read *Space and Time in the Modern Universe* (Cambridge University Press, Cambridge, 1977) by P. C. W. Davies.
* *The Cosmic Frontiers of General Relativity* (Little Brown, Boston, 1977) by W. Kaufmann deals in more depth with Einstein's ideas.
* *The Red Limit* (Bantam Books, New York, 1977) by T. Ferris gives good insights into development of modern cosmology.
* An excellent book on special relativity is *Space and Time in Special Relativity* (McGraw-Hill, New York, 1968) by N. David Mermin.
* A good recent exposition of relativity and cosmology is *The Big Bang* (Freeman, San Francisco, 1980) by Joseph Silk.

PART TWO
STARS: THE FOCUS OF COSMIC EVOLUTION

OUR SUN: HOME STAR

6

LEARNING OBJECTIVES

After studying this chapter, you should be able to:

1. Outline a contemporary method used to find the distance to the sun from the earth.
2. Show in detail how to determine the sun's mass, size, density, luminosity, and surface temperature.
3. Define a blackbody, describe the characteristics of blackbody radiation, and make use of these characteristics in astronomical situations.
4. Describe the difference in appearance of the three basic types of spectra—continuous, emission, and absorption—and the physical conditions under which each is produced.
5. Describe the physical meaning of the term *opacity* and how opacity affects the flow of radiation through the sun.
6. Use an energy-level diagram to explain how atoms absorb and emit light, with special emphasis on the hydrogen atom.
7. Describe the appearance of the sun's spectrum and what atomic processes produce this spectrum.
8. List the sun's two most abundant elements and describe how these and others have been found.
9. State the thermonuclear reactions that are thought to produce the sun's energy and describe the conditions for such reactions to take place.
10. Apply Einstein's mass-energy relation to the production of solar energy.
11. Discuss the results and consequences of the solar neutrino experiment.
12. Trace the flow of energy from the sun's core to the earth, and describe how the features of the quiet sun (photosphere, chromosphere, corona, and solar wind) result from this energy flow.
13. Identify and describe the major features of the active sun.
14. Describe the basic properties of an ordinary gas and make use of the relationship among pressure, temperature, and density for such a gas.

CENTRAL QUESTION:
How does the sun produce its life-giving energy, and how does its energy production and flow affect the sun's physical characteristics?

Sunlight gives us life. All creatures on the earth are children of the sun, for the sun provides essentially all the life-sustaining energy for this planet. The sun warms the earth and drives the weather cycle. In the past it raised the vegetation that produced our present fossil fuels. When these give out, we will likely turn to the sun directly for our usable energy.

The sun is basically a hot, huge ball of gas with fierce nuclear reactions firing its core. There, deep in the heart of the sun, the energy bound up in matter is released. Slowly, over tens of millions of years, that energy (as light) erratically flows to the sun's surface. Free of the sun's material, it flies into space. And 8.3 min later a very small part of that light strikes the earth.

For astronomers the sun is important for reasons in addition to these. Our sun is a star. It has the same basic construction as the other stars in the sky, which we can see directly only as pinpoints of light. We can find out the physical characteristics of these distant lights by using our sun as a guide. The sun serves as the local laboratory for testing our ideas about stars in general.

6.1 A SOLAR PHYSICAL CHECKUP

How large is the sun (Fig. 6.1)? How massive is it? To tackle such questions, we need to know how far the earth is from the sun. (You will find throughout the book that a critical question, perhaps *the* critical question in all of astronomy, is finding the distance to celestial objects.) Other facts we can infer from an analysis of the sun's light without having to know its distance. Let's see what we can find out more or less directly about the sun.

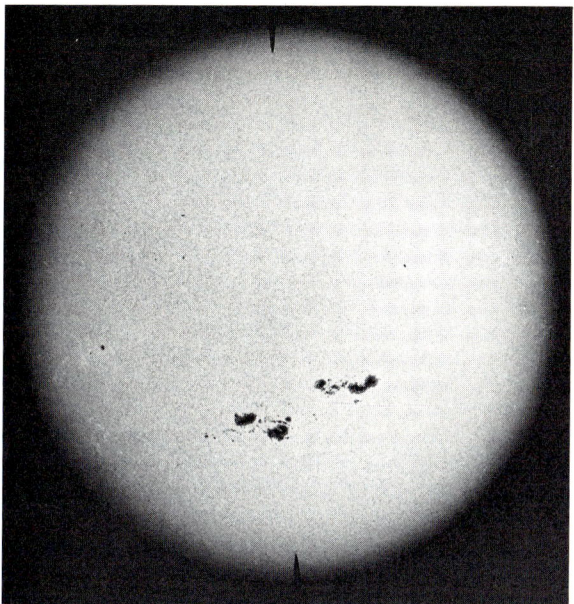

FIGURE 6.1 The sun's photosphere. Note that the photosphere has a granular look and fades out at the edge. The dark regions are sunspots. The dark spikes at top and bottom mark the sun's poles. [*Courtesy Mt. Wilson Observatory, Carnegie Institution of Washington.*]

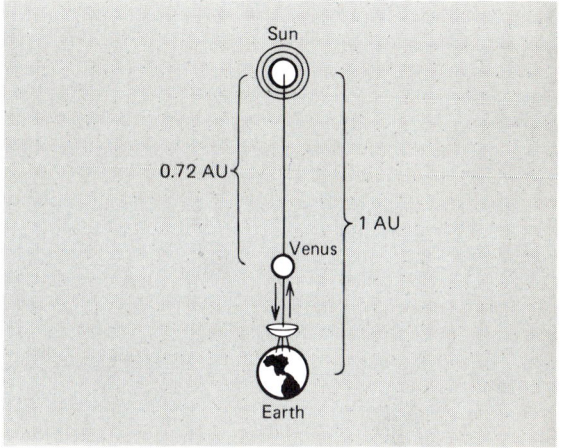

FIGURE 6.2 Bouncing radar signals off Venus to measure the Astronomical Unit.

How Far?

The earth's orbit about the sun is elliptical. The semimajor axis of its orbit, which is the average earth-sun distance, has a length of 1 AU. But how large is the AU in some basic physical units, such as kilometers?

Recall that simple angular measurements establish the scale of the solar system in terms of the AU (Section 3.1) and that Kepler's laws (Section 3.4) completely describe the orbital motions of the planets, with AUs as the measuring unit, so that we can draw a scale map of the solar system, with the planets' positions at any specific time all neatly laid out in AUs.

To work out the distance scale in kilometers requires that only one segment known in AUs be measured in kilometers. To illustrate, suppose you're given a map of your region of the United States with no distance scale in kilometers, but with all locations laid out correctly relative to one another. Drive your car from one point indicated on the map to another and keep track of the distance in kilometers. Measure the distance in centimeters between the points on the map, and you have found the distance scale for the map.

To do the same for the solar system, astronomers measure the actual distance to Venus using the speed of light, which is accurately known. Radar signals, which travel at light speed, are bounced off Venus, and the time between transmission and reception is accurately measured (Fig. 6.2). The Earth-Venus distance in kilometers at that time of observation is

$$d = \left(\frac{t}{2}\right)c$$

where d is the distance, c the speed of light (in km/sec), t the round-trip travel time of the signal, and $t/2$ the one-way travel time. Kepler's laws give the Earth-Venus distance at the measurement time in AUs. Since we know a segment of the AU in kilometers and the scale, we can easily calculate the AU in kilometers—1.496×10^8 km. (Remember it by rounding off to 150 million km.)

Why bother with bouncing radar off a planet? Why not bounce it off the sun and measure the AU directly? For two reasons: the sun is itself a powerful radio *source*, and it does not reflect radar signals very well.

How Big?

Viewed from the earth, the average angular size of the sun is 32'. (That's about $\frac{1}{2}°$, or half the tip of your finger at arm's length.) Because we know the AU in kilometers, we can calculate the sun's actual size from its angular size (Fig. 6.3). The sun's diameter is roughly 1.4 million km, or 109 times the earth's diameter (Fig. 6.4). Imagine the earth the size of a nickel (2 cm in diameter). Then the sun would be about 2 m in size and 200 m from the coin-sized earth.

How Massive?

One method of determining the sun's mass is to use Kepler's third law (Section 4.3). Here's another method, for which you need the earth-sun distance, the earth's average orbital velocity, and Newton's second law of motion and gravitation (Section 4.2). If

$$F = \frac{GM_{sun}}{R^2} M_{earth}$$

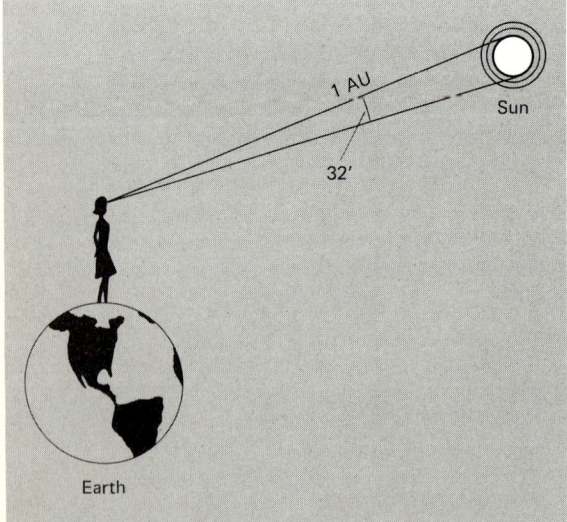

FIGURE 6.3 The angular diameter of the sun as viewed from the earth is 32'. From this angle and the distance, the sun's actual diameter can be determined.

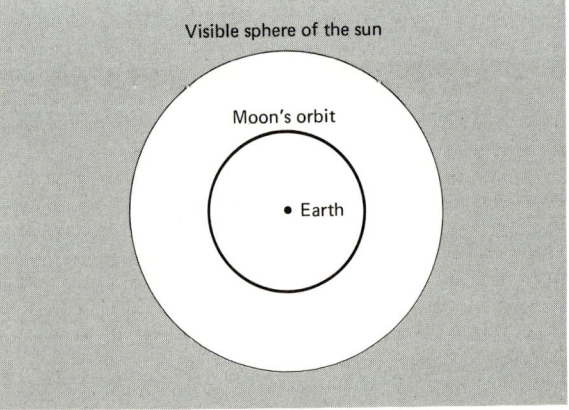

FIGURE 6.4 The relative sizes of the sun, the earth, and the moon's orbit.

6.1 A SOLAR PHYSICAL CHECKUP

and

$$F = M_{earth} a$$

then

$$M_{earth} a = \frac{GM_{sun}}{R^2} M_{earth}$$

and

$$a = \frac{GM_{sun}}{R^2}$$

This result is the centripetal acceleration of the earth around the sun. Since the earth's orbit is almost circular, we can write the acceleration (Focus 4.1) as

$$a = \frac{v^2}{R}$$

where v is the orbital velocity of the earth (30 km/sec = 3×10^4 m/sec). Then

$$\frac{v^2}{R} = \frac{GM_{sun}}{R^2}$$

or

$$M_{sun} = \frac{v^2 R}{G}$$

with R equal to the earth-sun distance. Here's the calculation:

$$M_{sun} = \frac{(3 \times 10^4 \text{ m/sec})^2 (1.5 \times 10^{11} \text{ m})}{(6.8 \times 10^{-11} \text{ m}^3/\text{kg} \cdot \text{sec}^2)}$$
$$= 2.0 \times 10^{30} \text{ kg}$$

This method implicitly assumes that the sun is so much more massive than the earth that the earth-sun distance is the same as the distance of the earth from the center of mass of the earth-sun system.

How Dense?

The sun is a massive object, but its radius, and hence its volume, is also very large. Is the average density, the mass per unit volume, high or low?

FOCUS 6.1 ENERGY AND WORK

Energy makes work possible and allows events to happen in the universe. It is no arbitrary phrase when you say you feel "energized" and demonstrate this statement by a flurry of activity. To scientists the concept of energy has fundamental importance because, although energy comes in many forms, the total amount of energy that an object has can be described and assigned a number. If you have a bunch of objects, the sum of their individual total energies is the total energy of the group. The fact that the total energy of a system of objects has a constant value is known as the law of *conservation of energy*. As a fundamental physical law, the conservation of energy helps to explain some of the complicated situations in the real world.

Energy exists in many different forms. Three common ones are (1) *kinetic* energy, due to motion; (2) *potential* energy, due to position under an applied force; and (3) *radiative* energy, which is carried by light.

Kinetic energy can be understood in terms of work. For example, to stop a moving object requires an obvious effort, and the more massive the object or the faster its motion, the greater the effort needed to stop it. Kinetic energy due to motion can be expressed in a simple equation:

$$KE = \tfrac{1}{2}mv^2$$

where m is the mass of an object and v its velocity. Note that the kinetic energy depends on the *square* of an object's velocity. So of two cars, one traveling twice as fast as the other, the faster car has *four* times the kinetic energy and will be four times harder to stop. In other words, with locked brakes, the faster car skids four times the distance of the slower.

Heat, as measured by temperature, is a measure of the kinetic energy of a large collection of particles. When you heat a gas, the average value of the random velocities of the molecules that make it up increases.

Potential energy can be defined as the capability of doing work. Any object to which a force is applied can move in the direction of the force, even if it is temporarily (or

---

Let's work it out. The sun's mass is about $M = 2 \times 10^{30}$ kg. Since the sun is spherical, its volume is

$$V = \frac{4}{3}\pi R^3$$

$$= \frac{4}{3}\pi (7 \times 10^8 \text{ m})^3$$

$$= 1.4 \times 10^{27} \text{ m}^3$$

permanently) restrained from motion. This possibility of motion for an object (which would require work to stop it) when some force is applied defines potential energy. An obvious example is the earth's gravitational attraction. All masses feel the earth's pull, although some, such as a ball held in your hand, do not immediately respond to the force. When you drop the ball, however, it gains kinetic energy, and the higher the point from which you drop it, the greater the kinetic energy it attains at the end of its fall. One form of energy is transformed to another (potential to kinetic), yet the total energy of the ball remains constant, for as it loses potential energy, it gains kinetic energy.

The energy-level diagram of an atom describes a situation analogous to that of objects pulled by gravity. Instead of gravitational force, however, we have the force of electrical attraction between the positively charged nucleus and the negatively charged electrons. When an electron falls from one energy level to a lower one, it loses potential energy, which must manifest itself in some form, in this case a photon (radiative energy). To reverse the process requires the electron to gain energy; the energy gain is produced by the absorption of a photon with the energy required to raise the electron to the next level.

The amount of energy carried by light depends on its frequency (or wavelength). The fundamental relation is

$$E = hf$$

where E is the energy in joules, f the frequency in cycles per second, and h a constant (called *Planck's constant*) with a value of 6.63×10^{-34} J·sec. Note that light with twice the frequency carries twice the radiative energy; three times the frequency, three times the energy, and so on.

Now back to the concept of the conservation of energy. Experiments (such as the dropping of a ball) have shown that energy is naturally transformed from one form to another, but that as long as all the energy is carefully accounted for, the total amount remains the same. One way to state the conservation of energy is: *Energy cannot be created or destroyed; it may be transformed, but the total does not change.* Don't forget that mass is a form of energy ($E = mc^2$), so that the transformation of mass into energy (such as in fusion reactions in stars) does not violate the conservation of energy.

So its mass per unit volume is

$$\text{density} = \frac{M}{V} = \frac{2 \times 10^{30} \text{ kg}}{1.4 \times 10^{27} \text{ m}^3}$$

$$= 1400 \text{ kg/m}^3$$

For comparison, water has a density of about 1000 kg/m^3, rocks about 3000 kg/m^3, iron about 7000 kg/m^3, and wood about 700 kg/m^3. Note that the sun is only 40 percent denser than water. This low average density, along with its high temperature, implies that the sun is a gas.

Ordinary Gases

To understand what happens inside the sun (and other stars), you must have some idea of how gases behave, because stars are huge balls of gas. Let's describe a simple model for ordinary gases. (You will come across extraordinary gases in Chapter 11.)

A gas consists of particles—atoms, molecules, ions (atoms with one or more electrons removed), or perhaps all of these. To simplify the model let's assume that all the particles are the same and that each particle is a sphere. Picture all the spheres trapped in a small box. Once set into motion, the spheres keep moving, bounding off the walls of the box and colliding with each other (Fig. 6.5).

Because of all these collisions, any one sphere will at times move faster or slower, but over time the sphere has a definite average speed. Also because of collisions, all the spheres in the box will have the same average speed. We use this average speed of the particles in a gas as a measure of its *temperature*. If all the particles were motionless, a gas's temperature would be zero. (In this book we will always use the Kelvin, or absolute, temperature scale, which is the same as the centigrade scale, but measured from absolute zero. On this scale the freezing point of water is 273.15 Kelvins (K).) At room temperature, about 300 K, the average speed in a gas of hydrogen particles is about 3 km/sec. Higher temperatures mean higher average speeds of the particles.

Suppose you put a partition into the gas container. The spheres bounce off both sides of the partition; each collision exerts a small force on it. The combined force of all collisions is the *pressure* of the gas. If you increase the temperature of the gas, the average speed of the particles increases, so they collide with each other and into the partition more frequently and with greater force. The pressure increases. For ordinary gases the increase is a direct one; if you double the temperature (in Kelvins) the pressure doubles.

What happens to the pressure if you increase the number of particles in the box? (You've done this if you've pumped up a bicycle tube, for then you forced more air particles into it.) For a gas the number of particles in a unit volume is called the *number density*. Suppose you increase the number density four times without changing the temperature. Each cubic meter now contains four times as many particles, so four times as many collisions occur, on the average, against the partition. Since each collision has the

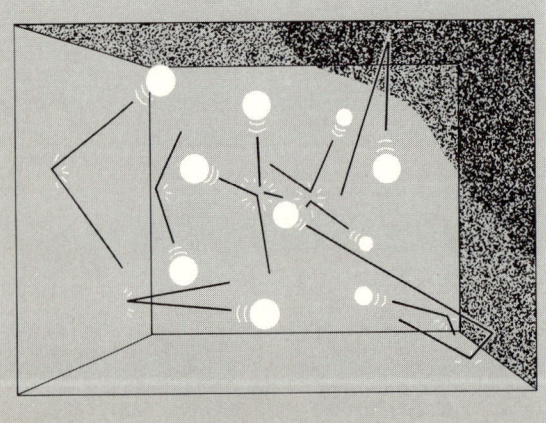

FIGURE 6.5 Hard-ball model of a gas in a box.

same average force as before, the pressure increases by a factor of four. Increasing the number density directly increases the pressure of an ordinary gas.

This hard-sphere model shows us that gas pressure depends directly on both the number density and the temperature. In algebraic form this relationship, known as the *perfect gas law*, is

$$P = nkT$$

where P is the pressure in atmospheres (sea level pressure on the earth, abbreviated atm), T is the temperature in Kelvins, n is the number density in a cubic meter, and k is Boltzmann's constant, which is equal to 1.38×10^{-28} in these units. For example, at the sun's surface the temperature (next section) is 5800 K and the number density roughly $10^{23}/m^3$. So the pressure there is approximately

$$P = (10^{23})(1.38 \times 10^{-28})(5800)$$
$$= 8 \times 10^{-2} \text{ atm}$$

Temperature and Heat

Note that temperature plays a role on a *microscopic* level for a gas; it's a measure of the average velocity of the particles that make up a gas. What, then, is the difference between temperature and heat?

Energy is the ability to do something useful, and usually manifests itself as either energy of motion or stored energy due to an object's location. *Heat* is a specific form of energy, the random energy of motion of the particles that make up some object. *Temperature* is a measure of the average random energy per particle in a body, which in turn depends on their average speeds. When you add energy to a system, some of that energy gets converted to random energy of motion or heat. The rise in temperature is a measure of the increased velocity of the particles. Those of you who in your youth were wont to pound slugs with a hammer have observed this; after a few blows with the hammer, the slug got hot. You were converting some of the kinetic energy of the hammer into thermal energy (heat) in the slug, speeding up the atoms in the slug, which made its temperature rise. (The rest of the hammer's kinetic energy went into deforming the slug.) If you used a bigger slug, you would have to pound it harder to get it as hot. Because there are more atoms in the larger object, you have to convert more of the hammer's kinetic energy into heat. So heat is the total random energy of motion of the particles in an object; temperature is a measure of the random energy *per particle*.

6.2 SUNLIGHT AND SPECTROSCOPY

Matter in the form of atoms makes up the sun. (Strictly speaking most of the sun is made up of ions but we will use the word atoms to mean both in this context.) Atoms can give off and absorb light; sunlight originates from atoms. Analyzing the light atoms emit and absorb tells about their structure and their physical environment. To understand more of the sun, you need to understand a little about atoms and the light they emit.

Atoms and Matter

In the eighteenth and nineteenth centuries chemists discovered that substances can be divided into two classes: chemical *elements* and chemical *compounds*. *Elements* cannot be broken by chemical reactions into simpler substances. (See Appendix G for the Periodic Table of the Elements.) They are the most basic substances, such as hydrogen (H), helium (He), carbon (C), and oxygen (O). Ninety-two elements occur in nature and 14 more have been created in the laboratory; many are rare, and only a dozen are common. Most substances you encounter are not elements, but *compounds*, substances made of two or more elements. For example, water (H_2O) is a chemical compound composed of the elements hydrogen and oxygen (Fig. 6.6). An *atom* is the smallest unit of an element that displays the chemical properties of the element, and a *molecule* is the smallest unit of a compound that still has the chemical properties of the compound. A compound is created when atoms join together to form molecules of the compound.

Scientists developed a useful model of the atom in the twentieth century. The modern concept of an atom pictures a tiny, dense *nucleus* surrounded by rapidly moving *electrons*. The study of electricity revealed that electrons are low-mass, negatively charged particles. *Protons* and *neutrons*, which are about 2000 times as massive as electrons, make up an atom's nucleus. The protons are positively charged, and the neutrons, as implied by their name, are neutral with no electrical charge. The nucleus of an atom, because of the protons it contains, is positively charged and attracts the negatively charged electrons. This attractive force binds the electrons to the nucleus and holds the atom together. (A strong *nuclear force* binds the protons and neutrons together in the nucleus, despite the mutual repulsion of the positively charged protons.) The electrons whiz around the nucleus in orbits, but their distance is very great in terms of the size of the nucleus. Most of an atom is empty space!

The difference between elements lies in the nucleus. Each element has a specific number of protons. The nucleus of a hydrogen atom, for example, contains one proton; helium, two; and carbon, six. In most elements the number of neutrons is approximately the same as the number of protons the atom contains. For example, atoms of the most common form of carbon contain six protons and six neutrons. Nuclei of an element which contain the same number of protons, but a different number of neutrons, are called *isotopes* of the element. For example, heavy hydrogen, sometimes called deuterium, has

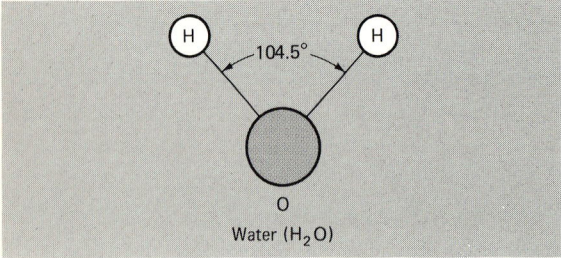

FIGURE 6.6 The chemical structure of water, H_2O.

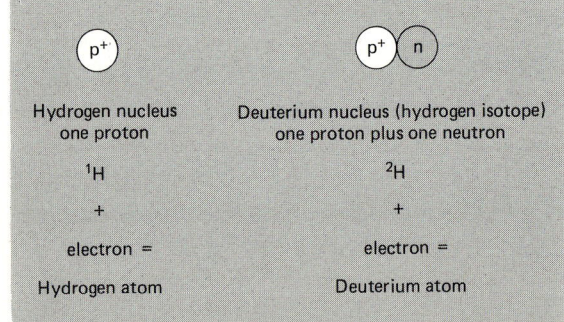

FIGURE 6.7 The difference between a hydrogen and a deuterium (an isotope of hydrogen) nucleus.

6.2 SUNLIGHT AND SPECTROSCOPY

one proton and one neutron, whereas ordinary hydrogen has one proton and no neutrons (Fig. 6.7). They are the same element, but an atom of heavy hydrogen has more mass than an atom of the ordinary form because of the extra neutron. To distinguish among isotopes of an element, we use a superscript attached to the chemical symbol, giving the total number of protons and neutrons in the nucleus; for example, ordinary hydrogen is ^{1}H, but heavy hydrogen is ^{2}H. Ordinary carbon, ^{12}C, has 6 protons and 6 neutrons in the nucleus, whereas ^{13}C (carbon 13) has 6 protons and 7 neutrons. (Sometimes a subscript giving the number of protons is added before the chemical symbol, such as $^{13}_{6}$C.)

A normal atom has the same number of electrons orbiting the nucleus as it has protons in the nucleus, so it carries no net electric charge. If the number of electrons is less than or greater than the number of protons, the particle is called an *ion*.

Simple Spectroscopy

The detailed analysis of light is called *spectroscopy*. You've probably done a little spectroscopy without knowing it. The next time you see a rainbow, look carefully to see that the colors run continuously from red to blue. What's happening? Raindrops are dispersing sunlight into a continuous band of colors.

You can use a prism to do the same at home or in a lab (Fig. 6.8). Use a piece of cardboard with a slit in it to pass a beam of sunlight through a prism. Let the light coming out of the prism fall on white paper. The sunlight spreads into an array of colors, with red light bent the least and violet the most. This sequence of colors is called a *spectrum*.

A spectrum with no breaks in it is called a *continuous spectrum* (Fig. 6.9a). White light produces a continuous spectrum with all colors running smoothly into one another and no colors missing. You can think of white light as a mixture of all colors.

Not all spectra are continuous. Suppose you put another slitted piece of cardboard in the light beam after it leaves the prism and let just a *single color* pass through. You'd then

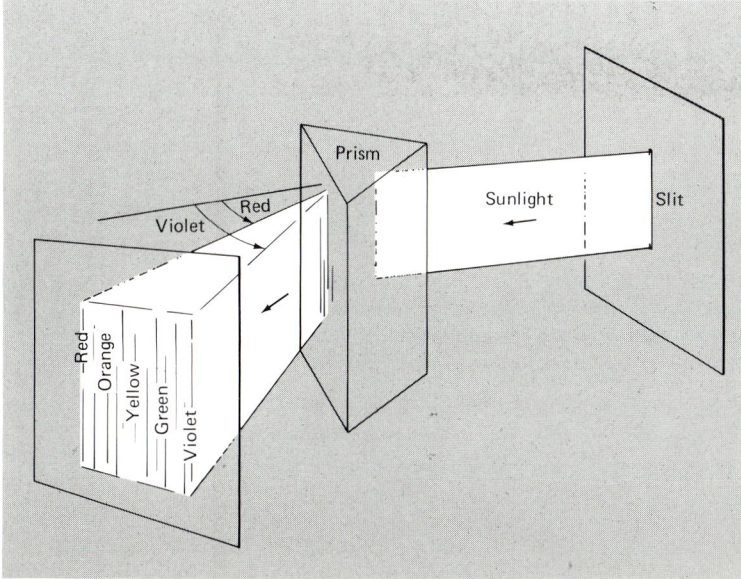

FIGURE 6.8 An experiment to display the sun's spectrum. The prism splits sunlight into the band of colors of the rainbow.

FIGURE 6.9 The three basic types of spectra. (a) A continuous spectrum, in which all the colors from red to blue run smoothly together with none missing. (b) A bright-line spectrum, in which individual bright lines of specific colors appear. (c) A dark-line spectrum, in which certain dark lines appear against a background of a continuous spectrum. [*Courtesy Bausch and Lomb, Analytical Systems Division.*]

see a bright line of that color (Fig. 6.9b), not a complete, continuous spectrum. This is what is meant by a *spectral line*. When light is present, or emitted, only at certain colors, as in the example above, we speak of *emission lines*.

Take your prism and slit and let sunlight pass through them; then magnify the spectrum with a telescope. You'll see a continuous spectrum crossed by dark lines (Fig. 6.9c), that is, there are some colors where light is *missing*. It appears as if something has removed, or absorbed, light of these colors. These dark lines are called *absorption lines*. (See Plate 1 for color pictures of spectra.)

The concept of spectral line may confuse you. Straight lines appear in a spectrum because the slit admitting the light to the prism is a straight line source of light. Each line is an image of the slit in the light of one color. It the slit, or other source of light, were curved in an S-shape, the lines in the spectrum would also be S-shaped. An example is the spectrum of the sun's chromosphere (Fig. 6.29, Section 6.5), in which the lines are curved because the visible crescent of the chromosphere is curved.

Analyzing Sunlight

To unravel the message of sunlight, let's first discuss a few experiments with a *spectroscope* (Fig. 6.10), an instrument for observing fine details in a spectrum. A simple spectroscope consists of a telescope, which feeds light from a slit through a prism. Another telescope allows viewing the spectrum produced by the prism.

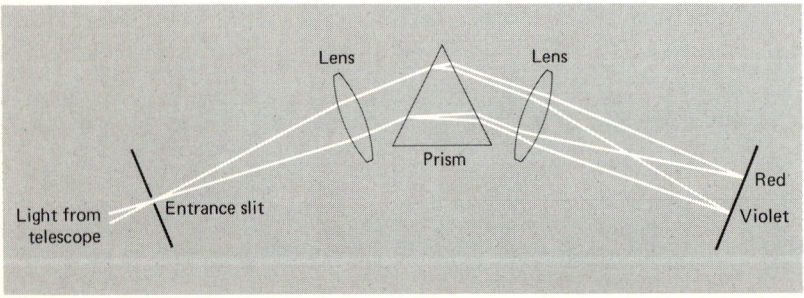

FIGURE 6.10 A schematic plan of a spectroscope. The light from a telescope passes through an entrance slit, through a lens that makes the light rays parallel, and finally through a prism to make a spectrum. The spectrum is focused by another lens and can be photographed or viewed directly.

6.2 SUNLIGHT AND SPECTROSCOPY

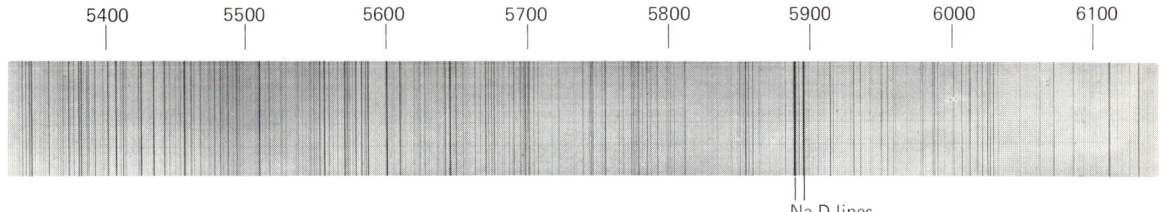

FIGURE 6.11 Dark lines in the sun's spectrum. Kirchhoff showed that the two especially prominent lines in the yellow region, called the D lines by Fraunhofer, are made by sodium. [*Courtesy Mt. Wilson Observatory, Carnegie Institution of Washington.*]

If you put sodium (in the form of table salt) in a hot, colorless flame, you will see the flame turn yellow. Through the spectroscope, you will see a series of bright lines, the brightest being a pair in the yellow region (Plate 1). A spectroscope shows that each chemical element displays a unique arrangement of lines. A particular spectral pattern demonstrates the presence of a particular element. For example, the brightest lines of neon gas lie in the red region of the spectrum, so a pure neon sign appears red. Hot mercury gas has the strongest bright lines in the yellow, blue, and green regions. These lines produce the blue-greenish color of mercury street lamps. Sodium street lamps appear a ghastly yellow because the brightest lines from a hot sodium gas fall in the yellow region.

If you look at sunlight with a spectroscope, you will see dark lines against a bright, continuous background of color (Plate 1 and Fig. 6.11). A pair of dark lines in the yellow region of the spectrum are particularly prominent; astronomers call this pair the *D lines*. (The dark lines in the solar spectrum were first studied by Joseph von Fraunhofer, 1787–1826, who labeled the most prominent lines A, B, C, D, etc.) Could these dark lines be related to the bright lines of sodium?

Suppose you try this experiment. Pass sunlight through a sodium flame into the spectroscope. You might expect that the added light of the sodium flame would make the lines less dark. But the lines do *not* become less dark! The pair of dark lines becomes even darker! Now pass light from a glowing solid such as a light bulb filament (which emits a continuous spectrum) through the sodium flame. This time a pair of dark lines appears in the spectrum at the same position in the yellow region as the pair in the sun's spectrum. So it must be sodium making these dark lines, *removing* yellow light from the spectrum.

From such experiments Gustav Kirchhoff (1824–1887) formulated his empirical rules of *spectroscopic analysis*, the determination of the composition of an unknown mixture of elements. Briefly put, Kirchhoff's rules are:

1. A hot and opaque solid, liquid, or highly compressed gas emits a continuous spectrum (Fig. 6.12a).
2. A hot, transparent gas produces a spectrum of bright lines (emission lines). The number and position of these lines (their wavelengths) depend on which elements are present in the gas (Fig. 6.12b).
3. If a continuous spectrum passes through a gas which is at a lower temperature than the source, the cooler gas causes the appearance of dark lines (absorption

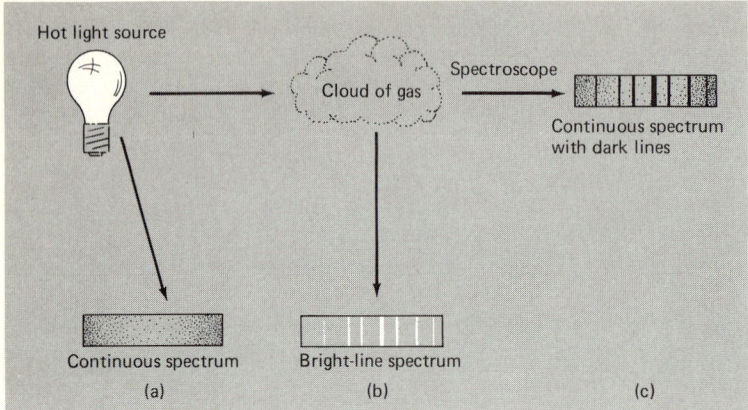

FIGURE 6.12 Kirchhoff's rules of spectral analysis. A hot, opaque solid or gas produces a continuous spectrum (a). A hot, transparent gas makes an emission spectrum (b). If a continuous spectrum passes through a cooler gas, dark lines appear (c). These dark lines are the same wavelengths as the bright lines produced by the same gas.

lines). Their position in the spectrum and their number depend on the elements in the cooler gas (Fig. 6.12c).

Note that these "rules" are really just summaries of Kirchhoff's experiments. They are not fundamental physical laws in the same sense as Newton's laws of motion.

The essence of the first rule is that an opaque, hot material produces a continuous spectrum. Recall that atoms usually emit and absorb at discrete wavelengths. However, when atoms are jammed together so densely that the gas is opaque to light, a continuous spectrum appears. For instance, the sun is so hot that it is gaseous, but it is still dense and opaque enough to produce a continuous spectrum.

In sunlight you see dark lines against a continuous background. What causes them? By Kirchhoff's third rule you expect that there must be a cooler and less dense gas between the visible surface of the sun and us. There are two places where this gas can be, in the atmosphere of the earth or the atmosphere of the sun. Although a few lines in the solar spectrum are produced by gases on earth (water vapor in particular), most lines are produced in the solar atmosphere. This atmospheric layer, cooler than the sun's surface, absorbs light from the continuous spectrum passing through it to produce the dark lines. (One point to keep in mind: Dark lines do not mean the complete absence of light at those colors; rather, the lines appear dark in contrast to the continuous spectrum.) The positions of the lines in the spectrum tell which elements are present. The solar composition determined from the dark lines relates only to the region of the sun that produces the absorption spectrum.

What happens when a continuous spectrum passes through a transparent gas that is *hotter* than that producing the continuous spectrum? A continuous spectrum with *bright* lines results. These lines are in the same positions as the bright lines seen when a thin, hot gas of the same composition is observed with a spectroscope.

A thin gas always emits some radiation in spectral lines. If the gas is cooler than the source of the continuous spectrum passing through, it absorbs more light in spectral lines than it emits, and the result is absorption lines. If it is hotter, it emits more than it absorbs, with the net result of bright lines overlaid on a continuous spectrum. So to Kirchhoff's three rules, we can add a fourth: A continuous spectrum passing through a transparent gas which is at a higher temperature than the source will have bright lines added.

6.3 SPECTRA AND ATOMS

With these ideas in mind, you should be able to understand what happens when sunlight passes through the sodium vapor to make darker D lines. The sodium in the flame is *cooler* than the sun's visible surface, so the sodium *absorbs* the light in the D lines, making them darker. This experiment proves that the missing colors in the continuous solar spectrum are due to absorption of those colors by the atoms in the cool gas of the solar atmosphere.

Here's the fundamental point concerning emission and absorption lines. *The lines absorbed by a gas from a continuous spectrum are the same lines emitted by the gas when energy is put into it.*

Whether the atoms in the gas emit or absorb depends on the physical conditions in the gas. Emission requires high temperatures in a transparent gas; absorption occurs when a continuous spectrum passes through a cooler thin gas. In either case the pattern of emission lines is the same as the pattern of absorption lines for a gas with the same chemical composition.

6.3 SPECTRA AND ATOMS

How can the physical conditions in the sun be discovered from its spectrum? The spectral code was cracked in a great revolution of physics in the twentieth century: the quantum theory and its explanation of the nature of the atom and of light. To understand the power of spectroscopy you must first know something about light.

Light and Electromagnetic Radiation

The nature of light has challenged many scientific geniuses, especially Newton, who pictured light as streams of small particles. This particle model of light hit many obstacles in the eighteenth and nineteenth centuries, when numerous experiments showed light had the properties of waves. (Today we know that light has both wave and particle properties.)

Waves have three fundamental properties: wavelength, frequency, and velocity. The *wavelength* is the distance between two successive crests of a wave (Fig. 6.13). The *frequency* is the number of waves that pass by each unit of time, for example, 1/sec. The *velocity* is the distance covered per unit of time, for example, 1 m/sec, by a crest traveling in a certain direction.

Be aware that waves do *not* involve the mass motion of material over long distances. If you're on the Pacific shore, for instance, the water in the waves pounding the sand is *not* the same water as that involved when the wave started at Hawaii. A wave carries energy of motion from one place to another, but does not transport material.

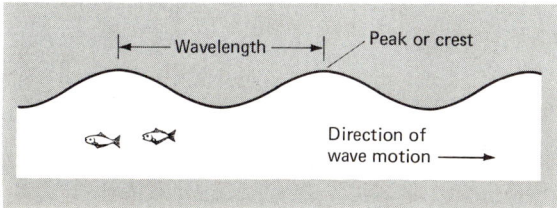

FIGURE 6.13 Waves. The distance from peak to peak of a wave is its wavelength. The number of waves that pass by each second is the frequency.

The three properties of waves are related. If the number of wave crests which pass by in a second is f (the frequency), the first crest to pass must move a distance of f wavelengths by the end of that one second. The distance traveled in a second is the velocity, so the velocity of the wave, V, must be

$$V = f\lambda$$

where λ is the wavelength.

This fundamental rule applies to all kinds of waves. It tells you that for waves traveling through some material, those of different frequencies must also have different wavelengths if the wave velocities are the same. For light the velocity is the same for all wavelengths. The speed of light is 299,793 km/sec in a vacuum. It is usually designated by c and is more easily recalled by its approximate value, 300,000 km/sec (3×10^8 m/sec). So for light waves

$$f\lambda = c$$

where λ is the wavelength and f the frequency. Because c is constant for all kinds of light, different wavelengths must have different frequencies.

Light is one example of electromagnetic radiation. The range of all different wavelengths of electromagnetic radiation is called the *electromagnetic spectrum* (Fig. 6.14).

Although the wavelength of electromagnetic radiation is sometimes measured in meters or centimeters, scientists often use special wavelength units for different regions of the electromagnetic spectrum. For example, the unit of measure in the region of visible light is the *Angstrom* (Å), equal to 10^{-10} m. In the infrared region the unit of length commonly used is the micron or micrometer (μm), equal to 10^{-6} m.

But wavelengths are not always used. For example, in the radio region astronomers commonly talk in terms of frequency, rather than wavelength. The unit of frequency is the *Hertz* (Hz), which is one cycle (or vibration) per second. When you see a wave go by, from peak to peak, it has gone through one cycle. Even in the radio region many cycles go by in a second. For example, the AM band of a radio covers the range from 540 to 1650 kilohertz (1 kHz = 10^3 Hz). The FM band ranges from 88 to 108 megahertz (1 MHz = 10^6 Hz). Radio astronomers work at such high frequencies that they use the unit of a gigahertz (1 GHz = 10^9 Hz). Police radars typically operate at a frequency of about 10 GHz.

Using frequency or wavelength is only a matter of convention and convenience, for the one can always be converted to the other, using the relationship $f\lambda = c$. For example, what wavelength corresponds to radar at 10 GHz? Since $f\lambda = c$, and 1 GHz = 10^9 Hz,

$$\lambda = \frac{c}{f}$$

$$= \frac{3 \times 10^8 \text{ m/sec}}{10 \times 10^9 \text{ /sec}}$$

$$= 3 \times 10^{-2} \text{ m} = 3 \text{ cm}$$

6.3 SPECTRA AND ATOMS

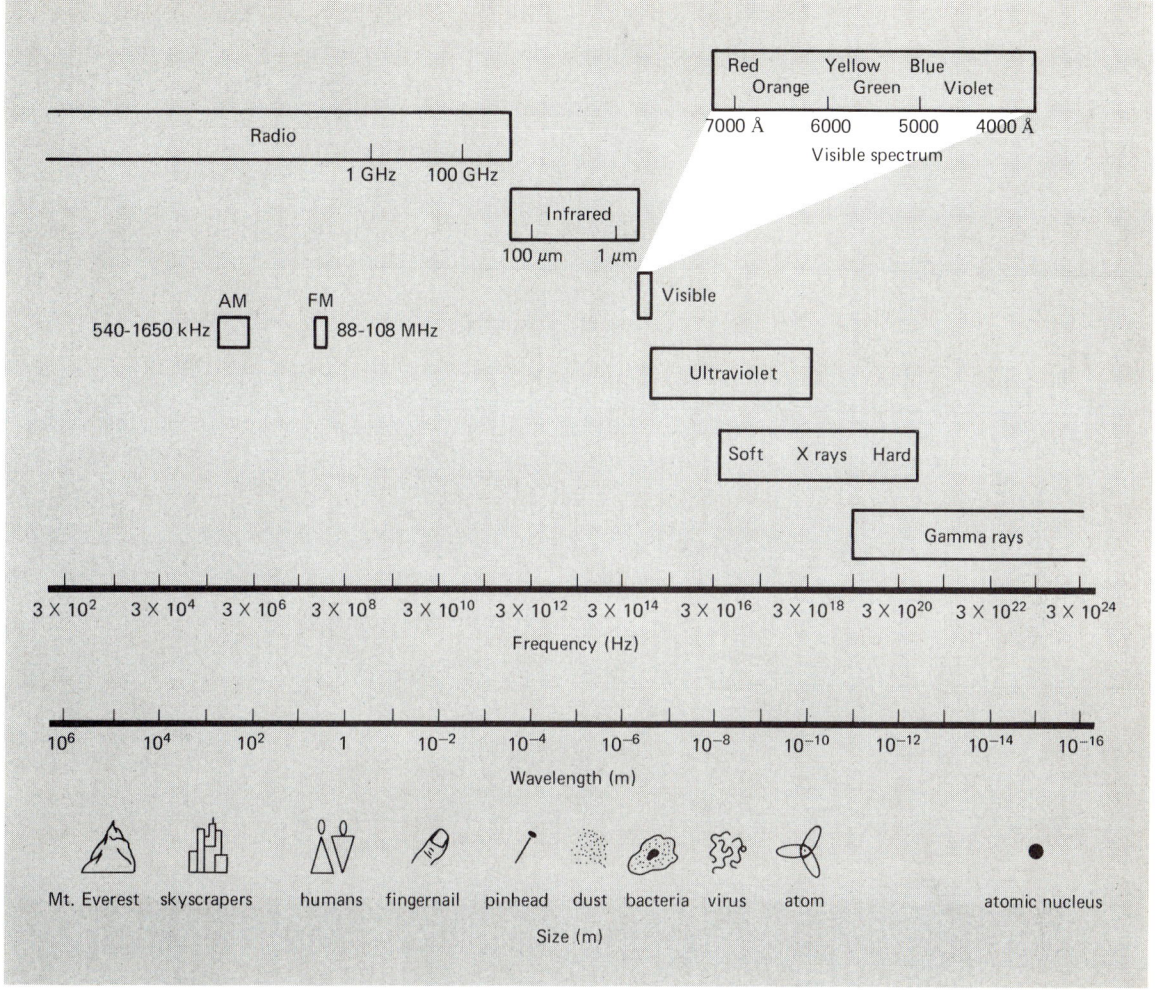

FIGURE 6.14 The electromagnetic spectrum. Wavelength in the visible region is measured in Angstroms (Å); one Å is 10^{-10} m. In the infrared region a micron (1 μm = 10^{-6} m) is the unit of measure. Radio waves are usually measured in cycles per second in a unit called a Hertz (Hz), 1 cycle/sec; megahertz (MHz), 10^6 cycles/sec; or gigahertz (GHz), 10^9 cycles/sec. [*Adapted from a diagram by E. J. Chaisson*].

Or what frequency corresponds to green light at 5000 Å? Since 1 Å = 10^{-10} m,

$$f = \frac{c}{\lambda}$$

$$= \frac{3 \times 10^8 \text{ m/sec}}{(5 \times 10^3)(10^{-10} \text{ m})}$$

$$= 6 \times 10^{14} /\text{sec}$$

$$= 6 \times 10^5 \text{ GHz}$$

The Atom, Quantum Theory, and Radiation

Although a wave model for light was championed during the nineteenth century, it did not explain how atoms produced waves. Physicists were puzzled by spectral lines, which for hydrogen appeared in a series in the visible region of the spectrum. This series of hydrogen lines was named the *Balmer series* (Fig. 6.15) for the Swiss mathematician Johannes J. Balmer (1825–1898), who in 1885 devised an empirical formula that described the regular sequence of lines. A modern form of Balmer's formula is

$$\lambda = 3.6470 \times 10^{-7} \frac{n^2}{n^2 - 4}$$

where λ is the wavelength (in meters) of the lines of the Balmer series, and n is an integer with a value of 3 for the first Balmer line, 4 for the second, 5 for the third, and so on. For example, the wavelength of the first Balmer line is

$$\lambda = 3.6470 \times 10^{-7} \frac{9}{9 - 4}$$

$$= 6.565 \times 10^{-7} \text{ m} = 6565 \text{ Å}$$

Because the Balmer series could be represented by such a simple formula, the inference was that light emission and absorption related to a simple structure of atoms. But what was that structure?

In 1911 the British physicist Ernest Rutherford (1871–1937) proposed a model for the atom in which all the positive charge resides in a tiny nucleus, with a surrounding negatively charged cloud of electrons. Rutherford theorized that the mutual attraction of

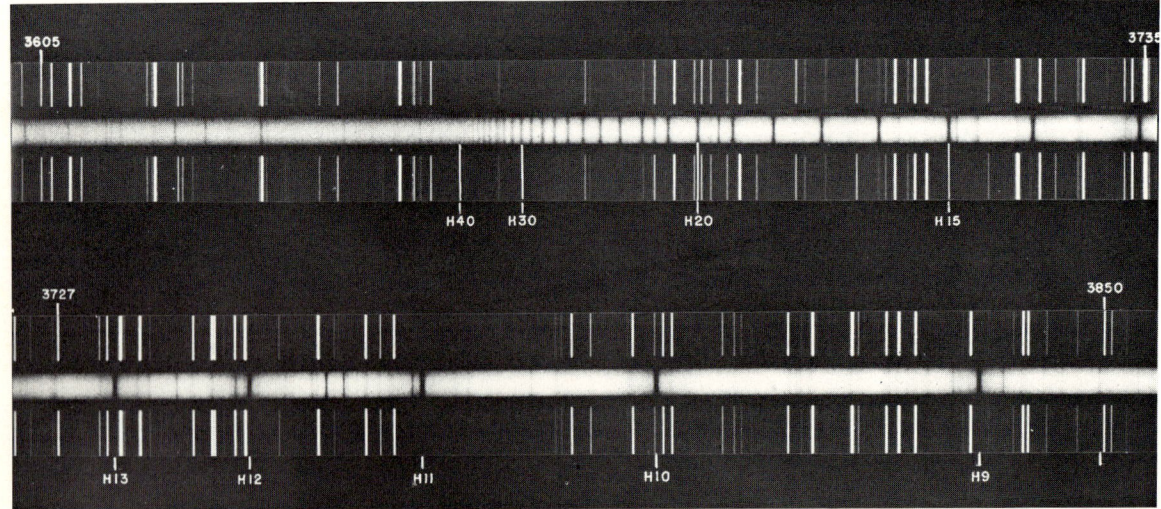

FIGURE 6.15 The Balmer series of hydrogen in the spectrum of a star. The lines are labeled H9, H10, etc. [*Courtesy, Palomar Observatory, California Institute of Technology.*]

6.3 SPECTRA AND ATOMS

unlike charges holds the atom together. If an atom loses one or more of the electrons, it becomes an ion. Physicists had noticed that an element's spectrum changes when it is ionized. For example, when hydrogen is ionized, it no longer produces the Balmer series of lines. They inferred that the arrangement of electrons somehow determines its light-emitting properties.

In 1900 Max Planck (1858–1947) announced a revolutionary idea about atoms and light. Perplexed by experimental work on spectra, Planck developed a startling new model about light: Radiating matter emits light in discrete chunks of energy that he called *quanta*. From this basis Planck explained some of the observed properties of radiation.

Taking up this quantum idea, Einstein showed that each quantum—usually called a *photon* when talking about light—carries an amount of energy that relates directly to its frequency. A photon of high frequency (short wavelength) transports more energy than one of lower frequency. For example, an X-ray photon carries much more energy than a visible light photon. This relationship between energy and frequency is a direct one: A photon with *twice* the frequency carries *double* the radiative energy. Similarly, a photon with one-half the frequency transports half as much energy. This crucial relationship between light's energy and frequency is

$$E = hf$$

where E is the energy in joules, f the frequency in Hertz, and h a constant called *Planck's constant* with a value of 6.63×10^{-34} J·sec. Now recall the relationship between frequency, wavelength, and speed for light:

$$\lambda f = c$$

Substituting for f,

$$E = \frac{hc}{\lambda}$$

which relates the energy of a photon to its wavelength.

How much energy does one photon have? Let's compute it for green light, 5000 Å wavelength. Remember that 1 Å = 10^{-10} m, so

$$E = \frac{hc}{\lambda} = \frac{(6.6 \times 10^{-34} \text{ J·sec})(3 \times 10^8 \text{ m/sec})}{5 \times 10^3 \text{ Å} \times 10^{-10} \text{ m/Å}}$$

$$= 4.0 \times 10^{-19} \text{ J}$$

That's pretty small: 1 J/sec = 1 watt. So the energy to light a 100-W light bulb for 1 second is equivalent to the energy in 2.5×10^{20} photons of green light.

With this idea in mind, note that the sun's continuous spectrum is, in fact, an energy spectrum, from lower (red) to higher (blue) energies. You have seen that atoms emit and absorb certain colors at specific wavelengths. This means that they are emitting or absorbing photons of specific energies. Atoms must absorb and emit energy in the form of whole photons, not $\frac{1}{3}$ photons or 1.67 photons. This quantum model emphasizes that in the emission or absorption of light the atom loses or gains energy in discrete amounts, the energies of individual photons.

Solving the Puzzle of the Hydrogen Atom

The Danish scientist Niels Bohr (1885–1962) meshed Planck's quantum and Rutherford's atomic pictures. The resulting scheme, known as the Bohr model of the atom, explained and predicted the absorption and emission of photons.

Bohr pictured atoms as having more than one possible arrangement of electrons. He imagined that the emission or absorption of light arises from a *transition* between electron arrangements. In line with Planck's theory and observed spectra, Bohr realized that only a limited number of specific electron arrangements are permitted, not an infinite number of arbitrary ones. Electron transitions can occur only between the special arrangements. You can visualize this atomic model by using concepts of energy (Focus 6.1). For a given nuclear charge (which is positive and varies from element to element), electrons have available a certain number of *energy levels*.

Consider the hydrogen atom (Fig. 6.16). It has one proton for a nucleus and one electron attached by the electric force between it and the proton. The electron has a certain total energy; the essence of quantum theory is that electrons remain in stable states of specific energies. For example, the electron in the lowest energy level (called the

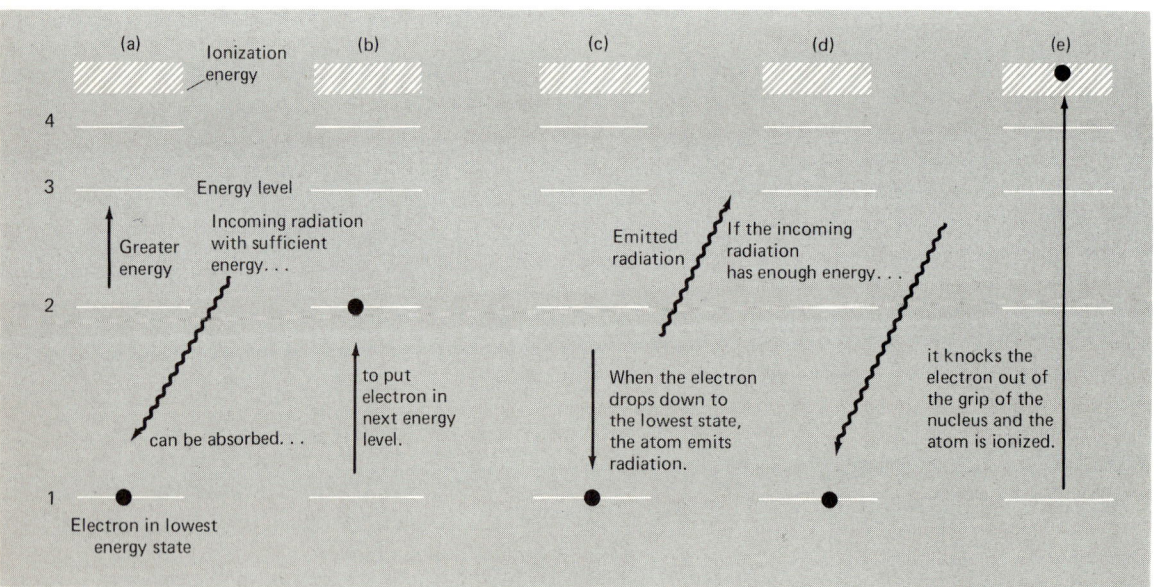

FIGURE 6.16 A schematic energy-level diagram for hydrogen atoms (the energy levels are not to scale). The electron is usually in the lowest energy level (a). The electron can gain energy by absorbing photons of certain energies, such as the difference in energy between levels 1 and 2. Absorbing such a photon, the electron jumps up to the second level (b); the atom is excited. If the electron drops back to level 1 (c), it gives off a photon with energy again equal to the difference in energy between levels 1 and 2. The electron can gain so much energy that it is no longer bound to the nucleus (d). The atom is then ionized (e).

ground state) securely orbits the nucleus. If enough energy is added to the atom, either by collision with another particle or by absorption of a photon with sufficient energy (Fig. 6.16a), the electron jumps up one or more energy levels (Fig. 6.16b). The atom is then *excited*. This condition does not last long; the electron drops to a lower level in about 10^{-8} sec. However, for the electron to descend to a lower energy level requires that it lose some energy. The electron does so by emitting a photon with an energy *equal* to the amount it needs to lose (Fig. 6.16c). This energy directly relates to the photon's frequency.

If the electron gains enough energy, it flies away from the nucleus (Fig. 6.16d, e). The atom is then *ionized*. The loss of an electron changes the energy arrangements available and so changes the atom's spectrum also.

As an analogy, imagine moving a bowling ball up and down the stairs. Each step is an incremental change in the ball's energy, and only full-step changes are permitted. If we try to add less than one step's worth of energy to the ball, it remains at its initial level. If we add exactly one step of energy, the ball moves up one step. When the ball loses energy, it descends in steps until it hits the floor at the bottom of the staircase; this is the ground level, the state of lowest energy.

Look at the energy-level diagram for hydrogen (Fig. 6.17). Every *upward* step requires the *absorption* of energy and every *downward* one the *emission* of energy. The greater the energy difference needed for a transition, the higher the frequency (and hence the shorter the wavelength) of the photon produced from that transition. All transitions to and from the lowest energy level involve large energy changes, so they correspond to wavelengths in the ultraviolet range of the spectrum. Although this set of lines, called the *Lyman series* (Figs. 6.17 and 6.18), cannot be seen by the eye, it can be detected by photography. The set of transitions down to and up from the second energy level is the Balmer series (Fig.

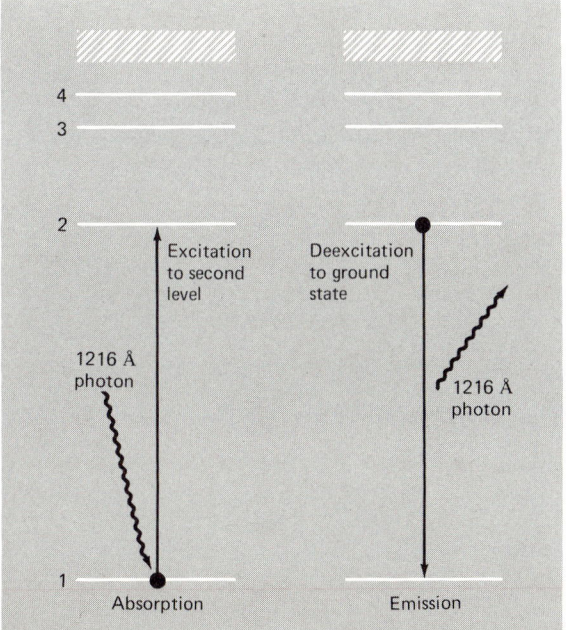

FIGURE 6.17 Energy-level diagram for the Lyman series of hydrogen.

FIGURE 6.18 A schematic of the Balmer and Lyman series of hydrogen as seen with a spectroscope. The Lyman series comes from the electron dropping into level 1; the Balmer series involves level 2. Note that the Balmer series lies in the visible region of the spectrum, but the Lyman series is in the ultraviolet.

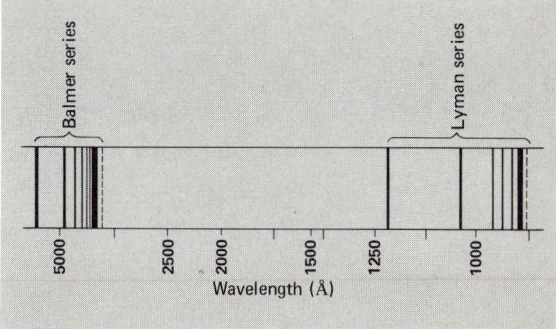

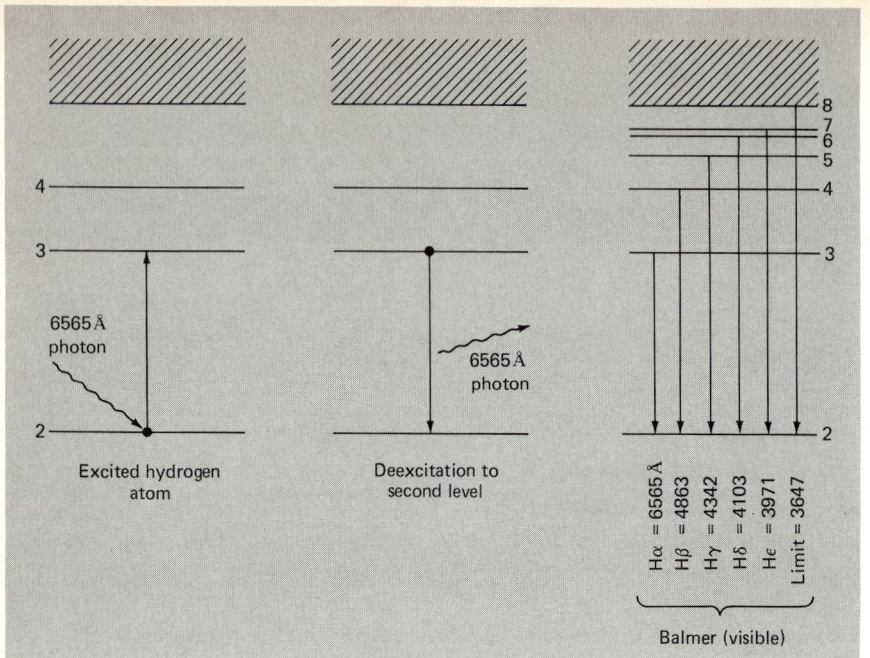

FIGURE 6.19 Energy-level diagram for the Balmer series of hydrogen. The energy level differences are drawn to scale.

6.19); it lies in the visible region of the spectrum (Fig. 6.18). The hydrogen absorption lines in the sun's spectrum are those in the Balmer series.

Using quantum and energy concepts, Bohr derived a formula for the wavelengths of all the lines of hydrogen:

$$\frac{1}{\lambda} = R\left(\frac{1}{n_B^2} - \frac{1}{n_A^2}\right)$$

where λ is the wavelength (in meters), n_A the number of an upper level from which an electron falls into a lower level, n_B the number of the lower level, and R the *Rydberg constant*, with a value 1.09678×10^7/m. (The Rydberg constant is a combination of other fundamental physical constants.) Now invert this equation:

$$\lambda = \frac{1}{R}\left(\frac{n_A^2 n_B^2}{n_A^2 - n_B^2}\right)$$

You see that this has the same form as the equation discovered by Balmer, if $n_B = 2$,

$$\lambda = \frac{1}{R}\left(\frac{4n_A^2}{n_A^2 - 4}\right)$$

$$= \frac{4}{R}\left(\frac{n_A^2}{n_A^2 - 4}\right)$$

$$= 3.65 \times 10^{-7}\left(\frac{n_A^2}{n_A^2 - 4}\right)$$

6.3 SPECTRA AND ATOMS

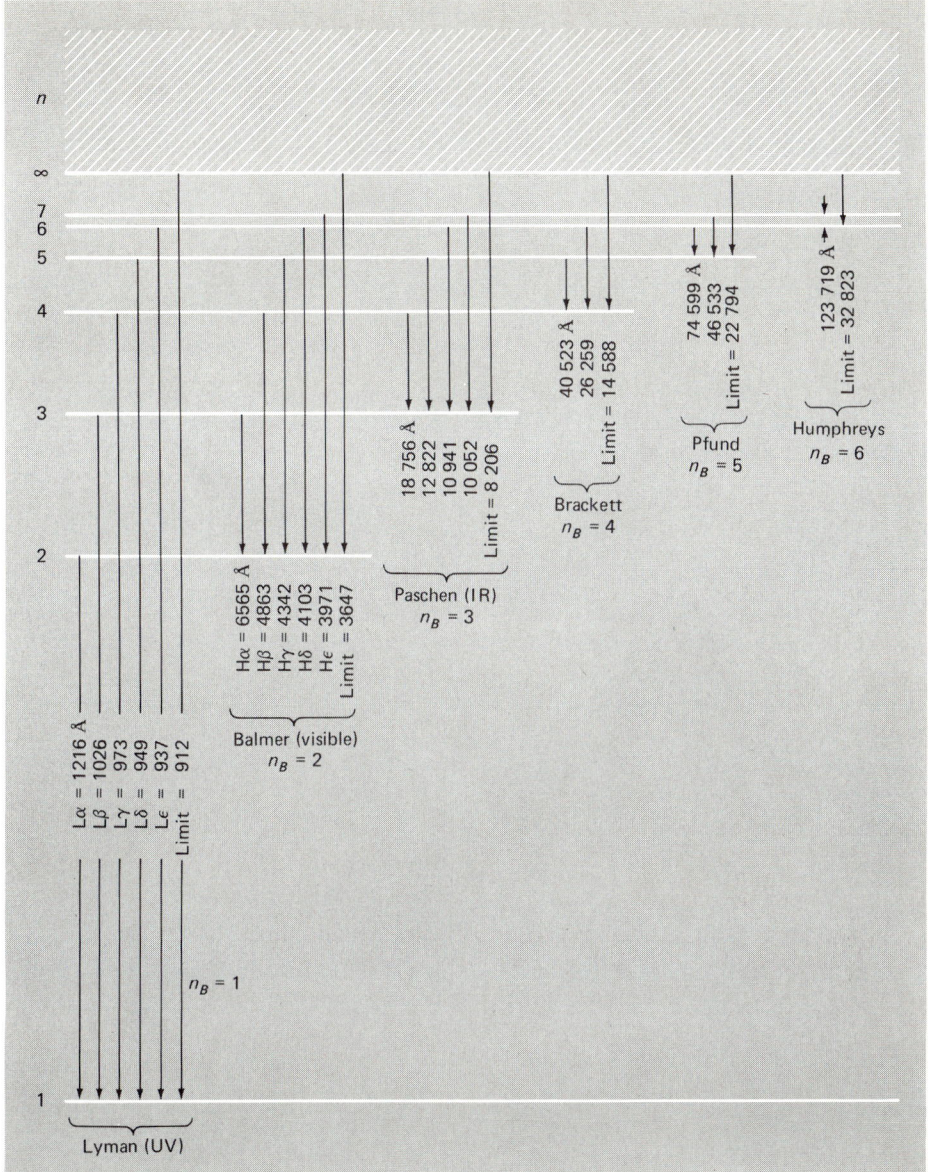

FIGURE 6.20 Energy-level diagram for the main spectral series of hydrogen. The energy level separations are drawn to scale.

The correspondence of the result predicted from the quantum model and the result discovered empirically marked a great triumph for quantum theory and also led to its acceptance.

If $n_B = 1$, you then have the formula for the lines of the Lyman series (Fig. 6.20), which arise from transitions down to or up from the ground state (level 1). For $n_B = 3$,

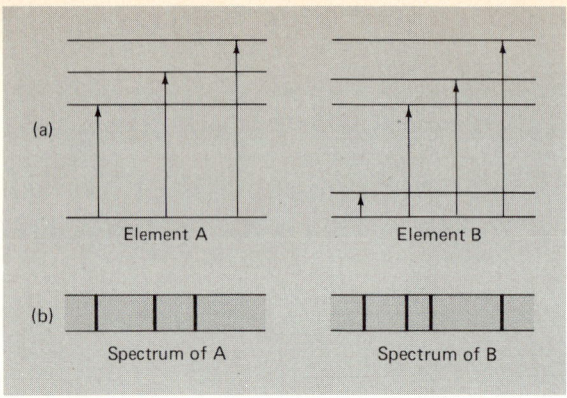

FIGURE 6.21 Different spectra (b) from hypothetical element A and element B which have different sets of energy levels (a).

you have the formula for a series of lines of low energy in the infrared. This series is called the *Paschen series*. Note that all these changes of upper and lower energy levels for the electrons reflect the quantum nature of atoms.

Other Atoms

We have focused so far on the hydrogen atom because it has the simplest characteristics of all the elements and plays a dominant role in the cosmos.

The simple Bohr quantum model needs drastic modification to work with other elements. Despite complications, one essential point remains: Each atom and each ion (even of the same atom, for more than one electron can be lost) has its own unique set of electron energy levels. So each has its unique energy-level diagram (Fig. 6.21a). Because spectral lines are produced by electronic transitions between energy levels, each element or ion has a unique set of spectral lines (Fig. 6.21b).

Spectra from Atoms

To sum up, electrons have only certain stable energies in atoms. Add sufficient energy, and an electron moves to a higher energy level. It shortly drops down to a lower energy level and emits a photon. The bigger the downward jump, the more energy the photon has. Any jump, however, produces or absorbs a photon with a specific energy and wavelength. This discrete jumping of electrons is how atoms absorb and emit photons, producing absorption and emission lines.

An electron transition between energy levels in which the electron is tied to the atom is called a *bound-bound transition*. When an atom is ionized and loses an electron from a bound state, the electron undergoes a *bound-free transition*. It is freed from a particular energy level and can have any energy over a wide, continuous range. When an ion captures an electron, we have a *free-bound transition*. The bound-bound transitions result in discrete lines. Bound-free or free-bound transitions produce absorption or emission over a continuous range of wavelengths.

Atoms have another important process of producing or absorbing radiation. Imagine an ionized gas, say of hydrogen. The electrons have been ripped away from the protons, leaving hydrogen ions, H^+. The free electrons and protons attract each other, and, since the electrons have much less mass than the protons, their paths bend as they speed by the protons. If a photon encounters an electron near a proton, the electron may jump suddenly to a path of higher energy. During this absorption the electron stays free from the proton, so we call it *free-free absorption*. Similarly, an electron skirting around a proton may lose energy and emit a photon. That is *free-free emission*. Because an electron has a continuous range of energies over which it can change, free-free transitions result in continuous (rather than discrete) emission and absorption.

What can excite atoms for emission or absorption? Imagine a box filled with hydrogen gas. The atoms of the gas collide with each other (and the sides of the box). Heat up the gas. The atoms then move faster and knock into each other harder. Collisions can bump electrons into higher energy levels; the harder the collision, the higher the level. This process is called *collisional excitation*. Some of the atom's energy of motion, called kinetic energy (Focus 6.1), transfers to an electron of another atom. Photons can also excite atoms when absorbed, but only certain photons, namely those with energies corresponding to the *difference* in energies of two energy levels of the atom. This process is called *photon excitation*. Whether excited by collisions or by photons, the excited atom usually quickly radiates.

Let's look at the sun and sodium vapor experiment with the quantum model in mind. When sodium chloride is placed in a flame, individual atoms of sodium are released from the salt. Some of these atoms collide with other atoms and are excited. As the electrons return to lower levels, they emit photons of yellow light at a wavelength of 5893 Å (Plate 1) when the outermost electron jumps from its lowest excited level down to the ground level. (Actually the lowest excited level for sodium is two levels of almost the same energy, so a pair of closely spaced lines results.)

Follow a 5893 Å photon on its way from the sun to the spectroscope through the sodium vapor cloud. If a 5893 Å photon encounters an unexcited sodium atom, the sodium atom *absorbs* the photon. In the gas there are enough sodium atoms to absorb almost all the 5893 Å photons that try to pass through. The sodium atoms absorb almost no other wavelength in the visual region. These other photons pass right through the gas preserving the continuous spectrum you see. There is a dark line at 5893 Å because these photons were absorbed by the sodium gas.

What happens to the sodium atom that has been excited by the photon absorption? An excited atom quickly emits a new 5893 Å photon. The original photon was headed directly for the spectroscope slit before it was absorbed by a sodium atom, but the brand-new photon can be reemitted in any direction (Fig. 6.22). Very rarely will it be emitted in the

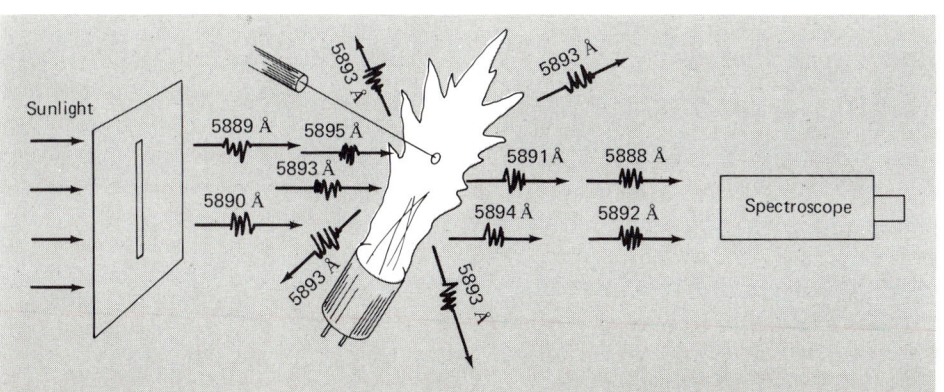

FIGURE 6.22
Paths of solar photons through a hot sodium flame.

same direction the original photon was traveling, so only very few of these new photons enter the spectroscope slit. That's how the sodium line in the sun's spectrum becomes darker when the sunlight is passed through the sodium vapor. Note that if you observe the sodium vapor without the sun behind it, you would see an emission line produced by the reemitted 5893 Å photons.

That's the basic idea of how atoms in transparent gases produce line spectra. With this background let's learn about the sun from its spectrum.

6.4 MESSAGES FROM SUNLIGHT

Most of what we know about the sun and other stars is information carried by light. To decode the message of sunlight and starlight takes up much of the time, energy, and ingenuity of astronomers.

Solar Energy

The sun's *luminosity* is its total output in radiative energy each second. By the time the radiation reaches the earth (in only 8.3 min), it has spread out over a large region of space—in fact, over a sphere whose radius is the earth–sun distance (Fig. 6.23). So you cannot measure directly the sun's luminosity.

How to find the sun's luminosity? Put a special detector in a satellite orbiting the earth. (You want to be above the earth's atmosphere because it absorbs some sunlight.) Point the detector directly at the sun, and measure the radiant energy absorbed. It amounts to 1370 W for each square meter of the detector's area. If every square meter on a surface 1 AU in radius (an imaginary sphere surrounding the sun) catches this amount of energy, the total energy radiated by the sun amounts to roughly 3.83×10^{26} W.

Let's do this calculation. The area of a sphere is

$$A = 4\pi R^2$$

where R is the radius. Here, R equals 1 AU or 1.496×10^{11} m, so

$$A = 4(3.14)(1.496 \times 10^{11} \text{ m})^2$$
$$= 2.81 \times 10^{23} \text{ m}^2$$

Then

$$\text{luminosity} = (2.81 \times 10^{23} \text{ m}^2)(1.370 \times 10^3 \text{ W/m}^2)$$
$$= 3.86 \times 10^{26} \text{ W}$$

That's enough power to light 4×10^{24} 100-W light bulbs!

The earth intercepts only about 10^{-10} of this total, but that is a large amount. On a clear day in New Mexico about 1 kilowatt (kW) falls on a square meter of the earth's surface. In a year the entire earth's surface catches about 10^{18} kilowatt-hours (kWh). At the rate of 5 cents per kWh, the annual solar energy would cost 5×10^{16} dollars! Fortunately, no one has a monopoly on this fantastic resource.

6.4 MESSAGES FROM SUNLIGHT

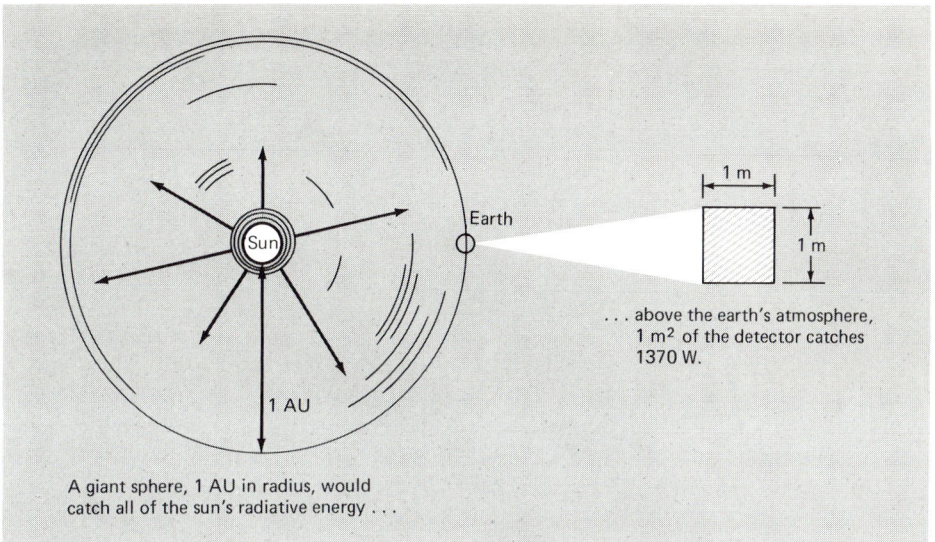

FIGURE 6.23 Measuring the sun's luminosity from the earth.

Geological evidence indicates that the sun's luminosity has not varied more than a few tens of percent over the past 3.5 billion years. Satellite measurements show it does not vary now more than a tenth of a percent. Long-term observations indicate a variation of only 0.25 percent in the past 50 years.

The Sun's Surface Temperature

The sun's color (yellow-white) provides a means of finding its surface temperature, that is, we can assign a temperature to the sun's surface by examining its continuous spectrum.

To see how, consider heating a piece of metal in a very hot flame. The metal first emits a dull red light. As it gets hotter, it glows bright red, then orange, yellow, white, and finally blue-white. So the overall color in the visible region of the spectrum relates to the temperature of the metal, which is another way of saying that a metal's continuous spectrum depends on its temperature.

How do color and continuous spectra relate? Let's look at the metal's emission in some detail (Fig. 6.24). Measure the intensity at a range of wavelengths (from ultraviolet to infrared) for the metal at different temperatures, that is, plot the metal's spectrum. Note three features in these spectra: (1) the emission *peaks* at some wavelength, (2) the peak shifts to *shorter* wavelengths as the metal gets hotter, and (3) the metal emits *more* intensely at *all* wavelengths at *higher* temperatures.

Your eye responds to only a small portion of the complete spectrum, the visible region from violet to red. (For example, your eye cannot sense any of the infrared emission.) Yet, for the temperature range considered here, your eye does reasonably well in discerning the shift in the balance of colors at the peak emission.

Radiation from Blackbodies

An object whose continuous spectrum has the shape shown in Fig. 6.24 and the variation with temperature described above is called a *blackbody radiator* or simply a

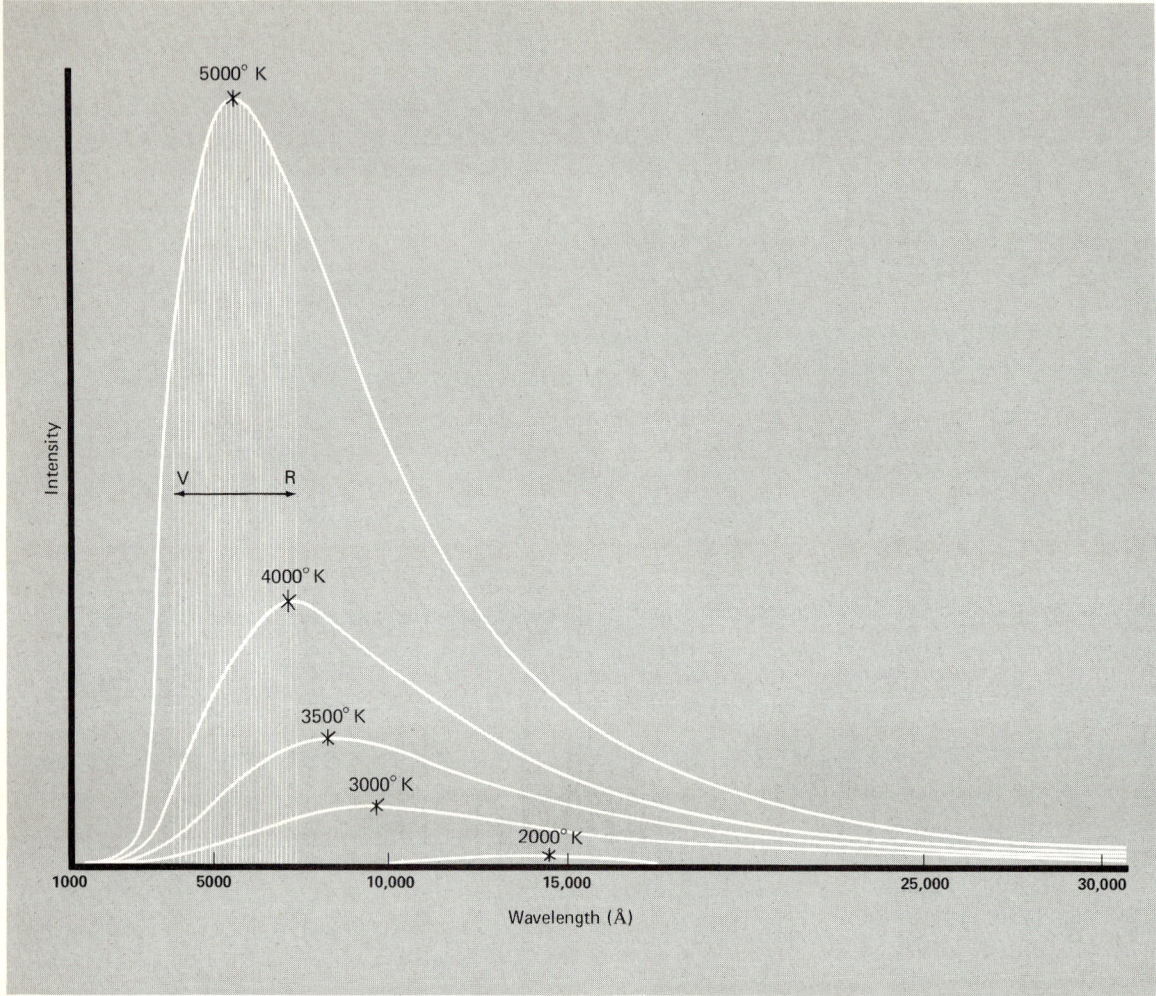

FIGURE 6.24 The spectrum of a hot metal (or opaque solid, liquid, or gas) as it varies with temperature (3000 to 5000 K). The visible region (from violet, V, to red, R) is indicated.

blackbody. Perfect blackbodies do not actually exist, but the sun (and other stars) emit radiation somewhat like an ideal blackbody.

One often confusing point about blackbodies is why they are called "black" when they give off light. Blackbodies are named from their light-absorbing abilities; they absorb light at any wavelength completely and reflect none. (If you beam light at the sun, none would reflect back.) When a blackbody absorbs radiative energy, it heats up and emits at all wavelengths, even though much of the emission may not be visible to our eyes. A good absorber, when heated, is a good radiator.

The radiation from a blackbody has a characteristic shape (Fig. 6.24) that depends only on the temperature of the body and not on any other property, such as its composition.

6.4 MESSAGES FROM SUNLIGHT

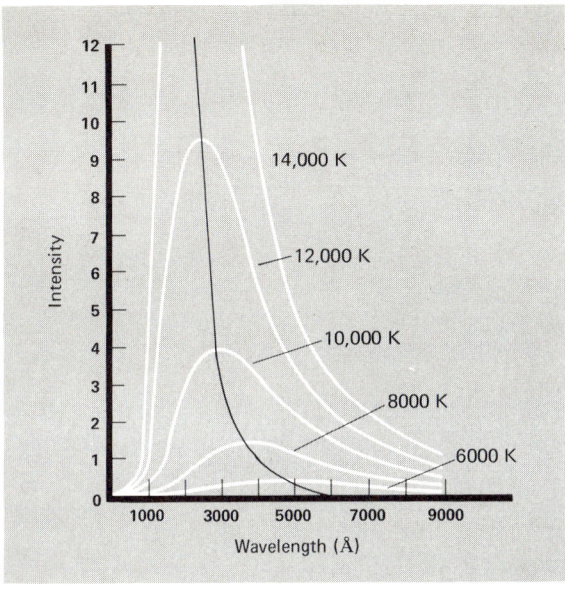

FIGURE 6.25 Spectra of blackbody radiators at different temperatures. Note that the higher the temperature, the shorter the wavelength of peak emission.

You can think of a blackbody as an energy homogenizer. It can absorb energy from many wavelengths, heat up, and emit radiation that has a characteristic shape with no trace of lines.

A blackbody radiator has a number of special characteristics. First, a blackbody with a temperature greater than absolute zero emits *some* energy at *all* wavelengths. Second, a hotter blackbody emits *more* energy at *all* wavelengths than does a cooler one. Third, a hotter blackbody emits *a greater proportion* of its radiation at *shorter* wavelengths than a cooler one. Fourth, the amount of radiation emitted by the surface of a blackbody depends on the *fourth power* of its temperature.

We can express these last two points algebraically. The wavelength at which the energy output from a blackbody peaks is

$$\lambda_{max} = \frac{2.90 \times 10^{-3}}{T}$$

where λ_{max} is the wavelength (in meters), at which the peak output occurs and T is the temperature in Kelvins. This is called *Wien's law* (Fig. 6.25).

The amount of energy emitted for every square meter of a blackbody at temperature T Kelvins is

$$E = 5.67 \times 10^{-8} \, T^4 \; W/m^2$$

This is called the *Stefan-Boltzmann law*.

Let's examine the sun's continuous spectrum as measured at the earth's surface (Fig. 6.26). Note that the earth's atmosphere absorbs parts of the spectrum, especially the ultraviolet (absorbed by ozone) and the infrared (absorbed by water vapor and carbon dioxide). However, very little of the visible light is absorbed. The spectrum peaks at about

0.5 µm (5000 Å). Without the atmospheric absorption the sun's spectrum follows a more or less continuous curve with this one peak in the yellow part of the spectrum.

Now to use Wien's law to estimate the temperature of the sun. The sun's emission peaks at 4.6×10^{-7} m (0.46 µm or 4600 Å). So the corresponding temperature is

$$T = \frac{2.90 \times 10^{-3}}{\lambda_{max}}$$

$$= \frac{2.90 \times 10^{-3}}{4.6 \times 10^{-7}}$$

$$= 6300 \text{ K}$$

The Stefan-Boltzmann law can also be used to infer the sun's surface temperature. To do so, you need to find out how much energy each square meter of the surface of the sun puts out. You can determine this quantity from the sun's luminosity (3.86×10^{26} W) and radius (6.98×10^8 m). Since the surface area of a sphere is $A = 4\pi R^2$, the energy radiated per unit area is

$$E = \frac{L}{4\pi R^2}$$

$$= \frac{3.86 \times 10^{26} \text{ W}}{4(3.14)(6.96 \times 10^8 \text{ m}^2)}.$$

$$= 6.34 \times 10^7 \text{ W/m}^2$$

Using the Stefan-Boltzmann law

$$E = 5.67 \times 10^{-8} T^4$$

$$T^4 = \frac{6.34 \times 10^7}{5.67 \times 10^{-8}} = 1.12 \times 10^{15}$$

$$T = 5780 \text{ K}$$

The temperatures determined by these two methods are not the same because the sun is *not* a perfect blackbody.

Opacity

If the sun's continuous spectrum has nearly a blackbody shape, then the region emitting must be a good absorber of light. What does the absorbing?

Recall (Section 6.3) that atoms can absorb light by three basic processes: bound-bound, bound-free, and free-free transitions. Bound-bound transitions produce only lines; but bound-free and free-free transitions can contribute to a continuous spectrum.

When an atom becomes ionized, it may absorb energy at *any* wavelength less than the minimum necessary for ionization. Hydrogen in the ground state can absorb photons with wavelengths 912 Å and shorter; excited hydrogen with electrons in level 2 absorbs photons

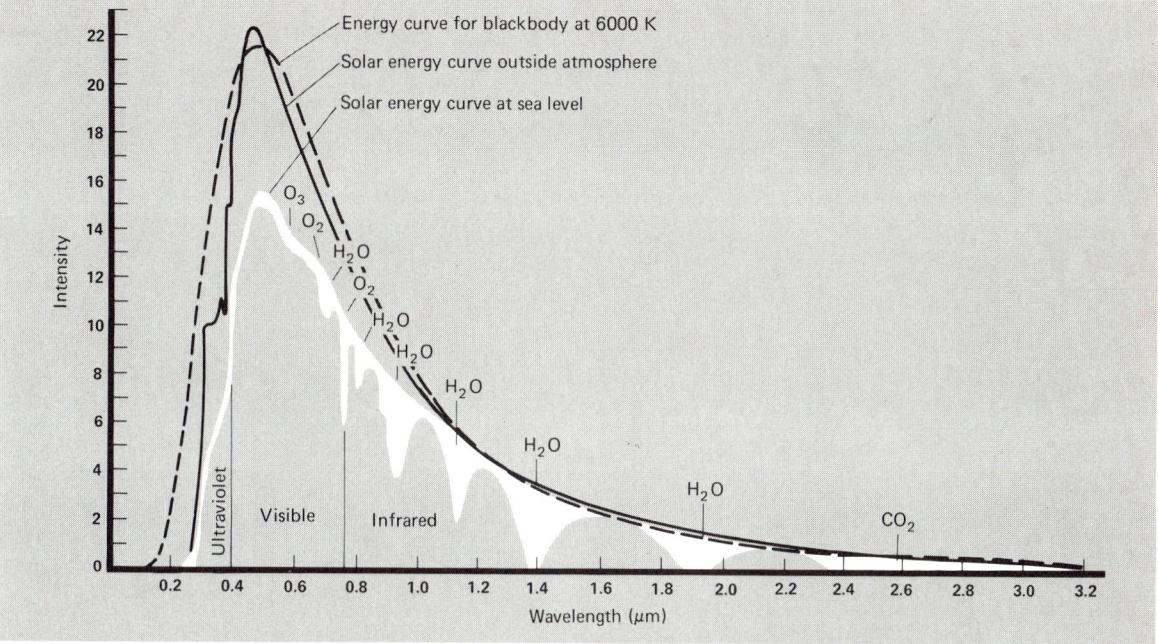

FIGURE 6.26 The absorption of the sun's radiation by the earth's atmosphere. The sun's radiation measured above the earth's atmosphere is indicated by the solid line. (Note how close this curve comes to a blackbody at 6000 K.) The shaded curve shows the actual measured spectrum at sea level. Indicated is whether the carbon dioxide (CO_2), water vapor (H_2O), oxygen (O_2), or ozone (O_3) are responsible for the absorption.

with wavelengths 3646 Å and shorter; in level 3, 8204 Å and shorter. Could these bound-free transitions from excited hydrogen plus free-free absorption from electrons absorb well enough to account for the blackbody spectrum? The answer is no. The excited hydrogen is there, but not enough of it to make the gas opaque.

To see this point, you need to understand the concept of *opacity*. On a clear day you can see many kilometers through the air. The air is transparent to light; its opacity is low. On a very foggy day you can see perhaps only a few meters. The air is opaque to light.

What makes a gas opaque? Interactions of light with atoms and electrons. When a photon is absorbed, it no longer exists and so can't carry energy any farther. The photon's energy is not destroyed; it has just been transferred to an electron (in a bound-bound, bound-free, or free-free transition). When the electron loses the energy, it emits a photon. But, and this is the key point, that photon can be emitted in *any* direction, including back in the direction from which it originated. It heads off and moves only a short distance before it is absorbed. When another photon is reemitted, it probably zips off in a different direction. So photons in an opaque gas travel very short distances before they are absorbed. In a gas of lower opacity, they travel greater distances.

So the opacity of a gas relates to how far photons can travel, on the average, between absorptions. The opacity depends on the density of the absorber and how effectively it absorbs. Although excited hydrogen atoms absorb fairly well, they are rare in the sun's atmosphere, only a few for every 10^8 atoms. The culprit for the sun's visible opacity is not neutral hydrogen, but a very strange type of hydrogen ion: a *negative hydrogen ion* (denoted as H^-). At a low enough temperature (roughly 6000 K) a hydrogen atom can acquire a second electron and still be stable. The proton binds two electrons, and the ion has a total negative charge. The second electron is not tightly bound—only 1.2×10^{-19} J will remove it. That energy corresponds to a wavelength of 16,000 Å (or 1.6 μm), which is infrared. So a negative hydrogen ion can absorb light (bound-free transition) at wavelengths of 16,000 Å or less, with a maximum absorption efficiency at roughly 8600 Å. The free electrons can then take part in free-free transitions. The absorption efficiency of these processes is so high that, even though only one H^- ion exists for every 10^8 H atoms in the sun, the opacity in the visible and infrared is very high. The visible surface of the sun is the layer where the gases become opaque. That layer, called the *photosphere*, is only 100 to 200 km thick and has a temperature of 5700 K.

The Solar Absorption-Line Spectrum

Recall (Section 6.3) that a spectroscope shows the solar spectrum to consist of a continuous spectrum crossed with dark lines. How is this spectrum produced?

To see how, let's use the concept of opacity for bound-bound transitions. Consider the transition that produces the 4383 Å line of iron (Fig. 6.27). When we measure the absorption or emission of this line, it has a finite width, centered on 4383 Å, where the iron atom absorbs very well. At somewhat shorter or longer wavelengths, say, 4379 Å or 4387 Å, the iron atom can still absorb light, but not as well. In other words, a gas containing iron has a much higher opacity for 4383 Å photons (at line center) than for those with wavelengths a few angstroms shorter or longer.

In the photosphere the opacity at the centers of lines traps the photons. They travel very short distances and so have little chance of escaping into space. The density of the photosphere drops rapidly from bottom to top. So the opacity decreases as fewer atoms per cubic meter are available to do the absorbing. Eventually the opacity falls enough for photons, even at the wavelengths of line centers, to escape.

As a result, absorption lines form at *different levels* in the photosphere. Take, as a specific example, the 4383 Å iron line (Fig. 6.27). At its center, where the opacity is high,

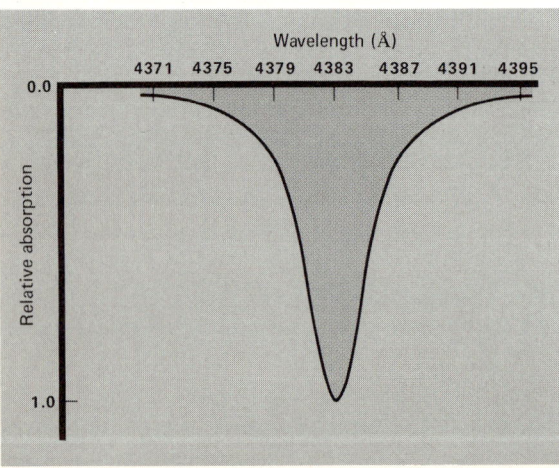

FIGURE 6.27 The relative absorption ability of the 4383 Å line of iron.

6.4 MESSAGES FROM SUNLIGHT

the light emerges from higher up in the photosphere, where the gas is cooler and emits less intensely. Off the line center, where the gas is more transparent, the light emerges from lower, hotter layers, which emit more intensely. So the spectrum is brighter away from the line center.

Note again that absorption lines are *not* perfectly black, containing *no* energy. They are only dark relative to continuous emission at neighboring wavelengths.

In a photosphere model (Table 6.1; one of the many available) the temperature rises quickly as you go down in the photosphere: a few thousand Kelvins in a few hundred kilometers. It is this sharp temperature change that results in the sun's dark-line spectrum. Also note that the photosphere's pressure amounts to about 10^{-2} that at the earth's surface, and its density is only 10^{-3} that of the earth's atmospheric density.

Astronomers have analyzed more than 20,000 lines in the solar spectrum to find the chemical composition of the atmosphere. The intensity of the dark lines from a particular chemical element relates to how much of that element is in the atmosphere. The line's intensity also depends on the temperature and density in the photosphere. Iron produces most of the absorption lines; others come from hydrogen, calcium, and sodium.

Identifying particular elements is a straightforward job—you compare the wavelengths of the solar lines with the wavelengths observed for various elements in the laboratory. But it's a harder task to find out an element's actual abundance. To determine it you need to know how atoms of a particular element absorb and emit light. Then you must make up a model of the sun's photosphere (such as Table 6.1). You can construct this model from theoretical calculations and basic physical concepts; it consists of a list of different temperatures, densities, and pressures at different depths in the photosphere. Then you add to the model some amount of the element, and you can calculate how intense its absorption lines must be. You try to match the calculated intensity to the observed intensity by changing the abundance. A match between observed and calculated intensities gives the correct abundance.

Warning: This procedure gives only the abundance in the *photosphere*; it does not tell directly the abundance in the *interior*.

TABLE 6.1 A Model of the Photosphere

Depth (km)	Temperature (K)	Density (kg/m^3)	Pressure (atm)
0	3054	4.6×10^{-6}	8.7×10^{-4}
56	4358	5.9×10^{-6}	1.6×10^{-3}
110	4448	1.0×10^{-5}	2.8×10^{-3}
206	4600	2.5×10^{-5}	7.0×10^{-3}
407	5069	1.4×10^{-5}	4.4×10^{-2}
500	5675	2.7×10^{-4}	9.5×10^{-2}
555	6723	3.4×10^{-4}	1.4×10^{-1}
605	8209	3.7×10^{-4}	1.9×10^{-1}

Note: The "top" of the photosphere, since it is a gas, is not sharply defined. Here the top is the lowest level from which all the visible light emitted at that level emerges without absorption.

Source: Theoretical calculations by R. L. Kurucz.

Of the 92 naturally occurring elements, so far 63 have been definitely identified in the solar spectrum. Most of the atmosphere's mass is hydrogen (78 percent) and helium (20 percent). All other elements (loosely called "metals" or "heavy elements" by astronomers) make up a mere 2 percent of the sun's mass.

The absence of an element's absorption lines in the visible spectrum is not a sure sign that it does not exist in the sun. Perhaps so little of the element is present that it does not produce detectable absorption lines. Or the strongest absorption lines may be in a region of the sun's spectrum unobservable through the earth's atmosphere. Still another possibility is that the temperature, pressure, and density of the sun's atmosphere may inhibit the formation of the element's spectral lines. If the conditions are too hot or too cool, the lines do not form.

Helium is a good example. You won't find any absorption lines of helium in the sun's visible spectrum. Why not? In the photosphere the temperature is too low to excite helium to levels which can absorb photons of visible wavelength. Above the photosphere the temperature rises to 40,000 K. Here the temperature is high enough to excite helium atoms, and during an eclipse you can find bright lines from helium in the spectrum of this high atmospheric region (the chromosphere). In fact, helium (from the Greek word *helios*, "sun") was discovered in the sun's spectrum before it was found on earth. However, there are too few excited helium atoms in this hot region to produce emission lines strong enough to be seen directly in contrast to the bright continuous spectrum.

6.5 THE QUIET SUN

The designation "quiet" relates to seemingly placid regions on the face of the sun. We use this word to describe those solar phenomena that are characteristic of large regions for long periods of time; for instance, the steady flow of energy out of the sun. This energy flow, from core to surface and beyond the atmosphere, controls the environment of the quiet sun.

The quiet sun that we can see directly consists of the sun's outer layers, together known as the atmosphere. Because the sun's atmosphere is a gas, it does not have distinct layers with sharp boundaries, but there are three substantially different zones: the photosphere, chromosphere, and corona.

Photosphere

The sun's photosphere has a granular structure (Fig. 6.28). The surface has a cobbled appearance, an individual "cobblestone" having an irregular shape about 1800 km across. (This is half the size of the United States!) Each granule has a lifetime of about 8 to 10 minutes (Table 6.2). In a time-lapse motion picture sequence the granulation looks like the top of a bubbling pot of oatmeal. You can visualize the photospheric granulation as the top layer of a seething zone where hot blobs of gas spurt to the surface, radiate energy, cool, and flow downward. The boiling photosphere has a region of convection just below it. Here the outward flow of energy heats the gas and makes it boil. The base of the convection zone is thought to lie at about 0.8 solar radius, though it may be as deep as 0.6 solar radius.

One curious and newly discovered fact about the photosphere is that it pulsates in

6.5 THE QUIET SUN

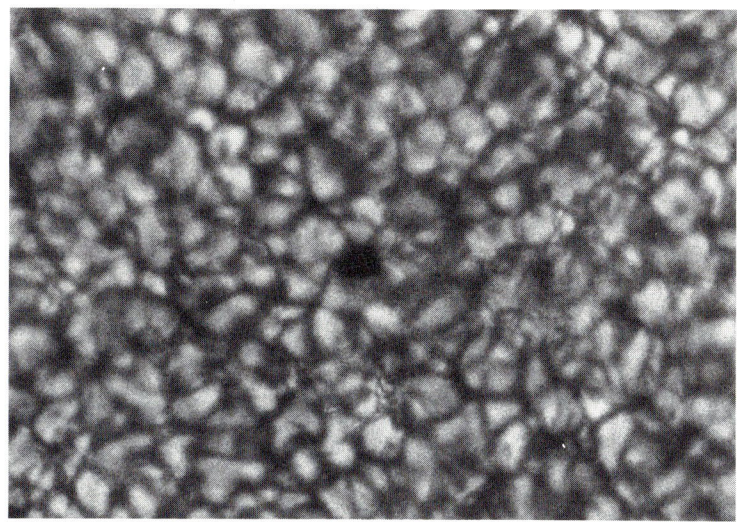

FIGURE 6.28 The granular structure of the photosphere. An individual granule is about 1500 km across. At the picture's center is an incipient sunspot. [*Courtesy Sacramento Peak Observatory, Association of Universities for Research in Astronomy, Inc.*]

waves with a period of about 5 min. In this time the photospheric gases rhythmically rise and fall some hundreds of kilometers over areas 2000 to 3000 km in diameter. These pulsations, like a solar heartbeat, are thought to be caused by low-frequency sound waves generated in the convection zone.

Chromosphere

Just before and after totality in a solar eclipse, a bright, pink flash appears above the edge of the photosphere. This is the chromosphere, the solar atmosphere just above the photosphere. The pink color comes from the emission of the first line of the Balmer series (Section 6.3), called H-alpha, in the red region of the spectrum (Plate 2).

A spectroscope directed at the chromosphere during its fleeting appearance shows a bright-line spectrum (Fig. 6.29). The temperature, density, and pressure in the chromosphere determine the intensities of its emission lines, so the line intensities provide clues to the physical conditions there.

The chromosphere begins just above the photosphere and extends only about 2500 km higher, where it merges into the corona. It is about 1000 times less dense than the photosphere but, surprisingly, gets much hotter. The temperature rises from 4300 K to

TABLE 6.2 Solar Granulation Characteristics

Characteristic	*Value*
Average intergranular spacing	1800 km
Dark lane width between granules	~350 km
Average granule lifetime	10 min
Temperature variation across granule	~100 K
Velocity of convective motion	~0.3 km/sec

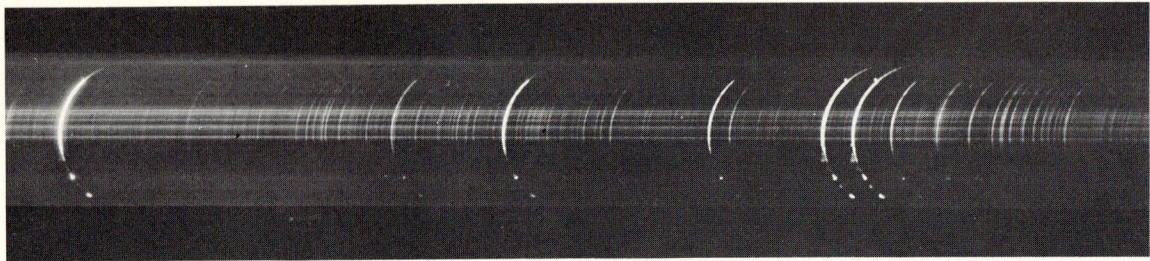

FIGURE 6.29 The emission spectrum of the chromosphere, visible just before total eclipse. The bright spectral lines are curved because the curved sliver of the chromosphere acts like a curved slit. [*Courtesy Mt. Wilson Observatory, Carnegie Institution of Washington.*]

FIGURE 6.30 Jets of hot gas along the edge of the sun. These jets, called spicules, transport material and energy from the photosphere to the chromosphere. [*Courtesy Mt. Wilson Observatory, Carnegie Institution of Washington.*]

10^6 K in the 2500 km of the chromosphere that extends above the photosphere. These high temperatures produce the emission lines from this region.

Why don't we see emission lines from the chromosphere when we look through it down to the photosphere? The chromosphere has such a low density that it is essentially transparent to the light passing through it. The photospheric spectrum (continuous and absorption lines) passes through the chromosphere, which adds only a little emission.

Why is the chromosphere so hot compared with the photosphere? A clue comes from observing it over the edge of the moon just before totality. Large spears of gas called *spicules* (Fig. 6.30) constantly pierce the chromosphere. These jets of hot gas can spurt up to heights of 10,000 km and fade away again in several minutes. They have diameters of about 1000 km. Their temperatures are about 8000 K. Because of the fountains of spicules, the chromosphere is not a uniform layer. The chromosphere ends jaggedly at the tops of the spicules.

The presence of spicules suggests the reason the chromosphere is hotter than the photosphere. The key here may be energy carried by sound waves—regions where the gas

6.5 THE QUIET SUN

is compressed alternating with regions where it is rarefied. These sound waves are generated in the convective zone in the sun's interior.

Convection causes this turbulence as it pumps material up through the photosphere. One model of how the chromosphere gets so hot views the chromosphere as the froth of the churning photosphere. Just as the spray whipped off an ocean wave travels faster than its parent wave, so the material spurting from the photosphere breaks away at high speeds, in fact, at supersonic speeds. (*Supersonic* means that the waves travel faster than sound waves move in the same environment.) This supersonic motion generates noise (sound waves), which carry energy. They give up their energy to the chromosphere when they crash into it. The energy they lose heats up the chromosphere so that most of the chromosphere is hotter than the photosphere. (*Note*: Not all solar astronomers accept this explanation, but many feel that it accounts for at least part of the heating of the chromosphere.)

Corona

You can see the sun's splendid corona directly during a total solar eclipse (Plate 3). Although almost as bright as the full moon, the corona is normally obscured by the sunlight scattered in the earth's atmosphere. During a total eclipse, the sky becomes dark enough for the corona to be visible. Stars and the bright planets can be seen near and in it.

Spectroscopic examination has found that the corona's spectrum contains bright lines. These were a mystery for many years. Some astronomers attributed the lines to an element, called coronium, not found on the earth. However, in 1940 the Swedish scientist Bengt Edlén demonstrated that highly ionized atoms of iron, nickel, neon, and calcium, rather than a strange new element, produced the emission lines. (Recall, Section 6.3, that when an atom loses an electron, its energy levels change.) Because it takes large amounts of energy to rip many electrons off an atom, the corona must be very hot. For example, lines of an iron ion with only 10 electrons are seen in the coronal spectrum. To strip iron of 16 of its normal 26 electrons requires a temperature greater than 2 million K. Although Edlén solved the problem of coronium, his solution presented another question: What makes the corona so hot?

Satellites such as Skylab that have observed the sun have provided some evidence for an understanding of the coronal heating. Most of the corona's emission lines arise in the ultraviolet region of the spectrum and so are not visible from the earth's surface. Recall that it takes a certain amount of energy and so a certain temperature to form ions of various elements (Section 6.3). Each ion of an element can serve as a thermometer to indicate the temperature in the region from which emission lines of that ion arise. These thermometers indicate that at the zone between the chromosphere and the corona, the temperature jumps sharply—roughly 500,000 K in just 300 km (Fig. 6.31). This abrupt temperature increase resembles the increase in the zone between the photosphere and chromosphere. So it may result from the same physical process: transport of energy by sound waves from the chromosphere to the corona.

This model of sound waves generated by the convection zone heating the sun's chromosphere and corona has recently hit a snag. It predicts that stars with substantial convection zones should have coronas; those without, should not. Because the sun's corona is so hot, it emits X rays; other stars with coronas should also emit X rays intensely. Those without will not. Observations by the Einstein Observatory, an X-ray telescope (Chapter

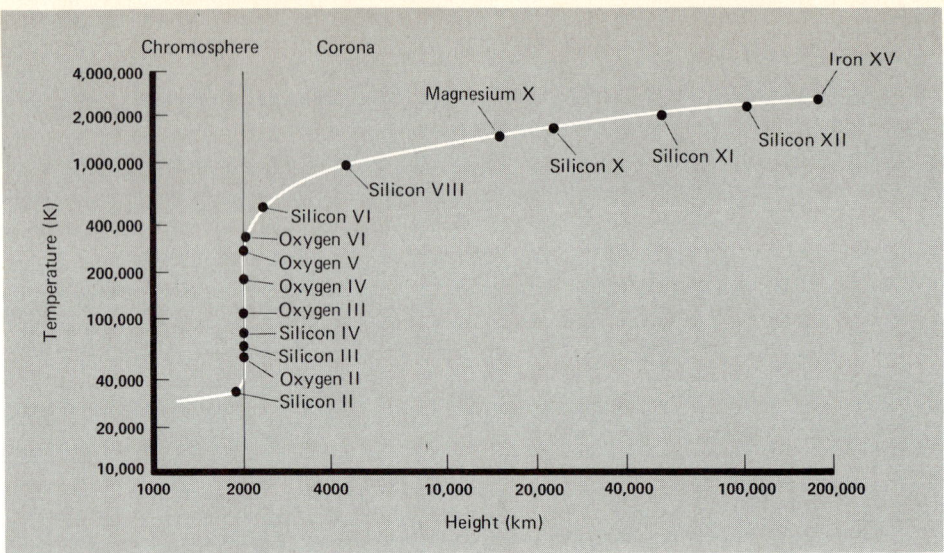

FIGURE 6.31 Ions used to flag temperatures in the corona. Since the corona is very hot, atoms in it are highly ionized. For a given element, it takes higher temperatures to ionize it more. So how much an element is ionized indicates the temperature where it is found. Stages of ionization are marked by a Roman numeral next to the element's name; the numeral II, for example, denotes that the atom has lost *one* electron. Note that the corona's temperature increases rapidly above the chromosphere and then levels out to about 2 million K. [*Data from Skylab.*]

7), show that *all* types of stars emit X rays strongly and so have hot coronas, even those stars that were not thought to have substantial convective zones.

Solar Wind

The corona does not end suddenly but extends into space to distances greater than indicated by its visual appearance. The corona moves; its particles continually flow out from the sun. Because of its high temperature, the coronal gas exerts an outward pressure greater than the inward pull of gravity. As a result, the gas from the corona streams away from the sun. Closer to the sun the expansion is slow because the sun's gravity is stronger. With increasing distance the flow speeds up because the gravity decreases. The gas stream becomes the solar wind, so named because of its high velocity. At the earth's orbit the solar wind whips by at roughly 450 km/sec. The particles in the solar wind (mostly protons and electrons, with a number density of 10 particles per cubic centimeter) travel from the sun to the earth in about 5 days. (In contrast, light takes 8.3 min.) The earth swims through the solar spray and catches some of the particles in its magnetic field. The sun does not lose much mass in the solar wind. The solar wind blows out about 10^{-14} solar mass per year; that's only 10^{-8} an earth mass in a year.

6.6 ENERGY FROM THE SOLAR INTERIOR

The interior of the sun contains the bulk of the sun's mass and the furnace that generates its energy. All the features of the quiet sun, from photosphere to solar wind, result from

6.6 ENERGY FROM THE SOLAR INTERIOR

the energy flow from the interior regions where it is produced. How does the sun generate its energy?

Energy Sources

During the nineteenth century, when the earth was found to be billions of years in age, the source of the sun's power became an embarrassingly difficult puzzle to explain. The crux of the problem is not the rate of energy production, about 4×10^{23} kW, but its longevity. The sun has been roughly the same luminosity for at least 3.5 billion years. How do we know? Geologists have found rocks containing fossils of living organisms which are at least that old. For life to exist, the earth must have been warm, and therefore the sun must have been roughly at its present luminosity.

The three major classes of energy source possible are chemical energy, gravitational energy, and nuclear energy. Energy from ordinary chemical reactions, such as burning (oxidation), could not provide the amount necessary. If the sun were composed entirely of oxygen and coal, it would have burned to a dark cinder in a few tens of thousands of years. Obviously, the sun could not be a coal-burning furnace!

In the middle of the nineteenth century Hermann von Helmholtz (1821–1890) and William Thomson, Lord Kelvin (1842–1907), proposed that the sun shone because it was releasing gravitational energy by shrinking, that is, gravitational contraction converted gravitational potential energy to radiative energy.

To understand this process of energy conversion, consider a ball held above the earth's surface (Fig. 6.32). Its velocity is zero; it has no kinetic energy. But it does have potential energy. Release the ball. As it falls, its velocity increases, so its kinetic energy increases. In fact, the more it falls, the more its velocity increases and the greater its kinetic energy.

Instead of a single ball, picture a cloud of gas that contains a large number of atoms (Fig. 6.33). Each can be thought of as the ball above. Imagine the cloud contracting gravitationally from the combined forces of all the atoms in it. As the cloud contracts, the atoms gain velocity. Though the velocities might initially all be directed generally inward, collisions among the atoms will soon distribute the velocities in random directions. So the atoms' kinetic energy of random motion increases. Because temperature is a measure of the average kinetic energy of random motion of the atoms (Section 6.1), the temperature of the gas increases. Meanwhile, the density is also increasing as the atoms come together.

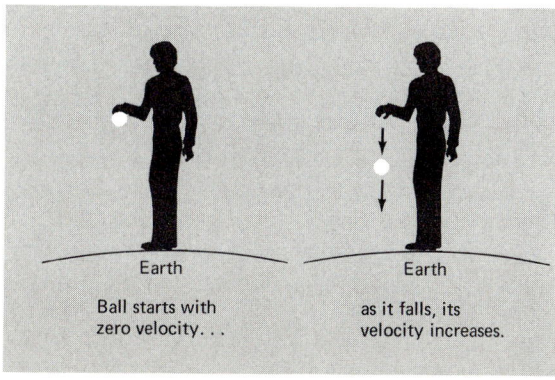

FIGURE 6.32 Converting gravitational energy into kinetic energy by dropping a ball.

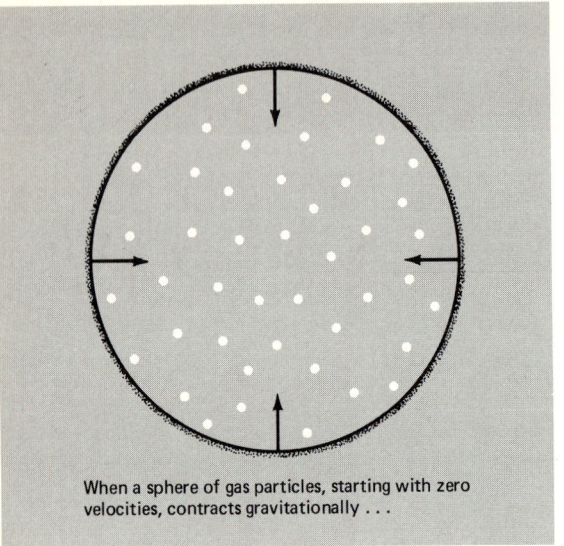 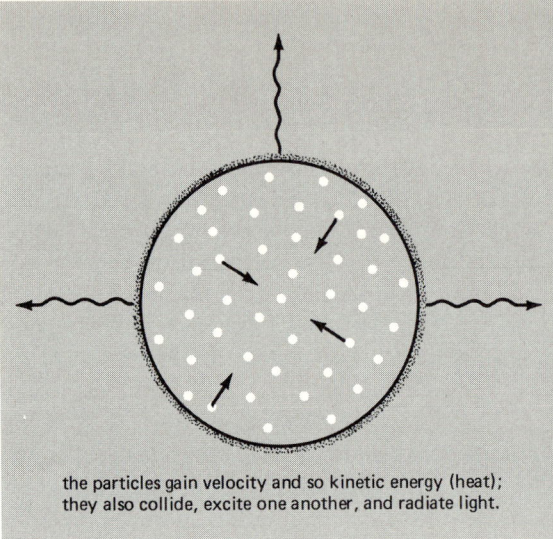

When a sphere of gas particles, starting with zero velocities, contracts gravitationally . . . the particles gain velocity and so kinetic energy (heat); they also collide, excite one another, and radiate light.

FIGURE 6.33 Heat and light produced by gravitational collapse of a cloud of gas.

With this increase in the velocity of the particles and the density of the gas, the number of atomic collisions increases. These collisions excite some atoms. When the excited electrons drop to a lower energy, radiation is emitted. The net result of gravitational contraction is that some of the initial gravitational potential energy is converted to thermal energy and some is emitted as radiation. In fact, exactly one-half goes into thermal energy and the other half ends up as radiative energy.

Note: This concept of the natural conversion of gravitational potential energy to heat and light by gravitational contraction is very important astronomically. Keep it in mind, as it will crop up again.

Because of the sun's substantial mass, a contraction rate of only 40 m/yr would liberate the required energy. The gravitational energy stored in the sun would last for about 20 million years, far longer than the earth's age as determined by the early geologists. However, under the impact of Darwin's theory of biological evolution (Chapter 25), later geological investigators expanded the earth's age, so the sun's resources of gravitational energy were no longer sufficient to account for the sun's shining.

Nuclear Transformations

Albert Einstein provided the key idea about the sun's energy at the beginning of this century. Grappling with the fundamental nature of electromagnetic waves, he demonstrated that mass and energy are related by the equation

$$E = mc^2$$

where E is the energy (in joules) released in the conversion of mass m (in kilograms), and c is the speed of light (in meters per second). Because c^2 is a large number (about 9×10^{16}), a minute mass stores enormous energy. For example, the conversion of 1000 kg of matter

6.6 ENERGY FROM THE SOLAR INTERIOR

into energy unleashes about 10^{13} kWh (9×10^{19} J), roughly equal to the total energy consumption of the United States in a year!

How to change matter into energy? Two operations in nature unleash the energy frozen in the nucleus of an atom: *nuclear fission* and *nuclear fusion*. In the process of nuclear fission a nucleus of a heavy element (such as uranium or plutonium) splits into two lighter nuclei. The mass of the remnants adds up to less than that of the original nucleus. The deficit in mass is released as energy. In nuclear fusion the nuclei of lighter elements are fused together to create a nucleus of a heavier element. However, the product has less mass than the combined mass of the original particles. This mass has been converted to energy.

It is unlikely that nuclear fission contributes to the solar luminosity because the sun has a small percentage of heavy elements. If fusion occurs, what atomic nuclei are involved? The natural candidates must meet stringent requirements. (1) Because like electric charges repel, and the greater the charge the greater the repulsive force, nuclei with small charges are more apt to combine. (2) An adequate supply of the nuclei that serve as fuel must be available. Hydrogen is the most abundant element with the smallest nuclear charge (one proton). The fusion of hydrogen nuclei results in the production of helium. To make a helium nucleus from hydrogen nuclei releases 4.2×10^{-12} J. That's a minuscule amount; it would raise the temperature of 1 g of water only 10^{-12} K.

Two sets of fusion reactions are possible processes for the transmutation of hydrogen to helium: the *proton-proton chain* (PP chain for short) and the *carbon-nitrogen-oxygen cycle* (CNO cycle). The latter is a minor contributor to the energy of the sun, but it is an important source in more massive stars (Chapter 8). Let's look at the steps in the primary proton-proton reaction. (There are two other PP reactions which we will describe later in Fig. 6.36.) A collision between two protons initiates it (Fig 6.34). If these nuclei collide with enough energy (a temperature of at least 8 million K), the protons stick together. A

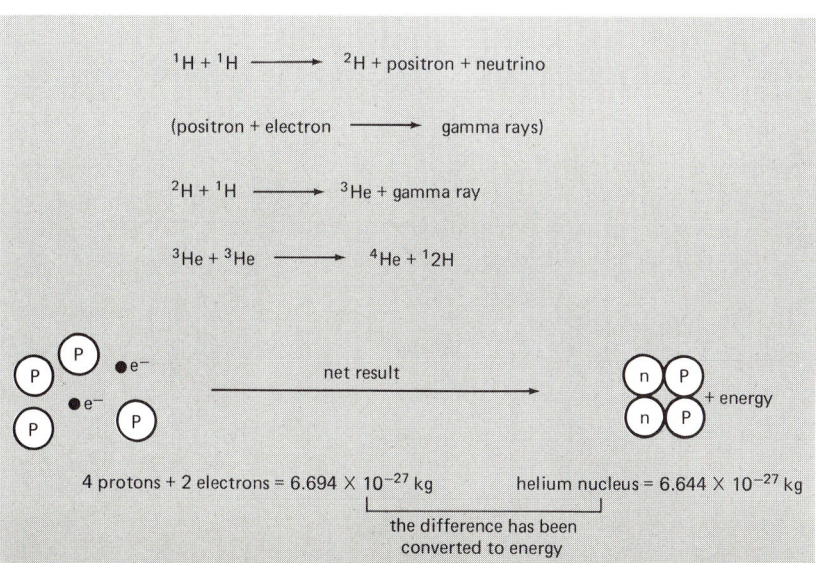

FIGURE 6.34 The primary proton-proton reaction (PPI). The net result is the formation of one helium nucleus that has a mass of about 5×10^{-24} kg *less* than the unbound particles that go into the reaction. The "missing" mass has been released as energy.

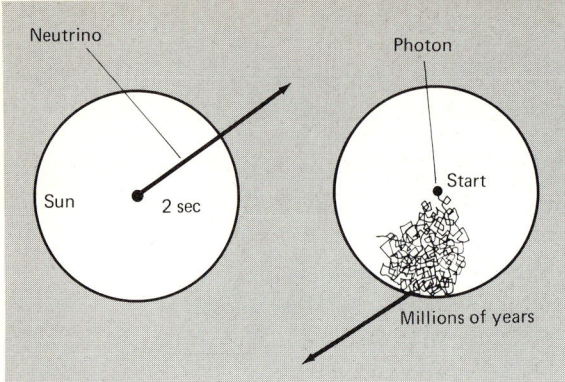

PPII

$$^1H + {}^1H \to {}^2H + e^+ + \text{neutrino}$$
$$^1H + {}^2H \to {}^3He + \text{gamma ray}$$
$$^3He + {}^4He \to {}^7Be + \text{gamma ray}$$
$$^7Be + e^- \to {}^7Li + \text{neutrino}$$
$$^7Li + {}^1H \to 2\,{}^4He$$

Note: Happens 8% of the time in the sun.

PPIII: First three steps same as PPII

$$^1H + {}^7Be \to {}^8B + \text{gamma ray}$$
$$^8B \to {}^8Be + e^+ + \text{neutrino}$$
$$^8Be \to 2\,{}^4He$$

Note: Happens 1% of the time in the sun.

FIGURE 6.35 The paths of neutrinos (left) and photons (right) from the core of the sun. A neutrino finds the sun completely transparent and zips straight out in about 2 seconds. In contrast, photons produced in the core find the sun opaque and take millions of years to make it from the core to the surface.

FIGURE 6.36 Other proton-proton reactions (PPII and PPIII) that are thought to take place in the sun's core. (Be is beryllium; Li, lithium).

heavy hydrogen nucleus (^2H) forms, consisting of a proton and a neutron. The other positive charge breaks away as a positron. (A *positron* is the antiparticle of the electron; it has the same mass, but carries a positive instead of a negative charge.) Almost lost in the shuffle is a neutral, supposedly massless particle called a *neutrino*, which zips away at the velocity of light. The dense solar interior offers no barrier to the neutrino's escape, and in about 2 sec the neutrino breaks into space carrying away some energy.

In the solar interior the positron quickly collides with an electron, and the two antiparticles are annihilated to form two gamma rays. Meanwhile the heavy hydrogen crashes into another proton and forms light helium (^3He) and another gamma ray. If this chain occurs again, another light helium is created, and it collides with the one previously formed. In this final reaction of the PP chain, the usual result is an ordinary helium nucleus (^4He) plus two protons and the release of energy as heat.

Keep in mind that very high temperatures and high densities are needed for protons to collide with enough energy to fuse and to collide frequently enough to generate as much energy as comes out of the sun. This requirement restricts the energy production to the sun's *core*, about the inner 25 percent of its radius. Only here is the temperature at least 8 million K. The density ranges from about 160,000 kg/m^3 at the center to about 20,000 kg/m^3 at the core's edge.

In summary, the input for the PP chain is six protons and two electrons, and the usual output is one helium nucleus, two protons, two neutrinos, and miscellaneous gamma rays. The gamma rays are absorbed by the surrounding gas and converted to heat. Any kinetic energy given to the various particles also ends up as heat through collisions with other particles of the gas. The net result of this nuclear cooking is the creation of helium and energy from four protons and two electrons. Each completed PP chain unlocks about 4.2×10^{-12} J. To account for the sun's luminosity, 1.4×10^{17} kg of matter must be converted to energy each year. The neutrinos fly off into space with about 2 percent of the energy; but the rest of the energy goes into heat. The energy flows to the surface primarily by radiation. In the center, where the temperature is high, the radiation is mainly at the wavelength of gamma rays. The gamma ray photons find the sun's interior opaque (Fig. 6.35). The photons bounce along random paths as they are absorbed and reemitted.

Slowly, the photons move out toward the sun's surface, from regions of higher to lower temperatures. As a consequence of this temperature decline, the average photon's energy declines. The original gamma rays are degraded by interactions with the sun's material into lower energy photons. In a few million years the photons break out of the photosphere in the form of less energetic, visible radiation.

The sun is a fusion furnace, forging helium from hydrogen in its core. The energy you see now was produced there many millions of years ago, when our ancestors first walked the earth.

How long can the sun survive at this rate? The sun has enormous hydrogen supplies—78 percent of its mass. However, only the hydrogen close to the core can burn. This is about 10 percent of the total. Also, the PP chain transforms only 0.7 percent of the mass into radiant energy. Even with these restrictions the sun's fusion energy can last about 10 billion years. Certainly we do not have to worry about a decline in solar power in the near future, for the sun is only a middle-aged star.

Solar Neutrino Experiment

How to probe the sun's interior when we can't see below its surface? We have some idea of the interior conditions because astrophysicists can make theoretical models of the sun. We can do this by applying simple physical laws to a ball of hot gas at whose center proton-proton reactions take place. Electronic computers calculate these models, which match the observed characteristics of the sun. The models show, for example, how the sun's temperature increases from edge to center, and how much the chemical composition in the core differs from that of the surface. (It differs because in the core the hydrogen is fused to helium while at the surface no such conversion takes place.)

Although these models of the sun's interior have been constructed as carefully as possible, an ongoing experiment casts some doubt on them. Remember the neutrinos produced in the PP chain? These come directly out of the sun's core and in about 8.3 min reach the earth. So if we could detect these neutrinos, we could "see" into the sun's interior. Unfortunately, the neutrinos in the primary PP reaction (PP I) do not have enough energy to be easily detected by present means. This reaction occurs about 91 percent of the time in the sun. The rest of the energy is thought to come from two other PP reactions (Fig. 6.36), which produce neutrinos with enough energy to be detectable. These two other reactions (PP II and PP III) are supposed to take place about 9 percent of the time.

Raymond Davis and his colleagues have developed a strange telescope to catch the neutrinos that are supposed to come from the sun's core (Fig 6.37). This telescope consists of about 378,000 liters of a chlorine compound (C_2Cl_4) placed in a huge tank located about 1.6 km below the earth's surface in Lead, South Dakota. A chlorine atom can absorb a neutrino and change to radioactive argon: $^{37}Cl + \nu = ^{37}A + e^-$. By a very delicate procedure the argon gas is flushed out of the tank and the amount of radioactivity measured, and so the number of neutrinos captured in a given time span is known.

From nuclear theory and models of the solar interior we can predict how many argon atoms should be produced from neutrinos each day. Then we can compare the prediction with the actual experimental results. To date the detected solar neutrinos amount to at most *one-third* the number predicted from standard models of the sun's interior and our current knowledge of the nuclear reactions involved! To put the result another way, the

FIGURE 6.37 The solar neutrino telescope. [*Courtesy Brookhaven National Laboratory.*]

experiment implies that some other source of energy besides PP reactions may be needed to account for the sun's present luminosity.

What's wrong? First, the experimental equipment may contain unknown problems. The chlorine capture experiment is the only one to have run for a long time. Its uniqueness demands results from another, independent experiment.

Second, the standard solar model may be incorrect. For example, solar models made with much lower abundances of heavy elements produce far fewer neutrinos, but they contradict our general understanding of the manufacture of elements in stars. Solar models can be fixed to agree with the experiment, but then they contradict other astrophysical data or the earth's climatic history.

Third, the experiment may be telling us that something is wrong with our understanding of the properties of neutrinos. For example, some models of particle physics suggest that the neutrino may actually have a small, but finite mass. If it does, then these models predict that the neutrinos produced in the sun will change into other types of neutrinos as they travel to the earth. There are three types of neutrinos known, and the type detectable by the chlorine experiment would exist only one-third of the time. If these models are correct, they would neatly account for the factor of three discrepancies in the experiment. Among all proposals, however, no really satisfactory explanation exists.

The result poses a great puzzle in contemporary astrophysics. The large discrepancy between theory and observation means that we lack an understanding of some critical part of the reasoning chain. When the solution is found, we will know a little bit more than we do now about the sun's interior and also about the interiors of other stars.

Energy Transport in the Solar Interior

Thermonuclear fires blaze in the sun's center, creating energy. Eventually that energy makes it to the earth. How does the energy get transported from deep in the sun to us? In general, energy is carried by the processes of *conduction*, *convection*, or *radiation*.

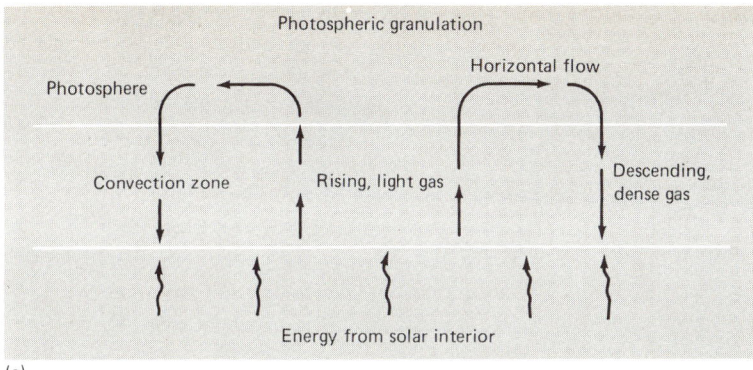
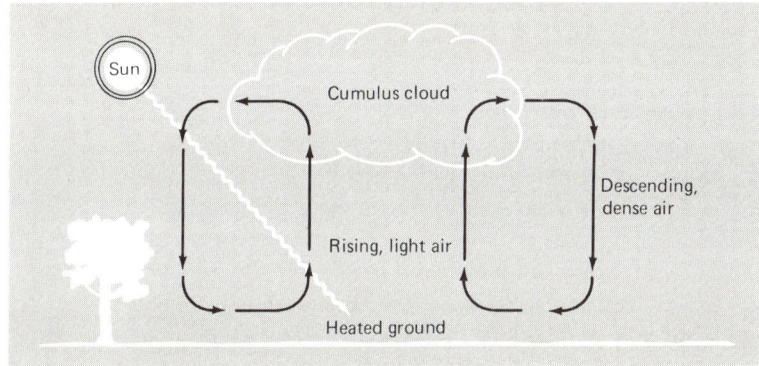

FIGURE 6.38 Flow of gas in the convective zone of the sun (a). Hotter, less dense gas rises to the top of the photosphere. There it radiates energy, cools off, and descends. The same process makes cumulus clouds on the earth (b).

When you relax in front of a roaring fire, you experience all three. You are directly warmed by the fire's heat, infrared radiation that travels directly from the fire to you to be absorbed by your skin. That's energy transported by radiation. Much of the energy from the fire, however, is wasted up the chimney. The fire heats the air just around it. Air is gas, and the air expands and becomes less dense. Cooler, denser air flows in and pushes the hotter air up—up and out of the house, where it cools off. This transport of energy by mass motions of a gas (or liquid) is convection. Finally, you may have by mistake left the poker in the fire. If you grab the handle without thinking, you yell upon finding it very hot. The poker's atoms that were actually in the fire were heated and so moved around at high speeds. They banged into their neighbors, agitating them. These collided with their neighbors, and so on throughout the poker from one end to the other. The transport of thermal energy by direct collisions of atoms within a gas, liquid, or solid is conduction. Since conduction depends on direct collision, the denser the material, the greater the conduction.

In ordinary stars like the sun, radiation or convection can carry energy along. Conduction does not play an important role because the sun's material is a gas and not very dense. (However, conduction does become important in extremely dense stars.) Local conditions determine which of the other two processes is at work. The general rule is that the transport process that can work most efficiently is the one that does operate. For most of the sun's interior, about 0.8 of the radius, energy flows most efficiently by radiation. Only in the outer 0.2 of the radius does convection come into play. We see the top of this convective zone as the turmoil of the photosphere (Fig. 6.38). The hot gas bubbling up here radiates out to space.

For a large part of the sun's radius, energy is carried by radiation. Why does the situation change? How well radiation transports energy through a material depends on its opacity. If the opacity is too high, radiative transport is inefficient, and convection takes over. In the sun's core the gas is fairly opaque (photons travel about 1 mm between absorptions), but radiation is still more efficient than convection for transporting the energy. As photons journey toward the surface, they run into a region where the temperature is low enough so hydrogen atoms form. These suddenly make the gas so extremely opaque that convection can transport energy more efficiently, and a convection zone forms.

6.7 THE ACTIVE SUN

In contrast to the easy-going processes of the quiet sun are localized, short-lived phenomena on or near the solar surface. The most important of these are sunspots, prominences, and flares. In general, these areas of solar activity are called *active regions*. Active sun phenomena are all associated with locally intense solar magnetic fields (Focus 6.2). Another solar property that contributes to them is the sun's nonuniform rotation. Because the sun is a gas, different parts can spin at different speeds; it spins fastest at the equator (1 rotation every 25 days) and slowest at the poles (1 rotation every 31 days). So as you travel from pole toward equator on the sun, the photosphere's velocity increases. This differential rotation distorts solar magnetic field lines and promotes the development of active regions.

Sunspots

If you view the sun through even a small, properly filtered telescope, you can see sunspots as dark blotches on the sun (Fig. 6.39). But they are not completely black. With a temperature of about 4200 K a sunspot is relatively cooler than the photosphere and so appears dark in contrast. Recall that the energy emitted from every square meter of the surface of a blackbody is proportional to the fourth power of the temperature ($E \propto T^4$). So

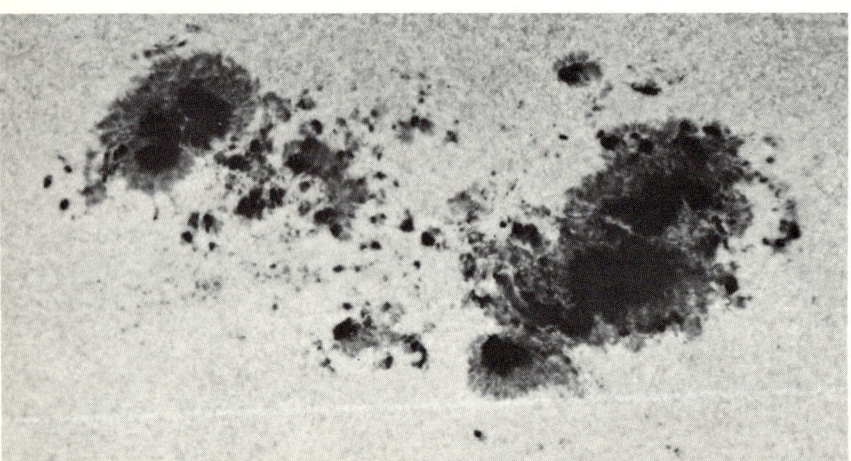

FIGURE 6.39 A large sunspot group, containing hundreds of sunspots. [*Courtesy Mt. Wilson Observatory, Carnegie Institution of Washington.*]

6.7 THE ACTIVE SUN

the brightness of a sunspot at 4200 K compared to that of the photosphere at 5800 K is $(4500/5800)^4 = 0.27$. The sunspot is almost four times fainter than the photosphere. If the sun could be whisked away, leaving its sunspots behind, you would see them shine with a bright orange glow against the darkness of space.

Sunspots have a strong tendency to form in groups. They are born in an active region where photospheric granules separate, and a tiny spot appears between them as a dark pore (see Fig. 6.28). Such pores have magnetic field strengths of roughly 2500 gauss (the magnetic field at the surface of the earth is about $\frac{1}{2}$ gauss). Usually more soon become visible and coalesce over a period of several hours, and a sunspot is formed. A large, single sunspot group may contain 100 individual spots and persist for two or more solar rotations.

Sunspot Cycle

In 1843 Heinrich Schwabe, a German amateur astronomer, noted that the variations in sunspot numbers are periodic. About 11 years pass between sunspot *maxima* (times of greatest numbers of sunspots) or sunspot *minima* (times of least numbers of sunspots). Later the British astronomer E. W. Maunder (1851–1928) discovered that not only do sunspot numbers vary during a cycle, but also that the sunspots' positions change. Sunspots usually appear only in the zone between the solar equator and 35° north or south latitude. At the start of a sunspot cycle a few spots emerge at high latitudes. As the cycle progresses, the sunspot zone migrates down toward the equator. As the survivors of one cycle expire near the equator (about 5° north or south latitude), new spots from the next cycle blossom at the higher latitudes.

In 1908 George Ellery Hale (1868–1938) detected intense magnetic fields associated with sunspots. The strongest magnetic sunspots have field strengths over 4000 times as great as the sun's normal field, that is, about 4000 gauss. Hale also noticed that sunspot groups contain spots of opposite magnetic polarity. For example, the leading spot of a twin group might be north polarity, the trailing spot south polarity. If this arrangement is true at a given time in the Northern Hemisphere of the sun, the situation in the Southern Hemisphere at the same time is reversed (Fig. 6.40). In 1913 Hale and his research

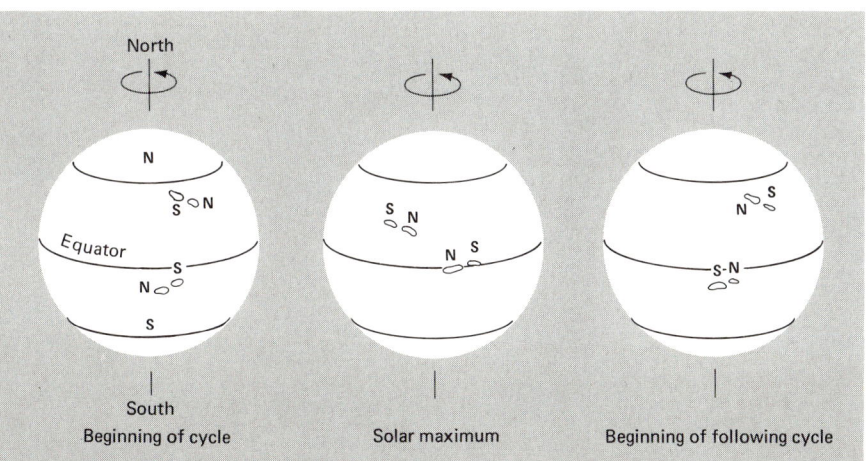

FIGURE 6.40 The magnetic cycle for sunspots. At the beginning of a sunspot cycle if spots with north polarity are leading groups (to the west) in the northern hemisphere, then south polarity leads groups in the southern hemisphere. At the beginning of the next cycle the polarity of the leading spots is reversed for each hemisphere.

FOCUS 6.2 MAGNETIC FIELDS AND FORCES

A field is another scientific model; it is a way of describing space that is somehow modified by the presence of matter. Magnetic fields are regions of space modified by electrical charges.

All magnetic fields come from electrical charges in motion. A bar magnet (Fig. F.8a) has no outward sign of motion, but the circulation of electrons around iron nuclei sets up its magnetic field (Fig. F.8b). You have probably seen an electromagnet in operation. Here an electric current, a flow of charged particles, passing through a loop of wire creates a magnetic field (Fig. F.8c). The earth's magnetic field is thought to be generated by the circulation of charged particles in its core. The magnetic fields around the earth and the sun have two poles, and so are called *dipole fields;* this characteristic allows us to think of these fields as arising from giant bar magnets buried in the earth and the sun (Fig. F.9), even though they really come from circulating electric currents.

To help visualize a magnetic field, take a very small compass and move it around a magnet (Fig. F.8). The changing direction of the compass needle shows the direction of

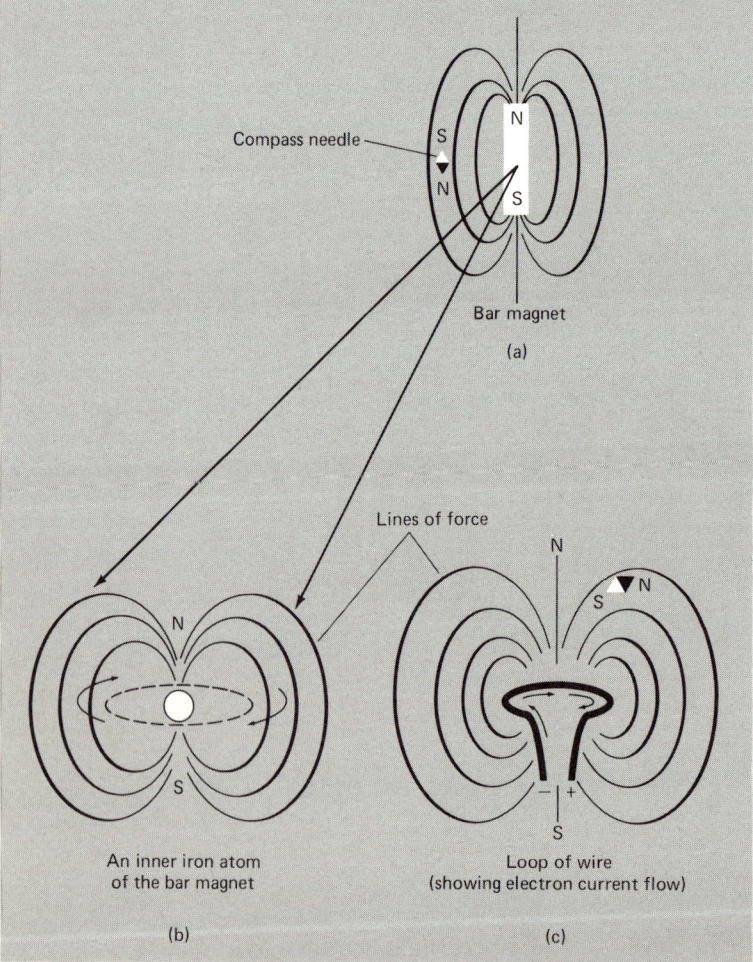

FIGURE F.8 The similarities behavior of magnets. At top (a) simple bar magnet; a small com needle (essentially another magn shows the direction of the magn field lines. These are produced flow of electrons around the nu iron atoms (b). The flow of elec in a wire (an electric current) m an electromagnet (c).

the magnetic lines of force. We usually draw the lines of force so that the spacing of the lines indicates the relative strength of the magnetic field: the closer the spacing, the stronger the field. Note that the farther away you are from the magnet, the weaker the field. Far from a dipole, the field intensity drops off as $1/R^3$—more rapidly than the inverse square law for gravity.

Charged particles and magnetic fields interact in such a way that the particles find it difficult to cross the field lines. Instead, the charged particles tend to spiral along the field lines, the direction of the spiral twist depending on whether the particle is positively or negatively charged (Fig. F.10). As a concrete analogy, imagine the magnetic field lines as elastic bands. If a charged particle attempts to plow across the bands, it encounters a resistance and stretches the field lines. So charged particles and magnetic fields are linked together by their interactions.

This linking is important for understanding what happens to a magnetic field that is immersed in an ionized gas (such as the sun). If the ionized gas moves, it carries the magnetic field lines with it. For example, if the gas is moving turbulently, it tangles and jumbles up the direction of the magnetic field lines.

To sum up, moving charged particles produce magnetic fields. In turn, magnetic fields affect the motions of charged particles. The linking of magnetic fields and charged particles has important astrophysical consequences.

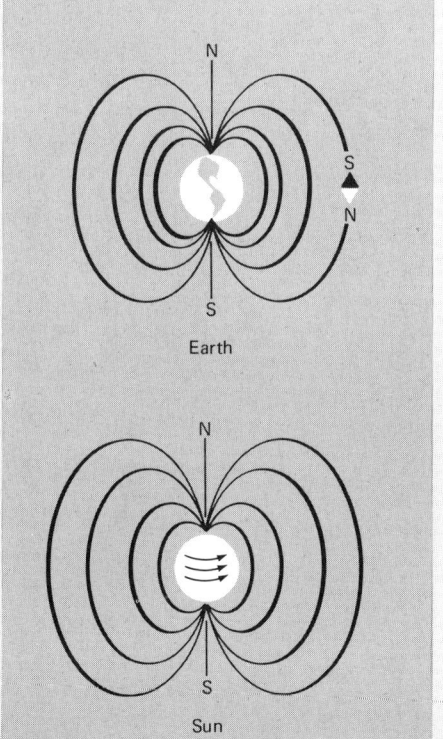

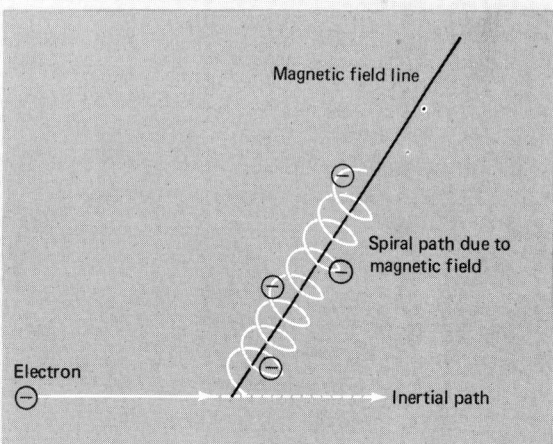

FIGURE F.9 The magnetic fields of the earth and the sun. Note that their shapes are similar to that of a bar magnetic.

FIGURE F.10 The path of an electron in a magnetic field. As the electron moves across a field line, a force acts on it, causing it to gyrate around the field line in a spiral path.

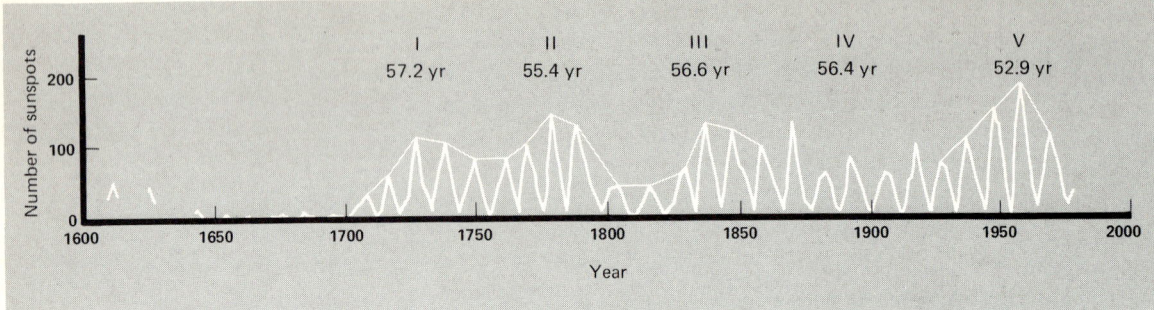

FIGURE 6.41 Sunspot cycles from 1600 to the 1970s. Plotted here are the number of observed sunspots during a year. The peaks (sunspot maxima) come roughly 11 years apart. Note that the peaks seem to come in larger cycles about 55 to 57 years long. Also note how few sunspots were seen prior to 1700, compared to modern times. [*Adapted from a diagram by H. Yoshimura,* Astrophysical Journal, *vol. 230, p. 905, copyright 1979 by the American Astronomical Society.*]

associates were surprised to find that the magnetic polarities for the trailing and leading spots had reversed for both hemispheres. At the end of that sunspot cycle in 1924 the polarities reversed again and reestablished the 1902 configuration. So the magnetic polarity cycle lasts for about 22 years, just double the 11-year cycle of sunspot numbers.

Recent investigations, especially by astrophysicist John Eddy, indicate that little historical evidence exists to show an 11-year cycle in sunspot activity before 1700. Sunspots were discovered by Galileo with his new telescope in about 1613; but, as first noted by Maunder, hardly any sunspots were seen in the 60-year period from 1645 to 1705 (Fig.

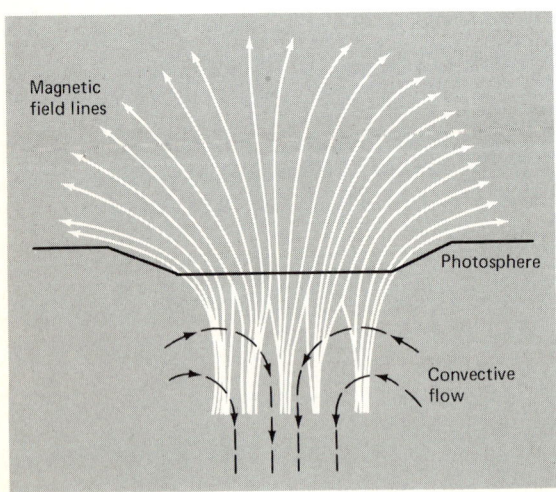

FIGURE 6.42 One model of the distribution of magnetic field lines in a sunspot. [*Adapted from a figure by E. N. Parker,* Astrophysical Journal, *vol. 230, p. 905, copyright 1979 by the American Astronomical Society.*]

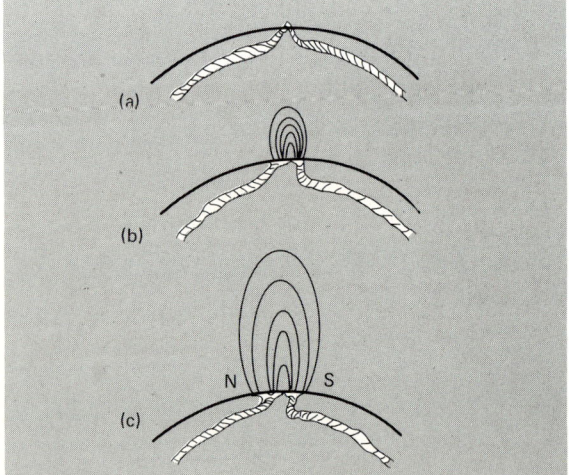

FIGURE 6.43 A possible model for sunspots. The sun's nonuniform rotation may wind the magnetic field into tight ropes (a). Kinks develop in these regions and burst through the surface (b). Sunspots develop in pairs of opposite polarity (c).

6.7 THE ACTIVE SUN

6.41). The relative consistency of the cycle in modern times may be a short phase of changes that take place over longer times.

Physical Nature of Sunspots

Sunspot magnetic fields are generated by enormous electric currents, much as the field of an electromagnet is produced (Focus 6.2). The hurricane of currents and fields in turn creates sunspots. Exactly how this formation happens is unclear. One possibility is that the sunspot's magnetic fields suppress the hot gas rising from the convective zone. The hot gas runs into a magnetic thicket and has trouble breaking the surface in the sunspot's center (Fig. 6.42). A convective downflow may draw away some of the hot gas, removing heat from the region. As a result, the sunspot is cooler and darker than the surrounding photosphere. It is also depressed about 700 km.

Here is one possible model for the origin of sunspots (Fig. 6.43). Magnetic field lines are attached to the solar gases. The nonuniform rotation winds up the lines of force. Turbulence twists the trapped field lines into tubes with a ropelike structure (Fig. 6.43a), and as the tubes are squeezed together, kinks appear. Eventually a critical point is reached when the kinked parts of the magnetic field lines burst through the solar surface (Fig. 6.43b), much as water breaks through a garden hose at its weak spots. One side of the broken kink exhibits a north magnetic polarity, the other side south polarity (Fig. 6.43c). The magnetic tube model for sunspots predicts a reversal of polarity as rotation unwinds the field lines every 11 years.

Warning: The magnetic tube model is just one possible model for sunspots. None of the models to date deals with the origin and cycle of sunspots in a completely satisfactory way. E. N. Parker, who has studied sunspots for over 20 years, states simply, "we do not yet have a satisfactory explanation of the sunspot phenomenon."

Prominences

Magnetic fields also play a major role in the production of *prominences*, huge red clouds of hydrogen gas above the photosphere. When viewed along the sun's edge, prominences often loop and surge up into the corona (Fig. 6.44a). When seen against the photosphere, a prominence resembles a dark snake winding across the solar disk (Fig. 6.44b). Because they are cooler than the photosphere, prominences absorb some of the photospheric light and appear dark in contrast.

Prominences are almost always found near a sunspot group in an active region. Frequently, a young prominence disappears, only to reappear in a few days relatively unchanged. This observation suggests that the basic underlying structure of the prominence is the magnetic field, made visible when gas is present at the right temperature and density to radiate visible light. A prominence may persist for weeks, with material raining down into the photosphere from it while, simultaneously, an infall of gases from the corona replenishes the prominence.

Flares

Sunspots are floating islands of electromagnetic storms, which generate short-lived, violent discharges of energy called *solar flares* (Fig. 6.45). These energetic bursts some-

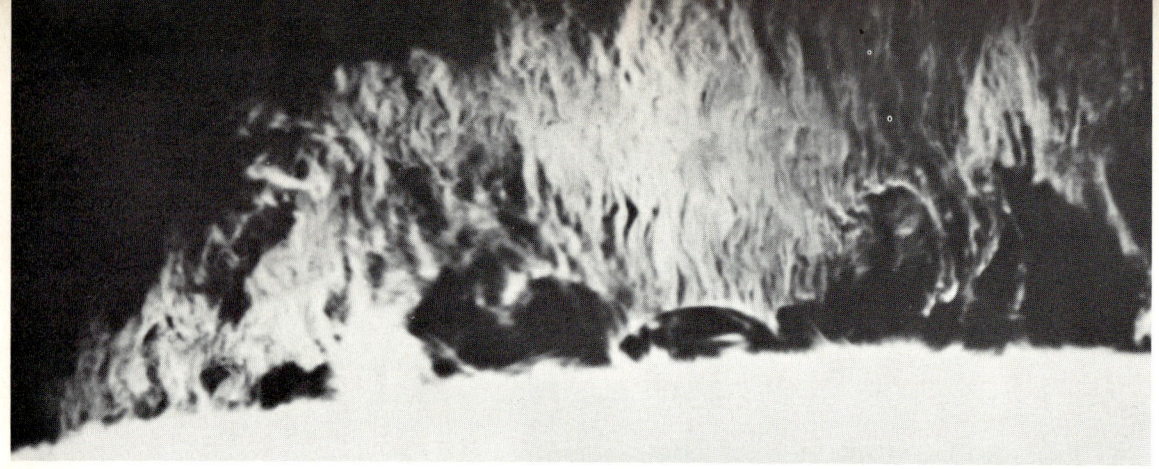

(a)

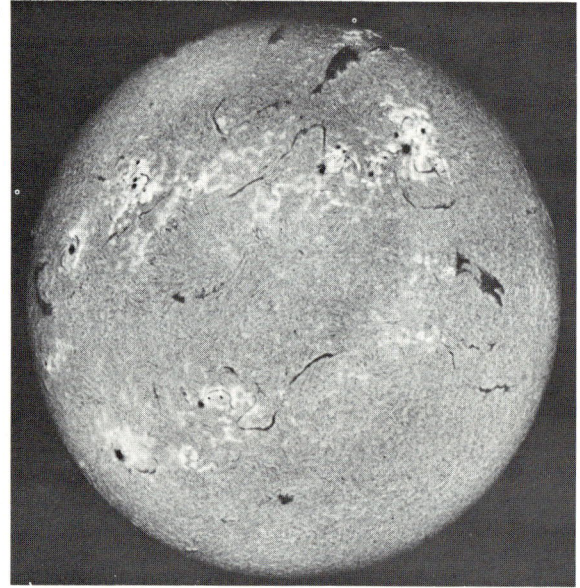

FIGURE 6.44 Prominences. (a) A prominence rising (b) over 150,000 km above the photosphere. [*Courtesy Mt. Wilson Observatory, Carnegie Institution of Washington.*] (b) Prominences, visible in a photo taken in H-alpha light, appear as dark, cloudlike and snakelike objects. [*Courtesy Sacramento Peak Observatory, Association of Universities for Research in Astronomy, Inc.*]

FIGURE 6.45 Outburst of a small solar flare (white region in photos) in a solar active region taken in H-alpha light. The time span (left to right) was about 45 minutes. [*Courtesy A. Maxwell and Harvard College Observatory.*]

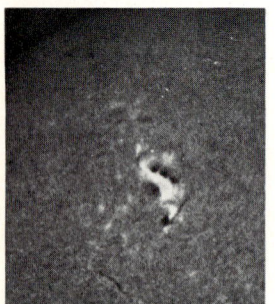 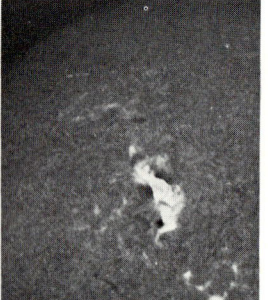 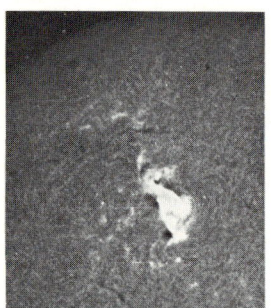 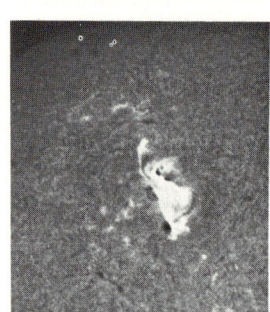

6.7 THE ACTIVE SUN

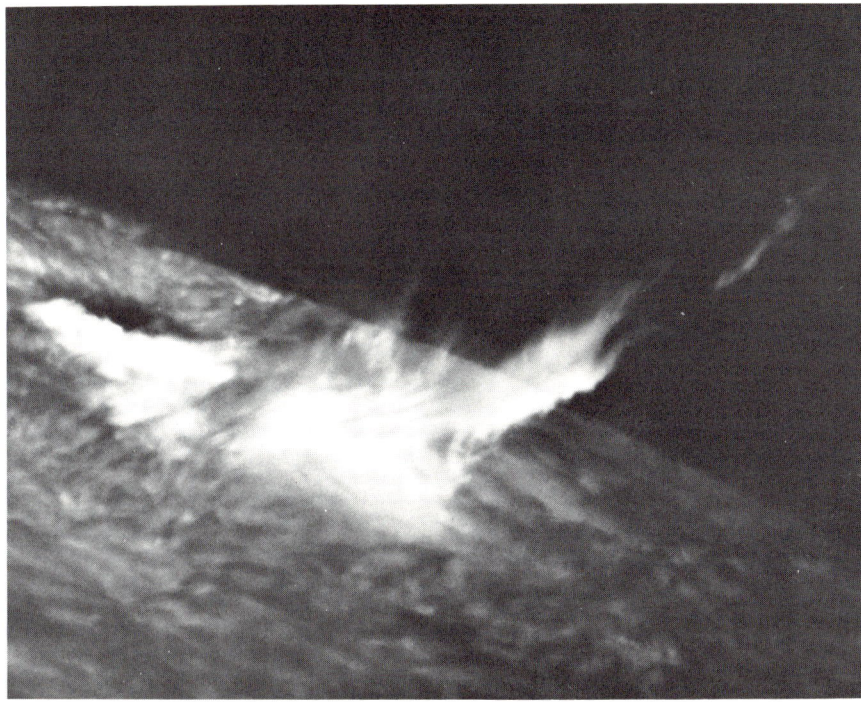

FIGURE 6.46 Material shot off the sun by a solar flare. Along with high-energy radiation and visible light, a solar flare ejects charged particles (mostly protons) into space. [Courtesy NASA.]

times bridge the gap between two close spots. Near large sunspots about 100 small flares occur each day. The elapsed time between the birth of a flare and its rise to peak intensity is only a few minutes, even for a large flare, and the decay time is about an hour. Emitting myriad forms of energy—X rays, ultraviolet and visible radiation, high-speed protons, and electrons—a large flare blows off about 10^{25} J, the equivalent of the energy released by a 2 billion megaton bomb. A typical flare has a temperature of 20 million K, a density of a few times 10^{16} particles/m^3, and a volume of some 1000 m^3.

Because closely packed sunspot groups are the most frequent locations for large solar flares, astronomers think that the concentration of flare energy is probably associated with local kinks in the magnetic fields. Flares often recur in the same location in active regions and may act as escape valves that release unstable accumulations of magnetic energy.

Flares blast energy into the corona and also shoot out energetic particles, mostly protons and electrons, into space (Fig. 6.46). A majority of the flare's energy, however, escapes as short wavelength (X-ray and ultraviolet) radiation. Arriving at the earth about 8 min later, the flare's radiation rips through the upper atmosphere and tears electrons from neutral atoms. The increase in local ionization in the atmosphere disrupts shortwave, long-distance radio communication. A few days later the lagging protons and electrons approach the earth, but they are usually trapped by the earth's magnetic field.

Occasionally the earth's magnetic reservoirs overflow with charged particles, particularly at times of sunspot maxima when flare activity is at its peak. As the particles spill into the earth's upper atmosphere, the swift electrons bump into atmospheric atoms and excite them. When the atoms deexcite, they emit visible radiation, causing a faint glow in the sky, often changing rapidly in shape and color, which is called the *aurora* (Fig. 6.47).

FIGURE 6.47 An aurora in the earth's atmosphere, taken by a NASA jet at an altitude of 30 km. [*Courtesy NASA.*]

Coronal Holes

Because the coronal gas is so hot, it emits low-energy X rays and shows up in X-ray photos of the sun (Fig. 6.48). These pictures reveal that the coronal gas has an irregular distribution above and around the sun. Notice the large loop structures; these show where the ionized gas flows along magnetic fields that arch high above the sun's surface and return to it. The hot gas is trapped in these magnetic loops (Focus 6.2). In fact, solar physicists now view the corona as consisting primarily of such loops, which provide some of the coronal heating.

Note also in Fig. 6.48 that some regions of the corona appear dark, especially at the top pole and down the middle part of the sun. Here the coronal gas must be much less dense and cooler than usual; these regions are called *coronal holes*. The coronal holes at the poles do not appear to change very much, but those above other regions seem somehow related to solar activity.

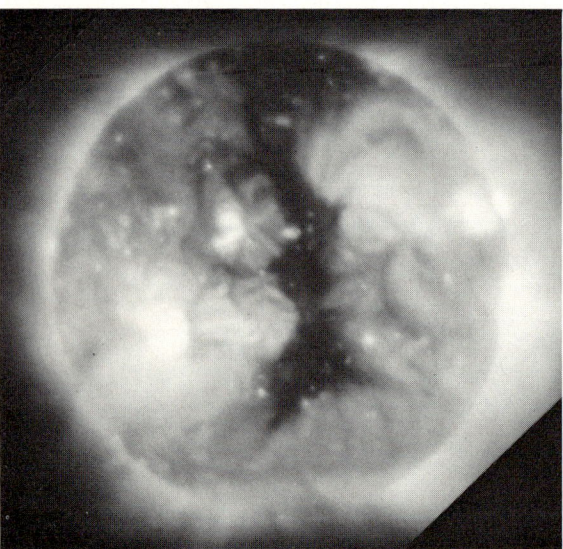

FIGURE 6.48 The sun seen in X-ray light. Very hot regions (a million Kelvins or so) appear bright in this picture, so most of these regions lie in the sun's corona. The looped appearance of streamers in the corona arises from strong magnetic fields. The dark region running down the middle of the disk is a coronal hole. [*Courtesy G. Vaiana and Harvard College Observatory.*]

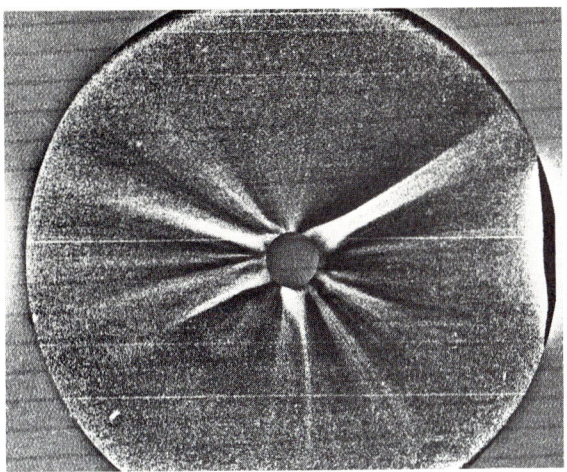 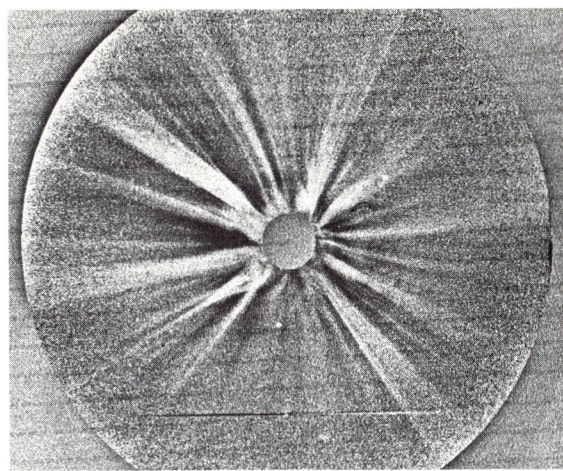

(a) (b)

FIGURE 6.49 Streamers in the sun's corona. (a) Photo taken during a total eclipse in July 1972. The streamers are visible out to a distance of about 12 solar radii. They show the magnetic field configuration and solar wind flow out beyond the region near the photosphere. (b) A similar photo taken in February 1979. Note how the flow shape has changed over this time. [*Courtesy C. Keller and the Los Alamos National Laboratory.*]

What makes a coronal hole? Solar astronomers believe that coronal holes mark areas where magnetic fields from the sun continue outward into space rather than flow back to the sun in loops. So the coronal gas, not tied down in these regions, can flow away from the sun out of the coronal holes; this is the source of the solar wind (Fig. 6.49).

The Evolution of Active Regions

All of the phenomena discussed above occur within active regions on the sun. They are all related by having at their roots intense, local magnetic fields. To link together these seemingly diverse phenomena, look at three phases in the lives of active regions: (1) the emergence of a local, intense magnetic field in the photosphere, (2) the response of the atmospheric structure to the magnetic field, and (3) the dissolution of the concentrated field.

The local magnetic field makes its photospheric debut as the dark pores that precede sunspots. These pores join to make sunspots of opposite magnetic polarity. Meanwhile, the chromosphere and corona above the developing active region form arches and loops that, within a few hours, connect to neighboring active regions. Once the active region is established, its sunspots change slowly, while flares and prominences punctuate the relative calm. An active region takes about 10 days to develop; about 10 to 15 days later it reaches a maximum activity; and a few months later it has essentially died out as the magnetic field dissipates and decays.

SUMMARY

Our sun is a star: a huge ball of gas that's a thermonuclear reactor in the sky. To understand the sun is to understand sunlight. Its production involves the nuclei of atoms. Its

TABLE 6.3 Summary of Important Physical Characteristics of the Sun

Property	Value
Distance from earth	1 AU = 1.496×10^{11} m
Radius	6.966×10^{8} m
Mass	1.991×10^{30} kg
Luminosity	3.86×10^{26} W
Surface temperature (photosphere)	5780 K
Average density	1410 kg/m³
Age	4.5 to 5 billion years
Composition (surface) by mass	Hydrogen: 78%
	Helium: 20%
	Other elements: 2%

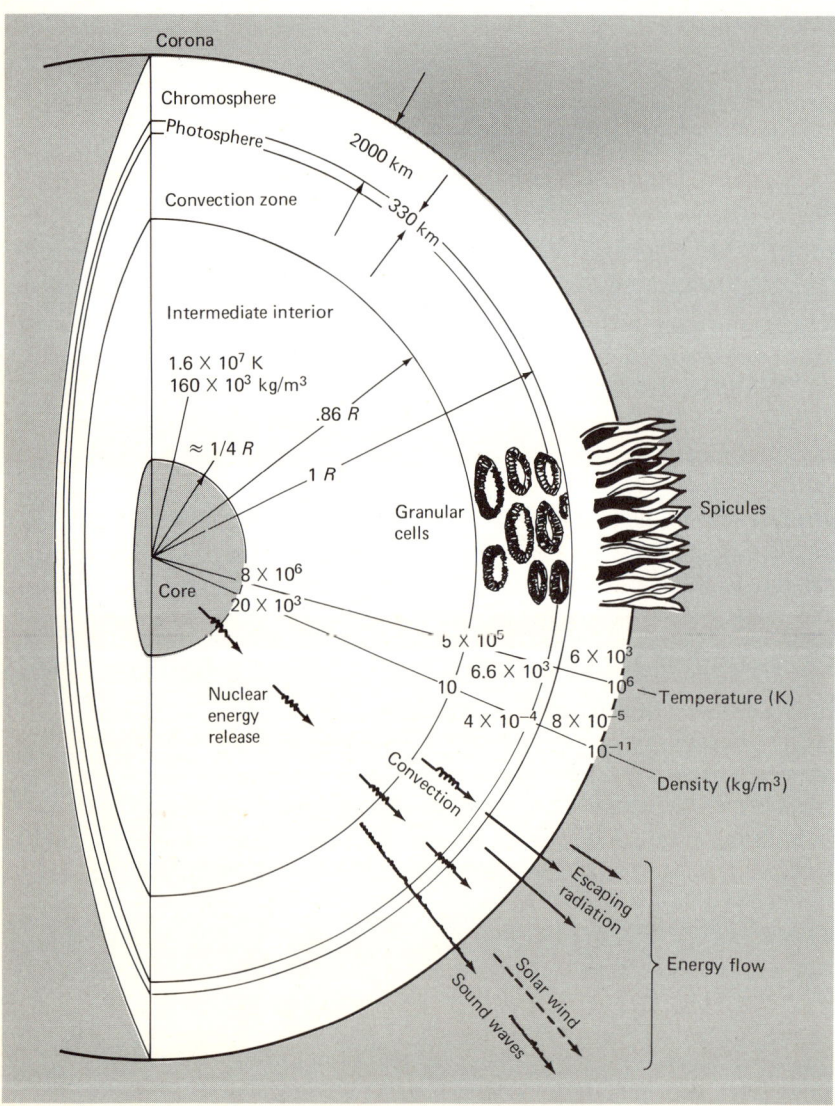

FIGURE 6.50 The outward flow of energy from the sun's core and the sun's general structure.

character involves the electrons orbiting atoms. The nature of light, as understood by quantum theory, and the analysis of sunlight, as viewed through a spectroscope, inform us of the physical properties of the sun (Table 6.3). These properties, we believe, are typical of stars in the Galaxy.

In the sun's core hydrogen fuses to helium, releasing energy. All general solar features result from the interaction of matter with energy generated in the core, flowing outward (Fig. 6.50). The energy-producing core occupies a quarter of the sun's radius. Here the awful crush of gravity packs the material to a density of 160,000 kg/m^3, ten times denser than the densest materials in the earth, but because it is ionized and hot, it is still a gas. The fury of the release of thermonuclear energy by the PP chain keeps the core at about 16 million K. Gamma-ray photons from the hot center journey to the surface and lose energy. At about 0.8 of the sun's radius the opacity is high enough to create a convection zone. This region produces the granular structure of the photosphere. The photons flee into space at the photosphere, creating the continuous and absorption-line spectrum of the sun. Most of the escaping photons are in the wavelength range of visible light, but some X-ray, ultraviolet, and radio waves emerge in the radiative scramble.

Supersonic waves generated in the convection zone may carry energy out from the photosphere to heat the upper chromosphere and corona to high temperatures. As a result of the high temperatures, the gas in the corona expands from the sun to create the solar wind. This wind blows past the earth, and our planet catches some of the particles from the sun.

Magnetic fields, twisted and stretched by the solar rotation, occasionally break through the surface, forming active regions. Here sunspots, which come and go in an 11-yr cycle, appear. Flares release excess magnetic energy from active regions in the form of high-energy particles and radiation, producing auroras on earth and disrupting radio communication.

Key Words

atom
molecule
ion
proton
neutron
nucleus
electron
positron
neutrino
element
compound
   nuclear force
isotope
temperature
pressure
number density
perfect gas law
Kelvin (absolute)
   temperature scale

Boltzmann's constant
heat
spectroscopy
spectrum
spectroscope
continuous spectrum
bright (emission) line
dark (absorption) line
D line
Kirchhoff's rules of
   spectroscopic analysis
electromagnetic radiation
electromagnetic spectrum
gamma ray
X ray
frequency
wavelength
angstrom
Hertz

Balmer series
Lyman series
Paschen series
quantum
photon
Planck's constant
electronic transitions
energy level
ground state
excitation
ionization
Rydberg constant
bound-bound transition
bound-free transition
free-bound transition
free-free transition
conservation of energy
kinetic energy
potential energy

radiative energy
luminosity
blackbody radiation
Wien's law
Stefan-Boltzmann law
opacity
negative hydrogen ion
quiet sun
photosphere
granule

chromosphere
corona
spicule
supersonic
solar wind
fission
fusion
proton-proton (PP) chain
conduction
convection

radiation
active region
sunspot
prominence
solar flare
aurora
coronal hole
magnetic field
dipole field
gauss

Review Questions

1. Which of the following do we need to know or to assume in order to determine the sun's luminosity?
   a. That the sun's output never varies
   b. That the sun radiates the same amount in every direction
   c. That our radiation detector absorbs all energy incident upon it (or at least that we know what fraction is absorbed)
   d. The distance to the sun
   e. The sun's surface area (derived from it's angular size)
   (Objective 2)

2. The energy generated in the sun (mark all correct answers):
   a. Ultimately comes from the conversion of mass into energy.
   b. Emerges from the photosphere mostly in the form of neutrinos, which are easily detected with special telescopes.
   c. Comes from the fission of one helium atom into four hydrogen atoms.
   d. Is produced in regions which are at temperatures of several million degrees.
   e. Must come from nuclear reactions, because it would be impossible for the sun to keep shining for more than a few hundred years if it derived its energy from either chemical reactions or gravitational collapse.
   (Objectives 9 and 10)

3. Mark all statements true or false.
   a. The photosphere is the layer in the sun's atmosphere from which most of the radiant energy is emitted to space; in other words, it's the layer of the sun that we see.
   b. The corona may be heated by sound waves generated from below which dissipate their energy as they move into the thinner, outer atmosphere of the sun.
   c. The chromosphere, according to Skylab observations of highly ionized species of iron, has a temperature of at least 2×10^6 K.
   d. The solar wind is an example of the force of gas pressure dominating gravitational force.
   (Objective 12)

4. The fact that the D lines in the solar spectrum get darker when sunlight is passed through a sodium flame shows that (mark all correct answers):
   a. Sodium is present in the atmosphere of the sun.

b. The sodium flame, though hot, is not as hot as the sun.
c. The enhanced brightness of a flame on the introduction of sodium is an illusion.
d. Fraunhofer had been wrong in identifying these lines as due to deuterium.
e. Sodium, not hydrogen, is the most abundant element in the sun.
(Objectives 4 and 8)
5. For what reason do astronomers bounce radar signals off Venus to find the distance to the sun? (Objective 1)
6. Suppose you examined sunlight with a spectroscope. Describe, in general, what you would see. (Objective 7)
7. Explain the appearance of the sun's spectrum by simple atomic processes. (Objectives 4 and 6)
8. For what reason do you not see helium absorption lines in the sun's visible spectrum, even though helium is the sun's second most abundant element? (Objectives 4 and 6)
9. Do you expect the chemical composition of the sun's core to differ from that of its photosphere? Why or why not? (Objectives 8 and 9)
10. Describe how to estimate the sun's surface temperature from its continuous spectrum. (Objective 2)

Problems
1. Calculate the light travel time from the sun to the earth. (Objective 1)
2. Calculate the energy released by one proton-proton chain. (Objectives 9 and 10)
3. Calculate the sun's surface temperature from the wavelength peak in its continuous spectrum. (Objective 2)
4. At the top of the photosphere the gas pressure is about 9×10^{-4} atm and the temperature is about 3100 K. Find the number density of the gas. (Objective 14)
5. What is the wavelength of the peak emission from a typical sunspot? (Objective 3)
6. What is the wavelength of a typical photon at the sun's center? (Objective 3)
7. From the fact that 0.7 percent of the mass is converted to energy in the PP chain, calculate the maximum age of the sun, if its luminosity (4×10^{26} W) has remained constant. (Objective 10)

BEYOND THIS BOOK . . .

* E. N. Parker paints a contemporary picture of the nearest star in "The Sun," *Scientific American*, September 1975.
* Some aspects of the active sun are treated in Part V of *The New Astronomy and Space Science Reader* (Freeman, San Francisco, 1977), edited by J. C. Brandt and S. P. Maran.
* A technical, but readable description of the sun is *The Quiet Sun* (NASA Special Publication S-271, 1970) by E. G. Gibson.
* For some popular articles on the sun, all by J. Pasachoff, look at "The Fiery Sun," *Natural History*, May 1972, p. 48; "The Sun: Still at the Center of Our Thoughts," *Popular Astronomy*, vol. 1, no. 1 (1975), p. 36; and "Our Sun," *Astronomy*, vol. 6, no. 1 (1978).
* For more details on the sun's outer atmosphere, read "The Solar Corona" by J. Pasachoff, *Scientific American*, October 1973, p. 68.
* A space-age view of the sun is found in *A New Sun: The Solar Results from Skylab* (NASA SP-402, 1979) by J. Eddy.

GATHERING LIGHT FROM SPACE

7

LEARNING OBJECTIVES

After studying this chapter, you should be able to:

1. Describe the impact of observations on scientific models.
2. Outline the main functions of a telescope—light-gathering, resolving, and magnifying power—and relate each to specific optical characteristics of a telescope's design.
3. State and make use of expressions for a telescope's light-gathering, resolving, and magnifying power.
4. Compare reflecting and refracting telescopes, including a sketch of the optical layout of each.
5. Compare a radio telescope to an optical telescope in terms of function, design, and use.
6. Cite a key drawback of a radio telescope compared with an optical telescope and describe how radio astronomers cope with this problem.
7. Describe how a radio interferometer operates and discuss its advantages over a single-dish radio telescope.
8. Describe what is meant by the term "invisible astronomy."
9. Compare an infrared telescope to an optical telescope in terms of function, design, and use.
10. Discuss at least two advantages a space telescope has over a ground-based telescope.

CENTRAL QUESTION:
How do astronomers collect the light
from celestial objects?

This chapter turns from astronomical concepts to astronomical tools. Telescopes and other equipment, such as spectroscopes, spring from advances in technology. The technological development of astronomical tools affects not only *what* we observe, but also *how* we observe. It expands, deepens, and sharpens our perceptions of the cosmos. What and how we observe act as prelude to and confirmation of our models of astronomical objects—and of the universe itself.

The image of an astronomer that many people have is a person who sits out all night in the cold peering through a telescope. That was accurate at one time, but it is rare these days for an astronomer to actually *look* through a telescope, other than to confirm that it is set on the desired object of study (Fig. 7.1). Instead, he or she uses many instruments to study the radiation from the stars—instruments which can provide much more accurate and detailed information than can the eye alone.

In the course of this century the technical advances have become so great that the labor of observing and explaining has been divided somewhat between the observational astronomers, who work with telescopes, and the theoreticians, who work with pencil and paper and computers and may never use a telescope in their lifetime. The technology has even become too complex for any one observer to master, and he or she now often works with teams of engineers and technicians. This chapter will not discuss all of the recent technological advances, but it will look at some of the developments providing new ways of observing the universe and the influence these have had on our astronomical models.

Observations accelerate the evolution of astronomical ideas. By them, the effectiveness of models is judged. From this judgment some models are discarded, new ones proposed, and a few finally adopted. Not all new observations have dramatic effects. In many instances the change of view brought on by the change of vision is slow and subtle. Slowly or swiftly, new observations compel new conceptions of the cosmos.

7.1 OBSERVATIONS AND MODELS

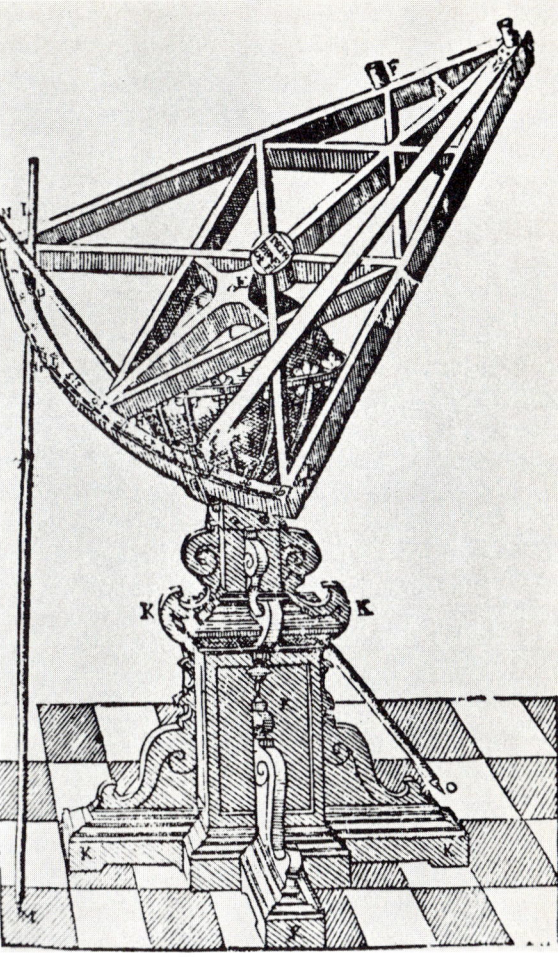

FIGURE 7.1 (a) The 61-cm telescope at Capilla Peak Observatory, New Mexico. A student, Gary Henson, is at the eyepiece. [*Photo by M. Zeilik.*] (b) One of Tycho Brahe's exquisite instruments—an astronomical sextant used at Hveen.

7.1 OBSERVATIONS AND MODELS

How do observations and models interact? The process usually goes like this (Section 2.1): Models spring from observations, whether straightforward or subtle. Basic astronomical observations naturally drove people to create models of the cosmos. The earliest of these were mythical (Section 2.2). The Ptolemaic model marked the first detailed attempt at a scientific explanation of these observations (Section 2.4). This model had two key functions: (1) It used geometrical, physical, and aesthetic ideas to *explain* what was seen, and (2) it also *predicted* planetary positions. A key aspect of the prediction is that it must be quantitative, that is, have definite numbers attached to it. Then how well a model corresponds to actual observations becomes crucial to its acceptability.

The issue of "how well" revolves around the techniques of observation, as well as some gut judgment of how good is "good enough." For example, astronomers knew for centuries that Ptolemaic predictions were frequently degrees off from the actual planetary positions. This discrepancy did not bother astronomers at the time, for a few degrees was considered "good enough," and their observations did not aim at any greater accuracy.

Not until Tycho Brahe (Section 3.3) did nontelescopic observations reach the limits imposed by the human eye. Brahe's technical achievement compelled Kepler (Section 3.4) to take a discrepancy of only 8 arcminutes seriously. The failure of the Copernican system to fit the data with circular orbits resulted in Kepler's devising a model with elliptical orbits. Their success formed a critical link in the evolutionary chain to Newton's idea of gravitation and a model of a cosmos tied by an invisible force (Sections 4.2 and 4.3).

Prior to Brahe's observations and astronomical use of the telescope by Galileo, both the Ptolemaic and Copernican models explained the motions of the planets equally well. And both made predictions just about as badly. Since both models were equally confirmed observationally, astronomers had to rely on aesthetic and philosophical beliefs to make a choice between the two. In fact, the crucial observation to distinguish between the simple heliocentric and geocentric models—that of heliocentric parallax (Chapter 8)—was not made until the 1830s, about two and a quarter centuries after the introduction of the Copernican model! Only by then had the techniques of measuring stellar positions become accurate enough to detect heliocentric parallax, which amounts to less than 1 arcsecond for even the nearest star. Copernicus guessed correctly that the stars were very far from the earth compared with the earth-sun distance.

To sum up, observations form the building blocks for model making, and they can also act as the driving force that causes models to be discarded. Their destructive effect often comes from new observations that don't fit the scheme of accepted models.

One more point. Astronomy, in comparison with physics and chemistry, is not an *experimental* science, but an *observational* one. We cannot physically bring a star or planet into a terrestrial lab to investigate its physical characteristics. We can work only with what we are given—for the most part, the light from celestial objects.

You should not, however, get the impression that astronomy totally lacks experimentation. We astronomers experiment in two basic ways. First, we can make different kinds of observations of the same objects. That's the importance of technological innovation, for it provides new tools for new experiments. Second, we can play with theoretical models in light of whatever observations we have. Theoretical model making may also be tied to technology; for example, electronic computers make it possible to manipulate quickly very detailed models of astronomical objects. And, of course, we make use of experiments in physics and chemistry (and may even do the experiments ourselves) that provide basic data on properties of matter and radiation.

7.2 VISIBLE ASTRONOMY: OPTICAL TELESCOPES

In the nineteenth century large telescopes began to peer beyond the solar system and local stars. Meteorites were not yet believed to be objects from space. No moon rocks had been ferried to the earth. Astronomers looked, but did not (because they could not) touch.

As extensions of the human eye, telescopes amplified the power of detection without extending the spectral range of our vision. Today we can sense much more than the visual

7.2 VISIBLE ASTRONOMY: OPTICAL TELESCOPES

part of the electromagnetic spectrum (Section 6.3). This section restricts itself to optical telescopes—those that manipulate light detectable by the eye. Before we deal with telescopes, let's discuss a little about *optics*—how light is controlled. You need to understand some basic optics to understand how telescopes work.

Refraction, Reflection, and Images

When traveling through space or a uniform medium, light moves in a straight line. Although light has characteristics of both waves and particles, in optics we think of it in terms of particles moving along straight-line paths, which are called *light rays*. Using lenses, mirrors, and prisms, we can change the direction of light rays or even break up white light into its component colors (light of different wavelengths). How light rays are affected by bouncing off or passing through materials is the essence of optics.

When light crosses the boundary from one transparent material to another (from air to glass, for example), its direction generally changes (Fig. 7.2). This bending of light rays is termed *refraction*. Refraction occurs because light travels *more slowly* in any substance than it does in a vacuum. Consider what happens when a beam of light (a bunch of rays) hits glass from air at some angle. The first part of the beam to strike the glass enters it and slows down. The rest of the beam continues to move at a faster speed and so gains on the light already in the glass. This catch-up results in the front of the beam turning toward the glass.

Here's an analogy. Imagine a line of a marching band turning a corner. To do this and keep a straight line, the people on the inside march more slowly than those on the outside. The line turns around some angle and remains straight.

The amount of refraction depends on the wavelength of light. In glass, and most other substances, blue light is bent more than yellow, and yellow more than red. The shorter the wavelength, the greater the amount of refraction.

You have all seen yourselves in a mirror, so you know that smooth surfaces return light by bouncing back the rays. This process is *reflection*. A light ray bounces off a polished surface the same way a ball bounces off a smooth wall: The ball rebounds at the same angle at which it hits. Reflection does not depend on the wavelength of the light; red and blue light are reflected the same way at the same angle.

The point of optics is to make *images* by refraction and reflection. An image occurs

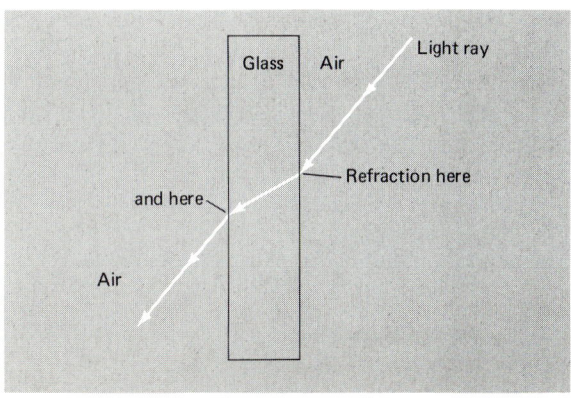

FIGURE 7.2 The refraction of light. When light rays cross the boundary between two different materials, their paths are bent. The ray here bends when it enters the glass from the air and also when it exits from the glass into air.

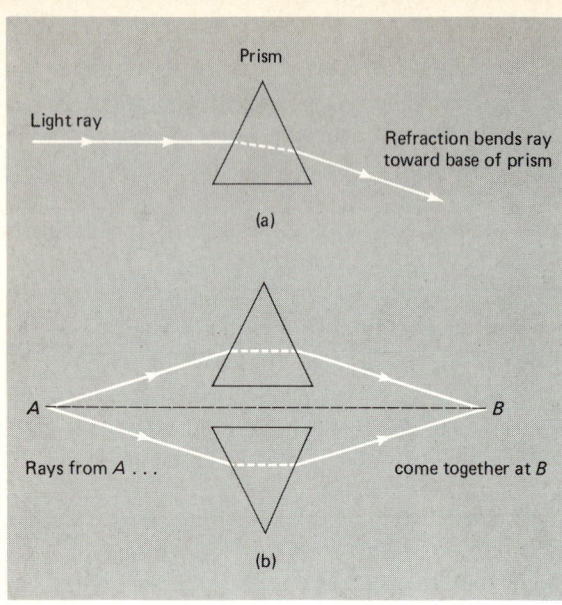

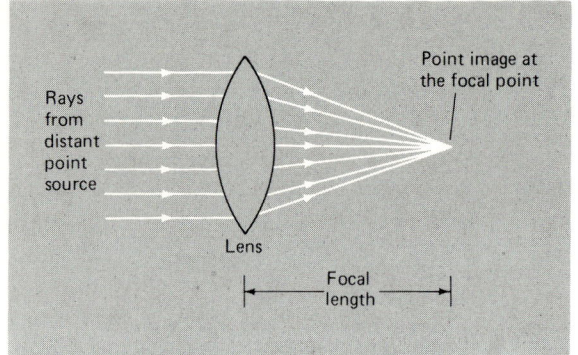

FIGURE 7.3 Prisms as crude lenses. (a) Because its sides are not parallel, a prism refracts a light ray toward its base. (b) Two prisms placed base-to-base refract rays from A to form at B a crude image.

FIGURE 7.4 A lens forming a point image from a distant point source. A lens acts like base-to-base prisms (Fig. 7.3). Because its surface is smoothly curved, it forms a sharp focus at the focal point. The distance from the lens to the focal point is the focal length, for a distant source.

when light rays are gathered together in the same relative alignment as when they left an object. You can recognize an image of an object as a visual representation of the object itself.

How can refraction form an image? Suppose light travels through a glass prism (Fig. 7.3a). Because the sides of a prism are not parallel, a light ray does not come out along its original path, but is bent toward the prism's base. Now place two prisms base to base (Fig. 7.3b). Then two light rays from a point source converge to a point. However, rays entering the prisms at different angles converge at different points. To get all rays to come to the same point requires a smoothly curved surface. Such a piece of glass is a *lens* (Fig. 7.4).

A lens brings rays from a point source to a point image at its focus. From an object of finite size, the lens makes an image of the object by focusing rays from each point of the source onto a separate point in the image. This image is generally smaller than the object and upside down (Fig. 7.5). For objects at large distances, the distance from lens to the image is approximately the same for all objects. This distance is termed the *focal length*.

How can a mirror make an image? You know that an image from a flat mirror is undistorted (but reversed right to left). An irregularly curved mirror, such as one in a

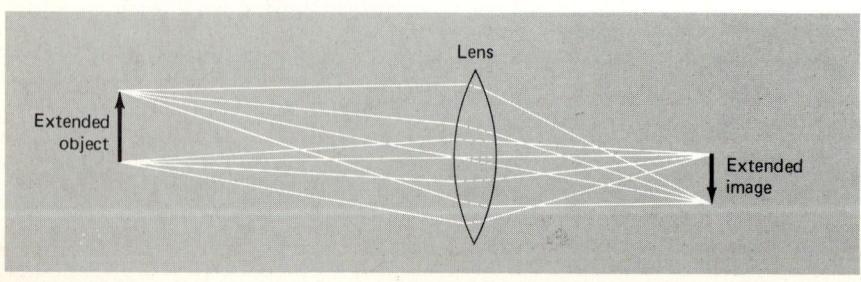

FIGURE 7.5 A lens forming an extended image, upside down, of an extended object.

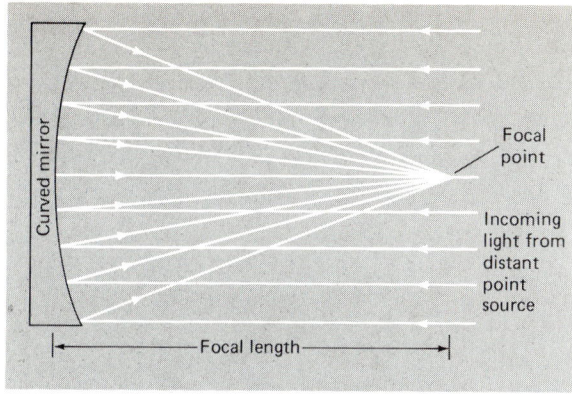

FIGURE 7.6 Making an image by reflection. A smoothly curved mirror reflects incoming light to a focus. Astronomical mirrors have their reflecting material on their front surfaces and follow a parabolic curve to make a clean focus.

funhouse, creates a distorted image. A smoothly curved mirror—whose surface, for instance, follows the curve of a parabola—brings all the light to a focus (Fig. 7.6).

Telescopes

Basically, a telescope is an instrument that gathers light and makes an image at a focus. A lens or mirror, called an *objective*, brings light to a focus. A lens at the focus, called an *eyepiece*, allows visual examination of the image.

There are essentially two types of telescopes, distinguished by their objectives: *refracting telescopes* (or *refractors*) that use a lens (Fig. 7.7) and *reflecting telescopes* (or *reflectors*)

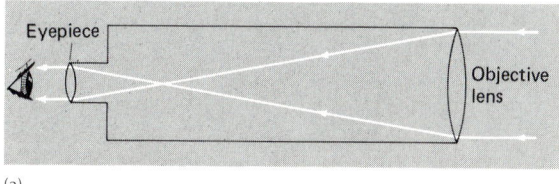

(a)

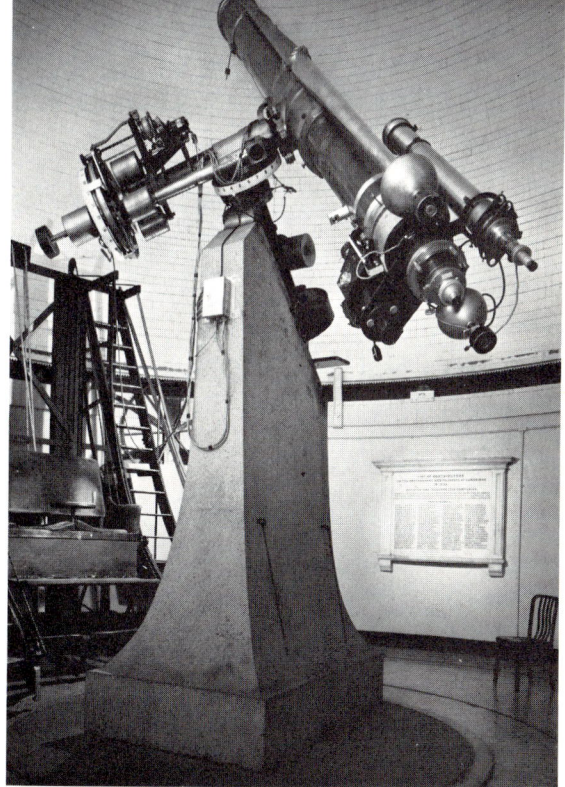

(b)

FIGURE 7.7 The refracting telescope. (a) The design of a simple refracting telescope. An objective lens gathers the incoming light and brings it to a focus. An eyepiece allows viewing of a magnified image. (b) The "Great Refractor" of Harvard College Observatory. Built in the middle of the nineteenth century, this refracting telescope has a 15″ objective lens. In its day it was one of the largest telescopes in the world. [*By permission of Harvard College Observatory.*]

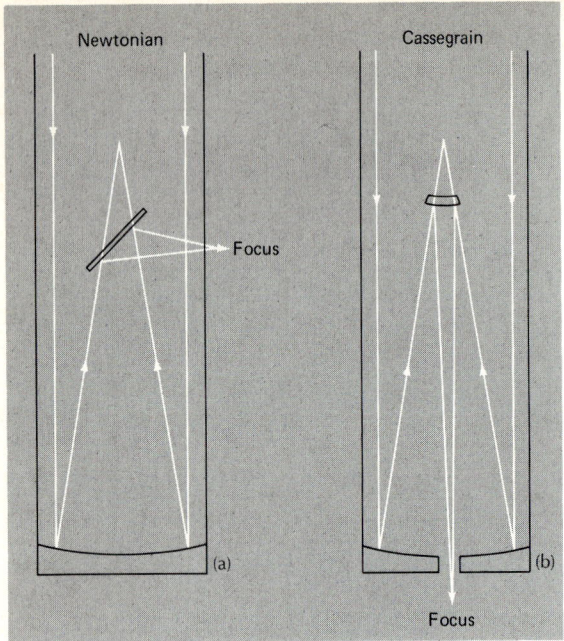

FIGURE 7.8 The design of reflecting telescopes. The main problem is viewing the image made by the objective mirror. Newton placed a small, flat mirror at an angle to reflect the light out to one side, where an eyepiece is positioned (a). This is called a Newtonian reflector. A common modern design uses a small convex mirror to reflect the light back down through a hole in the objective mirror for viewing (b). This is called a Cassegrain reflector.

FIGURE 7.9 (a) The Kitt Peak 4-m Mayall telescope, a modern large reflector with both Newtonian and Cassegrain foci. [*Photo by M. Zeilik.*] (b) An observer in the prime-focus cage of the Hale 5-m telescope on Palomar mountain. [*Courtesy Palomar Observatory, California Institute of Technology.*]

that use a mirror. Galileo's telescope (Section 4.1) was a refractor, Newton's a reflector (Section 4.2).

Newton designed the reflector because of a critical drawback of the refractor of his day. Since simple lenses act essentially as prisms, light of different colors has different focuses. Images viewed through a refractor may therefore have color halos (when one color is in focus the other colors are out of focus, and fuzzy). Newton noted that reflection does not make light break up into colors, so the image formed by a reflector does not have color halos. (Later lenses were designed to eliminate most of the refractor's color problems.)

One problem with a reflector is that the image is formed in front of the mirror, where it is difficult to examine without blocking the incoming beam. Newton directed the image out the side of the telescope tube by placing a small mirror, tilted at 45°, in the path of the reflected light (Fig. 7.8a). Such a design is called a *Newtonian reflector*. Most telescopes today use a different optical design called a *Cassegrain* in which a convex mirror brings the reflected light to a focus through a hole in the objective (Fig. 7.8b). All large telescopes built now are reflectors (Fig. 7.9a). In the very largest, the observer can get at the focus directly; he or she sits in a cage at the center of the telescope (Fig. 7.9b).

Reflectors Versus Refractors

In addition to affecting all colors alike, reflectors have other advantages over refractors. They are much cheaper to make, for, since the light does not actually pass through the mirror, the glass can be of lower quality (in fact, since its only function is to support a reflecting surface, the mirror need not be glass at all). Reflectors can be made much larger, for they can be supported from the back and sides, whereas a refractor can be supported only from its edges. On the other hand, the use of some kind of secondary mirror to bring the light to an accessible focus results in distortion of the light, and so images produced by a reflector are not as sharp and round as those produced by a refractor. When large light-gathering power is desired, a reflector is always chosen, but for precise positional measurements and observation of fine detail refractors are still very useful.

Functions of a Telescope

Whether it is a reflector or refractor, the primary function of a telescope is to gather light. A telescope is basically a light bucket, collecting photons. That's the reason astronomers want large telescopes—the bigger the bucket, the greater its light-gathering capacity.

A telescope's *light-gathering power* (LGP) is directly proportional to the square of its diameter; LGP $\propto d^2$. The amount of light a lens or mirror catches depends on its surface area, and the area of a circle with diameter d is $(\pi/4)\,d^2$. Light-gathering power is a relative, not an absolute measure; it states how two instruments compare to one another, not how much light is gathered. Since the factor $\pi/4$ is a constant, we need use only the diameters of the instruments to arrive at a comparative figure. For example, compared to your eye, which has a diameter of about 0.5 cm, a telescope with a 50-cm objective would have a light-gathering power of

$$\text{LGP} = \left(\frac{50}{0.5}\right)^2 = (100)^2 = 10{,}000$$

Similarly, the 5-m Hale telescope at Mount Palomar outdoes the 4-m Mayall telescope at the Kitt Peak Observatory by more than 50 percent:

$$\text{LGP} = \left(\frac{5}{4}\right)^2 = \frac{25}{16} = 1.56$$

A second important function of a telescope is to produce an image in which objects that are close together in the sky can be seen as clearly separate. This ability is called *resolving power* (RP). It is sometimes expressed as the inverse of the minimum angle between two points which can be clearly separated:

$$\text{RP} = \frac{1}{\theta_{\min}}$$

The resolving power and the minimum angle depend on the diameter of the objective and also on the wavelength of the light. For the same wavelength the resolving power depends directly on the objective's diameter. So a mirror *twice* the size of another has *double* the resolving power, that is, it can resolve objects that are half as far apart.

The minimum resolvable angle depends on both the diameter of the objective of the telescope and the wavelength of light being observed. In general,

$$\theta_{\min} = 206265 \frac{\lambda}{d}$$

where λ is the wavelength and d the diameter of the objective in the same length units, and $\theta_{\min}$ is the minimum resolvable angle in seconds of arc (the number 206265 is the number of arcseconds in one radian). For example, a 10-cm (0.1-m) telescope working at a wavelength of 5.6×10^{-7} m (5600 Å, yellow light) has a minimum resolvable angle of

$$\theta_{\min} = 206265 \left(\frac{5.6 \times 10^{-7}}{0.1} \right) = 1.4 \text{ arcseconds}$$

This result means that if this telescope is pointed at two stars which are more than 1.4″ apart, you will see two separate star images. If the angular separation of the stars is less than 1.4″, you will see a single star image.

The example above gives the *theoretical* resolving power of a 10-cm telescope. But such performance rarely, if ever, is attained by grounded-based telescopes. The resolving power of a big telescope is limited not by the telescope's optics, but by the earth's atmosphere (Section 14.6). You have probably noticed that stars twinkle. The twinkling comes from turbulence in the air that makes the atmosphere act like a huge, nonuniform lens. The motion of blobs of air, like the shimmering above a hot road, distorts and blurs images seen through a telescope. Even on the best of nights, the 5-m Hale telescope does not resolve better than a 10-cm telescope. At Kitt Peak, for example, it is a rare night when star images are smaller than 2″.

The limit which the earth's atmosphere sets on the resolving power of big telescopes makes a strong case for placing a large telescope in space. Here a telescope's resolving power would be limited by the optics, not the atmosphere.

The third function of a telescope is to magnify the image. *Magnifying power* (MP), the apparent increase in the size of an object compared with visual observation, depends on the ratio of the focal length of the objective to the focal length of the eyepiece.

$$\text{MP} = \frac{F}{f}$$

where F is the focal length of the objective and f is the focal length of the eyepiece in the same length units. For example, a 5-cm focal length eyepiece used with a 1-m (100-cm) focal length objective gives a magnifying power of

$$\text{MP} = \frac{100}{5} = 20$$

If you put in an eyepiece with *half* the focal length, you *double* the magnifying power.

You could choose eyepieces to produce as high a magnifying power as you like, but there is little point in using a magnifying power any greater than necessary to see clearly the smallest detail in the image. The degree of detail is determined by the resolving power. Extremely high magnification of an image produced by a telescope with a low RP merely makes the fuzziness larger.

Note: These days astronomers rarely, if ever, use their eyes directly for observations. Some light-sensing device, called a *detector*, is usually placed at the focus. The detector may be a photographic plate (light-sensitive materials on glass rather than film); that's how many of the photos in this book were made. Or it may be an electronic detector similar, for example, to a television camera.

To sum up, the three principal functions of a telescope are (1) to gather light, (2) to resolve fine detail, and (3) to magnify the image. Of these, we cannot overemphasize the importance of light-gathering power. Most astronomical objects are extremely faint. Without a telescope you can see about 6000 stars, but with even a small, 6-in telescope you can see some half million stars. The real power of a telescope is to enable us to see objects that we would otherwise not know existed.

Your Own Telescope

This book, or the course you're taking, may spark or develop your interest in observing. If so, you're probably wondering what telescope you should buy.

We won't recommend any specific brands, but here is a little general advice. First, decide how much money you can spend. Then, because light-gathering power is paramount, look for the *largest diameter* telescope with decent mechanical parts that you can find within your budget. On a cost basis this requirement pretty much forces you to consider only reflecting telescopes of Newtonian design if your budget is limited to a few hundred dollars.

Warning: Stay away from the small, cheap refractors that are typically sold in discount stores. Their objectives are so small (usually 50 mm) that they simply do not have the light-gathering power for serious astronomical observations.

You might be surprised to learn that neither of us owns a telescope. Instead, we apply for access to telescopes at observatories such as Kitt Peak (Fig. 7.10) or Lick Observatory (Fig. 7.11), where we can undertake our research projects.

7.3 INVISIBLE ASTRONOMY

Your eye senses only a tiny sliver of the electromagnetic spectrum (Section 6.3). When you have your teeth X-rayed at the dentist, the X-ray machine does not glow brightly

FIGURE 7.10 (a) Mike Zeilik setting up the 1.3-m telescope at Kitt Peak National Observatory for daytime infrared operation. (b) Controlling the 1.3-m telescope from a terminal linked to a computer. [*Photos by G. Henson.*]

FIGURE 7.11 Lick Observatory of the University of California on Mt. Hamilton. The dome of the Shane 3-m telescope is in the foreground. [*Courtesy Lick Observatory.*]

when it's on, but the film placed in your mouth senses the X rays and gives an internal picture of your teeth. When you stand next to an almost-dead fire, the coals look black, but your skin senses heat—infrared radiation—from the coals.

Invisible astronomy involves techniques that enable us to go beyond the observational limits of optical astronomy. (The development of detectors is a technological enterprise, often with no direct drive from astronomy. For instance, sensitive infrared detectors basically evolved for military purposes.)

There's also a less obvious aspect to invisible astronomy. How can we detect radiation from space that ordinarily does not reach the earth's surface? Our atmosphere effectively absorbs large blocks of the electromagnetic spectrum, especially ultraviolet light, X rays,

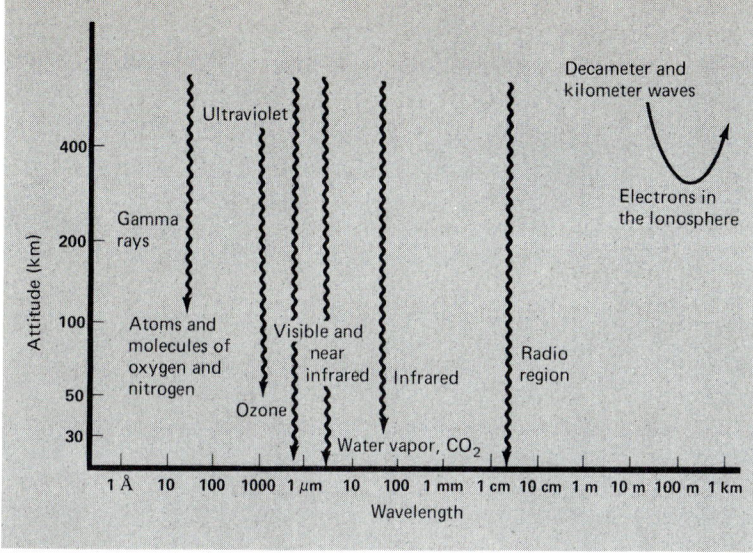

FIGURE 7.12 The transparency of the earth's atmosphere from short wavelength gamma rays to long wavelength radio waves. The scale at the left indicates the altitude down to which the radiation at various wavelengths penetrates.

some infrared wavelengths, and short-wavelength (millimeter) radio waves (Fig. 7.12). Infrared radiation is primarily absorbed by water vapor, which is found concentrated in the lower portions of the atmosphere, below 20 km. The ultraviolet and X-ray radiation is primarily absorbed in the ionosphere, at an altitude of 100 km, well above the levels which can be reached by balloons and airplanes. Light may have journeyed through space for millions or billions of years, only to be snuffed out in the last 0.001 sec of its trip, never making it to the earth's surface.

The obvious way to get around atmospheric absorption is to go above it. This is space astronomy. (In this book space astronomy includes rockets, balloons, and airplanes, as well as satellites and spacecraft.) So invisible astronomy has two natural divisions—that which can be done from the ground and that which must be accomplished in space.

Ground-Based Radio Astronomy

Radio astronomy was born in 1930 when Karl Jansky (1905–1950) undertook a study for the Bell Telephone Company of sources of static affecting transoceanic radiotelephone communications (Fig. 7.13). Jansky identified one source of noise as a celestial object: the

FIGURE 7.13 Karl Jansky with the radio antenna in Homdel, New Jersey, that he used to discover radio waves from space. [*Courtesy Bell Labs.*]

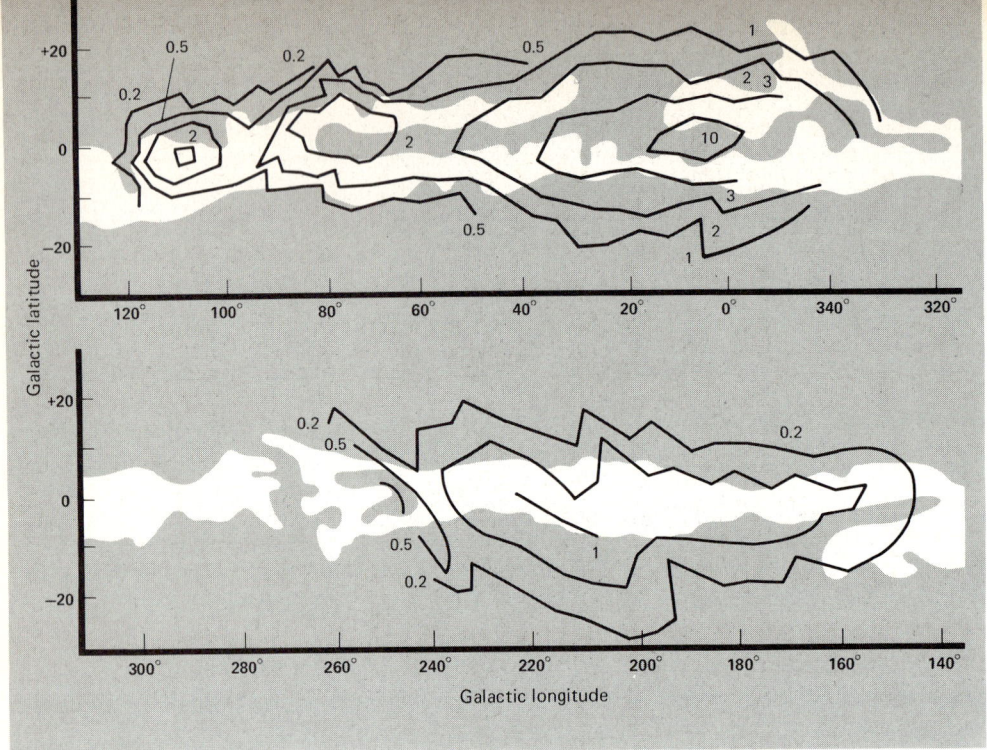

FIGURE 7.14 Grote Reber's contour map of radio emission from the Milky Way. The dark lines connect points of the same intensity of radio emission (see Focus 7.1). The shading indicates the brighter parts of the Milky Way visible optically. The galactic longitude and latitude (similar to terrestrial longitude and latitude) indicate the position in the sky. [*Observations from Reber's* Radio Data.]

Milky Way in Sagittarius. Jansky's discovery was published in 1932 but had little impact on the astronomers of the day.

However, an American radio engineer, Grote Reber, read Jansky's work and decided to search for cosmic radio static in his spare time. By the 1940s Reber had made detailed maps of the radio sky (Fig. 7.14). He sensed that a new astronomy was in the making and took an astrophysics course at the University of Chicago to learn more about astronomy and discuss his discoveries with astronomers—only a few of whom were impressed.

World War II forced technical developments in radio and radar work. John S. Hey in Britain accidentally discovered that the sun strongly emitted radio waves. After the war Hey continued his astronomical pursuits at radio wavelengths. So did other groups in Britain, the Netherlands, and Australia. Radio astronomy was reborn as a technological fallout from research by scientists forced to deal with the practical problems of war.

A common type of radio telescope, a radio dish (Fig. 7.15), functions like a reflecting telescope. Essentially, it's a radio wave bucket with a detector (a radio receiver) at the focus of the dish (Fig. 7.16) which reflects and concentrates radio waves in much the same way a mirror does in a reflecting telescope. The radio receiver translates incoming radio waves into a voltage that can be measured and recorded (Fig. 7.17). Finally, the measurements are made into a contour map (Focus 7.1).

Our atmosphere allows some millimeter, centimeter, and longer wavelengths to reach the ground. They can be observed both day and night. Radio telescopes can even observe on cloudy days at the longer wavelengths; clouds are transparent to these radio waves. Because large radio dishes are easier to construct than large mirrors, radio telescopes are

FIGURE 7.15 One antenna of the Very Large Array (VLA). It has a diameter of 25 m. Note the tracks for moving the antenna to different places along the array. [*Photo by M. Zeilik.*]

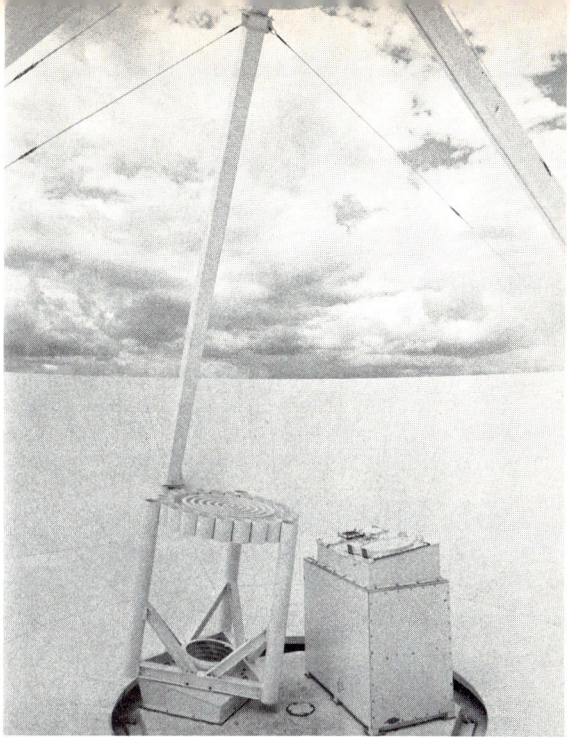

FIGURE 7.16 Surface of a radio dish, the antenna of Fig. 7.15. The metal surface reflects radio waves to a small convex dish (at top center) that reflects the radio signals down through a hole in the main dish (at left side of bottom center platform) to a radio receiver below. [*Photo by M. Zeilik.*]

typically much larger than optical telescopes, and therefore more sensitive because they can catch more radiation. These are a few of the advantages of radio compared to optical astronomy.

But radio telescopes have one major drawback: low resolving power. Resolving power depends on both the size of the objective and the wavelength of the gathered light. Radio waves are much longer than visible light, typically 100,000 times as long. So if an optical and a radio telescope had the same diameter, the radio one would have 100,000 times *less* resolving power. For example, for a radio telescope to have the same resolving power as the 5-m Hale telescope, it would have to have a diameter 100,000 times as great, about 500 km! Obviously, a single dish of this size cannot be built on earth.

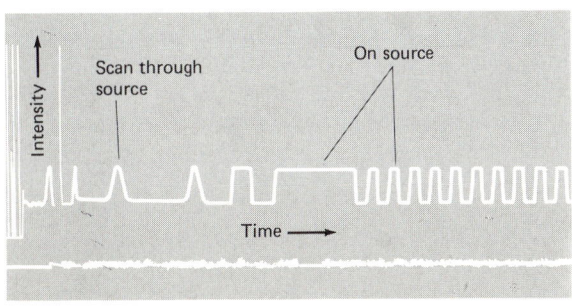

FIGURE 7.17 A radio observation. The intensity of the signal here is measured on a strip-chart recorder. The source observed here is called DR 21. [*Observations by M. Zeilik at Haystack Observatory.*]

FOCUS 7.1 UNDERSTANDING CONTOUR MAPS

Many of the figures in this book are *contour maps*, pictures of how the intensity of some kind of radiation (radio, visible, infrared, etc.) varies over some region of the sky. Such maps show a lot of wavy, connected lines labeled by numbers. What do they mean?

Here's an analogy you may be familiar with: a weather map (Fig. F.11). This is a map of atmospheric pressure across the United States, with high and low pressure systems indicated. What do the contours here tell you?

First, consider how a pressure contour is drawn. The weather stations around the United States report their local pressures. Each is put on the map. A line is drawn that connects all stations giving the same reading, *providing* that it does not cross a station of higher or lower reading. Then another contour (of higher or lower pressure) is drawn in, and so on. Contour lines cannot cross because if they did, it would mean that the *same* place has two *different* pressures, and that's impossible!

Second, note that there are places where the pressures hit a maximum (high) or a minimum (low). At the center of each is a last contour surrounding the region of highest or

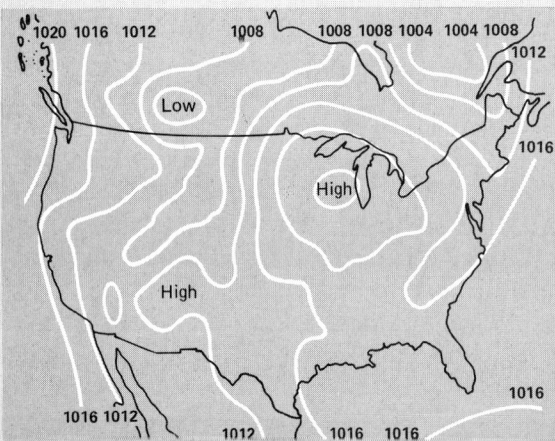

FIGURE F.11 A weather map showing surface pressure over the United States. The contour lines connect regions with the same pressure readings. A "High" marks a peak in the surface pressure; a "Low" a local minimum.

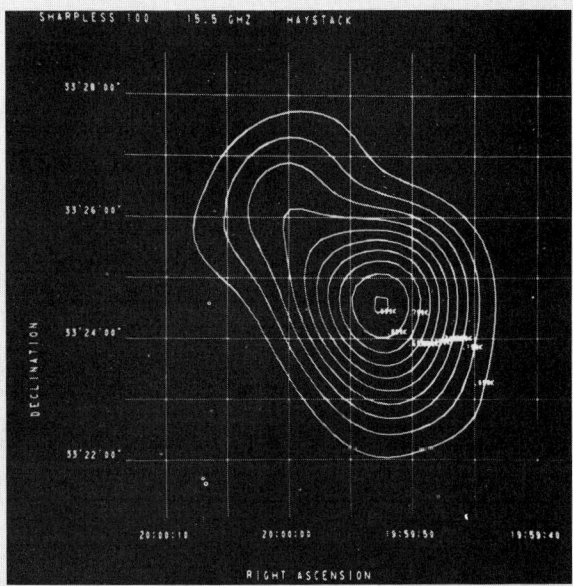

FIGURE F.12 An intensity contour map of radio emission from a region of hot, ionized gas. This cloud is called Sharpless 100; its position in the sky is indicated by the coordinate grid, one axis labeled "right ascension," the other "declination." Each contour line is 0.1 unit lower than the one interior to it. [*Observations by M. Zeilik at Haystack Observatory.*]

lowest pressure. You can imagine the center of a high as the peak of a pressure mountain. The contours around the peak tell you how the pressure falls away from the peak. If the contour lines are close together, the pressure drops quickly over a short distance; the fall-off is steep. If the contour lines are spread out, the pressure drops off slowly.

The same analysis applies to an intensity contour map, but instead of pressure, the map shows the intensity of radiation. For example, Fig. F.12 shows the intensity of radio waves of 2 cm wavelength from a cloud of ionized hydrogen. Note that there is one main peak where the radio intensity hits a maximum. To the left and above it lies a secondary peak. Between the peaks the intensity does not fall off as quickly as it does moving away from either one. Each contour line is a set amount of intensity higher or lower than adjacent ones.

To help you see this better, the same intensity map is plotted in a three-dimensional way (Fig. F.13). The height of the contours here indicates the intensity. Notice that the region of the radio source looks like a mountain! The peaks and steep sides stand out. This kind of image should be in mind when you see an ordinary contour map.

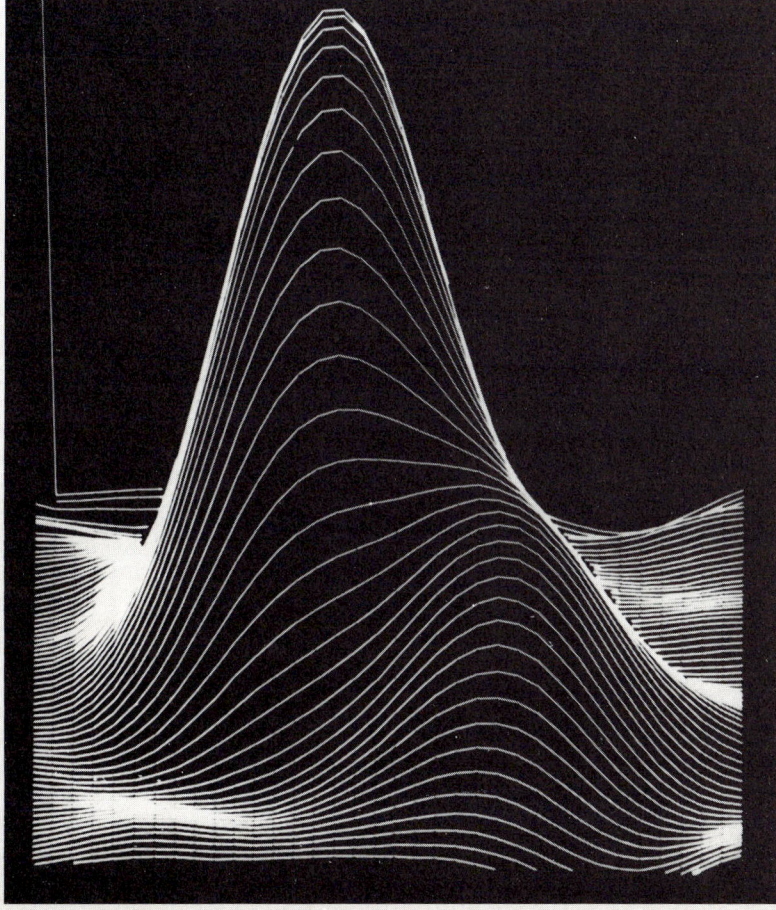

FIGURE F.13 The same observations as in the previous figure plotted so that the height of a contour line represents its relative intensity. The view is from the top of Fig. F.12 looking down at an angle of 25°. [*Map made by M. Zeilik at Haystack Observatory.*]

FIGURE 7.18 The VLA. (a) Central section of the "Y" of the VLA. The antennas are spread out and electronically linked together to achieve high resolving power. [Courtesy NRAO] (b) A view down the southwest arm of the VLA. [Photo by M. Zeilik.]

Let's examine this resolving power drawback of radio telescopes another way. Suppose you have a 5-m radio telescope operating at a wavelength of 1 m. Then its minimum resolvable angle is

$$\theta_{min} = 206265 \left(\frac{1 \text{ m}}{5 \text{ m}}\right)$$

$$= 41253 \text{ arcseconds}$$

or about 11°. That's 22 times the moon's angular diameter!

There's a method for making a small radio telescope function as if it were a large one. Imagine two radio telescopes placed, say, 10 km apart. By synchronizing the signals received by both, they can be made to act like a single dish with a diameter of 10 km, but only for a strip across the sky. (That's because they act like two small pieces at the opposite ends of a large dish.) To get good resolving power in a small, more or less circular region of the sky requires an array of coordinated radio telescopes. The newest such telescope, called the Very Large Array (VLA), is in New Mexico (Fig. 7.18). Completed in late 1980, the VLA has a resolving power at centimeter wavelengths equivalent to that of a moderate-size optical telescope. Such devices are called *radio interferometers*.

How does an interferometer work? Take the simplest example, that of two antennas separated by a few kilometers. The radio signals arrive in waves, with alternating peaks and dips. Both receivers transform the radio waves into electric signals that are fed into an electronic device called a mixer. With the radio source overhead, peaks hit both antennas at the same time (Fig. 7.19a). The mixer combines the two signals, and the signals add to each other to make a composite signal stronger than that from one telescope.

In a short time the radio source moves a little west of overhead, just enough so that when the waves arrive a peak hits one antenna and a dip the other (Fig. 7.19b). When the two signals are fed into the mixer in this case, the dip from one antenna cancels out the peak from the other, so the mixer does not produce any signal.

As the source continues to move westward, the mixer alternates between putting out a signal, zero signal, a signal, and so on (Fig. 7.19c). As a result, the two-antenna interferometer sees a striped sky: equal regions of signal and no signal. The stripes appear at right angles to the baseline of the antennas. The separation between stripes depends on the separation of the antennas; the farther the antennas are set apart, the closer the stripes.

That's the pattern for a point source of radio waves. But suppose there are two sources close together. The radio telescope will see a combination of the striped pattern for each source. Analysis of the combined radio signal can tell how far apart the sources are and how much they differ in intensity. If there are many sources, or if the source is an extended object, the analysis gets complicated and has to be done with computers.

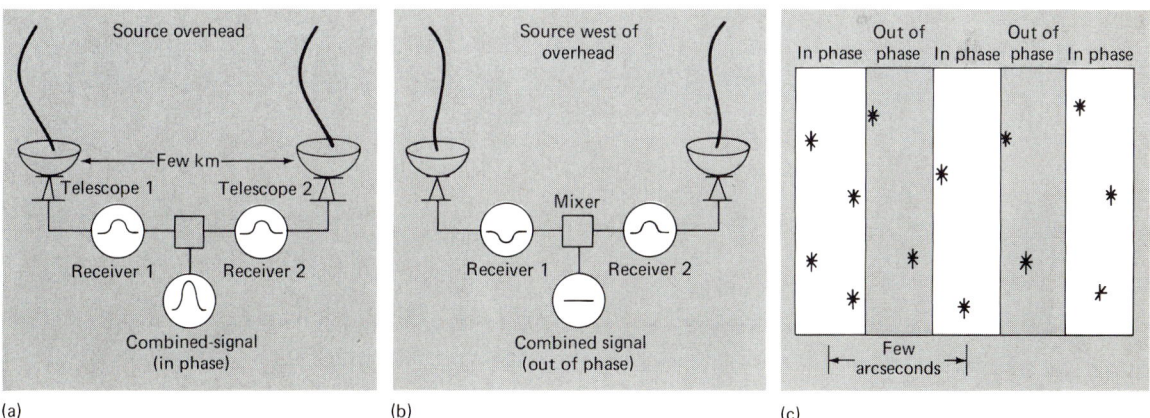

FIGURE 7.19 The operation of a radio interferometer. (a) When a radio source is directly overhead, the waves strike the two antennas in phase, that is, their peaks and dips arrive at both antennas at the same time. An electronic device called a mixer adds the two signals together to get a composite signal. (b) As the source moves west of overhead, the waves arrive out of phase; a peak arrives at one antenna at the same time that a dip arrives at the other. The combination of the signals at the mixer gives a zero net signal. (c) The interferometer sees the sky as a series of in- and out-of-phase bands, each a few arcseconds wide on the sky.

A two-antenna interferometer gives only one part of the separation between points in the source, that across the stripes (in the same direction as the baseline). To completely map a section of the sky with uniform resolving power requires a little work. The trick: twist the object with respect to the interferometer's striped pattern. This twisting can be accomplished naturally by observing the source move across the sky; its motion will twist its orientation with respect to the interferometer's baseline. To help this process along, the baseline of the interferometer can also be twisted or the antennas can be arranged in a "Y" (as in the VLA). Eventually, you end up with a set of maps with the stripes crossing it at many different angles. A computer combines the maps to give a composite picture of the source with high resolution.

How much resolution is possible? If the antennas are 1 km apart and operate at a wavelength of 1 cm, then

$$\theta_{min} = 206265 \frac{0.01 \text{ m}}{10^3 \text{ m}}$$

$$= 2.1 \text{ arcseconds}$$

This resolution is almost as good as optical telescopes and much better than either single dish could achieve.

In some cases even higher resolution can be obtained. In the technique known as Very Long Baseline Interferometry (VLBI), the signals received by very distant antennas (even located on different continents) are recorded on magnetic tape and combined later in a computer. The maximum baseline is the diameter of the earth, 12,000 km, so (for the example above) a resolving power of 2×10^{-4} arcseconds is possible. In the future radio telescopes in space, separated by larger baselines, will give even better resolution.

To construct a complete map of a radio source with a two-antenna interferometer is a tedious task that may take months. The VLA and other interferometer arrays are designed to overcome this difficulty. They consist essentially of many two-antenna arrays, at different angles to each other, operating on the same source at the same time. A complete radio picture can be constructed in about a day or less, depending on the resolution required.

Ground-Based Infrared Astronomy

Carbon dioxide and water vapor in the earth's atmosphere absorb much of incoming infrared radiation. The infrared astronomer can observe at only a few restricted wavelength ranges: 2–25 μm, 30–40 μm, and 350–450 μm. Such observations are best made from high sites in dry climates, where the least amount of water vapor is in the atmosphere above the telescope.

An infrared telescope differs from an optical one in the detector at the telescope's focus. Because our eyes and photographic film sense infrared radiation poorly, special infrared detectors are required; sensitive ones suitable for astronomical work have been around for only about 10 years. A common infrared detector is a *bolometer*; it is a tiny chip of germanium (about the size of the head of a very small nail) cooled to very low temperatures, about 2 K. When infrared radiation strikes a bolometer, it heats up, and its resistance to an electric current changes. Such changes can be measured electronically, and the amount of variation indicates how much infrared energy the bolometer is absorbing.

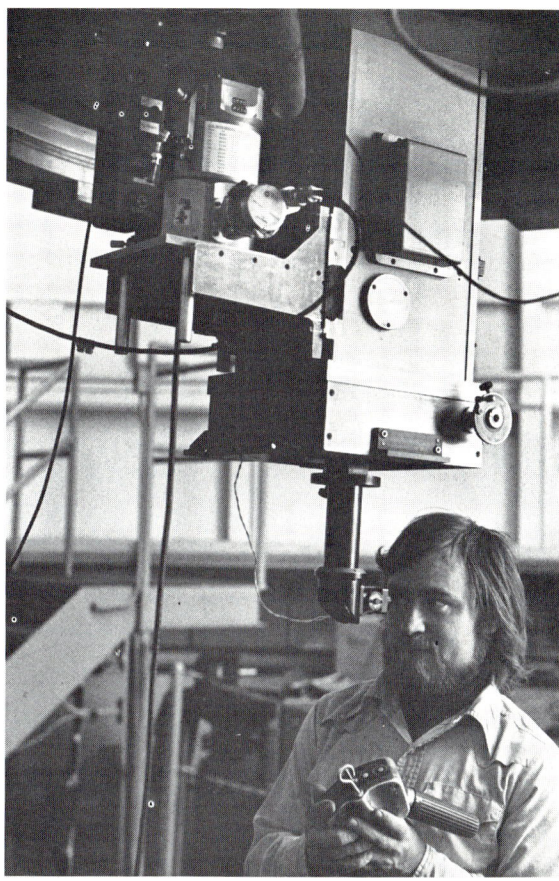

FIGURE 7.20 M. Zeilik observing in the infrared with the 1.3-m telescope of Kitt Peak. The infrared detector, in this case a chip of indium antimonide, is inside a dewar (the metal can in the upper left). The dewar contains liquid nitrogen that keeps the infrared detector cold, so it functions most effectively. [*Photo by P. Heckert.*]

Infrared observing has at least two distinct advantages over optical observing. First, infrared radiation is hindered less by interstellar dust (Chapter 9) than visible light, so you can see through dust clouds more readily. Second, cool celestial objects (3000 K and cooler) give off most of their radiation in the infrared. Typically, such cool objects cannot be seen in visible light, but can be detected in the infrared. So infrared astronomy brings the cold universe into view.

Another practical advantage is that much infrared observing can be done during the day, when crowded telescopes are not wanted by optical astronomers. This is because very little sunlight at infrared wavelengths is scattered by air molecules (Section 14.6), leaving the infrared sky dark both day and night (Fig. 7.20).

Space Astronomy

Parts of the infrared spectrum, ultraviolet light, and X rays can be detected only above the earth's atmosphere from airplanes, balloons, rockets, satellites, or spacecraft.

For example, most of the far infrared radiation (wavelengths longer than about 40 μm) does not make it to the ground. But at altitudes of 15–20 km or so, very little of the earth's atmosphere remains. Far infrared observations can be made at these altitudes from airplanes or balloons (Fig. 7.21). The equipment is simply a reflecting telescope equipped with a bolometer.

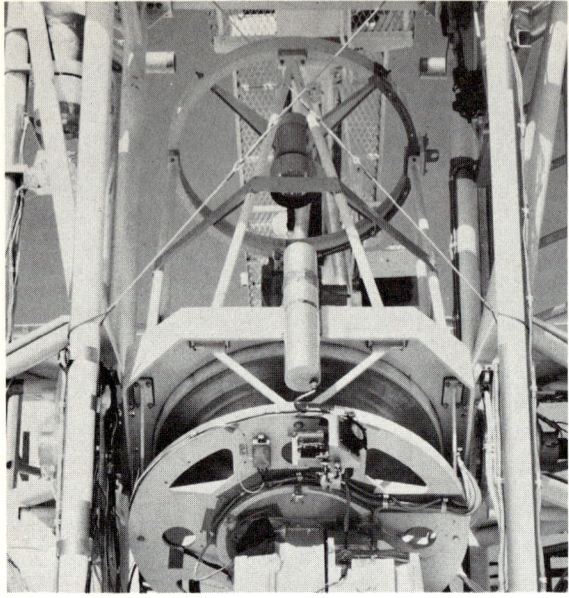
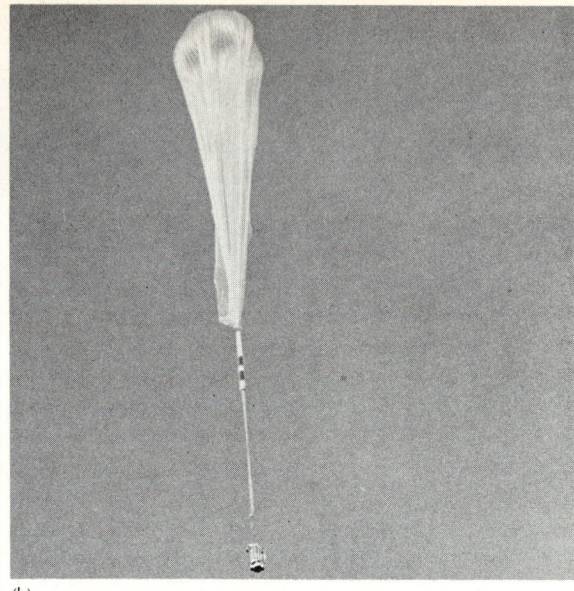

FIGURE 7.21 (a) The 1-m far-infrared telescope of the Center for Astrophysics and the University of Arizona. (b) The telescope at launch, carried aloft by a helium balloon to a height of about 25 km. [*Photos by M. Zeilik.*]

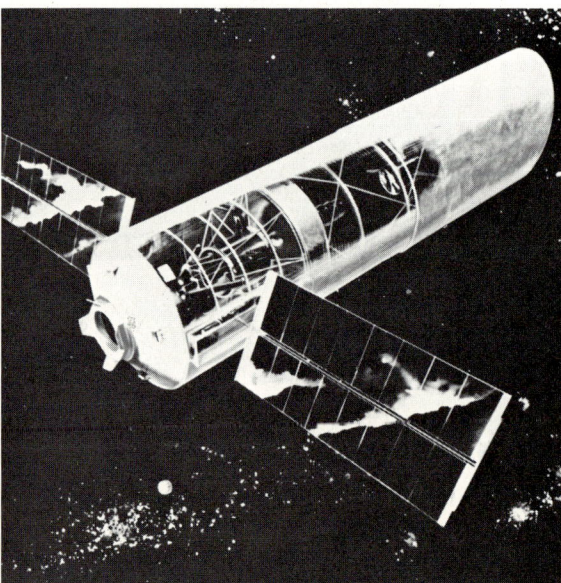

FIGURE 7.22 Artist's conception of the Space Telescope in earth orbit. This telescope will have a 2.4-m mirror and be placed 400 km above the earth by the shuttle. [*Courtesy NASA.*]

For ultraviolet astronomy methods of light gathering and detection are similar to optical astronomy. Photographic plates and some television tubes respond well to ultraviolet light, so detection presents no serious problem. Because glass absorbs ultraviolet light, refracting telescopes cannot be used, but reflectors work perfectly well. All ultraviolet astronomy must be done from space, for the absorbing layer of the atmosphere (the ionosphere) is higher than balloons or aircraft can reach.

For the high-energy realm of X-ray and gamma-ray astronomy, gathering and focusing the light require special techniques. X rays pass through most ordinary matter and so are almost impossible to reflect. In recent years X-ray telescopes have used the fact that X rays can be reflected from certain surfaces if they strike at very small angles, almost parallel to the reflecting surface. Such reflections produce crude, but reasonable images.

Gamma rays are so energetic that to focus them is an impossible task. Gamma-ray telescopes don't yet produce images as such, but they do indicate from what general direction in the sky gamma rays originate.

Finally, even optical astronomy in the near future will involve large space telescopes (Fig. 7.22). The main advantage in space (besides the lack of bad weather) is that a large telescope can be used at its theoretical limit of resolving power rather than as limited by the atmosphere.

SUMMARY

New technology produces new telescopes. These in turn provide new views of the universe (1) by gathering and detecting light too faint to be sensed by the eye, (2) by resolving details in the structure of astronomical objects, and (3) by providing magnified views of what's in the sky.

These new observations influence astronomical models in at least two ways. They provide the grand vision that sets the birth of models, and they confirm or refute the predictions of models. So observations have a push-pull function in astronomy; they pull along new ideas and push them to confirmation or over the edge to oblivion.

Contemporary astronomy has revealed a new face of the universe quite different from that perceived by optical astronomy. Both low-energy (radio and infrared) and high-energy (ultraviolet, X-ray, and gamma ray) astronomy provide complementary views of the newly sensed objects and familiar friends—a rich, sparkling cosmos.

Key Words

optics	objective	magnifying power
refraction	eyepiece	detector
reflection	refracting telescope	contour map
light ray	reflecting telescope	radio telescope
image	Newtonian reflector	radio interferometer
lens	Cassegrain reflector	Very Large Array (VLA)
focus	light-gathering power	infrared telescope
focal length	resolving power	bolometer

Review Questions

1. Reflecting telescopes, in contrast to refracting telescopes (mark all correct answers):
   a. Can be built of very large size.
   b. Focus all colors at the same place.
   c. Form sharp, round images.
   d. Have greater light-gathering power for a given diameter of objective.
   (Objective 4)
2. Which of the following are true?
   a. Radio telescopes generally have greater resolving power than do optical telescopes.
   b. Infrared telescopes differ from optical telescopes primarily in the type of device used to detect the incoming radiation.
   c. X-ray telescopes are usually placed in space satellites because of the danger to human life of exposure to radiation.

d. Light-gathering power, resolving power, and magnifying power all depend on the diameter of the telescopic objective, but in different ways.
e. Changing the eyepiece of a telescope changes the apparent size of an image, but has no effect on the detail which can be resolved or the faintness of the objects which can be seen.
(Objectives 2, 5, 9, and 10)
3. What is the most important function of a telescope? How do refracting telescopes accomplish this function? Reflecting telescopes? (Objectives 2 and 4)
4. Suppose you were presented with optical, infrared, and radio telescopes, all with the same objective size. List them in order of *increasing* resolving power. (Objectives 3, 5, and 9)
5. Imagine that you are going before a congressional hearing to justify the expense of putting a telescope in space. What arguments would you use to persuade the committee? (Objective 10)

Problems

1. Compare the light-gathering power of a 25-cm telescope (a large amateur's telescope) to that of the 5-m Hale telescope. Then compare their theoretical resolving powers. (Objective 3)
2. You buy a 10-cm telescope with a 100-cm focal length. What magnification do you get with a 25-mm eyepiece? 10-mm eyepiece? 5-mm eyepiece? (Objective 3)
3. The Haystack radio antenna has a diameter of 30 m. What is its theoretical resolving power at an operating wavelength of 2 cm? (Objective 3)
4. The maximum separation of two antennas of the VLA is about 30 km. What is the resolving power at a wavelength of 1.5 cm (the shortest it can operate at)? (Objective 3)
5. The transmission of the earth's atmosphere at 3.1 μm is 10 percent. How large a ground-based telescope is needed to gather as much infrared radiation at this wavelength as a balloon-borne telescope of 1-m diameter? (Objective 3)

BEYOND THIS BOOK . . .

* For an intriguing analysis of the development of radio astronomy in Great Britain, read *Astronomy Transformed* (Wiley, New York, 1976) by David Edge and Michael Mulkay.
* For a more American view on radio astronomy, try *The Invisible Universe* (Springer-Verlag, New York, 1974) by Gerrit Verschuur.
* *The Evolution of Radio Astronomy* (Science History Publications, New York, 1973) by J. S. Hey traces the development of techniques and ideas.
* An insight into astronomy as an experimental science can be found in *Experimental Astronomy* (Springer-Verlag, New York, 1970) by Jean-Claude Pecker.
* Colin Ronan in *Invisible Astronomy* (Lippincott, Philadelphia, 1972) makes a good case for how new techniques in astronomy lead to new views of the universe.
* Articles on invisible astronomy, all in *Scientific American*, are: "Infrared Astronomy" by G. Murray and J. Westphal, August 1965; "Intercontinental Radio Astronomy" by K. Kellerman, February 1972; "Ultraviolet Astronomy" by L. Goldberg, June 1969; "X-Ray Astronomy" by H. Friedman, June 1964; "The X-Ray Sky" by H. Schnopper and J. Delvaille, July 1972; and "Gamma-Ray Astronomy" by W. Krauschaar and G. Clark, May 1962.
* *Sky and Telescope* and *Astronomy* magazines contain ads by telescope manufacturers and also classified ads by individuals selling used telescopes (generally good buys because they don't depreciate).

THE STARS: OTHER SUNS

8

LEARNING OBJECTIVES
After studying this chapter, you should be able to:

1. Outline the methods astronomers use to find the following physical properties of stars: (a) surface temperature, (b) chemical composition, (c) radius, (d) mass, (e) luminosity, (f) density.
2. Describe the relationship between a star's color and surface temperature.
3. State and make use of a relationship among a star's luminosity, surface (effective) temperature, and size, assuming it radiates like a blackbody.
4. Use the Balmer series of the hydrogen atom to show that the spectral classification scheme for stars is a temperature sequence relating to their colors.
5. Explain the difference between apparent magnitude and absolute magnitude, between apparent magnitude and flux, and between luminosity and absolute magnitude.
6. State and make use of a relationship among flux, luminosity, and distance.
7. State and make use of relationship among apparent magnitude, absolute magnitude, and distance.
8. Use a star's absolute magnitude to infer its distance and luminosity.
9. Show by a simple diagram the relationship between a star's distance and its parallax; write an expression for this relationship and make use of it.
10. Sketch a Hertzsprung-Russell diagram for stars, indicating the positions of the sun, main sequence, giants, supergiants, and white dwarfs.
11. Use the Hertzsprung-Russell diagram to infer the relative luminosities, surface temperatures, and sizes of stars represented on it.
12. Use the Hertzsprung-Russell diagram to determine the distance of a star by the method of spectroscopic parallax.
13. Describe the three observational classes of binary stars: visual, spectroscopic, eclipsing.
14. Describe what is meant by a Doppler shift and make use of a relationship between Doppler shift and radial velocity.
15. Outline the steps by which astronomers determine the masses of stars.
16. Sketch the mass-luminosity relation for main-sequence stars and use it to infer relative lifetimes for stars of different mass.

CENTRAL QUESTION:
How do astronomers determine the
physical properties of stars?

What is a star? Have you ever asked yourself that question as you've scanned the sky in the perfect glory of a clear, dark night? People have posed it many times for many years and arrived at many answers. The answer astronomers have now is recent, born at the end of the nineteenth and the beginning of this century: stars are suns, enormous fusion reactors in space.

Prior to this understanding, astronomers had for centuries charted the positions of celestial bodies, both stars and planets. The planets, because of their strange retrograde motion, grabbed the astronomers' attention. The stars played a passive role, a placid backdrop to the curious activity of the planets. Many saw no hope of finding out about the physical properties of the stars, but developments in atomic physics at the beginning of the twentieth century proved them wrong.

Stimulated by an attempt to understand light, physicists probed atoms. Atoms in stars emit light, so an analysis of starlight provides information about the physical conditions and chemical compositions of stars. This information transformed the astronomer's view of the cosmos. Stars were no longer regarded as unknowable, eternal points of light. Instead they were seen as suns, made of the same stuff and having the same basic anatomy as our sun, which could be studied, analyzed, and understood.

8.1 SOME MESSAGES OF STARLIGHT

Go out on a clear January night to view the constellations (see the constellation chart for January). Face south. Orion immediately catches your eye (Fig. 8.1). Two stars in Orion shine the brightest: Betelgeuse, which looks reddish, and Rigel, blazing bluish-white. To the south and east of Orion lies Canis Major (the Great Dog), Orion's hunting companion. Sirius, in Canis Major, you can easily notice as the brightest star in the sky. How do these stars compare with our sun? Are they larger? More luminous? Hotter?

Starlight carries information about the physical properties of stars. By deciphering the messages of starlight, we can infer the nature of stars. These physical properties include:

8.1 SOME MESSAGES OF STARLIGHT

(1) chemical composition, (2) surface temperature, (3) radius, (4) luminosity, and (5) mass. For some of these properties we do not need to know the distance to the stars; for others we do. In all cases, we must observe, measure, and analyze the light from the stars as we do for the sun.

Brightness and Flux

If you have even casually looked at the night sky, you know that all stars do not appear to have the same brightness. You make this judgment with your eyes. The lens of your eye focuses the light onto your retina, which detects the light by transforming the radiative energy into electrochemical impulses that travel to your brain. How bright a star appears to you depends on how much energy, each second, strikes your retina. The more energy received, the greater brightness you perceive.

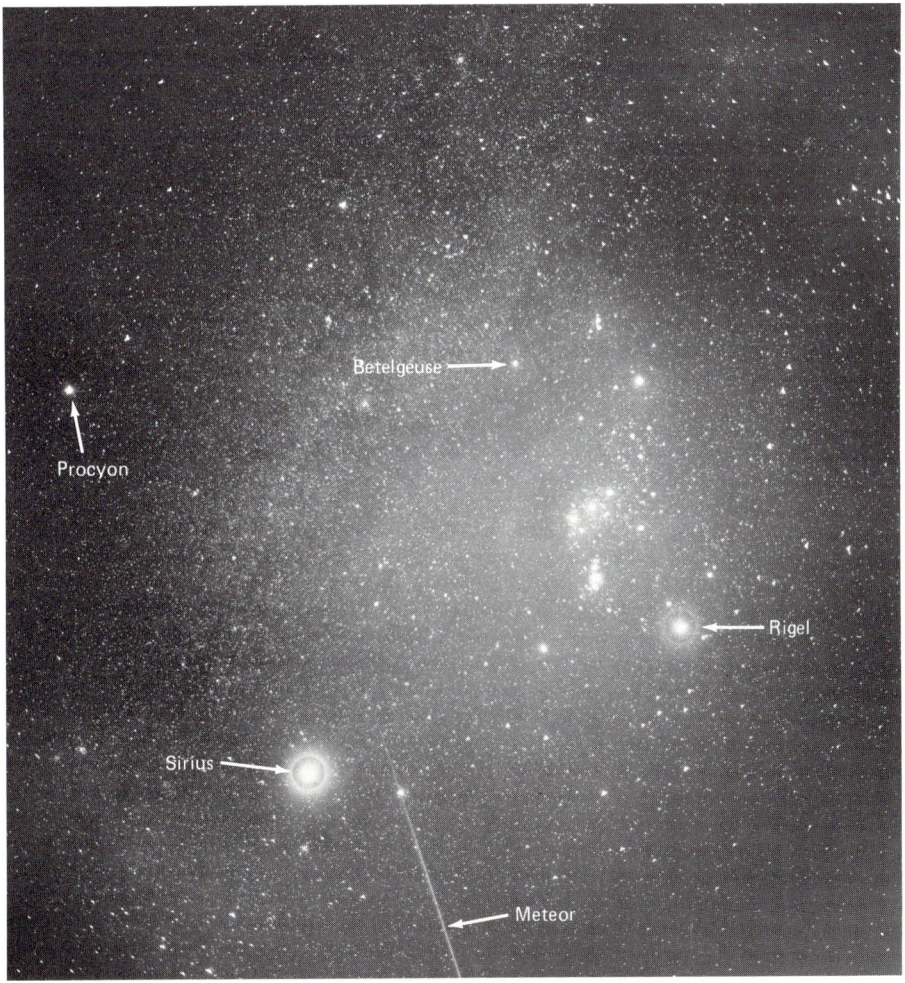

FIGURE 8.1 The sky around the constellation Orion. [*Courtesy F. Wright and the Smithsonian Astrophysical Observatory.*]

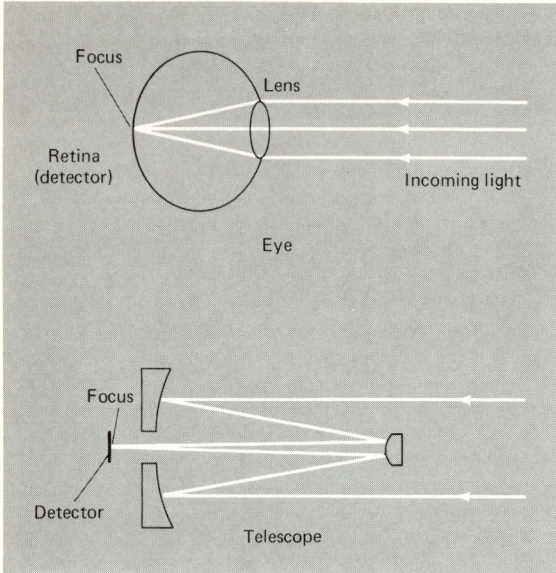 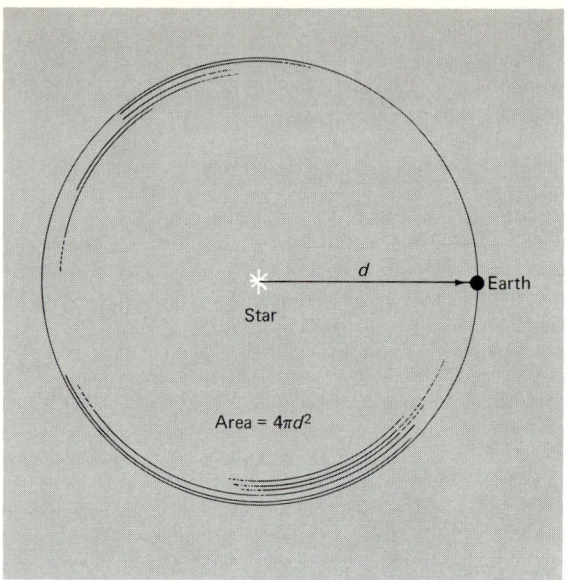

FIGURE 8.2 A comparison of the optics of a human eye and of a telescope.

FIGURE 8.3 Determining the luminosity of a star from its distance and the flux at the earth.

A telescope works the same way as your eye (Fig. 8.2). Its objective gathers light to a focus, and a detector senses the radiative energy. How bright a star appears in a telescope depends on how much energy, each second, arrives at the earth from the star, but also on how large the objective of the telescope is. Just to determine the brightness of the star, then, we need to get rid of the dependence on aperture size. We do this trick by considering only the energy crossing each square meter of the telescope's objective.

The amount of energy striking each unit of area each second is called *flux*. Since watts (Appendix A) are a measure of energy per second, flux can have units such as watts per square meter. For example, the flux at the earth from the sun is 1370 W/m^2.

Flux and Luminosity

Note that brightness actually means flux. "Brightness" is a more common word, but lacks the specific physical meaning of "flux" and is sometimes confused with a different quantity—luminosity. Luminosity is a star's total output in radiative energy each second, while flux is the amount of energy per square meter received at the earth.

Remember how to find the sun's luminosity (Section 6.4)? Measure the sun's flux, find the earth-sun distance in kilometers, construct an imaginary sphere with the earth-sun distance as the radius, and total up all the energy hitting that sphere. You can find a star's luminosity (L) by the same procedure (Fig. 8.3). Measure the star's flux (F) at the earth; knowing its distance (d), find the area of a sphere surrounding the star. Then calculate the flux over the total area, that is,

$$F = \frac{L}{4\pi d^2}$$

8.1 SOME MESSAGES OF STARLIGHT

so that

$$L = F \times 4\pi d^2$$

The formula is simple enough, but we need to know the distance to the star and that may not be easy to come by (see next section).

Warning: To get a star's total luminosity requires that we measure its flux over the complete range of the electromagnetic spectrum. That might not be possible. First, our detector may work only over a limited wavelength range. Second, the earth's atmosphere may absorb some of the energy. Third, matter in interstellar space may also do some absorbing. If the flux is measured over a limited range, say, the visible spectrum, astronomers report this fact by calling it the *visual flux*. The luminosity calculated from it is only the *visual luminosity* of the star. Stars like the sun emit most of their light in the visible part of the spectrum, so their visual luminosity is almost their total luminosity. This is not true, however, for very hot or very cool stars, and some way must be found to transform the measured visual luminosity to total luminosity.

The Inverse Square Law for Light

The square of the distance is a key term in the equation relating flux to luminosity because the brightness of a light source relates in a very specific way to your distance from it.

Imagine the following experiment. Put a light bulb in a socket and turn it on. Place a device used to measure the intensity of light (such as a photographer's light meter) 1 m from the bulb (Fig. 8.4). Note the reading on the light meter, and call that one unit. Move the light meter 2 m from the bulb; the reading will be one-fourth that at 1 m. At a distance of 3 m, the reading will be one-ninth that at 1 m. Note that the light intensity decreases as the *inverse square* of the distance.

To visualize this inverse-square relation easily, imagine a light source placed in the center of two spherical shells, one having twice the radius of the second (Fig. 8.5). The radiation from the source spreads out more by the time it reaches the larger shell than when it hits the smaller sphere. Both spheres receive the same total amount of radiation because the light source is the same, but the radiation is spread over a larger area in the larger sphere. How much larger? The area of a sphere is directly related to the radius

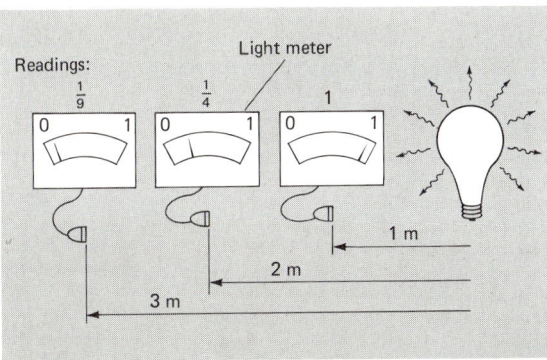

FIGURE 8.4 Light intensity and its change with distance measured with a light meter.

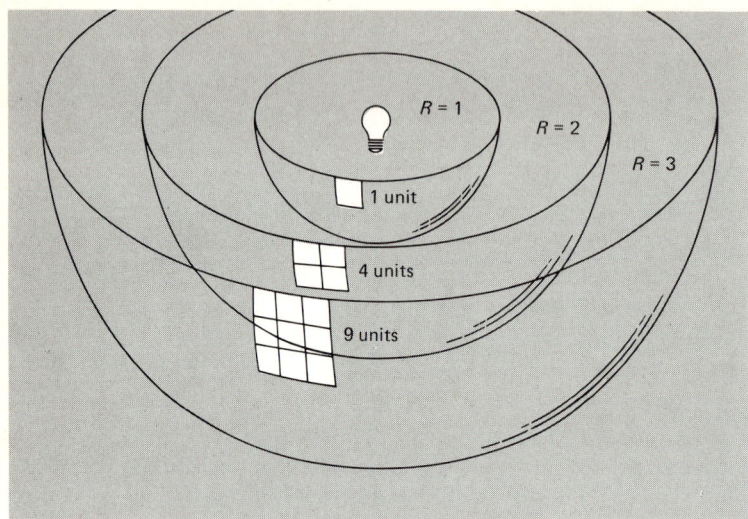

FIGURE 8.5 The geometry of the inverse square law for light intensity. [Adapted from a figure by J. Pasachoff, *Contemporary Astronomy*, Saunders, Philadelphia, 1981.]

squared; so the larger sphere has four times the area of the smaller one, and the light spreads out four times as much. Because of the radiation's dilution over a larger area, any unit of area on the larger sphere has only one-fourth as much light striking it as the same area on the smaller sphere. If the ratio of radii is increased to 3:1 (third shell), the brightness decreases by 9; if the ratio increases to 4, the brightness decreases by 16. In equation form,

$$\frac{\text{near brightness}}{\text{far brightness}} = \left(\frac{\text{far distance}}{\text{near distance}}\right)^2$$

or

$$\frac{b_2}{b_1} = \left(\frac{d_1}{d_2}\right)^2$$

where b_1 is the far brightness, b_2 the near brightness, d_1 the far distance, and d_2 the near distance. This is the *inverse square law* for light (or any electromagnetic radiation).

Suppose you observe that the apparent brightness of one star is 100 times that of another. If you assume that both stars have the same luminosity, how do their distances compare? The brighter one must be 10 times as close as the fainter one, that is, if $b_2 = 100 b_1$, then

$$100 = \left(\frac{d_1}{d_2}\right)^2$$

$$10 = \frac{d_1}{d_2}$$

where d_1 is the distance to the fainter star and d_2 is the distance to the brighter star.

8.1 SOME MESSAGES OF STARLIGHT

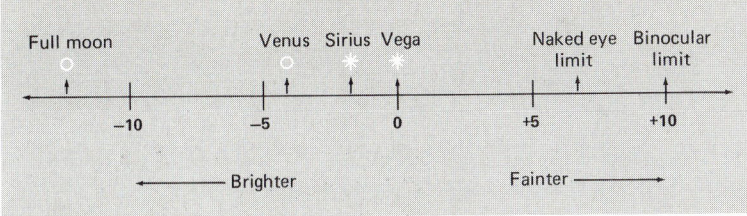

FIGURE 8.6 The astronomical magnitude scale with some common objects indicated. Note that the larger the *negative* magnitude, the *brighter* the object; the larger the *positive* magnitude, the *fainter* it is.

Flux and Apparent Magnitude

For reasons of history and convention, astronomers generally use the term *apparent magnitude* to talk about stellar brightness or flux. Flux and apparent magnitude are two different methods used to describe the same property.

The magnitude scale on which stars are rated has evolved from a convention first established by Hipparchus (160–127 B.C.). In his catalog of stars Hipparchus classified their apparent magnitude by rating the brightest star he could see as magnitude 1 and the faintest as magnitude 6. As this system evolved, some stars were found to be brighter than magnitude 1; for example, Vega is magnitude 0, and Sirius is magnitude -1.4. Note that in this scale the *larger* the magnitude of a star (the more positive), the *fainter* it is; conversely, the smaller the magnitude (the more negative), the brighter it is. A star of magnitude $+6.5$ is fainter than one of magnitude $+4.2$, and one of magnitude -1.3 is fainter than one of magnitude -2.1; a star of magnitude $+12.5$ is brighter than one of magnitude $+17.9$, and one of magnitude -1.8 is brighter than one of magnitude -0.9. Think of magnitude as measuring amount of *faintness*; larger numbers mean fainter stars (Fig. 8.6).

How the human eye judges light affects the magnitude scale. Suppose you tried to estimate the relative brightness of a 100-W and a 200-W light bulb at the same distance. The second bulb does *not* appear twice as bright as the first. And a 400-W bulb will not appear four times as bright. But the *difference* in brightness between the 100-W and 200-W bulbs will appear the same as the difference between the 200-W and 400-W bulbs. The eye sees equal *ratios* of brightness as equal *intervals*, that is, the eye perceives brightness on a logarithmic scale.

On Hipparchus's scale stars of first magnitude were about 100 times brighter than stars of sixth magnitude. The modern system therefore *defines* a difference of 5 magnitudes as corresponding to a brightness ratio of 100. A difference of 1 magnitude then amounts to a brightness ratio of 2.512, the fifth root of 100. Some magnitude differences and corresponding brightness ratios are given in Table 8.1.

Let's look at this mathematically. Take b_1 as the brightness of an object of first magnitude and b_6 as that of an object of sixth magnitude (b_1 and b_6 are physically the fluxes from the objects). Since by definition the difference is 5 magnitudes,

$$\frac{b_1}{b_6} = 100$$

Since $m_6 - m_1$ equals 5, 100 equals $100^{(1/5)(m_6-m_1)}$. So

$$\frac{b_1}{b_6} = 100^{(1/5)(m_6-m_1)}$$

$$= (10^2)^{(1/5)(m_6-m_1)}$$

$$= 10^{(2/5)(m_6-m_1)}$$

Take the log to the base 10 of this equation:

$$\log\frac{b_1}{b_6} = \frac{2}{5}(m_6 - m_1) = 0.4(m_6 - m_1)$$

In general, for any two objects, the brightnesses (b_A and b_B) are related to the magnitudes (m_A and m_B) by the equation

$$\log\frac{b_A}{b_B} = \log b_A - \log b_B = 0.4(m_B - m_A)$$

Note that for a difference of 1 magnitude, say, between first and second magnitude,

$$\frac{b_1}{b_2} = 100^{(1/5)(m_2-m_1)}$$

$$= 100^{(1/5)(1)} = 100^{(1/5)}$$

$$= 2.512$$

So a one unit difference in magnitude equals a brightness or flux ratio of 2.512, and in general

$$\frac{b_A}{b_B} = (2.512)^{(m_B-m_A)}$$

Here's an example. The apparent magnitude of the sun is -26.7. Sirius, the brightest star in the sky, has an apparent magnitude of -1.4. The difference in magnitude is roughly 25. For every 5 magnitudes the brightness ratio is 100 (10^2); so Sirius is approximately five factors of 10^2 (because $25 = 5 \times 5$) or 10^{10} times less bright than the sun. More exactly,

$$\log\frac{b_{Sirius}}{b_{sun}} = 0.4(m_{sun} - m_{Sirius})$$

$$= 0.4[(-26.7) - (-1.4)]$$

$$= -10.1$$

$$\frac{b_{Sirius}}{b_{sun}} = 9.1 \times 10^{-11}$$

8.2 STELLAR DISTANCES

TABLE 8.1 Conversion of magnitude differences to brightness ratios

A magnitude difference of:	Equals a brightness ratio of:
0.0	1.0
0.2	1.2
1.0	2.5
1.5	4.0
2.0	6.3
2.5	10.0
4.0	40.0
5.0	100.0
7.5	1000.0
10.0	10,000.0

In terms of flux, Sirius delivers to the earth about 10^{-10} as much flux as does the sun. The flux received from the sun is 1370 W/m^2 or about 10^3 W/m^2, so the flux from Sirius amounts to only about 10^{-7} W for every square meter of the earth's surface.

Differences in apparent magnitude provide a way of comparing fluxes if you convert to brightness ratios. But note that this comparison tells you nothing directly about a star's luminosity, only about the flux hitting the earth. For the luminosity, we have to know the distance to the star.

8.2 STELLAR DISTANCES

The key to finding a star's luminosity from its flux is knowing its distance. For nearby stars we have a fundamental, direct method to find distances: triangulation, similar to that done by surveyors on the earth. Stellar triangulation is called *heliocentric* or *trigonometric parallax*.

Parallax is actually a common phenomenon that you have observed even if you didn't know its name. Hold your hand out with one finger extended. Now alternately open and close each eye. Your finger will appear to shift back and forth relative to the distant background. Move your finger closer and the shift increases; move it farther away, the shift decreases. This angular shift in your finger's apparent position with respect to the background is the parallax of your finger.

Imagine that your finger is a nearby star, the background more distant stars, and your eyes the sighting positions of the earth in orbit around the sun separated by a time of six months. Parallax occurs when you view a relatively close object from each of two ends of a baseline. As the earth moves from one place in its orbit to another (Fig. 8.7a), the nearby star's position seems to change relative to the more distant stars (Fig. 8.7b). The maximum shift occurs when you view the star six months apart, so you are sighting from opposite sides of the earth's orbit. The angular shift (actually half the shift) is the star's *parallax*. Because you can measure the angular shift and you know the diameter of the earth's orbit, you can calculate the distance to the star.

Now suppose you travel out into space and look back at the earth-sun separation

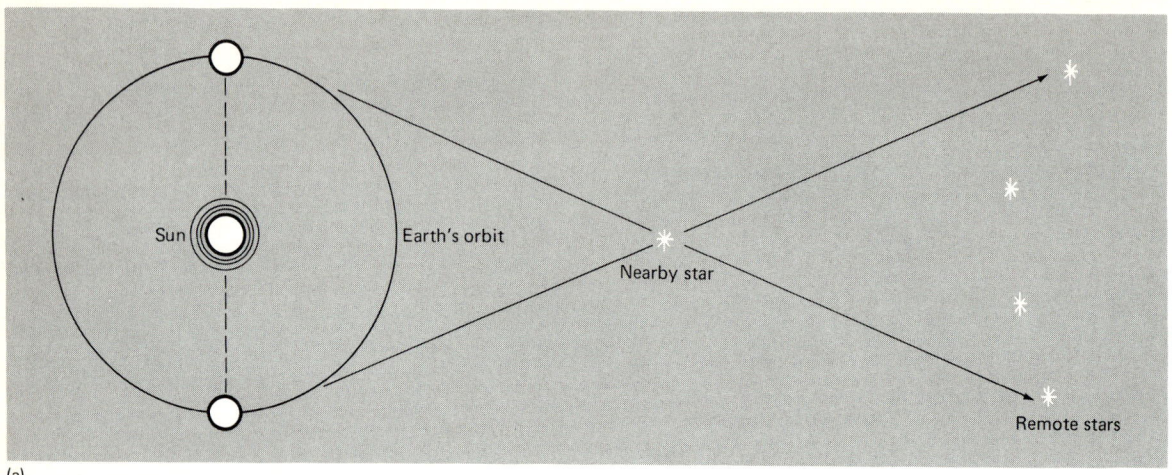
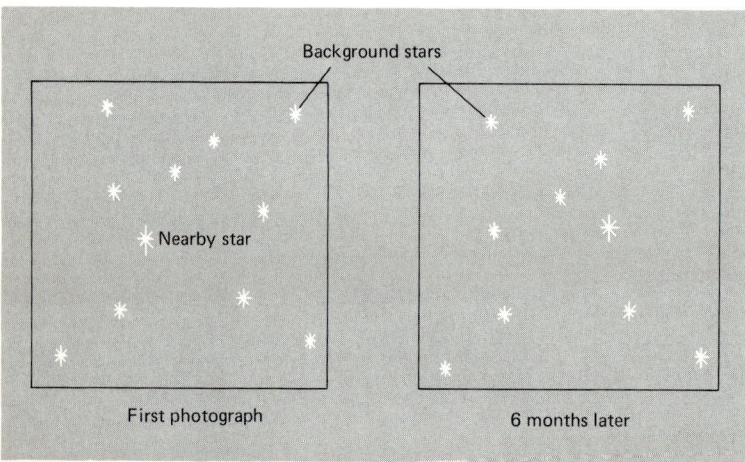
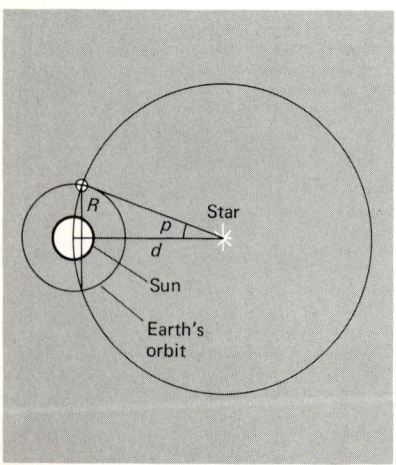

FIGURE 8.7 Heliocentric stellar parallax. As the earth goes around the sun (a), the positions of nearby stars shift relative to the positions of more distant stars (b). The *closer* the star, the *greater* the observable shift. The shift has a 1-yr cycle, the earth's period of revolution. (c) Geometry for heliocentric parallax.

8.2 STELLAR DISTANCES

(1 AU). You keep going until the angular size of the AU is 1 second of arc. Suppose you station a star at this point and hurry back to the earth. You observe this star for one year and measure its shift. One-half of the total shift for one year would equal 1 second of arc. We say the star is 1 *parsec* (pc) from the sun, that is, the distance at which the *parallax* is 1 *second* of arc, a distance roughly equal to 3.26 light years. Suppose the star were twice as far away; it would have half the parallax, $\frac{1}{2}''$. If it were half the distance away, its parallax would double and be $2''$. Note this simple inverse relationship of parallax and distance:

$$d = \frac{1}{p}$$

where d is the distance in parsecs and p is the parallax in seconds of arc.

As an example of the use of this relationship we can calculate the distance of the star 40 Eridani from the sun. The star has a parallax of 0.2 arcsec. Since the star's parallax is one-fifth that for a star 1 pc away, 40 Eridani must be 5 pc distant. Explicitly,

$$d = \frac{1}{0.2''}$$
$$= 5 \text{ pc}$$

Here's a simple geometrical explanation for the parallax formula. Draw a circle with a star at the center and the sun on the circumference (Fig. 8.7c). The circle's radius is d. Note that the parallax angle p (in degrees) is some fraction of the total circle (360°). Also, R, the earth-sun distance (1 AU), corresponds closely to an arc on the circumference of the circle; its fraction of the circumference is the same as the fraction p is of 360°. So

$$\frac{R}{2\pi d} = \frac{p}{360°}$$

and

$$d = \frac{360° R}{2\pi p}$$

Convert 360° to seconds of arc:

$$d = \frac{360 \times 60 \times 60}{2\pi} \frac{R}{p}$$
$$= 206,265 \frac{R}{p}$$

Now R is 1 AU; define 1 pc as 206,265 AU. Then

$$d(\text{pc}) = \frac{1}{p(\text{arcsec})}$$

or, since 1 pc = 206,265 AU = 3.086 × 10^{13} km = 3.26 ly,

$$d(\text{ly}) = \frac{3.26}{p(\text{arcsec})}$$

Heliocentric parallax works accurately only for close stars. The parallax of Sirius, a very close star, is only 0.37″, much less than 1 arcsec. One arcsec is about 1/2000 the angular diameter of the moon! That's the size of a U.S. quarter at a distance of a little more than 5 km. People who measure parallaxes have refined their techniques to an accuracy of about 0.01″, so they can get accurate distances out to about 300 ly (100 pc).

Many stars lie more distant than 300 ly. How are their distances measured? By ingenious, indirect methods. One, called spectroscopic parallaxes, comes later in this chapter. Others will crop up in Chapter 20.

Absolute Magnitude and Luminosity

If we know a star's distance by heliocentric parallax, we can find its luminosity from its flux (or apparent magnitude). The inverse square law, with the star's distance and flux known, gives the luminosity from the relation

$$L = F(4\pi d^2)$$

Since the small fluxes from stars are often difficult to measure in real units of W/m^2, astronomers often express luminosities by comparing the brightness of a star to that of the sun, with both stars imagined to be at the same distance. Astronomers do this comparison by a system called *absolute magnitude*. Suppose that you could place all stars the *same* distance from the earth. Then differing distances would not play a role in how bright the stars appeared. Any differences in magnitudes among them would arise only from differences in luminosities.

Astronomers do set stars, in an imaginary way, at a standard distance in order to compare their luminosities. We use a distance of 10 pc, or 32.6 ly, for the comparison.

Imagine that you could transport the stars in the sky, including the sun, to 10 pc from the earth (Fig. 8.8). Stars now closer than this distance would appear fainter, and those now farther would get brighter, as expected from the inverse square law. How bright a star would appear at 10 pc from us is termed its *absolute magnitude*.

Here are a few examples. The sun at 1 AU has an apparent magnitude of −26.7. At 10 pc it would have an apparent magnitude of about +4.8, and that is its absolute magnitude. Sirius is 2.7 pc away and its apparent magnitude is −1.4; move it farther away, to 10 pc, and it grows dimmer, until its magnitude appears to be +1.4. That's the absolute magnitude of Sirius. Now try Polaris, which is 240 pc (780 ly) from the sun. Its apparent magnitude is +2.3. Moving it to 10 pc brings it closer, so it would appear brighter, with an absolute magnitude of −4.6.

Compare the absolute magnitudes of these stars: the sun +4.8, Sirius +1.4, Polaris −4.6. These magnitudes now reflect the stars' luminosities. Note that Polaris is actually the most luminous, followed by Sirius, with the sun last.

Once we know a star's absolute magnitude, we can compute its luminosity by compar-

8.2 STELLAR DISTANCES

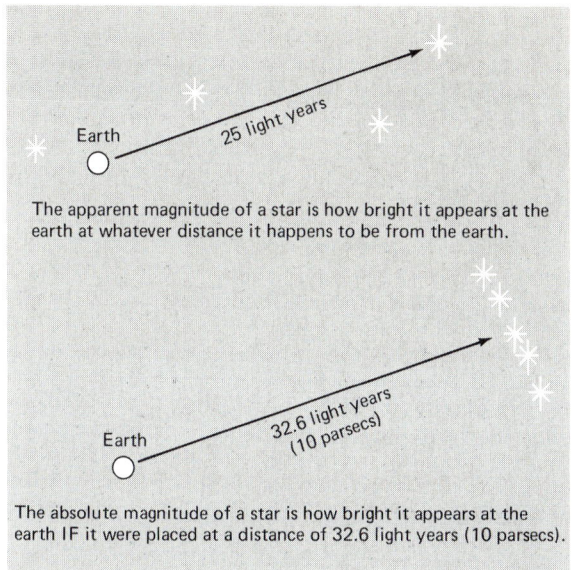

FIGURE 8.8 The concepts of apparent and absolute magnitude compared.

ing its absolute magnitude to that of the sun. The trick here is to convert magnitude differences into brightness ratios (Section 8.1). Let's compare the sun to Sirius. In the visual range the sun's absolute magnitude is +4.83, and that of Sirius +1.42. The difference is roughly 3.4 magnitudes, so the brightness ratio is about $2.512^{3.4} \approx 23$. So at visual wavelengths, Sirius has roughly 23 times the sun's luminosity.

Warning: This comparison does *not* give the luminosity ratio over *all* wavelengths, because it compares just visual magnitudes. The magnitude of the sun (or any star) including light at all wavelengths is called its *bolometric magnitude*. The absolute bolometric magnitude of the sun is +4.75, that of Sirius +1.37. So the magnitude difference is 3.38 and the brightness ratio 22. Including all wavelengths Sirius has 22 times the sun's luminosity.

The above examples indicate that apparent magnitude, absolute magnitude, and distance are related to one another. From a star's distance and apparent magnitude, you can work out its absolute magnitude. In fact, if you know any two of the three properties, you can find the third. Algebraically, the relationship is:

$$m - M = 5 \log d - 5$$

where m is the apparent magnitude, M the absolute magnitude, and d the distance in parsecs.

Let's see how to derive this relation. Since brightness ratios can be expressed as magnitude differences, and brightness depends on the inverse square of the distance, magnitude differences can be expressed in terms of ratios of distance. The inverse square law of light is

$$\frac{b_A}{b_B} = \left(\frac{d_B}{d_A}\right)^2$$

and the relationship of brightness ratio to magnitude is

$$\frac{b_A}{b_B} = (2.512)^{m_B - m_A}$$

where b_A, m_A, and d_A are the brightness, magnitude, and distance of star A, and b_B, m_B, and d_B are the brightness, magnitude, and distance of star B. Putting these two together,

$$\frac{d_B}{d_A} = (2.512)^{(m_B - m_A)/2}$$

This equation is useful enough as it is, but it can be written a little more neatly using logarithms:

$$\log d_B - \log d_A = 0.2(m_B - m_A)$$

Now suppose we say d_A is 10 pc, the distance at which absolute magnitude M is defined, and d_B and m_B are the distance and apparent magnitude of a star. Then this relationship becomes

$$\log d - 1 = 0.2(m - M)$$

A more common way to write it is

$$m - M = 5 \log d - 5$$

An example: The apparent visual magnitude of Sirius is -1.46 and its distance is 2.65 pc. What is its absolute magnitude?

$$M = m + 5 - 5 \log d$$
$$= -1.46 + 5 - 5 \log 2.65 = +1.42$$

To change absolute magnitudes to luminosities, you need to compare the star to the sun. The relation is

$$\frac{L_*}{L_{sun}} = (2.512)^{M_{sun} - M_*}$$

where the absolute bolometric magnitude of the sun is 4.75. Using Sirius as an example,

$$\frac{L_{Sirius}}{L_{sun}} = (2.512)^{4.75 - 1.37} = 22.5$$

8.3 STELLAR COLORS, TEMPERATURES, AND SIZES

Now return to the stars in the winter sky. We've dealt with one property you can easily observe: their apparent magnitude or flux. Another property you can see is color. Betelgeuse looks reddish, Rigel bluish-white, Sirius white, and Capella yellowish-white.

Color and Temperature

These different colors suggest that these stars have different surface temperatures. (Recall Section 6.4 pertaining to the sun's surface temperature.) Rigel is the hottest of the four (Fig. 8.9); it is so hot that it emits more blue light than any of the other visible colors. In contrast, Betelgeuse is the coolest, for it radiates mostly red light. The colors of stars give us clues to their surface temperatures.

Some words of caution: First, when we look at starlight with our eyes, we see only the visible part of its entire spectrum. Stars radiate very much like blackbodies, so their continuous spectra, from shortest to longest wavelengths, are similar to a blackbody shape (Fig. 8.10), called a *Planck curve*. The visible range covers but a small part of a blackbody's spectrum. Second, color indicates temperature only for light emitted by a body, not light reflected from it. A red shirt is not necessarily cooler than a pair of blue pants, for example.

Note in Fig. 8.10 how a change in temperature results in a change in the brightness of one color compared with another. Suppose you measured two colors' brightness, say, blue at 4400 Å and green at 5500 Å. From these measurements make a brightness ratio, in this case of blue to green. You expect this ratio to be larger for hotter stars and smaller for

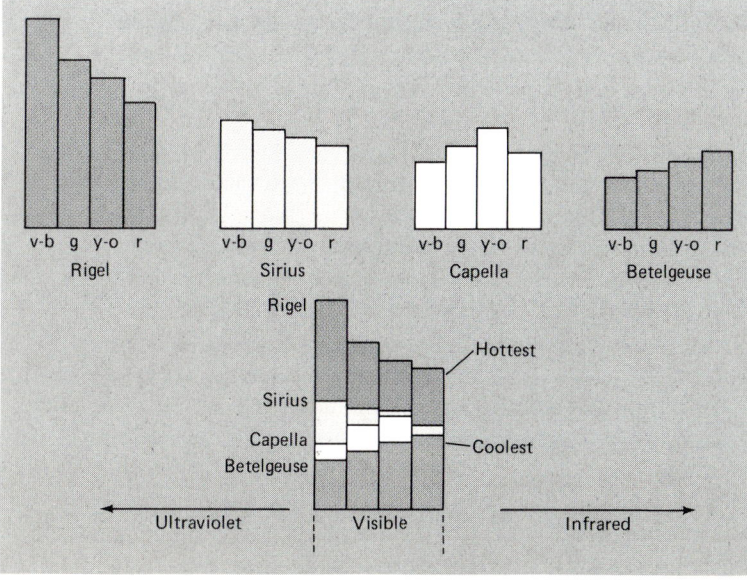

FIGURE 8.9 The intensities of Rigel, Sirius, Capella, and Betelgeuse compared in the color bands violet-blue (v-b), green (g), yellow-orange (y-o), and red (r).

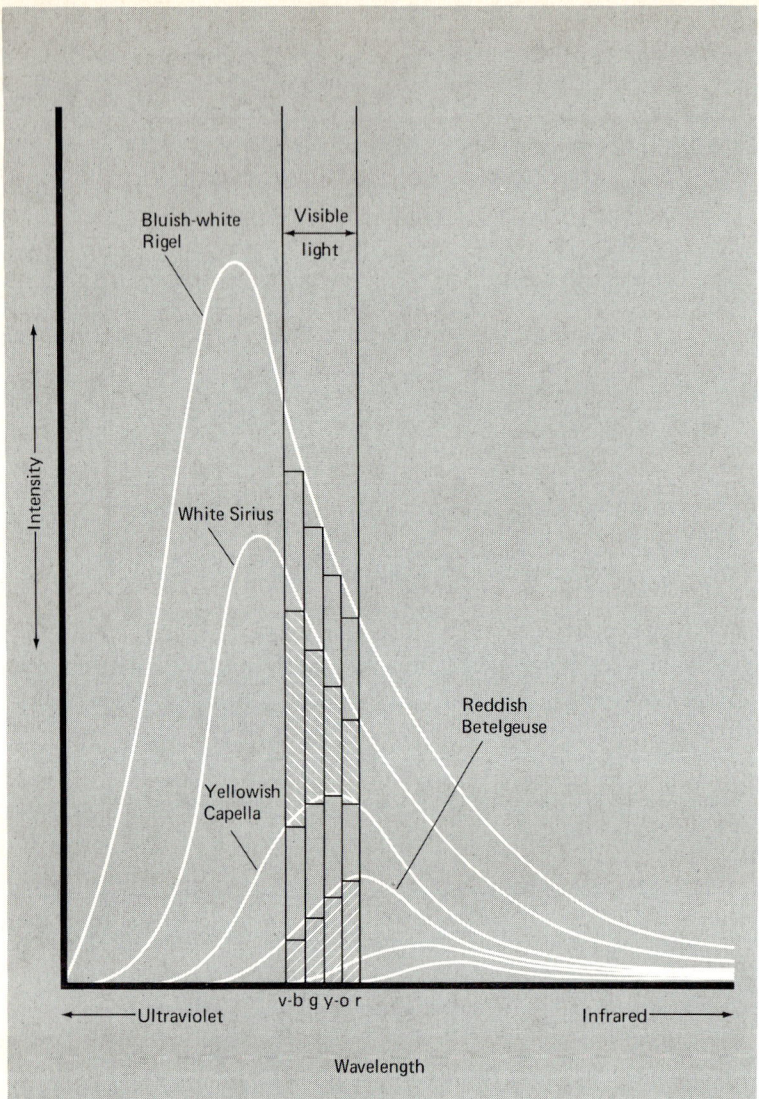

FIGURE 8.10 The continuous spectra of stars with different surface temperatures (those in Fig. 8.9). Their spectra have a distinctive shape known as a *Planck curve*. Note that the visible range is only a small part of the total spectrum.

cooler ones. So the two-color brightness ratio measures temperature. Astronomers call the temperature obtained by this method a star's *color temperature*.

Because stars differ so much in brightness, astronomers prefer to express their brightnesses as a magnitude. Similarly, they generally use magnitude differences to represent brightness ratios at different wavelengths. A filter that lets in light around 4400 Å is called a B filter; that which passes 5500 Å light a V filter; the magnitudes measured at these wavelengths are called B and V magnitudes. The difference between these magnitudes, B-V, is called the *color index* and is a way to measure a star's temperature (Table 8.2). The system is adjusted so stars of about 10,000 K temperature have a color index of 0.0. Very hot stars have negative color indexes, very cool stars positive ones. The sun has a color index of +0.66. Note that the color index is a magnitude difference, so it corresponds to brightness ratios at different colors.

Another way to measure a star's temperature is to find the wavelength of the peak in its

8.3 STELLAR COLORS, TEMPERATURES, AND SIZES

TABLE 8.2 Color index and surface temperature

Color Index (B-V)	Temperature (Kelvins)	Spectral Type
−0.32	47,000	O5
−0.29	30,300	B0
−0.16	15,300	B5
0.00	9,410	A0
+0.14	8,210	A5
+0.31	7,160	F0
+0.43	6,560	F5
+0.59	6,010	G0
+0.66	5,780	G5
+0.82	5,260	K0
+1.15	4,270	K5
+1.14	3,880	M0
+1.61	3,260	M5

Source: D. S. Hayes, "The Absolute Calibration of the H-R Diagram," from *The H-R Diagram*, edited by A. G. D. Philip and D. S. Hayes (Reidel, Dordrecht, 1978).

spectrum, and find what temperature blackbody would have the same peak in its Planck curve (Fig. 6.24 and Fig. 8.11). The hotter the star, the shorter the wavelength of the peak. Though simple, this technique has two drawbacks. First, we need to measure a wide range of the spectrum to find the peak. That's more work than measuring just two colors. Second, at the ground the earth's atmosphere may absorb the radiation at the wavelength of the peak (Section 7.2). For example, the hottest stars have spectra that peak in the ultraviolet, which doesn't make it through the atmosphere to a ground-based telescope.

We'll see later that the temperature can also be determined by examining the lines in a star's spectrum.

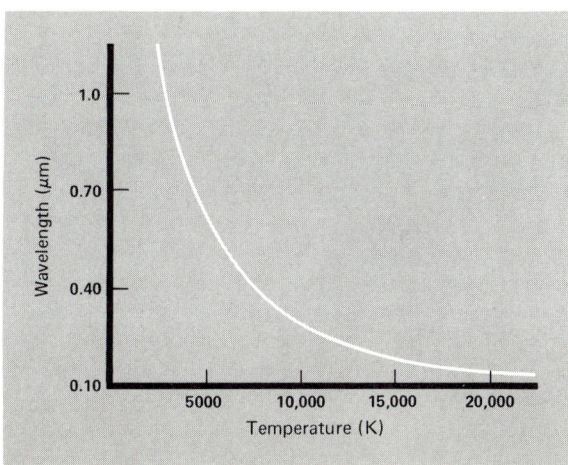

FIGURE 8.11 The peak wavelength of the continuous emission from stars as it relates to their surface (effective) temperature, for typical stellar temperatures.

Temperature and Radius

One reason we've concentrated on color is that we can use the stellar temperatures obtained from the colors to infer the sizes of stars, if we assume that the stars radiate like blackbodies.

Here is an example of how this estimate of size is done. In the constellation Scorpio, you can find the bright, reddish star Antares (see the June star map). If you view Antares through a telescope, you'll find it has a faint, bluish-white companion. Antares is a binary star (more to come in Section 8.6); it and the bluish-white companion revolve around each other, bound by gravity. The reddish star is called Antares A, the bluish-white companion Antares B. Antares A has a color index of $+1.81$ and thus a surface temperature of roughly 3000 K; its companion has a color index of -0.15 and thus a temperature of about 15,000 K.

Telescopic observations show that the flux from Antares A is about 40 times more than the flux from Antares B. Since both stars lie at the same distance from the earth, the difference in flux cannot arise from different distances; it must come from differences in the stars' luminosities. Antares A is 40 times more luminous than Antares B. But how can a cool star be so much more luminous than a hotter one?

Recall that the amount of energy emitted *by each unit of area* of a blackbody's surface depends only on its temperature, specifically, on the fourth power of the temperature. For example, Antares B is about 5 times hotter than Antares A, so each square meter of Antares B emits 5^4, or 625 times the energy of Antares A each second (Fig. 8.12). But the number of square meters on the surface of each star is not the same. So if Antares A had 625 times the surface area of Antares B, both stars would have the same luminosity. Because Antares A has 40 times the luminosity of Antares B, it must have 40×625, or 25,000 times the surface area.

From the example you can see that size can be estimated because a star's luminosity is related not only to its surface temperature (which determines how much energy each square meter emits), but also to its surface area (which determines the total number of square meters doing the emitting). For a spherical star, the surface area is simply related to its radius by $A = 4\pi R^2$, and the energy emitted per unit area is σT^4. So for a spherical blackbody there is a simple relation among its luminosity (L) in watts, its radius (R) in meters, and its temperature (T) in Kelvins:

$$L = 4\pi R^2 \sigma T^4$$

where σ is a constant with the value of 5.67×10^{-8} in the above system of units. You can apply this relation to stars since they radiate more or less like blackbodies. The T in the above equation is called the *effective temperature*, in essence, the temperature of a blackbody with the same radius and luminosity as the real star. The effective temperature is a measure of the total energy, over *all* wavelengths, radiated from each square meter of a star's surface.

Now you have a way to infer the radius of a star from its luminosity and surface temperature. Determine its luminosity and temperature, then plug these values into the relation above, and you have the radius. Note this procedure requires that you find the star's luminosity; to do this, you usually need to know the star's distance.

Take the star Betelgeuse in the constellation Orion as an example. Its surface tempera-

8.3 STELLAR COLORS, TEMPERATURES, AND SIZES

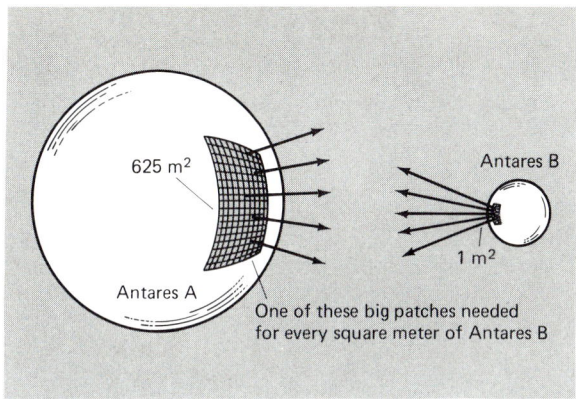

FIGURE 8.12 A comparison of the emission from the surfaces of Antares A and B.

ture is 3250 K and its luminosity 1.2×10^5 solar luminosities. Then, if Betelgeuse radiates like a blackbody, its luminosity is

$$L = 4\pi R^2 \sigma T^4$$

so its radius is

$$R = \left(\frac{L}{4\pi \sigma T^4}\right)^{1/2}$$

Since $L_{sun} = 3.83 \times 10^{26}$ W, the luminosity of Betelgeuse is $L = (1.2 \times 10^5) \times (3.83 \times 10^{26}) = 4.6 \times 10^{31}$ W, and T is 3250 K, so

$$R = \left(\frac{4.6 \times 10^{31}}{4\pi(5.7 \times 10^{-8})(3250)^4}\right)^{1/2}$$

$$= 7.6 \times 10^{11} \text{ m}$$

$$= 7.6 \times 10^{8} \text{ km}$$

Betelgeuse is about 1100 times larger in radius than our sun and about 5 times larger than the earth's orbit.

Direct Measurement of Diameters

If we could measure a star's angular diameter and its distance, we could calculate its actual diameter directly (using the small angle approximation). The problem with this procedure is that stars are so far away that their disks have very small angular diameters, on the order of a few milliarcseconds (1 milliarcsecond is 10^{-3} arcsecond)—far too small to resolve with earth-based telescopes. (Stellar images rarely have angular diameters less than about 1 arcsecond, even during excellent seeing conditions.)

One method of obtaining the high angular resolution required involves *lunar occultations*. The moon occults a star when it passes in front of the star as viewed from the earth. Since an occultation, like a solar eclipse, is visible only across a narrow path on the earth,

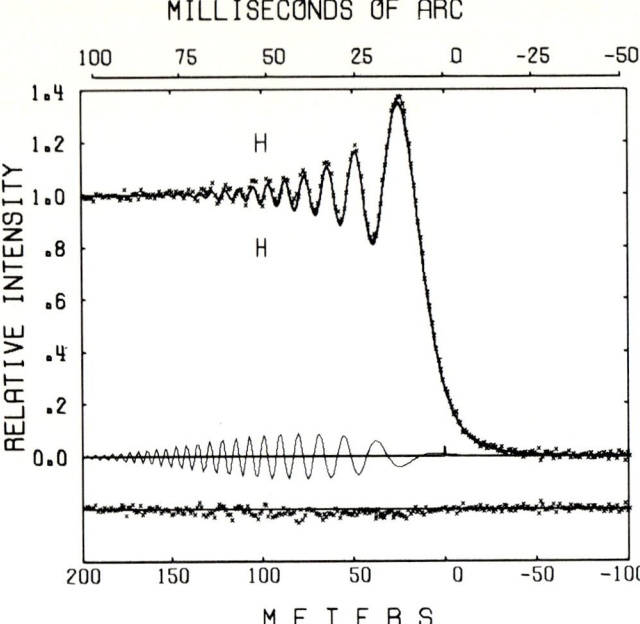

FIGURE 8.13 An occultation of a star (named BD +16° 625) by the moon, as observed at the Kitt Peak 4-m telescope. As the moon cuts in front of the star, its intensity makes a wavy pattern until it is completely covered. [*Observations by S. Ridgeway, D. Wells, R. Joyce, and R. Allen, Astronomical Journal, vol. 84, p. 247, copyright 1979 by the American Astronomical Society.*]

an observatory at a fixed location can see only a small number of all occultations that occur. Occultation observers make very high speed measurements of a star's light as the moon occults it (Fig. 8.13). The larger the star, the more slowly the light is dimmed. Because we know the moon's angular speed in the sky, we can find out a star's angular diameter from accurate timing of the occultation. There are complications (such as whether or not the star is uniformly bright across its surface), but several different groups have made measurements of the same stars and the results are consistent.

One star measured this way is Aldebaran (Alpha Tau), the red eye of Taurus the Bull. The results average about 21 milliarcseconds for the angular diameter (α). The distance (d) of Aldebaran is 67.8 ly (20.8 pc). Using the skinny triangle approximation (Section 1.5), we can compute its diameter D from the relation

$$\alpha_{radian} = \frac{D}{d}$$

Now 21×10^{-3} arcsecond = 1.02×10^{-7} radians, so

$$D = (1.02 \times 10^{-7})(67.8 \text{ ly})$$
$$= 6.92 \times 10^{-6} \text{ ly}$$
$$= 6.54 \times 10^{7} \text{ km}$$

So Aldebaran's radius is 3.27×10^{7} km, or about 50 times larger than the sun's radius.

8.4 SPECTRAL CLASSIFICATION OF STARS

The spectra of most stars resemble that of the sun (Fig. 8.14). From the sun's spectrum astronomers infer its photospheric composition (Section 6.2). The same procedure can be applied to stars. But there's one trick: A star's spectrum is affected by its surface temperature, as well as by its composition.

We now know that most stars have pretty much the same composition, but their spectra are *not* all alike. For example, stars cooler than the sun have spectra with weaker Balmer lines of hydrogen (Section 6.3), stars somewhat hotter show stronger Balmer lines, but stars *much* hotter again have spectra with weak Balmer lines.

Recall that the Balmer series is the name given to the set of spectral lines of hydrogen arising from transitions between the second and higher energy levels, that is, 2 to 3, 2 to 4, 2 to 5, etc. (in absorption) or 3 to 2, 4 to 2, 5 to 2, etc. (in emission). To absorb the visible light of the Balmer series, a hydrogen atom must be excited to level 2. In the sun

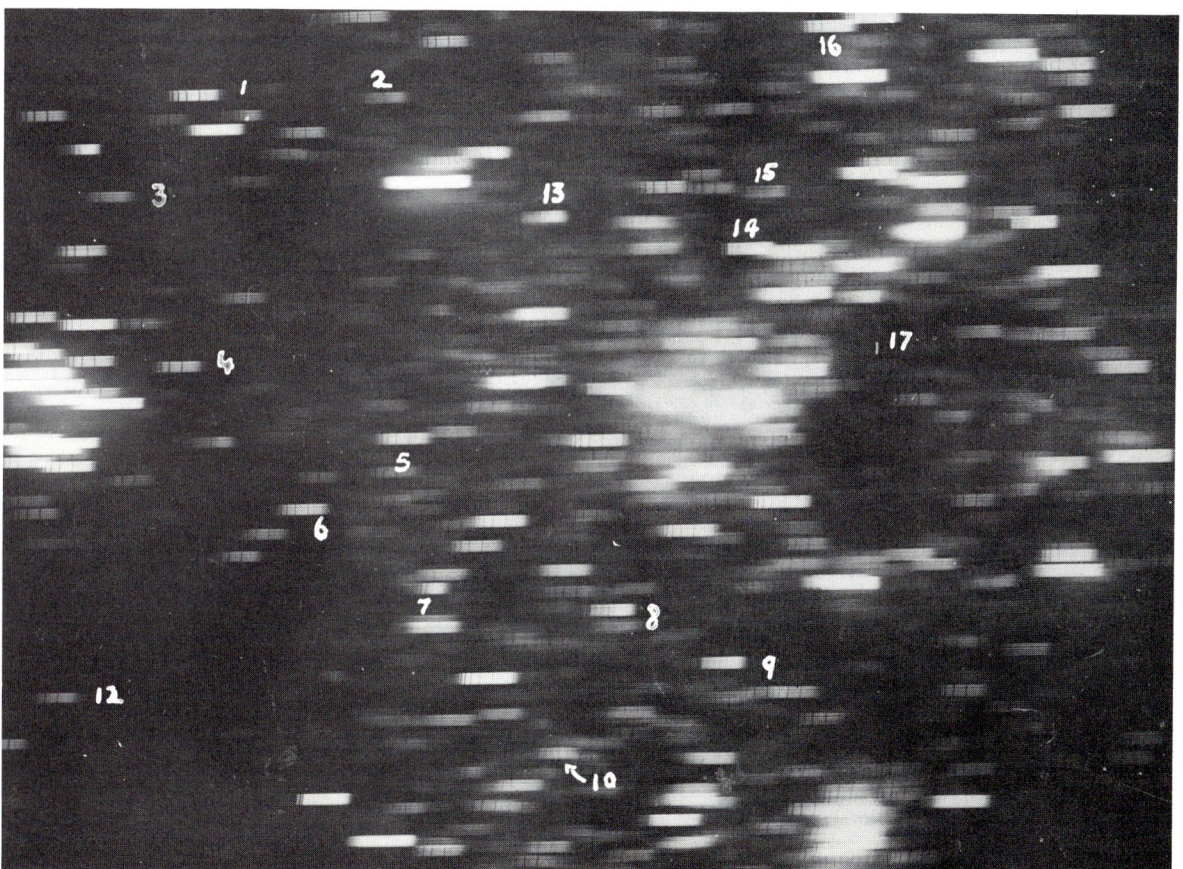

FIGURE 8.14 The spectra of many stars, made by placing a large prism in front of a telescope. Note that most show dark lines, as does the sun's spectrum. [*Courtesy Harvard College Observatory.*]

only 1 of every 10^8 atoms is excited; not enough energetic collisions occur to kick up many electrons from the ground level. So the Balmer lines in the sun's spectrum are not very strong. In a star cooler than the sun there are fewer energetic collisions, and fewer hydrogen atoms are excited; compared with the sun, the Balmer lines are much weaker. In a hotter star, say 10,000 K, there are more energetic collisions, more atoms are excited (about 1 out of every 10^6), so the Balmer lines are more intense (darker) in such a star. But if the star is very hot, say 20,000 K, collisions are so energetic that many electrons are knocked out of the atom entirely, leaving hydrogen *ions* (protons) behind. So, although a larger *fraction* of the atoms are excited, there are fewer atoms in total with excited electrons, and the Balmer lines are again weak.

This is the key point about the Balmer absorption lines: They are produced by atoms with an electron in the $n = 2$ energy level, that is, the atom must be *already* excited. If for some reason there aren't very many hydrogen atoms excited to the $n = 2$ state in a star, the Balmer lines in that star will be weak. This can happen in two ways: (1) The star may have a very high temperature, so that virtually all of its hydrogen is ionized, thus leaving very little hydrogen to produce absorption lines. (2) The star may have a relatively low temperature, so that even though there is much neutral hydrogen, there are very few atoms excited up to the $n = 2$ level. Once again, there aren't enough atoms in the right energy state to produce strong Balmer lines.

Let's compare the behavior of the Balmer lines to that of the Lyman lines originating in the $n = 1$, or ground state, of the hydrogen atom. (Recall from Section 6.3 that Lyman lines appear in the ultraviolet part of the electromagnetic spectrum.) At about 30,000 K essentially all the hydrogen will be ionized, and both the Balmer and Lyman lines will be weak or nonexistent. At 10,000 K the hydrogen is not hot enough to be mostly ionized, but it is hot enough to have a large number of atoms in the $n = 2$ state, so you will see strong Balmer lines in absorption. The Lyman lines will also be strong, because a large number of the neutral atoms will be in the ground state, no matter what the temperature. At cooler temperatures, say, about 4000 K, very few atoms can be excited up to the $n = 2$ state; most of them will be in the $n = 1$ state. The Balmer lines in this case will be weak, but the Lyman lines will be quite strong.

At the turn of this century workers at Harvard College Observatory classified stellar spectra by using absorption lines, especially Balmer lines. Much of this work was done by Annie Jump Cannon (1863–1941; Fig. 8.15), who single-handedly classified the spectra

FIGURE 8.15 Annie Jump Cannon (left) and Henrietta Swan Leavitt (right), two workers at Harvard College Observatory who did fundamental work in astrophysics. [*Courtesy Harvard College Observatory.*]

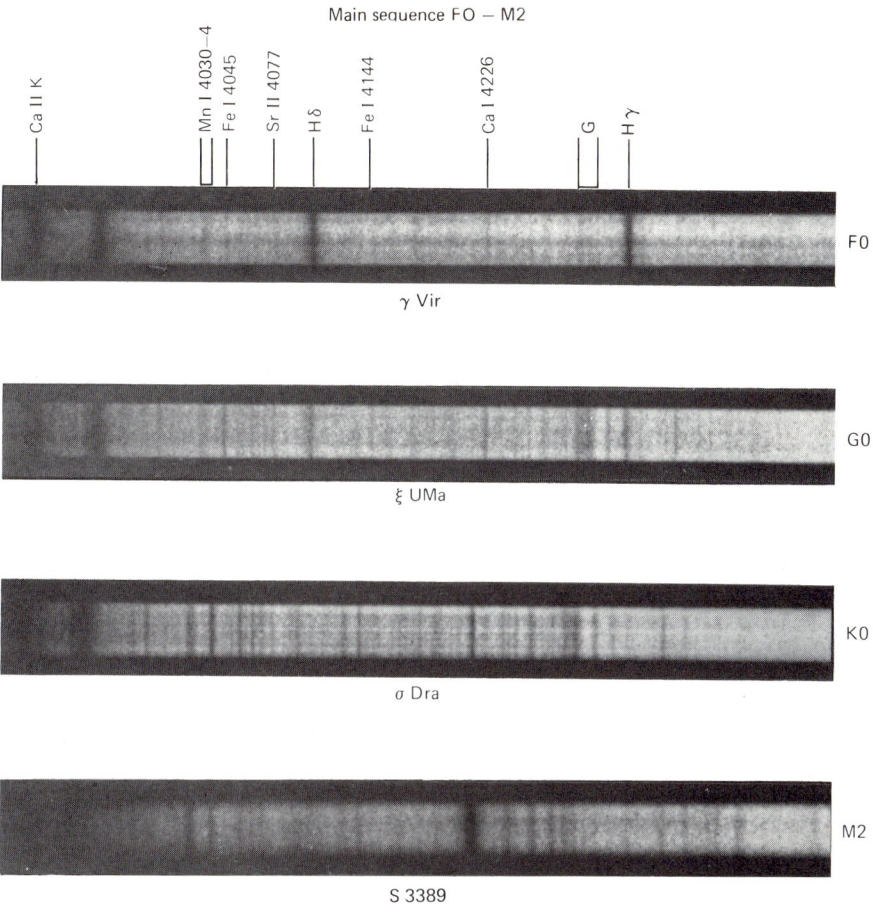

FIGURE 8.16 Stellar spectra classified on the Harvard system. Note these differences among the spectra: (1) the intensities of the hydrogen lines (H-δ and H-γ) and (2) the intensities of elements such as iron and calcium (Fe I and Ca I and II), which appear strong in some spectra and not at all in others. [*Courtesy Yerkes Observatory.*]

of over 250,000 stars! The original classification scheme was set up strictly on the basis of the strength of various lines (Fig. 8.16), well before there was any understanding of excitation and ionization effects produced by different temperatures. The Balmer lines played an important role in this scheme. Those stars with the strongest Balmer lines were called class A, those with slightly weaker lines class B, etc. Some classes were later dropped, because they contained too few stars or only very peculiar ones, and the order was rearranged to one of decreasing temperature, once the explanation of the line strengths in terms of temperature was understood.

The sequence of stellar spectra in the Harvard classification now runs O-B-A-F-G-K-M. (The standard mnemonic for the sequence is: Oh Be A Fine Girl! Kiss Me! You can avoid any sexist overtones by adapting the mnemonic to your own preference; *guy* or *girl* serves equally well.) Almost all stellar spectra fit into this sequence. Class O stars have spectra with weak Balmer lines of hydrogen and lines of ionized helium. Class A stars have the strongest Balmer lines. In class F stars the Balmer lines are weak and many other lines appear, mostly of metals. In the coolest stars, class M, we see lines of molecules.

The sequence from O to M, looking at the continuous spectra from the stars, is also a

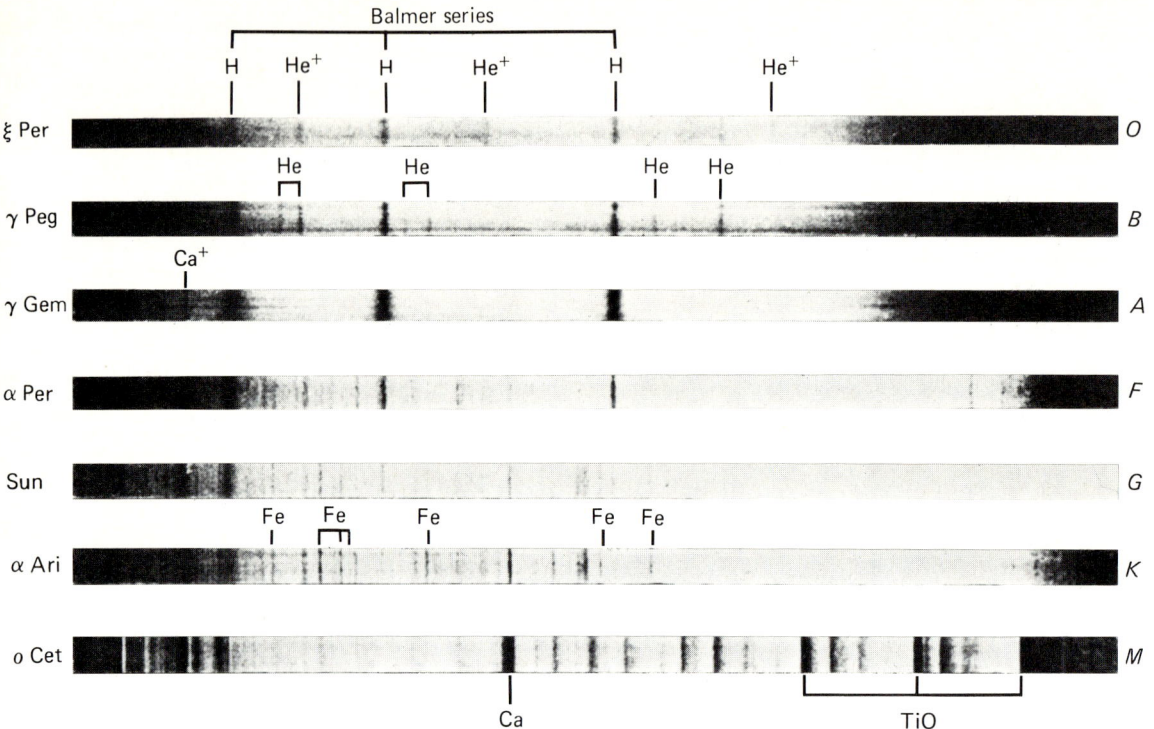

FIGURE 8.17 The basic spectral classification scheme from O (hottest) to M (coolest). The sequence is based on the intensities of the hydrogen Balmer lines and those of metals (everything except hydrogen and helium). Note that A stars have the strongest hydrogen lines; stars above and below type A have weaker hydrogen lines. [*Courtesy Yerkes Observatory.*]

color sequence. O stars appear bluish-white, G stars yellowish-white, and M stars reddish. Astronomers further divide each class into subclasses, labeled from 0 to 9, for example, G0, G1, G2, G3, and so forth; each subclass is distinguished by slightly different intensities of specific absorption lines.

The strengths of the Balmer lines give a clue that the differences in stellar spectra reflect primarily differences in *temperature* and not in the abundance of elements. These temperature differences lead to different degrees of ionization and excitation of the atoms in the star. How many atoms are excited and how many are ionized determine the strength of the atom's spectral lines.

Consider first O stars (Fig. 8.17). These have the hottest surface temperatures—30,000 K and higher. At such high temperatures atoms collide violently. The energies in such collisions can rip the electrons from hydrogen atoms so that most of the hydrogen is ionized. Very few neutral atoms remain to absorb at wavelengths corresponding to the Balmer series. Because the hydrogen is mostly ionized in O stars, the Balmer lines are weak.

Now turn to A stars, whose surface temperatures range from 8000 to 11,000 K. Collisions between atoms occur less violently, and most of the hydrogen is neutral. Although the collisions are not strong enough to ionize the hydrogen, they do possess the energy to

8.4 SPECTRAL CLASSIFICATION OF STARS

excite the electrons out of the ground state. In A stars many hydrogen atoms are excited by collisions, so their electrons are in the second level. These excited atoms readily absorb light at the Balmer wavelengths, and the lines appear strong.

In K stars surface temperatures are still lower, roughly 4000 K. Very few hydrogen atoms are ionized. In addition, the impacts between atoms do not have enough energy to excite very many. Most electrons are in the ground state. Such atoms cannot absorb at Balmer wavelengths. As a result the Balmer lines disappear almost completely.

The variation in Balmer line strengths arises from collisions that excite and ionize atoms. How much the collisions ionize or excite depends on temperature. So each spectral type corresponds to a restricted range of surface temperatures. These are listed in Table 8.3.

Other lines from other elements can be analyzed in a fashion similar to that for the Balmer lines (Fig. 8.18). In general, the lines of the neutral atoms are strongest in the coolest stars. At higher temperatures the atoms become more and more ionized, and the lines of the ions become stronger as those of the neutral atoms become weaker. Some atoms, iron for example, become doubly or triply ionized at the highest stellar temperatures, so we see lines of these ions in the spectra of very hot stars. At the very lowest

TABLE 8.3 Features of the stellar spectral classes

Spectral Class	Color	Approximate Temperature (K)	Principal Features	Examples
O	Bluish-white	30,000	Relatively few absorption lines. Lines of ionized helium and other lines of highly ionized atoms. Hydrogen lines appear only weakly.	Naos
B	Bluish-white	11,000–25,000	Lines of neutral helium. Hydrogen lines more pronounced than in O-type stars.	Rigel, Spica
A	Bluish-white	7,500–11,000	Strong lines of hydrogen. Also lines of singly ionized magnesium, silicon, iron, titanium, calcium, and others. Lines of some neutral metals show weakly.	Sirius, Vega
F	Bluish-white to White	6,000–7,500	Hydrogen lines are weaker than in A-type stars, but are still conspicuous. Lines of singly ionized metals are present, as are lines of other neutral metals.	Canopus, Procyon
G	White to Yellowish-white	5,000–6,000	Lines of ionized calcium are the most conspicuous spectral features. Many lines of ionized and neutral metals are present. Hydrogen lines are weaker even than in F-type stars.	Sun, Capella
K	Yellowish-Orange	3,500–5,000	Lines of neutral metals predominate.	Arcturus, Aldebaran
M	Reddish	3,500	Strong lines of neutral metals and molecules.	Betelgeuse, Antares

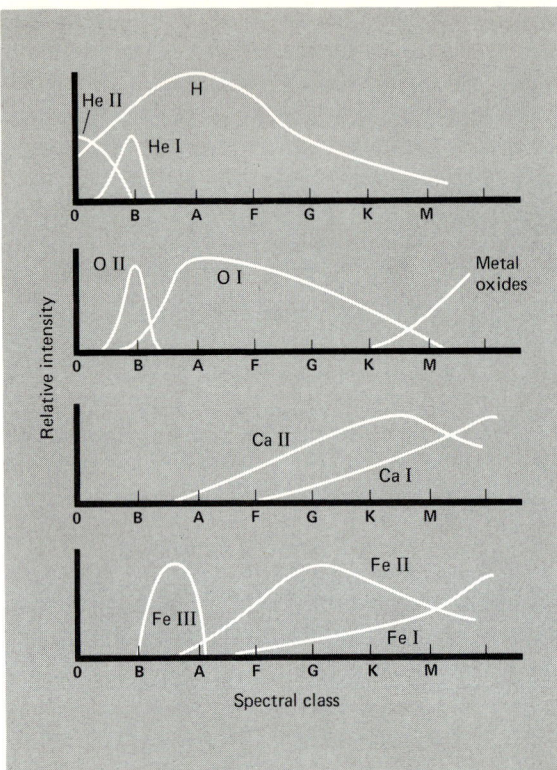

FIGURE 8.18 Relative intensities of dark lines of various elements in the different spectral classes. From top to bottom: hydrogen and helium, oxygen and metal oxides, calcium, and iron. The symbol I indicates a neutral atom, II is once ionized, and III twice ionized.

TABLE 8.4 Properties of Key Stars in the Winter Sky

Star	Apparent Visual Magnitude	Absolute Visual Magnitude	Distance (parsecs)	Distance (light years)
Epsilon Orionis	+1.50	−6.99(?)	500(?)	1600(?)
Rigel	+0.13	−7.03	270(?)	880(?)
Regulus	+1.35	−0.58	24.3	79.2
Sirius A	−1.46	+1.41	2.65	8.63
Procyon	+0.37	+2.65	3.50	11.4
Sol	−26.8	+4.77	5×10^{-6}	1.6×10^{-5}
Capella*	+0.09	−0.42	12.6	41.1
Epsilon Eridani	+3.74	+6.15	3.30	10.8
Aldebaran	+0.80	−0.79	20.8	67.8
Betelgeuse	+0.4 (variable)	−6.11(?)	200(?)	650(?)
Sirius B	+8.4	+11.27	2.65	8.63

*Capella is a double star. The magnitudes and colors are for the combined system; the temperature, radius, and luminosity are those of the brighter and cooler component. The distances of Epsilon Orionis, Rigel, and Betelgeuse are estimated, so the absolute magnitudes, radii, and luminosities of these three stars are approximate.

Source: Adapted from *The Restless Universe*, by H. L. Shipman (Houghton Mifflin, 1978), used with permission.

temperatures atoms combine into molecules, so bands of molecular lines appear in the spectra.

Caution: It is not sufficient that an ion exist in a star's atmosphere for it to be detectable in the spectrum. The ion must have energy levels that can be excited at that temperature and a spacing of the energy levels so that the lines will appear in the visible part of the spectrum. For example, neutral helium (He I) is abundant at low temperatures. (Roman numerals designate the ionization stage of an element: I for neutral, II for singly ionized, III for doubly ionized, etc.) But the energy levels of He I are so high that all transitions from the ground state are in the ultraviolet, and it requires a fairly high temperature to excite levels where transitions in the visible spectrum can occur. So we do not see helium lines in stars of spectral class A or cooler. Similarly, calcium is doubly ionized (to Ca III) in stars of class A and hotter, but again the strong lines are in the far ultraviolet.

The observation of stellar spectra coupled with an understanding of the atom gives astronomers information about the physical conditions in the atmosphere of a star. Here we've focused on the temperature; in Section 8.5 you'll see that spectra can also tell about the density. An analysis of spectral lines based on atomic theory also provides information about the abundance of elements in stars in the same way as for the sun (Section 6.4). Astronomers have found few surprises: just like the sun, stars consist mostly of hydrogen and helium.

8.5 THE HERTZSPRUNG-RUSSELL DIAGRAM

Let's focus again on the main stars in the winter sky. Table 8.4 summarizes their astronomical and physical properties. Examine this table for a minute. Can you find any

B-V Color Index	Spectral Class	Temperature (K)	Radius (solar radii)	Luminosity (solar units)
−0.18	B0	24,800	37	4.7×10^5
−0.05	B8	11,550	74	8.9×10^4
−0.11	B7	12,210	3.63	2.7×10^2
0.00	A0	9,970	1.67	2.5×10^1
+0.42	F5	6,510	2.07	7.0
+0.66	G2	5,780	1.00	1.00
+0.81	G5, G0	5,200	14.1	1.3×10^2
+0.87	K2	5,000 (est.)	0.7	2.8×10^{-1}
+1.55	K5	3,780	61	6.9×10^2
+1.85	M2	3,250	1100	1.2×10^5
−0.03	White dwarf	32,000	8×10^{-3}	5.8×10^{-2}

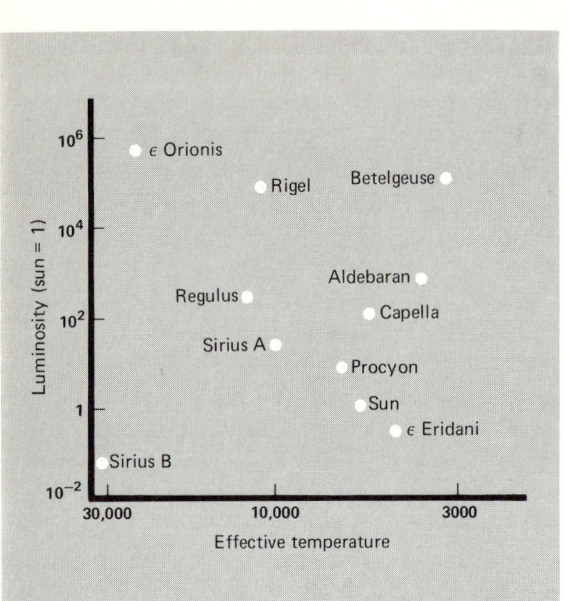

FIGURE 8.19 A temperature-luminosity diagram for important stars in the winter sky. [*Adapted from a figure by H. Shipman,* The Restless Universe, *Houghton-Mifflin, Boston, 1977.*]

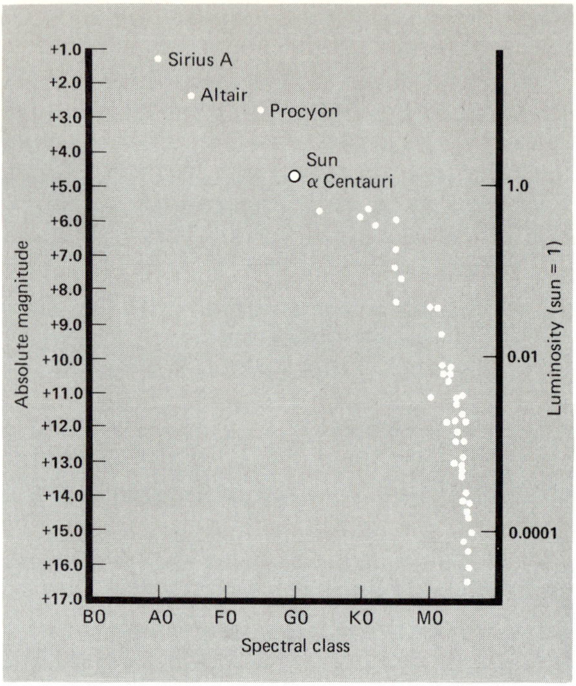

FIGURE 8.20 A Hertzsprung-Russell (luminosity-temperature) diagram for the stars nearest to the sun. Note that the vertical axis is in absolute magnitude, with the sun's absolute magnitude about +5.

pattern in it? Note the wide range of luminosities and sizes and the smaller range of temperatures. It's not obvious how these relate to the internal anatomy of the stars. Are they all like the sun? Or much different?

Let's turn the table into a picture—a temperature-luminosity diagram (Fig. 8.19). Such diagrams were set up early this century by Ejnar Hertzsprung (1873–1972) and Henry N. Russell (1879–1957). In their honor such plots are called *Hertzsprung-Russell diagrams* (commonly abbreviated *H-R diagrams*).

Examine Fig. 8.19. Do you see any patterns yet? Notice Sirius B in a corner by itself, and Betelgeuse alone in the upper right-hand region. Regulus, Sirius A, Procyon, the sun, and Epsilon Eridani fall along a sloping line close to each other.

Now inspect a different H-R diagram, one for the nearest stars, all those within 20 ly (6 pc) of the sun (Fig. 8.20). Notice that most of the stars are less luminous and cooler than the sun. (In fact, we can see them at all only because they are so close to us.) The star Alpha Centauri A has almost the same luminosity and temperature as the sun. This star is the sun's twin and is also the star nearest to us. (Alpha Centauri A is one member of a triple-star system with Alpha Centauri B and Proxima Centauri, and in this sense is unlike the sun.) Finally, note that the stars' properties clearly do not fall in a random scatter. Rather, there is a trend; if you draw a line through the points from luminous, hot, Sirius A to the coolest, faintest star in the lower right-hand corner, you have identified the *main sequence*. Most nearby stars fall on the narrow strip of the main sequence in the H-R

8.5 THE HERTZSPRUNG-RUSSELL DIAGRAM

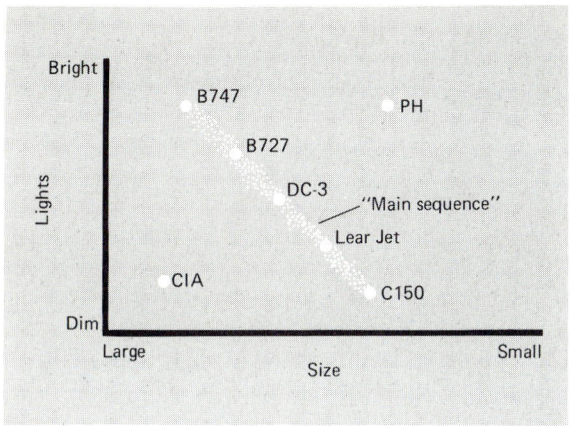

FIGURE 8.21 A concrete analogy to a Hertzsprung-Russell diagram.

diagram. Note the few stars in the lower, left-hand corner (Sirius B, shown on Fig. 8.19, is one of these.) These stars have very high surface temperatures but low luminosities, so they must be very small. (Remember the relation $L = 4\pi R^2 \sigma T^4$.) These peculiar stars are called *white dwarf stars* or *white dwarfs*.

Here's an analogy to the H-R diagram. Large airplanes have, in general, bigger, brighter lights than small airplanes. Suppose we draw a diagram plotting brightness of lights against size of airplanes (Fig. 8.21). Let's include planes ranging in size from a Cessna 150 to a Boeing 747. We also mark the position of a small police helicopter that has a big floodlight for disrupting lawn parties (PH), and the position of a CIA cargo plane secretly transporting tanks (CIA). Where do most airplanes fall on this diagram?

Note that size increases toward the left, just as on the H-R diagram temperature increases toward the left. (The reason for this is that astronomers use the color index of a star for determining temperature, and the index gets larger as the temperature gets smaller; an H-R diagram plotted in terms of luminosity versus color index would have the numbers getting larger toward the right.) Note also how most planes fall on a "main sequence," since size is generally related to brightness of lights. But there are exceptions (the police helicopter and the CIA transport), just as there are exceptions for the stars in our galaxy (the white dwarfs and the red giants).

Consider another H-R diagram (Fig. 8.22), one for the brightest stars in the sky. Compare it with the plot for the nearby stars (Fig. 8.20). What a difference! Almost all of these stars have a much higher luminosity than the sun. And many of them are also much hotter (spectral class B). The main sequence no longer appears so obvious.

What are the physical differences among these stars? Take the star Betelgeuse, whose properties put it in the upper right-hand corner of the H-R diagram. Here is a star whose surface is much cooler than the sun's. So if Betelgeuse were the same size as the sun, it would be much less luminous. But Betelgeuse has a luminosity 120,000 times that of the sun. To be so much cooler and more luminous than the sun, Betelgeuse must be much larger. In Section 8.3 we calculated Betelgeuse's diameter to be about 1100 times that of the sun—a star so big that it could engulf Mars if it were placed at the center of our solar system! Astronomers call Betelgeuse a *supergiant* star.

Similarly, a white dwarf of the same spectral class as Sirius (Fig. 8.20) will have the

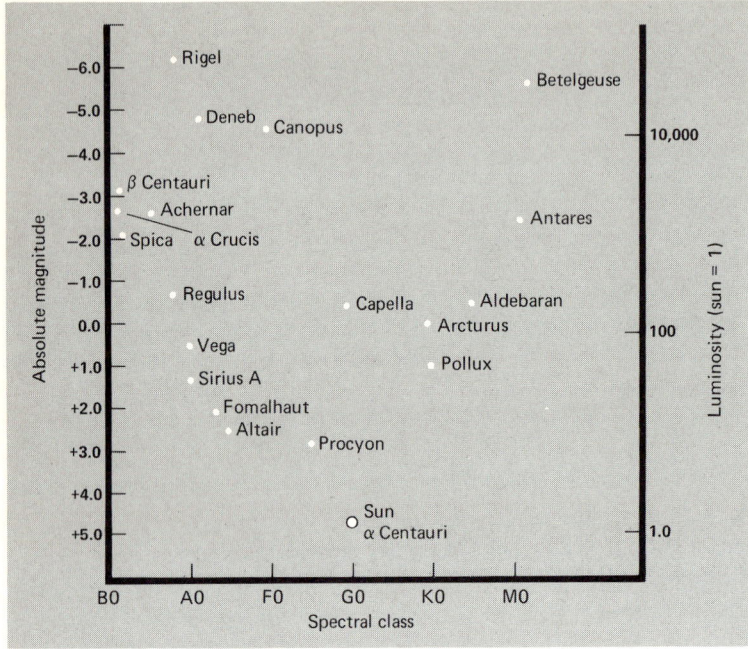

FIGURE 8.22 A Hertzsprung-Russell (luminosity-temperature) diagram for the brightest stars in the sky. Axes are the same as for Fig. 8.20. Note that the sun and Alpha Centauri have almost the same luminosity and surface temperature, and so they overlap on this diagram.

same temperature, but if it is 10 magnitudes fainter in absolute magnitude (a factor of 10^4 less in luminosity), it must be 100 times smaller in radius.

Here is the reason that the H-R diagram for the nearest stars (Fig. 8.20) differs from that for the brightest stars (Fig. 8.22). The first diagram contains a sample of the typical stars in our part of the Galaxy; no giants are among the nearest stars, for they are very rare. The sun is a main-sequence star. Most of the stars in the sun's vicinity are also main-sequence stars, of spectral class M. So they are cool and not very luminous. The second diagram contains many giant and supergiant stars, still visible among the brighter stars because of their high luminosity, even though they are scattered widely through space. At the distance of these highly luminous stars, most main-sequence stars are too faint to be seen.

Now piece these two diagrams together and add more stars (Fig. 8.23 and Plate 4). Most stars fall on the shallow "S" of the main sequence. A scattering of stars cuts across the tip of the diagram; these are the very luminous supergiants. A group of relatively cool, but luminous stars extends off the main sequence; these are the giants. Finally, note the white dwarfs in the lower left-hand corner of the diagram.

Is the Sun a Typical Star?

It is often said that the sun is a typical or average star. In some sense this is true, for the sun is neither the largest nor the smallest kind of star, it is intermediate in both mass and luminosity, and it has a composition like most other stars. However, it is not true that most stars in the Galaxy are like the sun. Examine Fig. 8.20 carefully. Counting up the stars on this diagram, there are 44 cooler than the sun and 9 as hot or hotter. There are 49 fainter than the sun and 4 as bright or brighter. Among the stars in the solar neighborhood, G stars like the sun are relatively rare. The most common kind of star in our immediate vicinity and in the Galaxy in general is a faint M main-sequence star, a cool, reddish star of very low luminosity, on the lower end of the main sequence.

Spectroscopic Parallaxes

An H-R diagram for many stars can be used to infer approximate distances to stars. How? First, find out the spectral type of a star. Suppose it's an M star. Then look at the H-R diagram to find the luminosity (or absolute magnitude) of an M star. Here a problem arises: M stars have a range of luminosities, from 1.6×10^{-5} solar luminosities for main-sequence stars to about 10^5 solar luminosities for supergiants. How do we decide what luminosity an M star is?

Fortunately, we have two ways to tell from a star's spectrum. The first involves the width of certain spectral lines, particularly the Balmer lines of hydrogen. Recall that to produce Balmer lines, hydrogen atoms must be excited by collisions with sufficient energy to raise the electrons one energy level. For gases at the same temperature the *rate* of collisions depends on the density of the gases. So in a denser gas collisions are more frequent than in a less dense gas at the same temperature. Collisions among atoms change their energy levels slightly, making it possible for them to absorb at slightly higher or lower frequencies. This means they can absorb over a range of frequencies around the average, and the line is broadened. Now giant stars, because they are so huge, typically have lower density atmospheres than main-sequence stars. The more frequent collisions in main-sequence stars make the absorption lines (particularly the hydrogen lines) in their spectra appear broader than the same lines in spectra of giant or supergiant stars (Fig. 8.24a). So a star's size, and hence its luminosity, is given indirectly by the widths of certain absorption lines when comparing the spectra of stars of the same spectral type.

The difference in density between main-sequence stars and giant stars creates a second difference in their spectra. The degree of ionization of an element, the ratio of number of ions to number of atoms, depends mainly on temperature; the higher the temperature, the greater the number of ions. But the ionization also depends on density. The greater the

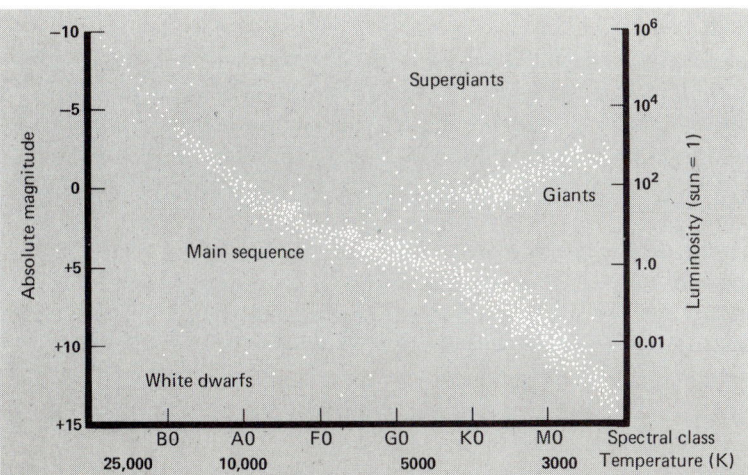

FIGURE 8.23 A Hertzsprung-Russell diagram for a large number of stars. Each star is represented by a point that denotes its absolute magnitude (luminosity) and spectral class (surface temperature). The stars fall into four general regions. Most lie along the main sequence. Many others fall into the giant region of luminous yellow to red stars. A small number make up the very luminous supergiants and the low-luminosity white dwarfs.

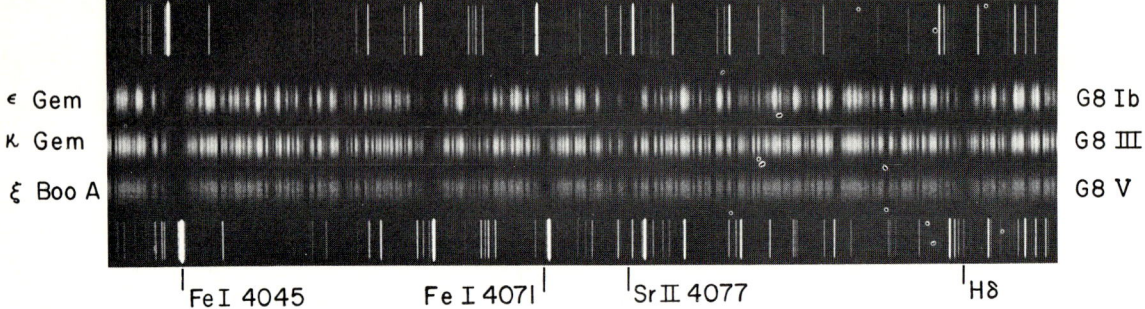

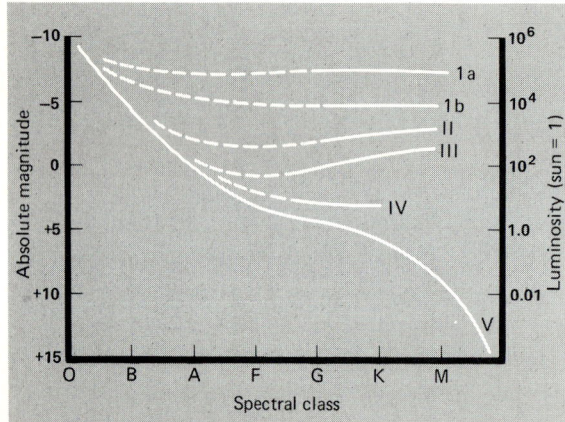

FIGURE 8.24a. Determining a star's luminosity class from its spectrum. Here are the spectra of three stars all with the same surface temperature (spectral class G8). The top spectrum is for a supergiant star, the middle for a giant, the bottom for a dwarf. The different intensities of identified dark lines permit astronomers to infer the luminosity class of the stars. [*Courtesy Lick Observatory.*]

FIGURE 8.24b. Luminosity classes of stars on the H-R diagram. The classes run from I, the largest supergiants, to V, stars like the sun. Classes Ia and Ib are supergiants; class II luminous giants, class III normal giants, class IV subgiants, and class V main-sequence stars.

density, the greater the rate of collisions between ions and electrons and the more easily will ions and electrons recombine to form neutral atoms. So for two stars of the same temperature, the one with the greater density will have more neutral atoms. The dependence on density is not the same for all elements; by examining the strengths of lines of ions and neutral atoms of different elements, we can determine both the temperature and density of a stellar atmosphere. Again, since the atmospheric density is less for the giant stars, we can estimate the size of a star and hence its luminosity from its spectrum.

Such as analysis reveals that stars fall into *luminosity classes*. The recognized luminosity classes are (Fig. 8.24b): Ia, most luminous supergiants; Ib, less luminous supergiants; II, luminous giants; III, normal giants; IV, subgiants; and V, main-sequence stars. The sun falls into luminosity class V.

So a star's spectrum allows it to be classified by spectral class *and* luminosity class. For a given spectral class, you can estimate, within a range of probable error (the width of a luminosity class on the H-R diagram), the luminosity or absolute magnitude (Fig. 8.25). If you know a star's luminosity and its flux, you can calculate its distance. Or, equivalently, if you know its apparent and absolute magnitude, you can also find the distance (Section 8.3). This procedure for working out distances from spectra is called the *spectroscopic distance method* or *spectroscopic parallaxes*.

Here's an example. Consider an M giant, luminosity class III. In Fig. 8.24b you see

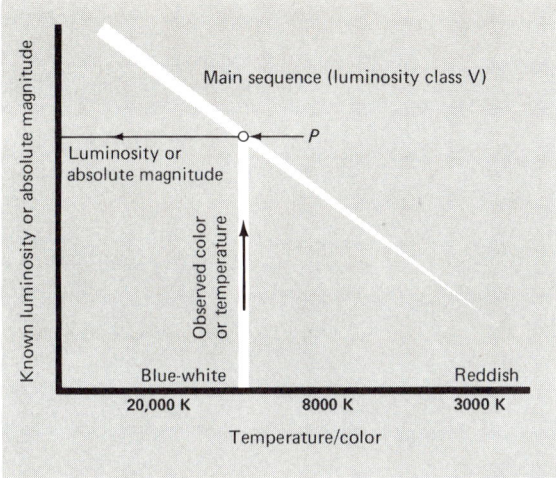

FIGURE 8.25 Using spectral class and luminosity class (in this case V, main sequence) to infer the luminosity and so the distance to a star, if its flux or apparent magnitude is known.

that an M III star has an absolute magnitude of about 0. You measure the star's apparent magnitude as 10. Then its distance is computed from the relation

$$m - M = 5 \log d - 5$$
$$10 - 0 = 5 \log d - 5$$
$$\log d = 15/5 = 3$$
$$d = 10^3 \text{ pc}$$
$$= 3.3 \times 10^3 \text{ ly}$$

In summary, to make an H-R diagram, you need to find stars close enough to the sun to measure their distances reliably by parallax. Calculate their luminosities from their distances and fluxes or absolute magnitudes from the distances and apparent magnitudes. The absolute magnitudes tell you how luminous the stars are compared with the sun. Next, take spectra of the stars to find out their spectral class. From the spectra you determine how hot the stars are. Then plot the luminosity against temperature or spectral type. Results: a calibrated H-R diagram—calibrated in the sense that you have the *luminosities* of the stars plotted against their *surface temperatures*. The H-R diagram graphically summarizes some of the important physical properties of stars. It also serves as a visual sorting tool to bring to light different classes of stars. And it can be used as a tool for obtaining (via spectroscopic parallaxes) the distances of other stars.

8.6 WEIGHING AND SIZING STARS: BINARY SYSTEMS

You have seen how stars differ in properties such as luminosity and radius. But what about mass? Some stars are larger than the sun, some smaller. But radii do not tell us directly if a star is more or less massive than the sun. How to find a star's mass?

Binary Stars

We have no direct way of knowing the mass of an isolated star. To find masses, we need to examine the gravitational effects of one star on another. Recall that to find the sun's mass (Section 6.4), we looked at the acceleration of the earth orbiting the sun. Similarly, we use the accelerations of two stars orbiting one another to find their masses. Two stars bound by their mutual gravity are called *binary stars*.

If both stars in a binary system are visible, we can trace out their orbital motion by observing them over a long time (Fig. 8.26), which gives us the angular size of the orbit and the orbital period. But that's not enough to get the masses! First, we need to find the distance to the binary system so that we can convert their angular separation into a physical one. Second, it is likely that the plane of the star's orbit is tilted from a direct face-on view; this orbital tilt needs to be accounted for. Then we have enough information to find the *sum* of the masses from Newton's revised form of Kepler's third law (Section 4.3).

Put in terms of the stellar masses, rather than the mass of the sun and a planet, this law is:

$$M_1 + M_2 = \frac{4\pi^2}{G} \frac{a^3}{P^2}$$

where a is the average separation of the two stars, that is, the sum of their average distances from the center of mass ($a = a_1 + a_2$). Write this equation for the earth-sun system:

$$M_S + M_E = \frac{4\pi^2}{G} \frac{a_E^3}{P_E^2}$$

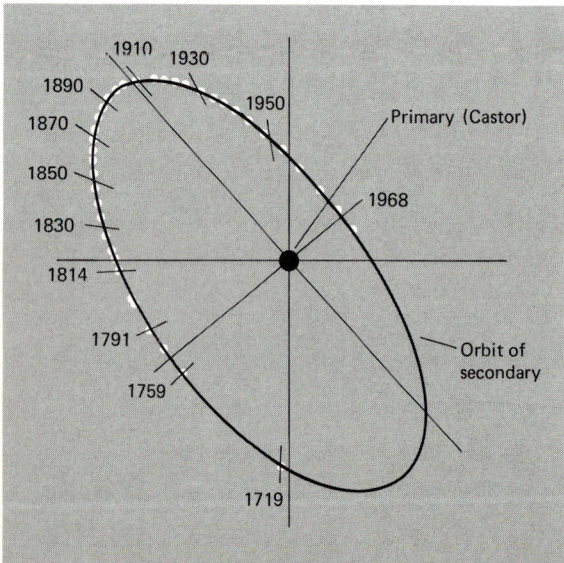

FIGURE 8.26 Motion of a visual binary star, Castor and companion. The orbit of the companion is drawn as if Castor were motionless. In fact, both stars revolve around a common center of mass with an orbital period of about 380 years. [Orbital data by K. A. Strand.]

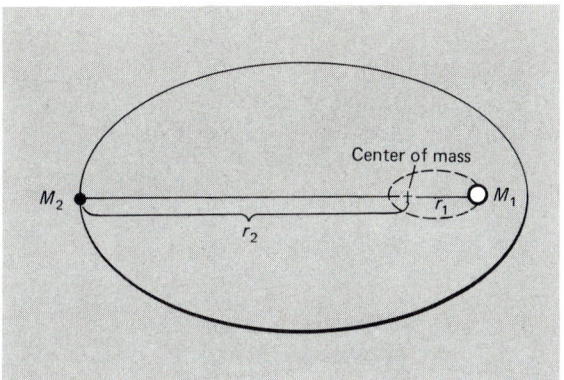

FIGURE 8.27 Center of mass in a binary star system. Both stars (M_1 and M_2) move in elliptical orbits (r_1 and r_2) around the center of mass.

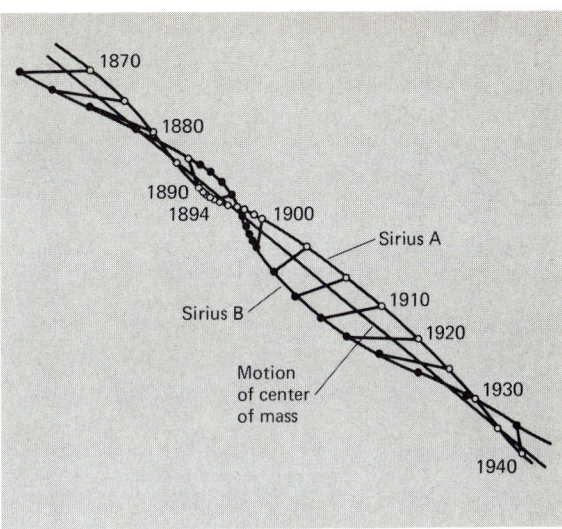

FIGURE 8.28 Motions of the binary star system Sirius A and its companion, Sirius B. This figure shows two motions: that of Sirius A and its companion about their center of mass and that of the center of mass of the two stars relative to background stars, as the center of mass moves through space. So the two stars make a corkscrew motion against the sky.

The mass of the earth is so small compared to that of the sun that $M_S + M_E \approx M_S$. Divide the two equations to get

$$\frac{M_1}{M_S} + \frac{M_2}{M_S} = \frac{(a/a_E)^3}{(P/P_E)^2}$$

If we express masses in solar masses, periods in years, and distances in AUs, then

$$M_1 + M_2 = \frac{a^3}{P^2}$$

Use of Kepler's law gives only the sum of the masses. To find the *individual* masses, we must have one more piece of information: how far each star is from the center of mass of the system. In a binary system each star orbits the center of mass at a distance inversely proportional to its mass. Algebraically, the relationship of masses and distances of the stars from the center of mass is (Section 4.3)

$$\frac{m_2}{m_1} = \frac{r_1}{r_2}$$

So the more massive star lies closer to the center of mass (Fig. 8.27).

As the system travels through space, and so across our line of sight, the center of mass traces a straight line, while the two stars spiral around it (Fig. 8.28). This corkscrew motion identifies the stars as binary and locates the center of mass.

Newton's revised form of Kepler's third law provides a way for calculating the sum of the masses in a binary system. But often a little thought about the physics of the situation will allow you to compare the masses of two systems without resorting to the formula. Remember that the force which keeps stars from flying out of orbit is gravity. Given two binary systems with the same distance between the two stars, the binary system with the shorter period of revolution (the one with the faster moving stars) must be the system with the greater gravitational force acting between the two stars. The force of gravity acting on the two stars depends on the masses of the stars and the distance between them. Since we assumed that the distances between the stars of each of our binary systems are the same, the binary system with the shorter period of revolution must have the greater total mass.

Now consider two binary systems with the same period of revolution, but different distances between the stars of each system. The pair with the greater separation must have higher velocities, since they have to cover a greater distance in the same period. The force of gravity decreases with distance. So in order to make up for the increased distance over which gravity must act, the system with greater distance between its two stars must have the greater total mass. In such simple cases it is sufficient to know that the shorter the period of revolution or the greater the distance between the stars in a binary star system, the greater the sum of the masses of the two stars.

Types of Binary Systems

All binary systems are physically the same, but we observe them in different ways. This fact of astronomical life has prompted astronomers to divide binaries into three general classes: visual, spectroscopic, and eclipsing.

In a *visual binary* a telescope clearly shows both stars. With enough observations we can trace the orbital path of the fainter star (called the *secondary*) around the brighter one (called the *primary*). Because a supposed pair may be only an accidental line-of-sight juxtaposition of two physically separate stars, we must sometimes wait for many years to confirm that the two stars are really bound by gravity. As you expect from Kepler's third law, binary stars with large separations will have large periods. For instance, Castor and its companion have an orbital period of 380 years (Fig. 8.26).

One problem with visual binaries is the determination of the angle of tilt. We can only observe the projection of the orbit onto the plane of the sky. How to determine the angle of projection? According to Kepler's first law, the orbit of one star about the other is an ellipse with the second star at its focus. Now an ellipse projects into an ellipse, so the observed orbit will also be elliptical. But the focus of the original ellipse does *not* project into the focus of the observed ellipse. So the observed orbit will *not* have the other star at its focus. Here's the observational clue we need to get the tilt. We need to find the angle by which the observed orbit must be untilted to get the other star back to the focus: a tricky, but feasible, problem in geometry.

Suppose two stars are so close together that we cannot resolve them with a telescope. Their orbital periods will be short, so the stars move quickly in their orbits. We can identify this binary by looking for the effect of their orbital motion in the spectrum. This effect is a shift in wavelength of the lines for one star toward the red and a shift of the lines for the other star toward the blue (Fig. 8.29). This shift, called a Doppler shift, identifies the star as a *spectroscopic binary*.

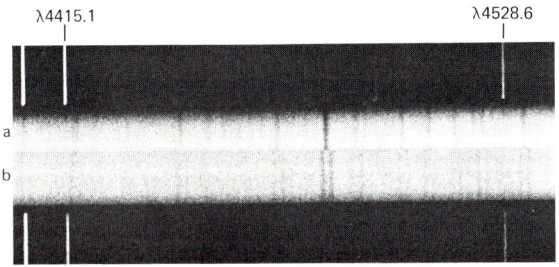

FIGURE 8.29 Spectra of a spectroscopic binary system (Mizar). The two spectra were taken at different times. In the upper spectrum (a) only single dark lines are visible. In the lower one (b) the lines are double because the secondary and primary are moving with different line-of-sight velocities. So the lines are Doppler-shifted relative to each other. The bright-line spectra above and below the dark-line spectra serve to mark the reference wavelengths of the spectral lines. [*Courtesy Palomar Observatory, California Institute of Technology.*]

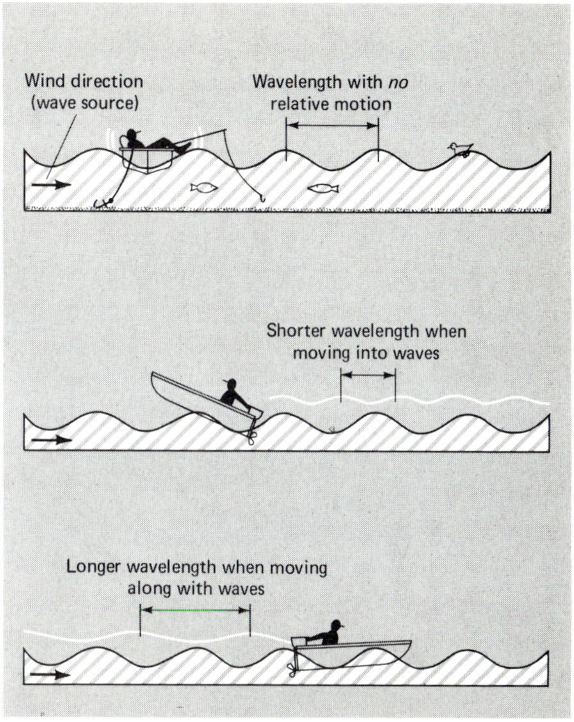

FIGURE 8.30 Wavelength and relative motion for the case of water waves.

The Doppler Shift

The Doppler shift is a powerful tool which allows astronomers to find the line-of-sight velocities of luminous objects without having to know their distance. The Doppler shift is named after Christian J. Doppler (1803–1853), who first noticed the effect in sound waves. Later the French physicist Armand Hippolyte Fizeau (1819–1896) applied the Doppler shift to light waves and recognized its importance in astronomical applications. In honor of these two men some scientists like to call it the *Doppler-Fizeau shift*. Most people call it simply the *Doppler shift*, and that is the name used in this book.

The Doppler shift occurs with all kinds of waves—light, sound, and even water waves. Here's an example with water waves. Imagine you are out fishing in a small motorboat (Fig. 8.30). You have been sitting in one spot for a while with little luck, and the rhythm of the waves, generated by a gentle wind, has lulled you almost to sleep. You decide to move for better fishing. First you go into the wind (the wave source). You notice that you bob up and down *more frequently* than when you were at rest; the wavelength appears to have gotten *shorter*. You drive the boat with the wind. You discover that your bobbing is less frequent; the waves seem to you to have a *longer* wavelength. The explanation is simple: When you moved in the direction of the waves, they had to catch up with you; when you went into them, you went to meet them.

That's the Doppler shift: When you are moving *toward* a wave source, the waves appear more frequent and shorter in wavelength; in contrast, when you move *away* from a wave source, the waves appear less frequent and the wavelength longer. It's only the

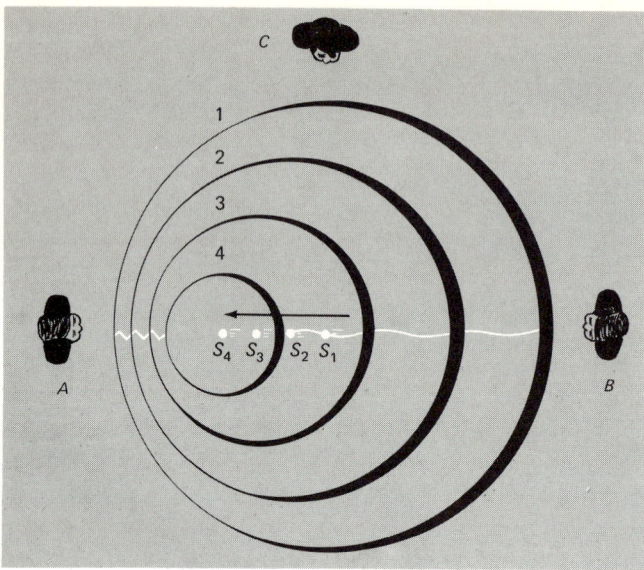

FIGURE 8.31 The Doppler shift for a source moving to the left and emitting light of a single wavelength.

relative velocity along the line-of-sight—called the *radial velocity*—that causes the Doppler shift (for speeds much slower than that of light). Since velocities are relative, it makes no difference if you're moving, or the source is moving, or both.

Let's look now at light waves. You can't see the waves directly, but you can see colors of light, which relate directly to wavelength. Keep in mind that red light has a longer wavelength than blue.

Imagine a stationary light source giving off just one particular wavelength every second. Each wave travels outward with velocity c, the speed of light. When the source is not moving, all the waves are concentric and separated by the wavelength of emission.

Now imagine that the source moves from point S_1 to point S_2 in 1 sec, and so on (Fig. 8.31). At each point (S_1, S_2, S_3, S_4) it emits a wave (1, 2, 3, 4) that travels outward at c. In the direction of its motion the source catches up a bit with the wave it has just emitted, so for an observer at A the wavelength appears shorter. This observer sees the distance between the waves as compressed and so observes a Doppler shift to the short-wavelength

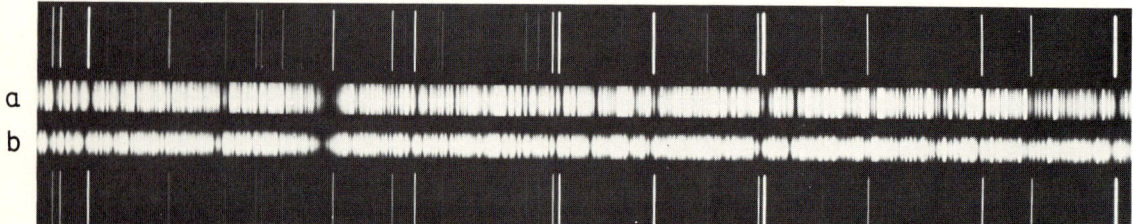

FIGURE 8.32 The Doppler shift in the spectrum of the star Arcturus. The two spectra were taken six months apart; above and below them are bright-line spectra made at the telescope to provide the wavelength comparison (longer wavelengths are to the right). The shift results from the orbital motion of the earth and the motion of Arcturus in space. [*Courtesy Palomar Observatory, California Institute of Technology.*]

8.6 WEIGHING AND SIZING STARS: BINARY SYSTEMS

end of the spectrum. In contrast, an observer B sees the waves as spread apart, a Doppler shift to the long-wavelength end of the spectrum. An observer at C, at right angles to the direction of the source's motion, measures no change in the wavelength and consequently no Doppler shift. *Note that only the velocity along the line of sight contributes to the Doppler shift.* It is also important to remember that the Doppler shift does *not* depend on the *distance* between the observer and the source, but only on their relative radial *velocities*.

The astronomer uses the Doppler shift to determine the line-of-sight velocities of celestial objects (such as stars) relative to the earth by using spectral lines to measure wavelength shifts. He or she takes a spectrogram of the object and at the same time superimposes a comparison spectrum from a local laboratory source (Fig. 8.32). The comparison source is at rest with respect to the telescope and provides the normal (zero relative velocity) placement of the lines with respect to which the astronomer measures the shift. With the measured shift and the value of c, the astronomer calculates the relative velocity between the source and earth by the expression

$$V_r = \left(\frac{\Delta\lambda}{\lambda_0}\right)c$$

where V_r is the relative radial velocity, λ_0 the rest wavelength of the observed line, $\Delta\lambda$ the observed shift in the line ($\Delta\lambda = \lambda_{observed} - \lambda_0$), and c the speed of light.

Here's an example. A strong dark line from absorption by calcium has a rest wavelength of 3933.68 Å. Suppose you measure this line in the spectrum of a moving object and find it shifted to 3934.07 Å. Is the object moving away or toward you? Away, since the wavelength is longer. How fast?

$$\Delta\lambda = 3934.07 - 3933.68 = 0.39 \text{ Å}$$

$$V_r = \frac{0.39}{3933.68}(3 \times 10^8 \text{ m/sec})$$

$$= 3.0 \times 10^4 \text{ m/sec}$$

$$= 30 \text{ km/sec}$$

Let's see how to use the Doppler shift to analyze spectroscopic binaries. Imagine the more massive star to be stationary, with the secondary revolving about it. As the secondary recedes from the earth, you see its spectral lines red shifted compared with those of the primary; as the secondary approaches, you see its lines blue shifted (Fig. 8.33a). At the intermediate points, when the secondary travels across the line of sight, you see no shift. If the two stars do not differ greatly in apparent magnitude (in practice, no more than about 1 magnitude), the shifts of both spectra can be observed. (The smaller mass star will have the higher velocity, and hence the larger shift, because it is farther from the center of mass.) Sometimes the spectrum of the secondary is too faint to be seen. We then use the Doppler shift in the primary's spectrum alone.

These relative wavelength shifts can be turned directly into relative velocities by using the Doppler relation (Fig. 8.33b). We then use the velocities and the period to get the

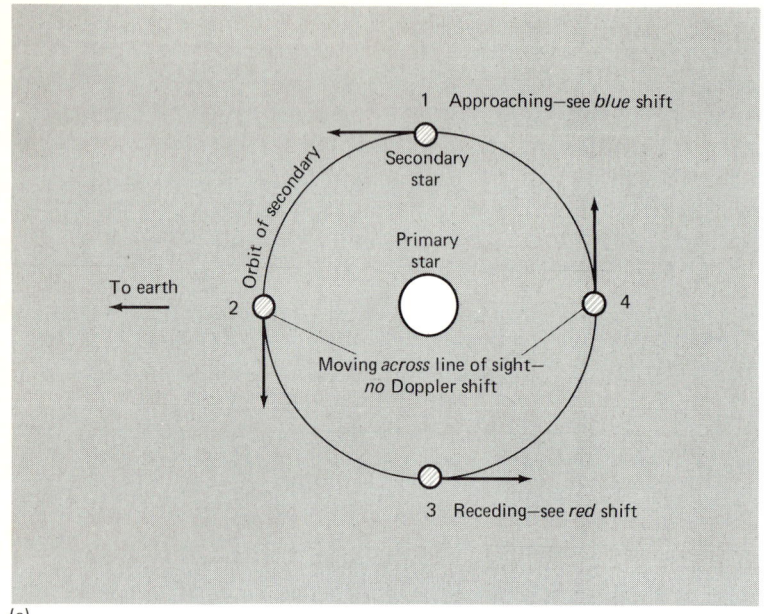

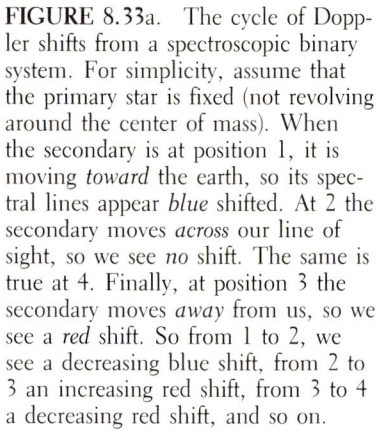

FIGURE 8.33a. The cycle of Doppler shifts from a spectroscopic binary system. For simplicity, assume that the primary star is fixed (not revolving around the center of mass). When the secondary is at position 1, it is moving *toward* the earth, so its spectral lines appear *blue* shifted. At 2 the secondary moves *across* our line of sight, so we see *no* shift. The same is true at 4. Finally, at position 3 the secondary moves *away* from us, so we see a *red* shift. So from 1 to 2, we see a decreasing blue shift, from 2 to 3 an increasing red shift, from 3 to 4 a decreasing red shift, and so on.

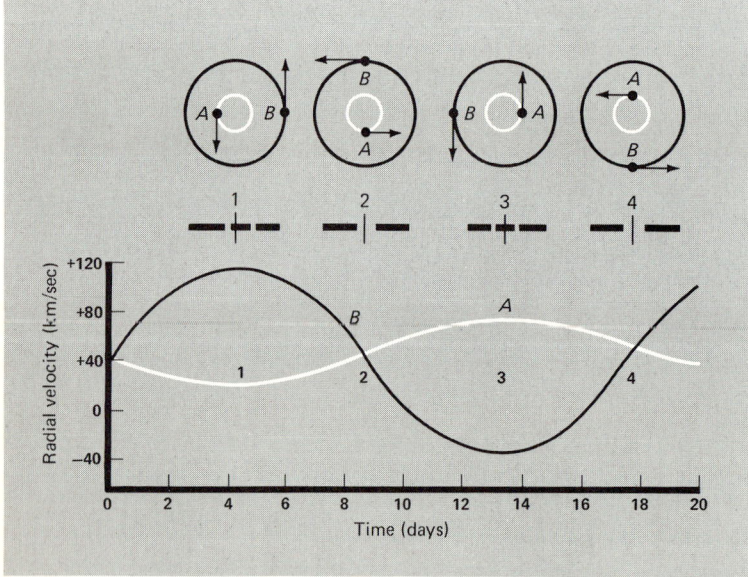

FIGURE 8.33b. Velocity curves for a hypothetical spectroscopic binary. Top view of orbits, shift in spectral lines, and velocity curves are shown, from top to bottom. Star A is more massive than star B. Note that when the stars move across the line of sight (2 and 4), the radial velocity observed is that of the entire system (in this case, 40 km/sec).

circumference of the orbits. Then we work out the radii of the orbits and so the separation of the two stars. So a spectroscopic binary gives us direct information on the system's orbit.

To see this point in detail, imagine a binary in which the stars have circular orbits. Suppose Doppler shift measurements show that the secondary has an orbital velocity V_2. Then the circumference of the orbit is

8.6 WEIGHING AND SIZING STARS: BINARY SYSTEMS

$$C = V_2 P$$

where P is the orbital period. The distance of the secondary from the center of mass (a_2) is

$$a_2 = \frac{C}{2\pi}$$

An example: Suppose V_2 is 200 km/sec and P is 2.87 days. Then C is 4.96×10^7 km, and a_2 is 7.89×10^6 km. The same analysis can be applied to the primary to find its distance from the center of mass (a_1). The sum $a_1 + a_2$ gives the average separation of the stars and thus the sum of the masses $m_1 + m_2$. The distance ratio a_1/a_2 gives the mass ratio m_2/m_1.

Note that in this method, because we can measure the actual velocity of the stars in km/sec, we get the actual radius of the orbit (in kilometers), not just the angular radius. So unlike the case for a visual binary, we do not need to know the distance of a spectroscopic binary in order to determine the stars' masses.

However, the difficulty here is that we cannot get the stars' masses unless we know the *tilt* of the orbit with respect to our line of sight, for we measure only the velocity along the line of sight, not the total velocity. Generally, we don't know that tilt. But in a few cases, the orbits are tilted so that one star passes in front of the other, producing an eclipse. That's an *eclipsing binary system*.

Algol, the "demon star" in the constellation of Perseus, is the prototype of eclipsing binaries. It has a period of 2.87 days and in mid-eclipse plummets 1.2 magnitudes in brightness (Fig. 8.34). (You can, with a little practice, observe Algol's light variation. *Sky and Telescope* provides the dates and time of eclipses.) Algol A, the brighter star (260 solar luminosities) is a B8 V (surface temperature 13,000 K); its companion, Algol B, is a fainter (5 solar luminosities) G-type star (surface temperature 4600 K). Recent spectroscopic observations of Algol B find that it orbits at 201 km/sec, Algol A at 44 km/sec, if the orbit is inclined 82.4°. (A 90° inclination would put both orbits directly in our line of sight.) The velocities and the period show that Algol A orbits 1.71×10^6 km from the center of mass, Algol B at 7.9×10^6 km. So Algol A has a mass 4.6 times that of Algol B. The total separation is 9.6×10^6 km. From Kepler's third law the total mass of the system

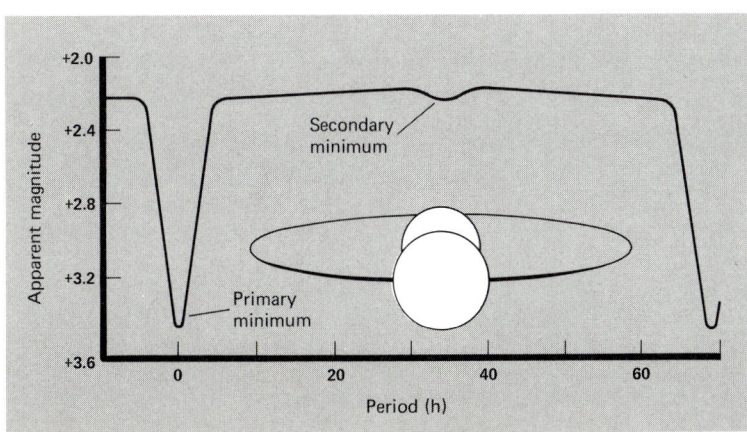

FIGURE 8.34 Algol, an eclipsing binary system. The orbits of the two main stars lie almost edge on as seen from the earth, so eclipses can occur. Algol A, the primary, is a B star with a diameter about three times that of the sun. Its dimmer companion, Algol B, is cooler; it is a G star with a surface temperature similar to the sun's. The eclipse of Algol A by its cool companion produces the largest dip (primary minimum) in the light curve, a plot of how the observed brightness changes with time. When the companion circles around Algol A, the loss of light is less, so a smaller dip (secondary minimum) occurs in the light curve.

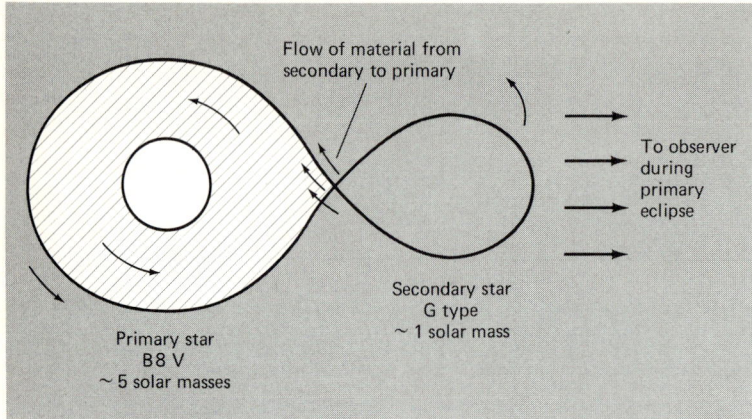

FIGURE 8.35 A model for Algol A and its main companion, Algol B. The two stars are so close that they distort each other by their mutual gravitational forces. Material probably flows from Algol B to Algol A, so both stars are shrouded by a gaseous disk.

is 4.51 solar masses. Combining this with the spectroscopic mass ratio, we find that Algol A is 3.7 solar masses, Algol B 0.81 solar masses.

Algol turns out to be a triple-star system. Algol C orbits Algol A and B once every 1.86 years. It has a mass of almost 2 solar masses, a surface temperature of 8700 K, and a luminosity 13 times that of the sun. Algol C has little effect on the orbits of A and B, for they are very close to each other compared to the distance to Algol C. In fact, these stars are so close together that their gravitational forces distort each other into ellipsoidal shape (Fig. 8.35) and matter flows from Algol B to Algol A.

Eclipsing binaries give directly one other property of the stars: their diameters. Let's see how. For simplicity, consider a pair of stars with orbits exactly in the line of sight (Fig. 8.36a). Then the smaller star (A) moves exactly across a diameter of the larger star (B). The light curve for the eclipse (the graph of brightness as a function of time) is shown in

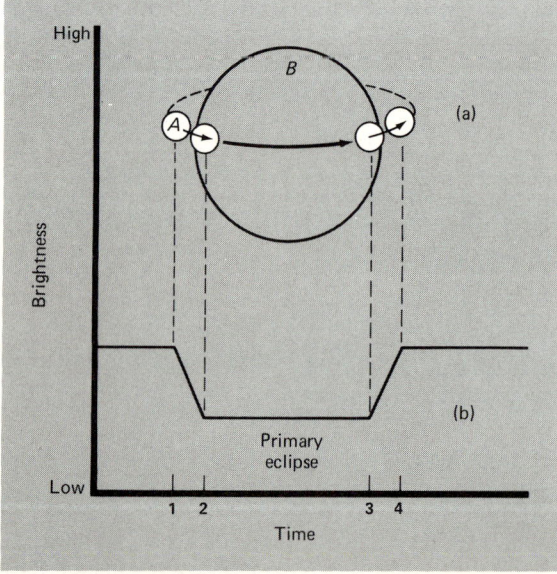

FIGURE 8.36 (a) An eclipsing binary system (primary star A, secondary B) and (b) its light curve as B eclipses A.

Fig. 8.36b. The light starts to diminish at point 1, when the star A first starts passing in front of star B. When it is completely in front of the star, at point 2, the light remains constant, for no more area is blocked off as it moves across. At point 3, star A starts to move off from in front of the star B, and the light begins to rise, reaching the original level at point 4, when the larger star B is completely uncovered.

From the diagram you can see that the star A moves its own diameter from point 1 to point 2, and it moves the diameter of star B from point 1 to point 3. The relative velocity of the stars, which we can get from the Doppler shifts in the spectrum, is $V_A + V_B$. Therefore

$$R_A = \frac{(V_A + V_B)(t_2 - t_1)}{2}$$

and

$$R_B = \frac{(V_A + V_B)(t_3 - t_1)}{2}$$

So we can get the radii of both stars by measuring the light curve and the changing Doppler shifts. (If the orbits of the stars are not exactly in the line of sight, the analysis is geometrically more complicated, but the principle is the same.)

Sirius: A Binary System

Finding stellar masses from binary stars is so important we will do another example in detail for you. Sirius is a binary star; the main star, Sirius A, is the one you see in the sky. Its companion, Sirius B, is much fainter, and you need a moderately large telescope to see it. The orbital motion of Sirius A and B is plotted in Fig. 8.37. From it you can infer that the orbital period is close to 50 years. The distance to the stars is 8.64 ly (2.65 pc); from this and Fig. 8.37, we find the separation (semimajor axis) is about 20 AU.

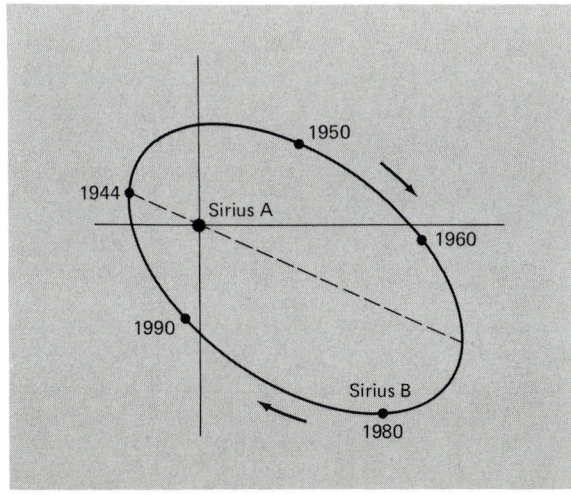

FIGURE 8.37 Orbit of Sirius B relative to Sirius A. Note that Sirius B was closest to Sirius A in 1944. The orbital period is 50 years.

If we want the masses to come out in *solar masses*, we use Kepler's third law in the form

$$M_1 + M_2 = \frac{a^3}{P^2}$$

with P in years and a in AUs. For the Sirius system,

$$M_A + M_B = \frac{20^3}{50^2}$$

$$= \frac{8000}{2500} = 3.2 \text{ solar masses}$$

Now Sirius A is about twice as close to the center of mass as Sirius B (Fig. 8.37). So from

$$\frac{M_A}{M_B} = \frac{a_B}{a_A}$$

with $2a_A = a_B$, we get

$$\frac{M_A}{M_B} = \frac{2a_A}{a_A} = 2$$

and

$$M_A = 2M_B$$

We now know the sum and ratio of the masses, so

$$2M_B + M_B = 3.2 \text{ solar masses}$$
$$3M_B = 3.2 \text{ solar masses}$$
$$M_B = 1.1 \text{ solar masses}$$

and

$$M_A = 2.1 \text{ solar masses}$$

Sirius A has about twice the mass of the sun, while Sirius B is about the sun's mass.

George Gatewood and Carolyn Gatewood have carried out a careful study of Sirius A and B. They find that the system has a period of 50.09 years and a total mass of 3.196 solar masses. Sirius A has a mass of 2.143 solar masses, a radius of 1.678 solar radii, and an effective temperature of 9970 K. In contrast Sirius B has a mass of 1.053 solar masses, a radius of 0.0073 solar radii, and an effective temperature of 29,500 K. Note the small radius—Sirius B is a white dwarf star.

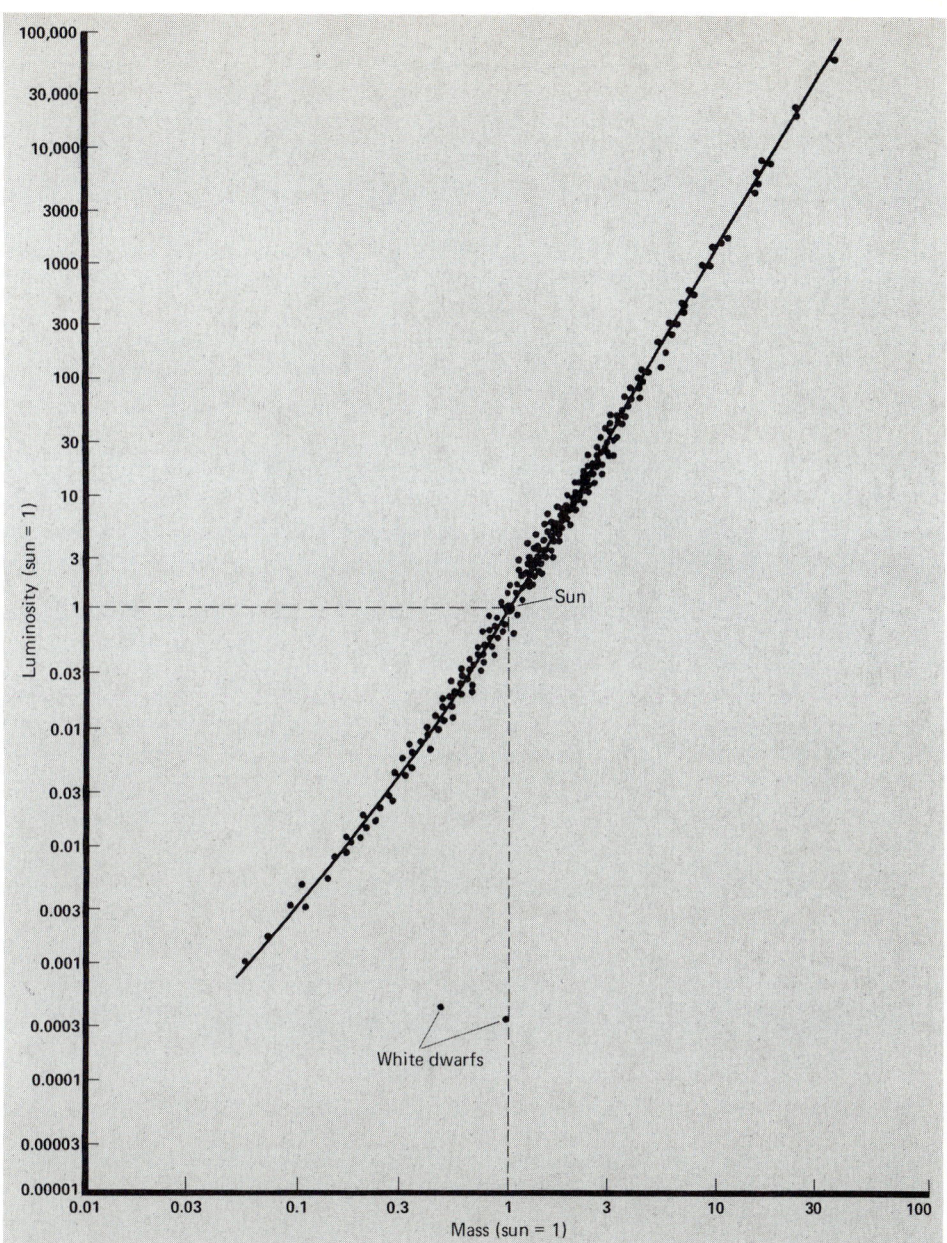

FIGURE 8.38 The mass-luminosity relation for stars, as determined from binary systems, in which the individual masses can be determined. These stars are also those whose distances can be measured, so that their luminosities can be calculated. Note that stars that have 10 times the mass of the sun shine with about 1000 times the luminosity.

The Mass-Luminosity Relation for the Main Sequence

Binary systems provide the only direct measure of stellar masses. In most cases the luminosities of the two stars can also be determined. When the luminosities are plotted against the stars' masses, the points fall into a definite pattern (Fig. 8.38). For most stars

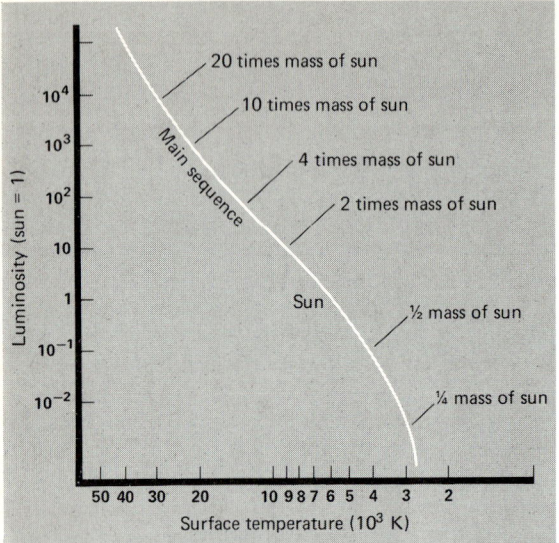

FIGURE 8.39 The approximate mass for stars on the main sequence of the Hertzsprung-Russell diagram. Note that the stars with the highest surface temperatures are the most massive; these are O and B stars.

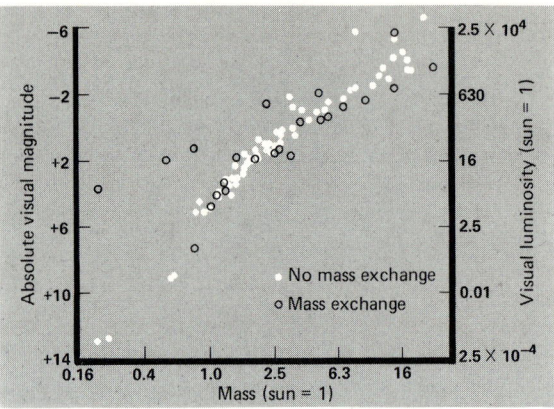

FIGURE 8.40 A mass-luminosity relation for eclipsing binary stars. The open circles represent systems which, like Algol, have mass flowing between the two companions. [C. Lacy, from the Astrophysical Journal, vol. 213, p. 458, copyright 1977 by the American Astronomical Society.]

the mass determines the luminosity, and the resulting correlation is called the *mass-luminosity relation*. The mass-luminosity relation shows that a star's luminosity is *roughly* proportional to the third power of its mass. For example, a star with a mass 10 times that of the sun has about 10^3 times the sun's luminosity. For main-sequence stars the mass-luminosity relation is given more exactly by

$$\frac{L_*}{L_{sun}} = \left(\frac{M_*}{M_{sun}}\right)^{3.3}$$

Main-sequence stars follow the mass-luminosity relation fairly well; hence *the upward swing in luminosity of the main sequence from M to O stars reflects an increase in the stars' masses* (Fig. 8.39). So main-sequence O stars are more massive than main-sequence M stars. Astrophysicists had predicted the mass-luminosity relation theoretically, and its confirmation came from investigation of binary stars. (In Chapter 11 you will see the importance of the mass-luminosity relation for stellar evolution.)

Claud Lacy has investigated the masses and luminosities of eclipsing binary systems of short period. He finds his mass-luminosity diagram (Fig. 8.40) has a large number of scattered points. These deviants turn out to be binaries in which the two stars are so close that, as with Algol, mass has been exchanged between them. We shall discuss some other properties of close binary stars in Chapter 11.

From the mass-luminosity diagram you see that the masses of other stars do not differ widely from the sun's mass. According to theoretical ideas, stars with masses greater than about 100 solar masses are unstable, and bodies with masses less than roughly 0.1 solar mass cannot become hot enough to start nuclear reactions and become stars. The pre-

8.6 WEIGHING AND SIZING STARS: BINARY SYSTEMS

dicted narrow range of stellar masses is borne out by the H-R diagram and the mass-luminosity relation.

Note that the white dwarf stars plotted in Figure 8.36 do *not* fall along the same line as do main-sequence stars. This fact hints that white dwarf stars are different beasts, with different interior structures. This is borne out by consideration of their densities. Let's check.

Stellar Densities

Although the masses of stars do not vary over a very wide range, their sizes do. So stellar densities must vary widely.

The sun's average density is 1400 kg/m³—40 percent more than that of water. Let's compare the sun to Sirius B, a white dwarf. For a spherical star the density (ρ) is

$$\rho = \frac{M}{(4/3)\pi R^3}$$

According to the Gatewoods, the mass of Sirius B is 1.05 solar masses and its radius is 0.0073 solar radius. So

$$\rho_{Sirius\ B} = \frac{(1.05)(2.0 \times 10^{30}\ \text{kg})}{(4/3)\pi[0.0073 \times (7 \times 10^8\ \text{m})]^3}$$
$$= 3.8 \times 10^9\ \text{kg/m}^3$$

Sirius B is about 4 *million* times denser than water! This fact tells you that Sirius B and other white dwarfs cannot be ordinary stars and cannot have an internal structure like the sun. That's the reason they do not fall along the main sequence in the H-R diagram or along the mass-luminosity relation with other stars. You'll find out more about the peculiar white dwarfs in Chapter 11.

Stellar Lifetimes

The existence of the mass-luminosity law provides a way of comparing stellar lifetimes. The argument goes like this. The total amount of energy available to a star from the conversion of hydrogen to helium is proportional to its mass. The rate at which it loses energy is given by its luminosity. Therefore, the time over which a star can radiate before its energy is used up is proportional to the mass divided by the luminosity (the energy store divided by the rate at which the store is used up). Relative to the sun,

$$\frac{t_*}{t_{sun}} = \frac{M_*/M_{sun}}{L_*/L_{sun}}$$

Now let's put in the mass-luminosity law, $L_*/L_{sun} = (M_*/M_{sun})^{3.3}$. Then

$$\frac{t_*}{t_{sun}} = \frac{M_*/M_{sun}}{(M_*/M_{sun})^{3.3}} = \frac{1}{(M_*/M_{sun})^{2.3}}$$

The lifetime is *inversely* proportional to the 2.3 power of the mass. More massive (more luminous) stars have a shorter lifetime. Though the massive stars have more fuel to burn, they use it up at such a fast rate that they last a shorter time.

For example, the lifetime of the sun is about 10 billion years, but a star of 30 solar masses has a lifetime of

$$t_* = 10^{10} \text{ yr} \times \frac{1}{(30)^{2.3}}$$

$$= 4 \times 10^6 \text{ yr}$$

Very massive stars last only a few million years.

SUMMARY

Everything we know about stars comes from analyzing their light. This analysis can be done only if we understand how atoms and light interact. It seems remarkable that from a knowledge of the smallest parts of the universe we can find out about the largest.

We know that the stars are other suns, big balls of hot gas burning by fusion reactions. Most of the stars we see are main-sequence stars and closely resemble the sun. But giants, supergiants, and white dwarfs do not follow the main-sequence pattern. These stars must have insides much different from those of main-sequence stars. Here's one clue that their *evolutionary* status differs.

An H-R diagram provides a graphical summary of the physical properties of the stars. You can think of an H-R diagram as a temperature-luminosity map of the stars in the sky. As we see the stars on the sky, they are arrayed randomly. The H-R diagram puts the stars into a pattern that gives us clues about stellar evolution.

Key Words

brightness
flux
inverse square law for light
apparent magnitude
heliocentric parallax
absolute magnitude
bolometric magnitude
  and luminosity
color temperature

color index
effective temperature
lunar occultation
spectral class
Hertzsprung-Russell diagram
main sequence
giant
supergiant
white dwarf
luminosity class

spectroscopic parallax
binary star system
visual binary
primary, secondary
spectroscopic binary
eclipsing binary
Doppler shift
radial velocity
mass-luminosity relation

Review Questions

1. The absolute magnitude of Procyon is +2.8 and its apparent magnitude is +0.5, much smaller. On the other hand, the absolute magnitude of Rigel is −6.2 and its apparent magnitude is +0.3, much larger. These data (mark all correct answers):
    a. Make no sense; one of the comparisons must be wrong.
    b. Mean simply that Procyon is closer than 10 pc, whereas Rigel is farther than 10 pc.

REVIEW QUESTIONS

   c. Mean simply that Procyon is farther than 10 pc, whereas Rigel is closer than 10 pc.
   d. Mean that Procyon appears slightly fainter to the naked eye than does Rigel. (Objective 5)
2. Look at the Hertzsprung-Russell diagram below.

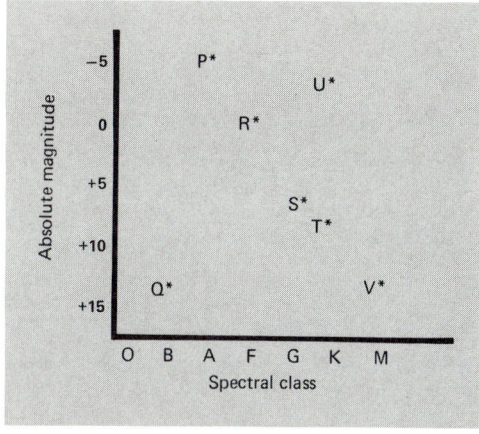

   a. Which two stars have the same temperature, but differ in luminosity, and which is more luminous?
   b. Which two stars have the same luminosity, but differ in temperature, and which is hotter?
   c. Which star has the smallest radius?
   d. Which star is reddest in color?
   e. Which star is similar to the majority of stars in the neighborhood of the sun? (Objective 11)
3. Would astronomers on Mercury be able to use trigonometric parallax more effectively, less effectively, or as effectively (compared to us on earth) for measuring distances to stars? (Objective 9)
4. The main spectral classes for stars, in order of decreasing strength of the Balmer lines of hydrogen, are A, B, F, G, K, M, O.
   a. Write the spectral sequence from bluest to reddest.
   b. Which spectral class is hottest? Which is coolest?
   c. Are most lines in the spectra of stars emission lines or absorption lines?
   d. Why are there no Balmer lines of hydrogen in O stars?
   e. Why are the Balmer lines very weak in M stars? (Objective 4)
5. a. In the winter sky Capella appears yellow, Betelgeuse red, and Sirius blue. List these stars in order of *increasing* surface temperature.
   b. Estimate the surface temperature of Betelgeuse. Of Sirius. Do they differ much? (Objectives 1 and 2)
6. Consider three stars with the following spectral and luminosity classes: M I, G III, and A V. Which star is the largest? Which the most luminous? (Objectives 3 and 10)

7. What considerations limit the accuracy of heliocentric parallax measurements? (Objective 9)
8. Refer to Figs. 8.19 and 8.22 to answer the following questions.
   a. Capella and the sun have roughly the same surface temperature. Which star is larger?
   b. Regulus and Capella have about the same luminosity. Which star is larger?
   c. Vega and Sirius A have about the same surface temperature. Which star is more luminous?
   d. Which star would appear redder: Vega or Pollux? (Objectives 10 and 11)

Problems
1. Sirius is 22 times more luminous than the sun; Betelgeuse 120,000 times more. Compare the sizes of the two stars to each other and to the sun. (Objective 3)
2. The heliocentric parallax of Alpha Centauri A is 0.76″. How far is this star in parsecs? In light years? (Objective 9)
3. Capella has an apparent visual magnitude of +0.09 and Regulus +1.35. Which star is brighter? How much brighter is it, in terms of a brightness ratio? (Objective 5)
4. Capella has a distance of 41.1 ly, Regulus 79.2 ly. Calculate the absolute visual magnitude of each star. Which is more luminous? By what factor? (Objective 7)
5. The sun has an absolute visual magnitude of +4.75. How luminous are Capella and Regulus compared with the sun? (Objectives 7 and 8)
6. The flux from Polaris at all wavelengths is 5.9×10^{-17} W/m^2, and it is roughly 390 ly from the sun. Calculate the luminosity of Polaris. (Objectives 5 and 6)
7. From the mass-luminosity relation, about how much more luminous than the sun do you expect a 10-solar mass star to be? (Objective 15)
8. The angular diameter of Antares is 41 milliarcseconds. Its distance is 130 pc. What is Antares' diameter? How does it compare with the sun? (Objective 1)

BEYOND THIS BOOK . . .

* *Atoms, Stars, and Nebulae* (Harvard University Press, Cambridge, Mass , 1971) by L. Aller has more details about spectra and the physical properties of stars.
* For original papers on spectra and stars, see Sections V and VIII in *Source Book in Astronomy* (Harvard University Press, Cambridge, Mass., 1960), edited by Harlow Shapley.
* Otto Struve and Velta Zebergs trace the development of ideas about stellar spectra in Chapters X and XI in *Astronomy in the 20th Century* (Macmillan, New York, 1962).
* For an update on the Hertzsprung-Russell diagram, see "The H-R Diagram as an Astronomical Tool," *Sky and Telescope*, vol. 55 (May 1978), p. 395, by A. G. D. Philip and L. Green; and *The HR Diagram* (Reidel, Dordrecht, 1978), edited by A. G. D. Philip and D. S. Hayes.

STARBIRTH AND THE INTERSTELLAR MEDIUM

9

LEARNING OBJECTIVES
After studying this chapter, you should be able to:
1. Present observational evidence for the existence of gas and dust between the stars.
2. Describe the different forms in which the interstellar gas is found and how each form is observed.
3. Describe the effects of interstellar dust on starlight.
4. Describe possible physical properties of interstellar dust.
5. Indicate how interstellar molecules and dust might be formed.
6. Describe the basic physical ideas of gravitational collapse, and make use of an expression for the free-fall collapse time.
7. Sketch a scenario for the formation of massive stars from molecular clouds, and indicate what observations support this model.
8. Outline possible processes for the birth of stars like the sun.
9. Sketch a model for the structure and evolution of the interstellar medium with emphasis on the role of supernova explosions.
10. Argue that starbirth must be occurring now in our Galaxy, with a focus on infrared and radio observations.
11. Define angular momentum, describe its conservation, and make use of an algebraic description in astronomical applications.

CENTRAL QUESTION:
How are stars born out of the interstellar medium?

Where do the stars come from? How and where are they born? You have found in Chapter 8 that stars have finite lives—long by human standards, but still limited. How long a star lives depends on its mass. A star like the sun will survive for some 10 billion years. More massive stars live scant millions of years. The fact that we observe so many stars now means not only that starbirth was happening in the past, but also that it must be going on now. Otherwise we would see fewer stars, especially the massive ones that die off relatively quickly.

Where are the wombs of starbirth? In clouds of gas and dust between the stars. Contrary to first impressions, interstellar space is *not* empty. It contains both gas and dust, thinly spread out and in clumps. Hydrogen makes up most of the gas, which in numbers of particles exceeds the dust by about 10^{12} to 1. In mass, however, the gas to dust ratio is only 10^2 to 1. The interstellar gas is not simply made up of neutral atoms; some is ionized, and some is bound up in molecules. A large fraction of the interstellar gas is locked up in short-lived clouds. From these interstellar clouds, stars are born. The interstellar medium is the main topic of this chapter. Here, stars form in vast stellar nurseries.

9.1 THE INTERSTELLAR MEDIUM: GAS

What do we know, in general, about the interstellar gas? First, it is made mostly of hydrogen. Second, it tends to clump in clouds. Third, a hot, dilute gas exists between the clouds. Fourth, the gas in different locations contains neutral atoms, ionized atoms, free electrons, and molecules. Fifth, the interstellar gas is very tenuous. It contains on the average roughly one hydrogen atom in a cubic centimeter—a vacuum by earthly standards! The distance between interstellar atoms is about 10^8 times larger than the size of the atoms themselves. If two people were separated by a proportional distance relative to their size, they would be about 10^8 m apart, about the distance between the earth and moon. Sparse as it is, the interstellar gas occasionally clumps and forms stars. Let's look at the variety of forms of the interstellar gas.

Bright Nebulas

On a winter's night you can easily spot the constellation Orion (see Fig. 1.1 and the January star chart). Dangling from Orion's belt is a short sword; if you look closely, the middle star appears fuzzy. A small telescope pointed at this fuzzy patch shows a diffuse, convoluted cloud surrounding a small cluster of stars (Fig. 9.1). This bright cloud is called the *Orion Nebula*. (*Nebula* is the Latin word for "cloud.") Only 1500 ly (460 pc) from the earth, the Orion Nebula, roughly 16 ly (5 pc) in diameter, is typical of *bright nebulas* (Fig. 9.2).

At the end of the nineteenth century spectroscopic analysis demonstrated that these bright nebulas did consist of gas. The bright nebulas showed spectra of emission lines. Recall from Kirchhoff's rules for spectra (Section 6.2) that an emission-line spectrum indicates a hot, diffuse gas. The Orion Nebula, for example, has lines of hydrogen, helium, and oxygen predominating in its bright-line spectrum.

In the 1930s Edwin Hubble demonstrated that nebulas such as that in Orion do not shine by their own light, but rather absorb the energy from hot O or B stars located in or

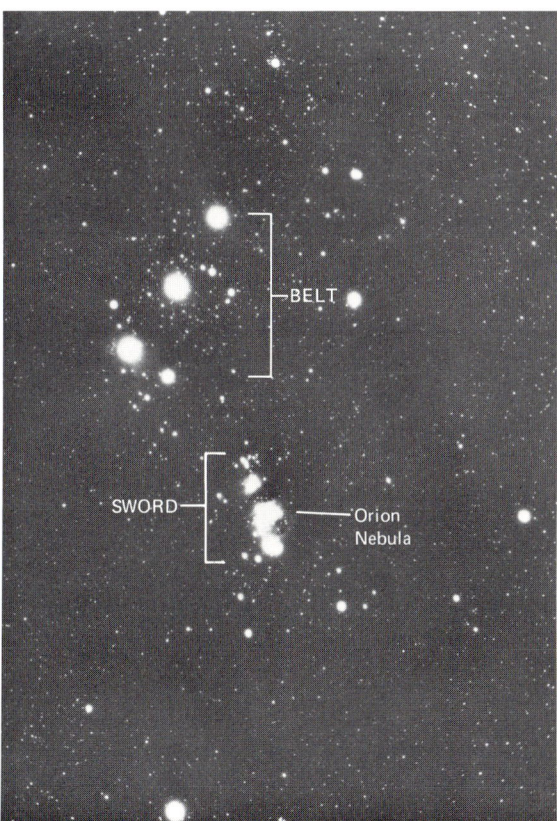

FIGURE 9.1 The belt and sword region of Orion (compare to Fig. 1.1). The Orion Nebula lies in the middle of the sword. [*Courtesy, Yerkes Observatory.*]

FIGURE 9.2 A close-up of the Orion Nebula, a typical bright, diffuse nebula. [*Courtesy, Lick Observatory.*]

near the nebula. The gas absorbs high-energy ultraviolet photons given off by the central star (or stars) and gives off photons in emission lines at lower energies. For a hydrogen gas the visible light comes out mainly in the red region of the spectrum as the Balmer line at 6563 Å. So emission nebulas glow predominantly with a red light (Plate 5). This process of splitting high-energy photons into photons of lower energy is called *fluorescence*; the same process occurs in a fluorescent light, which, like a bright nebula, also has an emission-line spectrum.

The star (or stars) in a bright nebula is quite hot, about 30,000 K, and it emits many photons with enough energy to ionize hydrogen. Most of these photons are absorbed by the gas surrounding the star so that, out to a considerable distance (a few tens of light years), the gas is almost totally ionized. This zone of ionized hydrogen is in its second state. It is called an *H II region* (Fig. 9.3). (H I stands for neutral hydrogen and H II for ionized hydrogen.) Astronomers use the terms H II region, bright nebula, and emission nebula interchangeably.

Keep these general ideas in mind as we describe in more detail the process by which a bright nebula emits light. Virtually all the atoms in an H II region are ionized. Most of these atoms are hydrogen. When a naked hydrogen nucleus (a proton) captures an electron (a process called *recombination*), the reformed atom gives off light. The details of this process are complicated by all the possible energy levels of the atom (Section 6.3).

To simplify matters, consider the lowest three energy levels in hydrogen. Imagine a hydrogen atom with an electron in the lowest level. The atom absorbs a high-energy ultraviolet photon (wavelength 912 Å or shorter). This addition of energy kicks the electron out of the nucleus's grip entirely and ionizes the atom. The electron zips around by itself with any one of a large possible range of energies. Ions and free electrons flying

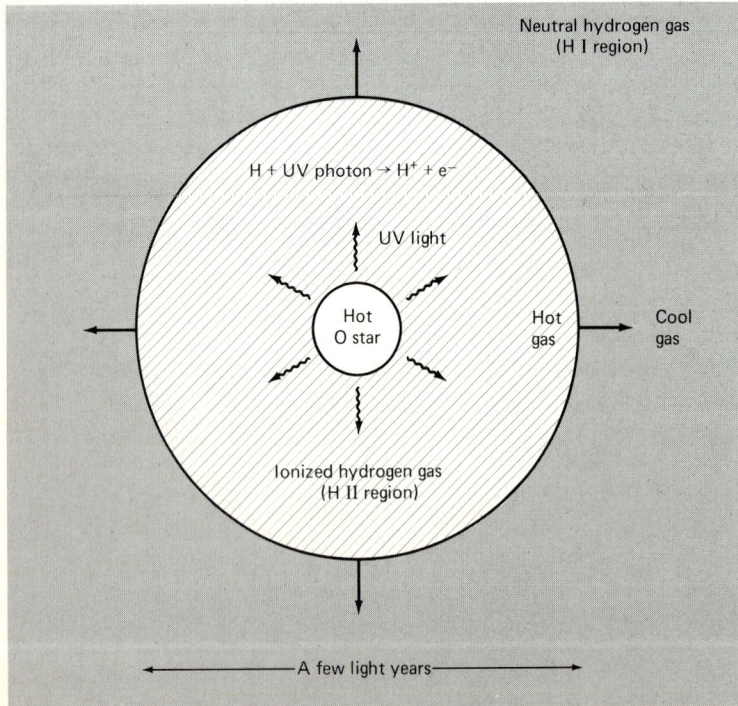

FIGURE 9.3 A schematic diagram of an H II region. A hot O star (or stars) emits ultraviolet light that can ionize hydrogen for a few light years around. The absorption of the ultraviolet heats up the ionized gas to a temperature of about 10,000 K. The hot, ionized gas expands into the cooler, neutral gas surrounding it.

9.1 THE INTERSTELLAR MEDIUM: GAS

around are the typical state of the gas in an H II region. The ions travel around until they encounter an electron with which to combine. This does not happen often because the gas in the nebula is sparse, only about 1000 atoms/cm^3.

Suppose the ion does pick up an electron. (Remember that they have opposite charges and so attract each other.) The electron does not necessarily end up in level 1, but if it does, the atom emits an ultraviolet photon that can ionize another atom. Quite quickly, another ultraviolet photon zaps the atom and ionizes it again.

Imagine that the electron is caught in one of the higher levels, say level 3 (Fig. 9.4a). As it is caught, it emits a low-energy photon. Since the free electrons may have any energy, the photons emitted when the electrons recombine with an ion also can have any energy, as long as it is greater than the energy difference between the final level (3 in this case) and the fully ionized state. Hence this free-bound transition results in a continuous spectrum of emission.

The electron does not stay in level 3 for long, because it seeks the lowest energy state (level 1). It can get there by dropping through the levels one by one or by skipping over some. However, as the electron drops down to the lowest level, it must emit a photon with each drop.

Suppose the atom, having caught the electron in level 3, then drops to level 2. It emits a Balmer photon with a wavelength of 6562.8 Å; the emission line that results is called the *H-alpha line* (Fig. 9.4b). It is the most common emission line from a nebula and makes it appear red.

You may wonder why the next drop, from level 2 to level 1 does not produce the brightest emission line. This drop produces an ultraviolet photon at a wavelength of 1216 Å, the Lyman-alpha line. Few of these photons make it out of the nebula, because they are constantly absorbed by other hydrogen atoms. Most of the hydrogen atoms are in their ground state, and so are capable of absorbing these ultraviolet photons of the Lyman series of lines. Even if these photons did get out, we would never see them on earth because the ultraviolet light does not penetrate the earth's atmosphere.

Optical astronomers don't have a monopoly on viewing bright nebulas; radio astronomers can see them also, in two ways. First, they can be viewed by photons of radio emission, in what are called *radio recombination lines*. The process resembles that for optical recombination lines, described above. Hydrogen atoms have some energy levels very close to the ionization limit. These levels lie very close together; their energy differences correspond to photons with energies in the radio range of the spectrum. Suppose a

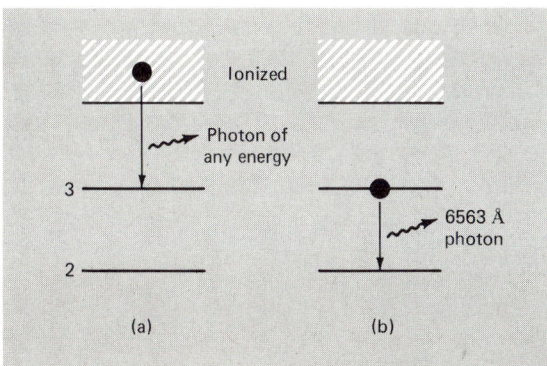

FIGURE 9.4 Emission of photons by hydrogen. (a) When a hydrogen ion captures an electron, it can end up in any energy level below the starting level; so a photon of any energy can be emitted. (b) If it falls between bound levels, it can emit photons with certain energies, such as the H-alpha (6563 Å) photon emitted when the electron drops from level 3 to 2.

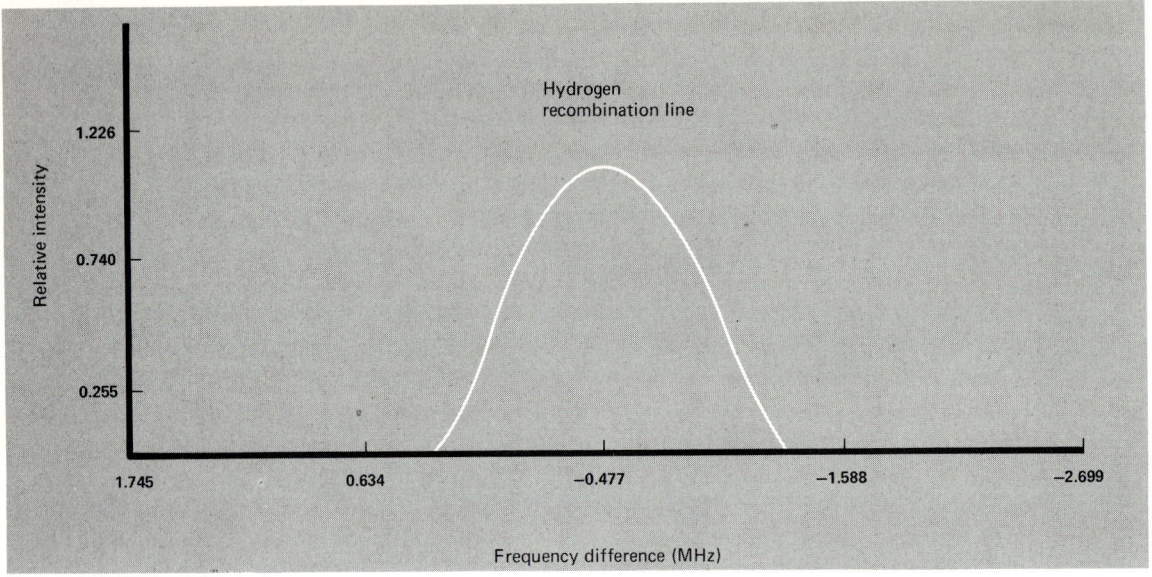

FIGURE 9.5 Observation of a hydrogen radio recombination line from the emission nebula M 17. The horizontal scale gives the difference in frequency from the rest frequency for the transition. [*Data taken by N. L. Cohen at Haystack Observatory.*]

proton captures an electron in one of these levels. If the electron drops down to the next level, it emits a radio photon that can be picked up by a radio telescope operating at the right wavelength, usually a few centimeters (Fig. 9.5). Second, H II regions are visible by means of their continuous radio emission, produced by a free-free process. In an H II region free electrons speed past protons. As the opposite charges attract, the electrons are bent from their straight-line paths. The electrons are accelerated, their energy changes, and they emit electromagnetic radiation, mostly at millimeter and centimeter wavelengths (Fig. 9.6). Since many different electrons undergo different energy changes at the same time, a continuous spectrum results, rather than an emission line. This continuous radio emission can be mapped by a radio telescope (Fig. 9.7). From such maps astronomers can infer how much ionized gas is contained in an H II region (the Orion Nebula,

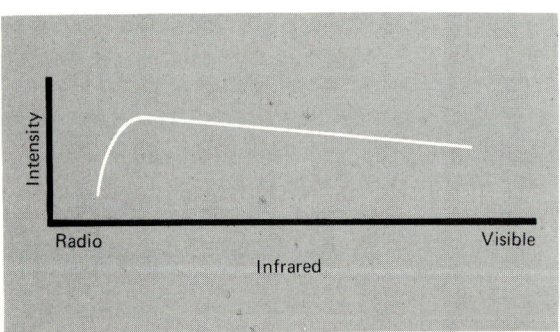

FIGURE 9.6 A schematic of the spectrum of free-free emission from a hot, ionized gas.

(a)

(b)

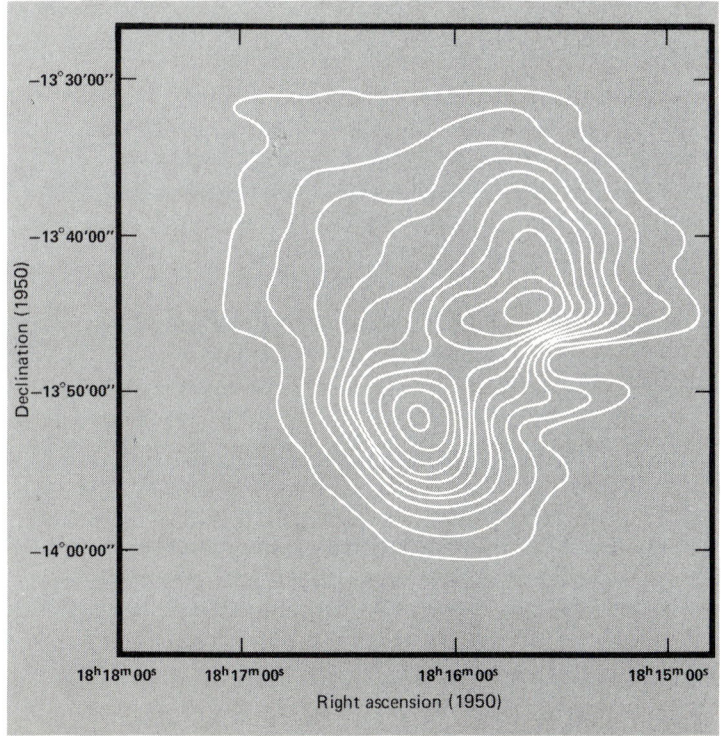

FIGURE 9.7 (a) A photo of the H II region Messier 16 (M 16). [*Courtesy Lick Observatory.*] (b) An intensity contour map of the free-free emission at 8 GHz from M 16. [*Observation by M. Zeilik and C. Lada at Haystack Observatory.*]

for example, contains about 300 solar masses), even if the nebula is not visible in an optical telescope.

In summary, H II regions emit in almost all possible ways that atoms produce light: bound-bound transitions (such as H-alpha and the radio recombination lines), free-bound transitions (as electrons are captured by ions), and free-free transitions (of electrons in the ionized gas).

Interstellar Atoms

Investigations of bright nebulas show that they are composed almost entirely of hydrogen. So the neutral gas ionized to form them must also have contained mostly hydrogen, as neutral atoms (designated H I).

Until the 1950s astronomers surmised that hydrogen atoms populated interstellar space, but they had not observed the H I gas. The Dutch astronomer Hendrick C. van de Hulst had suggested in 1944 that interstellar hydrogen atoms might be so abundant as to be detectable by radio telescopes. How can this simple atom give off radio radiation?

Recall that the hydrogen atom has one proton in the nucleus and one electron in orbit around it. Both the proton and electron spin like miniature tops. According to the rules of quantum physics the electron and the proton can be oriented in the atom so the two spins either align or oppose each other. If the spins oppose, the total energy of the atom is just a bit less than if the spins align. As usual, the atom prefers to be in the lower energy state. If the spins are aligned, eventually the proton flips over and emits a low-energy photon— energy that corresponds to a wavelength of 21.11 cm (Fig. 9.8).

How are the protons and electrons in hydrogen atoms aligned in the first place? It can happen by collisions with electrons and other atoms. The gas in interstellar space is very sparse, and collisions between two atoms occur only once every few million years. On the other hand, once the spins in a hydrogen atom are aligned, about 10 million years must pass before the proton flips and the atom drops to its lowest energy state. So several collisions occur before the atom radiates. Some will misalign the spins again. But the final result is that an equilibrium is established in which three times as many atoms have spins aligned as have spins opposed. Every so often, once in 10 million years on the average, an aligned atom flips over spontaneously, without a collision, and emits a 21-cm photon. This is a rare event for any one atom, but because so many hydrogen atoms exist in

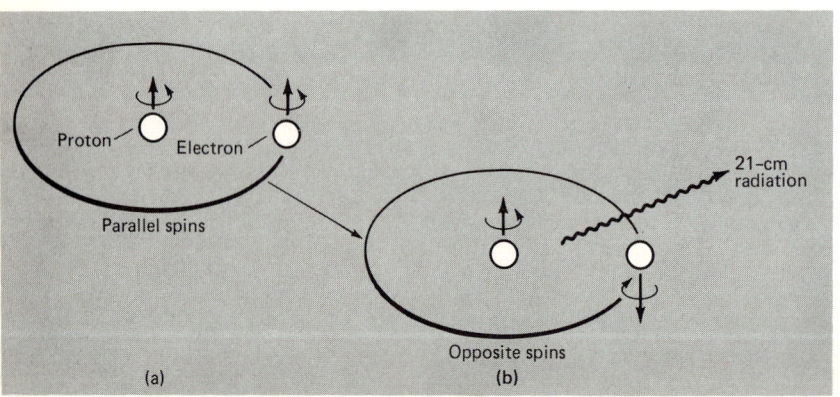

FIGURE 9.8 21-cm emission from hydrogen atoms. Collisions can line up the spins of the proton and electron so that they are parallel. After some time, if the atom does not hit another, the electron flips so that the spins are opposed, and the atom emits a 21.1-cm photon.

9.1 THE INTERSTELLAR MEDIUM: GAS

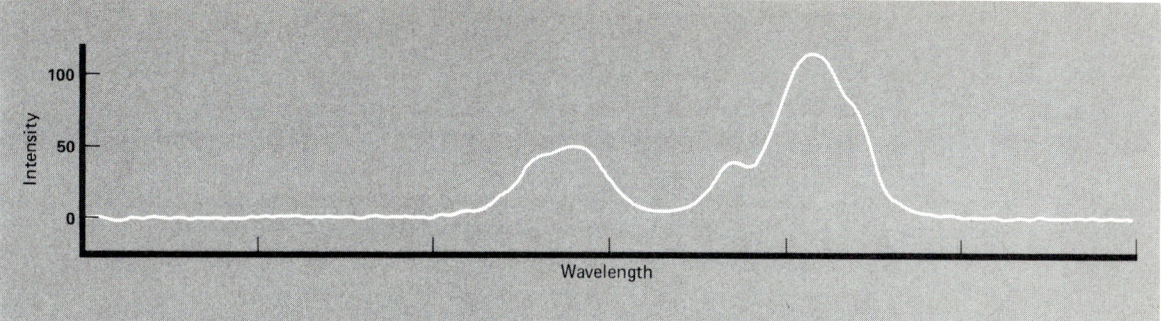

FIGURE 9.9 Observations of the 21-cm emission from atomic hydrogen in interstellar space. [*Observations by W. Burton with the Dwingeloo radio telescope.*]

interstellar space, at any given time enough are emitting 21-cm radiation that the interstellar gas radiates strongly at this wavelength and can be detected with radio telescopes (Fig. 9.9).

Surveys of interstellar hydrogen atoms with radio telescopes operating at 21 cm find that most neutral hydrogen is concentrated in the plane of the Milky Way. (The *Milky Way* is a faint, broad band of light in the sky; as Galileo discovered (Chapter 4), it consists, when viewed with an optical telescope, of a multitude of faint stars, all belonging to our Galaxy. When we talk about *our* galaxy, as opposed to all the others, we will often spell it with a capital letter, the Galaxy. See Chapter 20 for more on the Galaxy.) On the average the hydrogen atoms have a temperature of 70 K and a density of $0.33/cm^3$. So in a volume of space equivalent to the volume of your body, you'd find only 3×10^4 hydrogen atoms, whereas in fact your body contains some 10^{27} atoms.

The hydrogen atoms are not distributed uniformly in space. Rather, they tend to clump in small clouds. A typical H I cloud has a density of some 20 atoms/cm^3, a diameter of roughly 16 ly (5 pc), and a distance of approximately 1000 ly (300 pc) from the next nearest cloud. These clouds move at about 6 km/sec relative to each other.

As you might expect, there are also other gases in interstellar space. Even before atomic hydrogen was observed with radio telescopes, optical observations had revealed the presence of several other kinds of atoms. Superimposed on the spectra of some stars, astronomers found sharp, dark lines of elements such as sodium (Fig. 9.10). These absorption lines are produced when starlight passes through cool regions of the interstellar gas. Recall Kirchhoff's rules (Section 6.2): An absorption spectrum results when light from a continuous source passes through a cooler gas. But how do we know that these lines are formed by the interstellar gas rather than in the atmosphere of the star? For one

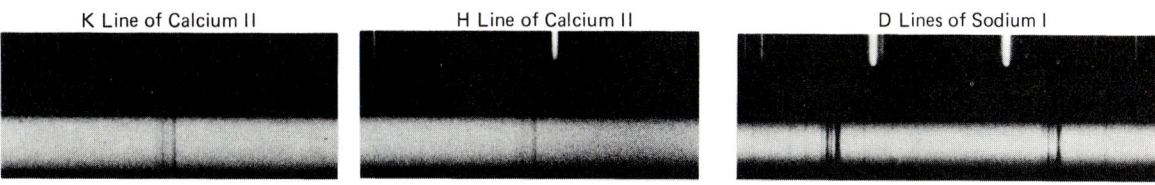

FIGURE 9.10 Absorption lines of sodium and calcium II in the interstellar medium. These elements in cool gas clouds absorb some of the starlight that passes through them. [*Courtesy, Palomar Observatory, California Institute of Technology.*]

thing, the lines are very narrow, an indication that the gas is cool. But the strongest evidence came when these lines were found in the spectra of some binary stars. As the stars move around each other, all the stellar absorption lines shift back and forth in wavelength, due to the Doppler shift caused by their changing velocities. But the interstellar lines remain steady at a constant wavelength, indicating that the gas producing them has nothing to do with the stars themselves. These kinds of optical observations uncovered atoms of sodium, potassium, ionized calcium (Fig. 9.10), and iron in the interstellar medium.

These interstellar absorption lines were the first direct indication of the existence of a pervasive interstellar medium. They led to the idea that cool regions in the medium exist in the form of small clouds.

You may be wondering why hydrogen, by far the most abundant atom in the interstellar gas, wasn't first found optically. Interstellar space is so empty and the temperature of the gas so low that most hydrogen atoms remain in their lowest energy state. In their ground state hydrogen atoms can absorb only ultraviolet light (in the Lyman series). But ultraviolet radiation cannot penetrate the earth's atmosphere, so these absorption lines cannot be observed with ground-based telescopes. With the advent of earth-orbiting ultraviolet telescopes (such as the Copernicus satellite), we can now observe hydrogen atoms optically.

Ultraviolet observations by the Copernicus satellite have enriched and complemented the radio picture of the interstellar gas. As hydrogen atoms absorb ultraviolet light, they leave their absorption lines in stellar spectra. Such observations by Copernicus have shown that the neutral hydrogen gas has a very patchy distribution in clouds with diameters from tenths of parsecs to tens of parsecs. The average density of neutral hydrogen is somewhat less than 1 atom/cm^3. However, the observations also show directions in space where the neutral hydrogen density is much less, 0.01 atom/cm^3 and lower. Now it doesn't seem likely that these regions have no gas at all, but rather that the gas is in a different form. For example, if the hydrogen is hot and ionized, it will no longer have any ultraviolet absorption spectrum and so will not be detectable by Copernicus observations. Here is evidence that some interstellar gas is (in part at least) ionized.

Intercloud Gas

What about the space between the interstellar clouds? It cannot be empty. As you'd expect, it must consist mostly of hydrogen. Is it ionized or neutral? Well, to make life complex, it seems to be both! Radio observations at 21 cm indicate a thin, neutral gas with an average density of only 0.17 atom/cm^3. Radio recombination lines of hydrogen point to an even thinner, hotter (about 10,000 K) ionized gas also between the clouds. The density of the H II gas averages a mere 0.03 ion/cm^3.

To make matters worse, ultraviolet observations by Copernicus provide direct evidence for a *very* hot gas permeating the intercloud regions. The observations show absorption lines (at 1032 Å and 1037 Å) of oxygen stripped of five electrons (Fig. 9.11). (The symbol for five-times ionized oxygen is O VI.) To rip so many electrons from an oxygen atom requires a very high temperature, about 2.5 to 7×10^5 K, or perhaps even 10^6 K. Because this hot gas has about the same temperature as the sun's corona, it is called the *coronal interstellar gas*. Evidence for it comes also from the X rays it emits. It must occupy a large fraction of the volume of interstellar space, because the O VI absorption lines and the X rays have about the same intensity in whatever direction we look from the earth.

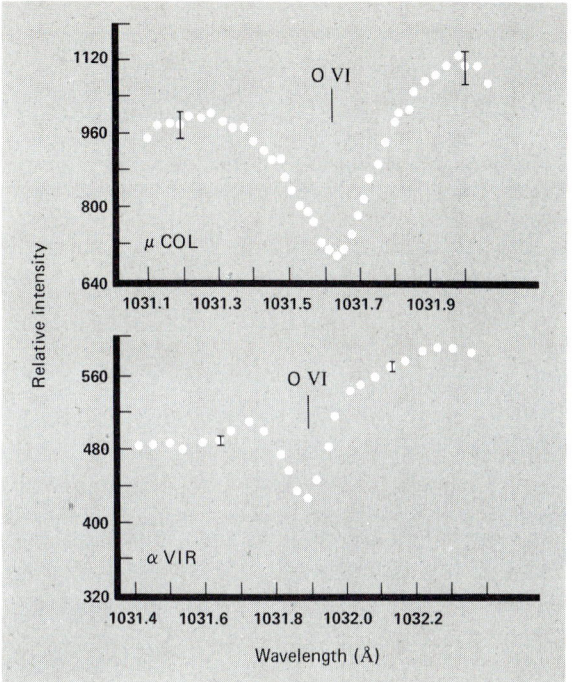

FIGURE 9.11 Interstellar absorption lines from O VI (five-times ionized oxygen) looking toward two different stars. [*Observations with the Copernicus satellite by D. York*, Astrophysical Journal (Letters), *vol. 193, p. L127, copyright 1974 by the American Astronomical Society.*]

Interstellar Molecules

So much for hot stuff. What about cold material? You might expect it to be in the form of simple molecules. Optical astronomers made the first discoveries of such molecules: CH (methylidyne radical), CH^+ (methylidyne ion), and CN (cyanogen radical).

But the optical part of the spectrum is not the most fruitful region to search. A molecule consists of atoms linked together in particular arrangements by electron bonds. A molecule can have different energy states if the atoms vibrate or the molecule spins in various ways. The energy difference between vibrational states amounts to only 10^{-21} J and that for rotational states only 10^{-23} J. (This contrasts with about 5×10^{-19} J for the typical energy difference between electronic states of atoms or molecules.) As with changes in electronic states in an atom, when a molecule changes vibrational or rotational states, it can emit or absorb a photon. For changes in vibrational states, the photons are infrared ones; for rotational states, radio ones. In the cold regions where molecules can exist in interstellar space, occasional collisions between molecules (or perhaps with atoms) kick the molecules and get them spinning. Eventually, the molecules emit radio photons that can be observed as a radio emission line, generally at millimeter wavelengths.

The radio search for molecules began in earnest in the 1960s. The first molecule discovered was the hydroxyl radical, OH. Later, molecules with more than two atoms—ammonia (NH_3) and water (H_2O)—were found, much to the surprise of astronomers who had not suspected that three (or more) atoms could get together in the sparse conditions in space to make a large molecule. To date, more than 50 molecules have been found (Table 9.1), some with as many as thirteen atoms.

You'll notice some familiar molecules in Table 9.1. Carbon monoxide (CO), one of

TABLE 9.1 Some Interstellar Molecules Observed to Date

Complexity	Inorganic		Organic	
Diatomic	H_2	hydrogen	CH	methylidyne radical
	HD	deuterized hydrogen	CH^+	methylidyne ion
	OH	hydroxyl radical	CN	cyanogen radical
	SiO	silicon monoxide	CO	carbon monoxide
	SiS	silicon monosulfide	CO^+	carbon monoxide ion
	NS	nitrogen monosulfide	CS	carbon monosulfide
	SO	sulfur monoxide	C_2	carbon
	NO	nitric oxide		
Triatomic	H_2O	water	CCH	ethynyl radical
	HDO	heavy water	HCN	hydrogen cyanide
	N_2H^+	imidyl ion	HNC	hydrogen isocyanide
	H_2S	hydrogen sulfide	DCN	deuterium cyanide
	SO_2	sulfur dioxide	DNC	deuterium isocyanide
	HNO	nitroxyl	HCO	formyl radical
			HCO^+	formyl ion
			HCS^+	thioformyl ion
			OCS	carbonyl sulfide
4-atomic	NH_3	ammonia	H_2CO	formaldehyde
			HNCO	hydrocyanic acid
			HNCS	isothiocyanic acid
			H_2CS	thioformaldehyde
			HC_2H	acetylene
			$HOCO^+$	protonated carbon dioxide
			C_3N	cyanoethynyl radical
5-atomic			CH_4	methane
			H_2CNH	methyleneimine
			H_2NCN	cyanamide
			HCOOH	formic acid
			HC_3N	cyanoacetylene
			H_2C_2O	ketene
			C_4H	butadiynyl radical
6-atomic			CH_3OH	methyl alcohol
			CH_3CN	methyl cyanide
			$HCONH_2$	formamide
			CH_3SH	methyl mercaptan
7-atomic			CH_3NH_2	methylamine
			CH_3C_2H	methylacetylene
			$HCOCH_3$	acetaldehyde
			H_2CCHCN	vinyl cyanide
			HC_5N	cyanodiacetylene

(continues)

TABLE 9.1 (continued)

Complexity	Inorganic	Organic	
8-atomic		$HCOOCH_3$	methyl formate
		CH_3C_3N	methyl cyanoacetylene
9-atomic		$(CH_3)_2O$	dimethyl ether
		CH_3CH_2OH	ethyl alcohol
		HC_7N	cyanotriacetylene
		CH_3CH_2CN	ethyl cyanide
11-atomic		HC_9N	cyanotetraacetylene
13-atomic		$HC_{11}N$	cyanopentaacetylene

the most common interstellar molecules, is an atmospheric pollutant given off by automobiles. Sulfur dioxide (SO_2) is another common pollutant that comes from burning fossil fuels with high sulfur content. Water (H_2O) is water; strange to think of water molecules floating in space! Ethyl alcohol (CH_3CH_2OH) gives the kick to beer and wine. Note that many of the molecules in Table 9.1 are *organic*, that is, compounds with carbon playing a central role. The most abundant atoms in these molecules—carbon, hydrogen, nitrogen, and oxygen—are also the most abundant in living creatures on the earth. (More to come in Chapter 25.)

By far the most abundant molecule is, of course, hydrogen (H_2), simply because hydrogen is the most abundant element in the universe (Table 9.2). Even though radio telescopes can easily observe atomic and ionized hydrogen, they cannot detect molecular hydrogen, because it does not emit or absorb at radio wavelengths. The hydrogen molecule does absorb and emit ultraviolet and infrared wavelengths, however. Infrared emission lines of H_2 have been observed from heated interstellar clouds, and ultraviolet absorption lines from cool clouds have been detected by the Copernicus telescope in the spectrum of hot stars (Fig. 9.12).

Warning: Do not confuse H_2, molecular hydrogen, with H II, ionized hydrogen, even though they may be pronounced the same.

TABLE 9.2 Fractional Abundance of Some Molecules in Dense Clouds (Hydrogen = 1.0)

Greater than 10^{-6}	10^{-6}	10^{-7}	10^{-8}	10^{-9}	10^{-10}
CO	HCN	OH	CH	HNCO	H_2CS
H_2O	HNC	CS	CN	NH_2CN	HCOOH
	NH_3	SO_2	H_2CO	CH_3C_2H	CH_2NH
		CH_3OH	HC_3N		$HCONH_2$

Source: Adapted from E. Herbst, "The Current State of Interstellar Chemistry in Dark Clouds," in *Protostars and Planets* edited by T. Gehrels (University of Arizona Press, Tucson, 1978).

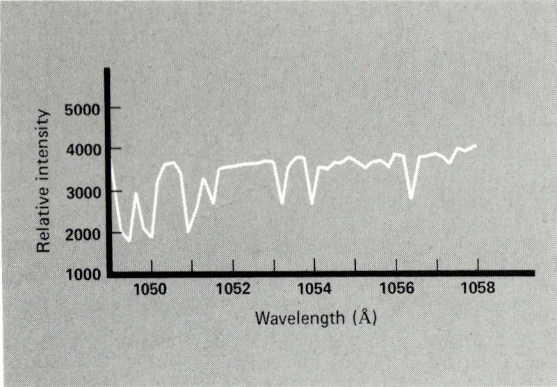

FIGURE 9.12 Interstellar absorption lines from molecular hydrogen, observed by the Copernicus satellite. [*From J. Rogerson, L. Spitzer, J. Drake, K. Dressler, E. Jenkins, D. Morton, and D. York,* Astrophysical Journal (Letters), *vol. 181, p. L97, copyright 1973 by the American Astronomical Society.*]

Molecular Clouds

Although some interstellar molecules, such as carbon monoxide, pop up in almost every direction in the sky, most are concentrated in dark, dense, cold conglomerates called *molecular clouds*. These clouds often lie near H II regions. One of the closest molecular clouds sits behind the Orion Nebula, as we view it from earth. The molecular cloud here consists of a large, low-density cloud (inferred from carbon monoxide emission) surrounding a dense, small core (inferred from formaldehyde emission). The low-density cloud has an enormous extent (Fig. 9.13a). It is at least 30 ly (10 pc) across, has a peak density of 10^3 hydrogen molecules/cm^3, and contains at least 10^4 solar masses of material. The core (Fig. 9.13b) is only 0.5 ly (0.15 pc) in size, has a peak density of 10^5 hydrogen molecules/cm^3, and a mass of only 5 solar masses.

The Orion region presents an excellent example of a *giant molecular cloud*. Observations so far indicate that the bulk of the material of the interstellar medium is bound up in complexes of giant molecular clouds. These immense globs of molecules, held together by gravity, typically have the following properties.

1. The cloud complexes consist mostly of molecular hydrogen. Many other molecules are present, but make up only a small fraction of the mass (Table 9.2).
2. They have average densities of a few hundred molecules per cubic centimeter; the individual clouds are slightly denser, with a few thousand molecules per cubic centimeter.
3. They have sizes of a few tens of light years.
4. The total masses of the complexes range from 10,000 to 10 million solar masses; 100,000 solar masses is typical. Masses of individual clouds are about 1000 solar masses.

The cores of these clouds are unusual places compared with the average interstellar medium. Here the temperatures are a frigid 10 K and the densities get as high as a *million molecules per cubic centimeter*. That's an immense concentration by interstellar standards, yet it is only 10^{-13} the density of molecules in the air at the earth's surface. Giant molecular clouds are so huge, though, that they contain an enormous number of molecules in total.

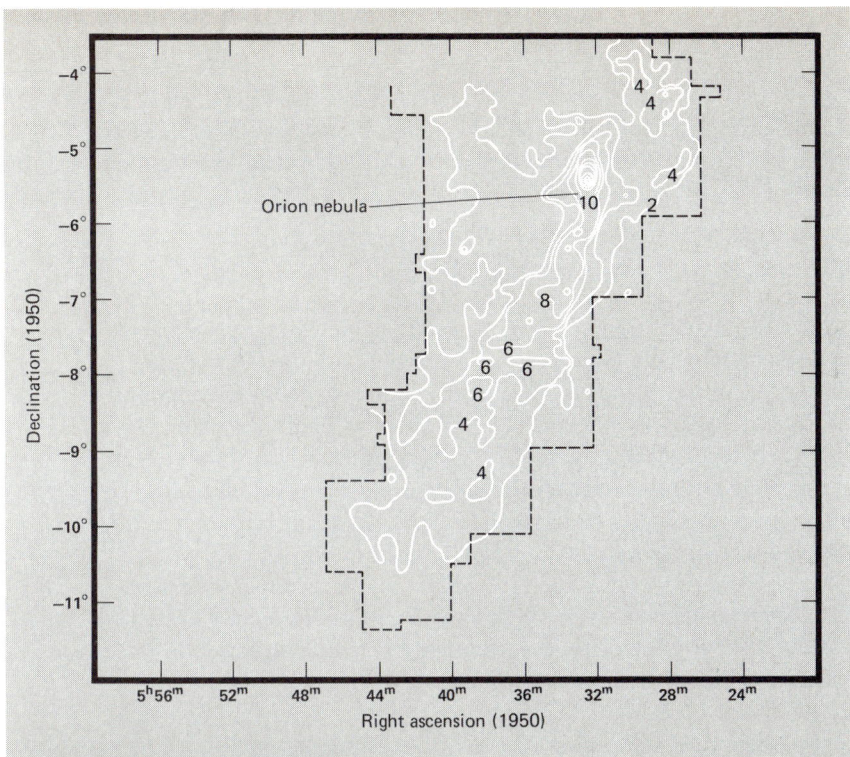

FIGURE 9.13a. A carbon monoxide intensity map of the giant molecular cloud associated with the Orion Nebula. The hottest, densest part of the cloud lies just above the contour marked "10"; the Orion Nebula lies just below and to the right of this peak. This map was made in New York City by the Columbia University millimeter-wave antenna. [*Adapted from a figure by M. Kutner, K. Tucker, G. Chin, and P. Thaddeus,* Astrophysical Journal, *vol. 215, p. 521, copyright 1977 by the American Astronomical Society.*]

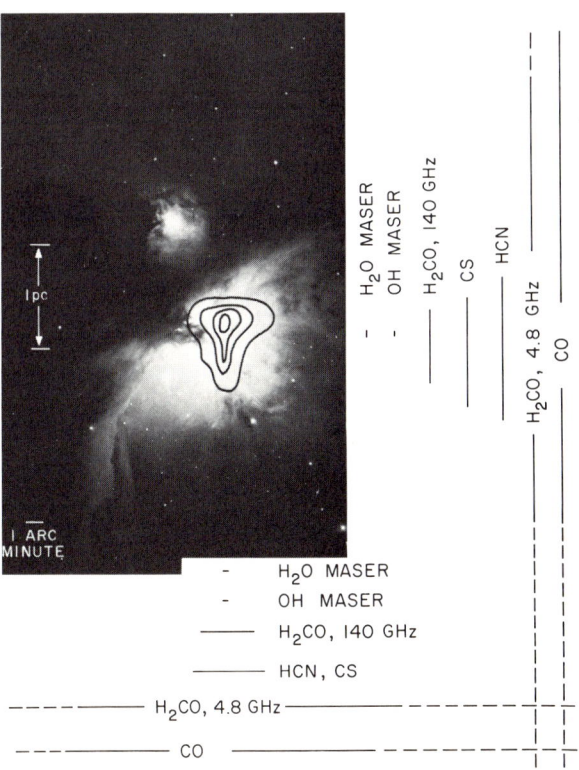

FIGURE 9.13b. A map of the central part of the giant molecular cloud that lies *behind* the Orion Nebula. The dark contour lines show the dense core of cloud; the line labeled "1 pc" gives the scale. The bars on the sides of the photo mark the extent of the core when mapped in the emission from carbon monoxide (CO), formaldehyde (H_2CO), and hydrogen cyanide (HCN). Very small regions of water and hydroxl emission come from the "H_2O maser" and "OH maser" regions. [*Figure courtesy of E. Chaisson; photo courtesy of Kitt Peak National Observatory.*]

311

Another important property of these clouds is their location in space. Giant H II regions, which surround young, massive stars, are always found near molecular cloud complexes. This proximity suggests that giant molecular clouds play a key role in the process of star formation.

Molecular Masers

The normal emission from interstellar molecules usually comes from excitation by collisions with other particles (typically H_2) nearby. For example, we can draw a simple energy-level diagram for the hydroxyl radical (Fig. 9.14), similar to those for electrons around atoms. This diagram represents schematically the different rotational states: four closely spaced energy levels sit near the ground state of OH (labeled 1, 2, 3, and 4 in terms of increasing energy). Collisions excite OH to level 3 or 4. When OH deexcites, it can do so by changing from level 4 to 1, 4 to 2, 3 to 1, or 3 to 2. With each transition the atom emits a radio photon of corresponding energy and frequency: 4 to 1, 1720 MHz; 4 to 2, 1667 MHz; 3 to 1, 1665 MHz; and 3 to 2, 1612 MHz. (The corresponding wavelengths are very close together, ranging from 17.4 cm to 18.6 cm, so this set of transitions is often referred to as the 18-cm line of OH, even though four different transitions are involved.) The most intense of these lines usually is that at 1667 MHz, the next intense at 1665 MHz, and then 1612 and 1720 MHz. (The ratio of intensities is 9:5:1:1.) So under normal conditions, you expect the 1667 MHz line to be the strongest.

In 1965 a radio astronomy group at the University of California discovered hydroxyl while observing the Orion Nebula. But the emission was not that expected from a hot gas at 10^4 K. The 1667 MHz line wasn't apparent, and the 1665 MHz line was the strongest. For OH to emit in this fashion, if it were simply excited by collisions in a hot gas, the temperature of the gas would have to be 10^{12} to 10^{13} K! Plainly, this high temperature was impossible, for the hydroxyl molecule couldn't exist under such conditions; it would be dissociated into hydrogen and oxygen and highly ionized as well.

What was happening in Orion? Clearly, the 1665 MHz emission was somehow ampli-

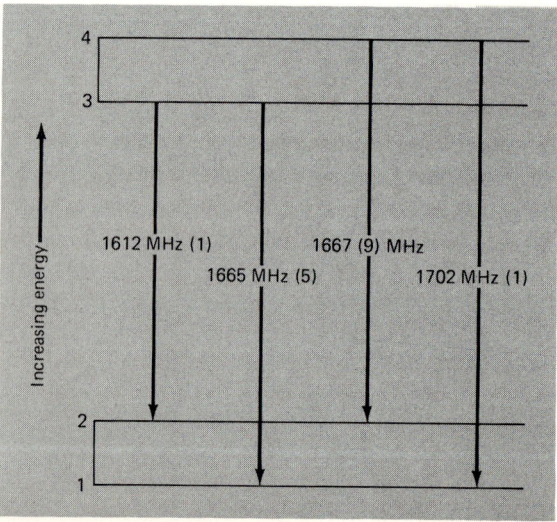

FIGURE 9.14 An energy-level diagram for the hydroxyl radical (OH). The transition from level 4 to level 2 is the most likely to occur.

fied over its normal intensity. The amplification results from the very same process as operates in a laser, but working at microwave frequencies. *Laser* is an acronym for *light amplification by stimulated emission of radiation*. *Maser* is the same, with "microwave" substituted for "light." The process involved is the same in both cases; the difference is the frequencies affected. Atoms or molecules in a gas are excited to some particular energy state and then stimulated to fall to a lower energy state at a more rapid rate than normal.

Let's illustrate the maser process with a three-level molecular laser (Fig. 9.15), with levels numbered 1, 2, and 3 in increasing energy. A molecule in the ground state, 1, is excited by collisions or by absorbing radiation to the highest level, 3. This process is called *pumping* a maser or laser. Assume that from level 3 the molecule has the greatest probability to give off a photon and drop to level 2, which is relatively stable, that is, the probability of dropping further, to level 1, is relatively small. Thus the pumping puts many molecules into level 2, primed for action. Now suppose a photon with energy equal to the difference between levels 1 and 2 comes close to such a primed molecule. It triggers the molecule's drop from level 2 to 1, sending off another photon of equal energy and in the same direction as the original photon. This process is called *stimulated emission*. The two photons now can stimulate two more molecules to radiate. The resulting four photons can trigger four more molecules to radiate, making a total of eight photons, and so on. The chain reaction amplifies the original photon millions of times as the photons travel through the gas. As a result, a maser's light is intense, narrowly directed, and at a single frequency.

Let's apply this maser process to hydroxyl (Fig. 9.16). The radical is pumped to some

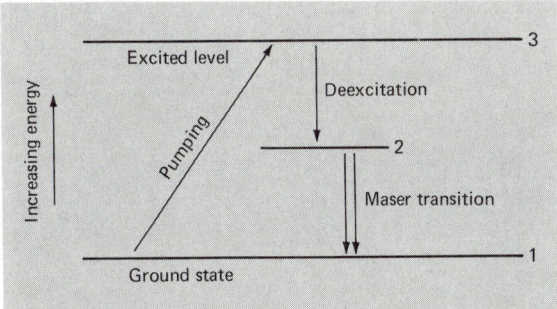

FIGURE 9.15 An energy-level diagram for a hypothetical three-level maser.

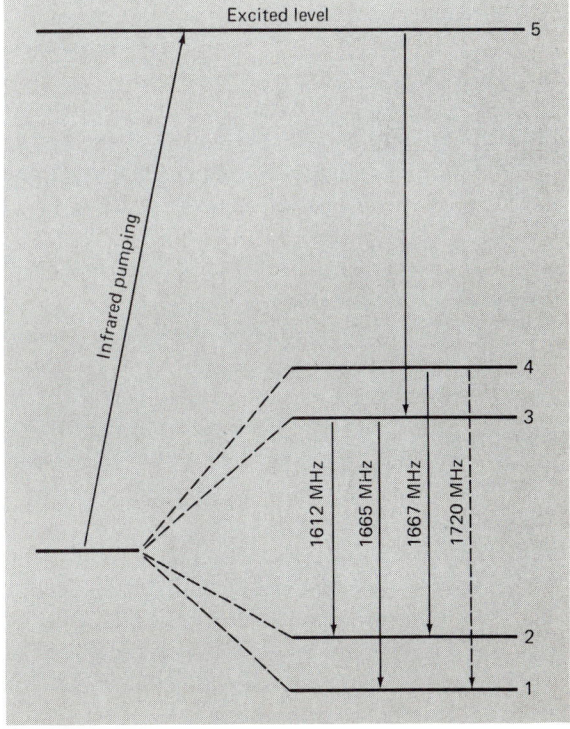

FIGURE 9.16 A simplified energy-level diagram for the maser transition of OH.

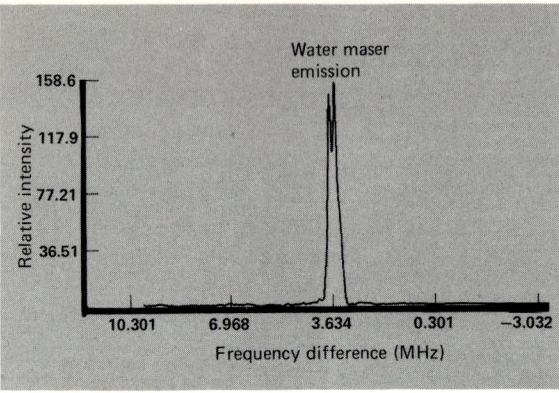

FIGURE 9.17 Radio emission from a water maser. The horizontal scale is the difference in frequency from the rest frequency of 22.2 GHz. [*Observations by N. L. Cohen at Haystack Observatory.*]

excited state (level 5). (It is believed, but not yet firmly proven, that the absorption of infrared photons pumps OH; a process involving ultraviolet photons has also been suggested.) The probabilities for transitions to lower levels are such that the natural tumble down in energy leaves most molecules in level 3. The molecule can then be stimulated to drop to level 1 and emit a 1665 MHz photon; another possible, but less likely drop is to level 2, with the emission of a 1612 MHz photon. This, in fact, is what is seen in sources such as the Orion Nebula. (In some other OH sources, particularly those associated with cool red giant stars, the 1612 MHz line is strongest. The physical reasons for this difference are rather complicated.)

Hydroxyl is not the only possible cosmic maser. Water has maser emission (Fig. 9.17) at one frequency—22,235 MHz or 1.35 cm. Silicon monoxide (SiO) can be a maser at 43,122 MHz (6.95 mm) and 86,243 MHz (3.47 mm). We are not yet sure how these interstellar masers are pumped.

Despite our lack of a complete understanding of masers, we do know that many hydroxyl and water interstellar masers tend to be found near giant molecular clouds. The regions of maser emission are extremely compact, only a few tens of AUs across, and very dense, at least 10^8 atoms/cm^3. They turn out to be signposts of incipient starbirth.

The Components of the Interstellar Gas: Summary

By now you may be wondering how all the forms of the interstellar gas fall together into a coherent picture of the interstellar medium. We wish we could present you a final picture now, but we cannot. Those who have been involved with research on the interstellar medium for a number of years have reached one firm conclusion: Our models for the interstellar medium have changed rapidly and will continue to do so.

Let's sum up what we know generally of the interstellar gas (Table 9.3):

1. *H II regions.* Zones of glowing, ionized hydrogen surrounding young, hot stars (spectral types O and B); contain a minor amount of the interstellar gas, perhaps 10 million solar masses total in the Galaxy.
2. *H I regions.* Clouds of cool, neutral hydrogen roughly 16 ly in diameter and each containing about 50 solar masses of material; total mass in the Galaxy may be 3 billion solar masses.

3. *Intercloud medium.* Contains two components: a hot, ionized, thin hydrogen gas between the clouds with some in the form of coronal gas and a cooler, neutral gas.
4. *Molecular clouds.* Small to giant in size, containing mostly molecular hydrogen (H_2); total mass of a few billion solar masses.

It's not clear now how these parts relate. Giant molecular clouds and H II regions appear to be tied together by the birth of massive stars (Section 9.3). The connection of these objects to H I regions and the intercloud gas is not obvious. But H I clouds and the intercloud gas do seem to be related by one piece of interstellar medium we have not yet mentioned, but will now discuss—supernova remnants.

The Violent Interstellar Medium

A supernova is a catastrophic explosion of a massive star at the end of its life that almost destroys the star completely (more details about supernovas in Chapter 11). A supernova blasts a tremendous amount of energy (about 10^{44} J) and material (1 to 50 solar masses) into space. The material blown off by a supernova expands as a shell into the interstellar medium. The expanding shell compresses and heats up the interstellar gas; behind the shell the gas is left hot and rarefied. It is so hot, in fact, that not only is the hydrogen all ionized, but also very highly ionized species, such as O VI, are formed. The hot gas can sometimes be seen optically (Fig. 9.18).

Christopher McKee and Jeremiah Ostriker have developed one model of the interstellar medium in which supernova explosions play a critical role. They picture the interstellar medium as containing three major elements: (1) expanding supernova shells, (2) small cool clouds, and (3) a hot intercloud gas (Fig. 9.19). The supernova shells are large structures, some hundreds of parsecs in diameter (Fig. 9.19a). As these sweep through the interstellar gas, they heat it to about 450,000 K and thin it out to a density of about 0.003 atom/cm^3. In the process they distort and destroy interstellar clouds (Fig. 9.19b) that have cold cores and hot edges (Fig. 9.19c). These clouds, which are about 13 ly (4 pc) in diameter, have core temperatures of roughly 80 K and densities of 40 to 50 atoms/cm^3. At their edges the clouds are hotter and thinner, with temperatures about 8000 K and densities of 0.3 atom/cm^3.

TABLE 9.3 Major Parts of the Interstellar Medium

Component	Indicator	Temperature (K)	Density (no./cm^3)	Fraction by mass (percent)
Molecular clouds	CO	10–50	10^2–10^7	40
H I regions	21-cm	50–100	1–50	40
Intercloud gas	21-cm	7000–10,000	0.2–0.3	20
Intercloud coronal gas	O VI	10^6	10^{-3}–10^{-4}	0.1
H II regions	H-alpha, continuous radio	10^4	10–10^4	Little

Note: The fractions by mass are very approximate and should be taken only as a guide.
Source: Adapted from D. D. Clayton, "The Cloudy State of Interstellar Matter," in *Protostars and Planets*, edited by T. Gehrels (University of Arizona Press, Tucson, 1978).

FIGURE 9.18 Optical emission from a supernova remnant in Cygnus. Note the filamentary structure. [*Courtesy Mt. Wilson Observatory, Carnegie Institution of Washington.*]

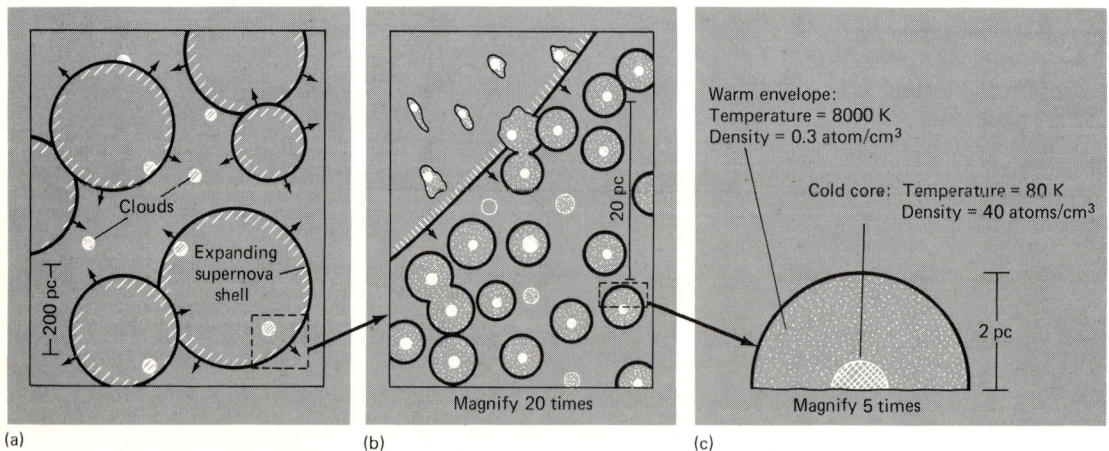

FIGURE 9.19 (a) An overall view of the dynamic interstellar medium, based on a model by C. McKee and J. Ostriker. The largest shells are expanding supernova remnants; these heat the gas to about 450,000 K. Only a few small clouds are shown; about 9000 would fill this region, which is 2000 by 3000 ly in size. (b) A view of (a) magnified 20 times to show the distribution of the clouds and the effects of a supernova remnant passing through them; the warm envelopes are squeezed and distorted by it. (c) A view of (b) magnified 5 times to show a "typical" interstellar cloud in this model.
[*Adapted from a diagram by C. McKee and J. Ostriker,* Astrophysical Journal, *vol. 218, p. 418, copyright 1977 by the American Astronomical Society.*]

This model is relatively new and does not cope with all aspects of the violent interstellar medium. For example, it does not deal with molecular clouds, which are certainly an important part. And it makes the interstellar medium look a lot more organized than it actually is. The model does emphasize what we do know about the interstellar gas; it is a dynamic, violent medium, blasted into action by supernova explosions.

Cosmic Rays

Most of the matter in the interstellar medium moves at modest speeds, at most a few tens of kilometers a second. But some particles that zip along at close to the speed of light permeate the interstellar medium; these are the *cosmic rays*. When they crash into the earth they form our only direct connection with matter from space beyond the solar system. And, although the relative number of cosmic rays in the interstellar medium is low, they carry a significant fraction of its total energy.

For historical reasons, cosmic rays are misnamed. They are *not* electromagnetic radiation, as you might infer from the use of the word *rays*. They are charged particles, usually protons (Table 9.4). In order of relative abundance after hydrogen, cosmic rays are made of helium nuclei (nuclear charge of 2), electrons, light nuclei (nuclear charges of 3 to 5), medium-mass nuclei (charges from 6 to 9), and heavy and very heavy nuclei (charges greater than 10). So, except for the electrons, cosmic rays are atomic nuclei, moving at near light speed. How do they get accelerated to such high speeds? From where do these nuclei come? Although cosmic rays have been studied since the beginning of this century, we do not yet have firm answers to these questions, but we do have a few clues and can make some good guesses. And we do know that cosmic rays must influence—perhaps strongly—the interstellar medium.

Observations indicate that, to within variations of only a few percent, cosmic rays stream to the earth *uniformly* from all directions in space. This observational clue really has only two possible simple explanations. (1) Cosmic ray sources are distributed uniformly throughout space. (2) Comic ray sources are *not* uniformly scattered, but the particles emitted by them are mixed up in direction by interstellar magnetic fields so that they arrive at the earth evenly from all directions. The second model seems most likely.

How can this directional mixing occur? Recall (Focus 6.2) that magnetic fields influence the direction of charged particles, causing the particles to gyrate around the field lines. The radius of a particle's corkscrew motion depends on the particle's speed and the

TABLE 9.4 Cosmic Rays

Type of Particle	Nuclear Charge (atomic number)	Abundance (relative to H)
Hydrogen nuclei	1	1000.0
Helium nuclei	2	68.0
Light nuclei	3–5	1.5
Medium nuclei	6–9	4.4
Heavy nuclei	≥ 10	1.5
Very heavy nuclei	≥ 20	0.41
Electrons	—	10.0

strength of the magnetic field; the lower the speed and the stronger the field, the tighter the spiral. We don't have a firm value for the strength of the interstellar magnetic field, but it is roughly 10^{-6} to 10^{-7} gauss. A light-speed proton in such a field spirals around a radius of some 10^{16} m (about a light year). Whatever the original direction of the proton, its path has been altered from that direction as it transits a light year—a short distance in the Galaxy.

Nuclei and electrons moving at close to the speed of light carry tremendous amounts of kinetic energy. Cosmic rays have energies ranging from 10^{-10} to 10^3 J. For comparison, the largest particle accelerators on earth boost matter to some 10^{-6} J, far below that from cosmic acceleration.

How do cosmic rays get accelerated to such high speeds? Frankly we're not really sure. Supernova explosions certainly drive particles to high speeds. Roughly, the total power of cosmic rays in the Galaxy is some 10^{33} W. If one supernova occurs in the Galaxy every 50 years, the accumulated explosions could accelerate particles to the observed energies and also provide the source of the particles. After an initial blast into space, a particle can be reaccelerated any place it runs into a concentrated magnetic field; one such place is in supernova remnants, which carry a tangle of magnetic fields through the interstellar medium.

Some low-energy cosmic rays come from the sun, accelerated in solar flares (Section 6.7). The highest-energy cosmic rays probably come from beyond our Galaxy, voyaging through millions of light years between the galaxies.

Whatever their source, the cosmic rays in the Galaxy heat the interstellar medium, especially the giant molecular clouds. These are so dense that nothing else has the energy to penetrate them. In fact, the energy density of cosmic rays near the earth is about the same as that for starlight, about 10^{-18} J/cm^3, even though the number density of cosmic rays is much less than that of photons.

9.2 THE INTERSTELLAR MEDIUM: DUST

Little was known of the interstellar gas until recently. Radio astronomy played a key role in its detection. But what else occupies interstellar space? Dust. There's not much out there; on the average, you find one dust particle in every million cubic meters (that's a cube with each side roughly the size of a football field). But the dust amounts to about 1 percent of the total mass of interstellar matter, and it can cut out light from distant objects or from those enshrouded in dense clouds. Piercing the dust veil has been an important goal of radio and infrared astronomers. That breakthrough has been critical in revealing the process of starbirth.

Cosmic Dust

Around 1930 Robert J. Trumpler at Lick Observatory was investigating the distances and sizes of small clusters of stars called *open clusters* (Fig. 9.20). His data seemed to reveal a strange feature in their size distribution: The farther a cluster was from the sun, the larger its diameter! Such an arrangement of sizes had no justification. If correct, it placed the earth in a special position to view such clusters, and special positions for the earth have been suspect since the days of Copernicus!

FIGURE 9.20 An open cluster of stars, New General Catalog 7510. [*Courtesy Lick Observatory.*]

To find the actual physical size of a star cluster, you need two pieces of information: its angular size and its distance from the sun. In order to get a cluster's distance you have to infer the absolute magnitudes of its stars, compare these to the observed apparent magnitudes, and from the difference compute a distance (Section 8.2). Essentially, this is the process of finding a spectroscopic parallax (Section 8.5) for the cluster. Interstellar dust cannot affect the angular size of a cluster, but it can dim the light from its stars. This dimming makes the cluster seem to be farther away than it actually is, and a cluster's size calculated from this wrong distance would be larger than its real size. For example, suppose dust makes the stars in a cluster one-quarter as bright as they would appear without the dust. If we didn't know the dust was responsible, we would think the cluster was twice as far away as it really is. From the angular size and this inflated distance we would calculate an inflated size; we would think the cluster twice as big as it really is.

Trumpler concluded that he could not take his cluster sizes at face value, but that there must be absorbing material in and near the plane of the Milky Way, where most open clusters are found. This material would reduce the light from distant clusters and would make their apparent distances greater than their actual distances. Most open clusters have about the same diameter; the effect noted by Trumpler reflected the absorption which led to a miscalculation of the clusters' distances. Trumpler's work confirmed the existence of interstellar dust.

Other observations hint at dust between the stars. *Dark nebulas*, such as the famed Horsehead in Orion (Fig. 9.21), display dramatic cutoffs of light due to dust. The dark rifts and lanes in the Milky Way (Fig. 9.22), once thought to be due to the lack of stars, are actually regions heavily obscured by dust. Some bright nebulas are not H II regions, but clouds of dust illuminated by nearby stars. One of the best examples of such a nebula is that around the Pleiades (Fig. 9.23). The spectrum of this nebula does not exhibit the bright lines characteristic of an H II region. It shows simply the reflected spectrum of the Pleiades' stars. Bright nebulas that arise from the reflection of starlight by dust are called *reflection nebulas*. (See Plate 5 for an example of a reflection nebula compared with an emission one.)

Aside from the dust nebulas, interstellar dust makes itself known by three primary effects: (1) *extinction*, the dimming of starlight, (2) *reddening*, the scattering of the blue

FIGURE 9.21 The Horsehead Nebula, a dark nebula in Orion. [*Courtesy Palomar Observatory, California Institute of Technology*]

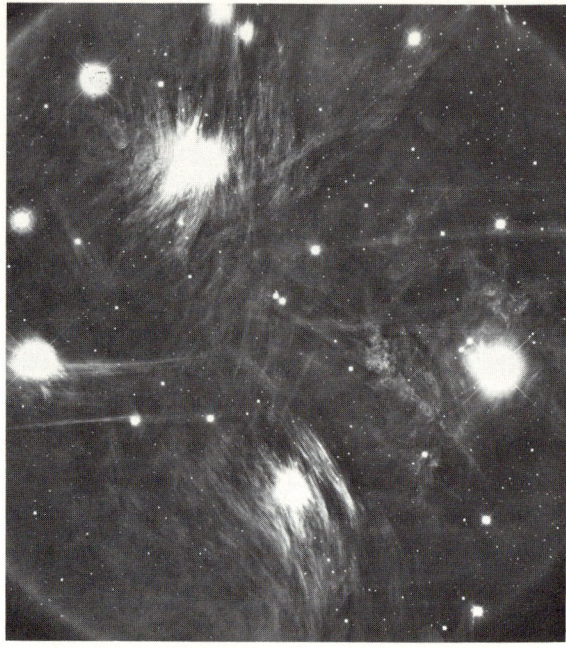

FIGURE 9.23 The Pleiades star cluster (Messier 45) immersed in a bright reflection nebula. [*Courtesy Kitt Peak National Observatory.*]

FIGURE 9.22 The Coalsack Nebula (left of center) in the southern Milky Way. [*Courtesy Harvard College Observatory.*]

9.2 THE INTERSTELLAR MEDIUM: DUST

wavelengths more than the longer wavelengths, and (3) *polarization* of the starlight. Let's look at each of these.

Imagine starlight traveling though a dust cloud (Fig. 9.24). The individual dust particles can absorb some of this light as it comes though. The dust particles can also scatter the starlight, so it goes off in a different direction from the original one. In either case less light exits from the dust cloud than enters it from the star. This dimming of starlight through absorption and scattering is called *extinction*.

Blue light is more affected by extinction than red light. Because red light penetrates the dust cloud more readily than blue, when you observe a star through the dust cloud, more of its red light reaches your eye than blue. This effect is called *reddening*. The part of the blue light that is scattered, and not absorbed, bounces around the dust cloud until it finally exits. So the cloud, a reflection nebula, appears blue (see Plate 5).

Except for obvious dark nebulas and effects such as that noticed by Trumpler, extinction is difficult to measure. Astronomers find it much easier to measure reddening, since this is a color effect, and then estimate the quantity of the dust from the amount of reddening. The H-R diagram states that a certain spectral classification of a star corresponds to a certain color. As long as a spectrum of a star can be obtained, its spectral class can be determined (from the strength of absorption lines) even if its light is reddened. We measure the star's color compared with that expected for its spectral class. The difference in color, the reddening, tells us how much dust lies along the line of sight to the star.

The third primary effect of dust, *polarization*, is a more subtle concept than extinction or reddening. Light, as you recall from Chapter 6, is an electromagnetic wave, which oscillates perpendicular to the direction of travel. If the oscillation is always in the same direction, that is, completely in one plane, the light is said to be completely polarized. If the oscillations are randomly oriented, equal in all planes, the light is said to be unpolarized. The wave drawn in Fig. 9.25a is polarized in the plane of the paper. Looked at along the direction of travel, the oscillation would be entirely along a single line, as shown in Fig. 9.25b. The other diagrams in Fig. 9.25b illustrate the direction of oscillation for unpolarized and partially polarized beams of light.

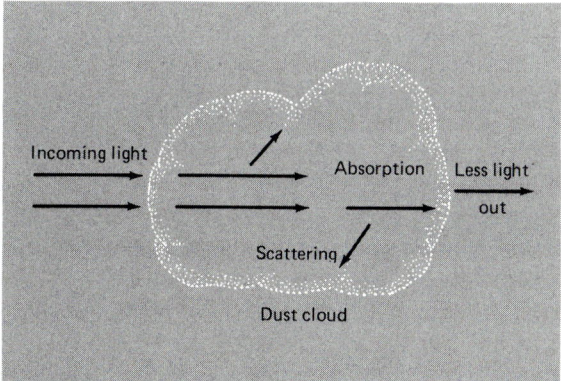

FIGURE 9.24 Starlight passing through a dust cloud is both absorbed and scattered. Both processes diminish the intensity of light in the beam finally leaving the cloud.

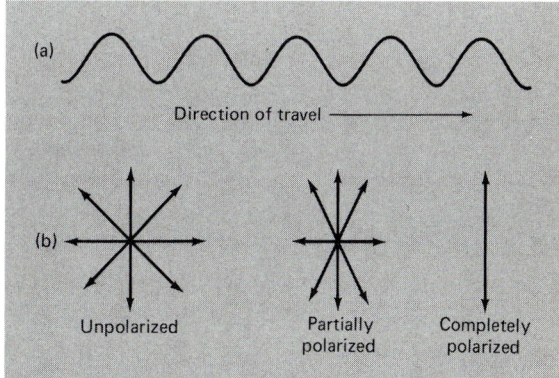

FIGURE 9.25 The polarization of light. (a) Side view of the electrical oscillations of a light wave. (b) Directions of oscillation for waves coming out of the page toward you. Note how the planes of oscillation line up for a fully polarized wave.

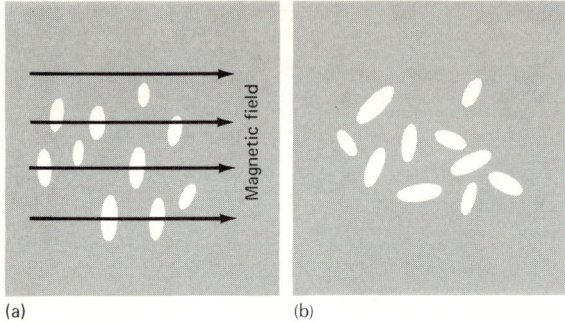

FIGURE 9.26 The alignment of dust grains by a magnetic field to cause the polarization of starlight, looking across the magnetic field lines (a) and along the field (b).

Light leaves a star unpolarized. It can become polarized upon passing through a dust cloud, *if* the dust particles are elongated, and *if* they tend to be oriented in the same direction. An elongated particle can more easily absorb and scatter light if the light wave is oscillating in the direction of the long dimension of the particle. Imagine a beam of unpolarized light coming in the direction perpendicular to the paper. If it hits the dust particles all lined up (Fig. 9.26a) the part of the light polarized vertically is absorbed (or scattered), and the part polarized horizontally gets through. However, if it hits a cloud of dust particles oriented randomly (Fig. 9.26b), there will be some particles absorbing (or scattering) light of all orientations, so the light that gets through will be less intense, but still unpolarized.

How are particles aligned in space? We think it is the effect of magnetic fields. So the observation of polarized light not only reveals that there is dust in space, but also that there are magnetic fields. It also tells us the direction of the magnetic field, for theoretical models predict that the particles should be lined up perpendicular to the magnetic field.

Dust and Infrared Observations

Interstellar dust blocks out visible light, which makes optical astronomers a bit unhappy, for it limits their view of distant stars and galaxies. Infrared astronomers feel rather happier, for infrared radiation penetrates dust. In fact, much infrared radiation from space comes from the dust itself. So infrared astronomers can both *see* dust and see *through* dust!

How does dust emit infrared? Basically, dust grains act roughly like (very small) blackbody radiators. Imagine a dust grain absorbing light from a nearby star. It must heat up until it emits (in the infrared) as much energy, each second, as it absorbs. When the energy input and output are balanced, the grain will have a certain temperature, say, 100 K. At this temperature, according to Wien's law (Section 6.4), its emission will peak roughly at a wavelength of 30 μm.

$$\lambda_{peak} = \frac{3 \times 10^{-3} \text{ m}}{T}$$

$$= \frac{3 \times 10^{-3}}{100}$$

$$= 3 \times 10^{-5} \text{ m}$$

$$= 30 \; \mu m$$

FIGURE 9.27 A photo in red light of the core of the Orion Nebula. The small group of stars in the center is the Trapezium cluster (arrow). [*Courtesy Lick Observatory.*]

In other words, the grain emits most strongly in the infrared. From the same calculation you can see that for the range from 3 μm to 300 μm, grain temperatures range from 1000 K to 10 K. (There's a limit to how hot a grain can get, depending on its composition; see below.)

The Orion Nebula marks a region of strong infrared sources. Let's examine the core of the nebula optically (Fig. 9.27). Here's the densest part of the visible nebula: hot gas (mostly hydrogen) ionized and excited by the O stars there. The brightest of these form a trapezoid figure (easily seen in a small telescope) called the *Trapezium*. Now let's look at an infrared map at 20 μm of the core region (Fig. 9.28). Quite a difference! The infrared emission does not peak around the Trapezium, but to the north and west. The infrared cluster here has two main parts: the Becklin-Neugebauer object, which emits most strongly from 3 to 10 μm, and the Kleinmann-Low nebula, which dominates the infrared emission at longer wavelengths (Fig. 9.29). The Kleinmann-Low source has a temperature of about 70 K, a size of 2000 AU, and a luminosity of 7×10^4 solar luminosities. Its position coincides with the core of the molecular cloud in Orion.

What are we seeing here? Probably the infrared emission from cool dust (about 70 K) located somewhere at or near the center of the molecular cloud—dust heated by something capable of putting out 7×10^4 solar luminosities. The visible Orion Nebula, illuminated and sustained by the Trapezium stars, lies in front of the molecular cloud like a hot bubble (Fig. 9.30). These stars have been born from a cloud of gas and dust; their placental material, altered by the light of the stars, is now hot and expanding away from the stars. Enough dust remains around one of the Trapezium stars to be visible as a weak infrared source.

The Becklin-Neugebauer object stands apart from the other infrared sources. It is less than 300 AU in diameter and emits about 1000 solar luminosities. Its infrared spectrum (Fig. 9.31) closely resembles that of a blackbody at 600 K. (Note that two absorption bands appear in the spectrum: one at 3.1 μm, the other near 10 μm. We'll get to these shortly.) The Becklin-Neugebauer object has played a key role in our understanding of starbirth.

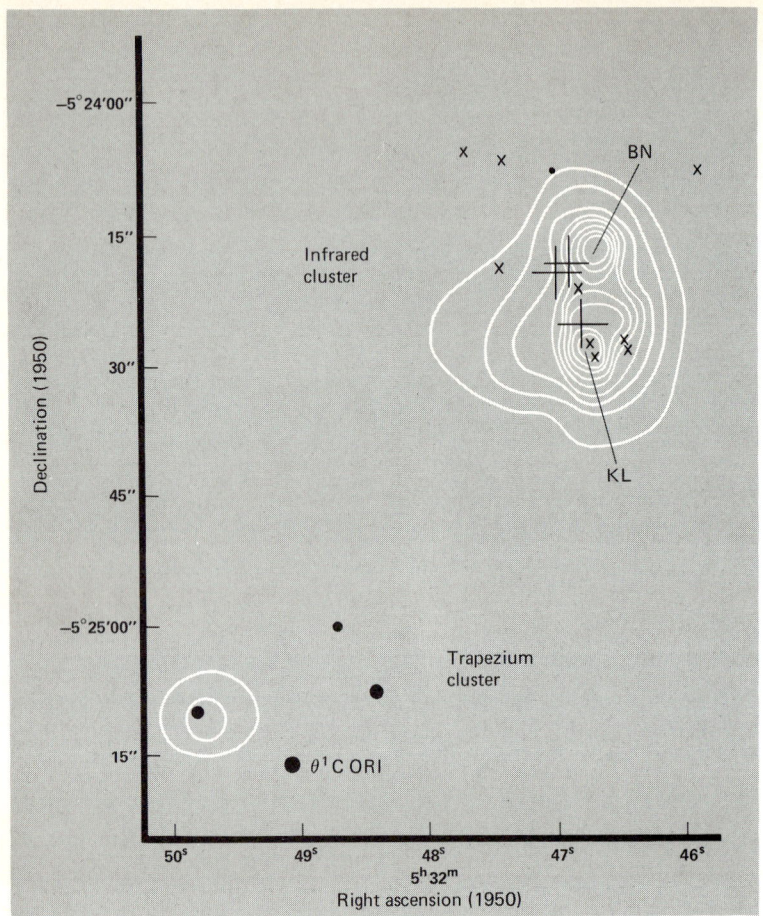

FIGURE 9.28 An infrared intensity contour map of the central part of the Orion Nebula. The infrared emission is concentrated in the infrared cluster (upper right). This cluster has two parts: the Kleinmann-Low (KL) source and the Becklin-Neugebauer (BN) object. The KL emission seems to come from the core of the molecular cloud. The BN object is suspected to be a protostar less than 100,000 years old. [Adapted from a diagram by E. Becklin, G. Neugebauer, and C. Wynn-Williams.]

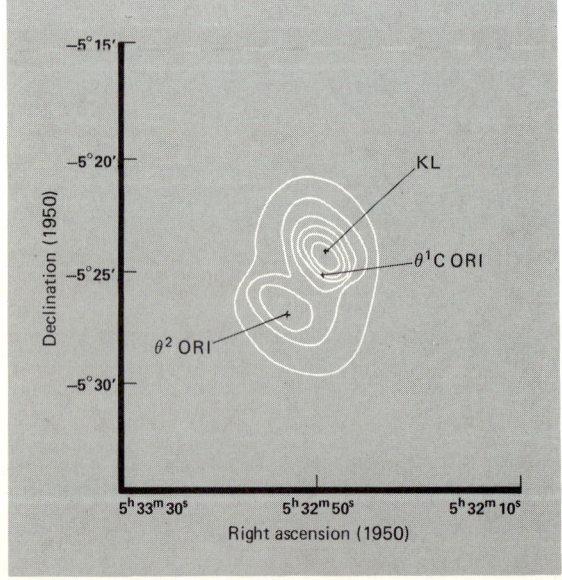

FIGURE 9.29 An infrared contour map of the Orion Nebula region at a wavelength of 69 microns. [Courtesy G. Fazio and the Center for Astrophysics.]

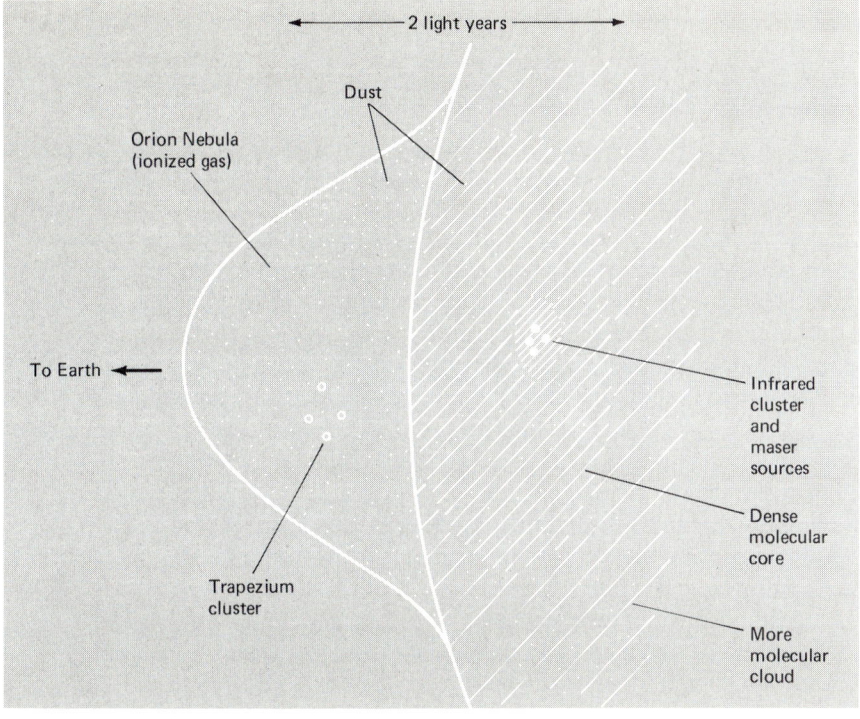

FIGURE 9.30 One model of the association of the Orion Nebula, its molecular cloud, and infrared sources. The Orion Nebula is a hot gas bubble, expanding outward, on the front of the molecular cloud. The infrared sources are embedded in the molecular cloud. [*Based on a model devised by B. Zuckerman.*]

The Nature of Interstellar Dust

What is the interstellar dust made of? Astronomers have been asking this question for a number of years, yet the answers are still a bit cloudy. We'll try to present the current state of knowledge without making it appear more certain than it actually is.

How do we find out about the physical properties of dust so far from the earth? The dust consists of small particles that absorb light in the infrared, visible, and ultraviolet. The particles heat up and emit in the infrared. We can study the spectra of absorption and

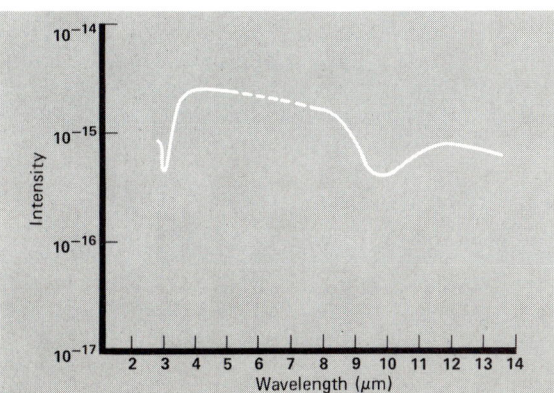

FIGURE 9.31 An infrared spectrum, from 2 to 14 microns, of the Becklin-Neugebauer source in the Orion Nebula. Note the absorption dips at 3 and 9.7 microns. [*Adapted from a figure by F. Gillett and W. Forrest, Astrophysical Journal, vol. 179, p. 483, copyright 1973 by the American Astronomical Society.*]

emission. Then we can try to match what we observe with the absorption and emission properties of grains made of various reasonable materials. (These properties are calculated from theory or measured in a lab.) This idea is indeed simple in theory, but in practice it's quite a mess. Our observations cannot be very selective. We usually end up looking at various kinds and sizes of dust grains all mixed together.

With this warning, let's examine the average *interstellar extinction curve* (Fig. 9.32) in the visible and ultraviolet. Note that the curve rises in the visible, has a funny bump in the ultraviolet (at about 0.2 μm), then, after a slight dip, rises again in the far ultraviolet (the data are limited at very short wavelengths by our observational techniques). What can be made of this curve? That no *one* type or size of grain can fit it. The extinction curve must be a composite resulting from the interstellar mixture. Calculations show that the bump and the rise in the ultraviolet must be caused by very small particles, 0.005 μm to 0.02 μm in radius. The rise in the visible can be the result of larger grains, 0.05 μm to 0.2 μm in radius.

What materials can make up these grains? One indirect, but valuable clue comes from considering the cosmic abundances of candidate elements (Table 9.5). Only those elements that make up an appreciable fraction of the interstellar material can contribute in a large part to the dust grains. Note that Table 9.5 contains elements that make up rather common substances: hydrogen and oxygen for water (H_2O), carbon and hydrogen for methane (CH_4), carbon and oxygen for carbon dioxide (CO_2), nitrogen and hydrogen for ammonia (NH_3), silicon and oxygen plus metals for silicates (compounds of Si and O commonly found in terrestrial rocks; for example, olivine, a common greenish mineral, is Mg_2SiO_4). Compounds like H_2O, CO_2, CH_4, and NH_3 are loosely called *icy materials* for these materials are solids at temperatures below about 100 K.

TABLE 9.5 Cosmic Abundances of Candidate Elements for Interstellar Grains

Element	Number of Atoms (relative to 10^6 H atoms)
H	1,000,000
He	120,000
C	370
N	120
O	680
Ne	630
Mg	34
Si	32
S	28
Fe	26
Ca	1.6
Na	1.3

Source: J. M. Greenberg, "Physics and Astrophysics of Interstellar Dust," in *Infrared Astronomy*, edited by G. Setti and G. Fazio (Reidel, Boston, 1978).

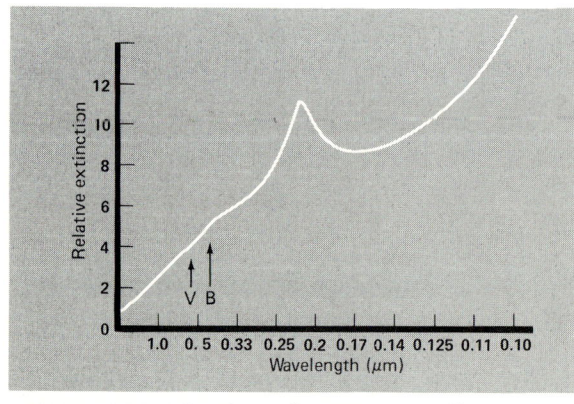

FIGURE 9.32 The observed average interstellar extinction curve from 0.1 to 1.0 microns.

9.2 THE INTERSTELLAR MEDIUM: DUST

Let's return to the extinction curve (Fig. 9.32). The bump at 0.2 μm can be explained by bare, small in radius (0.02 μm), graphite (pure carbon) particles. The bonds between the carbon atoms resonate and absorb at this wavelength. The rise in the ultraviolet must also come from very small particles; silicate particles with a radius of 0.005 to 0.01 μm can play this role. For the visible region of the spectrum, larger particles are needed; their radius must be about 0.2 μm. Such particles cannot be composed entirely of silicates, graphite, or pure iron; to account for how much extinction is seen requires more mass than that available in the carbon, oxygen, and other heavy elements in the interstellar medium (Table 9.5). We need to add some of the abundant hydrogen in the icy materials. To account for the shape and amount of the interstellar extinction curve, astronomers have developed *core-mantle grain models* (Fig. 9.33). The small core, about 0.05 μm in radius, can consist of silicates, iron, or graphite; silicates are plausible. The mantles are made of icy materials, likely some composite of them all.

Infrared observations bolster the idea that silicates and ices (at least water ice) make up part of the interstellar grains. Turn back to the spectrum of the Becklin-Neugebauer object (Fig. 9.31). Note the absorption bands at 9.7 μm and 3.07 μm. (These bands appear in the infrared spectra of other objects; the Becklin-Neugebauer source is a typical example.) Silicates in terrestrial rock, meteorites, and lunar rocks have absorption bands at about 10 μm; these involve changes in the energy of vibration in the Si-O bonds. Silicates also have another, but weaker, absorbing band at 18 μm that involves the energy of bending of the O-Si-O bonds. This absorption feature has been seen in a few other sources, strengthening the identification with silicates.

The band at 3.07 μm likely arises from water ice. But the amount of water ice in the grain is not enough to account for all of the extinction. Other icy substances are probably present, which have not yet been positively identified because their infrared bands are much weaker than that of water ice. It is commonly believed that water ice and silicates are some of the grain materials in a dust cloud surrounding a source such as the Becklin-Neugebauer object. As the infrared emission passes through the dust cloud, some of it is absorbed by the silicates and water ice to form the absorption bands.

Warning: Not all astronomers agree with the above interpretation of the materials

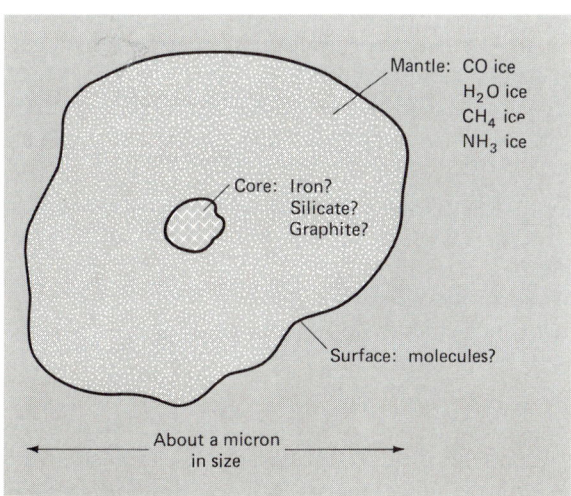

FIGURE 9.33 A simplified model of an interstellar dust grain. The composition is not really known; it may consist of any of the materials listed or some combination of them.

causing the infrared absorption bands. One vexing problem is that the shapes of the interstellar bands do not match well the shapes observed for terrestrial ices or silicates or the shapes calculated for various theoretical mixtures. One likely explanation is that since interstellar grains form and exist in environments unlike those on the earth, their crystalline structure is probably different from terrestrial substances. Therefore, the shape of their absorption bands is also different. Another possibility is that ices and silicates do not generate the bands. Some astronomers have proposed that organic polymers and polysaccharides of various types account for the infrared absorption bands. Organic compounds do have infrared absorption bands, and some organic materials may form in space. But we think an organic origin for the interstellar infrared bands in all cases is unlikely.

Dust and the Formation of Molecules

Dust and molecules are found intimately associated in space. They are almost always mixed together; wherever you find a molecular cloud, you usually find a concentration of dust. This association is not chance; grains probably play a role in the formation of molecules.

Consider the basic problem of forming an interstellar molecule. You have to get two or more widely separate atoms to come close together and chemically bond. That's no easy task in the dilute gas of interstellar clouds. Remember, the densities are about 10^6 atoms/cm^3, which means that the atoms are about 10^{-2} cm apart, a huge distance compared to their diameter of 10^{-8} cm.

Dust grains can mediate the chemical reactions. Infrared observations indicate that the grains in molecular clouds have temperatures of 20 to 40 K (some may be as cold as 5 K). If a hydrogen atom hits a grain, it will stick; if another hydrogen atom sticks, H_2 can form. Molecular hydrogen does not stick to a grain's surface as well as atomic hydrogen, so the molecule once formed eventually pops off into space.

Grain surface formation seems to work for hydrogen, but not for other small molecules. The main problem is how to get a molecule off the grain's surface without destroying the molecule. Such troubles have led astrochemists to investigate chemical reactions in the gas alone. Their work indicates that chemical reactions in the gas, especially ion-molecule reactions (the ions are formed by cosmic rays that can penetrate dense molecular clouds), seem to explain the formation of molecules other than hydrogen containing up to four atoms (about half of those in Table 9.1).

What about more complex molecules? We do not yet understand their formation. They may, like H_2, form on grains. If so, the relevant physical and chemical processes are not known.

To sum up what we understand so far about the formation of interstellar molecules:

1. H_2 forms on interstellar grains.
2. Molecules up to four atoms form in the interstellar gas.
3. The formation of more complex molecules is not understood.

The Formation of Cosmic Dust

So much for the molecules. What about the dust? Where does it come from? Remember that interstellar grains are made basically of two materials: ices of various types and

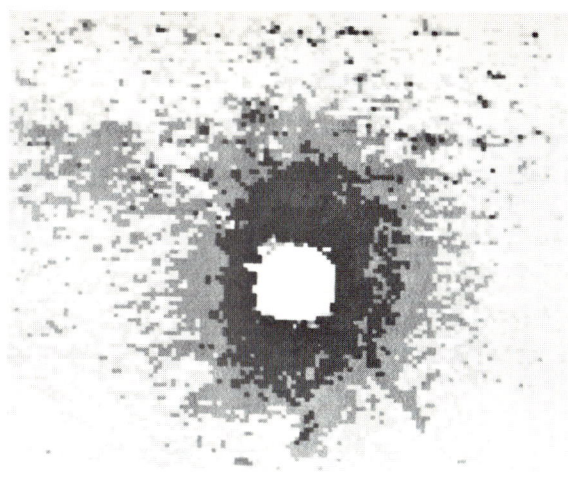

FIGURE 9.34 Material expanding in a shell from the red supergiant star Betelgeuse. This computer-processed infrared photo shows the star's image as a white hole in the center. The shell material appears dark; darker regions indicate denser parts of the shell. [*Courtesy R. K. Honeycutt.*]

denser solids. The ices solidify at a few hundred Kelvins, the denser materials at a few thousand Kelvins. The heavier grains are probably made in the atmospheres of cool supergiant stars. We know such stars are blowing mass into space at rates of about 10^{-6} solar mass a year (Fig. 9.34). The surfaces of these M stars have temperatures of only 2500 K or less. As gaseous material streams outward from them, its temperature drops, and solids can condense out of the vapor. In fact, spectra of some supergiant stars show the 9.7-μm silicate feature, indicating such dust exists around them. In a rarer class of stars, in which carbon is somewhat more abundant than oxygen, graphitelike particles and particles made of silicon carbide can form in the outflowing material. Infrared spectra of these stars are consistent with a cloud of carbon-containing particles around them.

The evidence is reasonable that dense grains form in the atmospheres and outflow from cool supergiant stars. This source probably provides many of the interstellar grains. Another source, probably minor, is condensation within material blasted out from stars in nova and supernova explosions (Chapter 11). A third stellar source of grains—but one not so well established—is mass loss from young, massive O stars. These spew out material at rates of 10^{-6} to 10^{-9} solar mass a year. Far out from these stars, some grains may condense from the blown-out material.

What about the ices that make up grain mantles (or perhaps entire grains)? These likely condense on cores in the deep interiors of dense molecular clouds. Here the temperatures are low and the gas densities high, so bare grains can grow crusts of ices. A core may have to grow a mantle once every 10^8 yr or so, since grains will lose their mantles when in an environment where temperatures range above a few hundred Kelvins, such as when they pass through an H II region or when clouds collide and heat up.

This discussion of dust completes our tour of the interstellar medium. We've dropped some hints about star formation along the way. Now to look specifically at the processes of starbirth.

9.3 STARBIRTH: THEORETICAL IDEAS

Stars are born out of interstellar molecular clouds. Because these clouds contain many times the mass of a single star, they must fragment into much smaller pieces during the process of star formation. That's the essential idea, but how *in detail*, stars are born remains a vexing question. We are just beginning to unravel the details. This question interests astronomers, if for no other reason than that the formation of planets intimately connects with the birth of stars.

To try to cast a clear light on our present understanding of starbirth, we have divided the topic into theoretical ideas (this section) and observational clues (next section).

Theoretical Collapse Models

We'll present here some highlights of the enormous amount of work done so far. Bear in mind, however, that these theoretical calculations (done on computers) represent highly idealized models based on uncertain assumptions about initial conditions and the relevant physical processes. The results may depend strongly on the assumptions, so the detailed correspondence to reality may be meager at best.

Newton first recognized the general outline of star formation: the process of *gravitational collapse* (Fig. 9.35). A cloud with enough mass and a low temperature will naturally contract from its own gravity. As long as there is little pressure in the cloud, so that it is contracting under the influence of gravity alone, it is said to be in *free-fall*. As gravitational potential energy is converted to kinetic energy, the material in the cloud heats up

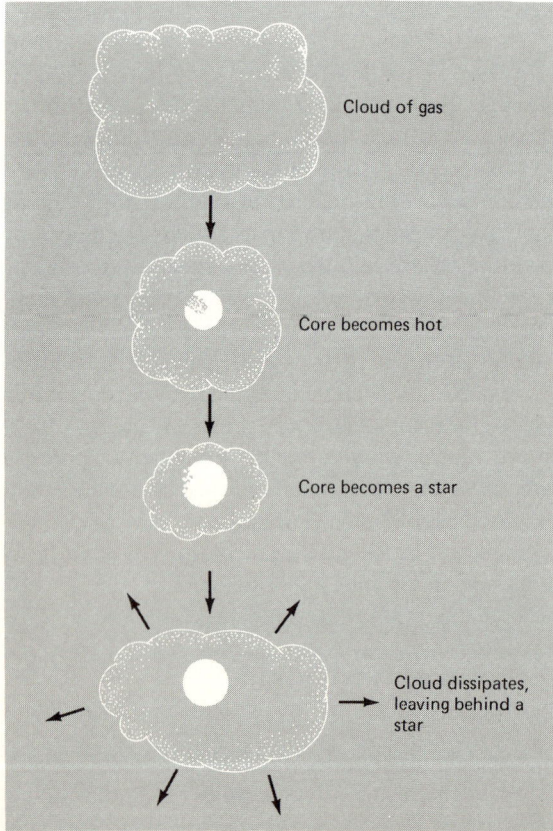

FIGURE 9.35 A very general scheme for the formation of a star by gravitational contraction. As an interstellar cloud contracts, the material at the core condenses faster than the envelope. When the core is hot enough, fusion reactions begin. The cloud dissipates, revealing the new star.

(recall Section 6.6). The heating up eventually has two results. First, the temperature (and density) will build up enough so that the outward pressure force halts the collapse and balances gravity; the cloud reaches a balance between gravity pulling inward and pressure pushing outward called *hydrostatic equilibrium* (Focus 9.1). Second, the temperature eventually reaches the kindling temperature of fusion reactions (Section 6.6); at that moment a main-sequence star is born.

Prior to the establishment of hydrostatic equilibrium, the contracting and heating cloud is called a *protostar*. Between that stage and the ignition of fusion reactions, while the incipient star gets its energy from gravitational collapse, it is called a *pre-main-sequence star*. As a star evolves, its luminosity and surface temperature change with time. So a point on the Hertzsprung-Russell diagram representing the protostar's conditions moves with time. The track traced out before the star hits the main sequence is called its *pre-main-sequence evolutionary track*.

The evolutionary tracks for protostars at different masses differ substantially. In part for this reason and in part for other evolutionary considerations, we'll divide the discussion into two parts: solar-mass protostars and massive protostars (which we'll define as 10 solar masses or more). Despite differences that come from different masses, the theoretical calculations have some common features:

1. The collapse starts out in free-fall, that is, it is controlled only by gravity (with negligible pressure).
2. It proceeds very unevenly. The central regions collapse more rapidly than the outer parts and a small condensation in hydrostatic equilibrium forms at the center. This core will become a star.
3. Once the core forms, it accretes material from the infalling envelope.
4. The star becomes visible to us, either by accreting all the surrounding material onto itself or by somehow dissipating it.

Free-Fall Gravitational Collapse

Before looking at specific cases, let's examine the notion of free-fall collapse in detail. Imagine a cloud of gas collapsing gravitationally so that the particles do not collide. The internal pressure is then zero, gravity is the only force acting on the particles, and the collapse takes place in free-fall for each particle.

We can describe this free-fall collapse and derive the free-fall time (t_{ff}) by applying Kepler's third law. Consider a particle of mass m at the edge of a spherical cloud with radius r (Fig. 9.36). All the matter interior to it acts gravitationally as if it were all concentrated in a point at the center; thus the particle moves as if it were attracted by an object at the center with the cloud's mass (M). Particle m falls on a straight line toward the center. Consider this path an orbit, an elliptical orbit with eccentricity $e = 1$. Use Kepler's third law to describe the motion, noting that M is much greater than m, so that $M + m \approx M$:

$$\frac{P^2}{a^3} = \frac{4\pi^2}{G(M + m)} \approx \frac{4\pi^2}{GM}$$

where a, the semimajor axis of the orbit, is $\frac{1}{2}r$.

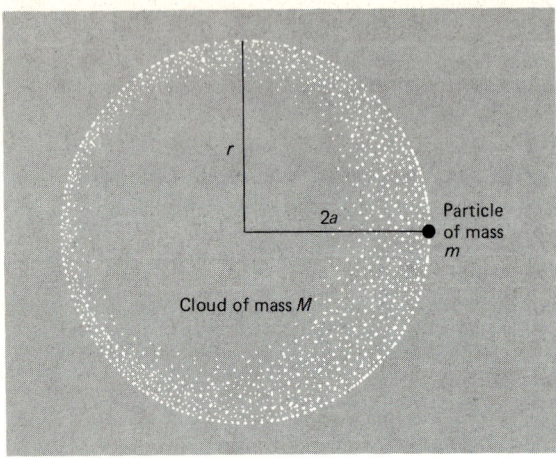

FIGURE 9.36 Geometry for the gravitational contraction of a spherical cloud.

Assume the cloud has an average initial density, ρ_0. Then

$$M = \frac{4}{3}\pi r^3 \rho_0$$

$$= \frac{4}{3}\pi (2a)^3 \rho_0$$

$$= \frac{32}{3}\pi a^3 \rho_0$$

and

$$\frac{P^2}{a^3} = \frac{4\pi^2}{(32/3)G\pi a^3 \rho_0}$$

or

$$P = \left(\frac{4\pi}{(32/3)G\rho_0}\right)^{1/2}$$

But P is the complete orbital period, the time to go distance $2a$ and $2a$ again (from the starting point and return). The free-fall time is only the time to fall r, which is $\frac{1}{2}P$, so

$$t_{ff} = \frac{P}{2} = \left(\frac{\pi}{(32/3)G\rho_0}\right)^{1/2}$$

$$= \left(\frac{3\pi}{32 G\rho_0}\right)^{1/2}$$

Note that the free-fall time depends only on the cloud's initial average density, not its mass.

9.3 STARBIRTH: THEORETICAL IDEAS

For ρ_0 in kilograms per cubic meter, the free-fall time in seconds is

$$t_{ff} = \frac{6.64 \times 10^4}{\rho_0^{1/2}} \sec$$

Solar-Mass Protostellar Collapse

With these general features in mind, let's now look at one model for the formation of a sunlike star, based mostly on work by Richard Larson.

Imagine a huge interstellar cloud of dust and gas, mostly in the form of molecular hydrogen. Also picture that a piece of this cloud has sufficient mass to contract gravitationally. Its initial density is 10^{-16} kg/m^3 (10^5 particles/cm^3) and the initial diameter 5×10^6 solar radii. The collapse begins! As it proceeds, material at the cloud's center increases in density faster than that at the edge. Because of the density increase, the collapse time at the center is decreased; it collapses faster, grows denser, and so collapses still faster. The rest of the cloud's mass is left behind in a more slowly contracting envelope. This part of the collapse takes place in free-fall.

With the rapid fall of material in the core, the hydrogen molecules gain kinetic energy. They bang into each other and also strike dust grains and so transfer kinetic energy to the grains. Heated by such collisions, the dust grains radiate at infrared wavelengths. As long as this heat radiation can escape into space, the kinetic energy is dissipated, the cloud stays cool, the pressure stays low, and the collapse continues in free-fall. But at some point the density of the core reaches a critical value at which the cloud becomes opaque and traps infrared radiation. The core heats up to a few hundred Kelvins, its pressure increases, and so the core's collapse slows down dramatically as hydrostatic equilibrium is established, 423,000 years after the collapse began.

Meanwhile, the envelope continues to fall inward, showering mass on the core. Where the infalling material bangs into the core, a *shock wave* forms. (You've experienced a shock wave every time you have heard a sonic boom; it's a piling up of matter in a gas or liquid whenever anything moves through it faster than the speed of sound.) As the matter piles up, it increases the core's mass and temperature. At about 2000 K the hydrogen molecules break up. In the process they soak up heat. The pressure doesn't rise fast enough to keep the star in equilibrium and gravity takes command again. The protostar again goes into free-fall. At the start of this stage of the collapse the core's radius is about 4 AU; at the end it will be about twice the size of the sun.

When all the hydrogen molecules have broken up, the pressure can rise again, and the collapse stops. The star slowly contracts. Gravitational energy heats the core and provides the energy to make the star luminous. A star is born. Its size is twice that of the sun, its luminosity a few times the sun's, and the total evolutionary time from the start of collapse to this stage is on the order of a million years.

But the envelope, still falling in, blocks the birth from the view of the optical astronomer. This infant star—still a protostar, because not all of its material is in hydrostatic equilibrium—hides in its womb. Dust in the envelope cuts out the protostar's light. However, the radiation absorbed by the dust heats it so that it gives off infrared radiation. So a sign of protostars should be small, intense sources of infrared radiation in or near known clouds of gas and dust.

FOCUS 9.1 HYDROSTATIC EQUILIBRIUM

When a fluid has all forces on it in balance, so that it does not move, it is in a state of *hydrostatic equilibrium*. In astronomy we encounter gases influenced by gravity, such as interstellar clouds and stars. Let's investigate the application of hydrostatic equilibrium to them.

Consider a coin-shaped piece of material in a star (Fig. F.14). This coin of gas has thickness Δr, surface area A, and density ρ. The gravity pulling the coin inward (F_{in}) at distance r from the star's center is

$$F_{in} = M \frac{GM_r}{r^2}$$

The mass of the coin (M) is equal to volume times density, and the volume is equal to surface area times thickness ($A\Delta r$). Substituting for M in the equation, we have

$$F_{in} = A\Delta r \rho \frac{GM_r}{r^2}$$

The symbol M_r indicates that it's only the mass *interior* to r that attracts the coin. As proved by Newton, the matter exterior to r has no effect as long as it has a spherical distribution.

The net outward force on the coin is the difference between the pressure on the top of the coin and that on the bottom, coming from the gas particles hitting it. Let's call this

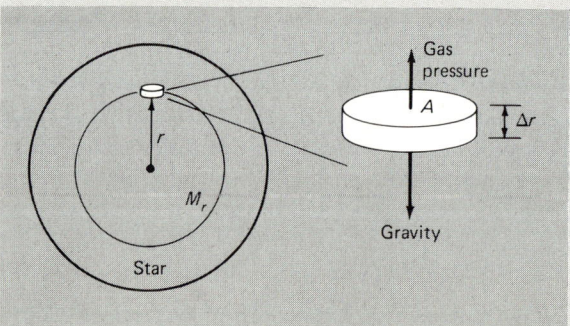

FIGURE F.14 Forces on a coin-shaped section of a star.

Eventually, the star rids itself of its cloaking cloud, which all falls onto the young star. As the cloud dissipates, we see a *pre-main-sequence star*, one that is larger and cooler than it will be in its final state on the main sequence.

Massive Protostellar Collapse

The collapse of a massive protostar follows the same general scheme as that for a solar-mass one. The two important differences are:

difference ΔP. Now, pressure is force per unit area, so the total net outward force (F_{out}) on the coin is its surface area times the pressure difference:

$$F_{out} = A\, \Delta P$$

This force pushes the coin outward because the density and temperature on the inward side of the coin are greater than the temperature and density on the outward side. So the pressure on the inward side is greater than that on the outward.

For the coin not to move, the net force must be zero,

$$F_{out} + F_{in} = 0$$

or

$$F_{out} = -F_{in}$$

and therefore

$$A\,\Delta P = -A\,\Delta r \rho \frac{GM_r}{r^2}$$

$$\frac{\Delta P}{\Delta r} = -\rho \frac{GM_r}{r^2}$$

The minus sign says that pressure and gravity act in opposite directions. This is the *equation of hydrostatic equilibrium*.

What does this equation mean? It tells you that for any point in the star the rate of change in pressure with distance along the radius ($\Delta P/\Delta r$) must balance the force of gravity for the gas to be in hydrostatic equilibrium.

Note from Newton's law of gravitation that GM_r/r^2 is just the acceleration due to gravity at distance r. Call this g_r. Then another way of writing the equation of hydrostatic equilibrium is

$$\frac{\Delta P}{\Delta r} = -\rho g_r$$

1. Fusion reactions begin *before* the accretion of the envelope stops. A massive star begins and spends part of its main-sequence lifetime obscured from view.
2. Massive protostars do not accrete all of their original cloud's material. Some is blown away (Table 9.6).

We'll describe the results of theoretical calculations done by H. Yorke and E. Krügel. Don't get lost in details; try to see the general trends as the action unfolds.

Take a dense cloud, hydrogen density of 10^{-15} kg/m^3 (10^6 atoms/cm^3), containing 20

TABLE 9.6 Some Properties of Massive Protostar Collapse

Mass of Cloud (solar masses)	Mass of Star (solar masses)	Main-Sequence Spectral Type	Time of Core Formation (yr)	Time of H II Region Formation (yr)
150	36	O6–O7	1.5×10^5	1.9×10^5
50	17	O8–O9	3.2×10^5	3.7×10^5
20	12	B0	4.7×10^5	6.4×10^5

Note: Based on theoretical calculations by H. Yorke and E. Krügel.
Source: P. Mezger, "Interstellar Matter," in *Infrared Astronomy*, edited by G. Setti and G. Fazio (Reidel, Boston, 1978).

solar masses of both gas and dust. Silicate and graphite grains (which evaporate at 1500 to 1700 K) are coated with ice mantles (which evaporate at 150 K) at the start. After 4.7×10^5 yr a starlike core 10^{10} m in radius (15 times the size of the sun) has formed. A shock layer, from accretion of material, surrounds the core (Fig. 9.37). Outside the shock is a region (about 10^{12} m in radius, 1500 times the sun) that is dust free; here the heating has evaporated all the dust. In the next region, about ten times bigger (10^{13} m in radius), which is somewhat cooler, graphite and silicate cores exist without ice mantles. These grains absorb the photons generated by the accretion and emit radiation in the infrared at a temperature of roughly 1000 K. So the protostar appears very red, with its emission peaking at about 3 μm.

Eventually fusion reactions turn on in the core, which becomes a massive-main sequence star (in this case, roughly a 12-solar-mass B0 star). Its surface temperature (28,000 K) is hot enough to emit photons shortward of 912 Å that ionize the hydrogen. A small H II region forms 6.4×10^5 yr after the start of the collapse. The H II region expands and helps to blow away some of the accreting material. Finally the star debuts to our optical view.

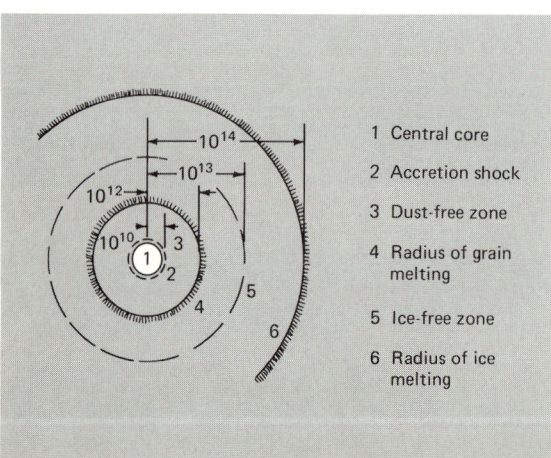

FIGURE 9.37 A schematic cross-section of a 20-solar-mass protostar about 470,000 years after the start of its gravitational contraction. [*Based on theoretical calculations by H. Yorke.*]

Collapse with Rotation

The results of theoretical calculations described above lack at least one fact of astronomical life: rotation. It is likely that interstellar clouds rotate at least a bit. Any rotating mass has *angular momentum*. Let's look at this concept and its application to star formation. Momentum is related to the inertia of matter. To get an idea of the meaning of momentum, consider the following. A bicycle and a truck are coming at you at the same speed. Which would be easier for you to stop? The bicycle, because it has less mass. Now imagine two bicycles coming at you, one with twice the speed of the other. Which is easier to stop? The one moving slower. These examples show you that momentum depends on both the mass and the velocity of the object involved. In fact, momentum is defined as the product of an object's mass and velocity:

$$\text{momentum} = \text{mass} \times \text{velocity}$$

If we let p be the momentum, m the mass, and v the velocity, then

$$p = mv$$

Note that since velocity has a direction, momentum does, too, and in the same direction as the velocity. You can think about momentum like this: Once you get a mass moving, you have to put out an effort to stop it.

Consider a spinning object, such as the earth rotating about its axis. Its inertia about its spin axis keeps it spinning. The faster it spins and the more mass spinning, the harder it is to stop. This spinning momentum is called *angular momentum*.

Angular momentum is the tendency for bodies, because of inertia, to keep spinning (rotating) or orbiting (revolving). In analogy to momentum, angular momentum is defined in terms of the products of the important quantities: the mass (but now the distribution of the mass about the center of motion complicates the picture), the velocity (around the center of motion), and the radius (the distance of the mass from the center of motion). For a single particle moving in a circle

$$\text{angular momentum} = \text{mass} \times \text{circular velocity} \times \text{radius}$$

Let L be the angular momentum, m the mass, v the circular velocity, and r the radius. Then

$$L = mvr$$

For a body such as the earth, which is made up of many particles moving at different velocities at different distances from the axis of rotation, we must add up the angular momentum of all the particles.

You can test whether this definition is reasonable by performing a simple experiment (Fig. 9.38). Tie a ball to the end of a string and whirl it around your head at a constant speed. Now with your free hand grab the string at half the distance you started with. The ball will move with double its circular speed. This is an example of the conservation of

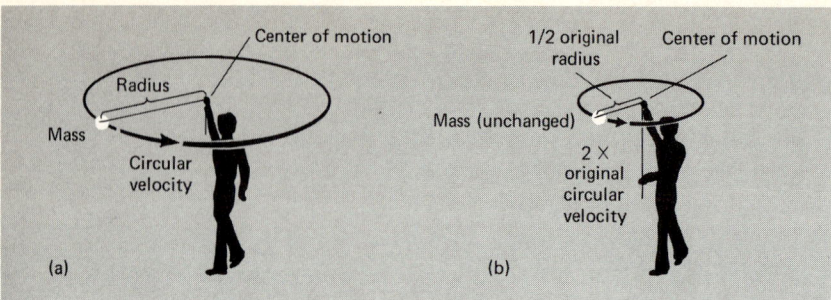

FIGURE 9.38 An illustration of the conservation of angular momentum. Twirl a mass around at the end of a string (a). Quickly grab the string at half its original length (b). The mass's circular velocity will double. In both cases the angular momentum is the same.

angular momentum. If $r_2 = \tfrac{1}{2} r_1$ and $v_2 = 2v_1$, then angular momentum stays the same:

$$L_2 = mv_2 r_2 = m(2v_1)(\tfrac{1}{2} r_1) = mv_1 r_1 = L_1$$

Another familiar example is the figure skater who spins faster by pulling in his or her arms.

Recall that the motion of an object moving on a straight line does not change unless a force is applied to it. Likewise, the angular momentum of a rotating or revolving object does not change unless a *torque* (a twisting force) is applied. In the experiment above no torques were applied, but the radius was reduced to half of its original value. Therefore the circular velocity had to increase so that the amount of angular momentum remained the same.

Kepler's second law is actually a statement of the conservation of orbital angular momentum. Compare a planet at the perihelion and at the aphelion points in its orbit (Fig. 9.39). Here the velocities (v_a, v_p) are at right angles to the radius (r_a, r_p). In a short time (Δt) the planet moves, according to Kepler's second law, so that the areas of the triangles A and B are equal:

$$\tfrac{1}{2} r_a v_a \Delta t = \tfrac{1}{2} r_p v_p \Delta t$$

or

$$r_a v_a = r_p v_p$$

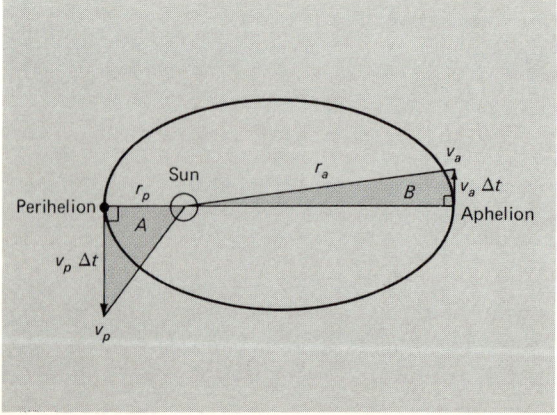

FIGURE 9.39 Kepler's second law as a form of the conservation of angular momentum.

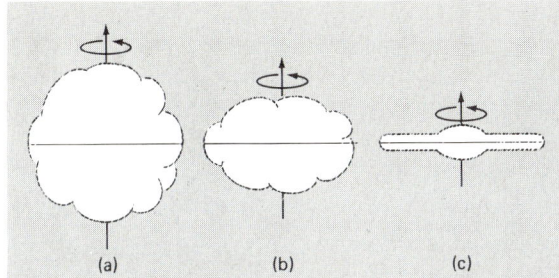
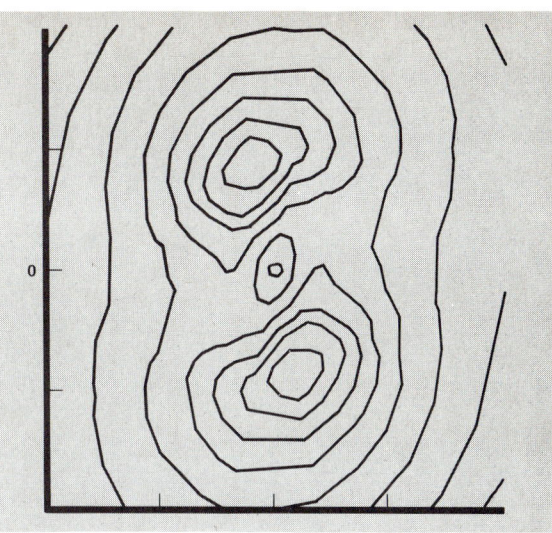

FIGURE 9.40 The gravitational contraction of an interstellar cloud. The cloud begins the collapse slowly spinning on its axis (a). As it contracts (b), the material falls in along the spin axis to form a disk with a central bulge (c).

FIGURE 9.41 Blobs forming in the disk of a cloud. Shown are the density contours in the equatorial plane 24,500 years after the start of the collapse of a rotating cloud. [*Adapted from a figure by* A. Boss and P. Bodenheimer, Astrophysical Journal, *vol. 234, p. 289, copyright 1979 by the American Astronomical Society.*]

Multiply by m, the planet's mass:

$$mr_a v_a = mr_p v_p$$

Note that $mr_a v_a$ and $mr_p v_p$ are the angular momentums at aphelion and perihelion. They are the same at these two points, and, in a similar way, it can be shown that the angular momentum is the same at all points in the orbit. So Kepler's second law is merely a consequence of conservation of angular momentum.

Now let's apply this concept to a collapsing protostar. An isolated, rotating cloud must conserve angular momentum. If the cloud collapses, as each particle moves closer to the axis of rotation, it must rotate faster, just like the ball on the string in the example above. So as a spinning cloud collapses gravitationally, it must spin faster. At some point the rotational velocity may become high enough that the centripetal acceleration (v^2/r) balances the gravitational force per unit mass, and the collapse halts. But note that it is the distance from the axis of rotation, not from the center of the cloud, that determines the angular momentum. So a particle which starts out near the rotation axis can fall a large distance toward the *center* without changing its distance from the *axis* much at all, and it will be able to move much closer to the center before its rotational velocity is high enough to stop it. Therefore, those parts of the cloud originally near the rotation axis will collapse more than those near the equator, and the cloud will flatten into a disk (Fig. 9.40).

The addition of spin to theoretical models of protostar collapse makes the calculations much tougher and the results less conclusive. The important point about calulations to date is that in some cases a disklike core of rings of material results. These rings turn out to be unstable in some instances, and they break up into several blobs (Fig. 9.41). Sometimes these blobs coalesce into fewer ones. If each blob eventually becomes a star, we then have a natural explanation for the fact that most stars in the Galaxy are in binary or multiple systems. A neat result!

9.4 STARBIRTH: OBSERVATIONAL CLUES

Enough of theoretical models. Let's now look at the real world to see how the models may (or may not!) be confirmed by observational evidence. It turns out that we have uncovered more information about the birth of massive stars than about the birth of solar-mass stars. There's a good reason. Massive protostars have greater luminosities than solar-mass ones and, once they reach the main sequence, massive stars ionize the gas around them. The ionized gas is detectable by radio telescopes. Since all this action takes place cloaked by dust, it's only infrared and radio observations that permit us to inspect stellar wombs. Nature seems a bit shy about exposing the secrets of starbirth.

Signposts for the Birth of Massive Stars

Before we turn to specific cases, let us outline what theory predicts should be the hallmarks, in the radio and infrared, of the birth of a massive star. First, because stars condense from molecular clouds, you need to find a molecular cloud; molecules emit at

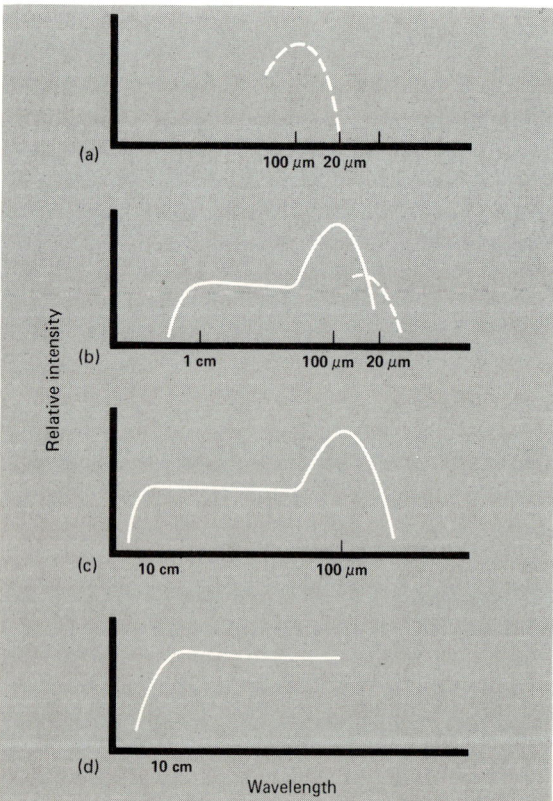

FIGURE 9.42 A schematic sequence of the evolution of the spectrum during the birth of a massive star: (a) dust at 30–50 K; (b) interior dust at 1000 K and exterior at 100 K; (c) ionized gas, H II region; (d) H II region after dust dissipated. [*Adapted from a figure by G. Fazio.*]

9.4 STARBIRTH: OBSERVATIONAL CLUES

millimeter wavelengths (such as carbon monoxide at 3 mm). Second, the free-fall collapse, at its early phases, heats the dust to low temperatures, roughly 30 to 50 K. This dust emits infrared radiation that peaks at roughly 100 μm (Fig. 9.42a). Third, as the protostar forms, the interior dust reaches about 1000 K, and therefore emits with a peak at 3 μm. The exterior dust is cooler, still about 100 K. So the spectrum will show the combination of two blackbody peaks, one near 3 μm, the other near 30 μm (Fig. 9.42b). Fourth, as the protostar reaches the main sequence, it ionizes the hydrogen gas, and a compact H II region develops, most easily observable at centimeter wavelengths (Fig. 9.42c). Fifth, the hot, ionized gas expands. The radio intensity decreases. The dust, pushed outward, is cooled; it emits with a peak in the far infrared, but less intensely. Sixth, when the dust has been pretty much dissipated, the H II region emits weakly a continuous spectrum of radio waves (Fig. 9.42d). Finally, the H II region expands enough to blow off its dusty cloak, and the star appears to optical view. (See Table 9.7 for a summary.)

With this scenario in mind, let's return to our old friend the Orion Nebula. The H II region around the Trapezium marks the oldest (most evolved in an evolutionary sense)

TABLE 9.7 Possible Sequence in the Birth of Massive Stars

Evolutionary Stage	Observational Signposts	Duration (yr)	Important Events
1. Collapse of prestellar clouds	Cool, dense molecular clouds; OH, H$_2$O sources	300,000	Gravitational collapse of unstable regions is set up by shock wave.
2. Very young, cool stars shining from gravitational contraction	Compact, far-infrared sources associated with molecular clouds	50,000	Stars form from gravitational collapse; go through cool, luminous phase. Stars increase a lot in surface temperature, a little in luminosity.
3. Start of nuclear burning in cores	Compact, near-infrared sources in or near molecular clouds		
4. Early, normal life nuclear burning	Infrared and radio sources in or near molecular clouds	30,000	Ultraviolet photons break up molecules and ionize gas to form an H II region around star.
5. Young, expanding H II region	Weak infrared emission, diffuse centimeter-wave radio emission, OB stars just visible	500,000	Heated gas in H II region expands at about 5–10 km/sec.
6. Old, very expanded H II region	No infrared emission, very diffuse centimeter-wave emission, OB stars plainly visible	2,000,000	Expansion dissipates both H II region and associated molecular cloud.
7. Naked OB stars	A single OB star or clusters of OB stars with no surrounding H II region	6,000,000	H II spreads into interstellar medium.

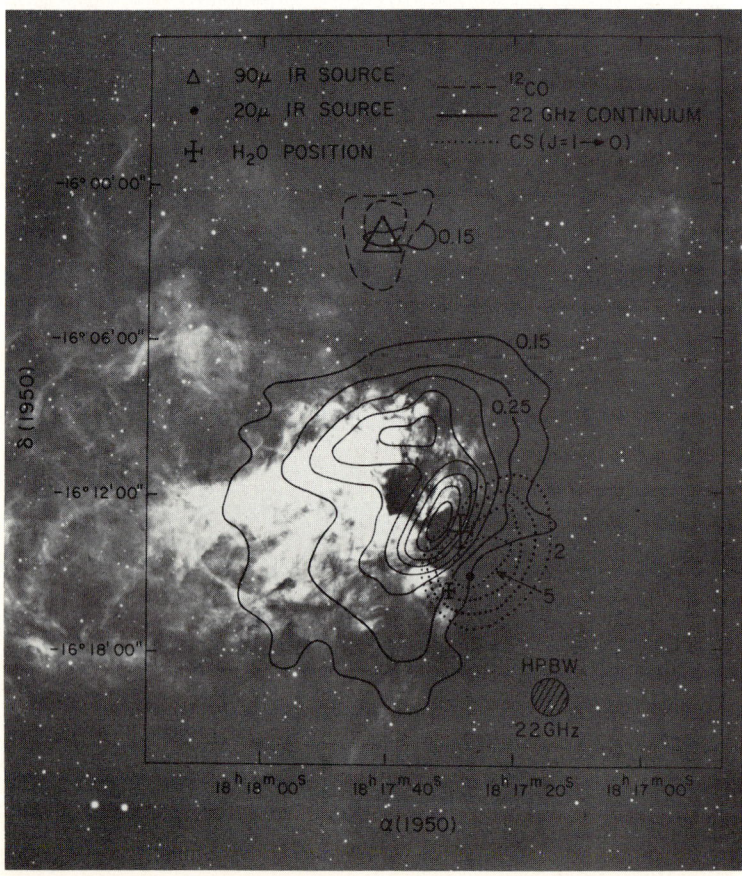

FIGURE 9.43 Messier 17, a nearby H II region. [*Courtesy Palomar Observatory, California Institute of Technology.*]

FIGURE 9.44 Infrared and radio emission from the M 17 region. The solid contour lines show the free-free emission at 22 GHz; dotted line, emission from carbon monosulfide (CS); dashed line, emission from CO; crosses, positions of water masers; large dot, 20-μm source; and triangle, 90-μm emission. [*Courtesy C. Lada.*]

part of the region. The Trapezium cluster consists, in fact, of a few hundred stars within 10^5 AU of each other; it's the O and B stars of this group that ionize the gas. These massive stars are no more than a million years old. Various observations show the gas flow here to be chaotic and swift, with velocities as high as 100 km/sec. The distance between the Trapezium stars and the front edge of the molecular cloud is roughly 1 ly (1/3 pc). In an evolutionary sense the molecular cloud core that lies behind the Orion Nebula is the youngest (least evolved) part of the region.

Where is starbirth happening? We see the embryonic material (the molecular cloud) and the end results (the Trapezium cluster). The most likely place for protostars lies between the two, in the infrared cluster associated with the Becklin-Neugebauer object and the Kleinmann-Low nebula. High-resolution infrared observations indicate that at least five sources lie in the cluster with separations of a few thousand AUs. These may be protostars. But most astronomers place their bets on the Becklin-Neugebauer object itself. Its observed characteristics match those expected from a massive protostar in its pre-main-sequence evolution.

Recently, infrared astronomers have carried out careful observations of the Becklin-Neugebauer object at the wavelength of an infrared line at 4.05 μm, called the *Brackett-alpha line*, which arises from a transition from level 5 to level 4 in hydrogen. From the viewpoint of protostellar evolution, this line can arise at two different evolutionary stages: (1) as a recombination line from a very small, newly formed H II region around a massive star approaching the main sequence or (2) as emission from infalling hydrogen gas excited by passing through the accretion shock around a protostar.

Observations to date indicate that the emission comes from a compact H II region (stage 5 in Table 9.7). So the Becklin-Neugebauer object, embedded in the molecular cloud, is, in this interpretation, a B0 star just making it to the main sequence, surrounded by dust and gas it has just begun to ionize. It is no more than a million years old, probably less.

Because the Orion Nebula region is so close to us, it's been heavily studied. But it's not the best place to look to unravel the processes of starbirth, for all the action occurs right along our line of sight, jammed together as we see it in the sky. By chance, Messier 17 (M 17), another bright H II region, presents a clearer observational view (because we happen to see it from the side) of one sequence of massive star formation.

You can easily see M 17 in a telescope (Fig. 9.43). It lies only 7200 ly (2500 pc) from us. The optical H II region marks a site of star formation some 10 million years old. Contained within it is a small cluster of O and B stars. To the west of the nebula nothing much appears optically, but millimeter radio observations reveal two pieces of a molecular cloud. The total mass within these clouds is about 10^5 solar masses and the peak densities may be as much as 10^5 molecules/cm^3. Both fragments have peak temperatures above that usually found in molecular clouds, so something there is heating up the gas. Infrared astronomers Douglas Kleinmann and Ned Wright have discovered in the south fragment a starlike infrared source that has some of the characteristics of a protostar. Near it lies a water maser (Fig. 9.44). Further to the west, connected to the two fragments next to M 17, lies a gigantic molecular cloud, some 70 by 280 ly (20 by 85 pc) in size and containing more than a million solar masses of material (Fig. 9.45). To give you some idea of how large this cloud appears in the sky, it would take 8 *full moons* touching (an angle as big as your fist at arms length) to span the cloud from its east to west end.

To round out the observational clues, infrared photos of M 17 (Fig. 9.46) show a small

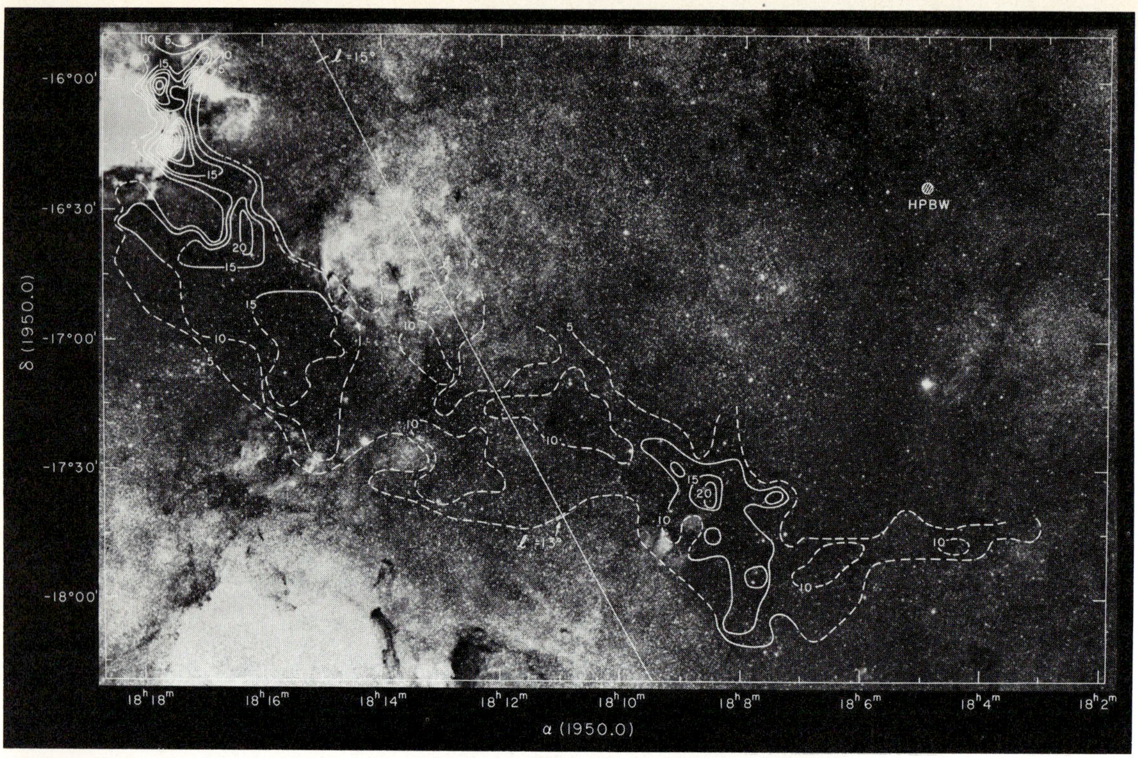

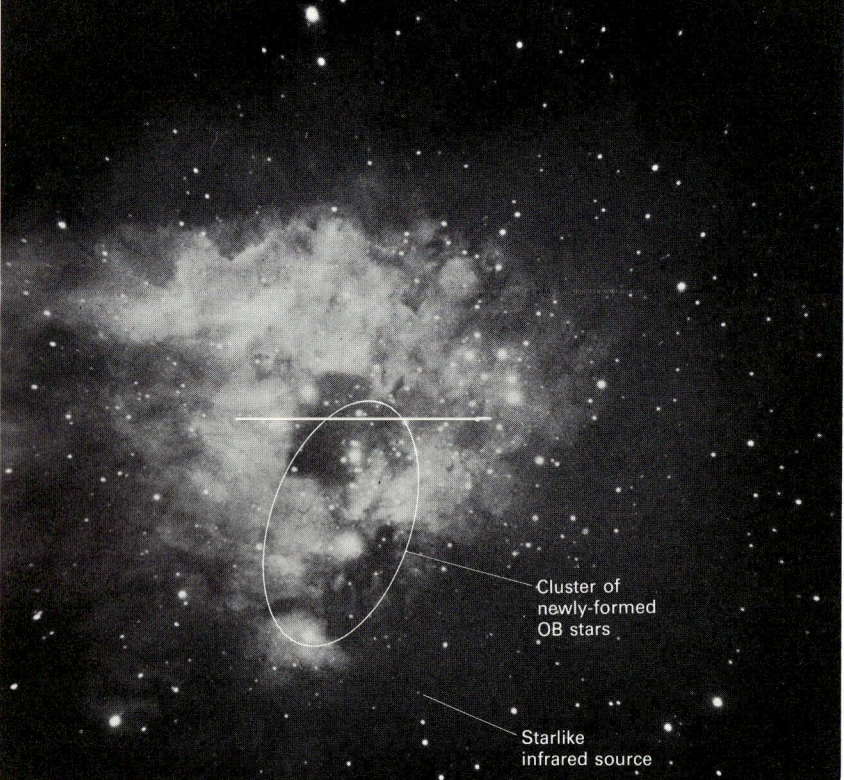

FIGURE 9.45 Carbon monoxide contour map of the giant molecular cloud extending to the west of M17. [*Courtesy, B. Elmegreen; based on observations by B. Elmegreen, C. Lada, and D. Dickenson.*]

FIGURE 9.46 An infrared photo of the region of M17 near the southern molecular cloud fragment. The white oval surrounds a cluster of young O and B stars. The line points to a starlike infrared source that may be even younger; it is located at the center of the cloud. [*Courtesy M. Beetz, H. Elsasser, C. Poulakos, and R. Weinberger; photo taken with the 123-cm telescope of the Centro Astronomico Hispano-Aleman Almeria on the Calar Alto in Spain.*]

cluster, previously undetected, of O and B stars in the zone between the H II region and the molecular cloud. The cluster's location coincides with the peak of continuous radio emission from M 17—an indication that this dust-obscured cluster now plays an important role in the ionization of the H II region.

These observations all add up to a composite picture of a sequence of massive star formation from giant molecular clouds (Fig. 9.47). On one side of the molecular cloud lies the oldest region of starbirth—M 17 and its immersed group of O and B stars formed out of the molecular cloud. These hot stars have heated up the gas around them, ionizing

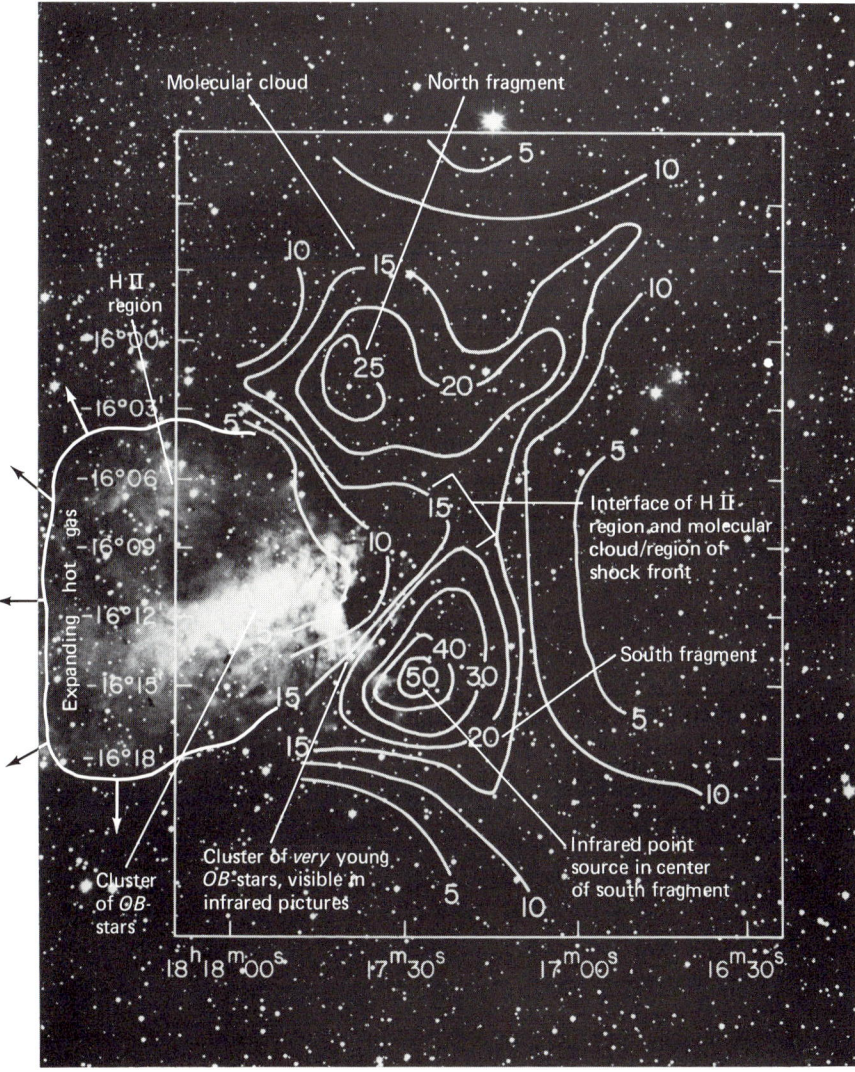

FIGURE 9.47 A map of the carbon monoxide emission of the M 17 region. It shows two fragments of a giant molecular cloud. An infrared source lies near the bottom peak (labeled "50"). The total mass in the two pieces is about 100,000 solar masses; the peak densities of molecules may reach $10,000/cm^3$. [*Courtesy C. Lada and Harvard College Observatory; based on observations by C. Lada and colleagues.*]

it and destroying the molecules there. The hot gas of the H II region slowly expands. On the west the expanding hot gas runs into the cold, dense molecular cloud. Here a shock wave forms. The shock wave, moving at about 10 km/sec, compresses the material and prompts gravitational collapse and star formation out of the molecular cloud. Moving more into the fragmented molecular cloud, the shock wave will probably trigger more star formation, and each group—if O and B stars—will develop an H II region and another shock wave. So the molecular cloud will finally self-destruct in an orgy of star formation. Material that doesn't make it into stars eventually is dissipated by the stars that have formed.

In this sequential model of massive star formation—recently revived by Bruce Elmegreen and Charles Lada—massive star formation begins at one end of a giant molecular cloud (Fig. 9.48). (Such clouds tend to be elongated, cigar-shaped objects.) A small group of about 10 O and B stars forms. They evolve to the main sequence. Their ultraviolet radiation then dissociates hydrogen molecules around them and ionizes the gas. The H II region, since it is hot (about 10,000 K), expands, pushing a shock wave into the molecular cloud. The gas behind the shock wave is compressed to densities sufficient to start gravitational collapse. A new group of O and B stars is born about a million years after the previous one. The process repeats. Small groups of massive stars are born in a sequence of bursts across the molecular cloud.

Note that this model predicts that the fossil remnants of a molecular cloud will be a string of small groups of O and B stars 30 to 100 ly (10 to 30 pc) apart, more or less in the same space as the parent molecular cloud. Have such strings been seen? Yes. In 1947 the Armenian astronomer V. A. Ambartsumian noted that loose groupings of O and B stars, called *OB associations*, must be very young. Since these stars do not live very long (a few tens of millions of years), the OB associations themselves cannot be more than 10^7 years old. Ambartsumian also noted that the associations were *not* held together by gravity; they simply did not have enough mass. Such star groups slowly expand until they lose their collective identity in about 10^7 years.

Later work, brought together by Adrian Blaauw, showed that many OB associations, which are 100 to 600 ly (30 to 200 pc) in size, consist of small clusters of stars called *OB subgroups*. These subgroups contain 4 to 20 stars (average about 10) and range in size

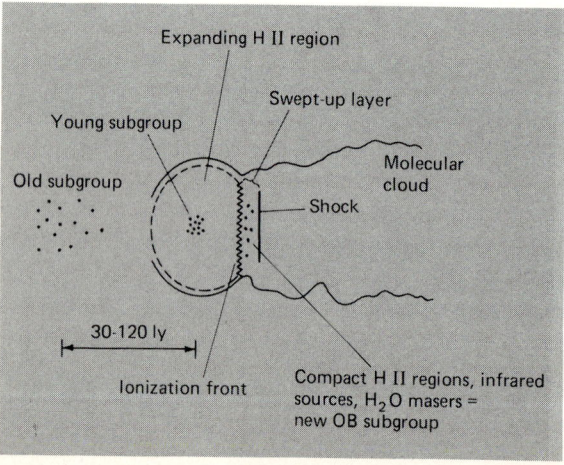

FIGURE 9.48 A schematic for the sequential model of massive star formation from a giant molecular cloud. [*Based on a figure by B. Elmegreen and C. Lada.*]

from a few to a few hundred light years across. Blaauw noted the remarkable fact that in an association OB subgroups are lined up in an evolutionary sequence. The most spread out, oldest subgroup lies at one end and the most compact, youngest subgroup at the other. For example, Orion contains a large association that has four OB subgroups; the smallest and the youngest subgroup (estimated by Blaauw to be 2 million years old) is the Trapezium cluster. Recall that the Trapezium is right up against the south Orion molecular cloud. Here we see the signposts of massive star formation.

You may have noticed in this model that once the star formation starts at an end of a molecular cloud, it propagates through it in a chain reaction. But what starts the first burst of star formation? The answer to that question is not known. Perhaps it starts from the collision of molecular clouds. Or, more likely, from the blast wave of a supernova remnant crashing into an end of a molecular cloud. This idea is aesthetically appealing: A supernova signals the death of a massive star; then the mark of its death ignites the birth of other massive stars. (But, of course, this idea must be checked against observations to be confirmed.)

Note also in this model that giant molecular clouds do not last long once starbirth begins—only a few tens of millions of years. Since we see many molecular clouds now, they must form rapidly, in order to balance their rapid destruction. Where and how these clouds form is a mystery now. Their origin likely has to do with the spiral structure of our Galaxy (Chapter 21).

The Birth of Solar-Mass Stars

We wish we could paint for you as neat an observational picture for the formation of stars like the sun. But we cannot. The observational evidence is skimpy and doesn't hold together in a comprehensive way, so this topic must be regarded as serious conjectures and working models. It does seem clear that, like massive stars, solar-mass stars are also born from molecular clouds. The questions are in which clouds and how.

Let's look to observations for clues. Some years ago Merle Walker investigated the H-R diagrams of clusters of stars he thought to be young because of their close association with gas and dust in the form of dark clouds. (Astronomers gave the name *dark clouds* to an interstellar cloud that contained so much dust that it blotted out the light of stars within it and behind it. We now know that dark clouds are one type of molecular cloud. They typically have temperatures of 10 K, densities from 100 to 1000 atoms/cm^3, and masses from a few tens to a few hundreds of solar masses.) One such cluster, New General Catalog (NGC) 2264, lies in front of and slightly embedded within a dark cloud. This cluster is about 2400 ly (700 pc) from us in the constellation Monoceros, just 15° east of Betelgeuse in Orion.

From his observations Walker constructed an H-R diagram of NGC 2264 (Fig. 9.49). He noted that the O and B stars of the cluster had already arrived at the main sequence. But stars from A to M in spectral type mostly fell *above* the main sequence by two magnitudes. A natural interpretation is that all the stars in the cluster formed at approximately the same time, but the more massive ones evolved faster. The more massive stars have already reached the main sequence, while the less massive ones (about three solar masses and less) are still contracting and evolving to it.

Among these pre-main-sequence stars Walker found a number of *T-Tauri stars* (named after their prototype, T Tauri). These are variable stars with spectral types from F to M. In

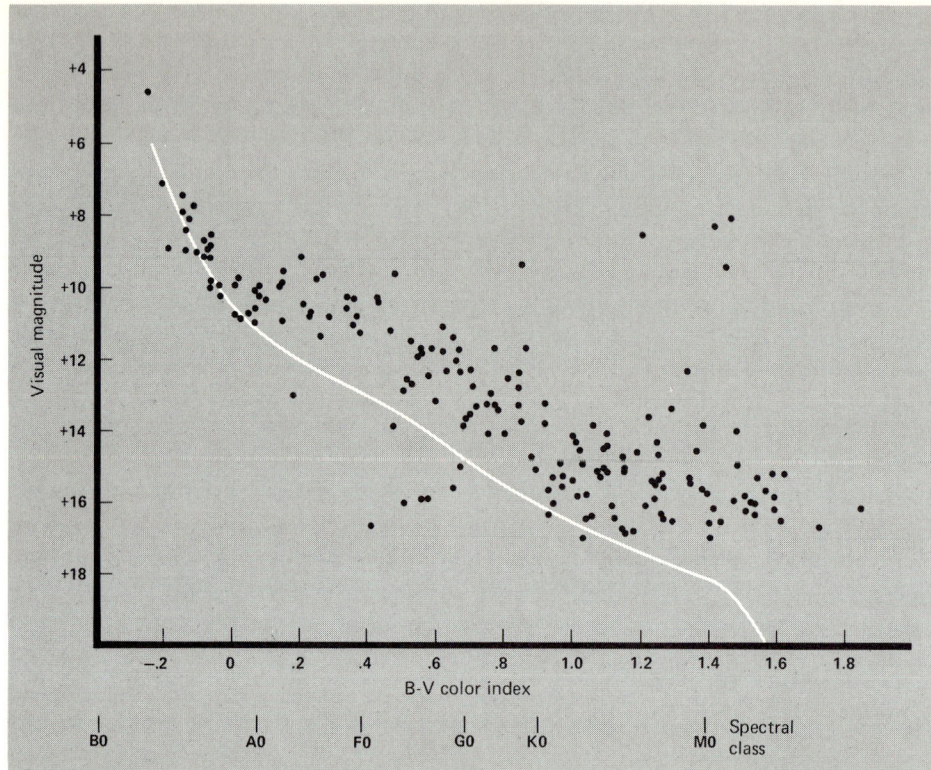

FIGURE 9.49 An H-R diagram for the open cluster NGC 2264. The solid line indicates the initial main sequence. [*Adapted from a figure by* M. Walker, Astrophysical Journal Supplement Series, *vol. 2, p. 365, copyright 1956 by the American Astronomical Society.*]

addition to dark-line spectra, they have emission lines of hydrogen, Ca II, Fe I, Fe II, and Ti II. Some T-Tauri stars have an excess of infrared emission, that is, more infrared than you'd expect from a main-sequence F to M star. A number of T-Tauri stars have been studied optically and in the infrared by Eric Rydgren, Steve Strom, and Karen Strom. They conclude that most T-Tauri stars are: (1) low-mass stars (0.2 to 2 solar masses) now approaching the main sequence, (2) young, with ages from 10^5 to 10^7 yr, (3) surrounded by hot (2×10^4 K), dense (10^9 to 10^{12} atoms/cm^3) envelopes, and (4) losing mass by stellar winds (analogous to the solar wind) at rates of 10^{-7} to 10^{-8} solar mass/yr that blow at speeds of a few hundred kilometers a second. (This last property of T-Tauri stars has been disputed recently.) Some of these have thin shells of hot (1000 K) dust, as evidenced by observations of the 9.7-μm silicate feature in emission and, in part, by the infrared excesses. From such observations T-Tauri stars are presumed to be characteristic of solar-mass pre–main-sequence stars.

T-Tauri stars can then be used as tracers of solar-mass pre–main-sequence stars. Hunted optically, T-Tauri stars appear almost always within dark clouds, sometimes within OB associations. They seem to occupy the surface layers of the dark clouds, but it may be that we just don't see the ones buried inside.

Fred Vrba, Steve Strom, Karen Strom, and Gary Grasdalen have made infrared sur-

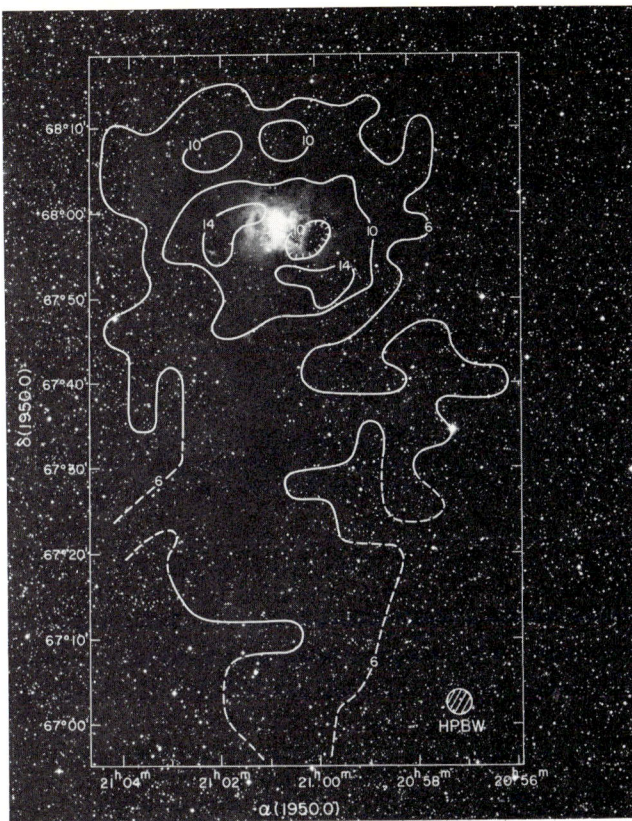

FIGURE 9.50 A carbon monoxide map of the nebula NGC 7023. [D. Elmegreen and B. Elmegreen, Astrophysical Journal, vol. 220, copyright 1978 by the American Astronomical Society. Background photo from the National Geographic Society Palomar Observatory Sky Survey.]

veys of a number of dark clouds to search for young stars embedded in them. They find numerous candidates. For example, in a dark cloud in Ophiuchus they detected about 70 infrared sources. These seem to make up a star cluster similar in many ways to NGC 2264, yet still contained, after millions of years, in its parent cloud.

These observations bring up the question of the mechanism for the dissipation of a dark cloud's material, which reveals its hatched stars. Right now we have no certain answer to this question. Some possibilities are (1) the material is blown away by stellar winds, (2) a nearby supernova blows off the cloud, (3) a supernova *in* the cloud blows it away, or (4) massive stars dissipate the cloud as they reach the main sequence and generate H II regions.

So one view of the birth of solar-mass stars is this. They are born in fairly massive dark clouds, perhaps in giant molecular clouds along with massive stars. They form from fragments throughout the cloud, rather than at the edges as massive stars do. The births of the massive stars or perhaps the death of one in a supernova, sweeps away the gas and dust to reveal the stars. In this picture most starbirth takes place in dark, massive clouds, out of which OB associations form. Thus the sun may have been born in an OB association, such as the one we described in Orion, perhaps blasted clean by a supernova. (More on the sun's formation in Chapter 13.)

An alternative idea that is also appealing is that solar-mass stars form out of small, isolated molecular clouds, not giant ones. These clouds are no more than a few light years in size, contain at most a thousand solar masses of gas, and are not near in space to any giant molecular clouds.

An example of such a cloud is NGC 7023 (Fig. 9.50), studied at millimeter wavelengths by Debra and Bruce Elmegreen. This cloud lies some 1400 ly (430 pc) from the sun and about 300 ly (100 pc) above the galactic plane. No large cloud is within 260 ly (80 pc) of it. NGC 7023 has overall dimensions of only 13 by 25 ly (4 by 8 pc) and contains some 300 solar masses of material in its densest region and at most 1000 solar masses overall.

Inspecting the carbon monoxide map you can see that the cloud is very clumpy. For example, the region to the east of the central star is much denser than that to the west, where you find a hole that allows you to look through the cloud. The region to the east has been found to contain a small, weak radio continuum source—an indication of a young, perhaps B-type star embedded in the clump next to the visible star.

Various observations show that the molecular cloud contains a number of relatively young stars spread throughout it. So it appears that star formation has taken place and continues in NGC 7023 in many places, not just at the edges, as appears to be the case with the formation of massive stars from giant molecular clouds.

From observations so far we are beginning to suspect that, in general, stars are born from molecular clouds; massive stars from massive clouds and less massive stars (the majority of those in the Galaxy) from less massive molecular clouds.

SUMMARY

The interstellar medium is a dynamic, violent place in continual evolution. It is filled with gas and dust, with the gas making up 99 percent of the mass. Most of the gas is tied up in clouds of various sizes, from the numerous H I clouds to the complex giant molecular clouds. Among these clouds streams a hot, dilute gas, perhaps heated up by supernova blasts.

Stars are born from these interstellar molecular clouds. The process of starbirth appears to follow two paths, depending on the mass of the protostar. Massive stars seem to condense in small groups in bursts of star formation at the edges of giant molecular clouds. When they reach the main sequence as O and B stars, massive stars quickly and violently disrupt their mother cloud. In contrast, solar-mass stars appear to condense throughout dark clouds. When formed, they affect their parent cloud minimally. Other influences must drive the dark cloud's matter away. In all, the process of star formation is not very efficient. It locks up at most a few percent of a cloud's mass as stable stars.

Key Words

interstellar medium	giant molecular cloud	Trapezium cluster
interstellar gas	molecular maser	Becklin-Neugebauer object
interstellar dust	stimulated emission	interstellar extinction curve
bright (emission) nebula	cosmic ray	icy material
H II region	open cluster	core-mantle grain
Orion Nebula	dark nebula	gravitational collapse
fluorescence	reflection nebula	hydrostatic equilibrium
recombination	extinction	free-fall time
radio recombination line	absorption	protostar
H-alpha line	scattering	pre-main-sequence star
H I region	reddening	shock wave
coronal interstellar gas	polarization	momentum

PROBLEMS

angular momentum
torque
Brackett-alpha line

Messier 17
OB association

OB subgroup
dark cloud
T-Tauri star

Review Questions

1. Describe how (one way for each) astronomers observe: (a) interstellar H I, (b) interstellar H II, (c) the coronal interstellar gas, and (d) interstellar molecules. (Objectives 1 and 2)
2. Outline two ways in which astronomers "see" interstellar dust. (Objectives 1 and 3)
3. What evidence do we have, if any, that dense materials make up part of the interstellar dust? Icy materials? (Objective 4)
4. List the observational evidence that leads to the guilt-by-association argument for the formation from giant molecular clouds of massive stars in small groups. (Objective 5)
5. What provides a protostar with its luminosity? (Objectives 5 and 6)
6. In a theoretical picture for the formation of a massive star, why do we not see a massive star directly (optically) until it reaches the main sequence? (Objective 5)
7. Once it reaches the main sequence, how does a massive star influence its parent cloud? (Objective 5)
8. What observational evidence do we have that solar-mass stars form from dark clouds? (Objective 6)
9. Present two ways in which supernovas can affect the interstellar medium. (Objective 7)
10. In *one short sentence*, argue that starbirth must be taking place *now*. (Objective 8)
11. Which of the following are evidence for the existence of interstellar dust?
    a. Dark clouds blotting out the light of distant stars
    b. Emission at 21-cm wavelength in parts of the Milky Way
    c. Nebulas with bright-line spectra
    d. Stars which appear redder than expected from their spectral type
    e. Polarization of starlight
    (Objective 1)
12. An infrared source is observed to be associated with a source of continuous radio emission, but there is no object visible at optical wavelengths. Why is this a possible indication of the presence of a newly formed star? What produces each kind of radiation? Why don't we see optical radiation from the star? (Objectives 7 and 10)

Problems

1. Using Bohr's formula for the wavelengths of the hydrogen lines (Chapter 6), calculate the wavelength and frequency of the radio recombination line with upper level $n_A = 93$ and lower level $n_B = 92$. (This is the line pictured in Fig. 9.5) (Objective 2)
2. If there is 1 dust particle/10^6 m^3 and 0.33 hydrogen atoms/cm^3, how massive must a dust particle be to account for 1 percent of the total mass of interstellar matter? If the density of a dust particle is taken to be 3000 kg/m^3, what is its radius? (Objective 4)

3. In the plane of the Galaxy dust dims stars by about 1 magnitude for every 1500 ly of distance. Suppose there is a cluster of stars at a true distance of 7500 ly. How far will it appear to be, if we fail to account for the dimming of the dust? (Objectives 1 and 3)
4. The Kleinmann-Low nebula in Orion has a temperature of about 70 K. Assuming it radiates like a blackbody, find the wavelength where it emits the most energy. The luminosity of the nebula is $7 \times 10^4 \, L_{sun}$. What is its radius? The Orion Nebula is about 1500 ly away. What is the angular size of the Kleinmann-Low nebula? (Objective 7)
5. What is the free-fall collapse time of a giant molecular cloud (density 10^3 hydrogen molecules/cm^3)? (Objective 6)
6. If a typical H I cloud ($r = 2.5$ pc, $n = 20$ atoms/cm^3) rotating once in 2×10^8 years collapses, conserving angular momentum, what will be its radius when the centripetal acceleration at its equator just balances the gravitational acceleration? (Objective 11)

BEYOND THIS BOOK . . .

* For good background material on star formation, try "The Birth of Stars" by B. Bok in *Scientific American*, August 1972. But be warned that some of his material is already dated.
* An advanced, technical article, but very readable, is S. Strom's "Star Formation and the Early Phases of Stellar Evolution" in *Frontier of Astrophysics* (Harvard University Press, Cambridge, Mass., 1976), edited by E. Avrett. Slightly more technical is "Young Stellar Objects and Dark Interstellar Clouds" by S. Strom, K. Strom, and G. Grasdalen in *Annual Reviews of Astronomy and Astrophysics*, vol. 13 (1975), p. 187.
* For more details on one aspect of starbirth, read "The Birth of Massive Stars" by M. Zeilik in *Scientific American*, April 1978. For another view, try "Bok Globules" by R. L. Dickman, *Scientific American*, June 1977.
* You can find a number of pertinent articles in *Protostars and Planets* (University of Arizona Press, Tucson, 1978), edited by T. Gehrels. Some of these articles are technical, but descriptive.
* For a nice summary of observational evidence for protostars, see the article by M. Werner, E. Becklin, and G. Neugebauer in *Science*, vol. 197 (1977), p. 723.
* For one view of the dynamic interstellar medium, read "The Structure of the Interstellar Medium" by C. Heiles, *Scientific American*, January 1978. A nice review of H II regions is "Gaseous Nebulas" by E. Chaisson, *Scientific American*, December 1978. For a look at molecular clouds, read "Molecules in Space" by B. Turner, *Scientific American*, March 1973.
* For a technical review on the evolution of interstellar dust, refer to "Formation and Destruction of Interstellar Grains" by E. Salpeter in *Annual Review of Astronomy and Astrophysics*, vol. 15 (1977), p. 267.
* For more on masers, read "Cosmic Masers" by D. Dickinson, *Scientific American*, June 1978.

THE LIVES OF STARS

10

LEARNING OBJECTIVES

After studying this chapter, you should be able to:

1. Show how the Hertzsprung-Russell diagram for many stars provides clues about the evolution of individual stars.
2. Describe the physical basis of a theoretical model of a star.
3. Trace the evolution of a 1-solar-mass star on a Hertzsprung-Russell diagram, describing the physical changes of the star that result from changes in the star's core.
4. State in one sentence why stars must evolve.
5. Compare the evolution of a 1-solar-mass star with that of a 5-solar-mass star.
6. Describe on an H-R diagram the evolution of a cluster of stars.
7. Back up theoretical ideas of stellar evolution with observational evidence.
8. List the sequence of thermonuclear energy generation reactions in stars of different masses.
9. Indicate how mass and chemical composition affect stellar evolution.
10. Describe how a degenerate gas differs from an ordinary one.
11. Describe the main physical differences between stars of Population I and Population II.
12. Locate on an H-R diagram and describe the evolutionary state of the major types of variable stars: RR Lyrae stars, cepheids, and red variables.
13. Show how to obtain the relative distances of clusters of stars by comparing their main sequences.
14. Compare the ages of star clusters by examining their H-R diagrams.

CENTRAL QUESTION:
How do the physical properties of stars change as they go through their normal lives?

The previous chapter explained that star formation occurs in the molecular clouds of interstellar space. Gravity has gathered at least 50 percent of the material in the sun's vicinity into stars. The rest floats as cool clouds trapped in a hotter gas that separates the clouds. Some of these clouds are gigantic; they contain thousands of solar masses of gas and dust. These huge clouds sometime in the future may condense into stars or into planetary systems.

Gravity controls the history of newborn stars. A star survives as long as it can counteract the relentless gravitational crunch. The story of this battle against gravity runs like the aging of a person from birth to death, but takes millions to billions of years. Let's try to put this span into a human perspective. Imagine time speeded up so that one year passes in one-fifth of a second. The sun would live only about 65 years. The sun's birth would be quick; only 4 months would pass from the start of the collapse of the sun's embryonic cloud to its establishment as an immature, but full-fledged star. For about the next 60 years the sun would shine calmly as it passed through middle age. Old age would gradually fossilize the energy production of the sun. In about 5 years the elderly sun would slowly expand to almost 100 times its present size; it would become a bloated red giant. Then a sudden burst would blow off the sun's atmosphere, leaving behind a hot core that would cool quickly.

How a star lives and dies depends mostly on how much mass it has at birth. This chapter first presents theoretical ideas about stellar evolution and contrasts the lives of stars like the sun to those of more massive ones. Then it offers some observational evidence for these theoretical concoctions.

10.1 STELLAR EVOLUTION AND THE HERTZSPRUNG-RUSSELL DIAGRAM

Contemporary ideas of how stars evolve focus on one general theme: As a star loses energy to space, it must change. Most objects cool off (their temperature goes down) when they lose energy to their environment. It's natural to think a star does the same, but just the opposite is the case. For most phases of its life a star does not cool off. Nuclear reactions generally supply the energy to keep a star hot. When one fuel (say, hydrogen) runs out, another (say, helium) can ignite at a temperature made higher by the compression of matter by gravity. Only when a star cannot contract any more does it finally cool off.

We approach stellar evolution from both theoretical and observational points of view. This chapter presents mostly theoretical ideas, because observations of a star's evolution do not come easily. The sun's anticipated lifetime is more than 100 million human lifetimes. There is no way you could watch a single star, like the sun, evolve. But you can see many different stars at one time. You can organize these stars on the Hertzsprung-Russell diagram (Fig. 10.1) if you know their luminosities and spectral classes. Then you can use the H-R diagram of many stars to guess at the evolution of one star. Let's see how.

Classifying Objects

Suppose you recorded the height and weight of all the 18- to 19-year-old males you met over a period of days. If you plotted your data as a graph of weight versus height (Fig. 10.2), you would find a trend; the points tend to fall along a line (call it a "main sequence") that shows that weight generally increases with height.

Now suppose you did the same experiment, but this time you recorded and plotted the height and weight of every person you encountered (Fig. 10.3). Compare this plot with the previous graph. You still have a main sequence, but you also have other groups that

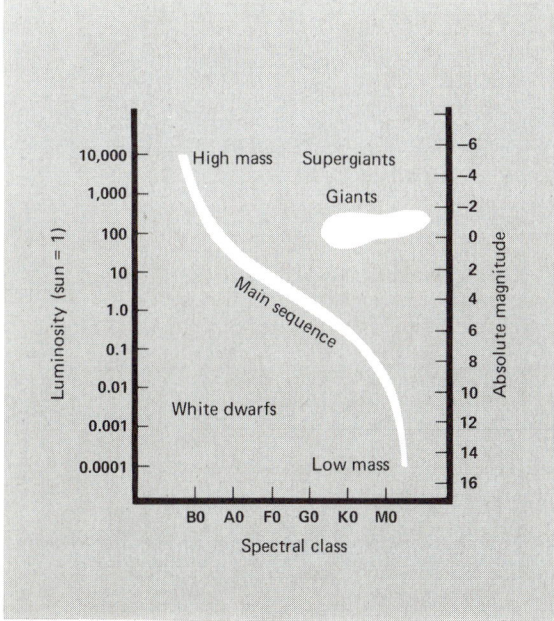

FIGURE 10.1 A schematic H-R diagram. A star's position on this diagram represents its surface temperature, its luminosity, and, indirectly, its size.

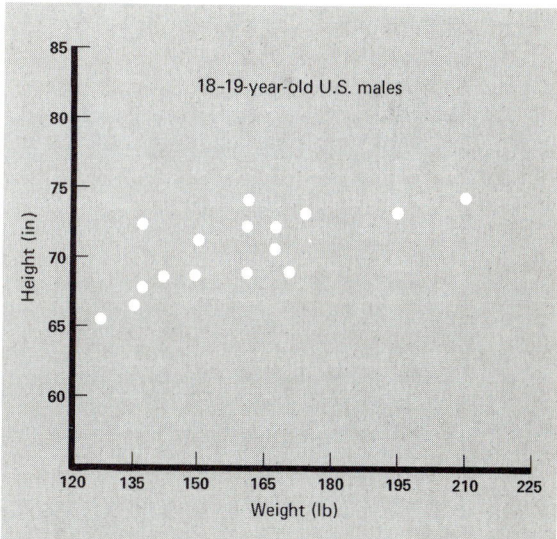

FIGURE 10.2 A height-weight diagram for a sample of 18- to 19-year-old U.S. males. [Data from Astronomy 10S class, University of California, Berkeley, Winter 1980.]

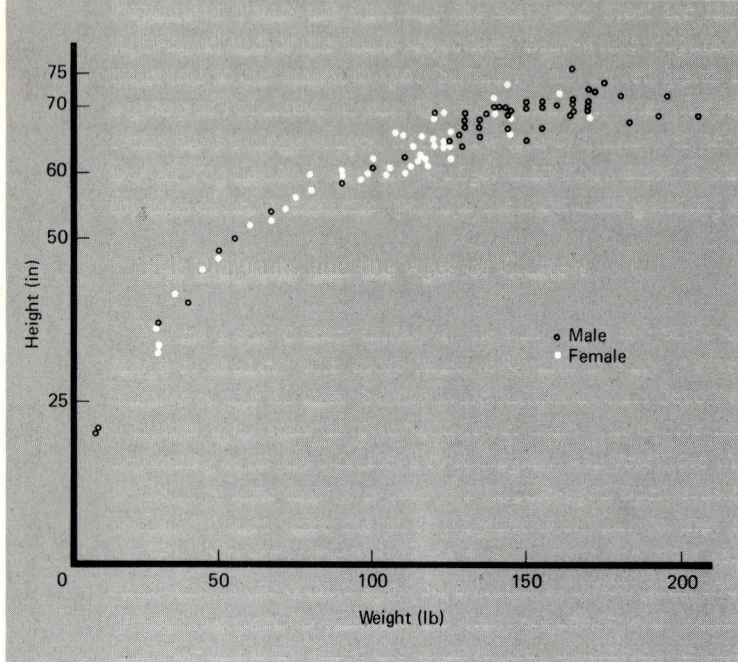

FIGURE 10.3 A height-weight diagram for a sample of U.S. males (open circles) and females (filled circles), ranging in age from 18 to 50. [*Data collected in Albuquerque, New Mexico.*]

don't follow the original main-sequence trend. The reason for the difference is probably clear to you. The first graph includes people of the *same* age, while the second includes people at *different* ages.

Now here's a third graph (Fig. 10.4). It's a plot of weight versus height for an average U.S. male at different times in his life from birth to age 20. It shows you how a *single* person's height and weight change as he ages. Note this evolutionary graph for one average person follows the trend in the height-weight graph of many different people. You can correctly interpret the graph for many people in evolutionary terms *if*, and *only* if, you know how one person evolves.

Time and age were implicit in the graph for many people chosen at random. In the same way, time and age implicitly play a role in an H-R diagram, in which are plotted two essential properties of stars: surface temperature and luminosity.

Time and the H-R Diagram

How does the H-R diagram tell you about the evolutionary sequence of a star? Imagine that you have a large family. They have gathered together, and you take a snapshot of them all, from the youngest infant to the great-grandparents. Most of the people in the picture are in their middle age (20 to 60 years old); you have a few infants, some children and teenagers, and a few old people. You have so many middle-aged people because most of your life you will be "middle-aged"; you spend relatively less time as an infant, teenager, or old person.

10.2 STELLAR ANATOMY

Now suppose you have a collection of any objects that evolve and you believe that the collection spans an evolutionary sequence. You can estimate the relative time spent in any evolutionary stage by the relative numbers you find at that stage compared with others. (This argument holds true only if the birth and death go on continuously; if no more people are born, eventually you will see only old people, then none.)

Go back to the H-R diagram for stars (Fig. 10.1). Recall that it shows the luminosity and surface temperature of the stars. Most of the stars fall on the main sequence, so stars found here are going through the longest, most stable stage in their evolution.

Section 8.1 pointed out that the main sequence, from O to M stars, represents a sequence from higher to lower masses. Section 6.6 explained the normal process of the conversion of hydrogen to helium in the sun, which is a main-sequence star. Here's the evolutionary meaning of the main sequence. It marks stars at the stage of converting hydrogen to helium in their cores; stars remain at this stage for the greatest part of their lives. That's why we see so many main-sequence stars now. Other stages, such as the red giant, must be shorter, because we see far fewer red giants now than main-sequence stars.

A star's mass determines how a star will evolve. So you have to examine the H-R diagram to find how stars of different masses evolve. But what is the correct interpretation of the H-R diagram for the evolution of stars? You need some hints from the physical nature of stars and theoretical calculations to make up the star models.

10.2 STELLAR ANATOMY

What is a star? A huge, hot ball of gas, mostly hydrogen, heated by thermonuclear reactions in its core. You can imagine a star as a controlled hydrogen bomb. A star does not fly apart, because gravity persistently pulls it together. A star withstands the inward squeeze of gravity by producing a compensating pressure force. For most of its life the compensating pressure comes from the heat of the star. A star consists of gas; a hot gas has

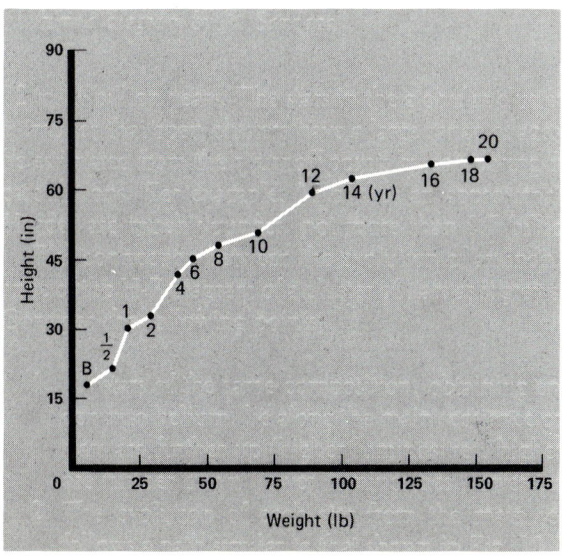

FIGURE 10.4 A height-weight diagram for a typical U.S. male from birth to 20 years of age. [*Data from the National Center for Health Statistics Growth Charts.*]

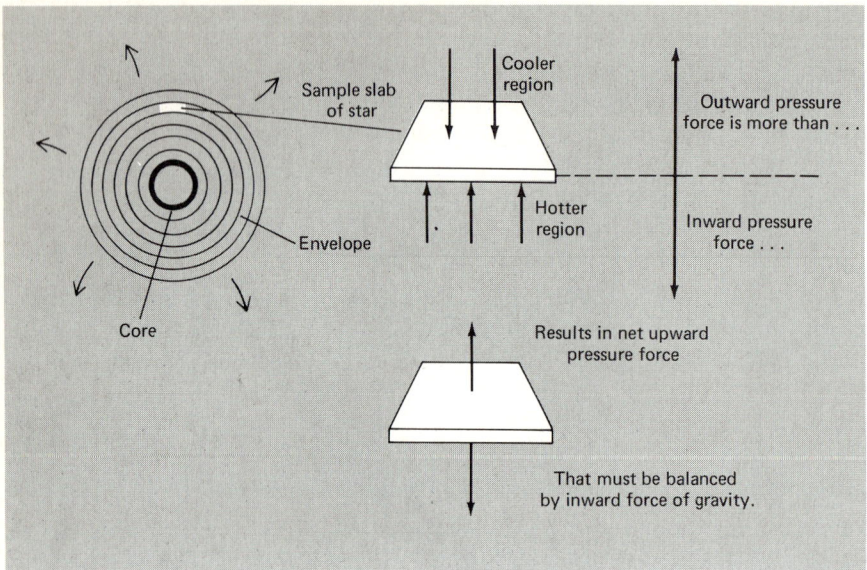

FIGURE 10.5 The balance of gas pressure and gravity in a star.

a high pressure; and this outward force balances the inward gravitational forces. This balance must hold true at every level throughout the star (Fig. 10.5); otherwise it would be unstable. So the first physical requirement for a star model is that it be in *hydrostatic equilibrium* (Focus 9.1).

Second, we need to know how the material of a star behaves at different temperatures, pressures, and densities. The physical relation, for a given material, among those quantities is called the *equation of state*. Luckily, the appropriate equation of state for most stars during most of their lives is a single one, that for an ordinary gas (Section 6.1).

Third, the star must generate energy internally. For most of a star's life thermonuclear fusion reactions operate as the internal furnace (details to come in Section 10.4). In general, the total energy produced inside the star must equal the rate at which the energy radiates away at its surface. This balance must be true not only overall, but also at each layer within the star. Otherwise the star will be unstable, heat up or cool down, and consequently expand or contract.

As an analogy, consider a factory assembly line on which cars pass through different assembly stages. The rate at which the cars enter and leave each stage must be equal, or the cars will pile up at one location. If the flow of heat through a star were uneven, the temperature of various layers would change. These temperature changes would result in pressure changes which would cause the star to expand or contract.

So, generally a star meets the condition of *thermal equilibrium*—energy lost equals energy produced. During brief phases of its life a star may not be in exact thermal equilibrium, but it always tries to reach equilibrium. To do so, it expands or contracts, but slowly, so that it is not totally disrupted.

Fourth, we need to look at how a star transports energy from its core to surface. Section 6.6 discussed the three methods available: conduction, convection, and radiation. Whichever method is predominant, an equation must be found relating the rate of flow of energy to the local conditions of temperature, density, and composition. In most regions of a star radiation is the important mechanism. The rate of flow depends directly on the *temperature gradient* (the amount by which the temperature drops in a given distance). The more rapidly the temperature falls with distance, the greater the energy flow. It also depends inversely on the *opacity* (Section 6.4). The more opaque a star's material is, the slower the flow of radiation through it. Opacity acts somewhat like insulation in a house in winter. The furnace is generating heat. The greater the house's insulation, the slower the flow of the heat to the cold outside. A slow flow keeps the exterior of the house cold and the inside hot. The outside of a poorly insulated house is warmer than the outside of a well-insulated house, and the inside is cooler. Similarly, if the opacity of a star were suddenly lowered, radiation would escape more easily; the star would become more luminous, but its inside would become cooler. So the internal pressure would drop, hydrostatic equilibrium would be upset, and the star would contract until equilibrium was regained.

Whether convection occurs at any region in a star is determined by the opacity of the material in that region. If the material is not too opaque, energy will flow easily by radiation. If it is so opaque that the radiative energy flow gets bottled up, convective transport will take over and operate instead.

Generally, a star's opacity depends on its chemical composition, density, and temperature (Section 6.4). Here are some rough guidelines to a star's opacity: (1) the greater the percentage of elements other than hydrogen and helium, the greater the opacity; (2) the greater the density, the higher the opacity; (3) the lower the temperature, the higher the opacity. You need to keep opacity in mind when considering stellar evolution, because the opacity determines how a star will transport energy (by radiation or convection) and that form of transport affects the internal structure of the star.

10.3 STAR MODELS

A theoretical model of a star incorporates all the physical conditions described above: hydrostatic equilibrium, the ordinary gas law, energy sources, thermal equilibrium, and energy transport. A star must produce energy, be in balance between gravity and pressure, and transport energy from its core to its surface evenly.

All these conditions combined are enough to develop a theoretical model of a star, if the astrophysicist also knows exactly what thermonuclear processes produce energy and the mass and chemical composition of the model star. Then the equations of stellar structure (Focus 10.1) formulate the problem, and these are solved in a consistent way to find the physical conditions (temperature, pressure, and density, for instance) in the star from the center to the surface. This catalog of values for important physical properties, such as temperature and pressure, for a specified mass and composition is called a *star model* or a *stellar interior model*. (This is not a real star; it's just a list in numbers!) One pair of numbers in this catalog, the luminosity and temperature at the surface, specifies a point on an H-R diagram, the *theoretical* point for a model star of given mass and chemical composition. If a series of models are computed for a star at several different times in its life, values of the luminosity and surface temperature are produced for each time, and we can see how the point representing a model star moves on the H-R diagram; that is, we can plot its *evolutionary track*.

The construction of a star model requires much tedious calculation. High-speed elec-

FOCUS 10.1 THE EQUATIONS OF STELLAR STRUCTURE

To construct a mathematical model of a star, each physical principal is expressed as a mathematical equation. A computer then solves the whole set of equations to provide a table of values for the physical properties—temperature, pressure, density, etc.—at every point in the star from center to surface.

The equation expressing the condition of *hydrostatic equilibrium*, derived in Focus 9.1, is

$$\frac{\Delta P}{\Delta r} = -\rho \frac{GM_r}{r^2}$$

where ΔP is the change in pressure in distance Δr along the radius r, M_r is the mass interior to r, ρ is the density, and G is the gravitational constant.

The quantity M_r increases outward in the star. The amount by which it changes (ΔM_r) in distance Δr is just the mass in the spherical shell of inner radius r and outer radius $r + \Delta r$. The volume of that shell is its surface area ($4\pi r^2$) multiplied by its thickness (Δr), and its mass is the volume multiplied by the density (ρ). So $\Delta M_r = (4\pi r^2 \Delta r)\rho$ or

$$\frac{\Delta M_r}{\Delta r} = 4\pi r^2 \rho$$

To use the above equation requires the relation among pressure, temperature, and density—the *equation of state*. For an ordinary gas this relation is simple

$$P = nkT$$

where P is the pressure, n the number density, T the temperature, and k Boltzmann's constant. If the gas has density ρ and consists of particles of average mass m, the number density is $n = \rho/m$, so this equation can be written as

$$P = \frac{k}{m}\rho T$$

In extreme conditions of temperature or density, the gas may not be ordinary, and a different equation of state applies.

To express the condition of thermal equilibrium mathematically, we define the energy flowing across the spherical surface of radius r as L_r. (The energy flowing out from the surface of the star, at $r = R$, is the star's luminosity L, so L_r could be called the local luminosity.) The change in this quantity (ΔL_r) in distance Δr must equal the energy produced within the shell between radius r and $r + \Delta r$. Suppose we define ϵ as the energy generated per gram of matter. The amount of mass in the shell is $\Delta M_r = 4\pi r^2 \rho \Delta r$. The energy generated in the shell is this quantity multiplied by ϵ ($\Delta L_r = \epsilon \Delta M_r$), so

$$\frac{\Delta L_r}{\Delta r} = 4\pi r^2 \rho \epsilon$$

This is the *equation of thermal equilibrium*. The quantity ϵ is different for different nuclear reactions, but in general it will depend on temperature, density, and chemical composition. For example, at temperatures around 14 million K, ϵ for the PP chain is given by the expression

$$\epsilon = \text{constant} \times X^2 \rho T^4$$

where X is the fractional abundance by mass of hydrogen.

The derivation of the *equation of radiative transport* is too complex for this book, but we can describe and make plausible some of its major factors. First, we expect that the energy flowing through a spherical shell of radius r should be proportional to its surface area $4\pi r^2$. Second, we expect this flow to be proportional to the difference in temperature between the two sides of the shell, ΔT; the greater the temperature difference, the more energy will flow, just as in conduction of heat along a rod, the greater the temperature difference between the two ends, the more heat flows. Third, the thinner the shell, the easier it will be for energy to get through, so we expect the flow to be inversely proportional to the thickness (Δr). Fourth, we might expect that the more opaque the shell, the *less* energy will flow. The more matter in the shell, that is, the higher the density (ρ), the more opaque it will be. But, depending on the process producing the absorption or scattering of radiation, the opacity will also depend on other factors, such as composition and temperature. We lump all these other factors into one, called the *coefficient of opacity*, κ. We expect the energy flow to be inversely proportional to the product $\kappa\rho$. When the derivation is done in detail, some other factors enter, and the final result for the equation of radiative transport is

$$L_r = 4\pi r^2 \frac{4}{3} \sigma \frac{T^3}{\kappa\rho} \frac{\Delta T}{\Delta r}$$

where σ is the Stefan-Boltzmann constant.

In order to use this equation, we must be able to relate the coefficient of opacity to the temperature, density, and composition. This relation will be different for different processes. One approximate formula used in a certain range of temperatures and densities, known as Kramer's law, is

$$\kappa = \text{constant} \times Z(1 + X) \frac{\rho}{T^{3.5}}$$

where X is the fractional abundance of hydrogen and Z is the fractional abundance of heavy elements.

These equations involve the variables P, T, M_r, L_r, ρ, ϵ, and κ along with compositional parameters and physical constants. There are seven variables, and seven equations, so it is possible to solve them. A computer makes the work easier. The solution of the equations provides values for all the variables at each point along the radius r. This set of numbers is the mathematical model of the star.

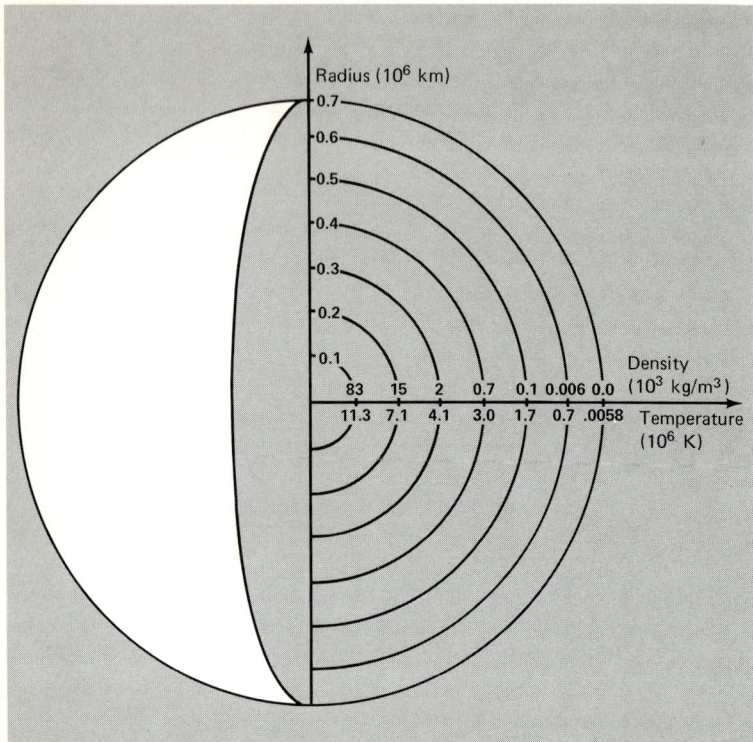

FIGURE 10.6 A theoretical model for the sun, for an age of 4.5 billion years. The values for the physical properties shown here have been calculated on an electronic computer. The central temperature is about 16 million K, the central density 1.58×10^5 kg/m^3.

tronic computers can complete the calculations for one star model in a few minutes. (See Figure 10.6 for a model for the sun.) A series of models used to describe the evolutionary sequence of one star may take hours of computer time.

The study of stellar evolution rests on the construction of physically reasonable models of stellar interiors. Model making is how astronomers get evolutionary tracks on the H-R diagram for stars of a specific mass and chemical composition. Models tell us how time enters into the H-R diagram for real stars. The goal of studying stellar evolution is to understand the physical cause of a star's evolutionary track on the H-R diagram.

Almost everything said in this chapter rides on the validity of stellar models. But one fly swims in the ointment: the solar neutrino experiment, which casts some doubt on our models for the sun (Section 6.6). Most of us ignore that problem, hoping that there is some good explanation for the missing solar neutrinos, such as a finite mass for the neutrinos. We do not think, or would not like to admit, that we have not understood the basic physics of stellar models.

Star models show that as a star evolves, its radius, temperature, and luminosity change in complicated ways. Such changes result in the change in a star's position on the H-R diagram. For example, if a star's surface temperature increases while its luminosity remains the same, the star's position on the H-R diagram moves horizontally from right to left, with no vertical change (line A in Fig. 10.7). If the luminosity increases while the surface temperature remains the same, the star's point on the H-R diagram moves vertically from bottom to top, with no horizontal change (line B in Fig. 10.7). Both motions on the H-R diagram represent changes in the physical properties of the star. (*Warning:* What moves on the H-R diagram is *not* the star itself, but a point *representing* the luminosity and temperature of the star.) Note in the first case that to keep the luminosity constant at a higher temperature requires that the star's surface area and so its radius decrease. In the second case, with no change in temperature, the star's luminosity can go up only if its surface area and so its radius increase. A star's surface temperature, luminosity, and radius change together.

One key aspect of the H-R diagram—the mass-luminosity law—can be understood fairly simply from the basic physical ideas determining the structure of a star. The condition of hydrostatic equilibrium tells us that a more massive star must have a higher pressure to support itself against gravity. For an ordinary gas a higher pressure means a higher central temperature. The higher the central temperature, the greater the temperature gradient, that is, the more rapidly the temperature must decrease with distance in order to reach a small value at the surface. Furthermore, the higher the temperature, the lower the opacity. The equation of radiative transport then tells us that the higher temperature star will have a greater flow of energy by radiation and hence a greater luminosity. The net result is that a higher mass star has a greater luminosity. (A more detailed mathematical derivation is given in Focus 10.2.) It is important to notice that nowhere in this discussion was it necessary to mention the source of energy in a star. The mass-luminosity law holds for many pre-main-sequence stars, which are obtaining energy from gravitational contraction, as well as for main-sequence stars, which are heated by nuclear burning.

10.4 ENERGY GENERATION IN STARS

Once an astrophysicist has established a model of a star, what mathematical gears must be turned to have the model *evolve*? The answer lies in the heart of the star. Here thermonuclear reactions cook lightweight elements into more complex ones by fusion. This change

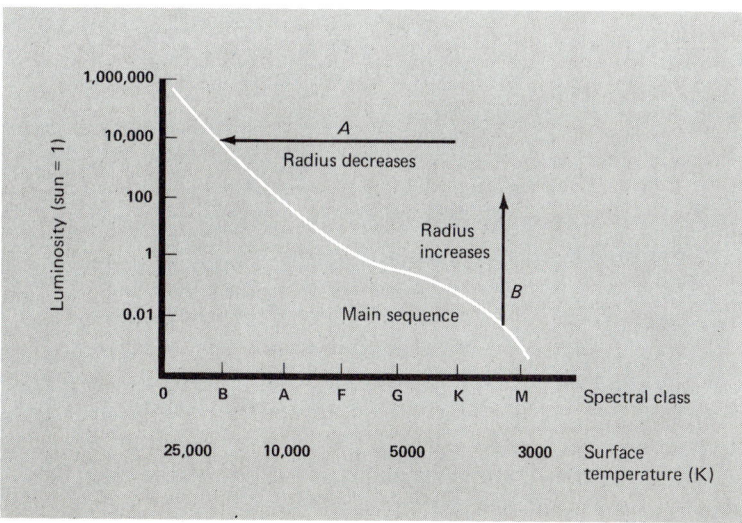

FIGURE 10.7 Hypothetical evolutionary tracks on an H-R diagram. If a star's luminosity remains constant while its temperature increases, its point moves horizontally from right to left (along line A). If its temperature remains constant while its luminosity increases, it moves vertically (along B). In the first case the radius decreases; in the second it increases.

FOCUS 10.2 PHYSICAL BASIS OF THE MASS-LUMINOSITY LAW

The equations of stellar structure (Focus 10.1) allow us to derive a *theoretical* mass-luminosity law. Start with the equation of hydrostatic equilibrium:

$$\frac{\Delta P}{\Delta r} = -\rho \frac{GM_r}{r^2}$$

Though this must hold for each shell in the star, let's solve it in a very approximate way, taking ΔP to be the difference between the central pressure (P_c) and the surface pressure (which is zero), and Δr to be the entire radius of the star, R. Then

$$\frac{P_c}{R} = \frac{GM\rho}{R^2} \quad \text{or} \quad P_c \propto \frac{M\rho}{R}$$

Now for an ordinary gas

$$P \propto \rho T$$

so

$$\rho T_c \propto \frac{M\rho}{R}$$

which gives

$$T_c \propto \frac{M}{R}$$

Let's do the same sort of thing with the equation of radiative transport:

$$L_r = 4\pi r^2 \frac{4}{3} \sigma \frac{T^3}{\kappa \rho} \frac{\Delta T}{\Delta r}$$

---

in chemical composition and its effects on a star's structure mark another major theme of stellar evolution.

The sun generates most of its energy by the proton-proton (PP) reaction (Section 6.6). Essentially, four hydrogen nuclei combine to form one helium nucleus, releasing 4.2×10^{-12} J per helium atom formed. A minor amount of the sun's energy comes from another reaction sequence, called the carbon-nitrogen-oxygen cycle (CNO cycle). In more massive stars this cycle becomes more important than the PP chain. Let's look at it in a little detail. The complete CNO cycle has six reaction steps (Fig. 10.8). (1) A proton (^1H) collides with a carbon nucleus (^{12}C) to convert it to a radioactive nitrogen isotope (^{13}N). (2) The nitrogen nucleus emits a positron and neutrino to become a carbon isotope (^{13}C). (3) A oton blasts into this particle and, with the emission of a gamma ray, turns it into stable nitrogen (^{14}N). (4) A proton bangs into the nitrogen nucleus to form a radioac-

Take ΔT to be the central temperature (T_c) minus the surface temperature (which is so small compared to T_c that it can be neglected). Then

$$L \propto R^2 \frac{T_c^3}{\kappa \rho} \frac{T_c}{R} \propto \frac{RT_c^4}{\kappa \rho}$$

The density (ρ) is the mass divided by the volume, so

$$\rho \propto \frac{M}{R^3}$$

and

$$L \propto \frac{R^4 T_c^4}{\kappa M}$$

Putting in the relation for T_c from the equation of hydrostatic equilibrium

$$L \propto \frac{R^4 (M/R)^4}{\kappa M} \propto \frac{M^3}{\kappa}$$

which is close to the observed relation, $L \propto M^{3.3}$. The difference has to do with the opacity and how it varies with temperature and density.

Note this derivation made use of only the equation of hydrostatic equilibrium, the equation of state for an ordinary gas, and the equation of radiative transport. We have *not* had to say anything about what it is that keeps a star hot—nuclear reactions or gravitational contraction.

tive oxygen isotope (^{15}O) and a gamma ray. (5) The oxygen isotope decays into a nitrogen isotope (^{15}N), a positron, and a neutrino. (6) A proton collides with the nitrogen isotope and splits it into ordinary carbon (^{12}C) and helium (^{4}He). Note that the carbon comes back unscathed; it acts only as a catalyst, helping to glue four protons together to make

^{12}C + ^{1}H → ^{13}N + gamma ray
^{13}N → ^{13}C + positron + neutrino
^{13}C + ^{1}H → ^{14}N + gamma ray
^{14}N + ^{1}H → ^{15}O + gamma ray
^{15}O → ^{15}N + positron + neutrino
^{15}N + ^{1}H → ^{12}C + ^{4}He

FIGURE 10.8 The carbon-nitrogen-oxygen (CNO) cycle. Note that the net result is the conversion of four hydrogen nuclei to one helium nucleus; the carbon entering the first step returns in the last.

helium. The net result of the CNO cycle is the same as the PP reaction: four hydrogen nuclei are converted to one helium nucleus, with the release of energy. Some 4.0×10^{-12} J for each helium nucleus formed goes to heating the star; an additional small amount goes off as neutrinos.

The PP chain and the CNO cycle contribute differently in different mass stars. Their rates depend on temperature in different ways, and the more massive stars have higher central temperatures. In the PP chain the colliding particles have relatively low charge (1 for a proton, 2 for a helium nucleus). A large fraction of the colliding particles have sufficient energy to overcome the repulsive force of these charges and stick together, and the rate of reaction varies as the fourth power of the temperature. But in the CNO cycle some of the reacting particles have higher charge (6 for carbon, 8 for oxygen), the repulsive force is greater, and only the small fraction of particles which have acquired high-energy can get close enough to stick together and react. The fraction of such high energy particles changes rapidly with temperature and the resulting rate of reaction is proportional to the twentieth power of temperature. This difference in temperature dependence of the reaction rates ($r_{PP} \propto T^4$, $r_{CNO} \propto T^{20}$) means that at the temperature of the sun the PP chain is dominant, but at slightly higher temperatures the CNO rate is higher. The two rates are equal at about 18 million K (Fig. 10.9). At 10 million K, $r_{PP}/r_{CNO} \approx (18/10)^{16} = 10^4$. At 25 million K $r_{PP}/r_{CNO} \approx (18/25)^{16} = 0.005$. So in the massive stars, which have higher central temperature, the CNO cycle is dominant, but in the less massive stars like the sun, which are cooler, the PP chain is more important.

These reactions take place only in the star's core, where temperatures are high enough to keep them going. Since a star has only a limited amount of hydrogen to burn, the core eventually is all converted to helium, and the CNO and PP reactions cease. What next? The core contracts and heats up. When the core's temperature gets up to roughly 100 million K, another reaction can take place: the *triple-alpha reaction*, so named because three helium nuclei (also known as alpha particles) fuse to form one carbon nucleus, with the release of energy (Fig. 10.10).

What happens when the helium runs out? The core contracts and heats up again. If the temperature increases enough, carbon can be fused into heavier elements. Such processes require extreme temperatures, at least 600 million K. Iron, the most stable of all nuclei, ends the sequence of fusion reactions; to form elements heavier than iron, energy must be added and absorbed. Such reactions occur only under special conditions. (When

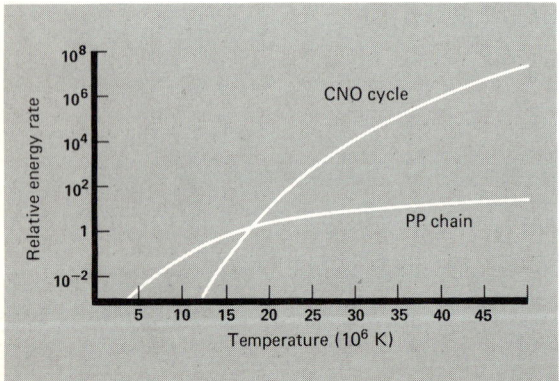

FIGURE 10.9 A comparison of the energy generation rates of the PP chain and the CNO cycle over a range of temperatures typical in the cores of stars. Note that the CNO cycle dominates at temperatures more than 18 million K. The energy generation has been calculated assuming solar (Population I) abundances. [*Adapted from a diagram by D. Clayton.*]

FIGURE 10.10 The triple-alpha process, a high-temperature fusion reaction.

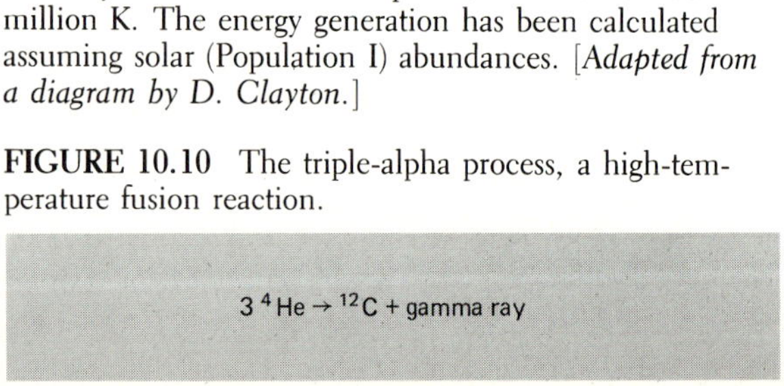

$$3\ ^4He \rightarrow\ ^{12}C + \text{gamma ray}$$

they do, they soak up energy from the core.) The steady climb from hydrogen to iron in fusion reactions in stellar cores is called *nucleosynthesis*. This nuclear cementing process makes heavy elements in a universe that otherwise would consist only of hydrogen and helium made in its Big Bang origin (Chapter 24).

You may be wondering why a star's temperature goes *up* when fusion reactions *stop*. The cause is gravitational contraction. A star is always losing energy to space by radiation. If that energy is not replaced by nuclear reactions, it must come from somewhere else, namely from gravitational contraction. As the star contracts, some of the gravitational potential energy of its particles transforms into kinetic (thermal) energy and some is radiated away. So the temperature goes up until the ignition temperature of the next set of fusion reactions is reached. Fusion turns on again, and the core contraction stops. During a star's life short periods of gravitational contraction alternate with long spells of fusion burning. (*Note*: It is typically the *core* which is contracting, not necessarily the star as a whole.)

10.5 THEORETICAL EVOLUTION OF A SOLAR-MASS STAR

Gravity instigates the birth of a star. The details of the subsequent evolution are controlled by the star's mass. If the mass is less than about 0.1 solar mass, the central temperature never reaches the 10 million K needed to start the PP reaction. (This happened to the planet Jupiter.) If the mass is greater than approximately 100 solar masses, the outward force of radiation pressure exceeds that of gravity, making the star so unstable that it blows itself apart (or perhaps never completely forms). Between roughly 0.1 and 100 solar masses, a stable main-sequence star can form. Let's look at the evolution of a star like the sun, a 1-solar-mass star, with the sun's chemical composition.

Evolution to the Main Sequence

Let's briefly review a solar-mass star's pre-main-sequence evolution (Sections 9.3 and 9.4). The difference here is that we'll look at the star *without* its surrounding cloud of gas and dust.

Gravity pulls together a spherical interstellar cloud of gas and dust. The cloud's density increases; collisions between particles occur more frequently. During the collapse the cloud heats up. Eventually the temperature reaches about 2000 K, and the collapse slows down (point 1 in Fig. 10.11).

The collapsing cloud makes its debut as a protostar. It has a larger radius than it will as a main-sequence star, and the surface temperature is lower (point 2). However, the protostar has a higher luminosity (point 3) than it will when it reaches the main sequence. The luminosity is higher despite the lower temperature because the protostar is larger and so has more surface area to radiate energy (Section 8.3).

Eventually, the rising temperature and density raise the pressure sufficiently to bring the star into hydrostatic equilibrium (point 4 in Fig. 10.11). Now it is a *pre-main-sequence star*. At this stage of its life the star's temperature is so low that its opacity is relatively high (even though its density is low). Convection rather than radiation transports the energy outward. So the star is completely convective—a huge, bubbling ball of gas. The effective transport of energy by convection makes the star very luminous.

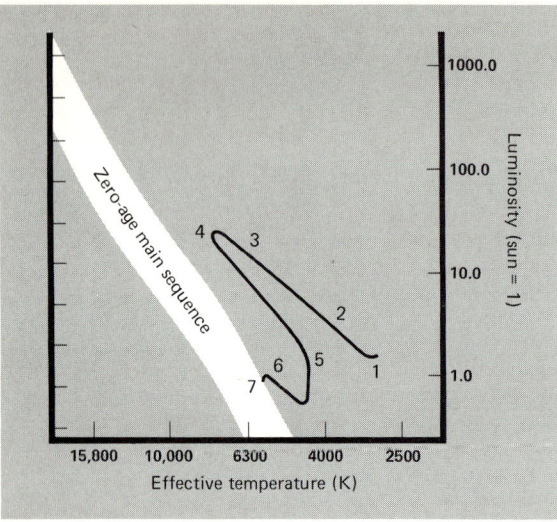

FIGURE 10.11 An evolutionary track for one model of the pre-main-sequence evolution of a solar-mass star. [*Based on theoretical calculations by R. Larson.*]

As the star radiates energy, it contracts in order to maintain its hydrostatic equilibrium. If the star did not contract, the loss of energy would decrease the internal temperature and therefore the pressure, throwing it out of equilibrium. Gravity takes the upper hand and squeezes the star. But this contraction converts gravitational potential energy into thermal and radiative energy. So the pre-main-sequence star shines by slowly shrinking.

As the star shrinks in size, its luminosity decreases (points 4 to 5 in Fig. 10.11). Its point on the H-R diagram moves downward. Meanwhile, the core continues to heat up. As it does, its opacity decreases. Eventually, the opacity drops enough for radiation to transport energy more efficiently than convection. The zone of radiative transport starts at the core and slowly creeps outward as the inner layers heat up. When a substantial fraction of the star's interior carries energy outward by radiation, its path on the H-R diagram kinks sharply to the left (points 5 to 6). The core eventually heats up to a few million degrees Kelvin, high enough to start thermonuclear reactions.

When the star gets most of its energy from thermonuclear reactions (PP reactions in the case of the sun) rather than gravitational contraction, it achieves full-fledged stardom (point 7 in Fig. 10.11). It no longer contracts to provide energy; the heat from fusion reactions keeps it in hydrostatic equilibrium. The star is now called a *zero-age main-sequence (ZAMS)* star. It settles down to the longest stage in its life, calmly converting hydrogen to helium in its core. Most of the interior transports energy by radiation; only the outer region of the envelope is convective. The total time elapsed from initial collapse to arrival as a star on the main sequence (from point 1 to point 7) is only 50 million years.

Evolution on the Main Sequence

The star's position on the main sequence depends mainly on its mass. The more massive the star, the hotter and more luminous it is; the less massive the star, the cooler and less luminous it is. Recall the mass-luminosity law (Section 8.6). The main sequence consists of a series of stars of different mass (but similar chemical composition) from the upper left-hand corner (O stars with high mass) to the lower right-hand corner (M stars

10.5 THEORETICAL EVOLUTION OF A SOLAR-MASS STAR

with low mass). A star, like the sun, spends about 80 percent of its total lifetime on the main sequence as it slowly transforms its hydrogen core to helium.

How the star evolves further also depends on its mass. Because massive stars have higher luminosities, the hydrogen in them must burn faster than in low-mass stars. Because they use their fuel at faster rates, massive stars spend less time on the main sequence, even though they have more fuel to burn. Such stars are spendthrifts compared with the miserly energy generation of a star like the sun.

In Section 8.6 we derived an approximate formula for a star's main-sequence lifetime: $t_{MS} = 10^{10}$ yr $(M/M_{sun})^{-2.3}$. For example, the sun's life on the main sequence lasts about 10 *billion* years, while the same phase of life for a 15-solar-mass star lasts 20 *million* years, only two-thousandths as long.

The main-sequence phase ends when almost all of the hydrogen in the core has been converted to helium. During this time the temperature in the core increases gradually and the star contracts slightly. This results in a greater flow of energy to the surface, and the star's luminosity increases (points 1 to 2 in Fig. 10.12; see also Table 10.1).

Evolution off the Main Sequence

When the hydrogen in the core is used up, the thermonuclear reactions cease there. However, they keep going in a shell around the core, where fresh hydrogen still exists. At the end of fusion reactions in the core, gravity takes over and the core contracts. This heats up the layer of burning hydrogen, so the reactions go faster and produce more energy. The luminosity increases. But the layer of burning hydrogen heats up the surrounding part and causes it to expand. So the radius of the star increases and its surface temperature decreases (points 2 to 3 in Fig. 10.12). The decrease in temperature increases

TABLE 10.1 Evolutionary Phases of a Solar-Mass Star

H-R Position (Fig. 10.12)	*Physical Processes*
1	Hydrogen core burning begins (zero-age main-sequence star).
2	Hydrogen core burning ceases; hydrogen shell burning begins.
3	Hydrogen shell burning continues; convection dominates the energy transport.
4	Helium flash occurs; helium core burning begins (red giant).
5	Helium core burning continues along with hydrogen shell burning.
6	Thermonuclear reactions in core end; helium and hydrogen shell burning continues.
7	Expansion and contraction throw off outer layers.
8	Planetary nebula.
9	All thermonuclear reactions stop white dwarf; slow cooling.

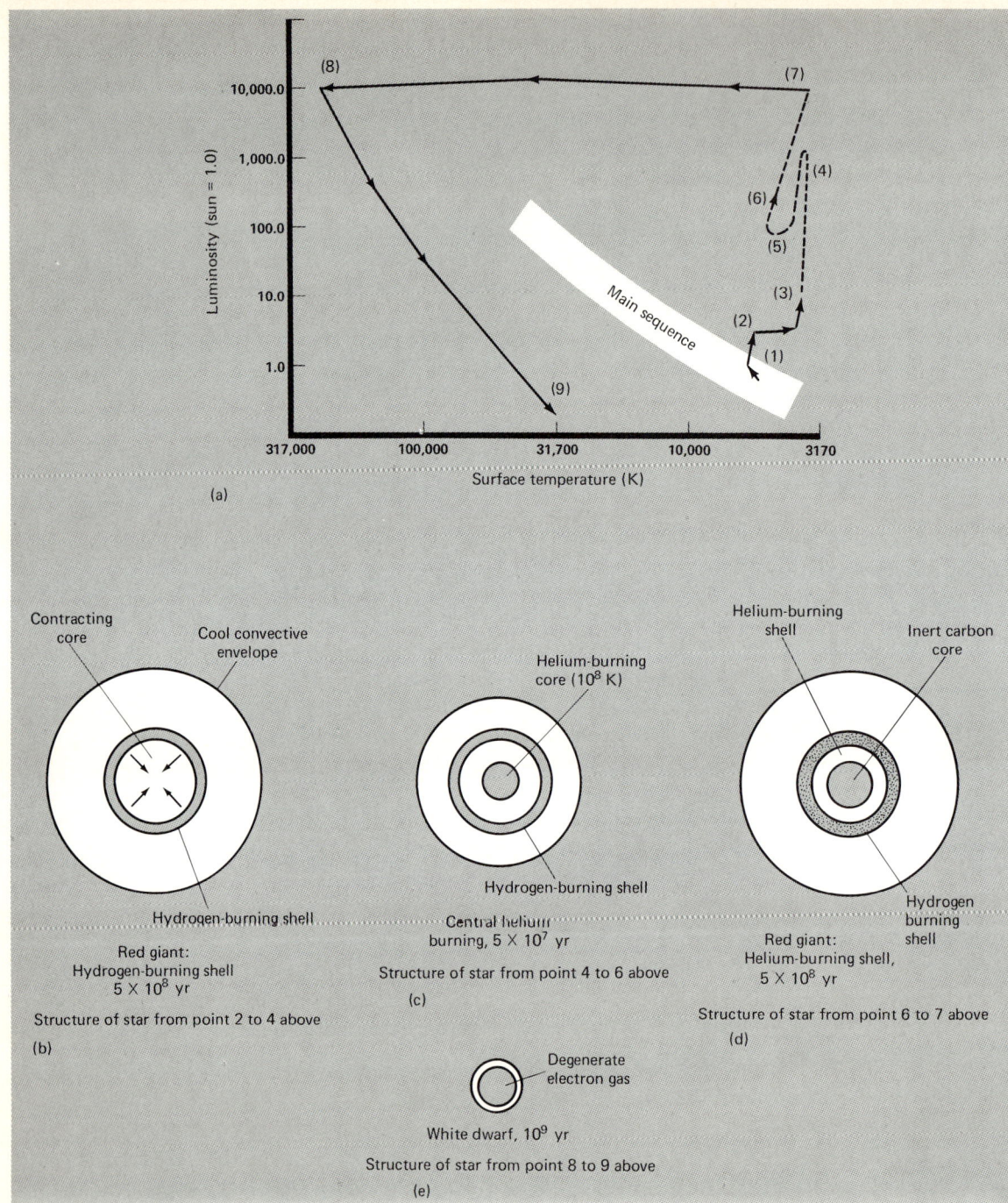

FIGURE 10.12 A theoretical evolutionary path of a solar-mass star off the main sequence (point 1). As the hydrogen is depleted in the core, the luminosity increases (1 to 2). When the core runs out of hydrogen, it burns hydrogen in a shell (2 to 4). Gravitational contraction heats the core until it gets hot enough to burn helium (4). When the core runs out of helium (6), it burns hydrogen and helium in a shell and becomes a red giant (6 to 7). The star throws off its outer layers (7 to 8) and becomes a white dwarf (8 to 9).

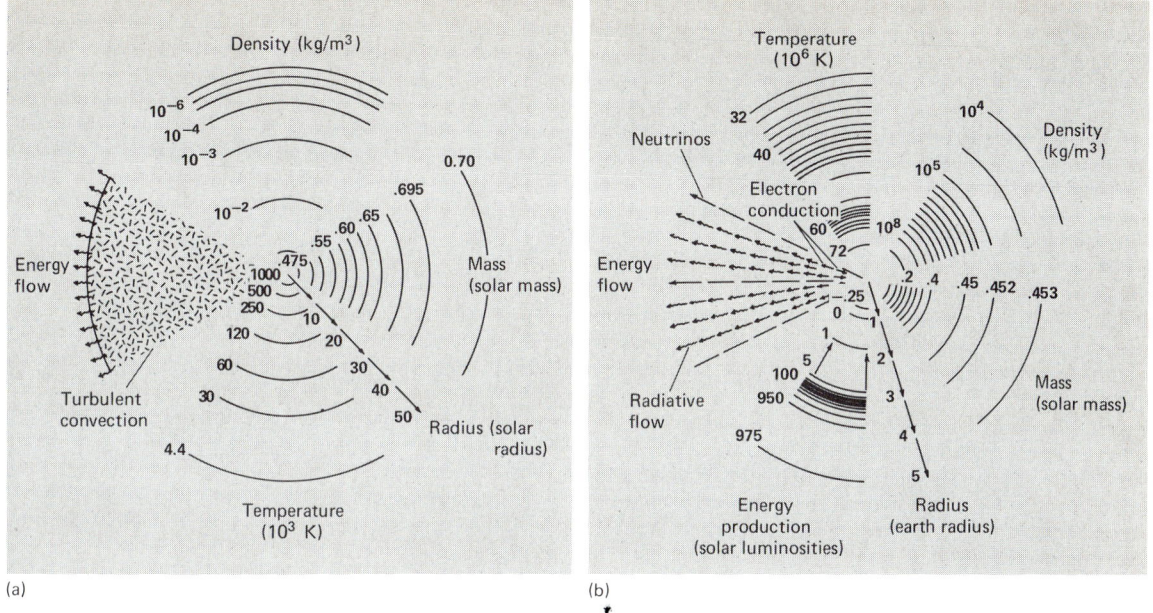

FIGURE 10.13 (a) Cross-section of the physical properties of the envelope of a 0.7-solar-mass red giant. (b) The same for the core of the red giant. [*Based on calculations by I. Iben, Jr., Publications of the Astronomical Society of the Pacific, vol. 183, p. 697, 1971.*]

the opacity, so that eventually convection carries the energy outward in most of the star's envelope (points 3 to 4).

The star now becomes a red giant; its position on the H-R diagram moves to the region of lower surface temperatures and higher luminosities. In an additional 500 million years after the main-sequence phase, a solar-mass star ends up with a luminosity of about 1000 solar luminosities, a surface temperature of about 3000 K, and a radius of about 100 solar radii. The sun at this stage would engulf Mercury!

A red giant star is really a strange beast compared with a main-sequence star. For example, a 0.7-solar-mass red giant (Fig. 10.13a), according to computer models by Icko Iben, has a radius of 50 solar radii, a surface temperature of 4400 K, and a surface density of only 10^{-6} kg/m^3. Most of the red giant's mass is concentrated in a tiny core of only 5 earth radii (Fig. 10.13b). At the core's center the temperature is 70 million K and the density is 0.51 *billion* kg/m^3. At such a high density the matter no longer behaves like an ordinary gas. It has become *degenerate*.

Degenerate Gases

For most of their lives most stars consist of ordinary gases. But when matter is packed to very high densities (greater than 10^8 kg/m^3), it no longer behaves in ordinary ways. In a normal gas the particles are widely separated and rush helter-skelter into one another and rebound away. In highly compressed material the matter is so jammed together that the electrons on the outside of the atoms are, in a sense, touching each other. The nuclei no longer can hold electrons in their usual energy levels, and the electrons move among the nuclei. But there's not much space for moving about.

Electrons abide by a quantum property called the *Pauli exclusion principle*. It states that no two electrons can be together in exactly the same energy state. Consider an electron confined to a very small volume, like a small box. Imagine adding another electron to the box. The electron already in it probably occupies the lowest energy state possible. So the next electron must occupy the second level available, and so on. In a low-density state many energy levels are available for occupation; a high-density gas has far fewer levels, which get filled up quickly.

What happens if we try to cool this dense gas by letting electrons give up kinetic energy? They can't lose very much kinetic energy, for only high energy levels are open. So no heat can be extracted from these electrons (in a sense, their temperature is zero). Yet they exert a large pressure because they move with high speeds.

A gas in this state is called a *degenerate gas* and the pressure from the exclusion principle in action is called the *degenerate gas pressure*. Unlike an ordinary gas where the pressure is directly proportional to temperature, degenerate gas pressure is independent of the temperature. The equation of state is

$$P = \text{constant} \times \rho^{5/3}$$

The fact that the pressure does not depend on temperature means that a star can preserve hydrostatic equilibrium in a degenerate core even though no nuclear reactions are going on there.

Electrons become degenerate at densities of about 10^8 kg/m^3. Such densities occur in stellar cores after main-sequence hydrogen burning and also in white dwarfs (Chapter 11). When electrons are degenerate, they conduct heat very efficiently and temperature variations are quickly smoothed out. So degenerate cores have the same temperature throughout. A high enough temperature can relieve the electrons of their degenerate condition; this requires a temperature of some 350 million K.

The Helium Flash

Let's return to the evolution of a red giant. As the bloated star attains its red giant status (point 4 in Fig. 10.12), the core temperature, which has been steadily increasing as the core contracts, hits the minimum necessary to start helium burning by the triple-alpha process. Because the helium core is degenerate, it conducts heat rapidly. Once part of it ignites in the triple-alpha reaction, the heat generated by the fusion quickly spreads throughout the core. The rest of the core quickly ignites. If the core were an ordinary gas, this explosive ignition would expand it from the rapid increase in temperature and pressure. But since the core is degenerate, increased temperature does *not* increase the pressure, and so the core does not expand. Instead, the increased temperature runs up the rate of the triple-alpha process, generating more energy, further increasing the temperature, and so on. This out-of-control process in the core is called the *helium flash*. When the core temperature finally reaches about 350 million K, the electrons become nondegenerate. The core expands and cools. The whole process of helium core ignition in the helium flash takes place in a very short time—perhaps only a few minutes. It may reach a peak of some 10^{11} solar luminosities!

10.5 THEORETICAL EVOLUTION OF A SOLAR-MASS STAR

But no one has ever seen this helium flash in a star, for all this action takes place deep in the core. The surface is hardly affected. The radius and luminosity decrease a little, and the star's point on the H-R diagram moves slightly downward and to the left. The star quietly burns helium in the core and hydrogen in a layer around the core (points 5 to 6 in Fig. 10.12). This phase of helium core burning is the analog to the star's main-sequence phase of hydrogen core burning.

Evolution to the End

Eventually the triple-alpha process converts the core to carbon. The reaction stops in the core, but continues in a layer around it. This situation—core shut down, but the thermonuclear reactions going on in a shell—resembles the situation when the star first evolves off the main sequence. The physical processes force the same evolution; the burning layer makes the star expand (points 6 to 7 in Fig. 10.12). The electrons in the core—this time carbon rich—become degenerate again.

Because the rate of the triple-alpha reaction is very sensitive to changes in temperature, the helium shell burning causes the star to become unstable. Here's how. Suppose the star contracts a little. The temperature and energy production in the layer increase; the pressure also increases. However, the increase in pressure more than compensates for gravity; the outer parts of the star expand. The expansion decreases the temperature, the pressure, and, more dramatically, the energy generation rate. Gravity takes command, the star contracts, the energy generation increases a lot, the star expands, and the cycle repeats. The star pulsates slowly, once every tens of thousands of years.

The pulsations gradually grow larger. A final violent one ejects the cool outer layers of the star (point 7 in Fig. 10.12). A hot core is left behind (point 8 in Fig. 10.12). The expelled shell forms a *planetary nebula* (Fig. 10.14), and the leftover core becomes its

FIGURE 10.14 A typical planetary nebula (NGC 7293 in Aquarius). The central blue star was once the core of a red giant. The nebula, which looks like a ring but actually forms a spherical shell, was the envelope of the red giant. [*Courtesy Palomar Observatory, California Institute of Technology.*]

central star. The nebula keeps expanding until it dissipates in the interstellar medium. The core, because it is degenerate and cannot contract and heat up, never reaches the ignition temperature of carbon burning. In about 75,000 years it forms a white dwarf star (point 9 in Fig. 10.12), composed mostly of carbon. Without energy sources the white dwarf cools to a black dwarf, the dark culmination of 10-billion year biography. The degenerate electron gas pressure supports it against gravity even as it cools down.

Note: If stellar evolution, especially the "motion" of a star on the H-R diagram, seems a bit confusing, keep in mind that all the turns happen for some physical reason. Look again at the track of a 1-solar-mass star in Fig. 10.12 and the phases listed in Table 10.1. Why does the star make that sharp turn to the right at point 2? That's where it ran out of hydrogen in its core. What about at point 4, where the star suddenly stops going up the red giant branch and moves down (fainter) and slightly to the left (hotter)? It started burning helium at this point, so it had a whole new energy source to play with and had to rearrange its entire internal structure to allow for that. Your turn now. Why do you think the star makes that sharp turn back up to being much brighter and slightly cooler at point 6? What happens at point 7 so that the star gets much hotter afterwards, but doesn't change in luminosity? (And what must be happening to the star's radius while this is going on?)

Stars of Less Than Solar Mass

The evolution of stars of lower mass than the sun is similar to that for the sun, with two exceptions. First, few stars less massive than the sun have had time to evolve off the main sequence. A 0.74-solar-mass star, for example, will have a luminosity about 0.37 that of the sun, according to the mass-luminosity law and hence a main-sequence lifetime of 20 billion years, longer than most estimates for the age of the universe (recall that $t_{MS} = 10^{10}$ yr $\times (M/M_{sun})^{-2.3}$).

Second, if the mass of a star is less than about 0.1 solar mass, it will not even reach the main sequence. Gravity is weak in such a low-mass star, and gravitational contraction does not heat it very effectively. Before it gets hot enough to start nuclear reactions, the density rises so high that the matter becomes degenerate. Then the pressure of the degenerate electrons supports the star and keeps it from contracting any further. If gravitational contraction is prevented from heating the star, the nuclear fires can never be lit, and the star simply cools off to become an invisible black dwarf or (if the mass is *very* small) a planet.

10.6 THEORETICAL EVOLUTION OF A 5-SOLAR-MASS STAR

To examine the history of a star much more massive than the sun, we've picked a 5-solar-mass star because these are fairly common, compared to, say, 50-solar-mass stars. (Regulus, the bright star that marks the heart of Leo the Lion, is a 5-solar-mass star; see the spring constellation chart.) A larger mass does not change the general flow of stellar evolution, but the details do change. Massive stars differ in their evolution from less massive stars because they reach higher temperatures in their cores. The greater temperatures have important consequences. (1) While on the main sequence the star burns hy-

10.6 THEORETICAL EVOLUTION OF A 5-SOLAR-MASS STAR

drogen by the CNO cycle. (2) The main-sequence lifetime is shorter. (3) The higher temperatures kindle carbon and heavier element fusion in the core. (4) The helium-rich core does not become degenerate.

Evolution to and on the Main Sequence

During most of its pre-main-sequence life a 5-solar-mass star evolves along a track of roughly constant luminosity. The star is hot enough so that its opacity is low enough for radiative transport to be more effective than convection. The conversion of gravitational energy powers the star; it contracts and heats up. As it does, its opacity falls and energy flows out more easily. So its luminosity increases a bit as the star contracts and its core temperature climbs. In about 600,000 years the star's point on the H-R diagram moves almost horizontally to the left (Fig. 10.15).

The PP reaction ignites first in the core. Then when the core's temperature rises to about 20 million K, the CNO cycle produces more energy each second than the PP reaction. The consumption of hydrogen causes the core to shrink, its temperature to go up, and thus its luminosity to increase. Stoked by the high core temperature, the CNO cycle uses up the core's hydrogen in about 60 million years, while the star's luminosity becomes twice as large (points 1 to 2 in Fig. 10.16).

Compared with PP reactions, the CNO cycle depends more sensitively on the temperature. So massive stars have very concentrated energy sources compared with those that are fired by PP reactions. This strong concentration results in the core transporting energy by convection. But the envelope still carries energy by radiation. As the star evolves on the main sequence, its convective core shrinks in size.

Note that a 5-solar-mass star on the main sequence is more luminous and hotter than a 1-solar-mass star (see Table 10.2). Regulus, for example, is a B star.

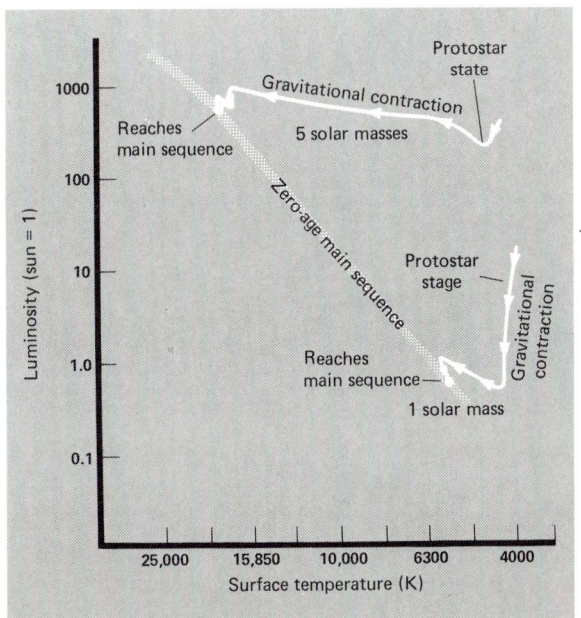

TABLE 10.2 Masses of Stars and Their Main-Sequence Type

Mass (sun = 1)	Main-Sequence Types
20–60	O stars
4–20	B stars
0.5–4	A, F, G, K stars
0.1–0.5	M stars

FIGURE 10.15 A comparison of the pre-main-sequence evolution of a 1-solar-mass star and a 5-solar-mass star. Note that the 5-solar-mass star approaches the main sequence at roughly constant luminosity. [Based on work by I. Iben, Jr.]

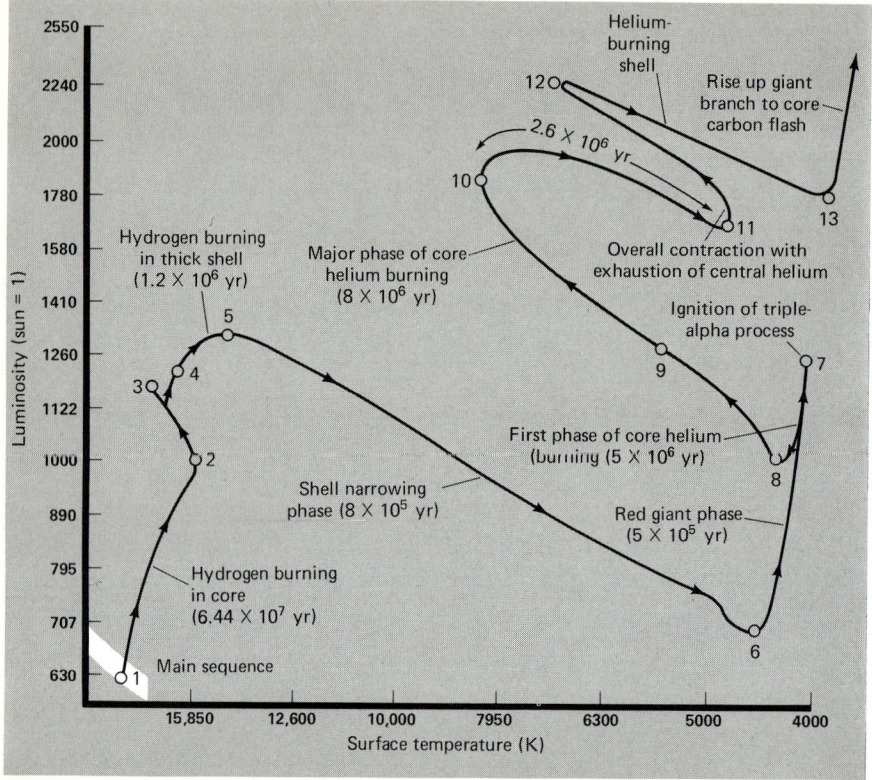

FIGURE 10.16 A theoretical evolutionary track for a 5-solar-mass star. The times given are those for the model star to evolve between the indicated points. Note that the star becomes a red giant every time it burns in a shell or shells around the core. Note also the alternation among core burning, shell burning, and gravitational contraction to reignite the core's fusion reactions. The details of the evolution after point 13 are not clearly known. [Based on work by I. Iben, Jr.]

Evolution to the End

When the central hydrogen fusion fires are exhausted, the star contracts. New hydrogen ignites in a shell around the burnt-out core (point 5 in Fig. 10.16). The luminosity increases but the radius expands, so the surface temperature drops. The star becomes a red giant. The core contracts until it gets hot enough to ignite the triple-alpha process (point 7). In contrast to a solar-mass star, the core has *not* become degenerate, so no helium flash occurs, just a relatively gentle triple-alpha ignition. (Stars greater than about 2 solar masses do not develop degenerate helium cores simply because they do not become dense enough before helium ignition takes place.)

The star burns helium in its core for about 10 million years. When the helium runs out, the star again contracts (point 11 in Fig. 10.16). Now fresh helium falls onto the core to make a thin, helium-burning shell around the core. The star expands again (beyond point 13). Note that whenever fusion reactions take place in a shell rather than in the core, the star expands in size.

We're really not sure about the details of what happens next. We expect that a 5-solar-mass star will develop a degenerate carbon core of about 1.4 solar masses. When the core gets hot enough to ignite carbon burning, it should do so explosively in a carbon flash. This reaction may detonate the carbon in the core so swiftly that the star blows apart. Such a cataclysmic detonation may result in a *supernova*. The outer layers of the star blast into space. The core is crushed to immense densities. Whatever the physical

10.7 THEORETICAL EVOLUTION OF VERY MASSIVE STARS

reasons for the detonations, a supernova explosion marks the end of the life of a massive star (more in Chapter 11).

10.7 THEORETICAL EVOLUTION OF VERY MASSIVE STARS

What about very massive stars, the O stars? Do they evolve in the same manner as the 5-solar-mass star? For 10- to 20-solar-mass stars, probably yes.

Here's a general outline of such a star's evolution (Fig. 10.17). Start with a 20-solar-mass star on the main sequence (point 1 in Fig. 10.17). When the core runs out of hydrogen, it shrinks and hydrogen burning starts in a shell (points 2 to 3). The star increases somewhat in luminosity and surface temperature. As hydrogen shell burning continues, the core contracts and the envelope expands (points 3 to 4). Helium burning soon begins in the core (point 4). When the helium burning in the core ends, fusion continues in a double shell—an inside one burning helium, the outside one hydrogen (point 5). With the establishment of a shell-burning structure, the star expands even more and becomes a red giant (points 5 to 6). The luminosity of these stars is so large, we should really call them *supergiants*.

As the star develops its red giant properties, the envelope forms a deep convective zone. Why? As the star's outer layers cool off, their opacities increase, and convection becomes the efficient method to transport energy.

The rest of the star's evolution is not reliably known in any detail. At some point (perhaps 6 in Fig. 10.17), carbon burning commences in the core. Then, as before, the core expands and the envelope contracts; the star evolves to the left and downward on the H-R diagram (point 7). When the core's carbon is depleted, burning continues in three

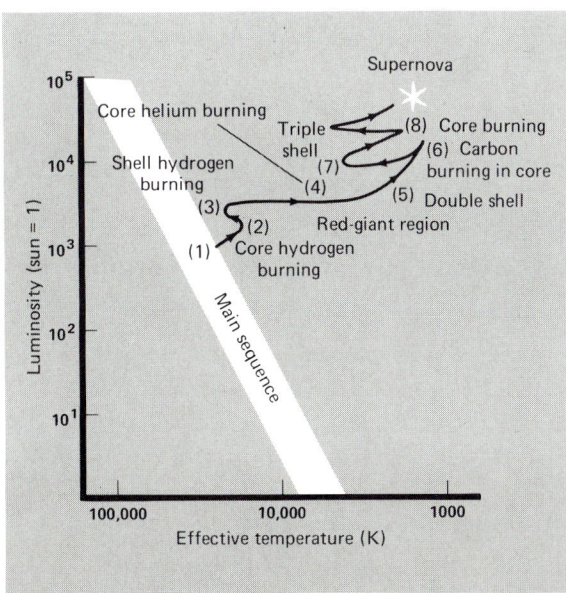

FIGURE 10.17 A very schematic evolutionary track for a 20-solar-mass star off the main sequence to its demise in a supernova.

shells—carbon, helium, and hydrogen (from inside out). Shell burning causes the star to expand again (point 8).

The star's evolutionary track keeps dancing back and forth on the H-R diagram as core sources turn on (motion to the left and down) and turn off so burning takes place in shells (motion to the right and upward). At each stage more mass is added to the core. In all these stages much energy is lost via neutrinos, which increases the rate of evolution. Eventually, the star becomes a supernova.

With this outline and summary in mind let's look at recent theoretical work on the evolution of *really* massive stars—50 to 100 solar masses. This work points out how one effect, which is usually ignored, dramatically changes the evolution of certain stars. That effect is *mass loss*. The sun loses mass, at a rate of about 10^{-14} solar mass per year, by the solar wind (Section 6.5). Other stars are known to lose mass at much greater rates. These outflows are called *stellar winds*. Red giants, for instance, blow off their envelopes at rates of 10^{-7} to 10^{-6} solar mass per year. Massive O stars, according to ultraviolet observations from the Copernicus satellite made by Ted Snow and others, also have stellar winds. Their rates are not yet worked out, but the best estimates are also 10^{-7} to 10^{-6} solar mass per year for the strongest winds.

Do you see the complication? What primarily determines a star's evolution? Its mass. The evolution described at the beginning of this section and in the previous two sections assumes that a star's mass remains the same throughout its life. For our sun, at least while on the main sequence, that's true. At its present loss rate it would take about 5×10^{13} years for the sun to blow off half of its mass. Contrast this to an O star with a stellar wind of 10^{-6} solar mass per year. Its main-sequence lifetime is a few million years, and in that time it will lose a few solar masses of material!

Workers at the Free University of Brussels have constructed computer models of massive stars with large mass loss. These models show that such stars lose about 50 to 60 percent of their initial mass by the end of their main-sequence lifetime. For example, a star which begins its main-sequence life with 100 solar masses ends up, 3 million years later, with only 42 solar masses. The outer layers of the stars are stripped off—so much, in fact, that the core is revealed and the products of the CNO cycle (Fig. 10.9; such as extra nitrogen) lie exposed at the surface. Such stripped cores may never become red giants because the layers above the core, where shell burning takes place, have been removed. (The peculiar objects known as *Wolf-Rayet stars*, which are hot stars with strong emission lines in their spectra, may be examples of these stars. They have abnormally large abundances of nitrogen and carbon and are losing mass.)

The evolutionary tracks that result from these calculations (Fig. 10.18) start with the stars on the initial main sequence. Take the 100-solar-mass star as the extreme case. From the initial main sequence (point 1 in Fig. 10.18), the star evolves to slightly lower luminosities and lower effective temperatures (from point 1 to 2). After 2.93 million years, the star has only 43 solar masses, a luminosity of 8.6×10^5 solar luminosities, a surface temperature of 4.3×10^4 K, and a radius of 16.7 solar radii. The evolutionary track then swings blueward; 60,000 years later (point 3) the star has crossed the initial main sequence! At this point it has a mass of 42 solar masses, a luminosity of 9.7×10^5 solar luminosities, a surface temperature of 5.6×10^4 K, and a radius of 10.5 solar radii. The evolutionary track then swings redward until helium ignition occurs (point 4), 2.99 million years from the initial main sequence.

Note that all the star models show this red-to-blue zig-zag in their evolution. And all

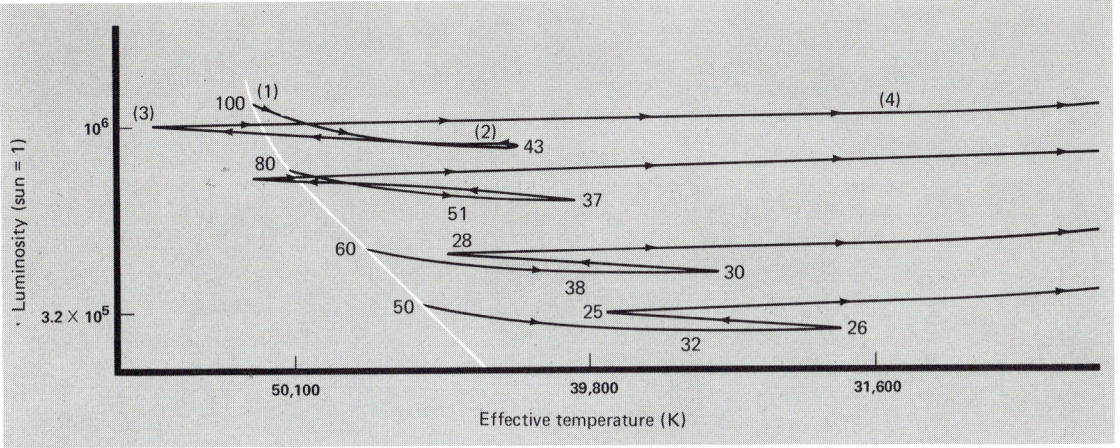

FIGURE 10.18 Theoretical evolutionary tracks for 50-, 60-, 80-, and 100-solar-mass stars undergoing mass loss as they evolve. The decreasing mass is indicated along the track. [*Based on calculations by C. de Loore, J.-P. De Greve, and D. Vanbeveren.*]

lose substantial amounts of their original mass. The lesson: In making evolutionary models of very massive stars, mass loss *cannot* be ignored.

10.8 CHEMICAL COMPOSITION AND EVOLUTION

So far we've focused on how a star's mass affects its evolution. What influence does its chemical composition have? For example, if a star model had a much smaller fraction of heavy elements than the sun—0.01 percent instead of 1 percent—would its evolution track take much different gyrations?

The answer is no. According to computer calculations the general trend is the same: evolution off the main sequence to the red giant region, a helium flash, then a move to the left and downward on the H-R diagram during helium core burning, a second trip to the red giant region, ejection of outer layers to create a planetary nebula, and then the final plunge to the white dwarf region.

Though the overall track is the same, these stars exhibit a significant difference in the position on the H-R diagram where they bide their time during core helium burning. Let's examine what happens to a star with a heavy element abundance of 0.01 percent that leaves the main sequence with mass of 0.7 solar mass (Fig. 10.19). Almost 16 billion years after the start of hydrogen core burning, PP reactions have consumed the core's hydrogen fuel. Shell burning takes over the energy production, and the star rises up to the red giant region, at first slowly and then swiftly. While a red giant, the star blows off some mass by a strong stellar wind. After the helium flash, the star settles down to core helium burning; it then has about 0.625 solar mass because of the mass lost via the stellar wind.

The star's interior at this stage looks much like that of other red giants: a dense core engulfed in a tenuous envelope. The star has a surface temperature of about 7300 K, a radius of 4 solar radii, and a luminosity of 43 solar luminosities. The energy production

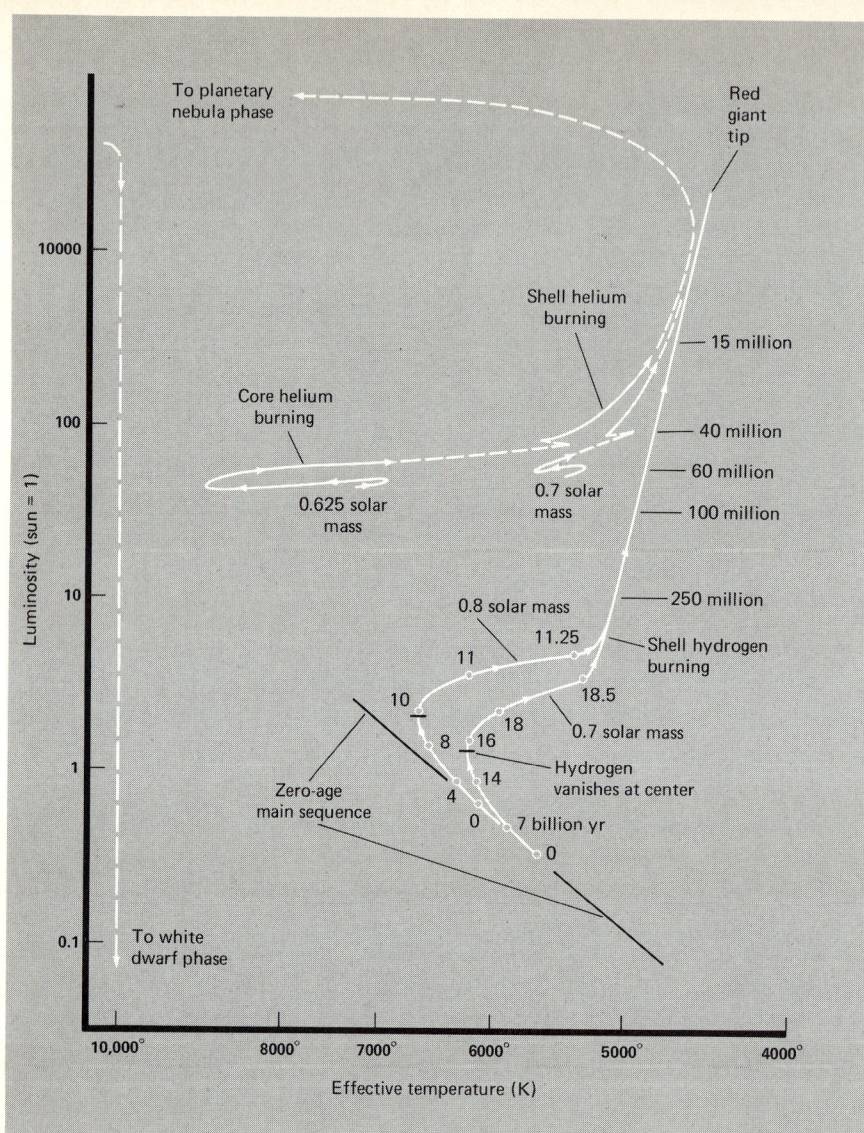

FIGURE 10.19 Theoretical evolutionary tracks for 0.8- and 0.7-solar-mass stars of low heavy element abundance (0.01 percent) as they move off the main sequence to become red giants. These stars lose mass during their red giant phase, and so have masses of 0.7 and 0.625 solar mass during their core-helium-burning phase, just before they eject their envelopes to make planetary nebulas. [*Adapted from a diagram by I. Iben, Jr., Publications of the Astronomical Society of the Pacific, vol. 83, p. 697, 1971.*]

goes on in a hydrogen-burning shell and a helium-burning core; there the density is roughly 10^7 kg/m³ and the temperature about 100 million K. As this star evolves, its luminosity stays roughly constant and its surface temperature changes, first to higher temperatures, then cooler ones before the core burning stops. The star's evolutionary track makes a zig-zag on the H-R diagram.

Helium-core-burning stars of the same low abundance of heavy elements but different masses occupy a locus on the H-R diagram which astronomers refer to as the *horizontal branch* (Fig. 10.20). The less massive stars have somewhat lower luminosities than those with greater mass, but there is a much greater variation in surface temperatures: from roughly 14,000 K for a 0.56-solar-mass star to some 5000 K for a 0.8-solar-mass star. (Note that the larger masses have lower surface temperatures.) Hence the stars are spread out in a more or less horizontal line on the H-R diagram.

10.8 CHEMICAL COMPOSITION AND EVOLUTION

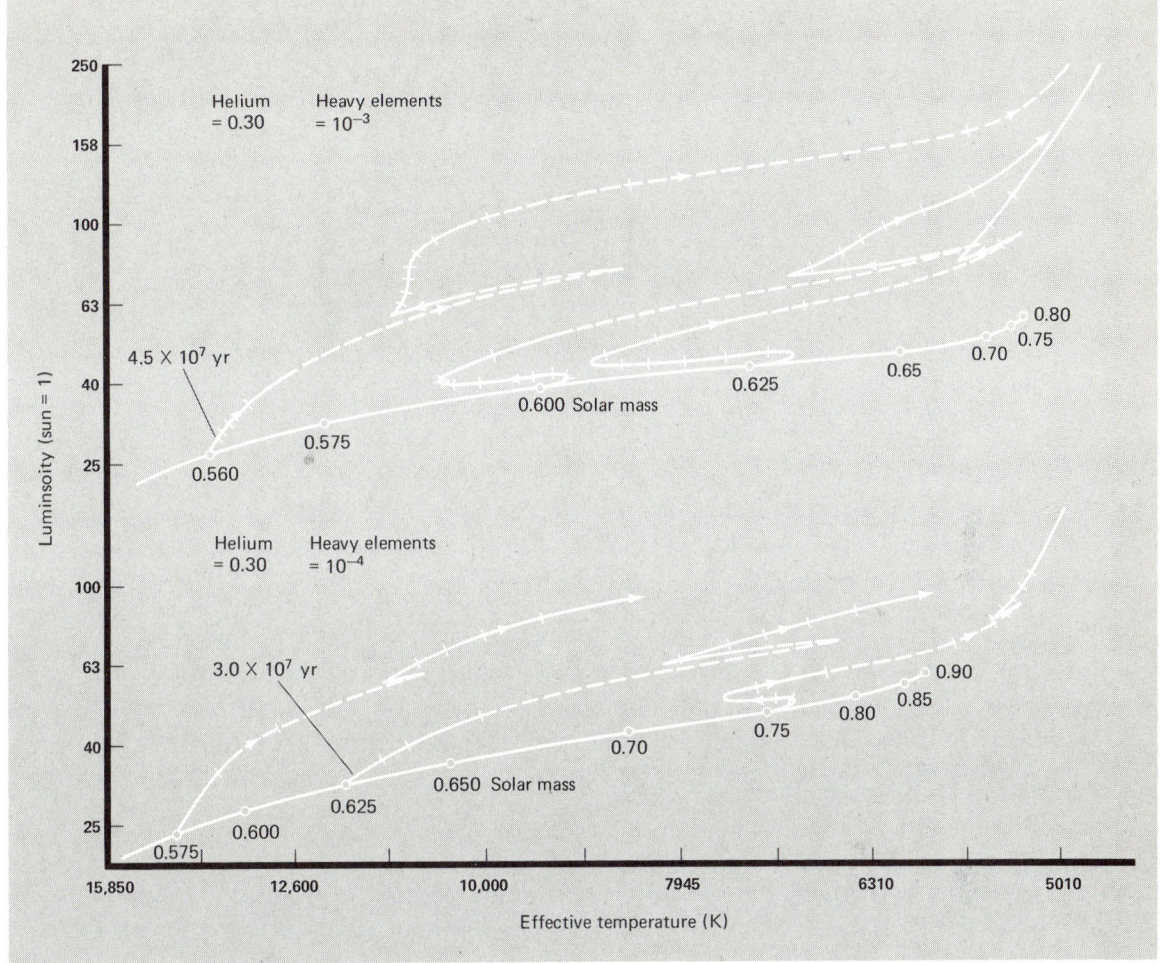

FIGURE 10.20 The helium-core-burning horizontal branch for low-mass stars of low heavy element abundance. In the upper diagram, the heavy element abundance is 10^{-3} of the mass; in the lower 10^{-4}. The helium abundance, 0.30, is the same in both cases. Note how the lower mass stars fall to the left. [*Adapted from a diagram by I. Iben, Jr., Publications of the Astronomical Society of the Pacific, vol. 183, p. 697, 1971.*]

Stars with fewer heavy elements (only 0.001 percent) and a range of masses also form a helium-core-burning horizontal branch. But compared to the same mass stars of higher heavy element composition (0.01 percent as above), these stars are arranged shifted to the left (higher temperatures) on the H-R diagram (Fig. 10.20).

In real clusters of stars the reason for the range of masses along the horizontal branch is the different amounts lost while the stars are red giants. All the horizontal branch stars in such a group start out with about the same mass and end up with the same size helium cores. But some lose more mass than others as red giants. Those that lose a lot of mass end up with a very thin hydrogen envelope around their helium cores. They are very much like main-sequence stars composed of pure helium and lie far to the left on the H-R diagram. Those that lose very little mass as red giants end up with thicker hydrogen envelopes. They are much like ordinary red giants, ending up farther to the right on the H-R diagram. The spread from left to right (hotter to cooler temperature) of the horizontal branch thus depends on mass loss during the red giant stage.

To sum up, stars with smaller percentages of heavy elements and less mass than the sun line up horizontally on the H-R diagram to make a helium-core-burning sequence. Increasing mass runs from higher to lower surface temperatures (left to right).

10.9 OBSERVATIONAL EVIDENCE FOR STELLAR EVOLUTION

The description of evolution given in the previous sections is based on theoretical calculations by electronic computers. The results presented for early and late stages of evolution are subject to change. Only evolution on and just off the main sequence seems well in hand.

Before looking at the observational evidence, let's summarize the theoretical conclusions again. A star's mass is the primary factor which determines its evolution. Massive stars are hotter in the cores; they spend less time on the main sequence and overall have shorter lives. A 1-solar-mass star first heats up from gravitational collapse and appears as a cool, luminous pre-main-sequence star. It contracts until the core is hot enough to ignite the PP process. When the core's hydrogen is gone, the star continues to burn hydrogen in a shell around the core. The shell burning causes the star to expand and become a red giant. The core heats up enough to burn helium. When the helium runs out in the core, the helium burning continues in a shell. The star expands again, but it is unstable. It pulsates more and more violently, until the outer layers are thrown out to form a planetary nebula, and a white dwarf remains.

A 5-solar-mass star has a pre-main-sequence stage of more or less constant luminosity. When it becomes a main-sequence star, the CNO cycle, rather than the PP cycle, converts hydrogen to helium. Because it reaches higher temperatures in its core, this massive star can fuse heavier elements, up to iron. Then it probably dies in the fierce demolition of a supernova explosion.

So much for theory. How well do these ideas connect with the real astronomical world? To find out, we first must look in some detail at clusters of stars.

Stars in Groups

Astronomers have found in the Galaxy three main types of star systems: *open clusters* (sometimes called *galactic clusters*), *globular clusters*, and *associations* (discussed in Chapter 9).

The Hyades and the Pleiades (Fig. 10.21) in the constellation Taurus (see the constellation endpapers for winter) are good examples of *open clusters*. Note how loosely the stars are arrayed. The Pleiades lie 414 ly (127 pc) from the sun. The cluster contains about 120 stars within a diameter of 13 ly (4 pc), for an average density of about 0.1 star in a cubic light year. These statistics are pretty typical of open clusters. They contain up to a few hundred stars in a space a few tens of light years in size, so their star densities are not more than a few stars per cubic light year. Astronomers have cataloged some 1000 open clusters to date.

FIGURE 10.21 The Pleiades, an open star cluster in Taurus. [*Photo taken with a Celestron 8 telescope; courtesy Celestron International.*]

Let's look at the H-R diagrams of open clusters. The one for the Pleiades is pretty typical (Fig. 10.22). Note that the stars below spectral type A0 fall squarely on the main sequence; above A0, the stars lie above and to the right of the main sequence. Other clusters show similar properties. The lower-mass stars fall on the main sequence, but at

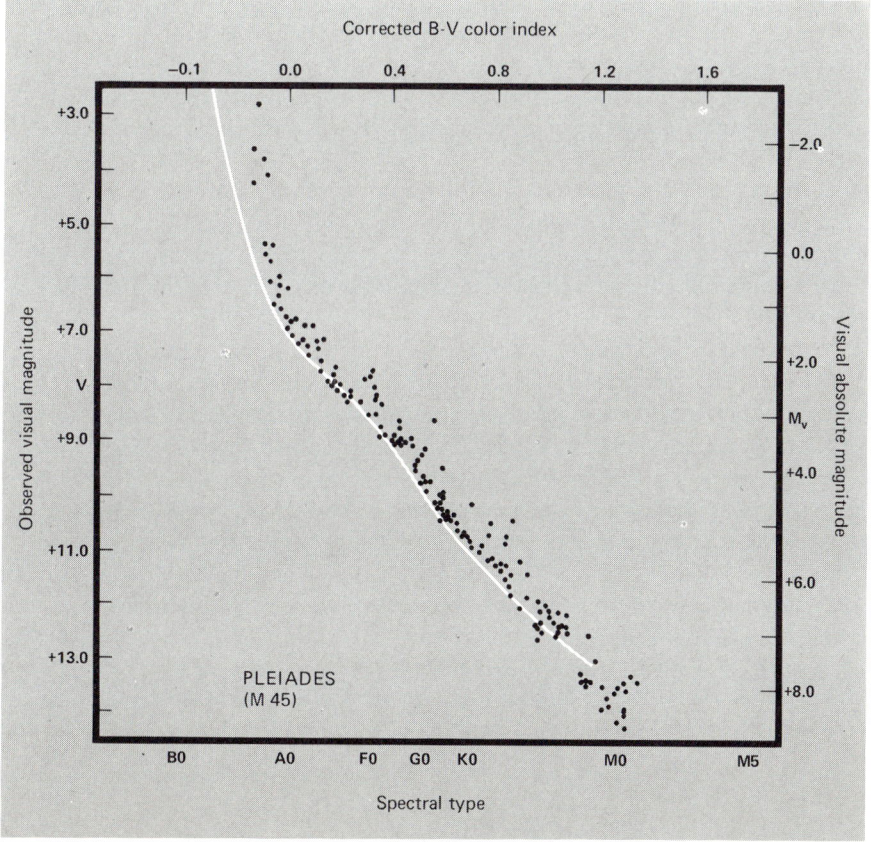

FIGURE 10.22 A H-R diagram for the Pleiades. The solid line indicates the zero-age main sequence. [*From "An Atlas of Open Cluster Colour-Magnitude Diagrams" by G. L. Hagen, David Dunlop Observatory, 1970.*]

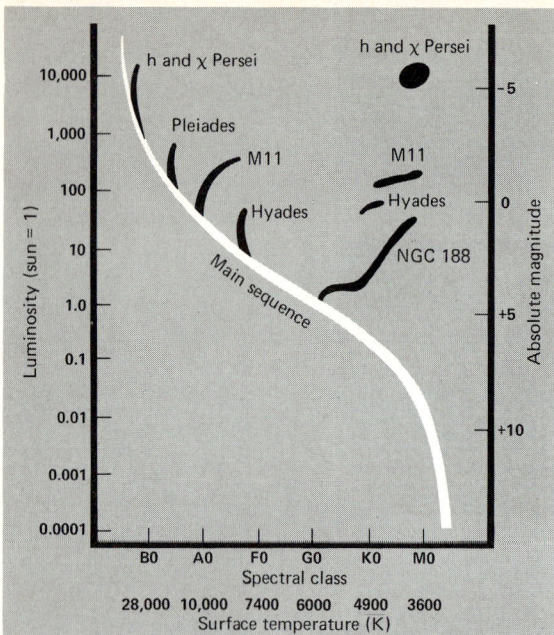

FIGURE 10.23 A schematic H-R diagram for some open clusters whose distances are known. Each cluster turns off the main sequence at a different point. The stars above it have evolved away from the main sequence.

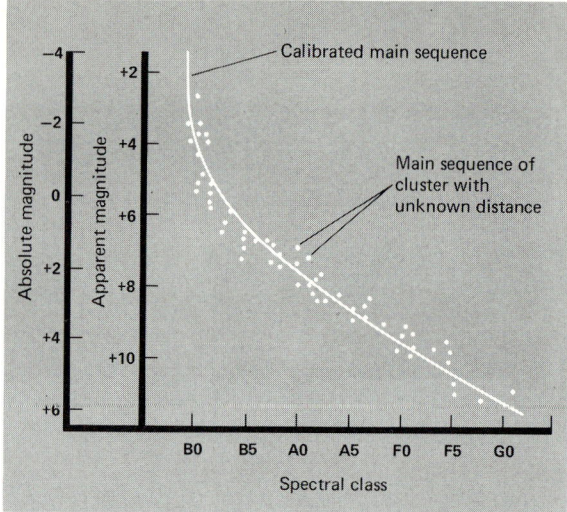

FIGURE 10.24 Using the H-R diagram to find the distances to open clusters. Here are two diagrams overlayed: one of a cluster of known distance (line) and one whose distance is to be found (dots).

some point the higher-mass stars turn off it (Fig. 10.23). In some clusters there are luminous blue stars on the main sequence. In other clusters the main-sequence stars are less luminous and the brightest stars are red giants.

Stars of the same composition and mass should all have the same luminosity during their main-sequence stage of core hydrogen burning. Only when the core hydrogen becomes exhausted do they evolve off the main sequence. By assuming that the stars in the lower part of the main sequence of all open clusters are essentially the same, we can find the distances to clusters.

We first select a cluster whose distance we find by independent means. The Hyades is used as the standard cluster. We know its H-R diagram in terms of luminosity (or absolute magnitude) versus spectral type. We then select an open cluster whose distance we don't know and determine the fluxes (apparent magnitudes) of its stars and their spectral types. From these data we can plot an H-R diagram and line up the main sequence of the standard calibrated cluster with that of the distant cluster (Fig. 10.24). For each spectral type we note the apparent magnitude of the stars in the cluster at unknown distance corresponding to the absolute magnitude of stars in the cluster at known distance. So we can find the average difference between the apparent (m) and absolute (M) magnitudes and then the distance. In Fig. 10.24 $m - M$ is 5.5. So

$$m - M = 5 \log d - 5$$
$$5.5 = 5 \log d - 5$$
$$2.1 = \log d$$
$$d = 126 \text{ pc or } 410 \text{ ly}$$

FIGURE 10.25 (a) The globular cluster 47 Tucanae, first photo of the Cerro Tololo 4-m telescope. [*Courtesy, Cerro Tololo Interamerican Observatory.*] (b) Messier 13 (M 13), a globular cluster in Hercules. [*Photo taken with a Celestron 8 telescope; courtesy Celestron International.*]

Globular clusters contrast dramatically with open clusters. Just compare Fig. 10.25a (the globular cluster 47 Tucanae) with Fig. 10.21 (the open cluster Pleiades). As the name implies and as you can see in the photos, globular clusters have a distinct spherical shape. You can see this shape easily in a small telescope. One of the best clusters to observe is Messier 13 (Fig. 10.25b) in the constellation Hercules. Binoculars will show it as a miniature, fuzzy sphere; a small telescope reveals some of the brightest stars.

Note how much more densely packed the stars are at the center of a globular cluster compared to its edges. The average density of stars in a globular is about 0.01 star in a cubic light year (or, roughly, 0.4 star in a cubic parsec; 1 star in every 100 cubic light years). In the center the stars are packed as high as 3 to 30 per cubic light year (100 to 1000 per cubic parsec). If the sun were a star in the core of a globular cluster, its nearest neighbors would be a few light months away. From the earth you would see thousands of stars scattered evenly over the sky. (See the story "Nightfall" by A. Clarke about a world within a globular cluster.) In all, a globular cluster contains 10^4 to 10^5 stars of approximately solar mass.

An H-R diagram of a globular cluster is dramatically different from that of an open cluster (Fig. 10.26). The main sequence turns off to the giant branch, and the upper end of the main sequence is missing. A horizontal line of stars stretches from the giant region to the region of the absent upper main sequence. This slash across the H-R diagram, the *horizontal branch*, is the special signature of a globular cluster.

One other difference between open and globular clusters relates to their stability. Because of their large stellar content and small size (30 to 100 ly in diameter), the globular clusters have average densities of stars about 1000 times greater than in the sun's neighborhood. The high concentration of matter ensures that the lifetime of a globular cluster

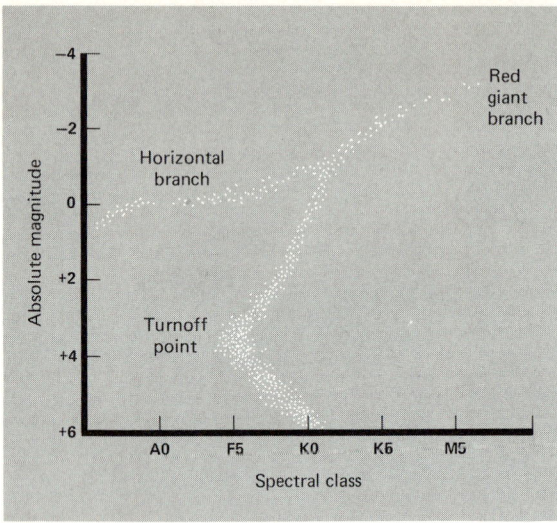

FIGURE 10.26 A schematic H-R diagram for a globular cluster. The stars along the lower part of the main sequence have masses less than 1 solar mass.

is long, on the order of tens of billions of years. (The average velocity of stars in a globular cluster is much less than the escape velocity from the cluster.) In contrast, open clusters dissipate in only a few hundreds of millions of years. *OB associations*, which you met in the last chapter, dissipate even faster. Since their star densities are even less than those of open clusters, most of the stars have velocities greater than the escape velocity. They last merely a few tens of millions of years. Table 10.3 compares and contrasts the three types of groups.

Stellar Populations

As the striking difference between the H-R diagrams of open and globular clusters seems to indicate, the stellar types in the two kinds of clusters are, in fact, quite different. Astronomers call the kinds of stars found in open clusters and associations *Population I stars* and those in globulars *Population II stars*. This distinction in stellar type was first discovered by Walter Baade in the 1940s.

TABLE 10.3 A Comparison of Star Groups

Characteristic	*OB Associations*	*Open Clusters*	*Globular Clusters*
Mass (solar masses)	10^2–10^3	10^2–10^3	10^4–10^5
Diameter (ly)	100–600	6–50	60–300
Color of brightest stars	Blue-white	Red to blue-white	Red
Density of stars (solar masses/cubic parsec)	Less than 0.01	0.1–10	0.5–1000
Examples	Orion (Fig. 9.27)	Pleiades (Fig. 10.21)	Hercules (M 13) (Fig. 10.25)

10.9 OBSERVATIONAL EVIDENCE FOR STELLAR EVOLUTION

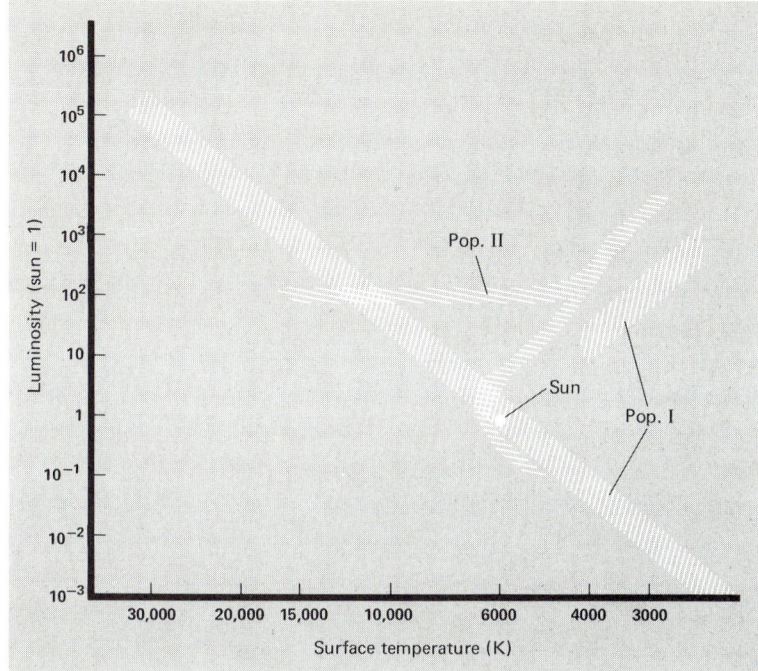

FIGURE 10.27 Baade's original divisions of Population I and II on an H-R diagram. [*Based on data by* W. Baade, Astrophysical Journal, *vol.* 100, 1944.]

You can see some of the obvious differences between Population I and II stars by examining Baade's original H-R diagram for them (Fig. 10.27). Note that the brightest Population I stars are blue-white; the brightest Population II stars, red. Also the brightest Population I stars (O stars) have about 100 times the luminosity of the brightest Population II stars (red giants). The luminous Population I stars must be relatively young, since they are O and B supergiants.

A really crucial distinction cannot be seen directly in the H-R diagram: chemical composition. Spectroscopic observations show that Population I stars have essentially the same chemical composition as the sun—1 to 2 percent, by mass, of heavy elements. (Remember that astronomers call all the elements beyond hydrogen and helium "heavy.") Population II stars contain about 1/100 as much heavy elements—0.01 to 0.02 percent of the mass.

From the description of stellar evolution so far, you should see that Population I stars must be *younger* than Population II stars, in the sense that they were born later. Population I contains some luminous blue stars on the main sequence, which have very short lifetimes, and thus must have been formed very recently. A clue to age also comes from the composition. According to the Big Bang model for the origin of the cosmos (Chapter 24), the original stuff of the creation was mostly hydrogen and helium, essentially no heavy elements. The heavy elements must be made in stars. When a massive star dies, it spews back into the interstellar medium a lot of material that has been enriched in heavy elements; from the medium, new stars are born. So we know that Population II stars are *older* (formed *earlier*) than Population I stars because they have fewer heavy elements. Population I stars have been formed out of enriched, recycled material.

Warning: Not all Population I stars are luminous blue-white stars. In fact, many

Population I stars are stars like the sun, and most are faint red dwarf stars (the lower right-hand end of the main sequence). Blue-white Population I stars are the most luminous and so the easiest to spot. Similarly, not all red giants are Population II stars, but in a group of Population II stars the most luminous ones will be red giants. Also, Population I stars have a range of ages, from a few tens of millions to perhaps 8 or 10 billion years old. But all Population I stars are younger than Population II stars, and they all have a larger fraction of heavy elements, perhaps the most distinguishing characteristic of Population I.

H-R Diagrams and the Ages of Clusters

When a star cluster forms, all its stars have the same chemical composition, but a range of masses. The more massive O and B stars evolve more rapidly than the less massive K and M stars. So the more luminous stars become red giants more quickly. As a cluster ages, stars of lower and lower mass will evolve off the main sequence. At the beginning of its life a cluster's H-R diagram will resemble that of an OB association (Fig. 10.28a). A little later it is that of a middle-aged open cluster, such as the Pleiades (Fig. 10.28b). Much later the H-R diagram is similar to that of an old open cluster (Fig. 10.28c). Note that the turnoff point away from the main sequence moves down to lower mass stars as the cluster ages. So a cluster's turnoff point indicates its age. (To get the age in years we need to calibrate the main sequence with stellar models. Without a calibration, we can only judge the *relative* ages of different clusters.)

Let's line up the H-R diagrams for a variety of star clusters (Fig. 10.29). The stars in the galactic clusters do not peel off the main sequence at the same point. In some clusters (such as the Pleiades) only a few stars at the upper end of the main sequence have evolved away from it. In others the turnoff point lies farther down the main sequence, but none is below absolute magnitude +4. In contrast to galactic clusters, all globular clusters (such as M 3 placed in Fig. 10.29 for comparison) have remarkably similar H-R diagrams, with roughly the same turnoff points.

What implications does this comparison have for stellar evolution? First, it says that globular clusters are older than open clusters. (M 3 *is* actually older than M 67; it appears slightly higher up on the H-R diagram because of its Population II composition.) Second, it implies that the ages of globular clusters are roughly the same; calibrated with theoretical models, the turnoff points for globular clusters indicate that they range from 10 to 16 billion years old, with about a 2-billion-year uncertainty. M 3, for instance, has an age of

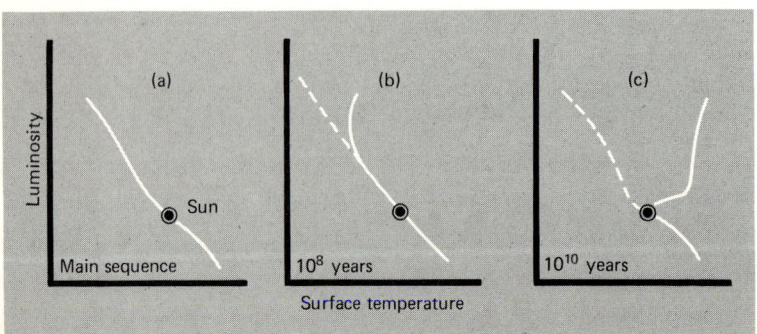

FIGURE 10.28 Theoretical evolution of a cluster of stars with the same chemical composition, born at the same time, but having different masses. "Sun" indicates the position of a solar-mass star.

10.9 OBSERVATIONAL EVIDENCE FOR STELLAR EVOLUTION

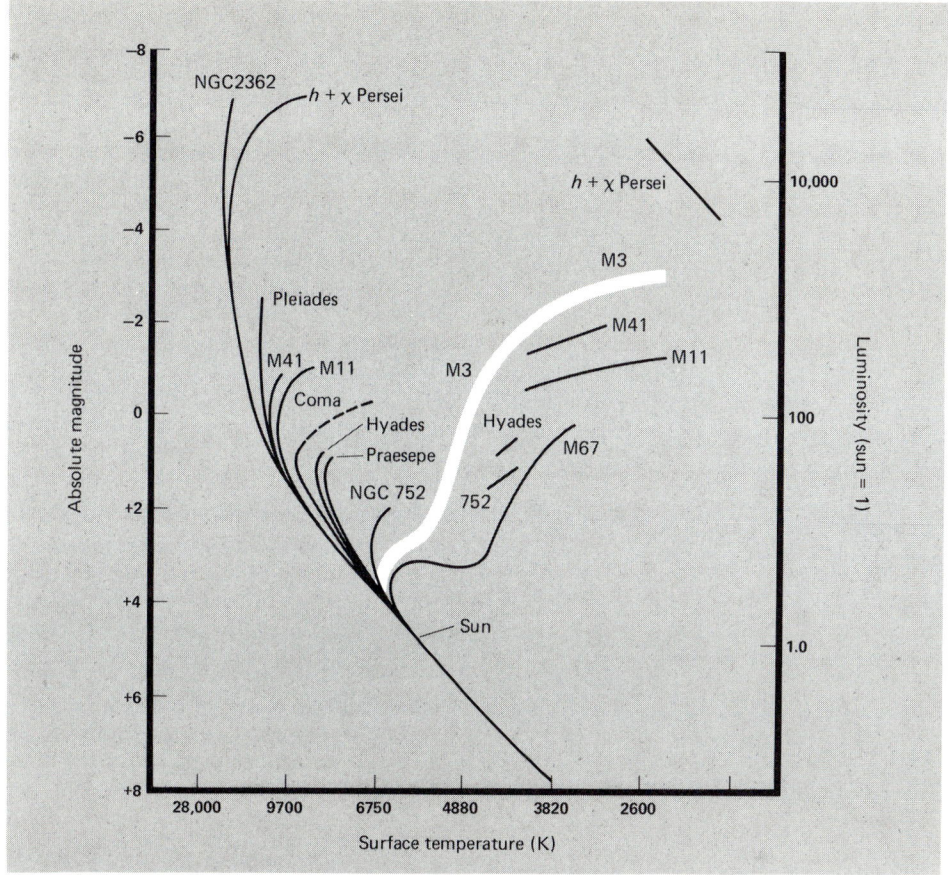

FIGURE 10.29 A composite H-R diagram for some galactic and globular clusters. The further down the main sequence the turnoff point occurs, the older the cluster. [*Adapted from a diagram by A. Sandage.*]

12 to 14 billion years. The stars in the globular clusters are the oldest stars known. Third, it implies that open clusters have a large range of ages. The ages of open clusters can also be estimated from their turnoff points by comparing them to those found from theoretical calculations. For example, M 67 has an approximate age of 3.2 billion years, and the Hyades are much younger, not more than 0.7 billion years old.

The connection between the evolution of an individual star and the changing appearance of the H-R diagram of a cluster of stars as it grows older merits a detailed explanation. Consider the following points.

1. All stars of the same composition, obtaining energy from the burning of hydrogen into helium, lie along the main sequence on the H-R diagram. Higher-mass stars are further up (higher luminosities, higher temperatures); lower-mass stars are further down (lower luminosities, lower temperature).
2. The evolution of each star from the main sequence to the red giant region follows a path similar to that shown in Fig. 10.16 for the high-mass stars or Fig.

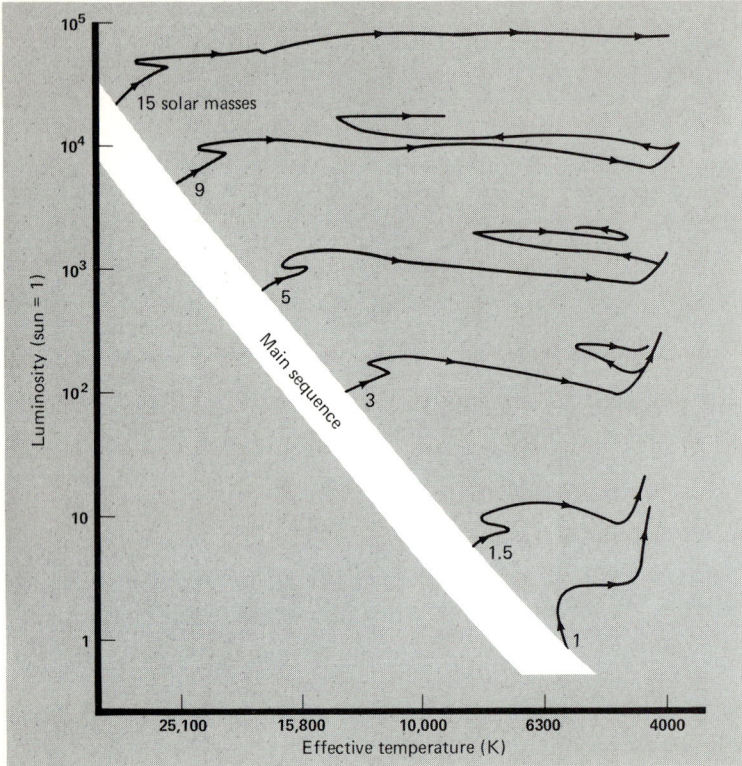

FIGURE 10.30 Theoretical evolutionary tracks off the main sequence for stars from 1 to 15 solar masses. [Based on calculations by I. Iben, Jr., *Astrophysical Journal*, vol. 140, p. 1631, copyright 1964 by the American Astronomical Society.]

10.12 for the low-mass stars, but each star starts from a different place on the main sequence, depending on the mass (Fig. 10.30).

3. Although the evolutionary paths are similar, the times to complete the paths are not. Those stars higher up on the main sequence (higher mass) evolve to the red giants in a shorter time.

4. In a very young cluster all the stars will be on the main sequence. As the cluster ages, the most massive (and luminous) main-sequence stars will evolve into red giants first, then the next most massive (next most luminous) ones, and so forth. The main sequence will gradually shorten as the stars peel off in order of mass and evolve over into the red giant region.

The stars last a relatively short time in the red giant phase. The more massive ones become supernovas. The less massive ones become planetary nebulas and leave white dwarf remnants behind. In either case they stop being red giants and disappear from that part of the H-R diagram. So, in any given cluster the red giants are those stars which have just recently evolved from the main sequence; they come from the region just above the main sequence turnoff point for that cluster and hence have masses just a little larger than those stars still remaining on the main sequence. Those stars which started out with still larger masses have already passed through the red giant phase and are now supernova remnants or white dwarfs.

We can see this evolution clearly by examining the theoretical paths on the

10.9 OBSERVATIONAL EVIDENCE FOR STELLAR EVOLUTION

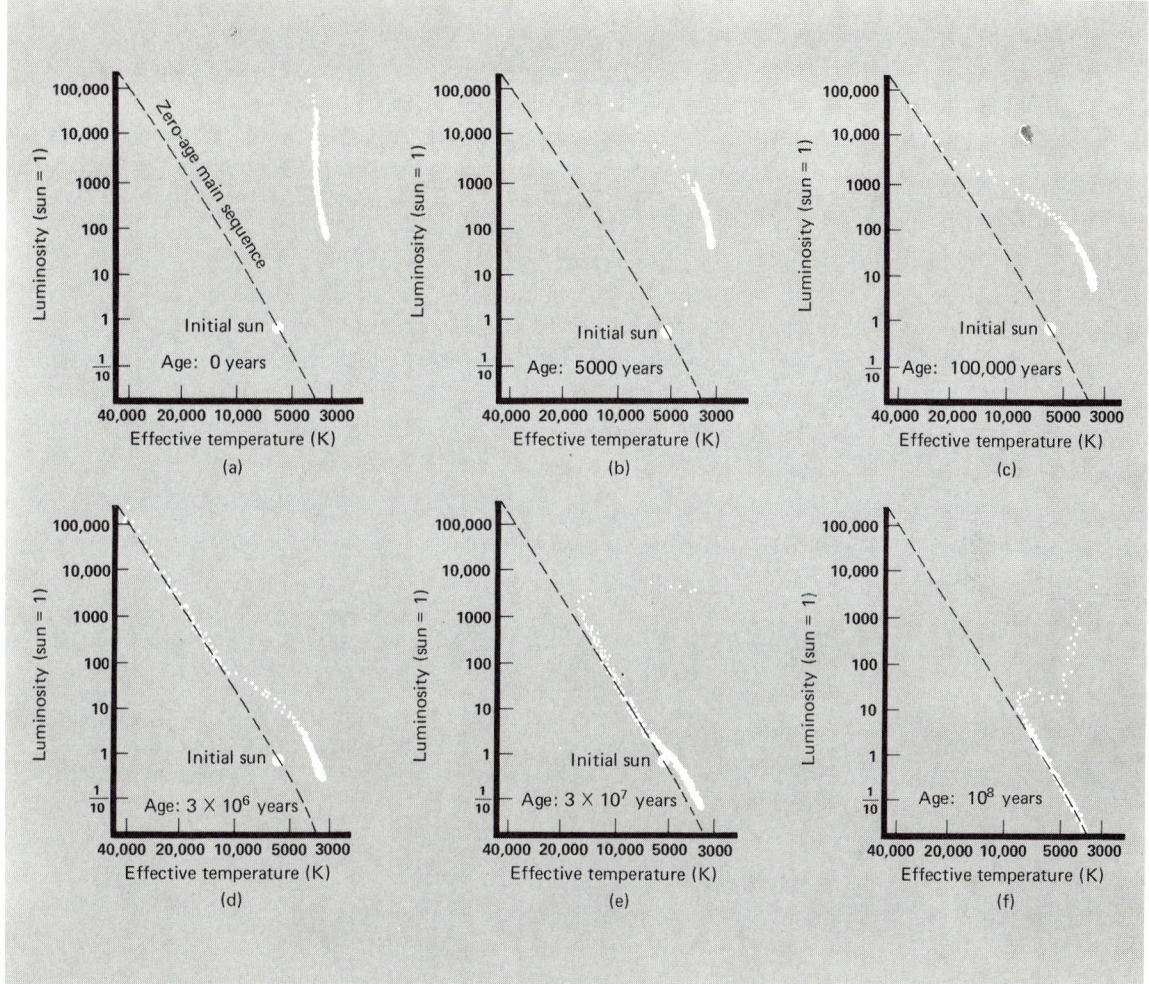

FIGURE 10.31 Theoretical evolution of Population I stars from 0.5 to 24 solar masses making up a hypothetical cluster with all the stars born at the same time. The times indicate that elapsed since birth. [*Based on calculations by R. Kippenhahn.*]

H-R diagram for an artificial cluster of stars. Let's look at a model cluster calculated by R. Kippenhahn. It consists of 190 stars with masses ranging from 0.5 to 24 solar masses, with a Population I chemical composition. At birth (Fig. 10.31a) the stars are protostars, approaching the main sequence. Just 5000 years later (Fig. 10.31b) the massive stars have already moved toward the main sequence, leaving the lower-mass stars behind. After 100,000 years (Fig. 10.31c) the massive stars have hit the main sequence and commence hydrogen core burning. At 3 million years (Fig. 10.31d) the upper main sequence is established. Here the cluster's H-R diagram resembles that for the young cluster NGC 2264 (Fig. 9.49). By 30 million years (Fig. 10.31e) the most massive stars are in the helium core burning stage and so are red giants. At 6.6 million years the upper main sequence is losing stars. By 10^8 years (Fig. 10.31f) many stars are red giants. The cluster's H-R diagram resembles that of M 67 (Fig. 10.29). Finally, at an age of 4.2×10^9 years the turnoff point has rolled down to about a solar mass.

5. The end result is that we can tell the age of a star cluster by drawing its H-R diagram and comparing it to theoretical ones. Those clusters with a long main sequence and luminous red giants are young; those with a shorter main sequence and less luminous red giants are older. In Fig. 10.29, NGC 2362 is the youngest galactic cluster shown, M 67 the oldest. (As we noted before, the higher luminosity of red giants of the globular cluster M 3 compared to the open cluster M 67 is a function of chemical composition, not age.) It is only by comparisons of theoretical evolutionary models with actual H-R diagrams for clusters that we can find their ages. The comparison also confirms the general validity of the models.

Variable Stars

A star's luminosity varies little during its main-sequence sojourn. That's about 80 percent of its lifetime. During the remaining 20 percent its luminosity varies dramatically, but in times too long for us to see it change for any one star. However, astronomers have observed stars whose luminosities change over periods of a few hours to a few years. These *variable stars* (so called because their luminosities change rapidly with time) lie above the main sequence on the H-R diagram (Fig. 10.32). What is their evolutionary status? We

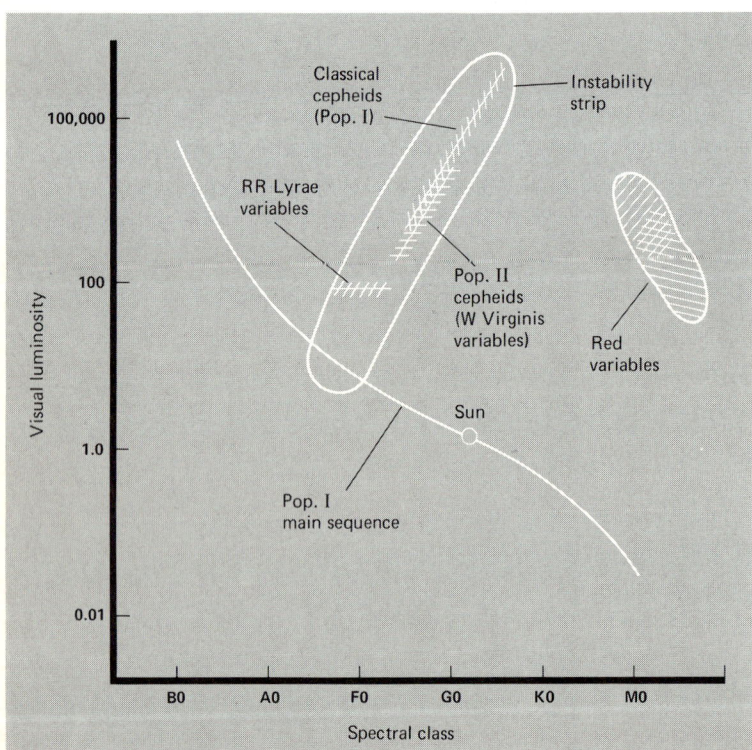

FIGURE 10.32 The positions of various variable stars on the H-R diagram. [*Adapted from a diagram by J. P. Cox.*]

10.9 OBSERVATIONAL EVIDENCE FOR STELLAR EVOLUTION

know from theoretical calculations that these variables are post-main-sequence stars and that those whose variations are regular are in the helium-core-burning stage.

There's a staggering array of variable stars; we'll limit discussion to just a few: RR Lyrae variables, cepheids, and red variables. RR Lyrae stars (named after their prototype, RR Lyrae, whose period is 13.6 h) vary in luminosity with periods of 1.5 to 24 h (typically 12). They are Population II, range in spectral class from A2 to F6, and have about 100 times the sun's luminosity. About 5000 RR Lyrae stars are known. Cepheids (named after their prototype Delta Cephei) fall into two groups: Population I, called classical cepheids, and Population II, sometimes called W Virginis stars. Classical cepheids have periods from 1 to 50 days (typically 5 to 10 days) and range from F6 to K2 in spectral class. Population II cepheids vary in periods from 2 to 45 days (typically 12 to 20 days) and range from F2 to G6 in spectral class. The RR Lyrae stars and Population I and II cepheids are *regular* or *periodic variables*; their change in luminosity with time follows a regular cycle.

In contrast, the *red variables* have irregular cycles of light variation that range from 100 to 700 days. They contain both Population I and II stars of spectral class K and M and have luminosities roughly 100 times that of the sun. They are red giant and supergiant stars.

The regular variables have a key physical characteristic in common: pulsation. Doppler shift observations of their spectra show that as they vary, these stars expand and contract. (When the star is expanding, its surface moves toward us, producing a blue shift in the spectrum; when it contracts, the surface moves away, producing a red shift.) Population I cepheids, for instance, expand and contract about 10 percent of their radius at speeds of about 30 to 40 km/sec.

The cepheids and RR Lyrae stars lie in a region of the H-R diagram called the *instability strip*. As low-mass Population II stars (0.5 to 0.7 solar mass) traverse this strip during their helium-core-burning phase, they become unstable and pulsate; these are the RR Lyrae stars. Population I stars of 3 to 18 solar masses cross the upper region of this strip also during their phase of core helium burning, and they also pulsate, becoming the cepheids.

What drives cepheids to pulsate? Models of the interior structure of Population I cepheids indicate that a layer where the helium changes from singly ionized (He II) to doubly ionized (He III) exists at a temperature of about 42,000 K. The opacity of the gas in the zone can change drastically upon compression or expansion as the level of ionization changes. When compressed, the opacity goes up. This dams up the outward flow of radiation, so the temperature and pressure increase. The higher pressure pushes the outer layers (30 to 50 percent of the star's radius, containing less than 1 percent of its total mass) and the star expands. The expansion decreases the opacity. Radiation then flows more easily; the pressure and temperature drop. Gravity contracts the star. Then the cycle repeats. So the change in opacity in the He II–He III zone drives the pulsations. The zone must be at the right depth for the mechanism to work. If it is too far in, the overlying layers are too massive to move much; too far out and the matter is too thin to impede the radiation flow. So these stars pulsate only when their evolution has produced just the right structure in their outer layers, which occurs when they evolve to the instability strip on the H-R diagram.

The red variables, on the other hand, fall in a region of the H-R diagram where the stars undergo *shell* rather than core burning. The cause of their pulsation is unknown.

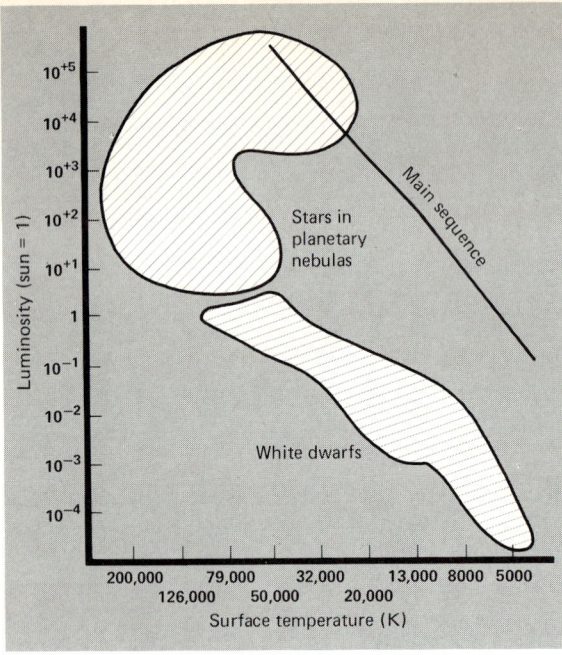

FIGURE 10.33 A schematic diagram for the location of the central stars of planetary nebulas on the H-R diagram.

Central Stars of Planetary Nebulas

In the scenario for the evolution of a solar-mass star, when a red giant pops off its outer layers, forming a planetary nebula, it leaves behind a hot, dense core. This cinder cools to form a white dwarf.

If this picture is correct, you'd expect that the central stars of planetary nebulas should fall along the evolutionary track after the red giant stage and before the white dwarf stage (points 8 to 9 in Fig. 10.12). Well, they do (Fig. 10.33)! Some central stars are extremely hot and luminous; others are hot, but not as luminous. Their positions on the H-R diagram fall neatly above that for white dwarfs. So the stars of planetary nebulas mark a transition between the core of a red giant and a white dwarf. This observation nicely confirms that stars of about the sun's mass evolve from red giants to white dwarfs.

10.10 THE SYNTHESIS OF ELEMENTS IN STARS

In order to survive, a star must fuse lighter elements into heavier ones and so generate energy. Gravitational contraction provides the initial heat to get fusion reactions going. The more mass a star has, the greater the central temperature produced by gravitational contraction and the heavier the elements it can fuse. From the ignition temperatures needed for fusion reactions, we can set limits on the heaviest elements that a star of a certain mass can fuse (Table 10.4). For example, our sun can burn helium to carbon, but will never get hot enough to fuse carbon.

Table 10.4 summarizes the principal stages of nuclear energy generation and nucleosynthesis in stars. Note that the products (or ashes) of one set of reactions usually become the fuel for the next set of reactions. What a beautiful scheme for energy production in the universe!

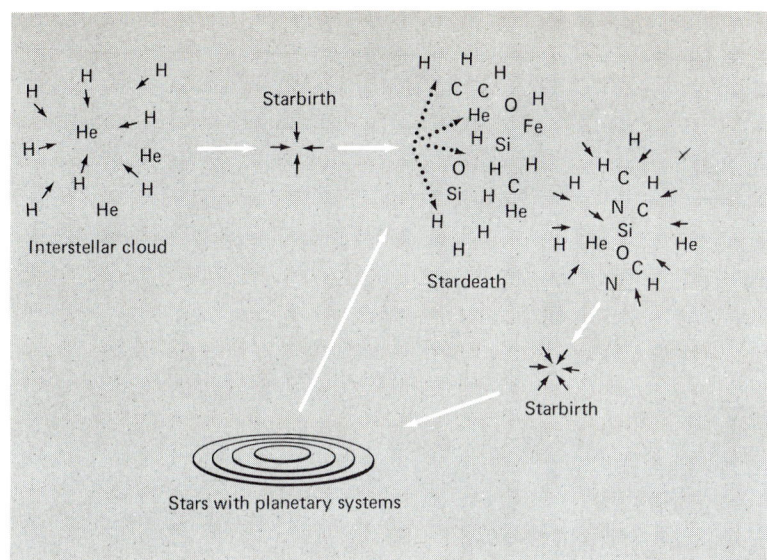

FIGURE 10.34 The cosmic recycling scheme involving starbirth, death, and the interstellar medium.

Also note in the table that only stars with masses greater than about 5 solar masses can produce elements heavier than oxygen, neon, and sodium. Few stars have this much mass, and so many stars come to the end of their nuclear evolution without having manufactured some important elements. This fact emphasizes the importance of massive stars in the scheme of cosmic evolution—they fuse heavy elements and throw some back into the interstellar medium (Fig. 10.34).

TABLE 10.4 Stages of Thermonuclear Energy Generation in Stars

Process	Fuel	Major Products	Approximate Temperature (K)	Approximate Minimum Mass (solar masses)
Hydrogen burning	Hydrogen	Helium	1–3×10^7	0.1
Helium burning	Helium	Carbon, Oxygen	2×10^8	1
Carbon burning	Carbon	Oxygen, Neon, Sodium, Magnesium	8×10^8	1.4
Neon burning	Neon	Oxygen, Magnesium	1.5×10^9	5
Oxygen burning	Oxygen	Magnesium to sulfur	2×10^9	10
Silicon burning	Magnesium to sulfur	Elements near iron	3×10^9	20

Source: Adapted from a table by A. G. W. Cameron from "Endpoints of Stellar Evolution" in *Frontiers of Astrophysics*, E. H. Avrett, editor (Harvard Univ. Press, Cambridge, Mass., 1976).

We hope you now see where part of the periodic table of the elements (Appendix G) comes from—that part up to iron. What about the rest? We'll find out in the next chapter.

SUMMARY

This and the previous chapter have described the contemporary concepts in our understanding of the life histories of the stars. Here's a summary of the major themes.

1. Stars are born when a dense interstellar cloud fragments and these pieces collapse gravitationally. Different fragments of the cloud form stars at different times. As a star forms, the dust in the cloud hides it from view.
2. A star's evolution depends mainly on its mass; more massive stars have higher core temperatures and so higher luminosities. They evolve faster and live shorter lives. Their higher core temperatures allow them to fuse heavier elements in later evolutionary stages. They are likely to die in supernova explosions.
3. Stars like the sun will evolve to become red giants, white dwarfs, and then black dwarfs. Most stars will go through this evolution because most stars contain a solar mass or less of material.
4. During its life a star constantly struggles against gravity. It resists gravitational collapse by pressure from heat in its interior. Fusion reactions in the core provide this heat; they are first ignited by the heat from gravitational collapse. A star must fuse heavier and heavier elements in order to withstand gravity.
5. Massive stars recycle some material back into the interstellar medium, but its composition has been changed by the addition of heavier elements. Some of the material stays locked in a star's corpse, never to participate again in cosmic evolution.

The stars form the crucial evolutionary links in the chain of cosmic evolution. They produce light and warmth vital to life on any planet around them. They create the elements out of which planets are made by fusion reactions. What ignites these reactions? Gravitational contraction. Gravity, the driving force of the astronomical universe, squeezes matter into heavier elements in the hearts of stars.

Key Words

hydrostatic equilibrium
equation of state
thermal equilibrium
opacity
temperature gradient
star model
evolutionary track
PP chain
CNO cycle
triple-alpha reaction

nucleosynthesis
protostar
pre-main-sequence star
zero-age main sequence (ZAMS)
degenerate gas
Pauli exclusion principle
helium flash
planetary nebula
black dwarf
mass loss
stellar wind

open (galactic) cluster
globular cluster
horizontal branch
Population I,
    Population II star
turnoff point
RR Lyrae variable
cepheid variable
red variable
instability strip

Review Questions

1. The fact that most stars we see are main-sequence stars and only 1 percent are red giants tells us that (mark all correct answers):
   a. Stars evolve from the main sequence to the red giant region.
   b. The main-sequence stage is about 100 times longer than the red giant stage.
   c. Red giants are 100 times more massive than main-sequence stars.
   d. Only about 1 star out of 100 ever becomes a red giant.
   e. Red giants are more difficult to discover than are main-sequence stars.
   (Objective 1)
2. State at which point in the H-R diagram (Fig. 10.35) for a 1-solar-mass star each of the following events occurs:
   a. Star loses envelope.
   b. Triple-alpha process in core begins.
   c. Energy is derived principally from gravitational contraction.
   d. Star becomes white dwarf.
   e. Hydrogen burning in core begins.
   f. Star arrives on main sequence.
   (Objective 3)

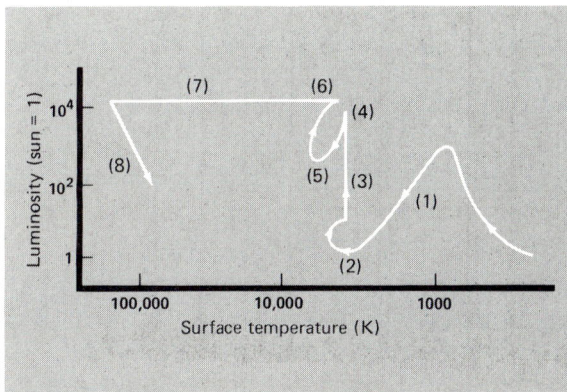

FIGURE 10.35 A H-R diagram for review question 2.

3. Which of the following statements apply to a 1-solar-mass star, a 5-solar-mass star, both, or neither?
   a. Its life is ended in a supernova explosion.
   b. On the main sequence it burns hydrogen in its core mostly on the PP chain.
   c. It never goes through a phase of helium shell-burning.
   d. The core gets hot enough to initiate carbon burning late in its life.
   e. It expands and becomes a red giant while hydrogen burns in a shell.
   f. When helium ignites in the core, it expands.
   (Objective 5)
4. A star like the sun consists completely of an ordinary gas. Why doesn't it suddenly collapse gravitationally? (Objective 2)
5. Calculations indicate that a solar-mass protostar is much more luminous than the sun and much cooler at the surface. How could the protosun be much cooler and yet more luminous than the present sun? (Objective 4)

6. How can you tell from an H-R diagram that the stars in the Pleiades cluster are younger than those in the Hyades? (Objective 6)
7. Why are massive stars able to fuse heavier elements than less massive stars? (Objective 8)
8. What evidence do we have that red giants become white dwarfs? (Objective 7)
9. Compare the star types in open and globular clusters. (Objective 11)
10. What kind of star is our sun likely to become in the future? What kind of corpse will it eventually leave behind? (Objective 3)

Problems
1. What is the approximate reaction rate (number of hydrogen atoms changed to helium per second) needed to account for the luminosity of a 10-solar-mass star? (Objective 2)
2. Estimate the size of the sun when it will be a red giant. (Objective 3)
3. For a 5-solar-mass pre-mass-sequence star, compare its size when its surface temperature is 6000 K and when it is 15,000 K. (Objective 5)
4. Compare the apparent and absolute magnitudes of the stars in the Pleiades cluster (Fig. 10.22) and estimate its distance. (Objective 13)
5. If the reaction rates for the PP chain and the CNO cycle are equal at 18 million K, at what temperature will the PP rate be 10 percent of the CNO rate? At what temperature will the CNO rate be 10 percent of the PP rate? (Objective 8)
6. By what factor does the pressure of an ordinary gas change when the density increases by a factor of 2 (the temperature staying constant)? By what factor does the pressure change when the temperature increases by a factor of 2 (the density staying constant)? What are the factors if the gas is a degenerate gas? (Objective 10)

BEYOND THIS BOOK . . .

* "Stellar Populations" by M. and G. Burbidge in *Scientific American*, November 1958, p. 44, is a good introduction to the subject.
* To find out more about cool stars, read "Red Giant Stars" in *Astronomy*, December 1976, and "Red Dwarf Stars" in *Astronomy*, July 1978, both by B. Johnson.
* A somewhat technical article for more details of thermonuclear reactions is "Energy Production in Stars" by H. Bethe, *Physics Today*, September 1968.
* For more details on planetary nebulas, look at "Recent Findings about Planetary Nebulas" by Y. Terzian, *Sky and Telescope*, December 1977, p. 459.
* I. Iben describes the evolution of Population II stars in "Globular Cluster Stars," *Scientific American*, July 1970, and gives a comprehensive review in "Stellar Evolution Within and Off the Main Sequence," *Annual Reviews of Astronomy and Astrophysics*, vol. 5 (1967).
* "Planetary Nebulas" by J. Miller in *Annual Reviews of Astronomy and Astrophysics*, vol. 12 (1974), gives more details about these objects.

STARDEATH

11

LEARNING OBJECTIVES
After studying this chapter, you should be able to:

1. Compare the physical natures of white dwarfs and neutron stars, and describe their place in stellar evolution.
2. Describe the basic physical properties of a degenerate star.
3. Indicate how white dwarfs are used to test general relativity.
4. Argue, with observational support, that pulsars are neutron stars and cannot be white dwarfs, and describe the model of a pulsar as a rapidly rotating, highly magnetic neutron star.
5. Compare the observed features of a nova and a supernova.
6. Describe a method for obtaining the distance to a nova.
7. Outline a possible model for a nova explosion that involves a binary star system.
8. Outline a possible model for a supernova explosion.
9. Cite observational evidence that the Crab Nebula is a supernova remnant, and describe the effect of the pulsar on the nebula now.
10. Contrast thermal and nonthermal radiation and describe how synchrotron radiation is emitted.
11. Describe how nucleosynthesis can occur in a supernova.
12. Place supernovas in the grand scheme of cosmic evolution.

CENTRAL QUESTION:
How do stars die, and what corpses do they leave behind?

Stars die more or less violently. Imagine, as in the previous chapter, that a year equals one-fifth of a second, so the sun would live 65 years. In this time scale you would see about 25 stars in the Galaxy wink out every second. These deaths are signaled by an ejection of a star's outer layers. The discarded shells replenish the interstellar medium, previously depleted by the formation of stars and planets. The dead star's remnant core cools. Locked tight by gravity, it forms a cinder in space. What are these remains like? In some instances, the burned-out core becomes a white dwarf star, a solid carbon crystal. In others the core becomes a neutron star, a smooth, spinning sphere of nuclear matter. In still others the core may disappear through a warp in spacetime as a black hole.

What paths do stars take to these strange deaths? Astrophysicists do not all agree on the details of the final stages of stellar evolution. But how a star comes to its final fate does depend on its mass at the end of its life. With less than 1.4 solar masses, a star's corpse will be a white dwarf. Between 1.4 and 5 solar masses, the end is a neutron star. With a greater mass, a star falls into a grave marked by a black hole. This chapter deals with white dwarfs and neutron stars, the next with the bizarre black hole.

Although the details are uncertain, observations do imply that almost all stars throw off mass before they meet their ends. A supernova is the most destructive example of mass loss. But a supernova is constructive too; in its immense explosion many heavy elements of the universe are made and thrown to the currents of space. Supernovas spice the interstellar medium and provide the impulse for the birth of new stars.

11.1 WHITE DWARF STARS: COMMON CORPSES

The evolution of a 1-solar-mass star (Section 10.5) illustrates the constant battle of pressure and gravitational forces. Because gravity never lapses the way thermonuclear reac-

11.1 WHITE DWARF STARS: COMMON CORPSES

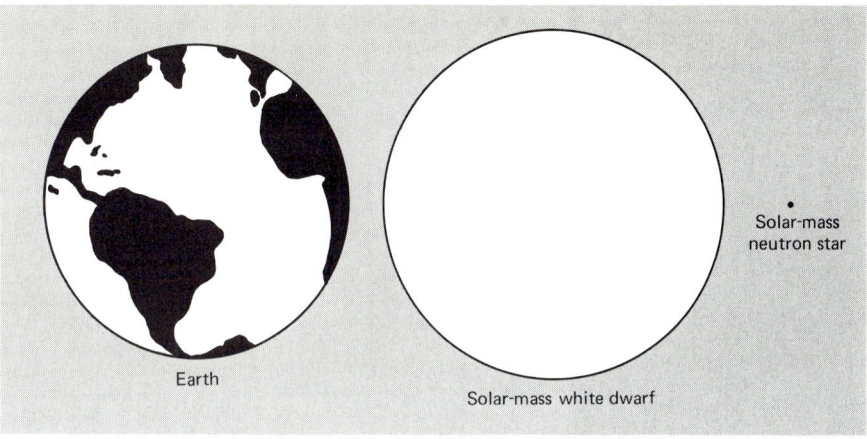

FIGURE 11.1 A comparison of the relative sizes of the earth, a solar-mass white dwarf, and a solar-mass neutron star.

tions do, the final state of any star depends only on the physical properties of matter at high densities and the total mass of the star. When all the thermonuclear reactions cease, what pressure can support the star? Only the pressure of a degenerate gas (Section 10.5).

In an ordinary gas the pressure depends directly on the temperature, but in a degenerate gas the pressure is *not* tied to the temperature. It depends only on the number of particles and how tightly they are crammed together. So this pressure will still exist even after nuclear reactions cease to keep the star hot. The density at which a gas switches to a degenerate state is about 10^8 kg/m^3. A degenerate star of 1 solar mass has an average density of about 10^9 kg/m^3. If the sun were compressed to this density, it would be approximately 7000 km in radius, about the size of the earth (Fig. 11.1).

White Dwarfs

In 1935 Subrahmanyan Chandrasekhar applied the physics of a degenerate electron gas to stars. He found that the pressure exerted by the electrons could resist the force of gravity only for stars less than 1.4 solar masses, and that such stars would have a density of 10^8 to 10^9 kg/m^3. Such a star, at the endpoint of its thermonuclear history, is a *white dwarf*. No thermonuclear reactions go on, no heavier elements are fused, and no energy is produced. How does a white dwarf fend off gravity? By the outward pressure from the degenerate electron gas. What about the nuclei? They form a crystal structure embedded in the degenerate gas. If the expulsion of the outer layers of the star leaves a carbon core, the white dwarf becomes a solid carbon crystal.

How is a white dwarf star both a degenerate gas and a crystalline solid? Both descriptions are correct, at least for the cooler white dwarfs. The *electrons* in the white dwarf always are free to move around from place to place; they have the properties of a gas. But the *ions* in the white dwarf, though they may be able to move around freely when the white dwarf is hot, soon get locked into a fixed location as the star cools. A white dwarf is both a crystalline solid (the ions) and a gas (the electrons) at the same time. The same is true of a metal. The electrons are free to move around (that is why metals are good conductors of electricity), but the ions are locked in place, giving the metal its hardness.

The small radius and high density of a white dwarf can be traced to the behavior of a degenerate electron gas in a gravitational field. Chandrasekhar found that the more mas-

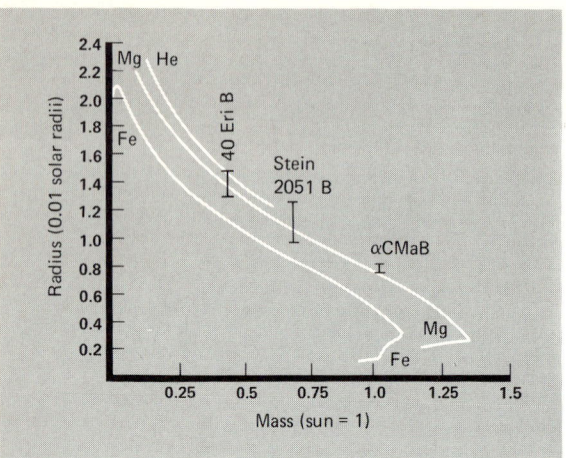

FIGURE 11.2 A mass-radius diagram for white dwarfs. The short bars indicate three white dwarfs whose radius and mass are inferred from observations. The solid lines are theoretical calculations for white dwarfs made by helium, magnesium, and iron. [Adapted from a diagram by J. Liebert, Astrophysical Journal, vol. 210, p. 715, copyright 1976 by the American Astronomical Society.]

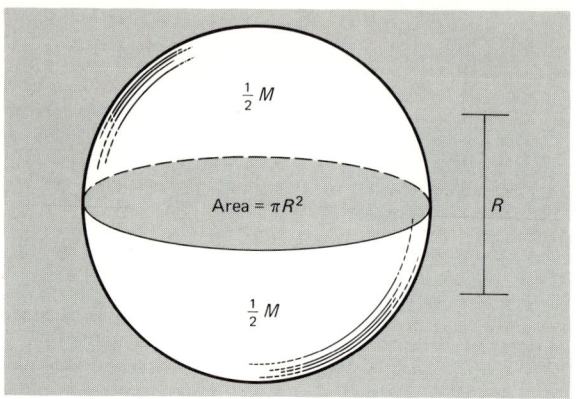

FIGURE 11.3 Cutting a star in half to estimate its internal pressure.

sive the white dwarf, the *smaller* its radius (Fig. 11.2). (For main-sequence stars the more mass a star has, the larger its radius.) How does this come about? More mass means more gravity. To balance gravity requires internal pressure. In a white dwarf the pressure does not come from internal heating from thermonuclear reactions. Instead it arises from the nature of a degenerate gas, where greater pressures are a response to closer packing of the materials (greater density).

The exact relationship between pressure (P) and density (ρ) in completely degenerate matter is

$$P = K\rho^{5/3}$$

where K is a constant number.

How much pressure is needed to support a star? To estimate this, imagine a star divided by a plane down the middle. The two halves each have a mass of $\frac{1}{2}M$, where M is the mass of the star, and the centers of these two masses are approximately a distance R apart (Fig. 11.3). The force pulling them together is, according to Newton's law,

$$F = G\frac{(1/2\,M)(1/2\,M)}{R^2}$$

The force keeping the two halves apart is the pressure (P) times the area of the hemispheres (πR^2). For the gravitational force to equal the pressure force,

$$\frac{GM^2}{4R^2} = P\pi R^2$$

11.1 WHITE DWARF STARS: COMMON CORPSES

or

$$P = \frac{GM^2}{4\pi R^4}$$

So far, we have said nothing about degenerate matter; the above relationship holds for all stars.

Now use the relationship between pressure and density for degenerate matter ($P = K\rho^{5/3}$) and express the density ρ in terms of the mass and radius:

$$\rho = \frac{M}{(4\pi/3)R^3}$$

Then

$$P = K\rho^{5/3} = \frac{K}{(4\pi/3)^{5/3}} \frac{M^{5/3}}{R^5}$$

But the pressure must also be equal to $GM^2/4\pi R^4$. So

$$\frac{GM^2}{4\pi R^4} = \frac{K}{(4\pi/3)^{5/3}} \frac{M^{5/3}}{R^5}$$

Solving for R in terms of M,

$$R = \frac{4\pi K}{G(4\pi/3)^{5/3}} \frac{1}{M^{1/3}}$$

that is, the radius of a degenerate star is proportional to the *inverse cube root* of the mass. As the mass gets larger, the radius goes down. (The actual observed relation between radius and mass differs a bit from this theoretical expectation.)

The mass of the white dwarf hits a crucial point at about 1.4 solar masses. At very high densities the equation of state changes, and the above relation between mass and radius does not hold. At 1.4 solar masses the radius would be *zero*. At this mass and above the gravitational forces overwhelm the degenerate electron gas pressure. The star collapses. It cannot be a stable white dwarf. This amount of mass, 1.4 solar masses, is called the *Chandrasekhar limit* and signals the point at which electron degenerate matter, crushed by gravity, is no longer able to support the star.

Observations of White Dwarfs

In 1862 the American optician Alvan Clark observed Sirius B (Fig. 11.4), the faint companion to Sirius A (Section 8.6). Later, this star was found to be a white dwarf. Because Sirius B is part of a binary star system, its mass (in terms of the sun's mass) can be calculated using Kepler's third law. The result of this calculation is about 1 solar mass for Sirius B. This star has a low luminosity of about 3×10^{-3} times the sun's luminosity and

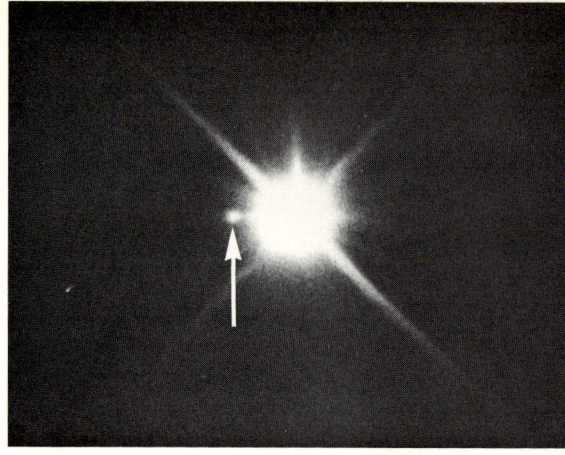

FIGURE 11.4 Sirius B (arrow), a white dwarf star, companion to Sirius A (brighter image). The sizes of the images in this photo relate only to the brightnesses, *not* to their sizes. [*Courtesy Lick Observatory.*]

a high surface temperature of about 29,500 K, and so it has a small radius, around 7×10^{-3} the sun's radius. Given its size and mass, Sirius B must have an average density of about 3 *billion* kilograms per cubic meter!

By coincidence, the brightest star in Canis Minor (the Little Dog, near Canis Major), Procyon, also has a white dwarf companion. This companion was predicted in 1862 from the motion of Procyon and was observed in 1882. Called Procyon B, it has a mass of about 0.65 solar mass.

Most white dwarfs seen so far are actually white, but a few are yellow, and some are red. Over 300 stars have been identified as white dwarfs. Because their low luminosities make them hard to detect, it's not easy to estimate the number of white dwarfs in the Galaxy, but they may make up as much as 10 percent of the stars.

White dwarfs tend to fall into two general catagories: those whose spectra show strong hydrogen lines and those which have strong helium lines. White dwarfs with the strongest hydrogen lines are put in class DA—D for dwarf and A to indicate that the spectra resemble those of A stars (strongest hydrogen Balmer lines).

An H-R diagram of DA white dwarfs for which decent observational data are available (Fig. 11.5) shows that (1) the DA white dwarfs actually fall along a temperature sequence from 6000 to 31,000 K, and (2) they lie parallel to lines of constant radius drawn on the H-R diagram. (Recall the formula $L = 4\pi R^2 \sigma T^4$. If R is kept constant, L will be high where T is high, low where T is low.) The average radius of these stars is 0.013 solar radius. Typical values for the physical properties of white dwarfs are 0.7 solar mass for the mass, 0.01 solar radius (7×10^6 m) for the radius, and 10^8 kg/m^3 for the density.

How often do white dwarfs form in the Galaxy? The ones we see probably originate from stars that have 1.4 to 6 solar masses while on the main sequence. Such stars produce hydrogen-exhausted cores of about 0.7 solar mass. If they lost their envelopes, these stars would have just about the average mass of known white dwarfs. If so, white dwarfs are born from their more massive progenitors at the rate of 1 every 4 years.

White Dwarfs and Relativity

The extremely high densities of white dwarf stars provide a natural laboratory in which to test the general theory of relativity, for their surface gravities are high enough to produce a detectable *gravitational red shift* in their spectra. Put simply, a gravitational red

11.1 WHITE DWARF STARS: COMMON CORPSES

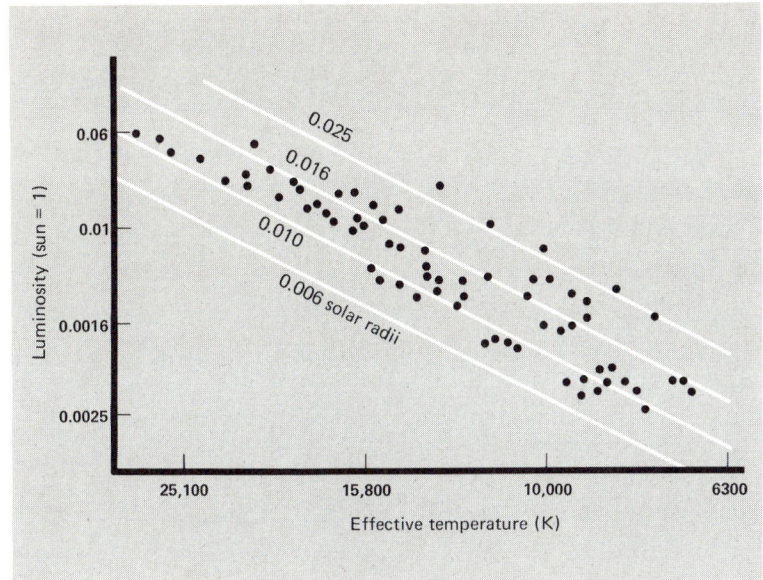

FIGURE 11.5 A Hertzsprung-Russell diagram for well-observed DA white dwarfs. The lines indicate a constant radius. [*Adapted from a diagram by V. Weidmann.*]

shift occurs whenever light moves from a stronger to a weaker gravitational field. As it does so, it must do work, since a photon has an equivalent mass ($E = mc^2$) and a gravitational field can affect it. In such a situation an ordinary particle loses kinetic energy (as it gains gravitational potential energy) and slows down. But photons cannot slow down; they travel only at the speed of light. Instead of slowing down, a photon's loss of energy shows up as a decrease in its frequency (an increase in its wavelength), that is, as a red shift.

The gravitational red shift produced by a star depends on the ratio of its mass to radius. The larger this ratio, the larger the gravitational red shift. The amount of red shift is

$$\frac{\Delta\lambda}{\lambda_0} = \frac{GM}{c^2 R}$$

where $\Delta\lambda$ equals $\lambda - \lambda_0$, M is the object's mass, R its radius, G the gravitational constant, and c the speed of light.

The actual observation is complicated by the motion of the star relative to the earth. This produces a Doppler shift (Section 8.6) to the red if the star is receding. The astronomer sees both shifts (gravitational and Doppler) together. The two can be separated only if the star's velocity through space can be found, a determination possible for binary systems because their velocities in space can be found from the spectrum of the primary star. With the velocity known, the Doppler red shift is subtracted from the total red shift to leave any gravitational red shift.

For a typical white dwarf with $M = 0.7$ solar mass and $R = 0.01$ solar radius, the gravitational red shift amounts to about

$$\frac{\Delta\lambda}{\lambda_0} = \frac{(6.67 \times 10^{-11})(0.7 \times 2 \times 10^{30})}{(3 \times 10^8)^2 (0.01 \times 7 \times 10^8)}$$

$$\approx 10^{-4}$$

The measured red shift for Sirius B is 3.0×10^{-4}. The predicted theoretical value is 2.8×10^{-4}. Within experimental error, the red shift observation confirms general relativity.

Magnetic White Dwarfs

In recent years astronomers have discovered that some white dwarf stars have intense magnetic fields—10^6 to 10^8 gauss at their surfaces. (Recall that the magnetic field of the sun is about 1 gauss.)

These strong fields are probably relics from the time before the star became a white dwarf. The basic physical concept that supports this idea is called *conservation of magnetic flux*. Consider a star with a magnetic field. The *magnetic flux* is essentially the number of field lines (field strength) times the surface area through which they thread. Imagine compressing the star to a smaller size. The number of field lines remains the same, but the surface area decreases, so the field lines draw closer together. The magnetic field strength increases, since the separation of the field lines indicates the intensity of the field. Simply put, the gas and the magnetic field are tied together (Focus 6.2). When the gas is compressed, the magnetic field has no place to go; it must stay trapped within the body of the star.

Because a star's surface area depends on the square of its radius, its magnetic field strength must (if flux is conserved) depend on the *inverse* square of its radius. Let's do an example. Start with a star like the sun: field about 1 gauss, radius 7×10^5 km. Imagine collapsing it to the size of a white dwarf, radius about 7×10^3 km. What does the conservation of magnetic flux predict for the white dwarf's field strength? We have

$$\frac{B_{wd}}{B_{sun}} = \left(\frac{R_{sun}}{R_{wd}}\right)^2$$

where B_{wd} is the white dwarf's field strength, B_{sun} the sun's field strength, R_{wd} is the white dwarf's radius, and R_{sun} the sun's radius. Then

$$B_{wd} = B_{sun}\left(\frac{R_{sun}}{R_{wd}}\right)^2$$

$$= 1 \text{ gauss} \left(\frac{7 \times 10^5}{7 \times 10^3}\right)^2$$

$$= 10^4 \text{ gauss}$$

Although lower than the strongest observed fields, this result shows that this simple collapse idea is plausible. The field may also be intensified by the internal changes in the star during its evolution.

To sum up, a white dwarf is a star with roughly the mass of the sun and size of the earth. A degenerate electron gas supports it against the crush of gravity. No thermonuclear reactions go on in a white dwarf; it has come to the end of the line of energy production. Very slowly (in billions of years), its stored internal heat radiates into space. Eventually, it becomes a black dwarf (not to be confused with a black hole). Our sun will become a white dwarf, then a black dwarf—a cold corpse in space.

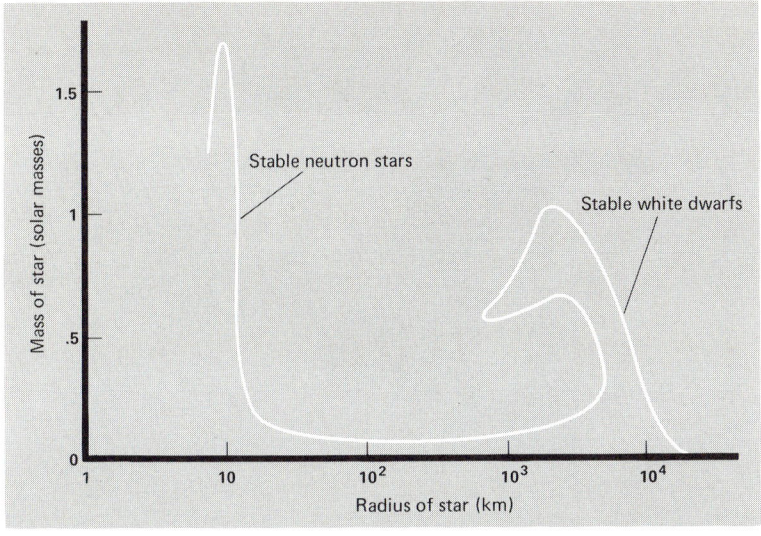

FIGURE 11.6 A mass-radius diagram for cold, nonspinning matter—the results of theoretical calculations. Given is the radius for a mass held together by gravity. Note a region for stable white dwarfs and neutron stars. In these parts of the curve internal pressure from the degenerate state of matter supports the star. [*Based on calculations by K. Thorne and colleagues.*]

11.2 NEUTRON STARS: COMPACT REMAINS OF MASSIVE STARS

What happens to stars which have more than 1.4 solar masses of material at the end of their evolution? Strange things! Since degenerate electron pressure cannot support them, gravity crushes them to higher and higher densities (Fig. 11.6). At about 10^{13} kg/m^3 the inward pressure has increased to such a value that *inverse beta decay* occurs, the process by which an electron and proton join together to form a neutron and neutrino (an electron is sometimes called a beta particle). This happens both to free protons and to protons that are part of the nucleus of a heavy element. At around 10^{15} kg/m^3 the neutrons begin to drip off the nuclei and to form a separate gas. At 10^{17} kg/m^3 the nuclei suddenly fall apart into a gas with proportions 80 percent neutron, 10 percent electron, and 10 percent proton. At this density the neutrons become degenerate in the same manner that electrons do at white dwarf densities. The neutrons provide a degenerate gas pressure and so balance the inward pull of gravity. This pressure allows the formation of a stable *neutron star*, a star composed mainly of neutrons. Its diameter will be about 10 to 20 km, depending on its mass (Fig. 11.1).

In a typical neutron star (Fig. 11.7) with a radius of about 15 km, the inner 12 km consists of a neutron gas at such high densities that it is a fluid. The outer 3 km is a mixture of the neutron super fluid and neutron-rich nuclei, which are arranged in a solid lattice. The structure is a crystalline solid similar to the interior structure of a white dwarf. In the outer few meters, where the density falls quickly, the neutron star has an atmosphere of atoms, electrons, and protons. The atoms are mostly iron.

Because a neutron star is so dense, it has an enormous surface gravity. For example, a solar-mass neutron star with a radius of 12 km has a surface gravity 10^{11} times greater than that at the earth's surface! Such an enormous pull means that mountains on a neutron star won't be very high, a few centimeters at most. This intense gravitational field also results in a huge escape velocity (Section 5.5), as much as about $0.8\ c$. Also, objects falling onto

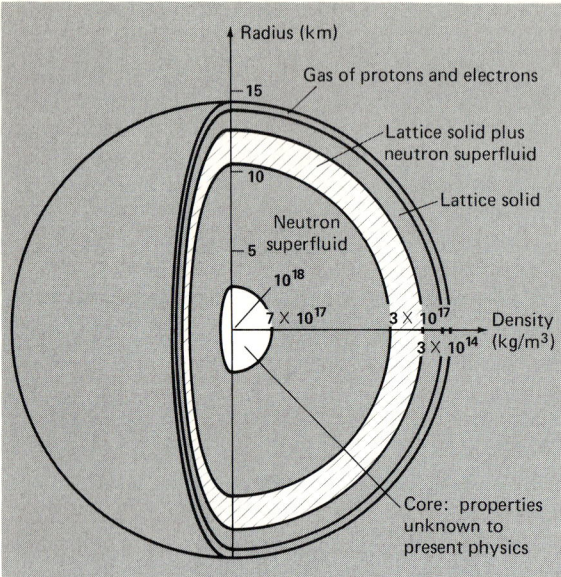

FIGURE 11.7 A theoretical model for the cross-section of a neutron star. The solid crust is mostly iron, topped off by an atmosphere of protons and electrons only a few meters thick.

FIGURE 11.8 Nova Herculis 1934. These photos show the star before (top) and during (bottom) its outburst. [*Courtesy Lick Observatory.*]

a neutron star from a great distance have at least the escape velocity when they hit. That means that even a small mass carries a fantastic amount of kinetic energy. For example, a marshmallow dropped onto a neutron star from a few AU's out will knock into the surface with a few *megatons* (TNT equivalent) of kinetic energy!

The gravitational red shift from a neutron star is substantial. For a solar-mass neutron star about 7 km in radius,

$$\frac{\Delta\lambda}{\lambda_0} = \frac{GM}{c^2 R} = \frac{(6.67 \times 10^{-11})(2 \times 10^{30})}{(9 \times 10^{16})(7 \times 10^3)}$$
$$\approx 10^{-1}$$

This means a 500 Å shift at 5000 Å (green light), a good fraction of the entire visible range, enough to change green light into yellow. A neutron star has these bizarre properties simply because its gravitational field is so intense.

An ordinary star with a mass at the end of its evolution greater than 1.4 solar masses and less than 5 solar masses probably ends up as a neutron star. Theoretically, a stable neutron star with a mass less than 1.4 solar masses can also form. These low-mass neutron stars could be made in the pile-driver compression of a supernova explosion, as are most neutron stars.

Because the neutron gas is degenerate, a neutron star has the same kind of mass-radius relation as a white dwarf star: the greater the mass, the smaller the radius (Fig. 11.6). In an analogy to the Chandrasekhar limit, a mass limit for neutron stars is reached when the gravitational forces overwhelm the degenerate neutron gas pressure. This limit—not known exactly, but about 5 solar masses—signals the next crushing point of matter by gravity. Stars greater than 5 solar masses at the end of their evolution will end up as black holes (Chapter 12).

Notice that we haven't presented evidence yet for the existence of neutron stars. What we've sketched out above are *theoretical* ideas about neutron stars, which were first worked out in the late 1930s. Evidence for the reality of neutron stars didn't crop up until almost 40 years later. That evidence involves the cataclysmic explosions of stars called supernovas (Section 11.4) and rapidly pulsing radio sources called pulsars (Section 11.5).

11.3 NOVAS: MILD STELLAR EXPLOSIONS

Aristotle asserted that the heavens were unchanging. Throughout the Middle Ages this precept required that the number of observed stars remain constant. This deeply ingrained principle received a hard knock in 1572 with the discovery of the new star, or *nova stella*, that was observed extensively by Tycho Brahe. Just a few years later, in 1604, Kepler kept a close watch on another nova that burst into view in the constellation of Ophiuchus.

With the advent of photography and large telescopes in the nineteenth century, astronomers discovered large numbers of novas scattered throughout the sky. By the beginning of this century a nova was no longer considered a new star, but rather the sudden eruption of light from an existing star (Fig. 11.8). Also in this century astronomers recognized that some of these outbursts took place with extraordinary violence. These special flare-ups are now called *supernovas* to distinguish them from ordinary novas. The novas of 1572 and 1604 are now known to have been supernovas.

Both novas and supernovas represent explosions of stars. Ordinary novas involve only the outer layers of the star, while supernovas involve the interior regions as well.

Ordinary Novas

In a typical nova outburst a star rises about 60,000 times in brightness in just a few days. It stays up at peak brightness for several hours. Then the nova's light slowly declines in a few hundred days to a level of brightness that is low, but usually slightly brighter than the star's prenova brightness. A plot of a nova's rise and fall in brightness (or luminosity) is called its *light curve* (Fig. 11.9). The general shape of the light curve is the same for all novas: a sharp rise with a gradual decline.

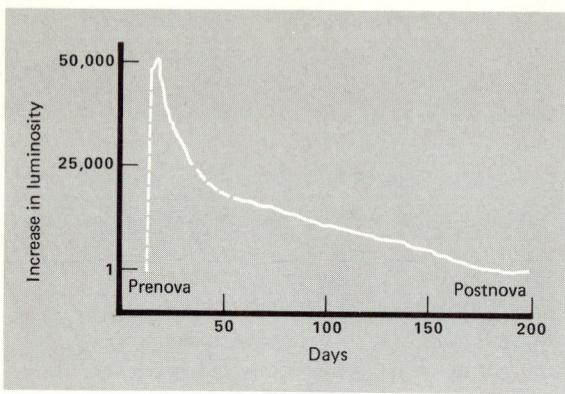

FIGURE 11.9 A light curve for a typical nova. [*Based on a compilation by C. Payne-Gaposhkin.*]

FIGURE 11.10 Nova Herculis 1934, photographed in red light in 1951. Note the shell of blown-off material, which is expanding into the interstellar medium. [*Courtesy Palomar Observatory, California Institute of Technology.*]

A typical nova hits a peak brightness of greater than 10^5 solar luminosities. It emits during its hundred days of flare-up and demise some 10^{37} to 10^{38} J or about as much energy as the sun generates in about 100,000 years.

The visual light curve displays only a part of a nova's activities. A nova in the constellation Serpentis in 1970 was observed by an ultraviolet telescope. It showed that the ultraviolet output increased for 60 days after the visual peak and just about compensated for the decline in visual luminosity. Later infrared observations demonstrated that the nova's infrared plus visual energy output—1.6×10^4 solar luminosities—roughly equaled the earlier output in the visual plus ultraviolet. Although it is dangerous to generalize from one nova, it appears that a nova's total energy output in all parts of its spectrum remains roughly constant for a few hundred days.

A nova's spectrum undergoes pronounced and complicated changes during its outburst. We won't describe them in detail, but generally the prenova star's spectrum has broad, dark absorption lines with weak or no bright emission lines. At maximum the nova has absorption lines similar to those of an A or F supergiant star. Some time after maximum the nova's spectrum develops emission lines similar to those from H II regions (Chapter 9). The Doppler shifts of these lines range from a few hundred kilometers per second in some novas to a few thousand in others.

The evolution of a nova's spectrum, the changing strengths and Doppler shifts of the lines, indicates several things. First, the star's photosphere dramatically expands in size to 100 to 300 solar radii. Second, the photosphere then collapses back onto the star. Third, a shell of material is blown off the star and rapidly expands away from it (Fig. 11.10). Early on, the density of this shell is about 10^6 particles/cm^3. Overall, a nova spurts off about 10^{26} kg of material, about 10^{-4} solar mass. Because the prenova star has about 1 solar mass, the ejected material makes up only a small fraction of the total.

Once the shell appears, its expansion can be used to find out the distance to the nova. The basic idea is to compare the *angular* expansion velocity, as measured on photographs, with the *spatial* expansion velocity, as determined from the Doppler shift. To do this, you have to assume that the expansion is symmetrical, that is, that the velocity perpendicular to the line of sight, which determines the angular expansion, is the same as the velocity along the line of sight, which determines the Doppler shift. Details of the method are in Focus 11.1.

11.1 DISTANCES TO NOVAS

If we can see the shell blasted out by a nova, measure the rate at which it expands (in arcseconds per year), and use the Doppler shift to find the actual rate of the shell's expansion (in kilometers per second), we can then determine the nova's distance. Let's look at the details.

The front side of the expanding shell results in a blue shift of spectral lines (Fig. F.15). In fact, the part of the shell directly along our line of sight to the star produces absorption lines. From their blue shift, we know the expansion rate of the shell by the Doppler formula:

$$v = \frac{\Delta \lambda}{\lambda} c$$

In a year, the shell expands some amount, D km ($D = v$ km/sec $\times$ 3.15 $\times$ 10^7 sec/yr). From the earth we measure an expansion of α arcsec/yr. Then we can apply the skinny triangle formula, with α in *radians* (1 radian = 206265 arcsec),

$$\alpha = \frac{D}{d}$$

to find the distance, d, to the nova:

$$d = \frac{D}{\alpha}$$

With the distance in hand, we can then work out the nova's luminosity from its light curve.

As an example, suppose we measure a Doppler shift of 1000 km/sec and an expansion of 0.1 arcsec/yr. Then

$D = (1000 \text{ km/sec})(3.15 \times 10^7 \text{ sec/yr})$

$= 3.2 \times 10^{10}$ km

and, since $0.1'' = 0.1/206265$

$= 4.8 \times 10^{-7}$ radians,

$$d = \frac{3.2 \times 10^{30} \text{ km}}{4.8 \times 10^{-7}}$$

$= 6.6 \times 10^{16}$ km

$= 2.1$ kpc

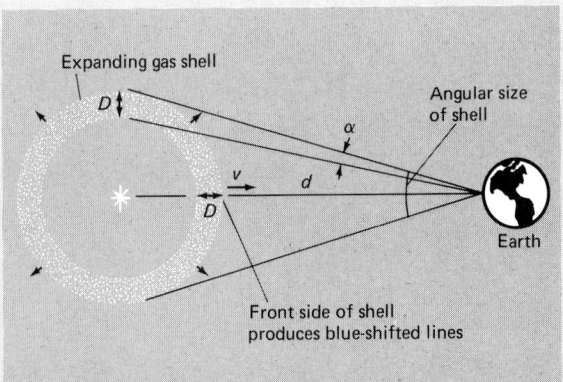

FIGURE F.15 Geometry for an expanding shell.

A Nova Model

What brings a nova to the point of explosion? One major clue is that observational studies indicate that almost all novas occur in close binary systems, that is, binary stars with short periods, such as spectroscopic binaries (Section 8.6). In such systems the two stars are so close that matter may flow from the more massive companion to the less massive one.

A nova may occur as part of the natural evolution of binary stars consisting of a main-sequence star and a less massive companion. At the end of its life the main-sequence star (which has more mass than its companion and so evolves faster) becomes a white dwarf. Remember that a white dwarf consists of a degenerate electron gas in a lattice of nuclei without hydrogen. In order to ignite hydrogen fusion again, it needs fresh hydrogen fuel.

Where might such material come from? In a binary star system the material can come from the companion star. Around each star is a region of space where its gravitational force dominates (Fig. 11.11). The edge of this region is called the *Roche lobe*; any matter within the Roche lobe is gravitationally bound to that star and cannot escape to the other star. But one point exists where the Roche lobes touch and join, and at that point the gravitational pull from one star just cancels that from the other. In many binary star systems the stars orbit so close together that this common point is very close to one or the

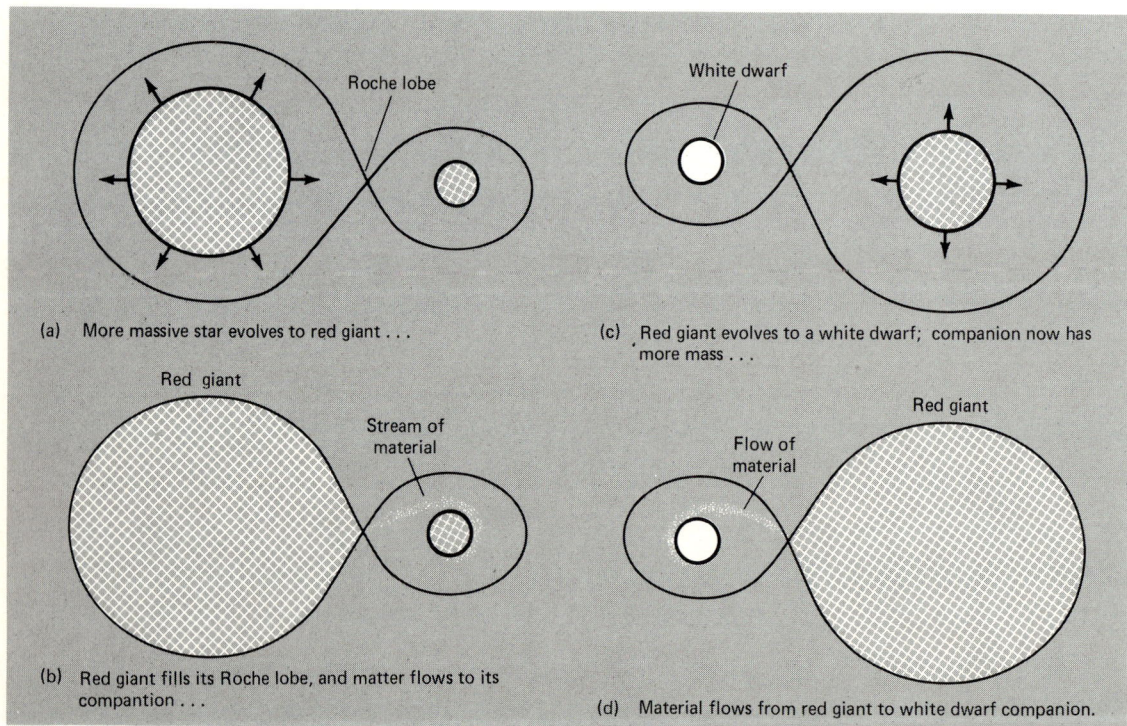

(a) More massive star evolves to red giant . . .

(b) Red giant fills its Roche lobe, and matter flows to its companion . . .

(c) Red giant evolves to a white dwarf; companion now has more mass . . .

(d) Material flows from red giant to white dwarf companion.

FIGURE 11.11 A model for the evolution of a red giant–white dwarf binary system to produce a nova outburst.

other star. This means that a gravitational highway exists for the flow of matter between the stars.

In order for the mass exchange to happen, material from one star has to get out to the Roche lobe. How? When stars like the sun reach old age, they become red giants before ending up as white dwarfs. Suppose two such stars are in a binary system. One of them evolves to a red giant and then a white dwarf (Fig. 11.11a–c). When the companion star evolves, it also bloats up as a red giant, so its atmosphere swells up and reaches its Roche lobe. Material then flows from the red giant to the white dwarf (Fig. 11.11d).

As the matter falls toward the white dwarf, it forms a disk around it, called an *accretion disk*. The material gathers in a disklike structure, rather than falling directly onto the other star, because of the conservation of angular momentum. The matter in the accretion disk gradually loses angular momentum by collisions with other particles and by the action of radiation pressure, and spirals onto the white dwarf's surface. Fresh hydrogen thus gradually accretes onto the white dwarf, forming a virgin envelope. Additional material piles on, compressing and heating it. When the temperature at the bottom of the accreted layers reaches 10^6 K, hydrogen fusion reactions ignite. Because the gas here is degenerate, the ignition is explosive (just like the helium flash in red giant stars). The runaway fusion reactions heat the layer to some 10^8 K; it loses its degenerate state. Then the material expands. Both explosive shocks and pressure blow the accreted material into space. A buildup of only some 10^{-6} solar mass prompts the nova outburst.

It doesn't take a large rate of mass infall to set up and ignite runaway reactions in a reasonable time. For a typical nova in a binary system, just 10^{-8} to 10^{-7} solar mass a year suffices. Some studies have shown that a mere 10^{-13} solar mass a year will do the trick. This low rate, only 10^8 kg/sec or so, can even be supplied by accretion from the general interstellar medium rather than from a companion star. That's one way a single white dwarf can become a nova.

Nova Cygni 1975: A Recent Nova

I (MZ) moved to New Mexico toward the end of August 1975. My department had a picnic in Corrales, on the west bank of the Rio Grande. It was a beauty of an August day, with the Sandia Mountains turning watermelon-pink at sunset. As I gazed up at the darkening sky, I sensed that something was wrong with the stars. Then I saw it. A new star in the constellation Cygnus near Deneb (Fig. 11.12). For a moment I was stunned. Then I recalled hearing about the discovery of this nova. But in the hustle and hassle of getting set up in a new place at a new job, I had forgotten about it. The shock of my personal discovery brought home to me the strangeness of a *nova stella* in the sky.

I (JG) too, was startled by Nova Cygni 1975. Comfortably asleep in my Berkeley brown-shingle house that August night, I was surprised by a midnight phone call from a former student, Luis Carrasco, who was observing at San Pedro Martir Observatory in Baja California. A bad connection, my stupor, and his Spanish accent hindered communication, but I understood the words nova, Cygnus, and Lick. I realized he had seen a star in Cygnus which wasn't on his star charts and wanted me to check it out with Lick Observatory. I thought he meant he had seen it through his telescope, and since faint novas (or wrong charts) aren't all that rare, I wasn't too excited. But then I walked out on my deck and was suddenly confused.

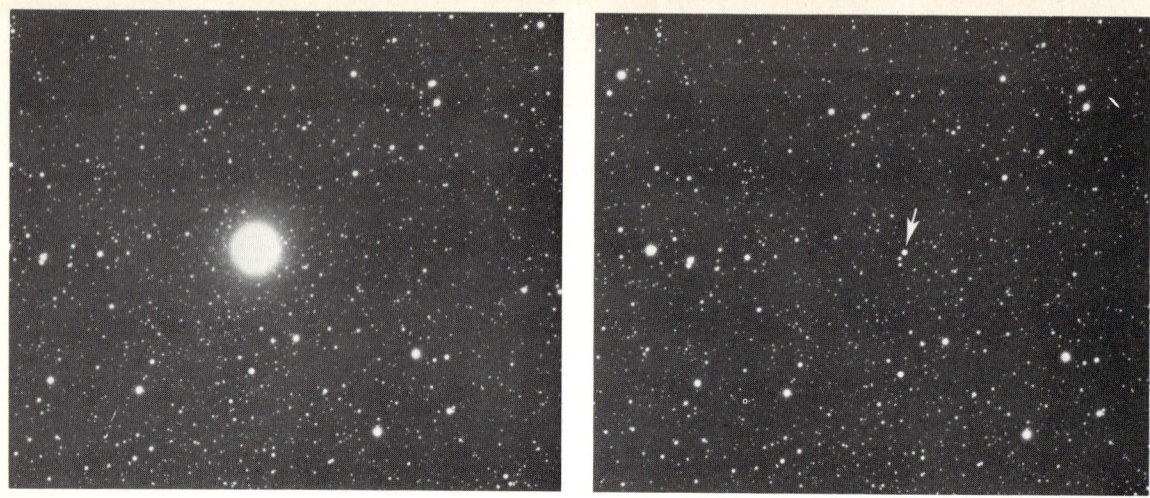

FIGURE 11.12 Nova Cygni 1975 during its maximum brightness (left) and about four months later (right). [*Courtesy Lick Observatory.*]

Cygnus, the Northern Cross, no longer looked like a cross! How disconcerting (but exciting) to see the familiar order suddenly disturbed by the interloper.

Nova Cygni 1975 peaked at magnitude +1.8. Its light curve (Fig. 11.13) showed a nova's typical sharp rise and decline. But note a less steep rise before the nova's light shot up to its peak. This part of a nova's light curve had never really been seen before. Fortunately, some photos of Cygnus, such as those by Los Angeles amateur astronomer Ben Mayer, caught the star before it went nova (Fig. 11.14).

Prior to its nova outburst, the star had a magnitude of less than 20 on old photos. So

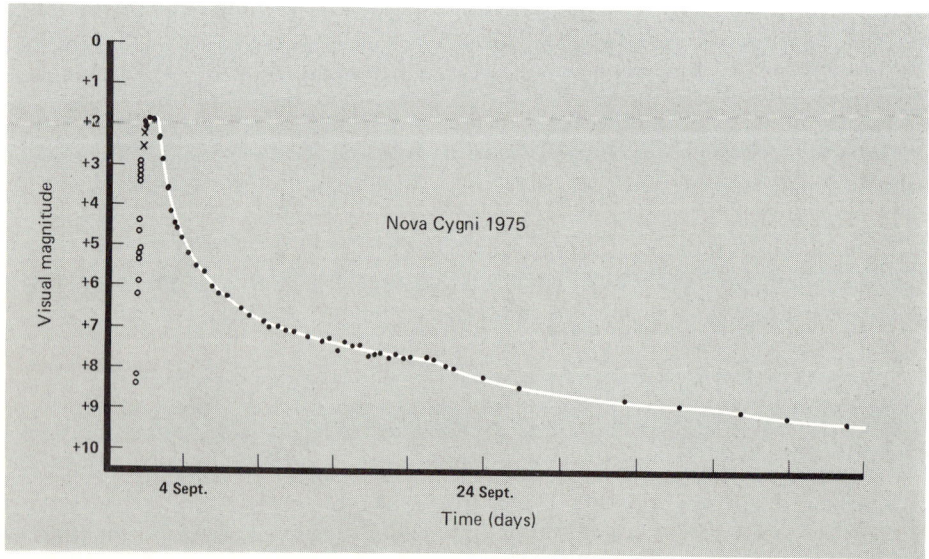

FIGURE 11.13 A visual light curve for Nova Cygni 1975. [*From work by P. Young, H. Corwin, J. Bryan, and G. de Vaucourleurs,* Astrophysical Journal, *vol. 209, p. 882, copyright 1976 by the American Astronomical Society.*]

FIGURE 11.14 A unique set of photos showing Nova Cygni 1975 before (arrow in 1 and 2) and during its outburst (3, 4, and 5). [*Courtesy Ben Mayer.*]

it increased at least 18 magnitudes, or 16 million times in luminosity. That's unusually luminous for a nova. Yet, Nova Cygni 1975 had a spectral evolution that followed the typical nova's pattern (Fig. 11.15). So this outburst was a regular nova, but an extremely violent one.

We have no indication to date that Nova Cygni 1975 is a member of a binary system. It may be one of those infrequent cases where a star goes nova from the accretion of interstellar material.

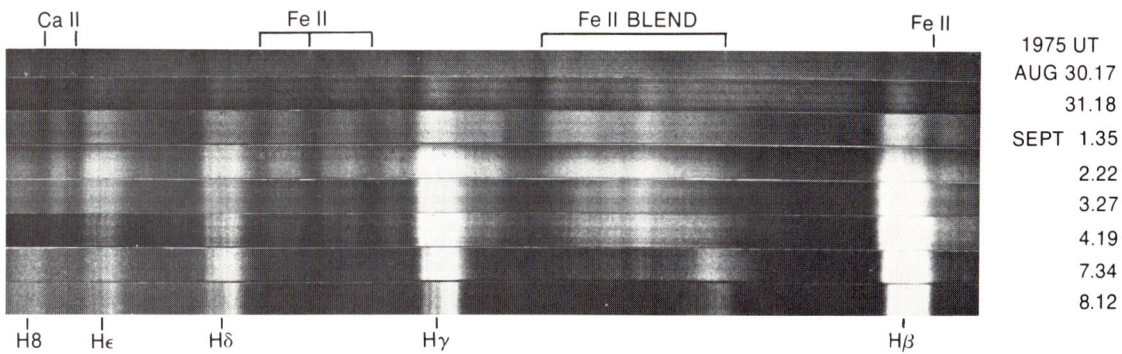

FIGURE 11.15 Evolution of the red part of the spectrum of Nova Cygni 1975. Note how very intense lines of the Balmer series develop. [*Courtesy B. Bohannon.*]

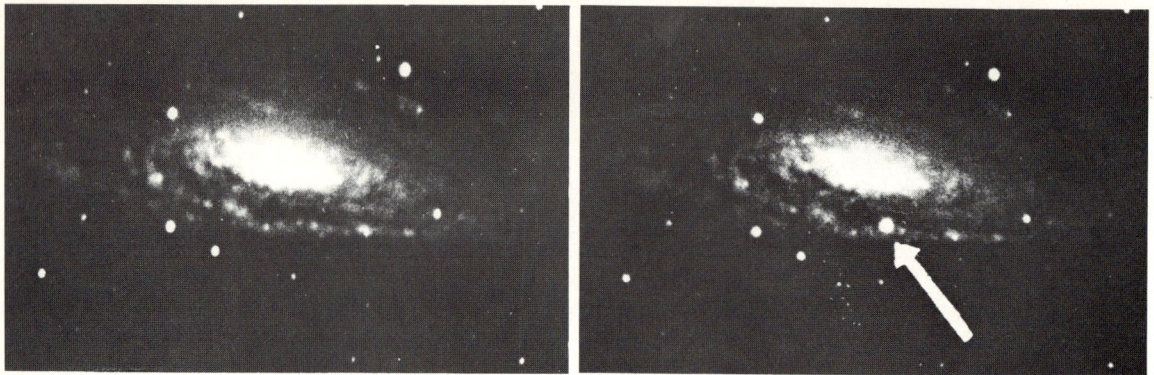

FIGURE 11.16 A 1959 supernova in the galaxy NGC 7331, before the supernova (left) and during the supernova's maximum brightness (right). [*Courtesy Lick Observatory.*]

In summary, novas may occur when material accretes onto a white dwarf. The infall heats the material, igniting runaway thermonuclear reactions that blow off the outer layers. Most novas are members of binary systems, with a red giant companion. Matter flows from the red giant to the white dwarf to set up the nova explosion. A few novas may be white dwarfs by themselves accreting interstellar matter.

11.4 SUPERNOVAS: CATACLYSMIC EXPLOSIONS

As violent as novas may appear, they cannot match the fierce destruction of a star in a supernova. These cataclysmic explosions spew out energy in extraordinary amounts—

TABLE 11.1 Supernovas Observed in the Galaxy by the Naked Eye

Date (A.D.)	Constellation	Apparent Brightness	Distance (kpc)	Observers
185	Centaurus	Brighter than Venus (-6)	2.5	Chinese
369	Cassiopeia	Brighter than Mars or Jupiter (-3)	10	Chinese
1006	Lupus	Brighter than Venus (-5)	3.3	Chinese, Japanese, Korean, European, and Arabian
1054	Taurus (Crab Nebula)	Brighter than Venus (-5)	2	Chinese, Southwestern Indian, and Arabian
1572	Cassiopeia	Nearly as bright as Venus (-4)	5	Tycho and many others
1604	Ophiuchus	Between Sirius and Jupiter (-2)	6	Kepler, Galileo, and many others

Source: Adapted from a table compiled by W. C. Straka.

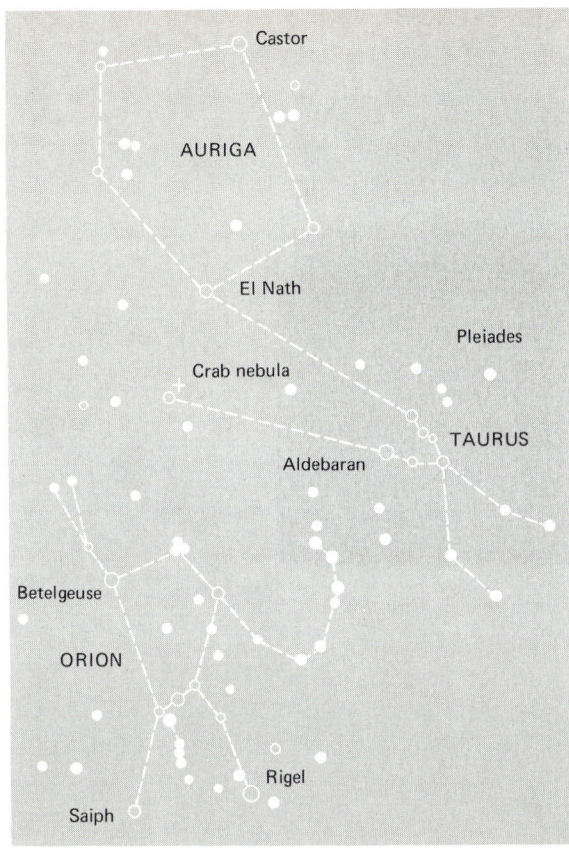

FIGURE 11.17 The position of the Crab Nebula in the constellation of Taurus.

about 10^9 to 10^{10} times the sun's luminosity at their peak (compared to a nova's 10^5 solar luminosities). According to contemporary ideas, a supernova usually signals the death of a massive star.

The name *supernova* was coined by Fritz Zwicky and Walter Baade for the extraordinary novas discovered in our own and other galaxies. Over 300 supernovas have been found in other galaxies (Fig. 11.16). Both the new stars observed by Tycho and Kepler were supernovas. Since the supernova of 1604, no such grand explosion has been seen in the Galaxy. Supernovas are such rare events that only six have been noted in the Galaxy during recorded history (Table 11.1). Of course, over the history of the Galaxy, about 10 billion years, hundreds of millions of supernova explosions have occurred. From this point of view, they are not rare at all.

The supernova in Taurus in 1054 marks an event of continuing interest since its sighting. Chinese astronomers termed temporary celestial objects, such as novas or comets, "guest stars." In the Chinese history *Shung-Shih* we find the report of a guest star that did not move, so it must have been a star rather than a comet. The dynasty dates the event well. The guest star entered the sky on July 4, 1054. Close study of Chinese and Japanese accounts of this visitor confirms that the star remained visible to the unaided eye for over 650 days in the night sky. It was visible in daylight for 23 days! The position noted by the oriental astronomers placed the event in the constellation Taurus.

In 1731 the amateur astronomer John Bevis discovered a faint nebulosity in Taurus just above the Bull's horns (Fig. 11.17). Much later, in 1928, Edwin Hubble measured the

FIGURE 11.18 The Crab Nebula, photographed to emphasize its filamentary structure. The starless areas have been artificially blocked out. [*Courtesy V. Regener, Capilla Peak Observatory, University of New Mexico.*]

expansion rate of this nebula, which had become known as the Crab. He deduced that its expansion began about 900 years earlier. Hubble concluded that since the Crab Nebula was near the position given for the Chinese guest star, that explosion was the source of the nebula. The Crab Nebula became the first identified supernova remnant in our Galaxy (Fig. 11.18).

Classifying Supernovas

Astronomers classify supernovas by the shape of their light curves into two general categories (Table 11.2). Type I exhibits a sharp maximum, about 10 billion solar luminosities, and dies off gradually; Type II has a less sharp peak at maximum, about 1 billion solar luminosities, and dies away more sharply (Fig. 11.19). Studies of other galaxies have revealed that Type II supernovas typically occur in association with Population I stars (Section 10.9) and Type I in association with Population II stars.

Supernova types also differ in their spectra (Fig. 11.20). At maximum brightness a Type II supernova shows a pretty nondescript spectrum; the only prominent line is an emission line at 6563 Å, the H-alpha line from hydrogen. About a month later the

TABLE 11.2 Properties of Supernovas

	Type I	Type II
Ejected mass (solar masses)	0.5	5
Velocity of ejected mass (km/sec)	10,000	5000
Total kinetic energy (J)	5×10^{43}	10^{44}
Visual radiated energy (J)	4×10^{42}	10^{42}
Maximum absolute magnitude	-19 to -20	-17
Frequency	1 in 60 yr	1 in 40 yr

Source: R. A. Chevalier, "The Interaction of Supernovae with the Interstellar Medium," *Annual Reviews of Astronomy and Astrophysics*, vol. 15 (1977).

11.4 SUPERNOVAS: CATACLYSMIC EXPLOSIONS

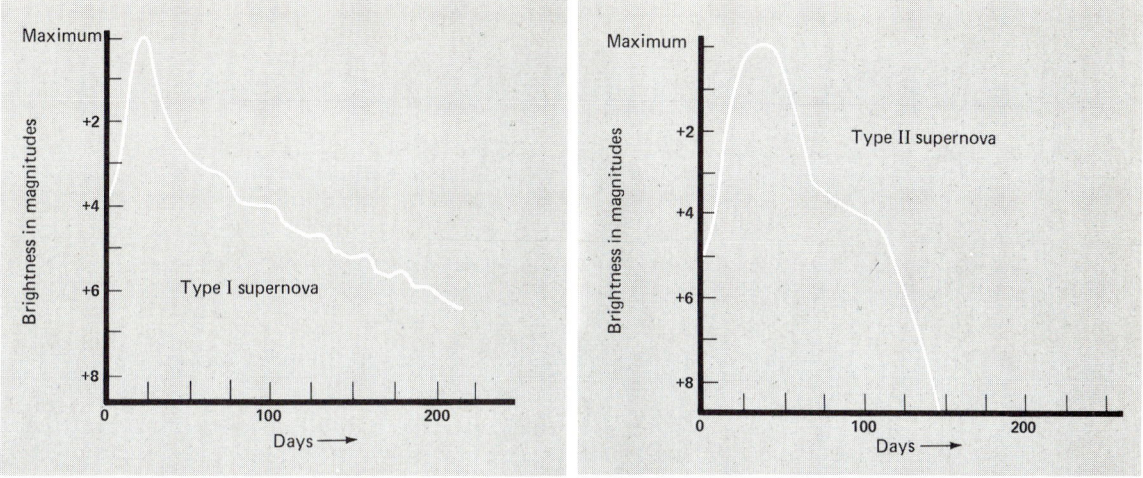

FIGURE 11.19 Typical light curves for Type I and Type II supernovas. [*Based on a compilation by W. Straka.*]

spectrum has evolved to show more emission lines and a few weak absorption lines. In contrast, Type I supernovas exhibit messy spectra. At maximum light broad emission lines appear along with some strong, dark lines. Later, four emission lines, whose origins are not yet known, dominate the spectra, along with emission lines H-alpha, Na I, and Ca II.

The total energy output from any supernova is stupendous: 10^{44} J, or approximately as much energy as the sun produces in its entire lifetime of 10 billion years. At its brightest a supernova shines with a light of *10 billion* suns!

How often this kind of cosmic violence takes place is still debated. The rate of occur-

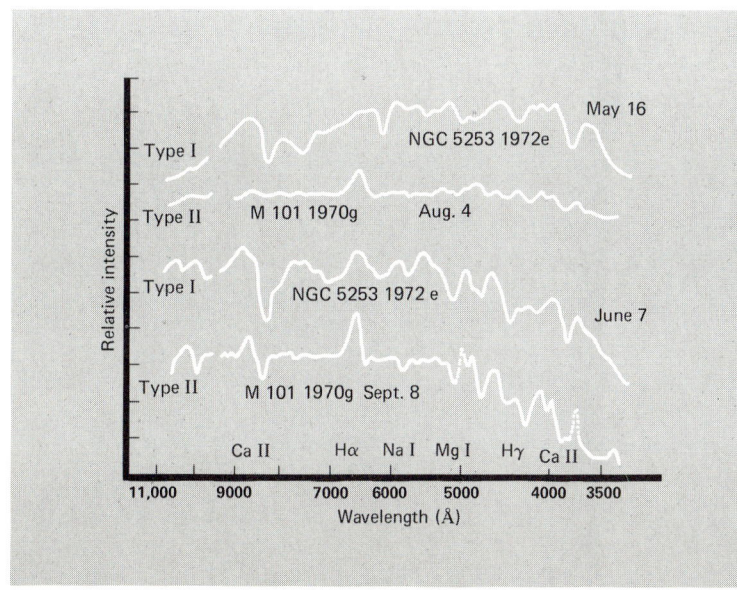

FIGURE 11.20 Changes in the spectra of a Type I (NGC 5252 1972e) and Type II (M 101 1970g) supernova. Lines from hydrogen, calcium, sodium, and magnesium are labeled. [*From a diagram by R. Kirshner, J. Oke, M. Penston, and L. Searle,* Astrophysical Journal, *vol. 185, p. 303, copyright 1973 by the American Astronomical Society.*]

rence in any one galaxy is low, but the vast number of visible galaxies ensures that a few supernovas will be observed every year. From the rate in other galaxies, we estimate that a Type I supernova bursts forth in a galaxy once in 60 years. Type II supernovas may be more frequent, once in roughly 40 years. Some astronomers argue that the true frequency must be greater because we do not observe all supernovas. In any case, 1 supernova every 50 years in a galaxy is a pretty reasonable estimate of average supernova frequency.

If supernovas occur in a galaxy once every 50 years, why haven't we seen one in our own Milky Way Galaxy since 1604? There probably have been some, but in distant regions of the Galaxy, where their light was cut off by the interstellar dust. Nevertheless, a supernova in our part of the Galaxy is almost due. Sidney van den Bergh has estimated that a supernova will be seen in our Galaxy sometime during the next 50 years. However, it may not present a spectacular stellar show since its light may also be cut down by interstellar dust.

Both types of supernova violently eject a large fraction of the original star's mass at speeds of about 5000 to 10,000 km/sec. At maximum brightness a supernova can reach a size about that of the solar system—a few light hours in diameter. For about the first month after the time of maximum light the supernova's size increases dramatically. For example, one Type II supernova observed in 1970 had a radius at maximum light of roughly 3×10^9 km, as large as the orbit of Uranus. The star expanded at about 5000 km/sec for 30 days, until it attained a radius of 2×10^{10} km, approximately three times larger than the solar system or one light day in diameter. After that the star's photosphere shrank. During the expansion the photosphere decreased in blackbody temperature from 12,000 K to 6000 K.

The Origin of Supernovas

What kinds of stars become supernovas? Type II are thought to be stars much more massive than the sun (10 to 100 solar masses), stars that live their normal lives as O and B stars. Current theoretical computer models for Type II supernovas ignore the details of the actual cause of the explosion, but assume it occurs in the core of a red supergiant star (Sections 10.6 and 10.7). They then investigate only the effects of the shock wave traveling through the atmosphere of such a star.

Such models mimic two important observed characteristics of supernovas. First, red supergiants are so large that they do not cool much when they expand to a size of roughly that of the solar system. Second, the atmosphere of a red supergiant has nearly constant density, so a shock wave traveling through it moves at almost constant velocity and transmits energy efficiently to the star's surface. The models predict that at peak brightness the supernova should have a photospheric temperature of roughly 10,000 K and a surface speed of 5000 km/sec. These numbers are in the right ballpark for what is observed.

Because Type I supernovas occur in Population II, these stars are roughly the mass of the sun. (Very massive stars cannot be old stars.) Type I supernovas are really a puzzle, for it is hard to see how a solar-mass star can detonate as violently as a supernova. One idea resembles that for binary novas. Imagine a binary system containing a white dwarf and a normal star in which the white dwarf has a mass very close to the maximum limit (1.4 solar masses). If enough mass flows onto the white dwarf to push it over the limit, it will collapse violently to a neutron star. This collapse may release sufficient energy to make a supernova. (This model is very tentative; don't accept it without reservation.)

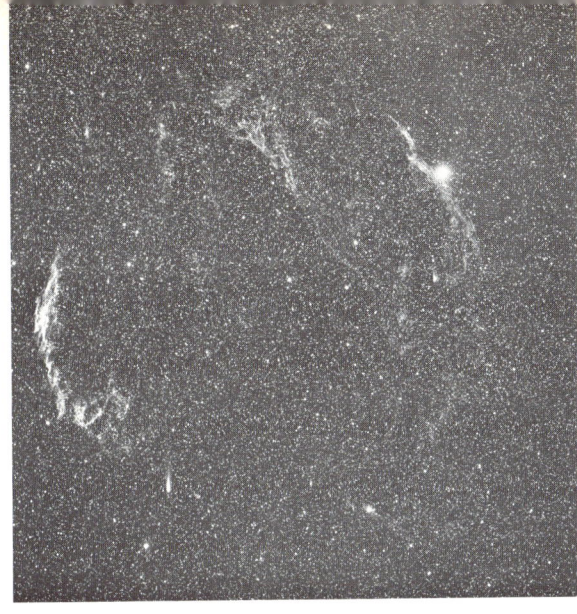

FIGURE 11.21 A supernova remnant in Cygnus. Note how the bright wisps almost make a circle, as if blown out from a central point. [*Courtesy Palomar Observatory, California Institute of Technology.*]

Supernova Remnants

A supernova bangs out a blast wave into the interstellar medium. Traveling at supersonic velocities, the shell of material creates a shock wave that plows through the interstellar gas and dust. The shock wave's collisions with the cool clouds of the interstellar medium excite the interstellar material so that it glows. This luminous material marks a *supernova remnant*. The Loop Nebula in Cygnus is such a remnant (Fig. 11.21). Note that it looks spherical—a shell produced by the interaction between the interstellar medium and a supernova shock wave.

A similar nebula in the southern sky—the Gum Nebula (Fig. 11.22)—extends over

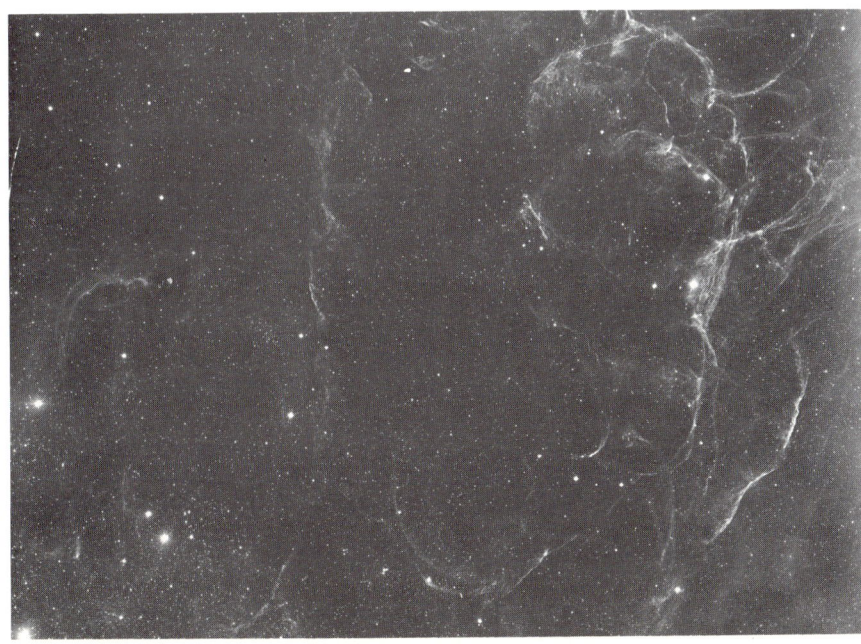

FIGURE 11.22 A part of the Gum Nebula, a supernova remnant. Note how the filaments form circular arcs. The entire nebula is more than 2000 light years in diameter. [*Courtesy B. Bok and Steward Observatory, University of Arizona.*]

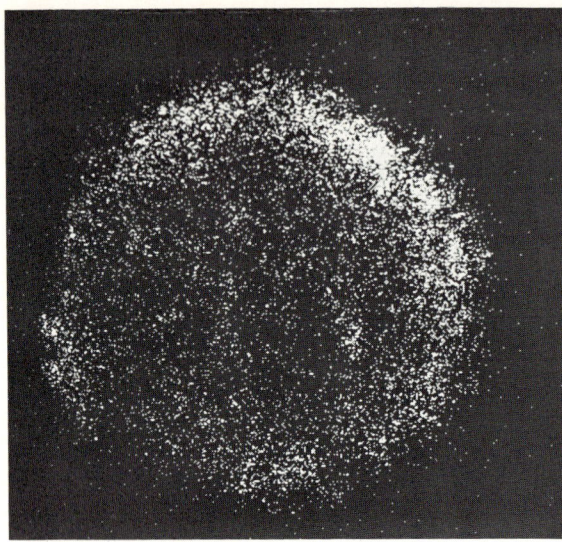

FIGURE 11.23 An X-ray image of Tycho's supernova remnant, taken by the Einstein X-ray Observatory. [*Courtesy P. Gorenstein and F. Seward, Center for Astrophysics.*]

50° in the sky. The Gum Nebula has a diameter of about 2300 light years, its closest edge being only 300 light years from the sun. Stephen Maran and coworkers believe that this nebula was created by the pulse of ultraviolet radiation and X rays generated by a supernova some 11,000 to 20,000 years ago. An X-ray source, named Vela X, lies almost in the nebula's center. It is a prime suspect as the supernova site. The discovery of a pulsar near the location of the Vela X source supports its nature as a supernova remnant. (Pulsars, see Section 11.5, are believed to be the neutron stars, Section 11.2, formed as the by-product of a supernova explosion.)

Radio and X-ray astronomers have a significant advantage over optical astronomers in the hunt for galactic supernova remnants. X-ray astronomers can observe young supernova remnants directly; over 30 have been observed so far with the Einstein X-Ray Observatory. The huge shock waves plow through the interstellar medium at speeds of hundreds of kilometers per second; they compress and heat the interstellar gas to temperatures of at least a few million Kelvins in the zone just behind the blast wave. This gas emits X rays. The X-ray pictures of Type I remnants, such as Tycho's supernova (Fig. 11.23) typically show symmetrical shells with variations in brightness around their rims—a possible indication of the patchy structure of the interstellar medium.

Radio astronomers can observe low-density excited gas that has no detectable optical emission (Fig. 11.24). For example, the radio astronomers were the first to detect Cassiopeia A, believed to be a supernova remnant. Later work with optical telescopes revealed faint patches of nebulosity, the debris from the supernova, at the same location.

Another advantage of radio astronomy is that it can separate supernova remnants from other radio sources by a special property of their radio emission. The intensity plotted versus frequency displays a *nonthermal spectrum*. What does this mean?

Thermal and Nonthermal (Synchrotron) Emission

The spectrum of blackbody radiation has a characteristic shape. The energy output increases from longer to shorter wavelengths, peaks at a wavelength that depends on the

11.4 SUPERNOVAS: CATACLYSMIC EXPLOSIONS 423

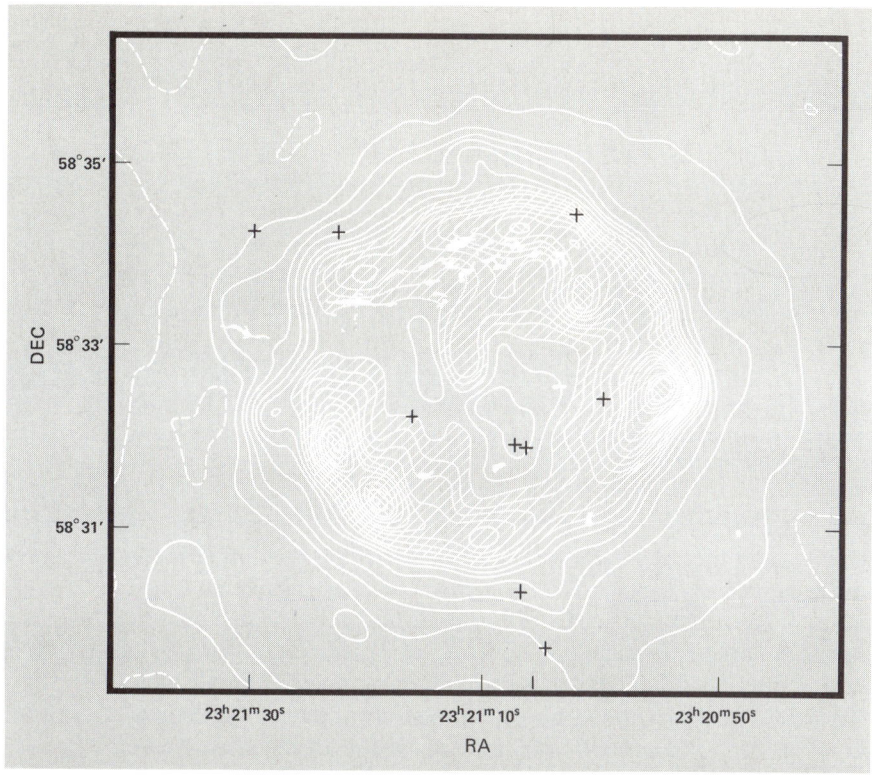

FIGURE 11.24 (a) A radio intensity map of a supernova that was the "nova" observed by Tycho Brahe in 1572. It shows a shell of hot gas expanding into the interstellar medium. [*Based on observations by M. Ryle, B. Elsmore, and A. Neville.*] (b) A visual photo, taken in red light, of the same region of the sky. The only indication of the supernova remnant is a few faint wisps visible at top center. [*Courtesy Palomar Observatory, California Institute of Technology.*]

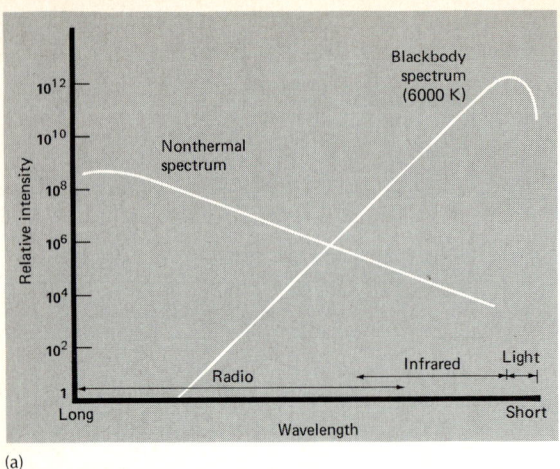
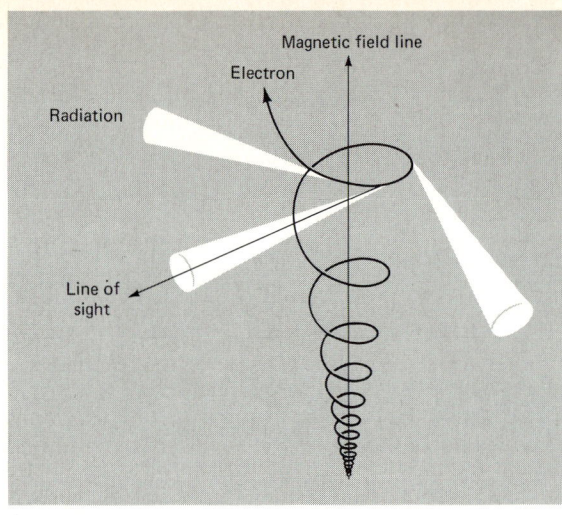

FIGURE 11.25 (a) A comparison of the spectra of a thermal (blackbody) source at 6000 K and a nonthermal (synchrotron) source. (b) Synchrotron radiation from an electron spiraling in a magnetic field.

blackbody's temperature, and then decreases at shorter wavelengths (Section 6.4). The spectrum of blackbody emission is the archetype of *thermal emission*, which arises basically from the motions of the particles involved. The greater the motions, the hotter the source, and the higher the output of radiation.

Nonthermal emission does not follow the characteristic signature of blackbody radiation. In general, the nonthermal spectrum increases in intensity at longer wavelengths, rather than decreases (Fig. 11.25a). *Synchrotron emission* is a frequently found example of nonthermal emission; it arises from the acceleration of charged particles (usually electrons) moving at relativistic speeds (near the speed of light) in a magnetic field. Moving charged particles interact with magnetic fields so that they spiral around the magnetic field lines rather than traveling across them (Fig. 11.25b). The spiral paths are curved, so the particle is continually accelerated and thus emits electromagnetic radiation. (The term *synchrotron emission* derives from the fact that such radiation was first observed from the General Electric synchrotron, which used magnetic fields to contain the high-speed electrons.)

The frequency of emission is directly related to how fast the particle spirals; the faster the spiral, the higher the frequency. Increasing the magnetic field strength tightens the spiral and so increases the frequency. High-energy particles require strong fields to keep them in a tight spiral.

The velocity of the charged particle also affects the frequency, in the sense that more energetic particles can produce higher-frequency emission. In most cases there are more particles of low energy than high energy, so the radiation is more intense at low frequencies (long wavelengths).

As a particle radiates, it loses energy and generates lower energy (longer wavelength) radiation. So a synchrotron source needs a continually replenished supply of high-energy electrons in order to keep emitting at relatively short wavelengths (high-energy photons).

One important point about synchrotron emission: It is usually *polarized*. Recall (Section 9.2) that light waves are polarized if the planes of their vibration are preferentially oriented in some direction. Thermal emission, such as from the sun, has no preferred alignment, that is, the planes of wave vibration are found in all directions in equal numbers. Synchrotron-emitting electrons, when viewed side on in their spiral motion, appear to be moving back and forth along almost straight lines. Their synchrotron emission has its waves more or less aligned in the same plane. Polarized synchrotron radiation at visible wavelengths can be observed with Polaroid filters.

If a radio source is observed at a variety of frequencies, the shape of its spectrum distinguishes between a possible supernova remnant (nonthermal) and an ordinary H II region (thermal). A nonthermal spectrum indicates that the radio emission comes from very high-speed electrons accelerating in magnetic fields. The radiation's intensity wavelength range depend on the intensity of the magnetic field and the kinetic energy of the electrons. The fact that supernova remnants display nonthermal spectra means that the remnants contain magnetic fields and high-energy particles.

The Crab Nebula: A Supernova Remnant

The systematic observation by Chinese and Japanese astronomers of the supernova that produced the Crab Nebula contrasts sharply with the utter lack of comment by European astronomers at the time. However, a Near Eastern account of the supernova appears in a journal kept by a physician named Ibu Bultntan, who lived in Constantinople. His record implies that the sudden appearance of a "spectacular star" took place in the summer of A.D. 1054.

The supernova may also have been observed and recorded in the southwestern part of North America. One good example is a pictograph (painting on a rock) in Chaco Canyon, New Mexico, which may represent the predawn conjunction of the waning crescent moon and the supernova (Fig. 11.26). The Anasazi (Section 1.7), who lived in Chaco then, may have made this painting to commemorate the event.

FIGURE 11.26 A painting in Chaco Canyon, New Mexico, that may represent the A.D. 1054 supernova. The view is looking up at the painting which is on the underside of a rock ledge. A starlike symbol lies next to a crescent symbol. The hand indicates that something extraordinary is shown here. On the morning of July 5, 1054, the supernova, brighter than Venus, rose with the waning crescent moon. [Photo by M. Zeilik.]

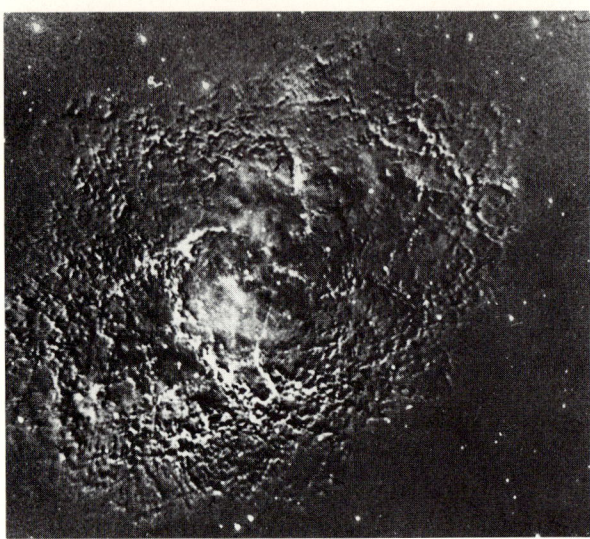

FIGURE 11.27 Expansion of the gas filaments in the Crab Nebula from 1959 to 1964. Here a positive print has been overlaid on a negative one, so that expanding filaments look white on their outer edges, black on their inner ones. [*Courtesy V. Trimble.*]

The Japanese and Chinese astronomical chronicles indicate that the Crab explosion was as bright at maximum as Venus. The material blown off in the explosion should be still expanding today. Indeed, photographs of the optically visible filaments in the Crab Nebula demonstrate that the gas *is* expanding (Fig. 11.27). Virginia Trimble has studied the Doppler shift of the expanding filaments and concludes, assuming a constant expansion rate, that they began expanding at around A.D. 1132–1148. That's close to the actual date of the explosion. The fact that it is later may mean that the velocity now is higher than the average, that is, that the expansion has been accelerating. From the expansion rate measured by the Doppler shift (1450 km/sec) and the observed angular expansion of 0.15 arcsec/yr (along its short axis—the Crab Nebula is shaped like a U.S. football), we can infer that the distance to the remnant (Focus 11.1) is about 6500 ly (2000 pc).

In 1953 I. Shklovsky resolved in part the enigma of the Crab Nebula's radio and optical emission when he suggested that the synchrotron process produced it. Synchrotron radiation requires a magnetic field and a source of energetic charged particles (such as electrons). The emission is polarized and has a nonthermal spectrum. Shklovsky's argument was clinched in the following year, when Russian astronomers Mikhail Vashakidze and Viktor Dombrovskii found that the optical emission was strongly polarized (Fig. 11.28).

If the radiation from the Crab Nebula is generated by the synchrotron process at all observed wavelengths, then this emission should be polarized at all wavelengths. The optical emission is definitely polarized. The Crab Nebula also emits X rays. Are they polarized too? The problem in making such an observation is that the pulsar in the Crab Nebula also gives off X rays, so the two different sources of the X rays—the pulsar (next section) and the gas of the nebula—are mixed together (Fig. 11.29). Recently, the polarization of the X rays from the Crab Nebula has been measured without interference from the pulsar's X rays. The X rays are polarized about 19 to 20 percent, which confirms that they must be produced by the synchrotron process.

The solution to one puzzle posed another one even more vexing: What is the source of the energetic electrons? As these electrons spiral through the magnetic field emitting

11.5 PULSARS: NEUTRON STARS IN ROTATION

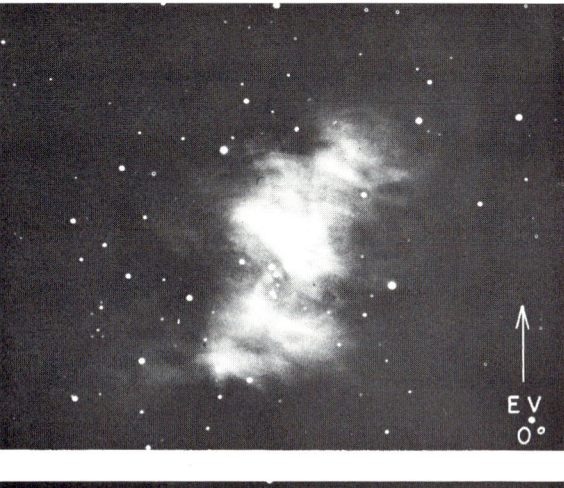

FIGURE 11.28 The Crab Nebula viewed through polarizing filters. The arrows in the corners indicate the orientation of the polarization of the filters. The differences in the images show that the light is polarized. [*Courtesy Palomar Observatory, California Institute of Technology.*]

FIGURE 11.29 An X-ray image of the Crab Nebula, taken by the Einstein X-Ray Observatory. The bright patch at the center is the pulsar. [*Courtesy F. Harnden, Jr. and H. Tananbaum, Center for Astrophysics.*]

synchrotron radiation, they rapidly lose energy. Half the energy of the electrons producing the optical emission would be drained off in only 70 years. So the supply of electrons must be continuously replenished. The problem of the electrons' source became even more acute when X-ray emission was discovered in 1963. The electrons which produce synchrotron X-ray emission have higher energies than those which produce optical emission. They also deplete their energy faster, losing half in only 7 years. The Crab Nebula emits about 100 times more energy as X rays than as radio or optical emission, so a large amount of energy must be added to the nebula over a time of only a few years. The energy problem disappeared in 1968 with the discovery of a pulsar in the Crab Nebula.

11.5 PULSARS: NEUTRON STARS IN ROTATION

Models of supernovas (Section 11.6) suggest that a neutron star may remain as the corpse of the exploded star. A neutron star found in a supernova remnant would clinch this

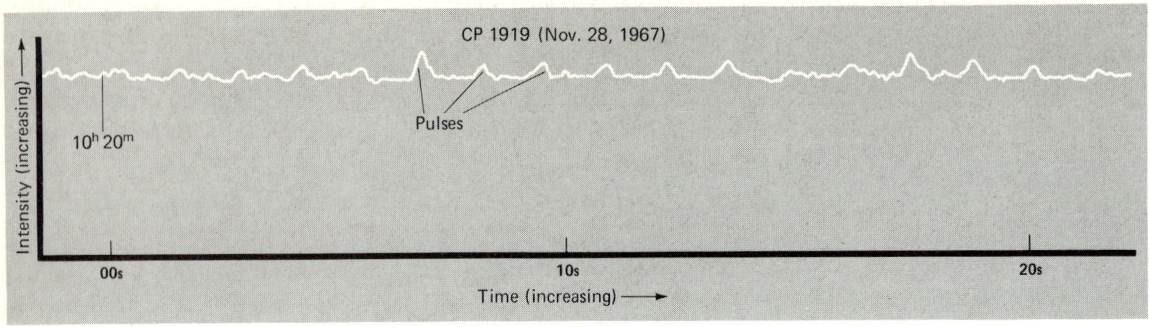

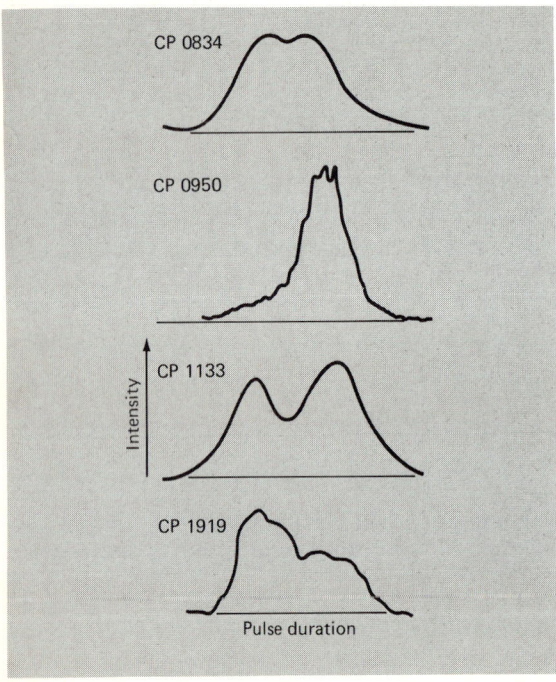

FIGURE 11.30 (a) Radio pulses from the pulsar CP 1919, the first one discovered. Note the peaks vary in intensity, but come at regular time intervals. [*Adapted from a diagram by S. J. Bell Burnell and the Mullard Radio Astronomy Laboratory.*] (b) Shapes of radio pulses from four different pulsars. The pulse durations are only a few milliseconds.

argument. But how would a neutron star be detectable? The answer came in a way not anticipated by astronomers: *pulsars*, accidentally discovered in the summer of 1967 by an English radio astronomy group headed by Anthony Hewish.

The pulsar detection ranks high on the list of those marvelous accidents of scientific discovery. The Hewish group was mapping the sky at radio wavelengths in an attempt to find quasars. A search of the entire sky necessitated a special radio telescope: 2048 antennas covering 4.5 acres of a field.

Jocelyn Bell Burnell, then a graduate student in charge of the preliminary data analysis, noticed a strange signal that suddenly disappeared, only to reappear three months later. The Hewish group concentrated on this unusual signal and found radio pulses occurring at a regular rate, once every 1.33730113 sec (Fig. 11.30a). Flushed with excitement, the radio astronomers searched the sky for any similar signals and discovered three more objects emitting radio bursts at different rates. The Hewish group concluded that the objects must be natural phenomena and named them *pulsating stars*, or *pulsars*.

Observed Pulsar Characteristics

To date, a total of about 150 pulsars have been studied in detail. About an equal number have been recently discovered in a special pulsar survey, for a total of roughly 300.

For a given pulsar the period between pulses repeats with very high accuracy, better than 1 part in 10^8. The amount of energy in a pulse, however, varies considerably; sometimes complete pulses are missing from the sequence. Although the intensity and shape vary from pulse to pulse, the average of many pulses from the same pulsar defines a unique shape (Fig. 11.30b). The average pulse typically lasts for a few tens of milliseconds.

For the well-studied pulsars periods range from 0.03 sec to 4.0 sec, with an average value of 0.65 sec. In the cases where accurate radio observations have been made, periods have been noted to increase in a regular fashion. The rates of change have typical values of 10^{-15} sec/sec, or about 10^{-8} sec/yr. Such small increases can be measured only with atomic clocks, whose stability is better than 10^{-10} sec/yr.

The distances to pulsars are difficult to measure. One way to do it involves observing the difference in arrival times of pulses at different radio wavelengths. If interstellar space were a perfect vacuum, pulses at all wavelengths would travel to us at the same speed, the speed of light. But interstellar space contains some gas, in particular, free electrons and ions, with a density of about $3 \times 10^4/m^3$. Radio waves travel through this ionized gas at speeds which depend slightly on wavelength, the longer wavelengths traveling slower. So if pulses of two different wavelengths λ_1 and λ_2 are emitted at time t_0 at the pulsar, the times at which they arrive at the earth, t_1 and t_2, are given by the expressions $t_1 - t_0 = d/v_1$ and $t_2 - t_0 = d/v_2$. We don't know t_0, but we can measure $t_2 - t_1$ which is equal to $(1/v_2 - 1/v_1)d$. The velocities depend on the electron density, so if we know that, we can determine the distance d. This method gives distances for pulsars ranging from 300 to 60,000 ly (90 to 18,000 pc).

These properties of pulsars provide clues to the physical character of these precise cosmic clocks. The pulse duration indicates an upper limit to the size of the bodies emitting the radiation. How? Suppose that you could switch off the sun. Because of the finite velocity of light, it would take slightly more than 2 sec for the entire solar disk to appear dark, and then another 2 sec to regain its original brightness if quickly turned on again. This is because the part of the sun that we see when we look at the edge of the visible disk is at a greater distance (halfway around the sun) then the part at the center of the disk. So if the sun is turned off, we won't know that the edge is dark until about 2 sec after we see the center become dark. So an object the size of the sun, an average star, is too large to be a pulsar, as the pulse durations amount to a few hundred milliseconds or less. The region emitting these pulses must be less than roughly 30,000 km in size (the light travel distance during the pulse's duration). The more typical pulse durations of a few tens of milliseconds imply sizes of roughly 3000 km or less.

What stellar objects exist at this size or smaller? Also, what objects can contain the large quantities of energy which pulsars emit? A dense object in some way moving rapidly acts as a storehouse of kinetic energy. This clue points to either white dwarfs or neutron stars as the candidates for pulsars.

Physical Characteristics of Pulsars

Immediately after the discovery of pulsars, some theoretical astrophysicists were inclined to consider white dwarfs as the source of their radiation. The reason was quite simple: Many white dwarfs had been observed, whereas no neutron star had ever been sighted. But problems arose in trying to fit characteristics of the white dwarf models to the observations. These ideas did not match well the observed characteristics of pulsars.

Pulsar models must account for the precise clock mechanism of pulsars, that is, the extremely regular repetition of pulses. Basically three clock mechanisms are available: revolution, pulsation, and rotation. Let's consider each in turn.

First, revolution. Two very close, compact bodies can orbit with short, regular periods. But a pair of white dwarfs even in *contact* cannot orbit each other faster than once every 1.7 sec; many pulsars are known with periods shorter than this. Two neutron stars can whip around each other with shorter periods, but here a different problem crops up. According to the general theory of relativity, two massive bodies in short-period orbit should lose energy in the form of gravitational radiation. With this energy loss, the two objects come closer together and their period *decreases* (following Kepler's third law). Eventually, the two bodies would collide. But pulsar periods are known to increase, not decrease. Scratch revolution as the clock mechanism.

Second, pulsation. Imagine a white dwarf or neutron star expanding and contracting regularly, with one expansion and contraction equal to a pulse period. Sounds good, but it doesn't work out. The time in which a spherical mass pulsates depends on its density, just as the free-fall collapse time depends on density (Section 9.2). More dense objects can pulsate more rapidly. That should seem reasonable. Imagine standing above the surface of a white dwarf star and pulling it outward. Release it; gravity pulls it in, and it pulsates about once a second. Now try the same imaginary experiment with a neutron star. Because it is denser, its surface gravity is greater and so its surface is pulled in more rapidly than a white dwarf's surface. The least massive, least dense neutron stars can pulsate no slower than once every 0.01 sec. That's faster than the fastest known pulsar, which has a period of 0.033 sec. Pulsation doesn't work. White dwarfs can't pulse as fast as the fastest pulsars, and neutron stars pulsate too fast.

Third, rotation. Consider one rotation equals a pulse period. Can rotation of white dwarfs explain the fastest pulsars? Let's consider two arguments. For any spherical object, the speed at its equator equals its circumference divided by its rotational period, that is,

$$V_{eq} = \frac{2\pi R}{P}$$

where R is the radius and P the period. For $R = 7000$ km, a typical white dwarf radius, and $P = 0.033$ sec,

$$V_{eq} = \frac{2\pi(7000 \text{ km})}{0.033 \text{ sec}}$$
$$= 1.3 \times 10^6 \text{ km/sec}$$

11.5 PULSARS: NEUTRON STARS IN ROTATION

That's about four times the speed of light; special relativity tells us that's impossible!

Here's another approach: If a spherical mass rotates too rapidly, its gravity will not be able to hold it together. Mass will fly off tangent to its equator. Consider a small mass m at the equator of white dwarf with mass M. The force on the small mass is

$$F_{grav} = \frac{GMm}{R^2}$$

The centripetal acceleration of the small mass is

$$a = \frac{V_{eq}^2}{R}$$

and so the centripetal force is

$$F_{centripetal} = ma = \frac{mV_{eq}^2}{R}$$

These two forces must be equal for the small mass to stay on the surface:

$$\frac{mV_{eq}^2}{R} = \frac{GMm}{R^2}$$

So

$$V_{eq} = \left(\frac{GM}{R}\right)^{1/2}$$

is the maximum possible equatorial velocity. But as above

$$V_{eq} = \frac{2\pi R}{P}$$

so

$$\frac{2\pi R}{P} = \left(\frac{GM}{R}\right)^{1/2}$$

Then

$$P = 2\pi R \left(\frac{R}{GM}\right)^{1/2}$$

is the *fastest* possible period. We can simplify this equation a bit by substituting for the mass in terms of density and radius:

$$P = 2\pi \left(\frac{R^3}{GM}\right)^{1/2}$$

$$= 2\pi \left(\frac{R^3}{G(4/3)\pi R^3 \rho}\right)^{1/2}$$

$$= \left(\frac{3\pi}{G\rho}\right)^{1/2}$$

$$= 3.8 \times 10^5 \rho^{-1/2}$$

where ρ is the average density in mks units. Since white dwarfs have densities of about 10^{10} kg/m^3 at most, their maximum period is

$$P_{max} = 3.8 \times 10^5 (10^{10})^{-1/2}$$

$$\approx 4 \text{ sec}$$

If a white dwarf cannot rotate faster than once every 4 sec without losing mass, the rotation is too slow for the fastest pulsars.

We are left with one more possibility. Neutron stars have average densities of about 10^{17} kg/m^3, so they can rotate once every 0.001 sec without losing mass. That period is fast enough to account for even the fastest pulsar, and, of course, they can rotate more slowly, too. A rotating neutron star theoretically can provide the clock mechanism for pulsars.

Pulsars and Supernovas

The idea that pulsars are neutron stars would be clinched if some observation could be found that made the connection. Recall that theoretical models predict that neutron stars are made in supernova explosions. So we need to look at supernovas or their remnants to connect neutron stars to pulsars. Fortunately, we have two examples to date: a pulsar in the Crab Nebula and another in the Gum Nebula (Section 11.4), both supernova remnants.

David H. Staelin and Edward C. Reifenstein discovered the Crab Nebula pulsar (Fig. 11.31), called PSR 0531+21. (PSR stands for pulsar, and the numbers 0531+21 refer to

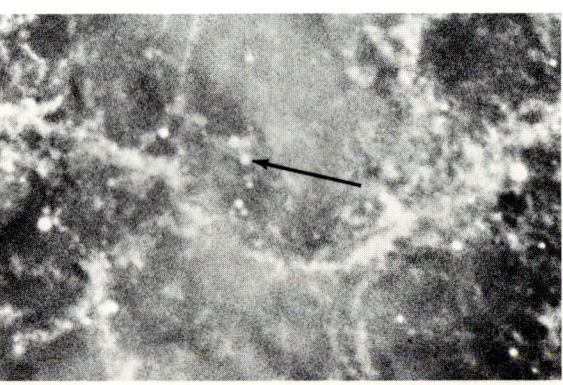

FIGURE 11.31 The location of the pulsar (arrow) in the Crab Nebula. [*Courtesy Lick Observatory.*]

11.5 PULSARS: NEUTRON STARS IN ROTATION

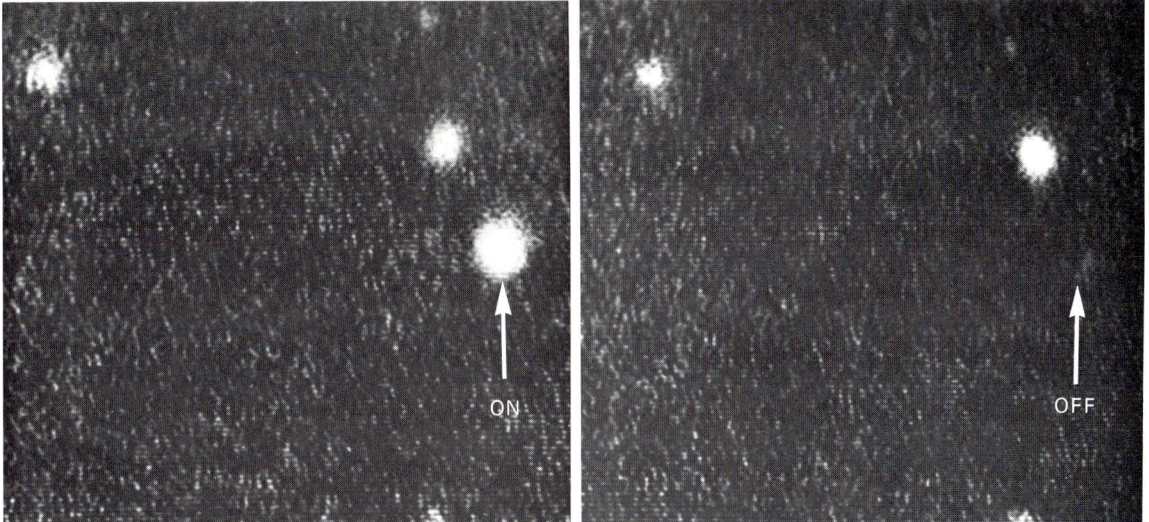

FIGURE 11.32 The Crab pulsar on (left) and off (right) at visual wavelengths. [*Courtesy Lick Observatory*]

its position coordinates in the sky.) PSR 0531+21 has the distinction of being the pulsar with the fastest period: 0.033 sec, or 30 pulses per second!

The Crab pulsar has other outstanding features. It was the first pulsar discovered to emit optical pulses as well as radio ones (Fig. 11.32); the optical and radio pulses were found to have the same period. Observations of these visible pulses showed a smaller interpulse between the main peaks (Fig. 11.33). Remarkably, the star emitting these pulses was picked out by Walter Baade and R. Minkowski in 1942 as a possible candidate for the stellar remnant of the supernova. Although this star is now known to be the pulsar, astronomers had observed it for years without noticing the optical blinking; a flicker of 30 times a second is just beyond the frequency which the eye can detect.

The Crab pulsar is the only pulsar so far discovered to pulse not only at radio and optical wavelengths, but also in the infrared, X-ray, and gamma-ray regions of the spec-

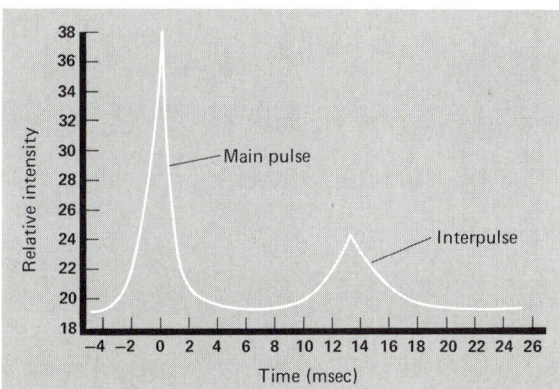

FIGURE 11.33 An intensity plot of the optical pulses from the Crab Nebula pulsar, showing the strong main pulse and the weaker interpulse. [*Observations by N. Visvanathan.*]

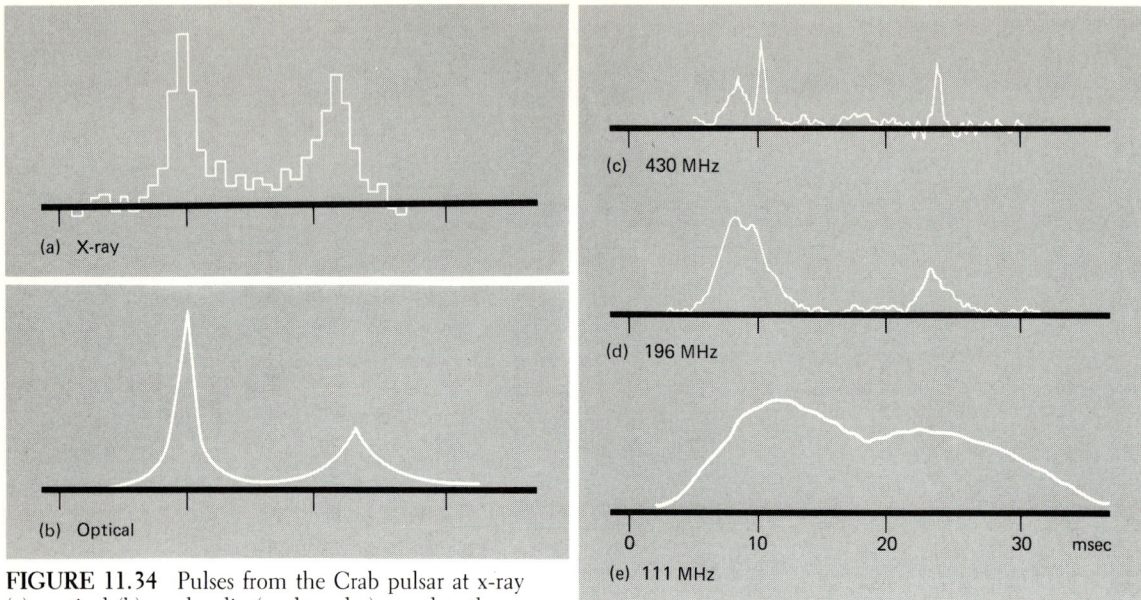

FIGURE 11.34 Pulses from the Crab pulsar at x-ray (a), optical (b), and radio (c, d, and e) wavelengths. Note the main pulse and interpulse appear in each.

trum (Fig. 11.34). That's a photon energy span from 10^{-25} J to 10^{-12} J! The total energy emitted in the pulses is about 10^{28} W.

Another important feature of the Crab pulsar is that it was one of the first to exhibit a definite slowdown in pulse period, at a rate of about 4×10^{-13} sec/sec. That's about 10^{-5} sec/yr—fast for pulsars, but less change than your electronic watch shows in a year.

The discovery of the Crab pulsar seems to solve the energy problem of the Crab Nebula. At all wavelengths the Crab Nebula emits about 10^{31} W. If the pulsar is a rotating neutron star, its slowdown in period gives a change in rotational energy of about 5×10^{31} W. That's enough to power the nebula—if the rotational kinetic energy of the neutron star can somehow be converted to kinetic and radiative energy of the nebula.

If the Crab pulsar were the only one associated with a known supernova remnant, it might be written off as a coincidence. But astronomers know of another one: the pulsar in the constellation Vela, near the center of the Gum Nebula (Fig. 11.35). This pulsar is called PSR 0833–45.

Although the Vela pulsar was known in 1968, it was eight years before astronomers finally observed its optical pulses because the pulses are very weak, with an average magnitude of only 25.2 (Fig. 11.36). The pulses come every 80 msec and have two peaks, separated by about 22 msec. Gamma-ray telescopes have also detected pulses from Vela (Fig. 11.36).

This pulsar resembles the Crab pulsar in many ways. Both are rapid, both emit pulses over a wide range of the electromagnetic spectrum, and their gamma-ray pulse profiles are very similar. The period of the Vela pulsar also slows down, but at a slightly different rate of about 1.3×10^{-13} sec/sec (4×10^{-6} sec/yr). This pulsar has also shown three sudden sharp increases in period; the Crab pulsar has done the same, but far less dramatically.

11.5 PULSARS: NEUTRON STARS IN ROTATION

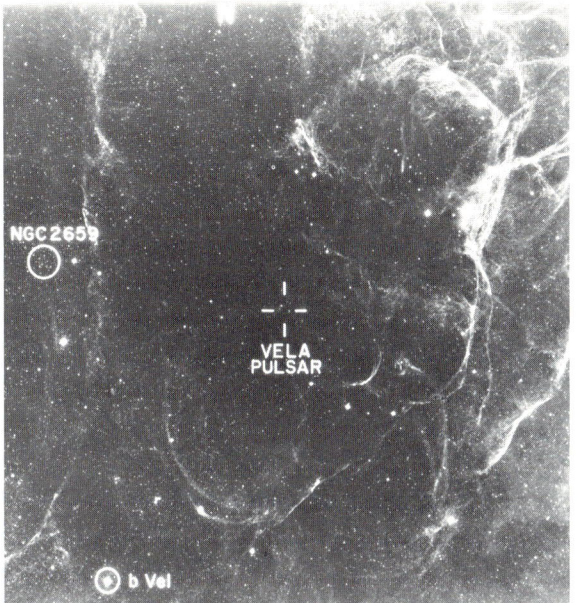

FIGURE 11.35 The location of the Vela pulsar in the Gum nebula (Fig. 11.22), a known supernova remnant. [*Courtesy B. Bok and Steward Observatory.*]

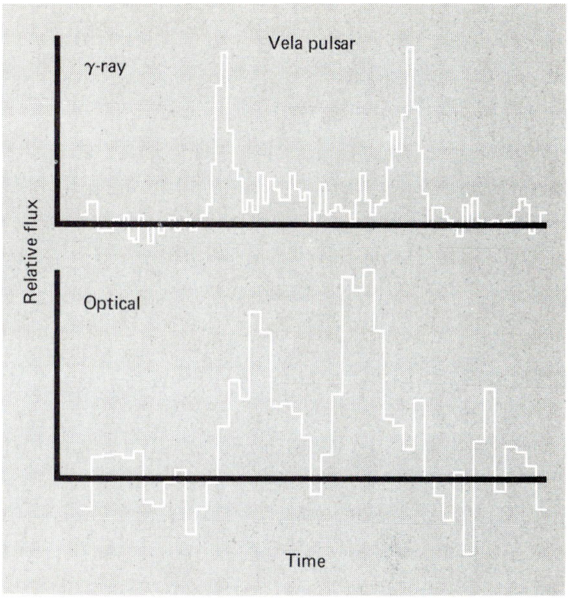

FIGURE 11.36 Gamma-ray (top) and optical (bottom) pulses from the Vela pulsar. [*Optical observations by P. T. Wallace and colleagues; gamma-ray observations by R. Buccheri and coworkers.*]

The Lighthouse Model for Pulsars

Let's try to tie these observations together in one model for pulsars—a rotating, magnetic, neutron star, otherwise known as the *lighthouse model*. The model has two key components: (1) the neutron star, whose great density and fast rotation insures a large amount of rotational energy, and (2) a dipole magnetic field that transforms the rotational energy to electromagnetic energy. (The basic idea that magnetic neutron stars could act as energy transformation devices was dreamed up by Franco Pacini one year *before* the discovery of pulsars!)

That neutron stars might possess extremely intense magnetic fields follows from the same conservation of flux argument applied earlier to white dwarfs (Section 11.1). (Recall that observational evidence supports this argument; some white dwarfs have surface magnetic fields of roughly 10^6 gauss.) Imagine our sun collapsed to a neutron star 7 km in radius. Calculating the field strength from conservation of magnetic flux, we have

$$B_{NS} = B_{sun}\left(\frac{R_{sun}}{R_{NS}}\right)^2$$

$$= 1 \text{ gauss}\left(\frac{7 \times 10^5 \text{ km}}{7 \text{ km}}\right)^2$$

$$= 10^{10} \text{ gauss}$$

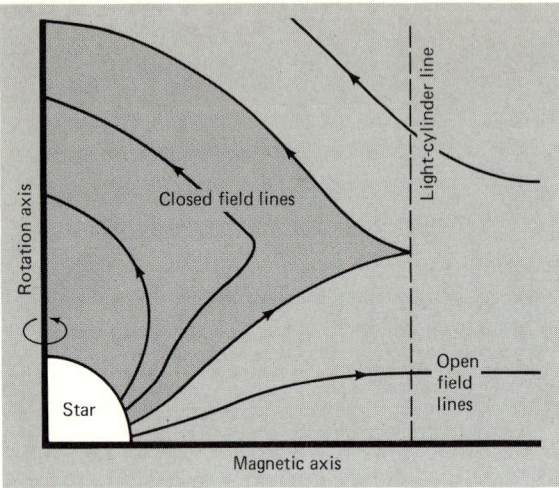

FIGURE 11.37 The magnetosphere around a rapidly rotating neutron star. The light-cylinder line is where the rotational speed of the magnetic field lines equals the speed of light. [Adapted from a diagram by R. Henriksen and J. Norton, Astrophysical Journal, vol. 201, p. 719, copyright 1975 by the American Astronomical Society.]

A magnetic field of that strength is probably found nowhere else in the universe. More careful calculations indicate that the fields are more typically 10^{12} gauss.

The region close to the neutron star where the magnetic field directly and strongly affects the motions of charged particles is called the pulsar's *magnetosphere*. Here all the energy conversion action takes place. In one model, in which the magnetic axis is tilted with respect to the rotational axis, the magnetosphere consists of two regions (Fig. 11.37): (1) a zone close to the neutron star where the magnetic field lines close on themselves, called the light-cylinder region; here rotation with the neutron star takes place at less than the speed of light; and (2) a region of open magnetic field lines beyond the light-cylinder region.

The overall picture of the lighthouse model is shown in Fig. 11.38. The magnetic axis of the pulsar is tilted with respect to the rotational axis. As the pulsar spins, its enormous magnetic field induces an equally enormous electric field at its surface. This electric field pulls charged particles (mostly electrons) off the solid crust of iron nuclei and electrons. The electrons flow into the magnetosphere where they are accelerated by the rotating magnetic field lines. The accelerated electrons emit synchrotron radiation in a tight beam more or less along the field lines.

You can now see how a pulsar pulses without actually pulsating. If the magnetic axis can fall within our line of sight, each time a pole swings around to our view (like the spinning light of a lighthouse) we see a burst of synchrotron emission. The time between pulses is the rotation period. The duration of the pulses depends on the size of the radiating region. As the pulsar generates electromagnetic radiation, the torque from accelerating particles in its magnetic field slows down its rotation. This slow-down is observed.

This is one possible model from the many available. A big problem with all models is the exact method by which a pulsar converts its rotational energy to electromagnetic energy. Models to date have limited success in explaining the abundant observational data. The basic clock mechanism seems right, but the detailed emission mechanism is not clear. The magnetic, spinning neutron star model is a good example of a basic idea that's been generally accepted, but whose details so far are somewhat vague.

11.5 PULSARS: NEUTRON STARS IN ROTATION

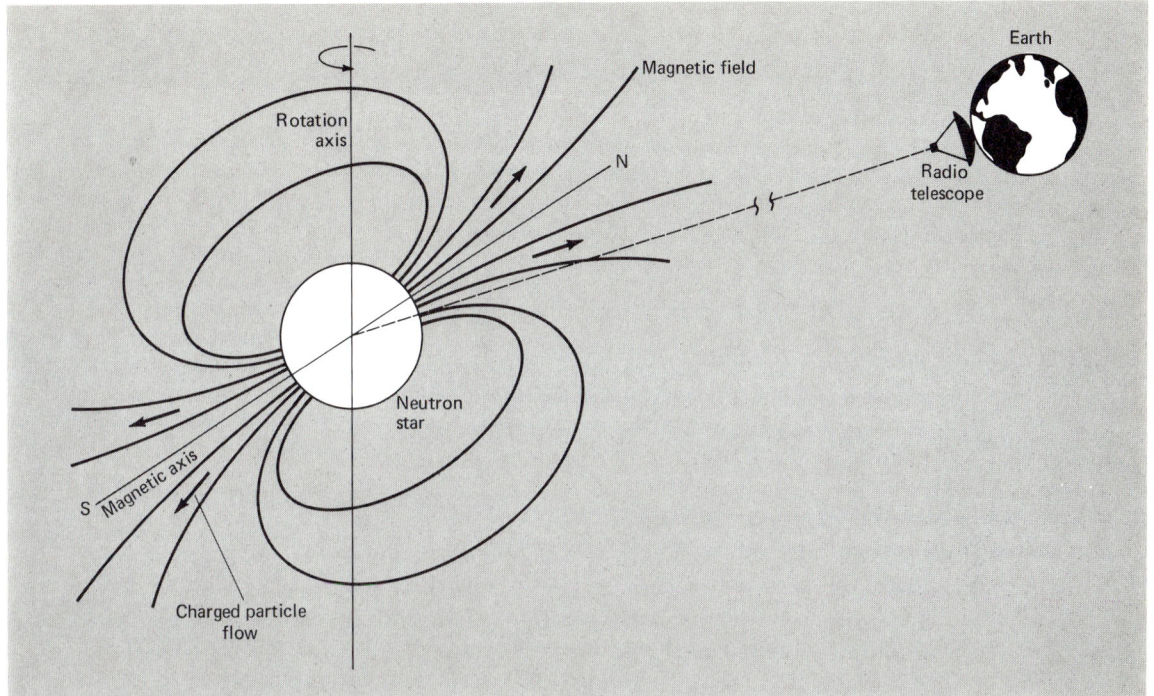

FIGURE 11.38 The lighthouse model of a pulsar. The magnetic axis is tilted with respect to the spin axis. Electrons from the neutron star's surface flow out along the magnetic field lines and so escape mostly at the north and south magnetic poles. These emit synchrotron radiation that we see as pulses when a pole spins across our line of sight.

Binary Radio Pulsars

Most stars in the Galaxy are members of binary or multiple star systems. Even after one member of a binary becomes a supernova, the system usually remains intact. So a radio pulsar could exist in a binary system.

The first one observed is called PSR 1913+16 and was discovered by Russell Hulse and Joseph Taylor in July 1974 during a search for new pulsars. This pulsar first attracted attention because its pulse period was only 0.059 sec, shorter than any known pulsar except the one in the Crab Nebula. When Hulse and Taylor reobserved PSR 1913+16 in September 1974, they found that its period went through a large cyclical change in only 7.75 h. What was going on? Hulse and Taylor recognized that such regular changes would naturally come about in a binary system of the pulsar and a companion with an orbital period of 7.75 h. What was seen was a Doppler shift in the signal produced by the orbital motion of the system (Fig. 11.39). When the pulsar is moving away from us, its pulses are spread out and come at longer intervals. When it is moving toward us, the pulses are pushed together and come at shorter intervals.

PSR 1913+16 lies about 15,000 ly (5 kpc) from us. Visual and X-ray observations have so far failed to detect either the pulsar or its companion. From radio observations alone we know that the pulsar and its companion have an orbital semimajor axis of only

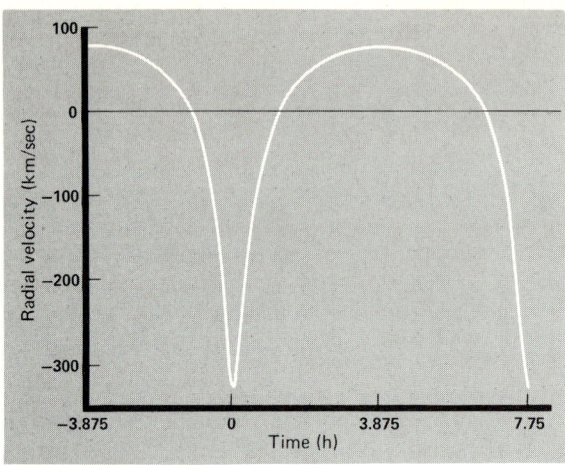

FIGURE 11.39 Doppler shift in the radial velocity of the binary pulsar PSR 1913+16. [Adapted from a diagram by R. Hulse and J. Taylor, Astrophysical Journal (Letters), vol. 191, p. L59, copyright 1975 by the American Astronomical Society.]

7×10^5 km—that's only the sun's radius! Their combined masses are 2.83 solar masses, so if the pulsar has a mass of about 2 solar masses (a typical neutron star), its companion has about 0.8 solar mass. The companion might be a white dwarf.

Other binary pulsars have been discovered. One called PSR 065564 has a period of 24 h 41 min; its orbit is almost perfectly circular with a radius of only 750,000 km, just a bit larger than the sun's radius. The companion to the neutron star must be small, perhaps also a white dwarf star. You'll meet other pulsars in binaries, seen with X-ray telescopes, in the next chapter.

SS 433: A Unique Binary System

We have presented evidence that neutron stars exist and that they exist by themselves and in binary systems. But the neutron star story does not end there. It pops up again with the bizarre radio, X-ray, and optical object called SS 433. That name comes from the fact that SS 433 is listed as the 433rd entry in a catalog compiled in 1977 by Bruce Stephanson and Nicolas Sanduleak of Milky Way objects showing strong emission lines. SS 433 had at first appeared to be a nondescript fourteenth magnitude star in the constellation Aquila. Then a serendipitous series of observations marked SS 433 as one of the strangest objects in the Galaxy. Radio observations showed that SS 433 lies near the center of a much larger radio source, called W 50, that may be a supernova remnant. The radio output of SS 433 varied by a factor of two to three over times of a few hours to days. X-ray observations revealed that the X-ray emission from SS 433 also varied.

These data prompted optical astronomers to take spectra of SS 433. The results presented a puzzle (Fig. 11.40). A very strong H-alpha line poked up, surrounded by less intense, but still rather strong lines. They were too strong to be lines emitted by oxygen or any of the heavier elements, yet they were not at the correct wavelengths to be lines from hydrogen or helium. But that's what they turned out to be—highly red- *and* blue-shifted lines of hydrogen and helium. Their displacements in late 1978 indicated radial velocities of more than 40,000 km/sec! Even stranger, during the first series of night-to-night observations (done at Lick Observatory) the Doppler-shifted lines moved a measurable amount, about 1 percent in a few days.

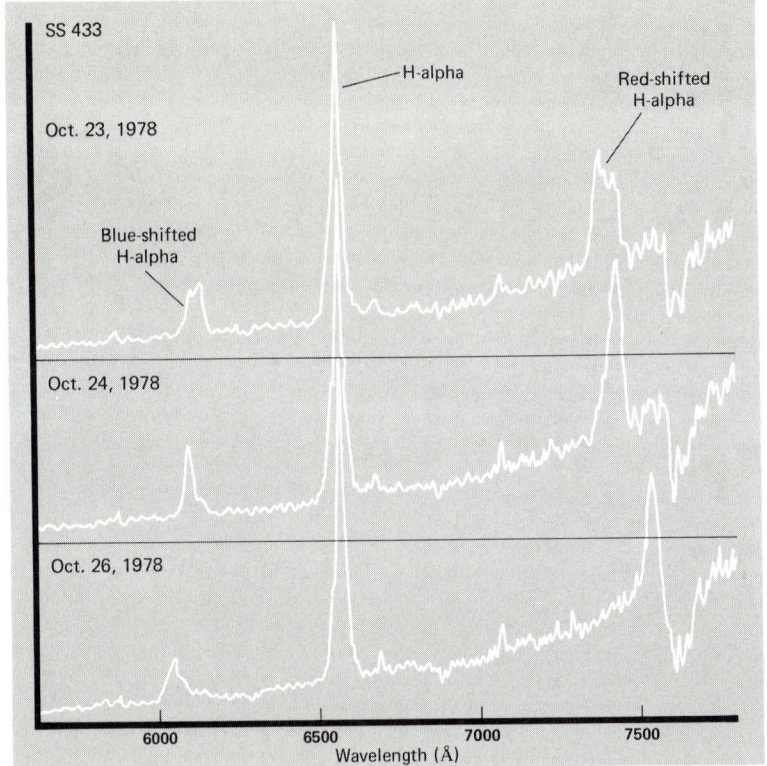

FIGURE 11.40 The red part of the spectrum of SS 433, observed by Remington Stone with the Lick 60-cm reflector. Note how much the red- and blue-shifted H-alpha lines move in this time. [Adapted from a diagram by B. Margon, H. C. Ford, J. I. Katz, K. B. Kwitter, R. K. Ulrich, R. P. S. Stone, and A. Klemola, Astrophysical Journal (Letters), vol. 230, p. L41, copyright 1979 by the American Astronomical Society.]

Two critical clues as to what was going on in SS 433 were yet to come. First, long-term observations by University of California astronomers showed that the Doppler-shifted lines moved periodically (Fig. 11.41a), with the red- and blue-shifted lines swinging back and forth in a 164-day cycle (Fig. 11.41b). The maximum red shift value corresponded to a radial velocity of a little over 50,000 km/sec. Second, very careful spectroscopic observations of the strongest emission lines, which at first seemed stationary, revealed a small Doppler shift (only 70 km/sec, about 0.1 percent of the main moving lines) with a period of 13 days.

A model developed by Bruce Margon and coworkers solves most of the puzzles presented by SS 433 (Fig. 11.42). It's a binary star system containing an ordinary (unseen) star and a neutron star orbiting every 13 days. Their proximity channels gas from the ordinary star onto the neutron star. This gas heats up as it falls in and produces the strongest emission lines. Meanwhile, the neutron star has two jets of material squirting out of it in opposite directions. The jet blasting away from us generates the highly red-shifted lines; the one pointing toward us, the blue-shifted lines. The line of the jets is tilted about 20° to the rotation axis of the neutron star; that, in turn, inclines about 80° with respect to the line of sight from the earth. Like the earth's axis, the neutron star's axis processes in space, with a period of 164 days. As the spin axis slowly whirls in space, it carries the jets around with it. This variation causes the Doppler-shifted lines to fluctuate between maximum and minimum displacements as the inclination of the jets changes along the line of sight.

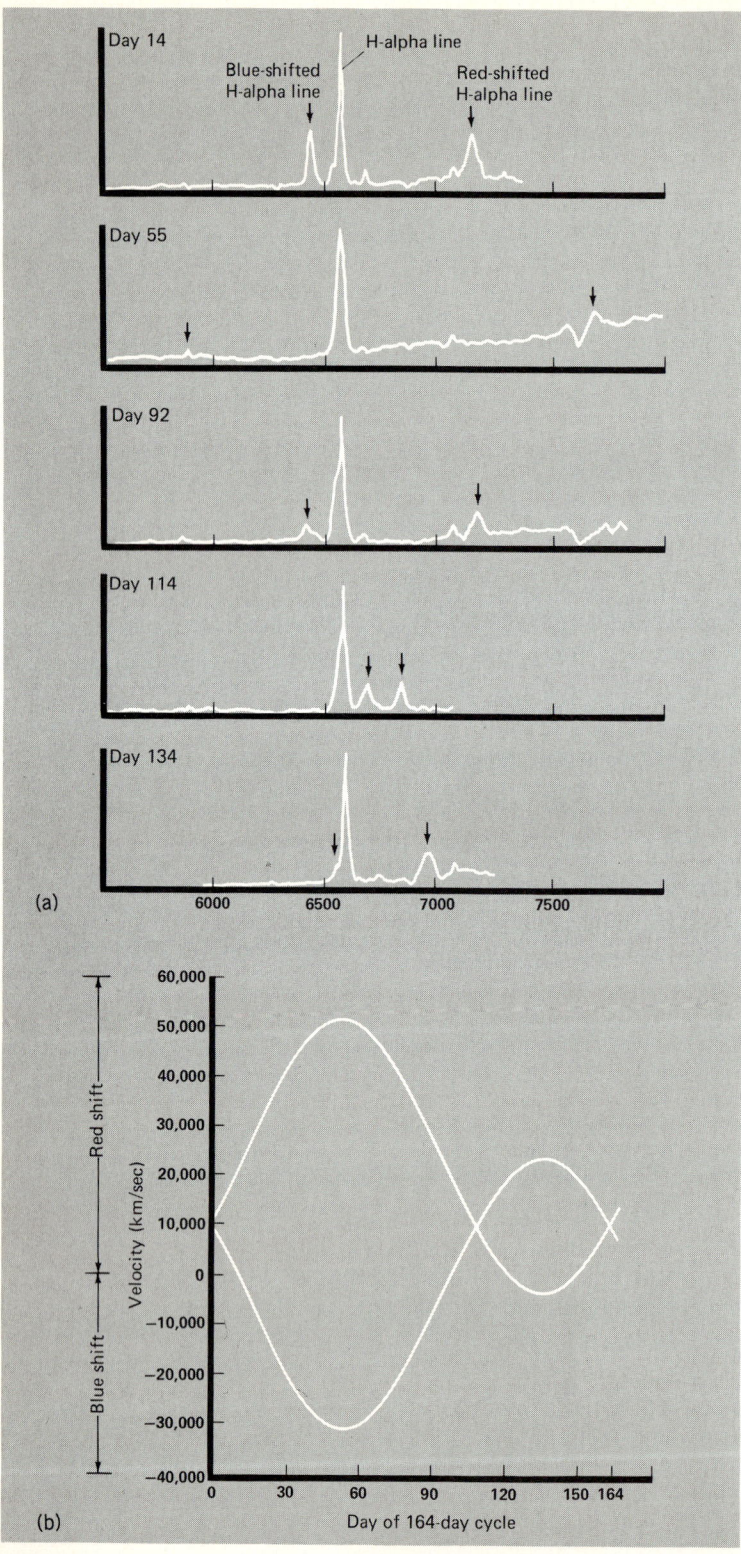

FIGURE 11.41 (a) Spectra of SS 433, showing the rapid motion of the Doppler-shifted H-alpha lines. Note that they cross between day 114 and 134. [*Spectra obtained by B. Margon and colleagues.*] (b) Theoretical curves of the Doppler shifts in the red- and blue-shifted emission lines for SS 433 for one cycle of 164 days.

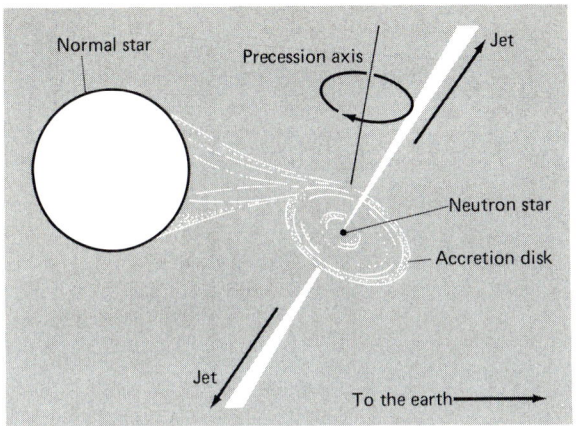

FIGURE 11.42 A model for the binary system that includes SS 433. Jets blast out from the face of the accretion disk, which precesses around the neutron star. [*Based on a theoretical model developed by B. Margon and colleagues.*]

What are these jets? Recent radio observations with the VLA imply that they consist of small blobs of ionized gas continuously blowing out of the source, like water from a hose. Incredibly, this material shoots out at some 78,000 km/sec, about 26 percent of c. The jets probably radiate by synchrotron emission—an idea backed up by the fact that they are highly polarized (some 15 to 20 percent). If so, they contain magnetic fields with strengths of about 10^{-2} to 10^{-3} gauss.

We've spent some time on SS 433 to show you how strange it is, to describe its binary nature, and to prepare you for even more energetic jet phenomena—those in the nuclei of galaxies (Chapter 23).

11.6 THE MANUFACTURE OF HEAVY ELEMENTS

A massive star dies in a supernova. The blast can leave behind a neutron star or a black hole. Does any good come of its death? Yes, for without massive stars dying violently, neither we nor the green, green grass would be here.

An extravagant claim? Not really, if you investigate where the elements in the universe are made. Start off with hydrogen which originates in the Big Bang beginning of the universe (Chapter 24). Recall in the discussion of nucleosynthesis (Chapter 10) that ordinary stars fuse hydrogen to helium by the PP reactions and CNO cycle in their normal lives. A solar-mass star converts helium to carbon by the triple-alpha reaction, but its core does not get hot enough to burn carbon and its thermonuclear reactions cease.

But other more massive stars do burn heavier elements. In the sequence of thermonuclear energy generation (look back at Table 10.4) the ashes of one set of reactions become fresh fuel for the next. Each step up in the fusion chain requires higher temperatures to overcome the greater repulsive electrical force of nuclei with more protons. These fusion reactions—carbon to oxygen, neon, sodium, and magnesium and from there to silicon—can fire only in massive stars. The end to the fusion chain comes with iron, the most tightly bound of the normal nuclei. Iron rests at the bottom of an energy well. To split iron into lighter elements or to fuse it to heavier ones takes energy. So even in stars greater than about 20 solar masses nuclear reactions naturally stop at iron.

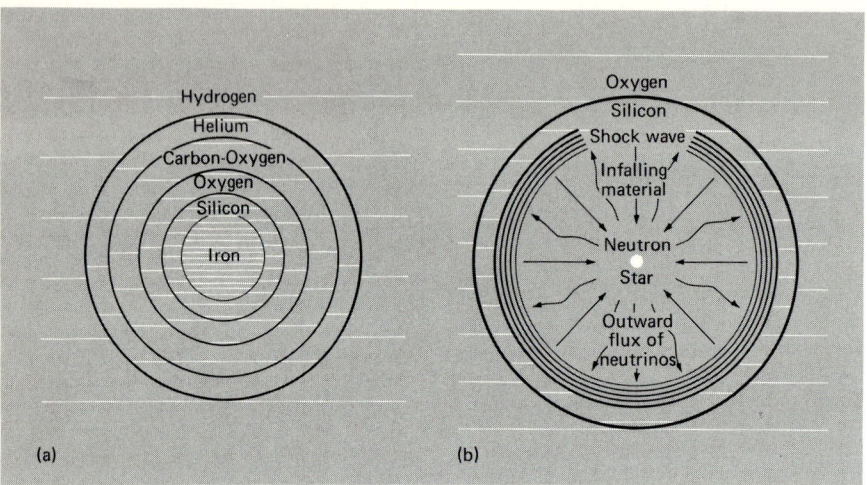

FIGURE 11.43 One model of the interior of a massive star going supernova. Fusion reactions in shells give layers of different elements (a). At high enough temperatures the iron core disintegrates into helium. This process soaks up heat from the core, and it collapses rapidly (b), sparking explosive ignition of fusion reactions. [*Adapted from a diagram by C. Wheeler.*]

But the elements that we know do not end at iron; many are heavier. Where and when are they manufactured in the course of cosmic evolution? A supernova explosion acts as nature's special workshop for forging some elements heavier than iron.

Here's one possible model of how a Type II supernova happens. Imagine a very massive star with a core of iron (Fig. 11.43a). Its interior temperature decreases from the core outward. Because of this outward temperature decline, the star's interior is layered like an onion. Around the iron core is a silicon layer; here temperatures did not get high enough to fuse iron. Around that layer is one of oxygen; here temperatures were too low to fuse oxygen to silicon. These shells are still burning: in the silicon core, oxygen is being fused; in the oxygen shell, carbon; and so on.

Once the core ends up as iron it must contract, for the fusion fires have failed. Relentlessly gravity squeezes the core to higher temperatures and densities. When the core gets to about 5 billion K, the photons there have so much energy that they can penetrate the iron nuclei and break them down into helium. As the iron disintegrates into helium, large amounts of heat are used up. The pressure in the core no longer supports the star and it collapses suddenly. The gravitational collapse rapidly pumps heat into the material.

Two important events occur in this collapse. First, protons and neutrons released by the disintegration of nuclei in the core pelt and penetrate remaining nuclei. These can capture neutrons and be transformed to heavier elements. Second, the layers above the core plummet inwards and heat up. Suddenly, ignition temperatures of many fusion reactions are reached. They turn on explosively. A blast shock wave from this explosive ignition bullies its way outward from the core (Fig. 11.43b). It carries material with it into the interstellar medium—material enriched with heavy elements.

Let's explain the synthesis of heavy elements in a little more detail. The most important process involves high-energy neutrons bombarding various nuclei. Because they have no charge, the neutrons have an easy time penetrating a nucleus. This nuclear capture of neutrons leads to a buildup of heavier nuclei.

Two processes partake in this buildup. Recall that a neutron under normal conditions of low density will disintegrate (in about 1000 sec) into a proton, an electron, and an antineutrino. This is called *beta decay*. The rate of beta decay naturally divides nucleo-

11.6 THE MANUFACTURE OF HEAVY ELEMENTS

synthesis into two processes. In one process neutrons are captured *faster* than the beta-decay rate, so several neutrons can be captured before any of them decay to protons, and neutron-rich nuclei are formed. This process is called the *rapid process* (or *r-process* for short). In the other process neutron capture is *slower* than the beta-decay rate, so a captured neutron normally decays to a proton before another neutron is added, and proton-rich nuclei are made. This process is called the *slow process* (*s-process*). A combination of both these processes leads to the manufacture of most of the elements and isotopes heavier than iron. In supernovas the time scales are short, and it is mainly the r-process which is effective.

A specific example: transforming lead 206 (^{206}Pb) to other isotopes of lead and finally to bismuth 209 (^{209}Bi). Sock ^{206}Pb with a neutron (Fig. 11.44); it becomes ^{207}Pb. Hit it with another neutron, and so on, up to ^{209}Pb. Each heavier isotope of lead is less stable than the previous one, and ^{209}Pb is so unstable that it decays rapidly (by beta decay) to ^{209}Bi (and an electron and an antineutrino). Then the ^{209}Bi can absorb neutrons, build up to other bismuth isotopes, finally beta decay to the next element, and so on.

We have emphasized nucleosynthesis in supernovas, but red giants also manufacture some heavy elements. When a red giant undergoes helium shell burning, small numbers of neutrons are produced and add at slow rates (s-process) to iron to fuse heavier elements. Although the s-process is indeed slow, the red giant stage lasts long enough (10^5 yr or so) to synthesize an appreciable amount of heavy elements. Some are also made during red giant helium flashes. But the s-process cannot synthesize the very heavy, radioactive elements, for the neutron addition is so slow that such nuclei decay by fission before more neutrons can be added.

In general, elements made by the s-process complement those made in the r-process to fill up the periodic table. Nucleosynthesis in red giants makes many of the elements heavier than iron and lighter than lead, but elements heavier than lead, such as uranium and thorium, are produced only in supernovas. (When you use electricity generated by a nuclear power plant, you're using the fossil remains of a massive star!)

Now to return to the claim about us and the green, green grass. Life as we know it (Chapter 25) builds on carbon atoms. A star like our sun can fuse carbon, *but* the carbon remains locked in the white dwarf corpse after the star's death. The material flung out by the star after it becomes a red giant comes just from its outer layers. Since no fusion reactions have gone on in these layers, the material has pretty much the same composition it started with. Only a massive star in a supernova spews into the interstellar medium newly made heavy elements. So if all stars had about the sun's mass, no carbon would get out into the interstellar medium to end up in living organisms on a planet. Your body and

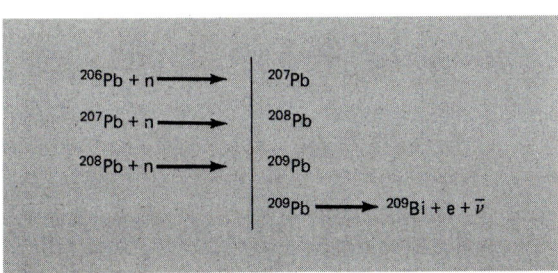

FIGURE 11.44 The synthesis of isotopes of lead and bismuth by neutron capture and beta decay.

the grass under your feet are mostly recycled stardust. Stars die to seed new life. That is a most incredible aspect of cosmic evolution.

SUMMARY

How a star dies depends on its mass at its time of death (Table 11.3). If its mass is less than 1.4 solar masses, its corpse takes the form of a white dwarf. If it is greater than 1.4, but less than roughly 5 solar masses, gravity crushes it to a neutron star. In either case gravity does not defeat the star completely. But if it has greater than about 5 solar masses, gravity wins out absolutely and forms a black hole (next chapter).

Violence marks the death of stars. The death rattle for a star of about 1 solar mass involves only a small fraction of the star's mass. Almost all of a massive star participates in the cataclysm of a supernova. This destructive violence has constructive ends. Heavy elements are made and ejected into the interstellar medium from which stars and their planets are born. So stars go from dust to dust again. And we partake of this cosmic recycling flow.

TABLE 11.3 Death of Stars

Main-Sequence Mass (solar masses)	Normal Life (main sequence)	Death and Final Corpse
0.1–0.5	M stars	White dwarf
0.5–4	K–A stars	Planetary nebula, white dwarf
4–10	A–B stars	Supernova, neutron star
10–20	B stars	Supernova, neutron star or black hole
20–60	O stars	Supernova, black hole

Source: Adapted from work by A. G. W. Cameron and J. C. Wheeler.

Key Words

Chandrasekhar limit
gravitational red shift
magnetic flux
conservation of magnetic flux
neutron star
inverse beta decay
nova

supernova
light curve
Roche lobe
accretion disk
Crab Nebula
Type I, Type II supernova
supernova remnant
synchrotron emission

nonthermal spectrum
pulsar
lighthouse model
beta decay
magnetosphere
rapid process (r-process)
slow process (s-process)

Review Questions

1. Which of the following are evidence that the Crab Nebula is a supernova remnant? (Mark all correct answers.)
   a. There is a pulsar at its center.
   b. The rate of expansion indicates the expansion began about 900 years ago.
   c. A new bright star was observed about 900 years ago in the present position of the nebula.

d. The star at the center is currently emitting far more light than a nova could.
e. The filaments, when viewed in polarized light, form an S.
(Objective 9)
2. Tell whether the following phrases would apply to a white dwarf, neutron star, both, or neither.
   a. Supported in large part by pressure from a degenerate gas
   b. Can have a mass greater than 1.4 solar masses
   c. Can have a mass greater than 15 solar masses
   d. Formed as a remnant of a supernova explosion
   e. Has a radius similar to that of the earth
   (Objective 1)
3. Pulsars are thought to be rotating neutron stars because (mark all correct answers):
   a. The pulse periods are about the same as the natural lifetime of the neutron.
   b. Theoretical calculations predict that neutron stars will be the remnant of some supernova explosions, and two pulsars have been discovered associated with supernova remnants.
   c. Some pulsars are gradually slowing down, and loss of rotational energy provides a natural explanation for this.
   d. They are always observed as members of short-period binary star systems.
   e. Only something so small can rotate rapidly.
   (Objective 4)
4. In a short paragraph describe the primary characteristics of a white dwarf. (Objective 1)
5. In a short paragraph describe to a friend who has not studied astronomy the chief features of a neutron star. (Objective 1)
6. What observational evidence do we have for the actual existence of neutron stars and white dwarfs? (Objectives 1, 2, and 3)
7. Look around you. Of the items you see, what would not be there if supernovas didn't occur? (Objective 11)
8. Assuming no loss of mass, what will be the final form of a 0.5-solar-mass star? A 2-solar-mass star? (Objective 1)
9. Make a list of the observational evidence that supports the idea of the Crab Nebula as a supernova remnant. (Objective 9)

Problems

1. Estimate roughly the amount of material consumed in fusion reactions and released as energy in a supernova. (Objective 8)
2. Imagine the sun shrunk to the size of a neutron star. If angular momentum is conserved in the process (no mass loss), calculate the rotation rate of the sun as a neutron star. How does this rate compare to that needed in the neutron-star model for pulsars? (Objective 4)
3. Radio observations show that the blobs in SS 433's jets expand with a proper motion of 3 arcsec/yr. Their actual speed is about 70,000 km/sec. Estimate the distance to SS 433. (Objective 6)
4. Compare the gravitational red shift of the sun to that of a white dwarf and a neutron star. (Objective 3)

5. Calculate the escape velocity from a solar-mass neutron star with a radius of 10 km. (Objective 1)
6. Imagine a hypothetical star of constant temperature composed of an ordinary gas, so that the equation of state is $P = C\rho$. How would the radius vary with mass for such a star? (Objective 2)
7. What is the fastest possible period with which the sun could rotate at its present radius? (Objective 4)

BEYOND THIS BOOK . . .

*For an advanced article on the deaths of stars, try "Endpoints of Stellar Evolution" by A. G. W. Cameron in *Frontiers of Astrophysics* (Harvard University Press, Cambridge, Mass., 1976), edited by E. Avrett. In the same book you'll find "Neutron Stars, Black Holes, and Supernovae" by H. Gursky; it's rather technical.

*We haven't had a supernova in our Galaxy recently, but we do see them in others. See "Supernovas in Other Galaxies" by R. P. Kirshner, *Scientific American*, December 1976.

*For a personal account of the discovery of pulsars, turn to "Little Green Men, White Dwarfs, or What?" by S. J. B. Burnell, *Sky and Telescope*, March 1978, p. 218. For a different view, read A. Hewish's account in *Science*, vol. 188 (1975), p. 1079.

*To find out more about SS 433, read "The Bizarre Spectrum of SS 433" by B. Margon, *Scientific American*, October 1980, p. 54.

BLACK HOLES

12

LEARNING OBJECTIVES
After studying this chapter, you should be able to:

1. Describe a black hole in terms of escape velocity and the speed of light.
2. Derive and make use of an equation for the Schwarzschild radius of a black hole using the concepts of energy and escape velocity.
3. Calculate the density of matter necessary to make a black hole of a given mass.
4. Describe what is meant by a *singularity* and its relationship to a black hole.
5. Describe what happens to an observer who falls into a black hole from the standpoint of the infalling observer and of an outside observer far from the black hole.
6. Explain the concept of tidal gravitational force and make use of an algebraic expression for it.
7. Indicate the place of black holes in the context of stellar evolution.
8. Describe and evaluate the observational evidence to date for the existence of black holes, concentrating on Cygnus X-1.
9. Describe how a black hole can emit X rays.
10. Describe how a binary star system of massive stars might evolve to a binary X-ray source.
11. Describe the observed properties of an X-ray burster and how it can be explained by accretion onto a neutron star.

CENTRAL QUESTION:
What is a black hole and how can it be detected?

A black hole is a region of spacetime (Chapter 5) in which gravity is so strong that nothing, not even light, can escape it. The idea of a black hole springs directly from Einstein's general theory of relativity. The prediction of black holes is one of the most profound in all of physics—and one of the most bizarre.

How are black holes made? Contemporary physics indicates that once a certain minimum mass gets together, it must eventually become a black hole, collapsing by its own gravity after all its nuclear fuel is exhausted. No known physical force can stop this self-destruction of mass that makes a black hole. And that mass is not large—about 5 times the sun's mass.

Have black holes been observed? Many astronomers would answer yes. We cannot see black holes directly, because they emit no light, but we can view the effects of black holes on nearby matter. They may orbit in a binary system and so affect the motion of a visible companion star. Or, if material falls into them, the conversion of gravitational potential energy heats that material sufficiently for it to emit X rays. X-ray satellites have detected emissions which may possibly come from material falling into black holes.

If only one black hole could be found for sure, then many must exist. Once formed, black holes are indestructible. They will exist until the universe ends—if it ever does!

12.1 BLACK HOLES: WARPS IN SPACETIME

Many stars in the Galaxy have masses greater than the neutron star limit of about 5 solar masses. Assume that such stars do not lose enough mass during their evolution to go below this upper mass limit for a neutron star. Further assume that all thermonuclear reactions have ceased, so that the star is cold, close to absolute zero temperature. Now ask this question. Is there any barrier to the collapse of this material with mass greater than 5 solar masses? The answer: There is none.

12.1 BLACK HOLES: WARPS IN SPACETIME

In our present understanding of physics, the crush of gravity overwhelms all outward forces, including the repulsive forces between particles with the same charge. No material can withstand this final crushing point of matter. The collapse cannot be halted; the volume will continue to decrease until it reaches zero. The density will increase until it becomes infinite. Neither of these events can be exactly true of a *real* object in this universe. This theoretical collapse to a singular point of zero volume and infinite density, called a *singularity*, marks a breakdown of the laws of physics as we know them.

Before a mass becomes a singularity, bizarre events occur near it. As its density increases, the paths of light rays emitted from the star are bent more and more away from straight lines going away from the star's surface (Fig. 12.1). Eventually the density reaches such a high value that the light rays are wrapped around the star and do not leave. The photons are trapped by the intense gravitational field in an orbit around the star. The escape velocity from the star is greater than the speed of light. Any additional photons emitted after the star attains this critical density can never reach an outside observer. The star becomes a *black hole*.

The Schwarzschild Radius

Let's consider the meaning of *black* in black hole in terms of escape velocity (Section 5.5). For an object to escape the earth permanently it must leave the surface at a velocity of at least 11 km/sec, the escape velocity from the earth. Imagine that you could squeeze the earth so that it would become smaller and denser. Its escape velocity would increase. Imagine the earth compressed until its escape velocity equaled the velocity of light. Then nothing, *not even light*, emitted at its surface could escape into space. Nothing gets away, so to an outside observer the earth would appear black.

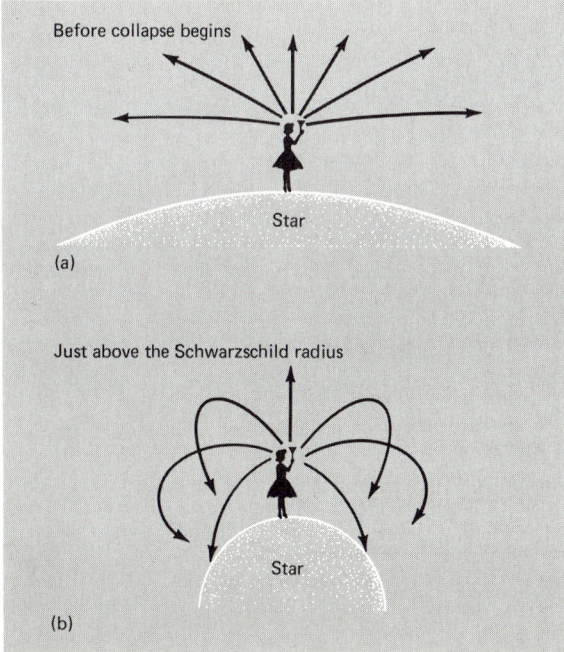

FIGURE 12.1 The trapping of light by the collapse of a mass into a black hole. Imagine a person standing on the star's surface with a flashlight. When much larger than its black hole size, the mass has a low enough gravity so that light leaves on straight-line paths (a). As the star contracts, the gravity at the surface increases so that light paths are bent. When the star is a little larger than its black hole size (b), all light paths, except for the one straight up, fall back to the surface. When the star is smaller than its black hole radius, even the straight-up photon returns. [Adapted from a diagram by W. Kaufmann.]

How small must an object become to be dense enough to trap light? Einstein's general theory of relativity provides an answer. Just after the general theory's publication, the German astrophysicist Karl Schwarzschild (1873–1917) calculated this critical size, now called the *Schwarzschild radius*. For the sun the Schwarzschild radius is about 3 km. If compressed to this size, the sun would have density of about 10^{19} kg/m^3. The mass of any object directly gives its Schwarzschild radius. For 1 solar mass, it's 3 km; for 2 solar masses 6 km; for 10 solar masses 30 km, and so on.

Here's a simple derivation of the equation for the Schwarzschild radius. The escape velocity is (Section 5.5)

$$V_e = \left(\frac{2GM}{R}\right)^{1/2}$$

Let V_e equal c, the speed of light. Then

$$c = \left(\frac{2GM}{R}\right)^{1/2}$$

$$c^2 = \frac{2GM}{R}$$

$$R = \frac{2GM}{c^2}$$

R here is the Schwarzschild radius. The equation can be simplified, by putting in numerical values for c and G, to

$$R = 3M \text{ km}$$

where M is the mass in units of solar masses (Fig. 12.2).

How can a star get as small as its Schwarzschild radius? Two ways are possible. The first is runaway gravitational collapse. More mass than about 5 solar masses must eventually squeeze itself into a black hole. Nothing we know about—not electron or neutron degeneracy pressure, not even the hardness of matter itself—can stop this final crushing. The second way is a supernova. A star's self-destruction can slam matter into a size smaller than its Schwarzschild radius.

Note: Any size object can be made into a black hole if a force is available which compresses it enough. But an object with a mass greater than the neutron star limit of 5 solar masses *must* become a black hole after thermonuclear reactions have ceased, for there is no known source of pressure that can support it.

Once a black hole forms, what happens to the matter that makes it? Einstein's general theory predicts that the matter keeps collapsing gravitationally until it has *no volume*. But it still has mass, so its density is infinite. This theoretical end to runaway gravitational collapse is called a *singularity* (Fig. 12.3). The matter has literally squeezed itself so that it occupies no space. Yet it's still there. How can matter *not* take up space? The general theory of relativity points to the formation of a singularity, cloaked in the center of a black hole, as the natural end of gravitational collapse. It is also where our present knowledge of physics ends.

12.1 BLACK HOLES: WARPS IN SPACETIME

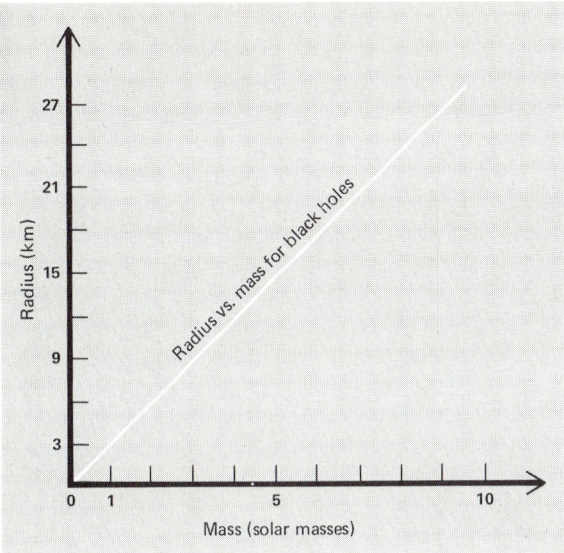

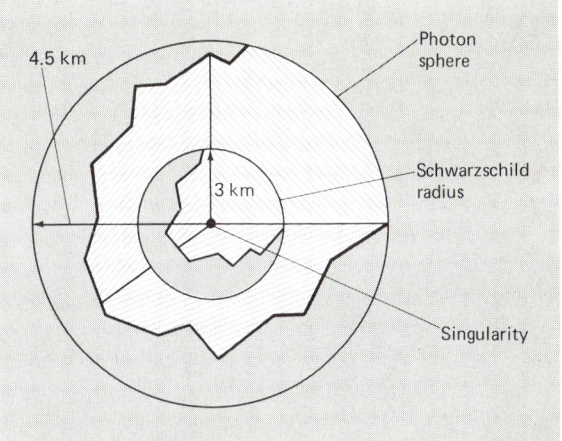

FIGURE 12.2 The relationship between mass and Schwarzschild radius for black holes.

FIGURE 12.3 A schematic of the structure of spacetime around the singularity that is supposed to form in the center of a black hole.

Journey into a Black Hole

Put aside the puzzle of the singularity for a moment and consider the theoretical properties of a black hole, both inside and outside. Keep in mind that a black hole marks a weird region of spacetime. A person falling into a black hole meets a fate an outside observer cannot even find out about—unless the outside person drops in too. (One way you can find out if a singularity exists in the center of a black hole is to jump in. Unfortunately, you'll never be able to come back to tell others what you've found.)

Let's take an imaginary journey into a black hole. We'll follow the adventures of a crazy astronaut who takes the plunge and compare this trip with what an outside observer sees of it.

You and a friend start out in a spaceship orbiting a 10-solar-mass black hole at a distance of 1 AU (Fig. 12.4). The ship orbits the black hole in accordance with Kepler's laws, as it would any ordinary mass. In fact, Kepler's third law and the spaceship's orbit permit you to measure the hole's mass. You know it is there, but you can't see anything. Your friend volunteers to hop in. She takes with her a laser light and digital watch. You and she synchronize watches. Once a second, according to her watch, she will send a laser flash back to you.

Down she goes! For a long time as she falls toward the black hole, nothing strange happens. But as she gets closer, stronger and stronger tidal gravitational forces (Focus 12.1) stretch her out (if she falls feet first) from head to toes. Also, another tidal force squeezes her together, mostly at the shoulders. (Such tidal forces exist on the earth, but they are so weak you don't feel them.) Near a black hole tidal forces grow enormously. Since ordinary human beings would be ripped apart about 3000 km from a 10-solar-mass black hole, we will have to imagine temporary indestructibility for our descending astronaut, at least until she enters the black hole.

FOCUS 12.1 TIDAL GRAVITATIONAL FORCES

You're familiar with gravitational forces pulling objects together. But they can also pull objects apart if you look at the *differences* between gravitational forces at different locations.

Consider two small balls, A and B, far above the earth (Fig. F.16). A is closer to the earth by distance d. If A is distance R from the earth, B is at distance $R + d$. Imagine the balls free-falling toward the earth. Because A is closer than B, it experiences a greater acceleration than B does. Some time later A will have fallen a greater distance than B. The distance between A and B will have increased. Viewed from either A or B, it looks like a force has pulled the two apart. This force, which arises from the differences between gravitational forces, is called a *tidal gravitational force*.

Let's compute the size of this force. The force on ball A is

$$F_A = \frac{GMm_A}{R^2}$$

The force on ball B is

$$F_B = \frac{GMm_B}{(R+d)^2}$$

Here M is the earth's mass. For simplicity, let's assume that m_A equals m_B and call them both m. Then the difference between the two forces is

$$\Delta F = F_A - F_B = \frac{GMm}{R^2} - \frac{GMm}{(R+d)^2}$$
$$= GMm\frac{(R+d)^2 - R^2}{R^2(R+d)^2}$$
$$= GMm\frac{d(2R+d)}{R^2(R+d)^2}$$

Because R is much larger than d, we can approximate

$$2R + d \approx 2R$$
$$R + d \approx R$$

and then

$$\Delta F = GMm\frac{d(2R)}{R^2 R^2}$$
$$= 2GMm\frac{d}{R^3}$$

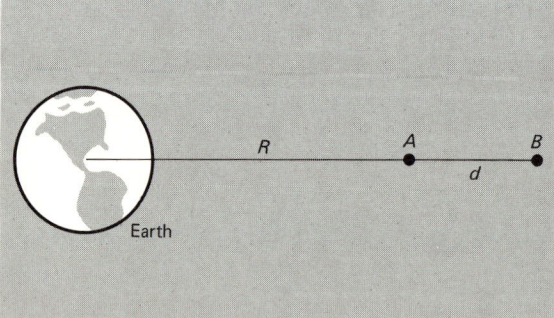

FIGURE F.16 Geometry for finding tidal forces on the masses A and B.

What a curious result: The tidal force depends on the inverse cube of the distance! Tidal forces become strong very quickly as you get closer to a mass. For example, at 3000 km from a 10 solar-mass black hole, the tidal forces can rip a person apart. At the edge of that black hole the tidal forces are $(3000 \text{ km}/3 \text{ km})^3$ stronger, or 10^9 times greater!

12.1 BLACK HOLES: WARPS IN SPACETIME

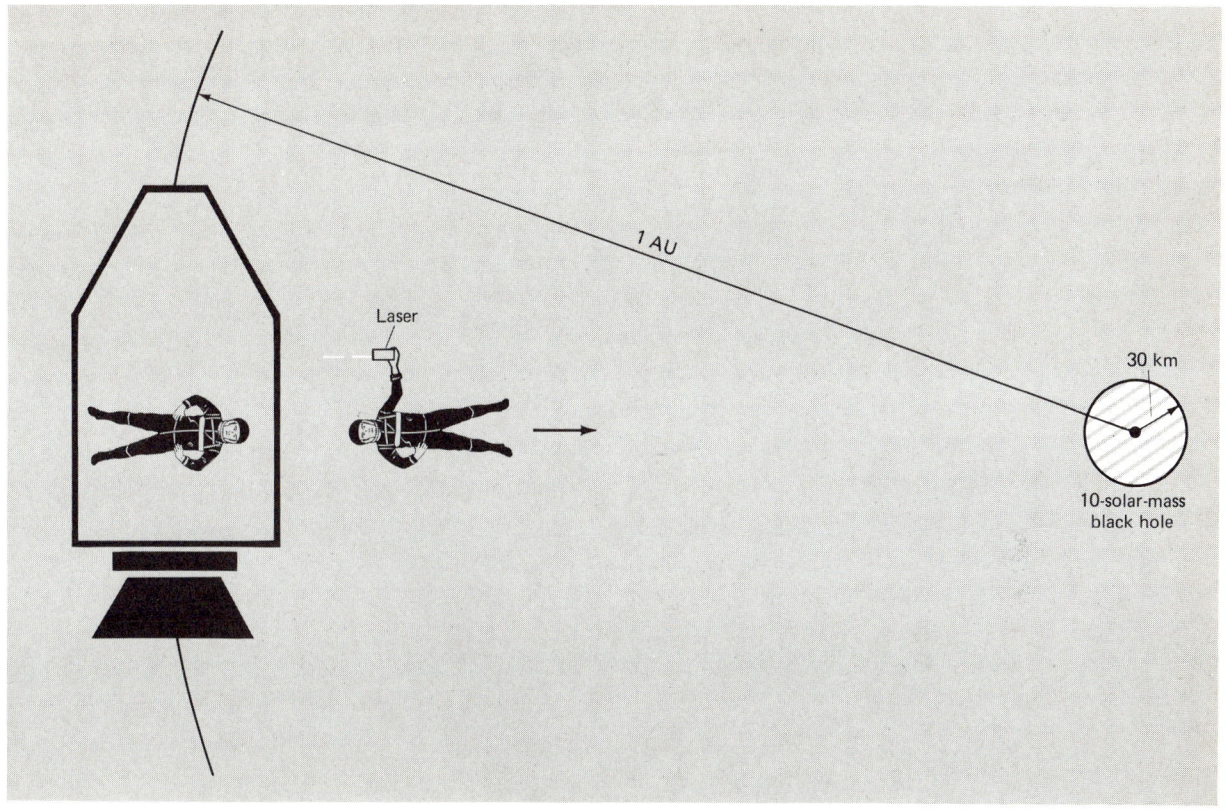

FIGURE 12.4 Orbiting a 10-solar-mass black hole. The black hole has a radius of about 30 km.

Down she drops. The tidal forces increase rapidly and make her more uncomfortable, but nothing else seems strange to her. Every second she sends out a blast of laser light. Peering down, she can just make out a small black region in the sky. (A 10-solar-mass black hole has a radius of only 30 km.)

Then it happens: she crosses the Schwarzschild radius! But nothing new happens to her. No solid substance, no signs mark the edge of the black hole. However, no amount of energy can push her out of the black hole; she has crossed a one-way gate in spacetime. The trip now swiftly ends for your unfortunate friend. In about 10^{-5} sec after she crosses the Schwarzschild radius, she crashes into a singularity. Crushed to zero volume, she is destroyed. Even if a singularity does not lie in the black hole's center, the mass that made the black hole probably does. So she would smash into it. Fatal end of friend's trip.

But what of your view, back in the spaceship? You do not see her final destruction; in fact, you do not even see her fall into the black hole. As she drops closer to the black hole, you notice that the light from her laser is red shifted, with the shift increasing the closer she falls. (The light must work against gravity to get to you, so it loses energy and increases in wavelength. That's the gravitational red shift; Section 11.1.) Also, you notice that the time between laser flashes increases. Compared with your watch, your friend's watch appears to slow down as she gets into regions of stronger gravity. Your watch and hers disagree about how long it takes her to travel to the black hole. (This slowing down of time is another prediction of general relativity; it has been confirmed by observations.)

As she comes closer to the Schwarzschild radius, the watches get more and more out of synch. The time between your reception of her flashes stretches out. In fact, a laser burst sent out just as she crossed the Schwarzschild radius would take an *infinite* time to reach you. It also would suffer an *infinite* red shift. To you, her fall seems to grow slower and slower as she gets closer to the black hole. Time slows down so much that it seems to be frozen. The light gets more and more red shifted until you can no longer detect it. A black hole practices cosmic censorship. It prevents you from seeing your friend even fall into it. Light—our only astronomical communication medium—is cut off by the black hole. So you cannot know what happens to your friend inside of it.

Warning: Black holes aren't cosmic vacuum cleaners. A popular misconception about black holes is that they have infinitely powerful gravitational fields that suck up everything that gets near them, scouring out the universe. Not quite. Suppose you'd never heard anything about the strange properties of black holes. Just thinking about Newtonian gravitation, what do you think would happen to the orbit of the earth around the sun if the sun suddenly shrank down to a ball 6 km across? That's right, nothing! The masses of the sun and earth haven't changed, and neither has the distance from the earth to the center of the sun. So the force of gravity on the earth hasn't changed and neither has its orbit. (Of course, things would get pretty dark.)

Now, it's true that the gravitational field of a black hole is very strong close to the black hole. In fact, that's the significance of the Schwarzschild radius; at that distance from the center of the black hole, the escape velocity is equal to the speed of light. Because no material object can go as fast as the speed of light, once you get closer than the Schwarzschild radius to the center of the black hole, you're stuck and you can't get out. But as long as you're safely in orbit outside the Schwarzschild radius, you'll stay in orbit; the black hole won't pull you in. You can certainly fall in, if you're not moving fast enough to stay in orbit. But that's just a property of gravity; the earth pulls on things, too, and pulls down the ones (like you) that aren't in orbit around it.

You can think of black holes as ghosts. The Schwarzschild radius is *not* a physical object. It's a region of spacetime so severely curved that weird things happen near it. Einstein's general theory actually describes the geometry of spacetime in detail; Fig. 12.5 shows a representation of this geometry. Note how badly warped spacetime is from flat (flat means no gravitational field, no mass). Likewise, the singularity predicted to be in the center of a black hole cannot be a real object. It's a region where space and time end in this universe. What a strange corpse from the death of massive stars!

The Collapse of a Star into a Black Hole

Imagine for a moment a nonrotating massive star (say, 10 solar masses) at the end of its nuclear-burning life. No internal pressure—not even the repulsive forces between elementary particles—can stop this mass from collapsing into a black hole. Once smaller than its Schwarzschild radius, the star gets crushed into the singularity. Once smaller than its Schwarzschild radius, no photon emitted by the star can ever get to an outside observer.

Do these predictions mean that the star instantaneously and completely disappears from the view of an outside observer? What would the outside observer see? At first, the

12.1 BLACK HOLES: WARPS IN SPACETIME

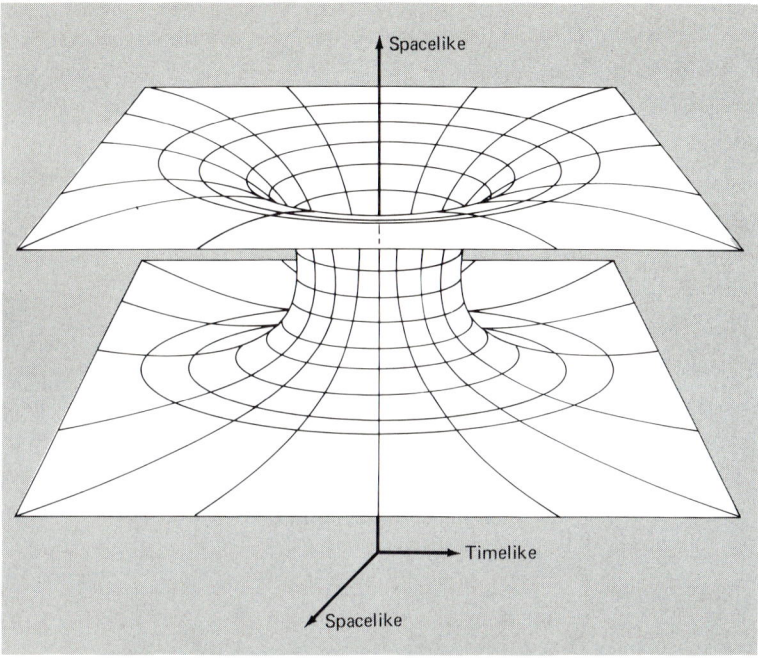

FIGURE 12.5 Representation of the geometry of spacetime around a black hole. This map is for a particular time (see Fig. 12.7).

collapse would go on rapidly. The light from the star would increase in red shift and also dim. In fact, the red shift makes the star go black. As the star's surface approaches its Schwarzschild radius, the collapse appears to slow. But the red shift and dimming rapidly increase. In only about 0.01 sec, the star winks out from view.

The Structure of Spacetime Around a Black Hole

Let's look at the geometry of spacetime outside of a black hole. To do this in detail, we will have to examine a spacetime diagram for the geometry there. We get the appropriate spacetime map by solving Einstein's equations of general relativity to find the geometry of spacetime for an empty region of space surrounding a nonrotating, spherical mass (the simplest case). To see what's going on will take a little concentration, so buckle down. The main point we want you to discover is this: *Spacetime is not static, but dynamic.* You will also see that spacetime does even stranger things than have been described so far.

Here's the spacetime diagram (Fig. 12.6). It's a bit different from the ones you've seen before (Chapter 5), because the coordinates used here are not space and time as you experience them. The horizontal axis has properties that are spacelike, the vertical one timelike, but they are not exactly the same as measured space and measured time. (This special set of coordinates, first used by M. Kruskal and G. Szekeres, has the neat property that the spacetime diagram around a mass is very easy to picture.) Past is at the bottom of the diagram; future at the top. Light follows a special path in this space-time diagram; it moves at 45° with respect to the axis. Any object moving slower than light has a path between the timelike axis and the light path; a path between the light line and the spacelike axis represents something moving faster than light, which is not normally possible.

You will note that the diagram has been divided into four regions (I, II, III, and IV),

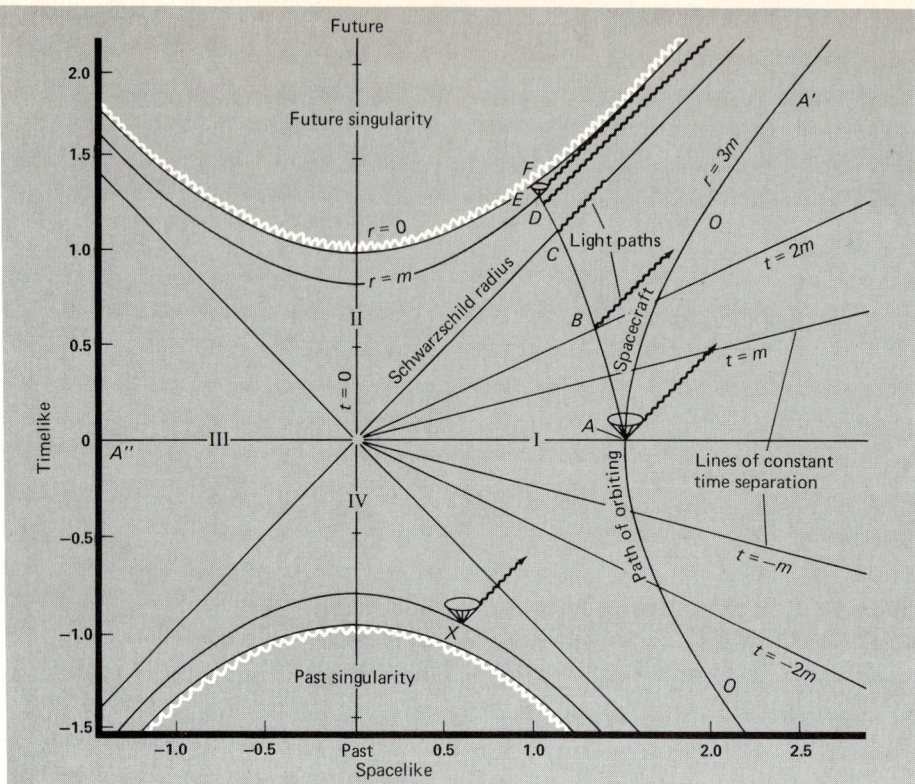

FIGURE 12.6 A spacetime map for the region around a black hole. The horizontal axis has spacelike properties, the vertical one timelike properties. The black hole (region II) is at the top; in it is a future singularity. Region I is our part of spacetime. Path AA' is that of an object orbiting the black hole. Region IV is a black hole and singularity in the past. [*Adapted from a diagram by R. Ruffini and J. Wheeler.*]

separated by *event horizons*. (An event horizon is another term to describe the Schwarzschild radius; it emphasizes the fact that any events taking place inside the Schwarzschild radius are cut off from outside view; they are beyond our visible horizon.) Note a singularity exists both at the top (future) and bottom (past) of the diagram. Also drawn in the diagram is the path of a spaceship orbiting the black hole (line O) and that for a person jumping into the black hole from the spaceship (path A to F).

With these preliminaries in mind, look at the journey into a black hole, described above, in the spacetime diagram. As the spaceship orbits the black hole, it moves along line O from past to future (to the upper right). Your friend jumps out of the spaceship at A. Her laser signals are indicated by wavy lines; note these lie at 45° with respect to the axes. The point at which the wavy line crosses line O is the point where and when you see it. The pulse emitted at B crosses line O, but the pulse emitted at C, that is, as your friend crosses the event horizon (Schwarzschild radius), does not intercept O until an infinite time has passed. All photons emitted after she crosses the event horizon (at D and E, for example) eventually get gobbled up in the singularity. At point F your friend plunges into the singularity. You cannot see any events at C or beyond.

This example shows that region I of the diagram is our region of spacetime, that is, our universe outside of the black hole. Region II is that part of spacetime within the Schwarzschild radius and containing the (predicted) singularity. What about region III? It's the mirror image of region I; another realm of spacetime outside of event horizons and singularities. Some people have called this region "another universe." (We don't believe that's the proper term, for as will be explained below, there's no way to get "there.") At the bottom of the diagram, region IV, lies a singularity that is the mirror image of the one at the top; that is, a singularity in the past or a time-reversed black hole. Notice that a permitted photon path (from X) can cross the event horizon into region I, our universe. This we would see as light erupting from an event horizon, a phenomenon sometimes called a *white hole*.

The so-called other universe of region III looks very inviting. But it's inaccessible, so we cannot demonstrate its existence. We are in region I. Suppose we tried to travel to region III along path AA''. To go that way means we must *travel faster than light*, because such a path makes less than a 45° angle with the spacelike axis. In fact, if you examine the figure, no path from region I to III requires less than light speed. Any slower than light path crashes into the top singularity. So we have no access to region III. Also, we cannot reach region IV, because it lies in the past, and we cannot travel back in time. We can get into region II, but we can't get out. Only in region I can we move in space at will.

Let's describe the inaccessibility another way. Suppose you would like to make a motion picture of the evolution of spacetime's geometry around a black hole. We can use the spacetime diagram to make such an imaginary movie. A horizontal slice across the diagram (like AA') gives the space geometry at a certain instant of time (Figs. 12.7a,b).

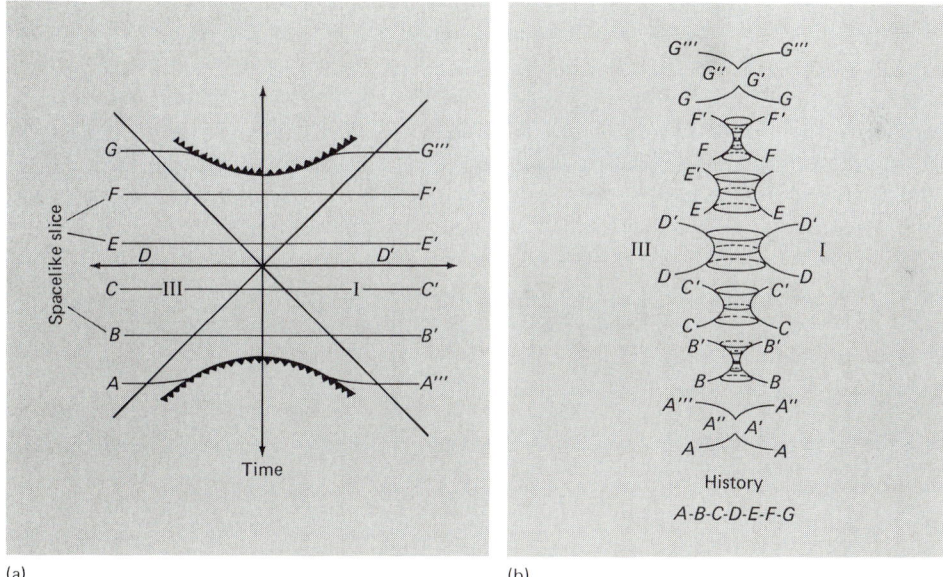

(a)   (b)

FIGURE 12.7 The evolution of spacetime around a black hole. Take slices through the spacetime diagram (a) at different times ($AA'A''A'''$, BB', and so on). These represent the form of spacetime at a sequence of times (b), showing how the throat of the geometry (Fig. 12.5) opens and closes, connecting regions I and II. [*Adapted from a diagram by C. Misner, K. Thorne, and J. Wheeler.*]

Notice that slice AA' has a geometry that bridges two flat regions of spacetime. This geometrical connection is called the *Einstein-Rosen bridge* or *wormhole*. Region I (top) is our universe; region III (bottom) is the mysterious "other universe." It looks like you could cross from region I to III along the surface of the wormhole. But you can't because the wormhole is not static and always open.

To see this point, make a movie by taking horizontal slices in the spacetime diagram (Fig. 12.7b) from bottom to top. This procedure gives a history of the evolution of spacetime's geometry. Notice that the throat of the wormhole opens (connecting regions I and III), reaches a maximum expansion (AA'), then closes again, disconnecting regions I and III. This opening and closing of the wormhole happens rapidly—so rapidly that if you tried to cross the bridge, it would shut before you got through, pinching you off into the singularity. Only by crossing the bridge faster than light (which you can't do) could you avoid being caught in the pinch-off.

This evolution of spacetime geometry comes right out of a special solution to Einstein's equations of general relativity. And it demonstrates one of the most profound insights of Einstein's theory: *Spacetime is dynamic; it evolves.* This insight will be extremely fruitful when we look at the evolution of the universe (Chapter 20).

12.2 OBSERVING BLACK HOLES

Enough for black holes as theoretical entities. How can you actually observe a black hole? Light emitted inside cannot get out; light sent out close to it is so strongly red-shifted that it's hard to detect; and a black hole is small, only a few kilometers in size. So you'll have a hard time seeing an isolated black hole.

But a black hole surrounded by clouds of material might be observable. Any matter falling toward a black hole gains energy and heats up. (It's also squeezed by tidal forces.) Heated enough, the atoms are ionized. Gravity accelerates the ionized gas, and it emits electromagnetic radiation. If heated to a few million Kelvins or so, the material gives off X rays. Before it is trapped in the gravitational gulf, infalling material can send X rays into space. So a black hole passing through an interstellar cloud or close to a star can sweep material into it and radiate. X-ray sources are good candidates for black holes.

X rays can't penetrate the earth's atmosphere, so X-ray astronomy can only be done from space. The *Uhuru* satellite, launched in 1970 and designed to observe X-ray sources, detected about 160 strong X-ray objects (Fig. 12.8). Some of these X-ray sources are prime candidates for black holes because they are binary—the X-ray source and a normal star (a potential source of infalling material) orbit a common center of mass.

Binary X-Ray Sources.

Why are binary X-ray sources most suspect? Imagine a black hole orbiting a supergiant star. Suppose that they are very close together so that their orbital period is a few days. Also, suppose their orbital phase is in our line of sight. The star has a huge, distended atmosphere, and material blown from this atmosphere (a stellar wind) can fall into the black hole (Fig. 12.9). Falling toward the black hole, the material gains kinetic energy, heats up to a million Kelvins, and emits X rays. Only a small region around the black hole gives off X rays, and, since the material falls in sporadically, the intensity of the X rays

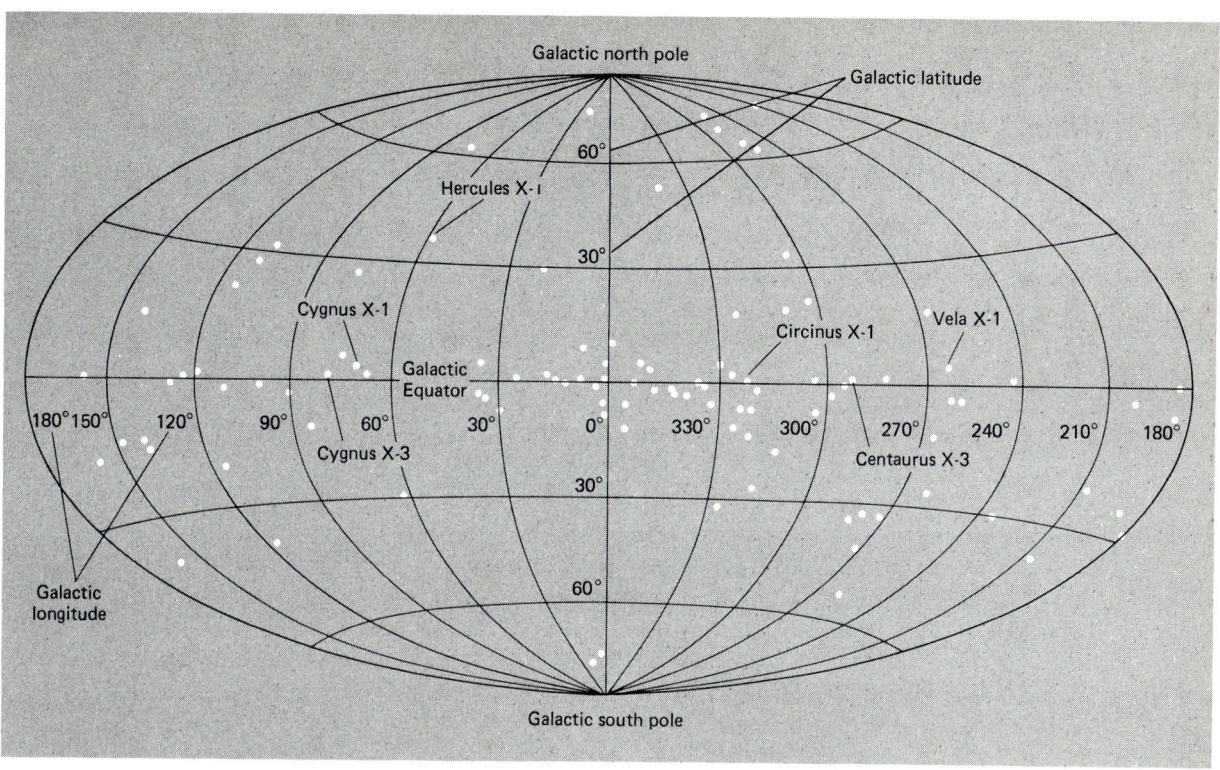

FIGURE 12.8 A map of the sky locations of strong X-ray sources detected by the Uhuru satellite. The strongest sources are named. Note the concentration to the Galactic Equator, an indication that the sources are in the Galaxy. [*Adapted from a diagram by H. Gursky.*]

varies quickly. When the black hole goes behind the star in our line of sight, its X rays are cut off. This is an *eclipsing* X-ray binary system.

So a sign of a possible black hole is a rapidly varying X-ray source, which may be eclipsed at regular intervals in a binary system. Have such variable X-ray sources been seen? Yes. Some are listed in Table 12.1. Each of these objects is believed to be a main-sequence or post-main-sequence star swinging around an X-ray source. They have X-ray luminosities in the range from 10^{29} to 10^{31} W. (That's 200 to 20,000 times the luminosity of the sun and all in X rays.) Three of the sources (Hercules X-1, Centaurus X-3, and Small Magellanic Cloud X-1) have short-period X-ray pulses (Fig. 12.10); they are X-ray pulsars. Searches for radio pulses from these objects have so far been unsuccessful.

Five of the systems (Centaurus X-3, Small Magellanic Cloud X-1, Vela X-1, Circinus X-1, and Hercules X-1) exhibit X-ray eclipses; the X-ray source passes behind the normal star as we view the system. Using spectroscopic analysis of the light from the visible star (*not* the X-ray source), we can observe the changes in Doppler shift, and so we can find out the orbital periods. They are typically a few days. These short periods indicate the orbits are only a few times larger than the primary stars. Then, if we can determine the separation of the two objects, we can ascertain, from Newton's form of Kepler's third law (Section 4.2), the sum of the masses (normal star plus X-ray source).

With an idea of the mass of the normal star from its luminosity (using the mass-

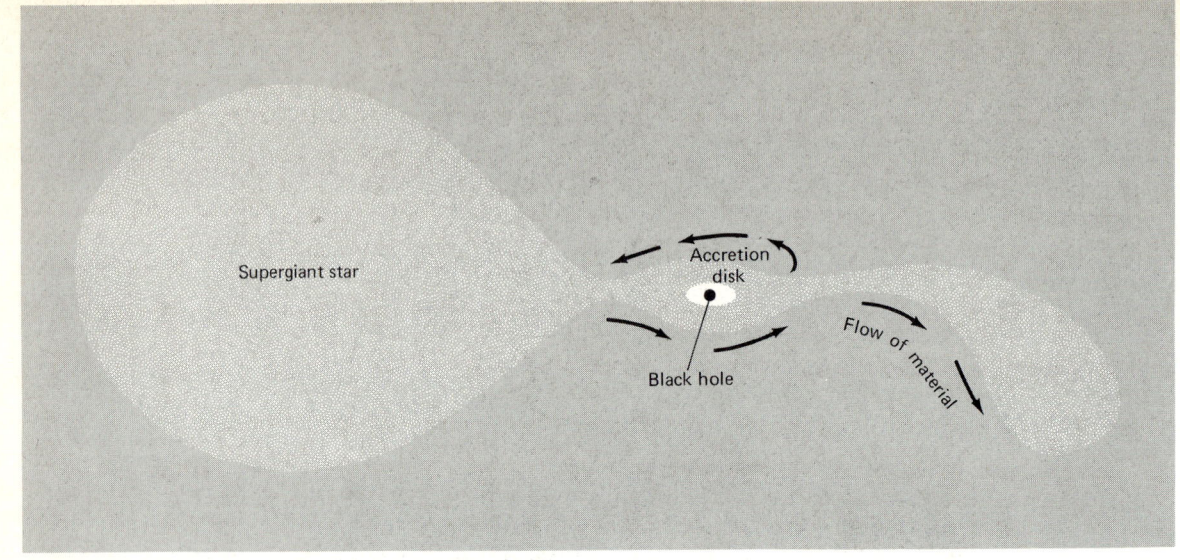

(a)

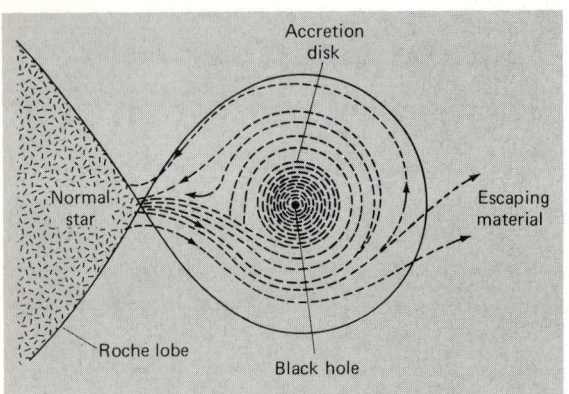

(b)

FIGURE 12.9 (a) One model for a black hole as an X-ray source. In a binary system the black hole is coupled with a giant or supergiant star. Material from the star flows to the black hole, where it falls into an accretion disk. As it falls, it heats up to about a million Kelvins and emits X rays. [*Adapted from a diagram by K. Thorne.*] (b) A close-up of the accretion disk around a black hole. [*Adapted from a diagram by D. Eardley and W. Press.*]

FIGURE 12.10 X-ray pulses from three binary X-ray sources. [*Adapted from a diagram by S. Rapapport and P. Joss.*]

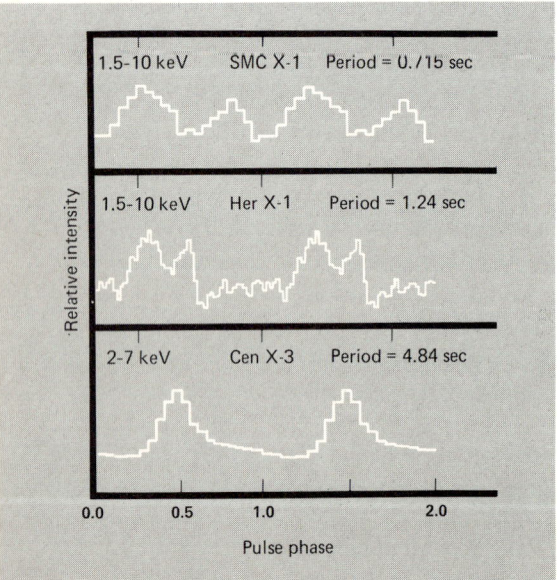

12.2 OBSERVING BLACK HOLES

luminosity law), we can also determine the mass of the X-ray source. And if that mass turns out to be large enough (greater than 5 solar masses, the upper limit for a neutron star), the X-ray source must be a black hole!

Let's investigate two of these binary X-ray sources in detail: Cygnus X-1 and Centaurus X-3. Might these be black holes?

Is Cygnus X-1 a Black Hole?

To prove the reality of black holes we need to observe one. So far, the most likely candidate is Cygnus X-1, a strong X-ray source in the constellation Cygnus (Fig. 12.11).

Cygnus X-1 emits about 4×10^{30} W in X rays. Observations have shown that Cygnus X-1 flickers rapidly, in less than 0.001 sec. (This is an intrinsic flickering of the source, not a "twinkling" like that of stars, which is caused by the earth's atmosphere.) In the same way as was done for pulsars (Section 11.5), we can interpret this observation as indicating that the X-ray emitting region must be less than 0.001 light second in size (less than 300 km). In 1971 radio astronomers discovered radio bursts from Cygnus X-1 and were able to pin down its location better than the X-ray astronomers could. In the most likely place for Cygnus X-1 lies an O supergiant star, that is, a hot massive star (Fig. 12.12). It is called HDE 226868 and is of spectral type O9.7 I, that is, a supergiant star of approximately 31,000 K surface temperature.

Optical observations show that the dark lines in the spectrum of the blue supergiant go

TABLE 12.1 Some Binary X-Ray Sources

Name	Distance (kpc)	Binary Period (days)	Characteristics of X Rays	Characteristics of Visible Star
Cygnus X-1	2.5	5.6	Vary in duration from 0.001 to 1 sec	Blue supergiant of about 20 solar masses
Centaurus X-3	5–10	2.087	Eclipses with 0.488-day duration; pulses every 4.84 sec	Blue giant of about 16 solar masses
Small Magellanic Cloud X-1	65	3.89	Eclipses with 0.6-day duration; pulses every 0.72 sec	Blue supergiant of about 25 solar masses
Vela X-1	1.4	8.95	Eclipses with 1.7-day duration; flares lasting a few hours	Blue supergiant of about 25 solar masses
Hercules X-1	2–6	1.70	Eclipses with 0.24-day duration; pulses every 1.24 sec	Companion HZ Her about 2 solar masses
Scorpio X-1	0.3–1.0	0.787	Vary in duration down to 1 sec; no eclipses	Less than 2 solar masses

Source: Adapted from tables by H. Gursky, E. P. J. van den Heuvel, S. Rappaport, and P. C. Joss.

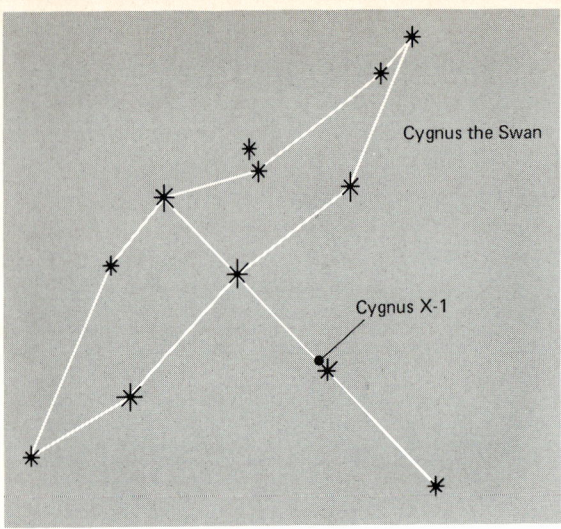

FIGURE 12.11 The position of Cygnus X-1 in the constellation of Cygnus.

FIGURE 12.12 The blue supergiant star HDE 226868 (arrow) about which Cygnus X-1 orbits. [*Courtesy J. Kristian.*]

through periodic Doppler shifts in 5.6 days (Fig. 12.13). So the supergiant orbits with the X-ray source about a common center of mass every 5.6 days. The supergiant has a massive but optically invisible companion—Cygnus X-1.

Recall that only for binaries can we find directly the masses of stars (Section 8.6). But we need to know the separation of the stars and their distance from the center of the mass. These distances can be worked out correctly only if we can observe *both* stars and know the orbital tilt with respect to our line of sight. In these two regards the mass of Cygnus X-1 is hard to determine. We can observe the Doppler shift in the spectrum of the visible companion, but we cannot obtain the velocity of the X-ray source. And because Cygnus X-1 has not been found to eclipse, we can't pin down its orbital inclination. So we don't have enough information to determine both individual masses.

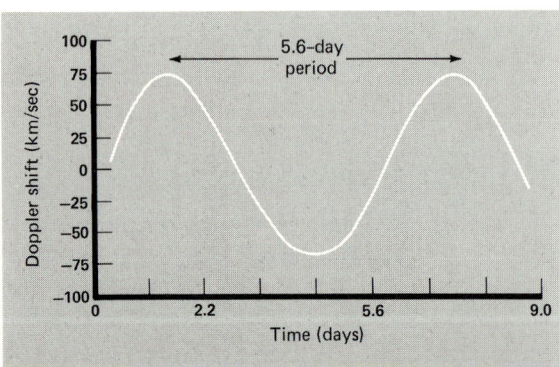

FIGURE 12.13 The orbital period of Cygnus X-1 inferred from the Doppler shift in the spectral lines of the blue supergiant star about which Cygnus X-1 orbits. [*Based on observations by C. Bolton.*]

12.2 OBSERVING BLACK HOLES

But we can make some reasonable estimates. Blue supergiant stars are typically 15 to 40 solar masses. Take 20 as typical. The orbital period and velocity of the supergiant give us a relation between the masses of the supergiant and the X-ray source, uncertain by the amount of orbital tilt. If the orbital tilt were 90°, the mass of the companion would be 4 to 5 solar masses. But the tilt can't be so high, because the star is not eclipsing. So the supergiant's orbital velocity must be higher than observed, and the mass of the companion must be higher. A study of the brightness of the supergiant indicates that it varies a little in 5.6 days. This fact leads to an inferred orbital tilt of 30°. If true, then Cygnus X-1 has a mass possibly as great as 14 solar masses, and most likely about 9 solar masses. If so, Cygnus X-1 must be a black hole, if the limit for a neutron star is 5 solar masses.

We feel that there is strong evidence here for a black hole. To sum up the chain of inference:

1. Because it emits X rays which vary in a short time, Cygnus X-1 must be a small, dense object onto which matter is falling.
2. The blue supergiant is the star about which Cygnus X-1 revolves.
3. From the Doppler shift of its spectral lines, the blue supergiant (and so the X-ray source) has an orbital period of 5.6 days.
4. The blue supergiant has a mass of 14 to 40 solar masses.
5. If the blue supergiant has a mass of 20 solar masses, Cygnus X-1 has at least 4 to 5 solar masses and more likely 9 to 14 solar masses.
6. A small object with a mass greater than about 5 solar masses is a black hole.
7. Therefore, Cygnus X-1 is probably a black hole.

Not all astrophysicists accept this conclusion. For example, some argue that the upper limit on a neutron star's mass is higher than 5 solar masses. In that case, if the mass of Cygnus X-1 is only 4 solar masses, it could be a neutron star rather than a black hole. Also, we have assumed that the supergiant is a normal star, but it may have been altered by the previous evolution of its companion. We wish we could report the existence of black holes as certain, but in all honesty we cannot. Yet, relativity predicts black holes. To believe in general relativity pretty much requires the acceptance of black holes.

Is Centaurus X-3 a Black Hole?

So far we've discussed binary X-ray sources in the context of black holes. But the X-ray emission could arise not only from accretion around black holes, but also onto neutron stars. (Recall from Section 11.2 that the escape velocity of neutron stars is close to the speed of light.) If one X-ray source, say Cygnus X-1, does get confirmed beyond reasonable doubt as a black hole, it does not follow that all binary X-ray sources must be black holes. What about the others?

Let's take a close look at Centaurus X-3 (abbreviated Cen X-3). The Uhuru satellite showed that this X-ray source pulses every 4.84 sec (Fig. 12.14). Also, long-term observations revealed that X-ray eclipses take place every 2.087 days and last about 0.5 day. So for Cen X-3 we know the orbit is tilted so that its plane lies in our line of sight. In the summer of 1973 Vojtek Krzeminski, a Polish astronomer, found a faint star at the X-ray source position that varies in light in the same period as Cen X-3. The star turned out to be a blue giant about 25,000 light years from us.

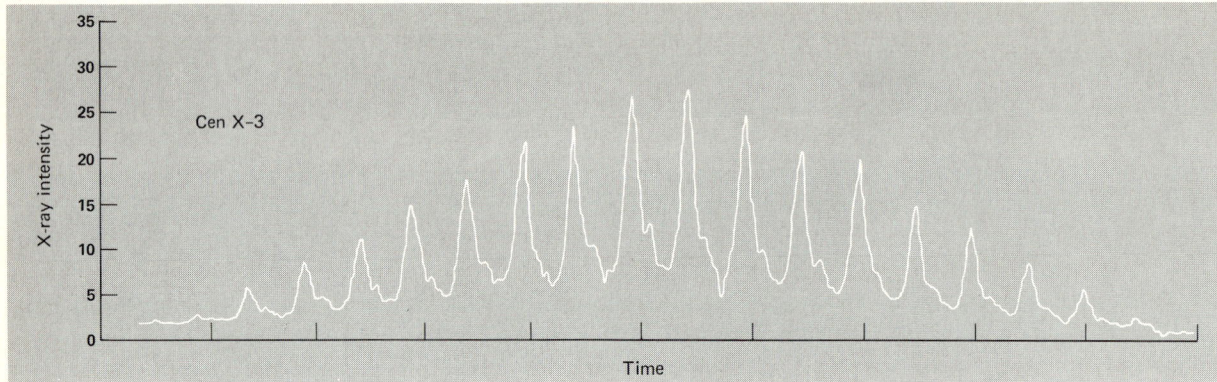

FIGURE 12.14 X-ray pulses, every 4.8 sec, from Centaurus A, as observed by the Uhuru satellite. Its rotation causes the variations in the intensities of the pulses. [*Adapted from a diagram by H. Gurskey.*]

This information all falls into place with a simple model for the Cen X-3 binary system (Fig. 12.15). Cen X-3 itself moves in an almost circular orbit around the blue giant star at 415 km/sec. Its orbit has a radius of about 11 million km, closer to the blue giant star than Mercury is to our sun. At this close distance mass flowing from the giant star is picked up by the X-ray source. About every 2 days the X-ray source orbits behind the giant star as seen from the earth, and an X-ray eclipse happens.

Is Cen X-3 a black hole or a neutron star? That depends on its mass. Since Cen X-3 eclipses, we can pin down its mass more definitely than we can for Cygnus X-1. Yoram Avni and John Bahcall have concluded that Cen X-3 has a mass between 0.6 and 1.1 solar masses, so it could be a neutron star rather than a black hole. Such a low-mass neutron star can be formed in a supernova (Section 11.2).

The fact that Cen X-3 is an X-ray pulsar also supports a neutron star model, in analogy with the model of radio pulsars as magnetic neutron stars. The X-ray pulses might arise from accreting matter channeled into the magnetic polar regions by the intense magnetic field (Fig. 12.16).

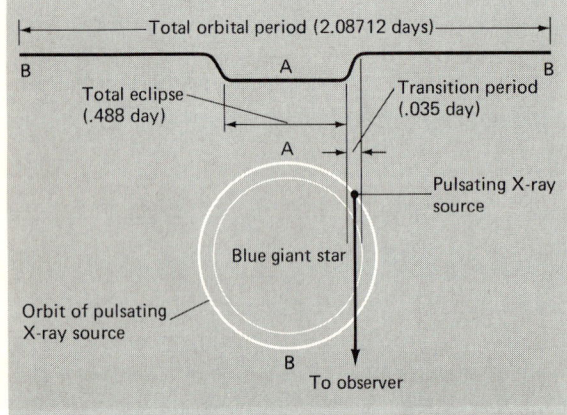

FIGURE 12.15 A model for the Centaurus X-3 binary system. [*Adapted from a diagram by H. Gurskey.*]

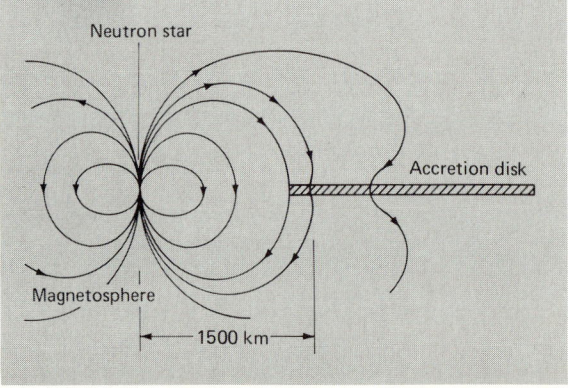

FIGURE 12.16 An accretion disk around a highly magnetic neutron star.

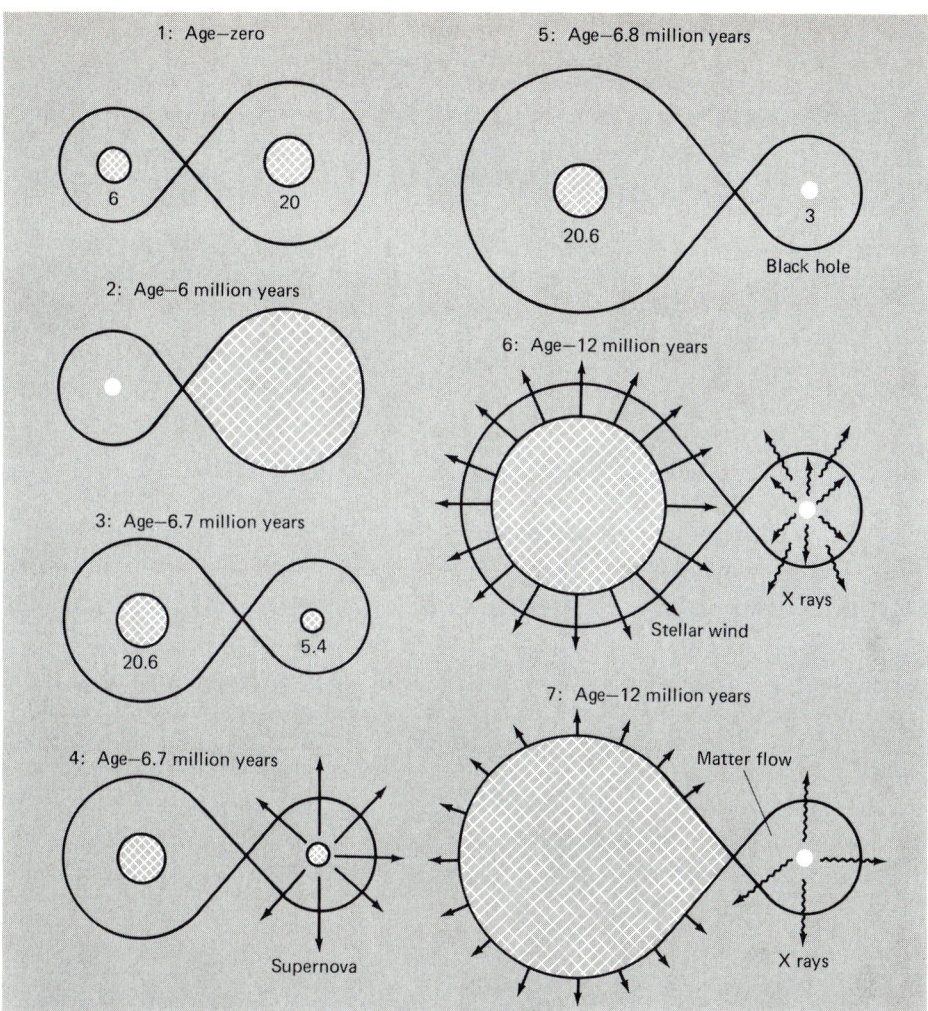

FIGURE 12.17 A model for the evolution of a close binary system with massive stars. Start out with 20- and 6-solar-mass stars. The more massive star evolves more quickly, expands in size, and material flows to the companion (1 and 2), which ends up with 20.6 solar masses (3). The core of the other star supernovas (4) to leave a black hole (5). The companion evolves, expanding and developing a strong stellar wind. Matter flows into the black hole and X rays result (6 and 7). [*Based on calculations by P. van den Heavel, C. de Loore, and J.-P. de Greve.*]

The Evolution of Binary X-ray Systems

How do you end up with an X-ray source (black hole or neutron star) in a binary system? One answer appears if you remember the basic fact of stellar evolution: The more massive a star is, the faster it evolves (and so the shorter its lifetime). With this fact in mind, let's examine some facets of one possible model for the evolution of a binary X-ray system.

Start with an ordinary binary containing a 20-solar-mass star and a 6-solar-mass star (stage 1 in Fig. 12.17). The more massive star evolves faster, becomes a red supergiant star, and fills its Roche lobe (stage 2). Matter streams from the more massive star to the less massive one. When the flow stops, the stars have switched roles (stage 3)—the 6-solar-mass star now has 20.6 solar masses, the other 5.4 solar masses. The star that is less massive at this stage is essentially the core of the former 20-solar-mass star. This core becomes a supernova (stage 4). It leaves behind a neutron star or a black hole (stage 5). The other star now evolves rapidly (because of its increased mass). It first loses matter by a stellar wind. Any of this material falling onto the black hole generates X rays (stage 6). Later, the star expands until it fills its Roche lobe (stage 7). Material can then flow along the gravitational bridge to the black hole and emit X rays.

What next? The massive supergiant should also become a supernova. That leaves a binary black hole system, a binary neutron star system, or a binary black hole–neutron star system. In all these cases, the matter is locked up. So the binary no longer emits X rays. Note that the X ray emitting stages don't last long, at most a few million years.

X-Ray Bursters: Active Black Holes or Neutron Stars?

X-ray astronomy played a key role in uncovering black hole candidates. But the story does not end with Cygnus X-1. In fact, it has gotten more curious and exciting since the discovery of *X-ray bursters*.

As you might guess from the name, X-ray bursters are set apart from other X-ray sources by their emission of brief but powerful bursts of X rays (Fig. 12.18). The bursts may peak for several seconds. Some bursters repeat their X-ray blasts at more or less regular intervals of a few hours or a few days. Others fire off in a rapid sequence like a machine gun, shooting off several thousand bursts in a day. A 10-sec burst carries as much X-ray energy as the sun gives off in a week at *all* wavelengths!

Here's the story of two X-ray bursters—both surprises with something in common that presents an astronomical puzzle.

The Uhuru X-ray sky survey detected a source, called 3U 1820–30 ("3U" stands for the third Uhuru catalog; the other numbers refer to the object's coordinates on the sky). The source lies near the center of the globular cluster NGC 6624. In 1975 Jonathan Grindley and Herbert Gursky discovered a brief burst of X rays from NGC 6624. Other

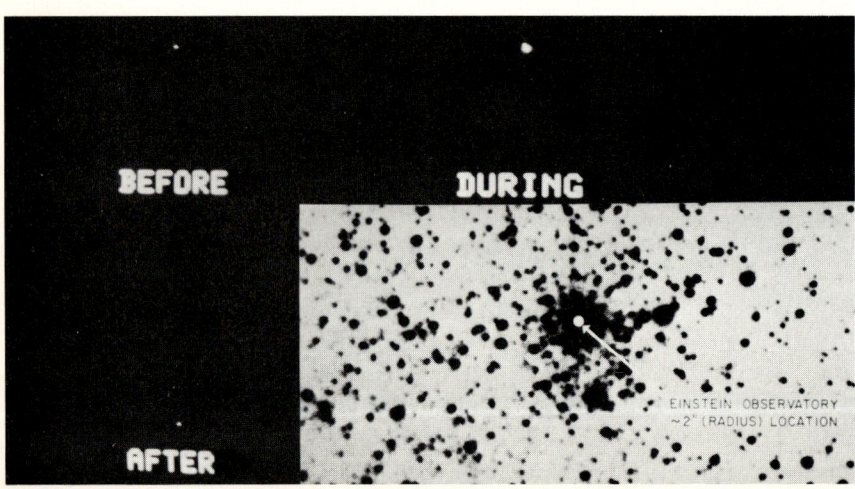

FIGURE 12.18 An X-ray burst (before at top left, during at top right, after at bottom left) from a globular cluster (called Terzan 2). Observations by the Einstein X-Ray Observatory. [*Courtesy J. Grindley.*]

12.2 OBSERVING BLACK HOLES

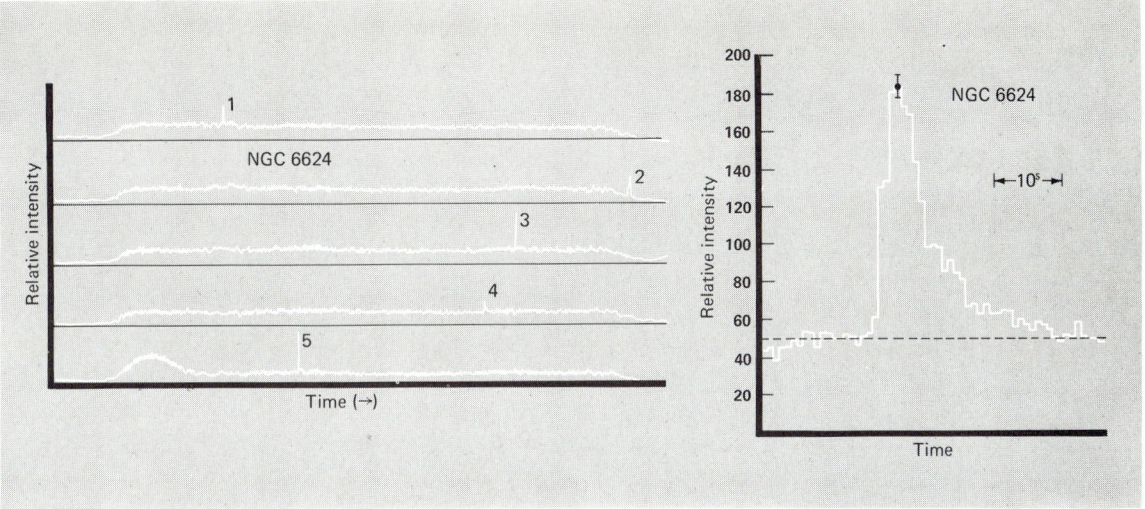

FIGURE 12.19 (a) X-ray bursts (numbered 1 to 5) from the globular cluster NGC 6624. (b) A profile of one burst. Note the sharp rise at onset. [*Observations by G. W. Clark and colleagues*, Astrophysical Journal (Letters), vol. 207, p. L105, copyright 1976 by the American Astronomical Society.]

X-ray satellites in operation at the time examined the source and confirmed its position very close to the core of NGC 6624. X-ray astronomers at MIT then examined some new and old data to find bursts that lasted about 10 sec (Fig. 12.19). Remarkably, the bursts seemed to recur every 4.4 h.

A photograph of the core of NGC 6624 taken by William Liller with the Cerro Tololo 4-m telescope shows a dozen or so red supergiant stars in NGC 6624 crowded into a region only 1 *light year* in diameter (Fig. 12.20). If you estimate that for every supergiant

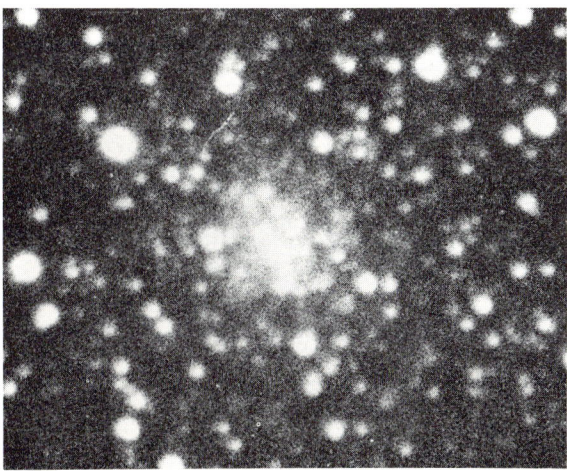

FIGURE 12.20 An infrared photo of the globular cluster NGC 6624. [*Courtesy W. Liller and J. Grindley.*]

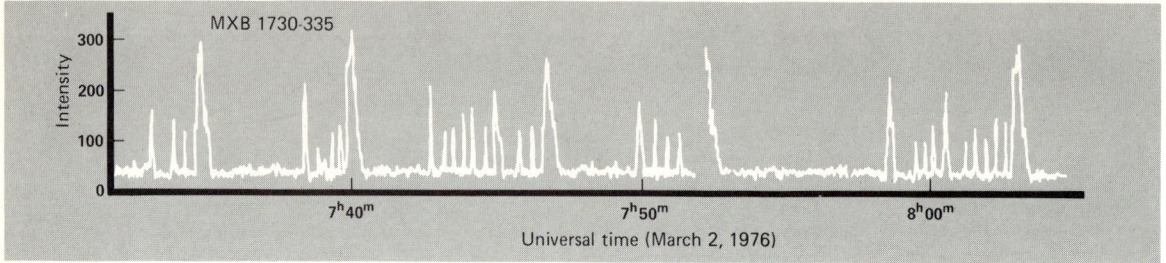

FIGURE 12.21 X-ray bursts from MXB 1730-335 (the "rapid burster"). Note how the burst sequence repeats about every 10 min. [*Observations by W. Lewin and colleagues,* Astrophysical Journal (Letters), *vol. 207, p. L95, copyright 1976 by the American Astronomical Society.*]

the cluster contains roughly 200 stars too faint to show in the photo, you end up with 2000 stars packed into a region 1 ly across—a million times the density of stars in the sun's neighborhood!

X-ray astronomers at MIT, headed by Walter Lewin, discovered a more curious burster, called MXB 1730–335 ("M" stands for MIT, "XB" for X-ray burster, the other numbers for the object's coordinates). This burster pumps out blasts in rapid-fire succession as fast as every 10 sec (Fig. 12.21). With the position of MXB 1730–335 known, William Liller used the Cerro Tololo 4-m again to take infrared photos of the region. (Infrared is used to cut through the dust in the Milky Way.) He found a vague blob of stars at the burster's position (Fig. 12.22). Meanwhile, Douglas Kleinmann, Susan Kleinmann, and Ned Wright made extensive infrared observations at the burster's position. They found that the source that Liller had photographed was, in fact, a distant (30,000 ly)

FIGURE 12.22 An infrared photo of the globular cluster whose position coincides with that of MXB 1730-335. [*Courtesy W. Liller.*]

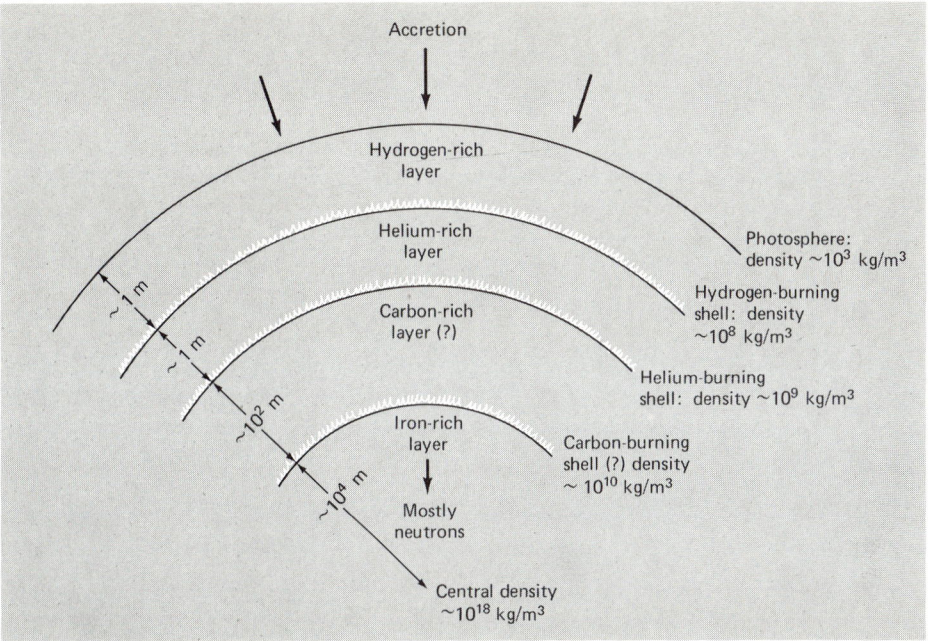

FIGURE 12.23 Surface layers of a model of an accreting neutron star. [*Adapted from a diagram by P. Joss.*]

globular cluster with a highly concentrated core. (At this distance each burst emits about 10^{31} W.) Liller then pointed out that the cores of NGC 6624 and the cluster containing MXB 1730–335 looked much the same.

What might these observations mean? The fact that both bursters lie in globular clusters presents a problem. Globular clusters contain little gas and dust compared with galactic clusters. The material needed to produce X rays by accretion around a condensed mass is not as available as it is in other parts of the Galaxy.

Not all X-ray bursters are associated with globular clusters, and those that are don't always fall close to a cluster's core. To date, it appears likely that neutron stars best explain bursters' properties. These are: rise to peak in less than 1 sec, duration of about 10 sec, interval between bursts of 10^3 to 10^4 min (about 2 h), a blackbody effective temperature at maximum of about 3×10^7 K. The bursters appear concentrated near the galactic center; if they are 30,000 ly from us, one burst has a luminosity at maximum of a few times 10^{31} W. As noted by J. Van Paradijs, an effective temperature of 3×10^7 requires a blackbody of roughly 9 km radius to produce 2 to 3×10^{31} W. This hints circumstantially that bursters are neutron stars.

Paul Joss of MIT has developed a model of helium-burning flashes on an accreting neutron star that accounts for a good number of burster properties. He starts out with a neutron star of 1.41 solar masses and 6.57 km radius. He lets hydrogen and helium fall onto the neutron star at a rate of roughly 10^{14} kg/sec to build an envelope on it (Fig. 12.23). As the matter accretes, the temperature rises to 2×10^9 K at the base of the helium layer. The helium ignites in a flash (Fig. 12.24), producing some 2.5×10^{31} W

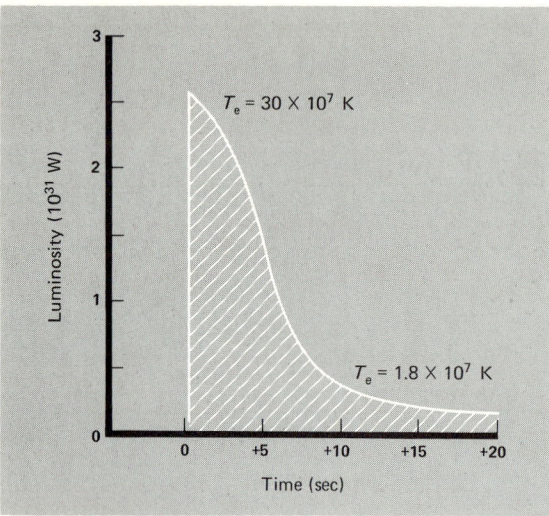

FIGURE 12.24 A theoretical model of X-ray bursts from the helium layer of an accreting neutron star. [*Adapted from a diagram by P. Joss,* Astrophysical Journal (Letters), *vol. 225, p. L123, copyright 1978 by the American Astronomical Society.*]

just 0.2 sec after the fusion reactions begin. With continual accretion, the flashes can recur every 15 h. These calculated characteristics are much like those observed. The model does not rely on any special source for the accreted material; it may come from a binary companion. This model is similar to that for novas (Section 11.3), the differences being that the material falls onto a neutron star, not a white dwarf, and the temperatures are higher, so that the energy comes out as X rays, rather than as visible light.

SUMMARY

A black hole forms when a given amount of matter is compressed within its Schwarzschild radius. Then the escape velocity from this region of spacetime exceeds the speed of light. The size of the Schwarzschild radius depends directly on the mass involved. For a 1-solar-mass star, it is roughly 3 km.

Because black holes are small and black, they are hard to observe directly. They can reveal themselves most clearly by their interaction with other matter, as in a binary star system where one star has evolved to a black hole. A black hole in a binary system can emit X rays by the flow and infall of material from the companion star to the black hole. So binary X-ray sources, such as Cygnus X-1, appear the most likely candidates for black holes. (But some may contain neutron stars.) The evidence to date builds a reasonable, but not airtight case for the existence of black holes. Those who are true believers of Einstein's general relativity feel strongly that black holes do exist.

Key Words

black hole
Schwarzschild radius
singularity
event horizon
white hole
Einstein-Rosen bridge or wormhole
Cygnus X-1
Centaurus X-3
binary X-ray source
X-ray burster
tidal gravitational force

PROBLEMS

Review Questions

1. To an outside observer, a light-emitting object falling into a black hole (mark all correct answers):
   a. Never makes it.
   b. Becomes more red shifted.
   c. Becomes very dim.
   d. Becomes distorted from tidal forces.
   (Objective 5)
2. What is the radius of a 100-solar-mass black hole? (Objective 2)
3. According to Einstein's theory of general relativity, a black hole (mark true or false):
   a. Cannot exist.
   b. Contains a singularity in its center.
   c. Is a severely curved region of spacetime.
   d. Sucks up matter from a great distance away.
   (Objectives 1 and 4)
4. A binary X-ray source (mark all correct answers):
   a. Must contain a black hole.
   b. Involves the evolution of two stars much more massive than the sun.
   c. Can emit X-rays for a relatively short time.
   (Objective 10)
5. In what sense does a black hole practice censorship? (Objectives 1 and 5)
6. What observational evidence leads some astronomers to believe that Cygnus X-1 is a black hole? (Objective 8)
7. If a black hole is "black," how can it give off X rays? (Objective 9)
8. a. To an outside observer, how long does it take an object to fall into a black hole?
   b. To an observer falling with an object, how long does it take (by the observer's watch) to fall from the edge of a black hole to its center?
   (Objective 5)

Problems

1. Calculate the Schwarzschild radius of (a) the earth, (b) the sun, and (c) a globular cluster. Then calculate the density of each mass if it just filled its Schwarzschild radius. Do you see a general trend? What is it? (Objective 2)
2. If an X-ray source flickers in intervals of about 0.01 sec, what is the maximum size of the X-ray emitting region? (Objective 11)
3. The orbital period of Cen X-3 is 2.1 days; the orbit has a radius of 11×10^6 km. Calculate the sum of the masses of the star and X-ray source. (Objective 8)
4. Calculate the blackbody radius of a spherical object with a luminosity of 2×10^{31} W and an effective temperature of 3×10^7 K. (Objective 11)
5. Find expressions for the tidal gravitational force of a mass M a distance R from two balls of mass separated by a distance d and for the attractive gravitational force between the two balls. Find an expression for the distance R at which these forces are just equal. At what distance from a 1-solar-mass black hole would two objects the mass and radius of the earth just be able to resist the tidal force by their internal gravitation? (Objective 6)

BEYOND THIS BOOK . . .

* K. S. Thorne describes "The Search for Black Holes" in *Scientific American,* December 1974.
* H. Gursky and E. P. J. van den Heuvel present a comprehensive picture of "X-Ray Emitting Double Stars" in *Scientific American,* March 1975.
* You can find more information about black holes in relativity theory in *The Cosmic Frontiers of General Relativity* (Little, Brown, Boston, 1977) by W. J. Kaufmann III.
* For a detailed detective story about one X-ray source, read "The Story of AM Herculis" by W. Liller, *Sky and Telescope,* May 1977, p. 351.
* W. J. Kaufmann III gives a general discussion of black holes in *Black Holes and Warped Spacetime* (Freeman, San Francisco, 1979).

PART THREE
THE EVOLUTION OF PLANETS

THE PLANETS: PAST AND PRESENT

13

LEARNING OBJECTIVES
After studying this chapter, you should be able to:
1. Outline a scenario for a supernova trigger for star formation.
2. Sketch a rough picture for the sun's birth from a molecular cloud.
3. Outline, in a general way, the origin of the planets with the sun.
4. Divide the planets into two general classes: terrestrial and Jovian.
5. Compare the terrestrial and Jovian planets, as classes, in terms of (a) size, (b) mass, (c) density, (d) period of rotation and revolution, (e) distance from the sun, and (f) atmospheric composition.
6. Describe the methods used to find each of the quantities listed in Objective 5.
7. Argue that Pluto is neither a terrestrial nor a Jovian planet.
8. Describe how radioactive decay (half-life) can be used to date rocks and the limitations of this procedure.

CENTRAL QUESTION:
How did the solar system form, and
what are the planets like today?

Our sun is a middle-aged star. Some 4.6 billion years ago, its material drifted as a cloud of gas and dust in interstellar space. Gravity pulled that cloud together to form the sun. As a spinoff of the sun's birth the planets of our solar system formed, too. That, in a nutshell, describes the contemporary model of the origin of the planets—a natural adjunct to the birth of the sun. Our home star, our home planet, and the neighboring planets grew from the same interstellar cloud.

Though most astronomers today would agree with this model in general, they still wrangle over the details of the processes of solar and planetary formation. No one model for the origin of the sun and the planets holds the lead among astronomers. In fact, few of the many models put on the firing line have actually been shot down by observations to date.

Despite our bewilderment we have a selfish interest to understand how our sun and solar system formed. This chapter serves as an introduction to that question. It first links the birth of the sun and planets to the death of a massive star. Then it portrays in general the planets as we know them today—the status of the solar system that any model of its evolution must explain.

13.1 STARDEATH AND THE SUN'S BIRTH

Recall the connection between the death and birth of stars (Chapter 9). Massive stars condense sequentially in small groups from giant molecular clouds. Expanding zones of ionized gas, generated by main-sequence OB stars, seem to trigger these bursts of star formation (Fig. 13.1). This chain reaction process moves through the molecular cloud, until it is used up in making groups of stars. This model raises the question of what sparks the *first* burst of star formation in a fertile molecular cloud. One possibility is the concussion of a shock wave from a supernova remnant. Charles Lada, Leo Blitz, and Bruce Elmegreen have argued that in the vicinity of OB associations a supernova blast within roughly 200 ly (60 pc) of a molecular cloud can induce the first episode of starbirth.

13.1 STARDEATH AND THE SUN'S BIRTH

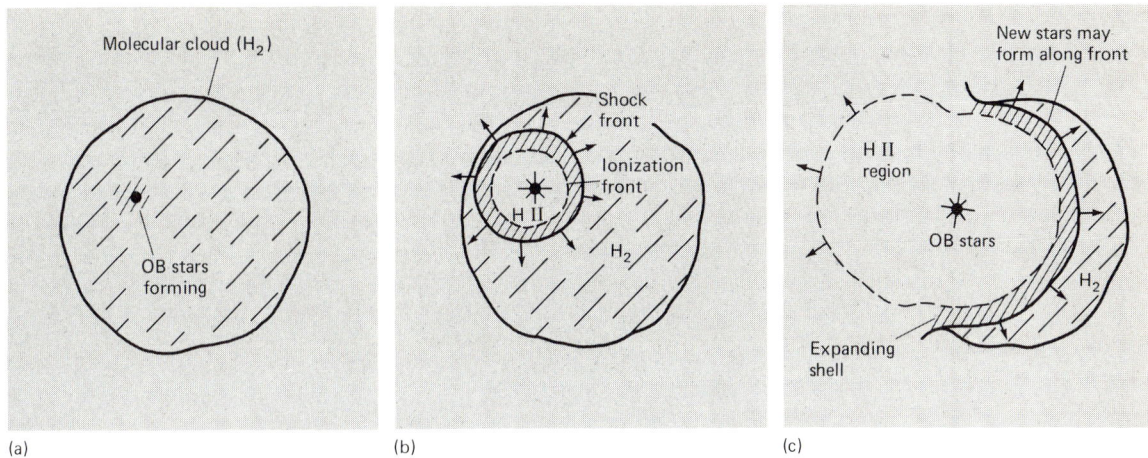

FIGURE 13.1 A schematic model of massive star formation from an interstellar cloud. (a) A small cluster of O and B stars is born in a giant molecular cloud. (b) As the stars evolve to the main sequence, they heat and ionize the gas around them, setting up an expanding shock front and ionization front. (c) Where the fronts plow into the molecular cloud, new stars may form. [*Adapted from a diagram by J. Bally and N. Scoville.*]

Does this interlocking of the birth and death of massive stars—their birth initiated by the touch of the ghost of a dead one—connect to the sun's birth? We think it likely, though the details still aren't clear. Here's one possible scenario and some evidence for it.

Recall the discussion of supernovas and supernova remnants in Chapters 9 and 11. A supernova flares in our Galaxy roughly once every 50 years (Fig. 13.2). It blasts matter into space at initial speeds of 1000 to 10,000 km/sec. As the ejected material hurls through the interstellar medium, it scoops up and pushes interstellar material ahead of it, like a snowplow. This interaction gradually slows the supernova remnant down to a speed of about 50 km/sec after some 10^5 yr. At this stage the supernova remnant spans about 100 ly (30 pc) and carries with it some 10^4 to 10^5 solar masses of material. It shows up as an expanding shell of neutral material that hauls along tremendous energy, some 10^{44} to 10^{45} joules.

Carl Heiles has observed wide areas of the interstellar medium at 21 cm and sees that expanding shells of H I, containing some 10^7 solar masses, pervade the interstellar medium (Fig. 13.3). He attributes these to supernovas, since the kinetic energies in these shells are 10^{45} to 10^{46} joules, about that available from one or more supernovas.

A supernova remnant carries with it more than energy. It also contains the products of nucleosynthesis (Chapter 11)—both explosive ones made in the supernova and ordinary ones made in the star's core during its normal life. Theoretical calculations show that as the shell expands and cools, dust grains condense from it (Fig. 13.4). These grains contain some heavy elements made in the supernova blast.

Clouds also pervade the interstellar medium. As a supernova remnant slams into a cloud, the cloud implodes. Very likely, star formation then begins. Imagine the supernova shell plowing into a small interstellar cloud. The grains penetrate the cloud like shrapnel, but the gas in the supernova remnant flows around the cloud, mixing little with it. The remnant's gas engulfs and squeezes the cloud, causing it to collapse. Perhaps the sun and planets formed from such a pressured cloud. (But remember, this is only one

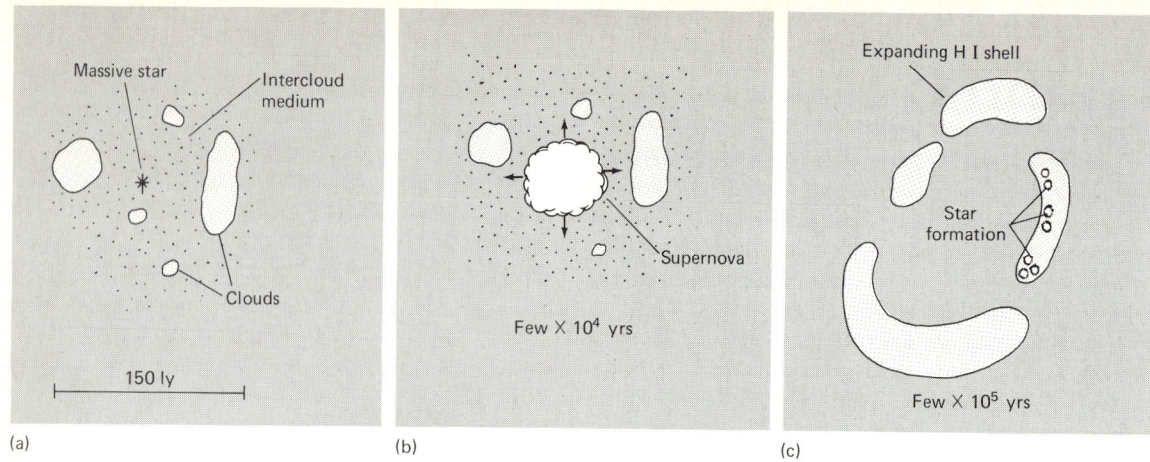

FIGURE 13.2 A schematic model for supernova-induced star formation. (a) A pre-supernova massive star lies among interstellar clouds. (b) The star explodes, blasting a shock wave out into the interstellar medium. (c) The expanding supernova remnant compresses H I clouds, triggering star formation. [*Adapted from a diagram by W. Herbst and G. Assousa, in* Protostars and Planets, *edited by T. Gehrels, University of Arizona Press, Tuscon, 1978.*]

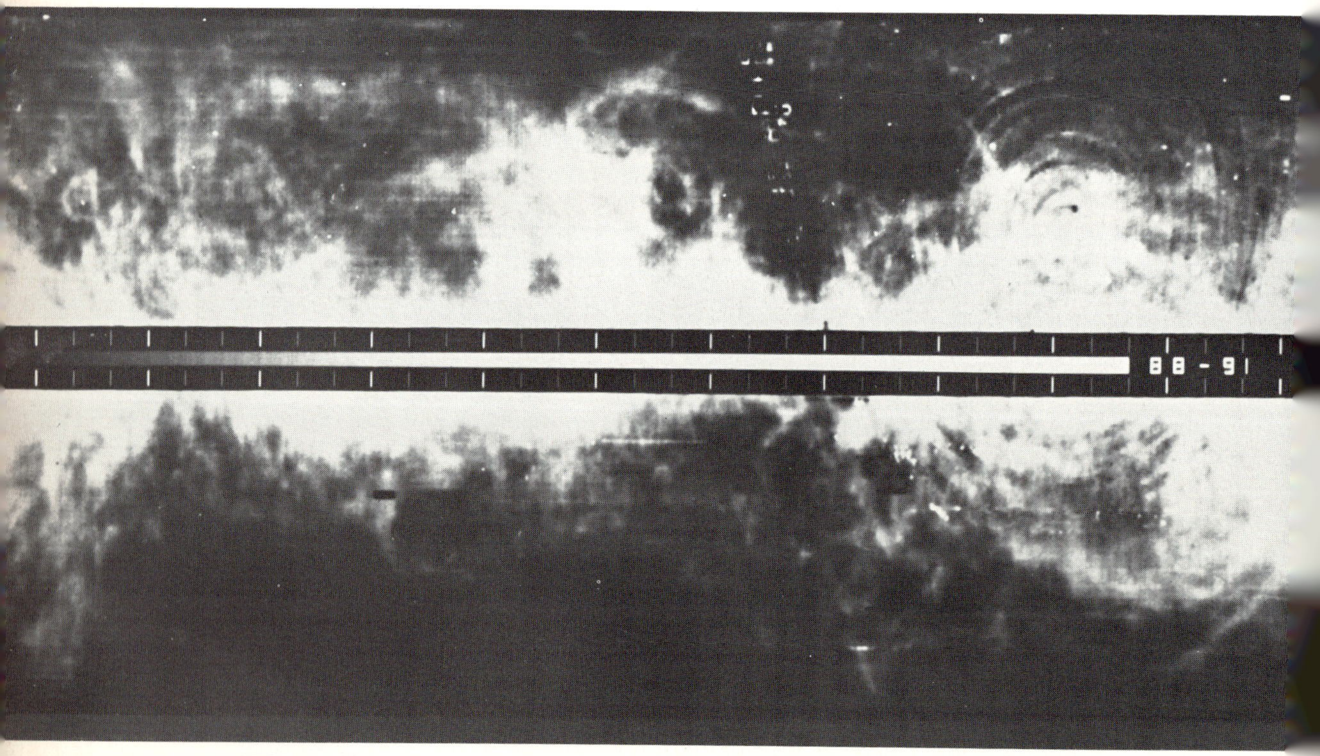

FIGURE 13.3 A 21-cm picture of giant filaments and shells in the interstellar medium. The largest of these, probably formed by supernovas, is hundreds of light years in size. The bar through the center is an artifact of the computer processing, which made this visual photo from 21-cm data; it also marks the position of the galactic plane. The view extends above and below the plane by about 60°. [*Courtesy Carl Heiles, University of California, Berkeley.*]

13.1 STARDEATH AND THE SUN'S BIRTH

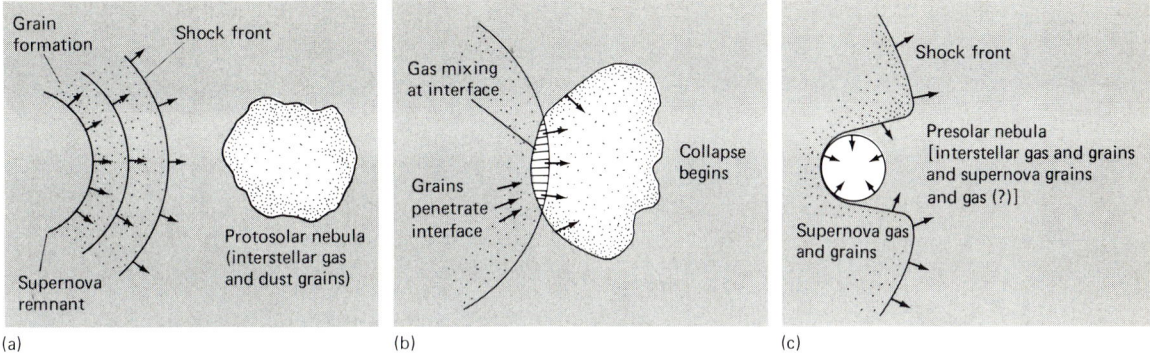

FIGURE 13.4 A schematic model for the injection of grains by a supernova into an interstellar cloud. (a) An expanding supernova remnant, in which grains are condensing, approaches the protosolar nebula. (b) Like shrapnel, the grains penetrate the cloud. (c) The supernova shock front flows around the nebula, compressing it. [*Adapted from a diagram by D. Schramm in* Protostars and Planets, *edited by T. Gehrels, University of Arizona Press, Tuscon, 1978.*]

model for the sun's formation; not all astronomers would agree that it is the correct one.)

Some evidence backs up this picture. First, it appears that star formation has happened at the expanding front of at least one supernova remnant. William Herbst and George Assousa have noted an association between young stars and an expanding shell of gas, possibly a supernova remnant, in the constellation Canis Major 3700 ly away. Here is a visible arc of fluorescing hydrogen. Radio observations at 21 cm show that the front side of this shell expands at about 30 km/sec. Assuming that this shell is a supernova remnant, Herbst and Assousa calculate its age as 800,000 years and the energy needed to form the shell as 10^{44} joules, about that released in a supernova. At one edge of the shell lie infrared sources that are probably young stars, only about 300,000 years old. Because this age is less than that of the suspected supernova remnant, the expanding shell may have compressed interstellar gas and dust enough to trigger the stars' birth.

Second—and this is a more subtle clue—is the unusual pattern of abundances of certain isotopes found in meteorites (more on meteorites in Chapter 18). Meteorites are the remains, after a plunge through the earth's atmosphere, of small bodies that orbit the sun. Along with moon rocks, they are the only samples on earth of extraterrestrial materials. Analysis in the lab provides detailed information of the chemical composition of meteorites.

Now to make a little digression, the reason for which will become clearer at the end of this section. As mentioned in Chapter 6, isotopes of an element all have the same number of protons (and so are the same chemical element), but they have different numbers of neutrons. For example, aluminum comes in only one stable isotope: aluminum 27, or $^{27}_{13}$Al, with 13 protons and 14 neutrons. Aluminum 26 ($^{26}_{13}$Al), another isotope, is unstable. It decays by the change of a proton into a neutron with the emission of a positron (antimatter electron, e^+) and a neutrino (ν):

$$^{26}_{13}\text{Al} \rightarrow e^+ + \nu + {}^{26}_{12}\text{Mg}$$

The ^{26}Al becomes ^{26}Mg because one positive charge has disappeared.

How fast an unstable isotope decays is indicated by its *half-life*, the time it takes for half of an initial sample to decay to something else. ^{26}Al has a half-life of 720,000 years. Suppose you start out with some amount of ^{26}Al. After 720,000 years have passed, you'd have $\frac{1}{2}$ the original amount, and the rest would be ^{26}Mg. Then 720,000 years later you'd have half of that, or $\frac{1}{4}$ the original amount. Another half-life later only $\frac{1}{8}$ the original ^{26}Al remains, and so on. After four half-lives (2,880,000 years), merely $\frac{1}{16}$ ($[\frac{1}{2}]^4$) of the original amount remains. In a few million years almost all the ^{26}Al is converted to ^{26}Mg.

Now the reason for this digression. Radioactive elements with known half-lives can be used to determine the age of meteorites and of the earth. On this basis 4.6 billion years ago is the date of the formation of the solar system. Now 4.6 billion years is many half-lives of ^{26}Al, so it all must have become ^{26}Mg by now. (All the aluminum on the earth is $^{27}_{13}$Al.) How can the ^{26}Mg made from ^{26}Al be detected? Magnesium has three stable isotopes: ^{24}Mg, ^{25}Mg, and ^{26}Mg. In a terrestrial sample of magnesium the normal isotopic composition is 78.99 percent ^{24}Mg, 10.00 percent ^{25}Mg, and 11.01 percent ^{26}Mg. Any magnesium from the decay of ^{26}Al would show up as an increased percentage of ^{26}Mg (more than 11.01 percent) in the isotopic abundances.

Where could the ^{26}Al come from? Grains from a supernova remnant penetrate the interstellar cloud that becomes the solar system. Models of supernova explosions (Section 11.6) indicate that ^{26}Al can be made in the star's carbon-rich shell near the core. If—and this is the critical if—a supernova went off near the cloud that became the sun and solar system not long before the sun's birth, it could have seeded the cloud with ^{26}Al. "Not long" means no more than about one half-life of ^{26}Al (700,000 years or so), because if a few million years or so had passed, the ^{26}Al would have turned to ^{26}Mg before being incorporated into meteorites. In the chemical and mineralogical processes involved in the formation of a meteorite, all the magnesium behaves the same way. The meteorites would then all have the same ratios of magnesium isotopes. We can get an excess of ^{26}Mg only in those meteorites which, for chemical reasons, are formed with an excess of aluminum minerals. In these the ^{26}Al decays to produce ^{26}Mg, but not ^{24}Mg or ^{25}Mg, hence the excess ^{26}Mg.

Has excess ^{26}Mg been found? Yes, in a meteorite called Allende, which fell in 1969 near the village of Pueblito de Allende in northern Mexico. Laboratory analysis has found some grains in the meteorite that contain 11.5 percent ^{26}Mg, rather than the normal 11.01 percent. That doesn't sound like much, but it's just the kind of difference expected if ^{26}Al had been incorporated in the primordial solar system cloud shortly after it was formed.

So we have some evidence that roughly 700,000 years before the sun was born, a supernova flared nearby, not more than about 200 ly (60 pc) away. Its remnant shell banged into an interstellar cloud, seeding it with grains and compressing it to collapse. This cloud became the solar system we know today.

Half-Life and the Dating of Rocks

The radioactive dating technique is crucial to finding cosmic ages. Let's look at it in detail. The technique works because of the natural instability of the nuclei of radioactive elements. When these nuclei decay, they break apart into simpler nuclei. Given just one atom, you cannot estimate when it will decay because the process is random; but given a

13.1 STARDEATH AND THE SUN'S BIRTH

large number of atoms, you can determine a gross rate of disintegration. (An analogous process is the popping of popcorn. You cannot predict which kernel will pop next, but you can estimate when the entire batch will be finished.) Half a piece of uranium 238 (^{238}U) decays to lead in 4.5 billion years, half again in the next 4.5 billion years, and so on (Fig. 13.5). So you can calculate the amount of uranium left at any time, even though the decay time for any one uranium atom cannot be specified. Conversely, given a rock sample containing ^{238}U and knowing the half-life of the uranium, we can calculate the age of the sample.

Elements in addition to uranium that can serve as radioactive clock include rubidium (^{87}Rb), which decays to strontium (^{87}Sr) with a half-life of 47 billion years, and potassium (^{40}K), which decays to the inert gas argon (^{40}Ar) with a half-life of 1.3 billion years. Whatever elements are used, the derived date is the time elapsed since the rocks last solidified.

How to get from knowing the half-life of radioactive elements to the age of a rock sample? You must know the original composition of the sample when it solidified. Often the mineralogical structure of the rock will reveal this. Suppose, for example, a sample contains a mineral which you know is formed with a certain fraction of ^{40}K. The potassium decays to ^{40}Ar. If the argon is trapped in the sample and doesn't escape, the amount of argon relative to the amount of potassium increases with time. This changing ratio tells the age of the sample.

For example, suppose a rock sample now contains equal numbers of ^{40}K and ^{40}Ar atoms. If there were no argon atoms in the rock originally, they must all have come from decay of ^{40}K. Exactly half of the ^{40}K nuclei have decayed (and half remain), so the rock must be one half-life old, or 1.3 billion years. How old is a rock which contains 7 times as many ^{40}Ar atoms as ^{40}K atoms? If all the argon came from decay of potassium, the remaining ^{40}K is $\frac{1}{8}$ of the original. So three half-lifes must have elapsed ($\frac{1}{8} = [\frac{1}{2}]^3$), and the rock must be $3 \times 1.3 = 3.9$ billion years old.

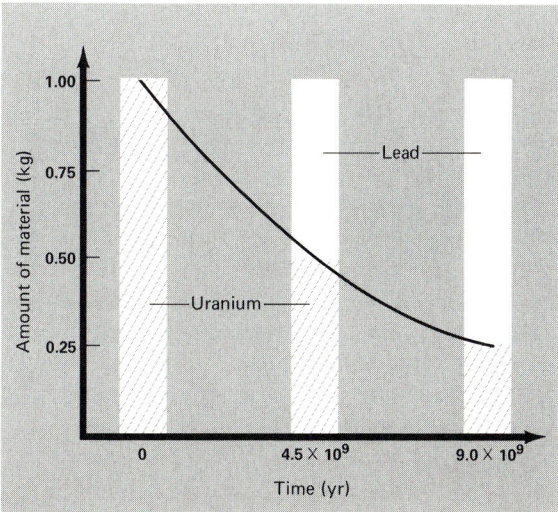

FIGURE 13.5 Radioactive decay of 1 kg of uranium 238. The half-life is 4.5 billion years, so 0.5 kg remains after 4.5×10^9 yr, 0.25 kg after 9.0×10^9 yr, and so on.

The above are somewhat simplified examples. Let's examine the more exact mathematical representation. Radioactivity is one example of an *exponential decay*. In such a process the number of remaining atoms is expressed by an exponential involving the time. If n_0 is the original number of atoms which exists at time $t = 0$, and n is the number of atoms remaining at time t, then the relation of the ratio of these numbers to the time is

$$\frac{n}{n_0} = 2^{-t/t_{1/2}}$$

where $t_{1/2}$ is the half-life for the decay process.

To find the age (t) of a sample from the measured value of the ratio n/n_0 we need to solve this equation for t. Take the logarithm to the base 2 of both sides:

$$\log_2 \frac{n}{n_0} = -\frac{t}{t_{1/2}}$$

$$t = t_{1/2} \log_2 \frac{n_0}{n}$$

To convert to the more familiar base 10 logarithms, we use the relation $\log_{10} x = \log_{10} 2 \, \log_2 x$. So

$$t = \frac{1}{\log_{10} 2} \log_{10}\left(\frac{n_0}{n}\right) t_{1/2}$$

$$= 3.32 \, t_{1/2} \log_{10} \frac{n_0}{n}$$

For example, suppose we find that in a certain rock the ratio of argon to potassium atoms is $\frac{1}{4}$. Then $n_0/n = 5$. The half-life is 1.3×10^9 yr. So

$$t = 3.32(1.3 \times 10^9) \log_{10} 5$$

$$= 3.0 \times 10^9 \text{ yr}$$

13.2 THE FORMATION OF THE PLANETS: A PREVIEW

A collapsed interstellar cloud gave birth to the solar system some 4.6 billion years ago. Since then, the sun and planets have evolved considerably. The crucial problem to understand in the origin of the solar system is how a dusty, smooth cloud became the sun and the planets.

Many competing models for solar system formation have been developed in the past, and more will develop in the future. This section will sketch a picture that's an amalgam of a few. We've synthesized the pieces that make a more or less clear story. Messy details will be ignored at this stage; we'll return to the topic in Chapter 19. Here we want to bridge the gap between the interstellar cloud (Section 13.1) and the overview of the planets today (Section 13.3). Then we'll look at the planets in detail (Chapters 14 to 18) to search for clues to their origin.

A little more than 4.6 billion years ago, a nearby supernova ignited. Not more than a few hundred thousand years later, its blast wave compressed an interstellar cloud of gas and dust. A small blob with perhaps twice the mass of the sun broke off and collapsed quickly. As it did, it formed into a disk, as required by the conservation of angular momentum, and heated up as the gravitational energy converted to kinetic energy. Eventually, rising temperatures resulted in pressures that halted the collapse.

The high temperatures near the center (about 2000 K) vaporized at least some dust there. As the nebula cooled, new dust grains condensed from the gas. The composition of the new grains depended on the temperature where they formed. Only metallic grains formed close to the center. A little farther out, both metallic and rocky grains could form. Even farther out, where the temperatures were much lower (about 100 K), icy grains formed along with rocky and metallic ones. These condensed grains were small. By hitting and sticking to each other they could form pebble-sized objects. These heavier particles quickly fell into the plane of the nebula and accumulated into boulder-sized objects forming a very thin disk of solids in a thicker disk of gasses (Fig. 13.6a).

Gravity then pulled these boulder-sized bodies into larger objects, some 1 to 100 (perhaps 1000) km in size. These became the basic building blocks of the planets—the *planetesimals* (Fig. 13.6b). Somehow (the how is not well known) the planetesimals congregated into larger bodies that would become the planets. These are called the *protoplanets* (Figs. 13.6c,d).

When the cloud collapsed, it formed a massive, central bulge, where the sun was born while the planets formed. After the sun's fusion fires ignited, it went through a rambunctious youth with a very intense solar wind, perhaps similar to that stage of evolution of T Tauri stars (Section 9.4). The solar wind blew out any free gas. Their process of birth finished, the sun and the planets gradually evolved to the state we know them now.

13.3 THE PLANETS TODAY

Each planet today has its unique mark. Mercury skirts closest to the sun, its marred surface seared by solar radiation. Venus hides a hellish surface behind a veil of clouds. Earth sports a surface covered mostly with water. Mars is rugged, with gigantic volcanoes and huge polar caps of water and carbon dioxide ice. Jupiter gives off more energy than it receives from the sun. Saturn spins rapidly in a setting of delicate rings. Uranus rotates on its side. Neptune sometimes orbits beyond Pluto. And distant Pluto has a surface frosted with methane ice.

The planets can be divided into two general classes. Mercury, Venus, Earth, and Mars make up the *terrestrial planets*, places like our home planet in terms of size, mass, density, and chemical composition. These planets are rocky globes with thin atmospheres. In contrast stand the *Jovian planets*: Jupiter, Saturn, Uranus, and Neptune. These planets are giant balls of gas and liquid with chemical compositions more like that of the sun than of the earth. The asteroid belt, which lies between the orbits of Mars and Jupiter, neatly separates the two clans in the family of the sun. Only Pluto, the most recently

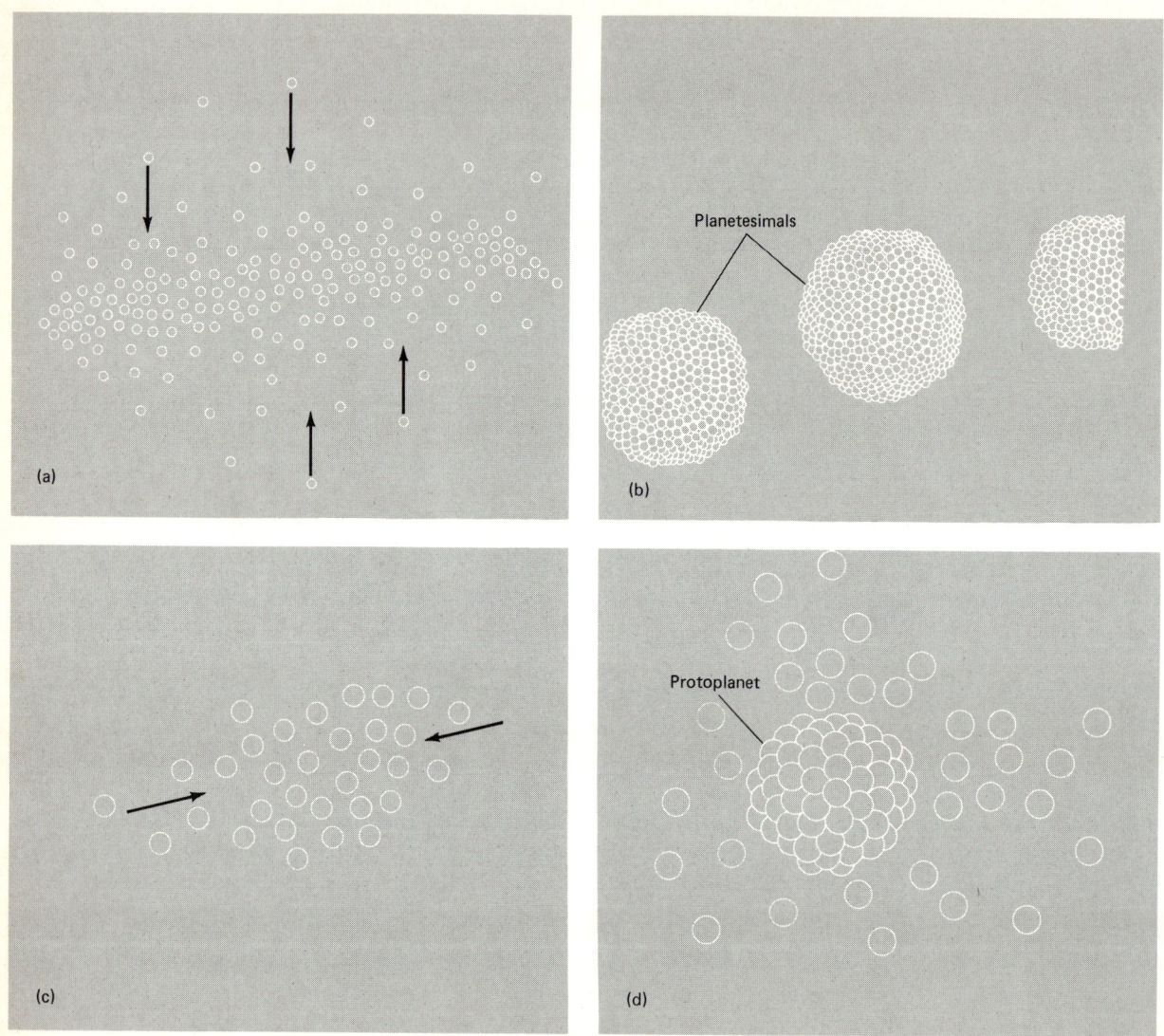

FIGURE 13.6 One model for the formation of the planets. (a) Dust grains in the nebula collide, stick, and clump. These fall into the nebula's midplane to form a thin disk. (b) Gravity pulls these clumps into asteroid-sized bodies, called planetesimals. (c) Planetesimals collect in clusters to make (d) the core of a protoplanet. [*Adapted from a diagram by A. G. W. Cameron,* Scientific American, *September 1975, copyright 1975 by W. H. Freeman and Co.*]

discovered and the most difficult to observe, does not seem to belong clearly to either branch of the sun's family.

More than the planets and the sun make up the solar system. The other members include satellites, comets, asteroids, meteoroids, and interplanetary gas and dust—material all kept here by the sun's gravity.

This section gives an overview of the planets as we know them today. Details will follow in Chapters 14 to 18.

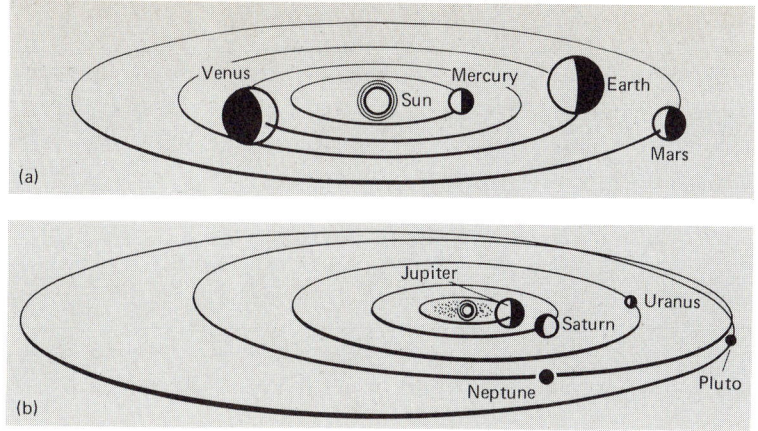

FIGURE 13.7 (a) The orbits of the terrestrial planets, drawn to scale. The view is from above the plane of the earth's orbit. The relative sizes of the planets are also drawn to scale, but not the same scale as the orbits. The size of the sun is not drawn to scale. (b) The orbits of the outer planets, drawn to scale, but one much larger than for the terrestrial planets. [*Adapted from a diagram by F. Whipple.*]

Terrestrial Versus Jovian

Any model of the planets' origin must explain the key physical differences between terrestrial and Jovian classes. The planetary characteristics that make clear the terrestrial-Jovian split are distance from sun, size, mass, atmospheric composition, density, and interior composition.

The terrestrial planets—Mercury, Venus, Earth (with its moon), and Mars—orbit within 2 AU of the sun (Fig. 13.7a). The Jovian planets—Jupiter, Saturn, Uranus, and Neptune—orbit in a zone from 5 to 30 AU (Fig. 13.7b).

In diameter the Jovian planets are 4 to 11 times larger than the earth (Fig. 13.8); the smallest terrestrial planet (our moon) is about ¼ the earth's size. (One way to look at this size difference is that the light travel time across a terrestrial planet is about 10^{-2} sec, for

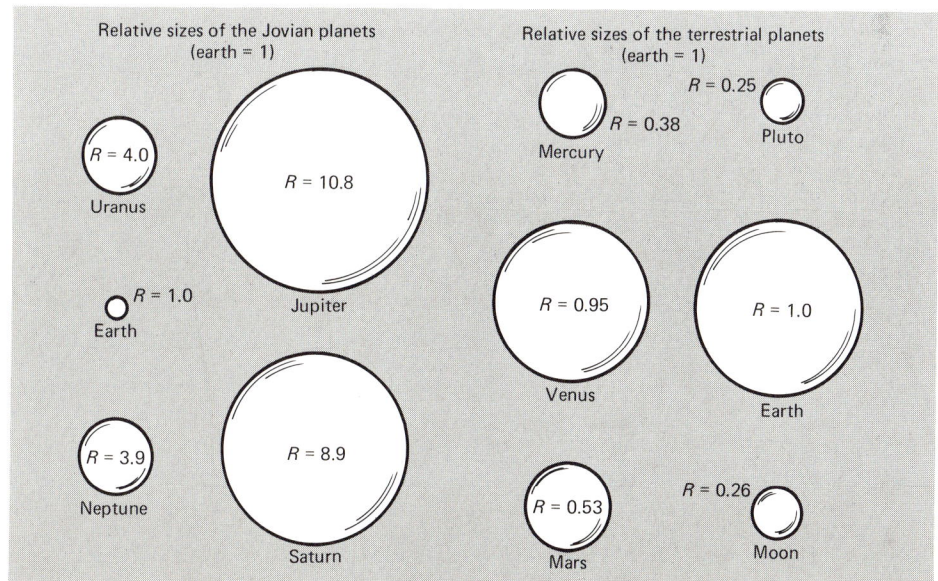

FIGURE 13.8 The relative sizes of the planets compared to the earth.

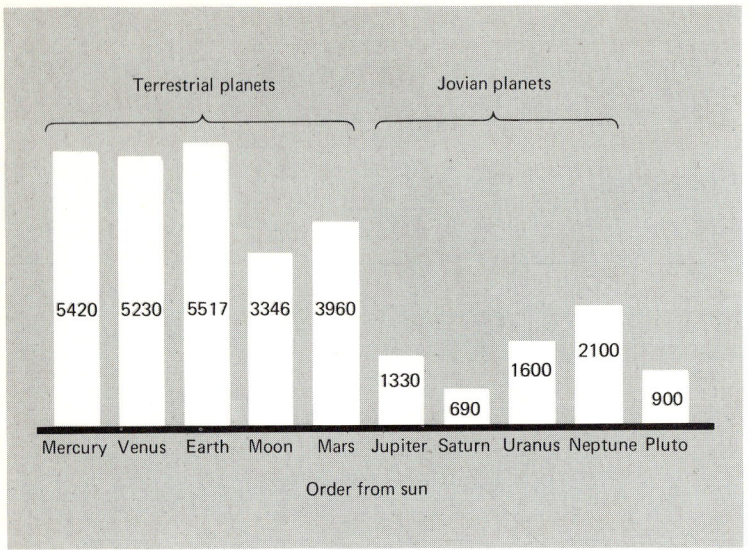

FIGURE 13.9 The bulk densities of the planets. Pluto's density is not well known.

TABLE 13.1 Approximate Planetary Data

	Mercury	Venus	Earth	Mars
Average distance from sun (AU)	0.39	0.72	1.0	1.5
Period of revolution (yr)	0.24	0.62	1.0	1.9
Inclination of orbit to ecliptic	7°	3.4°	0°	1.9°
Rotation period	59 days	243 days	23 h 56 min	24 h 37 min
Equatorial diameter (km)	4880	12,104	12,756	6787
Mass (earth = 1)	0.055	0.82	1.0	0.11
Density (kg/m^3)	5400	5200	5500	3900
Atmosphere (main constituents)	Helium, hydrogen	Carbon dioxide	Nitrogen, oxygen	Carbon dioxide
Surface gravity (earth = 1)	0.37	0.88	1.0	0.38
Satellites	0	0	1	2

13.3 THE PLANETS TODAY

a Jovian planet 10^{-1} sec.) The Jovian planets also have much more mass, from about 15 to 300 times the earth's mass.

The atmospheres of the terrestrial and Jovian planets present another dramatic contrast. Mercury and the moon essentially have no atmospheres. Venus and Mars have atmospheres consisting mostly of carbon dioxide, and the earth has a unique atmosphere rich in oxygen and nitrogen, but both types contain little hydrogen. The Jovian planets have much different atmospheres: mostly hydrogen and helium, with hydrogen compounds such as methane, ammonia, and water.

A very important difference is bulk density (Fig. 13.9). The terrestrial planets are denser than the Jovian ones: 3300 to 5500 kg/m^3 compared with 700 to 2300 kg/m^3. The bulk density of a solid body gives a direct clue to its overall chemical composition. Solid or liquid objects with densities of about 1000 kg/m^3 are made of icy materials (water, carbon dioxide, methane, ammonia); solid bodies with densities of about 3000 kg/m^3 are made of rocky materials; and those with densities of 7000 to 8000 kg/m^3 are made of metals. The terrestrial planets are rocks plus metals, the Jovian planets ices and gasses plus some rocks.

Finding Out Physical Properties

How do we know this physical information? Let's look briefly at the methods used for determining planetary data. Appendix B gives the data in detail; Table 13.1 gives approximate values of the main characteristics.

Jupiter	Saturn	Uranus	Neptune	Pluto
5.2	9.5	19.2	30.1	39.4
11.9	29.5	84.0	165	248
1.3°	2.5°	0.8°	1.8°	17.2°
9 h 50 min	10 h 39 min	12 h (?)	16 h (?)	6 days 9 h
142,800	120,660	51,800	49,500	3000
318	95	14.6	17.2	0.0022
1300	700	1200	1700	900
Hydrogen, helium	Hydrogen, helium	Hydrogen, helium, methane	Hydrogen, helium, methane	Methane
2.6	1.2	1.2	1.2	0.04
16	17	5	2	1

Planetary Distances and Diameters

The distance between a planet and the earth can be calculated by triangulation from various points in the earth's orbit (Section 3.2). The distances of the planets from the sun can be computed from observations of their synodic periods, transformed to sidereal periods (Section 3.2) and slipped into Kepler's third law. Both of these methods give distances relative to the Astronomical Unit. To find the distance scale in kilometers is a more difficult task, but it requires that only one segment be measured accurately in kilometers. Chapter 6 presented the modern method of finding the distance scale using radar beams bounced off Venus. Once we know the earth-Venus distance (a part of an AU) in kilometers, we can calculate the number of AUs per kilometer, and we have established the distance scale for the entire solar system map.

The best time to make this observation occurs when the earth and Venus lie closest together on the same side of the sun. Their separation is then about 0.28 AU. A radar signal takes roughly 280 sec for the round trip. The distance is then

$$d = \frac{(280 \text{ sec})(3 \times 10^5 \text{ km/sec})}{2}$$

$$= 4.2 \times 10^7 \text{ km}$$

This equals 0.28 AU, so

$$1 \text{ AU} = \frac{4.22 \times 10^7}{0.28}$$

$$= 1.5 \times 10^8 \text{ km}$$

or, more accurately, $1 \text{ AU} = 1.4959787 \times 10^8$ km.

So, from the relative distances determined by triangulation and by Kepler's law and from the calculated value of the AU we get the actual distances of the planets. Once the actual distance has been determined, we can calculate a planet's diameter from its apparent angular size.

Planetary Masses

To determine a planet's mass requires the use of Kepler's third law or some other form of Newton's law of gravitation. If the planet has a moon (or if a satellite can be placed in orbit around it) the period and the semimajor axis, incorporated into Kepler's third law, give the sum of the masses of the planet and satellite relative to the sun (Section 4.3). Since the mass of the satellite is always very small, we get the mass of the planet alone. (Why can't we use Kepler's law and the period and distance for the planet's orbit around the sun? We would get the sum of the planet's mass and the sun's mass, and it would be the mass of the *planet* that was too small to matter.) If no satellites are available, the accelerations that the planet produces on another planet, a passing asteroid, or a space probe are needed to derive the mass from Newton's laws.

This mass-finding procedure is important enough to do one example in detail. Let's find the mass of Jupiter using its satellite Callisto, whose orbital period is 16.689 days and

13.3 THE PLANETS TODAY

distance from Jupiter 1.884×10^6 km (1.884×10^9 m). Newton's version of Kepler's third law is

$$\frac{P^2}{a^3} = \frac{4\pi^2}{G(M+m)}$$

Let m be Callisto's mass, M Jupiter's mass, so $m + M$ approximately equals M. P is Callisto's period, and a the semimajor axis of its orbit. Then

$$\frac{P^2}{a^3} = \frac{4\pi^2}{GM}$$

and

$$M = \frac{4\pi^2}{G} \frac{a^3}{P^2}$$

$$= \frac{4\pi^2}{6.67 \times 10^{-11}} \frac{(1.884 \times 10^9)^3}{(16.689 \times 24 \times 60 \times 60)^2}$$

$$= 1.90 \times 10^{27} \text{ kg}$$

Note that a planet must have a satellite if we are to find its mass directly and easily.

Planetary Densities

Knowing the size and mass of a planet, you can calculate its density by dividing its mass by its volume. To work out the volume, it's good enough to assume the planet is a sphere. Then

$$V = \frac{4}{3}\pi R^3$$

and the density is

$$\rho = \frac{M}{(4\pi/3)R^3}$$

What is the density of Jupiter? From the angular size and distance, the radius of Jupiter is $R = 7.1 \times 10^4$ km (7.1×10^7 m). The mass is $M = 1.90 \times 10^{27}$ kg. The density is then

$$\rho = \frac{1.90 \times 10^{27} \text{ kg}}{(4\pi/3)(7.1 \times 10^7 \text{ m})^3}$$

$$= 1.3 \times 10^3 \text{ kg/m}^3$$

Jupiter is only slightly denser than water!

Planetary Atmospheres

The crucial data needed for the identification of atmospheric gasses come from a planet's spectrum. Because all planets reflect sunlight, the absorption lines of a planet's atmosphere will be superimposed on the solar spectrum. The absorption lines in the solar spectrum are produced under physical conditions different from those producing planetary absorption lines, and the two are easily distinguished. For the ground-based observer, the situation is confounded by the earth's atmosphere. Our atmosphere's absorption features blot out absorption lines of common gasses (such as carbon dioxide, water, and oxygen) in the light coming from another planet's atmosphere.

One way to avoid this masking effect involves making use of the Doppler shift. If a spectrographic study is made at the time a planet's radial velocity relative to the earth is greatest, the resulting Doppler shift will move the planet's absorption lines to a different wavelength from that of the terrestrial absorption lines. This technique has been used to detect water vapor in the atmosphere of Mars. The difficulties of ground-based planetary spectroscopy can also be overcome, with enough money, by sending out a planetary probe equipped with a spectrometer or sampling the atmosphere directly by a lander.

Motions of the Planets

We will describe the planet's characteristic motions in detail in Chapters 14 to 19, but we want to summarize here the key features to provide a framework for the information to come. In general, the planets all revolve in the same direction, counterclockwise around the sun, and the sun rotates in the same direction; most orbits lie pretty much in the same plane (within 17° of the ecliptic, Fig 13.10); most orbits are very close to circular; and most planets rotate counterclockwise, in the same direction that they revolve.

Basically, Kepler's three laws of planetary motion (Section 3.3) completely describe the dance of the planets. Table 13.1 includes the basic dynamical data.

13.4 THE NEW SOLAR SYSTEM: A PREVIEW

A spacecraft glides in the grip of the sun's gravity. As it approaches Mercury, an on-board computer, controlled from the earth, commands the cameras to focus in on the tiny planet's surface. The images travel as television signals back to earth. When they arrive here after a few minutes, a computer decodes the signals and paints, electronically, what the camera sees: a dead and cratered world, looking remarkably on its surface like our

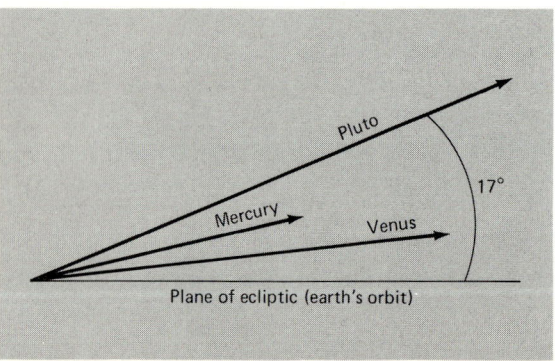

FIGURE 13.10 The range of inclinations of some planets' orbits relative to the plane of the earth's orbit.

COLOR PLATE CAPTIONS

Plate 1 Major types of spectra. This figure has short wavelengths (blue) to the left with wavelength increasing to the right (to the red). At top (1) is a continuous spectrum such as you would observe from a glowing solid. Below it (2) is the absorption-line spectrum of the sun. Here only the most prominent dark lines are indicated by the element that produces them. Below the sun's spectrum are the bright-line spectra of selected elements: sodium (Na), hydrogen (H), calcium (Ca), mercury (Hg), and neon (Ne). Note how the bright lines of sodium, calcium, and hydrogen line up with the dark lines of the same elements in the sun's spectrum. (From *General College Chemistry,* 5th edition, by C. Keenan, J. Wood, and D. Kleinfelter, Harper & Row, 1976, reproduced with permission.)

Plate 2 A close-up of the solar chromosphere, showing a small loop prominence. The colors here are not real, but are computer-generated to show the emission from lines in the ultraviolet. Red indicates emission from magnesium (9-times ionized), green from oxygen (5-times ionized), and blue from hydrogen (Lyman-alpha). The red region marks the corona above the (blue) chromosphere. (Skylab photos courtesy of R. Levine, Harvard College Observatory.)

Plate 3 A total solar eclipse, February 26, 1979, taken from an aircraft at 40,000 ft. The dark skies at this altitude permit details in the corona to be seen out to about 5 solar radii. Note the streams and knots in the corona. (Courtesy William H. Regan and Maxwell T. Sanford, Los Alamos National Laboratory.)

Plate 4 A color representation of the Hertzsprung-Russell diagram. The horizontal axis is surface temperature; the vertical one, luminosity (or absolute magnitude). For the vertical axis, the sun's luminosity is set as 1, and its absolute magnitude as roughly 5. Solid lines show where the luminosity classes fall: Ia supergiants at the very top, Ib supergiants below, III giants below them, and finally V main-sequence stars. The sizes of the stars are shown in the correct order (supergiants largest, white dwarfs smallest) but the stars are not drawn to scale. The colors are true to those perceived by eye.

Plate 5 Messier 20 (M 20), the Trifid Nebula in the constellation Sagittarius. Note that part of the emission is red (lower region) and part blue (upper region). The red light is from glowing hydrogen gas, heated by hot, young massive stars; the blue is starlight scattered by dust within the gas. The dark lanes mark regions where concentrations of dust block out light from the nebula. (Copyright by the Association of Universities for Research in Astronomy, Kitt Peak National Observatory, reproduced with permission.)

Plate 6 A laser shot to the moon from the 107-inch reflecting telescope at McDonald Observatory, Texas. (Courtesy McDonald Observatory and the University of Texas.)

Plate 7 Venus. Taken by the Pioneer Venus mission, this photo shows the circulation of the winds in the upper cloud layers of Venus's atmosphere. Note the yellowish color of the clouds—a color that is in accord with the idea that the clouds are made mostly of sulfuric acid. (Courtesy NASA.)

Plate 8 A view of the surface of Mars, taken by the Viking 1 lander. Note that most of the rocks have small holes in them, an indication of a volcanic origin. Also note that the sky is not dark blue in color but pink; the color comes from surface dust blown high up into the atmosphere by strong surface winds. (Courtesy NASA.)

Plate 9 A wide-angle view of Mars taken from the Viking Orbiter 1. Note the polar cap ice at bottom, covering a few craters; a volcano of the Tharsis ridge at top, with a cloud pluming away from its peak; and a section of Valles Marineris below the volcano. (Courtesy NASA.)

Plate 10 Jupiter, as seen by Voyager 1 from a distance of 20 million km. The two moons visible are Io the left, in front of the Red Spot, and Europa to the right. (Courtesy NASA.)

Plate 11 Volcanic eruption on Io, a Voyager 1 photo taken in March 1979. The eruption's plume stands out against the dark of space at the upper part of the picture. Computer processing has enhanced the brightness of the plume but preserved its greenish-white color. The plume reaches more than 200 km above Io's surface. (Courtesy NASA.)

Plate 12 Voyager 2 false-color photo of wind flows in the northern upper atmosphere of Saturn. The colors have been computer enhanced to emphasize fine details. The disturbance shaped like a 6 has been formed by wind shear. To the south (left), wind blows at 20 m/sec; to the north (right), in the white zone, at greater than 130 m/sec. Note the wavy pattern in the white, east-flowing zone. (Courtesy NASA.)

Plate 13 Voyager 2 close-up of Saturn's C ring and B ring (to the top and left), taken from a distance of 2.7 million km. This is a false-color image produced by a computer to emphasize the differences in color between the B ring (yellow) and the C ring (blue). These are not the actual colors of the rings. This difference indicates that the surface compositions of the particles making up these rings are also different. Over 60 ringlets are visible in this picture. (Courtesy NASA.)

Plate 14 Comet Bennett (1969) over the Swiss Alps. Photo taken by C. Nicolliet from Gornergrat Observatory, Switzerland. (Courtesy C. Nicolliet.)

Plate 15 Messier 31 (M 31), galaxy in the constellation Andromeda. M 31 is the nearest large spiral galaxy, only 2.2 million ly from the Milky Way Galaxy. Note the spiral arms and bright nuclear bulge. The streak above the galaxy is from a satellite passing by during the time exposure. (Copyright by the Association of Universities for Research in Astronomy, Kitt Peak Observatory, reproduced with permission.)

Plate 16 A VLA map of radio jets in the galaxy 3C 388, a typical double radio source. Extending from the nucleus (red dot in center) are two huge lobes of radio emission, with sizes of about 300,000 ly. Note in the bottom lobe a clear indication of a jet pointing out from the nucleus to the strongest part of the lobe. This VLA radio map has been processed by computer so that the different colors reflect different levels of the intensity of radio emission: dark blue the weakest, light blue the next stronger level, light yellow the next, yellow next, and red the strongest. (Courtesy J. O. Burns; observations with the VLA by J. O. Burns and W. A. Christiansen.)

Plate 17 The double quasar. This computer-processed photo shows the two images (left, looking black at their centers) separated by only 6 arcsecs. At right, the top image has been subtracted from the bottom one. What remains is the light from the galaxy acting as the gravitational lens. (Courtesy Alan Stockton, Institute for Astronomy, University of Hawaii.)

Plate 18 Messier 17 and its nearby molecular cloud fragments. This photo combines optical (black and white photo) and infrared (color) views. The color images were constructed to correspond to a visual image if the eye could see light at a wavelength of 40 microns. The infrared emission comes from dust in the molecular clouds to the west of M 17; the color changes from violet to red to illustrate the energy flow. (Courtesy A. Meyer, Ames Research Center, NASA.)

Plate 19 The Large Magellanic Cloud (LMC). This satellite galaxy to the Milky Way is visible to the naked eye in the southern hemisphere. Note how blue the stars appear—an indication that they are young Population I. At top of center is the huge H II region called 30 Doradus. (Copyright by the Association of Universities for Research in Astronomy, Inc., Cerro Tololo Inter-American Observatory, reproduced with permission.)

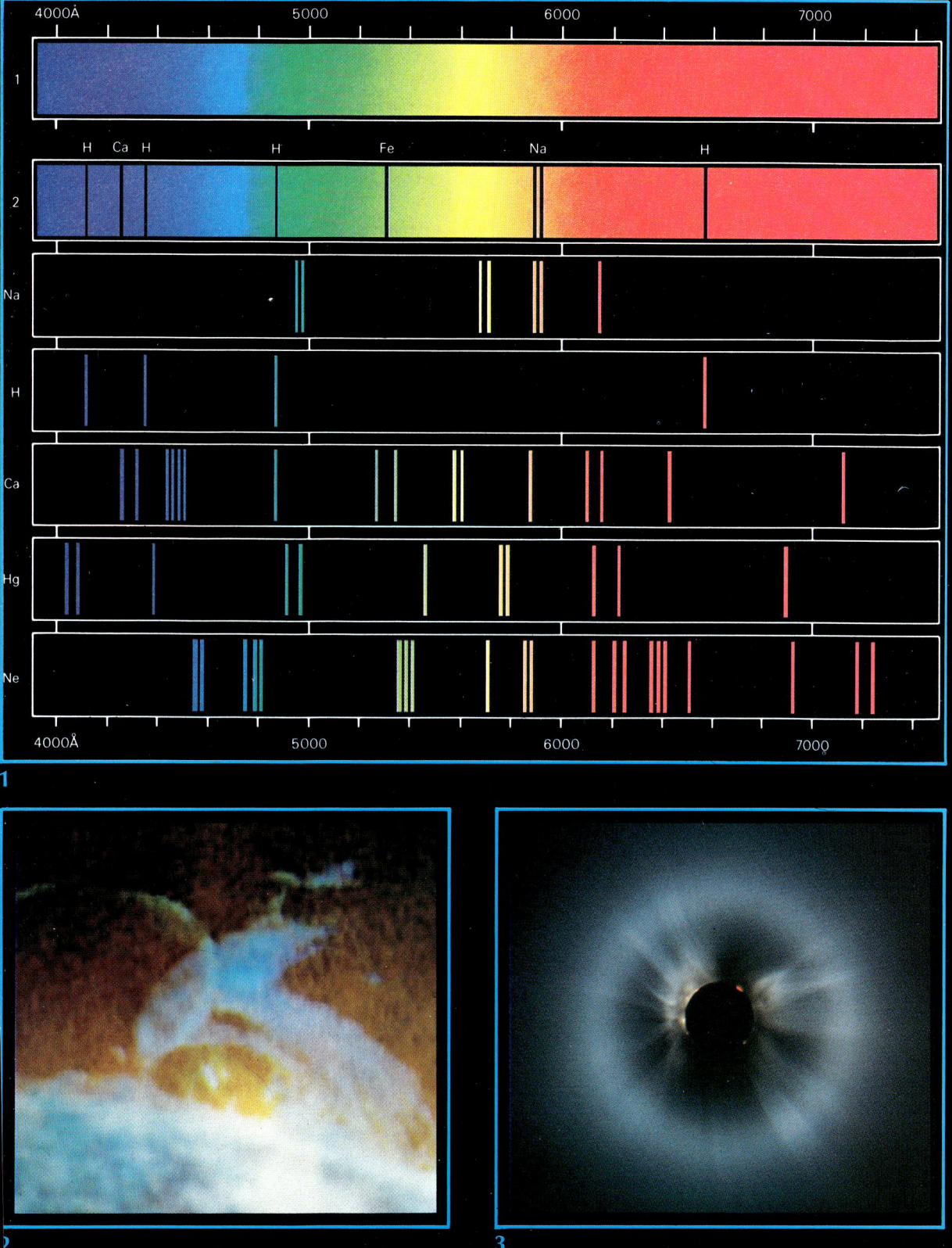

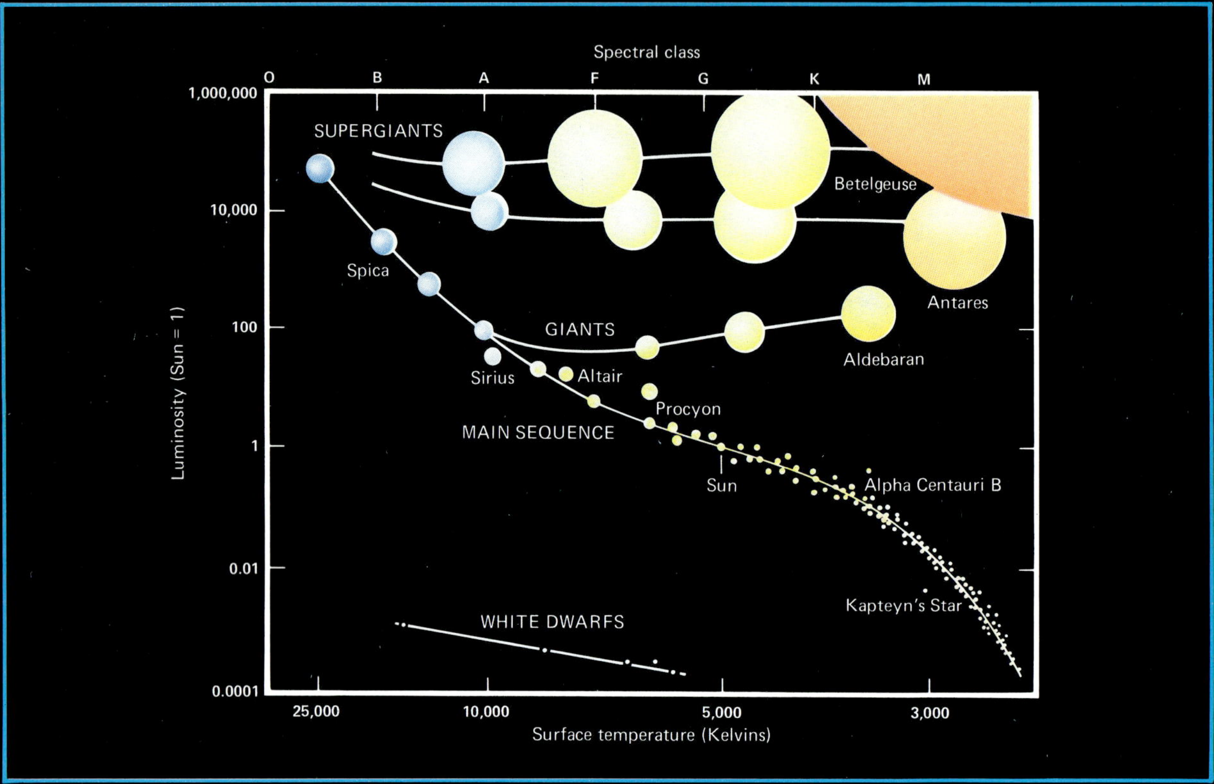
Spectral class
O
B
A
F
G
K
M
1,000,000
10,000
100
1
0.01
0.0001
Luminosity (Sun = 1)
SUPERGIANTS
Betelgeuse
Antares
Spica
GIANTS
Aldebaran
Sirius
Altair
Procyon
MAIN SEQUENCE
Sun
Alpha Centauri B
Kapteyn's Star
WHITE DWARFS
25,000
10,000
5,000
3,000
Surface temperature (Kelvins)

4

5

6

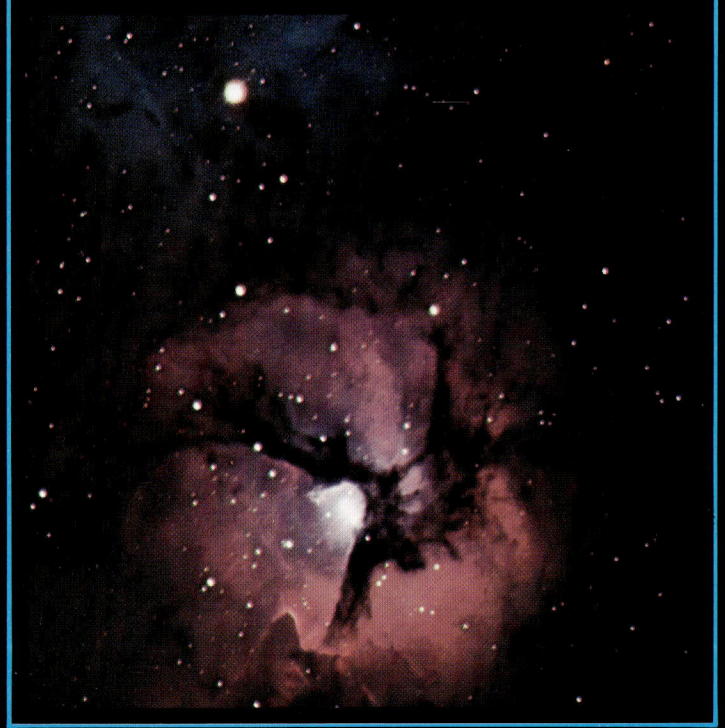

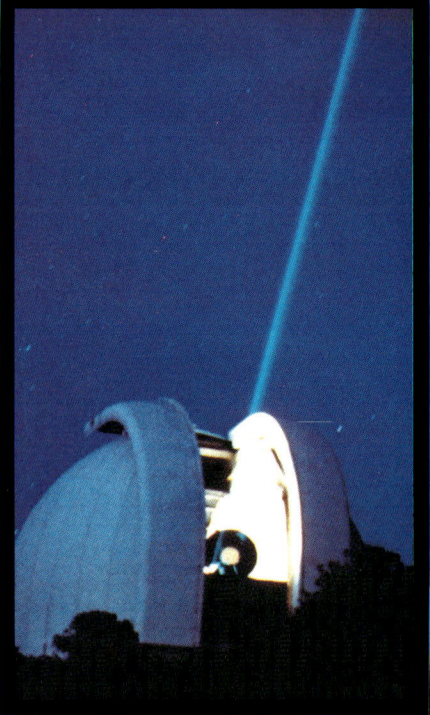

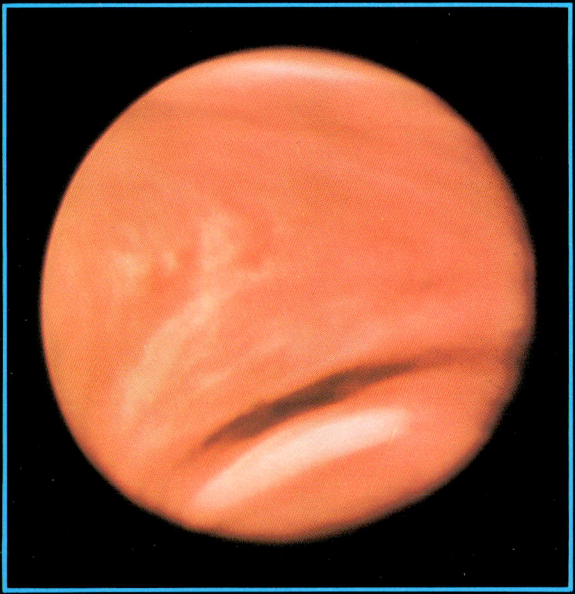

7

8

10

11

12

13

15

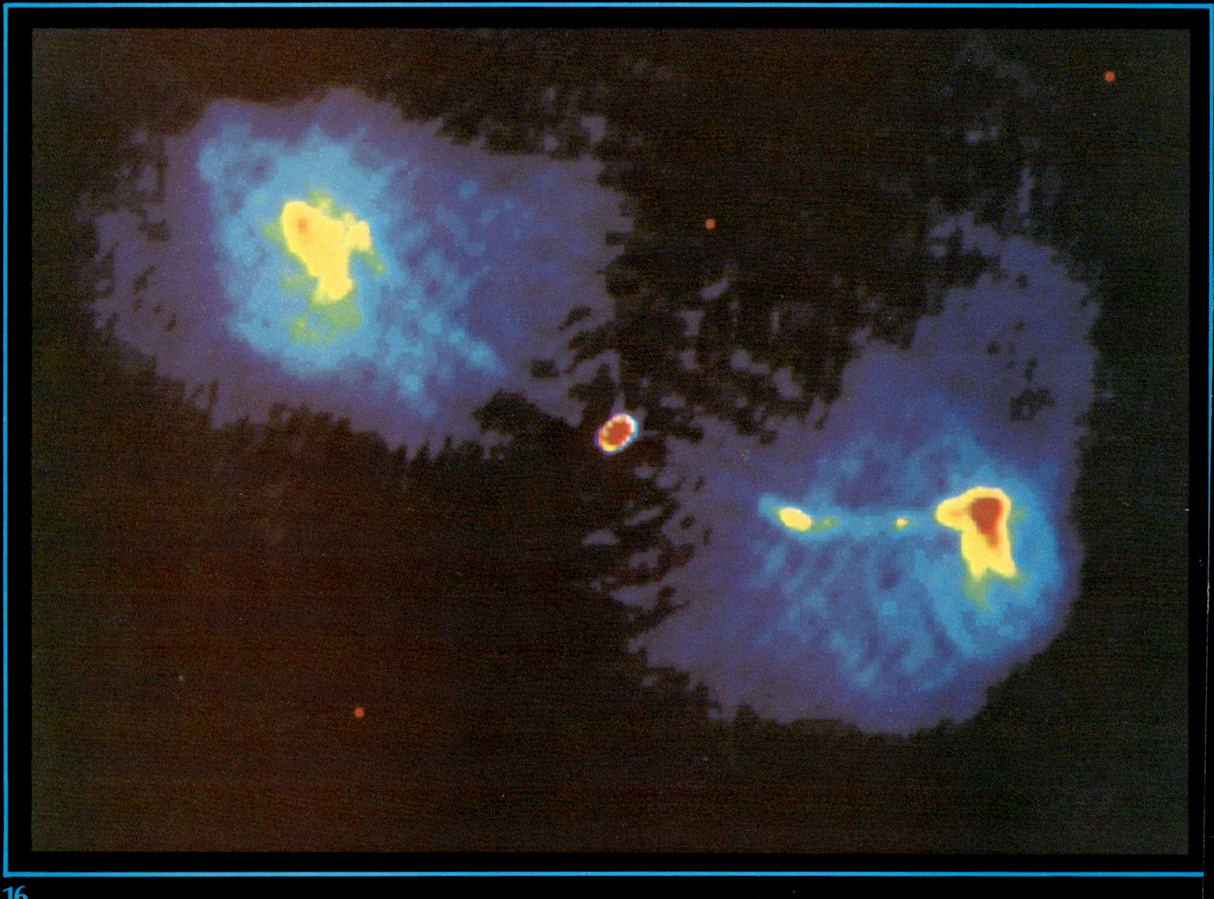

16

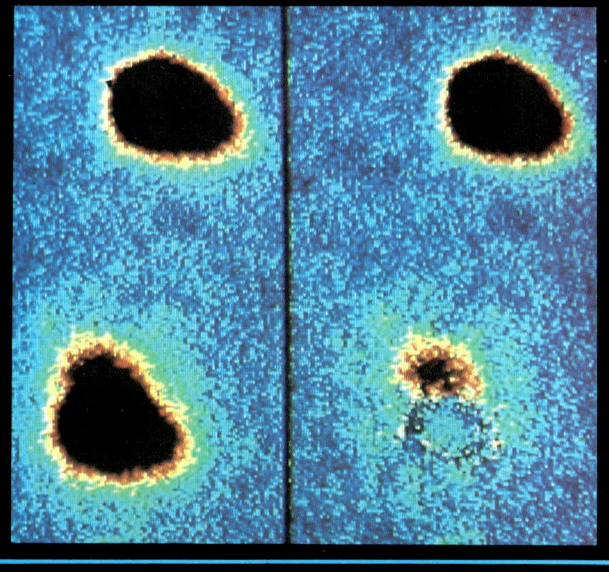

17

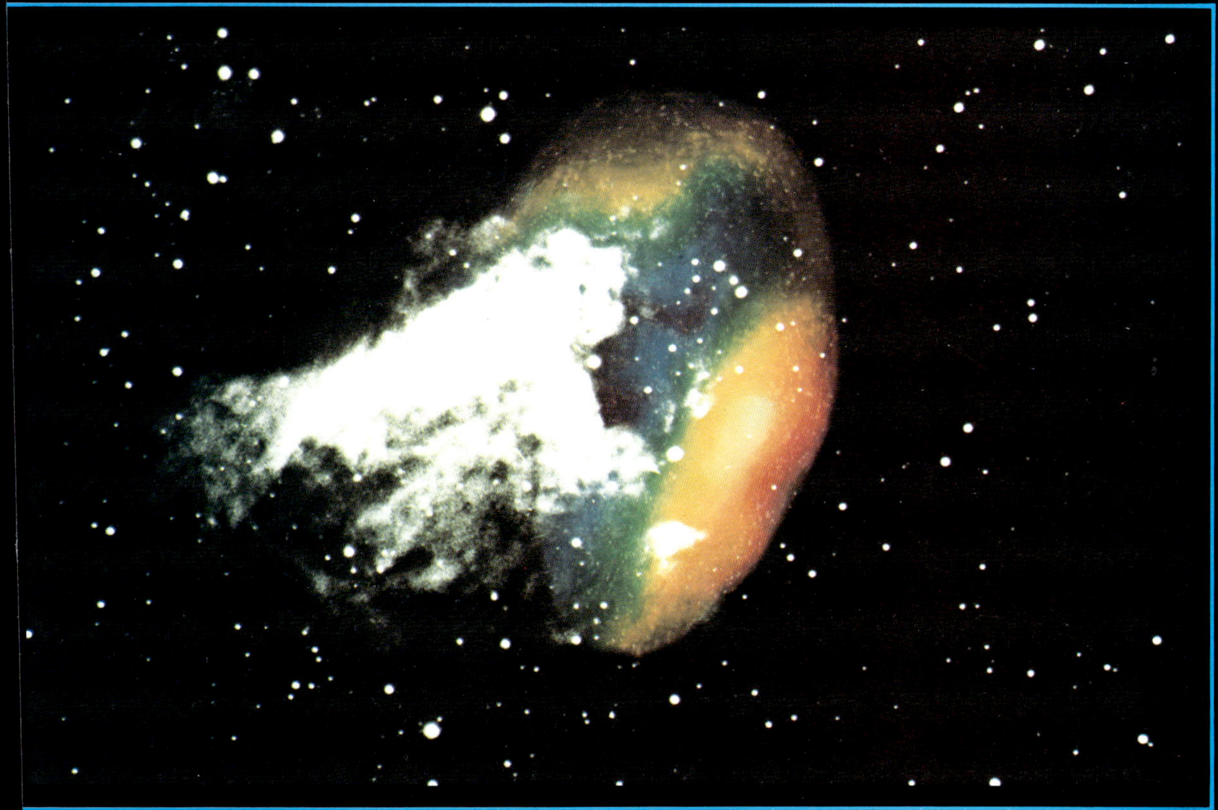

FIGURE 13.11 A close-up of Mercury's surface, taken by Mariner 10. Note the abundant craters. [*Courtesy NASA.*]

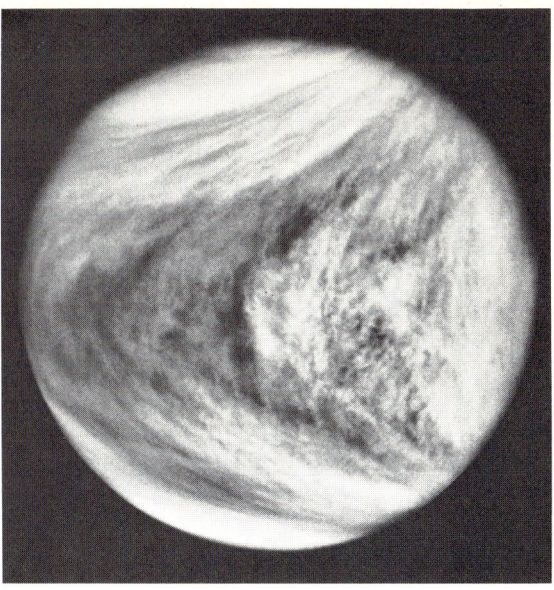

FIGURE 13.12 The clouds in the upper atmosphere of Venus, showing the turbulent flows and the caps above both poles. The mottled features near the center are convective zones from solar heating. The dark regions are *not* the surface, but places where we see deeper into the atmosphere. [*Courtesy NASA.*]

moon (Fig. 13.11). Mercury must have undergone planetwide violence in its past, a fierce bombardment by rocky chunks from space.

Before surveying Mercury, the same spacecraft had skirted past Venus. Its camera had scanned the upper regions of Venus's clouds. The photographs sent to earth revealed jet streams flowing around the planet at speeds of hundreds of kilometers per hour (Fig. 13.12). The clouds whirl completely around Venus once every four days.

Other spacecraft have plunged through Venus's clouds to its surface. These probes found Venus a hellish place: a surface temperature of 700 K, a surface pressure 100 times greater than on the earth, an atmosphere over 95 percent carbon dioxide, and the yellow clouds a mixture of sulfuric acid and water. Russian probes, which landed on the surface of Venus, took close-up pictures (Fig. 13.13) before they were destroyed in the heat and

FIGURE 13.13 A close-up of the surface of Venus, taken by Venera 10. Note the rock slabs in a darker soil.

FIGURE 13.14 The surface of Mars as seen from the Viking 1 lander. This 180° view shows many small rocks in the soil. [Courtesy NASA.]

crushing pressure. The photos show small rocks—some eroded, some not, and a few apparently volcanic in origin. It is a desolate scene of an empty landscape on a murderous world.

We have explored Venus also by bouncing radar signals off its surface. These silent probes have revealed many surprises: huge volcanoes (perhaps the largest in the solar system), possible impact craters, flows, giant plateaus, and a mammoth canyon, larger than any on the earth.

Almost as strange a world is ruddy Mars. The investigation of Mars began with the Mariner flybys and continued in earnest with the touchdown of the two Viking landers. Their primary mission was to probe the Martian surface (Fig. 13.14) to find its characteristics and to search for signs of life. The results: rusty sand covers the Martian surface; no life exists there. And no wonder. The Martian atmosphere is 95 percent carbon dioxide, the surface pressure is 1/200 that of the earth, water is scarce, and at night the temperature can fall to $-85°C$ (that's $-121°F$).

What of the fabled Martian canals? They don't exist. In fact, water cannot now flow on the Martian surface—the pressure is too low, and the temperature too cold. Mostly the water resides as ice in the core of the Martian pole caps.

Mars is a barren, desert world. But that doesn't mean its always been dull! Eroded craters pockmark the southern hemisphere of Mars. In the north enormous volcanoes, now dormant or extinct, point silently skyward (Fig. 13.15). Extensive lava flows coat the land. Here also a huge canyon scars the surface, its full length as long as the United States (Fig. 13.16).

As alien as Mercury, Venus, the moon, and Mars may seem, stranger still are the planets beyond the asteroids. Here, in the realm of a tiny sun and terrible cold, float worlds of liquid and gas, planets mostly without solid surfaces at all.

Jupiter lords over this realm. This massive, mighty planet carries with it an entourage of at least 16 moons (Fig. 13.17). Four of these, the ones discovered by Galileo, are worlds in their own right. Io, with its reddish-orange-yellow sulfurous surface, sports

FIGURE 13.15 Giant volcanoes on Mars. Four are visible in the upper part of this picture, taken by the Viking 2 orbiter. [*Courtesy NASA.*]

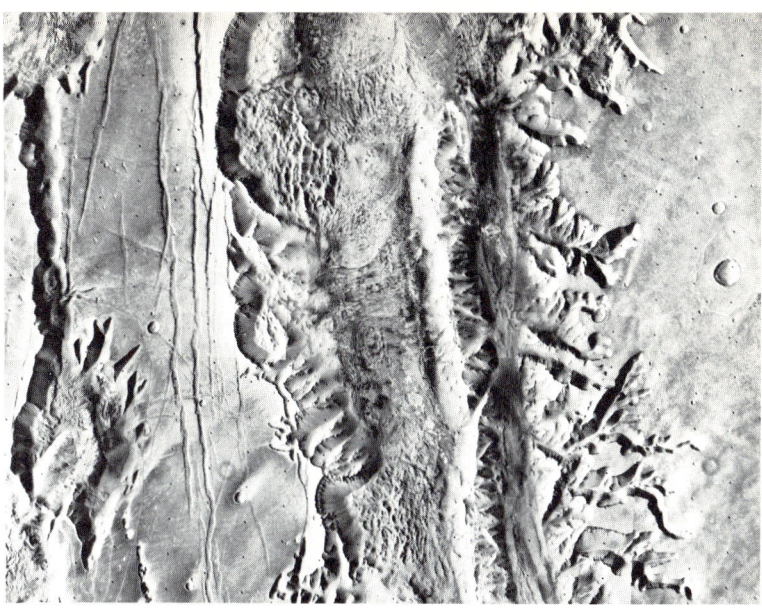

FIGURE 13.16 A section of Valles Marineris (the Valley of Mariner) taken by the Viking 1 orbiter. [*Courtesy NASA.*]

active volcanoes perhaps surrounded by lava lakes (Fig. 13.18). Europa shows a surface almost devoid of impact craters, crisscrossed with huge cracks across its face (Fig. 13.19). These cracks are thousands of kilometers long and tens of kilometers wide. Ganymede's surface displays impact craters (like those on Mercury) and long, complex cracks and ridges (Fig. 13.20). Callisto has a lunarlike appearance, with many old impact craters (Fig. 13.21).

What about Jupiter itself? Powerful jet streams zoom through its upper atmosphere,

FIGURE 13.17 Jupiter and its moons Io (left) and Europa (right), taken by Voyager 1 from 20 million km. The Great Red Spot is just below Io. [*Courtesy NASA.*]

FIGURE 13.18 Io, taken by Voyager 1 from 862,000 km. The circular feature near the center is an erupting volcano. [*Courtesy NASA.*]

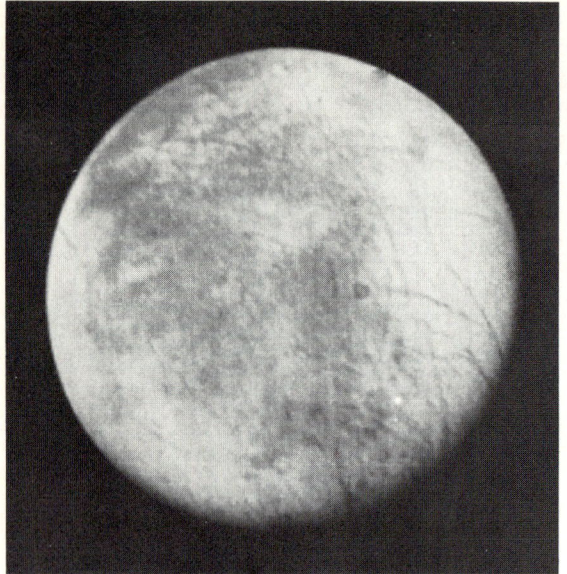

FIGURE 13.19 Europa, taken by Voyager 1 from 1.2 million km. Bright areas are ice. [*Courtesy* NASA.]

FIGURE 13.20 Ganymede, taken by Voyager 1 from 2.6 million km. The white spots are impact craters. [*Courtesy* NASA.]

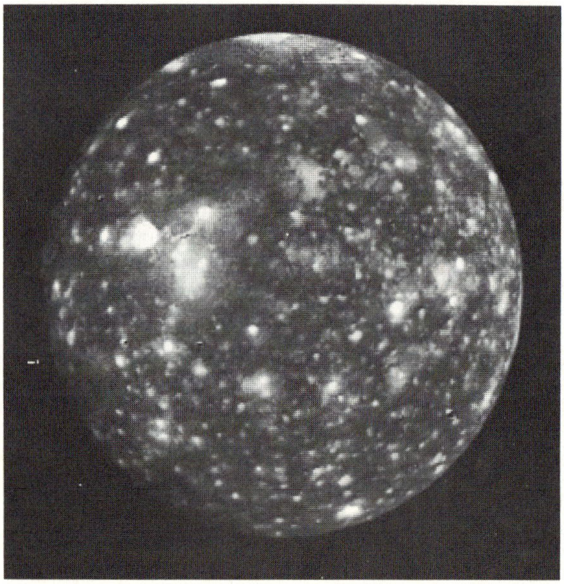

FIGURE 13.21 Callisto, taken by Voyager 2 from 2.3 million km. Note the bright rays extending from the impact craters. [*Courtesy* NASA.]

giving the planet its banded appearance. Huge storms swirl here. The most famous and longest-lived is the Great Red Spot, extending 10 km above the clouds like a mysterious eye (see Fig. 13.17). The Red Spot churns counterclockwise once every 6 days or so; all of its 30,000 km width and 10,000 km length whirls at more than 200 km/h. Thunderstorms rise high in the turbulent atmosphere, releasing immense lightning bolts more powerful than lightning on the earth. Surrounding all this violence swims a thin ring, some 57,000 km above Jupiter's active atmosphere (Fig. 13.22).

Saturn, of course, also has rings, five main ones with numerous divisions (Fig. 13.23).

13.4 THE NEW SOLAR SYSTEM: A PREVIEW

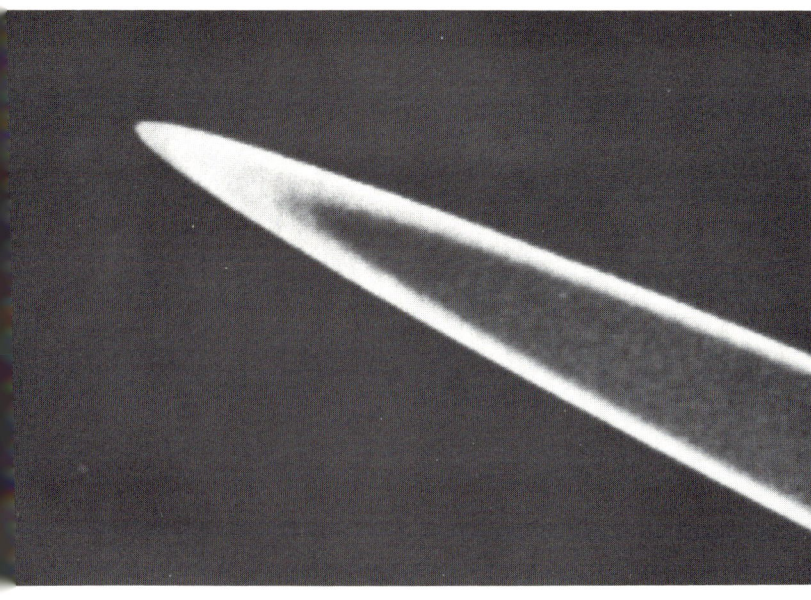

FIGURE 13.22 A close-up of the rings of Jupiter, taken by Voyager 2 from 1.5 million km. [Courtesy NASA.]

Saturn's atmosphere, like Jupiter's, shows a banded structure—an indication of fierce jet streams. And, also like Jupiter, Saturn has a retinue of many satellites, at least 17 and perhaps 22 or more. One, Titan (Fig. 13.24), has an atmosphere (it is the only moon in the solar system so far known to have a significant one) that contains nitrogen, argon, methane, and a stratospheric layer of yellow smog.

Far beyond Saturn orbits the giant, cold world of Uranus, a planet not yet well explored. Uranus also has rings, at least nine (Fig. 13.25). Unlike Saturn's bright, wide

FIGURE 13.23 A close-up view of the rings of Saturn, taken by Voyager 1 from 1.5 million km. [Courtesy NASA.]

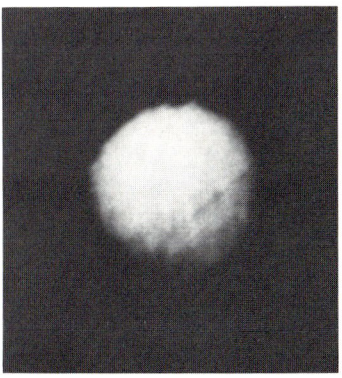

FIGURE 13.24 Titan, Saturn's largest moon, taken by Pioneer from 370,000 km. [Courtesy NASA.]

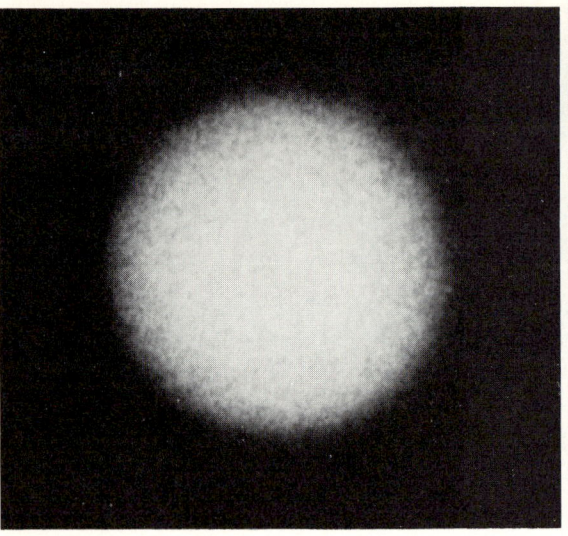

FIGURE 13.25 Uranus, photo taken by a balloon-borne telescope, Stratoscope II. Note the lack of any visible features. [Courtesy NASA.]

FIGURE 13.26 An infrared photo showing the structure of the clouds in the upper atmosphere of Neptune. The brighter regions are methane clouds. [Courtesy H. Reitsema, B. Smith, and S. Larson, the Lunar and Planetary Laboratory, University of Arizona.]

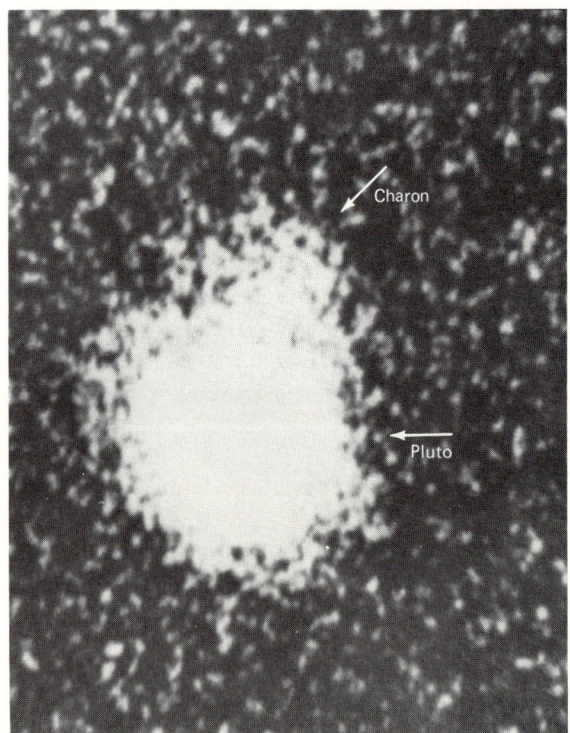

FIGURE 13.27 Pluto and its moon, Charon. The two bodies are so close together that their images merge. [Courtesy J. Christy.]

rings, Uranus's ring system consists of narrow (about 10 km wide) strips of dark particles (as black as the surface of coal).

Twin of Uranus is Neptune. We really don't know much about this distant planet. Recent observations imply that Neptune's atmosphere may contain high-altitude clouds of methane that come and go (Fig. 13.26)—a hint of some kind of weather pattern in the atmosphere.

Finally, in the severe cold at the outer reaches of the solar system lies puny Pluto. So cold is Pluto's surface that methane ice exists there. There is also a methane atmosphere, but it is very thin; most of it has fallen frozen to the ground. This barren world has a moon, called Charon (Fig. 13.27). The satellite, some 1500 km in diameter, orbits Pluto every 6.4 days. Using Charon's orbit, we have been able to find Pluto's mass: about 0.002 that of the earth or 0.2 that of our moon. With a small mass like a terrestrial planet, but composed of methane and other ices like a Jovian planet, how do we classify Pluto? It is unique, a bit like both, but not like either.

The solar system does not end at Pluto. Out beyond Pluto many comets slowly orbit, but probably no large planets inhabit that dark space. All searches for them to date have failed.

So spacecraft and new telescopes have imparted to us a new solar system, one more bizarre, curious, and mysterious than that known in the first half of this century. And, the inhabitants of one planet—a blue-white, water-covered world—have probed their neighbor worlds to find (much to our disappointment) no living creatures, not even simple microorganisms in the red sands of Mars.

SUMMARY

Nudged by a supernova remnant, an interstellar molecular cloud began to collapse some 4.6 billion years ago. Quickly, in a few tens of millions of years, this cloud, or a fragment of it, condensed to form a disk with a central bulge. The central area developed into the sun. In the disk dust grains hit and stuck to make pebble-sized objects; these aggregated into larger and larger bodies, until, finally, the young, unevolved planets materialized. Meanwhile, the sun became a main-sequence star, and its wind blew out the remaining gas and dust.

The planets today fall into two distinct categories. The terrestrial planets are small, rocky, dense bodies close to the sun; and the Jovian ones are large, massive, low-density bodies far from the sun. Pluto has some characteristics of both groups; it is small in diameter and low in mass like a terrestrial planet, but its density and composition resemble those of the Jovian planets.

Key Words
supernova-triggered star formation Allende meteorite   protoplanet
exponential decay   planetesimal   terrestrial planet
half-life   Jovian planet

Review Questions
1. What evidence do we have that a supernova exploded close to the solar system's location before its birth? Be sure to distinguish between the observations and their interpretation. (Objective 1)
2. What is a major problem in forming the planets with the sun? (Objectives 2 and 3)
3. Which is the largest terrestrial planet? Smallest? (Objectives 4 and 5)

4. Give a physical explanation for how Jupiter can be more massive than the earth, yet have a lower density. (Objective 4)
5. Mercury does not have a natural moon. How do we know its mass? (Objective 5)
6. What properties does Pluto have in common with the terrestrial planets? With the Jovian planets? (Objective 7)
7. Suppose that you had the following mineral samples, which a geochemist tells you were pure uranium when the rocks were formed. The numbers given are the present percentages of uranium and lead, respectively. The half-life of uranium is 4.5×10^9 yr.

    Sample A: 90, 10 Sample D: 12.5, 87.5
    Sample B: 25, 75 Sample E: 75, 25
    Sample C: 50, 50 Sample F: 60, 40

    a. Place the samples in order of increasing age.
    b. Which samples could not have been formed on the earth?
    c. Which sample is twice as old as Sample C?

Problems

1. a. Calculate the mass of Saturn using one of its moons. See Appendix B for the necessary data.
   b. From the diameter given in Table 13.1, calculate Saturn's density. (Objectives 5 and 6)
2. What is the maximum angular diameter Venus can have as seen from the earth? Minimum? (Objective 6)
3. How much of an original kilogram of pure ^{26}Al remains after 3.6 million years have passed? (Objective 8)
4. A rock known to have no ^{87}Sr when it was formed now contains 1/20 as much ^{87}Sr as ^{87}Rb. How old is the rock? (Objective 8)
5. How long does it take radio signals to reach Jupiter and return to earth when Jupiter is at opposition? When it is near conjunction with the sun? (Objective 6)

BEYOND THIS BOOK . . .

* *The Inner Planets* (Scribner, New York, 1977) by C. Chapman presents a comprehensive survey of the terrestrial planets.
* *The Solar System* (Viking, 1979) by P. Ryan and L. Pesek contains many excellent photographs.
* W. J. Kaufmann III gives the solar system a complete treatment in *Exploration of the Solar System* (Macmillan, New York, 1978). His later book, *Planets and Moons* (Freeman, San Francisco, 1979), is also a good summary.
* Check *Sky and Telescope*, *Astronomy*, and *Mercury* magazines for new information as it is discovered.
* The September 1975 issue of *Scientific American* was devoted to the solar system. It is available in reprint form from W. H. Freeman and Co., San Francisco.
* *The New Solar System* (Cambridge University Press, New York; Sky Publishing Corporation, Cambridge, Mass., 1981), edited by J. Kelly Beatty, Brian O'Leary, and Andrew Chaikin, covers everything up to the Voyager encounter with Jupiter (but not that with Saturn).

THE EARTH: HOME PLANET

14

LEARNING OBJECTIVES
After studying this chapter, you should be able to:
1. Describe one method for determining the earth's mass and density.
2. Sketch the interior structure of the earth, indicating the composition of each general region.
3. Argue simply that the earth's core must be denser than its crust and also that the core probably has a nickel-iron composition.
4. Argue that the earth's interior structure implies that it must have been molten at some time in the past.
5. Give the estimated age of the earth, and explain how we arrive at this estimate. (Be sure to give the assumptions and observations.)
6. Describe at least two ways in which the earth's atmosphere affects astronomical observations and two ways in which it affects the earth's surface environment.
7. Explain how the earth's atmosphere acts like a blanket that keeps the earth's surface relatively warm.
8. Outline a possible model for the evolution of the earth's crust and interior.
9. Outline a possible model for the evolution of the earth's oceans.
10. Outline a possible model for the evolution of the earth's atmosphere.
11. Outline a simple method for the escape of a planet's atmosphere into space and calculate an estimate of the escape time.

CENTRAL QUESTION:
What are the basic physical features of the earth, and how have they changed since our planet's formation?

How small the earth is: a tiny planet, whirling around one ordinary star. And, it is not even the center of the universe, as people believed for over 4000 years. But that change in cosmic position does not mean we should value the earth any less. For the earth is our delicate ship, protecting us on our dark passage through space.

With the aid of a modest telescope, astronomers on Mars would see terrestrial clouds as the most distinct features on the earth (Fig. 14.1). As is evident on satellite weather maps, the terrestrial clouds typically cover about 35 percent of the surface. They reflect back sunlight, so the earth appears bright. The swirls of cloud cover commonly obscure many surface features and certainly eliminate any hope that extraterrestrial astronomers could see directly evidence of human civilization.

This chapter looks at the physical makeup of the earth, the only planet we know to be inhabited. We live here, so we know this planet in more detail than any other. Our present understanding of the earth indicates that it is the most evolved of the terrestrial planets, that is, it has changed dramatically in its physical structure since its formation. We will use our home planet as the basis of comparison for understanding the makeup and evolution of the other terrestrial planets.

14.1 THE MASS AND DENSITY OF THE SOLID EARTH

How to find the earth's mass? Recall (Section 4.1) the discovery by Galileo that all bodies at the earth's surface have the same acceleration due to gravity (g). Newton's law of gravitation (Section 4.2) relates this acceleration to the earth's mass and radius and to the gravitational constant (G). So with G, the earth's radius, and g, we can figure out the earth's mass from Newton's law. The mass comes out to about 6×10^{24} kg.

FIGURE 14.1 The earth from space, taken by Apollo 11. [*Courtesy NASA.*]

Let's do this calculation in detail. At the earth's surface, the acceleration on any mass is g, or

$$g = \frac{GM}{R^2} = 9.8 \text{ m/sec/sec}$$

where M is the earth's mass, and R its radius (6371 km or 6.371×10^6 m). Then

$$M = \frac{gR^2}{G}$$

$$= \frac{9.8(6.371 \times 10^6)^2}{(6.67 \times 10^{-11})}$$

$$= 6.0 \times 10^{24} \text{ kg}$$

Knowing both the earth's mass and its volume (from its radius), we can compute its average density. Dividing the earth's mass by its volume, we obtain a density of 5500 kg/m³, or 5.5 times the density of water. This average density indicates that the earth, in bulk, consists of a combination of rocky and metallic materials. (Most rocks have a density between 2000 and 3000 kg/m³; iron has a density of 8000 kg/m³.)

Rocks near the earth's surface average 2400 kg/m³, about half the average density of the earth. This difference implies that the core of the earth must be denser than the average. Present estimates indicate that the density of the earth's core is perhaps 12,000 kg/m³.

Though the core is metallic, the weight of the overlying layers compresses it and creates this even higher central density.

14.2 THE INTERIOR OF THE EARTH

The earth's interior is *differentiated*. This means that it consists of layers with the most dense materials at the center and the least dense at the surface. Geologists generally divide the interior of the earth into three distinct layers: the core, the mantle, and the crust (Fig. 14.2).

The *core* makes up the central zone and extends more than halfway to the surface. The core's high density implies that it is probably composed of iron and nickel. (Some of the iron is probably combined with sulfur to make FeS, a compound called triolite.)

Above the core extends the *mantle*, roughly 2900 km thick. The mantle material is mostly rock made of iron and magnesium combined with silicon and oxygen (a silicate mineral called olivine). The mantle is plastic. Under slow, steady pressure such material flows like a liquid, but sudden changes in pressure make it snap and fragment like glass.

Encasing the mantle is the *crust*, the solid surface layer, which varies in depth from 16 to 40 km. Most of the crustal material consists of rocks that have solidified from molten

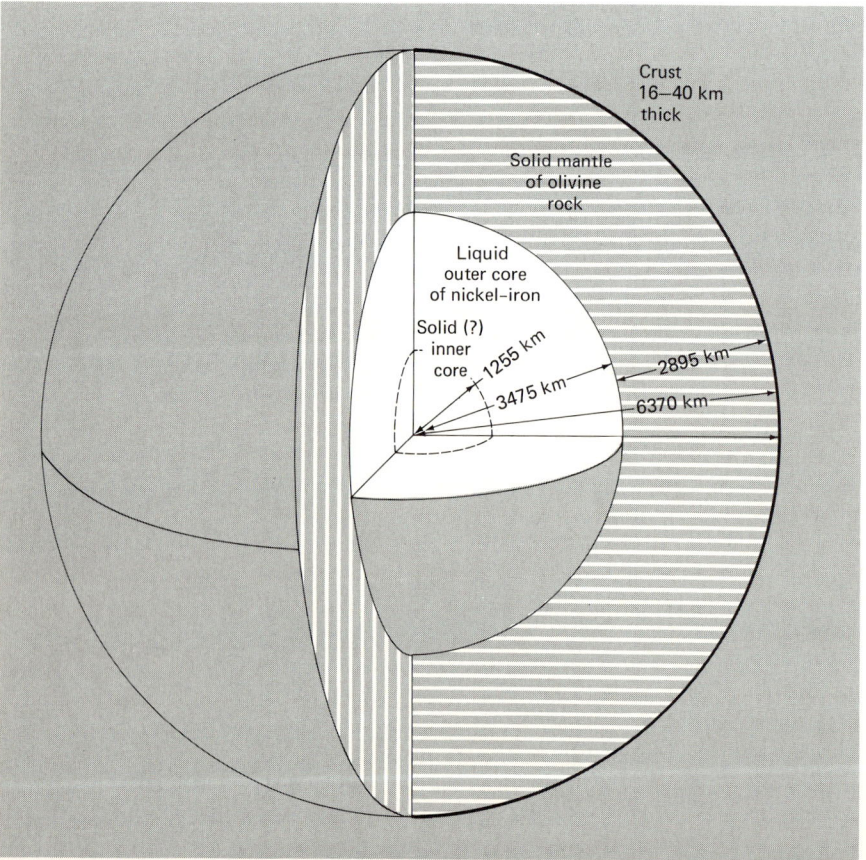

FIGURE 14.2 A model of the earth's interior, showing the core, mantle, and crust. [*Adapted from A. N. Strahler,* The Earth Sciences, *2nd ed., Harper & Row, New York, 1971.*]

lava, called *igneous* rocks. These rocks are *basalt*, a combination of oxygen, silicon, aluminum, magnesium, and iron. They comprise the ocean basins and the subcontinental sections of the crust. The continental masses are mostly *granite*, made of oxygen, silicon, aluminum, sodium, and potassium. Because the granite has a lower density than the basalt (granites contain less iron and magnesium), the continental plates float on the basalt. Also, because the mantle is denser than the basalt and granite, the entire crust floats on the mantle. (See Focus 14.1 to find out how geologists can "see" into the earth's interior.) Table 14.1 lists the ten most abundant elements in the three layers of the earth.

How did the earth become differentiated? Present theories of the solar system's origin see the earth as being formed from an aggregation of well-mixed material (Chapter 13). When the interior heated up enough so that it was mostly molten (or at least plastic), the dense materials (iron, nickel) settled at the core, and the less dense materials (silicates) formed a froth on top.

What heated the interior? One source is the heat generated by accretion. Another is radioactive decay. In the past the earth had about six times more radioactive material than now. The heat from accretion or the heat from the decay of so much radioactive material could melt the interior.

14.3 THE AGE OF THE EARTH

Geologists now estimate the earth's age at 4.6 billion years (with a range of error of 100 million years) on the basis of radioactive dating (Section 13.1), although the oldest known rocks on the earth's surface are not actually this old. The most ancient known rocks, found in West Greenland, have been dated by rubidium-strontium decay at 4.0 billion years. These rocks are igneous rocks; the material was first molten and then cooled

TABLE 14.1 The Ten Most Abundant Terrestrial Elements

Element	Symbol	Terrestrial Average* (% by mass)
Iron	Fe	34.6
Oxygen	O	29.5
Silicon	Si	15.2
Magnesium	Mg	12.7
Nickel	Ni	2.4
Sulfur	S	1.9
Calcium	Ca	1.1
Aluminum	Al	1.1
Sodium	Na	0.57
Chromium	Cr	0.26

*The averages listed include both the crust and the interior (mantle and core). The very high concentration of iron and nickel in the core contributes the largest amount to the averages of these elements. The crust itself is about 75 percent silicon and oxygen.

Source: Adapted from Brian Mason, *Principles of Geochemistry*, 3rd ed. (Wiley, New York, 1966).

FOCUS 14.1 SOUNDING OUT THE EARTH'S INTERIOR

Because none of our devices can plumb the earth's inner sanctum, geologists rely on earthquakes to map the lower depths. An earthquake occurs when two adjoining or overlying rock layers slip against one another because of built-up internal stresses. As the rocks move, they generate vibrations, called *seismic waves*, in the adjacent rock material. These seismic waves travel through the globe in two basic forms: *transverse* and *longitudinal* waves.

Transverse waves move up and down as they travel in a material. You can generate transverse waves by tying down a rope at one end and shaking the other end up and down. Transverse waves can travel through solids, but not through liquids. When they run into a liquid, they gradually dissipate. Longitudinal waves are push-pull waves; the most common example is sound. Longitudinal waves can travel through solids, liquids, and gasses.

The speed of a wave depends on the medium through which it passes. For example, the denser the material, the faster the waves go. When the medium suddenly changes, both the speed and the direction of the wave abruptly change. (The wave is *refracted*, just as light is refracted at a glass surface; Section 7.2.) When geologists record seismic waves, the changes in the wave speed and direction allow them to infer the physical properties of the earth's interior. In this technique the geologists look at the arrival of longitudinal waves (P-waves) and transverse waves (S-waves) from an earthquake. The P-waves are the first recorded by a seismograph; the S-waves arrive next. The S-waves do not penetrate the core, and the P-waves travel slowly through the core. Hence the core must be liquid rather than solid (Fig. F.17).

The reception of the earthquake waves provides information about the size of the core. The core shields out the S-waves from reception on the side of the earth opposite the earthquake's origin. It can also cause the P-waves to change direction sharply at the interface of the core and mantle. So some areas of the earth receive both P- and S-waves, some only P-waves, and some no waves at all. From this information we determine the size of the core compared with that of the mantle.

Put together, this information gives a sketch of the earth's interior properties (Fig. F.18). The P-wave velocity exhibits two abrupt changes, which mark the outer and inner core. The S-waves suddenly vanish at the outer core's boundary. The density of the material must rise rapidly in the core (c), but must not be much different in the inner core. The temperature (b), after a rapid increase in the crust and mantle, does not change much in the core, but it is high enough (about 2600 K) to melt the outer core. The inner core may be solid.

The seismograph is also an important tool for the astronomer, for it has been used to determine the interior structure of the moon and will be used in future studies of the other planets. Such information is valuable for any models of the origin of the solar system and the formation of its planets.

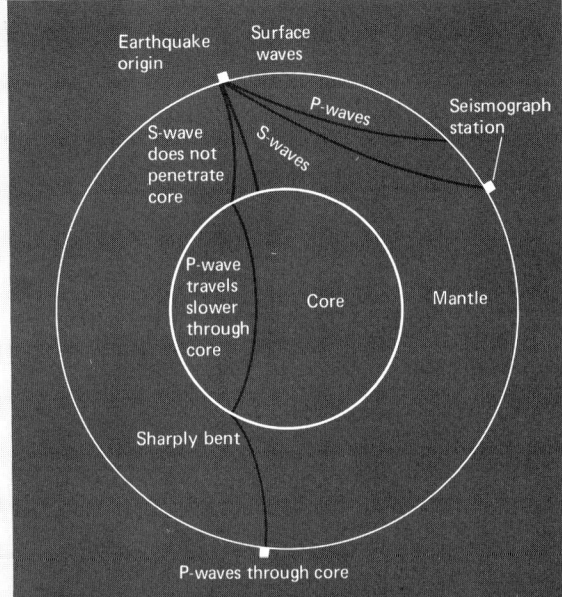

FIGURE F.17 Paths of seismic waves in the earth's interior. To a seismograph on the same side of the earth as an earthquake, S-waves and P-waves travel along the same path. To one on the opposite side, only P-waves arrive because S-waves cannot pass through the liquid part of the core. [*Adapted from A. N. Strahler, The Earth Sciences, 2nd ed., Harper & Row, New York, 1971.*]

FIGURE F.18 A model of the earth's interior inferred from seismic waves. The pressure (a) increases continually with depth. The temperature (b) at first rises quickly, then levels off. A sharp jump in the density (c) marks the division between the mantle and core. [*Adapted from A. N. Strahler, The Earth Sciences, 2nd ed., Harper & Row, New York, 1971.*]

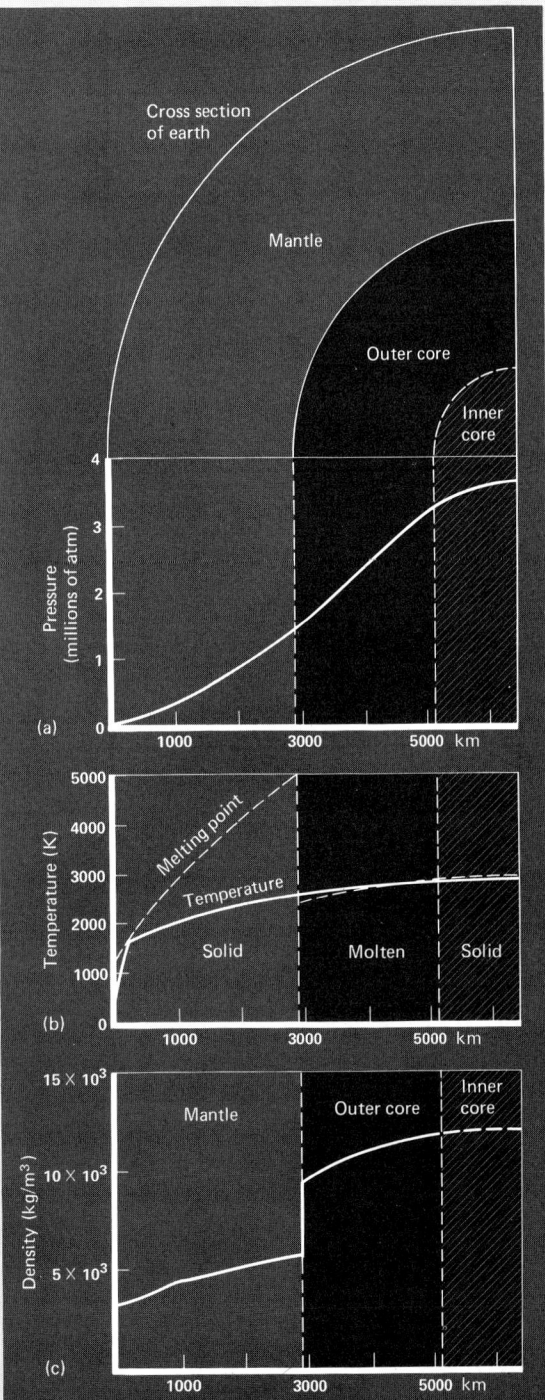

to form the present rocks. Geologists estimate that the time required for the crust's initial melting and cooling, to form the first rocks, was about 0.5 billion years. So by this estimate (they vary) the solid earth's age is approximately 4.5 billion years, the age of the rocks plus the estimated time to form them.

The estimate for the earth's age falls close to that for meteorite material (4.55 billion years) and lunar material (4.6 billion years), determined by the same radioactive dating techniques. The near-coincidence of these ages implies that the solar system formed in one single event about 4.6 billion years ago (Chapter 13).

Warning: Three assumptions are involved in making this estimate: (1) that the rocks have not been heated since they first formed, (2) that about half a billion years is a reasonable guess for the time between the earth's formation and the solidification of its crust, and (3) that the half-lives of radioactive decay were the same in the past as they are now. The exact times given above are still debated, but the earth's age is certainly a few billion years.

14.4 THE MAGNETIC FIELD OF THE EARTH

You can visualize the earth's magnetic properties by imagining a giant bar magnet located in the core (Focus 6.2). The magnetic lines of force protrude from the *south magnetic pole* in the Southern Hemisphere and return to the *north magnetic pole* in the Northern Hemisphere. The magnetic axis, which connects the magnetic poles, is inclined about 20° from the spin axis and does not pass through the earth's center (Fig. 14.3). The part of this magnetic field that is parallel to the earth's surface orients a compass needle along the lines of force so that the needle points to the north and south magnetic poles.

The earth's magnetic field changes with time in both direction and intensity. We have evidence that the magnetic poles have undergone actual reversals of polarity at least nine times during 3.5 million years and probably many more times in previous ages (see Section 14.7).

The source of the earth's magnetic field and the mechanisms for its changes are buried deep in the earth. They relate to the liquid nature of the iron-nickel core. The metal can conduct electrical currents, so it acts like a giant dynamo and electromagnet, generating electricity and creating a magnetic field. The earth's rotation supposedly stirs up the currents in the core. This *dynamo model* for the earth's magnetic field—if correct—has a major implication for other planets: Any planet that exhibits a strong magnetic field must have a substantial liquid conducting core and must rotate rapidly.

Warning: The dynamo model has yet to be worked out in detail. For instance, there is little agreement to date on how the fluid core flows, what drives these motions, and how these flows generate the complex field we measure at the surface. Recent work indicates that the flow may be driven by gravitational energy liberated as dense materials migrate to the center of the core and less dense materials flow outward.

In 1958 the early U.S. space satellites detected a region encircling the earth that contained a large number of protons and electrons. Later satellites revealed a similar, but larger region, farther from the earth's surface. These two doughnut-shaped belts of energetic particles trapped by the terrestrial magnetic field are called the *Van Allen radiation*

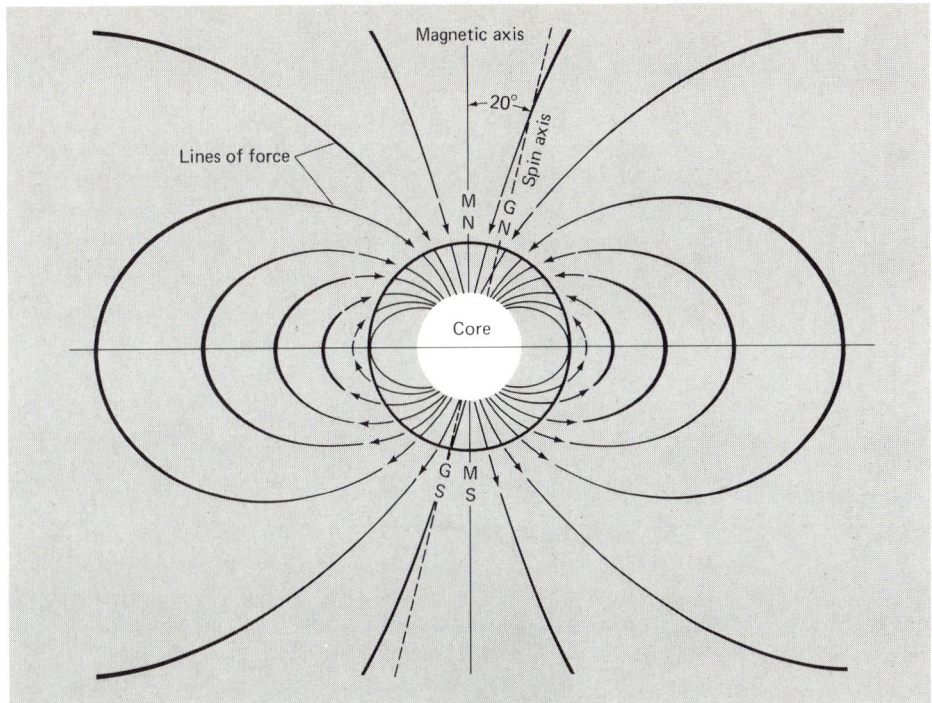

FIGURE 14.3 A model of the earth's magnetic field. Note that the magnetic axis (MN to MS) is *not* aligned with the spin axis (GN to GS), but is tilted about 20°. [*Adapted from* A. N. Strahler, The Earth Sciences, 2nd ed., Harper & Row, New York, 1971.]

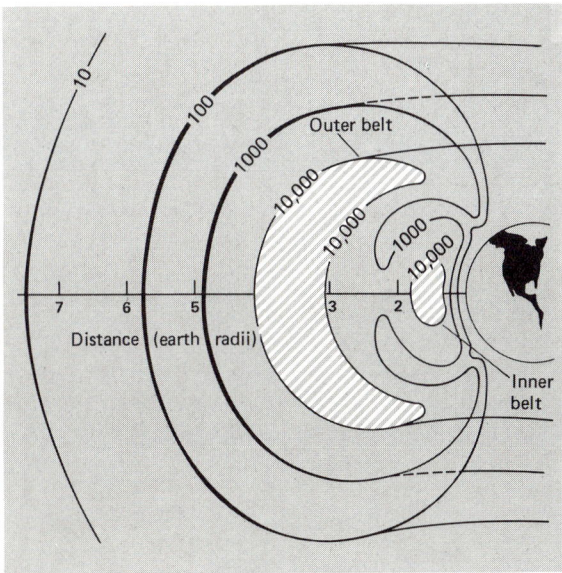

FIGURE 14.4 A model of the Van Allen radiation belts, which encircle the earth. These belts contain charged particles, mostly protons and electrons from the sun. The numbers on the contour lines indicate the average density of particles per cubic centimeter. [*Based on NASA data.*]

belts (Fig. 14.4), after their discoverer, James A. Van Allen. The particles trapped in the belts come from the solar wind and solar flares.

The Van Allen belts are only one aspect of the interaction of the earth's magnetic field with the charged particles that continually stream from the sun. In fact, the earth's

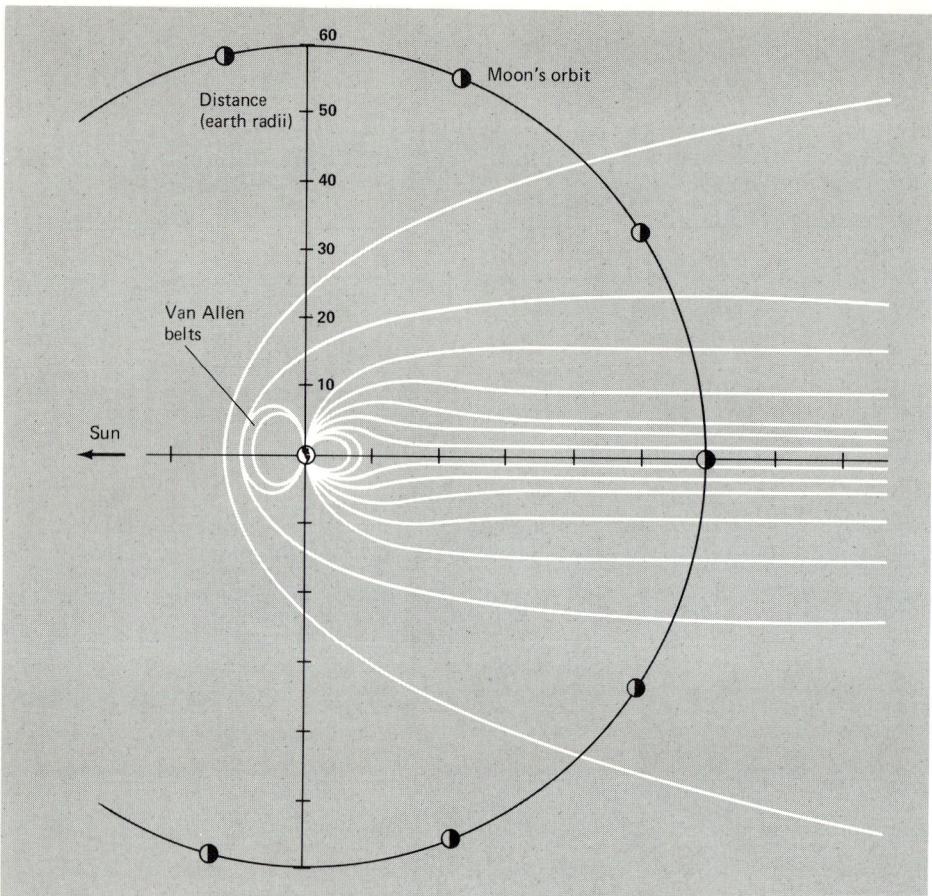

FIGURE 14.5 The earth's magnetosphere, created by the interaction of the earth's magnetic field with the flow of charged particles in the solar wind. [*Adapted from a NASA diagram.*]

magnetic field affects the flow of interplanetary charged particles for many tens of earth radii out into space. This region is called the earth's *magnetosphere* (Fig. 14.5). As the particles from the sun run into the earth's field, they cannot flow across the field lines, but are forced along them (Focus 6.2). Like the blunt prow of a boat in water, the earth's field deflects particles around it and leaves a wake in the direction opposite the sun. Most charged particles flow around the earth, but a few are caught in the field to make the Van Allen belts.

14.5 THE BLANKET OF THE ATMOSPHERE

The atmosphere provides oxygen for breathing, shields out the harmful radiation of the sun, and furnishes a thermal blanket to keep the surface warm. An understanding of the atmospheric composition and structure on earth provides information useful for the study of other planetary atmospheres.

14.5 THE BLANKET OF THE ATMOSPHERE

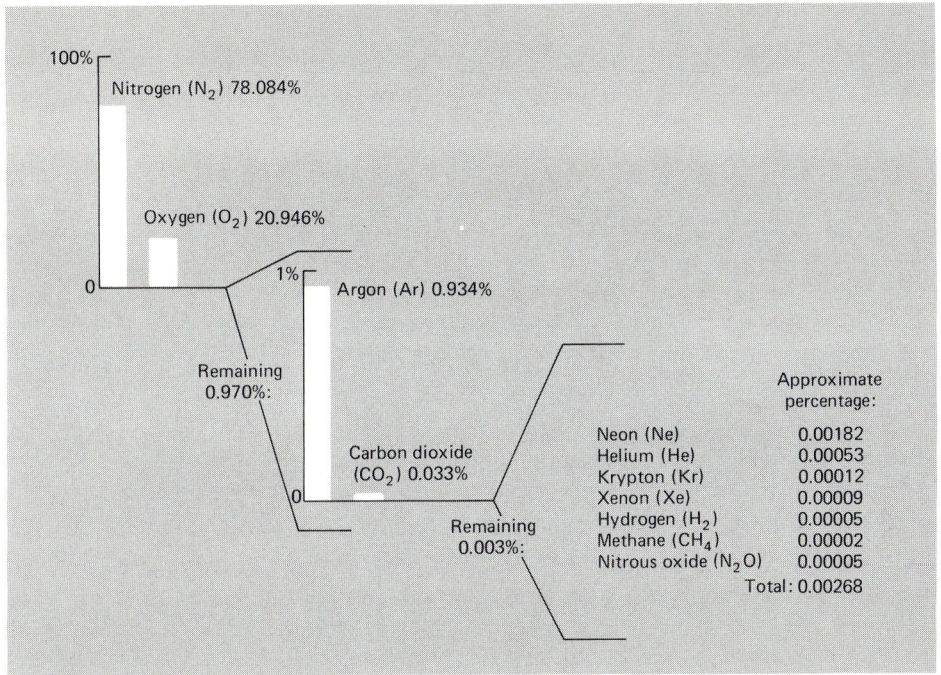

FIGURE 14.6 The composition of the earth's lower atmosphere. Not shown are gases whose percentage may vary considerably, such as water vapor. [*Adapted from A. N. Strahler, The Earth Sciences, 2nd ed., Harper & Row, New York, 1971.*]

Our atmosphere contains, relative to the total number of atoms and molecules available, approximately 78 percent nitrogen (N_2), 21 percent oxygen (O_2), 0.9 percent argon (Ar), 0.03 percent carbon dioxide (CO_2), and traces of other elements (Fig. 14.6). It also contains several constituents that are present in variable amounts: water vapor (H_2O), which sometimes may be as much as 3 percent near the surface, carbon monoxide (CO), sulfur dioxide (SO_2), and nitrogen dioxide (NO_2).

The weight of the upper atmospheric layers makes the lower portion denser than the upper, just as a sandwich at the bottom of a pile is squashed by the weight of those above it. The gas pressure at sea level on the earth's surface, where the entire atmosphere is piled above it, is called *1 atmosphere* (atm) of pressure. The atmospheric pressure and density decrease with height, rapidly at first, then more slowly (Fig. 14.7).

Although the atmosphere is a continuous fluid, with no definite boundaries, it is often convenient to discuss it in terms of various regions, or "spheres." The division between the regions is somewhat arbitrary and can be chosen in different ways, depending on the physical or chemical properties of interest. If we consider the way temperature changes with height in the atmosphere, we get four basic divisions: troposphere, stratosphere, mesosphere, and thermosphere (Fig. 14.7). In the *troposphere* the temperature decreases upward from the solar-heated surface. In the *stratosphere* the temperature stops decreasing and even rises somewhat, mainly due to heating by absorption of ultraviolet light in an ozone layer. The temperature falls again in the *mesosphere*, due to a decrease in heating from ultraviolet absorption and an increasing effectiveness of radiative cooling by carbon

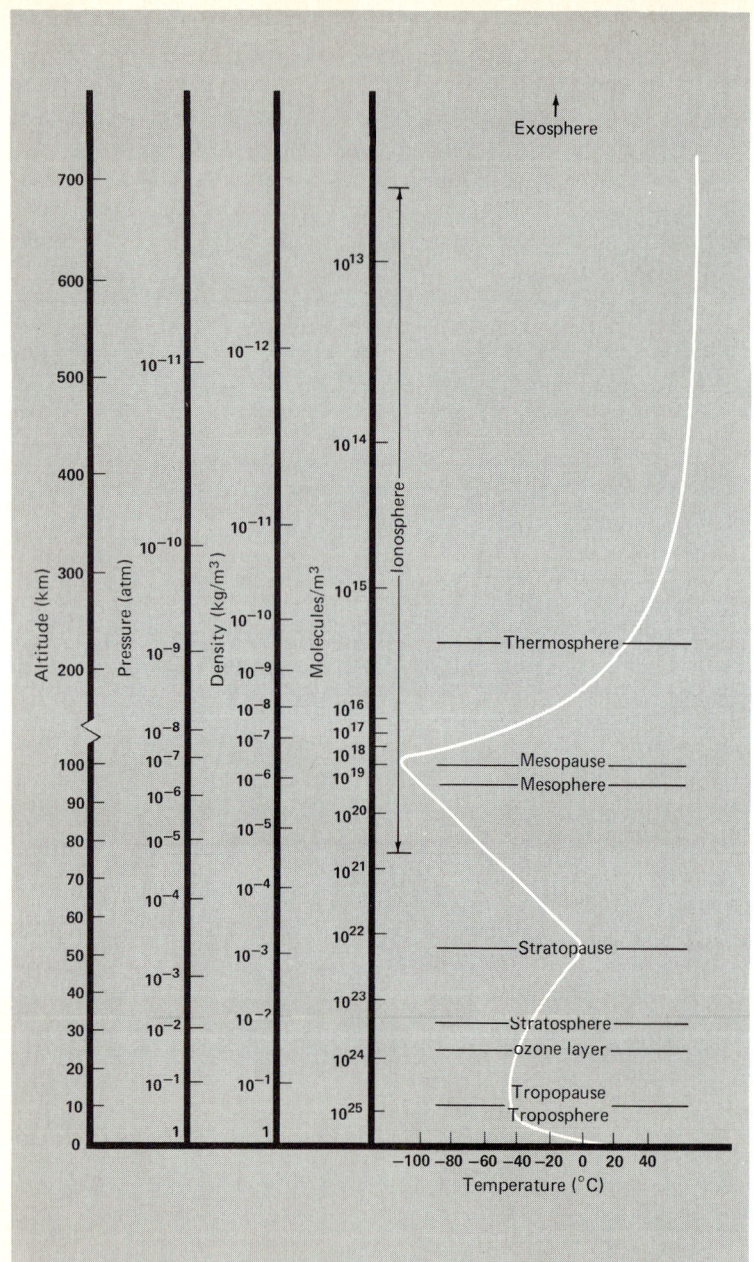

FIGURE 14.7 A cross section of the earth's atmosphere, showing the change of pressure, density, and temperature with altitude.

dioxide. Finally, heating by absorption of X-ray and far-ultraviolet radiation raises the temperature again in the *thermosphere*.

Troposphere

Meteorologists are most interested in the troposphere, the lowest layer, which extends to about 8 to 10 km above the surface. Practically all weather originates there, and few clouds form above it. The troposphere, particularly the boundary region near the top, also causes most of the twinkling of starlight, much to the astronomers' regret.

Stratosphere

The stratosphere extends to about 50 km above the troposphere (Fig. 14.7). It is so named because air movement in this region takes place in horizontal layers (or strata), in contrast to the vertical flow in the troposphere.

Within the stratosphere is an *ozone layer*, sometimes called the *ozonosphere*, at a height of about 25 km. It plays a critical role for the existence of life on the earth. Ozone is a combination of three oxygen atoms to form a molecule O_3. To create ozone from normal molecular oxygen requires energy. Ultraviolet light from the sun provides the energy. Normal molecular oxygen absorbs this radiation and dissociates into two oxygen atoms. An oxygen atom (O) can combine with an oxygen molecule (O_2), in the presence of another molecule or atom, to make ozone (O_3). The ozone is destroyed both by dissociation caused by more ultraviolet photons and by combination with atomic oxygen. In summary, the set of reactions governing the ozone abundance is

$$O_2 + \text{UV-photon} \rightarrow O + O$$
$$O + O_2 + M \rightarrow O_3 + M$$
$$O + O_3 \rightarrow O_2 + O_2$$
$$O_3 + \text{UV-photon} \rightarrow O + O_2$$

The balance of these reactions results in a peak concentration of ozone at an altitude of about 25 km.

The ozone layer of the stratosphere blocks out ultraviolet radiation. Life forms on earth have developed in an environment sheltered from ultraviolet light. We humans are so susceptible to this fairly lethal radiation that a mild exposure may result in a painful case of sunburn, and a large dose can kill. Even relatively mild doses over a long time promote skin cancer in fair-skinned people. For example, the per capita skin cancer rate among Anglo residents of New Mexico, which has clear skies and a high elevation, is one of the highest in the world.

Ionosphere

A region within the mesosphere and thermosphere, from about 75 to 700 km above the earth's surface, is called the *ionosphere* (Fig. 14.7) because it has large, variable amounts of ionized oxygen and nitrogen. Neutral oxygen and nitrogen absorb X rays from the sun and consequently become ionized. The presence of ions and electrons in the ionosphere creates layers that reflect shortwave radio waves. If the ionosphere did not exist, surface radio communication over long distances would be impossible because the radio waves would fly straight into space and never reach areas below the horizon.

The ionosphere has several sublayers with peak concentrations of ions at 90 km, 110 km, 200 km, and 300 km. Different physical processes govern the production and destruction of ions in these regions, some of which depend sensitively on the presence of sunlight. So some of the layers disappear after sunset, which accounts for the differences in shortwave radio propagation between day and night.

Exosphere

Because any free gas lacks a distinct boundary, the atmosphere has no sudden end. It gradually peters out to the interplanetary medium. However, as the atmosphere becomes thinner, collisions between atoms become less likely, and the distance between collisions increases. At a low enough density the chances for an atom to escape into space, because detouring collisions are few, become fairly large (as long as it is traveling faster than the escape velocity; Section 5.5). The region within the thermosphere above about 500 km, where atoms can escape, is called the *exosphere*. This layer effectively defines the top of the atmosphere.

Regions Based on Composition

A further division of the atmosphere can be made on the basis of composition. Below about 120 km, the atmosphere is well mixed and (except for water vapor) homogeneous in composition. This lower region is called the *homosphere*. The region above, the *heterosphere*, is relatively enriched in the lighter elements, such as helium and hydrogen. Rocket and spacecraft observations (Fig. 14.8) have shown that the earth has a *hydrogen corona* extending out to at least 10 to 15 earth radii (60,000 to 90,000 km).

Atmospheric Escape

How rapidly gasses leave a planet's atmosphere is important for its evolution. Let's look in detail at the atmospheric escape process.

To escape, an atmospheric particle must be moving outward with at least escape velocity (Section 5.5). It must not be knocked from its escape path by collisions, so it can escape only from the region where the atmosphere is thin (the exosphere). For an atmosphere at

FIGURE 14.8 An ultraviolet photo of a part of the hydrogen corona surrounding the earth. [*Courtesy NASA.*]

14.5 THE BLANKET OF THE ATMOSPHERE

a given temperature, low-mass particles move faster than higher-mass ones. So light elements like hydrogen and helium are more likely to escape than oxygen and argon.

Let's use the moon as an example. Its mass is 7.4×10^{22} kg, and its radius is about 1700 km (1.7×10^6 m). Its escape velocity then is

$$V_e = \left(\frac{2GM}{R}\right)^{1/2}$$

$$= \left(\frac{2 \times (6.7 \times 10^{-11}) \times (7.4 \times 10^{22})}{1.7 \times 10^6}\right)^{1/2}$$

$$= 2.4 \times 10^3 \text{ m/sec}$$

$$= 2.4 \text{ km/sec}$$

In a gas the temperature is a measure of the random thermal velocity of the particles. The specific relation is

$$\frac{mV_{th}^2}{2} = kT$$

where m is a particle's mass, V_{th} its most probable velocity, k Boltzmann's constant (1.4×10^{-23} J/deg), and T the temperature in Kelvins. Note that

$$V_{th}^2 = \frac{2kT}{m}$$

so the more massive particles in a gas move more slowly, on the average, than less massive ones.

During the day on the moon the temperature hits about 400 K. For hydrogen, $m = 1.7 \times 10^{-27}$ kg, so

$$V_{th} = \left(\frac{2kT}{m}\right)^{1/2}$$

$$= \left(\frac{2 \times (1.4 \times 10^{-23}) \times 400}{1.7 \times 10^{-27}}\right)^{1/2}$$

$$= 2.6 \times 10^3 \text{ m/sec}$$

$$= 2.6 \text{ km/sec}$$

So hydrogen atoms readily escape.

The temperature for particles of mass m to escape is computed by equating the most probable velocity to the escape velocity:

$$V_{th}^2 = V_e^2$$

$$\frac{2kT}{m} = \frac{2GM}{R}$$

$$T = \frac{GMm}{kR}$$

So the escape temperature goes up directly proportional to the mass of the particle.

Now for a complication. Particles in a gas at a given temperature do not all travel at the same speed. Most move near the most probable speed; some move faster, some slower. For atmospheric escape the faster particles play an important role, for even if the temperature is below that for the most probable velocity to equal the escape velocity, *some* particles will have greater than escape velocity and leave the atmosphere. But if the temperature is much below the escape temperature, it will take a very long time for the atmosphere to disappear.

How can the lifetime of different constituents in a planet's atmosphere be calculated? Basically, the lifetime depends on the temperature of the exosphere, the local acceleration due to gravity, the escape velocity, and the constituent's mass. The relationship is not simple. To get an approximate idea of the lifetime, you can use the following procedure. Calculate

$$r = \frac{V_e}{V_{th}} = \left(\frac{GM}{R} \times \frac{m}{kT}\right)^{1/2}$$

If $r = 1$, the lifetime is a few hours; $r = 2$, a few days; $r = 3$, a few years; $r = 4$, several thousand years; $r = 5$, about a hundred million years; and $r = 6$ or greater, essentially infinite.

In the case of hydrogen on the moon, $r = 0.9$, so the hydrogen's lifetime is only a few hours. Compare that to oxygen. The only difference is that oxygen's mass is 16 times that of hydrogen; the ratio r is 4 times larger, 3.6. The oxygen lingers for a hundred thousand years.

On the earth $V_e = 11.2$ km/sec (Section 5.5). The thermal velocity for hydrogen in the exosphere ($T = 1500$ K) is

$$V_{th} = \left(\frac{2 \times (1.4 \times 10^{-23}) \times 1500}{1.7 \times 10^{-23}}\right)^{1/2}$$

$$= 5.0 \text{ km/sec}$$

So $r = 2.2$ and hydrogen must escape in a few weeks. For oxygen $r = 8.8$, and so it does not escape.

14.6 THE ATMOSPHERE AND INCOMING RADIATION

The earth's atmosphere endlessly frustrates the ground-based astronomer. He or she must endure all the problems of bad weather that, in contrast, delight the meteorologist. Also, because of the ozone layer, the astronomer cannot view the ultraviolet radiation from celestial bodies, and the ionosphere filters out X rays and some radio waves.

Even on the clearest nights atmospheric turbulence causes telescopic images to flicker and so imposes the fundamental limitation on a telescope's effective resolving power (Section 7.2). The extent to which the atmospheric turbulence affects an object's image is termed *seeing*. When the seeing is very good, stellar images are sharp, steady pinpoints, about 1″ in diameter. At times of bad seeing, the images waver like candle flames in a gentle breeze. We have seen star images blow up like balloons to sizes of 10″ to 20″!

Extinction and Reddening

The atmospheric layers also absorb and scatter some of the light that penetrates them. This reduction of light is called *atmospheric extinction*. The closer an object appears to the horizon, the greater the atmospheric thickness through which the object's light must pass, and so the dimmer it becomes. Because of atmospheric extinction, the rising full moon has about half the brightness of the same moon overhead.

Why is the sky blue? Air molecules preferentially scatter blue light more than red. The atmosphere depletes a beam of light of its shorter (bluer) wavelengths, which are scattered uniformly through the sky. In any direction you look you see blue light, and so the entire sky is blue (Fig. 14.9). Light of longer wavelengths reaches you directly along the line of sight. The sinking sun appears a burning red because its light passes through a lot of atmosphere before reaching you. Along this path most of the blue light is scattered out, leaving mostly red light. This process is called *atmospheric reddening*.

Albedo

All the planets and their moons shine by reflected sunlight, and so does the earth when viewed from space. A celestial body's reflecting ability is called its *albedo*, the ratio of light reflected to the incoming light. If an object reflected all the light that struck its

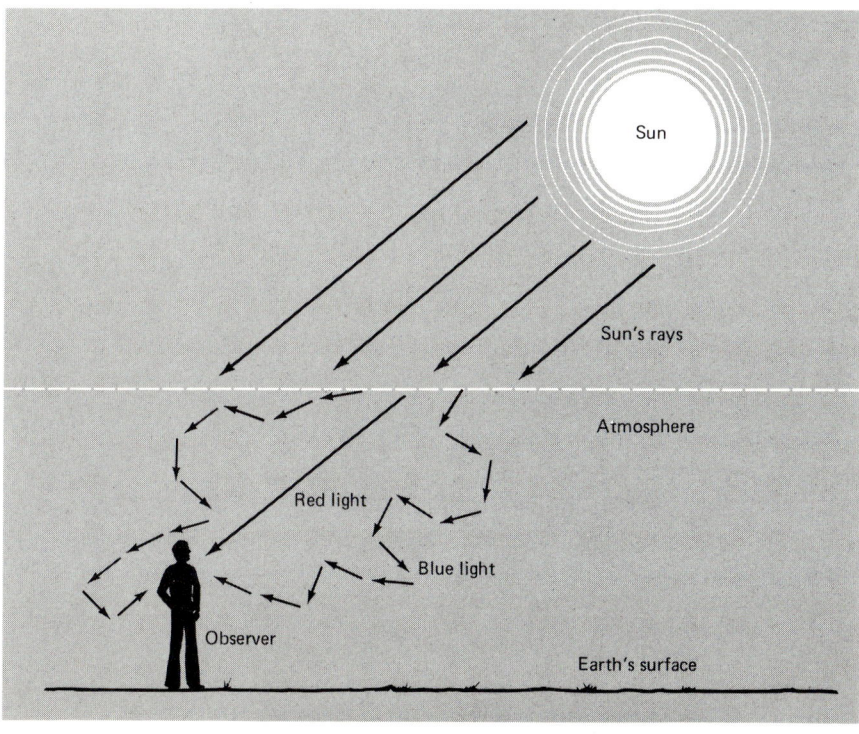

FIGURE 14.9 The blue sky. Air molecules let red light pass through relatively unhindered, but blue light is scattered in all directions. Looking toward the sun, you see red light directly. Since blue light scatters all around the air, in every direction you see blue light—and so a blue sky.

surface, its albedo would be 1.0. The clouds in the earth's atmosphere help to reflect visible light (see Fig. 14.1), and about 35 percent of the incident light reflects back into space; the earth's albedo equals 0.35. The atmosphere and surface absorb the other 65 percent.

The Greenhouse Effect

Of the incoming sunlight not reflected, 15 percent is absorbed in the troposphere, and the remaining 50 percent of the original strikes the ground and heats it and the air in contact with the surface. However, if this direct solar radiation were the only source of heat, the temperature at the ground would be a frigid 253 K (or −20°C), the temperature of a blackbody placed at the earth's distance from the sun. Water would always be frozen! The average temperature at the surface is actually much higher, about 293 K (+20°C).

How does this heating happen? Fortunately, the earth does *not* radiate like a blackbody. Visible light from the sun gets through to the surface and heats the earth. In turn the earth emits infrared radiation. *If* this infrared radiation simply escaped into space, the earth would be too cold for life. But this radiation *doesn't* escape completely. The infrared is absorbed by the earth's atmosphere (mostly by water vapor and carbon dioxide). The atmosphere heats up by absorbing this radiation. Some goes off into space (about 8 percent); the rest radiates back to the ground and so heats it. Both direct sunlight and infrared radiation from the atmosphere heat the earth's surface. The atmosphere acts like a blanket, insulating the ground from space and so helping to warm the earth.

If you have ever visited a high region with an arid climate, such as New Mexico, you know what a dramatic effect water vapor in the atmosphere has on the ground temperature. Water vapor absorbs infrared radiation, so the more humid the atmosphere, the more opaque it is to infrared and the better it insulates. For instance, in Albuquerque on a clear winter's day the high temperature can typically reach 15°C and, if the night is also clear, drop to −7°C at night. But if it's cloudy at night, the low may be only 0°C or so. Similarly, even in a low-lying region like Chicago, in the summer the temperature will go down to 20°C at night if it is clear. But if it's cloudy, it won't go below 30°C! Why the difference? At night the ground radiates in the infrared the energy it absorbed during the day. On a cloudy, high-humidity night the additional water vapor in the air traps the outgoing infrared radiation from the ground more than on a clear night. So the air temperature stays higher because not as much heat escapes to space.

This warming of the ground by the atmospheric trapping of infrared radiation is often called the *greenhouse effect* by analogy to one process that keeps a greenhouse warm. Glass is transparent to visible light, but opaque to infrared. So sunlight enters the greenhouse and warms the interior, which emits infrared. This heat can't radiate through the glass, so it stays to help warm up the interior.

To sum up, any planetary atmosphere that is more or less transparent to sunlight, but opaque to infrared, will act to keep the planet's surface warmer than if the planet had no atmosphere.

Note: The greenhouse effect is probably misnamed. Experiments have shown that the absorption of infrared radiation is not the main thing which keeps the greenhouse warm. When the glass is replaced by a rock-salt window (which transmits infrared), the greenhouse gets almost as hot. The air, heated by contact with the hot inside of the greenhouse, cannot escape. This inhibition of convective cooling occurs with *any* roof. However, the phrase "greenhouse effect" is so ingrained in the astronomers' vocabulary, we'll continue to use it in this book.

14.7 THE RESTLESS EARTH: EVOLUTION OF THE CRUST

When first studied, the unfolding of the earth's history seems rather uneventful. A fall of rain, a burst of wind, the transport of grains in a stream—all these erosive processes serve to flatten the earth's surface relentlessly and to wash materials into the sea. Gently, but inevitably, the mountains are leveled to plains and the ocean basins filled with sediments. But mountains exist and the continents and the ocean basins differ in average height by about 5 km. Given the great age of the earth (Section 14.3), the fact that these levels still stand apart implies that somehow the mountain heights and ocean depths are regularly replenished.

Earthquakes and Mountains

In the nineteenth century geologists noted that the zones of active volcanoes and frequent earthquakes are concentrated along the chains of young mountain ranges and submarine ridges. They argued that earthquakes and volcanic activity must be associated with mountain and island building. Modern research on the locations of earthquakes backs up this idea.

Another clue to the understanding of this activity shows up in the ocean basins, especially in the Atlantic. Modern sonar measurements have revealed a *mid-oceanic ridge*, an almost continuous submarine mountain chain that extends some 64,000 km through the ocean basins (Fig. 14.10). The mid-oceanic ridge indicates that important geologic processes take place in the ocean basins.

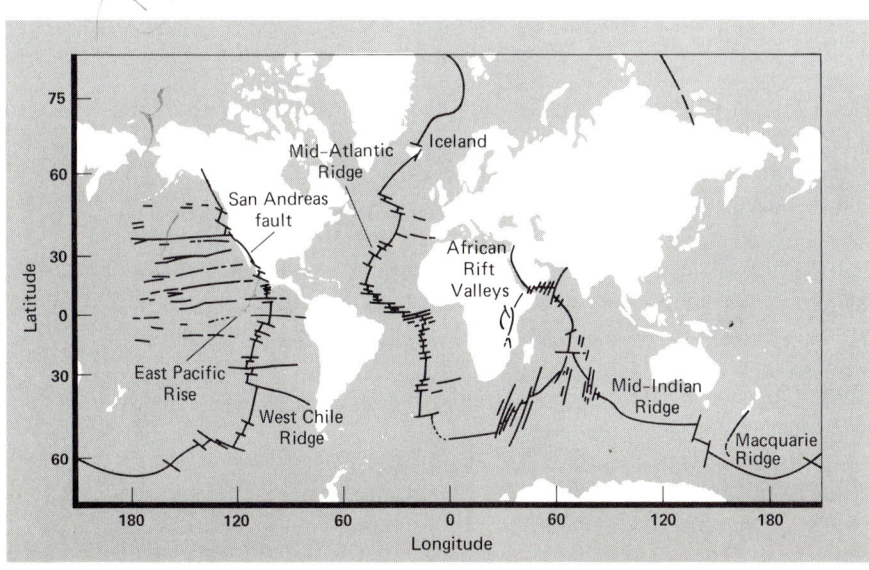

FIGURE 14.10 The system of mid-oceanic ridges (heavy lines). [*Adapted from A. N. Strahler*, The Earth Sciences, *2nd ed.*, Harper & Row, *New York, 1971.*]

FIGURE 14.11 Continental drift. The fitting of continental masses along the continental shelves, which are under water.

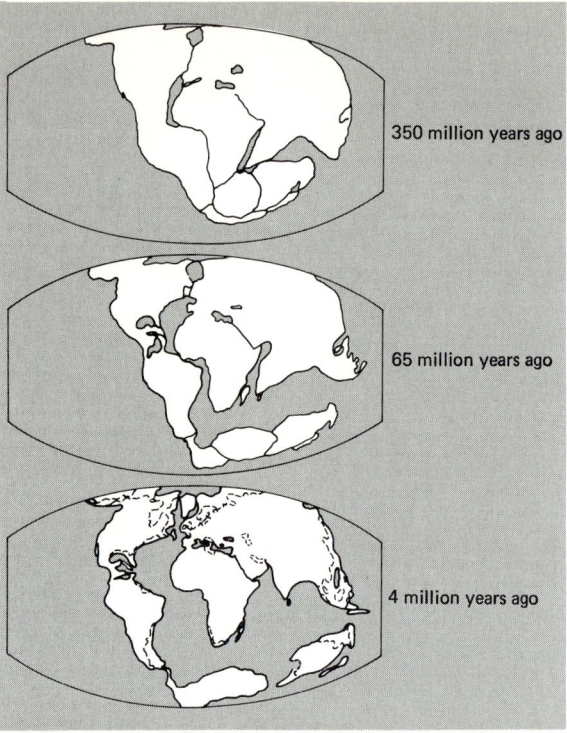

FIGURE 14.12 Wegener's original model for their separation from one large landmass. Present evidence suggests that the continents broke from two supercontinents: Laurasia (North America, Greenland, and Eurasia) and Gondwanaland (South America, Africa, India, Australia, and Antarctica).

Continental Drift

These facts fell together as a coherent picture during the 1960s with the revival of the model of *continental drift*, the idea that the present continents were at one time a unified landmass that fragmented and drifted apart. In 1910 Alfred L. Wegener (1880–1930) of Germany suggested that displacements of the earth's crust could shift the position of the continents. Wegener pointed to a number of remarkable geologic connections, such as similar rock formations and fossils, between the lands on opposite sides of the Atlantic (Fig. 14.11). In Wegener's picture the continents were originally joined in one vast land area, which broke up about 200 million years ago (Fig. 14.12). Today's evidence points to two primordial landmasses: one called Gondwanaland, in the Southern Hemisphere, the other called Laurasia, in the Northern Hemisphere. These may have broken from a single landmass called Pangaea.

Evidence from the magnetic characteristics of the ocean floors near ridges argues for the continental drift model. If the continents do move apart, the seafloor between them must be spreading. Oceanographic cruises across the Atlantic have found that the seafloor material contains remnants of ancient magnetism. When lava solidifies to form igneous

14.7 THE RESTLESS EARTH: EVOLUTION OF THE CRUST

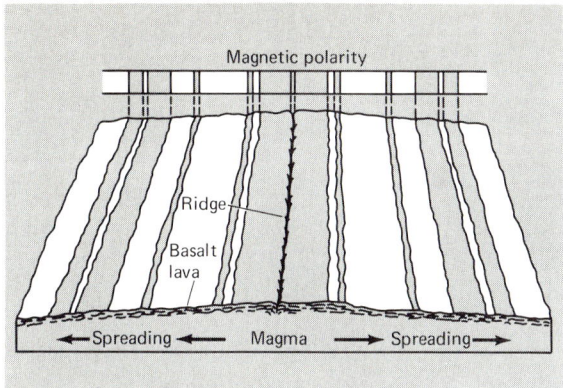

FIGURE 14.13 Magnetic evidence for seafloor spreading. As magma wells up along oceanic ridges, it cools and retains the direction of the magnetic field at the time of cooling. Rock samples from the ocean floors show the reversals of the magnetic field. The pattern of successive reversals (hatched areas) on both sides of the ridge for the past 10 million years is symmetrical. [Adapted from A. N. Strahler, The Earth Sciences, 2nd ed., Harper & Row, New York, 1971.]

rock, the iron minerals in the rock align with the earth's magnetic field. The directions and reversals of the direction of the earth's magnetic field in the past are preserved in the rock. They have a startling pattern. On both sides of the mid-Atlantic ridge the reversal patterns appear identical; one side is a mirror reflection of the other (Fig. 14.13). If the seafloor spreads, it needs a continuous supply of new material to add additional area. Lava flowing out pushes older material aside in both directions. When the lava solidifies, the rock on either side freezes in the magnetic field alignment of the time.

The alignment of magnetic field reversals not only indicates that new material emerges from a rift in the center of the ridge, but also gives the rate of expansion of the seafloor. The movement is about 2 to 4 cm/yr at its fastest speed across the mid-Atlantic ridge. The rate (if constant) amounts to more than 16,000 km in 400 million years, enough to push apart the Old and the New World.

Plate Tectonics

The new material oozing out from the earth's interior appears to account for the renewal of the ocean plains. The separate continental plates float like large rafts on the basaltic basin material, which forms another plate. Where one continental plate crashes into another, the impact raises up mountains. In some regions one plate may force another to fold under and descend into the mantle (Fig. 14.14). The plate's descent eliminates surface material essentially at the same rate it is created, so the earth's radius does not expand to accommodate the swelling plates. Because the plates' creation and

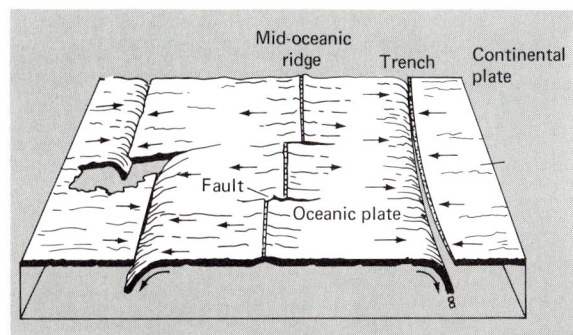

FIGURE 14.14 Interactions of oceanic and continental plates. The oceanic plates gain material from the outflow at oceanic ridges. As these plates expand, they crash into continental plates. Here mountain building and earthquakes occur. [Adapted from A. N. Strahler, The Earth Sciences, 2nd ed., Harper & Row, New York, 1971.]

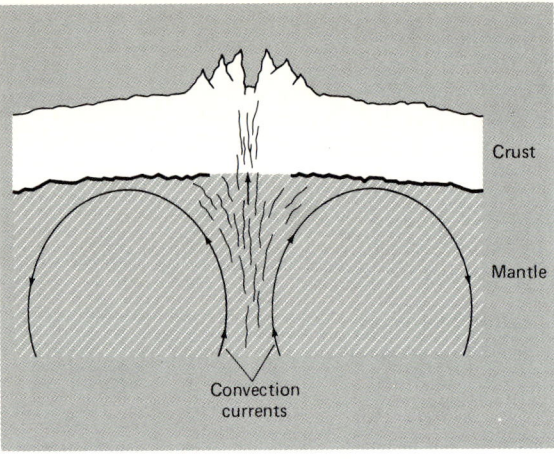

FIGURE 14.15 A convection model for the moving of crustal plates. Large convection currents in the plastic mantle carry along the plates that float on it.

destruction zones make natural fault areas, earthquakes and volcanoes occur mainly along the lines of plate collision. This contemporary model of the earth's crustal activity and evolution is called *plate tectonics*, the dynamics of the crust's continental and oceanic plates.

What moves the plates is not completely clear. One popular model pictures the upper part of the mantle as divided into large convection cells (Fig. 14.15). The mantle's plasticity allows a slow flow upward, horizontally, and downward. At the region of horizontal flow friction between the plate and the mantle drags the plate along with the mantle's flow. The upswelling magma supplies new materials to the plate. The energy source for such convection is still unknown. It may come from heating by radioactive decay or from internal currents persisting from the time when the earth's core formed.

Although many details are uncertain, the main point is clear: The earth's crust has evolved since its formation and is changing right now. When we investigate other terrestrial planets, we will look for evidence of their crustal evolution. Note that the earth's ocean basins, because they have been recently formed, are the *youngest* parts of the earth's crust.

14.8 EVOLUTION OF THE ATMOSPHERE AND OCEANS

No other planet in the solar system has the earth's combination of an extensive atmosphere plus oceans of water. The atmosphere and oceans have changed throughout geologic time, influenced by and influencing the earth's biological evolution. The future of life on the earth depends critically on the future of its fluid system.

Origin and Development of the Oceans

By the contemporary model of the formation of the planets (Chapter 13), the earth was probably not covered by oceans when it formed 4.6 billion years ago. The primeval surface may have been fairly hot, about a few thousand degrees Kelvin, if the earth formed

14.8 EVOLUTION OF THE ATMOSPHERE AND OCEANS

by the accretion of small masses. As these planetesimals banged together, their kinetic energy transformed, in part, into heat that melted rock. It was certainly too hot for water to be liquid!

When the surface had cooled to about 373 K (100°C), water could condense. At that time maybe one or two continents existed on the surface, and the rest of the surface comprised the initial ocean basin. This large tub contained very little water then, only a few percent of the present volume.

The rest of the oceanic water came from the earth's interior. When magma breaks through the crust, it carries a variety of gasses, such as carbon dioxide, and also a large amount of water vapor. The steam arises from water trapped in the solid earth when the planet formed. The present rate of water production gassing out from the interior, about 10^{11} kg/yr, if it has been constant for 4.5 billion years, accounts for the amount of water in the oceans today. Slowly that volume increases. In fact, in the past when the earth was hotter, we expect the outgassing to have been much faster.

When the new water steams up from the crust, it does not contain the salts that make up 3.5 percent of seawater. (The other 96.5 percent is pure water.) The solids that produce the saltiness come from two sources: (1) substances in the gasses from volcanoes, such as chlorine and sulfur, dissolved in rain and (2) minerals from the land, carried by runoff waters to the sea. Because both new salts and new water are added to the oceans, the saltiness remains fairly constant.

The earth's oceans now cover 71 percent of its surface with an average depth of 4 km. Together they contain some 10^{21} kg of material, mostly water. These figures have been roughly the same in the past billion years, but the configuration of the oceanic waters must have been very different in the past because of plate tectonics.

The Evolution of the Atmosphere

The outgassing from the earth's interior that created the oceans also influenced the development of the earth's atmosphere. It evolved from the chemical interplay of the solid and fluid earth (Fig. 14.16). It's important to understand that the earth's atmosphere does

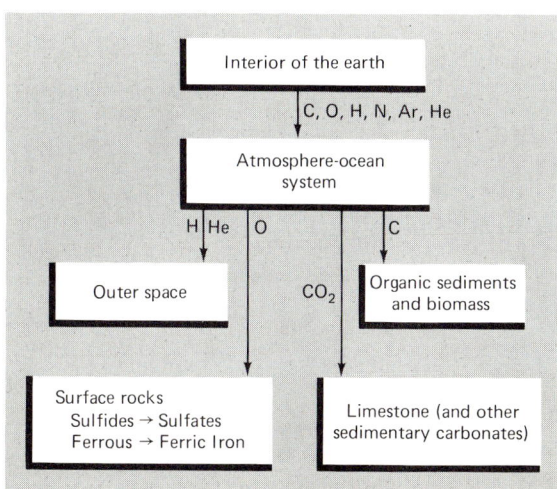

FIGURE 14.16 A schematic diagram of the interaction of the earth's interior, oceans, and atmosphere as affected by life.

not exist in isolation from the other parts of the earth with which it has contact. Let's look at its interaction with the oceans, using carbon dioxide (CO_2) as an illustrative example.

The carbon dioxide now in the atmosphere is removed by plant photosynthesis and returned by animal respiration, organic decay, and the burning of fuels. On the average, a carbon dioxide molecule spends 6 years in the atmosphere before it returns to the biomass—the surface reservoir of organic (living and dead) material. More carbon dioxide resides in the top, well-mixed 70 m of the oceans, dissolved in the form of bicarbonate ions (HCO_3^-), and even more is stored in the deep oceans. The bicarbonate ions flow into the oceans from rivers; they come from weathering of surface rocks. In the oceans organisms convert the bicarbonate into carbonate for shells. Part of this ends up as sediment on the ocean bottoms. Sedimentary rocks formed in oceans become exposed on continents in later eras.

The point of the above is not to tell you all about the earth's carbon dioxide budget (it doesn't!), but to show you the complicated interactions of atmosphere, oceans, and biomass on the earth now. These interactions must be understood in order to work out the evolution of the atmosphere. We do not yet clearly understand all the details, so our current models of the evolution of the earth's atmosphere are not the final one.

With this word of warning, here is one reasonable model of atmospheric evolution. If our ideas of planetary formation are correct, the earth's first atmosphere did not resemble the present one at all. It was an *accretion atmosphere* that may have contained hydrogen and helium. But these gasses, because they are so light, quickly escaped from the earth into space (Section 14.5). Our second atmosphere arose mostly from outgassing from the solid earth. Active volcanoes, for instance, spew out gases that were trapped in the earth when it formed—carbon dioxide, sulfur dioxide, hydrogen, nitrogen, water, methane, and ammonia. In addition, some gasses, such as helium and argon, come from the decay of radioactive materials. The outgassed materials (which are still entering the atmosphere now) interact with the oceans, surface materials, and biomass in complex ways. For example, carbon dioxide is now added to the atmosphere by volcanoes, organic decay, and combustion of fossil fuels. Carbon dioxide is taken out by plants and is being dissolved in the oceans, where much of it eventually ends up in rocks. The balance, however, has changed with time, so the atmospheric composition has evolved.

Michael H. Hart has made a computer simulation of how this evolution might have gone (Fig. 14.17). His calculations indicate that the earth's second atmosphere started out with a large amount of carbon dioxide. (The water vapor had quickly rained down to end up in oceans.) The carbon dioxide soon ended up in the oceans and rocks, so 3 billion years ago the atmosphere consisted mostly of methane (CH_4) and other hydrogen-carbon compounds.

At this time the atmosphere contained little free oxygen, so earth had no ozone layer. Ultraviolet light readily penetrated and broke up methane, ammonia, and water. The hydrogen from these molecules sped into space. Some of the oxygen freed from the water combined with some of the methane and gradually eliminated the carbon. The rest of the oxygen eventually created an ozone layer, cutting out ultraviolet light from interacting with most of the atmosphere. Nitrogen became the dominant constituent of the atmosphere.

The high abundance of atmospheric oxygen was produced (and is now maintained) by biological activity. Geologic evidence indicates that the transformation to an oxygen-rich atmosphere began roughly 2 billion years ago, when plant activity and photosynthesis

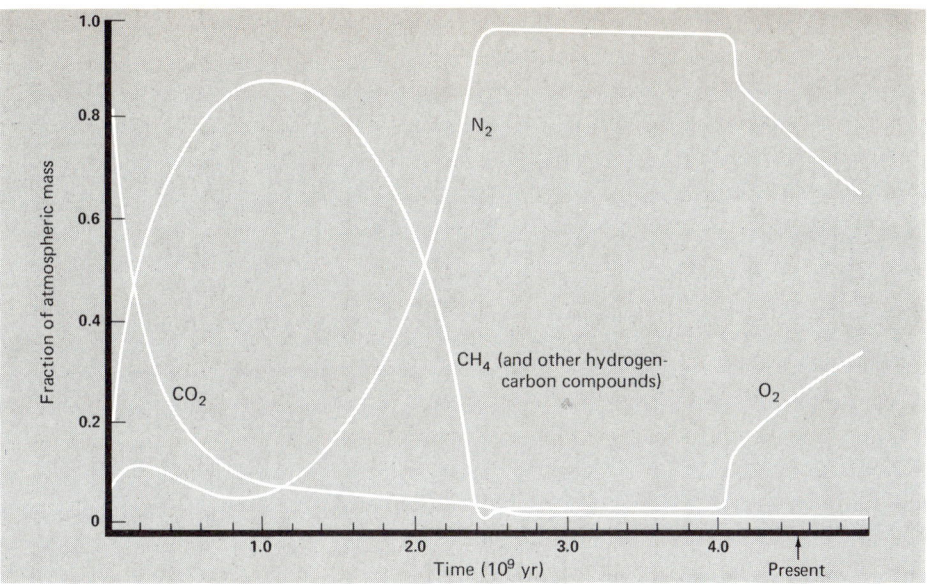

FIGURE 14.17 A model for the evolution of the earth's atmosphere. This graph shows, for a span of 5 billion years, the fraction of the earth's atmosphere in the form of carbon dioxide, molecular nitrogen, molecular oxygen, and methane. [*Based on theoretical calculations by M. Hart.*]

bloomed (Chapter 26). The increase in oxygen was probably a gradual, continuous one. About 1 billion years ago, the atmosphere may have contained only 10 percent of the present amount of free oxygen. A large increase occurred about 600 million years ago. The oxygen content suddenly increased to present levels, along with a sudden proliferation of large forms of life.

The Evolution of the Earth's Surface Temperature

How hot it gets at the earth's surface depends on how much energy it receives from the sun and how effectively the greenhouse effect operates. Less solar energy results in lower temperature. A better greenhouse effect (more carbon dioxide and water vapor in the atmosphere) delivers higher temperatures.

People have disturbed the natural carbon dioxide balance by extracting fossil fuels from the earth, burning them for energy, and so adding to the carbon dioxide in the atmosphere. In addition, our destruction of forests has eliminated a substantial part of the green plants that take in atmospheric carbon dioxide and has added some of their carbon to the atmosphere. The ocean can absorb only a part of the excess.

Our activities have a net result of increasing the percentage of carbon dioxide in the earth's atmosphere. This increase is big enough to have been observed; observations at Mauna Loa Observatory in Hawaii indicate an increase of 5 percent over the past 30 years (Fig. 14.18). That amounts to about 2×10^{12} kg of carbon added to the atmosphere in a year. Although there is still much controversy over what the impact of this increase will be, it may result in a temperature increase of about 2°C by A.D. 2020, if the overall water vapor and cloudiness do not change. This could have serious effects on climate and atmospheric circulation.

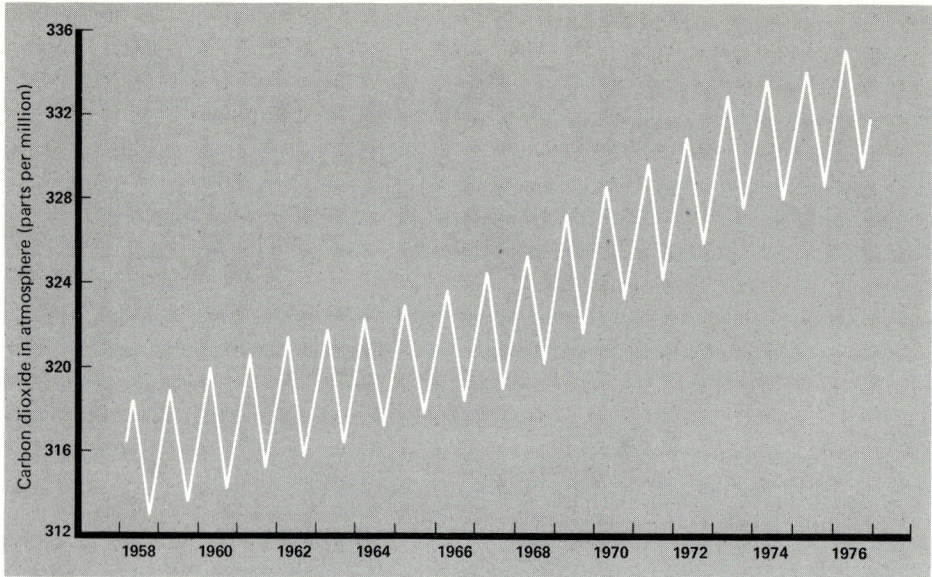

FIGURE 14.18 The measured carbon dioxide content of the lower atmosphere. The annual changes come from the removal of carbon dioxide by photosynthesis in spring and summer and its release in fall and winter. This variation occurs on top of a long-term trend of gradual increase. [*Based on data by C. D. Veeling, Scripps Institute of Oceanography.*]

The natural evolution of the sun's luminosity will in time also affect the global environment. The earth's climate may not be pleasant for our descendants. Models of stellar evolution (Chapter 10) predict that the sun's luminosity will slowly increase with time. As the sun grows brighter, the earth's surface temperature will increase. More water will evaporate and increase the atmospheric trapping of infrared radiation, which, in turn, will increase the temperature even more. About 4 billion years from now, according to some theoretical calculations, the increased temperature will completely evaporate the oceans and produce a hot, steamy atmosphere. If technology cannot reduce the steambath, the survivors, if any, will be forced to leave their superheated home. But 4 billion years is a long time, about as long as the whole past history of the earth. And remember that *Homo sapiens* has existed as a species for only a few *million* years. Interesting though it may be to speculate about it, we really needn't worry about such a distant catastrophe.

14.9 AN OVERVIEW OF THE EVOLUTION OF THE EARTH

The earth is the most evolved of the terrestrial planets. It's an active world now (powered by internal heat) and shows no sign of quitting. Geologic evidence implies it's been active since its birth 4.6 billion years ago. No other planet has been restless for so long! Let's step back and scan our planet's overall evolution (Fig. 14.19 and Table 14.2). It falls into four large-scale stages.

14.9 AN OVERVIEW OF THE EVOLUTION OF THE EARTH

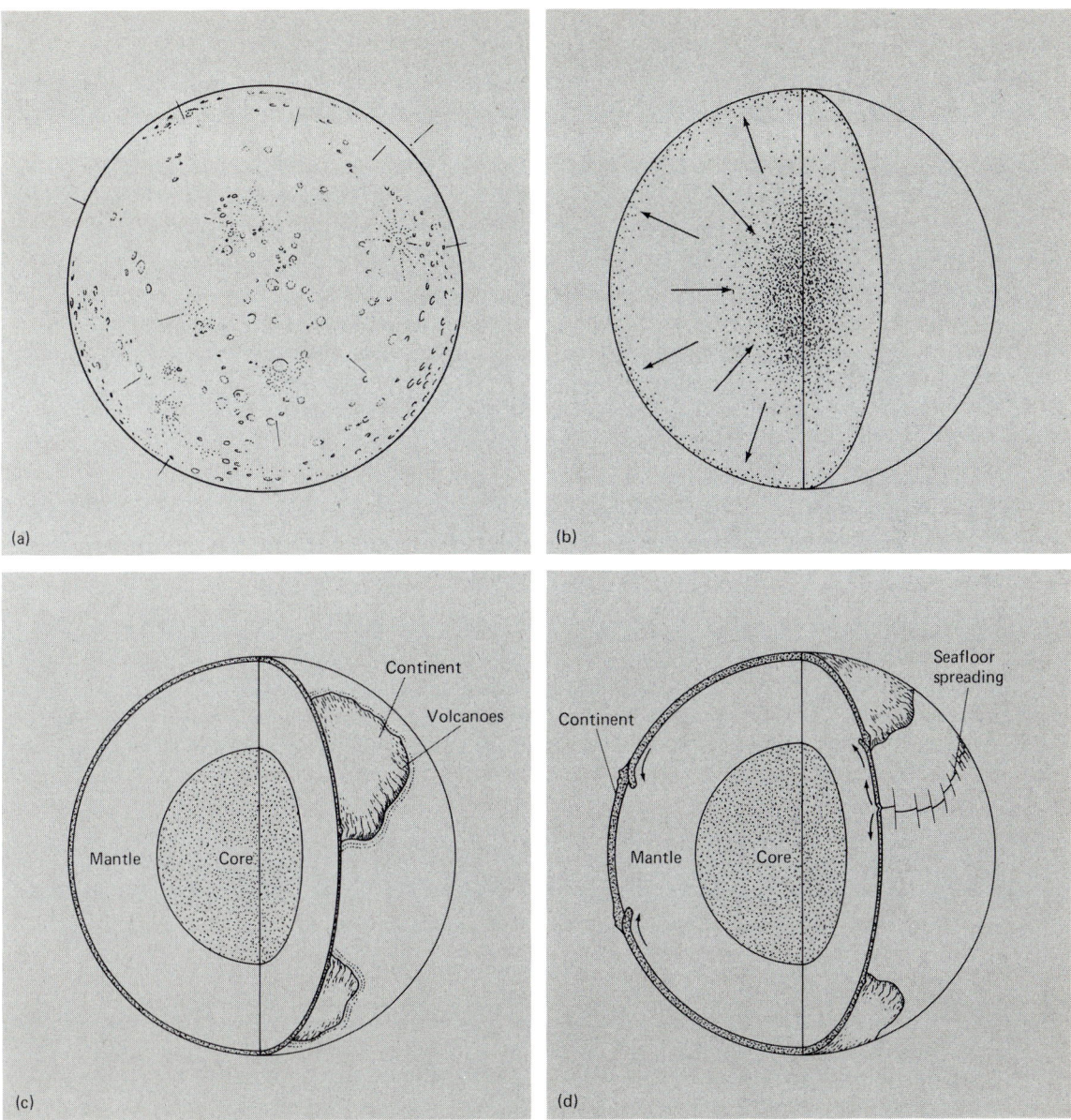

FIGURE 14.19 One model for the overall evolution of the earth. (a) The earth forms by the accretion of planetesimals. Late-arriving ones crater the surface. (b) The interior, heated by the accretion and radioactive decay, differentiates. (c) Water condenses, the first continents form, and intense volcanic activity occurs. (d) The crust thickens and crustal plates roam across the surface. [*Adapted from a diagram by R. Siever, Scientific American, September 1975, copyright 1975, W. H. Freeman and Co.*]

The first stage began 4.6 billion years ago. The accretion of the earth from planetesimals in the solar nebula probably took place quickly, in only a few million years. It left a pockmarked planet of more or less uniform composition, since each planetesimal was made of about the same combination of materials (Fig. 14.19a). The atmosphere at the time of accretion was rich in hydrogen and inert gasses. Accretion and radioactive decay produced rapid internal heating. The low mass gasses escaped into space.

In the second stage, beginning about 4.5 billion years ago, radioactive heating (and perhaps gravitational contraction) melted the interior. It differentiated to form a core of dense materials and a crust of light ones (Fig. 14.19b). Volcanic activity caused by interior heating created the second atmosphere, containing outgassed water, methane, ammonia, sulfur dioxide, and carbon dioxide. The infall of large objects continued, fracturing the crust. Ocean basins were formed and the earth's surface cooled enough for rain to fall to begin to fill the basins.

TABLE 14.2 A Possible Order for the Earth's Evolution

Stage	Events
I. 4.6 billion years ago	1. Formation of earth by accretion from the solar nebula 2. Rapid internal heating from accretion and radioactive decay 3. Blowing off of primeval atmosphere
II. 4.5 billion years ago	1. First differentiation of interior, formation of crust and core 2. Outgassing of carbon dioxide, water, methane, ammonia, etc. to make second atmosphere 3. Infall of large objects to fracture crust in places 4. Formation of ocean basins; begin filling with water
III. 3.7 billion years ago	1. First tectonic movements, seafloor spreading, formation of volcanoes and mountains 2. Cooling and thickening of crust 3. Ocean basins mostly filled
IV. 600 million years ago	1. Continuation of processes of stage III at slower rate 2. Enlargement of ocean basins 3. Formation and subsequent breakup of Pangea

Warning: The order, dates, and details in this table are speculative.

In the third stage, about 3.7 billion years ago, the first continents appeared, and plate tectonics began (Fig. 14.19c). Mountains grew, only to fall to the weathering of wind and rain. Slowly the atmosphere evolved. Roughly 2.2 billion years ago, crustal cooling thickened the crust enough to allow plate activity as we see it today (Fig. 14.19d).

By 600 million years ago the planet had entered the fourth stage of evolution. The processes that had begun in the third stage continued at a slower rate until the earth came to look much as it does today. And those processes continue to operate.

SUMMARY

This chapter has scanned the earth's structure from its dense core to the tenuous top of the atmosphere. Three of the four elemental divisions made by Aristotle (Section 2.3) fit the earth's structure well: earth, water, and air. Although the solid earth constitutes the bulk of our planetary environment, the atmosphere and oceans are as fundamental to our survival. The two fluid systems transport energy around the earth and continually erode the surface. The evolution of the atmosphere and oceans has directly shaped the geological and biological evolution of our home planet.

The rise of mountains yields finally to the onslaught of wind and rain. Mountains are scrubbed away, but the collisions of continental plates push up new peaks to replace the worn ones. The restless earth moves not only with the tremor of earthquakes, but also with the slow drift of continental masses. Viewed over a long enough time, the mountain ranges move like waves of rocks across the earth's surface. The fourth ancient element—fire—fuels our planet in two forms: solar energy (fusion) and radioactive decay (fission). Because many heavy radioactive elements are born in supernovas, that energy actually comes from stars, too. The energy for the motion of the solid earth derives from currents in the mantle, apparently generated from the heat of radioactive decay. Solar heating drives wind and water currents and waves. These two forms of stellar fire energize the evolution of earth.

Although our planet is 4.6 billion years in age, it is not static. Rather, the planet dynamically evolves. The earth's structure and evolution, which we know in more detail and depth than for any other planet, serves as a model for the investigation of similar bodies within the solar system.

Key Words

core
mantle
crust
igneous rock
basalt
granite
differentiated
seismic waves
   (transverse, longitudinal)
radioactive dating
dynamo model

magnetosphere
Van Allen
   radiation belts
troposphere
stratosphere
mesosphere
thermosphere
ozone layer
ionosphere
exosphere
homosphere

heterosphere
hydrogen corona
atmospheric escape
thermal velocity
seeing
albedo
greenhouse effect
mid-oceanic ridge
continental drift
plate tectonics
accretion atmosphere

Review Questions
1. Indicate which area of the earth (core, mantle, crust) each of the following best describes.
   a. Hottest
   b. Flows, yet brittle
   c. Highest pressure
   d. Made of basalt
   e. Likely source of earth's magnetic field
   (Objective 2)
2. Every good astronomer hopes to go to the moon when he or she dies, because, since the moon has no atmosphere (mark all correct answers):
   a. The astronomer's instruments could detect X-ray and ultraviolet photons, as well as visible light.
   b. All celestial objects would appear brighter than on earth.
   c. The astronomer's telescopes would be able to distinguish closely spaced objects more clearly than on earth.
   d. All celestial objects would appear with their true colors.
   e. The sunsets would be red and beautiful.
   f. Without an atmosphere and resulting weather patterns, every night would be warm and cozy. The astronomer would have no need to bundle up and sip coffee.
   (Objective 6)
3. What is the origin of the following?
   a. Oxygen in the earth's atmosphere
   b. Helium in the atmosphere
   c. Salt in the oceans
   d. Ozone in the stratosphere
   e. Magnetic field of the earth
   f. Mountains on the earth
   (Objectives 8, 9, and 10)
4. Explain how you could determine the earth's mass by jumping off a building. (Explain it; don't do it!) (Objective 1)
5. Contrast the composition of the earth's core to that of its crust. (Objectives 2 and 3)
6. Discuss uncertainties in the statement "The earth's age is 4.6 billion years." (Objective 5)
7. Make a simple argument to demonstrate that the earth's core must be denser than its crust. (Objective 3)
8. Describe two effects of the earth's atmosphere on sunlight passing though it. (Objective 6)
9. Suppose the amount of water vapor in the atmosphere suddenly increased by a large amount. What would happen to the earth's surface temperature? (Objective 7)
10. How can volcanoes affect the evolution of the earth's oceans and atmosphere? (Objective 8)
11. What was the composition of the earth's first atmosphere? What happened to it? What was the composition of the earth's second atmosphere? Where did it come from? What happened to it? (Objective 10)

12. Give two ways in which the oceans affect the atmosphere. (Objective 9)
13. Where did the oceans' water come from? (Objective 9)

Problems
1. Find the radius of the earth's core relative to the total radius if the core density is 10,000 kg/m^3, the mantle density is 2500 kg/m^3, and the average density is 5500 kg/m^3. (Objective 3)
2. Compute the escape temperature for the earth at the height of the exosphere for hydrogen, helium, oxygen, nitrogen, and argon. Compare with the exosphere temperature ($T = 1500$ K). Which elements are likely to escape? (Objective 11)
3. Compute the ratio of escape velocity to thermal velocity for the same elements as in problem 2, and estimate their lifetime according to the criterion given in Section 14.5. (Objective 11)
4. The light of a star directly overhead is dimmed by 0.2 magnitudes (Δm_0) due to extinction in the earth's atmosphere. Assume the atmosphere is a horizontal slab of constant thickness, and calculate the fraction f by which the pathlength is increased if the star is at an altitude of 45°, 30°, and 15°. At these altitudes the star is dimmed by the amount $\Delta m = f \Delta m_0$. Find the amount of extinction (Δm) at these altitudes. Estimate (by interpolation) the altitude of a star whose light has been dimmed by a factor 2 (0.75 magnitudes). (Objective 6)
5. The energy released within the earth by radioactive minerals is conducted to the surface as heat.
    a. The amount of heat conducted through unit area of a substance each second is equal to the thermal gradient (the change of temperature per unit length) times the heat conductivity (a property of the material). If the average thermal gradient is 0.07°C/m and the conductivity of rock is 0.84 J/m/sec/°C, what is the rate of heat loss per square meter? The heat input from the sun is of order 10^3 W/m^2. Which is more important in heating the surface of the earth: sunlight or internal heat?
    b. How much heat is conducted to the earth's entire surface every day?
    c. If the earth were to remain in a thermal steady state from internal sources alone, how much heat would have to be generated per kilogram of material in the earth?
    d. Compare your answer to (c) with the measured radioactive heat generation of basaltic rock, which is 2×10^{-10} W/kg. What does this tell you about the amount of basaltic rock in the earth? (Hint: If the entire earth were made of basaltic rock, would it be heating up or cooling down?)
    (Objectives 2 and 4)

BEYOND THIS BOOK . . .

*R. Seiver compares the earth with the other planets in the solar system in "The Earth," *Scientific American*, September 1975.
*There are many good geology books available. *The Earth Sciences* (Harper & Row, New York, 1971) by A. N. Strahler is a useful reference.
*You can find a technical, comprehensive view about our atmosphere in *The Evolution of the Atmosphere* (Macmillan, New York, 1977) by J. C. G. Walker.

* *Planetary Geology* (Prentice-Hall, Englewood Cliffs, N.J., 1975) by N. M. Short contains comparative information about the earth's evolution in Chapter 13.
* An excellent article on the fate and role of carbon dioxide in the atmosphere is "The Carbon Dioxide Question" by G. Woodwell, *Scientific American*, January 1978.
* For more on the dynamo model, see "The Source of the Earth's Magnetic Field" by C. Carrigan and D. Gubbins, *Scientific American*, February 1979.
* Read the details in "Plate Tectonics" by J. F. Dewey, *Scientific American*, November 1972.

THE MOON AND MERCURY: DEAD WORLDS

15

LEARNING OBJECTIVES
After studying this chapter, you should be able to:

1. Compare the moon and Mercury in size, mass, and density.
2. Explain the cause of tides and describe their effect on the moon's orbit.
3. Describe the moon's surface features and indicate a possible formation process for each.
4. Describe how Mercury's rotation period was discovered and explain its relation to the orbital period.
5. Describe Mercury's surface features and indicate a possible formation process for each.
6. Compare the surface environments (temperature, atmosphere, surface features, magnetic fields) of the moon and Mercury to each other and to the earth.
7. Explain what determines the surface temperature of an airless planet and make use of an equation for calculating that temperature.
8. Sketch the lunar interior, as inferred from Apollo experiments.
9. Sketch a model of Mercury's interior, and compare it to the interior of the earth and the moon.
10. Compare the evolution of the moon and Mercury to each other and to the earth.
11. Outline a possible history for the moon's evolution in light of Apollo results; present evidence for each of the major stages.
12. Compare models of the moon's origin using Apollo results to support or refute the models.

CENTRAL QUESTION:
What processes have driven the short evolution of the small, airless worlds of the moon and Mercury?

Through an earth-based telescope, the moon strikes you as a stark world, tantalizingly close (Fig. 15.1). Fascination with this neighbor bred tales of traveling to the moon. The stories range from men borne aloft by birds to the cannon-powered voyage described by Jules Verne. His astronauts, after their launch from Florida, circled the moon and returned home by plunging into the sea. Verne had the right idea! NASA carried out his vision with additions. In July 1969 Neil Armstrong's one small step imprinted an indelible mark in history (and on the moon's surface). The event fulfilled many earlier dreams: the first visit to an alien planet.

Another body in the solar system has similar mien—Mercury (Fig. 15.2). Like the moon, Mercury is a small, airless world pockmarked with many craters.

Both the moon and Mercury are now dead worlds. Their interiors are cooler than the earth; no heat drives the motions of crustal plates. No mountains rise, no volcanoes fume. Without atmospheres no wind or water wears down their landscapes. Their heyday of activity has passed.

This chapter compares the tiny worlds of the moon and Mercury to each other and to the earth to provide an insight into their evolution. The emphasis will fall on our moon because of the Apollo missions, which give solid clues to reconstruct the moon's history and to infer that of Mercury's in comparison.

15.1 ORBITAL AND PHYSICAL CHARACTERISTICS OF THE MOON

Greek astronomers knew that the moon circles the earth. They even had a fairly accurate determination of the moon's distance, in terms of the size of the earth. Hipparchus (Focus 2.2) estimated that the moon was 29½ earth diameters away, very close to the modern figure of 384,401 km, about 30 earth diameters (Fig. 15.3).

The moon revolves around the earth in an elliptical orbit. During a month the earth-moon distance varies. When the moon is closest to the earth, it is at *perigee*. The most

15.1 ORBITAL AND PHYSICAL CHARACTERISTICS OF THE MOON

(a)

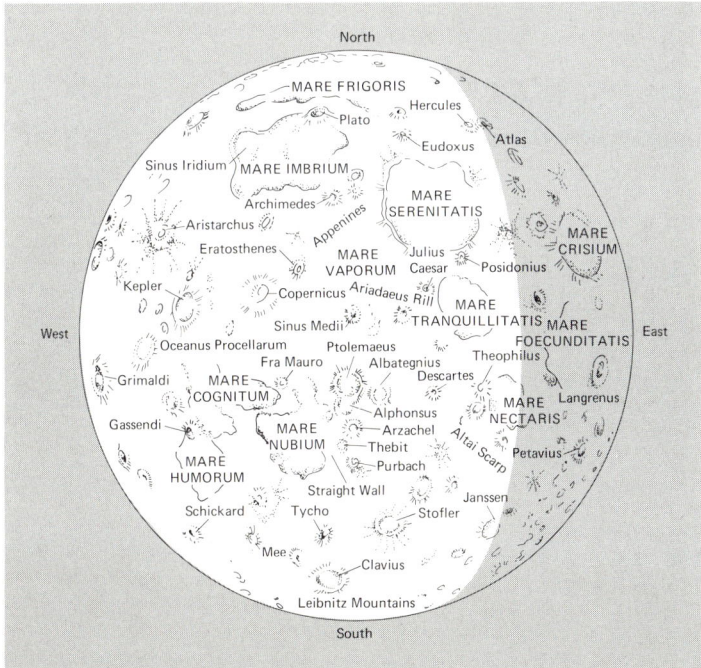

(b)

FIGURE 15.1 (a) The moon as seen through a telescope. [*Courtesy Lick Observatory.*] (b) Names of the major features on the moon's near side.

FIGURE 15.2 The cratered surface of Mercury. [*Courtesy NASA.*]

FIGURE 15.3 The moon's orbit relative to the earth. This view is down on the plane of the moon's orbit from above the north pole. The minimum perigee and maximum apogee distances are shown. The earth and moon are not drawn to the same scale as the distances, but they do have the correct relative sizes.

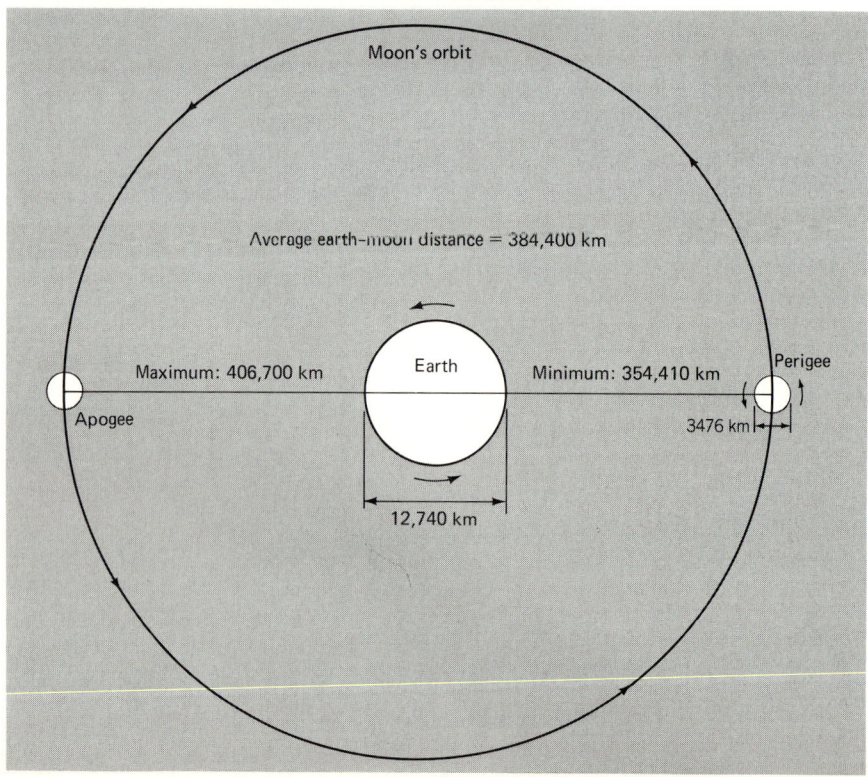

15.1 ORBITAL AND PHYSICAL CHARACTERISTICS OF THE MOON

distant point is called *apogee*. The difference between apogee and perigee is just over 50,000 km.

In 1969 the Apollo 11 astronauts deposited special reflectors on the moon's surface. Later Apollo missions put down other such reflectors. These devices reflect laser light bounced to the moon from the earth (Plate 6) back again to the earth. Timing the round trip permits measurements of the moon's distance from the earth with uncertainties as small as 8 cm!

Tides and the Moon's Orbit

The moon exerts tidal forces (Focus 12.1) on the earth. These result in ocean tides associated with the moon, with typically two high tides and two low tides occurring each day. (There are also tides in the body of the earth itself, but they are much smaller than the ocean tides.) We'll give you a general explanation of tidal dynamics using Newton's law of gravitation (Section 4.2), but be warned that it does not solve all the detailed problems about tides.

Imagine a smooth earth's surface covered completely with a layer of water. Consider for a moment the moon's gravitational attraction at three points lined up with the moon: point A closest to the moon, point B at the center of the earth, and point C on the side of the earth opposite the moon (Fig. 15.4). Recall that the force of gravity decreases as the inverse square of the distance between masses. So the moon's gravitational force must be greater at A than at B, and greater at B than at C. The greater the force acting on the same mass, the greater its acceleration. So a mass at A has a greater acceleration than a mass at B, and a mass at B has a greater acceleration than a mass at C. The *difference* between these accelerations is crucial. Because of the difference, the water at A bulges ahead of the

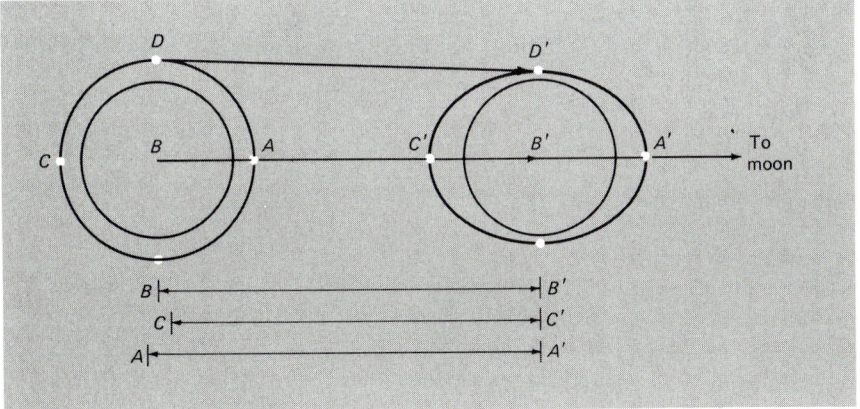

FIGURE 15.4 Tides. Imagine a smooth earth covered with water. Consider the water at A, C, and D pulled by the moon's gravity. How far the water falls in some time depends on its distance from the moon; water at A falls farther than that at C. Water at D and the earth's center (B) fall the same distance (because the acceleration on them is the same). So as the earth falls toward the moon (from B to B'), water on one side falls a bit more (from A to A') and on the other side a bit less (from C to C'). Water flows from D to supply the bulges; the depth at D' is less than at A' or C'.

earth (point B), and the water at C lags behind the earth and forms a bulge on the earth's side opposite the moon. So we have two high tides: one on the side of the earth toward the moon and one on the opposite side. These two high tides take place about one half-day apart.

Note that it is the *difference* in gravitational forces that accounts for the tides; this difference is what is referred to as a *tidal gravitational force* (Focus 12.1). It is proportional to the *inverse cube* of the distance.

History of the Moon's Orbit

These tidal forces affect the earth-moon system, causing the moon to gradually move away from the earth. Why? The answer lies in the notion of *tidal friction* and the conservation of angular momentum.

The total angular momentum of the earth-moon system consists of two parts: the spin angular momentum (of the earth and the moon rotating about their axes) and the orbital angular momentum (of the moon revolving about the earth). The conservation of angular momentum says that the sum of these must remain constant. Because the moon rotates slowly and its mass is small, the moon's spin adds very little to the total and can be ignored. So we need to consider only the spin of the earth and the revolution of the moon.

The earth is *not* a perfect sphere covered with a layer of water. Because there are continents separating the ocean basins, the tidal bulge cannot freely move around the earth as it rotates. The continents get in the way, producing a frictional drag which slows the rotation.

Because of tidal friction, the earth's rotation is slowing down, so its spin angular momentum decreases. For the system's total angular momentum to remain constant, the moon's orbital angular momentum must increase. This increase can happen if the moon moves away from the earth. What causes this outward movement? Gravity moves the moon outward. Picture the tidal bulges of the water on the earth (Fig. 15.5). The earth rotates faster than the moon revolves around it. Friction in the ocean basins carries the tidal bulges along in the direction of the earth's rotation. So the tidal bulge *nearest* the moon gets *ahead* of the moon. This mass of water pulls gravitationally on the moon, accelerating it in the direction of its motion. This gives the moon a higher velocity and hence a higher angular momentum (remember, $L = mvr$; Section 9.3). The bulge on the opposite side of the earth pulls in a direction opposite the moon's motion, but it is farther away and the force is weaker. The net result of this acceleration moves the moon outward in its orbit.

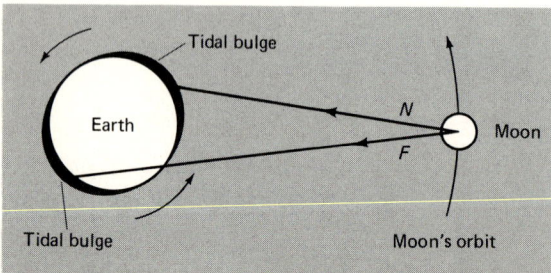

FIGURE 15.5 The tidal bulges on the earth and the moon's motion. The near bulge pulls the moon along N; the far one along F. The net force has a component in the direction of the moving motion. The bulges spin ahead of the moon because the earth rotates faster than the moon revolves and friction carries the bulges with the earth's surface.

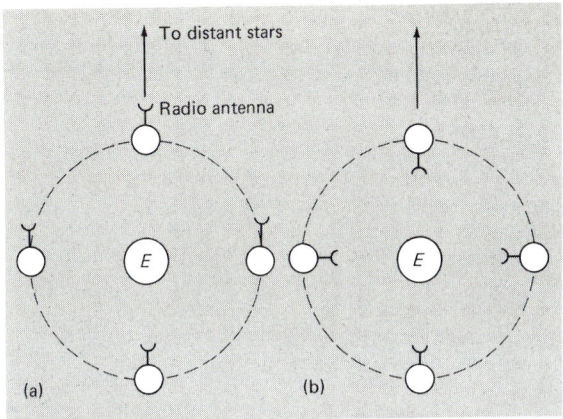

FIGURE 15.6 The moon's rotation. The moon keeps the same face toward the earth because its rotation period is the same as its period of revolution with respect to the stars. Imagine a radio antenna on the moon. (a) If the moon did not rotate with respect to the stars, the antenna would not always point at the earth. (b) But because it does rotate at a rate equal to its revolution, an antenna always points to the earth.

FIGURE 15.7 A close-up of the moon's far side. The smallest craters are about 1 km across. [*Courtesy NASA.*]

Tidal friction slows down the earth's rotation rate at about 10^{-3} sec a century. That's a deceleration of the rotation rate of about one part in 10^{10}. The corresponding acceleration of the moon results in an increase of the earth-moon distance. Kepler's third law tells us there will be a corresponding increase in the moon's period. So as the earth's rotation period increases, so does the moon's period of revolution. In the future the moon will be so far from the earth that the length of the month (longer than now) will equal the length of the day (also longer). Then the day and month will be 55 present days long.

Reversing this reasoning, we see that the moon must once have been much closer to the earth than it is now. One estimate is that the closest approach of the moon took place about 1.2 billion years ago. At that time, if the calculations are correct, the month was about 6.5 h long, the day 5 h long, and the moon only 18,000 km from the earth. The moon would have looked like an enormous defaced balloon in the sky! It would have covered 11°, which is 22 times its present angular diameter.

The One-Faced Moon

You know from having looked at the moon that we always see the same side. In order to keep the same side facing the earth, the moon must rotate on its axis with the same period, 27.3 days, as its period of revolution about the earth with respect to the stars (Fig. 15.6). Until satellites were placed in orbit around the moon, we had no knowledge of the appearance of the side turned away from the earth.

Lunar orbiters have allowed us to see the moon's once mysterious far side (Fig. 15.7). Their photos show that the far side looks quite different from the near side (Fig. 15.1). It is almost completely cratered, with little of the dark-colored, smoother areas that cover so much of the near side.

If you stood on the moon's near side, you would see the earth suspended, never rising or setting, against the stars. With the earth always in sight, astronauts on the near side can communicate directly here by radio. By contrast, any astronaut on the far side would never see the earth and would have to rely on a lunar orbiter to relay radio signals to earth.

The Lunar Day

The moon keeps the same face to the earth, but not to the sun. It rotates once with respect to the sun in 29.5 days. So the lunar day is 29.5 earth days long. This is also equal to the *synodic month*, the time between corresponding phases of the moon. Why? Suppose you were standing on the moon at what we on earth see as the center of the full moon. Where would the sun be? Directly overhead! When is the next time the sun is directly overhead? At the next full moon!

Note that the synodic month (29.5 days) is not the same length as the *sidereal month*, the period of revolution (and rotation) with respect to the stars. Why not? Because the earth moves around the sun. Consider the moon at full moon (M_1 in Fig. 15.8) when it is opposite the sun. After one sidereal month the moon is back at the same place in its orbit, with respect to the stars (M_2). But the earth has moved almost one-twelfth of the way around its orbit. So the moon has to move somewhat farther to get back to a place opposite the sun (M_3). Hence the synodic month, the time until the next full moon, is longer than the sidereal month. (The relationship between the two is similar to the relationship between the synodic and sidereal periods of the planets; Section 3.2.)

The Moon's Size

If you know the distance from the earth to the moon, you can find its physical diameter from an observation of its angular size. The result is 3476 km, about $\frac{1}{4}$ the earth's diame-

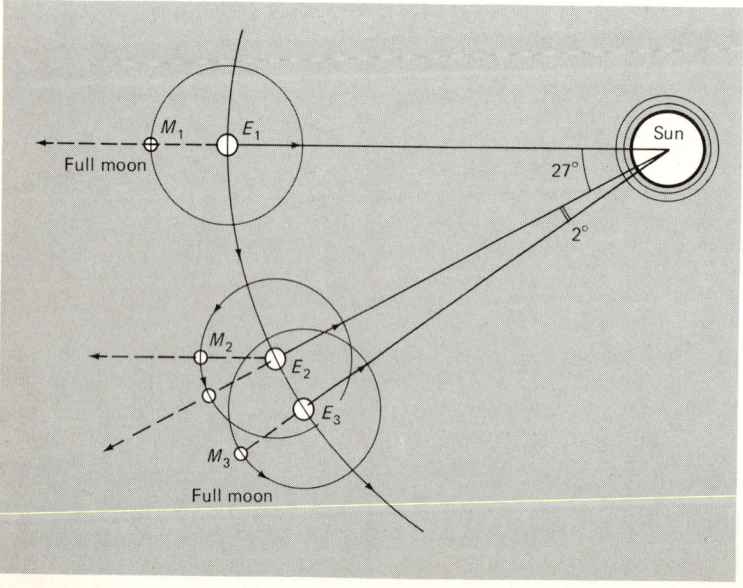

FIGURE 15.8 Geometry for the difference between a sidereal month (M_1 to M_2) and a synodic month (M_1 to M_3).

15.1 ORBITAL AND PHYSICAL CHARACTERISTICS OF THE MOON   • 539

ter. If the earth were the size of your head, the moon would be about the size of a tennis ball. On the same scale, the diameter of its orbit would be about 12 m (40 ft).

The Moon's Mass and Density

In order to find the moon's mass, we can't simply use Kepler's third law and the moon's period and distance; that would give the mass of the moon *plus* the mass of the earth. We need to look at the *center of mass* of the earth-moon system. (Recall Section 8.6, where the center of mass for binary stars was discussed.)

The earth and the moon balance at their center of mass, as a consequence of Newton's third law (Section 4.2). Each moves in an ellipse around the center of mass (Fig. 15.9). This motion around the center of mass can be measured by observing the monthly shifts in the positions of the other planets. From these observations we know that the earth-moon center of mass lies 4671 km from the earth's center in the direction of the moon. The distance of the moon from the center of mass is 379,730 km. The moon does *not* revolve around the center of the earth!

The center of mass of the earth-moon system is at a point such that the ratio of the distances from it to the earth and the moon is equal to the ratio of the masses of the earth and moon (Section 4.3):

$$\frac{r_{moon}}{r_{earth}} = \frac{m_{earth}}{m_{moon}}$$

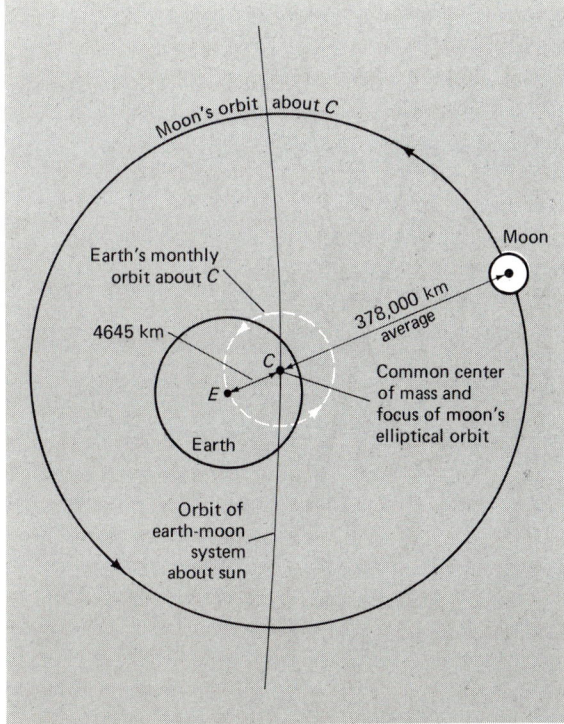

FIGURE 15.9 The center of mass (point C) of the earth-moon system. The distance between E and C varies; 4645 km is an average.

From this we can compute the mass ratio:

$$\frac{m_{earth}}{m_{moon}} = \frac{379{,}730 \text{ km}}{4671 \text{ km}} = 81.3$$

Because the ratio of the distances of the earth and moon from the center of mass of the earth-moon system is 1/81.3, the mass of the moon is 1/81.3 times the mass of the earth, or 7.2×10^{22} kg.

From the mass and radius, the average density comes out to be 3300 kg/m³, about the same density as the rocks in the earth's mantle.

15.2 THE SURFACE ENVIRONMENT OF THE MOON

Surface Gravity of the Moon

A planet's surface gravity is simply the acceleration of any object at its surface due to the planet's gravitational pull. Therefore from Newton's laws (Section 4.2)

$$g = \frac{GM_{planet}}{R_{planet}^2}$$

The moon has a mass 1/81.3 that of the earth and a radius 0.273 that of the earth, so

$$\frac{g_{moon}}{g_{earth}} = \frac{M_{moon}/M_{earth}}{(R_{moon}/R_{earth})^2}$$

$$= \frac{1/81.3}{(0.273)^2}$$

$$= 0.165 \approx \frac{1}{6}$$

The moon's surface gravity is $\frac{1}{6}$ that of the earth, and objects weigh $\frac{1}{6}$ as much on the moon as they do on the earth.

An easy way to lose weight is to travel to the moon! The astronaut who must lug around a 180-pound life-support system on the earth carries the equivalent of only 30 pounds on the moon. But the astronaut must be careful. Although the *weight* of the life-support system is much less, its *mass* is the same on the moon as on the earth (or anywhere). So to stop running, for example, requires the same effort. That's why the astronauts felt so clumsy as they moved around on the moon.

The Atmosphere of the Moon

The moon has no atmosphere to speak of. How do we know? When the moon moves in front of a star, the star suddenly vanishes without warning! Any substantial atmosphere on the moon would dim the star gradually before it disappeared behind the disk.

15.2 THE SURFACE ENVIRONMENT OF THE MOON

Why no atmosphere? The moon's escape velocity is 2.4 km/sec. At typical lunar temperatures gas particles of low mass at the surface have escape velocity (Section 14.5). Most gasses have escaped from the moon's gravitational grasp since its formation. Some material from the solar wind travels near the moon and stays briefly. But the surface density must be extremely low. The exhaust from the Apollo 11 landing dumped more gasses into the atmosphere than had previously existed there! But these gasses will not stay around forever: for example, oxygen dumped at the moon's surface escapes in roughly 100,000 years.

The Surface Temperature of the Moon

The earth's atmosphere acts like an insulating blanket (Section 14.6). During the night it retains much of the heat received in the previous day. Lacking such atmospheric insulation, the moon experiences a greater temperature range during a lunar day (which lasts a month of earth time). Under direct sunlight the moon's surface temperature exceeds 370 K, and at midnight it drops to 125 K.

Let's consider how to estimate the moon's surface temperature from its energy budget. Sunlight heats the moon's surface, but it does not absorb all the incoming sunlight (Fig. 15.10). Some is reflected back into space. The moon's albedo is about 7 percent, that is, it reflects 7 percent of the light that hits it and absorbs the remaining 93 percent. (Even though a full moon seems to shine brightly in the sky, the moon's surface is really quite black.) The absorbed sunlight heats the surface; suppose it radiates like a blackbody, emitting mostly infrared because of the low temperature. The balance between the incoming sunlight and outgoing infrared determines the surface temperature of the sunlit side. These must balance; otherwise the surface temperature would not be stable in sunlight.

Let's look at the mathematical details and also generalize to all planets by including the effect of distance from the sun. Imagine a planet with surface albedo A at distance D from the sun. The sun's luminosity, L, is spread out over the imaginary sphere of radius D, that is, over the area $4\pi D^2$. Each square meter of the planet intercepts a small fraction, $1/4\pi D^2$, of this energy. So the total solar energy hitting each square meter of the sunlit side of the planet is $L/4\pi D^2$. Of the total energy coming in, the planet reflects a fraction (its albedo, A) and absorbs a fraction $(1 - A)$. So the net solar energy absorbed is

$$E = (1 - A)L(1/4\pi D^2)$$

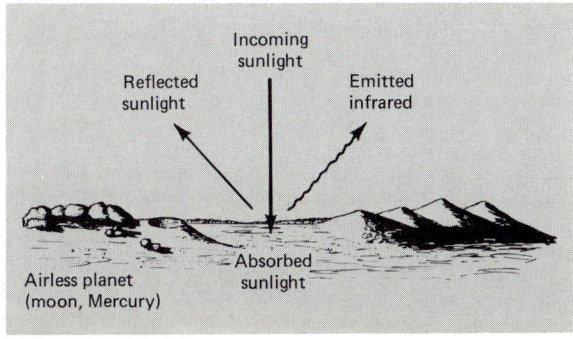

FIGURE 15.10 Energy budget for an airless planet.

This input heats the surface.

The surface loses energy by radiating infrared into space. Assume it acts like a blackbody. Each square meter of surface at temperature T emits σT^4 W (Section 6.4). This loss must balance the gain:

$$\sigma T^4 = (1 - A)L(1/4\pi D^2)$$

So

$$T^4 = \frac{1-A}{4\pi\sigma}\frac{L}{D^2}$$

Then, taking the fourth root,

$$T = \left(\frac{1-A}{4\pi\sigma}\frac{L}{D^2}\right)^{1/4}$$

Simplify this equation by putting in the luminosity of the sun (3.8×10^{26} W) for L, converting the distance to Astronomical Units, and using the proper values for the constants π and σ:

$$T\text{ (Kelvins)} = 394\left(\frac{1-A}{D^2}\right)^{1/4}$$

Let's apply this equation to the moon, for which $D = 1$ AU and $A = 0.07$. Then

$$T = 394\left(\frac{1-0.07}{1^2}\right)^{1/4}$$

$$= 386\text{ K}$$

This is the maximum temperature a solid blackbody can attain at the distance of 1 AU from the sun. The moon doesn't get quite this hot; it isn't a perfect blackbody, and some heat is conducted downward into the surface.

Note that, because of the greenhouse effect, this approach fails if the planet has a substantial atmosphere.

At night there is no heat from the sunlight, and the infrared, because no atmosphere exists to trap it, radiates away into space. Because the lunar night is so long (about 15 earth days), the surface has a long time to cool and the temperature plummets to 125 K. The large noon-to-midnight temperature difference, some 250 K, occurs because the moon rotates slowly and has no atmosphere.

Without a significant atmosphere, the moon has no shield from the lethal X rays and ultraviolet radiation from the sun, or from the small, solid particles coming in from space. The moon's surface presents a fierce, unfriendly place for people. The Apollo astronauts, imprisoned in their bulky life-support systems (Fig. 15.11), found the moon bleakly beautiful, but uninviting.

FIGURE 15.11 Astronaut Edwin E. Aldrin, Jr., on the moon. Note the bulky life-support system. A gold-plated visor filters out ultraviolet light from the sun; the surrounding scene is reflected in the visor. [*Courtesy NASA.*]

15.3 THE SURFACE OF THE MOON: PRE-APOLLO

Until the Apollo astronauts walked upon the lunar surface and sampled it, the study of the moon was confined to viewing it from the earth, from orbiting satellites, or from lunar landers. These investigations provoked many questions, such as how and when did the craters form? This section briefly surveys the moon's surface as a preview to a discussion of the results of the Apollo landings.

The consistent study of the lunar surface commenced in 1609, when Galileo turned his telescope to the moon (Section 4.1). His pioneer work sparked an explosion of careful observations. Galileo named the dark areas *maria* (Latin for "seas"; the singular form is *mare*.) We will use *maria* rather than *seas*, since the English word implies water more forcefully than the Latin one.

The moon viewed through a small telescope or binoculars is always fascinating (see

Fig. 15.1). Lunar craters give the moon a pockmarked face. Mountains and craters irregularly rim the moon's edge. Some mountains stand alone in the maria, while others link in long ranges. Bright rays flower from some craters. The moon impresses you as a rough, old world that has suffered significant violence—violence which has carved its splotched surface.

Maria and Basins

Photographs from satellites orbiting the moon show that most of the maria lie in the northern half of the moon's hemisphere on the side toward the earth; few are on the far side. The maria appear dark compared with the rest of the surface. (Their albedo is lower.) Their dark irregular stretches form the face of the "man in the moon."

Along with their darker appearance, the maria have other general features that give clues to their formation. Some maria look circular in shape, and some are interconnected. They have smooth surfaces compared with the brighter, cratered regions. Also, the maria have lower elevations, by about 3 km, than the rest of the surface. The maria are called the lunar *lowlands* and the lighter areas the *highlands*.

The vast flat extents of the maria look like solidified lava flows (and indeed they are). The regions around the maria have craters flooded by the dark material from the mare (Fig. 15.12). This indicates that the lava flow that formed the maria occurred *after* the lunar crust formed and after the formation of some craters. In other words, the lunar maria fill up large, shallow *basins* on the moon's surface. For example, you can easily see the Mare Imbrium basin on the moon's near side (Fig. 15.1). It's roughly circular in outline and has a diameter of about 1200 km. The moon's most striking basin lies on its far side: Mare Orientale (Fig. 15.13). The concentric rings of mountains make it look like a bull's-eye. The outer rim of mountains (the Cordillera Mountains) has a diameter of about 970 km and rises to a height of 7 km. Surrounding this basin for about 1000 km out lies a blanket of lighter material covering the older lunar surface. Mare Imbrium would look like Mare Orientale if most of its filling material were removed.

Planetary scientists have found concentrations of mass beneath most of the maria. These are called *mascons*. One mascon, for example, under Mare Imbrium, has a mass of 1.6×10^{19} kg and an average density of 3700 kg/m^3. (Note that this density is *greater* than the average density of the moon.) If this mascon were spread out to cover California, it would be 12 km thick! The fact that almost all maria have associated mascons implies that some process produced large amounts of dense material under the maria.

Craters

Craters (from the Greek word meaning "cup" or "bowl") litter the moon almost everywhere. They range widely in size, from many smaller than a coin to five with diameters greater than 200 km. (That's about the size of Connecticut.)

Craters are generally round. The heights of the rims of lunar craters are small compared with the diameters, and the floors are depressed compared with the surrounding landscape. The terrain just outside large craters has a wavy look as if shocked by an explosion. Some craters have light-colored rays emanating from them.

What formed the craters? After Galileo's observations, Robert Hooke compared lunar craters to those formed by bullets striking mud. This was the start of the *impact model* of

FIGURE 15.12 The surface of a mare (the Ocean of [Storm]s). Note how much smoother, darker, and lower [in hei]ght the mare is compared with the background [highl]ands. The baylike area near the center (where the [mare] meets the highlands) is a filled-in crater (arrow). [Cour]tesy NASA.]

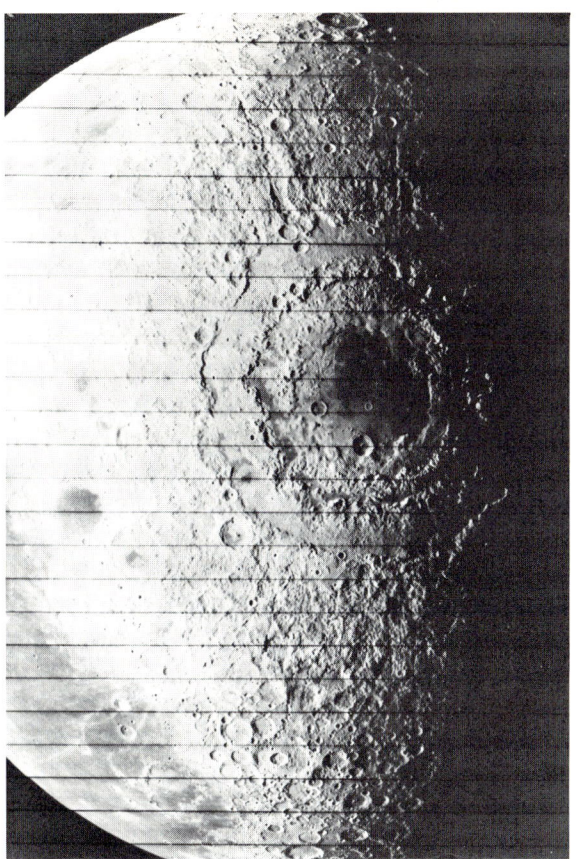

[FIGU]RE 15.13 Mare Orientale on the moon's far side. [The C]ordillera Mountains ring the basin like a bull's-[eye. Courtesy NASA.]

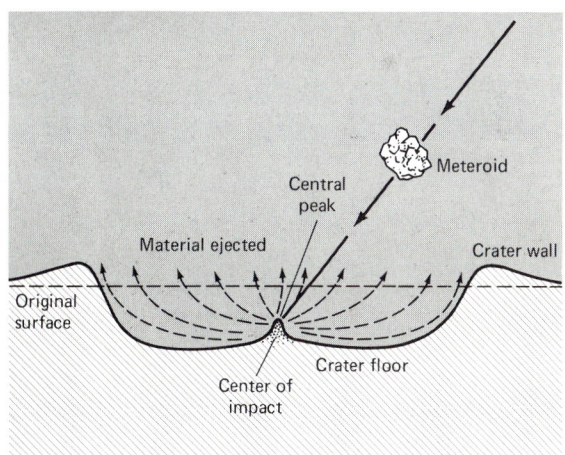

[FIGU]RE 15.14 Formation of a crater by impact of a [solid b]ody. Ejected material can form rays and smaller [craters] surrounding the main impact.

crater formation: Objects from space slammed into the moon to punch out the craters (Fig. 15.14). With the rise in the study of volcanoes in the nineteenth century, astronomers and geologists generally accepted the *volcanic model* instead: The craters were left over from a lava eruption. The volcanic model requires that the moon's interior be hot enough to provide lava.

On earth we find a few craters made by the impact of solid bodies from space and also

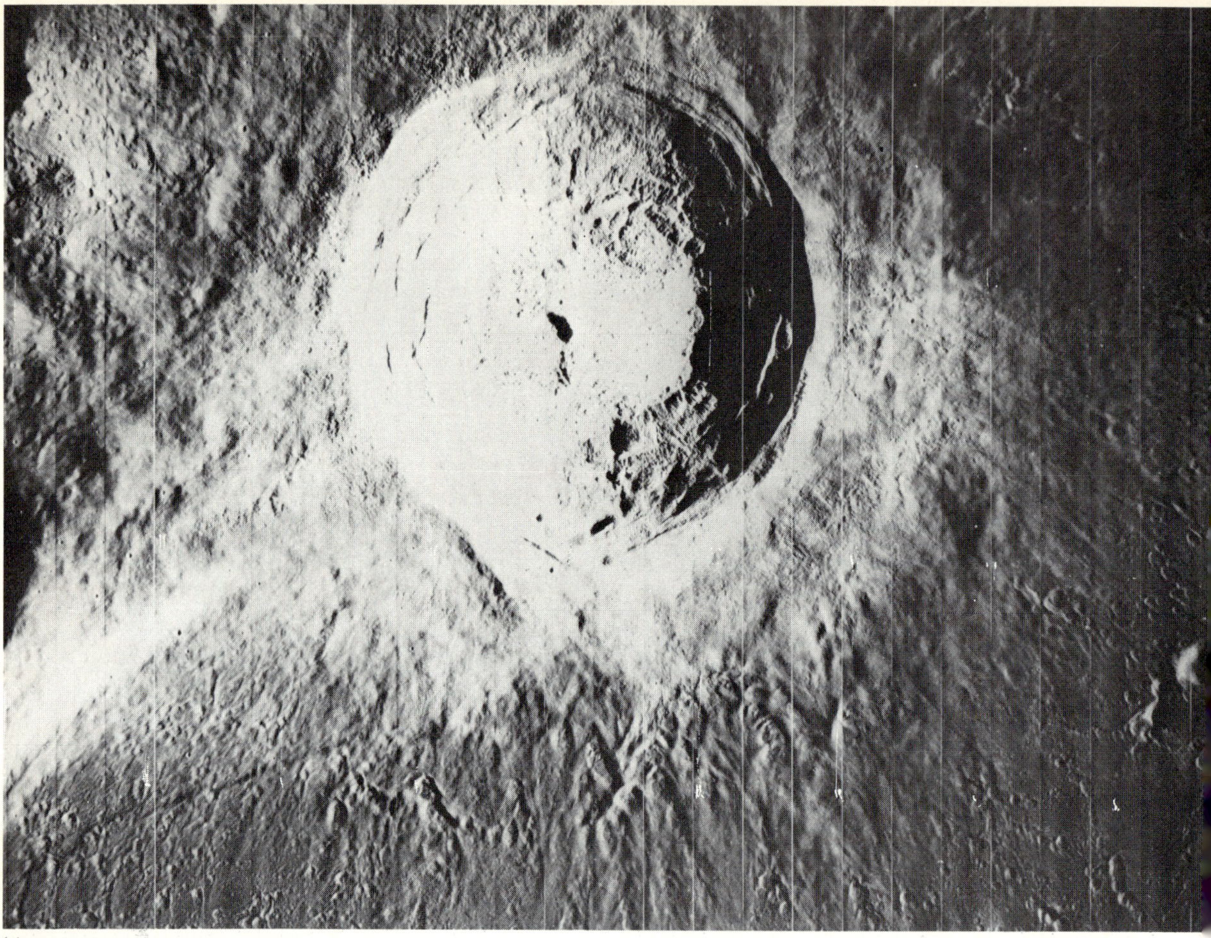

(a)

(b)

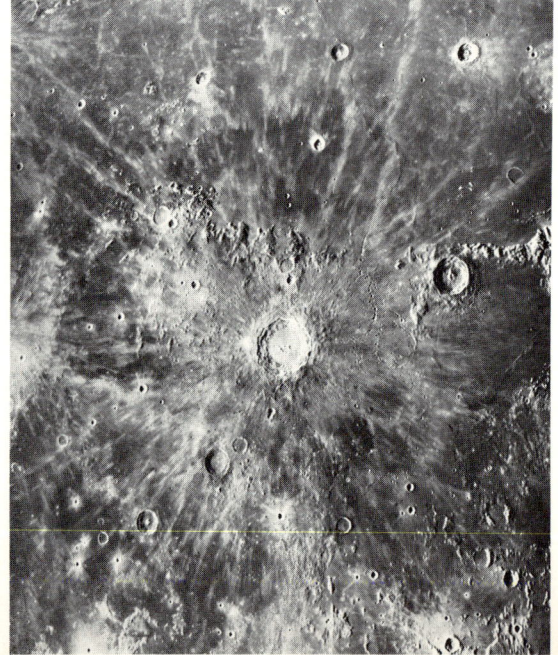
(c)

some made by volcanoes. Differences exist between the two. No volcanic crater on the earth is larger than 5 km, while some impact craters are larger. The outer slopes of a volcanic crater are smooth. An impact produces a crater with undulating slopes, usually covered by debris. If the debris and crater wall were put back into an impact crater, they would fill it up. Volcanic craters are smaller. Material blasted out by an impact falls in streaks, leaving a raylike pattern. Volcanoes eject material in a uniform rather than a streaked pattern. Large chunks of thrown-out material can create small secondary craters around impact craters. Volcanoes have no secondary craters. Since most lunar craters display impact characteristics, for example, ray systems (Fig. 15.15), we conclude that impacts sculpted most of the lunar craters.

What projectiles made these impacts? The small, solid bodies orbiting the sun, which are called meteorites when they strike the earth, will crash into the moon if they pass close enough to it. We have seen no new, large craters form recently, so the era of heavy impact cratering took place in the past. The Apollo missions aimed to find out when.

15.4 THE MOON CLOSE UP: APOLLO

Complex computers piloted the Apollo spacecraft to the moon. The Apollo 11 craft returned with a cargo of 22 kg of the moon. Those lunar samples, collected by astronauts Neil Armstrong and Edwin Aldrin in slightly less than three hours, were probably the most expensive scientific specimens ever gathered. They were also the most fascinating: the first pieces of another world brought back to the earth.

The Apollo program conveyed back to earth about 400 kg of lunar material from six different sites. Scientists have examined only about 10 percent in detail. Already these samples and other experiments in the Apollo program have provided the first deep understanding of a planet other than the earth. We can now sketch out many physical details of the moon and outline its history.

The Lunar Surface

The Apollo samples reveal the physical and chemical nature of the lunar surface. The very top layer is a porous, somewhat adhesive layer of debris (Fig. 15.16). It consists of fine particles (called lunar *soil*) and larger rock fragments. All soil samples contain a large amount of mostly round pieces of glass (Fig. 15.17). These glass spheres make the surface slippery.

The rocks returned from the moon mostly fall into four categories. (1) *Mare basalts* are dark, fine-grained rocks similar to terrestrial basalts (magnesium and iron silicates) (Fig. 15.18a). (2) *Anorthosites* are light-colored igneous rocks with visible grains (aluminum and calcium silicates). They are by far the most common rock on the surface (Fig. 15.18b). (3) *Breccias* are rock and mineral fragments cemented together (Fig. 15.18c). (4)

FIGURE 15.15 (a) The crater Aristarchus, about 60 km wide, with its ray system. Note the ripples in the surface just outside the crater; these formed from the shock of impact. [*Courtesy NASA.*] (b) A close-up of the southern wall of Aristarchus, showing evidence of impact and erosion. [*Courtesy NASA.*] (c) The crater Copernicus (center) with its ray system. [*Courtesy Lick Observatory.*]

FIGURE 15.16 An astronaut's footprint in the lunar soil, which has a consistency like that of wet sand. [*Courtesy NASA.*]

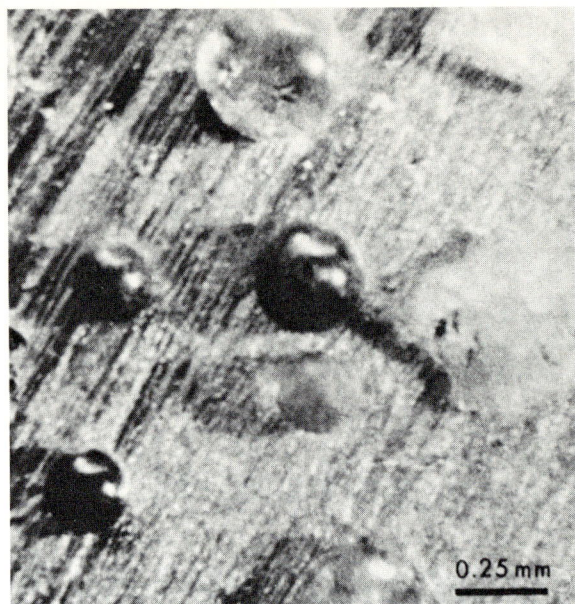

FIGURE 15.17 Small glass spheres, typically found in the lunar soil. [*Courtesy NASA.*]

A fourth type of rock, less common than the others, is similar to terrestrial basalts, but contains high amounts of potassium (K), rare earth elements (REE), and phosphorous (P). This material has been dubbed *KREEP*. The anorthorsites and KREEP make up the rugged lunar highlands.

What do these characteristics imply? First, because the moon rocks are igneous rocks, they formed from the solidification of lava. The rate at which lava cools determines the grain size of igneous rocks. Fast cooling results in small grains, slow cooling in large grains. So the dark rocks (found in the lowlands) cooled faster than the light ones (found in the highlands). In addition, the light-colored rocks are less dense than the dark ones. This difference occurs because anorthorsites contain calcium and aluminum silicates rather than the iron and magnesium silicates of the mare basalts. These compositional differences imply that the mare basalts formed from partial melting and slow cooling of silicates at a relatively low temperature (about 1200 to 1400 K) inside the moon and after some of the heavier elements had settled toward the interior. In contrast, the anorthorsites were hotter, cooled more rapidly, and formed from the less dense materials, probably on the surface.

Second, we can infer how the breccias formed. Imagine newly made igneous rocks on the moon's surface. Bodies from space pound into these rocks, fragmenting and heating them. The heat cements some fragments together to make breccias. The loose material left over makes up the lunar soil.

In a few important ways moon rocks are quite different from the earth's igneous rocks. First, they contain more titanium, uranium, iron, and magnesium. Second, compared with earth rocks or meteorites, they have a smaller fraction of elements that would con-

15.4 THE MOON CLOSE UP: APOLLO 549

FIGURE 15.18 (a) A typical mare basalt. The holes were air bubbles in the lava. The pit in the center resulted from an impact by a small meteorite. (b) A rock containing anorthorsite (white area). This is the sample from Apollo 15 known as the "Genesis Rock." (c) A lunar breccia. The dark areas are melted rock. Large pieces of other rock are visible embedded in it. [*Courtesy NASA.*]

dense at relatively low temperatures (1300 K) in the solar nebula (Chapters 13 and 19). The elements are called *volatiles* and include sodium, potassium, copper, argon, and chlorine. Third, a few lunar samples contain minute amounts of complex carbon (organic) compounds. (This discovery does not mean that life existed on the moon; organic compounds can be formed in nonbiological processes.) Some meteorites (Chapter 18) also contain carbon compounds. So the carbon compounds in the lunar soil may have been carried there by infalling bodies. Fourth, the moon rocks contain *no* water. Earth rocks always have some water locked up in their minerals. The moon rocks were found to be bone dry.

The difference in the bulk densities of the moon and earth, 3300 kg/m^3 versus 5500 kg/m^3, implies that the moon contains less metals and probably does not have a large nickel-iron core as the earth does.

In one critical way, on a nuclear level, the moon and the earth are similar. The isotopic composition of oxygen (relative abundances of ^{16}O, ^{17}O, and ^{18}O) in the lunar samples is the same as that for the earth. Studies of the oxygen isotopic compositions of meteorites show a distinct variation among samples. Some meteorites are thought to be primitive materials from the condensation of the solar nebula. Indirect evidence suggests that the variation in oxygen isotopes corresponds to condensation in different parts of the solar nebula. The identity for the earth and moon establishes that these bodies formed in the same general region.

Ages of Lunar Samples

Moon rocks are dated by the same radioactive-decay techniques used to date earth rocks (Section 13.1). One caution: These methods give the time since the rock last solidified. If a rock was heated enough to melt since its original formation, radioactive dating gives the age since that reheating.

Only a few fragments from the lunar soil have ages as great as 4.5 to 4.6 billion years. The light-colored rocks from the highlands generally are the oldest: 3.8 to 4.0 billion years. The KREEP basalts are younger, from 3.4 to 3.6 billion years old. The dark-colored rocks from the maria are the youngest, some only 3.2 to 3.3 billion years old and only a few as old as 3.8 billion years.

What do these ages reveal about the history of the moon's surface? The moon formed a little more than 4.6 billion years ago. After formation, the present highlands solidified, about 4 billion years ago. Then the lava flows that made the maria took place, about 3.5 billion years ago.

The Moon's Interior

The Apollo missions probed the moon's interior with the same seismic-wave method used to probe the earth (Focus 14.1). The astronauts placed seismometers on the moon to measure moonquakes. These instruments found that few moonquakes occur, and those few take place at least 800 km below the surface. Such shocks release only about 10^2 to 10^5 J, barely a tremble by our standards. (For comparison, a baseball pitch has an energy of some 10^2 J.) If you stood directly over the strongest moonquake so far recorded, you would not even feel your feet shake. Geologically, the moon is a quiet world.

This low seismic activity indicates that the moon is cold and solid down to a depth of about 800 to 1000 km (Fig. 15.19). This region makes up the moon's mantle, which very likely consists of silicates only a little more dense than the surface. Encasing the mantle is the crust, which is layered. On the very surface lie the rocks found in the Apollo missions: highland anorthorsites, lowland basalts, and KREEP. At a depth of about 25 km the crust probably consists of anorthorsite-rich silicates.

It's uncertain if the moon has a well-defined core. If it does, the core probably makes

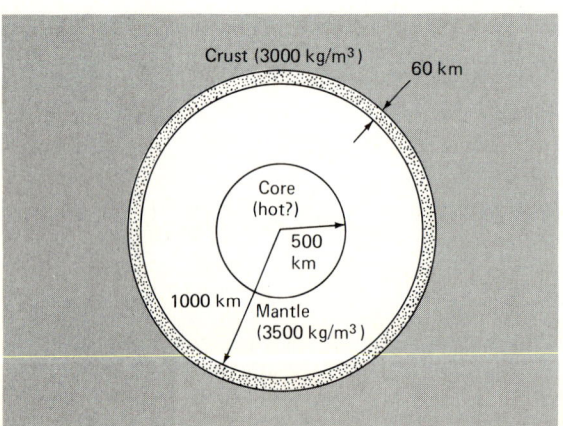

FIGURE 15.19 A model of the moon's interior inferred from seismic data. The core may be hot enough to be molten, about 1500 K.

up the inner 500 km of the moon. It may be hot, about 1500 K, and molten in whole or in part. Evidence for a hot core comes from the Apollo measurement of heat flowing up through the lunar surface. The amount of heat flow is three times as great as that for the earth. So some part of the moon's interior must be relatively hot. The mantle is solid and cool, so the core must be the source of the heat outflow.

But, the moon's core cannot be like the earth's. The moon's magnetic field is very weak, only 10^{-4} as strong as that of the earth. If the dynamo theory for the origin of a planet's magnetic field is correct (Section 14.4), then the moon's core cannot be completely molten or composed mainly of iron and nickel. (Also, the moon's density is too low for it to have a substantial nickel-iron core.) On the other hand, some surface rock samples are magnetized much more than you would expect from such a weak magnetic field. Recall (Section 14.8) that iron minerals in an igneous rock preserve the magnetic field present at the time of solidification. So in the past the moon's magnetic field must have been stronger than it is now.

Conjectured Lunar History

From the Apollo results we can concoct a scenario of the moon's history. The inferred sequence of events (Table 15.1) relies heavily on the dating of the lunar rocks.

About 4.6 billion years ago the moon formed. Whether the moon came together near the earth or not, it probably formed by the accretion of chunks of material. These pieces continued to plunge into the moon after most of its mass gathered. During the first 200 million years after formation, these projectiles from space bombarded the surface and, along with radioactive decay, heated it enough so that it melted. Less dense materials

TABLE 15.1 Evolution of the Moon: A Contemporary Model

Event	Time (billion of years ago)	Processes
Formation	4.6	Accretion of small chunks of material
Melted shell	4.6–4.4	Melting of outer layer by heat from infall of material and/or radioactive decay; volatile elements lost
Cratered highlands	4.4–4.1	Solidification of crust while debris still falls in to crater it
Large basins	4.1–3.9	Reduced infall, but formation of basins by impact of a few large pieces; outflow of KREEP basalts from lava below solid crust
Maria flooding	3.9–3.0	Flooding of basins by lava produced by radioactive decay
Quiet crust	3.0–now	Bombardment by small particles to pulverize and erode surface

Source: Adapted from H. H. Schmitt, "Evolution of the Moon: The 1974 Model," *Space Science Reviews*, vol. 18 (1975), p. 259.

FIGURE 15.20 The battered surface of the moon—an Apollo 14 photo of a field of boulders. [*Courtesy NASA.*]

floated to the surface of the melted shell; volatile materials were lost to space. The crust began to solidify from this melted shell about 4.4 billion years ago. From 4.1 to 4.4 billion years ago, the crust slowly cooled. The bombardment from space continued, but began to taper off. This falling debris made many of the craters now found in the highland areas.

Below the surface the moon's material remained molten. About 3.9 to 4.1 billion years ago, a few huge chunks smashed the crust to produce basins which later became maria. For example, the Mare Orientale basin formed some 4 billion years ago when an object about 25 km across smashed into the moon. Only later did the basins fill with lava. Though the crust lost its original heat of accretion, short-lived radioactive elements (which decay rapidly) reheated sections of it. From 3.0 to 3.9 billion years ago, lava formed by the radioactive reheating punctured the thin crust beneath the basins, flowing into them to make the maria.

The higher density basalts from the mantle probably produce the mascons detected beneath the lunar maria. They may be due in part, however, to remnants of the impact projectile buried beneath the surface or to compression produced by the impact. This is probably the case for a basin, such as Orientale, which contains little lava flow.

For the past 3 billion years the crust has been inactive. However, small particles from space have incessantly plowed into the surface since it solidified. These sand-sized grains scoured the surface, smoothed it down, and pulverized it. Continued bombardment by larger bodies churned the fragmented surface. Impacts melted the soil, which swiftly cooled to form breccias and glass spheres. The moon's surface today resembles a heavily bombarded battlefield—constantly fragmented, stirred up, and melted (Fig. 15.20).

To recap (Fig. 15.21), we suppose the moon to have formed about 4.6 billion years ago. The rocks fall into two main age groups: highland rocks about 4 billion years old and

FIGURE 15.21 A model for the moon's evolution. (a) 3.9 to 4 billion years ago. The surface has solidified and is cratered by the infall of small bodies; most of the surface is saturated with craters. (b) 3.0 to 3.2 billion years ago. Fractures in the surface allow lava from the interior to flow out and fill the lowland basins with material darker than the highlands. (c) Now. Many of the large rayed craters blossomed on the surface after the formation of the maria. Parts of the maria have grown lighter in color from the material blown out by crater formation. [*Paintings by D. Davis and D. Wilhelm, U.S. Geological Survey.*]

lowland rocks about 3.5 billion years old. The highland rocks are the oldest and must have melted and solidified before the lowland rocks of the maria. Melting may have been caused by the heat from radioactive decay and the energy from the impact of many bodies falling in from space. After the crust solidified, the region below the crust heated and then melted from radioactive decay. A few last large objects formed the maria basins and the mascons beneath. They later filled with lava. Everything cooled, and now the moon is a desolate, quiet planet.

553

15.5 THE ORIGIN OF THE MOON

Although the moon, like the earth, is thought to have formed from the accretion of planetesimals, the details of its origin still present major problems. The implications of the Apollo cargo so far have clarified a few aspects, but none of the rival models has claimed supremacy in explaining the moon's origin. These models fall into three broad categories: fission, binary accretion, and capture.

The *fission model*, the earliest of these ideas, was developed in 1879 by Sir George H. Darwin (1845–1912) and recently revived and revised. Noting that the earth was slowing down due to tidal friction, Darwin surmised that the earth must have been spinning more rapidly in the past than at present. A faster rotation rate would have created a greater equatorial bulge than now exists. If the earth were molten, its equatorial speed may have been so great that friction and gravitational attraction would not have been able to hold the bulge to the earth. A chunk of mantle may have detached, spiraled out from the earth, cooled, and formed the moon (Fig. 15.22).

In 1889 O. Fisher speculated that the protomoon had split off from the Pacific Ocean basin and so was made of mantle material. The fact that the average density of the moon and the average density of the earth's mantle rock (3300 kg/m^3) are roughly the same lends support to this model. Fisher's fission model was widely accepted for about 30 years. Then it ran into major difficulties. First, the angular momentum of the earth-moon system must be conserved. If the moon merged into the earth now, the combined mass would spin much faster than the earth does now, about 4 times faster (once every 8 hours). But

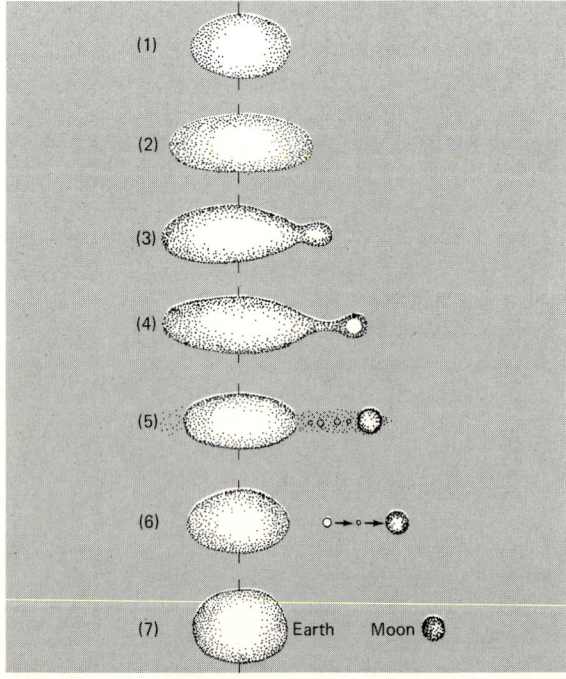

FIGURE 15.22 Formation of the moon by fission from the earth. Debris from the break-off rained down on the moon to crater it.

15.5 THE ORIGIN OF THE MOON

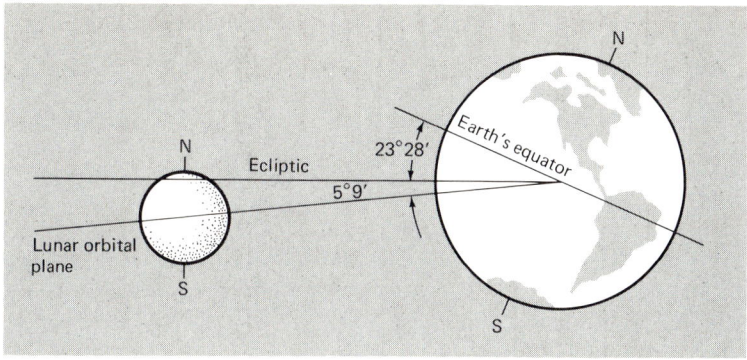

FIGURE 15.23 The tilt of the moon's orbit with respect to the earth's equator.

that rate is *not* fast enough to separate the lunar mass from the earth. Second, such a separation would take place at the earth's equator, where its rotation speed is the fastest. If angular momentum is conserved, the orbital plane of the broken-off mass must remain in the earth's equatorial plane. In fact, the moon's orbit is tilted with respect to the earth's equator (Fig. 15.23); it wobbles back and forth a bit, but the smallest angle it makes is 18.5°. Third, the evolution of crustal plates requires that the configuration of the ocean basins change constantly and rapidly, so existing basin features cannot be used to argue that the moon left the earth in the distant past. Fourth, the present rate of tidal friction would indicate that the moon separated from the earth only 1.2 billion years ago, but the earth was certainly solidified long before that.

Besides these problems, the Apollo results dealt the fission model a knockdown blow. The lunar basalts differ critically in chemical composition and water content from the terrestrial basalts that line the ocean basins. So the fission model, though once intriguing, is not now a promising idea.

The *binary accretion model* views the moon as created out of the same primitive cloud as the earth, rather than born from our planet. This approach connects to the condensation-accretion model of the formation of the solar system (Chapters 13 and 19). Dust particles grew from a gradual condensation of gas. These eventually accreted into planetesimals which accumulated to form the protomoon and protoearth. The moon formed so close to the earth that earth's gravity held the moon in a close orbit. The leftover bits and pieces from the formation of the two planets fell into the moon and heated its primitive surface.

Some lunar scientists argue that the composition of lunar minerals discredits this idea. Lunar and terrestrial rocks are somewhat *different* in composition. And the moon has no water. In addition, the moon has no dense iron core. If the embryonic environment and materials of the two bodies were the same, it is difficult to see how the earth accumulated iron and the moon did not. On the other hand, the fact that the oxygen isotopic composition of the earth and the moon are the same supports the condensation-accretion model. This model has the same difficulty as the fission model in accounting for the short evolution time of the lunar orbit.

Around 1955 the *capture model* was developed. According to this idea a vagabond moon did not form with the earth or from it, but rather in some other part of the solar system. By chance, the moon traveled close enough to the earth to be captured gravita-

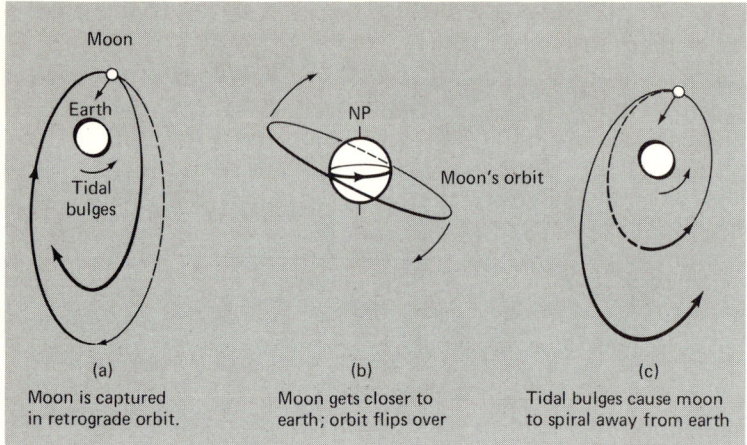

FIGURE 15.24 Capture of the moon. (a) The moon, formed in a part of the solar system away from the earth, passes close to the earth and is captured in a retrograde orbit. (b) Tidal forces flip the orbit over the earth's poles so the orbital direction is reversed. (c) Continual tidal interactions cause the moon to spiral from the earth.

tionally (Fig. 15.24). The earth caught the moon in a highly eccentric and retrograde orbit; that is, the moon orbited the earth in a direction *opposite* to the direction of the earth's rotation. In this circumstance the earth's tidal bulge acted to slow the moon down at its closest approach, so its orbit decreased in size. The tidal bulge also would act to change the inclination of the moon's orbit, tipping it over the earth's poles. When the orbit finally did flip all the way over, its direction was reversed relative to the earth's rotation. Once the moon was revolving in the same direction as the earth rotated, the tidal bulge worked to speed up the moon on its closest approach, and so the orbit became larger—as happens now at a rate of about 3 cm/yr.

The capture model agrees in part with the inferred dynamical history of the earth-moon system. If the moon is now moving away from the earth, it must have been closer in the past. The capture model requires a closer moon at the time of its capture. Theoretical calculations of the earth-moon dynamics further back up the capture model. These show that if the moon were captured 5 billion years ago, it would have come closest to the earth 1.2 billion years ago and then moved out to its present distance.

Some other pieces of evidence support the capture model. For instance, if the moon formed in another region of the early solar system, its early environment would have been different from that of the earth. So the differences between earth and moon—the moon's lack of water and iron, its lower average density, and its high abundance of uranium and rare earth elements—would arise from the different formation sites.

However, if the moon were formed far from the earth, it would not have the same isotopic composition of oxygen. Also, the capture model fails to come to grips with the details of the moon's formation. Instead it pushes them into the murky background of the formation of the rest of the solar system. Finally, the capture model requires that very special conditions must have occurred in a very particular sequence for the moon's capture by earth to have taken place. It's very hard for the earth by itself to capture a body as massive as the moon. In fact, it's impossible for a spherical body to capture another spherical body. The extra gravitational force of the earth's equatorial bulge, plus that of the sun, may have done the trick, but it's not easy to make it work; perhaps a third body was also nearby.

To date, none of the above models wins out as the best explanation for the origin of the moon. The results from the Apollo mission have complicated the situation with a wave of details, still to be put into a more general framework. Further exploration (it can be unmanned) of the moon might provide additional clues to support one of these models over another, but it might also generate a new set of questions.

15.6 CRATERING AND THE EVOLUTION OF THE TERRESTRIAL PLANETS

Cratering played a large role in the surface evolution of the moon. This chapter and the next two show you that every terrestrial planet and nearly every satellite in the solar system are scarred by craters. The fact that craters are so universal is fortunate, for their shape and number provide clues to the astronomical and geological processes that formed the terrestrial planets.

Impact craters are created by the infall of massive objects on a planet's surface. The great abundance of craters on the planets and satellites implies that a storm of ancient impacts blasted them long ago. Since that time, erosion and crustal evolution have modified the original pattern. These modifications tell us about the evolution of the planets' surfaces. In addition, if we can estimate the rate at which cratering occurred, the accumulation of craters gives a guide to the age of the surface on which they are found.

Here are a few major points about cratering. On any planet's surface small craters greatly outnumber the larger ones. The size of a crater relates to the mass of the projectile that formed it and so to the projectile's size. The largest of these must have been pieces of rock greater than 100 km in size. Craters, once formed, can be wiped out by a number of processes, even on an airless world: (1) by later impacts upon them, (2) by materials thrown up from younger, nearby craters, (3) by lava flows. On a planet with an atmosphere, erosion also plays a role.

The analysis of cratering leads to one key overall conclusion: The terrestrial planets have undergone different amounts and kinds of surface evolution. We can rank the planets and satellites according to the relative amounts of modifications of their surfaces. Our moon rates as the least modified, along with Deimos and Phobos (the moons of Mars), followed by Mercury, Mars, Venus, and the earth. Our planet has undergone by far the most surface evolution since its formation; the motion of crustal plates and weathering have completely obliterated the earth's original surface. The oldest regions of the continental plates are roughly 2 billion years old, and the oldest known impact crater is only about 1 billion years old.

Our moon stands in stark evolutionary contrast to our planet. The oldest regions of the lunar highlands have not changed much since their formation. Analysis of cratering and radioactive dating of Apollo samples tell us that a little over 4 billion years ago the cratering rate was 100 to 1000 times greater than now and that this intense influx tapered off, reaching the present rate about 3 billion years ago.

15.7 ORBITAL AND PHYSICAL CHARACTERISTICS OF MERCURY

Now let's look at another heavily cratered, little-evolved world: Mercury. Even though Mercury's surface reflects light poorly (its albedo, 0.056, is even a bit less than that of the

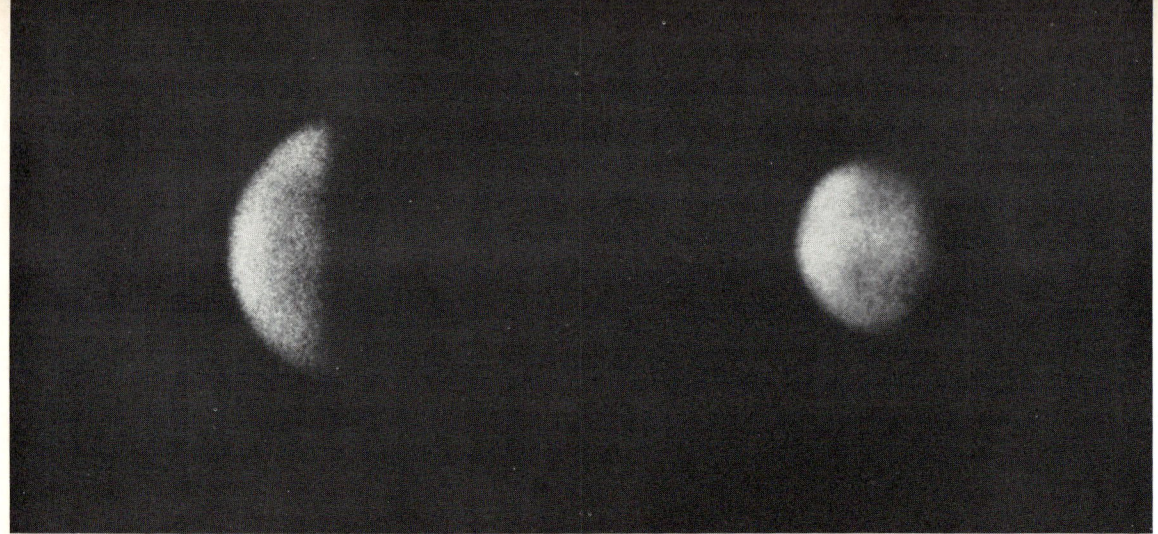

FIGURE 15.25 Two views of Mercury taken by earth-based telescopes. Note the vague, dusky features. [*Photo by C. Knuckles, New Mexico State Observatory.*]

moon), its closeness to the sun bathes it in such intense light that its brightness at maximum, as seen from the earth, can rival that of the star Sirius. However, because Mercury's maximum elongation from the sun averages only 23°, and at most is only 28°, the planet is hard to observe. This fact has resulted in limited and sometimes confusing information about this swift-moving world.

The Mariner 10 mission revealed that Mercury's surface resembles the moon in many ways; it opened up a direct comparison of these inactive, desolate bodies. This comparison hints at what processes drive the evolution of those planets without atmospheres or active interiors.

Mercury's Orbit

Mercury speeds around the sun once every 88 days. Its orbit is one of the most eccentric in the solar system. When closest to the sun, at *perihelion*, its distance from the sun is 46 million km; its greatest distance, at *aphelion*, is 70 million km. This difference means that sunlight at perihelion falls on the planet 2.3 times more intensely than it does at aphelion.

Mercury's Rotation

Mercury's small angular size and poor visibility render the study of the surface difficult (Fig. 15.25). But persistent and careful observation revealed faint, dark, apparently permanent markings. In 1877 Giovanni Schiaparelli (1855–1910) constructed the first map of the Mercurian surface. Since the position of the surface features apparently did not change, he concluded that the rotation rate of Mercury equaled its revolution rate, 88 days. Schiaparelli thought Mercury always presented one side to the blast of the solar radiation, and the other to the cold of space. He and other astronomers who believed that Mercury had the same rotation and revolution rate were in part influenced by our moon's example.

This long-accepted view changed radically in 1965, when radio astronomers Gordon

15.7 ORBITAL AND PHYSICAL CHARACTERISTICS OF MERCURY

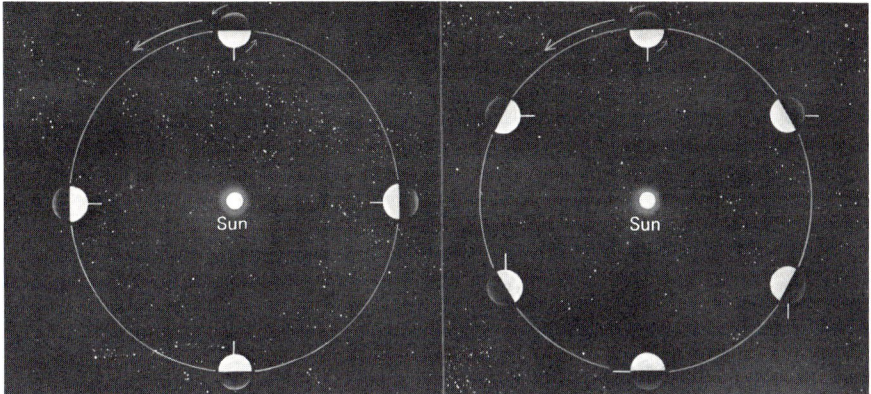

FIGURE 15.26 Mercury's rotation. Before 1965 astronomers believed that Mercury rotated in 88 days, so it kept the same side toward the sun (*left*). Radar measurements in 1965 showed the rotation period is 59 days (*right*). [*Courtesy NASA.*]

H. Pettengill and Rolf B. Dyce, using the mammoth radio telescope at Arecibo, Puerto Rico, measured the Doppler shift of radio beams reflected from opposite edges of Mercury's disk. They determined the rotational velocity and discovered that Mercury rotates in about 59 days (Fig. 15.26). This new result astounded many astronomers, for all maps drawn since Schiaparelli's time had supported the 88-day rotation period.

How could something believed to be correct for so long turn out so wrong? The trick is this: Mercury's rotation period is 2/3 its orbital period. An exact 2:3 ratio (rotation 58.65 days, revolution 87.97 days) implies that Mercury's spin and orbit are coupled, most likely by tidal interaction with the sun.

How did this 2:3 ratio mislead ground-based observers? First, because Mercury's closeness to the sun makes it hard to observe, there are relatively few observations. Second, it is difficult to get a rotation period from a small set of observations of surface markings. If we could see the planet all the time, it would be simple. We would simply wait until a particular mark comes around again. But if at the time the mark reappears Mercury is too close to the sun to observe or it is cloudy, we miss it. So we only get a fraction of the possible sightings of the mark, and we have to figure out what that fraction is, not knowing the period in the first place.

Look at Figure 15.27. Line A shows times at which a particular mark is sighted (the arrows). Lines B, C, and D show how three different rotation rates would fit those observations, the other appearances of the mark being missed for one reason or other. Which

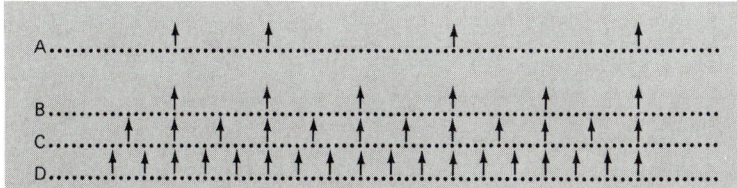

FIGURE 15.27 Rotation period and observational selection.

one is right? You can't tell without further observation. One of the periods that matched the observations of Mercury was 88 days, and since this was the same as the period of revolution, people jumped to the conclusion that Mercury kept the same face to the sun, just as the moon does to the earth. They completely ignored the fact that several other periods, such as 59 days, 29 days, etc., were possible. Sometimes an astronomer's preconditioned notions about something affect how it is seen!

Mercury got locked into a rotation period of $\frac{2}{3}$ its orbital period because of a curious coincidence arising from its eccentric orbit. When Mercury is at perihelion, it is at 0.794 of its average distance from the sun. By the conservation of angular momentum, the velocity then must be larger than average by $1/0.794$. The angular speed in its orbit must be larger than average by $1/(0.794)^2$, since both the distance from the sun is smaller and the velocity is larger. So the instantaneous period of revolution, that is, the period it would take to rotate once at that angular speed, is 88 days $\times (0.794)^2 = 56$ days. This is very close to the 2:3 ratio period of 58.7 days. Add the fact that Mercury is slightly bulged. Then with a 2:3 ratio of rotation period to revolution period Mercury always has its bulge toward the sun when it is at perihelion. The tidal forces acting on the bulge are strongest at this time (remember, they are proportional to $1/r^3$), and they act to keep Mercury synchronized as closely as possible to its instantaneous period of revolution at perihelion. In other words, near perihelion (but not in general) Mercury does keep the same face toward the sun.

How long is the solar day on Mercury? We can derive a relation between the solar day and the sidereal day in a way very similar to the derivation of the relation between synodic and sidereal periods of the planets (Section 3.2). Suppose you were on Mercury with the sun directly overhead (Fig. 15.28). Let T be the period of revolution of Mercury about the sun (88 days), P the sidereal period of rotation (58.7 days), and S the synodic period of rotation (the solar day). After one earth-day Mercury will have rotated $360°/P$ (angle A) with respect to the stars, but only $360°/S$ (angle B) with respect to the sun. The difference between these two angles (angle C) is equal to the angular distance Mercury has moved in its orbit, $360°/T$ (angle D). (Remember the theorem from geometry about alternate interior angles of parallel lines?) Stated algebraically,

$$\frac{360°}{P} - \frac{360°}{S} = \frac{360°}{T}$$

or

$$\frac{1}{S} = \frac{1}{P} - \frac{1}{T}$$

Putting in the numbers:

$$\frac{1}{S} = \frac{1}{58.7} - \frac{1}{88.0} = 0.00567$$

$$S = 176 \text{ days}$$

So Mercury's solar day is just twice the length of Mercury's year.

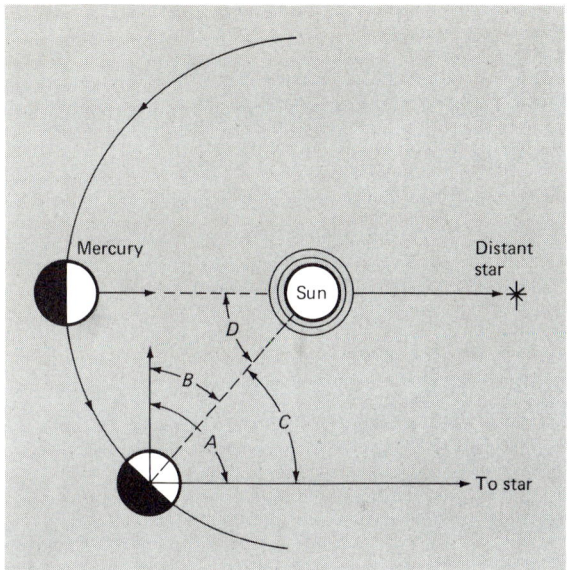

FIGURE 15.28 Geometry for finding the length of the solar day on Mercury.

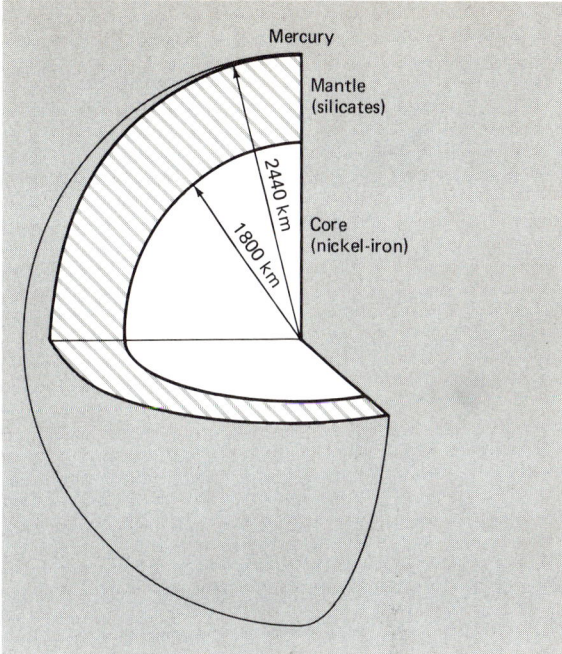

FIGURE 15.29 A model for Mercury's interior. Note the large nickel-iron core.

Mercury's Size

Once we know Mercury's orbit and thus its distance, we can find its physical size from its angular diameter. Mercury turns out to be a tiny world 2440 km in radius. That's only about 40 percent larger than our moon and roughly 38 percent the size of the earth.

Mercury's Mass and Density

Since Mercury has no known natural satellite, it's hard to determine its mass accurately. During the Mariner 10 mission the spacecraft sped past Mercury three times. Its acceleration due to Mercury's gravity was accurately measured, so we now have a good value for Mercury's mass: 0.055 that of the earth, or about 3.3×10^{23} kg.

For such a small planet Mercury has a fairly large mass. That means its bulk density must be high: 5420 kg/m^3, essentially the same as the earth's. But Mercury's gravity is smaller than the earth's, compressing it less, so it must contain more heavy elements. Comparing Mercury's interior with that of the earth, we expect a large nickel-iron core (Fig. 15.29) and a rocky mantle, but we don't yet have seismic information to confirm this interior model in detail.

15.8 THE SURFACE ENVIRONMENT OF MERCURY

Suppose that at perihelion at noon you stood on the equator of Mercury. You would need to be made of sturdy stuff, because the surface temperature would be about 700 K! The surface temperature drops to 425 K at sunset and reaches about 100 K at midnight. This range of temperatures is the widest known in the solar system.

The long, hot solar day and low escape velocity (4.2 km/sec) make it impossible for Mercury to have an extensive atmosphere. Gas molecules, even the more massive ones, would easily be heated to escape velocity. So any atmosphere could not be expected to last long (Section 14.5).

The Mariner 10 space probe that winged its way past Mercury had on board an ultraviolet spectrometer to search for an atmosphere. This device detected an atmosphere (of sorts) of helium and hydrogen, but the surface atmospheric pressure was very small, a little less than 10^{-15} that of the earth.

The following may give you some idea how incredibly thin Mercury's atmosphere is. At the earth's surface every cubic centimeter contains about 10^{19} particles. At Mercury's surface you'd find only about *10,000* particles in a cubic centimeter! Another way to see it: Imagine taking one cubic centimeter of earth's air and spreading it out over the whole state of Ohio; it would then have Mercury's atmospheric density.

No atmosphere means no insulation from space. That's why the range of noon-to-midnight temperatures on Mercury is so severe. Compare the temperatures to the moon's. Night on the moon and Mercury are essentially the same. Since neither has an insulating blanket, infrared radiation from the sunless side escapes directly into space during the long night. Some heat is supplied from the interior by conduction. So both bodies have about the same midnight temperature, about 100 K. At noon the surface temperature depends on the surface albedo and distance from the sun. The moon and Mercury have about the same albedo, but Mercury is 2.5 times closer to the sun than the moon. So you'd expect that Mercury would be $2.5^{1/2} = 1.6$ times hotter at noon, in the vicinity of 600 K (Section 15.2). That's just about what is measured.

With the day so hot and the escape velocity so low, why does Mercury have any atmosphere at all, especially of helium, a very low-mass gas? You'd expect helium by itself to last only about 10^5 sec (a month) before it is lost to space. One answer is the solar wind. Mercury's atmosphere does fly out into space, but the influx of material from the solar wind, which contains 2 to 20 percent helium, could possibly replenish the loss. A problem here is Mercury's magnetic field (Section 15.10), which deflects much of the solar wind, except when it is very strong. Another source, at least for helium, is decay of radioactive elements in Mercury's interior. This balance between loss and gain results in a thin, but detectable, atmosphere.

15.9 THE SURFACE OF MERCURY CLOSE UP

Television cameras onboard Mariner 10 scanned Mercury's surface as the spacecraft passed close by three times. The images sent back increased our resolution of planetary detail by 5000 times. The surface resembles our moon (see Fig. 15.2), but there are differences: fewer small craters (20 to 50 km in size); no mountains; many shallow, scalloped cliffs (called *scarps*; Fig. 15.30) reaching lengths of hundreds of kilometers and rising to 1-km heights; fewer basins and large lava flows; and more relatively uncratered flatlands (called *intercrater plains*) amid the heavily cratered regions. These differences are important, yet the general similarity jumps out. Mercury's highlands are riddled with

15.9 THE SURFACE OF MERCURY CLOSE UP

FIGURE 15.30 A scarp on Mercury's surface, more than 300 km long. In the photo it extends diagonally from upper left to lower right (arrow). [*Courtesy NASA.*]

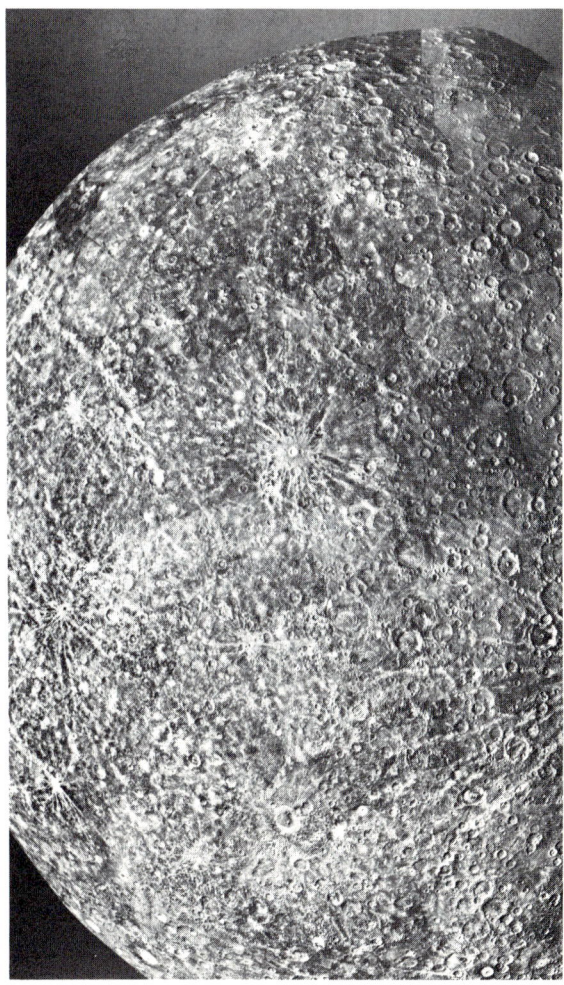

FIGURE 15.31 Impact craters in Mercury's south pole region. The craters with the brightest rays are the youngest. The largest craters in this photo are about 200 km across. [*Courtesy NASA.*]

craters like the moon's bleak highlands. Light-colored rays spring from some of the craters (Fig. 15.31), an indication that these were formed by violent impacts during Mercury's stormy past. Some craters are over 200 km in size, comparable to the biggest lunar craters.

One difference between the moon and Mercury results from the fact that Mercury has twice the surface gravity of our moon. Material ejected from an impact will cover a smaller area than from an impact with the same energy on the moon. So secondary impact craters cluster more tightly around their primary craters, and ejected material that forms rays does not spew out as far.

And what of large basins, such as found on the moon's near side? Mariner 10 found only one, the Caloris ("hot") Basin on the northwest part of the planet (Fig. 15.32). Since the Caloris Basin sat on the sunrise line, only about half of it was photographed. It probably has an overall diameter of some 1300 km. The basin is bounded by rings of mountains about 2 km high. In size and structure the Caloris Basin resembles the moon's Orientale Basin: the blasted area from the impact of a large celestial chunk (some tens of

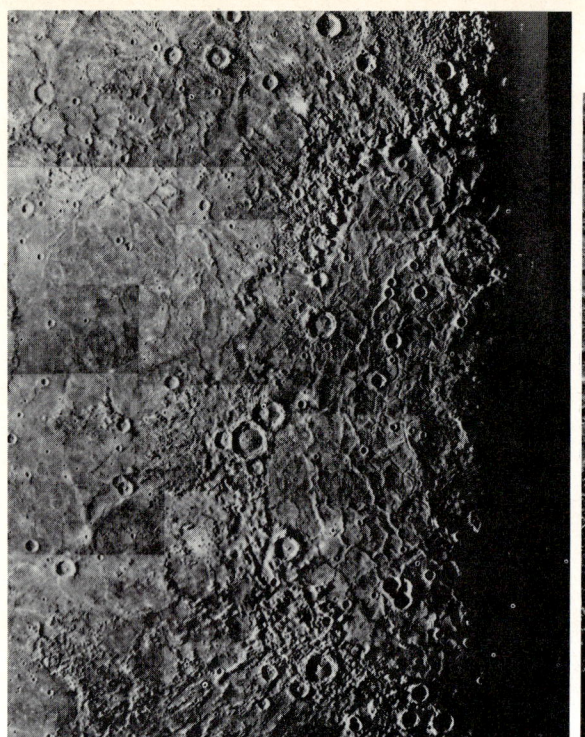

FIGURE 15.32 The Caloris basin. Only a part shows in the right half of this photo; the rest is in shadow. The basin is about 1300 km across and is rimmed by mountains 2 km high. [*Courtesy NASA.*]

FIGURE 15.33 Weird terrain on Mercury's surface diametrically opposite the Caloris basin. This rugged region may have resulted from shock waves from the Caloris impact that traveled through the planet. [*Courtesy NASA.*]

kilometers in size) that pushed up mountains, ejected material that fell over a large area, and sculpted the older surface.

The Caloris Basin has a crinkled floor, perhaps fractures from rapid cooling of lava. Older craters were flooded by the lava outpouring from the Caloris impact. The Caloris impact may have been so strong that shock waves actually traveled through Mercury to the other side of planet, disturbing the surface there. Opposite Caloris we see a jumbled region, unique on Mercury's surface (Fig. 15.33). Hills and ridges cut across craters and the intercrater plains. This region of weird terrain may be the cross-planet disruption powered by the Caloris impact.

On Mercury even the most heavily cratered regions are not completely saturated with craters, and the number of small craters is less than would be expected if the craters had formed at the same rate as on the moon. From these two observations we can infer either that the cratering material had a different range of sizes for Mercury than for the moon or that some process modified the surface after the cratering.

Mercury's scarps vary in length from 20 to over 500 km and have heights from a few hundred meters up to 1 to 2 km. Individual scarps often cut across different types of terrain. If the regions photographed by Mariner 10 represent Mercury's overall surface, the characteristics of the scarps imply that Mercury's radius has shrunk some 1 to 3 km, probably from cooling of its core or crust or both.

15.10 THE EVOLUTION OF THE MOON AND MERCURY COMPARED

The evidence is clear and direct that the force that shaped the surfaces of Mercury and the moon was the impact of projectiles from space that pitted the surfaces with craters. Both the moon and Mercury lack substantial atmospheres, so there is no erosion from weathering. And both are tiny worlds with relatively cool interiors compared to the earth's interior, so neither has had much, if any, volcanic activity and neither has undergone the continual surface evolution from the shifting of crustal plates such as the earth experiences (Section 14.7).

The lack of atmosphere and the short period of crustal evolution relates to the small masses of Mercury and the moon. Their surface gravities are so low that most gasses have thermal velocities greater than the escape velocity, and atmospheres are not retained. The small masses also imply that internal heating from radioactive decay would be less compared with that for the earth, and the flow of heat outward would be so fast that both bodies would cool off quickly. The earth is hot at its interior, and the outward flow of heat sets up currents in the plastic mantle; these power the evolution of the earth's crust. Both Mercury and the moon now lack this combination of a hot interior and plastic mantle.

Without wind and water erosion and crustal evolution, the moon and Mercury retain the evidence of their early years. The similarities of surface features suggest similar histories. For example, the Caloris basin on Mercury resembles maria on the moon. The lunar maria were probably formed by the flow of lava into large basins created by impacts of large bodies. Such processes most likely also formed the Caloris basin.

Lunar analogies imply a working model for the evolution of Mercury. (But keep in mind that we have a lot less information about Mercury than we do about the moon, so the details of the model are less certain. In particular, we know nothing about Mercurian rocks.) Mercury probably went through the following general stages: (1) heating and melting of the surface (by impacts or radioactive decay) followed by cooling and formation of a solid crust, (2) heavy cratering, (3) formation of impact basins, (4) filling in of basins by volcanism, and (5) low-intensity cratering. We can't date this sequence as we can for the moon because we do not have rock samples from Mercury's surface. But a lunar comparison suggests that the intense sculpting of Mercury's surface took place about 4 billion years ago (stage 1), not long after the planet formed.

The earliest phase of cratering (stage 2) must have been wiped out by later volcanism. We come to this conclusion because the intercrater plains seem to have covered up any scars of the earliest accretion. At about the same time, the scarps developed from global shrinking of the crust, interior, or both.

Basin formation (stage 3) must have come at the end of the heavy bombardment of the surface. A few large pieces crashed into the surface; one made the Caloris Basin. Not long after this time widespread volcanism (similar to that which made the lunar maria) created the broad, smooth planes such as those adjacent to the Caloris Basin (stage 4).

After stage 4 a few impacts punched out the rayed craters (stage 5), but basically Mercury has evolved little since its formative times. It is now a dead world.

15.11 THE MYSTERY OF MERCURY'S MAGNETIC FIELD

As Mariner 10 flew past Mercury, it detected, much to everyone's surprise, a weak planetary magnetic field, about 10^{-2} times the strength of the earth's surface magnetic field. Small as this sounds, it's sufficient to carve out a magnetosphere in the solar wind. Here the magnetic field deflects the charged particles (mostly protons) of the solar wind around the planet.

Mercury's field appears to be a dipole, more or less aligned with its spin axis. So, in general, Mercury's magnetic field is similar to the earth's, only weaker. And that's the problem. Mercury, like the earth, has a nickel-iron core. But it's presumed to be relatively cold and solid now because a small planet loses heat quickly. Recall (Section 14.4) that the earth's magnetic field supposedly arises from swirling motions in its hot, liquid nickel-iron core. The churning is thought to be driven by the earth's spinning. Because Mercury rotates much more slowly than the earth and is expected to have a cool core, no one really expected it to have a planetwide magnetic field.

What's the explanation? No one knows for sure. It may be that the dynamo model is incorrect; after all, Venus, which has a much larger and presumably hotter core than Mercury, has no detectable magnetic field. (But Venus rotates even more slowly.) Perhaps the field is left over from an older time, or maybe it originates from the planet's interaction with the solar wind. Or perhaps small planets with cool nickel-iron cores produce magnetic fields by a mechanism not yet imagined.

SUMMARY

The moon and Mercury are worn-out, inhospitable worlds. Essentially no atmosphere clings to their surfaces. Without a protective atmospheric blanket, temperatures range widely from noon to midnight. In addition, bodies from space have incessantly bombarded the surfaces. This bombardment has pulverized the surface layers.

The American and Russian space programs have brought back samples from the moon's surface, from which we can reasonably infer its history. Moon rocks fall into two broad categories based on age and composition. The samples from the maria are the younger, averaging 3.5 billion years in age. The aluminum-rich samples from the highlands are older, generally about 4 billion years old. Both types of rock show evidence of previous melting and rapid cooling. The first melting of the primeval surface took place over 4 billion years ago; it may have been caused by the infall of objects that also made many of the craters. Later, a few larger objects formed the mascons and maria, which then filled with lava. Since that time, 3.5 billion years ago, the moon has been an inactive planet, a silent companion to the earth.

We do not yet have samples from the surface of Mercury. But from a direct comparison of its surface and the moon's we infer that Mercury's history was much the same. We expect the timing of events to be roughly the same, too.

The moon and Mercury are dead worlds, their dramatic evolution ended. Compared with the earth, they are fossil planets rather than living ones. This difference arises mainly from the fact that the masses of the moon and Mercury are much less than that of the earth. Mass determines the extent of evolution of a terrestrial planet (more in Chapter 16).

PROBLEMS

Key Words

perigee
apogee
tides
tidal friction
tidal gravitational force
albedo
crater
mare
basin

Mare Orientale
Mare Imbrium
impact theory
volcanic theory
lunar soil
mare basalt
anorthorsite
breccia

KREEP
fission model
binary accretion model
capture model
perihelion
aphelion
scarp
intercrater plain
Caloris Basin

Review Questions

1. Suppose you stepped out of TWA flight 101 onto the moon's surface. You look around at both the ground and the sky. How does what you see differ from what you would see on the earth? (Objectives 3 and 6)
2. Suppose you stepped out of TWA flight 102 onto the surface of Mercury. How would the scene differ from that on the earth? (Objectives 5 and 6)
3. Argue from a comparison of average density that the moon *cannot* have a nickel-iron core like the earth's and that Mercury *must* have a nickel-iron core. (Objectives 1 and 10)
4. How were most of the craters on the moon formed? On Mercury? Back up your statement with specific evidence. (Objectives 3 and 5)
5. What specific evidence do we have that the moon's lowland regions (maria) formed after the highlands? (Objective 11)
6. You are writing a grant proposal to NASA to do research on the origin of the moon. Describe in the best light possible the model you plan to support. (Objective 12)
7. Compare the characteristics of the Orientale Basin on the moon to the Caloris Basin on Mercury. (Objectives 3, 5, and 6)
8. In one sentence, describe how the surfaces of the moon and Mercury *differ*. (Objectives 3, 5, and 6)
9. Neither the moon nor Mercury has a substantial atmosphere. Why not? (Objective 7)
10. In one sentence, describe the difference between the interiors of the moon and Mercury. (Objectives 8, 9, and 10)

Problems

1. For a magnetic dipole, the magnetic field strength is proportional to the inverse cube of the distance ($B = B_0/R^3$). The magnetic field of Mercury is about 1 percent as strong as that of the earth. At what distance from the center of Mercury would you find a field strength equivalent to that in the van Allen belts around the earth (which lie about 10,000 to 20,000 km from the center of the earth)? Does your answer suggest a reason why such belts are *not* detected around Mercury? (Objective 6)
2. Calculate the surface gravity on Mercury and compare with that on the moon. (Objective 1)

3. At what wavelength would Mercury radiate most strongly on the day side? On the night side? (Objective 7)
4. What would the surface temperature of the moon be if it were made of highly reflective material, with an albedo of 0.90? (Objective 7)
5. Few substances can be heated hotter than 2000 K without melting. How close could a spaceship (radiating and absorbing like a perfect blackbody) come to the sun and not reach this melting temperature? (Objective 7)
6. Geological and biological evidence shows that the average surface temperature of the earth could not have varied by more than about 20 K during the past 3.5 billion years. What does this fact imply about the variation of the Astronomical Unit and the solar luminosity over the same period? (Objective 7)
7. At a frequency of 10 GHz, calculate the difference (due to rotation) in the Doppler shift of a radio signal bounced off one side of Mercury compared to that bounced off the other edge. (Objective 4)

BEYOND THIS BOOK . . .

* The Apollo missions have produced a flood of data on the moon. You can find a short summary in "The Moon" by J. A. Wood, *Scientific American*, September 1975.
* For data about the moon as a planet, see *Planetary Geology* (Prentice-Hall, Englewood Cliffs, N.J., 1975) by N. M. Short. Another book with the same approach is *Geology of the Moon* (Princeton University Press, Princeton, N.J., 1970) by T. A. Mutch.
* For a personal story about Apollo 11, read *Carrying the Fire* (Farrar, Straus and Giroux, New York, 1974) by M. Collins.
* *Voyages to the Moon* (Macmillan, New York, 1960) by M. H. Nicolson is an amusing contrast to modern voyages to the moon.
* C. Chapman writes about the evolution of Mercury in *The Inner Planets* (Scribner, New York, 1977).
* For a pictorial tour of Mercury, look at *The Atlas of Mercury* (Crown, New York, 1977) by C. Cross and P. Moore.
* B. Murray discusses the results of Mariner 10 about Mercury in "Mercury," *Scientific American*, September 1975.
* A good recent text on the terrestrial planets, taking the approach of comparative planetology, is *Earthlike Planets* (Freeman, San Francisco, 1981) by Bruce Murray, Michael C. Malin, and Ronald Greeley.

VENUS AND MARS: EVOLVED WORLDS

16

LEARNING OBJECTIVES

After studying this chapter, you should be able to:

1. Compare Venus and Mars in size, mass, and density to each other and to the earth.
2. Compare the surface environments (temperature, atmosphere, general surface features) of Venus and Mars to each other and to the earth.
3. List the main constituents of the atmospheres of Venus and Mars and compare them to those on the earth.
4. Sketch a model for the interiors of Venus and Mars and compare them to the earth's interior.
5. List the major surface features of geologic importance on Venus and Mars and compare them to those on the earth.
6. Compare the planetary magnetic fields of Venus, Mars, and the earth.
7. Evaluate your chances of survival on Venus and Mars.
8. Discuss the role of cratering in shaping the surfaces of Venus and Mars.
9. Discuss the implications of the volcanoes of Mars for the evolution of the Martian surface and atmosphere.
10. Compare the inferred geologic histories of Venus, Mars, and the earth. Justify an ordering of these planets from the least to the most evolved.
11. Describe how radar is used to determine the rotation rate of Venus and to map its surface.

CENTRAL QUESTION:
What forces have shaped the evolution of Venus and Mars, planets with hot interiors and substantial atmospheres?

Venus, the brilliant light of love (Fig. 16.1); Mars, the red sign of war (Fig. 16.2). These two worlds come the closest to the earth in space. And these planets resemble the earth more closely than any other planets in the solar system. They are truly terrestrial planets.

Yet in many ways the similarities are superficial. The differences go much deeper. Venus is blistering hot at its surface—hotter even than Mercury at noon! A thick, dense atmosphere of carbon dioxide presses heavily on the barren ground. On Mars the atmosphere also consists of carbon dioxide. But it's thin, offering no protection from the incoming solar ultraviolet and no hindrance to the outgoing infrared. Mars is mostly a cold desert, where the water ceased to flow at least tens of millions of years ago.

How did Venus and Mars become so different from the earth? What forces shaped their evolution? This chapter deals with the comparative evolution of these worlds.

16.1 ORBITAL AND PHYSICAL CHARACTERISTICS OF VENUS

Viewed from the earth, Venus can outshine every celestial body except the sun and the moon. If you know where to look, you can spot Venus even during the day. Venus shines so brilliantly because its unbroken swirl of clouds reflects 77 percent of the incoming sunlight back into space. This cloud cover completely frustrates any attempt to view its surface features with optical telescopes.

But we are now piercing Venus's cloudy veil to uncover the mysteries of this world. Radar beams, bouncing off the surface, sense what the topography is like. Spacecraft have passed by the planet, gone into orbit around it, and plunged through the atmosphere to its forbidding surface. What a difference from the earth! Surface temperatures hit 750 K. The atmosphere presses down at 90 atm, and the atmosphere contains mostly carbon dioxide. The surface topology is mostly flat with a few highland masses, very little like the earth.

16.1 ORBITAL AND PHYSICAL CHARACTERISTICS OF VENUS

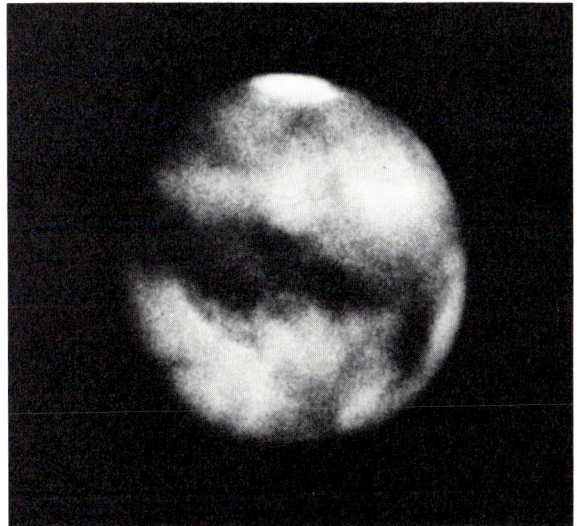

FIGURE 16.1 Venus, photographed in blue light. [*Courtesy Palomar Observatory, California Institute of Technology.*]

FIGURE 16.2 Mars as seen from the earth. Note the white polar cap at top and fuzzy dark regions. [*Courtesy Lick Observatory.*]

Venus has gained the reputation of being the earth's twin sister. In terms of size (the diameter of Venus is 12,104 km; earth, 12,756 km), average density (Venus, 5200 kg/m^3; earth, 5500 kg/m^3), and mass (Venus, 4.87×10^{24} kg; earth, 5.98×10^{24} kg) the sisterhood is appropriate. In most other ways Venus has a personality tremendously different from the earth. No oceans, streams, or lakes; no gentle, cool breezes. Instead, a veritable hell!

Revolution and Rotation

Second planet from the sun at an average distance of 0.72 AU, Venus completes one orbit in only 225 days. Venus's orbit is the most circular of all the planets, so Venus's distance from the sun varies by only 1.5 million km (1.4 percent) perihelion to aphelion.

Thwarted by the lack of a surface view because of the unyielding cloud cover, astronomers before 1961 had no real idea of Venus's rotation rate. Some proposed 24 hours (arguing that Venus is earth's twin), and others decided that the rotation period must equal the revolution period, 225 days. (These people were misled by the incorrect rotation rate agreed upon for Mercury.) Since 1961 radar echoes have been bounced off the surface of Venus. Measuring the Doppler shift in the returning signals, radar astronomers have found that Venus rotates once every 243.01 days, and the rotation is retrograde, that is, it

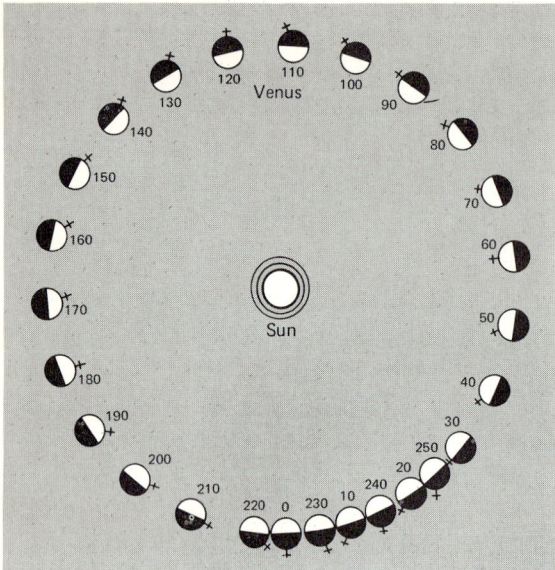

FIGURE 16.3 Rotation of Venus and the solar day. With respect to the sun, Venus rotates in 117 terrestrial days.

FIGURE 16.4 Tidal bulges and air flow in the atmosphere of Venus.

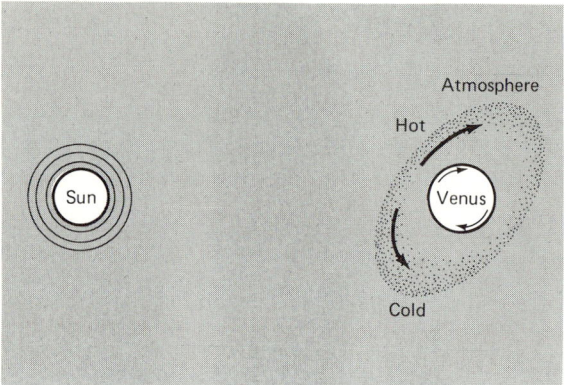

spins from east to west, rather than west to east as does the earth. So on Venus the sun (if you could see it through the clouds) rises in the west and sets in the east.

This long retrograde rotation results in a solar day 117 terrestrial days long (Fig. 16.3). But, because of the clouds, the sun's disk can't be seen directly. Like an overcast day on earth, it would be light during the day, but the sun's intensity on Venus is only 10 W/m^2, about 1/100 that at the earth's surface.

There is one curious fact about Venus's rotation. If Venus spun once every 243.16 days, it would rotate exactly four times during the interval between alignments of Venus and the earth on the same side of the sun. (This alignment is called *inferior conjunction*.) That value is very close to, but not exactly, the 243.01 day rotation period. So Venus almost presents the same face to the earth at successive inferior conjunctions—almost, but not quite!

Why is Venus's rotation period so close to being in resonance with the earth and its orbital period? One suggestion is that it arises from a balance between tidal forces acting on the core and mantle of Venus, which tend to slow down the rotation, and other tidal forces acting on the atmosphere, which tend to speed up the rotation. The slowing-down forces are similar to those produced by the moon on the earth, but on Venus it is the sun producing the force, and it acts on the bulk of Venus, not on oceans.

The atmospheric tidal force is a little trickier to understand. The sun heats the atmosphere of Venus, but because of its rotation, the hottest spot is not exactly on the side toward the sun, but displaced in the direction of rotation. The solar heating produces an increase in pressure, which causes some of the atmosphere to flow toward the cooler regions, producing a bulge (Fig. 16.4). The tidal force of the sun acting on this bulge tends to speed up Venus's rotation. According to computer models of this effect, the two tidal forces just happen to balance when the rotation rate is about 243 days. So the coincidence with the period between successive inferior conjunctions is just that, a coincidence. (Note that the *in*equality of the two periods was proven only when the measure-

16.2 THE ATMOSPHERE OF VENUS

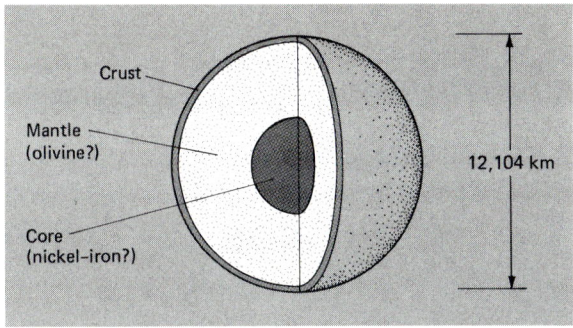

FIGURE 16.5 A model for the interior of Venus.

ment was made to five significant figures—an example of the importance of *accurate* observations.)

Size, Mass, and Density

We can measure the earth-Venus distance directly by radar. Then from the planet's angular diameter, we can calculate its physical diameter and radius. Venus turns out to have a radius of 6052 km, only 5 percent smaller than the earth.

Like Mercury, Venus has no known natural satellite. So we can accurately find the mass of Venus only when we send a spacecraft to it. Venus's mass comes out to be 0.815 times that of the earth, or 4.87×10^{24} kg.

With the mass and size in hand, we can find out Venus's bulk density: 5200 kg/m^3, almost the same as the earth. We guess (since we have no seismic data for Venus) that the interior of Venus should closely resemble the earth's interior (Fig. 16.5). We expect a rocky crust (which Venus landers confirmed), a mantle, and a nickel-iron core. Because Venus has a lower density than the earth, we imagine that it has a somewhat smaller core.

Where's the Magnetic Field?

But this model has a severe problem. A nickel-iron core, liquid in part, implies by comparison with the earth that Venus should have a planetary magnetic field. Because Venus rotates 243 times more slowly than the earth, we expect its internal dynamo to be weaker and so the magnetic field to be less intense than the earth's. But no probe to date has detected *any* magnetic field. If one exists, it must be at least 10^4 times weaker than the earth's magnetic field. That's *much* weaker than expected from a simple dynamo model. What a magnetic mess—Mercury has a planetary field, and Venus does not. Mars has a barely detectable magnetic field (Section 16.5). Perhaps the dynamo model is not the correct one to apply to all planets.

16.2 THE ATMOSPHERE OF VENUS

The atmosphere of Venus differs remarkably from ours, and that difference is a key to understanding the planet's evolution.

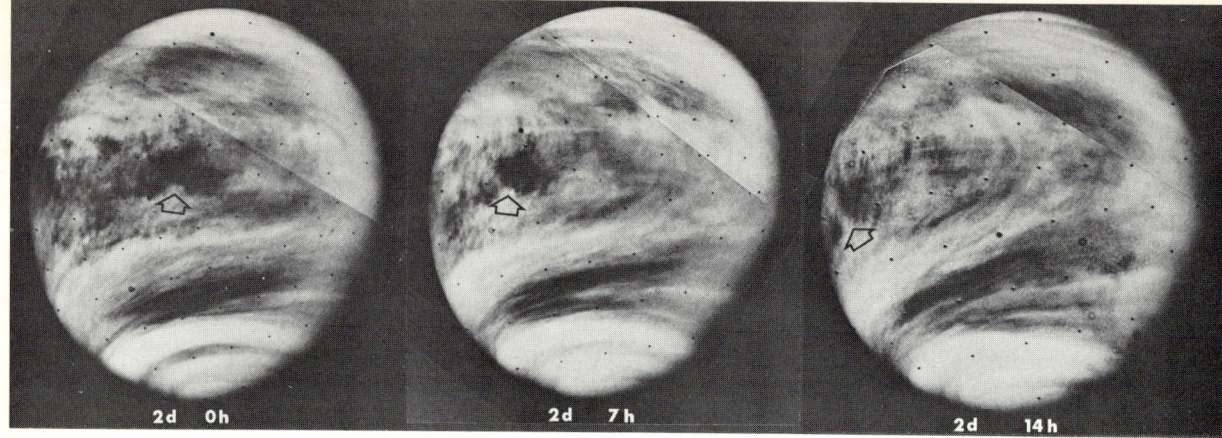

FIGURE 16.6 Rapid motions (arrow) in the upper cloud layers of Venus. The clouds circle the planet once in about 4 days; shown here are 14 h of that circulation. [*Courtesy NASA.*]

Clouds

The clouds of Venus, yellow-white in color, enshroud the planet completely (Plate 7). That's why we've known so little about the planet for so long.

Recent efforts have gathered much new information about the clouds of Venus. The cloud tops (which you see in a telescope) reach about 65 km above the surface. (The highest clouds on the earth go up to about 16 km; most are below 10 km.) Ultraviolet photography by Mariner 10 revealed that these cloud tops flow with the upper atmosphere in patterns similar to jet streams of the earth (Fig. 13.14). Ringing the equator, the clouds whiz around east to west at roughly 350 km/hr, fast enough to orbit the planet in only four days in a direction opposite its rotation (Fig. 16.6). So Venus's atmosphere circulates around the planet 60 times faster than the surface rotates! In contrast, the earth's atmosphere rotates west to east at the same rate as the surface. In addition to this planetwide circulation, winds also blow from the equator to the poles in large cyclonic cells 100 to 500 km in diameter. They culminate in two giant cloud vortices capping the polar regions. We don't know what drives such fierce winds. It's generally believed that solar heating does the trick, as on the earth, but details have yet to be worked out.

The Pioneer-Venus probes found that the clouds float in two broad layers (Fig. 16.7). The upper cloud deck tops at roughly 65 km and has a thickness of some 5 km. Here the liquid drops in the clouds have diameters of about 2 μm. Below the upper deck sits a thin haze layer. Below that, at 49 to 52 km, is the lower cloud deck, by far the densest layer. The cloud particles here are both liquid and solid. Below 49 km, the clouds gradually thin out. Below 33 km, the atmosphere is clear of any particles down to the surface.

In the densest part of the clouds the visibility is less than a kilometer, a fog which would be dangerous to fly through, but still not pea-soup. The clouds of Venus aren't particularly murky, just very thick. Because of the thick, cloudy atmosphere, the sunlight that reaches the surface has a yellow-orange color, giving the land an alien glow.

An infrared sensor on the Pioneer-Venus orbiter found the polar regions 20 K hotter than the 240 K average for the planet. The polar clouds are thinner and allow more infrared radiation to escape, but they are not completely transparent either. If they were, we could observe down to a region where the temperature was at least 290 K before the atmosphere itself became opaque.

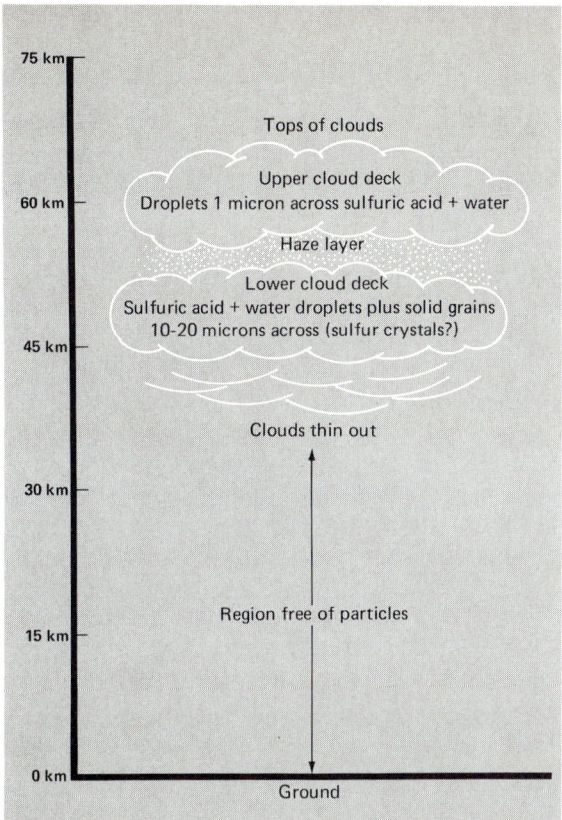

FIGURE 16.7 The structure of the clouds in the atmosphere of Venus. [*Based on NASA data.*]

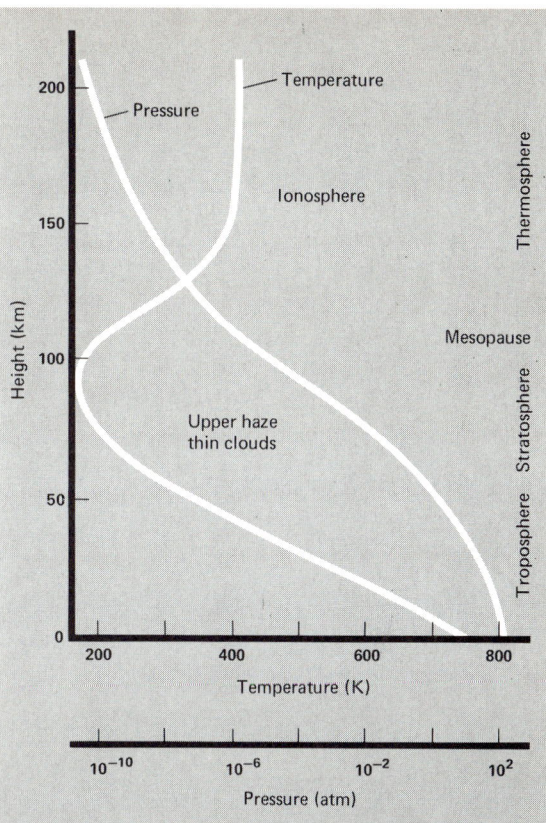

FIGURE 16.8 The overall properties of the atmosphere of Venus, based on data from USSR landers. [*Adapted from a diagram by M. Y. Marov.*]

That's the structure of the clouds. What are they made of? The best proposal to date sees the upper level clouds as concentrated solutions of sulfuric acid. Models of the clouds designed to match their observed infrared spectrum imply that they contain a solution of 85 to 90 percent sulfuric acid mixed with water.

Clouds of sulfuric acid and water mean that Venus does contain some water vapor, but it doesn't amount to very much compared to the total amount of water on the earth. If all the earth's water (in the atmosphere and oceans) were spread in a uniform layer over its surface, that sheet of liquid would be 3 km thick. All the water in the atmosphere of Venus (none exists on the surface because it's so hot) would amount to a layer only 30 *cm* thick. So our earth has about 10,000 times more water than Venus.

Atmosphere and Surface Temperature

Since 1932 we have known that the atmosphere contains carbon dioxide, but not how much. Interplanetary probes launched by both the United States and the Soviet Union indicate that the atmosphere contains about 96 percent carbon dioxide, 3 percent nitrogen, and traces of water vapor (0.1–0.4 percent), oxygen, argon, hydrogen chloride, hydrogen fluoride, hydrogen sulfide, sulfur dioxide, helium, and carbon monoxide.

The Russian probes found the surface pressure to be 90 to 100 atm, and the sunlit surface temperature about 730 to 750 K (Fig. 16.8). This high temperature probably results from the effective trapping of surface heat (Section 15.7), because carbon dioxide, water vapor, and sulfur dioxide absorb infrared radiation well. Venus is so hot because of an extreme greenhouse effect.

Stop for a second and think how hot it is on Venus: 700 K is about 430°C, or 800°F! (Your oven broiler operates at about 500°F.) Consider also the surface pressure: It's about that encountered at a depth of 1 km in the earth's oceans.

Atmospheric winds on Venus blow from the day to the night sides and from equatorial to polar regions. The wind flow carries heat. Along with the very effective atmospheric insulation, this helps to keep the temperatures fairly constant over Venus's surface. They vary about 10 K or less.

Besides composition, a key difference between the earth's atmosphere and Venus's is that the earth's is hotter at the top and colder at the bottom (Fig. 14.7), while Venus is colder at the top and hotter at the bottom!

16.3 THE SURFACE OF VENUS

We can investigate Venus's cloaked surface by two methods: landing probes on the surface to examine it close up and bouncing radar (which can penetrate the clouds) off the surface and analyzing the reflections. (That turns out not to be so easy.)

Surface Photographs

From 1972 to 1978 the Russians landed six probes by parachute to Venus's surface. These landers stopped working within an hour after landing because of the extreme pressure and temperature. Two landers—Venera 9 and 10—sent back close-up photos of the surface before they disintegrated (Fig. 13.15). These pictures show slabby rocks about 50 to 70 cm long and 15 to 20 cm wide. A few rocks have small holes that look as though they were once filled with gas; this implies a volcanic origin. Some rocks show jagged edges, indicating little erosion, and others show blunted, rounded edges, an indication of considerable erosion. The rocks rest on loose, coarse-grained dirt. A fine material fills the nooks and crannies of the rocks.

What kind of rocks are these? Measurements by the landers of the radioactivity in the rocks indicate that at one lander site they are basaltic, similar to those lining the earth's ocean basins (Section 14.3), but at the other site they are granitic, similar to those found in the earth's mountains. Both types are igneous rocks formed from lava.

In March 1982, two Soviet landers—Venera 13 and 14—touched down within 900 km of each other. Their views of the ground were very different. Venera 13 (Fig. 16.9a) photographed a rocky plain covered with clustered outcroppings. These are interrupted by patches of a dark, dusty material which may be Venus's soil. Some isolated rocks show evidence of erosion. An analysis of a rock sample revealed a composition that resembles that of basaltic rocks on the earth in continental rift zones. In contrast, Venera 14 sent back images (Fig. 16.9b) of fairly even layers (1 to 10 cm thick) of broken, rocky plates that extend to the horizon. No patches of dust are visible. The surface composition is basically volcanic, like the basalts found on the earth's ocean floor near mid-oceanic ridges.

Keep these views in mind and compare them to the surface pictures of Mars later in this chapter.

16.3 THE SURFACE OF VENUS

FIGURE 16.9 (a) Venera 13 photo of the surface. The horizon is visible in the upper corners and the lander at bottom center. Note the slabs of rock, patches of dark "soil," and pebbles close to the edge of the lander. (b) Venera 14 view of Venus. The flat expanses of rock are unbroken by any "soil" patches and fewer pebbles are visible.

Radar Mapping

The Venera pictures gave a worm's-eye view of the surface. What is the more general lay of the land? We now have a pretty good, overall idea from ground-based radar maps and radar mapping by the Pioneer-Venus orbiter. These maps reveal a varied terrain: mountains, high plateaus, canyons, volcanoes, ridges, and possible impact craters. Overall, Venus looks fairly flat. Elevation differences are small, with the exception of a few highland regions; 60 percent of the surface lies within 500 m of the average planetary radius. In the highlands the land reaches up to 10 to 12 km, compared with 4-km highland-lowland differences on the moon and Mercury, 25-km differences on Mars, and 19 to 20 km on the earth (Mt. Everest is 8.85 km high, and the Japanese trench 11.03 km deep).

The southern half and the northern half of the mapped face of Venus differ remarkably. The northern region is mountainous with uncratered *upland plateaus*. In contrast, the southern part consists of relatively flat *cratered terrain*.

Let's look at the radar maps in detail to see what hints they provide about the evolution of Venus. Be forewarned that these interpretations are very preliminary. Remember, because Venus rotates almost exactly four times between inferior conjunctions and the mapping is best done at that time (when Venus and the earth are the closest), we always examine the same half of Venus in the ground-based radar observations. Only the orbiters can see the "back" side.

The early radar images picked up two highland regions some 1000 km across, about the size of continental masses on the earth. The great northern plateau is called *Ishtar*

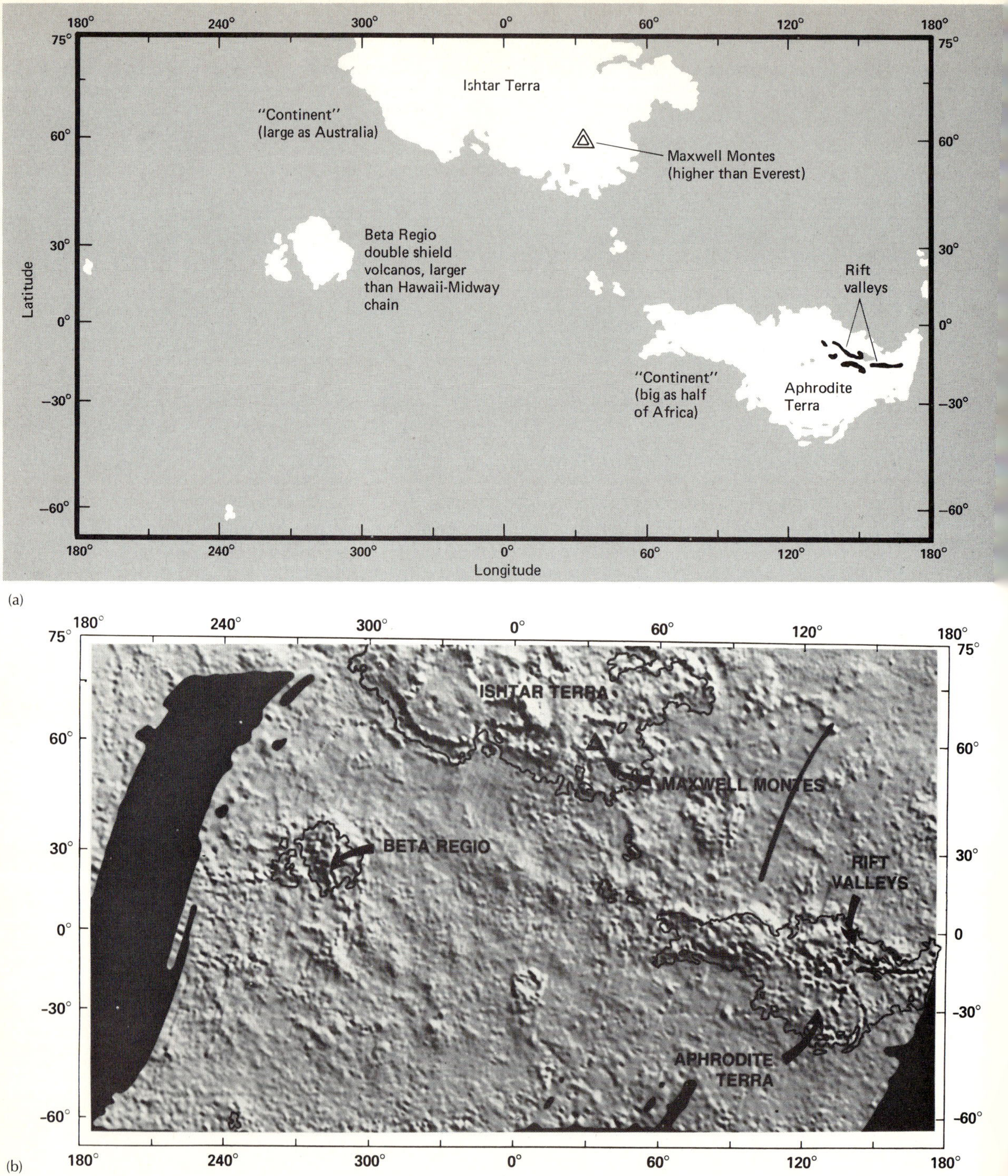

FIGURE 16.10 (a) A map of the highland regions on Venus's surface, as revealed by radar mapping. Because of the distortion of a spherical surface drawn as a flat map, Ishtar seems larger than Aphrodite, but it isn't. (b) A radar map of this region, taken by the Pioneer Venus orbiter. The map has been computer processed to emphasize changes in elevation; black areas are those not mapped. [*Courtesy NASA.*]

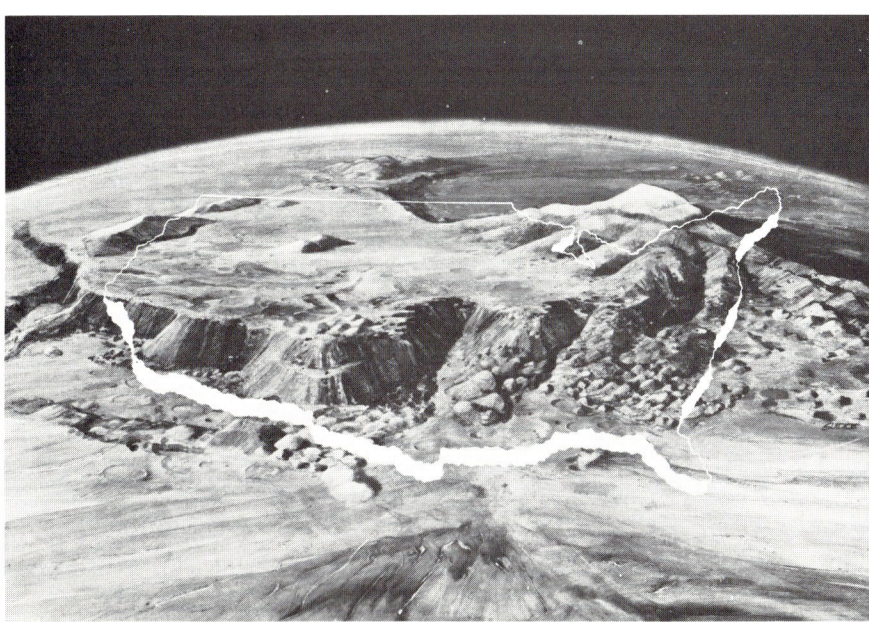

FIGURE 16.11 (a) A radar map (from earth) of the Ishtar region. The Ishtar plateau (upper center) is ringed by mountains. The bright area to the right is the Maxwell Montes. [*Courtesy D. Campbell, R. Dyce, and G. Pettengill; observations at Arecibo Observatory.*] (b) An artist's conception of the Ishtar plateau, comparing its size to that of the United States. [*Courtesy NASA.*]

Terra; the southern one is *Aphrodite Terra*. A third was discovered in a later map of the planet's northern region and is called *Beta Regio* (Fig. 16.10).

Upland Plateaus

Ishtar is huge, some 1000 by 1500 km (Fig. 16.11). (That's larger than the biggest upland plateau on the earth, the Himalayan Plateau.) Three mountain ranges border Ishtar on the west, north, and east. The eastern range, called *Maxwell Montes*, contains the highest elevations on Venus seen to date, some 12 km. Radar shows that this range is rough, as expected from a lava flow. And some radar views indicate a volcanic cone near Maxwell's center. The northern mountain range rises about 3 km above Ishtar; the western one reaches only 2 km above the plateau.

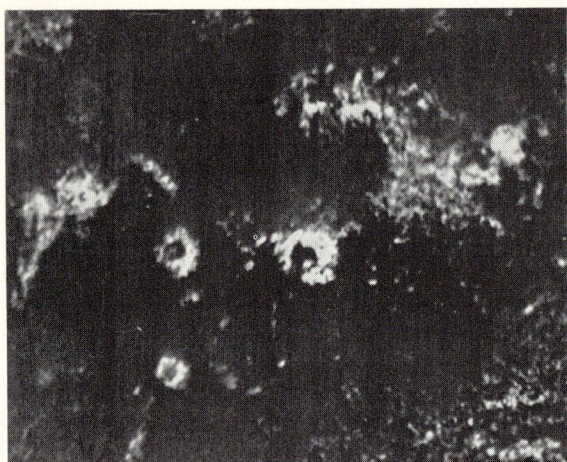 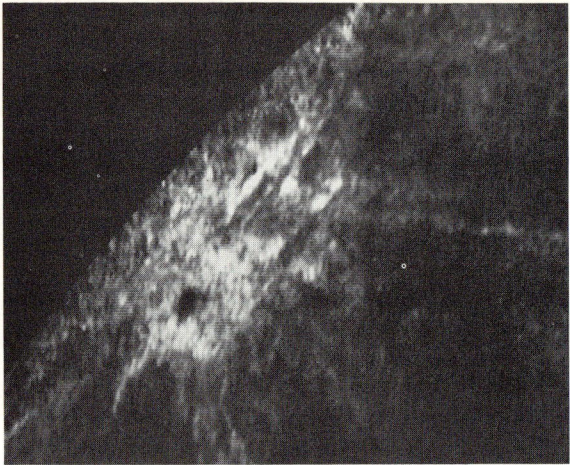

FIGURE 16.12 Ring-shaped features west of the Alpha region. They may be impact craters. The largest is about 100 km in size. [*Courtesy D. Campbell, B. Burns, and V. Boriakoff; observations at Arecibo Observatory.*]

FIGURE 16.13 A radar map of the Beta region, which appears to contain a large volcanic peak called Theia Mons. The central dark region is the volcano's crater, some 5 km high, and the streaks radiating outward may be lava flows. [*Courtesy D. Campbell, B. Burns, and V. Boriakoff; observations at Arecibo Observatory.*]

Cratered Terrain

The southern half of Venus's face consists of low, rolling plains apparently punctuated by large craters up to 800 km in diameter and smaller ones less than 1 km (Fig. 16.12). These are probably impact craters. (Craters smaller than this size do not form on Venus because the dense atmosphere completely burns up small bodies before they reach the ground.) The craters of Venus resemble those on the moon, Mercury, and Mars. They have peaks in their centers—a clue to their impact origin. In general, the craters are shallow from erosion (probably by wind), only some 500 m deep.

One Venera lander came down in the midst of the southern terrain. Its instruments indicated that the rock was more granitic than basaltic. This fact is one clue that at least some of the surface here is ancient. Later lava flows could have poured out basaltic rock. The fact we still see craters here also implies the surface must be old; otherwise volcanoes and mountain building would have obliterated all the craters.

One high region reaches up from the cratered terrain, Aphrodite Terra. Aphrodite is almost 3000 km across, about half the size of Africa. It also has a rough surface that hits 7 to 10 km in height.

Volcanoes on Venus

The Beta Regio, which seems to contain at least two separate volcanoes, appears to be an enormous volcanic complex that formed from a great north-south fracture zone (Fig. 16.13). The volcanoes are *shield volcanoes*. Instead of a sharply uplifted cone, shield volcanoes have gentle slopes and are relatively flat, like an armor shield. They often have a collapsed central crater at their summits.

One of the volcanoes in the Beta Regio has a diameter of 820 km, a height of 5 km, and a summit crater 60 by 90 km. In contrast, the island of Hawaii on the earth (a volcanic island) is 200 km across and 9 km high. The largest volcano on Mars, Olympus Mons, is 550 km in diameter and 20 km high. So Venus may have the largest diameter volcanoes in the solar system!

Are these volcanoes extinct, or could they still be active? Several arguments suggest that volcanic activity is still present on Venus. First, Venus probably has the same amount of radioactive substances in its interior as the earth. The heat generated by their decay must get out somehow. Venus doesn't appear to have mantle convection like the earth; there are no "mid-ocean" ridges, for example. So volcanic outpouring could be a primary source of heat loss.

Second, to support the shield volcanoes requires a crust which is quite thick, up to 160 km in some regions. To some geologists, this thickness seems unreasonable. The earth's crust is at most 40 km thick—why should Venus be so different? Perhaps expanding hot spots in the mantle support the raised masses instead.

Third, small gravitational disturbances in the orbit of the Pioneer orbiter suggest that the crustal masses are higher than would be expected if they were simply floating on the mantle. There seems to be some unusual process at work beneath them.

Fourth, the Beta Regio and the east end of Aphrodite Terra reflect radar well, as if they had relatively fresh lava coatings on their surfaces.

Fifth, analysis of radio radiation coming from beneath the clouds suggests that lightning crackles over the Beta Regio and the eastern part of Aphrodite, which could be generated by electrically charged dust propelled from the volcanoes.

Much of this evidence is circumstantial, but taken together, it suggests that perhaps some volcanoes on Venus are still quite active.

Rift Valleys

Venus not only has large mountains; it also sports enormous canyons. On the near side of Venus (the one facing the earth at inferior conjunctions) a huge canyon extends for more than 1300 km, has a width of 150 km, and a depth of 2 km. On the far side, an even larger canyon scars the landscape: it is 5 km deep, 320 km wide, and at least 1400 km long. (Its length has not yet been completely mapped.)

The canyons of Venus appear to be *rift valleys*, which form along fault zones rather than by water erosion. (A large rift valley splits New Mexico; the Rio Grande flows down its length.) Rift valleys appear on the earth where the crust is spreading apart. A similar process may have occurred on Venus in the past (Fig. 16.14), but the crust is most likely rigid now.

To sum up, the surface of Venus has mountains, plateaus, volcanoes, impact craters, lava flows, and rift valleys. For all this variety, the surface is remarkably flat. Only 18 percent of the mapped surface extends above 7 km, 11 percent above 10 km. In contrast, about 30 percent of the earth's surface reaches above 10 km.

Venus does not appear to have lunar-type basins, lowlands filled by lava flows. As indicated by the presence of craters, the lowlands of Venus must be older than the rest of the crust. That's just the opposite of the earth and moon, where the lowlands (the ocean basins) make up the youngest part of the crust. On Venus the highland masses are younger than the lowlands.

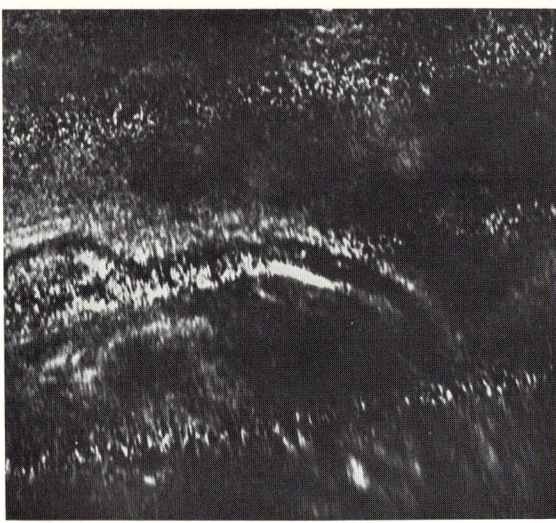

FIGURE 16.14 A radar map of parallel ridges near Venus's equator. They may be the result of crustal movement. [*Courtesy D. Campbell, B. Burns, and V. Boriakoff; observations at Arecibo Observatory.*]

Venus and Earth Compared

Overall, the surface of Venus resembles the earth but little, even though Venus has about the same size, density, mass, interior composition, and structure. The differences probably developed from the lack of liquid surface water. On earth rainwater reacts with atmospheric carbon dioxide to produce carbonic acid, which in turn combines with calcium to produce limestone. On Venus the carbon dioxide cannot be trapped this way. It stays in the atmosphere, keeping up the severe greenhouse effect.

On earth water helps the continents float on the mantle. It acts as a lubricant and lowers the melting point of the rock. The crust of Venus may have experienced some crustal plate movement. The evidence is its rift valleys, where plates may have pulled apart, and its mountain plateaus, where plates may have collided. But the widespread cratered terrain indicates that plate movement has not been a long-term planetwide process, as on the earth.

Without water large-scale erosion cannot carve landforms (such as river valleys) commonly found on the earth, sedimentary rocks cannot form, and, of course, life as we know it cannot develop.

16.4 THE EVOLUTION OF VENUS

What implications do these observations have for the geologic history of Venus? Right now we can only speculate. It does seem, though, that the early history of Venus (earlier than 4 billion years ago) must have been similar to the earth's early history, because both planets have similar mass, density, and size. The later history more closely resembled that of Mars.

We infer that Venus formed about 4.6 billion years ago with the other terrestrial planets (Chapter 13). As happened to the earth (Section 14.9), Venus's interior differentiated from internal heating. During the first 500 million years, a crust (part basalt and part granite) formed and solidified. About 4 billion years ago large masses bombarded the surface and fractured the crust. Volcanoes erupted. Bombardment by smaller bodies from space cratered the surface. This intensive bombardment ended suddenly, and erosion has since altered the surface of Venus.

In these early times plate movement may have pushed up some highland regions and created rift valleys. Huge volcanoes vented through cracks in the surface and may continue to do so; their cones form the shield volcanoes of today. Venus seems to have evolved in a sequence similar to the earth, but more slowly and not quite as far, as expected for a planet whose mass is a little less than the earth's.

16.5 GENERAL CHARACTERISTICS OF MARS

The ruddy spectacle of Mars has sparked interest for astronomers since recorded time. Its unique, bloody color and swift motions in the sky marked Mars as a special planet. Telescopes revealed polar caps, dust storms, and permanent dark features. Mars appeared to be the most likely home of extraterrestrial life in the solar system, as portrayed in the fantasies of H. G. Wells and other science fiction writers.

The Viking landers have dampened this expectation; Mars seems barren of life as we know it (more in Chapter 25). These and other spacecraft have presented us with a new Mars (Fig. 16.15): a planet that is a cross between the earth and the moon, with plentiful, ancient craters, giant canyons, and massive, extinct volcanoes.

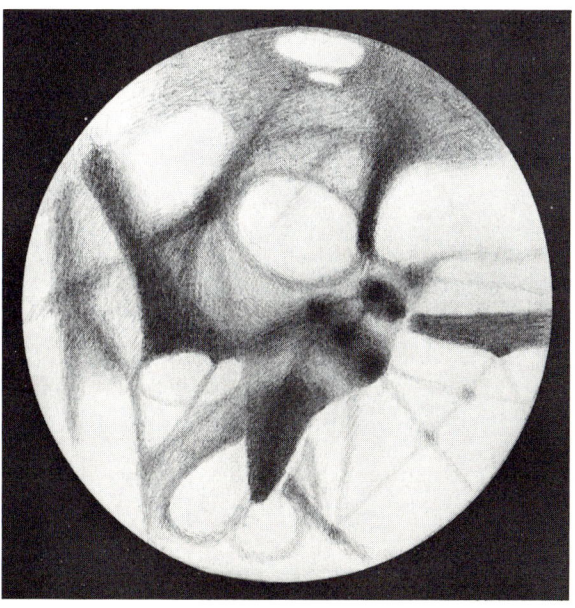
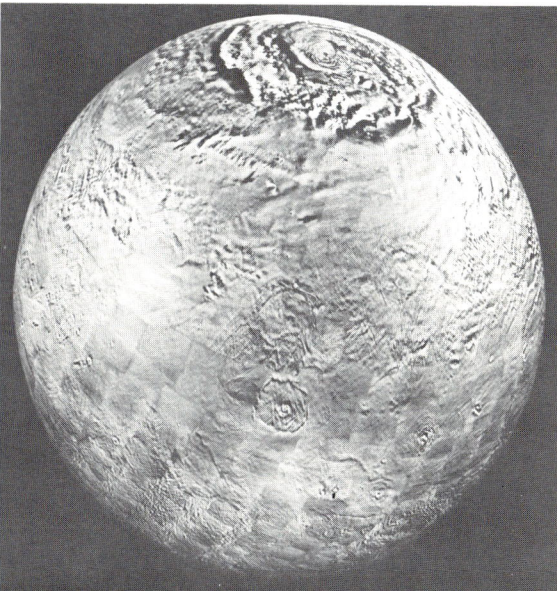

FIGURE 16.15 (a) Mars, a drawing made in 1926 of the surface features seen through a telescope. [*Courtesy Lick Observatory.*] (b) A mosaic of Mars compiled from Mariner 9 photos. The North Pole ice cap is at top. Just below the center is the giant volcano, Olympus Mons. [*Courtesy NASA.*]

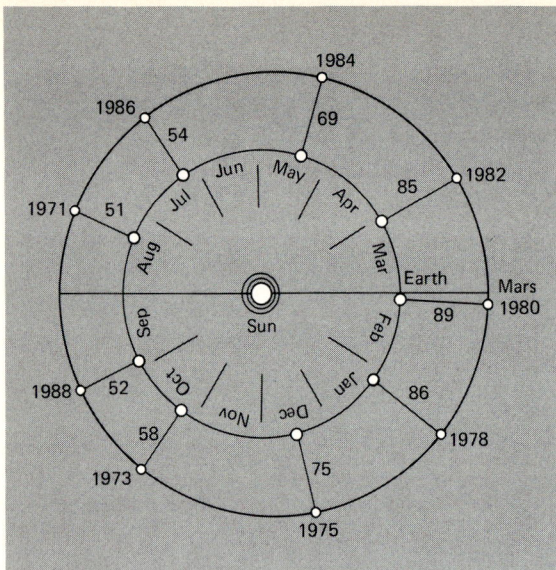

FIGURE 16.16 Oppositions of Mars from 1971 to 1986. The distances between the earth and Mars are in millions of kilometers. The opposition distances vary because of the eccentricity of the Martian orbit.

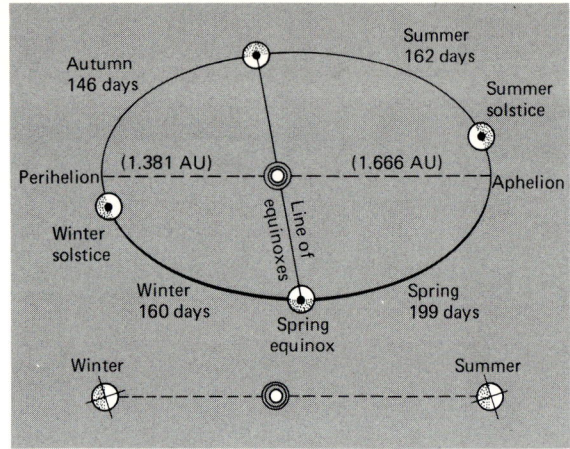

FIGURE 16.17 The Martian orbit and seasons for the northern hemisphere.

Martian Orbit, Day, and Seasons

Mars orbits the sun at an average distance of 1.52 AU. But its distance from the sun varies considerably (±9 percent) because its orbit is fairly eccentric. Consequently, the distance between Mars and the earth at opposition (when both planets lie on the same side of the sun) ranges from less than 56 million km to more than 101 million km (Fig. 16.16).

Surface markings visible through telescopes make the Martian rotation rate easy to measure. In 1659 the Dutch physicist Christian Huygens (1629–1695) observed the rotation rate to be close to 24 h. Modern measurements place the value at 24 h 37 min; a day on Mars lasts only a bit longer than on the earth.

How do the Martian seasons compare with ours? The equator of Mars inclines about 25° to its orbital plane. This angle is about the same as the earth's (23½°), so the Martian seasons vary like those of the earth, though Mars's spin axis points to a different direction in space. (Standing on the Martian north pole you would find Deneb, the brightest star in Cygnus the Swan, almost directly overhead.) Because Mars takes longer to orbit the sun than the earth does, Martian seasons last about twice as long as ours (Fig. 16.17). Also, the eccentricity of the orbit gives the seasons different lengths.

Size, Mass, and Density

With a known distance and measured angular diameter, we get Mars's size: 3394 km in radius. That's only 53 percent the earth's radius, but about 40 percent larger than Mercury's.

Mars has two satellites, so we can use Newton's version of Kepler's law (Section 4.2) to find its mass from their orbits. It is only 6.4×10^{23} kg, about 11 percent the earth's mass.

16.5 GENERAL CHARACTERISTICS OF MARS

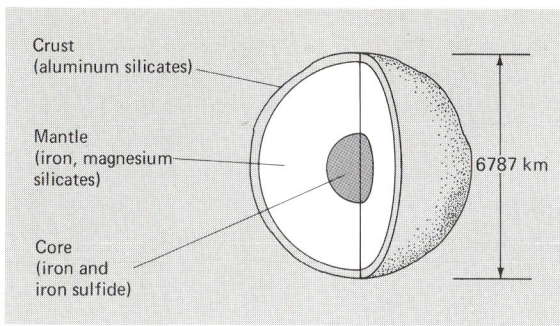

FIGURE 16.18 One model of the Martian interior. [*Based on calculations by J. S. Lewis.*]

The mass and size give Mars's density, 3900 kg/m³. It is only a bit higher than the moon's (3300 kg/m³) and much less than the earth's (5500 kg/m³).

This low density implies that Mars's interior must be different from the earth's (Fig. 16.18). In particular, its core must be smaller and probably consists of a mixture of iron and iron sulfide (FeS), which has a lower density than the materials in the earth's core. The Martian mantle probably has the same density as the earth's. The exact composition of the mantle is not known. Many different models have been developed. One has a mantle with olivine, iron oxide, and some water (0.3 percent).

The interior of Mars is probably much cooler than the earth's because of its smaller mass. Here's a general idea of how mass affects the evolution of an earthlike planet. For bodies of roughly the same density, the mass is proportional to the volume, and so to the cube of the radius: $M \propto R^3$. A planet's mass determines how much heat is generated during its original accretion and also the quantity of radioactive nuclei it contains that later will generate heat. The more mass, the more the body will heat up. How fast a planet cools off depends on its surface area (for bodies of the same composition) and so on its radius squared: $A \propto R^2$. The ratio of heating rate to cooling rate is therefore the ratio of mass to surface area, which is proportional to the radius or the cube-root of the mass:

$$\frac{\text{heating rate}}{\text{cooling rate}} \propto \frac{M}{A} \propto R \propto M^{1/3}$$

So larger terrestrial planets are more likely to get hot enough to melt internally and to retain their heat longer than smaller ones.

Planetary Magnetic Field

Mars has an extremely weak planet-wide magnetic field, only 2×10^{-3} the strength of the earth's. That small a value presents a puzzle if the dynamo model correctly describes the origin of planetary magnetic fields. Mars rotates just as fast as the earth. Though the core is smaller, it should contain a substantial amount of metals. We have no seismic evidence that the core is liquid, but the evidence for past volcanic activity implies a hot mantle and so a hot, somewhat liquid, core. So Mars should have a moderately strong field, but it does not.

16.6 THE MARTIAN ATMOSPHERE AND SURFACE TEMPERATURE

Astronomers have known for a long time that Mars has an atmosphere. From observations made from 1777 to 1784, William Herschel concluded that apparent changes in the surface features were actually due to "clouds and vapors floating in the atmosphere of the planet." Many astronomers afterward believed that the Martian atmosphere was like the earth's, but thinner.

Thin it is, but nothing like the earth's. The Viking landers found average surface pressures of roughly 1/200 the earth's surface pressure. (You'd have to travel 40 km up in the earth's atmosphere before the pressure falls that low.) This thin atmosphere consists of 95 percent carbon dioxide, 0.1 to 0.4 percent molecular oxygen (O_2), 2 to 3 percent molecular nitrogen (N_2), and about 1 to 2 percent argon, not really very different in percentage composition from the atmosphere of Venus.

The Viking orbiters measured the water vapor in the atmosphere and found the greatest amounts in the high northern latitudes. Peak concentrations were about 0.01 mm of precipitable water, that is, if all the water above that location rained to the surface, it would form a layer only 0.01 mm thick. On earth the atmospheric water vapor is typically several centimeters of precipitable water, and the oceans are several kilometers thick. The thick atmosphere of Venus contains about 3 cm of precipitable water. Mars is a very dry planet compared with the earth or Venus.

It cannot rain on Mars today because of the low surface pressure. Only in the deepest canyons, where the atmospheric pressure is higher than average, could water be liquid on the surface. However, it is common to have water ice on Mars, either on the surface or in the clouds. There is also some evidence for water in a permafrost layer beneath the surface.

Although the atmosphere contains mostly carbon dioxide, its low density does not provide much protection against temperature extremes. At the Martian equator, when Mars is closest to the sun, the difference between noon and midnight amounts to almost 100 K. The summer tropical high of 310 K (37°C) is exceptional. For a period of two Martian months the surface temperature remains below the freezing point of water both day and night.

At the Viking 1 site, 23° north latitude, the air temperature near the ground ranges from −85°C to −29°C (Fig. 16.19). At the Viking 2 site, which is farther north, the temperature falls even lower and water condenses on the surface (Fig. 16.20). The landers found that the layer of water ice that coated the rocks and soil was less than a millimeter thick and that the frost remained for about 3 months. One model for the development of the frost layer pictures solid water and carbon dioxide adhering to dust in the atmosphere. The icy dust settles to the surface, where the sun evaporates the carbon dioxide. The dust and water ice make up the thin frost layer.

As the Viking landers plunged through the Martian atmosphere, they measured the air temperature at different elevations. Using these and other data, we can compare the temperature profile of the atmosphere of Mars with that of the earth (Fig. 16.21). Both atmospheres have a steady drop in temperature up to an altitude of about 10 km; in this lowest region, the sunlit ground heats the air. Because of the thinness of the atmosphere,

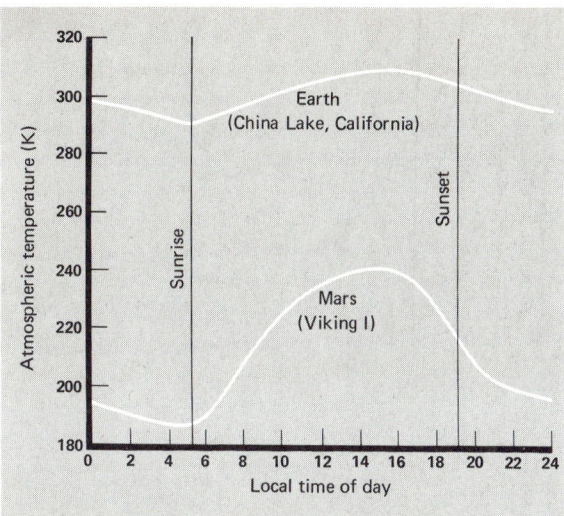

FIGURE 16.19 The daily variation of temperature on Mars compared to a desert site on the earth.

FIGURE 16.20 A Viking 2 lander photo of frost patches on Mars. The rocks are 10 to 80 cm across. The frost, made of water and dry ice, forms when the temperature drops to 160 K at night. [*Courtesy NASA.*]

we expected the atmosphere of Mars to be cooler than it is; a small amount of suspended dust absorbing infrared from the ground makes the difference. The temperature peak at 50 km on earth is caused by the absorption of solar ultraviolet in the ozone layer. Mars has no corresponding peak, so it probably lacks a definite ozone layer. In the earth's atmosphere the temperatures rise again above 100 km because of absorption of solar ultraviolet and X rays. No such heating takes place in the Martian atmosphere.

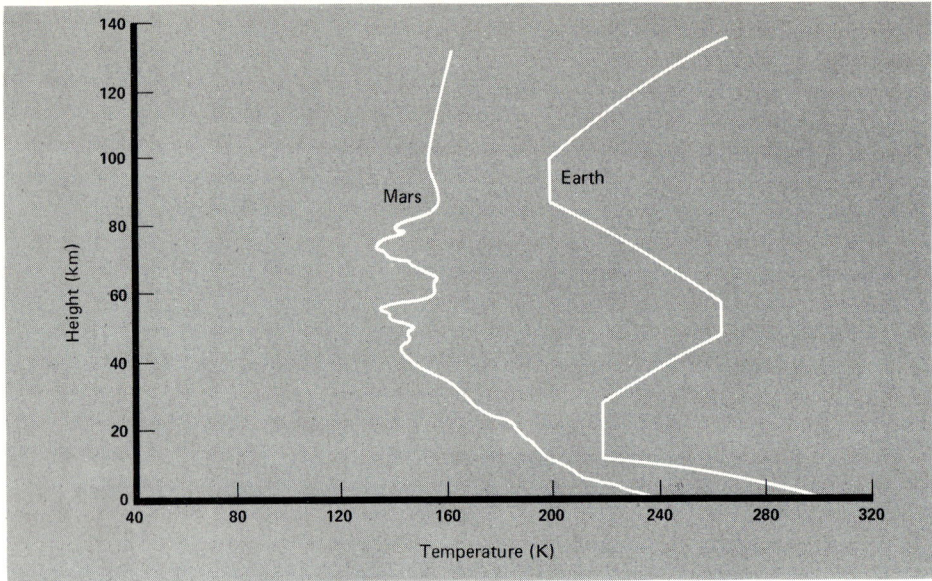

FIGURE 16.21 A comparison of the overall temperature profiles of the atmospheres of Mars and the earth. [*Based on NASA data.*]

16.7 THE MARTIAN SURFACE VIEWED FROM THE EARTH

Because an astronomer must look through two atmospheres, the view of Mars is usually blurry. A small telescope shows the main surface features: the white polar caps, light reddish-orange regions, and dark areas. Spacecraft visits to Mars improved dramatically the resolution of our vision.

Sands of Mars

Definite surface features on Mars are dark, apparently greenish-gray areas, in contrast to the reddish-orange color of the rest of the surface. Some early observers thought that these features were oceans. This idea was discarded when the surface pressure was found to be too low for liquid water. Others contended that the dark regions were green and so indicated vegetation. This greenness turns out to be an illusion caused by the contrast of the light and dark areas. The dark regions are not really green; they are actually red, but not as red as the lighter regions.

The light regions make up almost 70 percent of the Martian surface. They give Mars its striking reddish appearance. In 1934 the American astronomer Rupert Wildt suggested that these areas contain ferric oxide (rusted iron). Iron oxides come in many forms on the earth; all are characteristically brown, yellow, or orange. Infrared observations have added support to the idea that the surface contains substantial rusted iron combined with water; perhaps as much as 1 percent of the surface is water bound up with minerals.

The Viking landers' measurements were compatible with a surface composition of

FIGURE 16.22 Percival Lowell. "... the solidarity of the Martian land system points to an efficient government...." [*Courtesy Yerkes Observatory.*]

about 19 percent ferric oxide (Fe_2O_3). In addition, they measured about 44 percent silica (SiO_2), which leads to the conclusion that silicate minerals make up a major part of the surface. The Martian surface is covered with rusty sand (Plate 8).

Planetwide Dust Storms

The reddish sand, most of which is much smaller grained than that on earth's beaches, is blown by fierce winds greater than 100 km/h to create planetwide dust storms. They blow most violently when Mars is closest to the sun. Then the dust clouds, whipped up to heights of 50 km, shroud the entire planet. They cover Mars in a yellow haze for about a month. It takes many months for the fine dust to completely settle back to the surface. These global storms sandblast the surface and mix it up so much that the surface composition over the planet becomes essentially the same.

The winds are the major source of surface erosion. The wind-driven dust causes most of the changes in Martian surface features. For example, the wind storms blow dust into dunes or deposit it in streaks around mountains and craters.

The Martian Canals and Polar Caps

In 1877 Schiaparelli recorded Martian surface features in great detail. He charted a number of dark, almost straight features, which he called *canali*, Italian for "channels." This word was translated into English as "canals," which implied to some people that they were artificial structures.

Some observers could not see any canals. But others, especially those who regarded them as natural waterways, continued to find more. These so-called canals ignited the curiosity of the American astronomer Percival Lowell (1855–1915). To pursue his interest in Mars, Lowell (Fig. 16.22) in 1894 founded an observatory near Flagstaff, Arizona, to

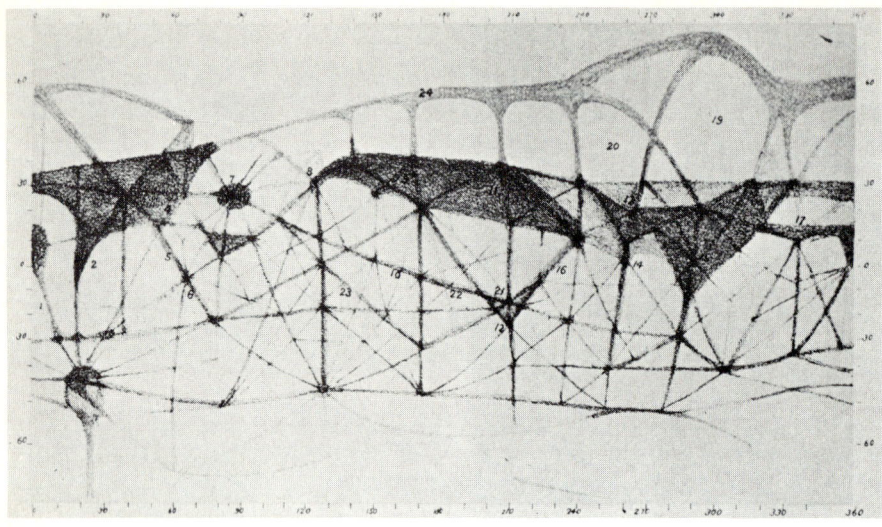

FIGURE 16.23 Surface features on Mars, including some "canals," drawn by Percival Lowell in 1896–1897. [*Courtesy Lowell Observatory.*]

take advantage of the excellent observing conditions there. Shortly afterward he published Martian maps showing a mosaic of over 500 canals (Fig. 16.23). In a series of popular books Lowell argued that the canals were artificial waterways, constructed by Martians to carry water from the polar caps to irrigate arid regions for farming. Lowell believed that the polar caps were water ice and that the dark regions were areas of vegetation that displayed seasonal growth, prompted by water from the polar caps.

The polar caps are indeed largest in winter and smallest in summer, and they do consist mostly of water ice, especially the residual cap left in the summer, which ranges in thickness from year to year from 1 m to 1 km (Fig. 16.24). The outer reaches of the caps,

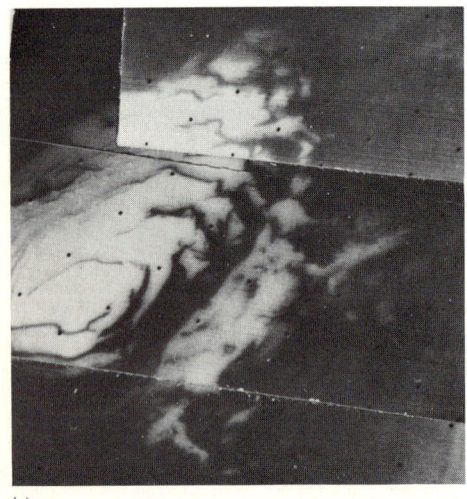
(a)

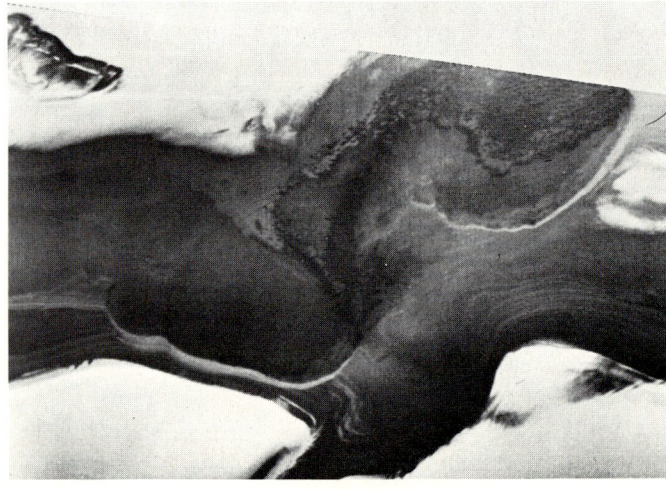
(b)

FIGURE 16.24 (a) The residual north polar cap in the summer. Note the layered appearance. The ice is water ice. (b) A close-up of part of the residual north polar cap in midsummer. Note the layered terrain beneath the ice. [*Courtesy NASA.*]

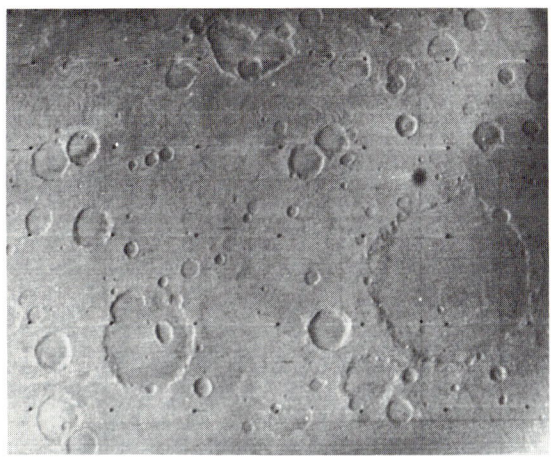

FIGURE 16.25 Craters in the Martian southern hemisphere. The largest is about 200 km in size, the smallest 10 km. Note the heavily eroded appearance; this erosion is caused by wind-blown dust. [*Courtesy NASA.*]

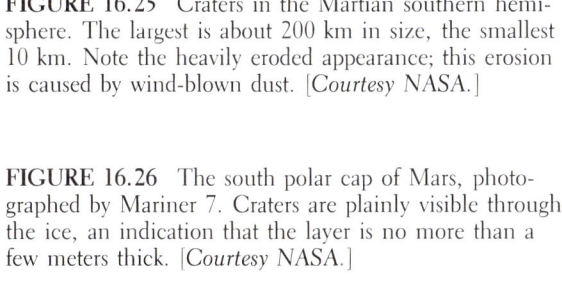

FIGURE 16.26 The south polar cap of Mars, photographed by Mariner 7. Craters are plainly visible through the ice, an indication that the layer is no more than a few meters thick. [*Courtesy NASA.*]

prominent in winter, consist of carbon dioxide ice, which condenses at a lower temperature than water ice. (At Martian surface pressures, water ice condenses at about 190 K, carbon dioxide ice at 150 K.)

Lowell was right in believing that water does exist on Mars, but it's not freely flowing on the surface, because the temperature is too cold and the pressure too low. (If all the water in the polar caps could cover the surface as a liquid, it would form a layer only about 10 m deep.) He was also wrong about the vegetation and the canals. Some astronomers now believe that wind-blown dust deposits might have created temporary features that were seen as the largest and fuzziest of the canals. The smallest ones were likely an illusion guided by wishful thinking.

16.8 THE INVASION OF MARS BY THE EARTH

Mars never had a chance to invade the earth—Martians don't exist. But we've invaded Mars, not to conquer, but to learn. We have found that Mars was once an active world, but now is a calm, cold desert world.

The Surface of Mars

The 1969 explorations of Mariners 6 and 7 reinforced the view (first developed from the Mariner 4 pictures of 1965) that the surface of Mars resembles that of the moon. They photographed abundant Martian craters (Fig. 16.25) visible even under the thinner regions of the polar caps (Fig. 16.26). The Martian craters do look like impact craters, but

FIGURE 16.27 A wide-angle view of Valles Marineris taken by the Viking 1 Orbiter. The canyon (arrow) is about 5 km deep at its west end and is shallower to the east. It runs parallel to and just south of the equator. [*Courtesy NASA.*]

tend to be shallower than their lunar counterparts. Their flat floors and low rims indicate that they have been strongly eroded.

Is Mars, like the moon, a dead world? No, for the photographs taken by Mariner 9 in 1971 also showed spectacular features of a geologically active planet. (See Plate 9 for a color view of Mars from space.) The extensive photographic survey by Mariner 9 showed that the two Martian hemispheres have different topological characteristics. The southern is relatively flat, older, and heavily cratered. The northern is younger, with extensive lava flows, collapsed depressions, and huge volcanoes. Near the equator, separating the two distinct hemispheres, lies a huge canyon, called *Valles Marineris* (Fig. 16.27; see also Fig. 13.18). This chasm is 5000 km long (about the length of the United States), some 500 km wide in places, and bordered by branching tributaries that look as if they have been carved by water.

On July 20, 1976, exactly seven years to the day after Neil Armstrong stepped out on the moon, Viking 1 touched down on Mars on the plains of Chryse (Fig. 13.16). Not long after, Viking 2 dropped to the Martian surface on the plains of Utopia. Seen close up, the Martian surface is bleak and dry. Large basaltic boulders are strewn about amid gravel, sand, and silt (Fig. 16.28). Some contain small holes from which gas has apparently escaped; the holes make the rocks look spongy. On earth such basalts originate in frothy, gas-filled lava; the Martian rocks probably had a similar origin.

Both landers presented indirect evidence for once-flowing Martian surface water. The Chryse region seems to be a flood plain where water sorted the smaller rocks into gravel, sand, and silt. The ground there also resembles the hardened soil of earth's deserts. Such soil forms when underground water percolates upward and evaporates at the surface. Upon evaporating, the water leaves behind minerals that harden the soil. Mineral analyses by Viking 1 and 2 indicated that the soil does contain such evaporated minerals, such as epsom salts.

FIGURE 16.28 A close-up view of the Martian surface near Viking 2. The rocks are 10 to 25 cm in size. Note that most contain holes. [*Courtesy NASA.*]

The Arroyos of Mars

The Mariner 9 mission discovered and the Viking orbiters confirmed a number of sinuous channels that appear to have been cut in the surface by running water (Fig. 16.29a). The largest ones have lengths up to 1500 km and widths as great as 100 km. (These channels are *not* the canals seen by Lowell and others; they are not straight, and are too small to be visible from the earth.)

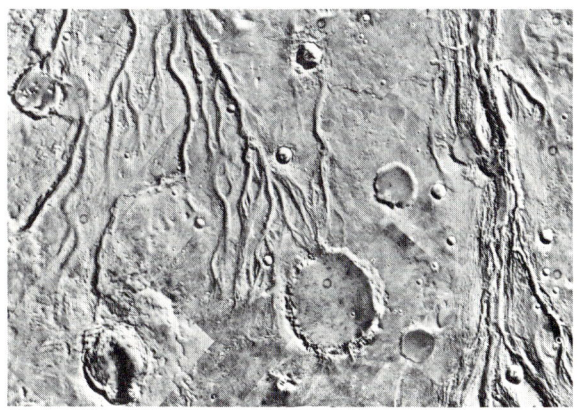

FIGURE 16.29 (a) Large arroyos on Mars. The terrain slopes about 3 km. Note how some arroyos cut through craters, an indication that they formed after the craters. [*Courtesy NASA.*] (b) An arroyo in New Mexico. Note the wavy patterns in the sand. [*Photo by M. Zeilik.*]

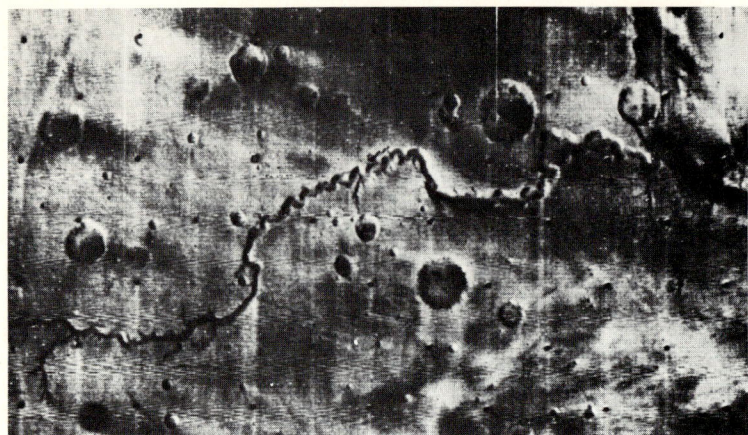

FIGURE 16.30 A single channel on Mars. It is about 5 to 6 km wide and some 400 km long. Note its meandering pattern. [*Courtesy NASA.*]

The channels resemble the arroyos commonly found in the Southwest of the United States. An *arroyo* is a channel in which water flows only occasionally. During the rainy season (the summer months) in New Mexico violent thunderstorms suddenly dump torrents of rain. The deluge of water, which is not soaked up by the hard soil, flows downhill to a river bed, such as the Rio Grande valley. The rush of water erodes the land into arroyos (Fig. 16.29b), some of which are quite large.

Certain characteristics of the Martian channels are fairly strong evidence that they were actually formed by flowing water. (1) The flow direction is downhill. (2) The flow patterns meander (Fig. 16.30). (3) Tributary structures indicate where several flows merged to form a larger one. (4) Sandbars are cut by smaller flow channels, as is commonly found in arroyos on the earth.

The formation of arroyos requires extensive running water for at least a short period of time. Because Mars does not have liquid surface water now, conditions for it must have occurred in the past.

The Martian Volcanoes

By far the most awesome Martian features are the shield volcanoes clustered on and near the *Tharsis ridge*. The largest is *Olympus Mons*, some 550 to 600 km across at its base (Fig. 16.31). Telephoto close-ups show a wavy texture of the cone's surface that is the result of lava flows. The cone reaches 25 km above the surrounding plain, and its base would span the bases of the islands of Hawaii, which are made of several volcanoes. If put down in California, the volcano would cover the territory from San Francisco to Los Angeles (Fig. 16.32). Olympus Mons soars more than $2\frac{1}{2}$ times the height of Mt. Everest above sea level! If this volcano grew at the same rate as terrestrial ones, it would have taken about 100 million years to develop. Indirect evidence from the lava flows implies that the volcanoes are about a billion years old.

The huge mass of Olympus Mons requires that the Martian crust beneath it be thicker than the crust beneath smaller such volcanoes on the earth. Geologists estimate the thickness of the Martian crust to be about 120 to 130 km, about three times that of the earth's.

Olympus Mons crowns a cluster of volcanoes situated on the Tharsis ridge. Thin ice clouds, sometimes seen decorating the tops of the volcanoes, might result from erratic

16.8 THE INVASION OF MARS BY THE EARTH

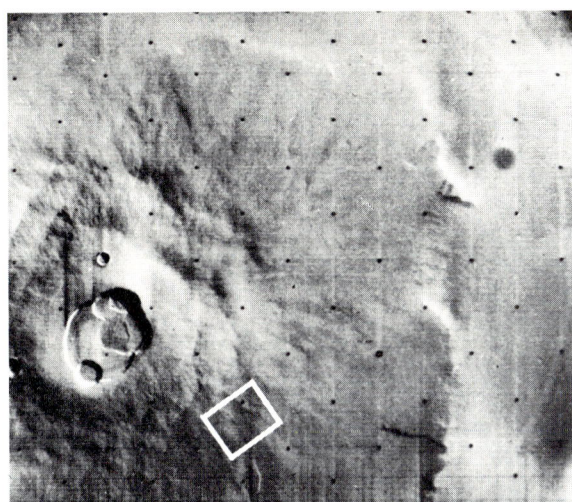

FIGURE 16.31 The top of Olympus Mons. (a) A wide-angle view shows the crater at the top of the cone, one side of the volcano, and the surrounding plateau. (b) A closer view of the area in the white box in (a) shows the surface texture which indicates a flow of material from the volcano's top. [Courtesy NASA.]

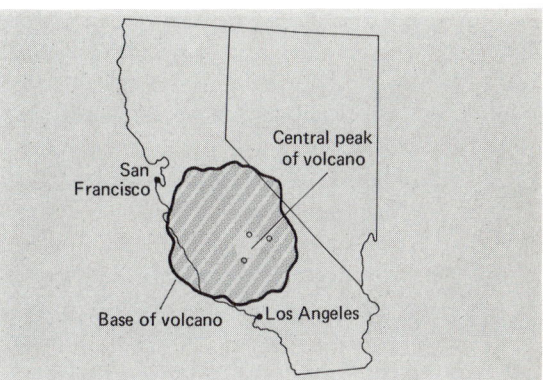

FIGURE 16.32 The size of Olympus Mons compared with the state of California.

spurts of outgassing. On earth volcanic activity spews forth gases (including water vapor) from the earth's interior (Section 14.9). Such outgassing by the giant Martian volcanoes in the past may have contributed significantly to the Martian atmosphere. The clouds may also result from the cooling of air that is being blown up the sides of the volcanoes.

The Tharsis ridge (look back at Fig. 13.15) is a hallmark of Mars's northern hemisphere, which differs so dramatically from the southern one. The ridge rises about 10 km above the average surface height for the planet and contains numerous volcanic structures. Very few impact craters are visible. In contrast, the southern hemisphere is basically a desert pockmarked by old, eroded craters; the oldest was made some 4 billion years ago. The geologic inference from this difference is that about 3 billion years ago in the northern hemisphere a huge mass of lava oozed out from under the surface, creating the volcanic plains and the volcanoes over a long period of time. This flow wiped out the older deserts and craters in this region. Other flows have taken place in this region since the first.

FIGURE 16.33 Argyre Planitia in the Martian southern hemisphere. Argyre is a large impact basin (left of center). Many relatively uneroded craters are on the surface. [*Courtesy NASA.*]

The Cratered Southern Hemisphere

The southern hemisphere of Mars has a cratered terrain that resembles the ancient highlands of the moon or the intercrater plains of Mercury (Fig. 16.33). The landscape contains impact craters that range in size from huge, lava-filled basins down to some only a few meters across. The Martian craters come in the same varieties as lunar and Mercurian ones: multi-ringed, centrally peaked, and bowl-shaped.

In general the Martian craters are shallower than the ones on the moon and Mercury. Many are filled with wind-blown dust. Wind scours the sides, rims, and terraces of the craters and piles the dust in dunes within the craters' bowls.

The Martian craters also do not usually have the rims of ejected material common to craters on the moon and Mercury. Instead, many Martian craters have bulges that protrude out from their rims, produced by the melting of frozen ice in the Martian soil during the impact; this melted water quickly turned the ground into mud that flowed in bulges away from the crater.

A Possible Evolutionary History of Mars

Putting this mass of new information together to infer the Martian past is a tough task. Many uncertainties crop up, especially with respect to the sequence of events. (Remember, we have not yet been able to make any measurements of Martian rocks.) But here is a preliminary, working model for the geologic evolution of Mars.

In the first phase Mars formed by accretion, and then impact craters covered the surface. Shortly afterward the planet differentiated to form a crust, mantle, and core. Regions of thicker crust rose to higher elevations.

In the second phase thin regions of the crust fractured, and the Tharsis ridge uplifted, cracking the surface around it. During this time a primitive atmosphere, denser and warmer than at present, held large amounts of water vapor from the volcanic outgassing. Rainfall may have eroded the surface in furrows and percolated to a depth of a few kilometers. Decreasing temperatures formed ice at shallow depths. When heated (perhaps by volcanic activity), this ice melted, leading to formation of collapse and flow features. Planetwide water erosion carved the surface.

In the third phase extensive volcanic activity occurred, especially in the northern hemisphere. The Tharsis region continued to uplift, generating more faults. Valles Marineris formed at this time. More recent volcanism, most of it concentrated on the Tharsis ridge, broke the surface and spewed out great flows of lava marking the fourth phase of evolution.

Since that last time of great eruptions, it is mainly wind erosion which has sculpted the Martian surface. A few small impact craters have probably formed from time to time. Recently, strange, small craft from the earth have touched down on the surface, finding a cold, windy, dusty, and rock strewn desert.

16.9 THE MOONS OF MARS

Two satellites circle the planet Mars; appropriately, they are named Phobos and Deimos ("Fear" and "Panic"), after the companions of the god Mars (see Appendix B). Asaph Hall (1829–1907) at the United States Naval Observatory discovered the two moons in 1877. The observations were preceded by predictions of the satellites' existence by Kepler in 1610. He reasoned that since earth has one moon, and Jupiter four (all that were known at the time), Mars ought to have the geometric mean, two. (Presumably he would have predicted eight for Saturn.)

Deimos and Phobos both lie close to Mars and orbit the planet rapidly. Deimos, the outer moon, circles Mars in $30\frac{1}{3}$ hours; Phobos, the inner moon, takes a mere $7\frac{1}{3}$ hours. Phobos is one of the solar system satellites which orbit their parent planet faster than that body spins (another is Jupiter's innermost satellite, 1979 J1, discovered by Voyager). So while Deimos rises in the east and sets in the west, as our moon does, Phobos rises in the west and sets in the east. Like the earth's moon, the Martian moons keep the same face to the planet.

Spacecraft observations show that Deimos and Phobos have the same general shape— an ellipsoid with three axes. Phobos, the larger, has axes about 27, 21, and 19 km long; Deimos's axes are only 15, 12, and 11 km. Photographs also show that Phobos and Deimos have cratered surfaces (Figs. 16.34 and 16.35). The sizes and numbers of these craters indicate that the surfaces of these satellites are at least 2 billion years old.

The surfaces of these satellites are very dark. The albedos of Deimos and Phobos are 0.022 and 0.018, respectively. Thus they reflect much less light than our moon (albedo 0.067) and are among the blackest known objects in the solar system. The surface characteristics resemble those of a certain kind of meteorite called carbonaceous chondrites (Section 18.1) and of dark asteroids such as Ceres (Section 18.2). Deimos and Phobos do not appear to be natural satellites formed by accretion at the same time as Mars. Rather they may have been formed in or near the asteroid belt and picked up by Mars later.

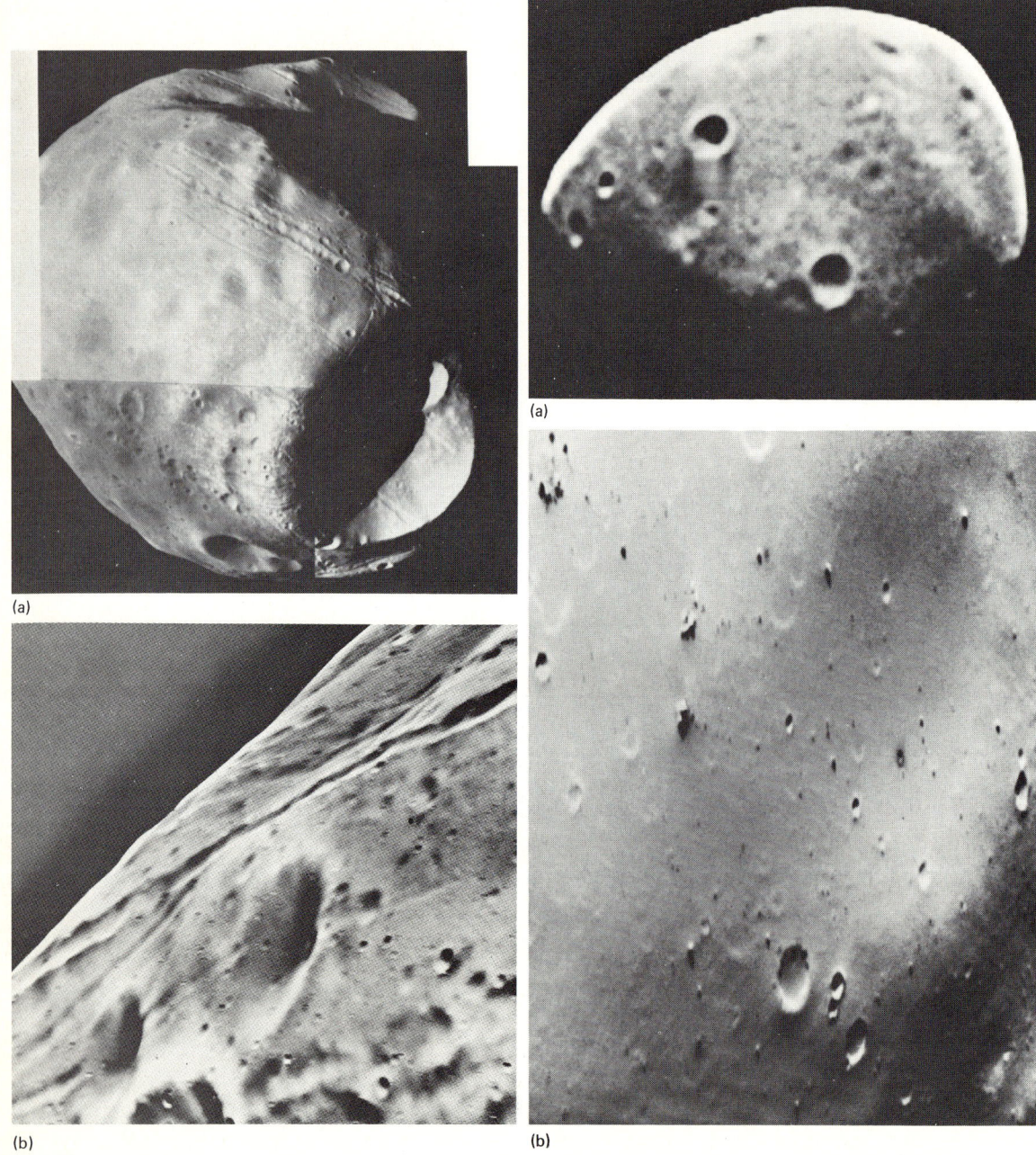

FIGURE 16.34 (a) Phobos photographed from 612 km by the Viking 1 Orbiter. This side of Phobos is the one that faces Mars. The largest crater is Stickney, 10 km in size. (b) Phobos from 120 km. The smallest craters visible are 10 m in size. The surface resembles that of the lunar highlands. [*Courtesy NASA.*]

FIGURE 16.35 (a) Deimos. The largest crater is 1.3 km in size. The illuminated part of the moon is about 12 by 8 km. (b) A close-up of Deimos. This photo shows a region 1.2 by 1.5 km and features as small as 3 m. The boulders are about the size of a house. [*Courtesy NASA.*]

16.10 VENUS, EARTH, AND MARS: AN EVOLUTIONARY COMPARISON OF KINDRED PLANETS

These planets share a special location in the solar system, have similar internal characteristics, and possess atmospheres. The atmospheres of Venus and Mars are about 95 percent carbon dioxide with traces of other compounds, while the earth's is only about 0.03 percent carbon dioxide (Section 14.5). This difference in atmospheric constitution is as striking as the earth's unique treasure of oceans of water compared with the arid wastes of Mars and Venus. Such profound differences among such otherwise similar planets relate to different evolutionary rates and processes on each planet.

The water question is a central one. Some perspective can be gained by asking how Mars and Venus may have lost their original water, while the earth has kept most of its original store. (The initial amount of water on the earth was probably much less than on Venus and more than on Mars, based on a contemporary model for the formation of the solar system; see Chapter 19.) Water may be trapped underground, but geologic activity such as volcanoes releases it. We have evidence for volcanoes on Venus and some regions of Mars. The Martian arroyos appear to be the work of water. Present conditions prohibit liquid water on the Martian surface, but it does exist as ice on the surface and maybe underground. On Venus it has always been so hot that volcanic-released water never flowed on the surface. So all three planets probably outgassed considerable quantities of water from their interiors during their history.

Water can, however, be removed from the atmosphere if it is split into hydrogen and oxygen. On Mars the exospheric temperature is high enough and the escape velocity low enough so that hydrogen escapes into space (Section 14.5). The active oxygen can combine chemically to form a stable compound, such as ferric oxide. What breaks up the water molecules? Ultraviolet radiation from the sun has enough energy to break the oxygen-hydrogen bonds; this process is called *photodissociation*. Mariner 6 and 7 found a substantial hydrogen cloud around Mars. (Earth also has such a cloud.) Because such a cloud could not last long, it must be supplied with hydrogen from photodissociation. Venus, because it is closer to the sun, and hence subject to stronger ultraviolet radiation, should have suffered an even greater rate of photodissociation and consequently a greater loss of hydrogen. (The ozone layer now protects earth's water from photodissociation, but some may have been lost before the layer formed.) So the earth initially had more water than Mars and has lost it less quickly, because of both a higher surface gravity and the protective ozone layer. Venus had more water than the earth originally, but lost it more quickly because of the higher flux of ultraviolet radiation.

Now consider carbon dioxide. The differences among the planets in amount of carbon dioxide are more apparent than real if the carbon dioxide in the earth's crust and oceans is added to the total. When immersed in water, silicates in rocks react with carbon dioxide to form carbonates. Also, shells of sea creatures are primarily carbonates formed from the carbon dioxide dissolved in the oceans. Added together, the crustal, oceanic, and atmospheric carbon dioxide total for the earth is about equal to the total in the atmosphere of Venus.

The carbon dioxide on Venus did not end up in the crust because of lack of water. The

TABLE 16.1 Comparison of the Terrestrial Planets

Planet	Diameter (earth = 1)	Mass (earth = 1)	Density (water = 1)	Surface Pressure (earth = 1)	Atmosphere's Main Constituents
Mercury	0.38	0.055	5.4	10^{-15}	Helium, hydrogen
Venus	0.95	0.82	5.2	100	Carbon dioxide
Earth	1.0	1.0	5.5	1.0	Nitrogen, oxygen
Mars	0.53	0.11	3.9	0.01	Carbon dioxide

early evolution of the atmosphere of Venus probably followed the early evolution of the earth's atmosphere (Section 14.9). Venus is so close to the sun that its initial temperature 4 billion years ago was high enough to vaporize all the primeval water. Outgassing from the interior added to the carbon dioxide, further raising the temperature. So, the surface temperature on Venus has always been high enough to boil water. Without liquid water not much carbon dioxide can be deposited in the crust as carbonates. And, the rate of deposition is slower at higher temperatures, so fewer carbonates would have formed on Venus, even if the liquid water were available.

What about Mars? Its small mass and size means that it has a small escape velocity and that, compared with the earth, it cooled off quickly. The low escape velocity allows hydrogen, oxygen, carbon, and nitrogen to escape rapidly from the upper atmosphere, even though the temperature is low. Photodissociation breaks up carbon dioxide and water in the Martian atmosphere, and the atoms from these molecules speed off into space. Calculations indicate that the total loss of water (since the time of the formation of Mars) amounted to roughly 10^6 times the present amount of water and that the amount of carbon dioxide lost about equals that in the atmosphere now. The rest of the water and some of the carbon dioxide spewed out by volcanism in the past has returned to the polar caps and the soil.

SUMMARY

This chapter has focused on Venus and Mars, the planets most like the earth. They are roughly the same size as the earth, lie relatively close to the sun, and have about the same average density as the earth (Table 16.1).

This last fact is a key one, for it implies that Venus and Mars have bulk compositions similar to the earth. The earth's interior consists of a core (nickel-iron) and mantle (dense rocky material) with an average density of 5500 kg/m³. Venus and Mars should have interior structures and compositions not wildly different, with a core, mantle, and crust. Until seismographs are placed on Venus's surface, we can only make reasonable guesses about the size of its core and mantle, but we expect them to be similar to the earth. Mars, because its average density is 3900 kg/m³, cannot have a large nickel-iron core. From theoretical considerations we expect a small core to be made of iron sulfide, with a mantle of silicates like the earth's.

Despite these similarities Venus and Mars look like alien worlds to us. The greatest difference among the three is how they have evolved since their formation. Mars is the least evolved (through stage IV in Table 16.2). Venus has changed somewhat (into the beginning of stage V); the earth the most (through stage V). In contrast, the moon and Mercury have only made it up to stage III. They are fossils of what the earth, Venus, and Mars were like 2 billion years ago. Venus lies between Mars and the earth with respect to the extent of its evolution. Because it's more massive than Mars, it should have been hotter in its interior and retained its heat longer.

TABLE 16.2 Main Stages in the Evolution of Terrestrial Planets

Stage	Processes
I	Formation by accretion, heating of crust and interior, crust formation
II	Crust solidification, intense impacts, cratering of surface
III	Basin formation and flooding, lowlands formation
IV	Low-intensity impacts, atmosphere formation by outgassing
V	Volcanoes, crustal movement, continents formation

Our earth is the most evolved of these planets. It now ferments with geologic activity. Volcanoes erupt, crustal plates crash, mountains rise and fall, and the wind and the rain carry the continents' surfaces to the seas. This world is an active, youthful planet—mainly because it has retained a hot interior for billions of years.

Key Words
inferior conjunction
retrograde rotation
radar mapping
upland plateau
cratered terrain
Ishtar Terra
Aphrodite Terra
Beta Regio
Maxwell Montes
shield volcano
rift valley
polar cap
canali
Valles Marineris
arroyo
Olympus Mons
Tharsis ridge
Deimos
Phobos

Review Questions
1. How is the Martian surface similar to that of Venus? (Objective 2)
2. In what respect is Venus most like the earth? Most different? (Objectives 1, 2, and 3)
3. How is Mars most like the earth? Most different? (Objectives 1, 2, and 3)
4. Suppose you are kidnapped by an evil alien creature who threatens to drop you on Venus or Mars with a limited amount of supplies. Which planet would you prefer, and what are the reasons for your choice? (Objective 7)
5. Under what conditions could the Martian arroyos have formed? (Objectives 2, 5, and 9)

Problems
1. What accuracy in timing of radio signals must be obtained in order to detect minimum height differences on Venus of the amount mentioned in the text? (Objective 11)
2. By what percentage does the surface temperature of Mars vary between its nearest and farthest points from the sun? (Objective 2)
3. From the statements given in the text about the amount of water in the polar caps of Mars, calculate the area of a cap if it is 1 km thick. (Objective 5)

4. Both Olympus Mons on Mars and Kilauea on the earth are shield volcanoes. This type of volcano is built up through a series of quiet eruptions of basalt, which, being very liquid when hot, produces a cone with a very low profile. Volcanoes like Mt. Fuji consist of alternating layers of ash and more viscous lava, and so have the more familiar high profile. Kilauea exudes 0.04 km^3 of basalt per year. Given that Olympus Mons is 500 km across at its base and 24 km high, and assuming that it has been erupting basalt at the Kilauean rate, how long has it taken Olympus Mons to reach its present size? Comment on the reasonableness of your result. (Objectives 5 and 9)
5. The Valles Marineris is a giant Martian gorge. Nearly 7000 Grand Canyons could fit inside it. It is not known what agent dug the canyon, but one possibility is water. The Colorado River carries a load of about 1.82×10^6 tons of material a day into the Gulf of California. Assuming this rate for a hypothetical river at the bottom of the Valles Marineris, how long would it take to carve the canyon? The valley is approximately 4000 km long, 200 km wide, and 5 km deep. Comment on the result. Is this a reasonable hypothesis? (Objective 5)
6. In using radar to detect rotation rates for planets, astronomers measure the Doppler shift of the radar beam. The difference in frequency between the signal reflected from one edge of the planet and that reflected from the other edge has a Doppler shift proportional to four times the rotational velocity (one factor of two because one edge is approaching at the rotational velocity whereas the other edge is receding, and another factor of two because there is a Doppler shift both on absorption and on emission). If the frequency of the radiation emitted by the radar transmitter is 430 MHz, what is the Doppler shift expected between the opposite edges of Venus? (Objective 11)
7. Calculate the length of the solar day on Venus. Hint: See Section 15.6 for the same calculation for Mercury, but remember that Venus's rotation is retrograde. (Objective 2)

BEYOND THIS BOOK . . .

* In the September 1975 issue of *Scientific American* you can find "Venus" by A. and L. Young and "Mars" by J. B. Pollack.
* The Viking landers and orbiters have produced a new vision of Mars. Look for articles in *Astronomy, Sky and Telescope, Science, Nature*, and the *Griffith Observer*. The August 27, 1976, issue of *Science* contains a number of technical reports from Viking 1. The September 1976 issue of *Sky and Telescope* has a report on Viking 1 and the October 1976 issue a report on Viking 2.
* A very technical, but intriguing report on Mars is "The Geologic Development of Mars: A Review" by T. A. Mutch and R. S. Saunders in *Space Science Reviews*, vol. 19 (1976), p. 3.
* For even more geology, read *Planetary Geology* (Prentice-Hall, Englewood Cliffs, N.J., 1975) by N. M. Short and *Earthlike Planets* (Freeman, San Francisco, 1981) by B. Murray, M. Malin, and R. Greeley.
* K. Weaver has written on "Mariner Unveils Mercury and Venus" in *National Geographic*, June 1975. To get the Russian view on Venus, read "Veneras 9 and 10 on Venus" in *Sky and Telescope*, December 1975 and "Report From a Torrid Planet," May 1982.
* Perhaps Mariner 9's most important discovery was the giant volcanoes on Mars. For details see "The Volcanoes of Mars" by M. Carr, *Scientific American*, January 1976.
* For more on the new mapping of Venus, see "The Surface of Venus" by G. Pettengill, D. Campbell, and H. Masursky, *Scientific American*, August 1980. For additional information on the atmosphere, read "The Atmosphere of Venus" by G. Schubert and C. Covey, *Scientific American*, July 1981.

THE JOVIAN PLANETS: PRIMITIVE WORLDS

17

LEARNING OBJECTIVES
After studying this chapter, you should be able to:

1. Compare the Jovian planets as a group to the terrestrial planets, emphasizing the greatest differences.
2. Compare the Jovian planets to one another in terms of (a) relative size, (b) relative mass, (c) average density, (d) atmospheric composition, (e) internal structure, and (f) special features.
3. Outline new information that space probes have provided about Jupiter and Saturn and their satellites.
4. Compare the rings of Saturn with those of Uranus and Jupiter in terms of size, shape, and possible composition.
5. Present the unique characteristics of Pluto that make it neither a Jovian nor a terrestrial planet.
6. Compare and contrast the general characteristics and surface features of the Galilean satellites of Jupiter—Io, Europa, Ganymede, and Callisto.
7. Compare the Galilean satellites to the earth's moon and to Pluto.
8. Compare the larger moons of Saturn to those of Jupiter.
9. Compare Jupiter's magnetic field and magnetosphere to that of Saturn and the earth.
10. Argue that, compared with the terrestrial planets, the Jovian planets have not evolved much since their formation, but that some of their moons have evolved.

CENTRAL QUESTION:
How do the Jovian planets differ from the terrestrial ones, and what do these differences tell about different evolutions?

When an amateur astronomer finishes constructing a telescope, after looking at the moon, he or she usually turns to Jupiter and Saturn. And no wonder! These are impressive, awesome planets (Fig. 17.1)—two giant worlds languidly circling the sun. Banded Jupiter drags along its coterie of satellites. Saturn is set in its rings like a prize gem of space. Once, at a public lecture given (by MZ) to young people, Saturn was voted by an overwhelming margin the most beautiful of all the planets.

It's hard to grasp the weird environments of these giant worlds; planets four to eleven times the diameter of the earth, mostly gasses and liquids, and without breathable atmospheres. And they are so distant from the sun and the earth. That gulf has now been bridged; Pioneers 10 and 11 and Voyagers 1 and 2 have skirted Jupiter and winged past Saturn. These probes have confirmed what was suspected: their awesome beauty cloaks an environment deadly to human beings.

This chapter investigates the features of the Jovian planets that set them apart from the terrestrial planets. The major differences between the Jovian and terrestrial planets furnish additional clues about the evolution of the planets and the formation of the solar system. The key point is this: The Jovian planets are primitive worlds, looking today very much the same as when they were formed. These giant worlds are little evolved compared with a planet like ours.

17.1 JUPITER: LORD OF THE HEAVENS

Jupiter is the largest and most massive body in the solar system (except for the sun). Its total mass is about $2\frac{1}{2}$ times that of all the other planets put together. Eleven planet earths placed edge to edge would stretch across Jupiter's visible disk. It's not farfetched to consider the solar system as a two-body system of just Jupiter and the sun. The center of mass of the Jupiter-sun system lies just outside the sun's surface.

17.1 JUPITER: LORD OF THE HEAVENS

Physical Characteristics of Jupiter

The contingent of satellites allows Jupiter's mass to be determined by using Newton's version of Kepler's third law. Its mass is the largest among the planets, 318 times that of the earth, and so is its diameter, almost 140,000 km. Yet for all this size, Jupiter's material is less concentrated than the earth's, for Jupiter's density is only 1330 kg/m^3.

All the Jovian planets have low densities compared with the terrestrial planets. This key difference between the two classes implies that the Jovian planets are made of fundamentally different stuff. The terrestrial planets are basically globes of rock and metal, made of elements such as iron, aluminum, oxygen, and silicon. Jupiter, in contrast, is made mostly of hydrogen and helium in liquid form.

The hydrogen and helium came together when Jupiter formed, and the giant planet has lost little, if any, since then. Jupiter's huge mass means that its escape velocity is high, about 57 km/sec. Remote from the sun, Jupiter's upper atmosphere is cold, only about 130 K. The average velocity (Section 14.5) for hydrogen molecules at 130 K is about 1 km/sec. So even low-mass hydrogen molecules here move around so slowly that they do not attain escape velocity. If the hydrogen cannot escape from the upper atmosphere, more massive atoms and molecules cannot either. Jupiter has retained its atmosphere for eons and will hold it for eons to come. What you see now is basically the atmosphere and mass with which Jupiter was born.

General Atmospheric Features

The visible disk of Jupiter is not the planet's surface, but its upper atmosphere. Notice in all the photos of Jupiter in this book that the upper atmosphere shows alternating strips of light and dark regions that run parallel to the equator. The Pioneer missions discovered that the light regions, called *zones*, have lower temperatures than the dark regions, called *belts*. So the zones rise above the belts. This difference in temperature implies that the zones flag the tops of rising convective regions of high pressure, and the belts must be the descending areas of low pressure. Jupiter's upper atmosphere contains continually rising

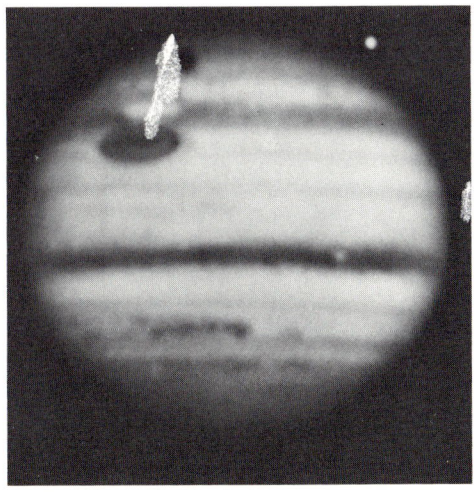
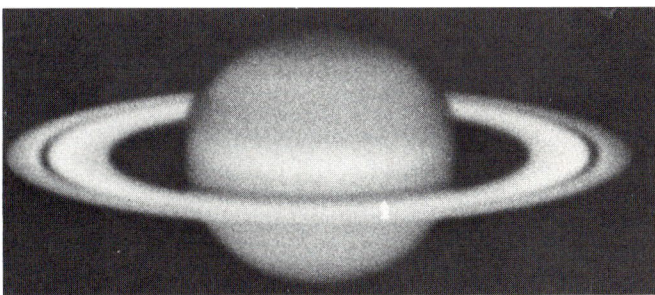

FIGURE 17.1 Jupiter and Saturn as viewed from the earth. *(Jupiter photo from Palomar Observatory, California Institute of Technology; Saturn photo for NASA by the Lunar and Planetary Laboratory, University of Arizona.)*

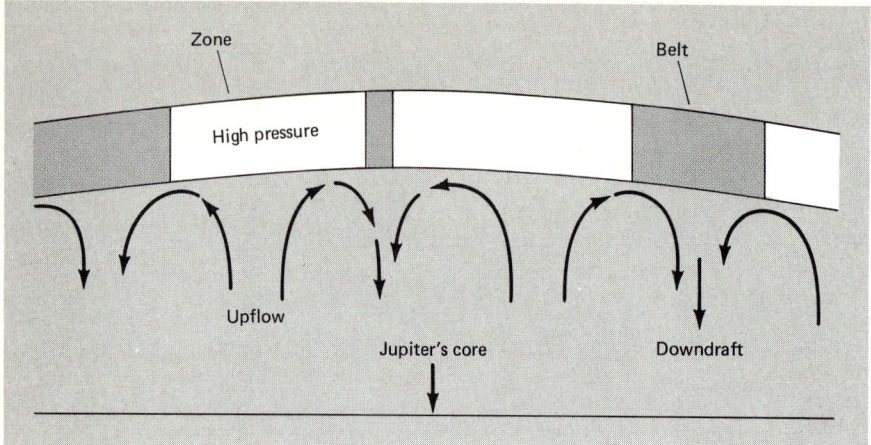

FIGURE 17.2 The circulation in Jupiter's upper atmosphere. Rising air creates high pressure regions (called *zones*) and the downflow makes low pressure areas (called *belts*). The zones generally appear lighter in color than the belts because they are higher up in the atmosphere.

and falling material (Fig. 17.2). This atmospheric circulation transports heat out to the surface from the planet's interior; from there it is radiated into space.

Semipermanent markings in the cloudy surface allow a measurement of Jupiter's rate of rotation. The rate varies with latitude: Jupiter spins in 9 h 50 min at its equator and 9 h 55 min at its poles. This rapid rotation and Jupiter's large radius produce an equatorial velocity in excess of 43,000 km/sec.

Such an enormous rotation speed affects the circulation in Jupiter's atmosphere. It causes the permanent high-pressure regions (zones) and low-pressure regions (belts) to stretch out completely around the planet instead of, as on earth, being somewhat localized. High-speed winds (jet streams) speed along at the boundaries between the belts and zones, creating atmospheric disturbances.

The Voyager spacecraft zoomed in on these complex streams, swirls, and gyrations of

FIGURE 17.3 A close-up view of turbulent patterns in Jupiter's atmospheric flow. Note the large white spots. (*Courtesy NASA.*)

17.1 JUPITER: LORD OF THE HEAVENS

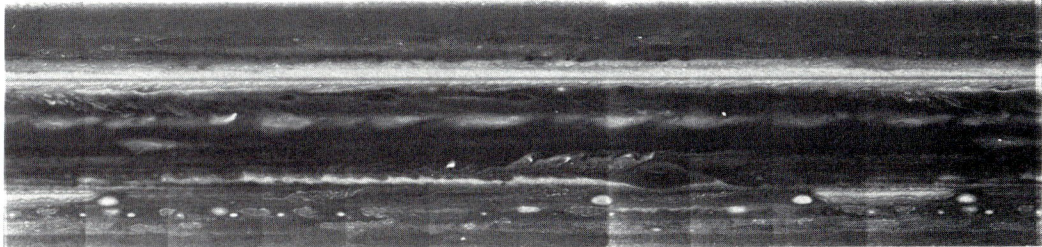

FIGURE 17.4 Changes in Jupiter's atmosphere between Voyager 1 (top) and 2 (bottom). Photos have been spliced together to show the whole planet. The top and bottom views are aligned along the same longitudes. West is to the left; east to the right. Note that the Red Spot has moved westward, and the white ovals eastward. (*Courtesy NASA.*)

Jupiter's upper cloud layer. The spacecraft sent to us a collage of color (see Plate 10) and a fantasy of flow (Fig. 17.3). These high-resolution photos show the turbulence and intricacy of the atmospheric flow.

Earth-based telescopes as well as Voyager have revealed complex changes in the belts and zones. Occasionally, dark blue, red, brown, and white ovals appear against the banded background (Fig. 17.4). These small oval spots last as long as a year or two.

The Red Spot

The most permanent and famous atmospheric disturbance is the Great Red Spot, first observed by Englishman Robert Hooke in 1630. The Red Spot changes in size; it is some 14,000 km wide and 30,000 to 40,000 km long—it could easily swallow the earth!

What is the Red Spot? The Pioneer flybys found that the Red Spot was a few degrees cooler than and poked about 8 km above the surrounding zone, so it is a rising region of high pressure. In ground-based observations astronomers at New Mexico State University (Fig. 17.5) found that the Red Spot rotates counterclockwise like a vortex—just what is expected from a high-pressure zone in Jupiter's southern hemisphere. The material around the Red Spot flows in a way to reinforce its circulation: to the west on the north end of the Red Spot and to the east on its south end. The Red Spot is like a rolling wheel turned by the surrounding atmospheric flow; in turn the Red Spot deflects nearby clouds and forces them around it.

The Voyager missions presented the Red Spot to us in dramatic detail (Fig. 17.6). They showed that the Red Spot's spin is so forceful that small spots roll around and get squashed in it. Behind the Red Spot runs a region of turbulent flow, like the wake of a

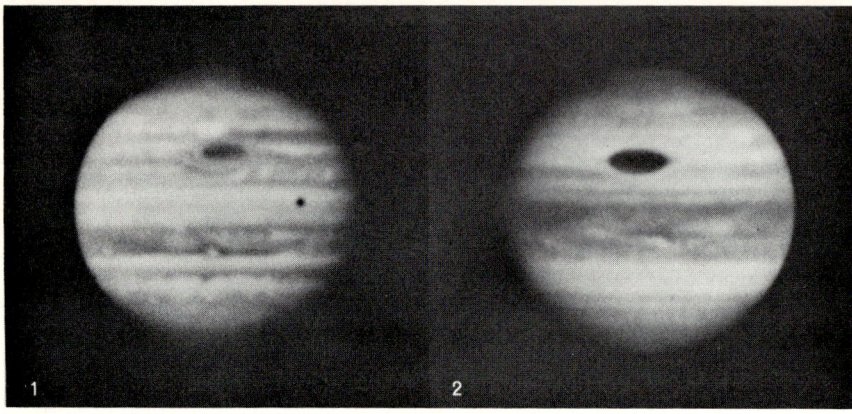

FIGURE 17.5 Jupiter's Great Red Spot viewed from the earth. These two photos were taken two years apart and show the changes over that time. (*Photos by A. Murrell and R. Minton; courtesy, New Mexico State University Observatory.*)

fast-moving boat. But in this case the turbulence is not caused by the Red Spot's movement, but by the flow of the atmosphere past it.

The contemporary model for the Red Spot pictures it as a high-pressure storm at least 300 years old that marks the top of an enormous, powerful convective column. Not yet clearly worked out from this model is how the Red Spot started and how long it will last.

Why is the Red Spot red? (Many other spots on Jupiter are white; see Plate 10.) We don't know for certain. The color may be caused by chemical reactions driven by the heat flowing up from below. These reactions may involve compounds that include phosphine (PH_3). Another possibility is that the color comes from organic compounds made lower down in the atmosphere and brought to the top in the upwelling region of the Red Spot.

Whatever the exact chemical explanation for the clouds' colors, we do know that

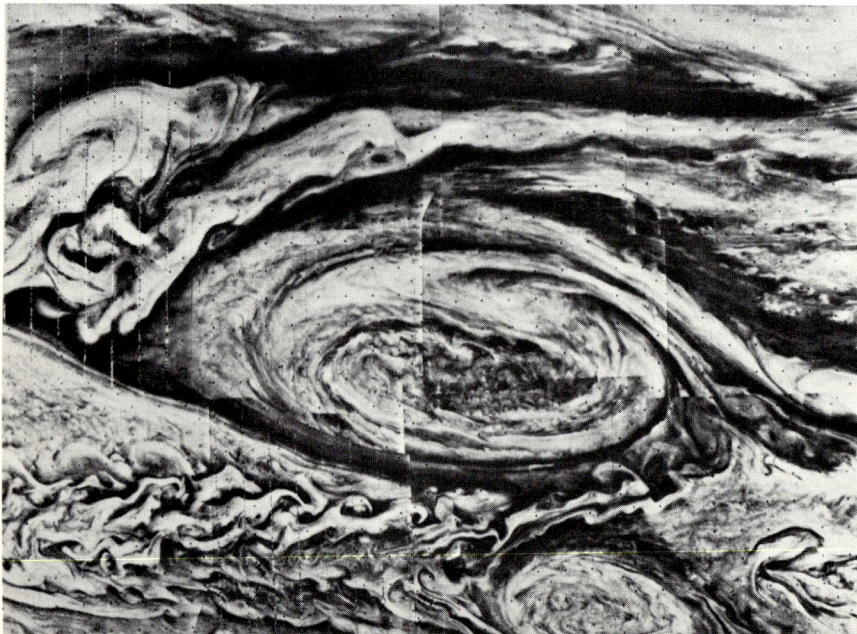

FIGURE 17.6 A close-up of the Red Spot and the turbulent region near it. The smallest details visible are about 30 km across. (*Courtesy NASA.*)

different colors correspond to different levels in the atmosphere. The highest clouds are red, the next highest white, lower ones brown, and the lowest blue.

Atmospheric Composition

We know the atmospheric composition above the clouds from infrared spectroscopy. In 1934 methane (CH_4) was discovered, the first molecule definitely identified in the Jovian atmosphere. About 10 years later ammonia (NH_3) was found. Molecular hydrogen (H_2) was discovered in 1960. The Pioneer flights measured atmospheric helium. Voyager found evidence for these and also for acetylene (C_2H_2), ethane (C_2H_6), phosphine (PH_3), water (H_2O), and germane (GeH_4). Some of these had been previously detected in spectra taken from the earth, as had CO and HCN. A few molecules containing deuterium are also known to be present.

Most of these compounds would *not* be present in Jupiter's atmosphere if all its chemical reactions reached their natural balance. The various carbon compounds would revert to methane and nitrogen to ammonia. So energy must be added to the atmosphere to keep it off balance. Possible sources for that energy are ultraviolet radiation from the sun, lightning, and charged particles raining down from the magnetosphere.

An analysis of Voyager spectroscopic data implies that Jupiter's upper atmosphere contains about 79 percent hydrogen, 20 percent helium, and 1 percent all other elements. It is essentially the same composition as the sun, but since Jupiter is not as hot as the sun, we are talking about molecules, not ionized gas.

The visible clouds at the tops of the zones are most likely ammonia ice crystals (Fig. 17.7). Below them, according to theoretical calculations, lies a layer of ammonia hydro-

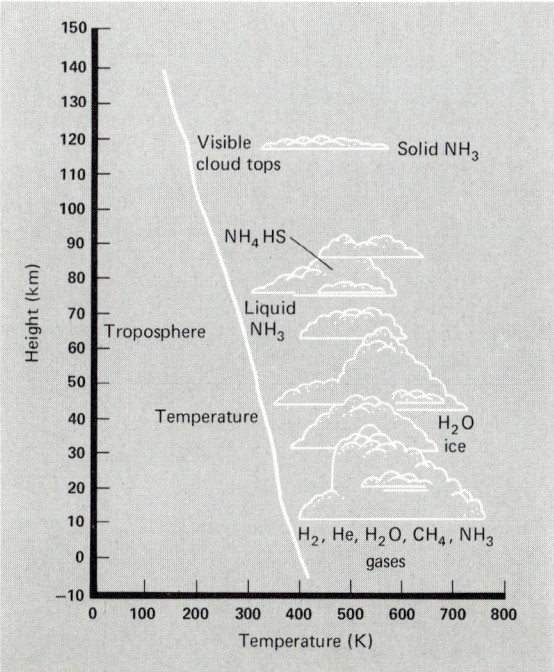

FIGURE 17.7 One model for the structure of Jupiter's atmosphere. Note that the temperature increases downward and hits room temperature (290 K) at about 60 to 70 km into the troposphere. (*Adapted from a diagram by W. Hartmann; based on calculations by J. Lewis.*)

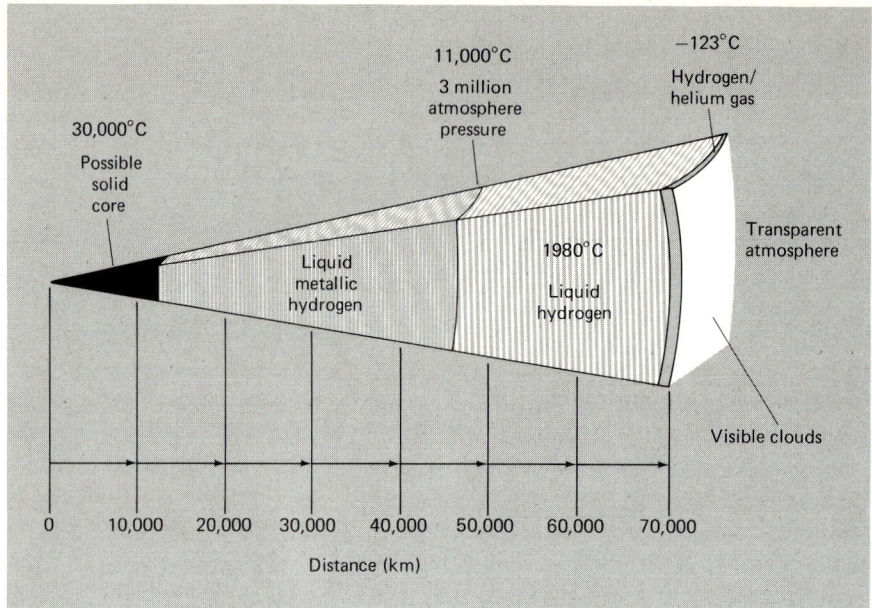

FIGURE 17.8 A model for Jupiter's interior structure. Below the atmosphere is a thick layer of liquid molecular hydrogen. The core is a dense material, perhaps silicates, that may be molten.

sulfide (NH_4HS) clouds. Below these float liquid ammonia and water ice clouds. This model has three separate cloud layers that make up the upper atmosphere.

Deeper down in the atmosphere the temperature rises more and the atoms and molecules are all in a gaseous state. In Jupiter's stratosphere the pressure is only about 0.1 atm and the temperature 100 K. Near the cloud tops, at a pressure of 0.7 atm, the temperature rises to 150 K. At a lower level, where the pressure hits 100 atm, the temperature reaches 300 to 400 K. The entire atmosphere may be 1000 km thick. In fact, there is no distinct boundary between atmosphere and interior. The atmosphere just gets denser and hotter further in, gradually merging into the liquid interior.

A Model of the Interior

Since we cannot see directly inside Jupiter, we can only infer its internal structure from calculated physical models, as is done for the sun and other stars. Two key pieces of information go into these models: (1) Jupiter's low density and atmospheric composition imply a solar mix of material throughout and (2) Jupiter radiates into space more energy than it receives from the sun, so it must be hot inside.

Infrared detectors on Pioneer 10 confirmed this peculiarity of Jupiter (discovered first by infrared observations from earth). The planet gives off as infrared radiation about twice as much energy as it receives from the sun. Jupiter has a luminosity about 2×10^{-9} the sun's present luminosity. So Jupiter must have its own internal heat. An essential difference between stars and planets is that stars produce their own energy by thermonuclear reactions in their cores, and planets do not generate energy. Jupiter has an essential attribute of a star, but it's a planet! This seeming paradox is resolved when we look at

Jupiter's origin and evolution. The internal heat is probably left over from Jupiter's formation. (Massive planets lose heat very slowly.)

Recent models of Jupiter, assuming a solar composition and internal heat, come up with the following picture (Fig. 17.8). The atmosphere covers the planet like a thin skin and consists mostly of molecular hydrogen. Going into the planet the density, temperature, and pressure increase, so the hydrogen exists in a liquid state. At a pressure of about 3 million atm the hydrogen is squeezed so tightly that the molecules are separated into protons and electrons that are free to move around and so can conduct electricity. This state is called *metallic hydrogen*. Although it has never been observed in a lab on the earth, its existence is predicted from atomic physics. The critical pressure is reached at a distance of about 0.75 to 0.80 Jupiter's radius from the center. This strange state continues to within about 14,000 km of the planet's center. Here, perhaps, if Jupiter does have a solar composition, lies a core of heavy elements. It may make up about 4 percent of Jupiter's total mass. Whether a heavy core exists or not, Jupiter's center must be hot—30,000 to 40,000 K.

Note that most of Jupiter is hydrogen and most of that hydrogen is liquid—quite a contrast to the earth's interior (Section 14.2) and that of the other terrestrial planets. The core temperature may be as high as 30,000 to 40,000 K, about 10 times hotter than the earth's core. It is the flow of the heat outward from the core that drives the circulation to produce the beautiful banded atmosphere of Jupiter.

Jupiter's Magnetosphere

Early radio observations implied that a strong magnetic field existed on Jupiter. The Pioneer 10 space probe confirmed that Jupiter has a magnetic field almost 1000 times stronger than the earth's. As the solar wind plows into this intense field, it creates an enormous shock wave that enshrouds the planet (Fig. 17.9). In addition the magnetic field traps energetic charged particles from the sun. The region within the Jovian magnetic field close to the planet contains so many such particles that it would be lethal to travel through it.

How does Jupiter generate such an intense magnetic field? Recall that a dynamo model (Section 14.4) pictures currents in the liquid nickel-iron core generating the magnetic field like an electromagnet. Jupiter certainly does not have a nickel-iron core, but in a large part of Jupiter's interior metallic hydrogen can conduct electric current. So the conditions for a dynamo to operate are available: a fluid able to conduct electricity and convective currents driven by heat and rapid rotation. A dynamo effect in the metallic hydrogen zone can produce Jupiter's magnetic field.

Radio Emission

Jupiter gives off intermittent bursts of radio noise of unexpectedly high intensity. At wavelengths of tens of meters, sharp bursts lasting about a second have been detected by radio telescopes. A strong burst generates approximately 10^{11} J; in comparison, an intense lightning bolt on earth discharges about 10^5 J. (It is fitting that the namesake of the ruler of the gods should unleash power equivalent to a million terrestrial lightning bolts in one burst!)

These radio bursts are likely generated by superbolts of lightning. Some Voyager pic-

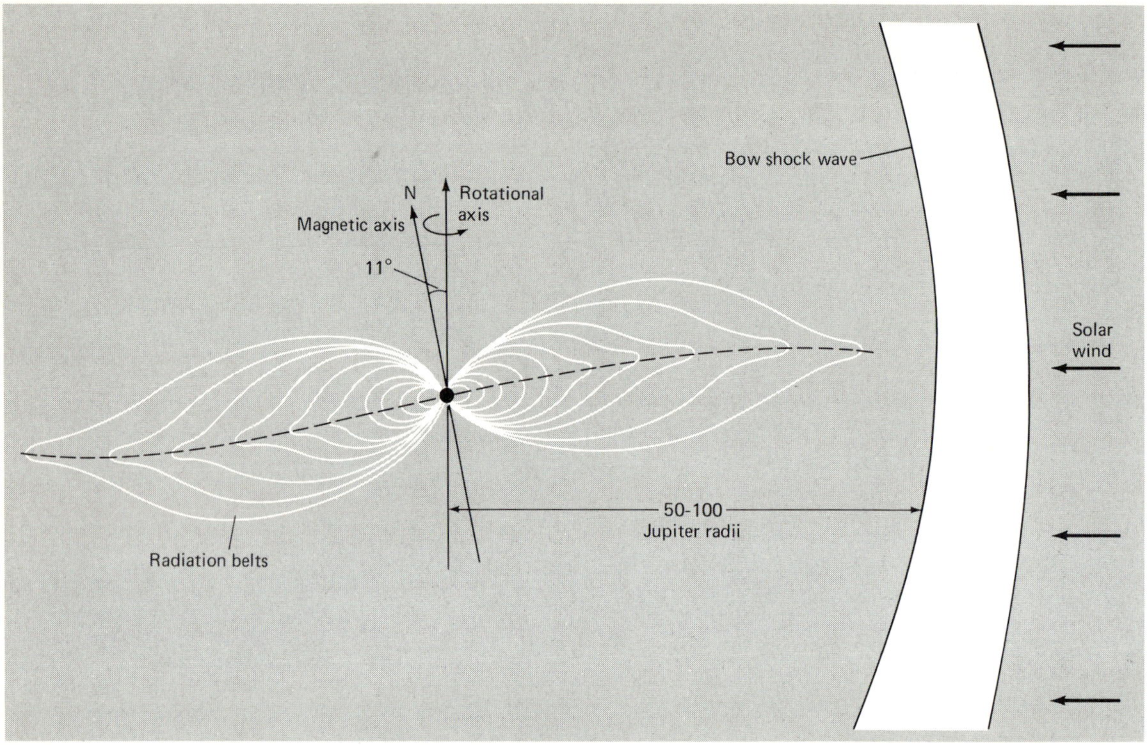

FIGURE 17.9 A model for Jupiter's magnetosphere based on spacecraft data. The size and shape varies according to the strength of the solar wind. (*Adapted from a NASA diagram.*)

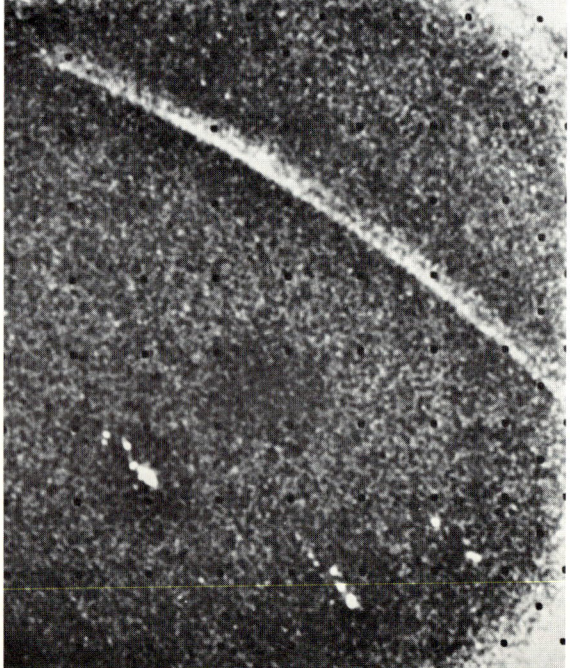

FIGURE 17.10 Jupiter's night side. The bright streak at the top is an aurora over the north pole. The bright spots below it are lightning flashes. (*Courtesy NASA.*)

17.1 JUPITER: LORD OF THE HEAVENS

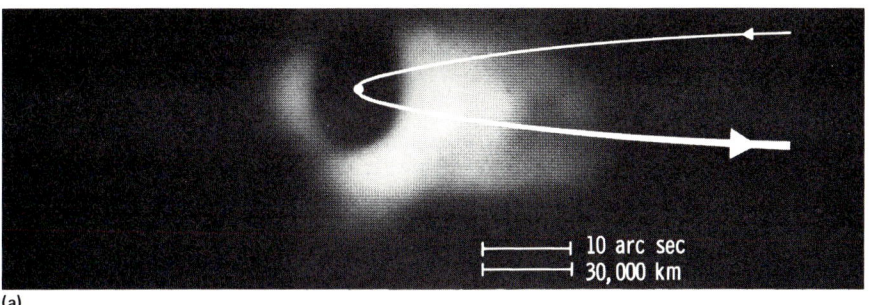

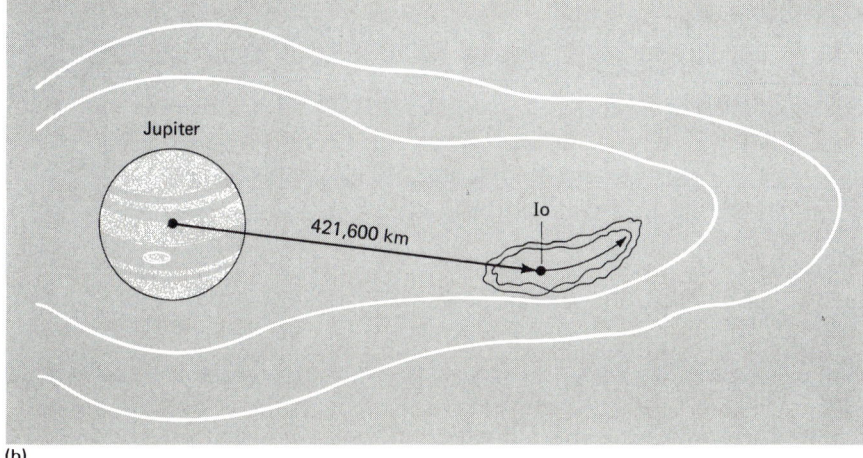

FIGURE 17.11 (a) The sodium cloud around Io, as seen from the earth. The size and position of Io have been drawn in. (*Courtesy B. Goldberg and the Jet Propulsion Lab.*) (b) The overall extent of Io's sodium cloud along Io's orbit.

tures showed bright regions on the planet's night side that may be lightning (Fig. 17.10). Violent updrafts and turbulence in Jupiter's atmosphere probably generate this lightning, in much the same way as thunderstorms do on the earth.

Radio radiation at a few tenths of a meter has also been detected. A surprising discovery was that variations in the radiation's intensity are correlated with the position of Jupiter's satellite Io. When Io passes in front of or behind the planet (as viewed from the earth), the radiation decreases. At Io's maximum elongation to either side, the intensity of the radiation peaks. How Io's position can trigger the radiation is unknown in detail, but the reason is probably related to belts of high-energy particles from the sun trapped by Jupiter's strong magnetic field.

Auroras

Voyager pictures of Jupiter's night side showed polar *auroras* for the first time (Fig. 17.11). The Voyager photos indicate that the auroras occur in at least three layers: at 700, 1400, and 2300 km above the cloud tops. We presume that these auroras happen for the same reason as on the earth: excitation of the upper atmosphere by energetic charged particles pouring in along the north and south magnetic poles (Section 6.7). These particles flow from Io and are trapped by Jupiter's magnetic field.

17.2 THE MANY MOONS OF JUPITER

Jupiter possesses the second largest satellite system of all the planets, an entourage of at least 16 moons (Appendix B). The Voyager missions discovered three moons close in to the planet (they are called 1979 Jl, 1979 J2, and 1979 J3). Charles Kowal, who discovered the thirteenth moon in 1974 (called Leda) discovered another possible one in 1975. That discovery has not yet been confirmed, but many people believe this moon exists. The brightest, largest, and first-known moons are the four discovered by Galileo: Io, Europa, Ganymede, and Callisto. Their orbits lie within 3° of Jupiter's equatorial plane, close to the ecliptic and to our line of sight.

All the Galilean satellites of Jupiter lie within the region found by Pioneer 10 to contain strong magnetic fields and energetic particles. The influx of these charged particles may charge the surfaces of the satellites. Such electric charges on Io interact with the Jovian magnetic field to produce the observed radio bursts.

The Galilean moons are large compared to our moon (Table 17.1). Ganymede, 5276 km in diameter, and Callisto, 4848 km in diameter, are both larger than Mercury. Io has a diameter of 3638 km, and Europa 3126 km. Only Europa is smaller than our moon (3476 km). These huge moons orbit within 2 million km of Jupiter in the following order: Io at 422,000 km, Europa at 671,000 km, Ganymede at 1,070,000 km, and Callisto at 1,883,000 km. As our moon does to the earth, each keeps one face toward Jupiter.

The other moons of Jupiter are quite small and difficult to observe. Amalthea orbits only 181,000 km (2.6 Jupiter radii) from Jupiter in a swift 0.5 day. The Voyager moons 1979 Jl and 1979 J2 are even closer, only 1.8 Jupiter radii out from the planet's center. The other satellites all lie far beyond the Galilean moons. There are two groups of four, one group at a distance of about 12 million km that orbit counterclockwise and another at about 23 million km that orbit clockwise (retrograde). These outer groups are thought to consist of captured asteroids.

The Voyager missions brought a bonanza of results about the Galilean moons. They resemble the terrestrial planets somewhat in size and composition, but each is a world of its own, different from the others (see Figs. 13.18–13.21). These differences are hinted at by the bulk densities of the moons: Io 3500 kg/m^3, Europa 3000 kg/m^3, Ganymede 2000 kg/m^3, and Callisto 1800 kg/m^3. This list is in order of increasing distance from Jupiter, and you can see the pattern. The density of the moons decreases as the distance from Jupiter increases. Such density differences tell us that the compositions of Io and Europa resemble that of our moon—mostly rock, with perhaps a little icy material. In contrast,

TABLE 17.1 Properties of the Galilean Satellites

Name	Diameter (km)	Average Distance from Jupiter (km)	Orbital Period (days)	Bulk Density (kg/m^3)	Mass (moon = 1)
Io	3638	4.22×10^5	1.77	3530	1.21
Europa	3126	6.71×10^5	3.55	3030	0.66
Ganymede	5276	1.07×10^6	7.16	1930	2.03
Callisto	4848	1.883×10^6	16.69	1790	1.45

Ganymede and Callisto must contain substantial amounts of water ice or other low-density icy materials and much less rock than do the inner moons. One explanation for this progression in density is that Jupiter was itself quite hot at the time the satellites were forming and prevented the icy substances from being included in the nearer satellites (Section 19.4).

Let's now look at the new data on each Galilean moon; these highlights open a new chapter in solar system research.

Io

Even prior to the Voyager flybys, we knew that Io must be a strange place. Compared with terrestrial planets, Io is a world in its own right. Three-fourths the size of Mercury, it orbits Jupiter every $42\frac{1}{2}$ hours—a giant satellite for a giant planet. Io has an atmosphere, but a very thin one. At the surface the pressure is about 10^{-10} atm. (Only two other satellites, Saturn's Titan and Uranus's Triton, are known to have an atmosphere.)

Io's atmosphere has a peculiar property: It gives off a bright, continuous glow of emission from sodium atoms. This sodium glow surrounds Io like a yellow halo out to a distance of about 30,000 km (Fig. 17.11a). The sodium cloud extends about 200,000 km along Io's orbit, forming a partial ring of gas around Jupiter (Fig. 17.11b). Gasses of potassium and sulfur also have been observed surrounding Io.

What produces Io's sodium cloud? Prior to Voyager we knew that the surface of Io is bright, reflecting sunlight well, the way ice does. But infrared observations showed that ice does not cover Io's surface. One proposal saw Io's surface coated with deposits of salt, rich in sodium and sulfur. Such materials (such as the gypsum salt of White Sands, New Mexico) reflect sunlight well. In this model the energetic particles trapped by Jupiter's magnetic field bombard the surface of Io, releasing sodium atoms that escape from the satellite's surface. These make up the sodium cloud.

We now know that this idea contains only a part of the picture. The other piece was revealed by Voyager: Io has active volcanoes! In fact, it is volcanically one of the most active places in the solar system; erupting volcanoes and fuming lava lakes cover its surface (Plate 11).

Voyager 1 picked up its first erupting volcano at the edge of Io, and at least eight were seen. All but one were still active when Voyager 2 came by four months later. Unless we have by accident caught Io at an unusually active time, spouting volcanoes must be common happenings on this world (Fig. 17.12).

Io's volcanoes eject plumes of gas and dust up to heights of 250 km at velocities of up to 1000 m/sec. (The earth's large volcanoes spit out material at about 50 m/sec.) On a nearly airless body like Io, the volcanic gas and dust crest like a fountain plume and then spread and fall in a dome shape (Fig. 13.12c). Io's escape velocity (2.5 km/sec) is greater than the speed with which the volcanic dust erupts, so little solid material is lost to space. However, about a ton of gasses each second is blown into Jupiter's magnetosphere. These eruptions provide the main source of particles there; the solar wind contributes relatively little.

It seems likely that Io's volcanoes are driven by hot liquid sulfur and sulfur dioxide in the interior. On Io's surface sulfur exists as crystals. Io's interior is hot; here sulfur is liquid. When the liquid sulfur combines with lava, the lava vaporizes, driving the lava up and out of volcanic vents.

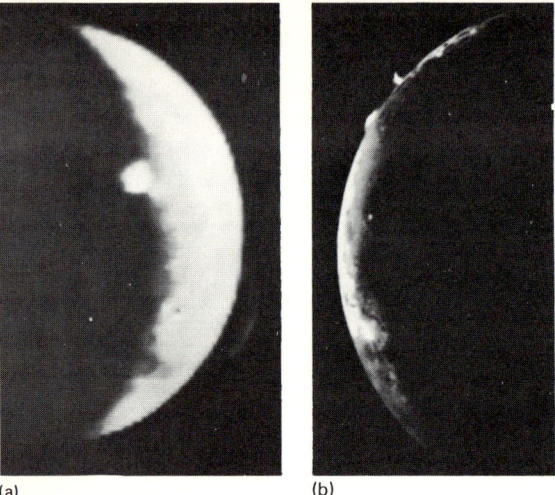

(a) (b)

FIGURE 17.12 (a) The discovery photo of volcanic eruptions on Io. One eruption is the faint plume at lower right, the other the bright spot just above the center. (b) Continuing eruptions are visible along the left edge in this photo taken 9 months later. (*Courtesy NASA.*)

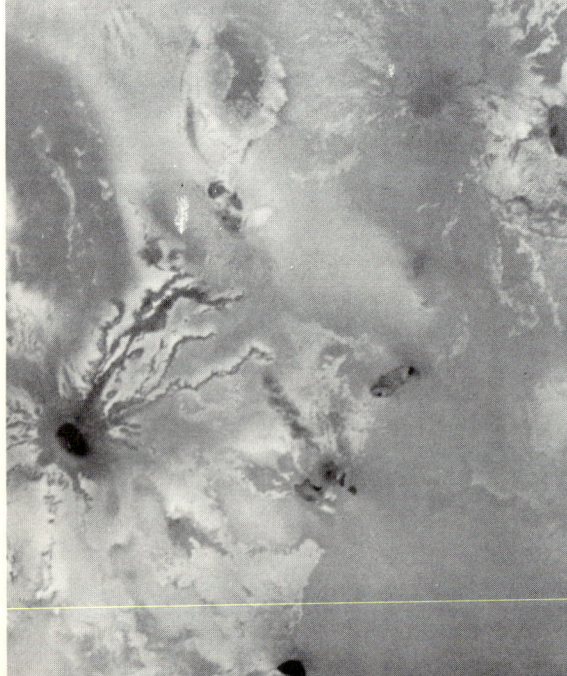

FIGURE 17.13 A close-up view of Io. The dark spot with the irregular radiating pattern (at the left) is a volcanic vent with lava flows. (*Courtesy NASA.*)

FIGURE 17.14 A view of Europa with a resolution of 4 km. Note the dark cracks along the surface. (*Courtesy NASA.*)

17.2 THE MANY MOONS OF JUPITER

Io's volcanoes are quite different from those commonly found on the earth, Venus, and Mars. Few appear as cones or shields; most resemble collapsed volcanic craters. Lava simply pours out of a crater vent and spreads outwards for hundreds of kilometers (Fig. 17.13), so dark-colored lava lakes surround many of Io's volcanoes. One of the largest is about the size of the island of Hawaii. The temperatures in these lava lakes are 40 to 80°C.

Because volcanic activity continually alters Io's surface, it must be very young. No impact craters are seen; volcanic flows have covered them up. Io's surface seems to be the youngest in the solar system—probably less than 1 million years old.

Why is Io so actively volcanic? One possible model views Io's interior as melted from tidal forces (Focus 12.1) produced by the other three Galilean satellites. The calculations indicate that the tidal heating of Io now amounts to about three times more than the current radioactive heating rate estimated for our moon. This model had in fact predicted "widespread and recurrent volcanism" on Io before the Voyager 1 photos were obtained.

Europa

The surface features of Europa consist of bright areas of water ice among darker, orange-brown areas. Europa's surface is crisscrossed by stripes and bands that may be filled fractures in the moon's icy crust (Fig. 17.14).

Europa appears much less active than Io; no volcanoes are seen here. Yet Europa's surface, too, is almost devoid of impact craters; only three possible ones have been identified. So Europa's surface cannot be a primitive one; it must have evolved since its formation. The crust must have been warm and soft sometime after formation to wipe out evidence of the early, intense bombardment.

On first look the most impressive features on Europa are the dark markings that crisscross its face, making it look like a cracked eggshell. Some of these cracks extend for thousands of kilometers, split to widths of 50 to 200 km, but reach to depths of only 100 m or so. Europa's surface is really incredibly smooth; compared to its size, its dark markings are no deeper than the thickness of ink drawn on a ping-pong ball.

Europa's cracked surface indicates that its solid, icy crust is thin and its interior hot and primarily molten. One tentative model proposes that its crust long ago may have been a slush kept partially melted by a molten interior (hot, perhaps, because of tidal stresses). As Europa cooled, its crust turned to smooth, glassy ice. Tidal stresses may then have cracked the icy surface into patterns that resemble pack ice on a frozen sea on the earth. Darker material from the interior welled up to fill in the cracks.

Ganymede

Largest moon of Jupiter, Ganymede is also the largest moon in the solar system. Its surface looks strangely like our moon's, with dark regions that resemble maria. Yet, it also has huge fault lines along its surface, like Europa.

Ganymede has two basic types of terrain (Fig. 17.15): *cratered* and *grooved*. Craters up to 150 km in size densely mark the cratered terrain. Their abundance indicates that the cratered terrain is old, some 4 billion years. Compared with those on the moon and Mercury, the craters are shallow and some have convex rather than concave floors. The craters of Ganymede also differ from those of the moon and Mercury in that they have central pits rather than central peaks (Fig. 17.15a). There are no extensive mountainous

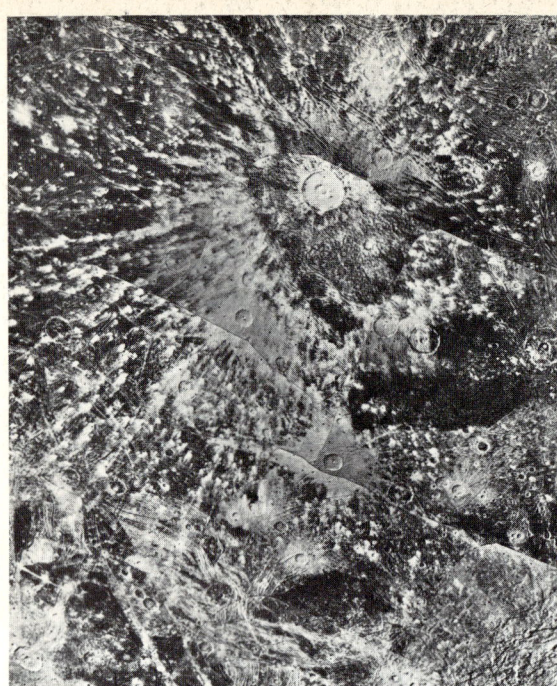 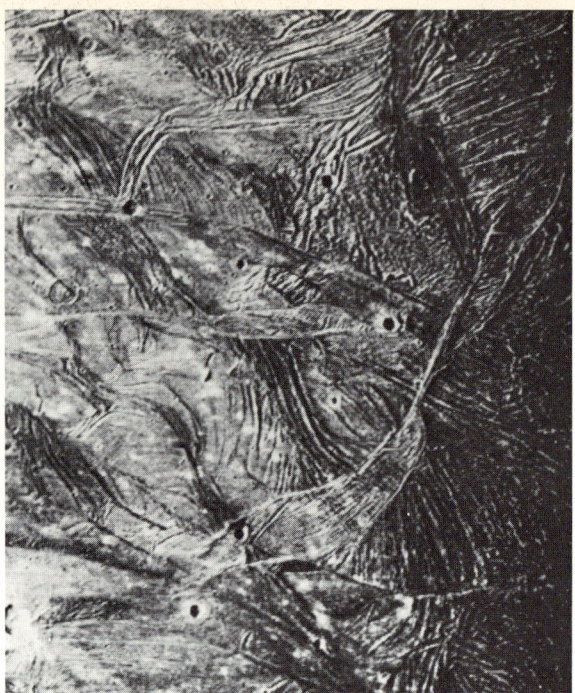

FIGURE 17.15 (a) Impact craters with bright rays on the surface of Ganymede. The large crater at the upper center is about 150 km in size. (b) Grooved and ridged terrain on Ganymede's surface. (*Courtesy NASA.*)

regions or large basins on Ganymede either; in fact nowhere on the satellite is there any relief greater than about 1 km. In some regions there are small bright patches with just a hint of surrounding walls, which look as if a crater has sunk into a soft surface. All these features suggest that the crust of Ganymede is somewhat plastic. Crater rims and mountains slowly sink back into the surface; crater floors gradually fill in. This plastic flow probably is due to the large amount of water ice in Ganymede's crust. Many craters on Ganymede have very bright rays extending from them, attesting to their formation by impacts on an icy surface.

In contrast, the grooved terrain separates the cratered terrain into polygon-shaped segments. From a distance Ganymede's grooved terrain looks like sand crossed by dune buggy tracks (Fig. 17.15b). The grooved terrain consists of a mosaic of light ridges and darker grooves where the ground has slid, sheared, and torn apart. The ridges have widths of 10 to 100 km, lengths of 10 to 1000 km, and heights of 200 to 300 km.

Ganymede's surface is covered with cracks called *transverse faults*, where the ground has moved sideways for hundreds of kilometers. Earth has many transverse faults, for example, in the ocean basins along mid-oceanic ridges where upwelling new rock forces the crustal plates apart.

Ganymede's bulk density is low, only about 2000 kg/m^3, so it must contain about half water and half rock. Occasional stresses on the water-rock crust have created the fracture patterns. Some ridges and grooves overlie others, an indication that many episodes of crustal deformation have happened in the past.

In some places craters cover the grooved terrain, while in others the ridges and grooves lie on top of craters. Near the south pole a large basin covers the land. In its general structure it resembles basins on the moon, so it probably formed the same way: by the violent impact of a large object from space. Craters on top of this basin imply it formed early in the satellite's history, perhaps within a billion years after Ganymede's accretion.

From studying the earth's moon, we believe that the era of intense cratering ended some 3 to 4 billion years ago. So Ganymede's surface is 3 to 4 billion years old, that is, Ganymede has been geologically inactive for over 3 billion years.

Callisto

Farthest out of the Galilean moons, Callisto (Fig. 13.21) has a surface that most resembles our moon and Mercury. It is riddled with craters of a wide range of sizes. Many have pitted centers, some have bright ice rays, and others are filled with ice. Callisto's craters are shallow, less than several hundred meters deep, because the surface is a mixture of ice and rock; the surface slowly flows, flattening out the ups and downs of the land.

Callisto has one huge and beautiful multi-ringed feature (Fig. 17.16). Its central floor is 600 km in diameter; 20 to 30 mountainous rings that have diameters of up to 3000 km surround it like a bull's-eye. The rings look like a series of frozen waves. They might have been formed in a stupendous collision that melted the subsurface ice, causing the water to spread in waves that quickly froze in the −180°C surface temperature. The ripple marks are preserved as rings—frozen blast waves.

The central floor of this ringed feature has fewer craters than the rest of the terrain. This difference indicates that the impact forming the rings occurred after much of the cratering, probably 3.5 to 4 billion years ago.

FIGURE 17.16 The giant ringed basin on Callisto (upper right). The bright spot at the basin's center is about 600 km across, the outer rings 2600 km. (*Courtesy NASA.*)

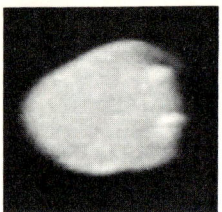

FIGURE 17.17 Amalthea, Jupiter's innermost moon. The indentations at top and bottom may be craters. (*Courtesy NASA.*)

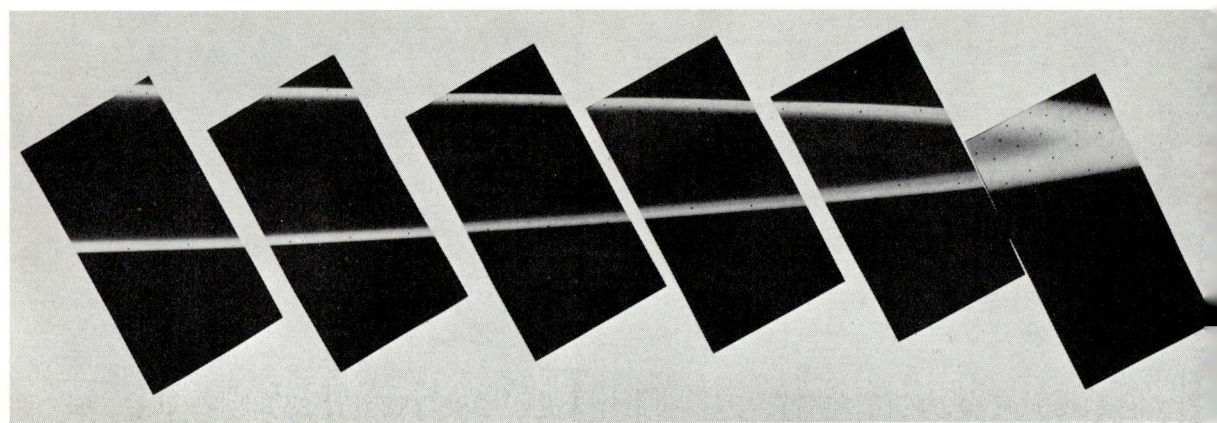

(a)

FIGURE 17.18 (a) Backlit view of Jupiter's rings. (b) Close-up of the rings. Note the bright, sharp edge and diffuse, inner region. (*Courtesy NASA.*)

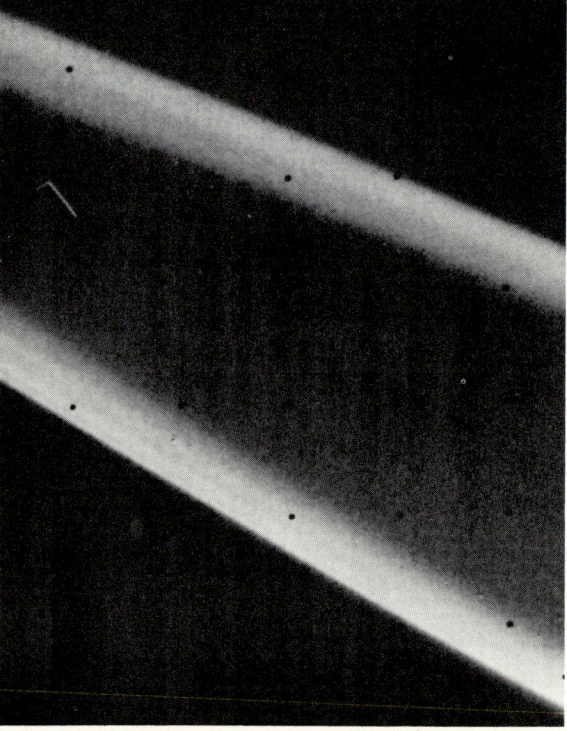

(b)

Asteroidal Moons

Orbiting closer to Jupiter than the Galilean satellites is Almathea. It whizzes around once every 12 h only 110,000 km above the cloud tops. Voyager 1 got a photo of this small, cold world (Fig. 17.17). About 10 times larger than Phobos, it is also elongated, 270 km by 155 km along its major and minor axes. The surface is cratered and has a dark red color. Almathea keeps its long axis pointed at Jupiter. Like the Galilean satellites, its rotation period equals its orbital period. The temperature on the surface is about $-120°C$. This moon's irregular shape, small size, and dark, cratered surface imply it is asteroidlike in character (more about asteroids in Chapter 18).

The surfaces of Jupiter's other moons were not photographed by the Voyager missions. These satellites are smaller than Amalthea. They also seem to be asteroidlike bodies, and we expect that they are indeed captured asteroids. (This is likely, for, after all, Jupiter lies just outside the asteroid belt.)

The Rings of Jupiter

Jupiter actually has millions of moons—tiny ones that make up the ring system discovered by Voyager (Fig. 17.18a). They are so thin (less than 30 km thick) that they are essentially transparent. The rings are best viewed when edge on and with back lighting; then the particles scatter light effectively.

A close-up view (Fig. 17.18b) shows the rings have a definite structure. The outer, brightest part is 800 km wide and lies about 128,500 km from Jupiter's center. Within it is a broader ring some 6000 km wide. And within that ring lies a faint sheet of material that extends from 119,000 km out from Jupiter's center down to the cloud tops.

Compared with Saturn's rings (Section 17.3) and those of Uranus (Section 17.4), the Jovian rings are closer to their parent planet (outer edge at 1.8 Jupiter radii) and between Uranus's and Saturn's rings in width.

What makes up the rings? They must contain some small particles. Voyager 2 photographs imply an average particle diameter of a few μm. They reflect blue light poorly, a characteristic of asteroids, meteorites, and the surfaces of the largest Jovian moons. Observations from the earth show the rings have a reflectivity in the infrared similar to that of the satellite Almathea. But small particles do not explain this observation well. Probably larger bodies, from a meter to a kilometer in size, are also present. The particles' composition is not yet known, but they are not icy, like those in Saturn's rings.

17.3 SATURN: JEWEL OF THE SOLAR SYSTEM

Outpost of the five ancient wandering stars, Saturn bears a marked resemblance to Jupiter, but its ring system outranks in splendor that of the larger planet. Saturn has a slightly smaller diameter (120,700 km) and less mass (95 earth masses) than Jupiter. It has the lowest density of any of the planets—only 680 kg/m^3, less than that of water.

Physical Properties

The atmospheric structure of Saturn is similar to that of Jupiter (Plate 12). Although not as conspicuous, it also has belts running parallel to the circles of latitude (Fig. 17.19).

Disturbances in the belts rarely occur (only ten spots have been observed to date from the earth), compared with their frequency on Jupiter. Voyager I discovered a reddish spot, but much smaller than the Great Red Spot on Jupiter; clouds only a few hundred kilometers across were detected at high latitudes. The atmosphere of Saturn probably has much the same composition as that of Jupiter. So far methane (CH_4), ammonia (NH_3), ethane (C_2H_6), phosphine (PH_3), acetylene (C_2H_2), methylacetylene (C_3H_4), propane (C_3H_8), and hydrogen (H_2) have been detected. The percentage of ammonia is less than that found on Jupiter; probably just as much exists, but at the lower temperature of Saturn ($-180°C$) it has frozen and fallen out of the upper atmosphere. Infrared spectroscopy has detected abundant molecular hydrogen, and a substantial percentage of helium. However, Voyager spectrometers measured only 11 percent helium by mass, compared to 20 percent for Jupiter. Some of the helium may have condensed and settled out toward the interior.

Saturn's clouds appear far less colorful than those of Jupiter. The predominant colors are a faint yellow and orange. Because of the low temperatures on Saturn compared to Jupiter, the clouds lie lower in the atmosphere, and a high-altitude haze subdues our view. However, the pictures obtained by Voyager showed much of the same complexity of cloud patterns seen on Jupiter, with wind speeds much higher, up to 500 m/sec near the equator. Voyager 2 photos showed that the weather on Saturn had changed enormously in a time equivalent to a week on the earth. Large storm systems changed shape, but still remained visible—a hint that Saturn's storms, like Jupiter's, are long-lived compared to the earth's storms.

The Voyager photos show that the upper atmospheric wind flow of Saturn is different

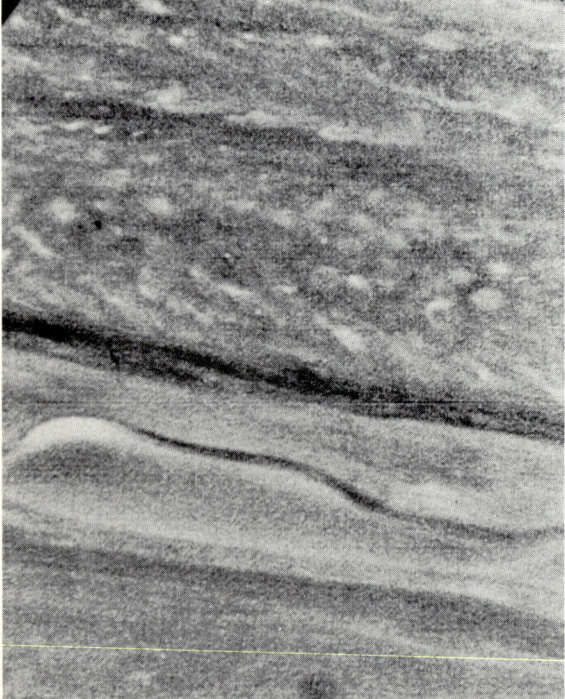

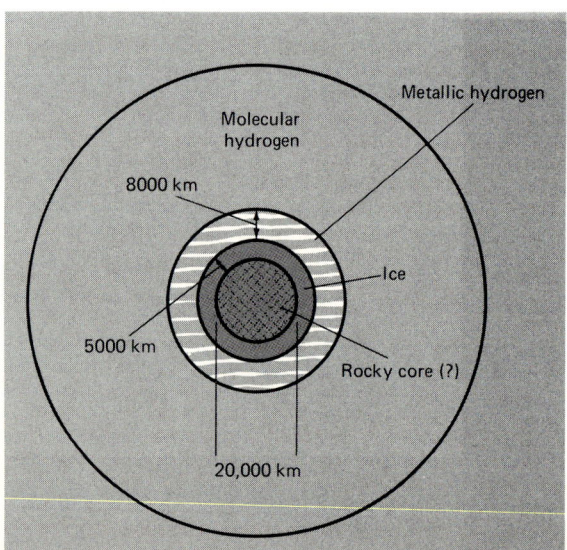

FIGURE 17.19 Jet streams and turbulence in the upper atmosphere of Saturn. (*Courtesy NASA.*)

FIGURE 17.20 A model for the interior structure of Saturn. Note how much smaller the metallic hydrogen zone is compared to that of Jupiter (Figure 17.8).

from Jupiter's. Near the equator the winds all blow eastward at equatorial speeds four times those found on Jupiter. At the higher latitudes the pattern follows an east-west flow alternation as found on Jupiter. The wind velocities for both planets fall off rapidly away from the equator, but Saturn's atmospheric bands do not mark jet stream flows as they do on Jupiter.

Saturn's interior (Fig. 17.20) probably reflects Jupiter's composition. Theoretical estimates are about 74 percent hydrogen, 24 percent helium, and 2 percent heavier elements. Again, this composition is roughly the same as that of the sun. Saturn may have a small, rocky core some 20,000 km in diameter and a mass of about 20 earth masses (about 25 percent of the total mass). Other models have the metallic hydrogen region extending right to the center. Note that the level at which hydrogen becomes metallic is much deeper in Saturn than in Jupiter. The change of state occurs at a pressure of roughly 3 million atm. This level is reached deeper in Saturn than Jupiter because Saturn has a smaller mass and density.

Saturn resembles Jupiter in two other important respects: (1) Infrared observations show that Saturn also emits about two times the energy it receives from the sun (about 10^{17} W). (2) Observations of radio bursts from Saturn imply that it has a magnetic field, but only 1/20 as strong as that of Jupiter. (That's still 40 times stronger than that of the earth.) Pioneer 11 and the Voyager spacecraft measured this field in some detail and found the magnetic axis aligned with Saturn's rotation axis and the field pattern much more regular than for Jupiter. The magnetic field is probably produced by a dynamo effect in the liquid metallic hydrogen zone of Saturn, in the same way as it is presumably produced in Jupiter. But since the metallic hydrogen region lies deeper in Saturn, the irregularities in the field lie buried beneath the surface, inaccessible to spacecraft measurement.

The excess heat from Saturn is somewhat of a puzzle. For Jupiter the emission can be accounted for as that left over from a period of gravitational contraction during its formation. A similar model for Saturn fails to account for the infrared excess. What is its source?

One suggested model uses helium rain to release energy. On earth, water condenses and raindrops fall when air is cooled enough. The condensation releases the heat added earlier to vaporize the water. Saturn's rain is helium droplets; as the raindrops fall through liquid hydrogen, they rub against it to produce heat.

If correct, this idea requires that the helium abundance in Saturn's atmosphere is less than that of Jupiter. Voyager measurements imply that Saturn's atmosphere has about half as much helium (in percent) as Jupiter's. The total is probably the same, so the rest must be hidden in the interior. If this difference does result from helium rain, it could have maintained Saturn's internal heat for the past 2 billion years.

Satellites

Saturn's band of moons totals at least 17 (Table 17.2). The largest one, Titan, has a mass of 1.37×10^{23} kg and a radius of 2575 km. Its density is about 1900 kg/m^3. Christian Huygens first noticed Titan, in March 1655, as a tiny star close to the planet. During an interval of about 20 years, no other satellites were discovered. Then Giovanni Cassini observed Iapetus and Rhea, and some years later he also found Tethys and Dione. More than a century passed before the four smaller satellites (Mimas, Hyperion, Phoebe, and

TABLE 17.2 Saturnian Satellites and Rings

Object	Radius (km)*	Density (kg/m^3)	Distance from Saturn (km)
Cloud tops			60,330
D-ring inner edge			67,000
C-ring inner edge			74,400
B-ring inner edge			91,900
B-ring outer edge			117,400
A-ring inner edge			121,900
Encke division			133,400
A-ring outer edge			136,600
1980 S28	10 × 20	?	137,670
1980 S27	70 × 50 × 40	?	139,350
F-ring			140,300
1980 S26	55 × 45 × 35	?	141,700
1980 S3	70 × 60 × 50	?	151,420
1980 S1	110 × 100 × 80	?	151,470
G-ring			170,000
E-ring inner edge			180,000
Mimas	196 ± 3	1190 ± 50	185,540
Enceladus	250 ± 10	1200 ± 400	238,040
1980 S13	17 × 14 × 13	?	294,670
1980 S25	17 × 11 × 11	?	294,670
Tethys	530 ± 10	1210 ± 160	294,670
Dione	560 ± 5	1430 ± 60	377,420
1980 S6	18 × 16 × 15	?	378,060
E-ring outer edge			480,000
Rhea	765 ± 5	1330 ± 90	527,100
Titan	2,575 ± 2	1880 ± 10	1,221,860
Hyperion	205 × 130 × 110	?	1,481,000
Iapetus	730 ± 10	1160 ± 90	3,560,800
Phoebe	110 ± 10	?	12,954,000

*Those satellites with two or three dimensions listed for the radius are not spherical.

Enceladus) were spotted. Another small satellite, named Janus, was detected in 1966, but its existence was in doubt for many years. Then in 1980 two satellites were found in about the same orbit ascribed to Janus. Voyager confirmed these, and discovered three more, one just outside the A-ring, and two which straddle the new F-ring. Two others share the same orbit with Tethys, leading and following it by 60°, and another co-orbits with Dione. Other possible satellites have been suggested, but not as yet confirmed, on the basis of Voyager data.

Except for outermost Phoebe and Iapetus, all the satellites stick close to Saturn's equatorial plane. They are all much smaller than Titan, less than 800 km in radius. Masses for some of the satellites were determined from their gravitational attraction on Pioneer 11

17.3 SATURN: JEWEL OF THE SOLAR SYSTEM

and Voyager. The corresponding densities range from 1200 to 1400 kg/m^3, all similar to the densities of the outer Galilean satellites of Jupiter.

Titan has given astronomers many surprises. It was the first satellite found to have an atmosphere. The ultraviolet and infrared spectrometers on Voyager showed that it consists of 82 percent nitrogen with about 6 percent methane and 12 percent argon. Several kinds of hydrocarbons have also been detected. The atmosphere's surface pressure, determined from the extinction of Voyager's radio transmissions when it went behind Titan, is about 1.5 atm. Taking into account the low surface gravity, this pressure translates into 10 times as much gas above each square meter of surface as on the earth. The corresponding surface temperature is 95 K.

Pioneer 11 took the first close-up pictures of Titan (look back at Fig. 13.24). They showed a fuzzy ball with more color variation than expected. A stratospheric layer of orange smog was directly visible and also a blue color along Titan's edge. The Pioneer flyby also found the light reflected by the atmosphere to be highly polarized—a direct indication that it contains smoglike particles. No surface features were seen.

The surface was completely obscured from Voyager's view also (Fig. 17.21), but similar variations in color were seen, including a definite contrast between the northern and southern hemispheres. Though Voyager arrived in Titan's "spring," with the sun directly above its equator, the atmosphere was still showing signs of "winter" in the north and "summer" in the south. (Terrestrial seasons similarly lag a month or two behind the position of the sun.) Between the arrivals of the two Voyagers, a dark polar hood had changed into a ringlike collar. So there are both season and weather changes on Titan.

In some ways Titan is a twin of the largest Jovian satellites, Ganymede and Callisto. They have about the same size, mass, and density. But Titan has an atmosphere, and the Jovian satellites do not. One reason for this difference is that Titan formed farther from the sun, in a colder environment, and so perhaps retained some methane and ammonia

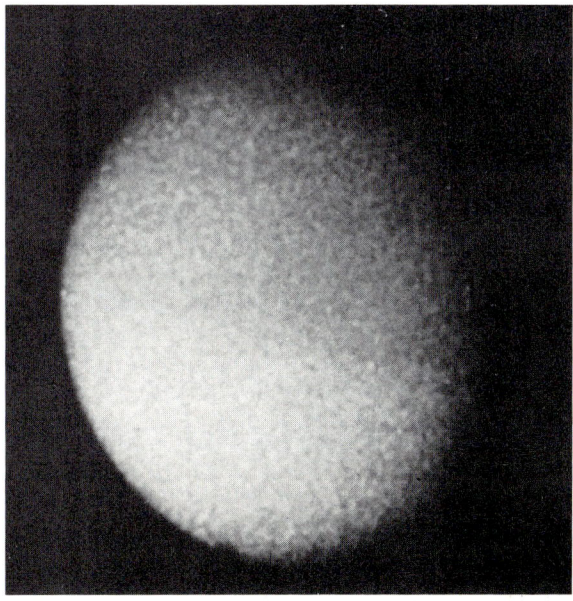

FIGURE 17.21 The clouds of Titan. Note they appear darker in the northern hemisphere (top) than the southern. (*Courtesy* NASA.)

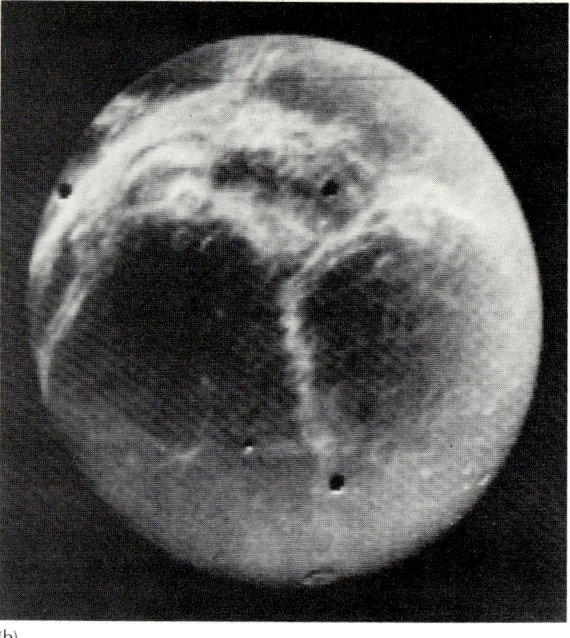

FIGURE 17.22 (a) The surface of Dione. Note the many impact craters. The irregular valleys are old surface faults eroded by impacts. (b) Overview of Dione, showing bright, wispy markings that are probably ice deposits. (*Courtesy NASA.*)

ices as well as water ice. Heating of Titan's interior would cause some of these ices to vaporize, forming an initial atmosphere. Then ultraviolet radiation from the sun might have dissociated the ammonia into nitrogen and hydrogen. The hydrogen would have escaped, and the more stable methane would have been retained (though some is being dissociated today, producing a thin ring of hydrogen along the orbit of Titan).

The presence of nitrogen in Titan's atmosphere allows some surprising chemistry to take place. One of the exciting discoveries of Voyager was the detection of compounds containing nitrogen as well as carbon, such as hydrogen cyanide (HCN), cyanoacetylene (HC_3N), and cyanogen (C_2N_2). These are the building blocks for amino acids and other compounds present in living matter. Their presence in the atmosphere of Titan, along with the pressure and temperature data, have led some to speculate that the surface is covered with organic tars along with pools of liquid methane. Titan is perhaps the best place in the solar system to examine the prebiotic chemical processes that might have given rise to life on the earth four billion years ago (Chapter 25).

Saturn's other satellites are much smaller than Titan. The four largest, next to Titan, are Iapetus, Rhea, Dione, and Tethys, with diameters ranging from 1160 km to 1530 km. They appear heavily cratered and covered with wispy white streaks (Fig. 17.22). In a few cases the streaks form rayed patterns around impact craters, but most do not. They are probably deposits of frozen ice, but whether from material emanating from the interior or from debris deposited by colliding cometary bodies is unknown.

Although all the satellites show brightness variations over their surface, Iapetus is the most extreme (Fig. 17.23). The light side of Iapetus is some 15 times more reflective than the dark side (equivalent to having clean snow on the light side and soot on the dark one). Although the reason for this difference is still unknown, two explanations have been suggested. Iapetus is locked into orbit with the same face toward Saturn at all times. It is the leading surface which is covered with dark material, which may be debris picked up

17.3 SATURN: JEWEL OF THE SOLAR SYSTEM

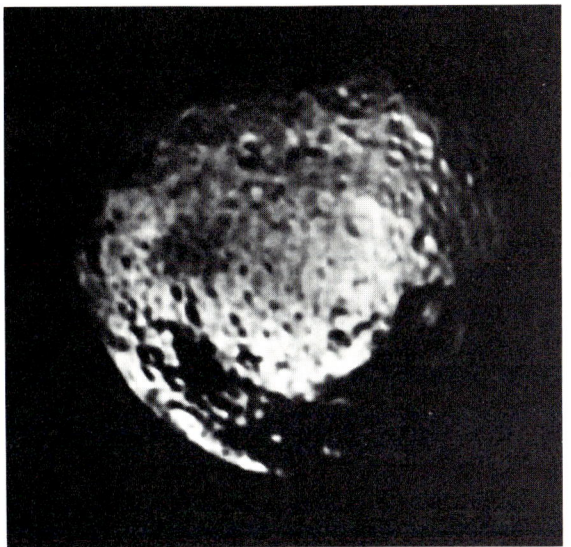

FIGURE 17.23 Iapetus, showing details as small as 20 km in size. Note the impact craters. The dark region at bottom covers the icy crust of the hemisphere that faces in the direction in which Iapetus orbits. (*Courtesy NASA.*)

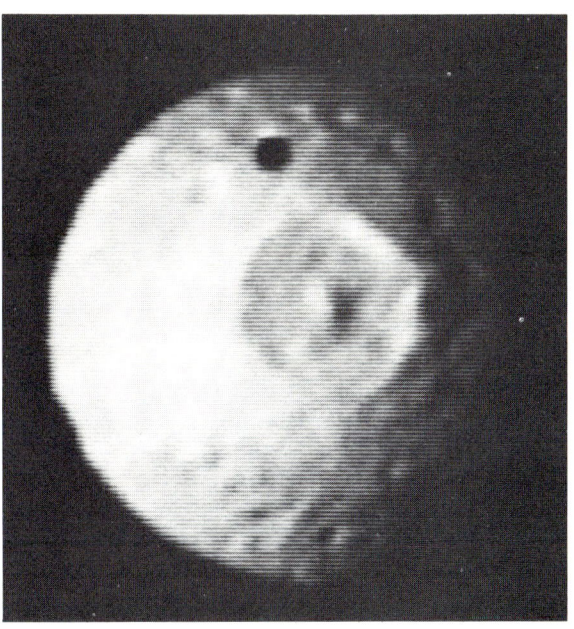

FIGURE 17.24 Mimas with its large impact crater, more than 130 km in diameter. Note the raised rim and central peak. (*Courtesy NASA.*)

during Iapetus's journey around Saturn. On the other hand, some dark material is seen on crater floors on the trailing side, suggesting an internal origin, perhaps the product of eruptions of methane from the interior. Voyager 2's pass of Iapetus permitted a measurement of its size and mass, and so its density. Its density of 1200 kg/m^3 implies that it is basically an icy body.

With one exception, the smaller of the nine named satellites are also cratered. On Mimas, for example, Voyager found a huge crater 130 km across (about $\frac{1}{4}$ the moon's diameter), with walls 9 km high, perhaps the deepest in the solar system (Fig. 17.24). A slightly larger impact probably would have shattered Mimas completely; cracks and troughs on the side opposite the crater were probably produced by the shock of the impact. The one exception to the general roughness is Enceladus, which appears remarkably smooth. One possible explanation is that since the orbital period is almost exactly half that of Dione, periodic tidal stretchings might have provided internal heating, keeping the surface soft until after the crater-making bombardment was finished.

The real surprise with Saturn's moons came with the images of Hyperion. They revealed it to have the shape of an oblate disk (something like a thick hamburger) 260 by 410 by 220 km thick (Fig. 17.25). Also, Hyperion rotates on an axis tilted some 45° to the plane of its orbit, a very unnatural, unstable situation. It may in fact be tumbling around in a cockeyed fashion.

A milder surprise came with the photos of Rhea, Dione, Enceladus, and Tethys. These icy bodies, on the basis of their small size, were thought not to have undergone internal heating or crustal movement. Not so! Especially for Enceladus (Fig. 17.26), the surfaces

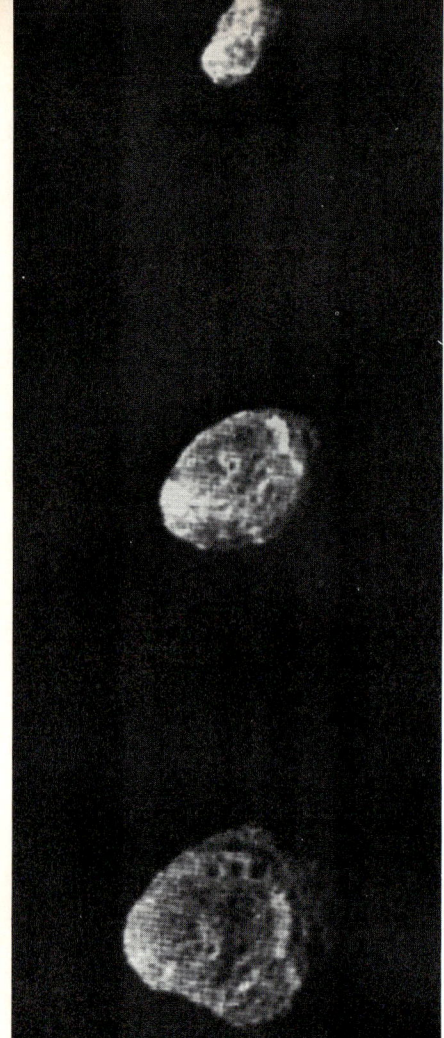

FIGURE 17.25 Three views of Hyperion, showing its unusual shape. Note the impact craters. (*Courtesy NASA.*)

FIGURE 17.26 A close-up view of Enceladus. The grooves and linear features imply that the crust was deformed from internal melting. The largest crater is about 35 km in size.

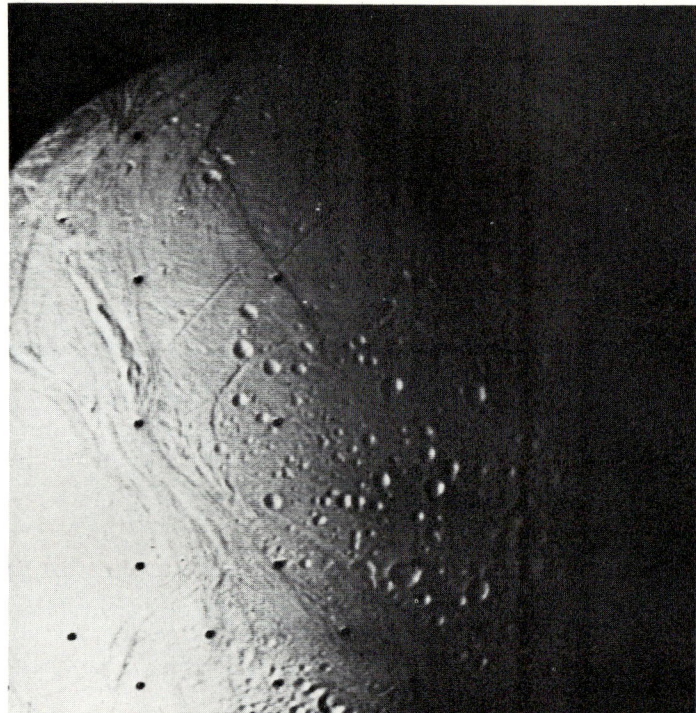

have fissures, canyons, and areas wiped clean of primeval craters—all indications of internal heating and crustal deformation.

The new satellites of Saturn are all small bodies, 100 km or less in radius. Their interesting characteristics are not their surfaces so much as their orbits. At 151,000 km from Saturn lie two bodies in almost exactly the same orbits, differing by only 50 km in semimajor axis. However, because the orbits and the periods are not *exactly* the same, the co-orbital satellites must pass each other from time to time. But since their sizes are bigger than the separation of their orbits, they *cannot* pass, unless they deviate from their average orbit. The gravitational dynamics of this collision avoidance maneuver has yet to be worked out. (The co-orbiting satellite following Dione is locked into a position 120° away along the orbit and never catches up.)

Close in to Saturn lie two satellites just inside and just outside of the narrow F-ring (see below). Nicknamed the shepherd satellites, their combined gravitational influence keeps the F-ring particles from wandering outside a narrow range of orbits. How does this work? The faster moving inside satellite accelerates the inner ring particles as it passes them,

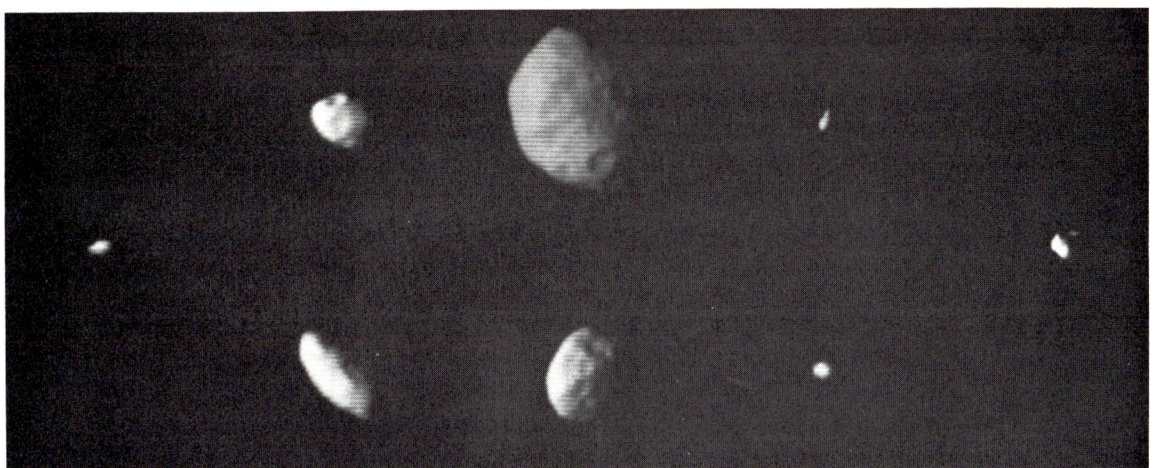

FIGURE 17.27 A family portrait of some of the smallest moons of Saturn. The images show the correct relative sizes. Top row, from left to right: 1980 S26, outer F-ring shepherd; 1980 S1, leading co-orbital; 1980 S25, trailing Tethys. Bottom row: 1980 S28, outer A-ring; 1980 S27, inner F-ring shepherd; 1980 S3, trailing co-orbital; 1980 S13, leading Tethys; 1980 S6, leading Dione. (*Courtesy NASA*.)

causing them to spiral out to larger orbits (just as the tidal acceleration of the earth on the moon causes it to spiral outward; Section 15.1). At the same time the slower moving outer satellite decelerates the outer ring particles as they pass by, causing them to spiral inward. A similar mechanism is thought to operate with the satellite just outside the A-ring, accounting for its very sharp outer boundary.

The Voyagers have opened up 17 new worlds to our view. Because it's easy to get lost in all the marvelous new information about Saturn's moons, here's a general summary of their key features and classifications. To help keep these in mind, consider them in three categories: Titan by itself; the 6 major icy moons (Mimas, Enceladus, Tethys, Dione, Rhea, and Iapetus in order out from Saturn); and the 10 smaller moons (Phoebe, Hyperion, and the asteroidal bodies).

First, all but one moon (Phoebe) keep the same face toward Saturn because of tidal forces. Second, all but two have nearly circular orbits in Saturn's equatorial plane. Third, measured densities are less than 2000 kg/m^3, and several are less than 1500 kg/m^3. So the moons must be made mostly of ice, roughly 60 to 70 percent with 30 to 40 percent rock. Only Titan appears to be about half ice and half rock. In contrast to Jupiter's largest moons, there is *no* trend of densities (higher to lower with increasing distance) among Saturn's satellites, and Saturn's moons generally are less dense than those of Jupiter.

Most of the moons are cratered from a massive bombardment, presumed to have occurred some 4 billion years ago. Surprisingly, for the internal heat produced by formation should not last long, these icy bodies have evolved since that time. For example, the giant canyons of Tethys and the ridged plain of Enceladus (where very few craters are visible) must be young. The driving power for this evolution is unclear; it may come from heating by the action of tidal forces.

A final word about the smallest and largest moons of Saturn. The 10 small satellites range in size from Hyperion (300 km diameter) to the A-ring shepherd (30 km in diameter). They are so small that they lack the self-gravity to pull themselves into a spherical shape (Fig. 17.27). Their rough-cut figures preserve the battering from impacts and the splits from fragmentation. We presume that they are made of icy materials.

Titan is a world of its own, even when compared to the Galilean moons of Jupiter, for it is the most earthlike of these bodies. That's not to say it isn't a strange place; you can think of it as a terrestrial planet in a deep freeze. It and the earth are the only worlds to have substantial atmospheric nitrogen and small amounts of carbon-nitrogen compounds, molecules found in the interstellar medium and essential to life as we know it.

Ring System

In 1659 Huygens observed that Saturn "is surrounded by a thin, flat ring" that does not touch the body of the planet. Further observations by Cassini uncovered a gap in Huygens's single ring; this gap is known as *Cassini's division* (Fig. 17.28). The rings lie tipped about 26° to the orbital plane, and because of their tilt they change their appearance as viewed from the earth during the course of Saturn's revolution about the sun.

The near disappearances of the edge-on rings indicate that they are very thin, no more than 5 km thick. The Voyager data imply a thickness of only a few hundred meters.

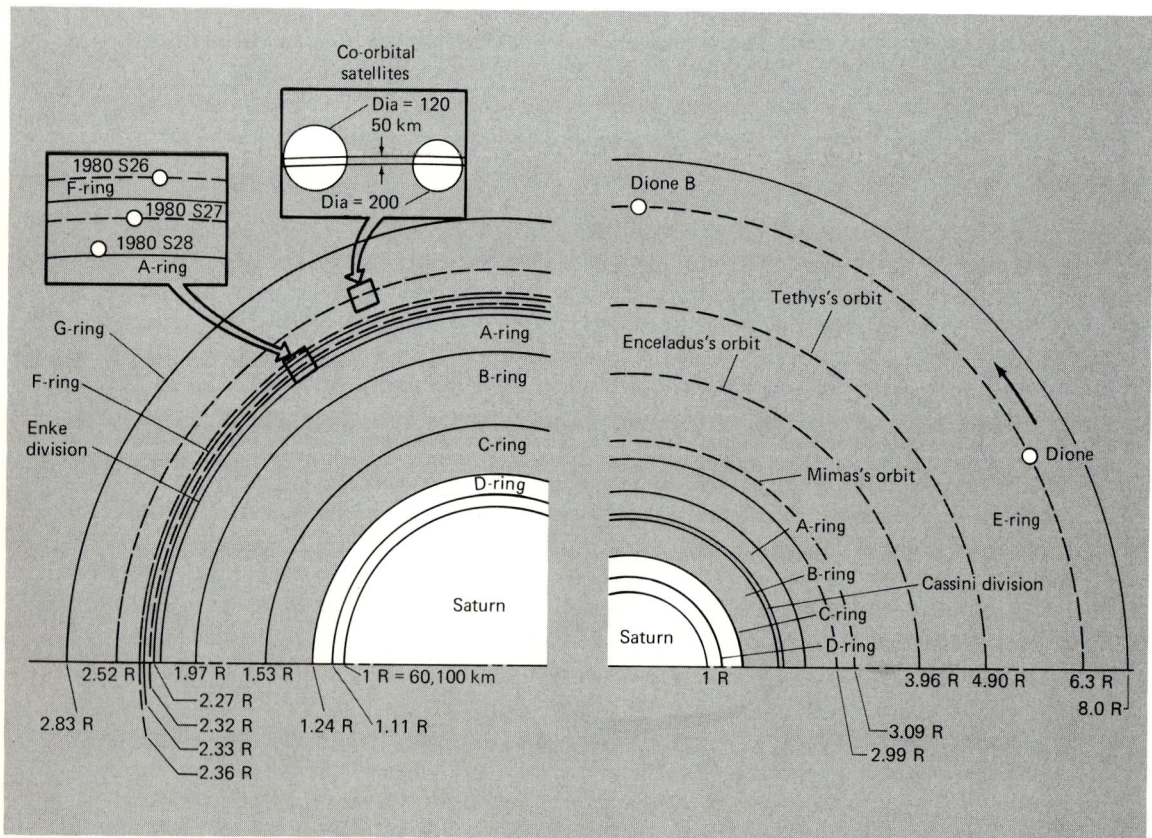

FIGURE 17.28 A schematic diagram of Saturn's ring system and the orbits of some of its moons. The view is from above Saturn's north pole. Note that the scales in the two halves differ. Distances are given in units of the planet's radius. (Adapted from a NASA diagram.)

17.3 SATURN: JEWEL OF THE SOLAR SYSTEM 631

FIGURE 17.29 A computer-processed photo of Saturn's rings, showing about 95 individual ringlets. *(Courtesy NASA.)*

Although thin, the rings are wide; the three main rings visible from the earth reach from 74,000 km to 137,000 km from Saturn's center.

The high-resolution photographs obtained by the Voyager spacecraft revealed spectacular detail in the ring system of Saturn. Although the A-ring is relatively smooth, the B- and C-rings break up into numerous small ringlets (Fig. 17.29), like grooves on a phonograph record. Many hundreds, perhaps a thousand, light and dark ringlets surround the planet, with widths as small as 2 km, the best resolution of the Voyager cameras. Some (in the C-ring) appear elliptical in shape, rather than circular. Even the Cassini division, apparently empty as seen from earth, was revealed to be filled with at least 20 ringlets (several of the brightest are visible in Fig. 17.29). Although the major gaps in the rings occur where particles orbit at some multiple of the period of an outer satellite, there are too many gaps and rings to explain with this idea. Some rings are possibly formed by spiral density waves, like those in spiral galaxies (Section 21.5). The true cause of all this complexity is as yet unknown.

A special instrument on board Voyager 2 (that had failed on Voyager 1) was able to record the light from a star passing through the many ringlets that make up the ring system (Plate 13). The flashes of starlight indicated that the ringlets number in the *hundreds* of thousands, not just in the thousands seen in regular photos. Also, the ringlets seemed to have a wavelike structure; that is, rather than each ringlet being a distinct entity that lasts some length of time, they may be short-lived structures formed by waves passing through the ring system.

Even more unexpected was the discovery of dark, spokelike features in the B-ring (Fig. 17.30). They are visible only when the sun is behind the camera, which implies that they

17 THE JOVIAN PLANETS: PRIMITIVE WORLDS

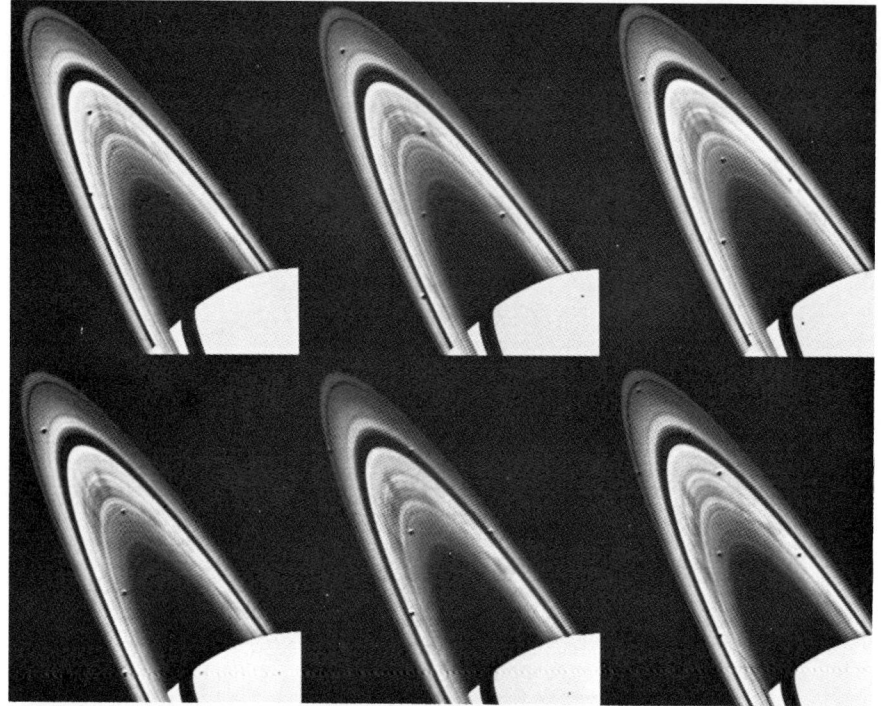

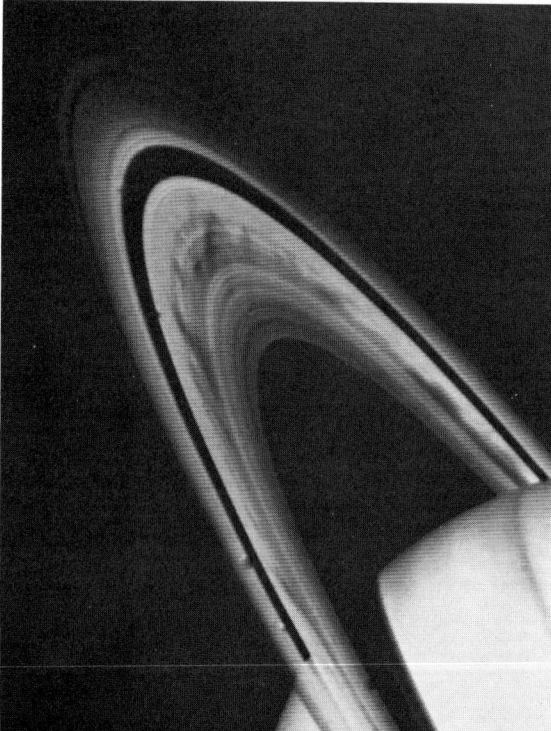

FIGURE 17.30 (a) Spoke-like features in the rings. These photos were taken in sequence (upper left to lower right) every 15 min. Note how the dark spokes revolve around. (b) Dark spokes in the B-ring. The markings are about 12,000 km long. Note their diffuseness. (*Courtesy NASA.*)

17.3 SATURN: JEWEL OF THE SOLAR SYSTEM

are made of *very* small particles. Typically, the spokes are about 10,000 km long and 1000 km wide. The particles in the spokes, just like the rest of the ring particles, follow Kepler's laws. Since the inner particles revolve more rapidly than the outer ones, the spokes, once formed, can last only a few hours before being pulled apart. One tentative explanation is that small particles acquire an electric charge and are lifted out of the ring plane by interaction with Saturn's magnetic field. However, once formed, the particles in the spokes seem to move only under gravitational forces.

Pioneer 11 discovered a new ring out beyond the previously known ones. Called the F-ring, it lies 3700 km outside the edge of the rings visible from the earth. The F-ring appears to be some 320 km wide and a mere 3 to 4 km thick. Voyager resolved this ring into a complex system of knots and a braided structure of at least three strands (Fig. 17.31). What produces such a noncircular, asymmetric structure is not yet understood. Evidence for another such very narrow ring (the G-ring), 30,000 km farther out, was obtained unexpectedly from the Voyager photograph of one of the co-orbiting satellites. A narrow shadow is visible which is not present on other photos.

Two other extremely faint rings are known. The E-ring, discovered photographically from earth and confirmed by Voyager, extends out beyond the F-ring to at least 8.0 Saturn radii (480,000 km). A ring inside the C-ring was photographed by Voyager 1 (Fig. 17.32). Called the D-ring, it extends at least halfway to the surface of Saturn. Such a ring had been reported earlier, but the ring seen by Voyager is much too faint to have been detected with ground-based telescopes.

Finding the rotation rate of the rings by measuring their Doppler shift was one of the earliest applications of spectroscopy to the planets (1895). The velocities range from 16 km/sec at the outer boundary of the A-ring to 20 km/sec at the inner boundary of the

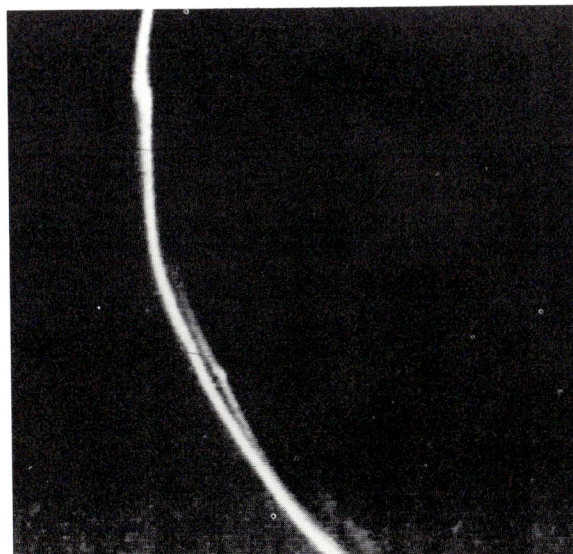

FIGURE 17.31 The F-ring, viewed by Voyager 1. Note the three strands of the ring, two of which are twisted around each other. Nine months later, the braided section was gone. (*Courtesy* NASA.)

FIGURE 17.32 The D-ring. The edge of Saturn is at the lower left, the inner part of the C-ring is at the upper right, and Saturn's shadow cuts through the middle of the photo. The D-ring extends down from the edge of the C-ring to about half the distance to Saturn. (*Courtesy* NASA.)

B-ring. The measured velocities agree with the theoretical velocities calculated from Kepler's third law for individual masses placed at the ring distances from Saturn; this agreement indicates that separate particles comprise the rings. (If the rings were solid, the velocity of the outer edge would be higher than that of the inner edge.)

Contemporary observations of Saturn's rings in the infrared region of the spectrum indicate that they are made of particles of water ice or rocky particles coated with water ice. The ice does not evaporate because the surface temperature of the particles is only about 70 K, and the particles are occasionally screened from direct sunlight by each other and by Saturn's sphere. Observations of radio signals from Voyager reflected and transmitted by the rings indicate that the particles are about 1 m in diameter, but a range of sizes, from centimeters to tens of meters, probably exists.

Although deceptively solid in appearance and covering a large area of space, the rings have a total mass estimated to be only 10^{16} kg, about 10^{-6} the mass of our moon and a mere 10^{-10} the mass of Saturn.

How did the rings form? Saturn's great mass creates strong tidal gravitational forces (Focus 12.1). If a satellite had approached close enough, it would have experienced tidal forces strong enough to tear it into pieces. Such may have been the fate of a former Saturnian satellite whose demise scattered the particles that then spread to form the rings. Another possibility is that the tidal forces may have kept close-in particles from ever forming into a satellite.

Magnetosphere and Radiation Belts

Pioneer 11 found (and Voyager confirmed) that Saturn's magnetosphere falls in size between that of the earth and Jupiter (Fig. 17.33). The magnetosphere acts like a magnetic bubble that expands and contracts as the variable solar wind slacks and gusts. The nose of the magnetosphere averages about 1.25 million km from Saturn; its width averages roughly 3.4 million km.

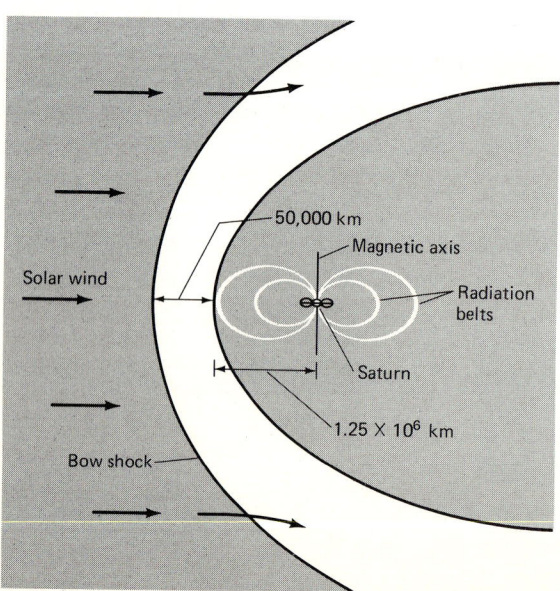

FIGURE 17.33 A schematic diagram of Saturn's magnetosphere, based on spacecraft data. (*Adapted from a NASA diagram.*)

17.4 URANUS: THE FIRST NEW WORLD

Within Saturn's magnetosphere lie intense radiation belts, similar to the Van Allen belts that encircle the earth (Section 14.4). These belts contain high-energy protons and electrons, but the particles' energies are hundreds of times less than those in Jupiter's belts. The probable reason is that Saturn's rings sit in the middle of the belts, and the ring materials absorb electrons and protons that hit them.

17.4 URANUS: THE FIRST NEW WORLD

On March 13, 1781, the then unknown amateur astronomer William Herschel perceived a star "visibly larger than the rest" in the constellation of Gemini and "suspected it to be a comet." Later observations in March and April revealed that the orbit was not like a comet's. Herschel concluded that he had discovered a new planet—the seventh in the solar system and the first to be discovered with a telescope.

Herschel first wished to name the planet *Georgium Sidus* (George's Star), after King George III; this suggestion pleased the king (who later gave Herschel a regal astronomical post) but irritated continental astronomers. To continue the long tradition of naming the planets for mythological characters, Johann E. Bode suggested Uranus, father of Saturn and god of the heavens. That name caused less controversy, so it was adopted.

At an average distance of 19.2 AU, it takes Uranus 84 years to journey around the sun. Far from the sun, the upper atmosphere must be very cold. A recent infrared observation puts the temperature at 58 K. In such a deep freeze all the ammonia has frozen out of the atmosphere and cannot be detected spectroscopically. Methane and hydrogen do appear in the spectrum. Helium may also have been detected, but this result has not been confirmed; the large quantity of helium in Jupiter strongly suggests that this element also should exist in the atmosphere of Uranus.

Viewed through a telescope, Uranus has a distinctive pale green color. This greenish color comes from sunlight that penetrates deep into the planet's atmosphere; some red light is absorbed in the atmosphere and much of the green is reflected back into space.

The low bulk density of Uranus, 1300 kg/m^3, implies that it has roughly the same composition as Saturn, but perhaps less of the lighter elements. Theoretical models support this expectation (Fig. 17.34): Uranus is thought to consist of roughly 15 percent hydrogen and helium, 60 percent icy materials (H_2O, CH_4, and NH_3), and 25 percent earthy materials (SiO_2, MgO, and Fe). The internal structure of Uranus differs from that of Jupiter and Saturn because of the differences in total mass, which result in different internal pressures.

For many years astronomers believed that Uranus rotated in about 10.8 h. In 1977 Michael Belton and Sethanne Hayes reported a 23-h rotation period. But observations by Robert Brown and Richard Goody indicate a period of 15.6 h, and other observers claim a period of 12 h. So we have not yet pinned down the rate of Uranus's spin.

How can we hope to determine the rotation rate of a planet so far away that we cannot see any features (look back at Fig. 13.25)? By the always useful Doppler shift. If the slit of a spectroscope is lined up with a planet's equator, the side approaching us blue shifts the line and the side receding red shifts it (Fig. 17.35). As a result the spectral lines are tilted. The amount of tilt tells us the velocity and hence the rotation rate, if we know the planet's radius.

The strange thing about Uranus's rotation is not the period, but the direction of the

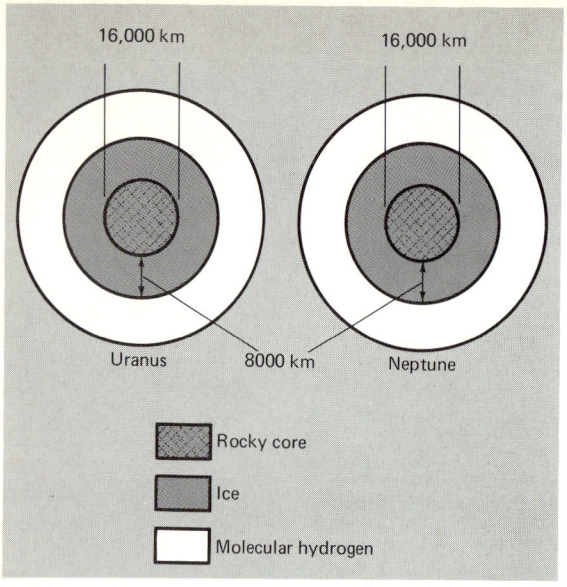

FIGURE 17.34 Models for the interiors of Uranus and Neptune. Both planets have the same mass and bulk composition, so their interiors are basically the same.

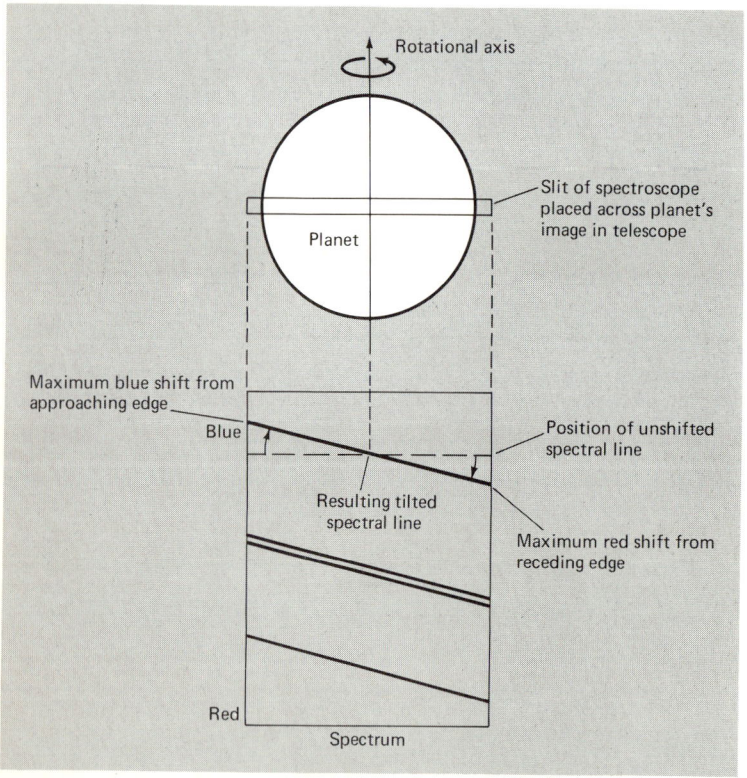

FIGURE 17.35 Measuring a planet's rotation rate with the Doppler shift. The slit of the spectroscope is placed at right angles to the rotation axis. The radial part of the rotational velocity increases from the center to the edge. At one edge the approach velocity is maximum and results in the largest blue shift. At the other edge the receding velocity is greatest and results in a maximum red shift. At the center the radial velocity is zero and no shift occurs. As a result of these Doppler shifts, the spectral lines are tilted. The greater the tilt, the faster the planet rotates.

rotational axis in space: It lies almost in the plane of the ecliptic; Uranus spins on its side! Journeying around the sun in this lopsided manner, Uranus exposes each pole to sunlight for 42 years at a time; night at the opposite pole lasts for an equally long time.

The tilt of the rotational axis contributes to the difficulty in measuring the rotation period. During the last decade the pole has been pointed somewhat toward the earth, so

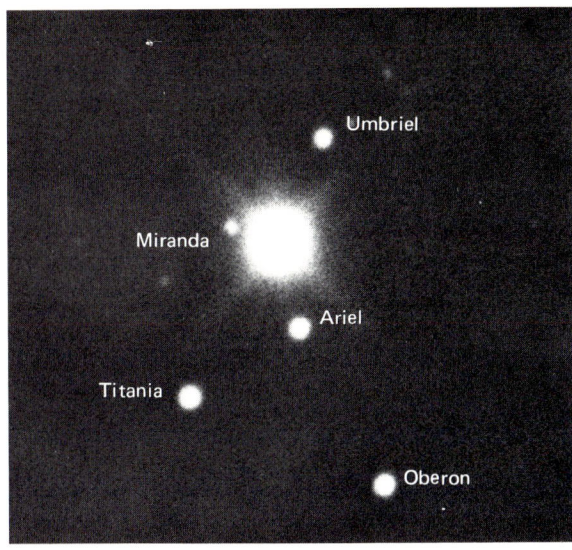

FIGURE 17.36 Uranus and its five moons. (*Courtesy, W. Liller; photo taken with the 4-m telescope at Cerro Tololo Inter-american Observatory.*)

the component of rotational velocity along the line of sight is small. In 2002 the poles will be perpendicular to the line of sight from the earth, so the measurement should be easier.

Five moons are so far known for Uranus (Fig. 17.36). They all move in the planet's equatorial plane and revolve in the same direction that the planet rotates (Appendix B). Askew to the ecliptic (because the moons lie in the same plane as Uranus's equator), their orbits as seen from the earth are alternately edge-on, then fully open, every 42 years; in 1966 they appeared edge-on, but in 1987 they will appear as circles.

The five moons are Miranda, Ariel, Umbriel, Titania, and Oberon. Miranda, the smallest (less than 100 km) and closest to Uranus, revolves in an eccentric orbit. The others are medium-sized (200 to 500 km) and revolve in circular orbits. Oberon and Titania seem to have bare, rocky surfaces, not at all like the ice moons of Jupiter and Saturn.

The key news about Uranus is that this green planet has rings, at least nine in all. Their discovery was accidental. In March 1977 three groups of astronomers—from Cornell, Lowell, and Perth Observatories and the Indian Institute of Astrophysics—were attempting to observe Uranus pass in front of a faint star in order to study its atmosphere. They were surprised to see the star momentarily dim even before Uranus covered it. The Cornell group, which got the most complete data, recorded the star dimming five times before going behind Uranus. Then, in a mirror-reversed fashion, the star dimmed five times again upon emerging from behind Uranus. This symmetry suggested a system of five rings around the planet (Fig. 17.37). A detailed analysis of the data and observations of Uranus passing in front of another star in 1978 showed that four other rings existed, for a total of nine.

The observations paint a picture of a ring system dramatically different from that of Saturn. The nine rings circle the planet in roughly four groups (Table 17.3): rings 6, 5, and 4 at about 42,500 km; α and β at 45,000 km; η, γ, and δ at 48,000 km; and the unique ϵ ring at 51,000 km from the center of Uranus. The smallest rings have widths of only about 5 km.

The ϵ ring is a curious one. It is elliptical in shape: 50,881 km from Uranus on one side, 51,439 km on the other. Its width varies around its circumference from 21 to 72 km. Some observations imply that this ring is slowly precessing around Uranus, like a hula-

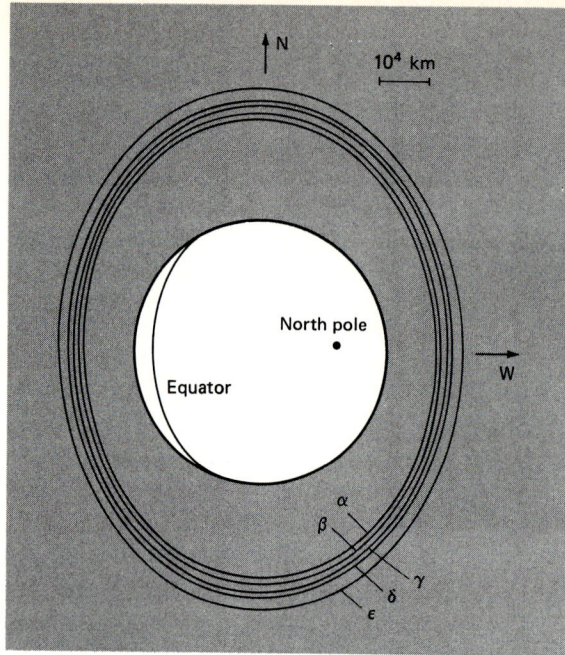

FIGURE 17.37 The five major rings of Uranus. In addition to the major rings (α, β, γ, δ, and ε) are rings 4, 5, and 6, just inside α, and the η-ring, between β and γ. (Adapted from a diagram by P. Nicholson, S. Persson, K. Matthews, P. Goldreich, and G. Neugebauer, in the Astronomical Journal, vol 83, p. 1240, copyright 1978 by the American Astronomical Society.)

hoop, that is, the major axis of the ring changes direction in space. This behavior is bizarre; for if tidal forces turn the ε ring, it should have spread out into a wider ring symmetrical about Uranus's equator. And it hasn't yet!

Why did astronomers not detect the rings of Uranus until now? (After all, we've known of Saturn's rings for about 300 years.) One reason is that the rings of Uranus are not very wide. More important, the material that makes up the rings is extremely black. Photographs of Uranus imply that the ring material can reflect no more than 5 percent of the light that strikes it. In contrast, because they are covered with water ice, the particles in the rings of Saturn reflect more than 80 percent of the light hitting them. So the particles

TABLE 17.3 The Rings of Uranus

Ring	Semimajor Axis (km)	Width (km)	Eccentricity
6	41,860	5	0.0014
5	42,270	5	0.0018
4	42,600	5	0.0012
α	44,750	9	0.0007
β	45,700	16	0.0004
η	47,210	5	0.0000
γ	47,660	7	0.0000
δ	48,330	5	0.0001
ε	51,180	20–70	0.0079

Source: J. L. Elliot, R. G. French, J. A. Frogel, J. H. Elias, D. Mink, and W. Liller, Astronomical Journal, vol. 86 (1981), p. 444.

in the rings of Uranus are probably bare and likely made of dark carbon materials. We have no good idea yet of how large the rings' particles are. Observations so far indicate that they cannot be larger than 5 to 6 km in size.

17.5 NEPTUNE: GUARDIAN OF THE DEEP

We don't really know very much about the cold world of Neptune. This planet is the eighth one from the sun (except for those times when Pluto comes closer) and is the most distant Jovian planet. It was discovered because of its gravitational effects on the orbit of Uranus. Soon after Uranus's discovery, astronomers observed it often and computed its orbit. To their surprise the planet consistently refused to follow the predicted path. With enough observations, the orbit and approximate mass of another planet, causing the perturbations of Uranus's orbit, could be calculated (Section 4.3). J. Galle at the Berlin Observatory was able to observe the new planet on September 23, 1846, near the spot predicted. Because of its remoteness from the sun, the new planet was named Neptune.

Neptune had actually been observed twice before. Joseph Lalande had seen it 55 years earlier, but he failed to identify it as a planet. Neptune was first seen by none other than Galileo, 234 years before Galle's discovery! Calculations of Neptune's orbit show that it should have been very close to Jupiter in the sky in January 1613 and perhaps was even eclipsed by Jupiter. Charles Kowal and Stillman Drake searched Galileo's journals, and, sure enough, there are entries showing that he observed an object in the vicinity of Jupiter near Neptune's predicted position on both December 27, 1612, and again on January 28, 1613. On December 27, Neptune was entering its retrograde motion (Section 1.4) and so was stationary with respect to the stars. But on January 28, Galileo detected a small motion of Neptune with respect to a nearby star. Inexplicably, Galileo never followed up on this discovery and thus failed to recognize the object as a new planet. Galileo's observation shows Neptune was not exactly in the predicted position. Kowal and Drake suggest that this may mean that Neptune's orbit is being perturbed by an as yet undiscovered planet.

In many ways Neptune is the twin of Uranus. Far from the sun, Neptune revolves once in 165 years. Like Uranus, Neptune shows a light green color. The upper atmosphere displays faint cloud bands. This cold atmosphere, −220°C, probably contains water ice and ammonia ice, mixed with gaseous methane, hydrogen, and helium. The internal structure of Neptune (Fig. 17.34) probably resembles closely that of Uranus, because the bulk densities are similar.

As for Uranus, the rotation period of Neptune has proven difficult to determine accurately. The value in common use has been 15.8 h. But in 1977 Sethanne Hayes and Michael Belton, using the same Doppler shift technique used on Uranus, found a value of 22 h. More recently, Dale Cruikshank found variations in the infrared brightness of Uranus that indicate a rotation period of 18 h 10 min or 19 h 35 min without a clear-cut way to tell which value is correct.

But all is not exactly the same with Uranus and Neptune. Recent infrared observations show that Neptune's temperature is about 56 K, compared with the expected value of 43 to 46 K if Neptune were heated only by the sun. So Neptune has internal heat. It gives off 2.8 times as much energy as it receives from the sun, which, as in the case of Jupiter, is most likely left over from its formation.

FIGURE 17.38 Neptune and its two moons. Triton is the bright spot just below Neptune; Nereid is indicated by the arrow. (*Courtesy Yerkes Observatory.*)

Does this mean that Neptune was formed with more internal heat than Uranus? No. Notice that the temperature of Neptune is now just about the same as that of Uranus (56 K vs. 58 K). Theoretical models show that, however they started, both planets should cool to about this temperature in 4.5 billion years. Uranus is close enough to the sun so solar heat can now keep it at this temperature. Neptune, farther away, will cool further before it comes into balance with sunlight.

Neptune's atmosphere appears to differ from Uranus's. For example, ethane (C_2H_2) has been detected in Neptune's atmosphere but not Uranus's. Infrared observations of Neptune by Richard Joyce, Carl Pilcher, Dale Cruikshank, and David Morrison indicate that Neptune's atmospheric reflectivity from 1 to 4 μm increased substantially over a 1-year period. Pilcher has interpreted this change as the formation of extensive cloud cover that then partially dissipated. So Neptune may have weather, in the sense of changing meteorological conditions (look back at Fig. 13.26). Uranus has not exhibited any such changes to date.

Neptune has two known moons, Triton and Nereid (Fig. 17.38). The orbit of the latter is extremely eccentric, and its distance from Neptune ranges from about 1 million to 10 million km. The larger satellite, Triton, revolves with a period of about 5 days in a retrograde (east to west) orbit that is inclined 20° to the plane of Neptune's equator. Triton has a diameter of about 3600 to 5200 km, which makes it one of the largest satellites in the solar system.

Triton now joins the short list of moons with an atmosphere. Dale Cruikshank and Peter Silvaggio have observed Triton in the infrared. They find an absorption band at 2.3 μm, which they attribute to gaseous methane. The pressure due to methane at Triton's surface is roughly 10^{-7} atm. Other observations indicate that the surface is mostly rocky. Cruikshank and Silvaggio picture the surface as mostly rocks with a few patches of frozen methane. These patches are probably concentrated on the moon's night side.

Neptune may also have a ring system very close to the planet. An international collaboration of astronomers observed two occultations by Neptune in May 1981 and found no observable rings. However, previously unpublished data on an occultation which occurred in 1968 have recently been uncovered, showing a dip in brightness 3 min after the star passed behind the planet and lasting for $2\frac{1}{2}$ min. If this dip was caused by a ring, its radius is 28,600 km (1.14 the radius of Neptune) and its width is 4300 km.

The 1981 observations hint at a third moon for Neptune. (They have not yet been confirmed.) If this moon does exist, it is at least 180 km in diameter and located about 3 Neptune radii from the planet's center.

17.6 PLUTO: GUARDIAN OF THE DARK

Prompted by the discoveries of Uranus and Neptune, astronomers searched the sky, hoping to find a planet beyond Neptune that would account for the slight perturbations remaining in Uranus's motions after the gravitational effects of Neptune were subtracted. In 1880 the United States Naval Observatory conducted the first serious search for a planet beyond Neptune, using the 26-in refractor. The search failed.

Early in the twentieth century Percival Lowell also became fascinated with the problem of a planet beyond Neptune and initiated a search program at Lowell Observatory. After Lowell died in 1916, the search for Planet X was terminated until after the completion of a 13-in refractor in 1929. This new telescope, built with funds provided by A. Lawrence Lowell (Percival's brother), yielded good star images over a large region of the sky. The number of stellar images per photo reached 1 million in the densest section of the Milky Way.

Clyde W. Tombaugh worked at the new search, which started on April 1, 1929. Because astronomers had assumed that Planet X would be similar to Uranus and Neptune, they had searched for a visible disk as the telltale sign of a new planet. Instead of searching for a disk, Tombaugh used a blink-comparator, which compares the stellar images recorded on two photographs exposed on different nights. By this method, any motion relative to the background stars could easily be detected, the motion of a possible planet as well as that of asteroids or comets.

The photographic search was trying and tedious, but on February 18, 1930, Tombaugh noted two images on different photographs, in the area near a star in Gemini, that had shifted by 3.5 mm (Fig. 17.39). The shift was such that the object had to be a body orbiting the sun. The detection was quickly confirmed as a new planet, and the discovery was announced on March 13, 1930, Lowell's birthday. (During this successful search and its continuation, Tombaugh "blinked" about 45 *million* stars.) The name Pluto was officially accepted by Lowell Observatory, and in honor of Lowell the trustees of the observatory chose ♇, a monogram of his initials, as the symbol for the planet.

We put Pluto in this chapter mostly because of its position just outside the orbits of the Jovian planets. Pluto is much smaller than the Jovian planets, but it may have a similar density. In fact, it resembles the large satellites of Jupiter and Saturn.

Pluto's average distance from the sun is 39.44 AU. Since it has a highly eccentric orbit, it ranges from 29.7 AU to 49.3 AU from the sun, and hence it is never closer to the earth than 28.7 AU at closest approach (opposition). Because of the vast reach of space, Pluto is difficult to observe and strains the resolving power of large telescopes. Tombaugh and his

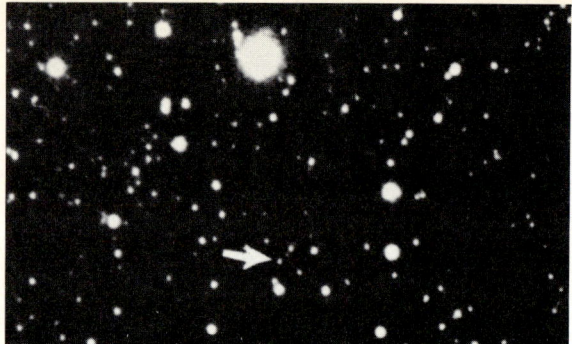

FIGURE 17.39 Discovery photos of Pluto (arrow). *(Courtesy Lowell Observatory.)*

FIGURE 17.40 Schematic diagram of the Pluto-Charon system.

TABLE 17.4 Properties of the Pluto-Charon System

Separation	17,500 km
Period of revolution	6.3871 days
Pluto	
Mass (earth = 1)	0.0018
Diameter	3000 km
Density	800 kg/m^3
Rotation period	6.39 days
Charon	
Mass (earth = 1)	0.0002
Diameter	1500 km
Density	800 kg/m^3
Rotation period	6.39 days
Pluto to Charon mass ratio	10

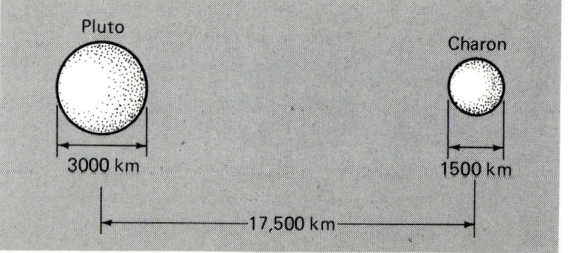

associates were extremely disappointed when the telescope at Lowell Observatory could not resolve Pluto's disk; they had anticipated uncovering a much larger planet.

More recent attempts to measure Pluto's diameter have also been frustrating. In 1965 Pluto was expected to pass in front of a faint star. If this event could have been observed from two separate stations on the earth, Pluto's radius could have been determined. Unfortunately, careful observations at Kitt Peak Observatory did not find any decrease in the combined light received from the star and Pluto, so all that could be concluded was that Pluto is less than 6800 km in diameter.

Later Dale Cruikshank, David Morrison, and Carl Pilcher discovered methane ice on Pluto's surface. The presence of methane ice means that the surface temperature is no more than 40 K. This discovery also leads to an estimate of Pluto's size. We know how much sunlight hits Pluto's surface from the inverse square law for light; if we knew Pluto's reflectivity (albedo) and its total area, we could work out how much sunlight Pluto reflects back into space. Then we could determine how bright Pluto should appear at a known distance from the earth. We can turn this argument around to find Pluto's size. If we measure Pluto's brightness from the earth, we can calculate how much surface area Pluto has and so its radius. The trick here is to know how much methane ice covers the surface, for that determines the reflectivity. If the surface is completely covered by ice, then the diameter is less than 3000 km, smaller than our moon.

Recently S. Arnold, A. Boksenberg, and W. Sargent have measured Pluto's diameter at the Hale 5-m telescope by a technique known as speckle interferometry. They determined a diameter of 3000 to 3600 km.

17.6 PLUTO: GUARDIAN OF THE DARK

Observations of Pluto's brightness have uncovered a slight increase about every 6.4 days. This cyclic variation is the only evidence of rotation, and the 6.4-day period is generally accepted as Pluto's rotation period. The observations also show that Pluto's axis of rotation, like that of Uranus, lies close to the plane of the ecliptic.

Something else has the same 6.4-day period—Pluto's moon, Charon. In June 1978 James Christy of the U.S. Naval Observatory in Flagstaff, Arizona (Lowell's old haunt), noticed what appeared to be a bump on Pluto's image in a photo (look back at Fig. 13.27). Checking photos taken in 1965 and 1976, Christy found seven showing the same bump, always oriented approximately north-south. He proposed that the bump was the faint image of a moon partially merged with the image of the planet. Christy named this moon Charon, after the mythological boatman who ferried souls across the river Styx to Pluto for judgment. An occultation of a star by Charon indicates it has a diameter of at least 1300 km. So Charon is about half the size of Pluto, and, if the same density, has about 1/10 its mass (Table 17.4).

The observations of Charon imply a revolution period of 6.4 days and a distance of 17,500 km from Pluto (Fig. 17.40). That's neat, for now we can use Charon to find Pluto's mass, using Kepler's third law.

The result is important enough to do the calculation in detail. Let's compare the earth-moon system to the Pluto-Charon system. For the earth and moon

$$M_E + M_M = \frac{4\pi^2}{G} \frac{a_{EM}^3}{P_{EM}^2}$$

and for Pluto and Charon

$$M_P + M_C = \frac{4\pi^2}{G} \frac{a_{PC}^3}{P_{PC}^2}$$

Divide these two equations to get

$$\frac{M_P + M_C}{M_E + M_M} = \left(\frac{a_{PC}}{a_{EM}}\right)^3 \left(\frac{P_{EM}}{P_{PC}}\right)^2$$

Because the masses of the moons are smaller than their parent planets, we can approximate $M_E + M_M$ by M_E and $M_P + M_C$ by M_P. Then

$$\frac{M_P}{M_E} = \left(\frac{1.75 \times 10^4 \text{ km}}{3.8 \times 10^5 \text{ km}}\right)^3 \left(\frac{27.3 \text{ days}}{6.4 \text{ days}}\right)^2$$

$$= (1.0 \times 10^{-4})(18.2)$$

$$= 1.8 \times 10^{-3}$$

So Pluto has a mass about 0.002 that of the earth. Our planet has a mass of 6×10^{24} kg. So

$$M_P = (6.0 \times 10^{24} \text{ kg})(1.8 \times 10^{-3})$$

$$= 1.1 \times 10^{22} \text{ kg}$$

FOCUS 17.1 IS PLUTO AN ESCAPED MOON OF NEPTUNE?

Because the orbits of Neptune and Pluto overlap, some astronomers have proposed that Pluto once was a moon of Neptune. We know that Pluto closely resembles a Jovian moon. But if Pluto once orbited Neptune, how did it move to its present orbit around the sun?

R. Harrington and T. Van Flandern have done some theoretical calculations to investigate this question. They note that Neptune's satellite system is unusual, which suggests possible disruption. Neptune has only two satellites; the other Jovian planets have more. Triton, the innermost, has an east to west orbit inclined 20° to Neptune's equator; no other planet has a close moon moving retrograde and so much inclined. Nereid, the outer moon, has an orbital eccentricity of 0.75, two times larger than that of any other solar system satellite. Nereid's orbital velocity of 3 km/sec at closest approach to Neptune is only 0.2 km/sec shy of escape velocity.

Harrington and Van Flandern feel that Triton's and Nereid's unusual orbital characteristics can be explained if the system was disrupted by an outside body that had more mass than the satellites.

Let's examine their scenario. It assumes that Neptune's moons originally had circular, equatorial orbits—modeled after the Galilean moons of Jupiter. Four moons are assumed. A hypothetical disrupting body, with a range of mass, is run through the satellite system (in their computer). Harrington and Van Flandern picked encounter conditions in ways to try to maximize the frequency of interesting results for a mass between 0.2 and 2 Neptune masses.

One trajectory produced especially amusing results (Fig. F.20). Here a 3-earth-mass planet comes in at a 30° angle, plunging through the system of four satellites. Satellite 1 gets captured by the perturbing planet and is pulled off Neptune. Satellite 2 also escapes and ends up in a Pluto-like orbit. Satellite 3 has its orbit flipped to retrograde and becomes Triton. Satellite 4, the outermost, falls into a Nereid-like orbit.

What happens to the disrupting planet? Harrington and Van Flandern conclude it ends up in an elliptic orbit with semimajor axis less than 100 AU and an eccentricity of less than 0.6—another planet beyond Neptune's orbit.

Of course, these calculations do not prove that such a close encounter produced Pluto. But they do show it's dynamically possible.

What a lightweight planet! Previous estimates of Pluto's mass, from its supposed gravitational influence on Neptune, came up with 0.1 that of the earth, about 50 times higher than the result above. It seems that Pluto does not have enough mass to affect Neptune's path. So Tombaugh's discovery may have been luck, the result of a diligent, comprehensive search rather than of an accurate prediction from Newton's laws of motion and gravitation.

With Pluto's mass and diameter in hand, we can figure out its density. The result is about 800 kg/m^3, which implies that Pluto consists mainly of ices and other frozen volatiles.

Pluto may have a very thin methane atmosphere, but a very peculiar one. Only where the sun is directly overhead is it warm enough to evaporate methane. Over the rest of the

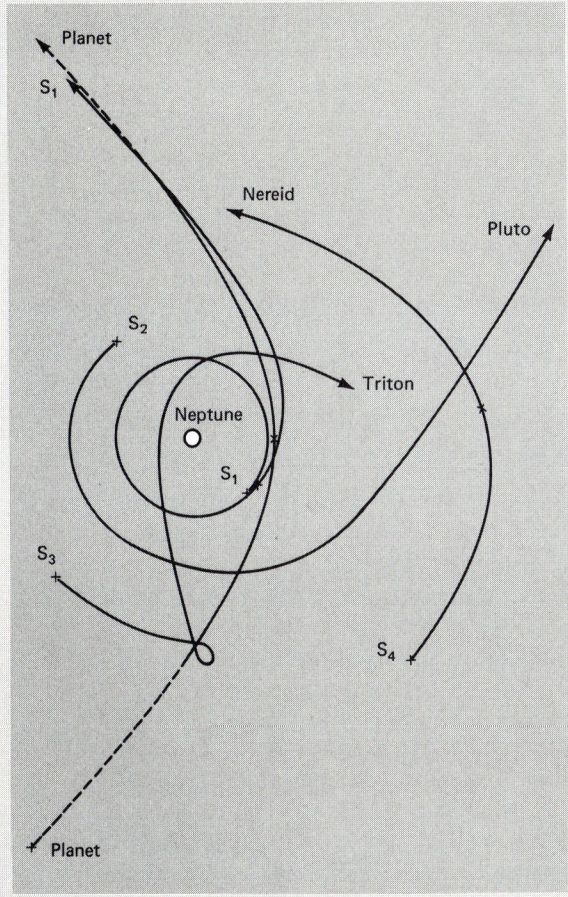

FIGURE F.20 A hypothetical encounter of Neptune with a body (labeled "planet") that has a mass of 3 earth masses Neptune's four moons are labeled S_1–S_4. (*Adapted from a diagram by R. Harrington and T. van Flandern,* Icarus, *vol. 39, (1979), p. 131.*)

planet it is frozen out. As Pluto slowly rotates, a mountain of methane gas forms at the noon point near the equator.

The idea that Pluto may be a small, icy planet, like a Jovian moon, fits with a speculation that Pluto is an escaped moon of Neptune. Why? The orbit of Pluto is so eccentric that it sometimes comes within Neptune's orbit, which means that at times Pluto is actually closer to the sun than Neptune. Recall that Neptune's Triton is a large satellite with an unusual retrograde orbit. Gerard Kuiper and Robert Lyttleton have argued that if Pluto were originally a moon of Neptune along with Triton, a close encounter between the two might have caused Triton to reverse its orbital motion and also thrown Pluto free of Neptune. However, there are many details that this model fails to account for. For example, it is hard to imagine a double satellite of Neptune. So most astronomers consider this only a speculation at the moment (Focus 17.1).

By the way, Pluto *is* now within Neptune's orbit. On January 21, 1979, Pluto edged closer to the sun than Neptune. It will orbit closer to the sun than Neptune until March 1999. Because of the high inclination of its orbit, Pluto is actually well above Neptune's orbital plane and Neptune is now about 60° around its orbit from Pluto, so there is no danger of a collision. Over the past ten years, as Pluto has gotten closer to the sun, it has gotten gradually darker, probably because some methane ice has evaporated from its surface.

The search at Lowell Observatory for other planets beyond Neptune ended in 1945 without yielding further positive results. A planet with the same characteristics as Pluto, placed at a greater distance, has little chance for discovery. However, one like Uranus or Neptune might be just visible if it were closer than 80 AU. Some astronomers have postulated a new Planet X orbiting retrograde beyond Pluto. If this planet exists—it has not been observed to date—it would bring the solar system's total to 10, the magic number of the Pythagoreans.

SUMMARY

The Jovian planets make up an aspect of the solar system worlds apart from the terrestrial planets. Here are huge, massive, low-density worlds with chemical compositions resembling the sun's.

Compared with the earth, the Jovian planets are primitive, relatively unevolved worlds, clinging to their pasts. If you could have observed the Jovian planets four billion years ago, they'd look much the same as they do now, although somewhat hotter. Their low atmospheric temperatures and large masses insure that they've kept their original atmospheres. They do not have rocky surfaces where volcanoes can vent and erosion can change the landscape. Infalling pieces from space don't make any dents.

Don't get the impression that the Jovian worlds don't change at all, for they certainly do. Just look at their stormy atmospheres, driven in part by internal heat left over from their formation. Yet these changes are ethereal; they leave no long-lasting marks. Today, the Jovian planets are places of change, but not of planetary evolution. In contrast, three terrestrial planets—the earth, Mars, and Venus—have evolved dramatically and continue to evolve today.

Key Words

belts and zones of Jupiter's atmosphere	grooved terrain	shepherd satellite
Great Red Spot	cratered terrain	planetary ring
metallic hydrogen	transverse fault	A-, B-, C-, D-, E-, F-, G-rings of Saturn
aurora	asteroidal moon	Cassini's division
Galilean moons: Io, Europa, Ganymede, Callisto	Amalthea	Triton
	Titan	Charon
	co-orbital satellite	

Review Questions

1. The Galilean satellites, in order outward from Jupiter, are Io, Europa, Ganymede, and Callisto.
   a. Which has active volcanoes?
   b. Which is the most heavily cratered?

PROBLEMS

c. Which has both grooved terrain and large dark cratered regions?
d. Which has the lowest density?
e. Which has a thin, fractured ice crust and few impact craters?
(Objective 6)
2. Which of the following are true?
a. The Red Spot is a large, solid island floating 8 km above the clouds of Jupiter.
b. Belts and zones arise from heat flowing out from the interior of Jupiter.
c. Like Jupiter, Saturn shows belts, zones, and a red spot.
d. The discovery of Pluto allowed astronomers to compute Neptune's mass.
(Objectives 2, 3, and 5)
3. Which of the following properties distinguish Jupiter from any of the other planets?
a. Largest mass
b. Lowest density
c. Large internal heat source
d. Shortest rotation period
e. Largest magnetic field
f. Largest orbital period
(Objectives 1 and 2)
4. In what one significant respect is Jupiter most different from the other Jovian planets? (Objective 1)
5. Suppose you flew very close to Jupiter. What would you look for? (Objectives 2, 3, and 4)
6. How do we know the density of Pluto? (Objectives 2 and 5)
7. How do we know that the rings of Saturn are thin? (Objectives 2 and 4)
8. What fact makes it relatively easy to find the masses of the Jovian planets? (Objective 2)
9. In one word state the greatest difference between the Jovian and the terrestrial planets. (Objective 1)
10. What are your chances of survival on any of the Jovian planets? (Objectives 1, 2, and 3)
11. In one sentence compare the rings of Saturn with those of Jupiter and Uranus. (Objective 4)
12. How is Jupiter's magnetic field similar to that of the earth? How is it different? (Objective 9)
13. What features does Pluto have in common with the Galilean moons of Jupiter? (Objective 7)
14. In what way is Saturn's Titan especially interesting? (Objectives 3 and 8)
15. What characteristics of the moons of Jupiter and Saturn indicate that some have evolved since their formation and others have not? (Objective 8)

Problems

1. The albedo of Jupiter is about 0.35 and its surface temperature is about 130 K. Calculate the amount of energy received from the sun, and compare it with the energy radiated (assuming Jupiter is a blackbody). What fraction of the radiated energy must be generated in the interior? (Objective 2)

2. Assume that the particles in Saturn's rings have circular orbits ranging from 87×10^3 km to 137×10^3 km from the center of Saturn.
   a. Calculate the orbital velocities at the inner and outer edges of the rings, assuming that they are made of individual particles obeying Kepler's laws.
   b. Calculate what the velocity at the outer edge would be if the rings were solid and rotating with the same period as calculated in (a) for the inner edge.
   c. What resolution (what accuracy in wavelength measurement) would a spectrograph have to have in order to distinguish between these two types of motion for the rings?
   (Objective 4)
3. The period of Mimas is 0.9 days. Its orbit has a radius of 185×10^3 km. Any particles with periods half that of Mimas have unstable orbits. This is the reason for the gap at one edge of Cassini's division. Calculate the distance of Cassini's division from the center of Saturn. (Objective 4)
4. Spectroscopic observations suggest that Pluto is covered with icy frost and thus has a high albedo (A = 0.5). The brightness of Pluto at opposition (d = 38 AU from the earth) is 15th magnitude, that is, 2×10^{-17} as bright as the sun (1 AU from the earth). From these two observations, calculate the radius of Pluto. Then, using the mass given in Table 17.4, calculate the density. (Objective 5)
5. Pluto's orbit has an eccentricity of 0.25 and a semi-major axis of 39.44 AU. What is its minimum distance from the earth? (Objective 5)
6. The total amount of thermal energy within Jupiter can be estimated by assigning an energy $\frac{3}{2} kT$ to each particle within it. Assuming that most of the mass is hydrogen molecules, calculate the thermal energy for an average internal temperature of 20,000 K. Then calculate how long it would take for Jupiter to radiate away this energy at its current luminosity. (Objective 2)

BEYOND THIS BOOK . . .

*Try "Jupiter" by J. H. Wolfe and "The Outer Planets" by D. M. Hunten in the September 1975 issue of *Scientific American*.
*You can read the technical report of the discovery of methane ice on Pluto by D. P. Cruikshank, C. B. Pilcher, and D. Morrison in the November 19, 1976 issue of *Science*, p. 835.
*Contrast D. Cruikshank and D. Morrison's "The Galilean Satellites of Jupiter" in *Scientific American*, May 1976, with "The Galilean Moons of Jupiter" by L. Soderblom in the same magazine, January 1980.
*The results from the Voyager missions have appeared in many places. Check *Sky and Telescope*, *Mercury*, *Astronomy*, and *Science* for recent articles.
*The January 1980 issue of *National Geographic* contains an excellent article by R. Gore on Voyager observations of Jupiter and its satellites.
*For the discovery papers on the rings of Uranus, look at *Nature*, vol. 267 (May 26, 1977), pp. 328–332. "Rings in the Solar System" by J. Pollack and J. Cuzzi, *Scientific American*, November 1981, p. 105, compares the ring systems of Jupiter, Saturn, and Uranus.
*For more details, read "The Moons of Saturn" by L. Soderblom and T. Johnson, *Scientific American*, January 1982.
*For a first-hand view about the discovery of Pluto, read *Out of the Darkness* (Stackpole, Harrisburg, Pa., 1980) by C. Tombaugh and P. Moore. For other information read *The Planet Pluto* (Pergamon Press, New York, 1980) by A. Whyte.
*A good survey of the Voyager discoveries is contained in *Voyager to Saturn* (NASA SP-451, 1982) by D. Morrison.

SOLAR SYSTEM DEBRIS: CLUES TO THE PAST

18

LEARNING OBJECTIVES
After studying this chapter, you should be able to:
1. Describe the physical properties of a typical asteroid.
2. State the location of the asteroid belt and its contents.
3. List the main types of asteroids as inferred from their reflectivities and infrared properties.
4. State the evidence for asteroidal moons.
5. Sketch the dirty-iceberg model of a comet.
6. Compare the orbits and physical properties of comets to those of asteroids.
7. Describe the difference between a comet's gas tail and dust tail.
8. Sketch a picture of the Oort comet cloud and argue that most of the comets must be far from the sun.
9. Describe the difference between a meteoroid, meteor, and meteorite.
10. Indicate the relationship between periodic comets and meteor swarms.
11. Describe the physical characteristics of the three main types of meteorites: irons, stones, and stony-irons.
12. Discuss the implications of chondrules and Widmanstätten figures found in meteorites.
13. Give observational evidence for interplanetary gas and dust.
14. Outline the evolution of each of the main types of solar system debris.

CENTRAL QUESTION:
What are the physical properties of the interplanetary debris, and what hints do they give about the solar system's origin?

Astronomers for centuries considered the flashes of meteors as terrestrial phenomena blazing high in the atmosphere (Fig. 18.1). Not until early in the nineteenth century did the extraterrestrial origin of meteorites become an accepted fact.

Meanwhile, comets had been making a bad name for themselves for centuries (Fig. 18.2 and Plate 14). Their passage had become identified with the deaths of kings and other dire historical events. Plutarch states, for example, that at Caesar's funeral a bright comet stretched across the sky. And Mark Twain, who was born in 1835 at one return of Halley's comet, died at the next return in 1910, as Twain himself had predicted.

FIGURE 18.1 A meteor—an unusually bright one, called a *fireball*. (*Courtesy F. Whipple and Harvard College Observatory.*)

18.1 ASTEROIDS: THE MINOR PLANETS

FIGURE 18.2 Comet Ikeya-Seki in October 1965. Note the structure in the tail. (*Courtesy Lick Observatory.*)

Comets blossom when they pass close enough to the sun to be illuminated by it and close enough to the earth to be visible. A delicate tail is a comet's emblem and the source of its name, a "hairy star." A comet has no tail away from the sun. A small, cold mass of icy materials, it glides invisibly through space. The small bodies that fall to the earth as meteorites also orbit among the planets. They are called meteoroids while in space. These small hunks become visible only when their orbits cross the earth and they dive through our atmosphere. Their luminous trail is called a meteor.

Comets and meteoroids are not the only interplanetary debris in the solar system. The asteroids, most of which inhabit a belt between Mars and Jupiter, slip almost unnoticed among the stars. These are rocky objects larger than meteoroids. Interplanetary gas and dust also drift among all the bodies of the solar system. Some solid material remains from the formation of the solar system. Most of the gas comes from the sun as the solar wind. The solar wind sweeps through this interplanetary material and pushes the gas and some dust out of the solar system.

The collected mass of the debris in the interplanetary junkyard probably amounts to less than the mass of the moon. Yet, archeologists have discovered that a city's trash pile provides direct clues to the history of a culture, even though it contains much less mass than the city. In the same way, the solar system's debris provides a fruitful hunting ground for clues to its past. In this chapter we hunt through this varied collection of leftovers. Here you will encounter important guideposts to the origin of the solar system.

18.1 ASTEROIDS: THE MINOR PLANETS

On January 1, 1801, the Sicilian astronomer Father Giuseppe Piazzi (1746–1826) noticed a small object through his telescope that appeared like a distant comet. Because

FOCUS 18.1 THE TITIUS-BODE "LAW"

The year 1972 marked the bicentenary of the Titius-Bode "law," a mathematical relationship applicable to the solar system which is neither a law nor a discovery by Titius and Bode together. In 1772 Johann E. Bode published the second edition of his popular astronomy book, in which he presented the relationship shown in Table F.1. Let the sun-Saturn distance be 100 arbitrary units; then Mercury is 4 such units from the sun; Venus is 7 units, or 4 + 3; the earth, 4 + 6 = 10; Mars, 4 + 12 = 16. Note that this sequence is the geometrical progression 3, 6, 12, . . . added to 4. If the numbers are divided by 10, they give the planetary distances in AU. Bode pointed out a gap in the progression at 28 units (2.8 AU) between Mars and Jupiter. Because he could not believe that the Creator could have left a hole in the planetary series, Bode considered that an undiscovered planet might reside there.

Independently, Johann D. Titius had written almost the same idea in a footnote of a book he had translated. Like Bode, Titius did not think that the gap between Mars and Jupiter was actually empty and postulated that the undiscovered satellites of Mars belonged in the space. Titius, however, did not compare the mathematically prescribed distance with the actual average distance (in AU) as Bode had done.

In 1781 William Herschel discovered Uranus in the next place in the progression after Saturn. Bode publicly announced that he was pleased that his "law" was proving so fruitful. The annoying Mars-Jupiter gap spurred the search that led to the discovery of Ceres in 1801. The host of asteroids, found later, tempered the apparent triumph of the "law." After the discovery of Uranus, both Adams and Leverrier used the next Titius-Bode position as a guide to predict the place of Neptune. However, the difference between the actual distance and the mathematical one was substantial for Neptune and even greater for Pluto (see Table F.1).

Piazzi was constructing a new star catalog, he carefully noted the object's position. On the next night the object had moved a bit to the east. This rapid motion indicated that it was a member of the solar system. Piazzi named the new object Ceres after the patron goddess of Sicily. He had discovered the first *asteroid*.

After a few weeks of observing, Piazzi became ill and lost Ceres's position. However, the German mathematician Karl Friedrich Gauss (1777–1855) had just developed a new technique for computing orbits from three observations and Newton's law of gravitation. With Piazzi's positions, Gauss calculated Ceres's orbit, and on December 31, 1801, Ceres was relocated near the predicted position.

Why are three observations necessary and sufficient for computing an orbit? To use Newton's laws to calculate the future motion of a body, we need to know its initial position and velocity in three dimensions, that is, we need to know three position coordinates and the parts of the velocity in these three directions—six numbers in all. In observing an astronomical body, we usually are unable to measure its position in space and can determine only its position on the sky. Any one observation yields two numbers, for example, the angular distance north or south of the ecliptic and the angular distance east of the vernal equinox. One observation yields two numbers, so three observations give six numbers. What Gauss did was to find a way of using these six angular measurements, instead of the six numbers giving initial position and velocity, to compute the entire orbit.

The Titius-Bode "law" is not a law in the same sense as, for example, Newton's law of gravitation, because, as far as we know, it does not rest on fundamental physical principles and its predictive powers are limited. However, the relationship does emphasize that the solar system has a regular structure, that is, the planets are not found at completely random locations. Similar spacing rules apply to the satellite systems of Jupiter, Saturn, and Uranus. The coincidence of the rules with actual distances implies that regular physical processes guide the formation of planetary and satellite systems.

TABLE F.1 The Titius-Bode "Law"

Planet	Titius Bode Distance ($\div 10$)	Actual Distance (AU)	Difference (%)
Mercury	0.4	0.39	+3
Venus	0.7	0.72	−3
Earth	1.0	1.00	0
Mars	1.6	1.52	+5
(Ceres)	2.8	2.7	+1
Jupiter	5.2	5.2	0
Saturn	10.0	9.5	+5
Uranus	19.6	19.2	+2
Neptune	38.8	30.1	+22
Pluto	72.2	39.4	+49

The discovery of Ceres also seemed to verify the Titius-Bode "law" of planetary spacing (Focus 18.1). The gap in this regular sequence, at 2.8 AU from the sun, prompted Johann Bode to suggest the possibility of a new planet at this position. His prediction spurred a search for the possible planet. Gauss computed that Ceres's distance was within 0.4 AU of the distance of the hypothetical planet.

Three more asteroids were soon discovered: Pallas in 1802, Juno in 1804, and Vesta in 1807. These (along with Ceres) are the brightest asteroids. They have orbits with semimajor axes close to 2.8 AU. Further discoveries of asteroids occurred only sporadically until 1891, when photographic search methods were introduced (Fig. 18.3). Over 2000 asteroids have been discovered and their orbits worked out. Astronomers estimate that over 50,000 more remain to be found.

Orbits

The "typical asteroid" moves around the sun in the asteroid belt at an average solar distance of 2.8 AU (Fig. 18.4). Most orbits have eccentricities that are less than 0.2 and inclinations that cluster around 10° and range from 0° to 40°. However, the orbital characteristics of some asteroids deviate widely from these typical values. Hidalgo, for

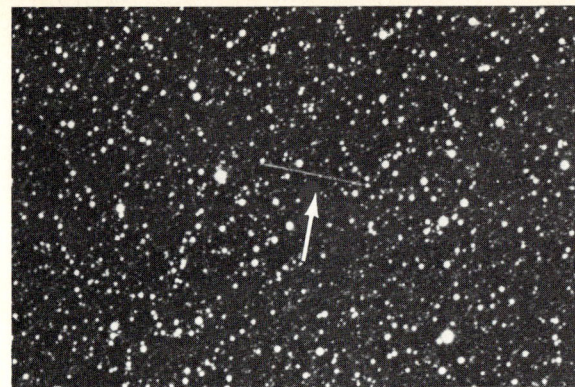

FIGURE 18.3 The asteroid Icarus appears as a streak (arrow) moving with respect to the stars. (*Courtesy Palomar Observatory, California Institute of Technology.*)

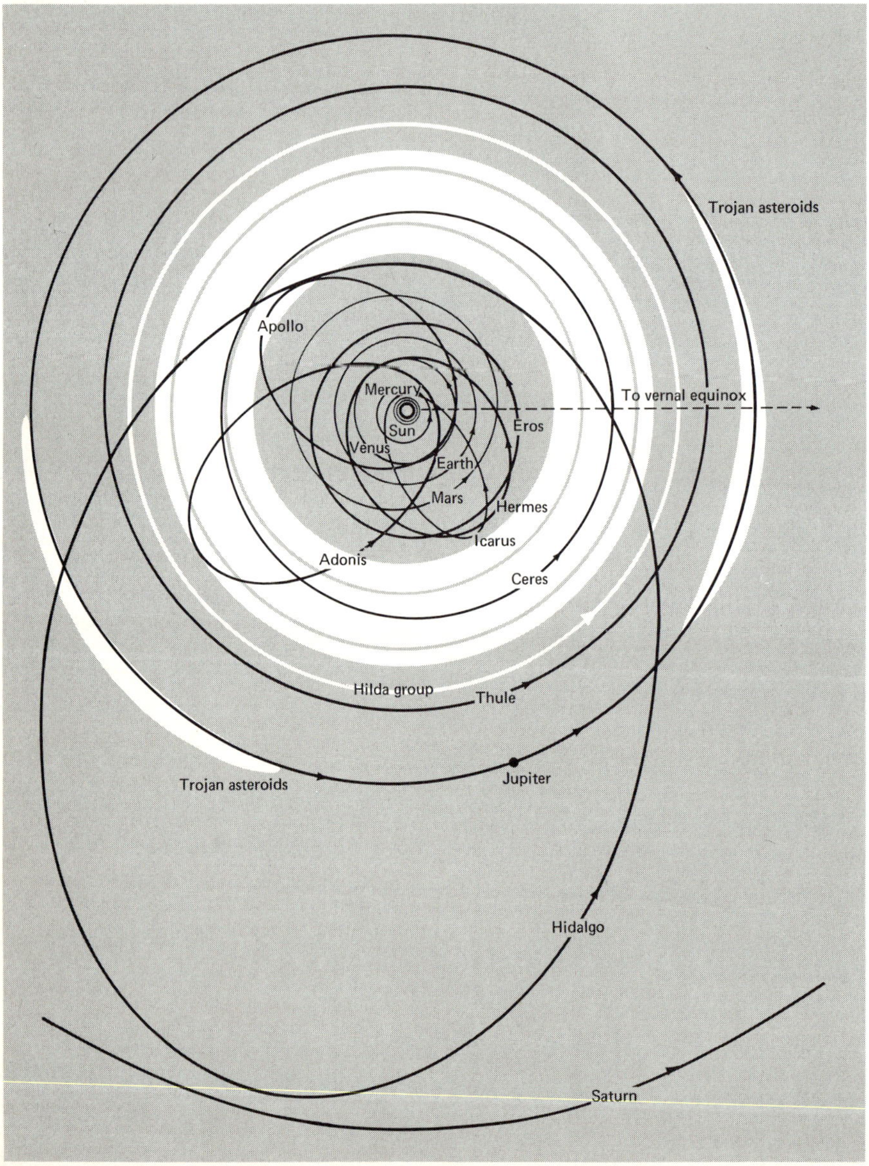

FIGURE 18.4 Orbits of asteroids. Most, like Ceres, are located in the asteroid belt (light area) between Mars and Jupiter. Some of these located outside the belt, such as Icarus and the Trojans, are also shown.

example, swings around the sun at an inclination of 43° to the plane of the ecliptic. Icarus actually skirts closer to the sun at perihelion than Mercury does and also approaches as close as 5.6×10^6 km to the earth (Fig. 18.4). On June 15, 1968, Icarus careened past the earth with a closest approach of 6.4×10^6 km. Some people, fearful of a collision, predicted that the world would end. It did not.

Physical Features

What is an asteroid? It is an irregular, rocky hunk, small both in size and in mass compared to a planet. (The word planetoid is a less frequently used, but perhaps more descriptive, designation for these bodies.) Ceres, the largest known asteroid, has a diameter of only about 1000 km (Table 18.1). (That's about $\frac{1}{3}$ the size of the moon.) Asteroid sizes, however, have not been directly measured. Instead, we measure an asteroid's reflectivity of visible light and its infrared emission (typically at 10 μm) when the asteroid is at a known distance from the sun. These two measures allow an estimate of the asteroid's size. Let's see how.

The amount of energy falling on an asteroid is

$$\frac{L_{sun}}{4\pi D^2} \pi R^2$$

where R is the radius of the asteroid and D is the distance from the sun. A fraction A (the albedo) is reflected back into space. If the earth is a distance d from the asteroid, the flux of reflected light at the earth is

$$F_{vis} = \frac{L_{sun}}{4\pi D^2} \pi R^2 \frac{A}{4\pi d^2}$$

We can measure this flux, and hence determine the quantity $R^2 A$ (the rest of the numbers

TABLE 18.1 The Largest Asteroids

Name	Diameter (km)	Color Class
Ceres	1000	C
Pallas	608	Unclassified
Vesta	538	Unclassified
Hygiea	450	C
Euphrosyne	370	C
Interamnia	350	Unclassified
Davida	323	C
Cybele	309	C
Europa	289	C
Patientia	276	C

Source: D. Morrison, "Asteroid Sizes and Albedos," *Icarus*, vol. 31 (1977), pp. 185–220.

in the equation are known). Now the fraction of the energy not reflected, $(1 - A)$, is absorbed, heats up the asteroid, and is reemitted into space as infrared radiation. We can observe this infrared flux at the earth. The ratio to the visible flux is

$$\frac{F_{vis}}{F_{IR}} = \frac{A}{1 - A}$$

So from this measurement we can determine the albedo A, then use it with the previous determination of R^2A to calculate the radius R.

Most asteroids regularly fluctuate in brightness. Such variations indicate that they may have a nonuniform albedo or an irregular shape or both. From the observed brightness variations of Eros, for example, astronomers estimate it to have a bricklike shape, 6 km by 22 km. Gravity alone could never hold such a spinning splinter together; the cohesiveness of rock keeps Eros in one piece.

Astronomers believe that they have directly observed two former asteroids: Deimos and Phobos, the satellites of Mars (Section 16.9). These bodies are rough, irregular, and pitted with craters. They reflect light very poorly, no better than a blacktop road. Some of the satellites of Jupiter and Saturn may also be captured asteroids.

Recent observations of the reflectivity of the larger asteroids' surfaces have given some new hints about their compositions. This work shows us that asteroids cannot be simply lumped together in one group—they have individual personalities!

For instance, asteroids have a wide range of reflectivities from Nysa (diameter 82 km), whose albedo is 38 percent, to Cybele (diameter 309 km), which reflects only 2 percent of the visible light that hits it. Nysa's surface reflects sunlight about as well as the icy satellites of Jupiter and Saturn; Cybele is darker than a lump of coal.

The reflectivities indicate that most asteroids fall into two major compositional classes. Some are relatively bright with albedos of about 15 percent, and others are much darker (albedos 2 to 5 percent), indicating that they contain a substantial percentage of dark compounds such as carbon or the black mineral magnetite (Fe_3O_4). These dark asteroids resemble a class of meteorites (Section 18.3), the carbonaceous chondrites, which are dark because they contain carbon compounds (roughly 1 to 5 percent carbon). The lighter class are dubbed *S-type* asteroids, the darker ones *C-type*. The S-type, in addition to having higher albedos, also show spectral absorption bands indicative of silicate materials (Section 9.2). A third class, called *M-type*, has characteristics suggestive of metallic substances. They have albedos of about 10 percent. Only 5 percent of asteroids belong to this class.

Recent investigations indicate that, based on albedos, compositions in the asteroid belt vary with distance from the sun. Near the orbit of Mars, almost all asteroids have S-type characteristics. Farther out are fewer high albedo ones and more dark ones. At the outer edge of the belt, 3 AU from the sun, some 80 percent of the asteroids are C-type. An explanation for this distribution is found in a current model for the formation of the solar system (Section 19.4).

Chiron: A Very Peculiar Asteroid

Late in 1977 the local newspapers trumpeted that a "new planet" had been discovered by Charles T. Kowal of Hale Observatories. Our students, eager and curious, bombarded

18.1 ASTEROIDS: THE MINOR PLANETS

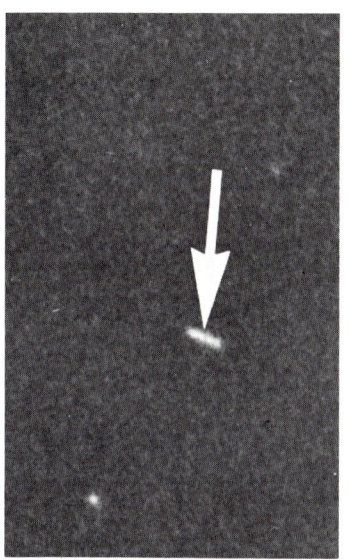

FIGURE 18.5 Discovery photo of Chiron (arrow). (*Courtesy C. Kowal.*)

FIGURE 18.6 The orbit of Chiron. The orbital period is about 51 years. (Based on calculations by B. Marsden.)

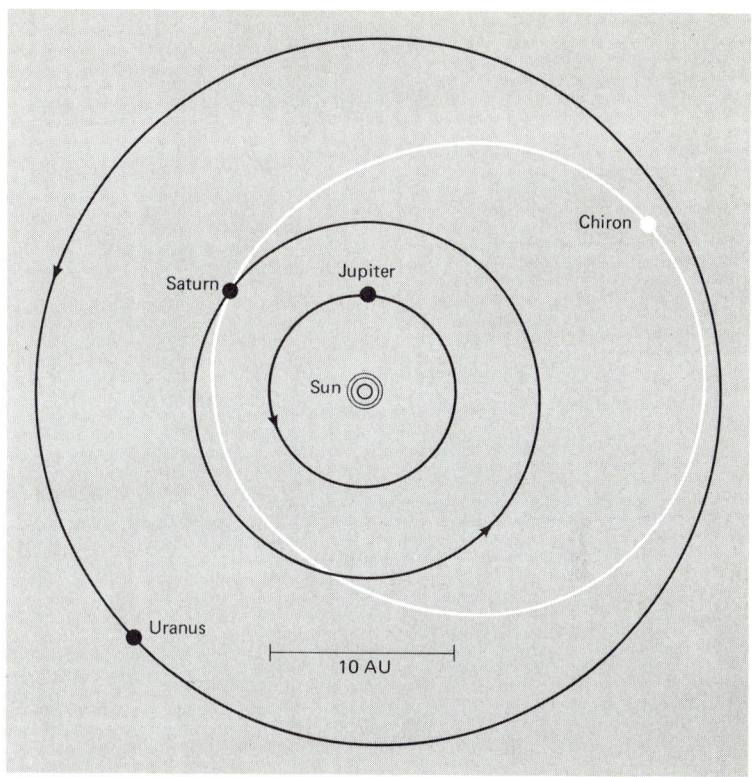

us with questions about this newly discovered object, now called Chiron. With regret we had to tell them that Chiron is not a major planet, but a strange object, nevertheless.

Kowal discovered Chiron while looking at photos taken in October 1977 (Fig. 18.5). Chiron appeared very faint at a distance of about 18 AU from the sun. From additional observations, Brian Marsden computed the orbit of Chiron (Fig. 18.6): a period of 50.7 years, a closest approach to the sun of 8.5 AU (inside the orbit of Saturn), and a far distance of 18.9 AU (almost out to the orbit of Uranus).

Some view Chiron as an unusual asteroid. Chiron, with its orbit lying between Saturn and Uranus, has at times passed close to Saturn, within 0.1 AU in 1664 according to calculations by Marsden. But where did Chiron come from? Chiron may be an asteroid gravitationally drawn out of the asteroid belt by a two-step process. First, Jupiter's pull draws some asteroids (about 300) out of the Mars-Jupiter belt and into a Jupiter-Saturn belt. Second, Saturn attracts some of these (about 40) into a Saturn-Uranus belt. Chiron is then one of the brighter of these displaced asteroids.

Recent infrared observations show that Chiron has a dark surface similar to the C-type asteroids. With an albedo of 3 to 5 percent, Chiron has a diameter between 310 and 400 km.

Asteroidal Moons?

Can asteroids have satellites? Do they? The answer to the first question is yes, to the second a qualified maybe.

Newton's laws certainly permit asteroids to have moons; small ones could orbit in accord with Kepler's third law. But asteroids have very small masses; Ceres, for example, if it were all rock, would have a mass of only about 10^{21} kg, only 1/10 that of our moon. A more typical asteroid is about 10 times smaller in size and so has a mass of only 10^{-3} that of Ceres. This small mass means that an asteroidal-moon system would be very unstable. The gravitational force between the bodies would be so weak that the tidal force of the sun or Jupiter could easily disrupt the binary system. Any existing system is likely then to have a very small separation between asteroid and moon, and so it would be hard to see the two as separate points of light.

Occasionally, as seen from the earth, an asteroid will pass in front of a star. The asteroid will block out the star's light for a few seconds; a moon would cut off the light for even less time. A few such observations have been reported. The most impressive occurred in 1978 with the asteroid Herculina (diameter 220 km); the suspected moon was 975 km away and had a size of only 50 km. Only one other observation looks as good as that for Herculina (it's for the asteroid Pallas); about 20 others hint at companions. (These observations may not all be valid evidence for asteroidal companions. None, for instance, has yet been repeated.)

If Herculina does have a moon, the system's lifetime, before tidal forces disrupt it, is only about 10^7 years. This short time implies that asteroidal moons are probably formed when collisions between asteroids break them up; some of the debris from the breakup may remain bound as moons for a short period of time.

18.2 COMETS: SNOWBALLS IN SPACE

When Edward the Confessor died in 1066, he left no direct heir to the English throne, and the nobles chose Harold II as their king. In the same year a bright comet streaked across the sky—the commonly accepted sign of a ruler's death and misfortunes to follow. In France William of Normandy cleverly interpreted the comet as a sign portending his victory. With this psychological support, he sailed to the British Isles and conquered the dispirited Saxon armies near Hastings. By the end of the year William the Conqueror was crowned king. The comet of 1066 was later to be known as *Halley's Comet*.

Orbits

Cometary orbits remained unknown until Edmund Halley (1656–1742), the Astronomer Royal and a friend of Newton, calculated the orbits of the comets of 1531, 1607, and 1682 by a method devised by Newton (Sections 4.2 and 4.3). Halley found that the orbits were almost identical. Noting that the comets appeared at intervals of approximately 75 or 76 years, Halley concluded that these several comets were in fact the same. He predicted it would return around 1758. Halley was right. His comet was sighted on Christmas night in 1758 and named in Halley's honor. Halley's Comet (Fig. 18.7) was the first comet to be recognized as a permanent member of the solar system, with an elliptical, periodic orbit.

Halley's Comet is the granddaddy of all periodic comets. Its passage near the sun has been recorded at least 29 times, as far back as 239 B.C., possibly even 466 B.C. In 1910 some people incorrectly believed the world might come to an end when the earth passed

FIGURE 18.7 The head of Halley's Comet in 1910. (*Courtesy Mt. Wilson Observatory, Carnegie Institute of Washington.*)

through the comet's tail. (A comet's tail is much too thin to have any effect on the earth.) Halley's Comet moves in an extremely elongated ellipse (eccentricity 0.97). Having passed aphelion beyond Neptune's orbit in 1948, it will return to the earth's neighborhood around 1986. Most comets have such highly elliptical orbits.

Comets are generally divided into two groups: long-period and short-period comets. The dividing line between the two groups is an orbital period of 200 years. Of the 566 comets whose orbits are well known, about 80 percent are long-period, and 20 percent short-period. The long-period comets have orbits that are such long ellipses that they return at very long intervals; the average is estimated to be 10 million years. The short-period comets have smaller elliptical orbits and return more frequently to the sun.

Although the dividing line is chosen somewhat arbitrarily, there are definite differences between the two groups. With a few exceptions, short-period comets have orbits with relatively small eccentricities, stay close to the plane of the ecliptic, and orbit in the same direction as the revolutions of the planets. (Halley's Comet, with a high eccentricity and retrograde orbit, is one of these exceptions.) The Mercury of the short-period comets is Encke's comet, which has a period of only 3.3 years and a perihelion distance of only 0.4 AU (Table 18.2). In contrast, the orbits of the long-period comets are not confined to the plane of the ecliptic and swing around the sun at random angles. About 200 have definite elliptical orbits with periods ranging from 250 years to 30 million years.

Some comets have open orbits that are not ellipses. This seems to indicate that the comet is not a member of the solar system, but an interstellar interloper, momentarily drawn in by the gravitational clutches of the sun. However, careful study, tracing back the orbital positions, shows that the original orbits were long ellipses that were transformed into open orbits by the gravitational influence of Jupiter on the comets. Jupiter's gravity sometimes throws a comet into an orbit that will take it out of the solar system forever. So

TABLE 18.2 Some Periodic Comets

Name	Period (yr)	Perihelion Distance (AU)	First Observed
Encke	3.30	0.399	1786
Giacobini-Zinner	6.42	0.729	1900
Biela	6.62	0.756	1772
Whipple	7.42	0.356	1933
Wolf	8.43	0.395	1884
Tempel-Tuttle	33.18	0.905	1866
Halley	76.03	0.967	239 B.C.

Note: These are all short-period comets, most of which have been observed at least once in the twentieth century.

far as we know, all comets are members of the solar system, held in orbit by the sun's gravity.

Where do the short-period comets come from? They also have orbits which come close to the orbit of Jupiter or some other major planet. Detailed calculations show that they were once in much larger orbits, like the long-period comets, but had their orbits changed by a close encounter with a planet.

Physical Characteristics

You probably associate comets with long, graceful tails. In fact, not all comets exhibit tails, even at perihelion, and no comet has a visible tail when far from the sun (Fig. 18.8).

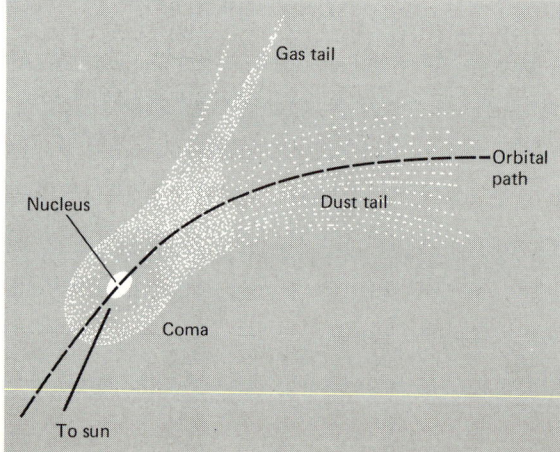

FIGURE 18.8 Comet Cunningham (1940). A comet far from the sun shows little or no tail. (*By permission of Harvard College Observatory.*)

FIGURE 18.9 The main parts of a comet. The diffuse coma encloses a bright, starlike nucleus. The gas tail streams out opposite the sun. The dust tail follows the orbital path of the nucleus.

18.2 COMETS: SNOWBALLS IN SPACE

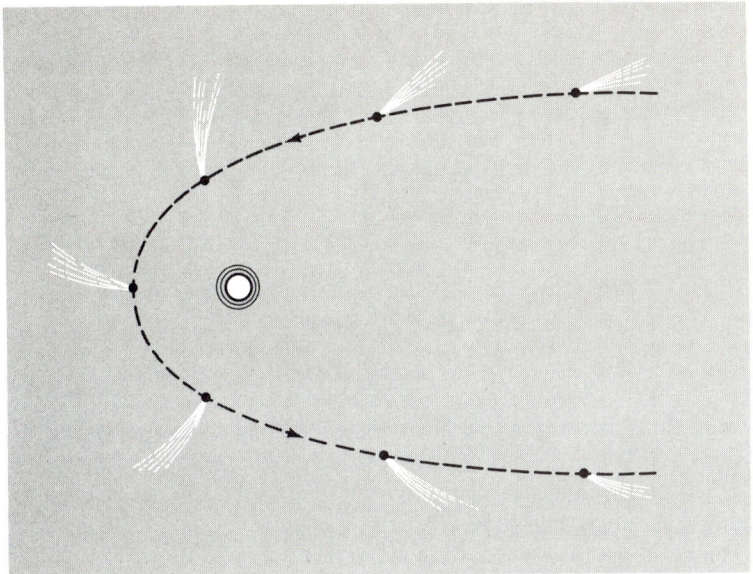 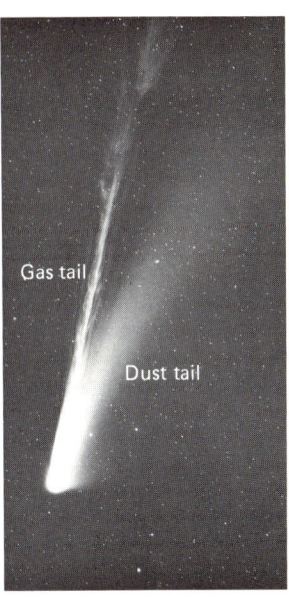

FIGURE 18.10 A comet's gas tail. As a comet swings by the sun, its gas tail, pushed by the solar wind, points away from the sun. The tail's size increases as the comet nears the sun.

FIGURE 18.11 Comet Mrkos (1957), showing both a gas and dust tail. (*Courtesy Mt. Wilson Observatory, Carnegie Institute of Washington.*)

When first sighted telescopically, a comet typically appears as a small, hazy dot. This bright head of the comet is called the *coma*. Sometimes the coma contains a small, starlike point called the *nucleus* (Fig. 18.9). Cometary nuclei are very small, not larger than 1 to 2 km across; none has ever been viewed as more than a point of light. As a comet moves toward perihelion, it grows brighter and sprouts a *tail*. A comet's tail may stretch for millions of kilometers. It always points away from the sun (Fig. 18.10).

Comets may have two types of tails: gas and dust (Fig. 18.11). The physical difference between the two shows up in their spectra. The spectrum of the gas tail has emission lines. The dust tail does not show emission lines, but rather a spectrum of sunlight, reflected from dust expelled out of the coma. The pressure from sunlight detaches dust from the coma and pushes it out to form a tail.

In the spectrum of the gas tail the most conspicuous spectral lines are those produced by carbon monoxide (CO), carbon dioxide (CO_2), nitrogen (N_2), and radicals of ammonia (NH_3) and methane (CH_4). Puffs of ionized gas sometimes shoot through the tail. Radiation pressure from sunlight is not strong enough to account for the fact that these gas tails point straight away from the sun. The German astronomer Ludwig Biermann has suggested that the solar wind has a major effect on the ionized tails. Measurements of the solar wind (Section 6.5) confirm that the magnetic fields carried by the wind's particles can indeed drag ions from the comet's coma.

At great distances from the sun the coma shows a reflected solar spectrum, so the heads must also contain solid particles that reflect sunlight. At about 1 AU from the sun, the

TABLE 18.3 Observed Composition of Comets

Head	Tail
H, C, C_2, C_3, CH, CN, HCN, CH_3, CN, NH, NH_2, O, OH, H_2O, Na, K, Ca, V, Cr, Mn, Fe, Co, Ni, Cu plus dust particles with silicates	CH^+, CO^+, CO_2^+, N_2^+, OH^+, H_2O^+, Ca^+ plus dust particles with silicates

Source: F. Whipple and W. Huebner, "Physical Processes in Comets," *Annual Review of Astronomy and Astrophysics*, vol. 14 (1976), p. 143.

head exhibits molecular emission bands of carbon (C_2), cyanogen (CN), oxygen (O_2), hydroxyl (OH), and hydrides of nitrogen (NH and NH_2). As the comet speeds nearer to the sun, emission lines of silicon (Si), calcium (Ca), sodium (Na), potassium (K), and nickel (Ni) appear. Table 18.3 summarizes the materials that have been observed to date in the heads and tails of comets. Note these two points about the composition. First, the head contains some of the same molecules as found in interstellar space (Table 9.1), such as hydrogen cyanide (HCN) and methyl cyanide (CH_3CN). Second, infrared observations confirm that comets contain considerable amounts of dust. Infrared spectra of some comets display the 10-μm and 18-μm bands characteristic of silicate dust.

For all their stunning length against the sky, comets have very small masses. Comet masses can be estimated only roughly because they are so small that they do not affect the orbits of other bodies. Halley's Comet, one of the largest, has an estimated mass of only about 10^{16} kg (10^{-8} that of the earth) and loses about 10^{11} kg during each perihelion passage. In 1910 the tail of Halley's Comet stretched about 100°, reaching from horizon to horizon; the lack of any noticeable effects as the earth traveled through the tail indicated that the gas was quite rarefied, less than about 10^{-17} kg/m^3. (The atmospheric density at the earth's surface is about 10^{-1} kg/m^3.) The density of the coma is estimated at 10^{-7} kg/m^3, much higher than the tail's density, but still a good vacuum. With so little mass, a comet achieves its spectacular display only by spreading itself thin.

The mass expelled from a comet, mostly as gas, flies off into space. The nucleus supplies this material, but what is the nature of the nucleus? In 1950 Harvard astronomer Fred. L. Whipple developed the *dirty-iceberg cometary model* (Fig. 18.12). In this concept the comet nuclei are compact, solid bodies made of frozen gasses (ices) of water, ammonia, and methane, embedded with rocky material. Beyond Jupiter low temperatures allow the ice-gravel conglomerate to persist unchanged for long periods of time. As the comet nears the sun, the icy material vaporizes. This released material enlarges the coma and creates the tail. As the ice evaporates, a thin coating of rocky material remains to form a solid, but fragile, crust on the nucleus. The heating of the subsurface material creates gas jets which blow off puffs of gas and act as small rockets that slightly change the comet's orbit.

As the comet rounds the sun, the semisolid nucleus can usually withstand the solar heat. Comet Ikeya-Seki, the great sungrazing comet of 1965 (Fig. 18.13), passed within 470,000 km of the solar surface and survived. But some comets passing close to the sun are not so lucky; Comet West (1976) split into at least four pieces after its perihelion

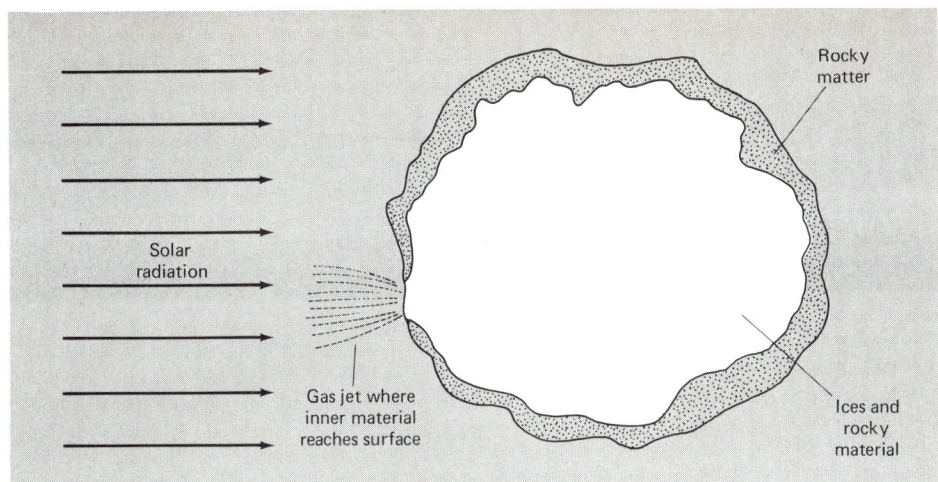

FIGURE 18.12 The dirty-iceberg cometary model. The nucleus is a pudding of ices and rocky material encrusted by a thin, rocky shell. Sunlight heats the nucleus and vaporizes the ices. Gas streams out to make the coma and tail.

FIGURE 18.13 Comet Ikeya-Seki (1965). This comet passed within 470,000 km of the sun and survived. (*By permission of Harvard College Observatory.*)

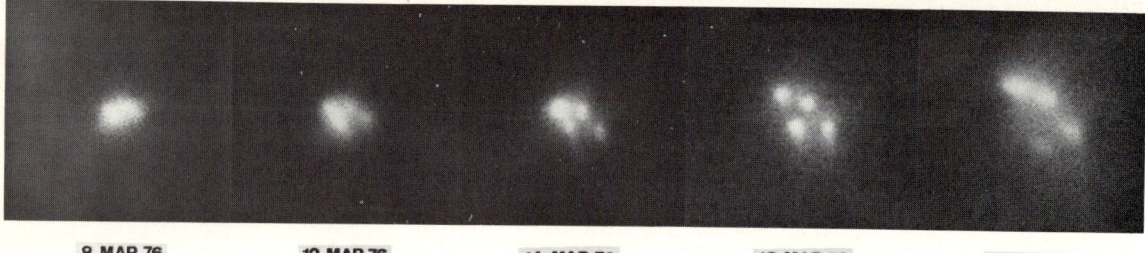

FIGURE 18.14 The breakup of Comet West (1975). After its perihelion passage, Comet West's nucleus split into at least four pieces. (Photos by C. Knuckles and A. Murrell; courtesy New Mexico State Observatory.) (*Courtesy R. Anderson.*)

passage (Fig. 18.14). Periodic comets lose a little material each time they pass the sun and eventually must be completely destroyed.

The Comet Cloud

Where do comets come from, and where are they all now? All comets are attached gravitationally to the sun. Most comets are long-period. From the observed orbits, we find that the average value of their semimajor axis is about 50,000 AU, and the corresponding orbital period is about 10^7 years. The orbits are highly elliptical; in accord with Kepler's second law the comets travel very slowly at aphelion, only a few kilometers per day. So, such comets spend most of their time coasting far from the sun. If you could step back and take a wide-angle photograph, you would find a large number of comets hovering in a cloud, far beyond Pluto.

This cometary cloud, proposed by the Dutch astronomer Jan Oort in 1950, is sometimes known as *Oort's Cloud*. It makes up the solar system's cometary reservoir. According to Oort's model, most comets never come very near the sun, and we never see them. But occasionally the gravitational action of passing stars pushes a comet into an orbit that does bring it closer. For example, Comet Kohoutek (Fig. 18.15) was probably a first-timer in the neighborhood of the sun. Comets are eventually lost, either by vaporization of the nucleus or by Jupiter's perturbation of their orbit. So the supply of sun approaching comets must be replenished with new comets from the cloud. To ensure sufficient input to make up the loses, Oort's picture requires at least 10^{11} comets to be clustered in the cloud.

Where did Oort's cloud come from? Is it possible for so many masses to be picked up gravitationally from interstellar space as the sun threads its way through the Galaxy? Calculations of capture say no, so the comets must have been formed along with the rest of the solar system. It is unlikely the density of the solar nebula (Chapter 19) was high enough for them to form in their present location. Some astronomers suggest they may have originated in the asteroid belt between Mars and Jupiter; others think they formed just beyond the orbit of Pluto, about 100 AU from the sun. Both of these suggestions run into the same difficulty: How did the comets get from their place of origin to their present location in the cloud? Only some speculations have been offered to date, for example, that the gravitational effects of Jupiter would kick comets out of an asteroid belt into long-period orbits.

FIGURE 18.15 Comet Kohoutek (January 14, 1974). Note the kinks in the gas tail. (*Courtesy Mt. Wilson Observatory, Carnegie Institute of Washington.*)

The 1986 Return of Halley's Comet

The greatest cometary show of all returns in 1985–1986: Halley's Comet. Only this comet has the combination of stunning brightness and relatively short periodic returns. Comet Halley is a human comet, marking off a human lifespan between returns.

Halley's Comet will be at perihelion on February 9, 1986, at a distance of only 0.59 AU (Fig. 18.16a). Before perihelion it will pass closest to the earth, at a distance of 0.62 AU, on November 27, 1985. After rounding the sun, the comet will again pass by the earth, closing to 0.42 AU on April 11, 1986. (The closest approach to the earth to date occurred in April 837 A.D.—a scant 0.04 AU!)

Unfortunately this pass of Halley's Comet will not be the best for observers in mid-northern latitudes (Fig. 18.16b). Before perihelion, when it will be visible in the evening sky, it will be fairly faint. After perihelion the comet will be much brighter, but low in the morning sky. Better views will probably occur at more southern locations.

18.3 METEORS AND METEORITES

On a good, clear night you can see about 5 to 10 meteors an hour. Some meteor experts estimate that about 90 million meteors visible to the naked eye occur each day.

A *meteor* is the flash of light associated with the entry of a solid particle into the earth's

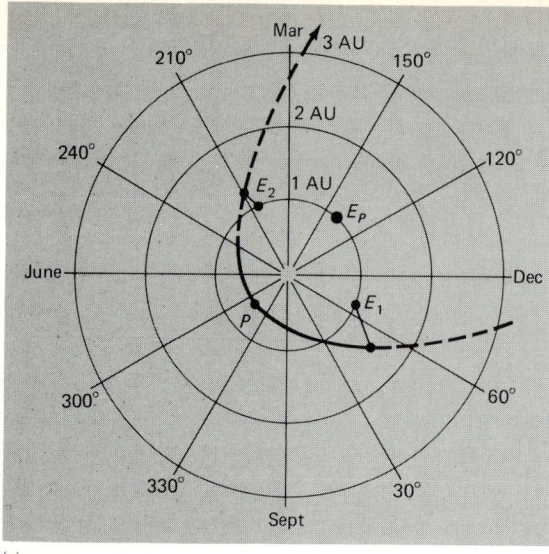

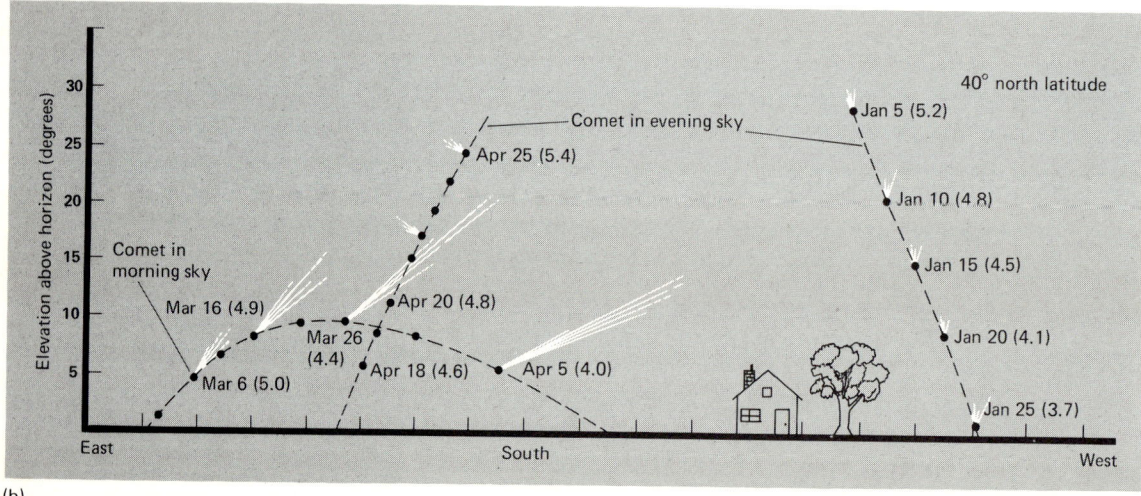

FIGURE 18.16 (a) Orbit of Halley's comet (1985–1986) near the earth. P marks the comet's perihelion point (February 9, 1986), E_1 the closest approach to the earth before perihelion (November 27, 1985), E_2 the closest approach after perihelion (April 11, 1986), E_p the earth's position at perihelion. (b) Observing conditions for Halley's comet in 1986 for 40° north latitude. Approximate visual magnitudes are given in the parentheses next to the dates. (*Adapted from NASA diagrams.*)

atmosphere and its reaction to the atmosphere (Fig. 18.1). As it plunges through the earth's atmosphere, the particle is destroyed by friction, and it may leave a bright *trail*, a glowing column of light, behind it. Before the particle meets its fiery doom in the upper air, it is called a *meteoroid*: a solid object traveling through interplanetary space. (Sometimes the word meteor is used to refer to the particle as well as to the train of light it produces; the context should make the meaning clear.) Of course, other objects such as comets, asteroids, and planets also travel through the interplanetary void; a meteoroid differs from these chiefly in its small size, no more than a few meters in diameter, usually much less.

If a meteoroid survives its plunge through the atmosphere and strikes the earth's surface, the body is then called a *meteorite*. A special kind of meteorite is termed a *micrometeorite*. This is a tiny particle which melts in its fiery fall, but does not vaporize completely. A micrometeorite is very small, about 0.1 μm in diameter. Because of this small size, it slows down enough to cool off and solidify before it reaches the ground.

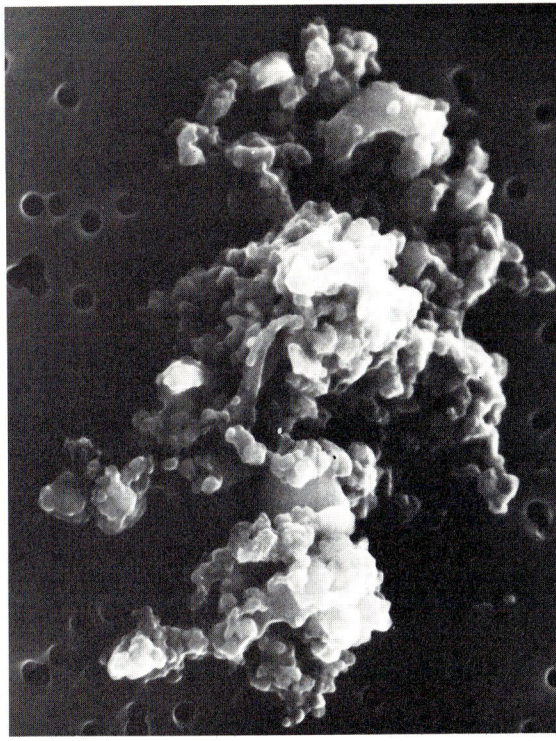

FIGURE 18.17 A meteoroid, a particle of interplanetary dust. It is about 10 microns in size. Note its fragile structure. (*Courtesy D. Brownlee, University of Washington.*)

Physical Characteristics of Meteoroids

From parallax observations of meteor trails made from two locations on the ground, meteoroids are found to disintegrate at an altitude of about 100 km. Most meteoroids quickly meet a fiery death, as their kinetic energy changes into heat in collisions with air molecules. The process resembles the heating of space vehicles during reentry, except that the meteoroid has no protection against the intense heating. Surface atoms boil off and collide with atmospheric atoms to produce ionization and excitation of the air; a glowing envelope forms around the moving meteoroid. As it interacts with the atmosphere, the meteoroid melts apart, layer by layer, and fragments into a scatter of pieces. The fragmentation enlarges the trail size to about 1 m in diameter, even through the meteoroid itself is much smaller.

Because the trail provides a brief record of the body's disintegration, astronomers have been able to determine the general physical characteristics of a meteoroid. Most meteoroids are fragile, delicate particles that crumble quickly in their contact with the air (Fig. 18.17). A block of meteoroid material a cubic meter in size would crumble under its own weight, for it is no stronger than cigarette ash. A meteoroid is typically a flimsy dust speck whose demise is its only remarkable asset.

Meteor Streams and Showers

On the night of November 12–13, 1833, thousands of meteors invaded the sky. People trembled with fear. Even scientists were bewildered by the sight of "stars falling as thick as snowflakes." An important discovery, however, arose from the confusion. Some astrono-

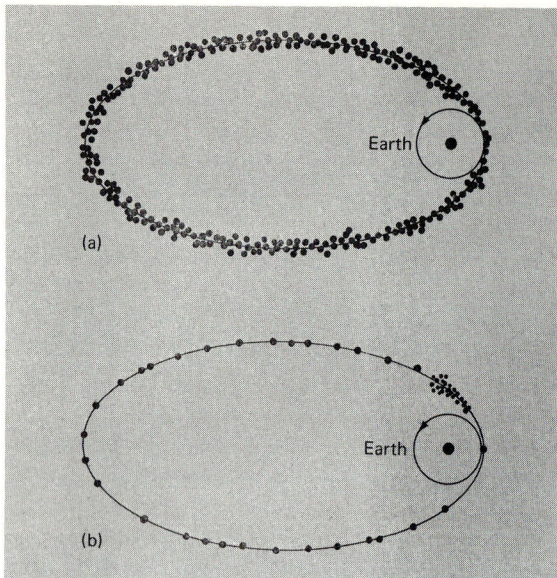

FIGURE 18.18 (a) A meteor stream. The pieces are distributed along the orbit. (b) A meteor swarm. The pieces are concentrated in one part of the orbit.

mers noticed that the meteors all seemed to be streaming from a point in the constellation Leo. This emanation point of the meteor trails is called the *radiant* and occurs because the meteoroids are hitting the earth's atmosphere while traveling parallel to each other. (Similarly, parallel railroad tracks or telephone wires seem to converge at a distant point.) A spectacular rain of meteors from a radiant is called a *meteor shower*. Showers are designated by the name of the constellation in which their radiant point seems to lie. The shower of 1833, when a radiant point was observed for the first time, apparently originated in the constellation Leo, and so is called the *Leonid shower*.

A Leonid shower was last seen in 1966. It recurs every 33 years, and we can therefore expect another spectacular display in 1999. The shower has a definite period because its meteoroids travel in an orbit around the sun. We see a shower only when the earth intersects the orbit of the meteoroids that make it. If the meteoroids are uniformly distributed along their orbits, they are called a *meteor stream* and the shower occurs every year (Fig. 18.18a). On the other hand, if the meteoroids are concentrated in one part of the orbit, the aggregation is called a *meteor swarm* and the shower may not be an annual event (Fig. 18.18b). For example, in one year the earth may intersect the center of a swarm, and a heavy shower will result. In the following year the earth may intersect the swarm's orbit after or before the swarm arrives at the intersection point; then no shower will occur in that year.

What is the source of the meteoroid material found in these orbits? In 1866 the French astronomer Wilhelm Tempel discovered a rather inconspicuous comet that now bears his name. (It is now called Tempel 1, for Tempel discovered a number of comets.) As Comet Tempel rounded the sun, astronomers calculated its orbital period as 33 years; they also noted that the orbit of Tempel's comet and the orbit of the Leonid meteor swarm were almost identical, which suggested that the comet might be the source of the meteoroid particles. Other meteor swarms are also now known to be associated with periodic comets (Table 18.4).

18.3 METEORS AND METEORITES

TABLE 18.4 Some Meteor Showers

Name*	Date for Maximum	Extreme Limits	Hourly Rate of Meteors	Associated Comet
Quadrantids	January 3	January 1–4	30	—
Perseids	August 12	July 29–August 17	40	1862 III
Aquarids	May 4	May 2–6	5	Halley
Orionids	October 22	October 18–26	13	Halley
Taurids	November 1	September 15–December 15	5	Encke
Leonids	November 17	November 14–20	6	Tempel 1
Geminids	December 14	December 7–15	55	—

*The name identifies the constellation in which the radiant point is found. The best time to observe most showers is in the early morning, although not all of the showers peak at those hours.

Whipple's dirty-iceberg comet model readily explains the affiliation between comets and meteor swarms. During a comet's successive passages by the sun, solar heating causes a continual loss of icy material from the cometary nucleus. The dust and solid particles interspersed in the ices flake off and scatter in an untidy array around the comet. This solid debris is very fragile and has a low density. The older the comet and the greater its number of passages by the sun, the greater the loss of icy material and release of meteoroid material. About 99 percent of meteors are of cometary origin. The remainder are probably associated with asteroids.

Not all meteors, however, originate in swarms or streams. The occasional meteors you can see on a clear night are termed *sporadic meteors* because we have no knowledge of their orbit or of the comets that may have given them birth.

No meteor, shower or sporadic, has been observed traveling at a velocity greater than 72 km/sec or less than 11 km/sec. Why these numbers? First, 11 km/sec is just the escape velocity of the earth; or, it is the velocity of a particle falling to the surface that was moving very slowly before its encounter with the earth. Second, 72 km/sec is just the sum of the earth's velocity about the sun, 30 km/sec, and the escape velocity from the solar system at the earth's distance from the sun, 42 km/sec. If a meteoroid were moving just at escape velocity and in the opposite direction to the earth's motion, its velocity relative to the earth would be 72 km/sec. So any meteoroid velocities greater than 72 km/sec would indicate that the meteoroid had been moving at greater than escape velocity before its encounter with the earth, and so must be an interstellar nomad caught by the sun's gravitational field. Since there are none observed with such high velocities, meteoroids cannot be nomads of interstellar space. They are all members of the solar system, orbiting the sun.

Meteorite Falls and Finds

Falling "stones from heaven" have been recorded since the seventh century B.C. But only since 1803 have astronomers generally accepted the idea that meteorites have an extraterrestrial origin. The French Academy investigated an extensive meteorite fall that occurred on April 26 of that year in L'Aigle, France. Careful detective work convinced the Academy that the strange stones did in fact fall from outer space, and this conclusion was announced to the world. Before this discovery, people commonly believed that the hot, dark stones were produced in violent thunderstorms.

If a meteorite is seen coming down from the sky and is recovered, it is called a *fall*. If a meteorite whose fall is not seen is picked up from the ground, it is termed a *find*. Once a year, on the average, a meteorite falls on continental North America and is recovered. Some four meteorites weighing at least 10 kg each hit the earth each day. There are many more of smaller mass. Every century, a meteorite weighing a few thousand tons plows into the earth. The largest known find struck near Grootfontein, Southwest Africa, and weighs about 55,000 kg (60 tons). The most massive stone meteorite fell in China in March 1976; it weighed in at 1770 kg.

A spectacular event that was once believed to be a meteoroid fall occurred in Siberia in 1908. It took place at 7:17 A.M. on June 30 in Tunguska, but was not investigated until 19 years later. Much evidence was obliterated in the interim. Observers at the time reported seeing a body moving downward for about 10 minutes. Then a bright, bluish-white flash leaped into the sky, followed by a column of black smoke. A tongue of flame was visible as far as 1000 km from the impact. The explosive shock knocked over the people and animals within a 60-km radius. The shock wave circled the world twice. Sunlight reflected from dust raised by the impact made the night sky so bright that newspapers could be read at midnight. When the site was investigated by Russian scientists, they found that a large forest and swamp area had been devastated by the blast. Strangely enough, no meteorite fragments were ever found, even when holes were drilled in the permafrost. The lack of solid remains has led some people to conclude that a comet's nucleus rather than a meteoroid was the object that rent the sky and smashed into the earth.

Meteorite craters are striking evidence of space visitors to the earth. Probably the most famous is the Barringer Crater, located a few kilometers from Winslow, Arizona. The crater is about 1200 m in diameter and 174 m deep, and its rim rises slightly above the local plateau (Fig. 18.19). The crater's origin was thought to be volcanic until it was

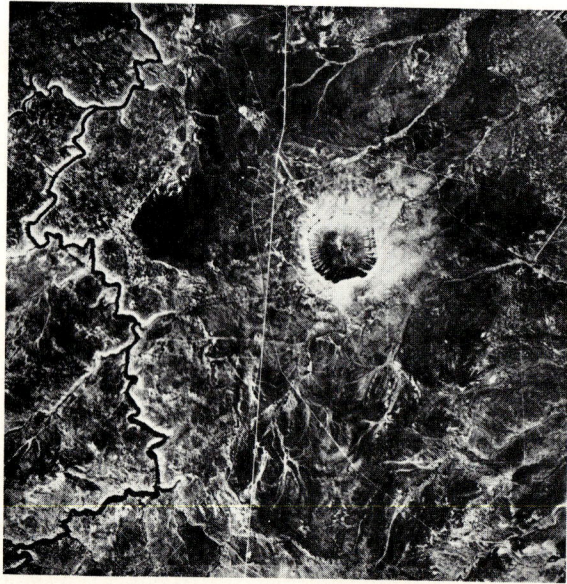

FIGURE 18.19 The Barringer Crater in Arizona. It is about 22,000 years old. (*Courtesy F. Whipple.*)

18.3 METEORS AND METEORITES

investigated in 1891. Many small meteorite fragments were uncovered, some of which contained small diamonds.

The Physical Nature of Meteorites

In terms of physical and chemical composition, astronomers divide meteorites into three broad classifications: irons, stones, and stony-irons. The *irons*, which are generally about 90 percent iron and 9 percent nickel with a trace of other elements, are the most common finds. They are easily identifiable because of their high density and melted appearance. The *stones* are composed of light silicate materials similar to the earth's crustal rocks. Though actually the most common kind of meteorite seen to fall, they are difficult to distinguish from an ordinary terrestrial stone and so make up a small fraction of the finds (Fig. 18.20). When examined under a microscope, many stones are seen to contain silicate spheres, called *chondrules*, embedded in a smooth matrix (Fig. 18.21). These stones are known as *chondrites*. *Stony-iron* meteorites represent a cross between the irons and the stones and commonly exhibit small stone pieces set in iron.

Irons are the most dense meteorites, with densities ranging from 7500 to 8000 kg/m^3. Stones are the least dense, averaging from 3000 to 3500 kg/m^3. Stony-iron meteorites, because they are a mixture of stones and irons, have an intermediate density of about 5500 to 6000 kg/m^3.

One of the most curious kinds of chondrites is the *carbonaceous chondrite*. The chondrules in these meteorites are embedded in material that contains a large amount of

FIGURE 18.20 A piece of a stony meteorite from Canyon Diablo, Arizona. (*Courtesy Institute of Meteoritics, University of New Mexico.*)

FIGURE 18.21 Chondrules—round silicate spheres found in chondrites. (*Courtesy F. Whipple and the Smithsonian Astrophysical Observatory.*)

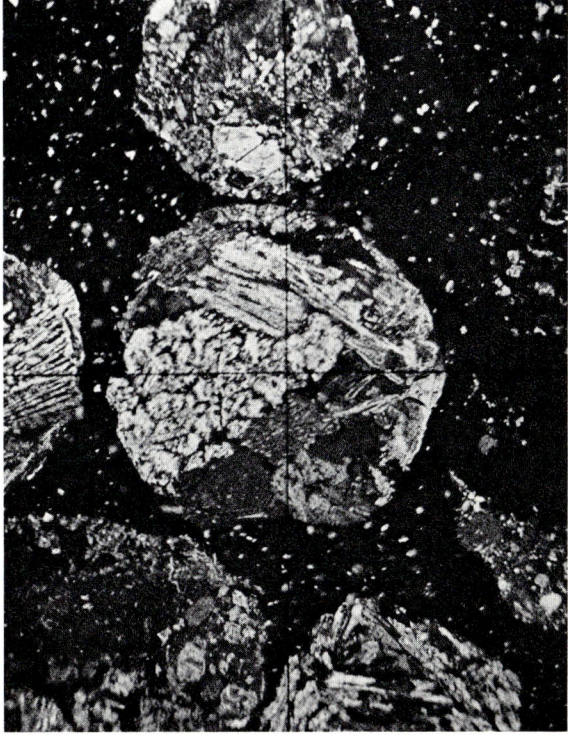

FIGURE 18.22 Widmanstätten figures in a nickel-iron meteorite from Glorietta, New Mexico. (*Courtesy Institute of Meteoritics, University of New Mexico.*)

carbon compared with that in other stony chondrites, typically from about 1 to 4 percent carbon by mass. Their carbon content gives these meteorites a dark appearance.

Carbonaceous chondrites also contain water (ranging from about 3 to 20 percent) and volatile materials. In addition, the relative abundances of condensable elements in carbonaceous chondrites more closely resemble those found in the sun's photosphere than those in the crust of the earth. That is, if some gas were extracted from the sun and cooled to below the freezing point of water, the condensed elements would be chemically quite different from the earth, but very similar to carbonaceous chondrites. This similarity suggests that carbonaceous chondrites formed out of the same primordial material as the sun and have suffered no major bulk heating or changes since that time.

Most meteorites have too great a density to come from comet-related meteoroids. They most resemble the inferred physical characteristics of asteroids. In cases where orbits of meteorites have been determined, they are like those of asteroids rather than of comets. Collisions between asteroids fragment them, and these pieces may be one source of meteorites.

What about the origin of iron meteorites? An important clue comes from etching with acid a polished surface of an iron meteorite. Large crystalline patterns, called *Widmanstätten figures*, become visible (Fig. 18.22). Terrestrial iron does not show such patterns when etched. Widmanstätten figures clearly distinguish meteoritic from terrestrial iron and also give a clue to the history of the meteorite material. A nickel-iron mixture, when cooled slowly under low pressures from a melting temperature of about 1600 K, forms large crystals. The cooling must be very gradual (about 1 K every million years). But metals conduct heat well, and in the cold of space a molten mass of nickel and iron would cool rapidly and not form large crystals. Nickel-iron meteorites could grow Widmanstätten figures only if they had protection from the cold. So it's likely that nickel-iron meteorite material solidified inside small bodies, termed *parent meteorite bodies*. To allow a cooling of only 1 K every million years these must have been at least 100 km in diameter.

Such bodies probably formed with the formation of the solar system. Parent meteorite bodies are envisioned to have been only a few hundred kilometers across. Once formed, they could be heated by the radioactive decay of short-lived isotopes, such as ^{26}Al. When heated to melting, a parent meteorite body differentiates; the densest material falls to the center and the least dense comes to the surface. So the object ends up with a core of metals and a cover of rocky material, which cools to form a crust. This insulates the molten metals and allows them to cool slowly and form large crystals. Much later, the parent meteorite bodies collide and fragment. Pieces from the outer crust make stony meteorites, pieces from farther down become stony-iron meteorites, and the core produces the iron meteorites.

Are any parent meteorite bodies around now? Possibly yes—as asteroids. Recall (Section 18.1) that there are three main types of asteroids, the dark C-type, containing much carbon, the lighter S-type, composed of silicate materials, and the intermediate M-type, with metallic characteristics. These are probably related to the carbonaceous chondrites, the stony meteorites, and the irons, respectively.

The view of meteorites outlined here implies that their parent bodies were among the first solid objects to form in the solar system. So the ages of meteorites should provide a direct indication of the age of the solar system. Radioactive dating techniques (Section 13.1) give ages very close to 4.6 billion years. Meteorites provide a direct, reliable estimate of when the solar system formed.

18.4 INTERPLANETARY GAS AND DUST

Interplanetary gas comes from many sources. Some may be left over from the solar system's formation, some has escaped from planetary atmospheres, and some has been released in the demise of comets. Most, however, comes from the sun. The solar wind, essentially an expanding extension of the solar corona, rushes past the earth at about 500 km/sec (Section 6.5). The ionized gas does not stop there; it sweeps onward beyond Pluto's orbit until it slows down and dissipates in interstellar space. The solar wind acts like a fan to sweep out most of the interplanetary gas.

However, the solar wind has only a minor effect on the heavier dust particles in the solar system. These particles, which lie approximately in the plane of the ecliptic, produce a phenomenon known as the *zodiacal light*, a faint cloud of light that extends in a roughly triangular shape above the horizon before sunrise and after sunset (Fig. 18.23). The zodiacal light shines so faintly that the lights of even a small town can completely obscure it.

The spectrum of the zodiacal light resembles that of reflected sunlight, a fact that indicates that dust is the source rather than gas. Satellite observations show that the zodiacal dust particles in the earth's neighborhood have a concentration of about 10^{-8} particles/m^3 and that they are composed mostly of silicates, iron, and nickel. When this cosmic dust falls to the earth, it is found as micrometeorites. It adds perhaps a few million tons to the earth's mass each year. The total mass of the zodiacal cloud is about 10^{16} kg.

FIGURE 18.23 The zodiacal light at Mt. Haleakala, Hawaii. This is a short time exposure; note that the star images are trailed. (*Courtesy A. Peterson and L. Kieffaber, University of New Mexico.*)

Along with falling into the planets and the sun, dust is eliminated by solar radiation pushing the smallest dust particles out of the solar system. These processes would quickly clean out the present interplanetary dust, so there must be some source of new dust to replenish the supply. The dirty-iceberg comet model sees comets as dispensing gas and dust between the planets. Asteroids may also be ground down to smaller pieces in random collisions, especially in the crowded asteroid belt. The fact that meteorites come from the interiors of small planetary bodies supports this idea. Also, the present flow of dust can sandblast small bits off asteroids and meteoroids.

SUMMARY

Different forms of matter fill interplanetary space. In orbits around the sun, we find asteroids, comets, meteoroids, and dust (Table 18.5). Comets, formed of low-density materials, make most of the meteoroids. This connection has been established by the association of meteor swarms and streams with cometary orbits and by the fluffy nature of meteoroids. In contrast, most meteorites have a dense composition of rocky and metallic compounds. These objects are probably stray fragments left over from the formation of the solar system. Collisions between asteroids, which resemble meteorites in density and composition (but are much larger in size), may produce chips and chunks that add to the interplanetary complement of meteorite material.

TABLE 18.5 Interplanetary Matter

Material	*Location*	*Composition*	*Origin*
Asteroids	Generally found between Mars and Jupiter in the ecliptic plane	Rocky material (silicates), iron, nickel	Uncollected debris from formation of the solar system or fragmented planet
Comets	Short-period: elliptical orbits in ecliptic plane with aphelions near Jupiter and Saturn. Long-period: orbits not confined to ecliptic plane, in Oort's cloud	Ices of water, methane ammonia, carbon dioxide, and rocky material (silicates)	Perhaps uncoalesced building blocks of Jovian planets and fragments from collisions between them
Meteors and meteoroids	Swarms or streams in elliptical orbits generally in ecliptic plane	Flaky silicate materials, possibly some ices	Remains of exhausted comets (if dirty-iceberg model is correct)
Meteorites	On earth's surface	Silicates, iron, nickel	Parent meteorite bodies, asteroids
Interplanetary gas and dust	Throughout the solar system, mostly in ecliptic plane	Gas: mostly hydrogen and helium. Dust: silicates, graphite	Solar wind, ashes of comets, fragmentation of asteroids, and meteoroids

Comets eventually disintegrate from heating and loss of volatiles after many passes of the sun. Some of their material remains in orbit as meteoroids. The gas and dust particles from comets add to the interplanetary medium. We see some of the interplanetary dust as the zodiacal light. Although comets lose material and gradually disappear, we still see many comets. So the supply of comets must be replenished. There may be a vast cloud of comets, far from the sun, containing the needed supply. Jogged by the gravitational force of a passing star, occasionally a few of these comets are forced into orbits that pass close to the sun. On arrival at the innards of the solar system, they may be trapped in shorter-period orbits by the gravitational force of Jupiter.

The interplanetary debris must be remnants of the original material that formed the solar system. Much of the original material became the sun and the planets. Some did not collect into these large bodies, but made the smaller meteoroids, asteroids, and comets. These objects hold important information about the original solar system material.

Key Words

asteroid	gas tail	sporadic meteor
Ceres	dust tail	meteorite fall
asteroid belt	dirty-iceberg cometary model	meteorite find
S-type asteroid	Oort's comet cloud	iron meteorite
C-type asteroid	meteor	stone meteorite
M-type asteroid	meteoroid	stony-iron meteorite
Chiron	meteorite	chondrule
comet	micrometeorite	chondrite
Halley's comet	radiant	carbonaceous chondrite
periodic comet	meteor shower	Widmanstätten figures
nucleus	meteor stream	parent meteorite body
coma	meteor swarm	zodiacal light

Review Questions

1. Suppose you hopped in your spaceship on Saturday night and flew to an asteroid. What would you see? (Objectives 1 and 3)
2. Then you speed off to a comet. What would you see? How would its appearance differ from that of the asteroid? (Objective 6)
3. After many perihelion passages of the sun, what happens to a comet in the dirty-iceberg model? (Objectives 5 and 10)
4. The Perseid meteor shower appears regularly at the middle of every August. What does this indicate about the particles which produce it? (Objective 10)
5. How can you tell an iron meteorite from a piece of terrestrial iron and nickel? (Objectives 11 and 12)
6. What direct observational evidence do we have for the existence of interplanetary dust? (Objective 13)

Problems

1. From the data given in Table 18.2, calculate the semi-major axis and aphelion distance of the orbit of Halley's Comet. (Objective 6)
2. The earth's atmosphere affords a certain amount of protection from meteoroids. A

meteoroid is slowed down significantly by the atmosphere only if it encounters more than its own mass of air. Over a square meter of the earth's surface there are about 10 metric tons of air. Making the approximation that meteoroids are spherical, what is the smallest meteoroid that will keep most of its incoming space velocity? Assume a density for meteoroids of 3200 kg/m^3. (Objective 9)

3. Radiation pressure (force per unit area) is given by the flux of radiation divided by the speed of light ($P = F/c$). The force of radiation on a particle therefore depends on the luminosity of the sun, the inverse square of the distance from the sun, and the area of the particle (proportional to r^2). The gravitational force on a particle depends on the inverse square of the distance from the sun and on the mass of the particle, which is proportional to r^3. Since the dependence of these two forces on particle size is different, there will be some size for which they just balance. Find this size for a particle of density 3000 kg/m^3. (Objective 15)

4. If the infrared flux from an asteroid is half the visible flux, what is the asteroid's albedo? (Objective 3)

5. Assume that Herculina does have a moon, which orbits 975 km from Herculina and has a diameter of 50 km.
    a. Calculate the moon's orbital period.
    b. Compare the tidal force of the sun pulling the system apart to the gravitational force of the two bodies pulling them together. (Objective 4)

6. In order to form the iron and stony meteorites, the meteorite parent bodies must get hot enough to melt and differentiate. One source of heat is radioactive decay. Below is a table of radioactive elements which might be present in the parent meteorite bodies. The quantity ϵ is the energy produced per second from each kilogram of meteorite material.

Isotope	$t_{1/2}$ (10^9 yr)	ϵ (J/kg/sec)
^{238}U	4.51	1.1×10^{-12}
^{235}U	0.71	1.9×10^{-12}
^{232}Th	1.39	6.5×10^{-13}
^{40}K	1.25	1.7×10^{-11}
^{26}Al	0.00072	6.0×10^{-6}

The effective heating time can be taken to be the half-life. The heat capacity of meteoritic material, the energy needed to heat it by 1 degree Kelvin, is about 10^3 J/kg/K. The body must be heated to about 1500 K to melt. Consider a parent meteorite body of 200 km diameter and density 2500 kg/m^3. Which of the radioactive elements, if any, can melt such a body (assuming the initial temperature is zero and no heat is lost from the surface)? (Objective 13)

BEYOND THIS BOOK . . .

* In *The New Solar System* (Sky, Cambridge, Mass., 1981), read "Comets" by J. C. Brandt, "Meteorites" by J. Wood, and "Small Bodies and Their Origins" by W. Hartmann.
* Part III of *The New Astronomy and Space Science Reader* (Freeman, San Francisco, 1977) deals with asteroids, meteors, and comets.
* A technical, but excellent book is *Comets, Asteroids, Meteorites: Interrelations, Evolution, and Origins* (University of Toledo Press, 1977), edited by A. H. Delsemme.

THE ORIGIN AND EVOLUTION OF THE SOLAR SYSTEM

19

LEARNING OBJECTIVES
After studying this chapter, you should be able to:

1. Identify at least two dynamical and two chemical properties of the solar system that any model of origin must try to explain.
2. Specify what clues asteroids, comets, and meteorites provide about the formation of the solar system.
3. In one sentence for each describe catastrophic and nebular models of formation, and use one specific example to illustrate each.
4. Describe the process of nebular collapse, and explain what is meant by the "angular momentum problem" for nebular models.
5. Describe one possible way out of the angular momentum problem in modern nebular models.
6. Evaluate how well catastrophic models explain present observations and point out why such models are not accepted now.
7. Describe briefly the contemporary condensation sequence, using one Jovian and one terrestrial planet to illustrate it.
8. Describe the role of accretion in the formation of the planets.
9. Outline a possible model for the formation of Jupiter and Saturn.
10. Sketch a modern scenario for the formation of the solar system, and evaluate how well it explains the known chemical and dynamical properties you listed for Objective 1.

CENTRAL QUESTION:
What physical processes resulted in the formation of the planets from a smooth cloud of gas and dust?

Most cultures have a story about the origin of the world. These myths tell of the world's development from some formless state—from chaos to cosmos. Having completed a grand tour of the solar system, you too are probably wondering: How did the solar system originate?

Because it is still unresolved, the puzzle of the origin of the solar system still arouses astronomers' curiosity. Many models have been proposed. None has been completely successful. One scientific justification of the space program rested on finding possible information to support or refute theoretical ideas.

To unravel the puzzle of the solar system's genesis requires more than astronomy. It demands the interplay of astronomy with physics, chemistry, and geology. The question refuses to be answered simply. This chapter investigates in some detail the contemporary tentative answer. In this picture the formation of the solar system arises as a natural consequence of the formation of the sun from an interstellar cloud of gas and dust. The general outline of this process reasonably explains the major features of the solar system. Many details and puzzles remain to be resolved; no one model yet fits them all together.

19.1 PIECES AND PUZZLES

The dynamical and chemical properties of the solar system impose crucial limitations on any theory of its formation. These features serve as broad templates for shaping more specific questions.

Chemically, the solar system falls into three broad categories of material: solar, terrestrial, and icy (Table 19.1). Each group is distinguished primarily by its melting point. The solar and icy materials together are sometimes called *volatiles*. These are generally gaseous under the conditions expected during the solar system's formation. The bodies of the solar system, including the comets and asteroids, are composed of various combinations of the three groups (Table 19.2). Note the kinship of Uranus and Neptune with the comets, of Jupiter and Saturn with the sun, and of the terrestrial planets with asteroids.

19.1 PIECES AND PUZZLES

TABLE 19.1 General Classes of Solar System Materials

Class	Elements	Melting Point
Terrestrial	Silicon, magnesium, iron oxide, etc., plus oxygen in oxides	1500–2000 K
Icy	Carbon, nitrogen, oxygen, and hydrogen in combination (methane, ammonia, water, etc.)	273 K
Gaseous	Hydrogen, helium, neon, argon, etc.	14 K

Though one class of materials may be predominant, the bodies of the solar system contain some of each group. For example, the sun contains terrestrial materials, but as gasses (because the sun is hotter than 2000 K), not solids. In fact, the sun contains, in terms of total mass, more terrestrial materials than all the terrestrial planets put together. Likewise, the sun also contains the icy group as gasses.

An object's density is a major clue to its composition. As a general rule, solid objects with densities of about 1000 kg/m^3 are made of icy materials, and those with densities of about 7000 kg/m^3 are made of metals, mostly iron. Bodies with densities between these are mixtures. For example, a density of about 2000 kg/m^3 implies a combination of icy and rocky materials, and that of about 5000 kg/m^3, a mix of rocky and metallic stuff.

The solar system also displays a regular structure in terms of its dynamical properties—those that relate to its motions. Viewed from above the sun's north pole, the solar system has the following regularities:

1. The planets revolve counterclockwise around the sun; the sun rotates in the same direction.
2. The major planets, except Mercury and Pluto, have orbital planes that are only slightly inclined with the plane of the ecliptic; that is, the orbits are *coplanar*.

TABLE 19.2 General Composition of the Major Bodies of the Solar System

Bodies	Composition (%)		
	Terrestrial	Icy	Gaseous
Sun	0.3	1.2	98.5
Terrestrial planets	70	30	0
Asteroids	70	30	0
Jupiter	13	5	82
Saturn	21	12	67
Uranus	25	60	15
Neptune	20	70	10
Comets	15	85	0

Source: Adapted from Fred L. Whipple, *Earth, Moon, and Planets* (Harvard University Press, Cambridge, Mass., 1968); John A. Wood, *The Solar System* (Prentice-Hall, Englewood Cliffs, N.J., 1979).

3. The planets, except Mercury and Pluto, move in orbits that are very nearly circular.
4. The planets, except Venus and Uranus, rotate counterclockwise, in the same direction as their orbital motion.
5. The planets' orbital distances from the sun follow a regular spacing rule, the Titius-Bode "law"; roughly, each planet lies twice as far out as the previous one (Focus 18.1).
6. Most satellites revolve in the same direction as their parent planet rotates and lie close to their planet's equatorial plane.
7. Some satellites' orbital distances follow a regular spacing rule.
8. The planets together contain more angular momentum than the sun (Table 19.3).
9. Long-period comets have orbits that come in from all directions and angles, in contrast to the orbits of the coplanar planets, satellites, asteroids, and short-period comets.
10. At least three of the Jovian planets have rings; the terrestrial planets do not.

The sun contains most of the mass of the solar system (Table 19.3). The rest lies close to the plane of the solar system. In terms of the layout of mass, the solar system is really very thin. If it were the diameter of an average pancake, it would be only the thickness of that average pancake, about 1 cm.

Although the sun holds 99 percent of the system's mass, it contains less than 1 percent of the angular momentum. The outer planets have the most, 99 percent of the total (Table 19.3). If all the planets with their present angular momenta were dumped into the sun, it would spin once every few hours rather than once a month.

A successful model must explain as many of these dynamical and chemical properties as possible. It must deal not only with the dynamical and chemical regularities of the system, but also with the interplanetary debris: comets, asteroids, and meteoroids (Chapter 18). These bodies are important relics of the solar system's early history. But a good model is not simply the one that accounts for the greatest number of these characteristics. It must explain them in some internally consistent, simple fashion (Sections 2.1 and 7.1). It must be "aesthetically pleasing."

Most theories of the system's genesis take one of two basic forms: (1) *nebular* models, in which the sun condenses from an interstellar cloud of gas and dust that also forms a disk, a solar *nebula*, out of which the planets condense; or (2) *catastrophic* models, in which

TABLE 19.3 Distribution of Mass and Angular Momentum in the Solar System

Object	Mass (% of total)	Angular Momentum (% of total)
Sun	99.86	0.5
Jovian planets	0.132	99.0
Terrestrial planets	0.003	0.2
Asteroids	0.00003 (?)	0.1 (?)
Comets	0.0000003 (?)	0.3 (?)

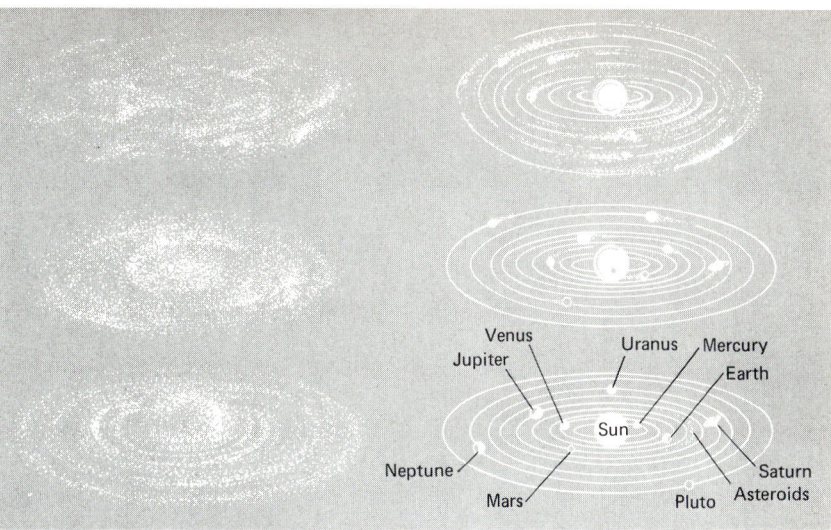

FIGURE 19.1 Marquis de Laplace. (*Courtesy Yerkes Observatory*.)

FIGURE 19.2 Laplace's nebular model for the formation of the solar system. A cloud of gas and dust contracts by its own gravity. It breaks into concentric rings, whose material forms the planets.

the occurrence of a catastrophic event involving the sun leads to the injection of gaseous material (which condenses to the planets) into solar orbit. Nebular and catastrophic models have battled for about 200 years, with fluctuating success, for the dominant position among astronomers. (Nebular models are now on top, and we think likely to remain so.)

19.2 EARLY IDEAS OF THE ORIGIN OF THE SOLAR SYSTEM

The nebular models interpret the existence of the solar system as a natural consequence of the sun's formation. So planetary formation should, perhaps, accompany any star's formation. If nebular models are correct, planetary systems may be very common in our Galaxy and in other galaxies. The opposing catastrophic theories require a rare, one-of-a-kind event. If they are right, planetary systems among the stars are rare. Here's a brief look at the origins of the two main views of the solar system's formation.

Nebular Models

Immanuel Kant (1724–1804) first developed a detailed picture of the dynamical evolution of the solar system consistent with Newton's law of gravitation. Arguing by analogy with Saturn's rings, he depicted the solar system as forming from a disk-shaped nebula containing a large percentage of dust.

In 1796, the French mathematician Pierre Simon, Marquis de Laplace (1749–1827), presented in *The System of the World* an elaborate Newtonian treatment of the history of the solar system. Laplace (Fig. 19.1) imagined that the sun in its primitive state condensed from a cloud of gas and dust and would have been surrounded by rings of material. Collisions and mutual gravitational attraction of the particles in these rings would have lumped the material into planets (Fig. 19.2). Then the planetary bodies would have contracted and shed material to form the satellites. Like Kant, Laplace pointed to Saturn as a concrete example of the process. (Today, he could also point to the rings of Jupiter, Uranus, and Neptune.)

The essential feature of the Kant-Laplace nebular models—that the sun and then the planets form from a nebula—is the foundation of the modern approach. The sun's formation takes place in the center of a flattened nebula. The planets grow from this nebula.

The growth of the planets occurs by three main methods: (1) gravitational collapse, (2) accretion, and (3) condensation. *Gravitational collapse* works only if regions in the nebula have enough mass so that they contract by their own gravity to form a planet. (This was Kant's view.) *Accretion* occurs when small particles collide and stick together to form larger masses that eventually grow into planets. (Laplace thought that accretion was important in forming planets.) *Condensation* involves the growth of small particles by the sticking together of atoms and molecules, just as water molecules combine in clouds to form raindrops.

The most compelling objection to the Kant-Laplace picture comes from the present *distribution of angular momentum* (Table 19.3). Imagine that the nebula out of which the sun and planets form is rotating slowly. (We expect it to be rotating a little, because the Galaxy itself is rotating.) As the nebula collapses, it spins faster (the conservation of angular momentum). Since the sun forms from the central part of the nebula, it should be spinning very rapidly, at least once every few hours. Actually, the sun spins 400 times more slowly than this rate. But, the sun would spin once every few hours *if* the present angular momentum in the planets were added to it. The angular momentum is there, but not in the right place! To adopt the Kant-Laplace nebular model requires a process to account for the peculiar angular momentum distribution, a mechanism not developed until this century.

Note that the conservation of angular momentum reinforces the most compelling feature of a nebular model: the sun and planets form out of a rotating cloud, so *naturally* the sun rotates and the planets revolve in the same direction. Also, the cloud finishes its collapse as a disk, and the planets' orbits do align in a thin disk.

Catastrophic Models

Early in this century the American scientists Thomas C. Chamberlain (1843–1928) and Forest R. Moulton (1872–1952) proposed a version of a catastrophic model that avoided the angular momentum problems of the nebular models. They envisioned a star skirting the sun and raising giant tides in the solar atmosphere. The tongues of material from the sun and star eventually condensed to form *planetesimals* (small bodies a few hundred meters across), which accreted to planetary-sized bodies (Fig. 19.3). Excess material either fell into the sun or left the solar system entirely. Because the planets obtained all their angular momentum from the passing star, they could easily pick up more angular momentum than the sun has.

All variations of catastrophic models run into four basic difficulties. First, the probability of an encounter or collision between the sun and another body is extremely low. The separation between stars and the small relative velocities of stars near the sun make even a distant encounter unlikely. The frequency of a sun-star encounter is about once in 10^{20} years, a period much longer than the age of the universe, which is estimated to be roughly 10^{10} years. A catastrophic model implies that the solar system must be unique, since it formed in a freak event.

19.2 EARLY IDEAS OF THE ORIGIN OF THE SOLAR SYSTEM

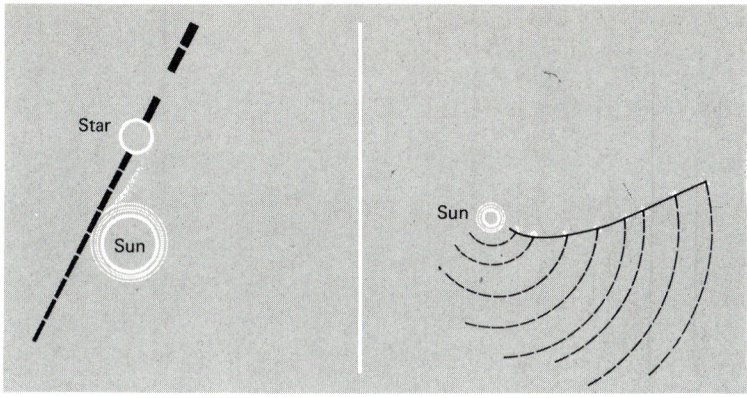

FIGURE 19.3 The Chamberlain-Moulton encounter model. A star passes close to the sun, and its tidal forces draw out a plume of solar material. A similar piece rips off the star. The material cools and forms the planets.

Second, the expansion of hot material from the sun must be rapid, so it would be difficult to form planets. Lyman Spitzer calculated in 1938 that a filament of solar material massive enough to form planets must come from deep in the sun, where it is quite hot (about 1 million K). Such hot gasses suddenly expelled into space would expand and dissipate rather than condense.

Third, the total amount of matter ejected from the sun is insufficient to form the planets. More than 99 percent of such material would fall back to the sun, even if the encounter were close enough to raise a sufficient mass. To create the planetary masses requires a filament mass about equal to the present mass of the sun.

Fourth, the detailed composition of the planets differs greatly from that of the sun in key respects. For example, the deuterium-hydrogen ratio on the earth is about 10^{-4}. In the sun the proton-proton chain (Section 6.3) quickly consumes deuterium at low temperatures, so the deuterium-hydrogen ratio is small (about 10^{-7}). If the earth condensed from solar material, after the sun formed, it should have the same relative abundance of deuterium to hydrogen. It does not. Similarly, observations have found the deuterium-hydrogen ratio in the atmosphere of Jupiter to be about 10^{-5}. Saturn and Uranus have about the same deuterium-hydrogen ratio as Jupiter. This value is also much larger than expected if Jupiter were made of material torn from the sun. So even though Jupiter has a composition like that of the sun, it cannot have been formed directly out of solar material. The deuterium-hydrogen ratio seen in the atmospheres of the Jovian planets is about the same as that in the interstellar medium, which supports the nebular model and suggests that it is the sun which has changed since its formation.

To summarize, catastrophic models were designed primarily to account for the angular momentum distribution in the solar system. They suffer from the rarity of close encounters between stars and from the difficulty of forming planets from hot gasses in space. Nebular models naturally account for most of the dynamical properties of the solar system, but fail to predict correctly the angular momentum distribution.

The Chamberlain-Moulton catastrophic model has been revived in modern form by M. Woolfson. He proposes a modification of the tidal encounter as follows. First, the encounter takes place in star clusters and associations and so the encounter probability is higher, perhaps as much as 1 chance in 100 in a cluster over its lifetime. Second, the encounter takes place between the sun and a protostar (Fig. 19.4), not a main-sequence star. A filament is more easily drawn off by tidal forces from a protostar than from a

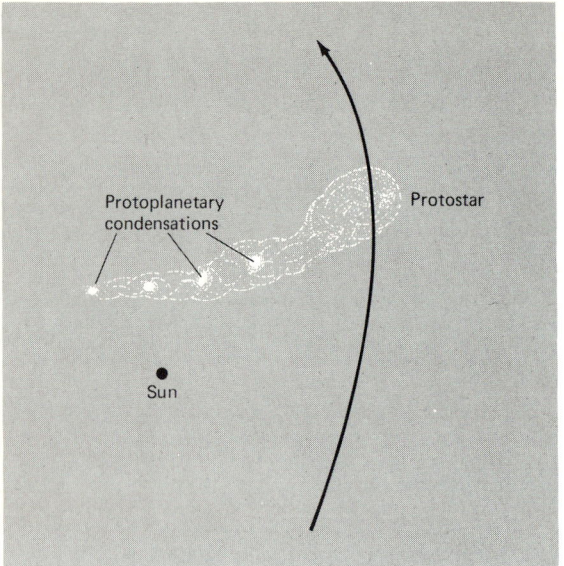 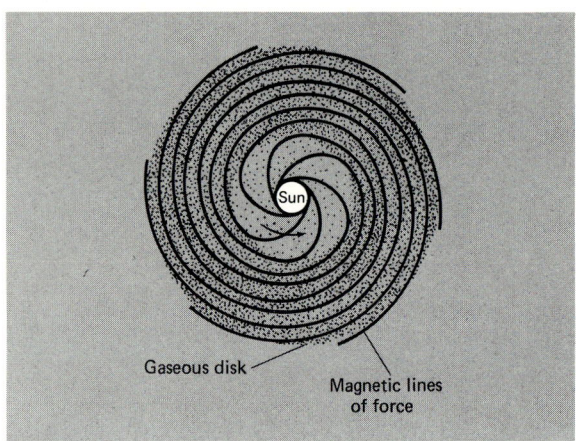

FIGURE 19.4 A modern catastrophic model. A protostar passes near the sun. Tidal forces rip material from it; this condenses into the protoplanets. (*Adapted from a diagram by M. Woolfson.*)

FIGURE 19.5 Top view of a possible primeval solar magnetic field. One end of the field lines rotates with the sun; the other end is dragged by the ionized material in the disk.

main-sequence star because the protostar is much less dense than a normal star and not as well held together. Also, since it is cooler, the material in the filaments will dissipate less rapidly. Third, the sun captures some material from the filament; these blobs form the planets. Their orbits must be highly elliptical. Some planet-planet collisions might occur before drag by the surrounding medium makes the orbits more circular. Such collisions could break up larger planets into terrestrial-sized ones.

There is at least one drawback to this encounter-capture model. Unless the sun-protostar encounter occurs very early in the protostar's lifetime, before its pre-main-sequence phase of evolution, it, too, will have consumed substantial deuterium. Then the same objection about the deuterium-hydrogen ratio applies.

19.3 AN OVERVIEW OF NEBULAR MODELS

A successful nebular model must account for four important stages in the solar system's evolution: (1) the formation of the nebula out of which the planets and sun originate, (2) the formation of the original planetary bodies, (3) the subsequent evolution of the planets, and (4) the dissipation of leftover gas and dust. Modern nebular models (there are more than one) give tentative explanations for all these stages, but many details are incomplete.

We have pointed out that the main objection to a nebular model is the present distribution of angular momentum in the solar system. Contemporary models cope with this problem by invoking the interaction of magnetic fields and charged particles to rearrange the distribution of angular momentum. The basic solution requires that the spin of the central part of the nebula be decreased and transferred to the outer regions.

Charged particles and magnetic fields interact so that the particles spiral along the magnetic lines of force (Focus 6.2). Such interactions can transfer spin from the young sun (in the center of the nebula) to the outer parts of the nebula. As the sun forms, it heats up the interior regions of the nebula. Here the gas is ionized, so charged particles are abundant. The magnetic field lines trap these particles. As the sun rotates, it carries its magnetic field lines with it; these drag along the charged particles that in turn collide with and drag along the rest of the gas and dust. So the magnetic field accelerates the material in the nebula near the sun. At the same time, the mass of the nebula resists the rotation. This drag on the magnetic field lines stretches them into a spiral shape (Fig. 19.5). The magnetic field links the material in the nebula to the sun's rotation. So the nebular material gains rotation (and angular momentum) at the sun's expense.

Whatever process transferred the angular momentum, the transfer must have taken place before large solid objects formed in the nebula. The transfer mechanisms described above work effectively only on gasses. With this in mind, here is a skeletal sketch of a modern nebular model. The main problem is how to get a lumpy solar system out of a smooth, diffuse cloud, that is, how to form the planets.

Prior to the formation of the planets, the solar nebula must condense from an interstellar cloud. We dealt with that collapse in Section 13.2, but let's review it briefly. From a large interstellar cloud a small blob broke off and contracted gravitationally into a flattened nebula with dimensions comparable to those of the present solar system. It heated up as well. Dust grains accreted into objects a few kilometers in size, called *planetesimals*. The swarms of planetesimals were swept up in zones to make larger bodies, called *protoplanets*, which would evolve into the planets we see today. The sun formed in the center of the nebula. These events took place quickly (in a hundred million years or so) about 4.6 billion years ago. The young sun blew off material that swept away large amounts of leftover gas and dust. This sweeping out requires that the nebula must have contained more mass than in the planets today—at least 0.1 solar mass and maybe as much as 1 solar mass.

Do we have any observational evidence for the formation of dusty nebulas around stars? Yes. Infrared observations of young stars (Section 9.5) indicate that dust surrounds them. From the conservation of angular momentum, we expect the dust will form a disk. Recent infrared polarization observations of suspected massive protostars confirm that the dust grains orbit the youngest of these in disks rather than in spherical shells.

The rest of this chapter focuses on the details of the processes that formed the planets. But, once again, be forewarned that there is a lot of confusion because no one of the competing models is completely satisfactory. To quote William K. Hartmann, a worker in the field: " . . . there is not yet a unified, well-accepted modern theory of planet formation. . . ."

19.4 FORMATION OF THE PLANETS

Two essential ingredients for contemporary models are: (1) the primeval nebula was turbulent, not quiescent, and violent motions helped the first steps in planetary formation;

(2) the formation of the planets was a multistep process proceeding through the formation of planetesimals and protoplanets. Much research work in the 1970s has focused on the complex problem of how to get planet-sized objects out of dust grains. That's no easy task!

One aid to the accretion of grains is turbulence in the nebula, because that makes grains collide more frequently. The solar nebula may have contained turbulent vortices, a characteristic of rapidly moving fluids or gasses. For example, you have probably seen that the flow of smoke out of a smokestack breaks up into numerous swirls. At the boundaries between swirls, particles would collide, accrete, and form the planets. Some recent work, especially that of A. G. W. Cameron and M. R. Pine, finds that the turbulence does not seem to last long enough to help the planets grow all the way.

A protoplanet model needs at least three stages for planetary growth. First, turbulence helps the growth of planetesimals by accretion of small particles. Second, the protoplanets then grow by direct collision and accretion of planetesimals. Third, when they gain enough mass, the protoplanets gravitationally pull in passing material. In a few million years to hundreds of millions of years a protoplanet's formation is completed.

The Chemical Condensation Sequence

The original protoplanet picture left a crucial detail unanswered: How did the protoplanets acquire differences in chemical composition? Recent research has developed the concept of a *condensation sequence*, an idea first suggested by chemist Harold Urey and worked out in detail by John Lewis and others. Urey noted that the nebula's center must have been hot, a few thousand degrees Kelvin. Here solid grains, even iron compounds and silicates, could not condense. Elsewhere, what materials would condense depended on the temperature. Just below 2000 K grains made of terrestrial materials would condense (Table 19.1); below 273 K grains could form of *both* terrestrial and icy materials.

Note that the temperatures reached in the nebula also affect interstellar grains in it. Icy interstellar grains vaporize wherever (and whenever) the temperature rises above 273 K; terrestrial grains (iron and silicates) evaporate at temperatures of roughly 2000 K. Wherever in the nebula temperatures never hit these points of vaporization, the appropriate interstellar grains remain.

Recent research has reached even more specific conclusions about the sequence in which compounds could condense from a heated nebula (Table 19.4). At different temperatures the gasses available and the solids present react chemically to produce a variety of compounds. Here's the key result of the condensation sequence: The densities and compositions of the planets can be well explained with the condensation sequence *if* the temperature in the nebula drops *rapidly* from the center outward. Then, at different distances from the sun different temperatures allowed different chemical compounds to condense and form grains that eventually made up the protoplanets. If a material could not condense because the temperature was too high, it would not end up in the protoplanet. For instance, the terrestrial planets lack the icy and gaseous materials common in the Jovian planets because at their close distances to the sun, the temperatures were too high for the condensation of icy and gaseous materials. In contrast, the Jovian planets pretty much reflect the original composition of the nebula (and cosmic abundances in general).

Because of its key role in the modern nebular model, let's examine the condensation sequence idea in detail. Start out with a temperature of about 2000 K in the solar nebula.

19.4 FORMATION OF THE PLANETS

(This is about the lowest temperature at which all the materials will be vaporized.) The first compounds to condense are of calcium oxide (CaO), aluminum oxide (AlO_3), and rare-earth oxides (Table 19.4). Next, starting at about 1500 K, an iron-nickel material similar to that found in iron meteorites would condense. In the next step, silicates, such as found in stony meteorites and the rock on the earth's surface, would form. In particular, various kinds of feldspars, a mineral commonly found in rocks, would be made. Where the temperature dropped to 680 K, hydrogen sulfide gas (H_2S) would act on iron to form the mineral triolite (FeS). Any leftover iron would combine with oxygen and silicon to make minerals such as olivine (Fe_2SiO_3), a dark mineral common in the rocks that make up the lunar maria. Below 170 K argon gas freezes. At 10 K neon and helium condense. It is unlikely that temperatures in the solar nebula were any lower than this.

The composition of carbonaceous chondrites supports this model. The chondrites are probably samples of the primordial solid stuff in the solar nebula. The mineral triolite (FeS) is found in almost all chondrites. The mineral olivine is also common. In addition, some chondrites contain large amounts of volatiles, especially water. These characteristics of chondrites are just those expected from the condensation sequence model.

TABLE 19.4 Equilibrium Condensation Sequence in the Solar Nebula

Temperature (K)	Reactions
1600	Condensation of metals combined with oxygen (refractory oxides), such as aluminum oxide (Al_2O_3) and uranium oxide (UO_3)
1300	Condensation of iron-nickel alloy
1200	Condensation of the silicate enstatite ($MgSiO_3$)
1000	Reaction of sodium with silicates and aluminum oxide to form common minerals such as feldspar; condensation of alkali metals
450–1200	Combination of iron and oxygen to form iron oxide (FeO) which reacts with enstatite to become olivine (Fe_2SiO_4 and Mg_2SiO_4)
680	Reaction of metallic iron with hydrogen sulfide gas (H_2S) to form the mineral triolite (FeS)
425–550	Combination of water vapor with olivine to make the mineral serpentine ($Mg_6Si_4O_{10} \cdot 8OH$)
175	Condensation of water ice
150	Reaction of ammonia gas (NH_3) with water ice to make solid hydrate ($NH_3 \cdot H_2O$)
120	Reaction of methane gas (CH_4) with water ice to make solid hydrate ($CH_4 \cdot 7H_2O$)
20–65	Condensation of inert elements, such as argon and neon, and any leftover methane
1	Condensation of helium into liquid

Source: J. Lewis, "The Chemistry of the Solar System," *Scientific American*, March 1974, p. 51.

Nebular Temperatures and Condensation

Some recent calculations by A. G. W. Cameron show how temperatures might differ in the solar nebula at different places and times (Fig. 19.6). His models indicate that the nebula starts out cool (less than 100 K), increases in temperature until a maximum is hit in about 50,000 years, then declines. The key point is how the maximum temperatures decrease with increasing distance from the sun: for the region of the formation of Mercury the temperature reaches about 1500 K, for Mars about 500 K, and for Uranus about 70 K. This dropoff with time and distance from the sun is necessary for the condensation sequence to work (Fig. 19.7). Note that this evolution of the nebula's temperature implies that the interstellar grains of terrestrial materials survive except within the orbit of Mercury. Interstellar grains of the ices survive outside the orbit of Mars.

The condensation sequence also predicts chemical differences among the terrestrial and Jovian planets. Let's take Mercury, the earth, and Mars for comparison. At Mercury's distance the temperature would be low enough for nickel-iron and silicates of magnesium to condense along with refractory oxides (including uranium, potassium, and thorium). The condensates of the radioactive elements provide one source of heating; accretion provides another. These heating sources drive the differentiation of the protoplanet. The resulting protoplanet would have a large nickel-iron core and a mantle of such silicates; we believe that Mercury does have such an interior structure (Section 15.1).

At the earth's distance the temperature is considerably lower, about 600 K. Here, silicates of both iron and magnesium could condense, along with iron oxide and iron

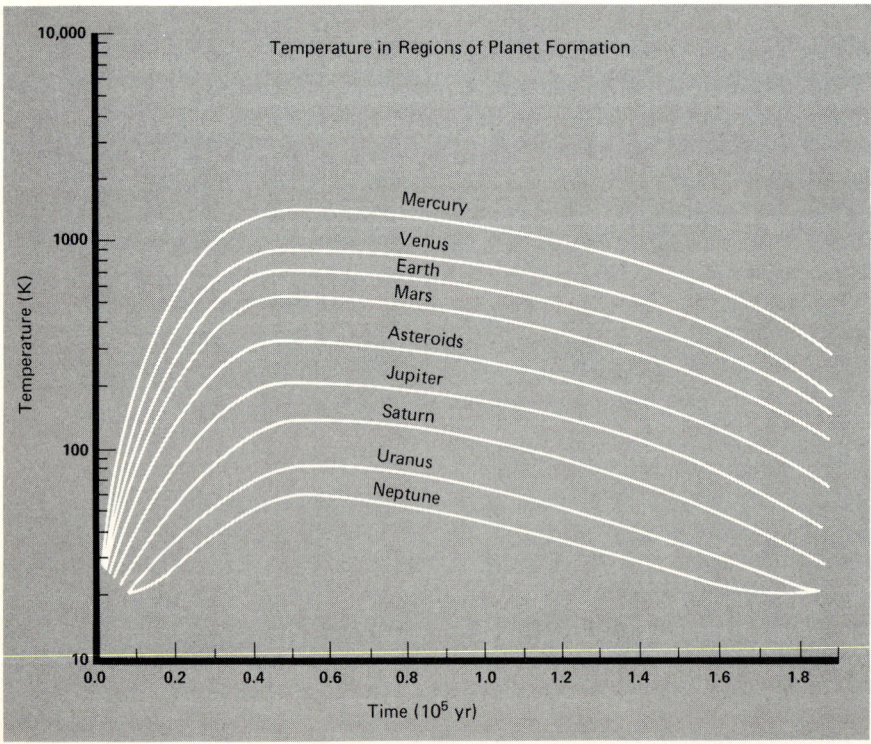

FIGURE 19.6 Temperatures in the solar nebula before the formation of the protoplanets. As the nebula contracts, temperatures rise, level off, and then fall. Each curve is for the distance from the sun for each protoplanet's formation. (*Adapted from a diagram by A. Cameron.*)

19.4 FORMATION OF THE PLANETS

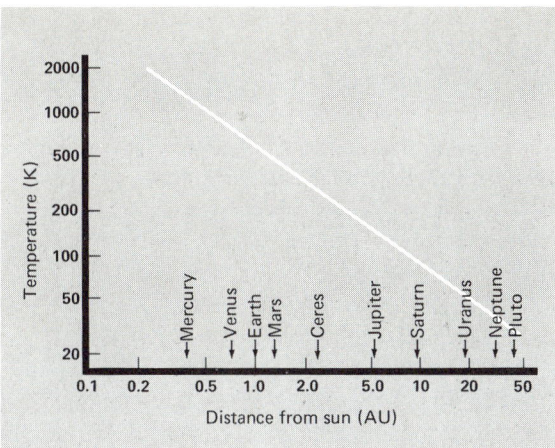

FIGURE 19.7 One model for how the temperature in the solar nebula might have decreased with increasing distance from the sun. At this stage the major constituents of the protoplanets are condensing from the solar nebula. (*Based on calculations by J. Lewis.*)

sulfide. Water would be bound up with minerals, perhaps as much as 5 percent of the total. So the theory predicts a planet with a smaller fraction of the mass in a nickel-iron core than for Mercury, containing sulfur (in the form of iron sulfide) and a larger mantle, rich in silicates and oxides of iron and magnesium. The earth does appear to have such an interior (Section 14.2).

What about Mars? Here the temperature was lower still. The resulting core would consist mostly of iron sulfide, not nickel-iron, and the mantle would be olivine, rich in iron oxide and water. We don't know the interior of Mars very well, but the prediction from the condensation sequence is compatible with our general inferences about the Martian interior (Section 16.5).

Note that it is how *low* the temperature falls that determines the matrix of reactions to produce condensates. For example, at the earth's position in the nebula, all the reactions in Table 19.4 down to 600 K can take place. The reactions at temperatures lower than 600 K do not happen.

In general, the condensation sequence requires a certain minimum temperature to be reached in order to account for the known chemical composition of the planets. Roughly, these temperatures are 1400 K for Mercury, 900 K for Venus, 600 K for the earth, 400 K for Mars, and 200 K for Jupiter.

The condensation sequence correctly predicts that Mercury has a larger, denser core than the earth and that Mars has a smaller, less dense one. Such successes imply that the compositions of the planets did arise from simple chemical reactions between the gas and dust in the nebula, reactions whose results depended on the local temperature.

The Aggregation of the Planets

How to get from small dust grains to large protoplanets? Turbulence makes the grains collide. They stick together and so accrete to form larger, perhaps pebble-sized, objects. These quickly fall into the plane of the nebula. These pebbles then accumulate into planetesimals.

Peter Goldreich and William Ward have proposed a model that gets the pebbles together into planetesimals. They have found that a thin disk of pebbles quickly fragments so that pebbles accumulate gravitationally into objects up to a few kilometers in size. Whatever materials happen to be available at a certain distance from the center of the nebula make up a planetesimal. So the planetesimals reflect the compositions established by the condensation sequence.

Once the planetesimals have formed, they might gather into larger bodies, perhaps as large as the moon. Somehow these objects finally end up in a few protoplanets. Here gravity would help. Once the planetesimals have gathered (in a few tens of thousands of years) into a few somewhat larger bodies (500 km in size), their masses would be enough to help pull in other smaller masses from a distance. So a growing planet will sweep clear a zone of the nebula in order to feed its mass. For the terrestrial planets to grow to their present sizes, calculations indicate an aggregation time of roughly 100 million years.

When the central regions of the nebula finally form the sun, solar energy evaporates any icy material in the nebula and around the inner protoplanets or planetesimals. Intense solar radiation pressure and a strong solar wind (a solar gale!) push leftover gasses out of the solar system.

The leftover planetesimals bombard the planetary surfaces, leaving remnant craters on the terrestrial planets. This bombardment heats the surfaces of the planets. Radioactive decay heats the interiors and melts them. Dense elements, such as iron and nickel, sink down to form a core. Less dense materials, such as silicates, float to form surface froth that cools to become the crust. So the planets become differentiated.

Jupiter and Saturn: A Different Story?

To make matters somewhat confusing, Jupiter and Saturn may have formed in a different way from the planetesimal accretion model described above. In analogy with starbirth, Jupiter and Saturn may have condensed gravitationally from single, large blobs of material in the nebula, rather than by accretion of planetesimals. If so, you can trace the evolution of proto-Jupiter and proto-Saturn by making the same kind of models as those used to work out the evolution of protostars. In fact, the chemical compositions of Jupiter and Saturn match those of stars fairly closely. The main difference arises from the fact that Jupiter and Saturn don't have enough mass to get hot enough to ignite fusion reactions. The heat they do gain comes from the conversion of gravitational potential energy into heat during gravitational contraction.

Harold Graboske and colleagues have made theoretical calculations of the evolution of Jupiter from a hot beginning, after the proto-Jupiter had come together (Fig. 19.8). They assume a solar mixture of material (74 percent hydrogen, 24 percent helium, and 2 percent everything else) and start the calculations with a proto-Jupiter 16 times Jupiter's present size, a central temperature of 16,000 K, a surface temperature of about 1000 K, and a luminosity of almost 10^{-2} the sun's present luminosity. Gravity quickly pulls in the proto-Jupiter. In only a million years the planet shrinks to just about twice its present size, its central temperature is about 40,000 K (the interior is heated by the gravitational contraction), and its luminosity drops to about 10^{-5} the sun's luminosity. (The luminosity drops even though the central temperature rises largely because of the large drop in surface area which can radiate.) The shrinking then slows down because the planet's interior is liquid, and liquids are difficult to compress. In the next 4.5 billion years, Jupiter contracts to its present size and its central temperature drops to 16,000 K as it loses some of its heat of formation to space at a rate of 1.8×10^{-9} times the solar luminosity.

19.4 FORMATION OF THE PLANETS

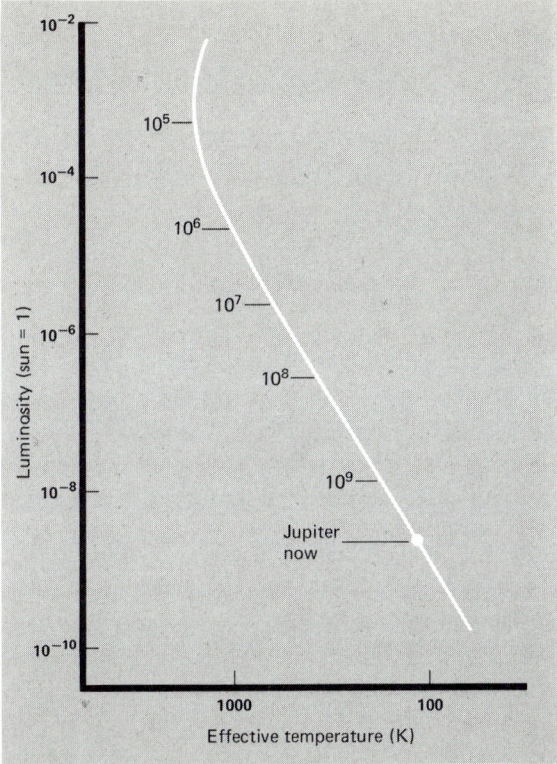

FIGURE 19.8 An evolutionary track for Jupiter. The horizonal marks indicate the time elapsed since the start of the contraction that formed the planet. (*Adapted from a diagram by H. Graboske, J. Pollack, A. Grossman, and R. Olness*, Astrophysical Journal, vol. 199 (1975), p. 265.)

These theoretical models indicate that soon after its formation the proto-Jupiter went through a brief phase of high luminosity. This period in its evolution resembles the protostar stage of stellar evolution. The early high luminosity of Jupiter may explain, in the context of a condensation sequence, why the Galilean satellites decrease in density going outward from Jupiter (Section 17.1). This density decrease implies, for example, that Callisto contains proportionally more icy materials than Io. At Io's closer distance, less ice condensed than at Callisto's distance. If Jupiter were hot at the time of the satellites' formation, the inner ones would not have accreted as much icy materials as the outer ones. So the Galilean moons may mimic the condensation and accretion of the terrestrial planets.

James Pollack and colleagues have made similar theoretical models for the early evolution of Saturn. As you might expect, the sequence resembles that for Jupiter. Their models begin 4.5 billion years ago with a planet ten times its present size, with solar composition and with no rocky core. Gravity quickly shrinks the planet. In about a million years Saturn is about twice its present size and has a central temperature of about 21,000 K. This high temperature serves to raise the internal pressure and slow down the gravitational contraction. It then takes 4.5 billion years for Saturn to contract to its present size.

During this time Saturn radiates away its internal heat. At first the energy is lost rapidly and then more slowly. The calculations show, for example, that while Saturn contracts rapidly, it emits about 10^{-6} the sun's luminosity. Now it emits only 10^{-10} the sun's luminosity, and its central temperature has dropped to 7000 K. This is enough to show up as an excess of infrared energy emitted by Saturn.

The evolution of Jupiter and Saturn may follow that of the solar system generally. Their rings may be unaccreted material.

The Asteroids: A Planet That Didn't Make It?

Contemporary research on asteroids has provided new clues about their origin. It now seems most likely that the asteroids are planetesimals that just didn't get together to make a planet, rather than remnants of a planet that exploded.

Recall (Section 18.1) the composition variation (determined from the albedos) across the asteroid belt. On the inner edge it contains mostly S-type asteroids, at the outer edge mostly C-type. These composition differences fit in nicely with the condensation sequence if the C-types contain more carbon than the S-types. At the inside of the belt the temperatures were low enough for silicates to condense, but too high for carbon-bearing materials to do so. Farther out both types of materials condensed to end up in planetesimals.

Why didn't the asteroids form a planet? Probably because of the gravitational influence of the proto-Jupiter. Recent theoretical calculations indicate that the proto-Jupiter (and the proto-Saturn) formed quickly, in a time as short as perhaps a few months. Their formation happened fast simply because a lot of material gathered together, which drew in more material, and so on. Once the proto-Jupiter had formed, it would tug on the planetesimals just within its orbit. Meanwhile the proto-sun would pull these planetesimals toward it. In this tug-of-war, the orbits of the planetesimals changed from circular to elliptical. Some crashed into others, shattering them into smaller pieces. Some of these pieces caromed into the inner part of the solar system and eventually rained onto the surfaces of Mercury, Venus, the moon, Earth, and Mars, forming craters. The remainder of the broken planetesimals stayed mainly in the region about 3 AU from the sun—these remnants are today's asteroids.

Meteorites and Planetesimals

The characteristics of chondritic meteorites (Section 18.3) support the condensation picture. Their chemical composition (similar to the sun) and unmixed structure suggest that they are the original condensed material of the nebula. The turbulence in the nebula may have created shock waves that swiftly melted and cooled grains, forming chondrules (Fig. 19.9). The glassy spheres that resulted accumulated in planetesimals (see Fig. 18.21). Chondrules suggest that the bulk of their condensation took place at temperatures around 500–700 K. (According to the condensation sequence, this range produces materials like those that make up the earth.) About a million years after formation, radioactive decay reheated some planetesimals, melting them to some extent, and allowing them to differentiate into iron cores and stony mantles. The planetesimals that were not gathered into a protoplanet possibly became the parent meteorite bodies. These bodies collided and fragmented to make the pieces that we pick up as meteorites after they fall to the earth.

What happened to other planetesimals? Some may have collided at high speeds with others and disintegrated into small pieces. A few may have passed close enough to a protoplanet to be captured as a satellite. Others may have experienced near misses. Their

19.4 FORMATION OF THE PLANETS 693

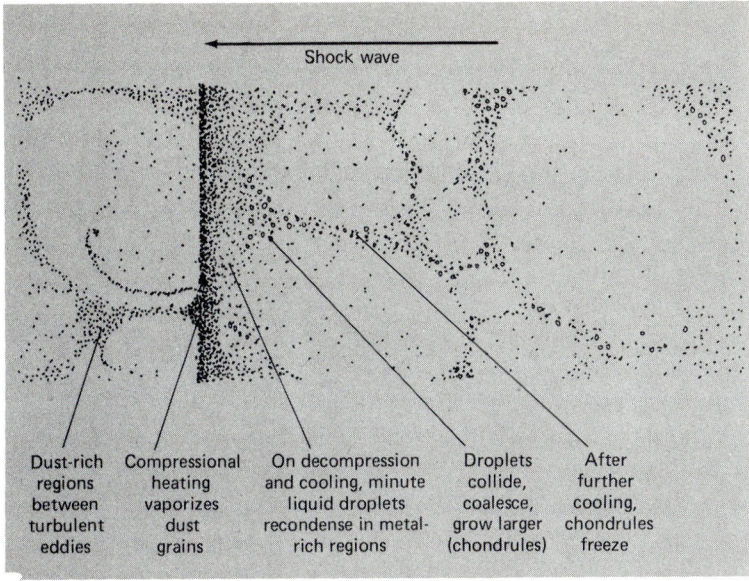

FIGURE 19.9 The formation of chondrules by shock waves passing through dust-rich regions in the solar nebula. The dust vaporizes in the shock and recondenses as droplets that quickly solidify.

orbits might have changed enough to throw them out of the solar system. Near Jupiter planetesimals would be mostly icy materials; these may have formed comets.

Here is the scenario in summary. The original cloud is homogeneous in composition; a clump of gas from anywhere in the cloud will be mostly hydrogen, but will have a few percent of heavy elements such as carbon, silicon, oxygen, or iron. Most original grains evaporate as the nebula heats up. When the nebula again cools and the grains reform, terrestrial grains form everywhere because of their high condensation temperatures, while icy grains, which contain hydrogen, form only farther out. Accretion makes protoplanets. The outer protoplanets are more massive because there is so much more hydrogen than anything else. They have terrestrial material, too. Then the sun becomes a protostar and blows most of the gas out of the solar system. There's a lot of gas close in because it had always been too hot there for any hydrogen to condense. So the interior of the solar system is left with only a few percent of its original mass—the heavy elements, the terrestrial material, all locked up in planets. The outer part of the solar system, in contrast, has locked up most of the gas in the planets, because the Jovian protoplanets are massive enough to retain the remaining gaseous hydrogen. Those planets, mostly solar and icy material, but still containing a few percent terrestrial material, remain large and cool. It's not certain why Uranus and Neptune have less hydrogen and helium than Jupiter and Saturn. Perhaps because they formed farther out in the solar nebula, where the density was lower, they were unable to accrete as much of these light gasses before the nebula was swept away by the rising luminosity of the sun.

The Nebular Model and Observations

How well does this amalgam of contemporary models match up with the chemical and dynamical properties of the solar system?

On the chemical side it does pretty well. The condensation sequence tied into the nebular model explains how the planets fall into two compositional classes (terrestrial and Jovian), and how planets in the same class do differ somewhat in chemical composition. On the dynamical side the model has some successes and a few failures. Let's compare the model's results to the features listed in Section 19.1.

1. Planets' revolution and sun's rotation: well explained as the original rotation of the nebula
2. Coplanar orbits: well explained by the conservation of angular momentum applied to the nebula's rotation
3. Circular orbits: explained fairly well by the interactions and sweeping up of planetesimals
4. Planets' rotation: expected from spin of the nebula, but not clearly worked out
5. Titius-Bode "law": explained in a general way by the sweeping out of zones of planetesimals by the protoplanets
6. Satellite systems: well explained if formed as miniature solar systems
7. Satellite spacings: works out in a general way, as for the planets, if satellites formed by sweeping up material
8. Angular momentum distribution: decent attempts, but no detailed solution
9. Comets: weakly explained as icy planetesimals that did not get caught by Jupiter, Saturn, Uranus, and Neptune, but were thrown out by them
10. Planetary ring systems: explained vaguely, like the asteroids, as unaccumulated debris

SUMMARY

The story outlined above amalgamates a few, but not all, contemporary ideas. Unmanned exploration of the solar system will provide new information for a more complete model. Also, the calculations needed to back up the details of the models are very difficult to make. Results so far have been only approximate, and some may be just plain wrong. For now, the scenario outlined in this chapter remains tentative and incomplete.

1. Process of Nebular Collapse

Time: about 4.6 billion years ago. Duration: 10 to 50 million years. A slowly rotating interstellar cloud of gas and dust collapses gravitationally from a diameter of about 50,000 AU. The total is perhaps a few thousand solar masses. The cloud fragments, and one piece with at least 1.1 solar masses (maybe 2 solar masses), continues to collapse. Pressure and density increase. Rotation rate increases. Cloud forms a disk about 60 AU across and 1 AU thick. Disk heats up more at center (about 2000 K) than at edges (about 100 K). Dust vaporizes near center.

2. Transfer of Angular Momentum

Duration: maybe as short as a few thousand years. The sun's magnetic field transfers spin from sun to disk.

3. Evaporation and Reformation of Grains; Formation of Planetesimals

Duration: 100,000 years. Grains condense in disk, composition being dependent on temperature. Generally, denser terrestrial materials are near the center, icy materials farther out. Grains hit and stick together, forming pebbles a few centimeters across. In a few years pebbles fall into the plane. Gravity brings them together as planetesimals, a few kilometers across.

4. Formation of Protoplanets
Duration: as much as a few hundred million years. Planetesimals accrete to form protoplanets. Sun forms and heats disk.

5. Dissipation of the Nebula
Duration: at most a few million years. Solar gales and radiation sweep out leftover gaseous material from nebular disk. Violent flares on the sun may contribute to the process.

6. Evolution of Protoplanets
Time: about 4.5 billion years ago. Duration: a few million years. Inner protoplanets heat up and differentiate. Jupiter and Saturn slow down in their contraction and sharply decline in luminosity.

7. Formation of the Earth-Moon System
Time: 4.6 billion years ago. Duration: perhaps a few hundred million years. Debris in space near the earth accumulates to form the moon at a distance of about half the moon's present distance. Leftover chunks of nebular material fall into the earth and the moon. Tidal interactions between the earth and the moon slow down the rotation of the earth and cause the moon to spiral away from the earth.

Key Words

solar, icy, terrestrial materials
volatiles
coplanar
distribution of angular momentum

Titius-Bode "law"
nebular model
catastrophic model
gravitational collapse
condensation
accretion

deuterium-hydrogen ratio
planetesimal
protoplanet
chemical condensation sequence
turbulence

Review Questions
1. What was the almost fatal flaw in the original Kant-Laplace nebular theory? (Objectives 4 and 5)
2. What planets do *not* fit well into the general chemical and dynamical properties of the solar system? (Objective 1)
3. What is the most important objection to any catastrophic theory? (Objectives 3 and 6)
4. Use the condensation sequence to explain the general chemical differences between the earth and Jupiter. (Objective 7)
5. Outline a currently accepted sequence of planetary formation. (Objectives 3, 7, and 9)

Problems

1. Calculate how fast the sun would rotate if all the orbital angular momentum in the planets now were put into the sun. (Objective 4)
2. Calculate the time between collisions of the sun with another star. The easy way to do this is to find out how long it takes the sun to sweep out a volume which contains one star. The sun's velocity is 25 km/sec and the number density of stars in the solar neighborhood is $0.1/pc^3$. Do the calculation again for the center of the Galaxy, where the stars move with velocities of 100 km/sec and the number density of stars is $10^5/pc^3$. Compare these times to the age of the Galaxy. Do you expect solar systems to be common in either part of the Galaxy? (Objective 6)
3. Suppose the planets do form from the sun during a collision. The mass of material pulled from the sun must have been at least the total mass of all the planets, which is about 600 M_{earth}, 3.6×10^{27} kg.
   a. Assuming the material has a density like that in the outer part of the sun, 100 kg/m^3, find the diameter of the ejected material, if it were a sphere.
   b. Calculate the thermal velocity for $T = 10^5$ K, the approximate temperature of material in the outer part of the sun. Compare to the escape velocity from the gas ball (see Section 14.5).
   c. Assuming the gas ball radiates like a blackbody, estimate how long it would take it to radiate its thermal energy ($E = \frac{3}{2} kT$ per particle).
   d. The answer to part (b) should suggest that the gas can expand freely at a speed equal to the thermal velocity. Calculate how long it takes to double its radius.
   e. Compare the expansion time to the cooling time, and discuss the relevance of your results to the encounter hypothesis for formation of the solar system. (Objective 6)
4. Calculate the free-fall time (Section 9.3) for the cloud properties in (1) in the summary and compare it to the duration given there. (Objective 4)

BEYOND THIS BOOK . . .

* Up-to-date nontechnical material on the origin of the solar system is hard to come by. We suggest you try, after reading this chapter, "The Origin and Evolution of the Solar System" by A. G. W. Cameron in *Scientific American*, September 1975. An explanation of the condensation sequence is "The Chemistry of the Solar System" by J. Lewis in *Scientific American*, March 1974. For details about the earth, see "The Formation of the Earth from Planetesimals" by G. W. Wetherill, *Scientific American*, June 1981.
* If you'd like to tackle some technical articles, read *On the Origin of the Solar System*, a book of the Centre National de la Recherche Scientifique de France, 1974 (ISBN 2-222-01512-X).
* W. Hartmann has a relatively nontechnical review of "The Planet-Forming State: Toward a Modern Theory" in *Protostars and Planets* (University of Arizona Press, Tucson, 1978), edited by T. Gehrels. Articles in Part V are also pertinent but more technical; some of their introductions are good.
* Read "Putting It All Together" by J. Lewis in *The New Solar System* (Sky, Cambridge, Mass., 1981) for one synthesis of the modern nebular model.

PART FOUR
GALAXIES: ISLANDS OF STARS

THE MILKY WAY: HOME GALAXY

20

LEARNING OBJECTIVES
After studying this chapter, you should be able to:
1. Make a sketch of the Milky Way Galaxy, showing the disk, nucleus, halo, spiral arms, position of the sun, and its distance from the galactic center.
2. Describe the technique for outlining the general shape of the Galaxy by counting stars in samples of the sky and describe the drawbacks to this technique.
3. Describe the two components of a star's motion in space: tangential velocity and radial velocity.
4. Define the proper motion of a star, and relate the proper motion to a star's velocity and distance.
5. Explain what is meant by the apex and antapex of the sun's motion.
6. Describe how the method of statistical parallaxes can be used to estimate the distances to stars.
7. Outline the method by which the distance to a cepheid variable can be determined.
8. Argue, as Shapley did, that the observed distribution of globular clusters in the sky leads to the conclusion that the sun is not located at the center of the Galaxy, and describe how he determined the distance to the galactic center.
9. Summarize the physical differences between young Population I, old Population I, and Population II stars, and indicate on a drawing of the Galaxy the regions where you find them.
10. List and describe the general contents of the disk, nucleus, and halo of the Galaxy.
11. Describe what X-ray, optical, infrared, and radio observations tell astronomers about the nucleus of the Galaxy.

CENTRAL QUESTION:
How did astronomers discover the layout of the Milky Way Galaxy, and what objects and material does it contain?

How are the stars arranged in space? On a moonless night you can trace out the soft, luminous band of the Milky Way wrapped like a fine scarf around the sky. Ever since Galileo discovered that the Milky Way consisted of a multitude of faint stars, astronomers have tried to determine their layout.

This chapter first deals with the quest for a detailed map of the nearby regions of space. We look at this question historically to see how astronomers have attacked this puzzle and to show the key importance of distance measurements and the properties of stars in solving it. The search culminates with the discovery of the Galaxy: a vast, flat pinwheel of stars, gas, and dust. Only one out of at least 100 billion stars, the sun sits about halfway out from the Galaxy's center and swings around it once every 250 million years.

The chapter then turns to the contents of the Galaxy. Three stellar populations mark out three distinct regions: the flat disk, the bulging nucleus, and the encompassing halo. Each part presents us with a different aspect of our complex Galaxy.

20.1 A TOUR OF THE MILKY WAY

We've mentioned the Milky Way a number of times without describing it in detail. So let us take a quick tour through the Milky Way as visible from the earth. Imagine that you are on the top of a mountain in the desert. A warm September day slowly cools as the sun vanishes behind the purple hills to the west. The brightest stars emerge from the darkening sky. Overhead you see the Northern Cross in the constellation Cygnus. As the darkness gathers, you pick out the wispy glow of the Milky Way. Scan down the Milky Way to the south. Before you reach Scorpius, you find that the Milky Way splits into two distinct bands (Fig. 20.1). The dark blot between the bands is called the *Great Rift*. Near it lie the brightest regions of the Milky Way visible from the Northern Hemisphere.

View this region with binoculars. Stars appear in countless numbers (Fig. 20.2). Slowly scan around the constellations Scorpius, Scutum, and Sagittarius. You get the

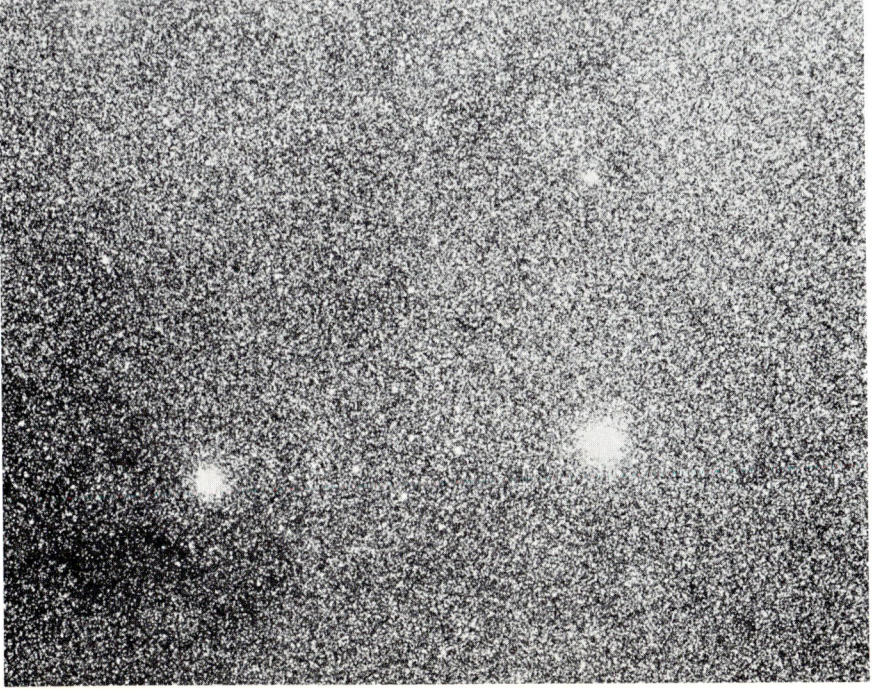

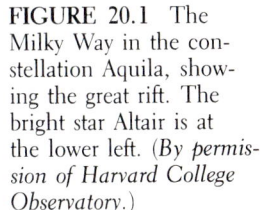

FIGURE 20.1 The Milky Way in the constellation Aquila, showing the great rift. The bright star Altair is at the lower left. (*By permission of Harvard College Observatory.*)

FIGURE 20.2 The Milky Way as seen through a large telescope. More than a million stars are visible in this photo. The two concentrations of stars are globular clusters. (*Courtesy Kitt Peak National Observatory.*)

FIGURE 20.3 An open cluster of stars (Messier 67). (*Courtesy Palomar Observatory, California Institute of Technology.*)

FIGURE 20.4 Two bright nebulas, Messier 20 (left) and Messier 8 (right), in the Milky Way. (*Courtesy Mt. Wilson Observatory, Carnegie Institution of Washington.*)

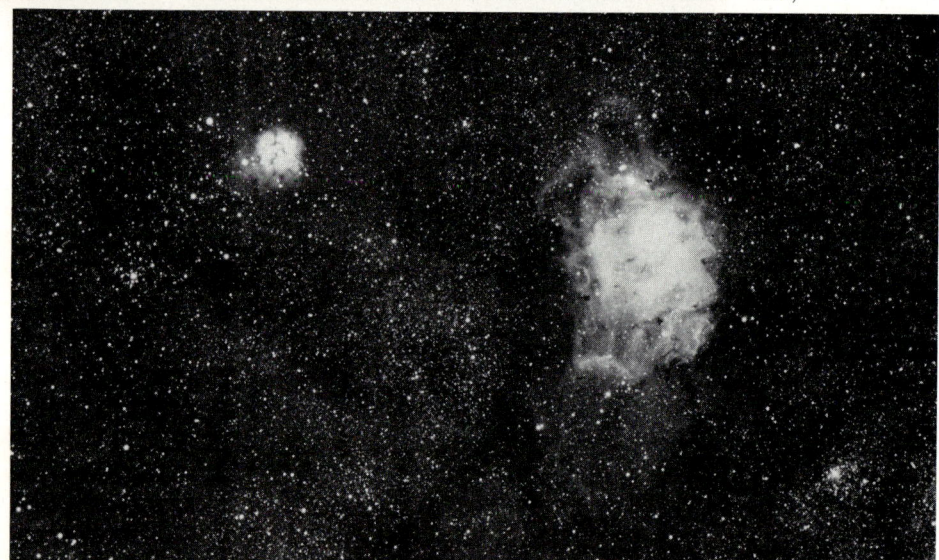

FIGURE 20.5 A globular cluster (Messier 15). Two globulars are also visible in Fig. 20.2. (*Courtesy Kitt Peak National Observatory.*)

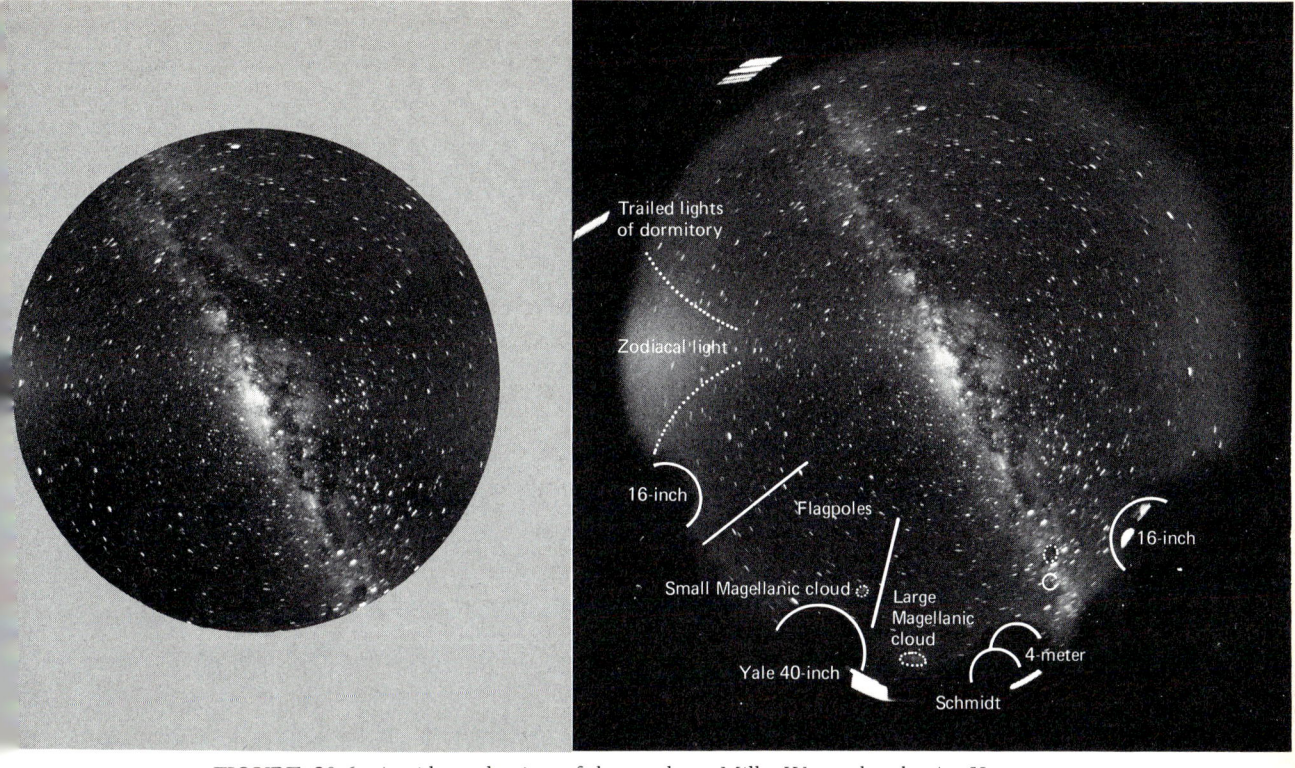

FIGURE 20.6 A wide-angle view of the southern Milky Way, taken by Art Hoag at Cerro Tololo Interamerican Observatory. (*Courtesy A. Hoag and Kitt Peak National Observatory.*)

impression that where you see more or fainter stars, you are looking deeper into space. Occasional areas seem swept clean of the stars. Are there no stars there or is dust blocking out the starlight?

If you scan more, you will come across small clusters of stars (Fig. 20.3). These groups are the *open* or *galactic clusters*. A few you see immersed in a faint, glowing cloud (Fig. 20.4). Such a cloud is a *nebula*. You can pick out with your binoculars very dense star clusters (Fig. 20.5). These appear more tightly packed than the galactic clusters and more symmetrical in shape; they are *globular clusters*. In other regions you come across loose groups of bright, blue-white stars, the *OB associations*.

Stars in loose and tight clusters, irregular clouds of glowing material, dark blotches, individual stars everywhere: These are the contents of the Milky Way viewed with binoculars or a small telescope. What is their layout in three dimensions? Seeing the Milky Way encircling the sky (Fig. 20.6), you get the impression that the sun is situated in a flat system of stars that is thick in the center and thins out toward the edges.

Earlier astronomers had similar impressions. Before we present the story of their detective work, we'll give away the story's end by sketching a rough model of the Milky Way Galaxy (Fig. 20.7). Astronomers picture the Galaxy as a vast disk of stars and gas with a bulge in the middle. The disk of stars has a diameter of 120,000 ly (40,000 pc); its thickness is about 1/60 of its diameter. The stars orbit around the center of the Galaxy bathed in a disk of gas that extends to a diameter of 160,000 to 200,000 ly (50,000 to 60,000 pc); its thickness is only 1/200 of its extent. The nuclear bulge is about 12,000 ly

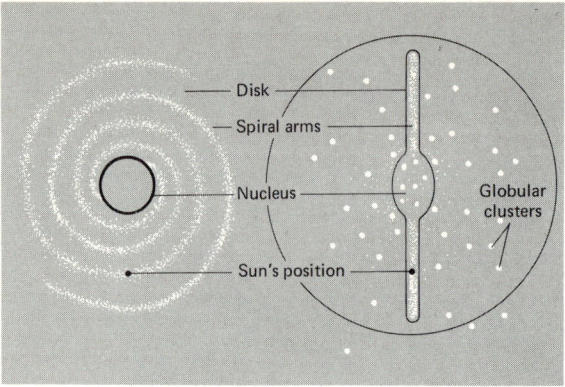

FIGURE 20.7 A schematic, contemporary model of the Milky Way Galaxy. Its diameter is about 120,000 ly.

(4000 pc) in diameter and 10,000 ly (3000 pc) thick. The sun lies in the Galaxy's suburbs, about 30,000 ly (10,000 pc) from the center, but almost directly in the Galaxy's plane. The sun and other stars whirl around the Galaxy at speeds of hundreds of kilometers per second. Such rapid motion accounts for the flatness of the disk of stars.

Note: You might want to stop here before you plunge ahead and review the general properties of stars, Chapter 8. Also, you must reorient your thinking to a scale of distances much vaster than the solar system (Section 5.4). In the Galaxy we deal with distances of thousands of light years, whereas light travels across the solar system in half a day. In this and following chapters we will often use the units kpc (10^3 pc) and Mpc (10^6 pc) to express these large distances.

20.2 HERSCHEL MAPS THE MILKY WAY

In the latter part of the eighteenth century, William Herschel (who also discovered Uranus) made systematic observations to attack the problem of the Galaxy's structure. Herschel had constructed telescopes that dwarfed other instruments of the day. With these powerful tools he set out to establish the extent of the Galaxy by counting stars of different brightness in selected regions of the sky. Herschel thought that directions where he found large numbers of faint stars were places in which the Galaxy extended the farthest.

This procedure required a number of assumptions. One can be stated as: *Brightness means nearness* (or *faintness means farness*). You assume that all stars are essentially the same and give off the same amount of light. So the ones which appear fainter must be farther away. (Recall from Chapter 8 the relation between apparent magnitude, absolute magnitude, and distance: $m = M - 5 + 5 \log d$. If all stars are assumed to have the same absolute magnitude, $\log d = 0.2 (m - M + 5) = constant + 0.2\, m$. So large apparent magnitude m, or faintness, means large distance d.)

From the number of stars counted at each apparent magnitude, Herschel estimated the number at different distances in various directions (Fig. 20.8). Because he did not know the luminosity of any star, he could determine distances only relative to some standard star (he chose Sirius as the standard). So Herschel obtained the relative shape of the Galaxy, but he could not determine its distance scale (Fig. 20.9).

Herschel concluded that the stars were arranged in a thin slab, with the sun not far from the center. The contours of the edge of the stars' distribution were irregular in his

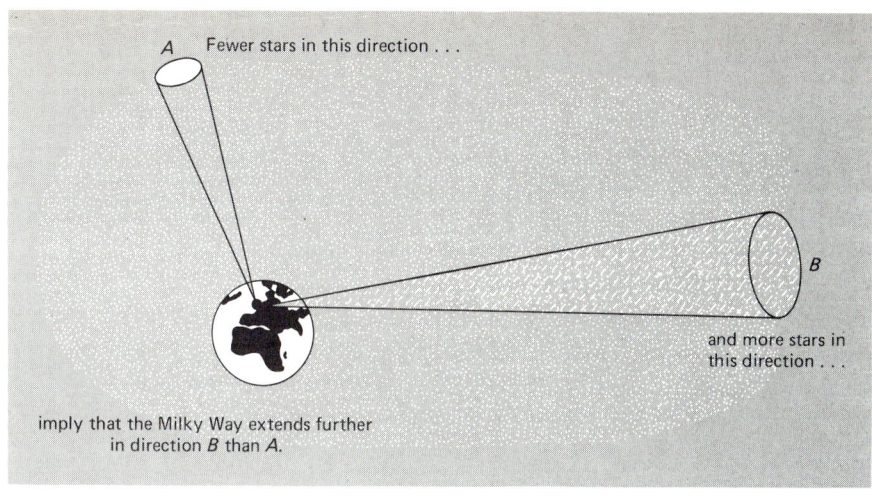

FIGURE 20.8 Herschel's star-counting technique to infer the shape of the Galaxy.

model to account for the differences he discovered from his star counting—where the number of faint stars was small, the edge of the Galaxy was nearby. The general conclusion of a flat system is right, but because of the simplifying assumptions Herschel used and because of his ignorance of interstellar dust and its dimming effect (Section 9.2), his model was wrong in detail.

From this picture of the Galaxy, Herschel made a bold mental leap. His telescopes showed him nebulous objects in the sky. He was able to resolve many of these into groups of stars. These observations led him in the 1780s to conclude that all spiral-shaped nebulosities were distant stellar systems that would ultimately be resolved into stars by large enough telescopes. So he inferred that these clouds (we still sometimes call them spiral nebulas) were other galaxies and that our Galaxy would appear like them, if viewed from a distance.

20.3 STARS TRAVELING THROUGH SPACE

After Herschel other astronomers tackled the problem of the detailed layout of the stars in space. To solve this problem requires the measurement of distances to stars. But that's the hardest measurement to make in astronomy!

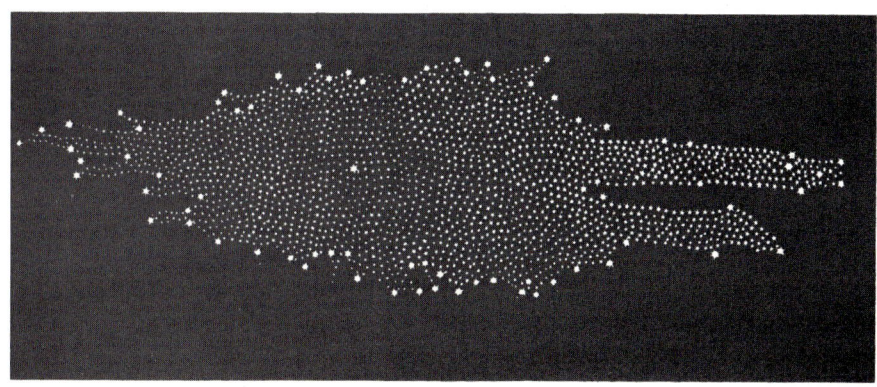

FIGURE 20.9 Herschel's model of a cross section of the Milky Way inferred from star counting. He described it as "a very extensive, branching, compound congeries of millions of stars". (*Courtesy Yerkes Observatory.*)

You have already seen how astronomers use *heliocentric (trigonometric) parallaxes* (Section 8.2) to find the distances to nearby stars. With a Hertzsprung-Russell diagram in hand, we can estimate stellar distances by *spectroscopic parallaxes* (Section 8.5). But we can also estimate the distances to stars by observing their motions in the sky. To understand how this is done, we first need to look at stellar motions in general.

Proper Motion

Early in 1718 Edmund Halley compared the positions in the sky he had found for Arcturus and Sirius with those given by Ptolemy. Halley found that these stars had moved a considerable amount in 1500 years; the "fixed" stars were not in fact fixed, but moved about in space. In one year the change in a star's position, as seen from the earth, amounts to a very small angular movement. Over many years, and certainly over the centuries between Ptolemy and Halley the changes add up and so become easy to observe (Fig. 20.10).

The change in position on the sky of a star with respect to other stars due to its motion in space is called its *proper motion* (Fig. 20.11). (The term arose historically to differentiate this real motion from the apparent motion of rising and setting.) Observing the proper motions of stars tells something about their distance, for if they were all moving with the same real speed through space, those with the highest angular speed must be the closest.

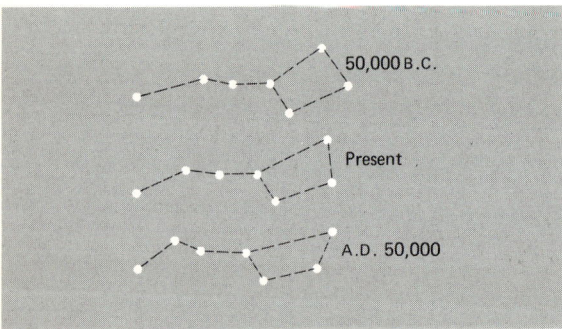

FIGURE 20.10 Proper motions of stars in the Big Dipper over 100,000 years.

FIGURE 20.11 The proper motion of Barnard's star (arrow) over 22 years. (*Courtesy Yerkes Observatory.*)

20.3 STARS TRAVELING THROUGH SPACE

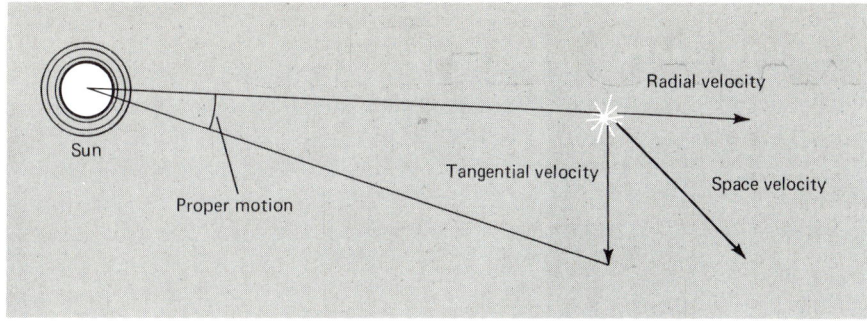

FIGURE 20.12 The motion of stars in space. The part of a star's space velocity across our line of sight is its tangential velocity; the part along the line of sight is its radial velocity.

So here's another rule of thumb to estimate stellar distances: *Swiftness means nearness* (or *slowness means farness*). Note that proper motion is an angular speed, not the actual velocity in kilometers per second.

Space Velocity

There is a complication in the swiftness means nearness rule. Seen from the earth, a star's motion in space has two parts (Fig. 20.12): a *tangential velocity* across the line of sight (which results in the angular proper motion) and a *radial velocity* along the line of sight. The actual motion of the star, in kilometers per second with respect to the sun, is a combination of its velocities across and along the line of sight. This total speed and direction is called the star's *space velocity*.

Let's look at this relationship in a little more detail (Fig. 20.13). Call the radial part of the star's velocity V_r, the transverse velocity V_t, and the space velocity V. Note that V_r and V_t are at right angles, and that V is the hypotenuse of a right triangle with sides V_r and V_t. By the Pythagorean theorem,

$$V^2 = V_r^2 + V_t^2$$

A star's proper motion is related to its tangential velocity. Though two stars may have the same space velocity, their proper motions could be quite different, if one is moving mainly in the tangential direction and the other mainly in the radial direction.

Astronomers possess a powerful tool for determining radial velocities: the Doppler effect (Section 8.6). Determining the tangential velocity is harder, for to do so requires both the proper motion and distance. The tangential velocity is proportional to the proper motion (the greater the angular speed, the greater the velocity), and it is also proportional

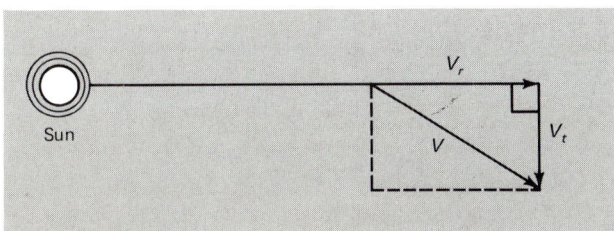

FIGURE 20.13 The components of a star's motion in detail.

FOCUS 20.1 PROPER MOTION, TRANSVERSE VELOCITY, AND DISTANCE

A star's proper motion relates to both its distance and its transverse velocity. The derivation of the relationship follows that for heliocentric parallax (Section 8.2).

Suppose a star at distance d moves in space along path AC in 1 yr (Fig. F.21). Then AB is its transverse motion in a year and AD its radial motion. From the earth (E), we see a proper motion (an angle) of μ in a year that results from the motion along AB. Note that AB closely approximates an arc on the circumference of the circle centered on E with

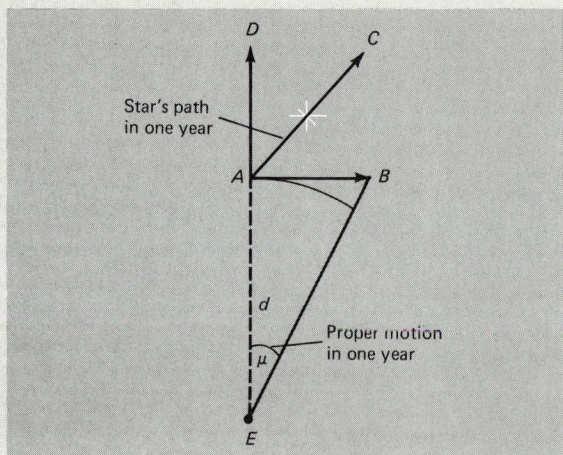

FIGURE F.21 Geometry for the relationship among proper motion, transverse velocity, and the distance to a star.

to the distance, for the farther an object is, the faster it must be moving to cover the same angle in the same time. The exact relationship is

$$V_t = 4.74 \mu D$$

where D is the distance in parsecs, μ is the proper motion in arcseconds per year, and V_t is the transverse velocity in kilometers per second. (See Focus 20.1 to find out where this equation comes from.) So if we know the distance, we can translate the star's proper motion into a velocity in kilometers per second. From the velocities along and across the line of sight we can find the star's motion through space.

Now reverse this procedure. If we knew a star's transverse velocity and its proper motion, we could calculate its distance:

$$D = \frac{V_t}{4.74\mu}$$

The trouble is, for an individual star, we don't know its transverse velocity. First of all, the

radius d. The ratio of this arc to the circumference is the same as the ratio of the angle μ to a full circle of 360° (which is equal to 1,296,000 arcsec). So

$$\frac{\mu}{360°} = \frac{\mu}{1,296,000''} = \frac{AB}{2\pi d}$$

$$AB = \frac{2\pi \mu d}{1,296,000}$$

But AB equals the transverse velocity V_t, in kilometers per second, times the number of seconds in a year (3.16×10^7 sec), so

$$V_t(3.16 \times 10^7) = \frac{2\pi \mu d}{1.296 \times 10^6}$$

or

$$V_t = \frac{\mu d}{6.52 \times 10^{12}} \text{ km/sec}$$

with d in kilometers, V_t in kilometers per second, and μ in arcseconds per year. To use this relationship with d in parsecs, note that one parsec equals 3.086×10^{13} km. Then

$$V_t = \frac{\mu d(3.086 \times 10^{13})}{6.52 \times 10^{12}}$$

$$= 4.74 \mu d$$

with V_t in kilometers per second, μ in arcseconds per year, and d in parsecs.

---

space velocities of stars are not all the same, though we do know that the average is about 25 km/sec. Second, we don't know the orientation of the space velocity, what fraction is along the radial direction and what fraction is along the transverse direction. Third, the proper motion may not all be due to the star's motion through space; some may be due to the motion of the sun itself.

The Sun's Motion

The above discussion assumes that the sun—the base from which we observe—is standing still. The proper motions of stars also provide valuable information on the sun's motion in space, relative to the nearest stars. In 1783 Herschel noted that the sun's motion should produce a part of the proper motions of stars, varying in a regular way with position on the sky. The direction in which the sun moves in space he called the *apex* of its motion. From the proper motions of only 13 stars, Herschel deduced that the apex of the sun's motion lies in the constellation Hercules, next to Lyra. He noticed that the direction of the proper motions of his dozen stars radiated outward from a point in that direction. They converged on a point in the opposite part of the sky, the *antapex*, in the

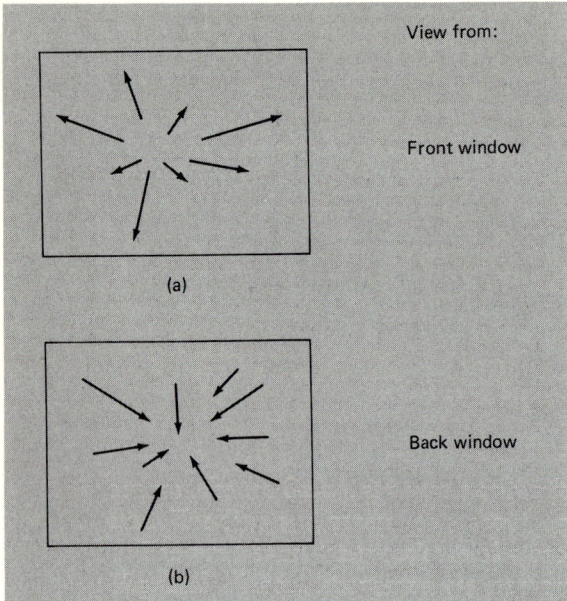

FIGURE 20.14 Motion of falling snowflakes (no wind) as viewed from inside a moving car through the front window (a) and through the back window (b).

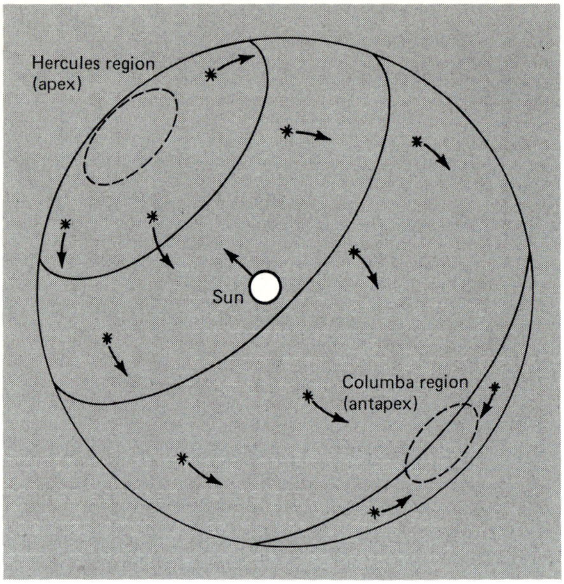

FIGURE 20.15 The sun's motion relative to local stars. The movement makes the stars appear to stream away from the apex and toward the antapex.

constellation of Columba, south of Orion. He guessed that this effect resulted from the motion of the sun through space.

Imagine this situation with the following analogy. Consider standing in a snowfall with no wind blowing. The snowflakes fall in their usual shifting patterns. Drive a car through the snowstorm. Peering through the front windshield, you note that the snowflakes appear to fan out from a point directly in front of the car (Fig. 20.14a). If you use the rearview mirror to look out the back window, the snowflakes appear to converge at a point behind the car (Fig. 20.14b). Their apparent motions reflect the car's motion through the swarm of falling snowflakes.

Similarly, the sun's motion at about 20 km/sec with respect to the local stars creates the effect of an apex and antapex of solar motion (Fig. 20.15). We know how fast we're moving by using the Doppler shift (Section 8.6). In general, stars in Hercules have blue-shifted spectra; those in Columba show red shifts. If we measure the blue shifts for a number of stars in the Hercules region and find an average velocity, it comes out to about 20 km/sec, which is 4.2 AU/yr. Note that when we talk of the sun's motion, we are talking about the motion of the entire solar system through space. We on the earth are included in that motion.

Statistical Parallaxes

Now to the point—using stellar motion to determine distances. In order to plot the structure of the Galaxy we need to reach beyond the local stars. One way to do this involves observing the average proper motions of selected groups of stars.

Consider this analogy. Suppose you are riding along a highway past a skyscraper-

20.3 STARS TRAVELING THROUGH SPACE

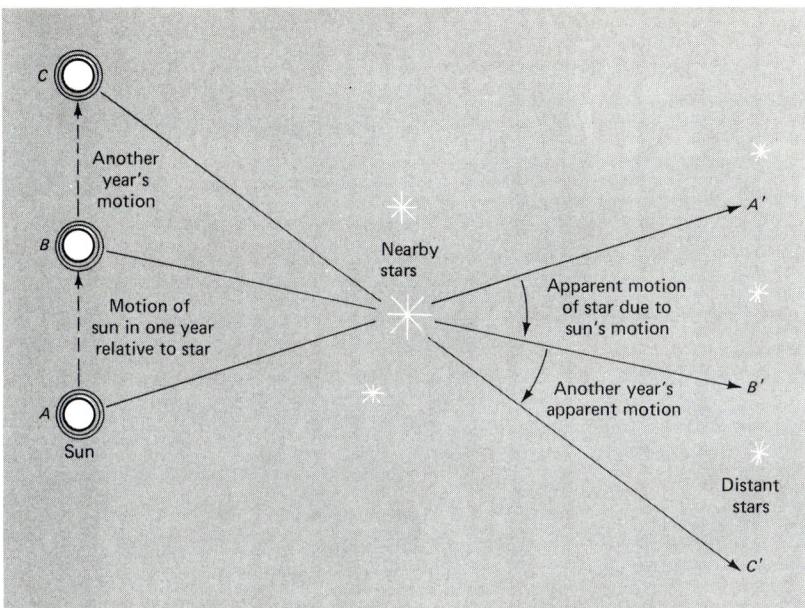

FIGURE 20.16 The sun's motion and the apparent motions of stars used in statistical parallaxes. Imagine only the sun moves (from A to B to C) relative to other stars. Consider a nearby star; relative to more distant ones, it will appear to move a certain angular distance (from A' to B') in a year because of the sun's motion. In another year it will appear to have moved from B' to C', for a total angle of A' to C' in two years. Each year adds to the baseline from which distances can be measured. For the technique to work, astronomers must use selected *groups* of stars, not the single star shown here to simplify the explanation.

studded city with a mountainous backdrop on your right. If you look at the scene with one eye and then the other alternately, no building shifts position relative to the background mountains. The city is too far away from you to show a trigonometric parallax using your eyes to set the baseline, because the baseline is too small. However, as your car speeds along past the city, you notice that the buildings appear to slowly move, relative to the mountains, and in the direction opposite your direction of motion. You can measure the amount of angular displacement they go through in a certain length of time (say a minute). The longer you drive, the larger the angular displacements become, and the farther away you can see measurable shifts. Because you know your speed, you know how far you traveled to see a certain angular shift. This information allows you to compute the distance to the buildings using the skinny triangle method.

Replace the car with the sun and earth (Fig. 20.16), the city with a group of stars and the mountains with background stars farther away than the selected group. The nearby stars will appear to move with respect to the background stars, in a direction opposite to the solar motion. Here's a complication: You must choose a group of stars whose *average* motions are zero so that their *average* proper motion reflects the sun's motion relative to them; they must also be at about the same distance. (The buildings in the city are at the same distance and do not move relative to each other.) If you took a single star, you wouldn't know whether its proper motion was due to its own velocity through space or to the velocity of the sun.

An example: Choose from all over the sky stars of spectral class A having apparent magnitude of +5. These stars have about the same luminosity, so this group of the same apparent magnitude must all lie at about the same distance from the sun. Subdivide the sky into small groups of these stars, and measure the average proper motion of each group. These groups seem to reflect the sun's motion: They stream away from the apex of the

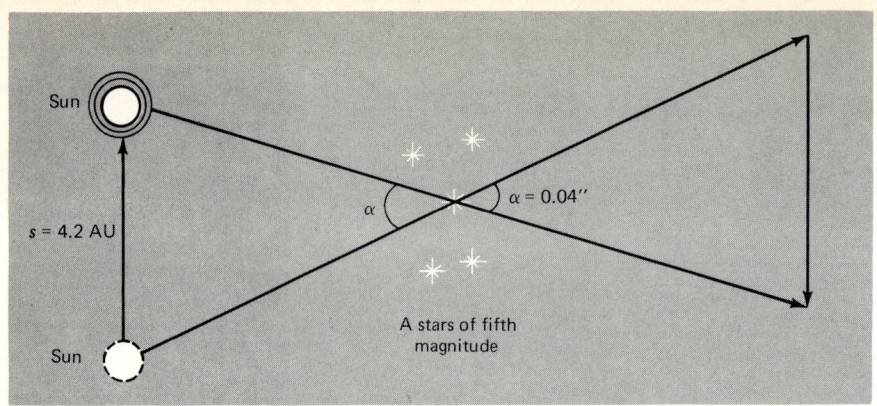

FIGURE 20.17 Statistical parallax for A stars of fifth magnitude.

sun's motion. Their average angular speed is 0.04″ a year. Now, in a year the sun moves about 4.2 AU toward the apex. So we have a skinny triangle, for which

$$\alpha_{rad} = \frac{s}{d}$$

with $\alpha = 0.04'' = 1.9 \times 10^{-7}$ radians and $s = 4.2$ AU (Fig. 20.17). Putting in the numbers:

$$1.9 \times 10^{-7} = \frac{4.2 \text{ AU}}{d}$$

$$d = \frac{4.2}{1.9 \times 10^{-7}}$$

$$= 2.2 \times 10^7 \text{ AU}$$

$$= 105 \text{ pc} = 342 \text{ ly}$$

So the *average* distance to A stars of fifth magnitude is 342 ly.

Here's another way to do this. Since all motion is relative, we can think of the solar motion in terms of the star having a transverse velocity with respect to the sun of 4.2 AU/yr (20 km/sec), and use the formula relating proper motion, transverse velocity, and distance (Focus 20.1)

$$V_t = 4.74 \mu D$$

with V_t in kilometers per second, μ in arcseconds per year, and D in parsecs. Then

$$D = \frac{V_t}{4.74 \mu}$$

$$= \frac{20}{4.74 \times 0.04}$$

$$= 105 \text{ pc} = 342 \text{ ly}$$

This method utilizes only the contribution to the proper motion of the sun's velocity.

The contribution of the stars' motions are averaged out; some will have proper motions in the same direction as the solar motion, some in the opposite direction.

Another method explicitly makes use of the intrinsic stellar proper motions. Suppose we again select a group of stars of the same type and same apparent magnitude, so that we can be assured they are all at the same distance. We measure their proper motions, but subtract that part due to the solar motion. Now we measure the radial velocities (V_r) of all the stars and find their average. If we assume that the directions of the stars' motions are random, the average transverse velocity (V_t) will be equal to this average radial velocity. Then we calculate the average of the stellar proper motions, disregarding their direction, and find the average distance from the relation:

$$D_{avg} = \frac{(V_r)_{avg}}{4.74 \mu_{avg}}$$

These techniques are called *mean* or *statistical parallaxes* to emphasize that they permit estimates only of average distances to groups of stars. They greatly extend the range of distance determinations, out to about 1500 ly (500 pc). This information provides a crucial leap beyond the direct measure of distances to the nearest stars, a leap necessary to discover the structure of the Galaxy.

20.4 THE DISCOVERY OF THE GALAXY

At the start of this century astronomers photographed a large part of the sky in order to plot the layout of the Galaxy. Such photographs provided a valuable resource for improving upon Herschel's star-count method.

Prelude: The Kapteyn Universe

In 1901 Jacobus C. Kapteyn (1851–1922) employed statistical parallaxes to find the average distances of stars in the Milky Way. Kapteyn's study provided the distance scale for an outline of the Galaxy given by the star counts, essentially Herschel's technique. He found the Galaxy to be 26,000 ly (8000 pc) in diameter and 6500 ly (2000 pc) in thickness, with the sun within 8000 ly (2500 pc) of the center. This is smaller than the true size of the Galaxy, but much larger than anyone had previously imagined.

Kapteyn did have one worry that later turned out to be important: the absorption of starlight by dust. If absorption did occur, distant stars would be so faint that they would not be seen. Kapteyn assumed the absorption was small, but he was right to worry; interstellar dust does absorb starlight significantly. The evidence for interstellar absorption by dust came after Kapteyn's work and served to reset the boundaries of the Galaxy. The cutting off of the starlight caused Kapteyn to ascribe a small, almost heliocentric character to our Galaxy (Fig. 20.18), which is neither heliocentric nor small.

Interlude: A New Distance Indicator

A novel technique of surveying the stars came from an unexpected quarter in 1908. While studying variable stars in the *Magellanic Clouds*—two large star systems in the

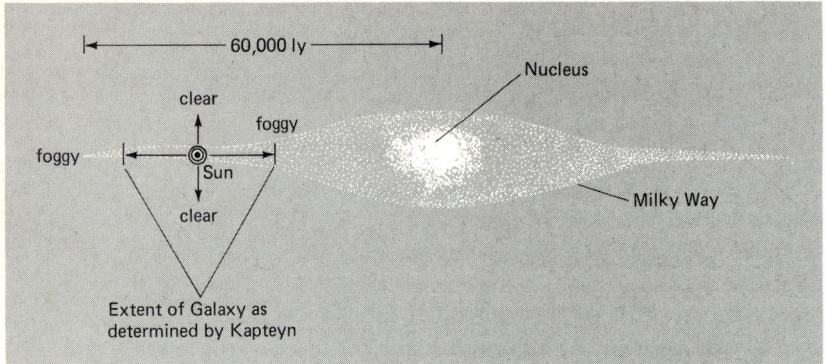

FIGURE 20.18 Star counts and interstellar obscuration. The sun is in the midplane of the Galaxy's disk, where interstellar dust is plentiful. This dust cuts out the light of distant star, especially in the direction of the Galaxy's center. This absorption leads to the impression that the Galaxy is small and heliocentric.

southern sky now known to be other galaxies physically connected with our own Galaxy—Henrietta S. Leavitt (1868–1921; see Fig. 8.11) discovered that when the apparent magnitudes of a certain type of variable star were plotted against their periods, a definite relationship appeared (Fig. 20.19). Since all the stars in the Small or Large Magellanic Cloud are essentially at the same distance from the sun, the plot of apparent magnitude against period can be translated into a plot of absolute magnitude against period, if the distance to the Clouds is known. (This calibration requires a chain of steps and is difficult to do.) Somehow the periods of these variables were related to their luminosities: the longer the period, the greater the luminosity.

The next year Hertzsprung pointed out that the variables Leavitt had discovered had light curves similar to that of the star Delta Cephei. (A *light curve* is a plot of a star's apparent magnitude against time; Fig. 20.20.) A star whose light varies in a regular

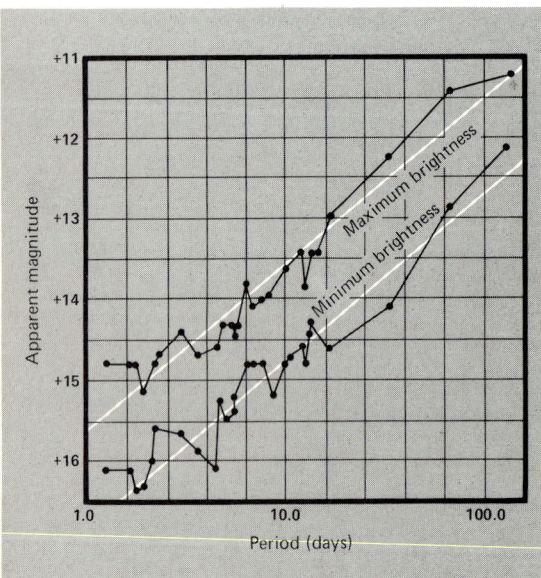

FIGURE 20.19 Leavitt's original period-luminosity relation for cepheids in the Magellanic Clouds. The stars are plotted for the maximum (upper line) and minimum (low line) brightnesses. (*Based on a diagram in the* Harvard Circular, *no. 173, 1912.*)

FIGURE 20.20 A light curve for a typical cepheid variable star—Delta Cephei, the prototype of the class. It shows how the star's brightness varies with time. Note that the rise in brightness is steep, but the decline less so. The period of brightness variation is about 5.4 days.

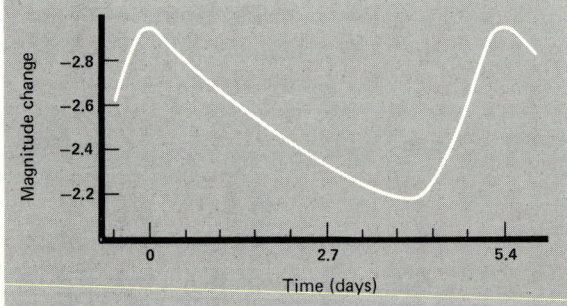

20.4 THE DISCOVERY OF THE GALAXY

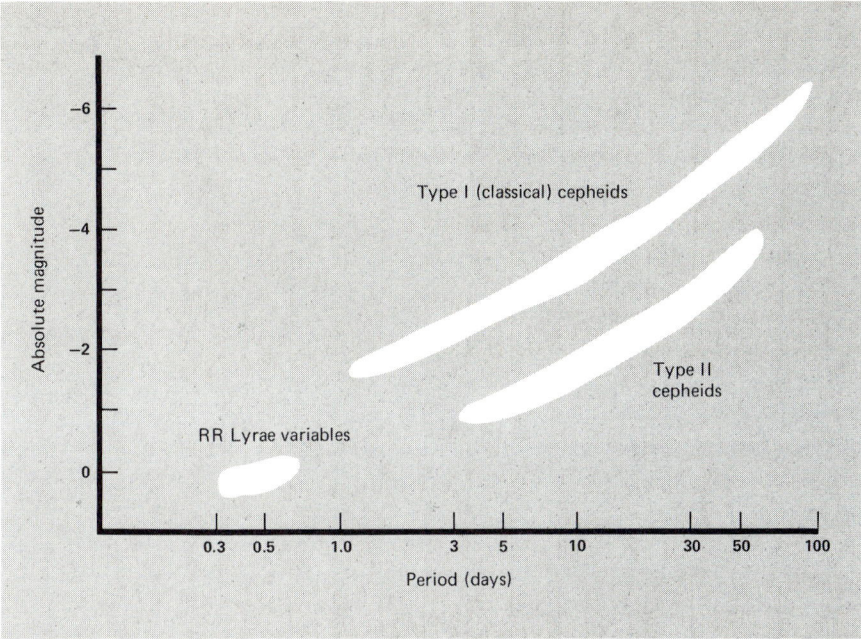

FIGURE 20.21 A simplified period-luminosity relation for cepheids, which fall into two groups—Types I and II. Also shown is the position for RR Lyrae stars, which are short-period variables usually found in globular clusters.

fashion is known as a *periodic variable*; the star Delta Cephei sets the standard for one such class of variables, called *cepheid variables* or *cepheids* (Section 10.9), by the special shape of their light curves.

Hertzsprung deduced from Leavitt's data that a unique relationship exists between the luminosity of a cepheid variable and its period. This connection is called the *period-luminosity relationship* (Fig. 20.21). It was a key discovery, for we can use it to find distances to cepheids. Here's the method. (1) Find a cepheid (identifying it by its light curve). (2) Measure its period (a relatively easy task that does not depend on any knowledge of the star's distance or spectral class). (3) Find the star's absolute magnitude from the period-luminosity relationship. (4) Measure the star's apparent magnitude. (5) Calculate its distance.

The period-luminosity relationship plays a vital role in surveying the Galaxy. With detection of only one cepheid we can find the distance to some object with which the cepheid is associated (assuming no absorption by dust). Note that only two measurements are needed: the star's apparent magnitude and its period.

Let's work out one example in detail. Say we discover a cepheid in a star cluster. We observe for awhile and find that the variable's period is 10 days and its average apparent magnitude is $+7$. Turning to Fig. 20.21, we find that for classical cepheids the absolute magnitude corresponding to a 10-day period is -3. The difference between the apparent and absolute magnitude is $+7 - (-3)$, or $+10$. The star is so far away that it is 10 magnitudes fainter than it would appear if it were at 10 pc. Ten magnitudes equals a brightness ratio of 10^4 (Table 8.1). To get a decrease by 10^4 requires a distance increase of $(10^4)^{1/2}$, or 10^2. So the star is 100 times farther away than 10 pc, or 1000 pc. Since the cepheid is part of a star cluster, we have determined the distance to the cluster to be also 1000 pc.

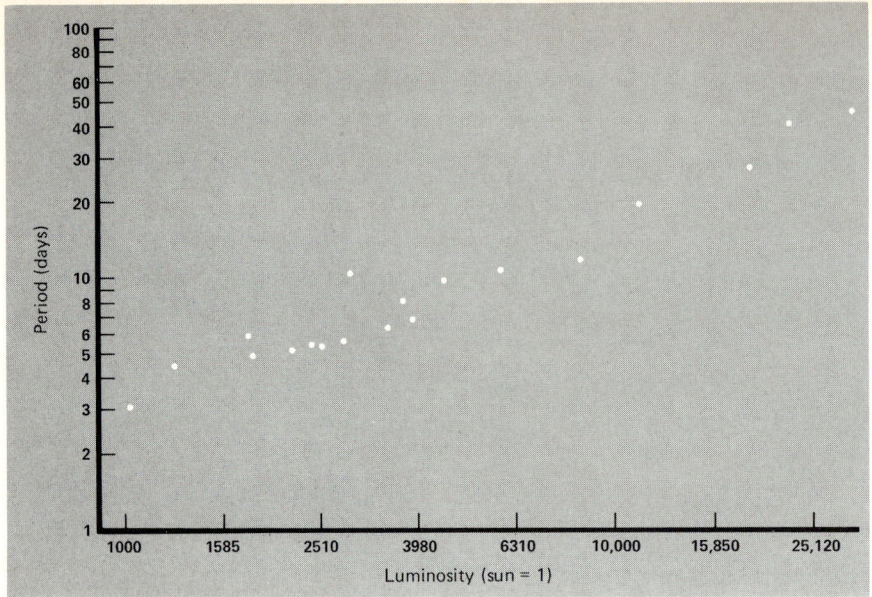

FIGURE 20.22 The period-luminosity relation for cepheids whose distances are known. These serve to calibrate the general period-luminosity relation (Fig. 20.21).

You can also work this out algebraically.

$$m - M = 5 \log d - 5$$
$$7 - (-3) = 5 \log d - 5$$
$$3 = \log d$$
$$d = 10^3 \text{ pc}$$
$$= 2.2 \times 10^3 \text{ ly}$$

There are some problems with this method. First, the technique assumes that dust does *not* cut out the light from the cepheid. If it did, the apparent magnitude would be fainter than it actually is, and you would estimate the star and the cluster to be farther away than its actual distance. Somehow, you must be able to correct for the effect of dust.

Second, today we know that there are actually *three* types of what used to be clumped together as cepheid variables: Type I (classical) cepheids, Type II cepheids, and RR Lyrae stars. Type I cepheids are Population I stars; Type II are Population II. RR Lyrae stars, also Population II, are commonly found in globular clusters. So we must not only identify a star as a cepheid, but also determine what type of cepheid it is.

Third, in order to use the cepheids' period-luminosity relationship, we must calibrate it, that is, we must find the luminosities of a few nearby cepheids. Unfortunately, no cepheids are close enough to measure their distances by heliocentric parallax. So we look for clusters that contain cepheids and find the distances to the clusters by statistical parallaxes (Section 20.3) or main-sequence fitting (Section 10.9). To date we have distances of 20 cepheids; these are the calibrators for the period-luminosity relationship (Fig. 20.22). But there are still some uncertainties in this calibration.

Shapley Dethrones the Sun

In 1917 Harlow Shapley (Fig. 20.23) proposed a radical idea about the size of the Galaxy and the sun's position in this island of stars. The observational foundation of his

20.4 THE DISCOVERY OF THE GALAXY

FIGURE 20.23 Harlow Shapley. "We are all brothers of the boulders, cousins of the clouds." (*By permission of Harvard College Observatory.*)

model rested on the period-luminosity relation of cepheids. His model evicted the sun from its central position in the Galaxy, the second (Copernicus producing the first) significant move of humankind away from special status in the universe.

In 1915 Shapley noted that globular clusters did not follow the distribution of stars along the Milky Way by sticking along the plane. Rather, they tended to be widely distributed, with an odd concentration in the direction of Sagittarius. No other objects were so strangely arranged relative to the Milky Way. In the closer globular clusters he could distinguish what he thought were cepheid variables (the stars which later turned out to be RR Lyrae stars). With his own calibration of the period-luminosity curve, Shapley found these closer clusters to be at a distance about 39,000 ly (12,000 pc) from the sun. Later he estimated M 13 to be 100,000 ly (30,000 pc) distant. This number was staggering, for it placed the globular clusters outside the boundaries of the Galaxy as mapped by Kapteyn.

The nonuniform distribution of globular clusters in the sky remained a nagging problem. Shapley thought that the globular clusters must be gravitationally allied with the Galaxy in a uniform distribution. Then the observed nonuniformity on the sky would indicate that the sun is not placed in the Galaxy's center.

Let's look at Shapley's argument, which rests on the observation of the concentration of globular clusters in the southern sky. From what vantage point do we, whirling around the sun, view these groups of stars? The key lies in a valid assumption about the distribution of the globulars in and around our Galaxy. Shapley knew that our Galaxy had the shape of a flattened disk. He could have assumed that the globular clusters followed the same distribution in space as ordinary stars. However, you would then expect that the

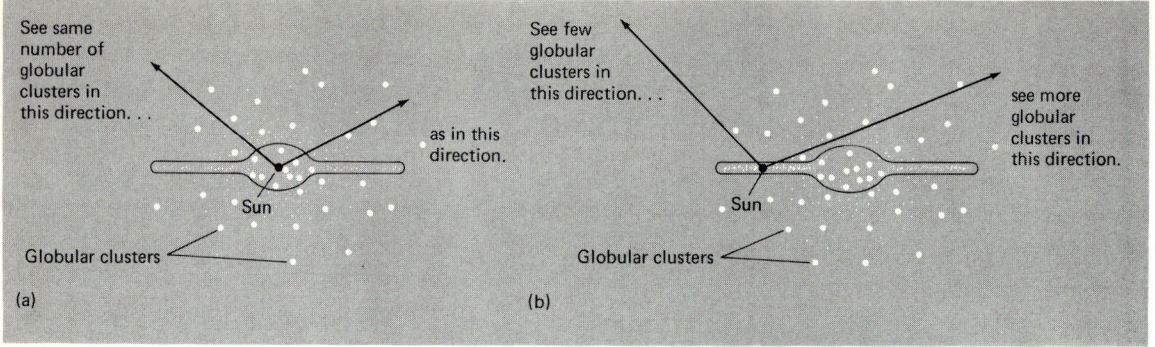

FIGURE 20.24 The position of the sun in the Galaxy inferred from the observed distribution of globular clusters in the sky. (a) The situation if the sun were in the center of the Galaxy. (b) The actual situation with the sun away from the center.

apparent globular cluster distribution in the Milky Way would more or less coincide with the distribution of stars. Such a coincidence does not exist. So Shapley assumed that the globular clusters had a uniform distribution around the Galaxy's center, but not the same distribution as most of the stars.

What are the observational consequences of this assumption? Suppose the sun were located in the center of the Galaxy (Fig. 20.24a). Trace lines of sight in a number of different directions. Because we have assumed a central vantage point in a uniform distribution of objects, every line of sight we choose should intercept the same number of globular clusters. But this uniformity does not in fact occur. Shapley therefore had the choice between dropping his assumption of the central location of the sun or explaining the observed distribution of the globular clusters on the sky by some nonuniform distribution in space. The latter choice would have cut across his aesthetic grain. The alternative was to move the sun away from the center (Fig. 20.24b). Now some lines of sight cut longer distances than others through the globular clusters, so more clusters are seen in these directions than along other lines of sight. The expected distribution is not symmetric, but is most concentrated in the direction of the galactic center. So when the sun was moved from the center, the predicted result matched the observational one.

During 1914, even before his determination of the position of the sun, Shapley had published two papers that foreshadowed his work on the size of the Galaxy. One paper dealt with variable stars in the globular cluster M 13, the other with the nature of variable stars. When he recognized that the variables he observed in globular clusters had the characteristics of cepheids, he adopted the technique of deriving the distance to the cluster by using a calibrated period-luminosity relation.

Here Shapley faced the difficult and crucial problem of how to get the calibration. No cepheid is close enough to the sun to have its trigonometric parallax measured, and in Shapley's day no cepheids were known to be members of galactic clusters. To overcome this lack, Shapley examined the proper motions of a number of nearby cepheids. He then applied the statistical parallax method to estimate the distance to the cepheids. With this estimate, he then could translate apparent magnitudes into absolute ones, and so calibrate the period-luminosity relation (as has been done for Fig. 20.21).

Shapley had measured the periods of many cepheids in globular clusters. Now with the calibrated period-luminosity relation, he could translate the periods into luminosities (or

20.4 THE DISCOVERY OF THE GALAXY

absolute magnitudes). Comparison with the apparent magnitudes of the cepheids gave their distances, and hence the distance of the globular cluster in which they were located. He then took the directions and distances to these clusters and made a three-dimensional model of their distribution. He assumed that the center of that distribution was also the center of the Galaxy, and, since he knew all the distances in the model, he could determine the distance of the sun from the center and the overall dimensions of the Galaxy.

Shapley made two mistakes in his procedure. First, he was wrong in assuming that the variables in globulars were the same as those in the disk of the Galaxy. (The variables in globulars actually are RR Lyrae stars, which have a different period-luminosity relationship.) Second, his calibration was off, so he underestimated the luminosities of nearby cepheids by about a factor of four. By coincidence, the first mistake, which overestimated the variables' luminosities by a factor of four, just about cancelled out the second error. But he also was unaware of another problem that caused him to overestimate the distance to globulars, interstellar absorption by dust.

So the figure Shapley finally arrived at for the galactic diameter—300,000 ly (100,000 pc)—was too large in the light of modern measurements, mainly because he did not account for interstellar absorption.

The Spiral Nebulas

What about Herschel's spiral nebulas (Fig. 20.25)? Were they, like globulars, part of the Galaxy? The question of whether or not they were connected to our Galaxy remained unsettled, because no distances had been found for them. If they were other galaxies and

FIGURE 20.25 A spiral galaxy (Messier 51). (*Courtesy Kitt Peak National Observatory.*)

if they were the same size Shapley proposed for our Galaxy, then their distances, as implied by their small angular diameters, must be immense—*millions* of light years!

In 1917 George W. Ritchey of Mt. Wilson accidentally discovered one of the strongest pieces of evidence for the distances and nature of the spiral nebulas. While photographing spiral nebulas to determine their motion in space, Ritchey caught the flash of a nova in one. Astronomers recognized the significance of this event. Since a nova could occur in a cluster of stars, but not in a cloud of gas, the spiral could not be just a nebula. Other plates at Mt. Wilson were searched, and more novas were discovered. Within two months of the discovery astronomers had found a total of eleven novas in spiral nebulas.

Heber D. Curtis (1872–1942) of Lick Observatory took part in the nova search. He regarded the novas as proof that the spiral nebulas were misnamed, that they were actually systems of stars rather than disks of gas. He devised a clever method of using novas to determine the distances of galaxies of which they were a part.

Let's see how Curtis estimated the distances to spirals. Focus 11.1 detailed how distances to novas in our own Galaxy can be determined from the expansion of their shells. This method shows that the absolute magnitude of a nova in our Galaxy at maximum brightness is roughly -8.

Curtis first assumed—using simplicity and the uniformity of nature as his philosophical foundations—that novas in another galaxy were the same as novas in our Galaxy. In particular, he assumed that they have about the same absolute magnitude at maximum as novas in the Galaxy. (Some support for this idea came from the fact that most novas in the same galaxy have the same apparent magnitudes at maximum.) Using -8 as the absolute magnitude at maximum, Curtis compared this value with the apparent magnitude at maximum to find the distance to the nova and hence to the galaxy.

Here's an example. The nearest galaxy similar to our own is M 31, also called the Andromeda galaxy, because of its location in the constellation Andromeda. Suppose a nova in M 31 shines at apparent magnitude of $+16$ at maximum brightness. We assume its absolute magnitude is -8. Then

$$m - M = 5 \log d - 5$$
$$16 - (-8) = 5 \log d - 5$$
$$\log d = \frac{29}{5} = 5.8$$

and

$$d = 6.5 \times 10^5 \text{ pc}$$
$$= 21 \times 10^5 \text{ ly}$$

That's the distance to the nova and hence to M 31.

Curtis used these measurements of distances to galaxies to infer the size of our own Galaxy. The distances and angular sizes of other galaxies gave their diameters, which were smaller than Shapley's diameter for the Galaxy. Curtis then argued that Shapley's measurements of the size of the Galaxy were incorrect and that Kapteyn's smaller value was the proper one.

The Shapley-Curtis Debate

The opinions of Shapley and Curtis represented the view of two different schools of thought among astronomers around 1920 concerning the spiral nebulas. The older view, upheld by Curtis, pictured the spiral nebulas as galaxies of approximately the same size and shape as our Galaxy (according to his estimate, roughly 30,000 ly, or 10,000 pc, in diameter). The opposing view, espoused by Shapley, pictured the Galaxy as outlined by globular clusters and about 300,000 ly (100,000 pc) in diameter. This size ruled out the spiral nebulas as galaxies, because their relatively close distances, as estimated by Shapley, were not consistent with the large diameters required if the spirals were the same size as the Galaxy. Shapley relegated the spirals to a place as minor groupings of stars at the fringe of the Galaxy. His view was simultaneously radical and conservative, for his large size for the Galaxy excluded similar systems, whereas Curtis envisioned a universe filled with spiral systems.

Shapley and Curtis debated their respective positions at the April 1920 meeting of the National Academy of Sciences in Washington, D.C. The kernel of the conflict between Shapley and Curtis was not the general shape of the Galaxy, the sun's position within it, or the fundamentals of any of the astronomical surveying methods. One difference was basically technical and revolved around points of detail (including an erroneous measurement of proper motions for some spiral nebulas). Another difference was philosophical: Did other galaxies exist in the universe?

Both men were right in their general scheme of the Galaxy, but neither had arrived at the correct size. Shapley's was two times too large, and Curtis's was too small by the same amount. (In the present view the Galaxy is about 120,000 ly, or 40,000 pc, across; the sun is about half of the Galaxy's radius from the center.) Curtis did score some points for the external galaxy theory, which Shapley later accepted after considering additional evidence. (One piece came when M 31, the nearest spiral, was resolved into stars by the 100-in telescope.)

However, Shapley had shaken one of our cosmic foundations, and people could no longer picture themselves proudly in the center of the Galaxy, but had to rather humbly accept their place in the outskirts.

20.5 STELLAR POPULATIONS AND THE STRUCTURE OF THE GALAXY

With the general outline of the Galaxy developed, astronomers turned to working out its structure in more detail. One new insight came by accident, helped by World War II. In 1944 Walter Baade was working at the 100-in telescope at Mt. Wilson. As it was wartime, the German-born Baade was restricted to staying in the Mt. Wilson–Pasadena area. Because his colleagues were off working for the government, Baade had access to plenty of time on the telescope. In addition, Los Angeles and neighboring towns were blacked out, resulting in an unusually dark sky.

Using film sensitized to red light, Baade photographed the Andromeda galaxy, M 31 (Fig. 20.26), and resolved the central region into stars. (See Plate 15 for a color photo of M 31.) When he made an H-R diagram of the stars in the central bulge, he found that most of the stars were red giants, and few were main-sequence stars. He was impressed by

FIGURE 20.26 (a) Messier 31, the spiral galaxy in Andromeda. The stars in the spiral arms in the disk are mostly Population I (*Courtesy Kitt Peak National Observatory.*) (b) Individual stars in the nucleus of Messier 31. These are Population II. (*Courtesy Palomar Observatory, California Institute of Technology.*)

the similarity of these stars to those in the globular clusters studied by Shapley. He also noted the differences between these stars and those found in the solar neighborhood or in galactic clusters. From those observations Baade suggested that the stars in a galaxy, like ours or M 31, could be put into two groups, which he called *Population I* and *Population II*. (Populations in the context of stellar evolution were introduced in Section 10.9.)

Baade noticed also that the two populations were found in different regions of the Andromeda galaxy and inferred the same situation in our Galaxy. Astronomers had found the Galaxy to be shaped like a thin *disk*. Viewing some other galaxies, such as Andromeda, they noted a bright *nuclear bulge* in the central region; its core is the nucleus. Globular clusters mark out the third general region of the Galaxy: a *halo* that is spherically distributed around the nucleus. The halo consists of globular clusters, some stars, and a little hot gas.

What kinds of stars populate these three regions? Baade concluded that the stars in the disk outside of the nuclear bulge are mostly Population I. The nuclear bulge he labeled as primarily Population II. And the halo stars, which are mostly in globular clusters, he also assigned to Population II.

Although it marked a first step toward working out the details of galactic structure, astronomers now recognize that Baade's division of stars into only two populations was too simple. We have since learned that the stars in globular clusters are not the same as those in the nuclear bulge. Although both groups contain numerous red giants, the globular cluster stars contain a much smaller abundance of heavy elements, whereas the nuclear stars have about the same abundance of heavy elements as in the sun. Astronomers today, therefore, refer to stars in the nuclear bulge as *old Population I*, those in the disk as *young Population I*, and only those in the halo as Population II. Some astronomers make even more divisions, for there is no strict boundary between groups, but for simplicity we will stick with these three.

In summary, young Population I stars are found in the Galaxy's disk. The most luminous are hot, blue stars; the youngest are only a few million years old. About 2 percent of their composition, by mass, is heavy elements. Astronomers call these stars *metal rich*. Old Population I stars are found in the nuclear bulge. They are also metal rich, but the most luminous are red giant stars. They are old, probably more than 10 billion years in age. Population II stars are found in the halo, especially in globular clusters. The most luminous are red giant stars. Typically, they have only about 1 percent of the heavy elements found in Population I, that is, 0.02 percent of the total. These stars are called *metal poor*. They are also old, the oldest stars in the Galaxy.

With these stellar populations in mind, let's look at the contents of the different parts of the Galaxy.

20.6 THE DISK OF THE GALAXY

The Galaxy's disk contains stars, gas, and dust. You've already been acquainted with these components in Chapters 8 and 9; this is a good place to review that material. Here we'll deal just briefly with the overall distribution of stars and the interstellar medium.

The Common Stars

Recall from Section 8.5 the H-R diagrams for the brightest stars (Fig. 8.15) and for the stars within 20 ly of the sun (Fig. 8.16). These groups complemented each other. The brightest stars contained many very luminous ones and the nearest stars many very low-luminosity ones.

One goal of studying the Galaxy is to find out, on the average, the properties of stars in a typical volume of space. For example, if we looked at a cube 1000 ly (300 pc) on each side, how many stars would we find with the sun's luminosity? In other words, what would a star census of our Galaxy look like?

To take such a census requires a combination of methods to get the distances and luminosities of many stars. For nearby stars, we use heliocentric parallax. For the distant stars (which turn out to be the most luminous ones), we employ both spectroscopic and statistical parallaxes. The final census, for stars of all spectral types, is called the *luminosity function* (Fig. 20.27). It is simply a compilation (but very hard to get) of how many stars of a given luminosity can be found in a typical volume of space in the disk of the Galaxy.

What does this census reveal? First, the bulk of stars are ones of very low luminosity. Note that even the sun has a greater luminosity than most of the stars in the disk. Second, although very luminous stars are rare, they contribute the most to the background starlight in the disk.

Once the luminosity function is calibrated in terms of mass, we can say even more. In a typical region of the Galaxy near the sun there is about 0.0014 solar mass of stars in a cubic light year (0.05 in a cubic parsec) emitting 0.0017 solar luminosity of visible light. The stars are spaced about 7 ly (2 pc) apart. That should give you an idea of how empty the Galaxy's disk is in terms of stars.

How thick is the disk? Since there is no sharp boundary to the Galaxy, the answer depends on what you mean by thickness. It also depends on the objects chosen to measure

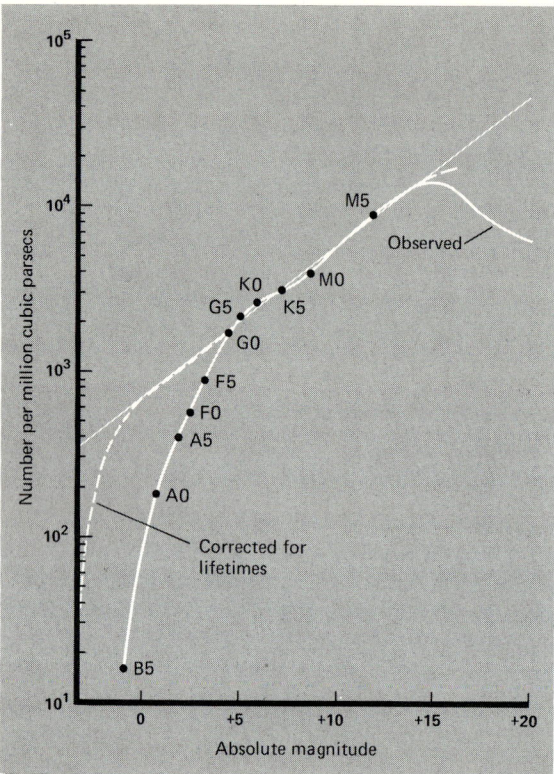

FIGURE 20.27 A luminosity function for stars in a typical volume of space in the Galaxy. The solid line indicates the observed numbers; the dashed line is what we would see of high-mass stars if their lifetimes were not relatively short.

it with. Here let's use stars, and we'll define the thickness as the distance from the plane where the density falls to half its midplane value. Stars of different spectral type do not congregate the same way above and below the galactic plane. For instance, O stars spread out about 160 ly (50 pc) above and below the plane; G main-sequence stars about 1100 ly (350 pc). Using G, K, and M stars to define the thickness gives a result of about 2300 ly (700 pc). The Galaxy's disk is razor thin compared with its diameter of about 120,000 ly (40 kpc).

Where's the sun in the direction perpendicular to the plane? It lies about 39 ly (12 pc) above the galactic plane.

The Interstellar Medium Revisited

The gas and dust of interstellar space have a distribution very different from that of the stars. Let's look first at how neutral hydrogen (H I) is arranged and then at molecular hydrogen (H_2).

The H I gas forms a thin layer that has a thickness of about 800 ly (250 pc) at the sun's location. But at greater distances from the center the thickness increases considerably, reaching 9000 ly (3000 pc) at the outer edges of the Galaxy. In addition to the gas widening out at the outer parts, it's warped; it bends "up" on one side and "down" on the other. The average density in most of the disk is about 0.3 atoms/cm^3, though much less in the region beyond the sun. The total mass of H I is about 3×10^9 solar masses.

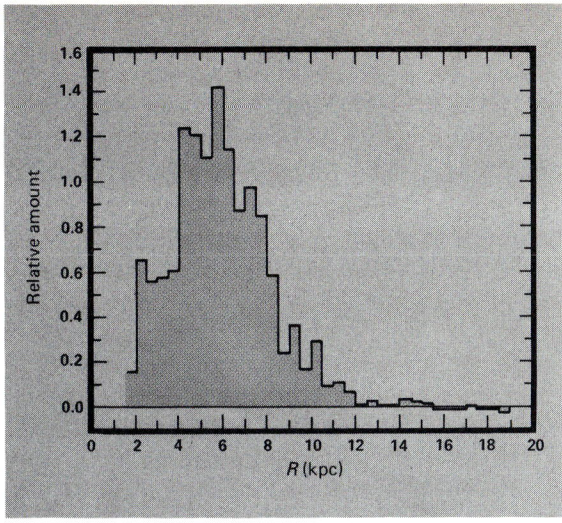

FIGURE 20.28 The relative density of CO in the galaxy from 2 to 19 kpc (6000 to 60,000 ly) from the galactic center (R). (*Based on data compiled by M. Gordon and B. Burton.*)

The inner part of the Galaxy, within 6000 ly (2000 pc) of the center, is also peculiar. Observations indicate that much of the gas here lies in a thin disk tilted some 24° with respect to the galactic plane. The disk has a thickness of 300 ly (100 pc) and a maximum rotational velocity of 360 km/sec. One model has the gas streaming on elliptical orbits, which gives the appearance of radial motions at some points, but there is no net expansion or contraction. If the model is right, and if the stars also have this elliptical distribution, then the center of our Galaxy may be like that of a barred spiral (see Chapter 22)! The disk contains some 10^7 solar masses of H I.

What about the distribution of H_2? Because we cannot observe H_2 directly with radio telescopes, we look at CO instead. There is considerable evidence that wherever we find CO, H_2 must also exist, that is, if the interstellar environment is suitable for making CO, it must also shelter H_2.

The CO distribution does not in general follow that for H I. Observations indicate that the CO is most dense at a distance of 20,000 ly (6 kpc) from the Galaxy's center (Fig. 20.28). Outward from 26,000 ly (8 kpc) the density of CO takes a slide, whereas the H I density stays roughly the same. Inside 13,000 ly (4 kpc) the CO density also decreases, but not as rapidly. The CO layer has a thickness of 410 ly (125 pc). Remember, the CO indicates the presence of H_2, molecular hydrogen. Within the sun's distance from the Galaxy's center, about 93 percent of the hydrogen exists as H_2. In contrast, outside of the sun the hydrogen takes the form of H I.

In one respect the molecular distribution is like that of the H I; it shows the same tilted disk (or bar) structure in the innermost part of the Galaxy. The disk contains some 10^9 to 10^{10} solar masses of H_2. The total mass, determined from the rotation of matter outside the disk, is only a few times 10^{10} solar masses. So this inner part of the Galaxy may be exceptionally rich in gas, rather than being predominantly composed of stars.

The interstellar dust is mixed in with the gas. We presume that it more or less follows along with the CO distribution because we find dust in molecular clouds. Surveys of the reddening of stars show that the amount of reddening is generally the greatest in and around the Milky Way, especially in the direction of the galactic center. It is the extinction from dust that prevents our getting a direct optical view of the galactic nucleus. If the dust were not present, the nucleus would appear brighter than the full moon!

20.7 THE NUCLEUS OF THE GALAXY

Although dust largely obscures the center of the Galaxy, there are a few "windows" of lesser absorption where optical astronomers can glimpse at least a few parts of the nuclear bulge. In addition, we can surmise the nature of the stars in the nucleus from observations of the nuclei of other galaxies, such as M 31. From these two kinds of observations, we picture the nucleus as containing mostly old Population I stars densely packed together.

Because dust scatters red light much less than blue light, infrared observations can penetrate into the nucleus. Radio waves, which are longer than infrared, are even less affected by the dust. X rays are so energetic that they go right through the dust. So radio, infrared, and X-ray astronomers can probe the nucleus more easily than can optical astronomers.

The non-optical observations show that the heart of the Galaxy is a bizarre place. Not only does it contain many very old Population I stars, but also some very young supergiant M stars and O stars. Motions of gas here suggest a high concentration of mass at the center, perhaps a black hole. In the very center of the Galaxy lies a radio source less the 140 AU (about 10 light *hours*) in size.

Radio Observations

Let's look first at the continuous emission of the galactic center (Fig. 20.29). An intense radio source lies smack in the direction to the center. It is called *Sagittarius* A, or *Sgr* A for short. Clustered around Sgr A and lying more or less along the galactic plane (or galactic equator) is a string of radio sources. When investigated at different radio wavelengths, these sources appear to have characteristics of H II regions—hot ionized gas around young OB stars. The total extent of this region is about 290 by 850 ly (90 by 260 pc); and the ultraviolet energy output from the OB stars needed to keep the region ionized is at least 2×10^{33} W, or 5×10^6 solar luminosities.

Sgr A itself is a bit different. Some of the radio radiation is from ionized gas, but some also is synchrotron emission coming from high-energy electrons traveling through a magnetic field (Section 11.4). A high-resolution radio map (Fig. 20.30a) shows that Sgr A actually consists of two separate radio sources. *Sgr A East* emits a nonthermal spectrum (synchrotron emission is likely); *Sgr A West* has a thermal spectrum like a giant H II region. Sgr A West is associated with an agglomeration of infrared sources (next section). Within Sgr A West is a pointlike radio source less than 0.1" in diameter that may mark the actual core of the Galaxy. The ionized gas here, which amounts to a few million solar masses of material, seems to be rotating at about a few hundred kilometers a second.

The VLA has recently mapped the galactic center (Fig. 20.30b). This map, which has been computer-processed to take out the nonthermal emission from the point source within Sgr A West and leave only the thermal emission of Sgr A West, shows the radio output from the inner 10 ly (3 pc) of the Galaxy. Note that the bulk of the radio emission lies along a ridge source at the galactic center. A more recent map shows Sgr A West clearly (Fig. 20.30c).

Now to check out the radio line emission. Luis Rodriguez and Eric Chaisson have observed radio hydrogen recombination lines (see Section 9.1) with the 36.5-m Haystack

20.7 THE NUCLEUS OF THE GALAXY

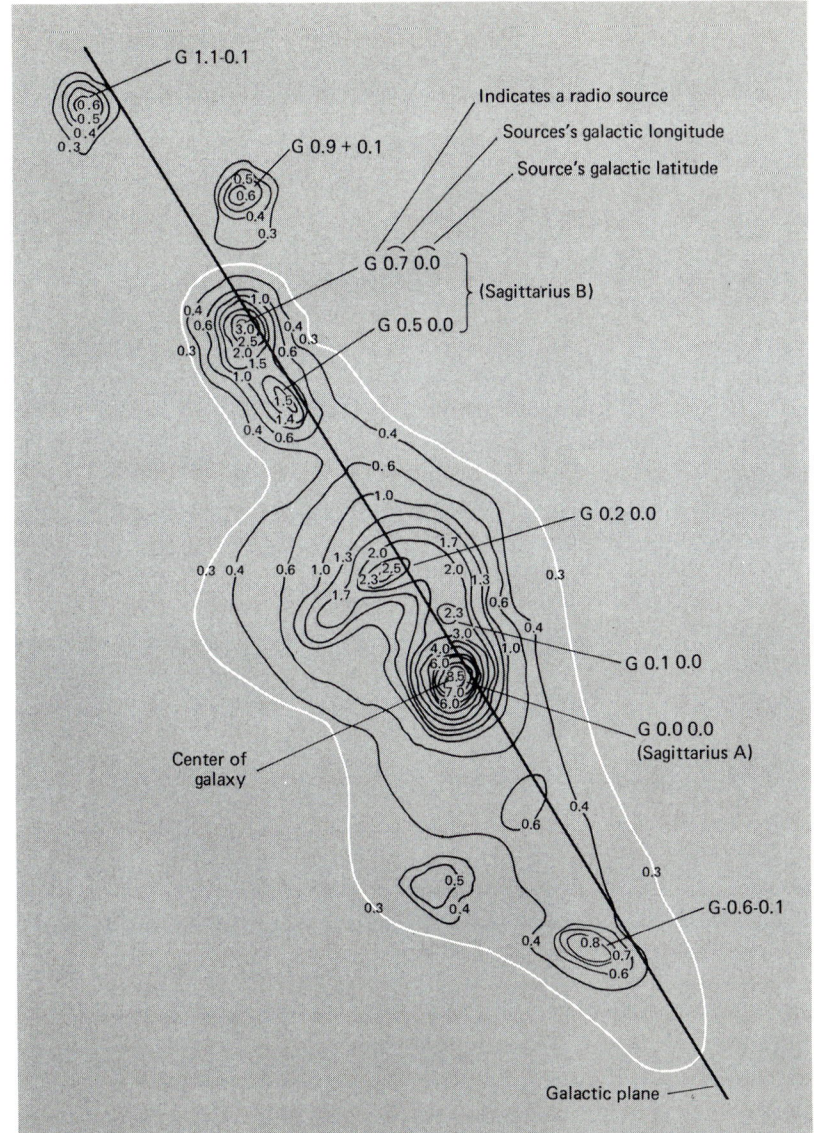

FIGURE 20.29 A radio map of the galactic center region at a wavelength of 3.75 cm. The intensity contour map shows where the emission is the strongest. One strong source, Sagittarius A, is associated with the Galaxy's center. Note that this view is edge on to the nucleus. (*Adapted from a diagram by D. Downes, A. Maxwell, and M. L. Meeks, who made this map with the Haystack Observatory antenna.*)

radio antenna. They find these lines from the ionized hydrogen in the galactic center to be very broad. Now one physical process that commonly broadens spectral lines is the Doppler shift from the rotation of a mass of gas. The regions of the gas approaching us give blue-shifted lines, those receding red-shifted ones. When the entire rotating mass is observed within the field of view of a radio telescope, all the various Doppler shifts get smeared together, resulting in a broad line. The lines observed by Rodriguez and Chaisson, if interpreted in this way, imply rotational speeds of 150 km/sec at a distance of 6 ly (2 pc) from the Galaxy's center.

Sgr A also gives off radio line emission from molecules such as CO. The observations indicate that this molecular cloud, which may contain as much as a million solar masses

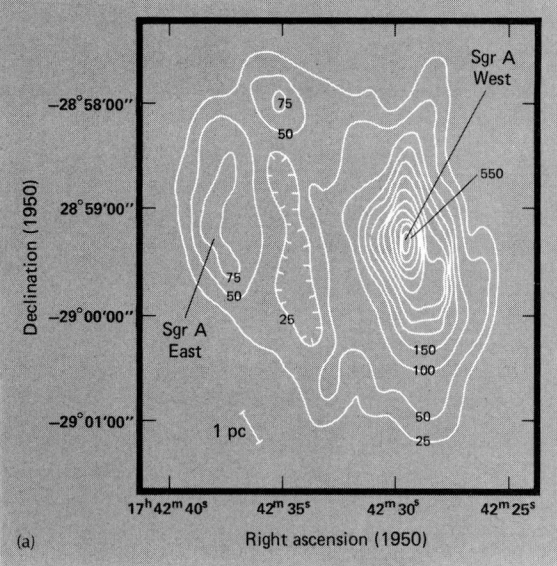
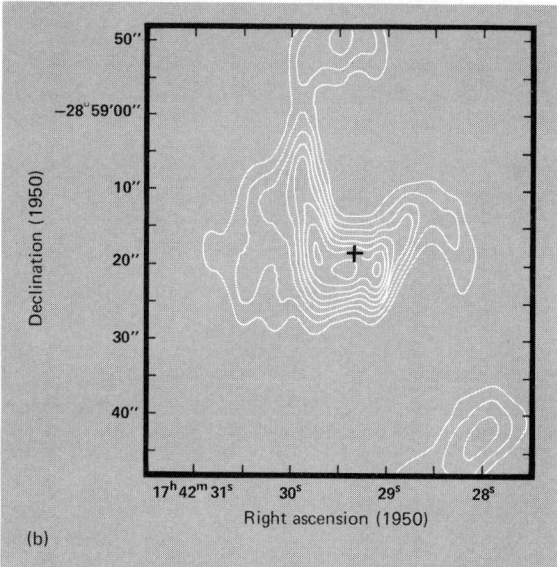
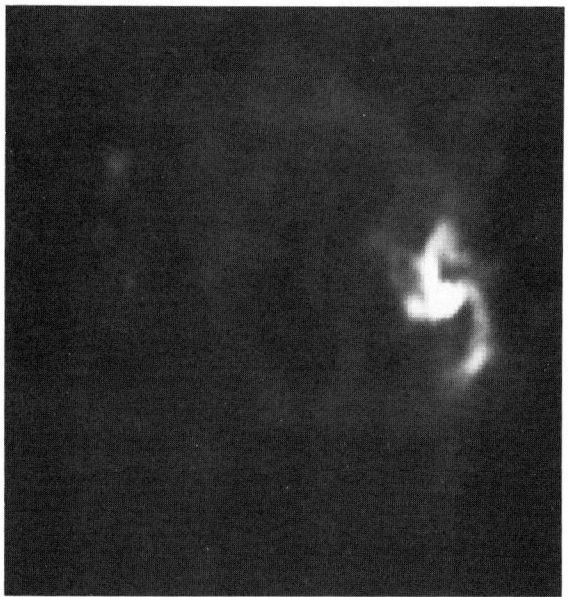

FIGURE 20.30 (a) A high-resolution map of the Sagittarius A (Sgr A) source. Sgr A West appears to be a giant H II region and is at the Galaxy's core. Sgr A East is a nonthermal source. (*Based on observations by R. Ekers, W. Goss, U. Schwarz, D. Downes, and D. Rogstad.*) (b) A map of the galactic center made with the VLA at 5 GHz. The cross indicates the position of the pointlike nonthermal source within Sgr A West. This may be the Galaxy's core. (*Adapted from a diagram by R. Brown, K. Johnston, and K. Lo, Astrophysical Journal, vol. 250, p. 155, copyright 1981 by the American Astronomical Society.*) (c) A computer-generated radio photo of Sgr A made from VLA observations at a wavelength of 6 cm. The area in this picture is about the same as that for (a). Sgr A West shows up clearly as the white region at the right. (*Courtesy R. D. Ekers; observations by R. D. Ekers, U. J. Schwarz, and W. M. Goss.*)

of material, is associated more with Sgr A East than with Sgr A West, so it is not right at the center of the Galaxy.

More interesting from a molecular viewpoint is the source east of Sgr A, known as *Sagittarius B (Sgr B)*. This source consists of at least seven compact H II regions embedded in a rich molecular cloud some 100 ly (30 pc) in diameter. Almost every interstellar molecule has been first discovered here, including ethyl alcohol (the drinking kind).

NE

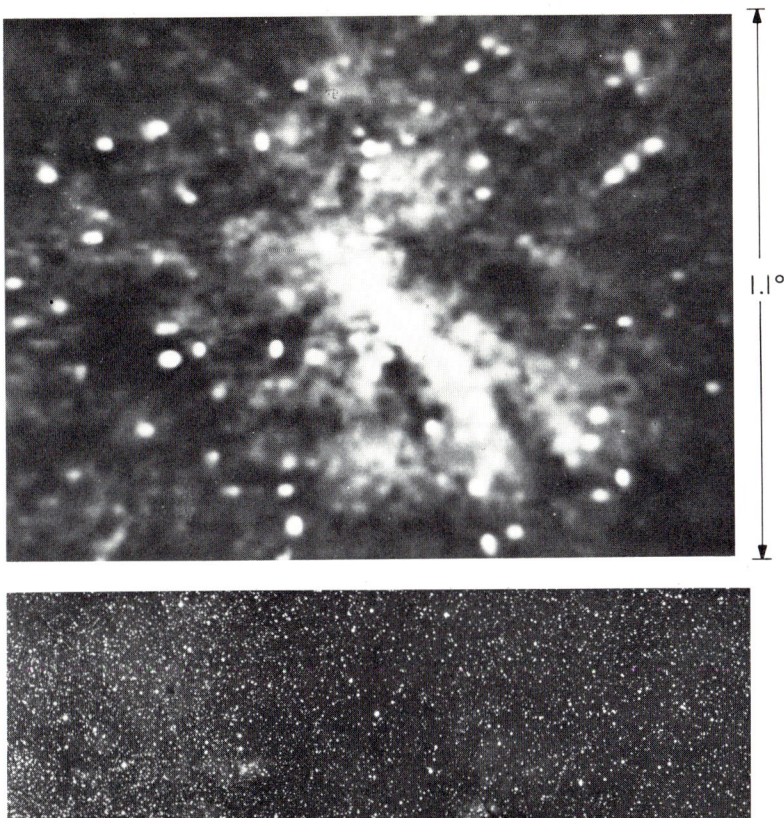

1.1°

FIGURE 20.31 A 2-μm map of the galactic center (top) compared to an optical photo (bottom) of the same region of the sky. (*Courtesy G. Neugebauer.*)

Infrared Observations

Early infrared observations by Eric Becklin and Gerry Neugebauer showed that the galactic center region emits strongly at 2.2 μm (Fig. 20.31). The most intense part of this emission coincides with Sgr A. The source of this radiation is simply the combined 2.2-μm emission from all the old Population I stars (probably mostly from K giants) that inhabit the galactic nucleus.

Higher angular resolution with the 5-m Hale telescope revealed that the region just around Sgr A was packed with 2.2-μm sources (Fig. 20.32a). One of these (A) coincides with Sgr A West. Another (B) is probably an M supergiant star. Others are stars or clusters of stars, most likely of spectral type M. (This infrared cluster coincides with the ridge of radio emission seen in Figure 20.30b.)

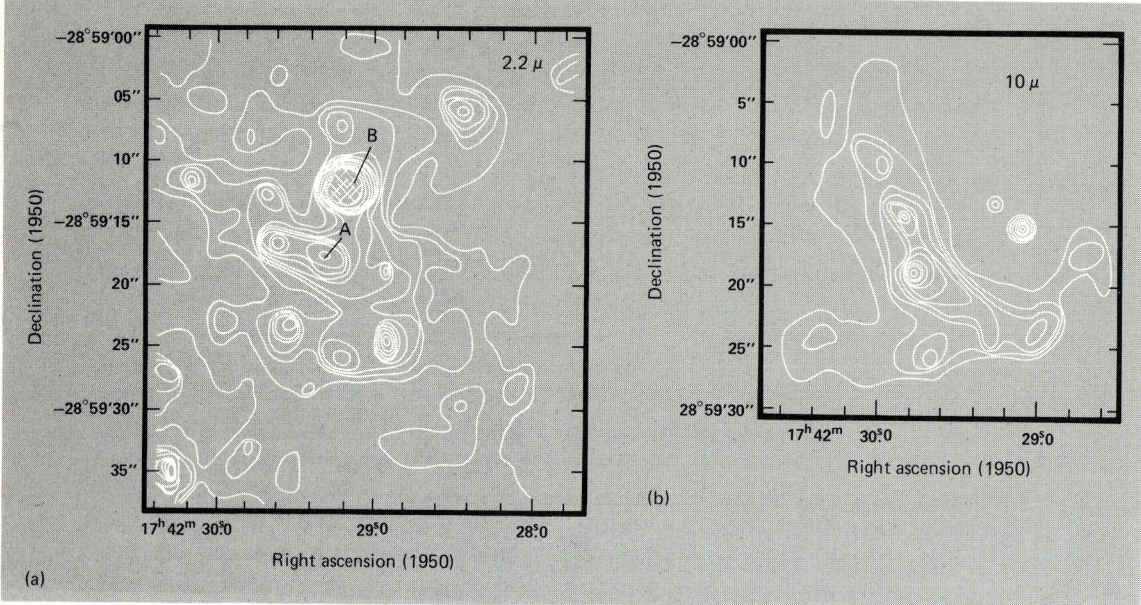

FIGURE 20.32 (a) A detailed 2-μm map of the galactic center region. Most of these infrared sources are thought to be stars or clusters of stars. Source A coincides with Sgr A West and so marks the center of the Galaxy. (*Adapted from a diagram by* E. Becklin and G. Neugebauer, Astrophysical Journal, *vol. 200, p. 71, copyright 1975 by the American Astronomical Society.*) (b) A 10-μm map of the Galaxy's center (region around Source A in the previous map). This infrared radiation from dust heated by stars. (*Adapted from a diagram by* E. Becklin, K. Matthews, G. Neugebauer, and S. Willner, Astrophysical Journal, *vol. 219, p. 121, copyright 1978 by the American Astronomical Society.*)

A map of the same region at 10 μm looks quite different (Fig. 20.32b). Here we are seeing the infrared emission from dust that is heated by the radiation from stars. Some of the heating radiation comes from the old Population I stars. But some also derives from high-luminosity O stars; the condensations in the 10-μm map are probably the locations of newly formed O stars. These regions have diameters of less than a few light years, the same size as small H II regions. The combined luminosity from them, in the range from 2 to 20 μm, is roughly a million times that of the sun.

Far-infrared observations (by airplane and balloon-borne telescopes) have found that Sgr A emits more intensely at 40 to 300 μm than at 10 μm (Fig. 20.33). (Sgr B is also a powerful source at these wavelengths.) Infrared astronomers estimate that about 100 million times the sun's luminosity pours out from the galactic center at these far-infrared wavelengths. This emission also probably comes from dust, heated by O and B stars in this region of space. Dust absorbs the light from these stars, so we cannot see them as stars optically.

A series of beautiful infrared observations of an infrared emission line at 12.8 μm produced by singly ionized neon (Ne II) has provided new insights about the core of the Galaxy. The emission comes from ionized gas in the nucleus and, as for radio recombination lines, information about the core can be inferred from the width and Doppler shift of the line.

20.7 THE NUCLEUS OF THE GALAXY

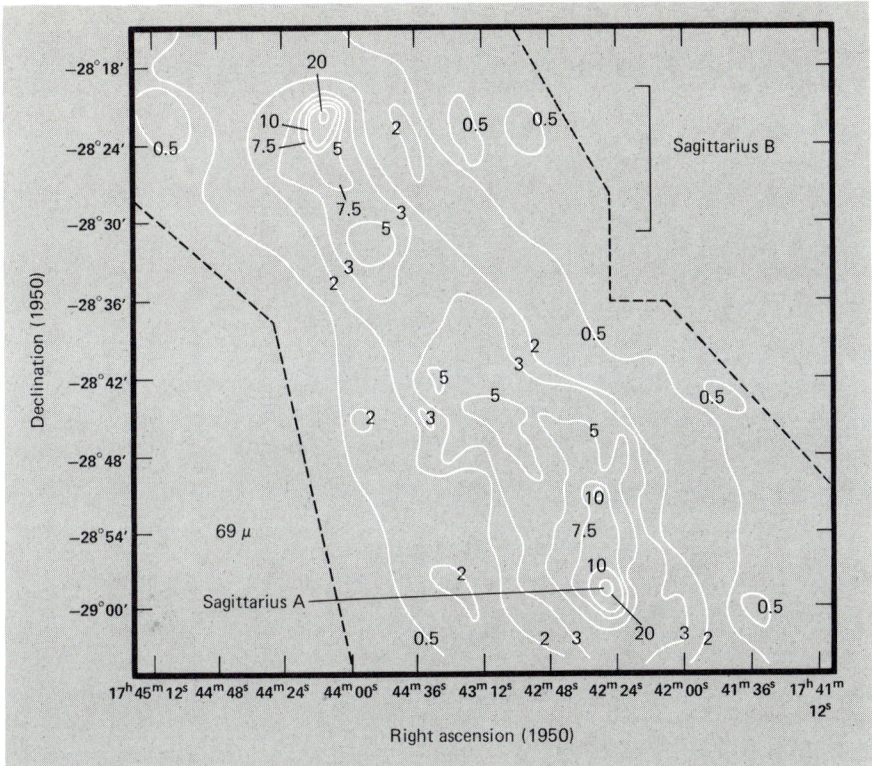

FIGURE 20.33 A 69-μm map of the galactic center region, including Sgr A and B. The view is over a much larger region than in the previous two maps. (*Courtesy G. Frazio and the center for Astrophysics.*)

These observations show that the ionized gas concentrates in a region about 5 ly (1.5 pc) in radius. The lines are very broad, an indication of random motions of around 200 km/sec. The lines show a systematic trend from blue to red shifts across the infrared core. This trend implies that the gas rotates at about 150 km/sec at 1.3 ly (0.4 pc) from the center. The core contains at least seven discrete, ionized sources, many of which coincide with the peaks in the 10-μm map. These ionized zones are small, less than 1.5 ly (0.5 pc) in diameter, and contain a few solar masses of ionized material. They appear to orbit around the galactic center on an axis tilted 45° to the main rotational axis of the Galaxy. So the galactic core contains, within the Sgr A molecular cloud complex, a disk of rotating ionized gas (Fig. 20.34).

X Rays from the Galactic Center

Astronomers have known for a number of years that the galactic center emits X rays. Early observations by the Uhuru X-ray satellite found an extended X-ray source about 2° in size. Later observations with other satellites discovered two pointlike, short-lived sources very close to the galactic center. More recent observations have detected bursts of X rays, with the amount of energy in each burst comparable to that from the short-lived sources. To date it is not clear if or how these sources relate to each other.

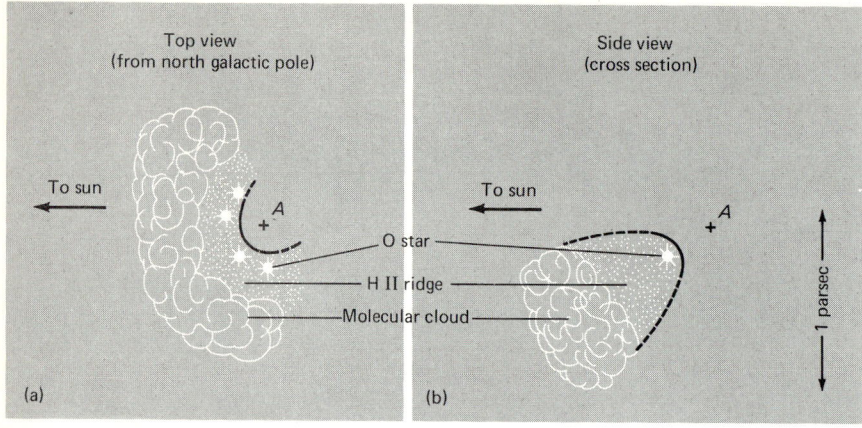

FIGURE 20.34 A schematic map of a top view (a) and side view (b) of the inner 3 ly of the Galaxy. The position marked A is the source in the infrared maps. (*Adapted from a diagram by E. Wollman.*)

The region within 2° of the galactic center consists of at least four, and probably six, separate sources. Assuming that these sources lie 33,000 ly (10 kpc) away, their X-ray luminosities range from 0.4 to 3.5×10^{30} W (1 to 9×10^3 solar luminosities). They all may be the sources of the X-ray bursts. None coincides with any of the main features seen in radio or infrared observations.

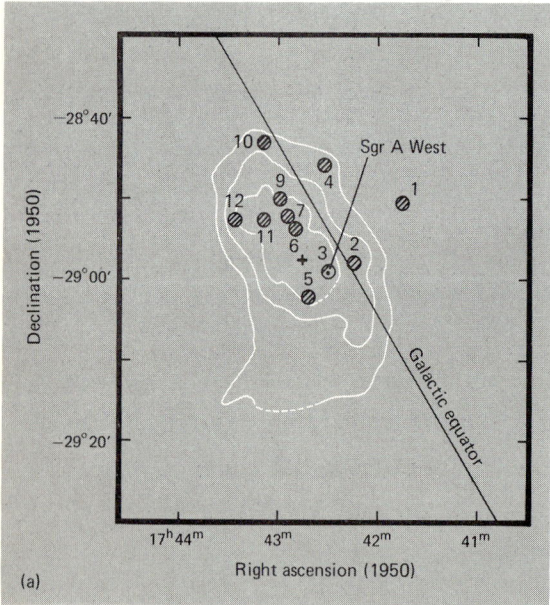 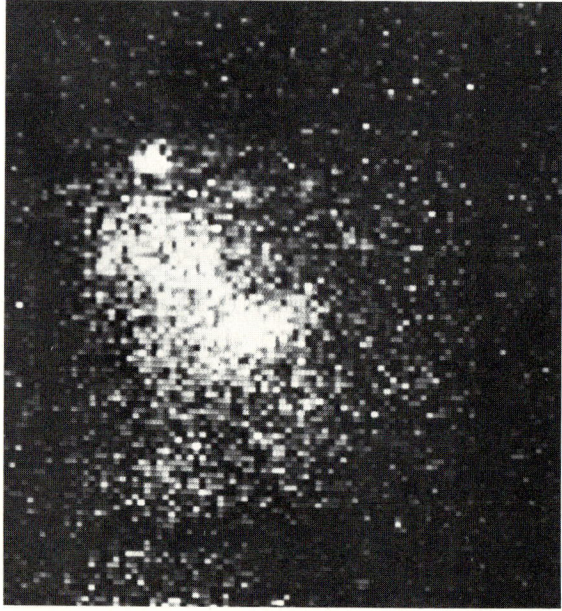

FIGURE 20.35 (a) A schematic map of the X-ray emission from the galactic center region. The shaded circles are discrete sources; the contour lines show the diffuse emission; the cross marks the center of this emission. (b) A computer-generated photo of the X-ray emission observed by the Einstein X-ray Observatory. (*Courtesy J. Grindley, Harvard-Smithsonian Center for Astrophysics; both figures from M. Watson, R. Willingale, J. Grindley, and P. Hertz,* Astrophysical Journal, *vol. 250, p. 142, copyright 1981 by the American Astronomical Society.*)

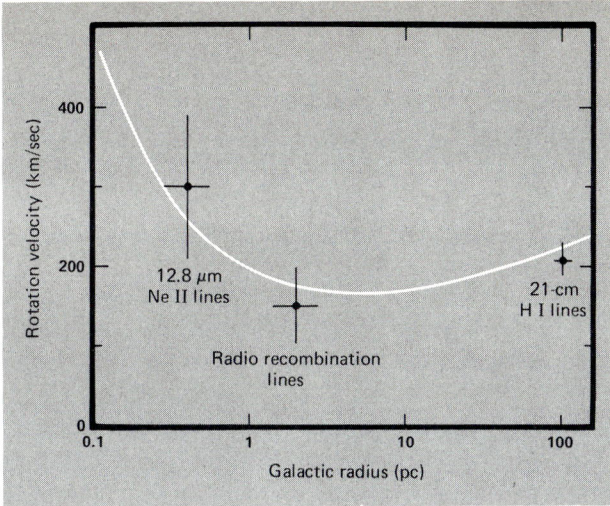

FIGURE 20.36 A rotation curve for the inner 300 ly of the Galaxy. The line fits both radio and infrared observations; it assumes a central mass of 5 million solar masses. (Adapted from a diagram by L. Rodriguez and E. Chaisson, Astrophysical Journal, vol. 228, p. 734, copyright 1979 by the American Astronomical Society.)

Recent observations with the Einstein X-Ray Observatory have finally revealed the X-ray image of the galactic center region (Fig. 20.35a). They show modest X-ray emission, at wavelengths below 0.6 nm (6 Å), within 300 ly (100 pc) of the galactic center. This emission consists of a complex of weak sources covering 200 ly by 150 ly embedded in a weaker halo of diffuse X-ray emission (Fig. 20.35b). At least three stronger and nine weaker sources make up the cluster of discrete sources.

One of these coincides with the position of Sgr A West (infrared source A); it has an X-ray luminosity of some 10^{28} W. Over a six-month span this source did not show any significant variation in luminosity. The other discrete sources also have X-ray luminosities of about 10^{27} to 10^{28} W and lie along a ridge just south of the galactic equator, in the same location as the cluster of infrared sources.

Does a Black Hole Lurk in the Core?

A puzzle generated by the radio and infrared line observations arises from the rapid rotational motions near the Galaxy's core. It appears that the rotational velocities increase closer to the core (Fig. 20.36).

Why is that a problem? The rotational velocities are so high that a huge concentration of mass is needed to hold all that rapidly moving gas together. For example, if the Galaxy's core simply contained a cluster of stars, you would expect the rotational velocities to decrease toward the core because as you get closer in, you have less and less mass to bind the moving materials gravitationally. To account for the rapid rotation requires a mass in the core of several million solar masses—all lumped together in a region only 0.13 ly (0.04 pc) in diameter.

An approximate calculation shows this point. The infrared line observations imply rotational velocities of about 200 km/sec at a radius of 1 ly (= 9.5×10^{15} m). Apply the same argument as was done for pulsars (Section 11.5), in which a rotating, spherical mass just holds together by its own gravity:

$$V_{eq} = \left(\frac{GM}{R}\right)^{1/2}$$

where V_{eq} is the velocity of the object's equator. Solve this equation for the mass, M:

$$M = \frac{RV_{eq}^2}{G}$$

$$= \frac{(10^{16} \text{ m})(200 \times 10^3 \text{ m/sec})^2}{7 \times 10^{-11} \text{ nt} \cdot \text{m}^2/\text{kg}^2}$$

$$= 6 \times 10^{36} \text{ kg}$$

$$= 3 \times 10^6 \text{ solar masses}$$

What form might this mass have? One possibility, and it's very hard to come up with another, is that the mass is locked up in a black hole. If it were in the form of, say, solar-mass stars, these stars would be separated on the average only 1 to 2 AU from each other. That seems unlikely, because stars so close, especially if many of them were red giants, would collide rather frequently.

The idea that a massive black hole lurks in the heart of the Galaxy has yet to be confirmed. Indirect support for the idea comes from observations that a few other galaxies may have a similar mass concentration in their nuclear region (Chapter 22).

To recap, infrared, radio, and X-ray telescopes can directly probe the galactic center. They have found that while the nucleus is very small, less than a few tens of light hours in diameter, it emits enormous amounts of energy, about 10 percent of the total from the Galaxy. Within it material orbits the center at a rapid rate. At the very center lies a massive object, perhaps a black hole.

20.8 THE HALO OF THE GALAXY

When Shapley investigated the distribution of globular clusters in space (Section 20.4), he found that they had a nearly spherical distribution around the galactic center. The globular clusters form a halo around the Galaxy. Little else is known to exist in the halo for sure. Stray stars are seen. The halo contains some gas that is hot and ionized. Cosmic rays whirl around, trapped by the Galaxy's magnetic field. The halo may also contain as yet undetected objects, such as very faint, low-mass stars, and extend at least 120,000 ly (40,000 pc) beyond the edge of the disk.

Globular Clusters

Section 10.8 presented the physical characteristics and stellar content of globular clusters (Table 10.3). To review briefly, a globular cluster has a spherical shape (some tens to hundreds of light years in diameter) and contains 10^4 to 10^5 Population II stars, each a little less than 1 solar mass on the average.

The globular clusters form a sphere around the Galaxy's center. Their elliptical orbits bring them out to extreme distances of 33,000 to 39,000 ly (10 to 12 kpc) from the Galaxy's nucleus. The clusters orbit at speeds of more or less 100 to 150 km/sec, diving into and shooting out of the disk (Fig. 20.37). These passages have helped to wipe globular clusters clear of any gas and dust they once had.

20.8 THE HALO OF THE GALAXY

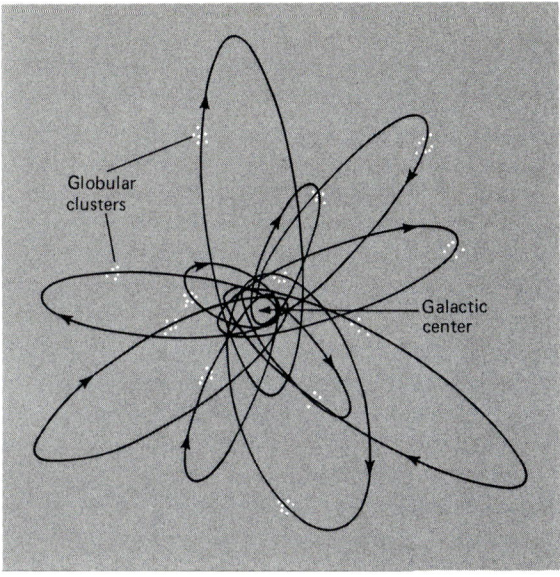

FIGURE 20.37 A schematic picture of the orbits of globular clusters around the galactic center.

Other Material in the Halo

The Galaxy's halo is thin and far from us, so it's hard to observe. Some indirect evidence and theoretical arguments indicate that the halo may actually contain considerable material, perhaps more mass than is in the visible part of the Galaxy. What might it be?

Astronomers have spotted a few RR Lyrae variables above and below the galactic plane that do not belong to globular clusters. In addition to these stars, the halo may contain a large number of low-mass, faint red stars that are difficult to observe directly. Recent observations of a few other nearby galaxies like the Galaxy imply that they may have extensive, massive halos of such faint red stars.

The halo also contains gas, but much less than the disk. Observations at 21 cm show hydrogen clouds traveling at high speeds above and below the galactic plane. So the halo gas has some H I in it. Most of the halo's gas, however, is probably ionized hydrogen. This gas could come from the disk, blown out by supernova explosions, expanding H II regions, and stellar winds. So the halo may be fairly hot and expanding into intergalactic space.

The halo may contain high-speed electrons and protons—cosmic rays trapped by the Galaxy's magnetic field. There is no solid, unambiguous evidence for a cosmic-ray halo, but observations so far, especially the fact that energetic cosmic rays striking the earth come from all directions, suggest that it may indeed exist.

But the halo may also contain other objects, currently unobservable. Low-mass main-sequence stars would be very hard to detect. Smaller objects, similar to planets and asteroids may also exist. Some astronomers have even suggested numerous black holes. We just don't know what might be there.

SUMMARY

This chapter narrated astronomers' efforts, in the span of two centuries, to piece together the construction of the Galaxy (Table 20.1). You have seen that this fitting together of the celestial architecture depended on the measurement of distances, which in turn is based on the fundamental properties of stars and electromagnetic radiation. Star counting, like that of Herschel and Kapteyn, gives only the relative shape of the Galaxy. Parallaxes were first used to set the distance scale. Later the H-R diagram was used to get more distances. The big breakthrough came with Leavitt's discovery of the period-luminosity relationship for cepheids. Once calibrated in terms of luminosity, this relationship could be used to find the distances to globular clusters. From these, Shapley found that the Galaxy was large, with the sun far from the center.

Three general regions make up the Galaxy: the nuclear bulge (containing the nuclear core), the disk, and the halo. Each region has stars that are classified in two general groups: Population I and Population II. Population I stars are metal rich. If young, the most luminous are blue and associated with regions of gas and dust. But an old Population I can have luminous red giants as its brightest members. Population II stars are red, old, and metal poor and are found in regions relatively free of interstellar material.

The halo consists of Population II stars, almost all in globular clusters. From infrared observations of the galactic nucleus and of other galaxies, astronomers infer that most of the stars there are old Population I. In the disk away from the nuclear bulge, such as in the neighborhood of the sun, young Population I stars abound.

Along with the Population I stars in the disk is an interstellar medium that consists of gas and dust. The gas is mostly hydrogen atoms and molecules, with some other molecules mixed in. A large fraction, perhaps as much as half, of the molecules appears to be concentrated in huge clouds that lie 1200 to 2500 ly (4 to 8 kpc) out from the Galaxy's center. The inner part of the disk seems to be tilted 24°. The outer parts are much thicker than the rest, and are warped up on one side, down on the other.

Radio and infrared observations of the Galaxy's nucleus have penetrated the dust veil there to find a place very different from the part of the Galaxy where we are located. The nucleus contains old Population I stars, singly and in clusters. It also contains molecular clouds, dust, ionized gas (all moving swiftly), and, surprisingly, young O and B stars, all concentrated in a small region of space. And the very heart of the Galaxy may harbor a massive black hole.

TABLE 20.1 Evolution of Ideas About the Galaxy

Astronomer	Sun's Position	Size of Galaxy
W. Herschel	At center	At least 900 times the distance to Sirius
J. C. Kapteyn	Near center	10 kpc diameter
H. D. Curtis	Near center	10 kpc diameter
H. Shapley	20 kpc from center	100 kpc diameter
Present	8–10 kpc from center	40 kpc diameter

Key Words
Great Rift
star counting
proper motion
space velocity
radial velocity
transverse velocity
apex
antapex
statistical parallax
Magellanic Clouds
cepheid variable
period-luminosity relationship
light curve
RR Lyrae variable
spiral nebulas
Population II
old Population I
young Population I
metal-rich star
metal-poor star
halo
disk
nucleus (nuclear bulge and core)
luminosity function
Sagittarius A East
Sagittarius A West
Sagittarius B

Review Questions
1. Radio observations show that some of the radio emission from the galactic center has characteristics of radio emission from H II regions. What can you guess about the physical characteristics (such as size and temperature) of these areas? (Objective 11)
2. Imagine that the sun *were* located in the center of the Galaxy. How would globular clusters be distributed in the sky? (Objective 8)
3. What is the main reason Herschel's star-count technique resulted in a very irregular shape for the Galaxy? (Objective 2)
4. Relative to the apex and antapex of the solar motion, in what direction in the sky would stars show the largest radial velocity toward the sun? Away from the sun? The largest proper motion? (Objective 5)
5. Why does the method of statistical parallax give distances to stars which are farther away than does the method of heliocentric parallax? (Objective 6)
6. Baade assigned the stars in globular clusters and in the nuclear bulge to the same population. Why do we now put them into different populations? (Objective 9)
7. Where in the Galaxy do we find young Population I stars? Old Population I stars? Population II stars? Interstellar gas and dust? (Objective 10)
8. Why is more information about the galactic center provided by radio, infrared, and X-ray observations than by optical observations? (Objective 11)

Problems
1. Suppose a group of stars has an average radial velocity of 25 km/sec and average proper motion of 0.03"/yr. How far away are these stars? (Objective 6)
2. For a star 1 pc from the sun, how much does the solar motion contribute, in arcseconds per year, to its proper motion? (Objective 6)
3. What is the absolute magnitude of a Type I cepheid with a period of 100 days? What is its distance, if its apparent magnitude is +15? (Objective 7)
4. Suppose we discovered that the absolute magnitudes of the RR Lyrae stars were 5 magnitudes brighter than had been thought. By what factor would that change the distance to the center of the Galaxy, as determined by Shapley's method? (Objective 8)
5. What is the orbital period of a typical globular cluster? Hint: Assume all the Galaxy's mass is concentrated in its nucleus. (Objective 10)

BEYOND THIS BOOK...

* Two excellent books about the evolution of ideas about our Galaxy and other galaxies are *The Discovery of Our Galaxy* (Knopf, New York, 1971) by C. A. Whitney, and *Man Discovers the Galaxies* (Science History Publications, New York, 1976) by R. Berandzen, R. Hart, and D. Seeley.
* "Stellar Populations" by M. and G. Burbidge in *Scientific American*, November 1958, is a good introduction to the subject.
* For more on cepheids, read "Pulsating Stars and Cosmic Distances" by R. Kraft, *Scientific American*, July 1959.
* *The Milky Way* (Harvard University Press, Cambridge, Mass., 1981) by P. and B. Bok presents an up-to-date summary of astronomical knowledge of our Galaxy.
* "Interstellar Smog" by G. Herbig in *American Scientist*, March 1974, reviews the content of the interstellar medium.
* For a look at the evolution of ideas about the Milky Way, read *The Milky Way: An Elusive Road for Science* (Science History Publications, New York, 1972) by S. Jaki.

THE EVOLUTION OF THE GALAXY

21

LEARNING OBJECTIVES
After studying this chapter, you should be able to:
1. Explain two astronomical difficulties in trying to figure out the structure of the Galaxy from our location in it.
2. Sketch the optically observable spiral arm structure of the Galaxy near the sun, naming the nearby arms and their positions and distances from us.
3. Describe what methods astronomers use to work out the positions of spiral arms.
4. Describe the sun's orbit around the galactic center and the orbits of nearby stars.
5. Sketch the rotation curve of the Galaxy and find from it the approximate mass of the Galaxy.
6. Explain how radio astronomers use 21-cm observations to trace spiral arms, indicate the limitations of their method, and explain its advantage over optical observations.
7. Describe the contents of a typical spiral arm.
8. Describe the evolution of spiral arms in terms of the density-wave model for spiral structure.
9. Outline a model for the evolution of the disk of the Galaxy.
10. Outline a model for the evolution of the halo of the Galaxy.

CENTRAL QUESTION:
What evolutionary processes induce the structure of the Galaxy?

Like a majestic cosmic pinwheel, our Milky Way Galaxy spins slowly in space. The fact that the main body of the Galaxy is a flat disk implies that it must rotate rapidly. Our sun, along with most of the stars in its neighborhood, orbits the center of the Galaxy at a speed of about 250 km/sec. Since the sun is about 33,000 ly (10,000 pc) from the nucleus, it completes one revolution roughly every 250 million years.

Our sun is but one of some 100 billion stars in the Galaxy. What is the structure of this enormous system of stars? Optical astronomers can probe the structure near the sun, and radio astronomers can study regions farther away. These optical and radio investigations show that the Galaxy has a spiral structure—a pattern inferred in part from spiral designs observed in other galaxies (Fig. 21.1). The maps of the Galaxy's spiral structure obtained to date are subject to revision. Though the details are not yet complete, astronomers have been able to establish the broad outlines of the Galaxy's structure, a remarkable achievement, considering that we ourselves are buried within the Galaxy.

How does the Galaxy evolve? Recent theoretical work, aided by the development of electronic computers, has produced new ideas about the physical cause of the Galaxy's spiral structure. The leading idea, called the *density-wave model*, links many of the galactic entities in the disk by the process of galactic evolution. The Galaxy's structure evolves, but at a rate so slow that we won't see any changes in our lifetimes.

In the past the Galaxy looked much different than it does now. This chapter ends with a brief excursion into our Galaxy's history to see how our spiral Galaxy came into being.

21.1 AN OVERVIEW OF THE GALAXY'S STRUCTURE

As discussed in the previous chapter, the Galaxy has three main parts: a nucleus (nuclear bulge plus core), a disk, and a halo. The disk, the main body of the Galaxy, has a

21.1 AN OVERVIEW OF THE GALAXY'S STRUCTURE

FIGURE 21.1 A typical spiral galaxy, Messier 101. (*Courtesy Kitt Peak National Observatory.*)

diameter of some 120,000 ly (40,000 pc) and a thickness of about 2300 ly (700 pc). Population I stars and interstellar clouds of gas and dust inhabit the disk, the gas extending out farther than the stars, but in a thinner layer. The sun resides a little above the central plane at a distance of approximately 33,000 ly (10 kpc) from the Galaxy's center. (Some recent studies suggest a lower value for the distance to the center, perhaps as low as 8.5 kpc; we will use the standard value of 10 kpc in this book.)

We are in a bad position to observe the Galaxy's structure. Imagine, for example, that you are watching a half-time show at a football game. The band has set up some elaborate formation. Up in the stands you can easily observe what the formation looks like, but down on the field at the edge of the formation it appears to be a jumble! You could eventually work out the shape if you could find the distances to all the band players. Then you could make a map of their positions by plotting the distances and directions of each person. But suppose you had to do this mapping in a fog so dense that only the closest people were visible. Then you would need some other method of estimating distances, perhaps by measuring the intensity of the sound of the instruments that pierce the fog.

Optical astronomers find their view blocked by interstellar dust. Radio waves are not stopped by the dust, so radio astronomers can pick up the radio emission from clouds of gas that mark the Galaxy's structure. But they have more difficulty in determining distances than the optical astronomers do. The results from the two techniques, though not agreeing in all details, have uncovered the general spiral-arm structure.

What defines a spiral arm? You find many O and B stars in a spiral arm. Also, the overall density of material—gas, dust, and stars—inside a spiral arm is roughly 10 times that in the region between arms. The spiral arms contain most of the gas, dust, and young stars in the Galaxy. A typical segment of a spiral arm has a width of 1500 ly (500 pc) and a height of 500 ly (150 pc). Near the sun, stars contain about 50 percent of the spiral-arm material; the other half forms the clouds of gas and dust in the interstellar medium. In

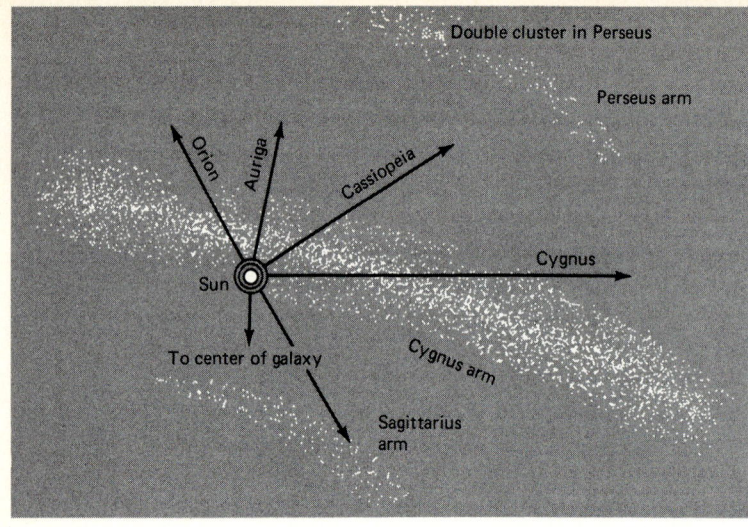

FIGURE 21.2 Spiral arm segments near the sun, which lies on the inner edge of the Cygnus arm in a branch called the Orion spur.

contrast, for the Galaxy as a whole only a few percent of the material consists of gas and dust. The dust is in general only about 1 percent as abundant as the gas.

The sun lies on the inner edge of a poorly defined structure called the *Cygnus arm* (Fig. 21.2). (Note: Astronomers use the word "arm" to indicate a well-defined *segment* of a larger overall spiral-arm structure. We'll use it the same way, but you should be aware that we are discussing pieces of larger structures.) Outward, about 10,000 ly (3 kpc) from the sun, lies an arm parallel to the Cygnus arm; since the most well-observed portion is in the direction of the constellation Perseus, it is called the *Perseus arm* (Fig. 21.3). At about 6000 ly (2 kpc) interior to the sun curves the *Sagittarius arm*. Some evidence indicates that another arm may lie 13,000 ly (4 kpc) from the sun toward the galactic

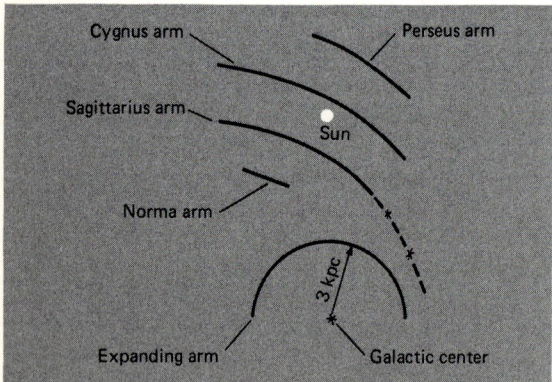

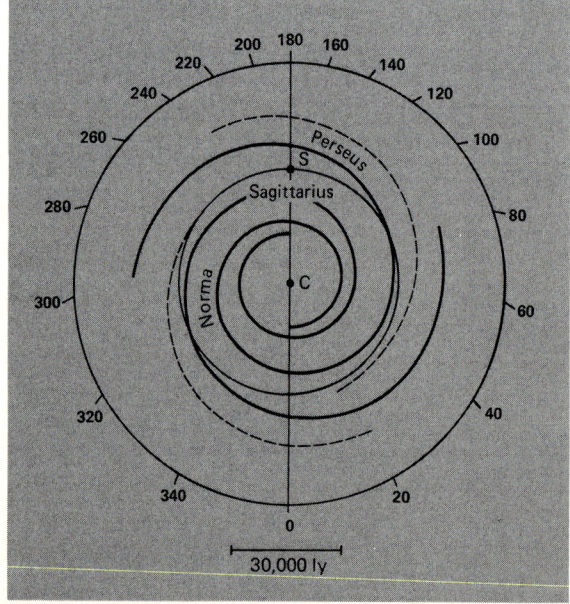

FIGURE 21.3 The spiral arms in the sun's vicinity. The arms are named after the constellations in whose directions they lie as seen from the earth.

FIGURE 21.4 One model for the overall spiral structure of the Galaxy. The sun is located at S. The numbers around the outer circle indicate the galactic longitude as viewed from the sun. Note that the two-arm pattern (solid lines) becomes four-armed (solid plus dashed lines) far away from the sun.

center; it is often called the *Norma arm*. Finally, encircling the galactic center at a distance of about 1000 ly (3 kpc) from the center is the *3-kpc arm*, or *expanding arm*, so called because this innermost arm appears to be moving toward us at roughly 50 km/sec, expanding away from the center.

We cannot yet see the whole scheme clearly. But some other spiral galaxies have two arms that wind around the nucleus. We guess that our Galaxy also has two major arms (and perhaps two less distinct arms) tightly wound around the nucleus (Fig. 21.4). Each arm starts near the galactic center about 180° away from the other and twists around, moving outward at an angle of 5° to 12° with respect to a circle. The arm segments mentioned above probably are parts of the two major arms.

21.2 GALACTIC ROTATION: STARS IN MOTION

Located in the galactic disk, far from the center of our Galaxy, the local stars appear to be moving in a variety of directions at a variety of speeds. Relative to other stars, the sun moves at a speed of about 20 km/sec. But how to find out what motion the sun and nearby stars share in relation to the center of the Galaxy?

One way would be to measure the proper motion, distance, and radial velocity of the many stars in our Galaxy, then calculate their average motion, and compare the average to the motion of stars in the solar neighborhood. But that is an impossible task. Most of the stars in the Galaxy cannot be seen even with the largest telescopes, and many are so faint and so distant that we cannot learn anything about their motions.

During the nineteenth century astronomers grew aware of the sun's motion in space as reflected in the relative motions of nearby stars. One pioneer astronomer, Bertil Lindblad (1895–1965), proposed that these relative motions occur partly because of the orbital motion of the stars around the Galaxy. So from those motions we can infer the character of the Galaxy's rotation.

Radial Velocities and Rotation

To see beyond our local region, we must find stars bright enough so that, even if they are far away, we can photograph their spectra to learn their radial velocities. The thousand or so brightest OB stars found all around the belt of the Milky Way are extremely luminous stars, so we can determine their radial velocities, even though they are thousands of light years away.

The results of this radial velocity survey are shown in a graph that plots radial velocity against *galactic longitude*, which is the angular position along the galactic equator in the plane of the Milky Way (Fig. 21.5). The point labeled 0° is the direction toward the galactic center in Sagittarius. In this direction the radial velocities of OB stars are about zero. As we move around the Milky Way to Aquila (45° longitude), the stars show red shifts, indicating velocities away from us. At 135° longitude the OB stars have blue shifts, indicating velocities toward us. The red and blue shifts show a regular cycle in the velocities, alternately away from and toward us, with the curve crossing zero at 0°, 90°, 180°, and 270° galactic longitude.

How to interpret this curve? In the neighborhood of the sun most stars revolve around

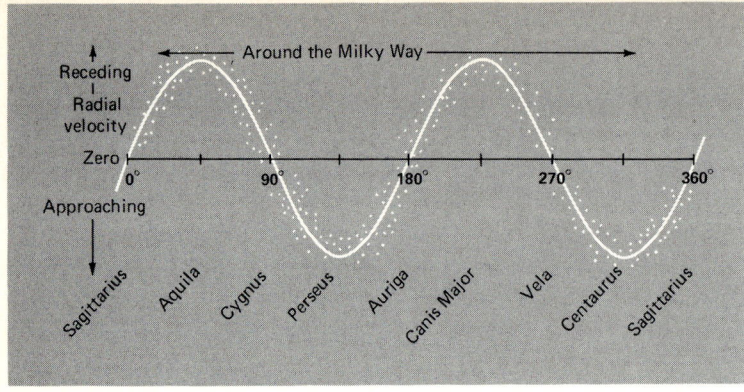

FIGURE 21.5 Measured radial velocities for stars located around the Milky Way. The constellations are indicated by their galactic longitude.

the galactic center in more or less circular orbits, but those close to the center move somewhat faster, and those farther from the center somewhat slower (Fig. 21.6). What are their radial velocities relative to the sun? Consider first the stars at the same distance from the center as the sun (longitudes 90° and 270°). They are moving at the same speed as the sun, so their radial velocity with respect to the sun is zero. Now consider the stars toward the galactic center and in the opposite direction (longitudes 0° and 180°). Though revolving at different velocities, they are moving parallel to the sun's motion, so again their relative radial velocity is zero.

Now consider stars on an inner orbit, but somewhat ahead of the sun, say at longitude 45°. They are moving faster than the sun, pulling ahead of it, so their radial velocity relative to the sun is positive. Stars at a similar position on an inner orbit, but following the sun (longitude 315° for example) are catching up, so their radial velocity relative to the sun is negative.

Lastly, consider stars on an outer orbit. They are moving slower than the sun. The sun is pulling away from those behind (such as at longitude 235°), so the relative radial velocity is positive, but it is catching up on those ahead (longitudes around 135°), resulting in a negative radial velocity.

Put it all together: 0°, zero velocity; 45°, positive velocity; 90°, zero velocity; 135°, negative velocity; 180°, zero velocity; 225°, positive velocity; 270°, zero velocity; 315°,

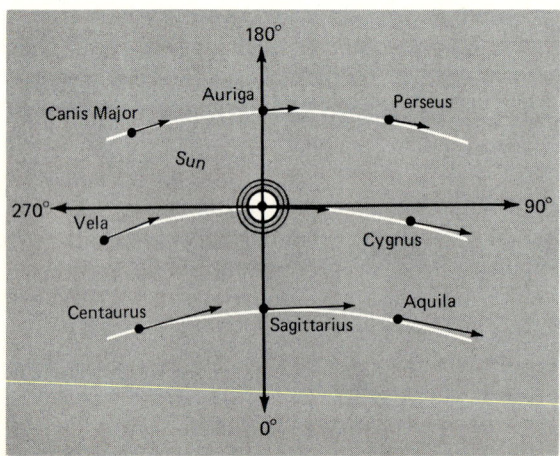

FIGURE 21.6 The motions of the sun and nearby stars orbiting around the Galaxy.

21.2 GALACTIC ROTATION: STARS IN MOTION

negative velocity; 360°, zero velocity. That's the same behavior as the observed radial velocity curve for OB stars (Fig. 21.5).

So the radial velocity curve can be explained if stars nearer to the center of the Galaxy are moving faster, and those farther out slower. Does this make sense? In the Galaxy most stars are concentrated in the galactic disk, which surrounds the dense central nucleus. The stars exert a gravitational attraction on one another. Out in the disk where the sun dwells, stars are attracted essentially toward the Galaxy's center.

This setup resembles the flattened system of planets forming our solar system. Although the planets attract each other, each orbits about the sun, because the gravitational attraction of the central sun is much stronger. The stars in the disk revolve about the nucleus of the Galaxy much as the planets revolve around the sun. The forces are similar, so the motions should be similar. If the Galaxy's mass were *entirely* concentrated in the center, the stellar motions would exactly follow Kepler's laws (Section 3.3), and the velocities of the stars would decrease with increasing distance from the Galaxy's center—just as, for example, the orbital velocity of Mars is less than that of the earth.

What would the velocity law be if the mass of the Galaxy were all concentrated at the center? From Kepler's third law,

$$P^2 = \frac{4\pi^2}{GM} R^3$$

where R is the distance from the Galaxy's center and M its mass interior to R. If the orbit is circular, the orbital velocity V is

$$V = \frac{2\pi R}{P}$$

and

$$\frac{4\pi^2 R^2}{V^2} = \frac{4\pi^2 R^3}{GM}$$

$$V = \left(\frac{GM}{R}\right)^{1/2}$$

So as R gets larger, V decreases, proportional to the inverse square root of R.

In fact, the velocities don't follow Kepler's laws exactly, and that reveals an important fact about the Galaxy. The major part of the mass is *not* concentrated at the center, which is quite different from the case of the solar system. But there is, of course, *some* relationship between orbital velocity and distance from the center, and these differences in orbital velocities show up as systematic differences in Doppler shifts, an idea first demonstrated in 1927 by Dutch astronomer Jan Oort. Here we have direct evidence that the Galaxy rotates in a special way—*differential rotation*, where differing orbital speeds mark differing distances from the center.

Oort showed that near the sun the variation of radial velocities (Fig. 21.5) arising from differential rotation can be described by the simple relation

$$V_r = Ad \sin 2l$$

where V_r is the radial velocity in kilometers per second, d is the distance to a star in kiloparsecs, l the star's galactic longitude in degrees, and A is *Oort's constant*, a number with a value of about 15 km/sec/kpc. Note that this relation has the proper characteristics. V_r is zero at $l = 0°$, $90°$, $180°$, and $270°$ and alternates positive to negative in the four quadrants.

Near the center of the Galaxy (inner 1000 ly or so), the orbital motion is different, because the Galaxy's mass is not all concentrated in a central body. Stars near the center do not feel as large an attraction as would be expected if all the mass were concentrated in the nucleus, for much of the mass is at greater distances from the center and exerts no force on them. So these stars travel more slowly in their orbits than might be expected (Fig. 21.7).

Proof that this picture is correct relies on radio observations of H I clouds, not of stars. (We can't use stars because we can't see distant ones through all the interstellar dust.) From observations of the variation of H I radial velocities in the galactic longitude range from 0° to 90°, radio astronomers have been able to confirm that the interior region of the Galaxy does *not* rotate in a Keplerian fashion. It rotates so that orbital velocity *increases*, rather than decreases, with distance from the center. The rate of increase is approximately like that expected for solid body rotation (though, of course, the Galaxy itself is *not* a solid body).

Let's work out this result. Assume a spherical mass with uniform density ρ throughout. Also, let the orbits be simple circles. Then the centripetal acceleration at distance R from the center is

$$a = \frac{V^2}{R} = \frac{GM_R}{R^2}$$

where M_R is the mass within radius R. Since

$$M_R = \frac{4}{3}\pi R^3 \rho$$

then

$$\frac{V^2}{R} = \frac{G(4\pi/3)R^3\rho}{R^2}$$

$$V^2 = \frac{4}{3}\pi G\rho R^2$$

$$V = \left(\frac{4}{3}\pi G\rho\right)^{1/2} R$$

Why is this "solid body" rotation? On a solid disk, like a phonograph record, each point goes around in the same time. But the distance traveled in one revolution, the circumference, is proportional to R. So the velocity must also be proportional to R.

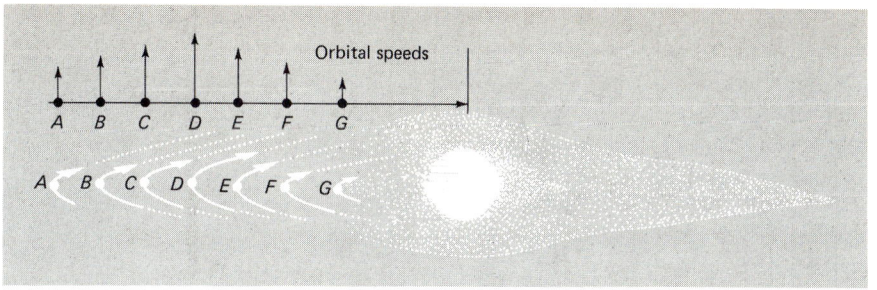

FIGURE 21.7 Galactic orbital speeds. From A to D the orbital speeds increase inward, just as the planets' velocities increase with decreasing distance the sun. But inward of D the orbital speeds decrease rather than increase.

The Sun's Orbit and the Mass of the Galaxy

How fast does the sun move around the Galaxy? We use a variety of indirect approaches to find out. One method utilizes the motions of globular clusters. The globulars orbit the Galaxy in random orbits, with a roughly spherical distribution around the nucleus. With respect to the nucleus, the average motion of all globular clusters is roughly zero. In other words, the system of globulars has no overall rotation about the galactic center (although individual clusters move rapidly, some in one direction, some in another). A study of the radial velocities of the globular clusters enables us to find the sun's motion with respect to the system of globulars. Because the system of globulars has no rotational motion with respect to the nucleus, the sun's motion found in this way is its motion with respect to the Galaxy's center. Such an analysis shows that the sun revolves around the center at about 250 km/sec. (Recent work, using other methods, has pushed this value down to 220 to 225 km/sec, but the adopted international standard is still 250 km/sec. In this book we'll continue to use this value.)

How far is the sun from the Galaxy's center? That's a tough question to tackle because we cannot see the center optically. However, we can look above and below the galactic plane, where the obscuration is less, to observe objects thought to be concentrated or symmetrical about the galactic center. That was Shapley's technique (Section 20.4).

Modern methods follow the same basic procedure. We use the RR Lyrae variables in globular clusters. These stars are Population II, have periods of light variation that range from $1\frac{1}{2}$ to 24 h (typically 12 h), and range in spectral type from A2 to F6. It turns out that all RR Lyrae stars have essentially the same average luminosity no matter what their period—48 solar luminosities or an absolute visual magnitude of +0.6. So once an RR Lyrae star is identified by the shape of its light curve, we can measure its apparent magnitude and its period and work out its distance. By this technique recent work has found the sun's distance from the galactic center to be 28,000 ly (8.5 kpc) with an uncertainty that ranges from 24,000 to 33,000 ly (7.5 to 10 kpc). We'll typically use 33,000 ly (10 kpc) in this book. This value is still the international standard, but the actual distance to the center *is* still uncertain. The true value is probably a bit less.

Knowing the sun's velocity and distance, we can apply Kepler's third law to deduce the mass of the Galaxy. The sun swings around the Galaxy at 250 km/sec at a distance of 10 kpc from the center. Assume that the sun moves in a circular orbit. Apply Kepler's third law in the form used for binary star systems (Section 8.6) to the Galaxy:

$$M_1 + M_2 = \frac{R^3}{P^2}$$

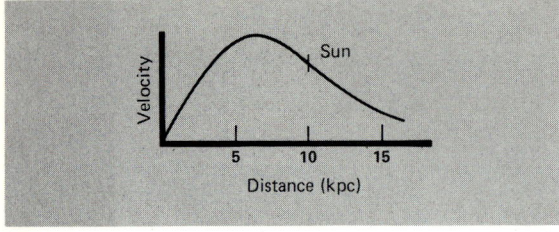
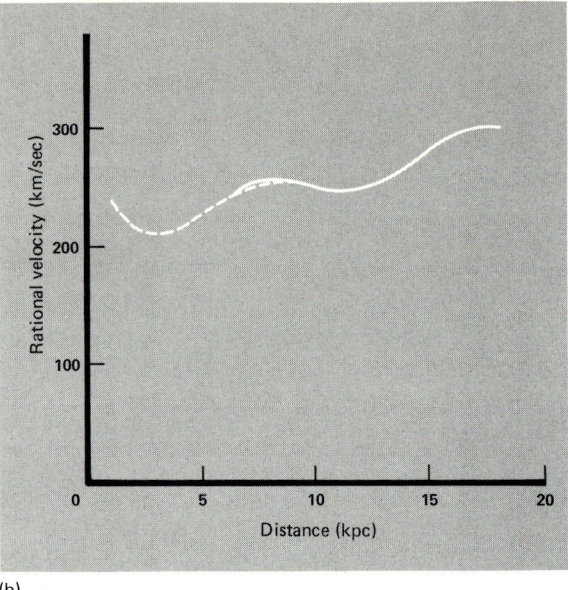

FIGURE 21.8 (a) A simplified rotation curve for the Galaxy, if most of the mass were concentrated in the center. (b) Observed rotation curve for the Galaxy. The dashed part of the curve is from atomic hydrogen data; the solid part from carbon monoxide observations of molecular clouds. (*Adapted from a diagram by L. Blitz, M. Fich, and A. Stark.*)

where R is in AUs and P in years. The mass then comes out in solar masses. At 250 km/sec it takes the sun about 2.5×10^8 yr to complete a circuit of the Galaxy. R is roughly 10 kpc, and 1 kpc equals 2.1×10^8 AU, so 10 kpc equals 2.1×10^9 AU. Then

$$M_1 + M_2 = \frac{(2.1 \times 10^9)^3}{(2.5 \times 10^8)^2}$$

$$= \frac{9.3 \times 10^{27}}{6.3 \times 10^{16}}$$

$$= 1.5 \times 10^{11} \text{ solar masses}$$

What is $M_1 + M_2$? It is the mass of the sun plus the mass of the Galaxy. The mass of the sun is so small compared with that of the Galaxy (just look at the result!) that we can ignore it. This result refers only to the mass interior to the sun's orbit. So the Galaxy's mass is at least 1.5×10^{11} solar masses.

With the sun's orbital distance and velocity known, we can use measurements of velocity and distance relative to the sun to find the *galactic rotation curve*, that is, how fast an object some distance from the galactic center revolves around it. In the simplest case we'd expect solid body rotation close to the center and the Keplerian motion in the outer regions (Fig. 21.8a), but the observed rotation curve (Fig. 21.8b) determined from 21-cm data does not follow this simple pattern. From close to the center out to 1000 ly (300 pc), the curve rises steeply, then drops, bottoming out at about 10,000 ly (3 kpc). It then rises slowly out to the position of the sun. Leo Blitz and colleagues have used CO observations to determine the rotation curve in the outer parts of the Galaxy. They find that the curve rises more steeply beyond the sun, reaching almost 300 km/sec at 59,000 ly (18 kpc).

What does this curve tell us? Since even the outer parts of the Galaxy do not revolve in a Keplerian fashion, much of the Galaxy's material must lie out beyond the sun's orbit.

21.2 GALACTIC ROTATION: STARS IN MOTION

From the rotation curve out to 59,000 ly (18 kpc), the Galaxy's mass is 3.4×10^{11} solar masses. So at least as much mass lies exterior to the sun as interior to it. Much of this matter is invisible. The Galaxy seems to have a massive halo of nonluminous matter.

High- and Low-Velocity Stars

The stars in the sun's neighborhood appear to be moving in all different directions at speeds of 20 to 30 km/sec with respect to each other. But the broad picture of our Galaxy has stars and gas clouds moving in circular paths around the center of the Galaxy. Are these two pictures consistent with each other?

If all the stars in our local neighborhood moved right along with the sun in perfectly circular orbits, then they would be traveling in essentially parallel paths along with the sun. Nearby stars and the sun would move for a long time like a band in a perfect parade; nobody would be getting closer to or farther from anybody else. In our local sample we do observe stars moving relative to the sun, so our picture of perfectly circular orbits is too simple.

Modify it by letting the stars have slightly elliptical orbits. Some stars will have orbits smaller than the sun's. They are near the sun only because they are at the apocenter (the point farthest from the focus) of their orbit. At apocenter, according to Kepler's second law, they are moving slower than their average speed. Similarly, some stars near the sun are at the pericenter (the point farthest from the focus) of orbits which are larger than the sun's. These stars are moving faster than average. The local random motions reflect the fact that the galactic orbits of the different stars are slightly different from one another.

For example, the sun, Sirius, and Procyon are moving about the center of the Galaxy at speeds of about 250 km/sec (Fig. 21.9), but each of their speeds is slightly different. Today, the three stars are within 12 ly of one another, but in 100 million years they will be separated by thousands of light years. During this time the motions of Sirius and Procyon

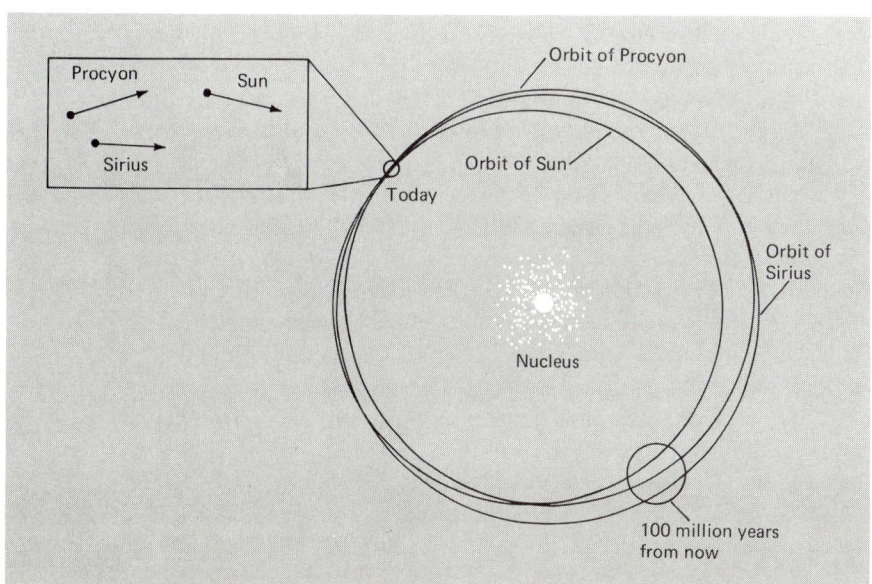

FIGURE 21.9 Schematic galactic orbits of the sun, Procyon, and Sirius.

will be almost, but not exactly, the same as that of the sun. The stars will move with low velocities relative to the sun. Stars observed to have speeds relative to the sun of less than 60 km/sec are called *low-velocity stars*.

Astronomers also see *high-velocity stars*, defined as those that move relative to the sun with speeds greater than 60 km/sec. These stars cannot be revolving around the Galaxy in the disk. What orbital paths are they following? Imagine a star moving in an orbit nearly perpendicular to the galactic plane, rather than in the plane (for example, a star in the halo of the Galaxy). It shares very little of the general motion around the Galaxy in its plane. (In fact, a large part of its velocity is up or down with respect to the galactic plane.) As seen from the center of the Galaxy, a star with a large orbital inclination has a small velocity in the plane compared to stars with small orbital inclinations. Suppose this star cuts through the plane close to the sun. The velocity of the star we see is the sun's velocity (around the Galaxy) minus the star's velocity (around the Galaxy). This difference is large, so we observe a high-velocity star.

The fact that the orbital motions of halo stars depart so dramatically from the disk stars gives a clue to the evolution of the Galaxy (Section 21.6).

21.3 GALACTIC STRUCTURE FROM OPTICAL OBSERVATIONS

In 1951 Walter Baade and N. U. Mayall studied the structure of M 31, the spiral galaxy nearest to our own. They found that the spiral arms stood out most strongly in blue-sensitive photographs. When examined in detail, the photographs showed that O and B supergiants, OB associations, H II regions, and Population I cepheids seem to trace out the spiral structure. These objects are called *spiral tracers*.

The next natural step was to observe these spiral tracers in our Galaxy to determine the location and extent of its spiral arms near the sun (Fig. 21.10). It's not a simple operation. First, it requires an accurate technique for measuring the distance to each of the tracers. Second, optical observations are restricted by the blotting out of starlight by dust. Most of the interstellar dust lies concentrated in the galactic plane, so the sun sits in the thick of the interstellar smog. Third, the sun's location in the plane gives us a poor vantage point for seeing the Galaxy's spiral structure, since we are forced to observe it edge on rather than face on.

H II regions and supergiant O and B stars trace the arms best because their high luminosities make them visible over large distances. In addition, cepheids have been used to delineate spiral features. The cepheids have the advantage that their distances are easy to determine by the period-luminosity relation.

You must take these optical maps with a bit of caution. The data extend only to about 14,000 ly (4 kpc) from the sun, and a larger range is needed to outline the spiral arms definitely. In addition, although the outline of spiral structure is probably correct, observations of other galaxies show that irregularities also commonly occur. It's futile to draw a master diagram from optical data alone.

Disagreements have arisen about the details, but most optical astronomers concur that their investigations have found at least three major arm segments spaced about 7000 ly (2 kpc) apart. The Galaxy appears to have a spiral structure with much irregularity in the general pattern.

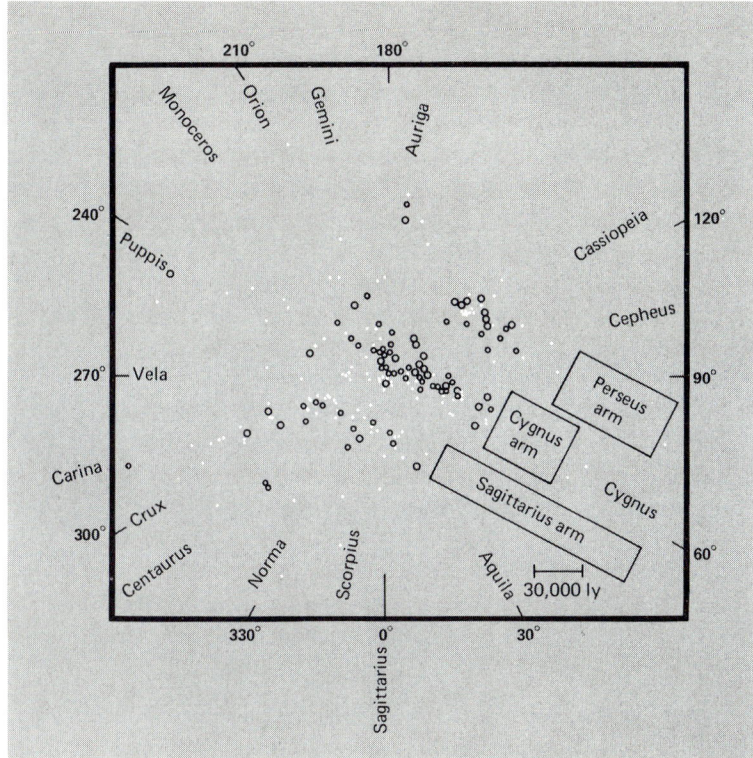

FIGURE 21.10 Spiral structure near the sun, as indicated by O and B stars, H II regions, and cepheids. Three arm segments (Perseus, Cygnus, and Sagittarius) are visible. (*Based on data compiled by W. Becker and T. Schmidt-Kaler.*)

21.4 EXPLORING GALACTIC STRUCTURE BY RADIO ASTRONOMY

Radio observations are not handicapped by obscuration by interstellar dust, so radio astronomers can reach far beyond the restricted range of optical astronomers, even to the other side of the galactic nucleus. Because hydrogen is so abundant, the 21-cm line (Section 9.1) plays a major role in diagraming the Galaxy. This radio emission comes from the neutral hydrogen clouds, which are concentrated in the spiral arms.

The mapping of spiral structure by 21-cm observations is an indirect, time-consuming task. Neutral hydrogen is concentrated in spiral arms, and the more hydrogen there, the more strongly it emits at 21 cm. We distinguish among spiral arms not only by looking in different directions (we are, after all, in the middle of the mess), but by looking at different wavelengths in the same direction. Not all 21-cm radiation arrives at the earth at exactly 21.11 cm. It is Doppler shifted to different wavelengths because of the different velocities of the hydrogen gas clouds. These differences in velocities come mostly from the rotation of the Galaxy. So *if* we know how the Galaxy rotates (and that's the tricky part), we can translate 21-cm observations into a map of spiral structure.

What is the technique radio astronomers use? They look for 21-cm radiation from some specific galactic longitude and latitude. (*Galactic latitude*, like terrestrial latitude, is the angular distance above or below the galactic equator.) If all the H I clouds in this

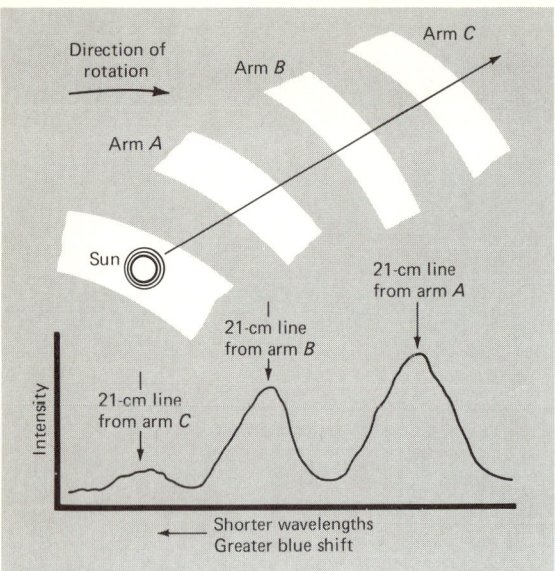

FIGURE 21.11 Using 21-cm emission to trace out spiral structure. Consider looking out of the Galaxy through three spiral arms (A, B, and C). The outer arms move more slowly than the inner ones. So the velocity difference between us and A is smaller than that between us and C, and the blue shift from A is less than that from C.

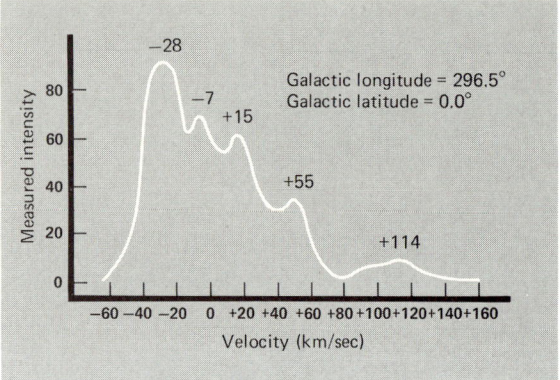

FIGURE 21.12 Actual observations of 21-cm emission at galactic longitude 296.5°, latitude 0.0°. Five distinct clouds appear at velocities of −28, −7, +15, +55, and +114 km/sec. (*Data from NRAO.*)

direction were at rest with respect to the sun (that is, moving at the same velocity as the sun), all of their emission would pile up at exactly 21 cm. However, because of galactic rotation, the clouds along the line of sight have different radial velocities, hence different Doppler shifts, so their signals are received at slightly different wavelengths.

Here's a particular case looking outward from the sun (Fig. 21.11). Our line of sight intercepts three clouds at successively greater distances. The outer clouds travel more slowly than the inner ones. They all travel more slowly than the sun. So the difference between the sun's velocity and the inner cloud's velocity is the least, and its Doppler shift is the least. The difference between the sun's velocity and the outer cloud's velocity is the greatest, so it has the greatest Doppler shift. Assume that each of these clouds corresponds to a piece of a spiral arm. How do we arrange them at different distances from the sun?

Recall Oort's ideas of a galactic rotation curve (Fig. 21.8), which assumes circular or nearly circular orbits. We know the sun's velocity around the Galaxy. When we observe 21-cm emission, we know the longitude at which we are looking and can measure a radial velocity (Fig. 21.12). From that radial velocity, and the assumption that the space velocity is perpendicular to a radius, we derive the rotational velocity of the cloud. We then look up this velocity on the galactic rotation curve to find the distance to which it corresponds. (See Focus 21.1 for details.)

In essence, the galactic rotation curve tells us that at a given galactic longitude a certain radial velocity corresponds to a specific distance from the sun. Scanning around the galactic equator, we make a series of 21-cm observations. Using the rotation curve to

21.4 EXPLORING GALACTIC STRUCTURE BY RADIO ASTRONOMY

translate velocities into distances, we plot the positions of the clouds. These are then connected to trace a spiral arm and so outline the Galaxy's structure.

Here's a simplified example. Suppose we are looking outward at galactic longitude 135°, and we find two H I peaks, one at a blue shift corresponding to a radial velocity of −10 km/sec, the other at a greater blue shift of −20 km/sec. (The minus sign attached to the velocity indicates a blue shift.) Now Oort's formula (valid only for small distances from the sun) relates the radial velocity (V_r), galactic longitude (l), and distance (d) by

$$V_r = Ad \sin 2l$$

where A equals 15 km/sec/kpc. For the −10 km/sec peak

$$-10 = 15d \sin 2(135°)$$

$$d = \frac{-10}{15 \sin 270°}$$

$$= \frac{-0.67}{-1.0}$$

$$= 0.67 \text{ kpc}$$

The other peak comes from a cloud at 1.3 kpc. So on a scale map of the Galaxy we can draw in the positions of these two H I clouds (Fig. 21.13). Then we move our radio telescope to a slightly different galactic longitude, 140°, for example, and make the same observations to find two more points. Point by point we build up our radio map of the spiral structure (Fig. 21.14). (Oort's formula is valid only if the distance is much less than the distance to the galactic center. For distances of many kpc, we need to use the more complicated formulas of Focus 21.1, but the general principle is the same.)

Conflicting radio maps have been drawn by different investigators. The heart of the problem is that the neutral gas clouds don't follow exactly the simple scheme of circular rotation; in addition to their circular motion, they have their own random motions.

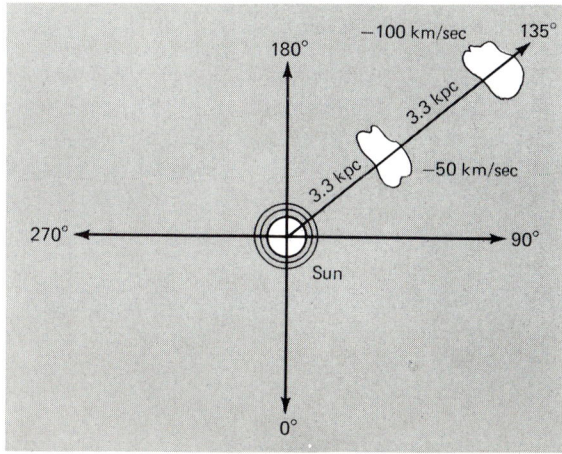

FIGURE 21.13 An example of finding distances to two H I clouds of known galactic longitude and Doppler shift.

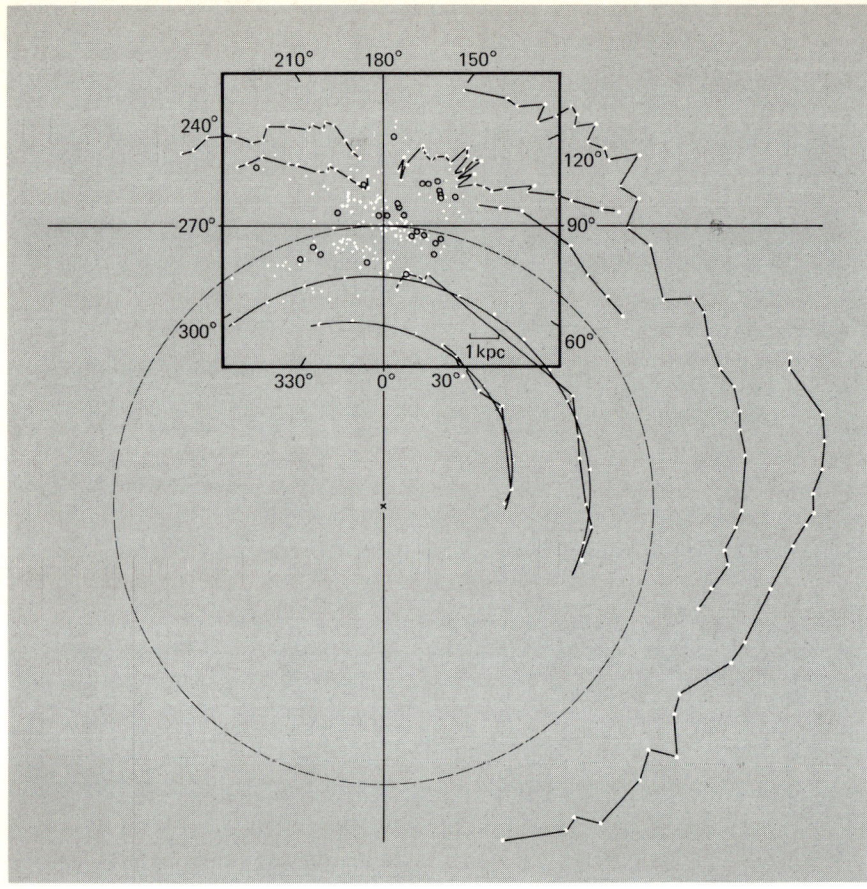

FIGURE 21.14 The Galaxy's spiral structure inferred from 21-cm data. The area in the box includes optical data (Fig. 21.10) for comparison. (*Based on a compilation by H. Weaver.*)

Unfortunately, such noncircular motions lead to incorrect distances and so disrupt the unity of the neutral hydrogen spiral-arm map. To quote Gerrit Verschuur, a researcher in the area of radio mapping,

> The problem is that we can determine how the Galaxy is rotating only if we know where all the matter is, but we only know where all the gas is if we know how the Galaxy is moving!

Despite such problems, the radio and optical maps coincide fairly well. The optical results are most reliable near the sun, where the radio approach is most subject to error. At distances greater than 13,000 ly (4 kpc), at least in the inner parts of the Galaxy, the radio astronomers probe regions inaccessible to optical astronomers. The two techniques complement each other. And the structure of the Galaxy derived from them shows through in its broad outline.

21.5 THE EVOLUTION OF THE SPIRAL STRUCTURE OF THE GALAXY

For many years astronomers supposed that the spiral arms in our Galaxy and others were material arms, a coherent bunch of objects—stars, nebulas, gas, dust—somehow physi-

FIGURE 21.15 A highway bottleneck as an analogy to a density wave. The slow-moving truck has cars jammed up behind it waiting to pass. The jam is always behind the truck, but it consists of different cars at different times. From above, the jam moves more slowly than the average speed of the cars; it is a density wave.

cally held together. Such a point of view faced two questions: What holds the material in an arm together, and how does an arm persist for a long time? Astronomers first surmised that a galactic magnetic field could shape the interstellar medium into spiral arms, but recent work has found the Galaxy's magnetic field to be too weak to keep stars and gas together.

Astronomers have been struck by the remarkable persistence of spiral arms despite the fact that the outer parts of the arm rotate differently from the inner parts. The Galaxy has turned about 60 times since its origin, but the arms have not wound up or spread out around the Galaxy. Of course, it could be that the arms are not as old as the Galaxy and that we are observing them at a very special time soon after their formation. But it is not likely that the same would be true for all galaxies. Of the brightest galaxies in the sky over 60 percent have a spiral form (Fig. 21.1). We cannot argue that all these galaxies have by chance just now formed spiral arms which have not yet had time to be wound up. And because it takes light a certain amount of time to reach us, we are seeing the more distant galaxies as they were in the past, not as they are now. Seeing so many galaxies to be spiral now implies that the structure lasts for at least several billions of years.

How can this persistence be explained? The contemporary attack pictures spiral arms not as material arms at all, but rather as the result of a *density wave*, a region of higher density moving through the Galaxy's disk.

What is a density wave? A common example is a sound wave. Push against air molecules to compress them together. This first group of molecules bangs into adjacent ones in the direction of their motion, which transfers the compression to the next bunch of molecules. As this compression (density wave) travels forward, it leaves behind a trough of lower density. The two important points here are: (1) A sound wave requires a source to start it, and (2) the high-density part of the wave persists and moves through the medium even though the specific particles that make it up are different at different points.

Here's another analogy. Suppose you are driving on a mountain road filled with traffic (Fig. 21.15). Everyone moves along happily at the speed limit. Ahead, an overloaded

FOCUS 21.1 GALACTIC STRUCTURE FROM 21-CM OBSERVATIONS

How can we derive the structure of the Galaxy from radio observations? Let's look at the geometry of the situation. Suppose we observe at galactic longitude l along the line of sight from the sun to point P (Fig. F.22). Let α be the angle between the line of sight and the tangent to the circle at P. Then the component of the space velocity at P along the line of sight is $V(R)\cos\alpha$, and the component of the sun's space velocity along the line of sight is $V_{sun}\sin l$. Their difference is the radial velocity:

$$V_r = V(R)\cos\alpha - V_{sun}\sin l$$

The angle SPC is $90° + \alpha$, and we can use the law of sines from trigonometry to relate this angle to the galactic longitude:

$$\frac{\sin(90+\alpha)}{R_{sun}} = \frac{\sin l}{R}$$

or

$$\cos\alpha = \frac{R_{sun}}{R}\sin l$$

Thus the radial velocity is

$$V_r = V(R)\frac{R_{sun}}{R}\sin l - V_{sun}\sin l$$

$$= R_{sun}\sin l\left(\frac{V(R)}{R} - \frac{V_{sun}}{R_{sun}}\right)$$

So if we measure the radial velocity at certain longitudes, we can obtain the quantity $V(R)/R$, and from the rotation curve we can determine the corresponding value of R and hence the position of the point P in the Galaxy.

But how do we get the rotation curve in the first place? Along any line of sight in the inner part of the Galaxy, there will be one point at which the space velocity is exactly along the line of sight (Fig. F.23), the point at which the line of sight is tangent to the circle around the galactic center. At this point the radial velocity of the gas relative to the sun will have its largest value, for at both greater and lesser distances along the line of

---

truck can go about half the maximum speed. Cars jam up just behind the truck as the drivers wait for a clear road ahead in order to pass. When they do pass, they move along again at the speed limit, leaving the truck behind. Imagine that you watched this situation from the air and concentrated on the motion of the cars. You'd see a denser region of cars just behind the truck, where they pile up for a short time; the truck moves down the road at half the average speed of the cars. You would also note that the jam persists, even

sight, only part of the space velocity is projected onto the line of sight, and in addition the space velocity is closer in absolute value to the velocity of the sun. So all we have to do is find the largest Doppler shift in the 21-cm emission at a given longitude, and that will give us the space velocity at the tangent point. A simple geometrical calculation ($R = R_{sun}$ sin l) tells us how far the tangent point is from the center, and so we have one point of the rotation curve.

The above method works only for points closer to the center than the solar distance. For points exterior to the sun, it's a little more difficult. One approach is to make a model of the mass distribution in the Galaxy, and then calculate from Newton's laws the space velocity of objects in circular orbits. Another approach is to measure the velocities of objects whose distance can be determined by other means. There are problems with both methods, and so the rotation curve for the outer part of the Galaxy is less well known than that for the inner part.

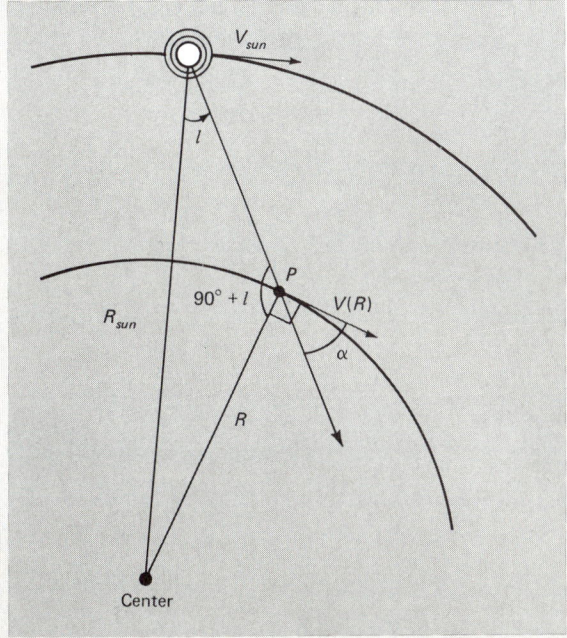

FIGURE F.22 Geometry for the rotation curve equation.

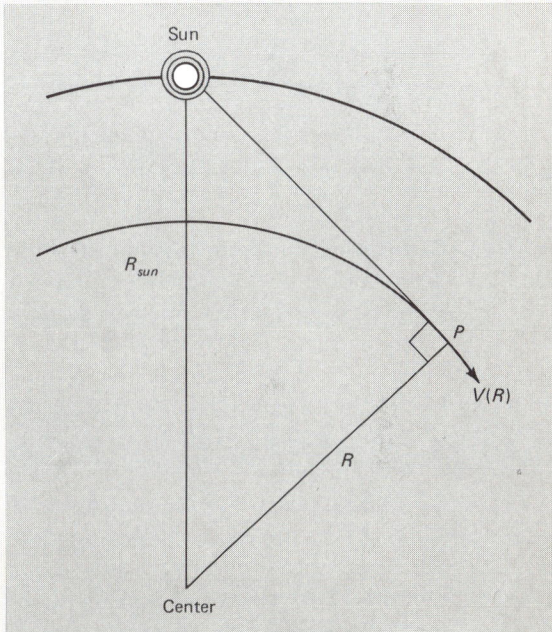

FIGURE F.23 Geometry for the observations used to find the rotation curve.

though it does not contain the same cars. New cars get caught up in it as other cars move out.

You can think of the cars as the stars and the interstellar medium in the galactic plane, and the jam as the visible effect of a moving density wave. This wave produces all the signposts of a spiral arm—young stars, H II regions, dust lanes. None of these objects lasts very long. As they die and the density wave moves, new spiral-arm tracers are born from

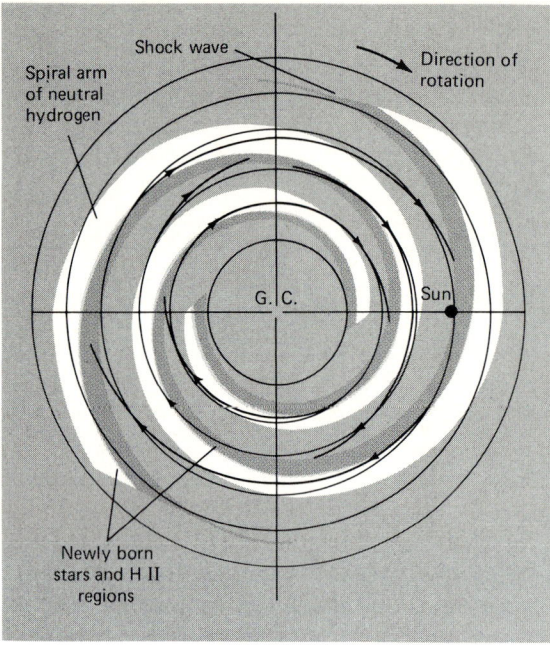

FIGURE 21.16 A model of the Galaxy's spiral structure made with the assumption that two spiral density waves (lines with arrows) ripple through the disk. Shock waves (dark regions) occur as the density waves plow through the interstellar medium. The spiral arms (white bands) form as a jam-up of material. (*Adapted from a diagram by* W. W. Roberts, Jr.)

the interstellar medium. So a spiral arm always contains the same *kinds* of objects, but not the *same* objects. Any particular arm is a transient phenomenon. Individual objects rotate at the speed appropriate for their distance from the center, but the *wave* pattern rotates with a constant angular speed, and does not wind up. This contemporary approach—pioneered by Lindblad and developed by C. C. Lin, Frank Shu, William Roberts, and Chi Yuan—is called the *density-wave model* of spiral structure. Of the models proposed to date to explain spiral structure, it best predicts and describes the overall scheme.

The density-wave model *assumes* that a two-armed spiral density wave sweeps through the galactic plane, but the model does not yet explain the wave's origin (Fig. 21.16). Although researchers can't explain its origin, they can calculate the wave's velocity and show that, once formed, it persists for a fairly long time. The gas in the disk piles up at the back of the wave. The buildup of pressure and density heats up the gas suddenly so that a shock wave forms along the front of the density wave. (You can make a shock wave by plowing through any medium faster than the speed of sound in that medium. For example, traveling through the air in a jet plane faster than the speed of sound creates a shock wave called a sonic boom.) The density wave travels through the gas faster than the sound speed, so a shock wave forms. The compression at the shock squeezes neutral hydrogen clouds together. This shock may initiate the collapse of the clouds to form giant molecular cloud complexes, which in turn form young stars and H II regions. Such squeezing also helps to make dust out of the gas, and a thin dust lane forms along the shock front. The compression of the interstellar medium by the density wave forms the features associated with a spiral arm.

During the short lifetimes of the newly formed O and B stars, the density wave moves only a short distance. So these stars, while they last, clearly mark the spiral arm. As the density wave moves on, it provokes the formation of more stars. These take the place of the ones that have rapidly faded out. So the spiral arms persist by a continual destruction and creation maintained by the density wave.

Warning: Just because we've been emphasizing luminous O and B stars (since they are good tracers of spiral arms), don't be misled into thinking that a density wave initiates the formation of only O and B stars. It probably prompts the formation of stars with all possible masses. The massive ones die quickly compared with the less massive ones that are greatest in number. The less massive stars persist after the density wave moves on, but they are less conspicuous because of their low luminosities.

How well does the density-wave model describe the observed spiral structure? First, it outlines the grand scheme of a two-armed spiral pattern that we observe in other galaxies and probably in our own. Second, it explains the persistence of the spiral arms in the face of galactic rotation. Third, it predicts the general features of a spiral arm. So the density-wave model succeeds fairly well in explaining the prominent features of spiral structure.

However, the model so far falls flat on a number of points. It does not explain the origin of the density waves. Nor does it clearly work out what keeps the density waves going. As the density waves ripple through the interstellar medium, they lose energy and should dissipate in about a billion years. But as evidenced by the abundance of spiral galaxies, the density waves—if they are the correct explanation—must last longer than a billion years. Some mechanism must keep supplying energy to maintain them. Finally, the model uses a two-arm spiral pattern, but does not explain how two arms form rather than, say, four. (Two arms *do* explain many of the observed spiral structures.) Alar Toomre, a spiral-wave researcher, states that "Recent observations leave little doubt that such spiral density waves exist and indeed are fairly common, but no one seems to know why." This is a mystery to be resolved in the study of our Galaxy and others.

21.6 A POSSIBLE HISTORY OF OUR GALAXY

The previous sections showed that we have a pretty fair idea of the architecture of the Milky Way Galaxy. What clues does this information provide about its birth and evolution? The crucial clues come from the chemical composition of galactic material and its dynamics. The process of galactic evolution links the chemistry with the dynamics. This linkage marks an important theme of cosmic evolution, because the evolution of the Galaxy results from the evolution of the stuff that makes it up.

Populations and Positions

We stated earlier that the chemical compositions of Population I and Population II stars differ considerably in their abundances of heavy elements. In general, Population II stars contain about 1 percent of the metal abundance of Population I stars. However, we do not find a simple division of metal abundances into just two groups. Rather, we find a continuous range of abundances, from about 3 percent to less than 0.1 percent for the ratio of metals to hydrogen. So, though the division into two populations is a useful first approximation, there is really a continuous range of populations. When objects are cataloged by metal abundance, a striking correlation emerges: *The lower the metal abundance of an object, the greater its height from the Galaxy's disk.*

Interpretation of this key observation relies on basic concepts of starbirth, stardeath, and the recycling of the interstellar medium. First, stars are born from clouds in the interstellar medium. Their atmospheric elemental abundance reflects that of the gas from which they formed. Second, stars inherit the orbital motions about the Galaxy of their parent gas and dust clouds. Third, massive stars evolve quickly and spew back into the interstellar medium heavy-element enriched material. So as long as new stars, especially massive ones, are born, the abundance of heavy elements in the interstellar medium of the disk gradually increases as the Galaxy ages.

With these basics in mind, let's try to make sense of the observations, which show that the youngest objects (highest heavy-element abundances) hug close to the disk, while the oldest objects (lowest heavy-element abundances) range far from the disk. Other objects fall in between these extremes.

The Birth of the Galaxy

We can estimate the Galaxy's age by finding the oldest stars in the halo. A comparison of theoretical models for globular cluster stars with their H-R diagrams (Section 10.9) indicates an oldest age of 17 billion years. That's when the Galaxy formed. How did it form?

Because globular clusters contain the oldest stars associated with the Galaxy, the halo marks the fossil remains of the Galaxy's birth. Within it globulars orbit the Galaxy on extremely elongated elliptical paths. The globulars move slowly through the halo at the outer extremes of their orbits, and only briefly do they whip in and around the nucleus. These stars exhibit the motions of the cloud from which they were formed. So the Galaxy must have been born from an initially huge gas cloud—at least 300,000 ly (100,000 pc) in radius.

Here's one model of the Galaxy's birth (Fig. 21.17). Imagine a tremendous cloud of gas roughly twice as big as the Galaxy's halo today. This proto-Galaxy cloud is probably turbulent, swirling around with random churning currents. Slowly at first, the cloud's self-gravity pulls it together, with its central regions getting denser faster than its outer parts. Throughout the cloud turbulent eddies of different sizes form, break up, and die away. Eventually, the eddies become dense enough to contain sufficient mass to hold

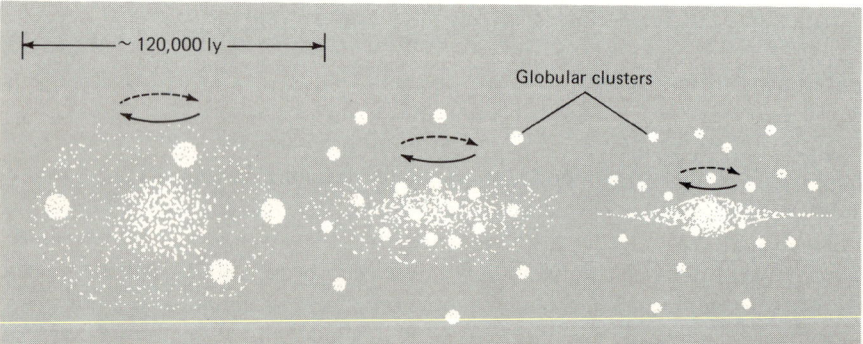

FIGURE 21.17 A schematic sequence of the collapse and condensation of a cloud to make globular clusters and the Galaxy's disk.

themselves together. These might be hundreds of light years in size—incipient globular clusters. Each blob then subdivides to form individual stars, all born about the same time with the same chemical composition. This happened 17 billion years ago, and it took place quickly. Recent evidence indicates no more than 2 billion years elapsed during the formation of halo stars.

Not all of the gas is consumed in this burst of globular cluster formation. As this material contracts more, it falls slowly into a disk. (Sound familiar? Look back at Section 19.4 on the formation of the solar nebula.) Because the original cloud had a little spin, the conservation of angular momentum requires that it spin faster around its rotational axis as it contracts. As the disk forms, its density increases, and more stars form. Each burst of starbirth leaves behind representative stars at different distances from the present disk. Finally, the remaining gas and dust settles into the narrow layer we see today. Somehow density waves appear and drive the formation of spiral arms.

During this time massive stars are manufacturing heavy elements and flinging them back into the cloud. So as stars are born in succession, each later type has more heavy elements. That enrichment continues today in the disk of the Galaxy.

There are two problems with this model. First, we have not discovered any star which contains *no* heavy elements. Where did the heavy elements contained in the oldest known stars come from? We shall see (Chapter 24) that it is unlikely that any heavy elements were made in the Big Bang origin of the universe. Was there then some previous generation of stars, formed before the globular clusters, which made the first heavy elements? Many astronomers think so, but then where are those objects today? They seem to have disappeared without a trace. Perhaps all such stars had high mass, so that they have all evolved to neutron stars or black holes.

Second, during the quick formation of halo stars, a rapid enrichment of metals occurred. How do we know? Globular clusters all appear to have the same age and so must have been formed at roughly the same time (within a span of two billion years). Yet, all globulars do *not* have the same metal abundances. For example, compared to the sun (a Population I star), some globular cluster stars have as little as 0.3 percent as much iron and others as much as 34 percent, that is, some have 100 times more than others. How did this difference come about in only two billion years? A related curiosity is that disk stars—the oldest of which match the ages of globular clusters—show a much more gradual metal enrichment. The reasons for this are not yet clear.

SUMMARY

Optical and radio astronomy complement each other to reveal the grand scheme of the Galaxy's structure. The placement of O and B stars, cepheid variables, H II regions, and dark nebulas marks the local domain of spiral arms out to 4 kpc from the sun. Beyond this distance the penetrating power of radio waves probes the interstellar gas to delineate the large extent of the spiral arms.

In spite of missing details, the overall picture that emerges shows the Galaxy as a tightly twisted, spiral structure (Fig. 21.18). Astronomers have plotted in some detail five segments of what are likely to be segments of two large-scale arms: the expanding arm (3 kpc out from the galactic center), the Norma arm (6.5 kpc out), the Sagittarius arm (8.5 kpc out), the Cygnus arm (10.5 kpc out), and the Perseus arm (13 kpc out). The Cygnus arm has a small branch, the Orion spur, in which the sun resides.

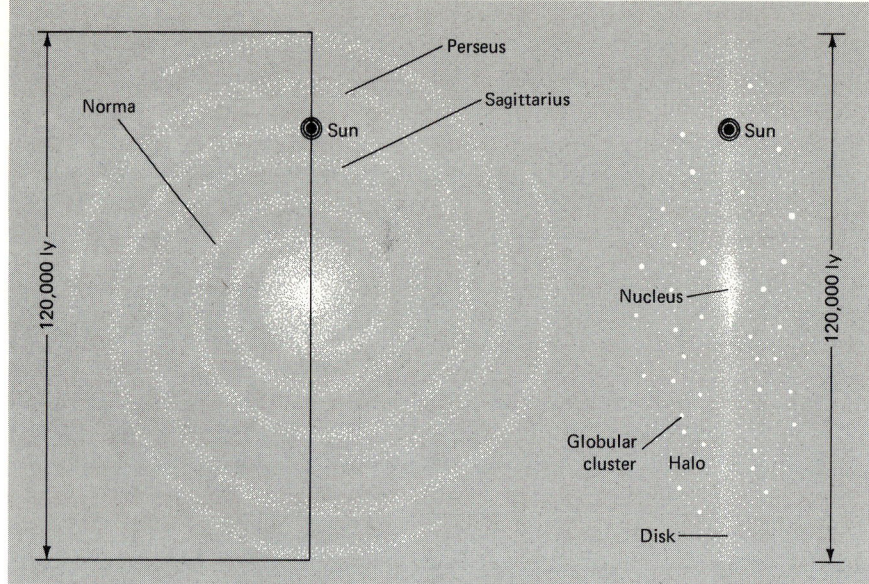

FIGURE 21.18 A simplified view of the possible overall form of the Galaxy. The spiral pattern is two-armed near the center and four-armed in the outer regions.

The flattened shape of the Galaxy has a pinwheel structure that may result from a two-armed density wave swinging through the interstellar medium. The wave compresses the interstellar gas to promote star formation. A spiral arm forms in the compressed region, marked by short-lived O and B stars. As these die out, new stars form, and the spiral arm retains its appearance, although replenished by new materials and new stars. So the arms of the Galaxy are both a short-lived and permanent phenomenon—structures continually becoming and dissolving.

Our Galaxy was born some 15 to 20 billion years ago from a gigantic cloud of gas. As this cloud contracted into a disk, successive stages of starbirth formed the different components of the halo, disk, and nucleus. The sequence can be judged from the differing heavy metal abundances of different dynamical subsystems of the Galaxy.

Key Words

spiral arm	expanding (3-kpc) arm	Oort's constant
Cygnus arm	galactic longitude	high-velocity star
Perseus arm	galactic latitude	low-velocity star
Sagittarius arm	differential rotation	spiral tracer
Norma arm	galactic rotation curve	density-wave model

Review Questions

1. What limits an optical astronomer's investigation of the Galaxy's structure? (Objective 1)
2. Why are Population I cepheids good spiral-arm tracers? (Objectives 3 and 7)

3. Why do radio astronomers need the rotation curve of the Galaxy in order to use 21-cm observations to establish its spiral structure? (Objective 6)
4. Argue that a spiral arm cannot be a material arm. (Objective 8)
5. What are the kinds of celestial objects found in spiral arms? (Objective 8)
6. What characteristics of spiral arms does the density-wave model account for? In what respects is the model at present inadequate? (Objective 8)
7. Relate the orbits of globulars and their chemical composition to the birth of the Galaxy. (Objective 10)

Problems
1. Calculate the mass of the Galaxy from the rotation curve in Fig. 21.8b. (Objective 5)
2. There's some dispute as to the actual distance of the sun from the galactic center and as to the actual velocity of rotation. Some astronomers favor $R = 8.5$ kpc rather than 10 kpc, and $V_{rot} = 220$ km/sec rather than 250 km/sec. Calculate the Galaxy's mass first using the alternate value for R, then the alternate value for V_{rot}, then both alternate values: (a) $R = 8.5$ kpc, $V_{rot} = 250$ km/sec; (b) $R = 10$ kpc, $V_{rot} = 220$ km/sec; (c) $R = 8.5$ kpc, $V_{rot} = 220$ km/sec. (Objective 5)
3. Looking at galactic longitude 60°, you find H I emission at +25 km/sec. Roughly how far away is the H I cloud? (Objective 6)
4. Calculate the amount of Doppler shift for a 21-cm line emitted by a cloud with a radial velocity of 100 km/sec. (Objective 6)
5. Calculate the free-fall time (Section 9.3) for material at the edge of the Galaxy's halo to collapse into the disk. (Objective 10)

BEYOND THIS BOOK . . .

* "The Arms of the Galaxy" by B. Bok in *Scientific American*, December 1959, is interesting to contrast to the information in *The Milky Way* (Harvard University Press, Cambridge, Mass., 1981) by B. and P. Bok.
* "The Morphology of Hydrogen and of Other Tracers in the Galaxy" by W. B. Burton in *Annual Reviews of Astronomy and Astrophysics*, vol. 14 (1976), p. 275, is a very technical, but recent review about spiral-arm tracers.
* An excellent exposition of 21-cm mapping is "Steps Toward Understanding the Large-Scale Structure of the Milky Way" by H. Weaver in *Mercury*, September/October 1975, p. 18; November/December 1975, p. 18; January/February 1976, p. 19.

BEYOND THE MILKY WAY: GALAXIES

22

LEARNING OBJECTIVES

After studying this chapter, you should be able to:

1. Compare spiral, elliptical, and irregular galaxies in terms of general physical characteristics such as size, shape, mass, color, types of stars, and amount of interstellar gas and dust.
2. Outline the methods used to find the bulk properties of galaxies and be able to estimate a galaxy's mass from its rotation curve.
3. Use the relative brightnesses and angular sizes of galaxies to estimate their relative distances.
4. Indicate what observations clinched the idea that the "spiral nebulas" were actually other galaxies.
5. Describe what is meant by standard candles and use them to find distances to galaxies.
6. Outline a contemporary method of finding distances to distant galaxies, starting with the Astronomical Unit and ending with the Hubble constant.
7. Evaluate the weaknesses in the procedure outlined in Objective 6 so you can estimate the possible errors in distances to galaxies.
8. Show how getting distances and radial velocities for galaxies results in a value for the Hubble constant, indicate present uncertainties in its value, and use this value to estimate distances to other galaxies.
9. Describe how galaxies are distributed around the sky.
10. Define a cluster of galaxies, describe the appearance of the different classes of clusters, and sketch the general properties of clusters.
11. Describe the effects of close encounters of galaxies in clusters.
12. Evaluate the evidence for intergalactic material between and within clusters of galaxies.

CENTRAL QUESTION:
What is the structure and content of galaxies, and how are they distributed throughout the universe?

No celestial object is quite as grand as a galaxy. In a large telescope a bright spiral galaxy is a stunning sight, a star-bright nucleus and misty swirl of spiral arms. Billions of stars are caught in a whirlpool spanning hundreds of thousands of light years.

The galaxies form the basic elements of our modern cosmological vista. Their sheer numbers are almost beyond our comprehension (Fig. 22.1), and their diversity of structure is also astounding. Even more surprising is the fundamental unity found in spite of their wide variety. The fact that galaxies can be sorted into broad divisions hints at a common evolutionary process.

Galaxies form the skeleton of the universe. Many conclusions of modern cosmology hinge on the correct establishment of their distances. This chapter notes that the notorious difficulties of surveying our own Galaxy are amplified when we try to appraise the vastness of the universe. In spite of present problems, our vision of the universe, underpinned by the theory of general relativity, has a coherence that allows us to draw conclusions about how it began.

22.1 THE RESOLUTION OF THE SHAPLEY-CURTIS DEBATE

Recall (Section 20.5) that Curtis opposed Shapley on the question whether the "spiral nebulas" were simply outer members of the Milky Way Galaxy, as Shapley claimed, or galaxies in their own right.

Curtis felt that the observations of novas in the spirals gave a clue to their nature. Only groups of stars, not clouds of pure gas, could have nova explosions.

Curtis also argued that the wide range of apparent angular sizes of spirals—approximately 2° for M 31, the nearest, to 10′ and less for the smallest—required a large range of distances and so they could not be part of our Galaxy. Starting from the principle of the

FIGURE 22.1 A group of galaxies in Leo. (*Courtesy Palomar Observatory, California Institute of Technology.*)

uniformity of nature, Curtis assumed that all spirals have roughly the same physical diameter. The range in observed sizes (about 10 to 1) implied that the spirals must be enormous distances from the Galaxy, for if they were the same in diameter, the range in apparent size meant that the ones 10 times smaller must be 10 times farther off, or about 10 times the radius of the Galaxy (Fig. 22.2). So they could not be members of the Milky Way Galaxy, as Shapley argued.

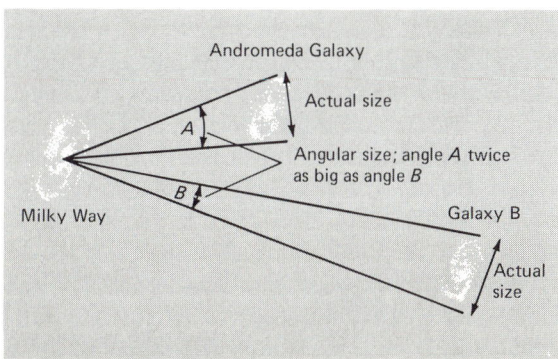

FIGURE 22.2 Distances to galaxies estimated by their angular sizes. If galaxies have roughly the same physical size, the more distant will have smaller angular sizes. In the case shown here, Galaxy B is about twice as far away as the Andromeda Galaxy.

767

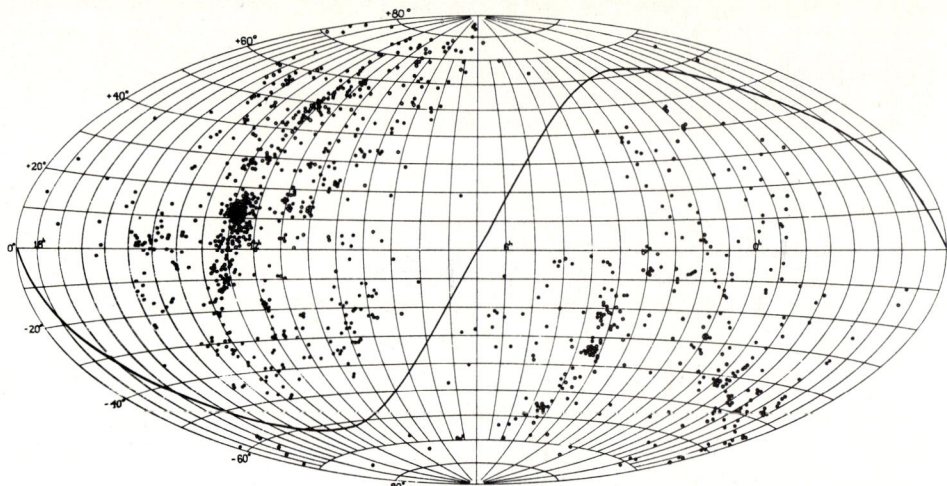

FIGURE 22.3 (a) The distribution of bright galaxies in the sky. The area on both sides of the heavy dark line is the region where few galaxies are visible. (*Courtesy, Yerkes Observatory.*) (b) Dust in the plane of a spiral galaxy (M 104). The dark band occurs from dust cutting out starlight. (*Courtesy Palomar Observatory, California Institute of Technology.*)

In addition, Curtis noted the so-called *zone of avoidance* (Fig. 22.3a), a region near the plane of the Galaxy where very few spirals are visible. Curtis argued that interstellar dust in the galactic disk cuts down the light from the spirals; we see fewer in the zone of avoidance because there we look through the disk of the Galaxy. If the spirals were actually associated with the Galaxy, they would be found concentrated in the plane along with the stars, rather than avoiding it. As evidence for this point of view, Curtis cited photographs of spirals showing dark lanes cutting through their planes (Fig. 22.3b). If the Galaxy and other spirals were similar in structure, Curtis reasoned, then our Galaxy must also have obscuring material collected in the plane.

Finally, Curtis pointed out that the spectra of spirals are not emission spectra like those of diffuse nebulas such as the Orion Nebula, but rather are absorption-line spectra. Such spectra are like those from a conglomeration of stars—the same spectrum the Galaxy would show if viewed from a great distance.

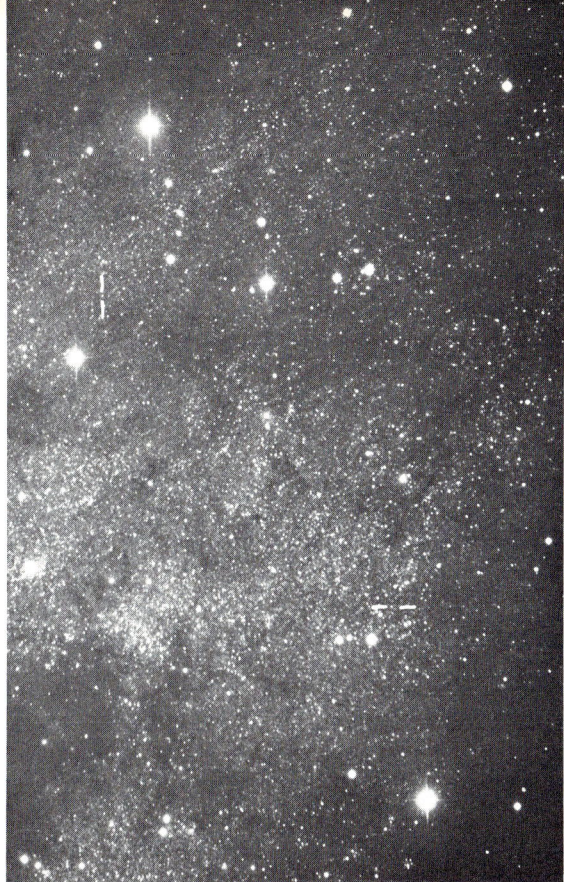

FIGURE 22.4 Two cepheid variables (stars between the white lines) in the Andromeda Galaxy (M 31). (*Courtesy Palomar Observatory, California Institute of Technology.*)

In 1924, after the Shapley-Curtis confrontation, Edwin Hubble settled the dispute conclusively by the discovery of cepheid variables in M 31. Although variables had been suspected as early as 1922, Hubble confirmed their existence in the outer arms of M 31 (Fig. 22.4). Using the period-luminosity relation (Section 20.4), he derived a distance of 490,000 ly (150 kpc) for M 31, far beyond the farthest globular clusters that Shapley contended marked the outer limits of the Galaxy. Even Hubble's estimate was too small, because he used an incorrect calibration for the period-luminosity relation for cepheids. Recent work on cepheids in M 31 establishes a distance of 2.22 million ly (680 kpc).

The procedure is a direct one. One of the cepheids found by Hubble has a period of about 45 days and an apparent average magnitude of about +18.5. From the period-luminosity relationship of Population I cepheids (Fig. 20.21), a period of 45 days implies an absolute magnitude of roughly −6. Then

$$m - M = 5 \log d - 5$$
$$18.5 - (-6) = 5 \log d - 5$$
$$\log d = 5.9$$
$$d = 780 \text{ kpc} = 2.6 \text{ million ly}$$

a reasonable value given the approximate choice of the absolute magnitude.

Warning: Several times so far we have referred to the principle of the "uniformity of nature." For example, Curtis assumed that the novas in M 31 were the same as the novas in our Galaxy, and Shapley later assumed that the cepheidlike stars in globular clusters were the same as the cepheids in the galactic disk. This is a very powerful principle, but it has to be used with caution. The novas in M 31 *are* pretty much the same as those in our own Galaxy, because M 31 is a spiral galaxy very similar to the Milky Way Galaxy. But the variable stars in globular clusters are *not*, it turns out, the same as the cepheids in the disk of our Galaxy, for they belong to quite different populations of stars. You can always make the assumption of uniformity everywhere, and that can lead to new insights, but you always have to keep looking for ways to check that assumption and be prepared to abandon it if inconsistencies develop.

22.2 THE GALAXIAN ZOO

Hubble pioneered in the field of extragalactic astronomy, first with the 100-in and then the 200-in telescope. He recognized that galaxies had different shapes. To sort out the differences in form, Hubble in 1926 proposed the first scheme for the classification of galaxies. His initial design is now considered too simple, and modern classification schemes have expanded his format, but astronomers still use the fundamental categories of *elliptical*, *spiral*, and *irregular* galaxies. Let's take a brief look at these basic galaxy types.

Elliptical Galaxies

Elliptical galaxies exhibit no spiral structure, but do show an elliptical shape with symmetry about a major and minor axis (Fig. 22.5). Very little gas (as evidenced by the

FIGURE 22.5 A giant elliptical galaxy, M 87. Note the symmetry of the shape and the lack of distinct structure. The small dots are globular clusters that orbit the galaxy's nucleus. (*Courtesy Lick Observatory.*)

22.2 THE GALAXIAN ZOO

lack of 21-cm emission) or dust appears in elliptical galaxies, and O and B stars are also absent. The ellipticals generally have a redder color than other types. They come in a range of sizes from giants to dwarfs.

Hubble subdivided the ellipticals in classes from E0 to E7, according to how elliptical they appear. Imagine looking at a circular plate face on; such is the appearance of an E0 galaxy. Now slowly tilt the plate so that it looks more elliptical and less circular. This flattening of shape presents the same views as the sequence from E0 to E7 galaxies. Be warned that Hubble based the classifications on the appearance of the galaxy, not its true shape. For example, an E7 is really a flat elliptical viewed edge on, but an E0 may be either a truly spherical galaxy or a flattened galaxy seen face on.

Spiral Galaxies

Spiral galaxies display obvious spiral structure, usually with two, but sometimes more, spiral arms (Fig. 22.6). One type has a prominent bar through the nucleus, the spiral arms winding out from the end of the bar (Fig. 22.7). Hubble called the spirals without a bar

FIGURE 22.6 A spiral galaxy with a large nucleus, M 81. It is Hubble type Sb. (*Courtesy Lick Observatory.*)

FIGURE 22.7 A typical barred spiral galaxy (NGC 1330). Note the bar crosses the nucleus and links the two spiral arms together. (*Courtesy Kitt Peak National Observatory.*)

normal and denoted them by the letter *S*. The others he called barred, denoting them as *SB*. Both groups are subdivided further into categories *a*, *b*, and *c* (Fig. 22.8). These are judged by how tightly the spiral arms wind around (a, tightest; c, most open), the relative size of the nucleus (a, largest; c, smallest), and the degree to which the arms are resolved into patches (a, most resolved; c, least resolved). For example, a galaxy of Hubble class Sa is a normal spiral with a large nucleus and tightly coiled arms. A few galaxies appear to have the disk of a spiral but no arms. Hubble dubbed those *S0*. These are now sometimes called *lenticular galaxies* because of their shape.

Irregular Galaxies

Irregular galaxies are those that are devoid of spiral structure or symmetry, but are resolvable into distinct patches of stars (Fig. 22.9). They fall into two groups. *Irr I* can be resolved into O and B stars and H II regions. Conspicuous dust clouds are usually absent. *Irr II* don't have visible H II regions, and usually do show prominent dust lanes.

Fairly recently astronomers have recognized that each Hubble type of galaxy, say Sb, comes in a range of luminosities (absolute magnitudes). In analogy to stellar luminosity classes (Section 8.6), galaxies also have *luminosity classes* of I, II, III, IV, and V, with I most luminous and V least luminous. An Sc I galaxy, for instance, is a very luminous spiral with small nucleus and spread-out arms. It turns out that luminosity class I galaxies

FIGURE 22.8 Examples of different types of spiral galaxies. Note that from type S0 to Sc, the nucleus is relatively smaller and the spiral arms more spread out. (*Courtesy Palomar Observatory, California Institute of Technology.*)

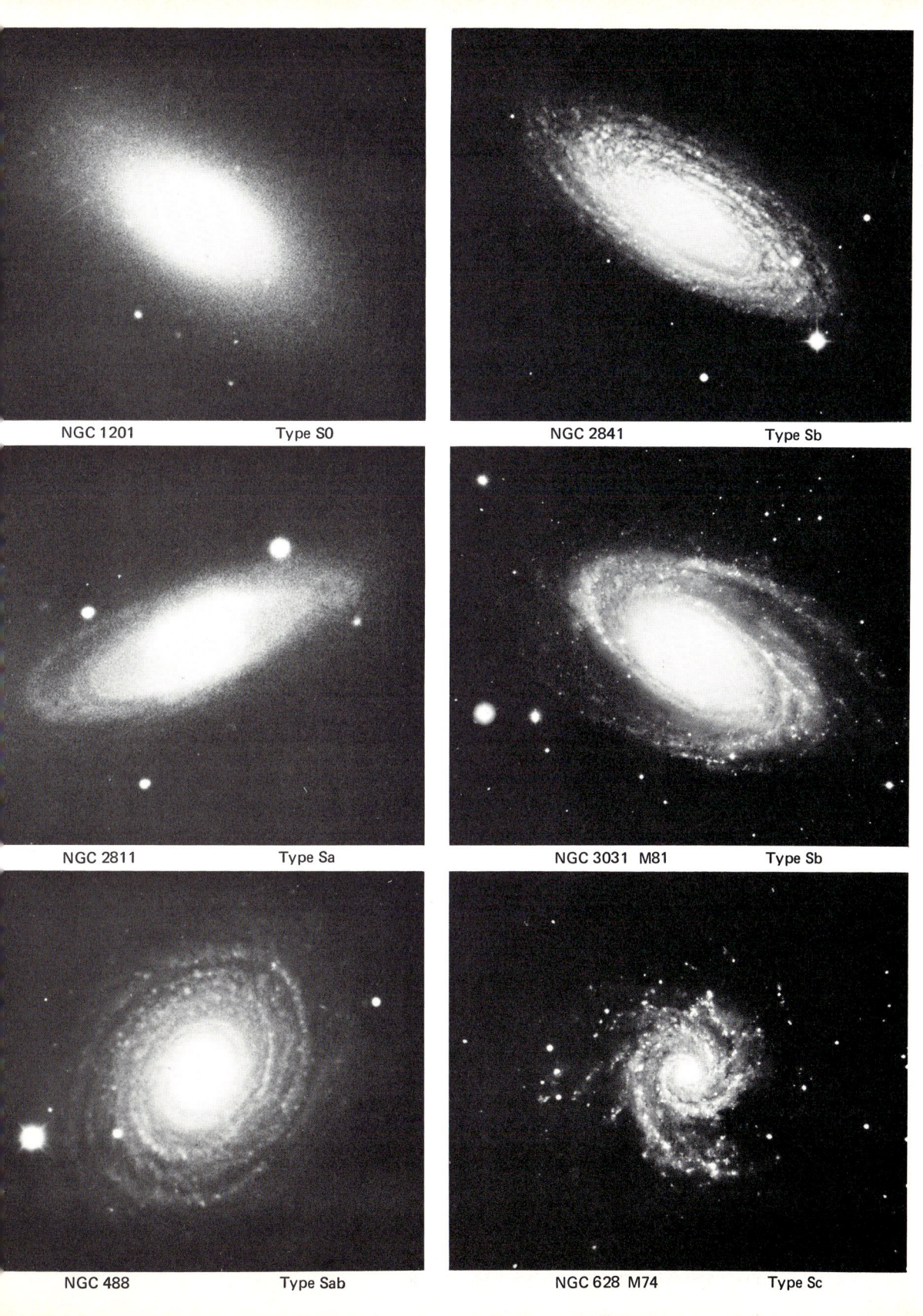

FIGURE 22.9 A dwarf irregular galaxy in the constellation Sextans. (*Courtesy Palomar Observatory, California Institute of Technology.*)

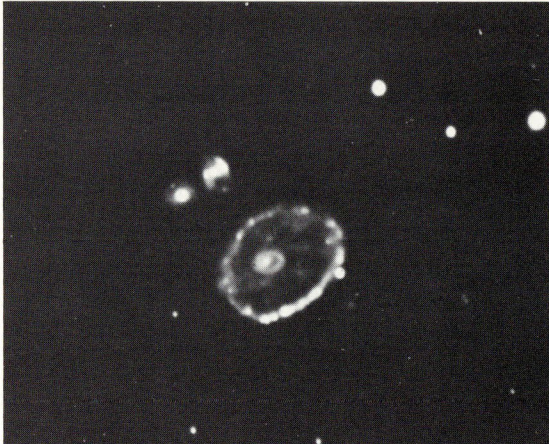

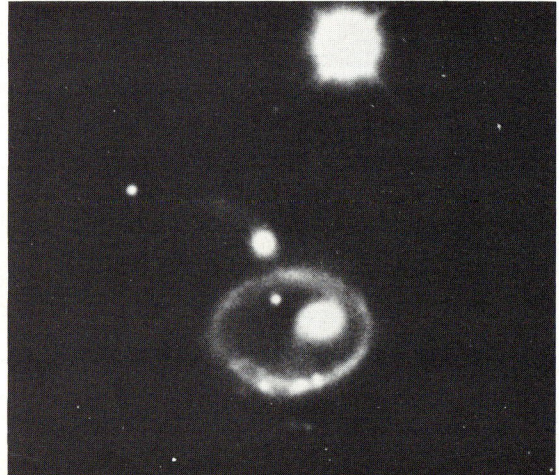

FIGURE 22.10 Two galaxies with peculiar shapes that do not fit the simple Hubble scheme. These are *ring galaxies*. (*Courtesy Kitt Peak National Observatory and Cerro Tololo Inter-American Observatory.*)

must be more massive than class II, and so on. So class I galaxies can be thought of as supergiant galaxies.

Warning: The Hubble scheme does not include all types of galaxies. Some galaxies stand out as peculiar in shape (Fig. 22.10) and do not fit into the three general Hubble categories. Many of these peculiar galaxies turn out to have evidence of unusual activity. Some appear to be pairs of galaxies close together interacting gravitationally (Section 22.5).

About 77 percent of known galaxies are spirals, 20 percent ellipticals, and 3 percent irregulars. This sample is dominated by the luminous spirals, however, which are visible at very great distances. The relative numbers in a given volume of space are quite different. A survey of the region of space out to 30 million ly (9.1 Mpc) finds that only 34 percent of the galaxies in this volume are spirals, 13 percent are ellipticals, and 54 percent are irregulars. Many of the irregulars are small galaxies of fairly low luminosity, similar to the Magellanic Clouds, which orbit our Milky Way Galaxy.

22.3 SURVEYING THE UNIVERSE OF GALAXIES

The grouping of galaxies by shape marked an initial step toward delving into the far depths of the universe. But to probe the physical properties of galaxies—their masses, sizes, luminosities—ultimately depends on a knowledge of their distances.

Only about 50 years have passed since we learned for certain that galaxies are faraway islands of stars. Given this short time, you should not be surprised to discover that we know the distances to galaxies only roughly. For the nearest galaxies we have measurements that are good to about 10 percent. For the most distant visible galaxies we are lucky if we know their distances within 50 percent of their actual value. It's a hard but essential astronomical task to survey the vast universe of galaxies.

Although the distances to galaxies have continually been revised, the essential techniques remain the same. The distance indicators are the old criteria of brightness and size: "brightness mean nearness" and "smallness means farness." The galaxies with the smallest angular size tend also to be the faintest and the most distant. By applying these simple criteria you can make rough estimates of the relative distances to galaxies. For example, if you look at two galaxies, one apparently half as large as another, then—if you have some reason to believe that they are actually the same size—the apparently smaller galaxy must be twice as distant as the apparently larger one. Similarly, if one galaxy is 5 magnitudes fainter than another, that is, 100 times fainter, and if you assume they are the same luminosity, then the fainter one must be roughly 10 times farther away.

The difficulties with this approach are, first, that we need to have a way of determining that the two galaxies, near and far, are really the same (in size or luminosity). Second, we need an independent technique to determine the distance of the nearby object. Theoretical models and careful observations of form, color, and spectrum provide the criteria of sameness. A step-by-step process, moving from nearby objects to more distant ones, paying attention to the calibration of each step, provides the distance to the nearby galaxies. Each step in surveying the universe applies to certain objects and over a certain range of distances; one piles step upon step to establish a *distance pyramid* (Fig. 22.11), which we will describe below. This structure is very much an astronomical house of cards, for it is only as strong as its weakest lower support. At present the degree of uncertainty in the distance determined to nearby galaxies is *at best* about 10 percent. As you move farther up the pyramid, the percentage of error increases.

In order to discuss the layers of the pyramid, we need to establish the scale of the AU, say in kilometers, from solar system observations (Chapter 13). To move out to the stars requires the use of heliocentric parallax (Section 8.2). Beyond the reach of heliocentric parallaxes astronomers rely on open star clusters (Section 10.9), statistical parallaxes (Section 20.3), spectroscopic parallaxes (Section 8.5), and the cepheid period-luminosity relation (Section 20.4). These techniques allow a reasonable assessment of the distance to the galactic and globular clusters of the Milky Way Galaxy, and so to the establishment of its size. More important, these local distances provide the luminosities of typical galactic occupants.

To bridge the distances to other galaxies, we must accept the assumption of the uniformity of nature: that the essential character of objects in our Galaxy (such as cepheid

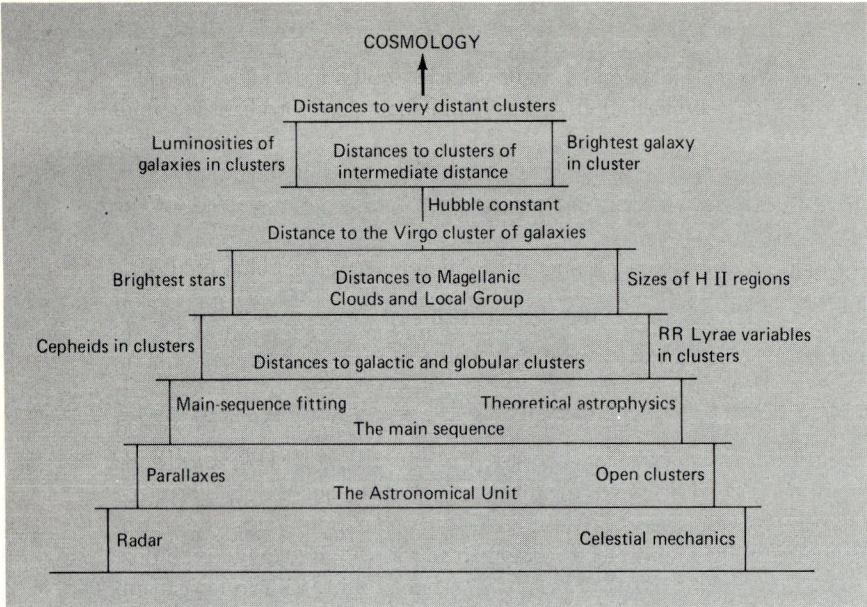

FIGURE 22.11 The distance pyramid. The various distance-measuring techniques are linked together to get finally the distances to galaxies. (*Adapted from a diagram by P. Hodge.*)

variable stars or supernovas) is the same for similar objects in other galaxies. Without this crucial assumption we couldn't get anywhere. However, it's not a blind, unsupported assumption. Observations made so far are consistent with it. For example, cepheids in other galaxies have the same spectra and the same shape light curves as cepheids of the same type in our Galaxy.

To find distances to galaxies, we must use identifiable objects whose luminosities we know. Astronomers often refer to these as *standard candles*. The trick here is to choose as standard candles objects whose luminosities we can determine reasonably well and relate to some other observable property. Then when we detect such an object in another galaxy, we can compare its flux (apparent magnitude) with its luminosity (absolute magnitude) to infer the distance. Unfortunately, some objects are too faint to be picked up by even the largest telescopes. So we want to choose the *most* luminous objects in galaxies to use as standard candles, as long as we know what their luminosities are.

Starting with close galaxies, we apply the period-luminosity relation to cepheids in other galaxies. The cepheids, however, are useful only over a very limited range, for they are not especially luminous. Also, E and S0 galaxies don't contain any cepheids. Of visible galaxies, only about 30 are of the right kind and close enough for us to detect cepheids within them. We can use cepheids as standard candles out to approximately 13 million ly (4 Mpc). At that distance cepheids are fainter than the limits of photographs taken on large telescopes (apparent magnitude of roughly +22).

To go beyond 13 million ly requires the establishment of other standard candles whose visibility is greater than that of the cepheids. Fortunately we can use other objects to make this leap, such as the brightest stars in galaxies (but again, elliptical and S0 galaxies are excepted, since they don't contain O and B stars).

Supergiant O and B stars appear to have an upper limit on their luminosities. In our own Galaxy the brightest blue-white star (Population I) has a luminosity of 5.2×10^5

22.3 SURVEYING THE UNIVERSE OF GALAXIES

solar luminosities (absolute magnitude of -9.5). In M 33, a nearby spiral galaxy whose distance we can measure by cepheids, the brightest blue-white star has a luminosity of 4×10^5 solar luminosities (absolute magnitude of -9.2). For other nearby galaxies the luminosities range from 10^5 to 10^6 solar luminosities (absolute magnitude -8.0 to -10.0), depending on the luminosity class of the parent galaxy.

Here's how to use this information to find distances. In the supergiant spiral galaxy M 101, you observe that the brightest blue stars have apparent magnitudes of $+19$. For a large Sc I galaxy like M 101, we expect the brightest blue stars to have an absolute magnitude of -10. Roughly how far is M 101 by this standard candle?

$$m - M - 5 \log d - 5$$
$$19 - (-10) = 5 \log d - 5$$
$$\log d = 6.8$$
$$d = 6.3 \times 10^6 \text{ pc} = 21 \times 10^6 \text{ ly}$$

A more precise calculation gives a distance of 24 million ly (7.2 Mpc). This method can be used out to a distance of 82 million ly (25 Mpc).

The designation of other distance candles follows the same strategy. Find fairly common objects, find their luminosities in our own or nearby galaxies, check other galaxies by methods known to be reliable, and then utilize the candle to the limits of its accuracy.

Another technique uses angular size of H II regions as a measure of distance. For nearby galaxies astronomers have found that the largest H II regions are the same size in galaxies of the same type (Hubble and luminosity class). A comparison of the angular sizes of H II regions leads to relative distances that are calibrated by the nearest galaxies whose distances are well established. If the largest H II region in one galaxy is $\frac{1}{3}$ the angular size of the largest found in another galaxy of the same type, then the first galaxy is three times as far as the second. If you know the distance of the nearer galaxy, you know the distance of the farther one. Or you can calculate the physical size of an H II region in a nearby galaxy, and use it as a standard ruler for more distant galaxies. This technique is useful out to 82 million ly (25 Mpc).

As an example, for Sc III galaxies close enough to find their distances by cepheids, we discover that their largest H II regions have a size of about 200 pc (650 ly). Suppose we find a distant Sc III galaxy whose largest H II region is resolvable with an angular size of 60". What's the distance? Use the small angle approximation, $\alpha_{rad} = s/d$, with 60" equal to 2.9×10^{-4} radians. Then, if the largest H II region in an Sc III galaxy has a size of 200 pc, the distance (d) to the galaxy is

$$d = \frac{200}{2.9 \times 10^{-4}}$$
$$= 6.9 \times 10^5 \text{ pc} = 2.2 \times 10^6 \text{ ly}$$

This method works as long as you can resolve the H II regions.

At distances greater than 82 million ly, we can't see individual objects in galaxies, at least with ground-based telescopes. (The Space Telescope will allow us to resolve stars at distances nearly 10 times farther out.) We then use the luminosities of the galaxies them-

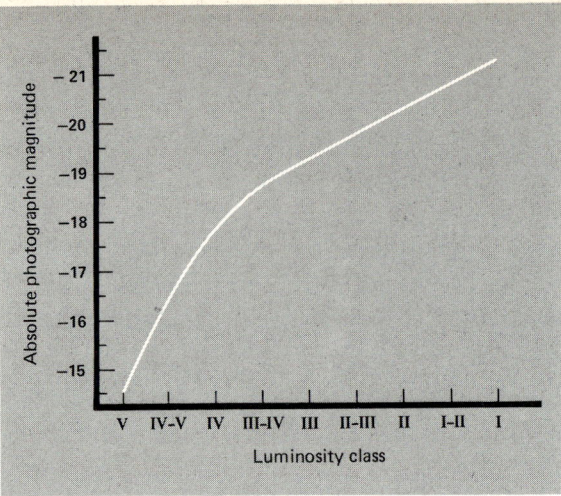

FIGURE 22.12 The Sandage-Tammann calibration of the absolute photographic magnitudes for different luminosity classes of spiral and irregular galaxies. A "photographic magnitude" is measured from a photographic plate. (*Adapted from a diagram by A. Sandage and G. Tammann,* Astrophysical Journal, *vol. 194, p. 539, copyright 1974 by the American Astronomical Society.*)

selves. Galaxies tend to lie in clusters (see Section 22.5). Unfortunately, we cannot simply choose any galaxy in a cluster as a standard candle, because a wide spread exists in their luminosities. To ensure choosing galaxies with the same luminosity, we select one of the brightest galaxies in a cluster rather than an average to one picked at random. The brightest galaxies in clusters appear to have about the same luminosity, *if* you stick to the same kind of galaxy (such as large spirals). In addition, we choose the brightest galaxies because for the most distant clusters only the most luminous galaxies can be seen easily.

Of course, for this method to work, we need a calibration for the galaxian luminosities. We get the calibration by using cepheid variables and the sizes of H II regions to find the distances to nearby spiral galaxies. Then we have a sample over a range of luminosity classes that we hope represents all galaxies (Fig. 22.12). (The uniformity of nature principle again!)

What kind of galaxies serve as useful standard candles? In the recent work of Allan Sandage and Gustav Tammann, the candidates are supergiant spiral galaxies with small nuclei and spread-out spiral arms, that is, Sc I galaxies. (M 101 is such a galaxy; turn back to Fig. 21.1.) Sandage and Tammann calibrate the absolute magnitude of such galaxies as about -21.2, or 25×10^9 times the sun's luminosity. Seen in distant clusters, these galaxies are relatively easy to identify because of their high luminosity and distinctive shape. With contemporary telescopes they serve as standard candles to distances of roughly 13 billion ly (400 Mpc). Giant elliptical galaxies are also used. Again, their luminosities must be calibrated by reference to some nearby galaxy or cluster whose distance is known by other techniques.

The Hubble Law Revisited

How can we measure the distances of galaxies if their standard candles are too feeble to be seen? We can rely on the distance-velocity relationship first found by Humason and Hubble—the expansion of the universe itself (Section 5.4).

Recall that compared to finding a galaxy's distance it is a relatively easy task to measure its red shift and from it determine the galaxy's radial velocity. A comparison spectrum made at the same time as the galaxy's spectrum affords a direct measurement of the shift in some prominent spectral lines (such as the H and K lines of calcium; Fig. 22.13). The

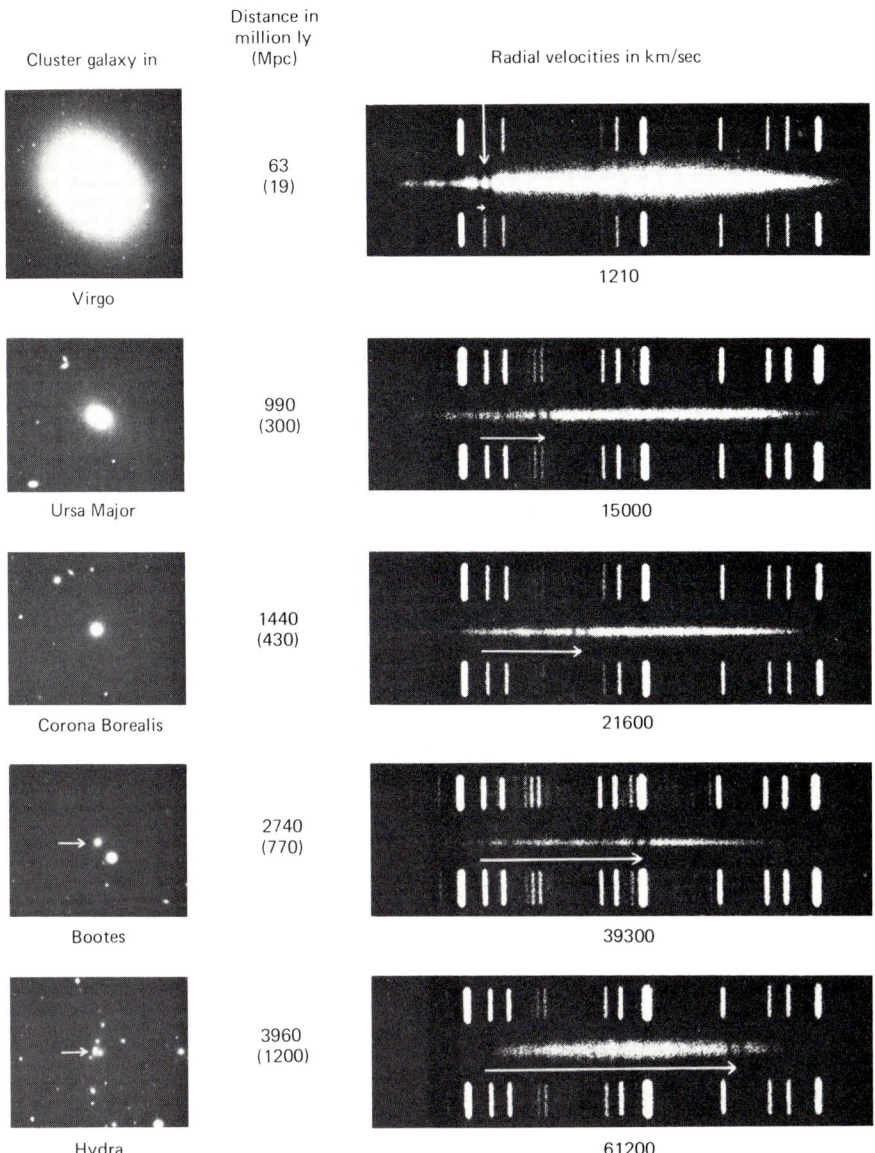

FIGURE 22.13 Red shifts and measured radial velocities and distances for selected galaxies in order (top to bottom) of increasing red shift and distance. The red shifts are visible in the spectra on the right, where a white arrow indicates the size of the shift in the H and K lines of calcium; these are the two darkest lines in the spectra. The radial velocities are given under the spectra. (*Courtesy Palomar Observatory, California Institute of Technology.*)

galaxy's distance is much tougher to find, as has been emphasized, but it is measurable out to the limits of the present standard candles.

To see how this method works, look at the galaxies pictured and at their spectra with red shift indicated (Fig. 22.13). Notice that as the red shifts get larger, the galaxies appear smaller and fainter. So the rules of thumb "faintness means farness" and "smallness

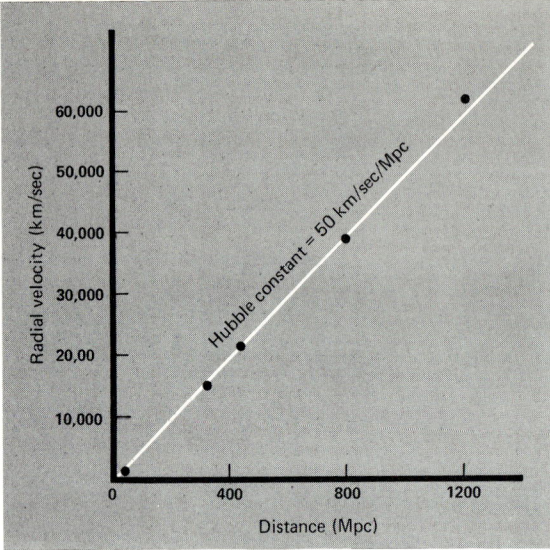

FIGURE 22.14 The Hubble plot for the data in Fig. 22.13. The straight line corresponds to a slope for a value of Hubble's constant of 50 km/sec/Mpc.

means farness" do work out as rough guesses. Now plot, for these galaxies, their distances versus their red shifts. You can draw a straight line that represents the trend of these points and so find a value for the Hubble constant, which is the slope of the line (Fig. 22.14).

Reverse this procedure to find distances from red shifts. Suppose you were given the red shift of a galaxy. You look up this value on the red-shift axis of your plot and see to what distance it corresponds, that is, you use the Hubble relationship relating a galaxy's radial velocity (V_r), distance (D), and the Hubble constant (H):

$$V_r = HD$$

In using this method, you have to assume that the line drawn for the plotted points applies to the galaxy you are considering. For relatively nearby galaxies for which you wish to use the method simply because they don't contain bright standard candles, it is fairly safe to assume that it does. But if you have to extrapolate beyond the last well-measured point, you cannot be certain the relation is still valid.

For example, suppose a galaxy's radial velocity measured from its red shift is 40,000 km/sec. What is its distance? If H equals 50 km/sec/Mpc, then

$$D = \frac{V_r}{H}$$

$$= \frac{40,000 \text{ km/sec}}{50 \text{ km/sec/Mpc}}$$

$$= 800 \text{ Mpc}$$

It looks simple, but we must inject a note of caution. This indirect method of distance measurement rests on two crucial assumptions: (1) that the galaxies with known distances give an accurate value for the Hubble constant and (2) that the relation is expressed by a

22.3 SURVEYING THE UNIVERSE OF GALAXIES

straight line that may be extrapolated. In the face of present evidence the first assumption seems good, but not totally secure. Estimates of the Hubble constant range from 50 to 100 km/sec/Mpc, with at least a 10 percent uncertainty. But the various estimates differ by a factor of two! The distances derived by this method may be wrong by at least 10 percent and perhaps by as much as 100 percent.

The validity of the second assumption depends on the geometry of the universe (Section 5.5). For example, if the universe is closed (so that it will eventually stop expanding), then the Hubble plot cannot be extended very far as a straight line. It must eventually curve sharply upward, because the expansion was much faster in the past than now. (The farther away an object, the longer the time required for its light to travel to us; so when we observe objects at greater and greater distances, we are also seeing farther and farther back in time.)

If we knew the overall geometry (open or closed) and the present density of our universe, we could apply relativity theory to draw the expected red shift–distance relation and use it to find the distances to galaxies. But we have a circular problem here, for the observed red shift–distance relation is one of the things used to determine the geometry of the universe. At distances beyond our standard candles we must assume a geometry in order to get a distance. So our second assumption has a weak foundation.

The Sandage-Tammann Procedure to *H*

Finding distances to galaxies is one of the toughest jobs for the astronomers. Yet it's one of the most important. Let us summarize the steps in a contemporary procedure followed by Sandage and Tammann, which yielded a value of the Hubble constant of 50 to 55 km/sec/Mpc. (To determine this value took *19* years of careful observing!)

1. Measure the AU using radar reflection from Venus and then the distances to nearby stars using trigonometric parallax with the AU as the baseline.
2. Determine the distance to the Hyades, a galactic cluster, using the motions of its stars, checked for consistency by comparing brightnesses of Hyades stars with those of nearby stars with distances known by trigonometric parallax.
3. Find the distances to cepheids in our Galaxy by searching out galactic clusters that contain cepheids, comparing the brightness of the stars in these clusters with that of the stars in the Hyades cluster, and so finding the distances to the clusters with cepheids (there are only a few). This calibrates the cepheid period-luminosity relation.
4. Use cepheids to determine the distances to nearby galaxies by the period-luminosity relationship.
5. Measure the angular sizes of H II regions in these nearby galaxies. We find that the actual sizes of H II regions in spiral and irregular galaxies depend on the luminosity of the galaxy. Use nearby galaxies to calibrate the relation between H II region size and luminosity.
6. Extend this calibration to Sc I galaxies (supergiant, very luminous spiral galaxies), for which the nearest, M 101, has its distance determined by different methods as a check. We now have the sizes of H II regions for Sc I galaxies. However, we must be careful because the calibration is based on only *one* galaxy, M 101.

7. Use the size-luminosity relation for H II regions to find the distances of galaxies to the limit of this method. (We can use this procedure on about 50 galaxies.) Now we know the absolute magnitudes of these galaxies and their relationship to luminosity classes.
8. Look at Sc I galaxies, the objects we can see distinctly at the greatest distances. Use the absolute magnitude calibration (step 7) to find their distances. Measure their red shifts to get their radial velocities. Divide their radial velocities by their distances. We then have the Hubble constant.

Warning: The current determinations of the distance scale of the universe contain one worrisome point, the assumption that the galaxies which are the standard candles do not change much with time. Light takes a finite amount of time to reach us, billions of years in the case of the most distant clusters of galaxies observed. But we know full well that since galaxies are composed of stars, they *will* change with time. The questions are how rapidly do galaxies evolve and how much are their luminosities and colors affected over the timespan which we can now observe. Many astronomers are currently working on this problem, both theoretically and observationally. It is an important unsolved problem, for it affects not only our understanding of distances, but also our measurements of the very nature of space itself.

Another Way to *H*: The Aaronson-Huchra-Mould Procedure

Not all astronomers agree with Sandage and Tammann that *H* is about 50 km/sec/Mpc. Marc Aaronson, John Huchra, and Jeremy Mould have developed a new technique—independent of the Sandage-Tammann steps—to find a value of *H* equal to 95 km/sec/Mpc (with a claimed uncertainty of 5 percent). Let's briefly look at their procedure, which relies on radio and infrared observations to infer the luminosities of galaxies.

The technique rests on a relation, found by R. Brent Tully and J. Richard Fisher, between the absolute magnitudes (in blue light) of spiral galaxies and the spread in frequency of their 21-cm H I emission: the larger the 21-cm line width, the greater a galaxy's luminosity. The 21-cm emission comes from the neutral gas in a spiral galaxy's disk. If we measure the 21-cm line from a spiral galaxy viewed edge on, the emission from the gas moving away will be red-shifted and that moving toward us will be blue-shifted. So the 21-cm line will be broader than expected if the galaxy had no rotational motion. In fact, the line width for edge-on galaxies measures the maximum rotational velocity in the disk. Most galaxies have maximum rotational velocities of 100 to 300 km/sec. So the Tully-Fisher relation is really a relation between rotational velocity and luminosity of spiral galaxies.

Why should there be *any* relation between the maximum rotational velocity and the luminosity of a galaxy? The rotational velocity at a given distance from the center relates to the mass of a galaxy (Section 21.2), and the luminosity of the stars in a galaxy is also related to their mass. So if we stick with one type of galaxy, if the galaxies have more or less the same mix of stars of different masses, and if the degree of concentration of mass toward the center is more or less the same, then we might expect the total luminosity to be related to the rotational velocity. But it is surprising that the relationship is so good, with relatively little variation from one galaxy to the next, at least among galaxies of the same type. (The relationship seems to be most exact for Sc galaxies.)

If properly calibrated, the Tully-Fisher relation provides a galaxy's luminosity from a measurement of the width of its 21-cm line. The main problem with the method, as originally developed, lay in its use of blue magnitudes. Dust scatters blue light well, so that blue-light observations of another galaxy suffer from an unknown amount of extinction produced by dust in our Galaxy and in the disk of the other galaxy. Aaronson, Huchra, and Mould noted that infrared light suffers virtually no extinction from galactic or extragalactic dust. So they made infrared observations of nearby galaxies (such as M 31 and M 33), whose distances are known from standard candles such as the cepheids. This gives the absolute infrared magnitudes (essentially the infrared luminosity at a wavelength of 1.6 μm). These infrared absolute magnitudes connect well with the width of the 21-cm lines from local galaxies.

Aaronson and colleagues then measured the apparent magnitudes and 21-cm line widths for some nearby clusters of galaxies. They again found a good connection between the magnitudes and line widths. Applying the calibration from local galaxies, they get the distances. The measured red shifts then give H.

For four clusters of galaxies Aaronson and workers find a Hubble constant of 95 km/sec/Mpc. For the nearby Virgo cluster, they get an H of 65 km/sec/Mpc. (We'll relate one possible explanation for this difference in Section 22.5.) Note that a value for H of 95 km/sec/Mpc implies a universe no older than about 10 billion years. That result poses a problem, for the oldest stars (in globular clusters) are now thought to be some 14 to 18 billion years old.

Who's right? Both the Sandage-Tammann and Aaronson-Huchra-Mould methods probably contain unknown systematic errors. But we can say confidently that H lies in the range 50 to 100 km/sec/Mpc. Several other investigations also give values in this range. A very recent one using supernovas as standard candles concluded that H was 68 to 76 km/sec/Mpc. If the actual value is less than about 75 km/sec/Mpc, it wouldn't cause serious problems with the ages of the oldest stars.

22.4 GENERAL CHARACTERISTICS OF GALAXIES

Let's step back a bit from our cosmological vista and study the characteristics of galaxies, now that we have methods to find out their distances. We'll discuss nearby galaxies, which have been carefully studied, as specific examples. From these we infer the properties of spiral, elliptical, and irregular galaxies in general (Table 22.1).

Size

Once you know the distance to a galaxy, you can find out its actual diameter from a measurement of its angular diameter. The hitch here is that the definition of the "edge" of a galaxy is more or less arbitrary; different definitions result in different diameters.

No matter how the angular diameter is defined, the procedure for finding the actual diameter from it relies on the relationship $\alpha_{rad} = s/d$, where α_{rad} is the angular diameter in radians, s is the linear diameter, and d is the distance (both in the same units). For example, long exposure pictures show that M 31, the Andromeda galaxy, has an angular

TABLE 22.1 General Properties of Different Types of Galaxies

	Spirals	Irregulars	Dwarf Ellipticals	Giant Ellipticals
Diameter (kpc)	30	6	3	50
Mass (solar masses)	10^{11}	10^6	Not well known	10^{13}
Luminosity (solar luminosities)	10^{10}	10^9	10^8	10^{11}
Color	Blue (disk), red (halo and nucleus)	Blue	Red	Red
Neutral Gas (fraction of mass)	5%	15%	Less than 1%	Less than 1%
Types of stars	Young (disk), old (halo and nucleus)	Young	Old	Old

Source: Adapted from a table by H. L. Shipman.

diameter of 4.5° or 7.85×10^{-2} radians. For a distance of 2.2 million ly (680,000 pc), determined using cepheids, the diameter is

$$s = \alpha_{rad} d$$
$$= (7.85 \times 10^{-2})(2.2 \times 10^6 \text{ ly})$$
$$= 1.7 \times 10^5 \text{ ly (52 kpc)}$$

Dwarf ellipticals and small irregulars tend to be the smallest galaxies, only 300 to 3000 ly (90 to 900 pc) in diameter for some. The typical diameter of galaxies of all types is about 50,000 ly (15 kpc). Giant ellipticals can range up to 200,000 ly (60 kpc) across. To put this size range into perspective, imagine your height (about 2 m) to be the size of a dwarf galaxy. Then an irregular galaxy would be about 2 times your size, a spiral 10 times your size, and a giant elliptical some 20 times your size! The very largest galaxies are the supergiant ellipticals, sometimes called *cD galaxies*. These can have diameters up to 6 million ly (2 Mpc). That's greater than the distance from our Galaxy to the Andromeda galaxy!

Mass

The most widely used methods of finding a galaxy's mass are rotation curves and binary galaxies.

Let's look first at the rotation curve method. Any object in a galaxy orbits the nucleus at a speed such that the centripetal acceleration is balanced by the net gravitational force, which is determined by the distribution of mass within the galaxy. So if we observe a galaxy's rotation curve, we can derive a model for that galaxy's mass distribution, assuming that the motions are circular, and then work out the galaxy's total mass, including dark masses that we can't see directly.

How to observe the rotation curve? Suppose we set our telescope on a spiral galaxy and position the slit of our spectroscope across the length of the galaxy (Fig. 22.15). Whereas a

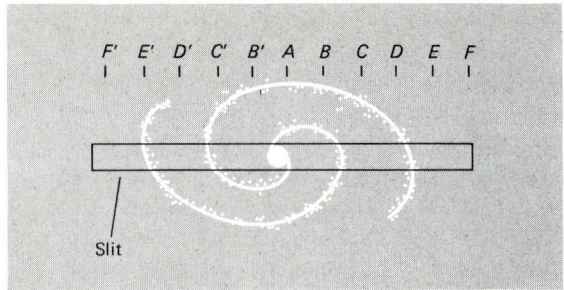

FIGURE 22.15 Observing the rotation curve. A spectroscope's slit is placed over the image of a spiral galaxy. The positions along the slit are marked.

spectral line from a nonrotating galaxy would look straight, the line from the rotating galaxy will be curved because the light coming from each position of the slit will be Doppler shifted according to the velocity of that point in the line of sight.

Assume that the galaxy is tilted with respect to the plane of the sky, as indicated in the figure by the fact that it is elliptical in shape, not circular. Let's say the upper part is moving away, the lower part toward us. In Figure 22.15, several points along the slit are marked (A, B, C, D, E, F); the points on the opposite side of the galaxy, marked with primes, will have the same velocities as the unprimed points, but of opposite sign. Suppose (Fig. 22.16a) we observe that point A shows no shift, points B, C, and D have progressively larger wavelength shifts, point E has the same shift as point C, and point F has a shift between that of points B and C. Each position we marked off along the slit

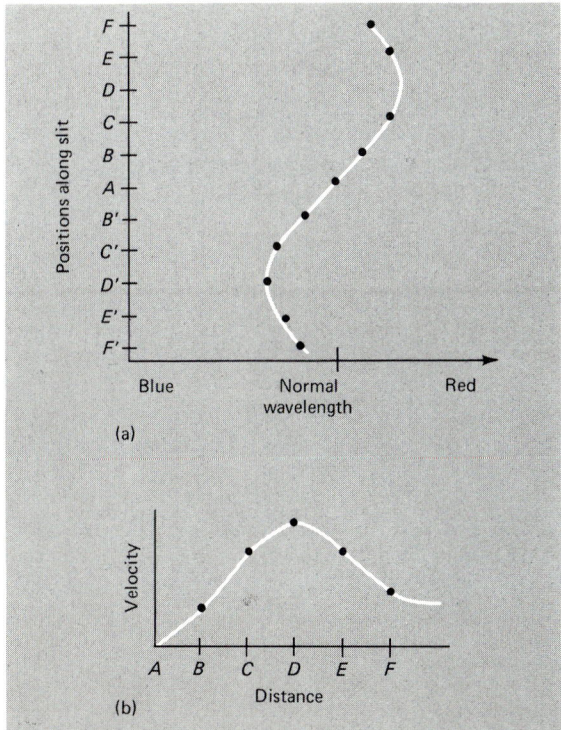

FIGURE 22.16 (a) Doppler shifts along the slit positions in Fig. 22.15. (b) Rotation curve inferred from the Doppler shifts.

TABLE 22.2 Masses of Some Galaxies

Galaxy	Type	Distance (10^6 ly)	Mass (10^{10} solar masses)
Milky Way	Sb	—	34
M 31	Sb	2.2	31–34
M 33	Sc	2.1	1.3–1.8
NGC 6503	Sc	16	0.13
NGC 3646	Sc	180	27
NGC 972	Sb	72	2
NGC 681	Sa	75	1.9
NGC 3623	Sa	30	20
SMC	Irr	0.20	1.2

Source: E. Burbidge and G. Burbidge, "The Masses of Galaxies," in *Galaxies and the Universe*, edited by A. Sandage, M. Sandage, and J. Kristian, University of Chicago Press, Chicago, 1975.

corresponds to a certain distance from the galaxy's center. The Doppler shifts give the line-of-sight velocities at each point; knowing the tilt of the galaxy we can calculate the true rotational velocities in the plane of the galaxy and hence translate the observed shifts into a rotation curve (Fig. 22.16b). Note we need to know the distance to the galaxy for this method to work, for we have to convert angular distances from the center to linear distances. With the rotation curve, we can calculate the mass in the same way as we do for our Galaxy (Section 21.2). Table 22.2 gives the rotation curve masses of some selected galaxies.

Note a remarkable feature of the rotation curves for spiral galaxies (Fig. 22.17). They rise steeply near the nuclear region and then flatten out at large distances from the nucleus. But they do *not* decrease, as they would if the mass were concentrated toward the center and the velocity followed Kepler's law. This fact implies that a large fraction of the

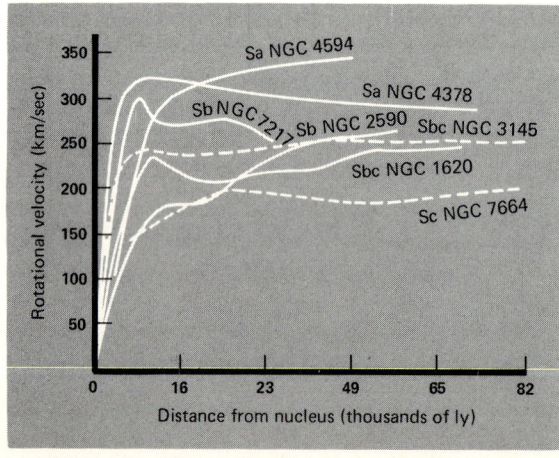

FIGURE 22.17 Measured rotation curves for seven spiral galaxies. (*Adapted from a diagram by* V. Rubin, W. K. Ford, Jr., and N. Thonnard, Astrophysical Journal (Letters), *vol. 225, p. L107, copyright 1978 by the American Astronomical Society.*)

mass lies not in the interior regions, but in the halo. If the rotation curve of the Galaxy resembles that for other spirals and so continues flat out to 180,000 ly, then its total mass is 7×10^{11} solar masses!

For binary galaxies we again make use of the versatile Doppler shift. Imagine two galaxies orbiting about the center of mass of the binary. Assume that the orbits are stable. Just as with visual binary stars (Section 8.2), we could apply Newton's form of Kepler's third law to find the masses, if we knew the distance, the angular size of the orbit, the period, and the position of the center of mass. However, galaxies revolve too slowly to see their actual orbits, periods, and relative centers of mass. So we cannot find the individual masses of binary galaxies. All we can measure are the present radial velocities and the separation; we don't know what part of the orbit the galaxies are on or what the inclination is, and so we don't know what the true orbital velocities are. But if we examine a large sample of galaxies and assume their orbits are nearly circular and randomly oriented to our line of sight, we can estimate from these data the *average* masses of the galaxies sampled.

A recent investigation of 279 binary systems, mostly spirals, uses the Doppler shift of the 21-cm line to get their velocities. It finds an average mass of 1.0×10^{12} solar masses for these spirals (for $H = 75$ km/sec/Mpc).

The most massive galaxies we know of are the supergiant elliptical (cD) galaxies. They can have up to 10^{14} solar masses, 100 times more mass than our Milky Way Galaxy contains.

Warning: These two methods—rotation curves and binary galaxies—do not always give the same results; masses from rotation curves are generally smaller than those from the binary galaxy method. This has led many astronomers to believe that galaxies have massive halos surrounding the bright disk of the galaxy. Such halos would not be detected in the rotation curve method, because rotation velocities can only be measured for the luminous material, and matter at large distances does not contribute to the net force at points within it. Our own Galaxy seems to have such a halo (Section 21.2).

From the study of our Galaxy we know that some of the material in the disk of a spiral resides in the form of H I. We detect H I in the Galaxy by the 21-cm line. We can observe H I in other galaxies the same way and, by adopting simple models for the distribution and temperature of the H I gas, we can estimate a galaxy's H I content.

Morton Roberts, analyzing these data for a sample of spiral and irregular galaxies, finds that the H I mass makes up a small fraction of the total mass, only 0.03 for lenticulars and 0.22 for irregulars. Roberts also notes that looking at the Hubble sequence from S0 to Irr, the different galaxy types contain increasing fractions of H I relative to their total mass. For example, Irr galaxies have more H I gas, relative to their total mass, than Sa galaxies.

Luminosity

If we know their distances and apparent magnitudes, we can work out the absolute magnitudes of galaxies and so their luminosities. One trouble here is that, because a galaxy thins out gradually at its edge, it's not easy to measure its apparent magnitude. You cannot be sure that you're catching all the light from the galaxy. In addition, corrections have to be applied for light absorption due to dust in our Galaxy and dust in the galaxy being measured (especially spirals).

The absolute magnitudes of galaxies range from -8 (2×10^5 solar luminosities) for dwarf ellipticals to −25 (10^{12} solar luminosities) for supergiant (cD) ellipticals. The cD ellipticals are very rare, however. The Galaxy, if we could see it all from space, has an absolute magnitude somewhere near −21 (2.5×10^{10} solar luminosities).

Mass-Luminosity Ratio

Divide the total mass of a galaxy by its luminosity. You then have its *mass-luminosity ratio* (abbreviated *M/L*), an indication of the average energy output per unit mass from the galaxy (Table 22.3). It is usually expressed in units of solar masses and solar luminosities.

Recent determinations using binary galaxy masses give 35 for the average mass-to-light ratio for spiral galaxies and about twice as much for giant ellipticals and lenticulars. (For comparison, the *M/L* ratio for stars in the sun's neighborhood is about 1.)

Why do ellipticals have a larger *M/L*? The simplest answer is that elliptical galaxies contain a greater percentage of low-mass stars with low light output—main-sequence stars of class M. If true, this extra abundance of M stars would mean that ellipticals should be redder in overall color than spirals (see below). Other possibilities are: neutron stars, black holes (including perhaps a giant one in the nucleus), and dark interstellar matter, which contribute to *M* but not to *L*.

One other trend relates to total luminosity and H I mass. Along the Hubble sequence from S0 to Irr, the ratio of hydrogen mass to optical luminosity increases. So the more H I gas a galaxy has, the greater its light output per unit mass. This relationship probably arises from the greater rate of ongoing starbirth in spirals with more gas (and dust).

Color

As for stars, we can measure the colors of galaxies using various filters. The color of a galaxy depends on the predominate stellar type in its mixture of stars. For example, a galaxy with many O and B stars is bluer than a galaxy with many red M stars.

A direct correlation exists between galaxy type and its color. Ellipticals tend to be much redder than spirals, and spirals redder than irregular galaxies. Within the spiral group the galaxies grow redder as their nuclear bulges grow larger and their spiral arms less extensive. One way to describe a galaxy's color is to specify the stellar spectral class whose color resembles that of the galaxy. The elliptical and Sa galaxies have the same color as K stars, whereas Sb galaxies resemble stars of class F to K, Sc and Irr galaxies show the color of class A to F.

The progression of color from the bluer irregulars to the redder ellipticals reflects a trend in the composition of the galaxy's population. In general terms, an old Population I predominates in ellipticals, whereas a much younger Population I stands out in the irregulars. The mixture in the spirals is determined by the size of the nucleus (old Population I) compared with that of the spiral arms (young Population I). (Population II is probably a minor contributor in all *large* galaxies, existing mainly in the globular clusters and galactic halo.)

Here is a way to remember the primary difference among galaxy types. Recall that our Galaxy has a halo and nucleus of red stars and spiral arms of blue stars. Imagine our Galaxy without the halo and nucleus. What remains (the arms) is like the stars, gas, and dust found in irregular galaxies. Now imagine our Galaxy stripped of its spiral arms. The remains (nucleus and halo) are typical of the composition of elliptical galaxies.

TABLE 22.3 Mass-Luminosity Ratios

Galaxy	Type	Mass-to-Light Ratio (solar units)
M 31	Sb	8.0–8.4
M 33	Sc	3–8
NGC 6503	Sc	0.7
NGC 3646	Sc	3.4
NGC 972	Sb	1.2
NGC 681	Sa	3.6
NGC 3623	Sa	7
SMC	Irr	2–4

Source: E. Burbidge and G. Burbidge, "The Masses of Galaxies," in *Galaxies and the Universe*, edited by A. Sandage, M. Sandage, and J. Kristian, University of Chicago Press, Chicago, 1975.

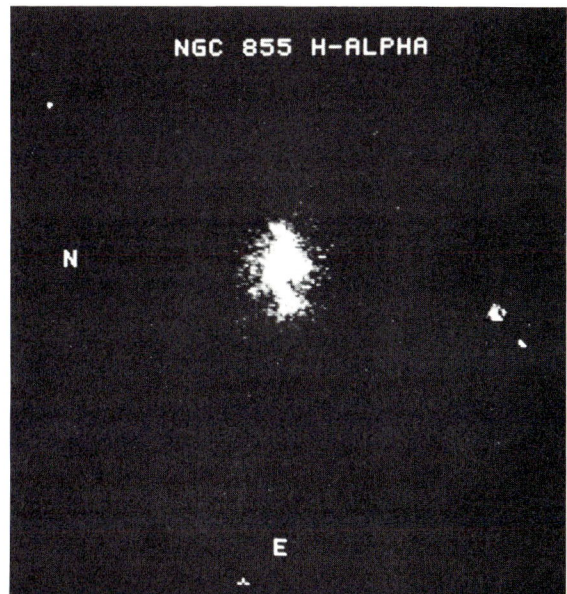

FIGURE 22.18 H-alpha emission from hydrogen gas in the nucleus of the elliptical galaxy NGC 855. (*Photo by H. Butcher and G. Gisler at Kitt Peak National Observatory; courtesy, G. Gisler.*)

A galaxy's color also relates to its overall content of gas and dust. The reddest galaxies, ellipticals, contain almost no gas and dust. The bluest galaxies, irregulars, contain the greatest percentage of gas and dust relative to their total mass. Why this progression? As the population comparison implies, irregulars have both old and young stars, but ellipticals contain only old stars. In irregulars there is still enough gas and dust so that star formation continues; in ellipticals it halted long ago. In spirals star formation from interstellar matter continues in the disk, but it has ceased in the nuclear bulge.

Note: Although gas and dust make up only a small percentage of the total mass of an elliptical galaxy, these galaxies are not completely devoid of gas and dust. Roughly 10 percent of ellipticals have detectable emission from gas—usually ionized and usually confined to the nuclear region (Fig. 22.18).

To sum up, galaxies contain stars, gas, and dust. The mixture relates closely to a galaxy's type and structure. Color and spectra measurements show that the nuclei of spirals contain an old stellar Population I, the disk much younger stars. Most elliptical galaxies have colors similar to those of spirals' nuclei, and irregular galaxies appear most akin in color and spectra to spirals' disks. Ellipticals contain very little gas and dust. In contrast, spirals and irregulars embrace extensive quantities of interstellar material. In spirals the gas is most obvious as H I clouds or as H II regions cloaking O and B stars. Although the masses of galaxies are difficult to measure for distant systems, we know that giant ellipticals are the most massive and irregulars the least massive.

Note that the sequence from ellipticals to irregulars marks a sequence from galaxies in which starbirth ceased long ago (ellipticals, with old Population I, some Population II, and no gas and dust) to those in which starbirth now is carried on very actively (irregulars,

FIGURE 22.19 The central region of the Virgo cluster of galaxies. (*Courtesy Kitt Peak National Observatory.*)

with young Population I and a large percentage of gas and dust). This trend gives a key clue to understanding the evolution of galaxies.

22.5 CLUSTERS OF GALAXIES

If you take a close look at a photographic survey of the sky (such as the Palomar Sky Survey), on some photos you will see extensive swirls of gas and many stars, but on the photos of regions where the stars thin out (away from the Milky Way) you can see the tiny forms of galaxies. If you look at enough photos, you will notice that if you find one galaxy, you're likely to see others nearby. Galaxies tend to come in clusters (Fig. 22.19), and it may be true that *all* galaxies belong to clusters, though many of these clusters may be a simple marriage of two galaxies. George Abell comments that on the Palomar Sky Survey "tens of thousands" of clusters of galaxies are "easily identified." Our universe is one of clusters of galaxies!

The Local Group

The nearest cluster of galaxies is the one to which the Milky Way Galaxy belongs. It is called the *Local Group*. This aggregation takes up a volume of space nearly 3 million ly (1 Mpc) across in its long dimension (Fig. 22.20). Our Galaxy is located near one end of the Local Group, and M 31 is near the other.

As the most massive objects in the cluster, the Milky Way Galaxy and M 31 dominate its motions and secure the other members gravitationally. The Galaxy and M 31 orbit each other. The other members of the Local Group come along for the ride. The Local Group contains at least 20 galaxies. They consist of 3 spirals, 4 irregulars, and 13 ellipticals (Table 22.4). Some of these ellipticals are quite faint and are called dwarf ellipticals to distinguish them from the giant ellipticals found in other clusters. The dust in the Milky

22.5 CLUSTERS OF GALAXIES

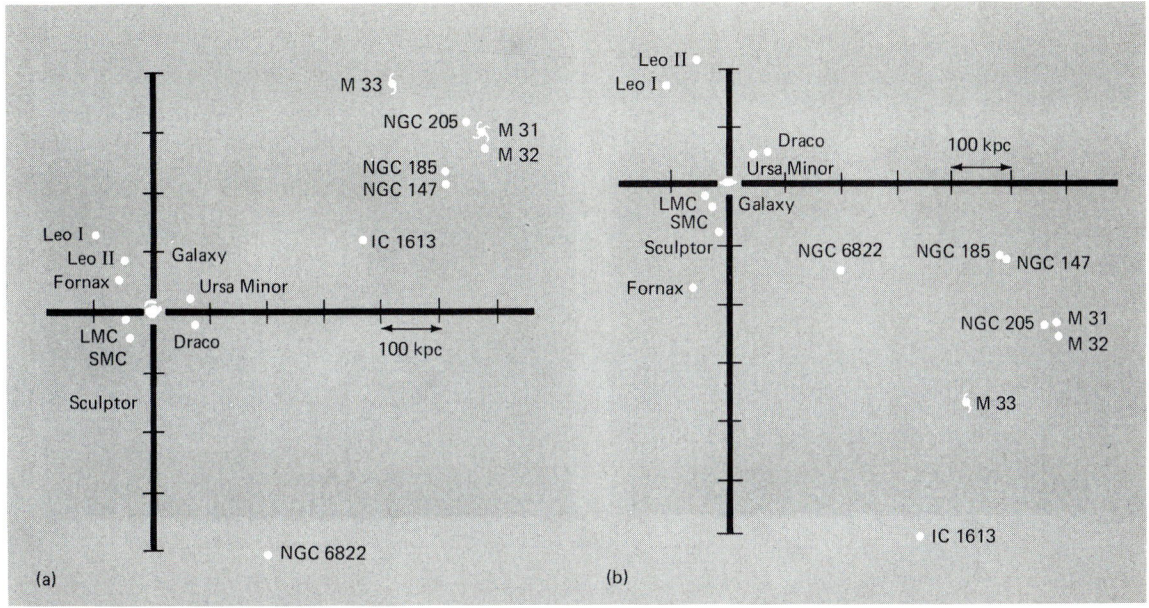

FIGURE 22.20 The most prominent galaxies in the local group of galaxies, as viewed (a) looking down on the plane of the Milky Way Galaxy and (b) looking edge on.

TABLE 22.4 Physical Properties of Some Galaxies in the Local Group

Name	Type	Diameter (kpc)	Distance (kpc)	Mass (solar masses)
Milky Way	Spiral	40	—	3×10^{11}
NGC 147	Elliptical	2.4	680	?
NGC 185	Elliptical	2.9	680	?
NGC 205	Elliptical	4.2	680	?
M 31	Spiral	52	680	4×10^{11}
M 32	Elliptical	2.1	680	2×10^9
Small Magellanic Cloud (SMC)	Irregular	5	50	?
Sculptor	Elliptical	2.4	86	3×10^6
IC 1613	Irregular	4	680	?
M 33	Spiral	18	700	2×10^{10}
Fornax	Elliptical	6.2	188	2×10^7
Large Magellanic Cloud (LMC)	Irregular	8	53	?
Leo I	Elliptical	1.8	230	3×10^6
Leo II	Elliptical	1.3	230	10^6
Ursa Minor	Elliptical	2.4	68	10^5
Draco	Elliptical	1.0	77	10^5
NGC 6822	Irregular	1.7	660	?

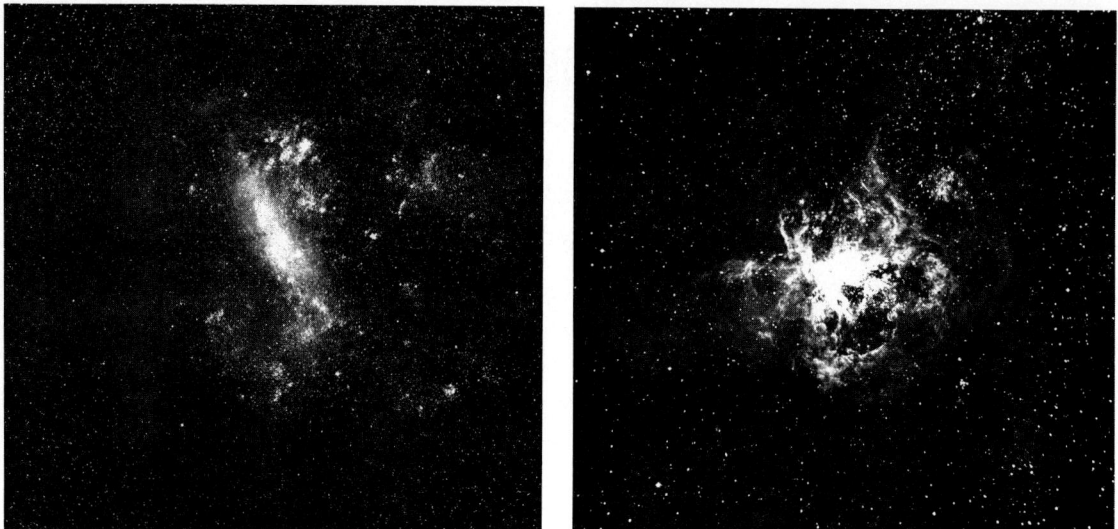

FIGURE 22.21 (a) The Large Magellanic Cloud (LMC). Note the many bright nebulas, especially the brightest (upper right). (b) The giant emission nebula, 30 Doradus, in the LMC. (*By permission of Harvard College Observatory.*)

Way probably clouds our sight of other members, especially faint dwarf ellipticals. A photographic survey of the southern sky has uncovered several new, nearby dwarf galaxies. Three of these, one elliptical and two irregulars, appear to belong to the Local Group.

Let's look briefly at some of the more important members of the Local Group.

The Large and Small Magellanic Clouds are closest to the Galaxy. They are, in fact, connected to our Galaxy by a bridge of hydrogen gas. The two clouds are physically connected to each other by a large, but thin envelope of neutral hydrogen. Both orbit our Galaxy and are distorted by tidal interactions with it.

The Large Magellanic Cloud (abbreviated LMC) contains stars totaling about 20 billion solar masses and orbits 170,000 ly (52 kpc) from the Galaxy (Fig. 22.21a). Spectra of a large number of stars in the LMC show that most stars are similar to those in the solar neighborhood, but there are more O and B stars. Over 2000 variables have been discovered in the LMC; about 600 of these are the famous cepheids which established the period-luminosity relation. Radio studies demonstrate that the LMC has large amounts of neutral hydrogen gas, a total of approximately 3 billion solar masses. The gas is also evident from the more than 400 emission nebulas found so far. One of these nebulas, called 30 Doradus, which is illuminated by a cluster of O and B stars, extends over 1600 ly (500 pc) in diameter and contains roughly 5 million solar masses of material (Fig. 22.21b and Plate 19).

Although very small compared with the LMC, the Small Magellanic Cloud (SMC) displays a similar stellar population (Fig. 22.22). It has a total mass of about 2 billion solar masses, 1/10 that of the LMC. Its distance from us is about 205,000 ly (63 kpc).

The Magellanic Clouds are not the only galaxies orbiting our Milky Way Galaxy. A swarm of six dwarf galaxies accompanies us in space. These dwarf galaxies have masses of only about 1 million solar masses and range from 160,000 to 980,000 ly (50 to 300 kpc) in distance from us.

22.5 CLUSTERS OF GALAXIES

FIGURE 22.22 The Small Magellanic Cloud (SMC). Note the lack of bright nebulas compared to the LMC. (*By permission of Harvard College Observatory.*)

M 31 is sometimes still called the Great Nebula in Andromeda. Actually a spiral galaxy, M 31 is easily visible to the unaided eye on a dark moonless night. With binoculars you find an elliptical, hazy patch of light. The spiral arms are too faint to see without a big telescope.

M 31 tilts 15° to the line of sight. Because of the tilt, dark lanes of obscuring material are plainly visible along with the spiral arms marked by O and B stars (Fig. 22.23). A halo of globular clusters surrounds M 31 like bees around a hive, in a distribution like that around our galactic system. The Andromeda galaxy also has companions, a total of seven dwarf ellipticals that orbit it.

M 33 is the only other large spiral in the Local Group (Fig. 22.24). It is the closest Sc II galaxy to us, only 2.7 million ly (0.72 Mpc) away. In the photo you can easily trace out wide, open spiral arms. Note that the arms are resolved into stars; most of these are blue supergiants. Because M 33 is so close, astronomers have been able to investigate in detail its dark dust lanes, H II regions, open star clusters, novas, and cepheid variables. M 33 has an overall diameter of about 60,000 ly (18 kpc), only about half the size of the Milky Way Galaxy.

It may seem strange, but we probably have more complete knowledge of other nearby groups and clusters of galaxies than we do of our own Local Group. It is quite likely that we have not yet even discovered all of its members. Some may be very faint and hard to see. Others may be located in space so that to see them from our Galaxy we have to look through the dust of the zone of avoidance, which can't be done optically. In the past decade two new galaxies have been found near the zone of avoidance: Maffei 1 and Maffei 2. Maffei 1 is a giant elliptical galaxy, and Maffei 2 a spiral. It's hard to get the distances to

FIGURE 22.23 A close-up of the central regions of the Andromeda galaxy. Note the dark, dusty regions. (*Courtesy Palomar Observatory, California Institute of Technology.*)

FIGURE 22.24 The spiral galaxy Messier 33 in Triangulum. It is Hubble type Sc. (*Courtesy Lick Observatory.*)

these galaxies, since so much obscuring dust is in the way. They may or may not be members of the Local Group. Other potential members of the Local Group have been found by radio astronomers. Although hidden by the dust of our own Galaxy, they are strong emitters of the 21-cm line of hydrogen and are therefore likely to be irregulars or spirals.

Other Clusters of Galaxies

Other clusters of galaxies range from compact groups (Fig. 22.25) to rather loose arrays. The Fornax cluster, one relatively close to us, displays a wide variety of types of galaxies, even though the total number is only 16. The huge Coma cluster (Fig. 22.26) spreads over at least 23 million ly (7 Mpc) of space and contains thousands of galaxies. Observations of even just the brightest galaxies show how common clustering is (look back at Fig. 22.3a). From these observations we find that a typical cluster contains about 100 galaxies and is separated some tens of millions of light years from its neighboring clusters.

George Abell has cataloged and studied 2712 rich clusters of galaxies that contain a number of bright galaxies. (*Rich* means having more than 50 galaxies within a radius of 2 Mpc.) Abell finds that the clusters tend to fall into two groups, which he calls simply *regular* and *irregular*. Regular groups have (1) a marked spherical symmetry, (2) a high concentration of galaxies at center, (3) many bright E and S0 galaxies, (4) 1000 or more

22.5 CLUSTERS OF GALAXIES

FIGURE 22.25 A small cluster of galaxies called Stephan's quintet. (*Courtesy Lick Observatory.*)

FIGURE 22.26 A cluster of galaxies in Coma Berenices. Note how the smaller galaxies appear to cluster around the two giant ellipticals at the center and to the left. (*Courtesy Kitt Peak National Observatory.*)

galaxies in the brightest 7 magnitudes, (5) masses on the order of 10^{15} solar masses, and (6) no subclustering. In contrast, irregular groups have (1) little symmetry, (2) no concentrated, single center, (3) mostly spiral and irregular galaxies, (4) 10 to 1000 galaxies in the brightest 7 magnitudes, (5) masses from 10^{12} to 10^{14} solar masses, and (6) subclustering into two or more groups.

An example of a regular cluster is the *Coma cluster* (Fig. 22.26). Note the large, bright elliptical galaxy near the center about which the others seem to concentrate; it is a supergiant elliptical galaxy. Examples of irregular clusters are the Local Group and the *Virgo cluster* (Fig. 22.19). Of the 205 brightest galaxies in the Virgo cluster, the four brightest are giant ellipticals, but ellipticals make up only 19 percent in all compared with the spirals' 68 percent. The Virgo cluster covers about 7° in the sky (14 times the diameter of the moon), which implies that its physical diameter is some 10 million ly (3 Mpc) at a distance of 51.2 million ly (15.7 Mpc).

Clusters and the Galaxian Luminosity Function

Section 16.2 showed that the local stars have a luminosity function, that is, a certain number of stars can be found in a given range of luminosity or absolute magnitude. The basic trend is that very few extremely luminous (OB) stars exist compared with the large numbers of low-luminosity (M) stars.

There is a similar hierarchy for galaxies. In the Local Group only three galaxies are very luminous (the Milky Way, M 31, and M 33); most galaxies in the Local Group are dwarf, low-luminosity galaxies. So galaxies have a luminosity function somewhat like that of stars.

Clusters provide a neat way to determine the luminosity function of galaxies, for you can see a wide range all at once. You can count the number of galaxies in specific apparent magnitude ranges, from the brightest to the faintest. Because every galaxy in a cluster lies at about the same distance from us, a plot of number versus apparent magnitude translates into number versus absolute magnitude once you know the distance to the cluster. The main drawback here is that dwarf galaxies are undercounted because they are so faint at great distances.

Abell has determined the luminosity functions of a few clusters. He finds that the number of bright galaxies falls off very rapidly as the luminosity increases (Fig. 22.27). This behavior tells us that there are many more low-luminosity galaxies in a cluster than high-luminosity ones. So much of a cluster's mass resides in very faint galaxies.

It's difficult to estimate the masses of these clusters. Not all the material in them can be seen, so adding up all the observable galaxies gives a lower limit to the cluster's mass. On the other hand, if the cluster is assumed to be gravitationally bound, its motions establish an upper limit on its mass. (The actual value lies between these two limits.) Masses range from 10^9 to 10^{15} solar masses. However, it is not possible to tell if all clusters are bound and stable or if they are unstable and expanding.

Superclusters

Are there clusters of clusters of galaxies, that is, do *superclusters* exist? Abell notes that in his catalog of 2712 clusters he sees evidence of roughly 50 superclusters, each containing about 11 clusters and having diameters of about 250 million ly (75 Mpc).

22.5 CLUSTERS OF GALAXIES

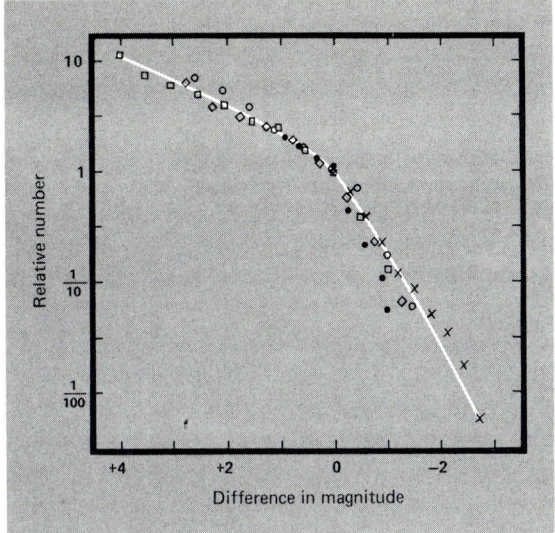

FIGURE 22.27 A typical luminosity function for a rich cluster of galaxies. Note that the least luminous galaxies are the most abundant. (*Adapted from a diagram by G. Abell.*)

Much work on superclustering has been done by Gerard de Vaucouleurs. He finds evidence for a *local supercluster* that has a diameter of some 100 million ly (30 Mpc). It contains the Local Group, Virgo cluster, and Coma cluster among others, for a total mass of some 10^{15} solar masses. The local supercluster appears to be flattened (about 6 million ly thick), which may imply it is rotating, but it may simply have been formed this way. The center of mass lies in or near the Virgo cluster (which contains about 20 percent of all the galaxies in the local supercluster); our Galaxy, which lies at the edge of the supercluster, moves at about 500 km/sec about the axis of rotation.

Recently, Tully and Fisher have completed a three-dimensional map of the local supercluster. (This project involved measuring the positions and red shifts of some 2200 galaxies and took nine years to complete!) They discovered a rich, convoluted structure that breaks into two main clouds with streamers—thin, cigar-shaped clouds—emerging above and below the central plane. Most of the supercluster is empty space: 98 percent of the visible galaxies are contained in just 11 clouds that fill a mere 5 percent of the overall volume. Yet, the clouds do delineate a disk structure, with a width about 6 times its thickness—a cosmic pancake of clusters of galaxies.

Now let's return to the difference Aaronson and colleagues found for the Hubble constant using the Virgo cluster only (65 km/sec/Mpc) and using other clusters of galaxies (95 km/sec/Mpc). One explanation of the difference is that the Local Group has a motion *toward* the Virgo cluster, resulting in a smaller Hubble constant in that direction. If so, the velocity would amount to some 480 km/sec. The local supercluster may have a large enough overall density to produce this motion, resulting in an apparent slowdown in the expansion of the universe in this direction.

A recent investigation of the Hercules supercluster covers a volume of some 60,000 Mpc^3 (Fig. 22.28)! The supercluster centers on a radial velocity of 11,000 km/sec (corresponding to a distance of 720 million ly, or 220 Mpc, if $H = 50$ km/sec/Mpc). In front of the supercluster is a void some 330 million ly (100 Mpc) deep that separates the supercluster from foreground galaxies. These voids near superclusters appear common.

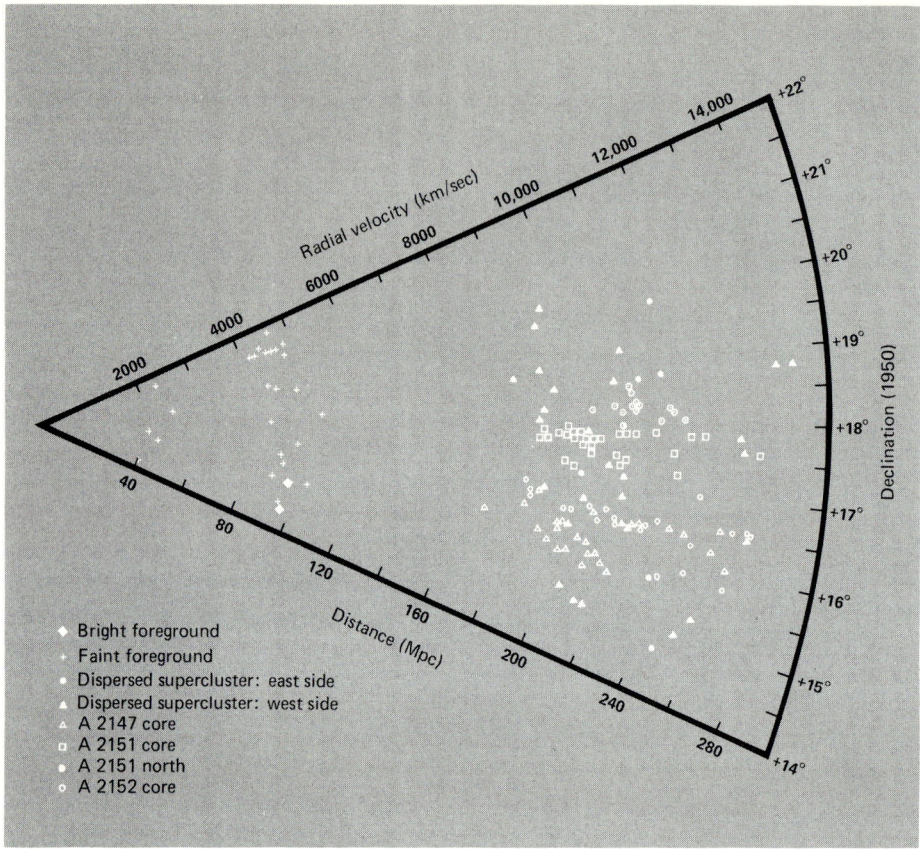

FIGURE 22.28 A slice of the Hercules supercluster. Shown here are the radial velocities (and so distances, for H = 50 km/sec/Mpc) and positions in the sky for the galaxies in the supercluster, which contains a number of clusters (labeled Abell 2147, 2151, and 2152). (*Adapted from a diagram by M. Tarenghi, W. Tifft, G. Chincarini, H. Rood, and L. Thompson,* Astrophysical Journal, *vol 234, p. 793, copyright 1979 by the American Astronomical Society.*)

They were first noted by Stephan Gregory and Laird Thompson in the Coma supercluster.

The Cosmic Tapestry

What does the universe of galaxies look like on the grand scale? P. James E. Peebles and coworkers have investigated in detail a survey of a million galaxies (Fig. 22.29) performed by Donald Shane and Carl Wirtanen of Lick Observatory. (This survey took 12 years to complete!) They find that galaxies cluster in knots and filaments in a hierarchical fashion. These are chains of galaxies and clusters (looking like a chain-linked fence). Nearly all galaxies are within these clusters, with huge holes between them devoid of luminous matter. Here we may be seeing the explosive imprint of the Big Bang—a filamentary texture similar to that of supernova remnants.

22.5 CLUSTERS OF GALAXIES

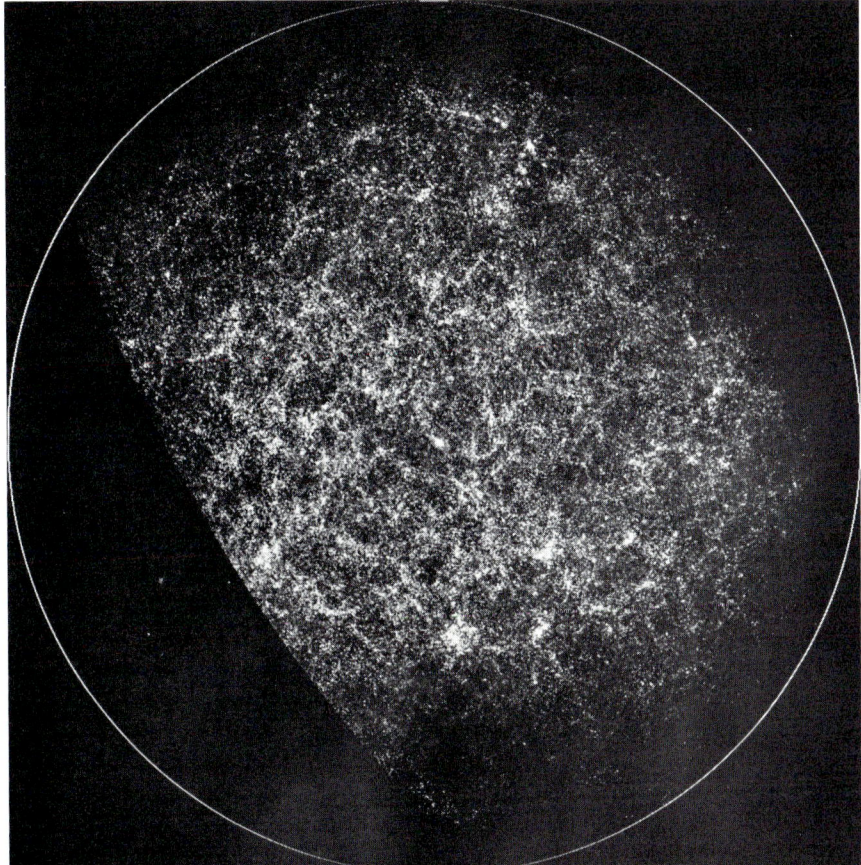

FIGURE 22.29 The universe of galaxies. More than a million galaxies brighter than apparent magnitude 19 are shown. The lighter a region, the more galaxies within it. Note the clumpy, chainlike structure. (*Courtesy P. J. E. Peebles, based on the Lick Observatory Catalog by C. Shane and C. Wirtanen.*)

Galactic Cannibalism?

A remarkable fact about clusters of galaxies is that—relative to the sizes of the galaxies in them—the spacing of galaxies is pretty close together. Compare them with planets and stars. In the solar system, planets are spaced out about 100,000 times their diameters. In the Galaxy, stars are spaced about a million times their diameters. But in a cluster of galaxies, the spacing is only about 100 times a typical galaxy's diameter. In a sense, the galaxies in a cluster are crowded together (astronomically speaking!).

Now, if in relative terms, galaxies are so close, consider the additional fact that the most massive galaxies (giant ellipticals) are at least 10^7 times more massive than the least massive ones (irregulars and dwarf ellipticals). Then it's not hard to imagine that the largest galaxies could disrupt the smaller ones by tidal forces—strongly enough to destroy their structure and then pull the pieces in. This devouring of a smaller galaxy by a larger one has been called *galactic cannibalism*.

FIGURE 22.30 The galaxies NGC 4038 and 4039, a tidally interacting pair. (*Courtesy Palomar Observatory, California Institute of Technology.*)

Why consider this idea? Some recent observations imply that supergiant elliptical galaxies (cD galaxies) are a special class of galaxies, not just larger elliptical ones. Their peculiar properties include: (1) extensive halos, up to 3 million ly (1 Mpc) in diameter, (2) multiple nuclei, and (3) location at the center of clusters. These observed properties plus theoretical calculations of their motion in clusters have suggested that the cD galaxies result from galactic cannibalism, that is, from close encounters at the center of clusters or the infall of material tidally stripped from other cluster members. As appealing as this idea appears, it has little strong observational support so far, mainly because of the difficulty of seeing directly the effects of one galaxy gobbling up another.

Although galaxies may not actually merge, certainly they undergo close encounters and interact by tidal forces. Such interactions would have some general effects. First, as illustrated by the earth's tidal bulges, matter would be pulled out in bulges on both sides of each galaxy. Second, because galaxies rotate, their material will conserve angular momentum after a tidal encounter and move off in arc-shaped streams. So we expect that tidal bridges may join two tidally interacting galaxies and tails may flow away from each in opposite directions.

Have such interacting galaxies been seen? Most likely. Many galaxies with peculiar shapes—those that do not fall into the standard Hubble form—show some of the characteristics of tidal interactions. An excellent example is the pair NGC 4038 and 4039 (Fig. 22.30). Here is visible a bridge of material between the galaxies and tadpolelike tails heading off in opposite directions.

22.6 INTERGALACTIC MATTER

Is intergalactic space empty, or is there an intergalactic medium similar to the interstellar medium? If an intergalactic medium is present, it may contain both gas and dust. The gas

22.6 INTERGALACTIC MATTER

(probably hydrogen) may be in neutral form, or it may be ionized. We can look for the intergalactic medium in two locations: *between* clusters of galaxies and *within* clusters of galaxies.

Matter Between Clusters

To get some idea of how much material might be in an intergalactic medium, imagine the following. Take the matter from all the galaxies we can see and spread it out over the entire volume of space we can observe. This spread-out material would have a density of about 4×10^{-28} kg/m^3. (That's about 2 hydrogen atoms every 10 m^3.) Recall that this density is about 10 times less than that needed to close the universe (Section 5.5). For the intergalactic medium to be significant, its density would need to be about this large. What is the evidence for such a density of dust, neutral hydrogen, or ionized hydrogen?

Consider the possibility of intergalactic dust. Such dust, if it resembled the interstellar dust in our Galaxy (Section 9.1), would extinguish and redden the light from distant galaxies. This extinction and reddening has been searched for but not found. It is less than 4×10^{-4} magnitudes per megaparsec distance. So intergalactic space cannot contain very much dust; the density must be less than 4×10^{-30} kg/m^3, which is two orders of magnitude less than the density of the spread-out material in galaxies.

How to detect neutral hydrogen? Hydrogen atoms are very good absorbers of ultraviolet radiation, especially at 1216 Å, the Lyman alpha absorption line produced when the electron is excited from its ground state to its first excited state (Section 6.3). Such ultraviolet absorption has been sought in the spectra of distant objects at both small and large red shift. It has not been detected. This lack of ultraviolet absorption implies that neutral hydrogen cannot have a density greater than about 10^{-9} atom/m^3 (10^{-33} kg/m^3). So if hydrogen is there, it must be ionized. Because these observations probe the universe's past, they also imply that any intergalactic gas must have remained highly ionized for most of the history of the universe.

These arguments leave ionized hydrogen (H II) as the most likely candidate for the intergalactic medium. Because intergalactic material would not have a high density, ionized hydrogen would take a very long time to find an electron and recombine. Unfortunately, detecting a low-density ionized gas is a tough job. If the gas is cool (a few thousand Kelvins), you can hunt for radio emission. If it is hot (a few tens of millions of Kelvins), you can search for X-ray or ultraviolet emission. X-ray observation of local superclusters finds 15 sources that are probably clustered in 7 superclusters. The sources consist of spots centered in rich clusters; this implies that hot gas in the superclusters is highly clumped.

Matter Within Clusters

For over 40 years astronomers have known about the puzzling characteristics of clusters of galaxies: Their masses estimated from the motions of the galaxies come out to about 10 times greater than those calculated by adding up the estimated masses of visible galaxies. In what form could this unseen material exist?

One possibility is that spiral galaxies have much larger and more massive halos than normally believed. These halos would be dark, about a million light years in size, and contain about 10 times the mass of the disk. The flat rotation curve of our Galaxy (Section 21.2) supports this idea.

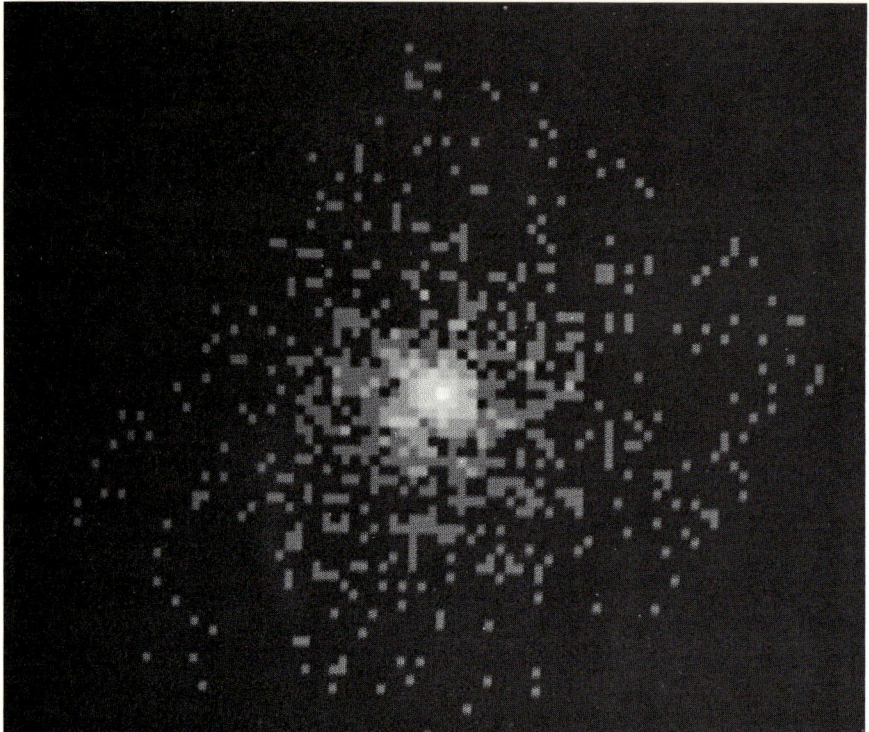

FIGURE 22.31 A map of the X-ray emission from the Perseus cluster of galaxies. The little squares are the size of the best resolution of the telescope. The brighter the square, the more intense the X rays. (*Observations by P. Gorenstein, D. Fabricant, K. Topka, F. Harnden, and W. Tucker; courtesy, P. Gorenstein and the Center for Astrophysics.*)

Another good candidate for matter within clusters is ionized hydrogen gas. Recent X-ray observations back up this idea. At least 40 clusters of galaxies are known to date to emit X rays (Fig. 22.31). The X-ray luminosities of clusters range from 10^{36} to 10^{38} W. The sizes of the X-ray–emitting cores range from 160,000 ly (50 kpc) to 5 million ly (1.5 Mpc). The richer clusters tend to be more luminous in X rays.

A reasonably confirmed model for this X-ray emission is that it comes from hot, ionized gas. This model requires typical temperatures of 10 to 100 million K and densities of about 1000 ions/m^3 to explain the X-ray observations. So we have reasonable evidence of intergalactic gas in clusters, about equal to the amount of mass in the galaxies themselves. But its density does not appear sufficient to bind the cluster gravitationally.

SUMMARY

This chapter has jumped out beyond our Galaxy to describe the galaxies found in the rest of the universe. The answers to the crucial questions about the physical properties of galaxies hinge on a knowledge of their distances. Beyond the limited range of present standard candles, we must rely on the calibration of the Hubble constant and on the assumptions of cosmology to estimate the distances to galaxies.

The distances to other galaxies are enormous compared with the distances between stars, but the distances between galaxies are small relative to their diameters. Suppose our Galaxy were shrunk down to the size of a dime, about 1.5 cm. Then the nearest galaxies—the Magellanic Clouds—would be two dimes away. The Andromeda galaxy (M 31) would be about 54 cm away, at the other end of the Local Group. The nearest cluster of galaxies—the Virgo cluster—would be almost 4 m away. The limits of the distance-measuring techniques would be about 1 km.

The fact that galaxies similar in form exist throughout the observable universe reaffirms the assumption of the uniformity of nature. So as we delve into the structure and evolution of our local Milky Way Galaxy, we are also investigating our ideas about the rest of the universe.

When we go beyond our Galaxy we run into the largest bound structures we can see in the universe: clusters of galaxies. These vast swarms of stellar systems mark the peak of the pyramid of gravitational groupings in the universe and probably are linked to the process in which the universe was created.

Key Words

zone of avoidance
uniformity of nature
Hubble classification
  scheme for galaxies
elliptical (E) galaxy
spiral (S) galaxy
irregular (Irr) galaxy
lenticular (S0) galaxy
barred (SB) spiral galaxy
supergiant elliptical
  (cD) galaxy

galaxy luminosity class
distance pyramid
standard candles
Tully-Fisher relation
rotation curve of a galaxy
binary galaxies
mass-luminosity ratio
cluster of galaxies
Local Group
Large Magellanic Cloud (LMC)
Small Magellanic Cloud (SMC)
M 31

M 33
regular cluster
irregular cluster
Coma cluster
Virgo cluster
luminosity function
  for galaxies
supercluster
local supercluster
galactic cannibalism
intergalactic medium

Review Questions

1. Which galaxies appear redder, ellipticals or spirals? (Objective 1)
2. Which galaxies contain more gas and dust relative to their total mass, spirals or irregulars? (Objective 1)
3. At the same distance from us, would irregular galaxies appear larger or smaller than spiral galaxies? (Objective 1)
4. How do astronomers know that other galaxies are made of stars? (Objective 4)
5. In recent years the value of the Hubble's constant has been revised from 100 km/sec/Mpc to 50 km/sec/Mpc. How does this change affect the distances to galaxies inferred from red shift and the Hubble's constant? (Objectives 6 and 8)
6. Describe how supergiant O and B stars can be used to estimate distances to nearby galaxies. State the assumptions and limits of this method. (Objectives 6 and 7)
7. Why must intergalactic gas be both ionized and hot? (Objective 12)

Problems

1. With the best techniques of 1980, galaxies can be detected to magnitude $m_V = 23$. There are about $10/\text{arcmin}^2$. How many are there over the entire sky? (Objective 9)

2. How close must a dwarf elliptical galaxy get to a cD galaxy in order for the smaller galaxy to be tidally disrupted by the larger? (Objective 11)
3. If the hot gas in clusters of galaxies has a temperature of 50 million K and emits like a blackbody, what is the wavelength of peak emission? (Objective 12)
4. Estimate the free-fall time (Section 9.3) for a cluster of galaxies and compare your result to the age of the universe inferred from the Hubble constant. (Objective 10)
5. If our Galaxy and M 31 are a binary pair, calculate their orbital period. (Objective 2)

BEYOND THIS BOOK . . .

* *Galaxies* (Harvard University Press, Cambridge, Mass., 1972) by H. Shapley, recently revised by P. Hodge, gives a comprehensive view of galaxies.
* *The Realm of the Nebulae* (Dover, New York, 1958) by E. Hubble was first published in 1936. Compare it with *Galaxies* by Hodge.
* *Galaxies and Cosmology* (McGraw-Hill, New York, 1966) by P. Hodge deals with the observed and physical properties of galaxies and their relationship to the universe.
* "Extragalactic Distance Scale" by S. Van Den Bergh in *Nature*, vol. 225 (1970), p. 503, is a fairly technical view of the problem.
* "The Content of Galaxies" by W. Baade in *Scientific American*, September 1956, is a dated article, but still useful for basic ideas.
* The volume *Galaxies and the Universe* (University of Chicago Press, Chicago, 1975), edited by A. Sandage, M. Sandage, and J. Kristian, contains a wealth of material about the topics in this chapter.
* Neta Bachall reviews "Clusters of Galaxies" in *Annual Reviews of Astronomy and Astrophysics*, vol. 15 (1977), p. 505.
* For a gorgeous tour of the galaxies, examine *The Hubble Atlas of Galaxies* (Carnegie Institution, Washington, D.C., 1961) by A. Sandage.

COSMIC VIOLENCE: ACTIVE GALAXIES AND QUASARS

23

LEARNING OBJECTIVES
After studying this chapter, you should be able to:
1. Outline the observational evidence for violent activity in our Galaxy and other galaxies, with special emphasis on synchrotron radiation and dust emission.
2. List the observational characteristics of quasars.
3. Sketch a possible model that accounts for the observed characteristics of quasars.
4. Discuss the red shift controversy for quasars.
5. Discuss the energy problem for quasars.
6. Apply the relativistic Doppler shift to quasars.
7. Compare the Milky Way Galaxy, active galaxies, and quasars.
8. Outline the method used to estimate distances to quasars, and discuss its uncertainties.
9. Discuss possible alternatives to quasars as distant objects.
10. Discuss the rationale and evidence for quasars as hyperactive, luminous nuclei of galaxies.
11. Describe superluminal motions in quasars and give one explanation for them.

CENTRAL QUESTION:
What observational evidence do we
have for violent activity in the Galaxy
and in objects beyond it?

In general, the universe appears calm, caught up in the well-controlled generation of energy in stars and the strict Newtonian dance of matter. Until the middle of the twentieth century we saw it as a gentle cosmos. Rare outbursts such as novas only occasionally shattered the stillness.

The advent of radio astronomy ripped off the veil covering a violent universe. Radio astronomers found the nucleus of our Galaxy to be a radio emitter. They also detected intense radio sources beyond the Galaxy. Such sources at enormous distances required a tremendous outpouring of energy at radio wavelengths. Later work by infrared astronomers revealed that some extragalactic radio sources emitted even more energy at infrared than at radio wavelengths. By the 1970s observational evidence forced us to recognize that violent events in the cosmos, especially associated with the nuclei of many types of galaxies, were quite common. The opening of the high-energy end of the electromagnetic spectrum in X-ray and gamma-ray astronomy has further enhanced our awareness of the violent aspects of the universe.

Quasars—objects that look like stars, with large red shifts—are one new and puzzling element in the range of cosmic violence. Originally discovered as radio sources, quasars have the largest known red shifts of any extragalactic objects. If the red shifts of quasars result from the general expansion of the universe, then the quasars must be the most energetic bodies in the universe. In one hour some quasars spew out energy equal to the amount generated during the sun's entire lifetime! At the fringes of the cosmos, quasars represent some of the first-formed objects in the visible universe—and the most powerful. The mystery remains of how quasars produce their energy.

23.1 VIOLENCE IN THE NUCLEUS OF THE GALAXY

As we have gained access to more regions of the electromagnetic spectrum, some galaxies—and the nuclei of many—have acquired the aspect of a compact arena of violent events. The kernel of the nucleus is too small to investigate directly in detail in other galaxies, except M 31, and too obscured to observe optically in our Galaxy. The nature of the physical conditions and processes at the heart of a galaxy remains mostly unknown.

Evidence of Violence in the Galaxy

Chapters 20 and 21 presented the structure of our Galaxy. In this chapter we place the results about the nucleus in the context of cosmic violence.

The nucleus lies at the center of the radio source Sgr A, which consists of Sgr A East and Sgr A West. Sgr A West appears to mark the actual core of the Galaxy. The central core of Sgr A West observed in the radio has a size of roughly 140 AU. A string of radio sources extend along the galactic equator from the Sgr A West position.

Infrared observations show that the galactic nucleus shines brightly at infrared wavelengths between 2 and 350 μm. The peak of the infrared emission comes from a region only 3 ly across. High-resolution maps show that the nucleus holds a cluster of infrared sources, many less than a few light years in diameter. The infrared luminosity of the nucleus is about 10^6 solar luminosities at short wavelengths and more than 10^8 solar luminosities at long wavelengths for a region some 2° around Sgr A.

The galactic nucleus also puts out high-energy photons: X rays and gamma rays. The Uhuru satellite discovered X-ray emission from an extended region around the galactic center, roughly coinciding with the source of the far-infrared emission; the Einstein telescope mapped it in detail. The energy emitted in X rays varies, but it averages about 10^{30} W (10^4 solar luminosities). Gamma rays are emitted strongly from galactic longitude 300° to 50° with no obvious peak at the galactic center itself. The energy emitted in gamma rays from 1° around the galactic center is 10^{27} to 10^{28} W (about 10 solar luminosities). In addition to continuous emission, the galactic center emits gamma-ray lines at 0.024, 0.026, and 0.028 Å; the first comes from electron-positron annihilation.

Emission at such a wide range of wavelengths (Table 23.1) indicates that the spectrum of the nucleus is in part nonthermal, that is, some of the emission must be by the synchrotron process. So high-energy electrons, spiraling in magnetic fields, must exist in the nucleus.

In summary, the nucleus of the Galaxy appears to have a different environment from the rest of the Galaxy. It is characterized by a distinct stellar population, rapid rotation, and violent activity. The support for the model of violence in the nucleus includes (1) the presence of ionized gas, (2) energy emitted from a very compact nucleus, (3) nonthermal

TABLE 23.1 Emission from the Galactic Center

Wavelength	Approximate Energy Output (W)
Centimeter continuous emission	10^{30}
100-μm extended source	10^{35}
10-μm sources in core	10^{32}
2-μm from stars	10^{34}
X rays (1–10 Å)	10^{30}
Gamma rays (10^{-4} Å)	10^{27}

radiation at a wide range of wavelengths, and (4) a strongly peaked concentration of rapidly rotating gas within 3 ly of the galactic center.

You'll see in what follows that most large galaxies have similar characteristics, and that there exist galaxies more active than ours which display energy output at a much higher scale of violence.

One Nuclear Model: A Supermassive Black Hole

What could explain the origin of the nuclear emission? One model relies on a black hole of a few million solar masses in the central 3 ly (1 pc) of the Galaxy. Recall (Section 20.7) that the nucleus contains several ionized clouds within 5 ly (1.5 pc) of the center, about half of which concentrate in the inner 2 ly (0.6 pc). These clouds have diameters of 0.3 to 1.5 ly (0.1 to 0.5 pc), masses of roughly a few solar masses, electron densities between $10^{10}/m^3$ and $10^{11}/m^3$, and an average radial velocity of 126 km/sec. The spectra show lines of H II, Ne II, and Ar II; other ionized species are probably also present. The gas in the clouds appears to be expanding at about 50 km/sec, so the clouds cannot last long, only a few thousand years. They must be continuously replenished; otherwise we would see none now. A nuclear model must account for the generation, ionization, and dissipation of these little clouds.

Consider cloud creation. Imagine a 1-million-solar-mass black hole, with a Schwarzschild radius of 3×10^6 km, in the center of a 3 ly (1 pc) radius cluster of some million old Population I stars. Any stars that orbit close to the black hole can be pulled apart by tidal forces (Focus 12.1) and become clouds.

Next, the ionization of the clouds. Recall (Section 12.2) that infalling material will form an accretion disk, because of its angular momentum, around a black hole. For a 10^6-solar-mass black hole, matter falling in at a rate of 10^{-5} solar mass/yr will heat the disk to a blackbody temperature of roughly 35,000 K, essentially the same as an O star. This hot accretion disk can provide the ionization. The total luminosity would be about 10^7 solar luminosities.

Finally, gas elimination. The gas from a dissipated cloud can simply spiral into the accretion disk and finally be gobbled up by the black hole. This process works out to provide the infall of the roughly 10^{-5} solar mass/yr required above.

Somehow this nuclear black hole may help power other forms of radiative emission. It doesn't have the energy to power all aspects of our Galaxy's nuclear output, but it does have the advantage of the efficient conversion of energy from gravitation into other forms.

Warning: This is just *one* possible model. It has seductive appeal; black holes are fashionable these days. But it could be wrong.

With supermassive black holes in the back of our minds, let's turn to active nuclei of other galaxies.

23.2 ACTIVE GALAXIES

Radio astronomers have found that many radio sources lie beyond our Galaxy and have nonthermal spectra. Many have turned out to be associated with some kind of galaxy. Those galaxies with nonthermal spectra have been lumped into the category of *active galaxies*.

23.2 ACTIVE GALAXIES

TABLE 23.2 Properties of Active Galaxies

1. High luminosity, greater than 10^{37} W
2. Nonthermal emission, with excess ultraviolet, infrared, radio, and X-ray flux (compared with normal galaxies)
3. Rapid variability and/or small size (a few light years at most)
4. Peculiar photographic appearance: high contrast in brightness of nucleus and large-scale structures
5. Explosive appearance or jetlike protuberances
6. Broad emission lines (sometimes) and nonstellar spectrum

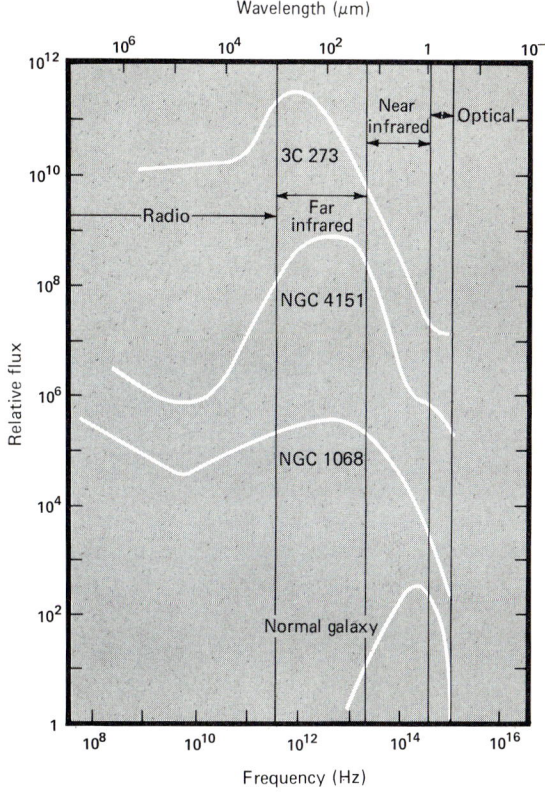

FIGURE 23.1 A comparison of the spectra in the radio, infrared, and optical for three active galaxies and a normal galaxy. (*Adapted from a diagram by R. Weymann.*)

But how active is "active"? We'll use the term here in contrast to "normal," which applies to our Milky Way Galaxy. Basically, an active galaxy's spectrum does not look like that of a collection of stars. It is mostly nonthermal and has infrared, radio, ultraviolet, and X-ray outputs greater than that in the optical. (Our own Galaxy does emit at all these wavelengths, but not nearly as strongly as an active galaxy.) Table 23.2 lists some of the typical properties of active galaxies, and Fig. 23.1 compares the spectra of some active galaxies with that of a normal galaxy.

There are many observations of active galaxies, but few successful models for explaining them. For this reason, we'll limit the discussion to the major classes of active galaxies.

Radio Galaxies

We find radio galaxies by comparing the positions of radio sources with the positions of optically observed galaxies. Two principal types have cropped up: *compact* and *extended*. *Extended* means that the radio emission is larger than a photographic image of the galaxy; *compact* means it is the same size or smaller. Compact radio galaxies often display *very* small (usually nuclear) radio sources, sometimes no more than a few light years in size. Extended radio sources, in contrast, sometimes show a double structure of two giant lobes, up to millions of light years in extent, symmetrically placed on opposite sides of the nucleus.

FIGURE 23.2 A short-exposure photo of the active galaxy Messier 87 (M 87). Note the jet in the upper right extending from the nucleus. (*Courtesy Kitt Peak National Observatory.*)

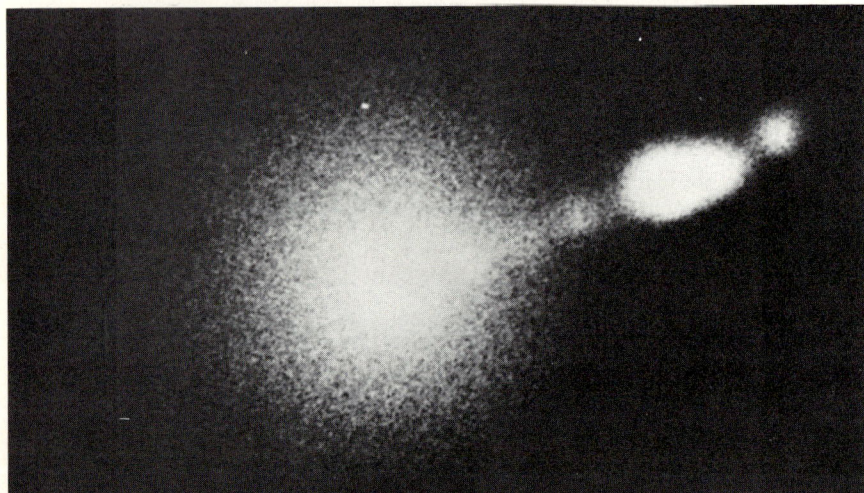

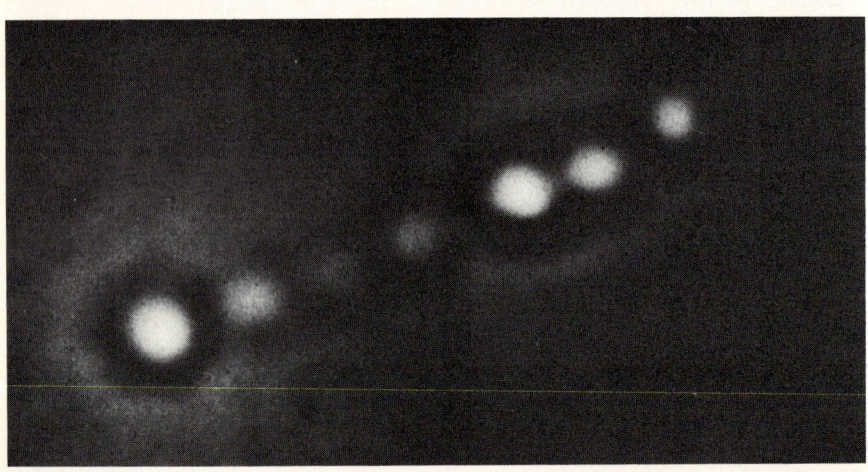

FIGURE 23.3 A close-up of the optical jet in M 87. The upper view is a conventional photo. The lower photo has been computer-processed to bring out the structure in the jet. Note that six blobs are clearly visible along with the nucleus. (*Courtesy H. Arp.*)

23.2 ACTIVE GALAXIES

Compact Radio Galaxies

M 87 is a fine example of a compact radio galaxy. A giant elliptical galaxy, M 87 dominates the Virgo cluster of galaxies. It lies about 65 million ly (20 Mpc) away and is the nearest elliptical galaxy that presents direct evidence for violent activity. One radio source only $1\frac{1}{2}$ light *months* in diameter appears in M 87's core along with a group of other compact radio sources. Poking out from the core, a remarkable, optically visible jet (Fig. 23.2) extends over a length of some 6000 ly. The jet has a luminosity of roughly 10^{34} W (10^7 solar luminosities); its emission is polarized.

A beautiful photograph taken by Halton Arp shows that the jet contains at least six blobs of material, each no more than a few tens of light years in size (Fig. 23.3). Over 22 years the blobs have changed slightly, but significantly, in intensity and polarization. In long exposure photos a fainter jet is visible on the opposite side of the galaxy from the brighter one—in other words, a counterjet.

M 87 also emits X rays with about 50 times more energy than its optical emission, about 5×10^{35} W (10^9 solar luminosities) in X rays from the whole galaxy. The jet itself is now also known to emit X rays. High-resolution observations with the Einstein Observatory show that the jet contains X-ray knots.

The VLA has mapped M 87's jet in detail and confirmed that its radio emission coincides with the optical and X-ray emission (Fig. 23.4). So the jet overall emits over a wide range of frequencies, from radio to X rays, and each knot of the jet generates this spectrum of energies.

What process can produce this fantastic radiative output? Observed at a wide range of wavelengths, the radio spectrum looks nonthermal. This fact led Russian astronomer Josif Shklovsky to propose that much of the emission comes from the synchrotron process, including the optical emission from the jet. The polarization of the jet's light (about 25 percent) confirms this proposal. This synchrotron emission presents a problem, however. It must result from very energetic electrons moving in magnetic fields. Such electrons emit radiation and lose energy quickly, only a few hundred years for those producing the radio emission, a few tens of years for those generating the optical radiation, and a few days for those responsible for the X rays. As the electrons spiral in the magnetic fields

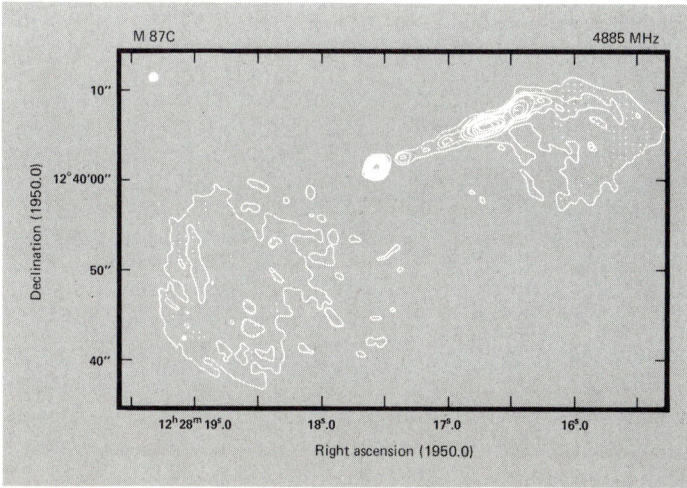

FIGURE 23.4 A high-resolution radio map of the central region of M 87, made with the VLA at a frequency of 5 GHz. Note the radio jet, which coincides with the optical one, extending to the right from the nucleus. (*Courtesy F. Owen; observations by F. Owen and P. Hardee.*)

(intensity about 10^{-5} to 10^{-6} gauss), they emit photons and so lose energy. Traveling slower, they then emit lower-frequency photons, and so on. The rate of loss of energy by the electrons is greatest at the highest energies and less at lower energies, if the magnetic fields are constant in strength. The problem is, what replenishes the electrons?

Other elliptical galaxies possess nuclear jets. For example, radio observations have detected a jet in the nucleus of the elliptical galaxy NGC 6251. The nuclear jet has a length of only 5.5 ly (1.7 pc), its material exceeds a temperature of 10^8 K, and it starts only 7 ly (2 pc) from the nuclear core. In fact, radio jets are common. Almost all radio galaxies of the lowest luminosities have jets (Plate 16).

Extended Radio Galaxies

Extended radio galaxies appear at first glance to be completely different beasts from compact radio galaxies. Radio astronomers find that many extended radio galaxies are double, with the lobes lined up with the galaxy's center. These radio clouds are huge: most are 150,000 to 3,300,000 ly (45 to 1000 kpc) in diameter. When classified by structure, extended radio galaxies fall into three main groups:

1. Doubles (example, Cygnus A): highest luminosities, lobes aligned through center of galaxy, bright hot spots at ends.
2. Bent doubles (example, Centaurus A): intermediate luminosities, bent through nucleus, taillike protrusions.
3. Narrow-tailed sources (example, NGC 1265): lowest luminosities, U-shaped, rapidly moving galaxies in a cluster. (Most extended radio galaxies fall into this group.)

Cygnus A, one of the strongest radio sources in the sky and one of the first discovered, provides an excellent example of the double structure typical of an extended radio galaxy. Its radio output, some 1.2×10^{38} W (10^{11} solar luminosities), comes from two giant lobes set on opposite sides of the optical galaxy (Fig. 23.5). Each lobe has a diameter of 55,000 ly (17 kpc), about half the size of the Milky Way Galaxy. They hang roughly 163,000 ly (50 kpc) away from the central galaxy. Each lobe contains a cloud of energetic electrons and magnetic fields that must be storing the 10^{53} J needed to account for the radio luminosity lasting 10^7 to 10^9 years. This is more energy than produced by all the stars in the Galaxy in 10^8 years.

The central galaxy of Cygnus A is a giant elliptical with a dust lane down its middle (Fig. 23.6). It has an active nuclear region, with a spectrum showing emission lines and a synchrotron (or at least nonstellar) continuum. But beyond 8 kpc from the center, the spectrum is just that of a mix of stars. This galaxy appears to have blasted out the two clouds some 10^7 to 10^9 years ago.

Centaurus A (NGC 5128) is another extended radio source somewhat similar to Cygnus A. It is a supergiant (cD) elliptical (E2) galaxy bisected by an irregular dust lane (Fig. 23.7). At a distance of 13 million ly (4 Mpc), Centaurus A is the closest active galaxy; it almost outshines M 33 visually. Viewed with a radio telescope, Centaurus A has two huge outer lobes, 650,000 ly (200 kpc) and 1,350,000 ly (400 kpc) in diameter. The double lobes form a bend through the nucleus and have taillike protusions (Fig. 23.8). Closer in, another pair of radio lobes sit on the edges of the optical galaxy; these are some 33,000 ly (10 kpc) in diameter. The inner and outer lobes are close to being in alignment.

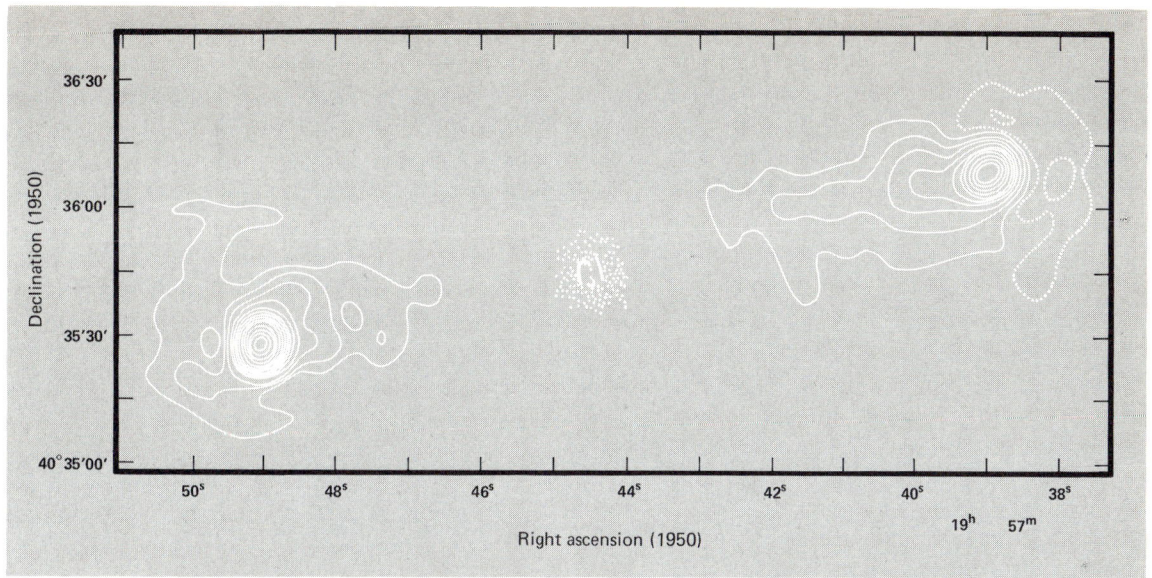

FIGURE 23.5 A radio map of Cygnus A, an extended, twin-lobed radio galaxy. The patch in the center represents the optically visible galaxy (next figure). On each side of the nucleus lie immense lobes of radio emission. (*Based on observations by S. Mitton and M. Ryle with the Cambridge Radio Telescope.*)

FIGURE 23.6 An optical photo of Cygnus A. Only the bright nucleus is visible, with a dust lane cutting through its middle. Compare with the next figure of a much closer elliptical galaxy, Centaurus A. (*Courtesy Lick Observatory.*)

FIGURE 23.7 Centaurus A (NGC 5128), a radio galaxy. Note the dust lane across the galaxy's middle, like that in Cygnus A. (*Courtesy Kitt Peak National Observatory.*)

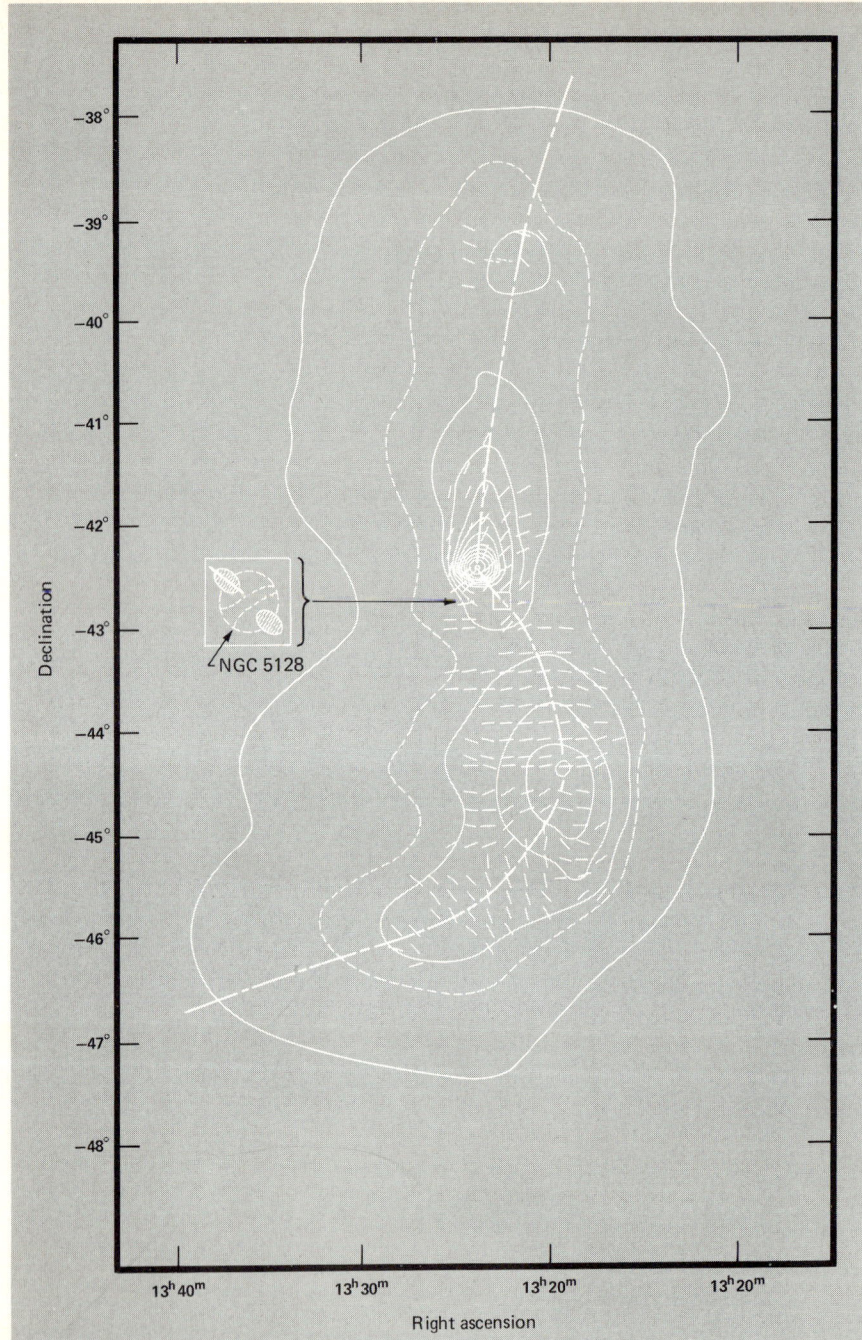

FIGURE 23.8 Radio emission from Centaurus A. This map shows the enormous extent of the two radio lobes. The short lines in the contours indicate the direction of the local magnetic fields. The optical galaxy is the size of the small square in the center. The enlargement of the optical galaxy in the insert outlines the galaxy, dust lane, and inner radio lobes. (*Courtesy R. M. Price; based on observations by B. Cooper, R. M. Price, and D. Cole.*)

Centaurus A also emits X rays intensely; the source is very small and coincides with the nucleus. Remarkably, the X-ray emission varies—by as much as a factor of 2. For instance, during July–August 1975 the X-ray luminosity was 1.3×10^{36} W; during July–August 1976 it was 6.9×10^{35} W. During the same time the radio flux also varied, but only about half as much. The nucleus of Centaurus A also emits infrared and radio

23.2 ACTIVE GALAXIES

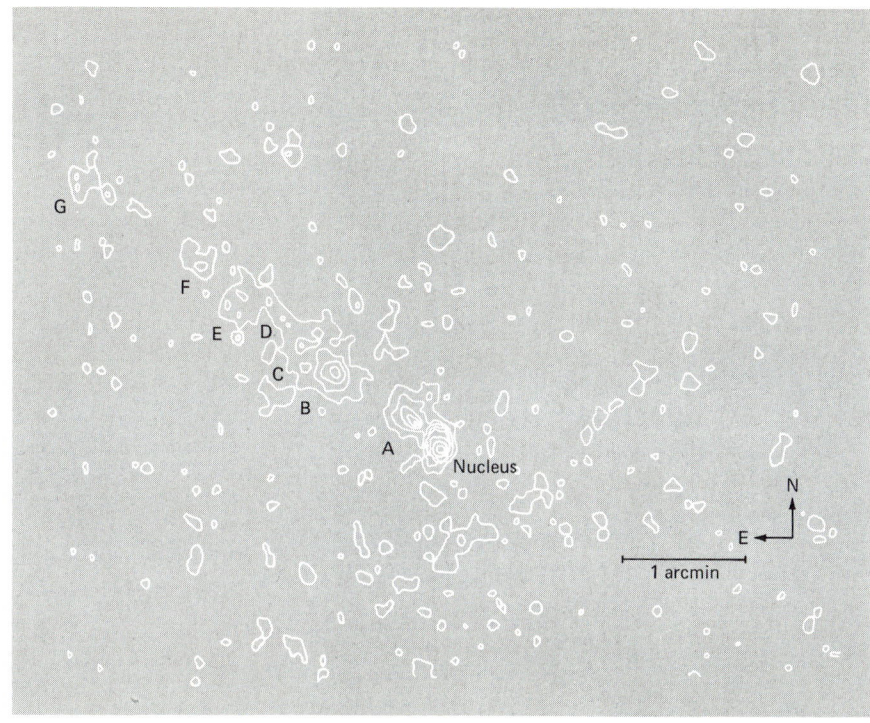

FIGURE 23.9 An X-ray map of the nuclear region of Centaurus A, taken with the Einstein X-Ray Observatory. Note the X-ray knots (A–G) extending from the nucleus. These make an X-ray jet. (*Courtesy P. Gorenstein, E. Schreier, and E. Feigelson.*)

radiation strongly. These observations, with those taken at X-ray energies, show that the nuclear emission is nonthermal.

The nucleus of Centaurus A has a direct connection to the inner radio lobes. The Einstein X-ray Observatory detected an X-ray jet streaming northeast from the nucleus and consisting of at least seven distinct blobs (Fig. 23.9). This discovery prompted Jack Burns, Ethan Schreier, and Eric Feigelson to observe Centaurus A with the VLA (a very difficult observation, because Centaurus A is so far south that it's above the horizon only a few hours a day as viewed from New Mexico). The VLA map at 20-cm wavelength shows radio emission along a jet that extends to one of the nuclear radio lobes (Fig. 23.10). The jet has a bloblike structure that coincides with the X-ray blobs. So Centaurus A and M 87 look similar in that both have nuclear jets that emit radio and X rays.

One of the largest radio galaxies, 3C 236, has outer lobes that extend 20 million ly (6 Mpc) end-to-end; each is about 3.3 million ly (1 Mpc) in size. One cloud could engulf the Milky Way Galaxy and the Andromeda Galaxy at the same time! This extended radio galaxy fills 10^{71} m^3 of space and is the largest known object in the visible universe found so far. 3C 236 also contains a radio core that extends some 2 kpc in length. This core source has a jetlike shape which aligns close to the axes of the giant lobes (Fig. 23.11).

Narrow-tailed sources make up the majority of extended radio galaxies, of which *head-tail galaxies* are one type. As implied by the name, head-tail radio galaxies have a radio head (a strong radio source around the visible galaxy) and an extended tail (a narrow source extending from the galaxy). NGC 1265 in the Perseus cluster of galaxies is a good

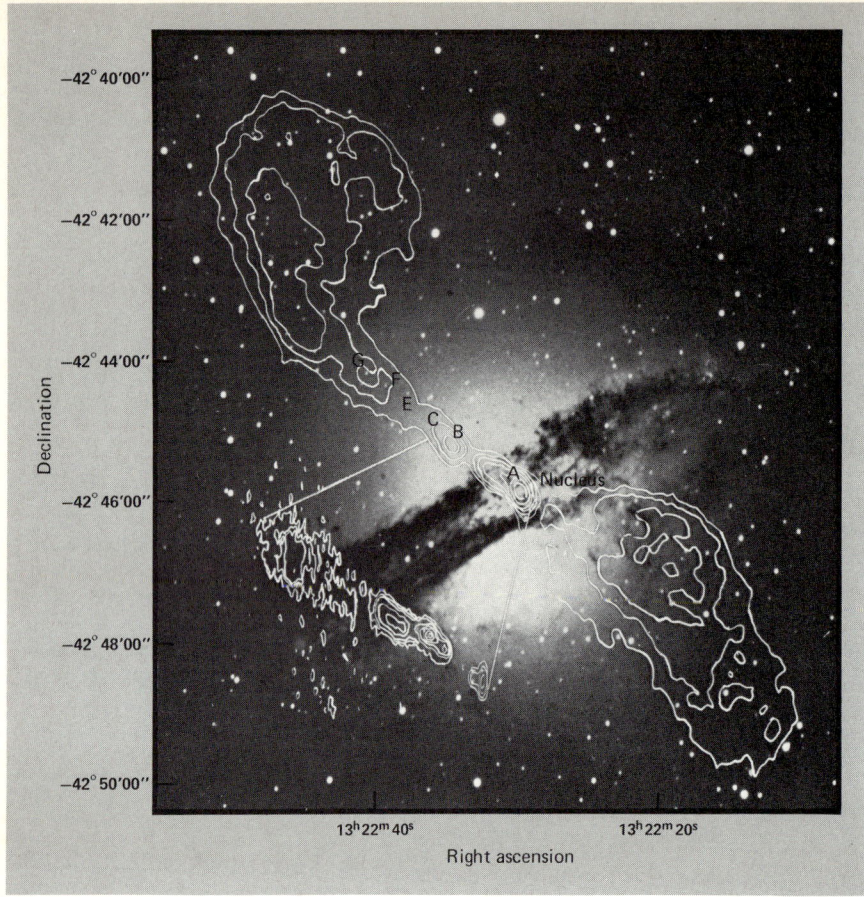

FIGURE 23.10 A VLA radio map of the nuclear region of Centaurus A overlaid on an optical photo. Note the two inner radio lobes and the radio jet extending from the nucleus to the lobe at the upper left. The jet consists of a series of blobs that correspond well to the ones in the X-ray map. (*Courtesy J. Burns; observations by J. Burns, E. Schreier, and E. Feigelson.*)

example (Fig. 23.12). Not all tails trail. Instead, we find a sequence in the amount of bending shown by the tails (Fig. 23.13), which ranges from 180° apart (double source; Figure 23.13a) to 0° (head-tail source; Fig. 23.13f). At high resolution, the heads often contain radio jets (Fig. 23.12).

How to explain this structural sequence? Recall (Section 22.6) that clusters of galaxies

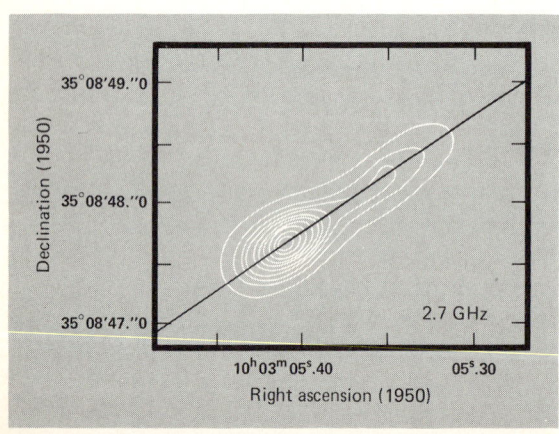

FIGURE 23.11 A VLA map of a radio jet in the nucleus of 3C 236. (*Courtesy A. Bridle; based on observations by E. Formalont, G. Miley, and A. Bridle.*)

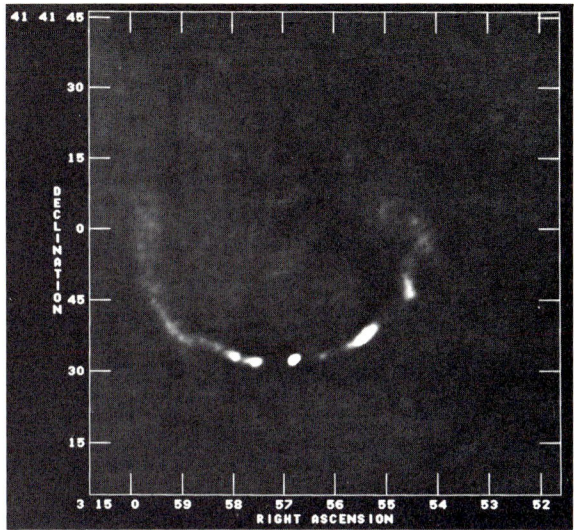

FIGURE 23.12 A radio picture, made with the VLA, of the nucleus of the head-tail galaxy NGC 1265. Note how the radio emission curves away and trails from the nucleus. (*Courtesy J. Burns; observations by F. Owen, J. Burns, and L. Rudnick.*)

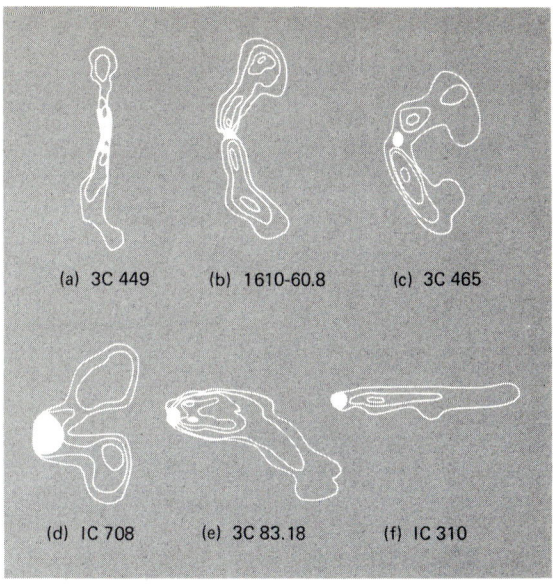

FIGURE 23.13 A sequence for the bending of radio tails around the nuclei of radio galaxies. (*Adapted from a diagram by G. Miley.*)

contain a hot, ionized intracluster medium. Imagine that a galaxy, moving rapidly through this medium, shoots out material (high-speed electrons, for instance) in a jet. The material flowing out of the galaxy is decelerated by the intracluster medium and the moving galaxy leaves it behind. As the galaxy travels along, it leaves behind a radio-visible trail—a fossil record of where it's been.

Head-tail galaxies interact with the surrounding gas and show that a significantly dense gas exists in clusters. The discovery of head-tail galaxies prompted the acceptance of intracluster gas *before* X-ray observations (Section 22.6), confirmed its existence (at a density of about 10^{-24} to 10^{-27} kg/m^3) and showed it to be very hot (about 10^7 K).

To sum up, one common type of active galaxy is a radio galaxy. Many radio galaxies have emission in the form of lobes or streams that extend far beyond the visible galaxy (Fig. 23.14). The lobes may be a few million light years apart and thousands of light years in size. The vexing problem with these extended radio lobes is the vast amount of energy

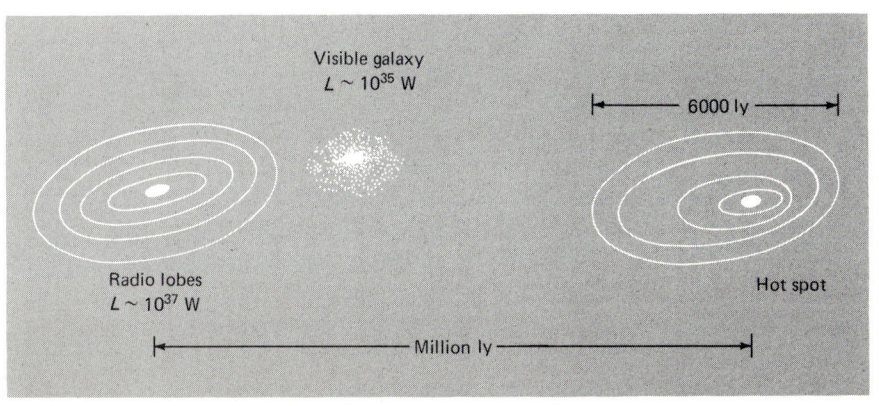

FIGURE 23.14 A schematic drawing of a typical twin-lobed radio galaxy. The visible galaxy is usually an elliptical one. (*Adapted from a diagram by A. Bridle.*)

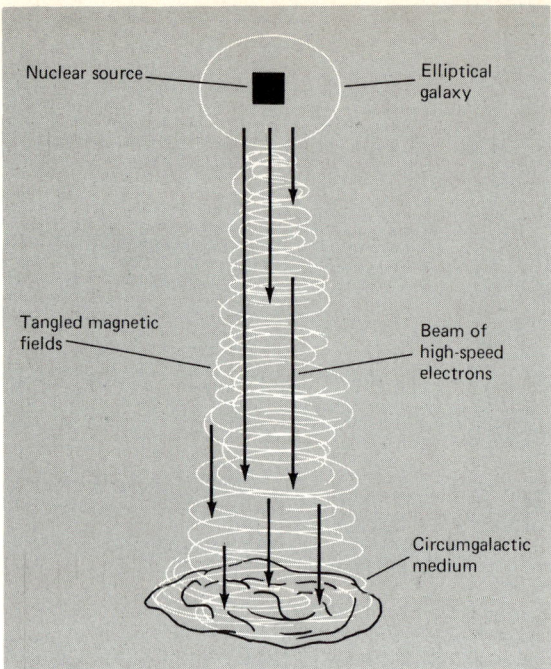

FIGURE 23.15 A schematic drawing of the possible connections among a radio galaxy's nuclear source, radio jet, and radio lobe. (*Adapted from a diagram by A. Bridle.*)

they contain: a typical lobe luminosity is 10^{36} to 10^{37} W, while the visible elliptical galaxy with which it is associated may emit only some 10^{35} W. Another way to consider it: the lobes are energy reservoirs that are a tangle of magnetic fields (strengths about 10^{-6} gauss) and high-speed electrons, if the emission is synchrotron radiation. Then a typical lobe contains more than 10^{52} J. (For comparison, the complete conversion into energy of a solar mass of material releases about 10^{47} J.)

What powers active radio galaxies? The clues are nonthermal emission, usually polarized, with nuclear jets that have a bloblike structure and often point to twin lobes of extended emission beyond the nucleus. That evidence suggests that the synchrotron process is operating. The nucleus provides the high-energy electrons. These are expelled either as a fairly constant beam of particles or a sequence of ionized blobs that are thrown out along a magnetic field, so half the particles fly in one direction, the other half in the opposite direction. If the nuclear machine is active and stable, extended lobes of ionized material build up at the end of the jets. Repeated bursts from the nucleus can account for variability.

We don't know all the details yet, but high-resolution radio observations have revealed the crucial clue: radio jets from the nucleus, aligned (more or less) with the lobes. These jets suggest that high-speed electrons are channeled, perhaps in bursts, from the nucleus into the circumgalactic medium, where they pile up to form a lobe (Fig. 23.15). What's the nuclear energy source? At the moment, we don't know.

Seyfert Galaxies

In 1943 Carl Seyfert collected the spectra of six spiral galaxies that showed unusual, broad emission lines. These are now called *Seyfert galaxies*. Later work has added to this collection; some 90 Seyfert galaxies are cataloged to date.

23.2 ACTIVE GALAXIES

FIGURE 23.16 The Seyfert galaxy NGC 1275, a strong emitter of X rays. Note how bright the nucleus appears compared with the rest of the galaxy. (*Courtesy Kitt Peak National Observatory.*)

What is a Seyfert galaxy? Most galaxies have a bright nucleus, which looks fuzzy when viewed with a telescope. In contrast, a Seyfert looks like a bright star surrounded by a faint haze (Fig. 23.16). In short-exposure photographs a Seyfert's nucleus could be mistaken for a star.

Along with this trademark optical appearance, a Seyfert has a particular spectrum. It shows very broad emission lines. We interpret the width of the lines as due to Doppler shifts produced by motion in the emitting gas. An astronomer observing the galaxy sees some blue-shifted photons from gas moving toward us, some red-shifted photons from gas moving away, and some unshifted radiation from gas at rest with respect to the average velocity of the galaxy. Taken together, the multitude of lines blends into one broad line. Other galaxies have emission lines with Doppler widths equivalent to a few hundred kilometers a second, but the lines from a Seyfert have Doppler widths equivalent to a few *thousand* kilometers a second, 10 times broader.

Could this broadening be due simply to a high temperature producing high velocities of the emitting atoms? Recall (Section 14.5) that the thermal velocity of a gas is

$$V_{th} = \sqrt{\frac{2kT}{m}}$$

or, solving for T,

$$T = \frac{mV_{th}^2}{2k}$$

with k (Boltzmann's constant) equal to 1.4×10^{-23} J/K. To produce a thermal velocity of

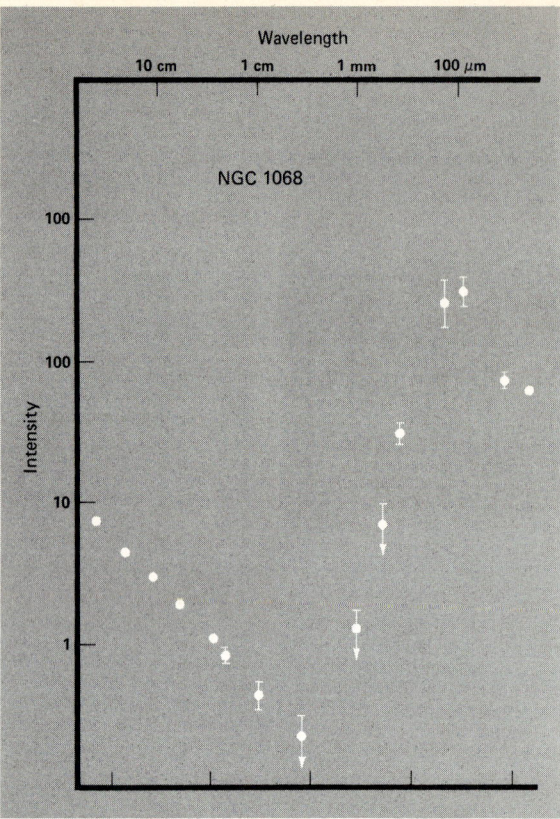

FIGURE 23.17 The spectrum of the Seyfert galaxy NGC 1068 in the radio and infrared. The decline in intensity from centimeter to millimeter wavelengths is typical of a nonthermal source. The rise of the emission in the infrared probably comes from heated dust. (*Adapted from a diagram by J. Elias and coworkers,* Astrophysical Journal, *vol. 220, p. 25, copyright 1978 by the American Astronomical Society.*)

1000 km/sec (10^6 m/sec) in a hydrogen gas ($m = 1.7 \times 10^{-27}$ kg) would require a temperature of

$$T = \frac{(1.7 \times 10^{-27})(10^6)^2}{2(1.4 \times 10^{-23})}$$

$$= 6.0 \times 10^7 \text{ K}$$

Gas at a temperature of 60 million K would be so highly ionized it would not produce any of the lines observed. So the line broadening must be due to high random velocities in the gas, not high temperature.

Seyferts are almost always spiral galaxies. A detailed survey of 80 Seyfert finds that only 5 to 10 percent might be ellipticals. (The small angular size of some Seyferts makes it hard to classify them.) Compare this with the fact that many extended radio galaxies are ellipticals. Overall, about 1 percent of all spiral galaxies (ordinary and barred) are Seyferts.

NGC 1068 is a good example of the peculiarities of Seyfert galaxies. Its continuous spectrum does not resemble that expected from a group of stars (Fig. 23.17). In fact, much (but not all) of the spectrum is nonthermal, and the thermal part peaks at relatively long wavelengths. These observed facts imply that at least three different sources contribute to the continuous spectrum: stars, synchrotron radiation, and infrared radiation from heated dust. The optical emission from NGC 1068 is polarized, as expected from synchrotron emission or light passing through thick dust.

23.2 ACTIVE GALAXIES

Infrared astronomers have found that the bulk of NGC 1068's luminosity, at least 10^{38} W, comes out in the infrared. Such infrared emission is most easily explained if the nucleus of NGC 1068 is embedded in dust. These particles can absorb ultraviolet and optical radiation from the nucleus and heat up to about 100 K. At that temperature the dust grains give off most of their radiation in the infrared part of the spectrum.

Daniel Weedman has found that Seyferts fall into two classes based on their spectra. Class 1 Seyferts have strong, wide Balmer lines but narrow lines of other elements, such as those from O III. Class 2 Seyferts have Balmer lines and O III lines of the same width, both a bit wider than the O III lines of Class 1 Seyferts. (NGC 1068 is a Seyfert 2.) The use of wide and narrow here is relative. For example, the average Balmer line width for 40 Seyfert 1 galaxies is 3400 km/sec. Their other lines have Doppler widths of "only" 200 to 500 km/sec. In some Seyfert 1 galaxies, the Balmer lines have Doppler widths as great as 10^4 km/sec (3 percent of c)! In contrast, the lines in Seyfert 2 galaxies are only 500 to 1000 km/sec wide.

As in diffuse nebulas (Section 9.1), the Balmer lines in a Seyfert are produced by recombination of hydrogen ions with electrons, which then cascade to lower levels. What makes the lines so broad? The simplest explanation: filaments or clouds of gas moving at speeds of a few thousand kilometers a second. To explain the Balmer emission from a Seyfert 1 requires a few tens to thousands of solar masses of ionized gas in the nucleus at densities of 10^{13} to 10^{15} ions/m^3 moving at such speeds. The narrow line emission then might come from a large, surrounding volume of lower density gas moving at slower speeds.

What ionizes the gas and moves it around? Probably the energy source in the nucleus that generates the synchrotron emission. That source is unknown. Also in the nucleus must reside a source of high-energy electrons and gas. Observations indicate that the nuclei of Seyferts are small, only a few light years in diameter. Gas moving at 10^4 km/sec would flow across such a tiny nucleus in only 100 years. So the gas must be replaced as it flows out, or somehow, even at 10^4 km/sec, the nucleus must hold it in.

To sum up, Seyferts are mostly spiral galaxies with extraordinary features. The most prominent are:

1. They have extremely small and bright nuclei.
2. Their nuclei have spectra that show emission lines not usually seen in the spectra of spiral galaxies. These bright lines do not come from stars.
3. The emission lines are very wide. Considered as Doppler shifts, the widths of the lines indicate gas motions of 500 to 4000 km/sec. Such high velocities could result from violent explosions.
4. Many Seyferts have compact, low-luminosity radio sources within them.

In addition, the continuous spectra of Seyferts have a combination of stellar, nonthermal, and infrared (from dust) radiation. The total energy output is some 10^{37} to 10^{38} W (about 10^{11} solar luminosities). Their luminosity varies, but only by small amounts, over time spans of a few days to a few months. Seyferts tend not to be strong radio sources; sensitive radio surveys have detected only about half of the Seyferts examined.

In some ways the characteristics of Seyferts resemble the Sgr A radio source in the center of the Galaxy. But Seyferts emit much more energy, at least a thousand times that of the nucleus of the Milky Way Galaxy.

BL Lac Objects

One more animal of the active galaxy zoo has some similarities to those described so far. These objects are named after their prototype, BL Lacertae, and so are called *BL Lac objects*.

As a group, the BL Lac objects have the following characteristics: (1) rapid variability at radio, infrared, and visual wavelengths, (2) *no* emission lines, (3) nonthermal continuous radiation with most of the energy emitted in the infrared, and (4) strong and rapidly varying polarization. Also, BL Lac objects generally have a starlike appearance—no structure is visible.

The greatest difference between BL Lac objects and other active galaxies is that their emission varies so frequently, erratically, and rapidly. For example, BL Lac itself, recorded on a long sequence of Harvard Observatory photos, fluctuates between visual magnitudes 14 and 16 with occasional bursts to brighter than 13. These fluctuations mean that BL Lac's optical emission varies by a factor of 20 times or so. Observers have noted night-to-night variations in luminosity of 10 to 32 percent. That doesn't sound like much, but imagine our Galaxy changing its light output by some 20 percent in a day. That's like 10^{10} suns turning on and off simultaneously. A few BL Lac objects have changed their brightness by as much as 100 times in luminosity.

The radio emission from many BL Lac objects is compact or only slightly extended. The extended radio structure is weak in contrast to the intense emission from the nucleus.

What puzzles astronomers most about the BL Lac objects is that their energy variations take place in objects that show almost no emission lines in their spectra! As discussed above, the standard model for active galaxies pictures synchrotron emission produced by steady injection or bursts of high-energy electrons. That synchrotron output in the ultraviolet (and even the electrons themselves) should ionize any gas near the nucleus and produce emission lines through recombination. But where are the BL Lac's emission lines if they are powered the same way?

Some 40 BL Lac objects have been classified to date. They may actually not all be the same kind of beast. A few are possibly the nuclei of galaxies. Some, like BL Lac itself, have a faint surrounding fuzz that might be a galaxy. Others look pointlike without a hint of enveloping material. A number of BL Lac objects are found in clusters of galaxies—indirect evidence that they are also galaxies.

Let us note finally that we don't have good distance determinations for very many BL Lac objects. Beverly Oke and James Gunn have reported a red shift of 0.07 in the weak absorption features of the nebulosity around BL Lac, which corresponds to a radial velocity of 2.1×10^4 km/sec. This has been confirmed by Joseph Miller and his colleagues at Lick Observatory, who in addition have shown that the nebulosity has a spectrum like that of a luminous elliptical galaxy. If the red shift is cosmological, this velocity corresponds to a distance of 1400 million ly.

To sum up, active galaxies put out some 10^5 times more energy than our Galaxy. Their spectra are basically nonthermal, probably synchrotron emission, but they also include some stellar emission and infrared radiation from dust. Often the emitting regions are small, only a few light years in size. Our models for the emission are uncertain and incomplete. In particular, we do not know what engine powers these intense sources.

FIGURE 23.18 3C 48, the first quasar to be discovered. (*Courtesy Palomar Observatory, California Institute of Technology.*)

FIGURE 23.19 3C 273, the first quasar to have its spectrum deciphered. Note the jet sticking out of the quasar at the upper right; compare it to the one in M 87 in Fig. 23.2. (*Courtesy Palomar Observatory, California Institute of Technology.*)

23.3 QUASARS: MYSTERIOUS ENERGY EMITTERS

About 20 years ago quasars made a rather meek debut on the astronomical scene. During the boom period of radio astronomy in the late 1950s radio astronomers, like modern Tycho Brahes, compiled catalogs replete with radio sources that were not identified with any familiar visible objects. Hunting for possible associations of radio and optical sources, Thomas Matthews and Allan Sandage in 1960 discovered a faint, sixteenth-magnitude starlike object (hence the name *quasi-stellar object*, or *quasar*) at the position of radio object 3C 48. (3C means the Third Cambridge Catalog.) This object had a spectrum of broad emission lines that could not be identified, and it emitted more ultraviolet light than an ordinary main-sequence star.

3C 48 (Fig. 23.18) remained a unique object until 1963, when the strong radio source 3C 273 was identified with a thirteenth-magnitude starlike object (Fig. 23.19). The emission lines of 3C 273 were just as puzzling as the emission lines from 3C 48: They coincided with no known atomic lines. Astronomers were baffled.

Quasar Red Shifts

Maarten Schmidt finally deciphered the spectral code of 3C 273 by recognizing the prominent emission lines as those of the hydrogen Balmer series, red shifted by 15.8 percent (Fig. 23.20). Recent ultraviolet observations of 3C 273 show strong emission with one intense peak at 1410 Å. This line turns out to be the Lyman-alpha line of hydrogen at 1216 Å red shifted to shorter wavelengths by 16 percent. So the red shift measured in the ultraviolet confirms that found optically.

After Schmidt decoded 3C 273, Jesse L. Greenstein applied the same analysis to 3C 48 and found that its spectrum was red shifted by 36.7 percent. If the red shift of 3C 48 is a Doppler shift, its radial velocity is 9.1×10^4 km/sec, 30 percent the speed of light!

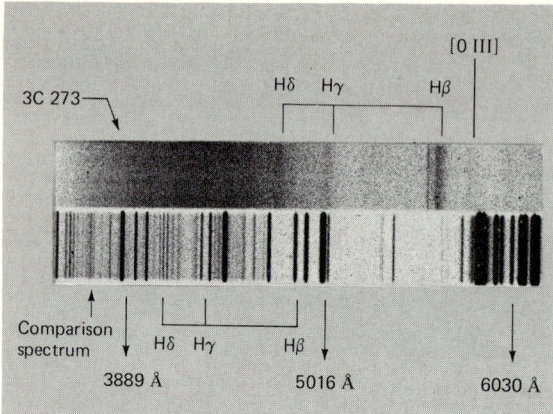

FIGURE 23.20 A spectrum of 3C 273, showing its large red shift. The upper spectrum is that of the quasar, the lower one a comparison spectrum that establishes the wavelength reference scale (at rest with respect to the observer). Compare the hydrogen Balmer lines. Note how the quasar's lines are shifted to the red end of the spectrum. This red shift amounts to about 16 percent. (*Courtesy M. Schmidt.*)

Following an initial elation at the discovery, the squabble about the nature of the red shifts began. Did the red shifts arise from the cosmic expansion, so that, according to the Hubble law, the quasars were at stupendous distances? Or were they Doppler shifts of masses expelled from the center of our Galaxy? Or from other galaxies? Or what?

Over 1500 quasars have been identified, and red shifts have been measured for most of these. On the average, one quasar appears in every 30 square degrees of sky—a patch about the size of the bowl of the Big Dipper. Some have red shifts that exceed 2.0, and a few even 3.0. For example, the quasar 4C 25.5 (4C means the Fourth Cambridge Catalog) has emission lines shifted by a factor of 2.358, or 235.8 percent. If interpreted as a red shift from the expansion of the universe, the light from 4C 25.5 comes from such a distance that it must have originated about 9 billion years ago. The strongly red-shifted quasars would then be the youngest objects we can see in the universe. Note by "youngest" we mean objects that existed far back in the past, closer in time to the Big Bang origin of the universe.

Warning: When the measured red shift approaches or exceeds 1, the simple Doppler formula $\Delta\lambda/\lambda_0 = v/c$ no longer gives the correct relative velocity. For example, if applied to 4C 25.5, the formula would indicate that this quasar was fleeing at 236 percent the speed of light! A modified formula, based on special relativity, must be used instead.

We won't derive the relativistic Doppler shift, but here's the result

$$\frac{\Delta\lambda}{\lambda_0} = \left[\frac{1 + v/c}{1 - v/c}\right]^{1/2} - 1$$

where $\Delta\lambda$ equals $\lambda - \lambda_0$, λ_0 is the original wavelength, v the radial velocity, and c the speed of light. Let's compare this expression with the classical Doppler equation by graphing them (Fig. 23.21). Note that the graphs are the same at low v, but as the red shift gets larger, v approaches, but never reaches c.

Let's use this relativistic Doppler shift formula to find v for a quasar with large red shift. Suppose the red shift is $\Delta\lambda/\lambda_0 = 2$. Then

$$2 = \left[\frac{1 + v/c}{1 - v/c}\right]^{1/2} - 1$$

23.3 QUASARS: MYSTERIOUS ENERGY EMITTERS

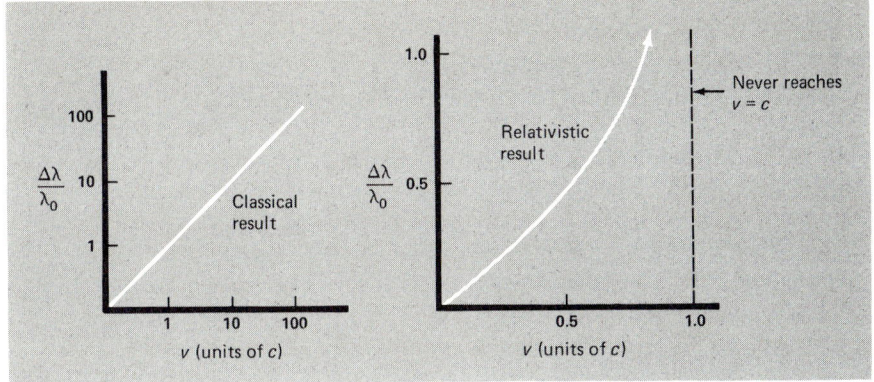

FIGURE 23.21 A comparison of the Doppler red shift versus radial velocity for the classical and relativistic cases. Note that for the relativistic one that v can never exceed c.

$$\frac{1 + v/c}{1 - v/c} = 3^2 = 9$$

$$1 + v/c = 9(1 - v/c)$$

$$10\frac{v}{c} = 8$$

$$\frac{v}{c} = 0.8$$

General Observed Properties of Quasars

Although most of the early quasars were detected because of their strong radio emission, many uncovered in later optical searches were found to be radio-quiet. So radio emission is not necessarily present in quasars. The observed properties that make quasars unique and that serve as identification tags are (1) starlike appearance with a large red shift, sometimes associated with a radio source (only about 10 percent are known radio sources), (2) broad emission lines in the spectrum with absorption lines sometimes present (usually the red shift of the absorption lines is less than that of the emission lines), (3) often variable luminosity, and (4) for those that are radio sources, aligned, double-lobed structures (like Cygnus A).

The main—and remarkable—feature of quasars is their large red shifts (compared to galaxies). At first glance the most natural explanation of the quasars' red shifts is a cosmological one; quasars participate in the universe's expansion. If so, their enormous red shifts indicate that they are very far from us. If they are as far away as indicated by their red shifts, but still bright enough to be observed, they must expend vast amounts of energy. For example, 3C 273's red shift of 16 percent, if due to the expansion of the universe, implies a distance of 3100 million ly (960 Mpc) for $H = 50$ km/sec/Mpc. At this distance, to appear at its observed apparent magnitude of 13, 3C 273 must emit about 10^{14} solar luminosities, or about 40 times as much as the most luminous galaxies. (Most of this energy is in the infrared.) This light left 3C 273 about 3 *billion* years ago—at the time simple life appeared on the earth.

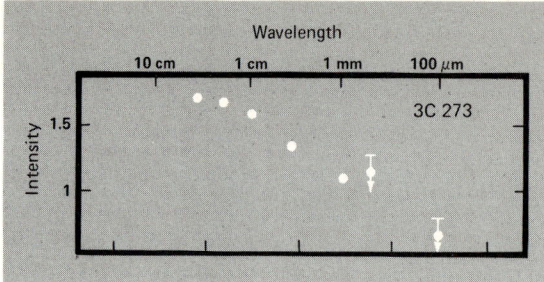

FIGURE 23.22 The spectrum of 3C 273 in the radio and infrared. The steady decline at shorter wavelengths is typical for a nonthermal source. (*Adapted from a diagram by J. Elias and coworkers,* Astrophysical Journal, *vol. 220, p. 25, copyright 1978 by the American Astronomical Society.*)

Quasars with smaller red shifts are closer in space and time. A recent candidate for the closest quasar is an X-ray source called 4U 0241+61, which has all the characteristics of quasars but a small red shift, only 4.4 percent. If this red shift arises from the expansion of the universe, this quasar is a mere 240 Mpc from us. It is also a radio source, and radio observations done at the VLA indicate that the size of the quasar, at least the core emitting the radio energy, is less than 120 pc. Its energy output at a distance of 240 Mpc is roughly 10^{11} times the sun's luminosity, or 10^{37} W—more like an active radio galaxy than a quasar.

The Light From Quasars

Some quasars emit radio waves intensely and all emit visible light. What produces this emission? A key clue comes from the spectrum of quasar radiation (Fig. 23.22). The spectrum does not resemble that for a blackbody emitter, but rather that for nonthermal, synchrotron emission. Such synchrotron emission is polarized if the magnetic fields are organized rather than chaotic. Note that synchrotron radiation requires a continuous supply of clouds of energetic electrons in the presence of magnetic fields. Any model of a quasar must include these two features.

Optical observations of quasars first discovered as radio sources have added more details to the general physical picture of a quasar. The continuous optical emission is in part polarized, so synchrotron emission produces some of the optical radiation. Given a common magnetic field, the electrons that produce the optical radiation must have higher energies than those that emit the radio radiation.

Line Spectra

All quasars have bright lines in their optical spectrum—the emission lines that are used to measure a quasar's red shift. The strongest emission lines in a quasar's spectrum are typically the Lyman-alpha line of neutral hydrogen (H I) at 1216 Å, a line of four-times ionized nitrogen (N V) at 1240 Å, one of triple-ionized carbon (C IV) at 1549 Å, another of single-ionized magnesium (Mg II) at 2798 Å, and the Balmer H-beta line of hydrogen at 4861 Å.

What do these emission lines reveal about the quasar? Recall (from Section 6.1) that emission lines are characteristics of a hot, low-density gas. For example, the Orion Nebula (Section 16.3) is heated by the ultraviolet radiation from young stars and glows with a characteristic bright-line spectrum. The techniques used to analyze the physical conditions in diffuse, glowing nebulas using their emission spectra have also been applied to quasars.

The emission-line spectra of quasars indicate that a low-density cloud of gas is irradiated with photons energetic enough to ionize hydrogen. The source of this radiation is probably the synchrotron emission from energetic electrons, which supplies ultraviolet and X-ray photons. So to the central synchrotron source we must add clouds or filaments of gas that convert high-energy radiation from the synchrotron source into visible emission lines.

What other information comes from an analysis of a quasar's spectrum? First, the composition of the radiating gas has no surprises. Lines of hydrogen, helium, carbon, oxygen, nitrogen, and other common elements have been observed. Second, these elements are in high ionization states, a fact that reaffirms the existence of energetic photons. Third, the emission lines are extremely broad. This fact implies that the filaments or clouds from which the emission lines originate move rapidly. Part of this broadening arises from the fact that the ions and atoms in the filaments are hot. However, the emission lines from quasars are a thousand times broader than the width predicted from the broadening effect of temperature. Most of the broadening, then, comes from motions of the clouds or filaments themselves at relative velocities from 1000 to 1500 km/sec. (Sounds like a Seyfert galaxy, doesn't it?)

Many (but not all) quasars also have absorption lines in their spectra. Quasars with emission-line red shifts less than 2.2 typically do not have absorption lines; those with greater red shifts have strong absorption lines. These lines are very narrow compared with the emission lines. Generally they are identified with ionized states of common elements, such as carbon, silicon, and nitrogen.

The absorption lines of quasars present a vexing puzzle. In general, their red shifts are *less* than that for the emission lines, and they sometimes show more than one red shift. For example, the spectrum of the quasar PHL 938 has an emission red shift of 1.955 and absorption red shifts of 1.949, 1.945, and 0.613. The difference between red shifts of 1.955 and 0.613 amounts to a relative radial velocity difference of $0.5\ c$!

What produces these absorption lines? They may be produced by gas clouds in or close to the quasar at temperatures cooler than the emission line regions. If the clouds are moving outward from the quasar, their velocity relative to us, and hence their red shift, will be less than that of the quasar. They might be produced by intervening (and otherwise invisible) intergalactic gas clouds. Another possibility is a halo of gas around an intervening galaxy. The intervening clouds or halos result in a smaller red shift because they are closer to us and so not moving away as fast. None of these possibilities has been well confirmed, but the discovery of the double quasar (see below) supports the third model.

Variability in Luminosity

Light variability was first observed for 3C 48 by Thomas Matthews and Allen Sandage, who reported variations of about 40 percent in 13 months. In 1963 Harlan Smith and Dorrit Hoffleit attempted the same task for 3C 273 and found an erratic variation over periods of about a year.

Most radio-emitting quasars also vary in radio output over periods of years (Fig. 23.23). In contrast, about 20 percent of the quasars exhibit rapid variations in light and radio output with periods on the order of days or weeks.

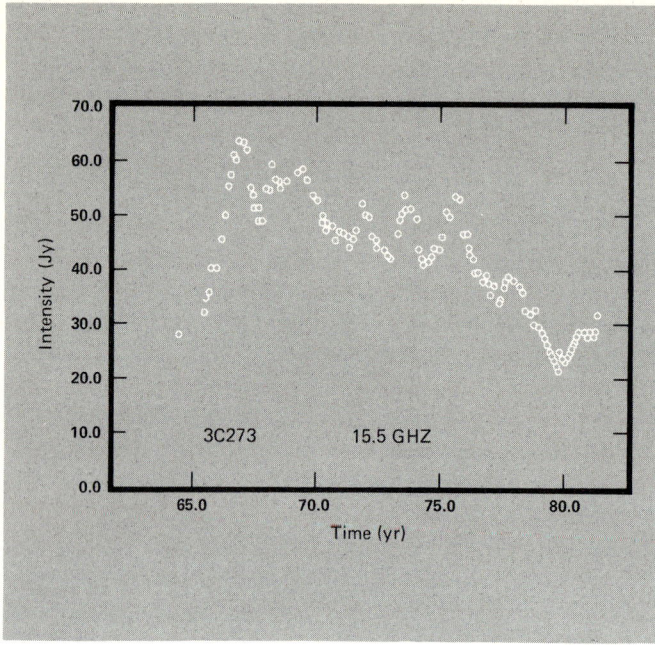

FIGURE 23.23 Variation of the radio emission of 3C 273 from 1965 to 1982 at a frequency of 15.5 GHz. (*Courtesy T. Balonek; observations by W. Dent and T. Balonek.*)

Problems of Quasars at Far Distances

Placing quasars at cosmological distances has the value of simplicity; it explains the observed red shifts without additional assumptions. As an additional bonus, quasars also serve as probes to the distant past. Two problems crop up, however: energy and size.

If quasars are actually billions of light years from the sun, they must be extremely luminous. The quasar 3C 273, for example, must emit a total radiative energy of about 4×10^{40} W, an amount of energy in one second equivalent to all the energy produced by the sun in about 3 million years. A typical quasar produces about 100 to 10,000 times as much energy as an ordinary spiral galaxy.

Not only do quasars blast out energy at enormous rates, but the energy comes from relatively small regions of space in the centers of quasars—from possibly light hours or light months to no more than light years in diameter. Two pieces of evidence point to small energy-emitting volumes. First, there is the variation of light output with periods of days to years. As an extreme example, the quasar 3C 446 has doubled in brightness in two days. Whatever the energy source, and whatever provoked the sudden flare-up, the size of the region that emits the energy can be no more than 2 light days across.

Why this restriction? The special theory of relativity sets the speed of light as the maximum velocity in the universe for the transportation of energy and information. Suppose that 3C 446 were larger than 2 light days in diameter, say 10 light days. Now suppose it suddenly doubles its luminosity. Radiation from the far side of the quasar will reach us 10 days later than radiation from the near side. So we would see the brightness of the quasar gradually increase, reaching twice its former value only after a period of 10 days. So if 3C 446 varies in 2 days, it cannot be larger than 2 light days across. This conclusion

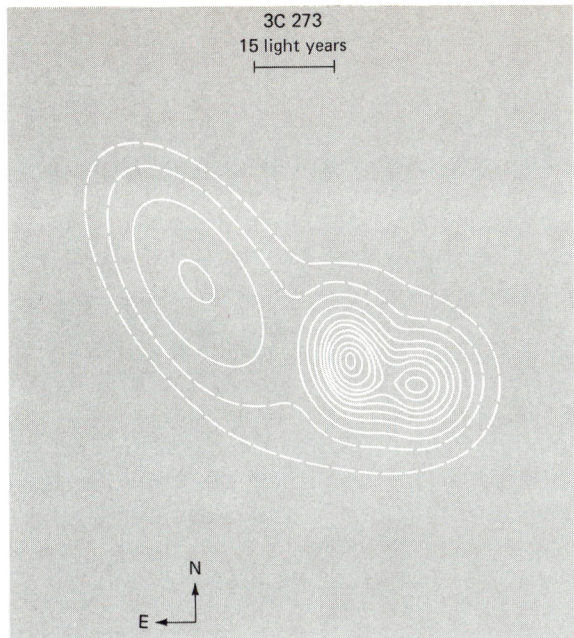

FIGURE 23.24 A very high-resolution radio map of 3C 273, which shows three distinct radio pieces. The size scale on this figure assumes a distance of 950 Mpc. (*Based on observations by K. Kellerman and coworkers, Astrophysical Journal, vol. 211, p. 658, copyright 1977 by the American Astronomical Society.*)

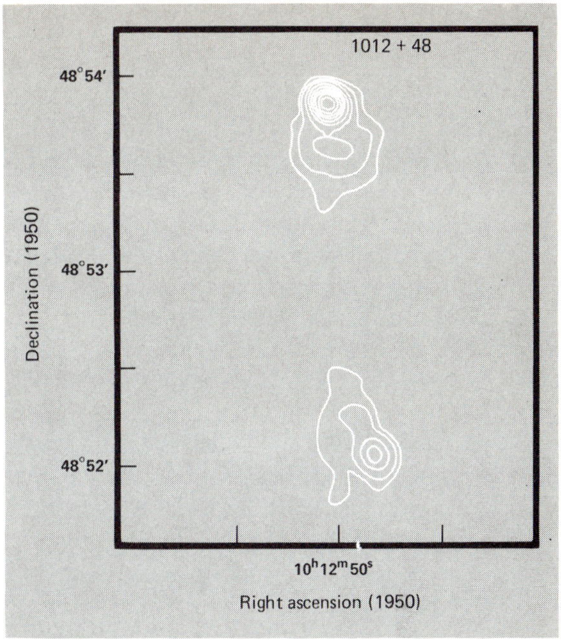

FIGURE 23.25 Radio emission from the quasar 1012+48. Note the symmetrical, twin-lobed structure. Compare Cygnus A in Fig. 23.5. (*Observations by G. Miley and A. Hartsuijker with the Westerbork radio telescope.*)

arises from the finite speed of light and does not rest on any derived property of quasars or their distances. Because quasars' light output varies over times of days to years, their energy-emitting regions cannot be larger than light days to light years—smaller than the distance from the sun to the nearest star.

Note that this same light travel time argument applies to any of the active galaxies described in Section 23.2. Their variability ranges from days to months to a few years, so their emitting regions cannot be more than a few light years in size.

Radio results obtained from international efforts of astronomers also support the conclusion that quasars must be small. With a maximum baseline of 10,536 km, intercontinental radio interferometers (Section 7.3) can resolve very small angular diameters. They show that 3C 273 consists of three separate pieces (Fig. 23.24). The smallest parts—if the quasar is at the distance of 950 Mpc demanded by a cosmological red shift—cannot be more than a few tens of light years across, a value consistent with the smaller size inferred from the light variations.

If these observations are correct, the energy of radiation and of moving particles trapped in a quasar's small volume is fantastically huge. Consider that for a moment. Quasars emit about 100 times the total energy output of our Galaxy from regions no more than a few light years in diameter!

High-resolution studies have also revealed the radio structure of quasars, especially those that are relatively nearby. These observations show that, for the quasars which can be resolved, the radio structures are generally symmetrical doubles (Fig. 23.25), similar to Cygnus A.

Faster Than Light?

Intercontinental radio astronomy permits radio astronomers to pick out fine details (roughly a few light years in size) even in distant quasars. These observations have detected change in the radio structures of six distant objects (five quasars and one galaxy) that lead to a remarkable conclusion, if taken at face value: Parts of these objects are moving with speeds greater than c! This apparent faster-than-light motion is called *superluminal motion* or *superluminal velocity*.

An example is our familiar quasar 3C 273. Recall it has an optical jet about 300 light years long. Its radio emission comes mainly from the body of the quasar (called 3C 273B) and near the end of the jet (called 3C 273A). So 3C 273 has a core with a radio jet protruding from it. A series of radio observations from 1977 to 1980 shows that a knot in the source has steadily moved away from the central peak (Fig. 23.26). The total separation has increased in the three years by 2 milliarcseconds (2×10^{-3} arcseconds). The expansion seems to have happened at a constant rate over this time.

Now, if 3C 273 is at the distance given by its red shift, the observed angular separation rate corresponds to a transverse velocity of almost *10 times the speed of light*! That, according to Einstein's special theory of relativity, is impossible.

What's going on here? The sources (such as 3C 273) that show superluminal motions all have a common feature: a radio structure of a strong central source with a weaker jet out one side. So they resemble nearby radio galaxies with single jets, such as Centaurus A (Fig. 23.10). These jets are thought to be electrons flowing outward at close to the speed of light from a nuclear source; they are *relativistic jets*. The same process may occur in the superluminal sources. In 3C 273 the jet is pointing almost directly at us (tilted only about 10° away from a perfect alignment). The moving knot is a blob of material streaming out along the jet. But the knot is *not* moving faster than the speed of light—it only *appears* to be doing so. The apparent superluminal speed is an optical illusion, caused by the almost head-on orientation of the relativistic jet and the finite speed of light.

To understand this effect, consider a jet emitting blobs of material at close to c. Suppose the jet opens at some small angle, say 8° with respect to our line of sight (Fig. 23.27). Suppose a blob ejected by the nucleus (point N) gets to point A in 101 years. Suppose the light emitted from the blob when it was at N reaches point B after 100 years. The separation between A and B is 14 light years (for an 8° angle). But the light at B is one year ahead of that emitted by the blob when it reaches A. (It's taken 100 years for the light from N to reach B, 101 years for the blob to reach A.)

Many years later, the light that was at B reaches us; only one year later, that emitted at A reaches us. The source seems to have moved from B to A—14 light years—in only one year. It seems to have a transverse speed of 14 times the speed of light. Yet no such physical motion has occurred. The superluminal velocity is only an apparent motion. And the smaller the angle of the jet to our line of sight, the more superluminal the motion will appear.

The Double Quasar: An Optical Illusion

Quasars are rarely close together, so astronomers were surprised in 1980 to discover two quasars only 6 arcsec apart. They are called 0957+561A and 0957+561B. Even more surprising, the emission-line red shifts of both are essentially the same—1.41. The ab-

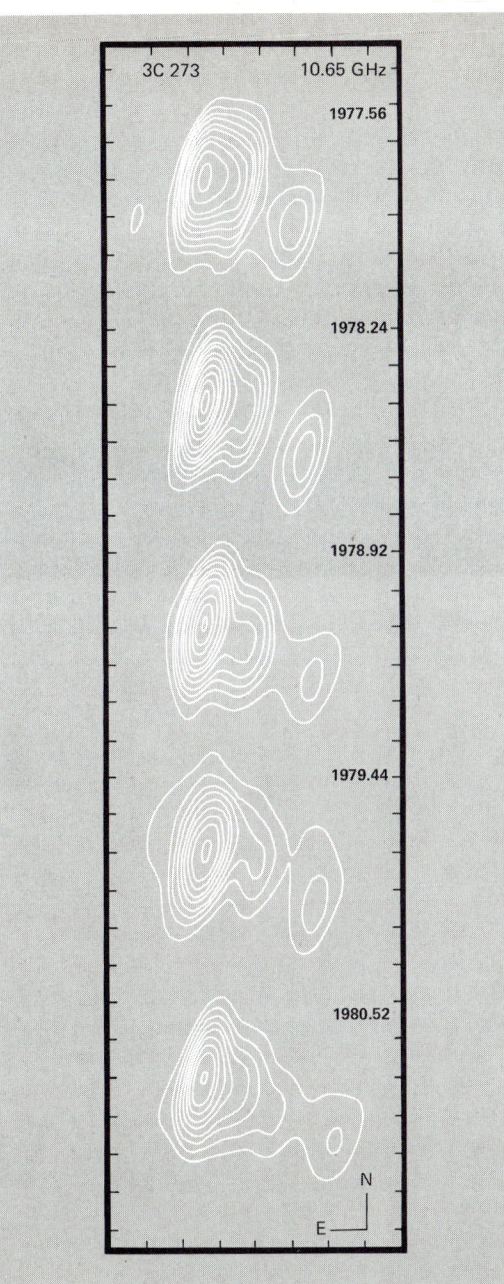

FIGURE 23.26 High-resolution radio maps of 3C 273 from 1977 (top) to 1980 (bottom) at a frequency of 10.65 GHz. The strongest peak is the main body of the quasar; the extension to the lower right is the radio jet. Note how a piece of this jet has moved away from the quasar. (*Observations by T. J. Pearson, S. C. Unwin, M. H. Cohen, R. P. Linfield, A. C. S. Readhead, G. A. Seielstad, R. S. Simon, and R. C. Walker of the Owens Valley Radio Observatory and the National Radio Astronomy Observatory.*)

FIGURE 23.27 Geometry for the illusion of faster-than-light speeds. NA is the axis of the jet, tilted 8° to our line of sight (NB).

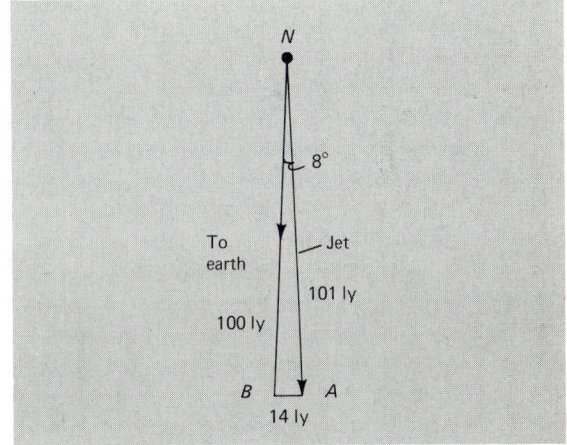

sorption line red shifts in both quasars are also the same and only slightly less than that of the emission lines. Could there be twin quasars with the same emission and absorption red shifts, 14 billion ly (4200 Mpc) from the sun, and separated by 200,000 ly? That's not likely.

This situation forced the conclusion that the quasars are not separate twins, but are optical images of the *same* quasar. Recall that general relativity predicts (Section 5.3) that masses deflect the paths of light rays, in essence, acting like a lens. If a very small, dense mass (a black hole, for instance) lies along our line of sight to a quasar, its image may be

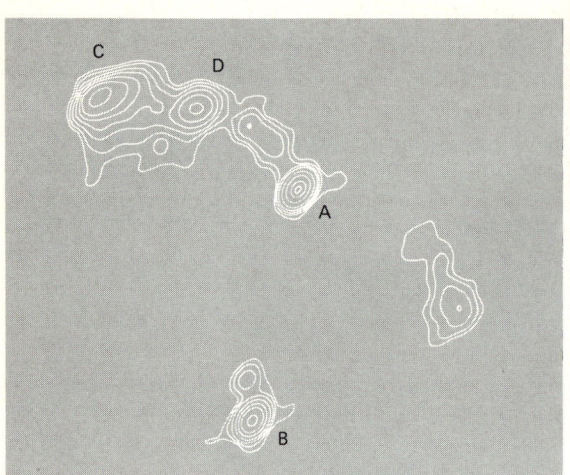 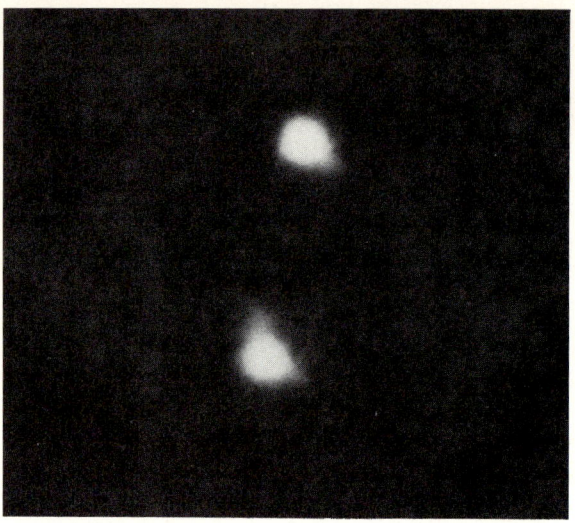

FIGURE 23.28 A VLA map of the radio emission from 0957+561 A and B. The double quasar consists of the two elliptical regions near the center (marked A and B). Note that there are blobs of emission (marked C and D) to the left of the upper quasar. (*Observations by P. Greenfield, B. Burke, and D. Roberts.*)

FIGURE 23.29 A high-resolution photo of the double quasar. Note the little fuzz sticking out of the top of B (the lower image). This turns out to be the poorly resolved image of the galaxy making the gravitational lens. The image of B is superimposed on it. (*Courtesy A. Stockton, Institute for Astronomy, University of Hawaii.*)

split into two, one above and one below the quasar's actual position. This phenomenon of image making by a mass is called a *gravitational lens effect*.

But radio astronomers at the VLA found data inconsistent with this model. The sources are not identical doubles, as expected from the simple gravitational lens model. Rather, there are radio blobs to the east of the northern quasar (A) not visible west of the southern quasar (B), and there is no indication of any object between A and B (Fig. 23.28).

The solution to the puzzle came when Alan Stockton, observing with the 2.2-m telescope on Mauna Kea, Hawaii, obtained photographs of the quasars on a night of exceptionally good seeing. The photos show the quasar B with a little bit of fuzz sticking out of it (Fig. 23.29 and Plate 17). This fuzz turns out to be the poorly resolved image of a faint galaxy—the gravitational lens! But since the galaxy is an extended mass, it acts like an imperfect lens and produces a complex pattern of up to three images. By a quirk of placement, we see only a part of the complete picture (Fig. 23.30).

So the twin quasar puzzle seems solved. We are seeing two of the three images formed by a gravitational lens, an intervening, probably elliptical, galaxy between us and the quasar. This discovery has three important implications. One, it provides another confirmation of general relativity. Two, it proves in this case that the quasar is more distant than the galaxy, and so the quasar's red shift is cosmological. Three, cool gas around the galaxy creates the quasar's absorption line spectrum; this situation may be the case for other quasars as well.

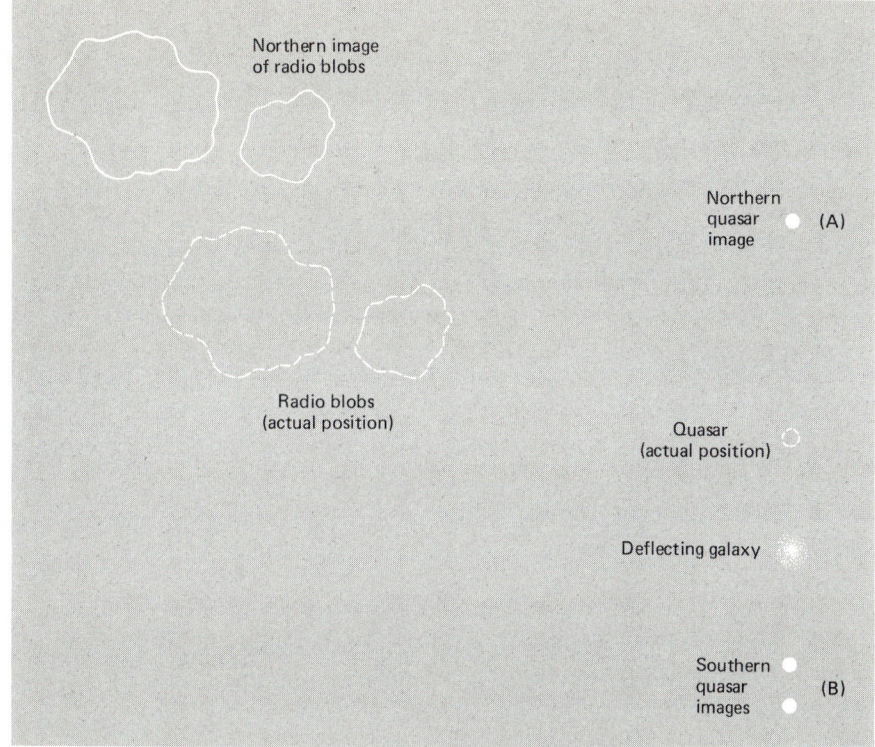

FIGURE 23.30 A schematic drawing of the optical illusion of the double quasar. The actual quasar and its radio emission (broken outlines) are invisible. An elliptical galaxy just below the actual quasar acts as the gravitational lens. It forms three images of the quasar: one above and two below the quasar's actual position. Only one image of the radio blobs is made, above and to the left of the actual position of the radio emission. The two lower images of the quasar lie almost on top of each other and so are not separately visible. (*Adapted from a diagram by F. Chaffee*, Scientific American, *November 1980; copyright by W. H. Freeman & Co., all rights reserved.*)

Models of Energy Sources in Quasars

What generates the vast energies of quasars? A model similar to that for Seyferts suggests that all or almost all of a quasar's continuous spectrum comes from synchrotron emission—high-speed electrons gyrating in a magnetic field. As these electrons emit electromagnetic radiation, they lose energy and move more slowly. So they emit radiation with lower and lower energy. This loss of speedy electrons implies that the supply of high-energy ones must be replenished at least about every year or so.

The central energy source of a quasar must yearly blast out clouds of high-energy electrons containing a total of at least 10^{43} J. The rest of the quasar acts like a transformation machine, trapping the energy of electrons and converting it into other forms.

What energy source lies in the heart of a quasar? We don't have any really well-confirmed ideas. Quasars are still a mystery after many years of study and imaginative theoretical work. We'll describe two ideas briefly to give you a flavor of the notions which have been kicked around so far. The basic concept usually involves some very massive object that can store rotational energy. (*Note:* Whatever the object is, it probably does not generate energy by thermonuclear fusion reactions, for they are very inefficient. The proton-proton reactions, for instance, unleash only 0.7 percent of the energy from the input mass.)

Perhaps a superstar—a star with a mass of 1 million to 100 million solar masses—lies in the core of a quasar. It may spin very rapidly and so store a tremendous amount of

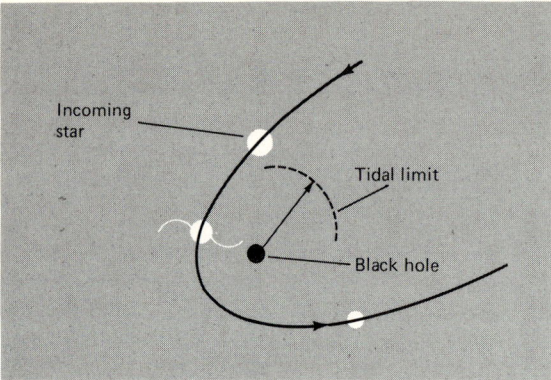

FIGURE 23.31 A supermassive black hole disrupting stars. As a star orbits within the black hole's tidal influence, material is drawn off it and eventually falls into the black hole. (*Adapted from a diagram by P. Young, G. Schields, and C. Wheeler.*)

kinetic energy. Physicist Philip Morrison calls this rotating superstar a *spinar* and compares it to a pulsar (Section 11.5), the neutron star corpse of a massive star. Even though we don't know the exact details, we do know that pulsars convert rotational energy into electromagnetic radiation at a rate of about 10^{30} W. At least some of this output is synchrotron emission.

Although the energy conversion mechanism is not yet understood, the same process may operate in quasars, but on a much grander scale. For example, a spinar of 10^9 solar masses rotating once a year could easily convert only a small fraction of its rotational energy into radiative energy and still emit 10^{40} W. The spinar would need to have a magnetic field, and electrons escaping from its surface in steady streams or bursts would accelerate through this field to produce synchrotron radiation.

The most developed quasar model to date involves supermassive black holes, objects of about 10^7 to 10^9 solar masses. This model takes off from that for binary X-ray sources (Section 12.2), in which material from a normal star forms an accretion disk around a black hole before the material falls into it. In the quasar model a supermassive black hole in a dense galactic nucleus is fueled by the tidal disruption of passing stars (Fig. 23.31). The stellar material forms an accretion disk and radiates as it spirals into the black hole, thus powering the quasar. The model calculations show that luminosities of 10^{12} solar luminosities, about that of bright quasars, are possible. To feed the black hole requires about a solar mass of material a year.

One part of this model that works is that the supermassive black hole can easily generate the level of quasar luminosity in a region of space only a few light years in size (the Schwarzschild radius of a 10^8-solar-mass black hole is only 3×10^8 km, or about 2 AU). And it does the energy conversion (from gravitational to radiative) with high efficiency. However, black hole models so far do not account well for the shape of a quasar's spectrum.

Quasars and Active Galaxies Compared

You've probably noticed that quasars and active galaxies share some observed characteristics. First, consider the radio galaxies. Most have only absorption lines in their spectra. Those that have emission lines come in two types. One type has *narrow* emission lines. Cygnus A is a good example; it has strong emission in the lines of O III at 5007 Å and N II at 6583 Å and other lines from such ionic states as S II, Ne V, and Fe X. These lines are all narrow; the Doppler widths are roughly 500 km/sec. Other narrow-lined radio galaxies have emission-line widths of 400 to 800 km/sec. Narrow-lined radio galaxies make up about $\frac{2}{3}$ of the total. The other $\frac{1}{3}$ have broad lines of hydrogen and helium, some as wide as 10^4 km/sec. Other emission lines for these galaxies tend to be narrow.

Quasars with low red shifts have optical spectra that resemble the broad-line radio galaxies in terms of the emission lines present, their widths, and also the shape of the optical continuous spectrum. For instance, the hydrogen and helium lines of such quasars have widths of 3000 to 6000 km/sec.

Quasars also resemble Seyferts in their emission-line spectra. Recall that Class 1 Seyferts have broad hydrogen lines and narrow lines of other elements, and Class 2 Seyferts have all broad emission lines. The Seyfert 1 galaxies, just like the broad-line radio galaxies, look like low-red-shift quasars in terms of their emission spectra. So the physical conditions in the regions producing the spectra must be basically the same in Seyfert 1 galaxies and in low-red-shift quasars.

In addition, Seyferts look like quasars. The nuclei of both are starlike. A few nearby quasars have been shown to have galaxylike disks surrounding them. The colors of the Seyferts with the largest nuclei resemble the colors of quasars. And, the nuclei of Seyfert 1 galaxies vary in light over periods of months, which implies, by the light travel time argument, that the emitting region cannot be larger than a few light months in size.

One other connection between Seyfert 1 galaxies and quasars has been pointed out by Weedman. He has compared the luminosities of Seyferts and quasars using the Balmer series H-beta line and the Lyman-alpha line. The Lyman-alpha luminosity is roughly 3 times that for H-beta. It can be used to infer the H-beta line intensity when the latter line is shifted too far to the red of the visible spectrum to be observed. (The 3:1 ratio comes from observational work; Weedman presumes, but does not prove, that it applies equally well to Seyfert 1 galaxies and to quasars.) When plotted against red shifts, the H-beta luminosities for Seyfert 1 galaxies and quasars form a continuous trend, that is, the most red-shifted Seyfert 1 galaxies (red shift of 0.06) have the greatest H-beta luminosities (10^{36} W), about the same as the low-red-shift quasars.

Comparing quasars to BL Lac objects, we first find a pronounced difference: BL Lac objects do *not* have strong emission lines. However, in both BL Lacs and quasars the nonthermal nature of the continuous emission stands out most clearly compared with other active galaxies. If the nonthermal emission is synchrotron, BL Lac and the quasar 3C 279 need about the same magnetic field intensity (roughly 3×10^{-3} gauss) to account for their synchrotron spectra in the radio range.

BL Lac objects and about 15 percent of radio-bright quasars show wide variations in optical output over periods of days, weeks, and months. These swings in luminosity often occur very abruptly.

One other possible connection relates to radio structure. Quasars have high luminosities and tend to exhibit symmetrical double structures. The higher luminosity active galaxies have similar radio shapes. So the physical processes responsible for both may be the same. The nuclear jets imply that this structure is somehow tied to violent activity in the nuclei of luminous active galaxies and quasars.

To sum up, active galaxies and quasars share some of the same *observed* properties and so, by inference, some of the same *physical* properties. The general aspect most striking is the nonthermal emission from a region a few light years in size.

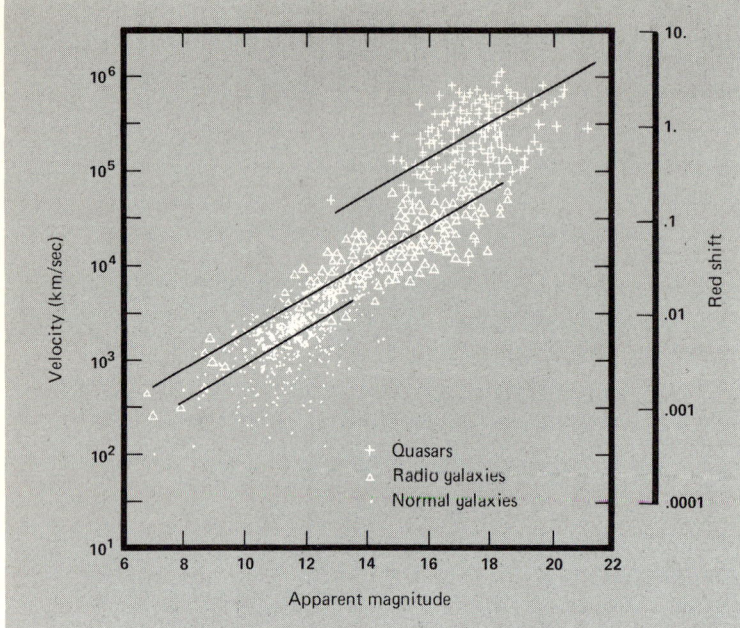

FIGURE 23.32 A composite Hubble diagram for quasars (crosses), radio galaxies (triangles) and normal galaxies (dots). Note how the three groups seem to merge together. (Courtesy K. Lang; from K. Lang, S. Lord, J. Johanson, and P. Savage, Astrophysical Journal, vol. 202, p. 583, copyright 1975 by the American Astronomical Society.)

The observations suggest, but do not prove, that quasars have some connection with active galaxies. One popular idea views them as similar objects at different stages of evolution, that is, the quasar phenomenon signals very violent activity in the nucleus of a galaxy at a very early stage in its life. Other active galaxies are older quasars, and so less active. The sequence would be roughly quasar, BL Lac, Seyfert, then radio galaxy stages, ending up with a normal galaxy. This sequence seems likely because, in general, quasars have the greatest red shifts, Seyferts less, and radio galaxies the smallest. If the red shifts are cosmological (see next section), then quasars are younger (less evolved) than Seyferts, which are younger than radio galaxies.

So quasars may be hyperactive galaxies. And most (or all) galaxies may go through evolutionary stages that resemble quasars, then active galaxies, and finally normal galaxies—which have some of the properties of active galaxies but display them less violently.

A composite Hubble diagram supports this notion (Fig. 23.32). When the red shifts and apparent magnitudes are plotted for normal galaxies, radio galaxies, and quasars, the three groups appear to join together. Quasar luminosities average 10^{38} W, radio galaxies 10^{37} W, and normal galaxies 10^{36} W. This decrease in luminosity and the merging of positions in the Hubble plot may indicate an evolutionary connection between the groups.

This evolutionary connection is a nice idea, but has one severe drawback. It implies that the energy source in quasars relates directly to that in active galaxies. We simply don't know the nature of that energy source—yet. If it is a black hole, then all (or most) galaxies need to have a supermassive black hole in their nucleus.

If we accept a black hole model, then the nucleus should show an intense, pointlike source of light (due to a concentration of stars around the black hole), stars orbiting the center should have high velocities, and emission lines with high Doppler shifts might be

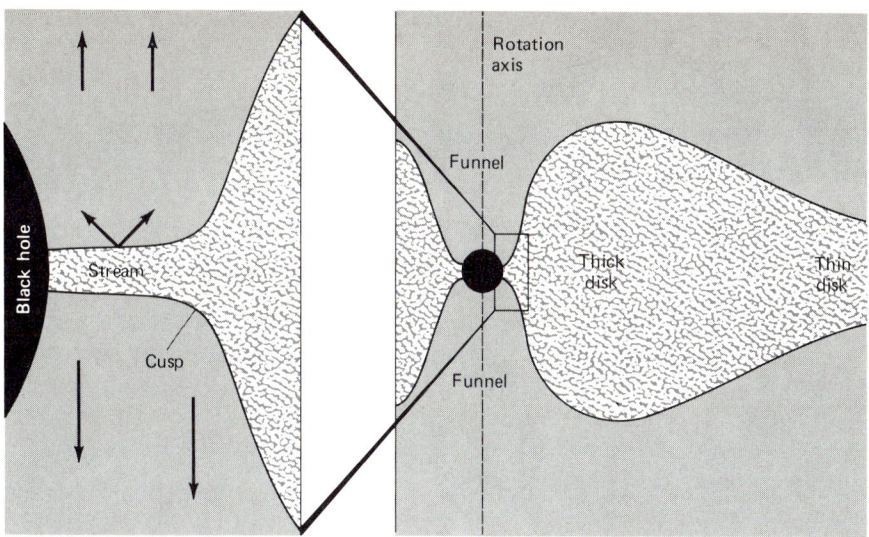

FIGURE 23.33 One possible model for a supermassive black hole powering a quasar or active galaxy. (The area within the rectangle in the drawing on the right is shown enlarged on the left.) An accretion disk surrounds the black hole; it is thick close to the hole and thins out away from it. A sharp funnel forms where the accretion disk streams into the black hole. The disk is hot, so gas blows off it. The funnel directs the gas up along the rotation axis of the black hole, where it streams out as a double jet. (*Adapted from a diagram by M. Abramowicz, M. Calvani, and L. Nobili,* Astrophysical Journal, *vol. 242, p. 772, copyright 1980 by the American Astronomical Society.*)

visible from the infalling matter. Observations of such effects have been reported by two research groups. One has found a high-luminosity point in M 87's nucleus (such a bright point is not found in the nucleus of normal elliptical galaxies), and the other has discovered dramatically high velocities for stars in the nucleus compared to those outside the nucleus. A model consistent with these observations is one of a 5×10^9-solar-mass black hole hiding in the inner 300 ly (100 pc) of the nucleus. The black hole would cluster the stars in the nucleus closely around it, and they would orbit at high velocities.

If supermassive black holes power active galaxies, what is their connection with the observed radio and optical jets that spout from some nuclei? That's not clear in detail, but here's one idea. Look at a black hole surrounded by an accretion disk (Fig. 23.33). Note that where the material actually flows into the black hole, it is pinched into a very narrow stream. Here the walls of the accretion disk rise sharply, forming a central funnel around the black hole. The funnel centers about the rotation axis of the black hole and accretion disk. The gas in the funnel is hot—so hot that material blows off of the accretion disk. The funnel acts to channel the blown-off material pretty much parallel to the rotation axis. (Material streaming at an angle will hit the sides of the funnel or fall into the black hole.) The oppositely directed streams of hot gas blow out the funnel and make the jets seen close to the nuclei of some active galaxies.

What about our Galaxy? In this model it, too, should have a nuclear supermassive black hole. Does it? Maybe. Recall (Section 20.7) that radio and infrared observations indicate rapid rotational motions near the Galaxy's core. To account for the rapid rotation requires a mass of several million solar masses all lumped in the core of a region only 0.13 ly in diameter! Possibly this mass is a supermassive black hole.

Note that this model resembles that for binary X-ray sources (Section 12.2). So you'd expect X rays to be emitted by the hot gas in the accretion disk around the black hole. Some active galaxies (such as Virgo A and Centaurus A) *do* have strong nuclear X-ray sources. And our Galaxy is now known to have an X-ray source at the position of the galactic core. Active galactic nuclei have X-ray luminosities of about 10^{33} to 10^{39} W; the Milky Way Galaxy's nuclear source emits only about 10^{28} W in X rays, so it is much weaker. But Sgr A West is also weaker than the radio emission from the nuclei of active galaxies, and the *ratio* of X-ray to radio luminosity is about the same. For the Galaxy, this ratio is about 10; for Centaurus A, for example, it is also roughly 10 (X-ray luminosity about 2×10^{34} W; radio, 3×10^{33} W). So the Galaxy's nucleus is similar to its more active counterparts, but less energetic. It may be a scaled-down version of the same process or in a relatively quiet state, which would agree with the evolutionary connection.

Be warned. Even though many astronomers consider the black hole models to be the current best buy in terms of conventional physics's understanding of quasars and active galaxies, the models are hard to test observationally. Martin Rees, a theoretician, has expressed the optimistic view: "The energy source in these galaxies will be shown to be a black hole, I think, even though it may take 100 years before we have proved it."

Alternative Views About Quasars and Their Red Shifts

Some of the problems of quasar energy production would ease if quasars were relatively close. (By close, we mean tens of millions of light years rather than billions.) For example, if 3C 273 were 100 times closer than it is, it would be emitting only 10^{10} solar luminosities (4×10^{36} W), about the luminosity of an ordinary galaxy.

But if quasars were not really far from us, we would have to change another part of our thinking, for then we could not interpret their large red shifts as cosmological. Instead we'd have to view all or part of the red shift as the result of another cause. Don't consider this an easy step to take, for it means abandoning the simplest explanation for quasars' red shifts. Perhaps our queasiness with quasars, mainly the failure to come up with a good model for the energy source, is strong enough to warrant it; at least a few astronomers think so. For 20 years, this small band has fueled the controversy about quasar red shifts.

Frankly, we believe that quasars' red shifts *are* cosmological (some further evidence is quoted below), and so do most astronomers. So we will treat the other side's views here very briefly. Their main thrust aims at discrediting at least part of quasars' red shifts as cosmological.

Quasars might acquire large, noncosmological velocities if they were small objects shot from galaxies by gigantic explosions. (Recall that we have evidence of explosions in our Galaxy and in active galaxies such as M 87.) Then, if the quasar is shot away from us, the red shift of the quasar would be the Doppler shift from its expulsion plus the Doppler shift of its parent galaxy. If the expulsion velocity is high, we could see a very large red shift from the combination of the two. If the quasar is shot toward us, its radial velocity is less than that of its parent galaxy; if it moves fast enough, we would see a blue shift.

Put aside for a moment the problem of what physical mechanism could shoot quasars from galaxies. From this model you would expect to find galaxies and quasars close together in the sky more often than by chance. But remenber, when you look at the sky you see both near and distant objects. Just because you see a galaxy and quasar close together does not mean that they are the same distance from us and physically associated.

23.3 QUASARS: MYSTERIOUS ENERGY EMITTERS

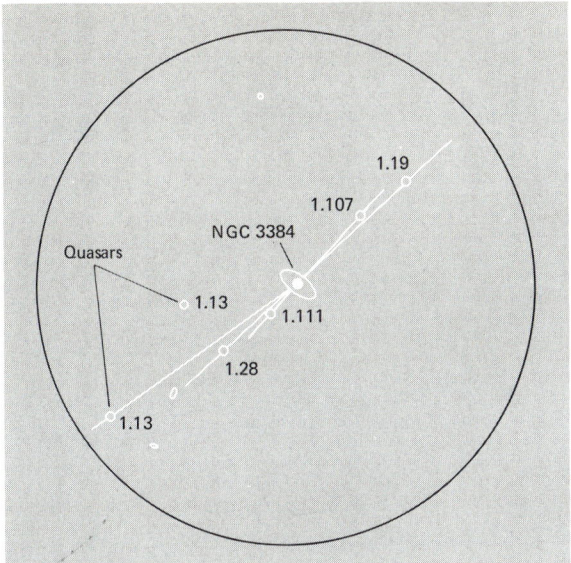

FIGURE 23.34 Orientation of the galaxy NGC 3384 and nearby quasars. The numbers indicate the red shifts. (*Adapted from a diagram by H. Arp, J. Sulentic, and G. di Tullio,* Astrophysical Journal, *vol. 229, p. 489, copyright 1979 by the American Astronomical Society.*)

FIGURE 23.35 A high red shift object (arrow) in front of the galaxy NGC 1199. (*Courtesy H. Arp.*)

The proponents of alternative models of quasars' red shifts believe that observational evidence supports the association of some quasars with galaxies. For example, Halton Arp and colleagues have investigated quasars near the S0 galaxy NGC 3384. They find eight quasars within 30′ of the galaxy; six of these have similar red shifts (from 1.11 to 1.28), much different from the galaxy's red shift of 0.003. Of these, five are on straight lines drawn through the nucleus of NGC 3384 (Fig. 23.34). Arp and colleagues argue that this spatial association is not chance, but indicates a physical association of the galaxy and the quasars, as if the quasars had been ejected from the galaxy. (Recall how radio galaxies tend to eject material symmetrically.)

In another peculiar case Arp has photographed a region in which he contends a quasar-like object (red shift of 0.044) lies in front of an elliptical galaxy, NGC 1199, with a red shift of only 0.009 (Fig. 23.35). How does Arp know the object with the higher red shift lies in front? He believes that he sees a dark ring around it and argues that the compact object is absorbing light from the galaxy. If the object was ejected toward us, it could be in front of the other galaxy, but then its velocity would be less than that of the galaxy, not more. Arp concludes that the compact object was ejected by the galaxy, and its red shift is *not* Doppler, but due to some other cause. (An alternative, which Arp rejects as unlikely, is that the compact object was ejected from *our* Galaxy, at a velocity of 13,300 km/sec, and just by chance lies in the direction of NGC 1199.)

Note that these alignment arguments don't conclusively *prove* physical association. There are about 300 bright galaxies and 60 bright quasars in the sky. Suppose you scat-

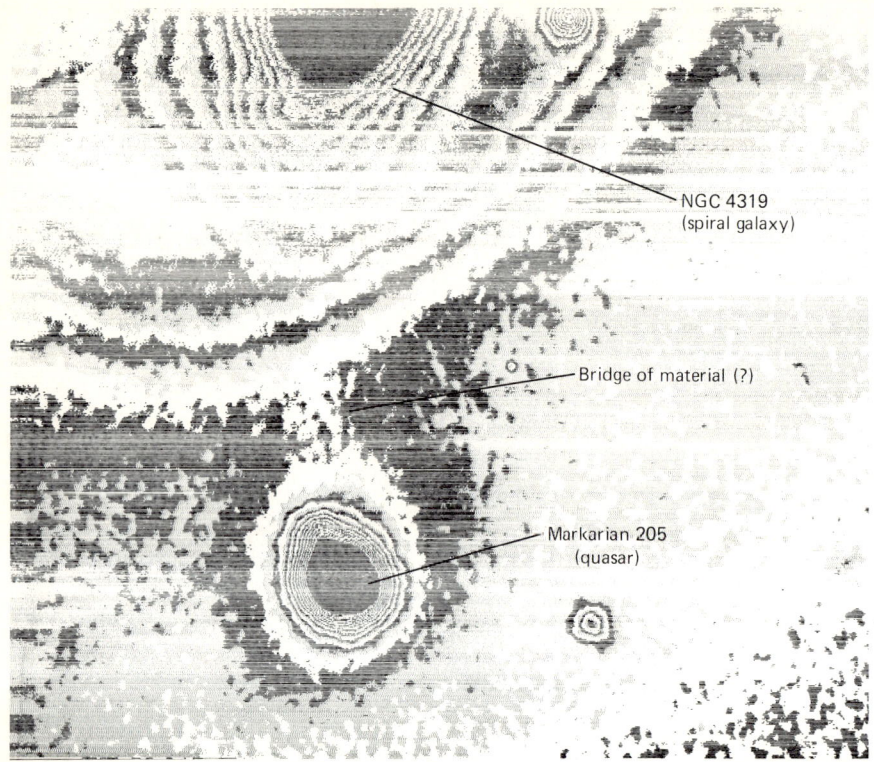

FIGURE 23.36 Possible evidence for a physical link between a galaxy and a quasar. This figure is a computer-generated contour map of a photo. The object at the top is the spiral galaxy NGC 4319. Below it appears the quasar Markarian 205. There seems to be a bridge of material linking the two images, but the supposed bridge may be a bending of the images as a result of the photographic process. (*Courtesy R. Lynds and Kitt Peak National Observatory.*)

tered 300 corn kernels and 60 tomato seeds over a huge map of the sky, and then asked yourself what were the chances of a corn kernel and a tomato seed being as close to or closer than a certain distance. You could calculate the chances of these occurrences, and if they were actually more frequent, you could argue for a physical connection that makes the associations more frequent than expected. But it is still possible for the associations to be simply chance.

A more compelling case would be made if evidence were found of an actual physical connection between a quasar and a galaxy, such as bridges of gas and dust. Arp thinks he has one strong case: the galaxy NGC 4319 and the quasar Markarian 205 (Fig. 23.36). The galaxy has a red shift of 0.006, corresponding to radial velocity of 1800 km/sec, and the quasar a red shift of 0.07, equivalent to a velocity of 21,000 km/sec. Arp contends that a bridge of luminous material, visible in the photograph, connects the galaxy and quasar—certainly a strong argument if true!

A recent reexamination of this notorious pair seems to refute Arp's arguments. First, the "bridge" may be an artifact of the merged images on the photo. Second, an object just 3 arcsec away from Markarian 205 turns out to be a galaxy with the same red shift. So it may be the companion to Markarian 205 and the object causing its distorted look. Finally, computer processing of Markarian 205's image reveals a fuzzy envelope around it. The spectrum of this envelope is that of a galaxy with the same red shift. So the quasar appears to be an extremely bright nucleus of a galaxy!

One strong argument against quasars as objects shot from galaxies is this: At least a few

23.3 QUASARS: MYSTERIOUS ENERGY EMITTERS

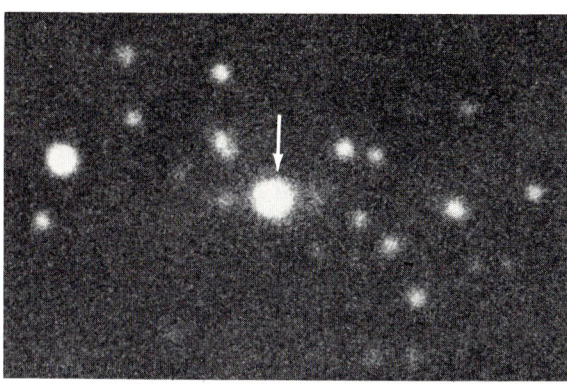

FIGURE 23.37 The quasar 3C 206 (arrow) surrounded by galaxies, which probably all lie in the same cluster. (*Courtesy H. Spinrad; observations by S. Wyckoff, P. Wehinger, H. Spinrad, and A. Boksenberg.*)

quasars should have blue shifts, since some should be aimed at us. But no blue-shifted quasars have been found to date.

Support for cosmological red shifts comes from a study by Alan Stockton. He chose an unbiased sample of 27 quasars; that is, he did not intentionally choose quasars known to appear close to galaxies. In this sample he finds 29 galaxies within 45″ of the quasars; of these, 13 have red shifts within 1000 km/sec of the quasar in the vicinity. (Why an allowable spread of 1000 km/sec? Because in a cluster of galaxies, 1000 km/sec is the typical spread in velocities. So two galaxies or a galaxy-quasar pair could belong to the same group, and so be the same distance from us, but have a 1000 km/sec difference in radial velocities.) The chances of eight such red shift agreements is less than one in a million for random associations, based on the red shift distribution of galaxies. The fact that 13 were found makes the cosmological nature of these red shifts virtually certain, and the study supports the concept that most quasars' red shifts do not arise in other ways.

In a few cases quasars appear in the same direction as clusters of galaxies. It's hard to prove that they actually are members of the cluster, rather than just foreground or background objects, because it is difficult to obtain spectra of the cluster galaxies, which are much fainter than the quasar. One good example is the quasar 3C 206, recently observed by Peter Wehinger and colleagues. It seems to be surrounded by about 200 faint galaxies (Fig. 23.37). The red shift of the quasar, from its emission lines, is 0.206. A nearby pair of galaxies has a red shift of 0.203. These observations support the idea of cosmological red shifts for the quasar; it is located in a cluster of galaxies with the same red shifts, and theirs are cosmological.

The red shift controversy still rages. It will probably take a tremendous amount of strong evidence to shake the simple cosmological view, which most astronomers find so compelling.

Are Quasars the Nuclei of Galaxies?

If the evolutionary connection among quasars, active galaxies, and normal galaxies is correct, then a quasar is the hyperactive nucleus of a galaxy. What about the rest of the quasar's galaxy? At the great distances implied by the large red shifts of quasars, the disk of a quasar's parent galaxy will be too faint and too small to see easily. For example, the Andromeda Galaxy spans an enormous 2.5° in the sky. But if it were placed at a red shift of 0.2 (small for a quasar!), meaning at a distance of 4 billion ly, its angular diameter would be a mere 4 arcsec. Keep in mind that the seeing of the earth's atmosphere limits a telescope's resolution to 2 to 3 arcsec at best. So a galaxy at a red shift of a few tenths would produce an image hard to distinguish from that of a faint star. With these difficulties in mind, let's look at some observations that support the idea that quasars are the luminous nuclei of galaxies.

First, consider again the quasar Markarian 205. We mentioned that it's surrounded by a faint fuzz that has the spectrum of a galaxy (rather than a nebula) at the same red shift as the quasar. This fuzz is probably the galaxy of which Markarian 205 is the nucleus.

Second, consider 3C 206, also mentioned above. A careful examination shows that the image is distorted and surrounded by a fuzzy envelope. It has the color of an elliptical galaxy and absorption lines with a red shift of 0.203. Within observational errors, that's the same as the quasar's red shift. Here again is the galaxy whose nucleus is the quasar.

Finally, consider 3C 273, the closest of the high-luminosity quasars. On photos processed to bring out faint features, 3C 273 has a fuzzy and asymmetrical appearance because of a surrounding envelope. Its spectrum shows emission lines with the same red shift as that of the quasar.

We have chosen the best examples to date backing up the quasar-galaxy connection. The case is not yet complete, but we are beginning to find that galaxies faintly perceived may be the key to unraveling the quasar tangle.

SUMMARY

This chapter has investigated the enormous range of cosmic violence in the spectacular outbursts from quasars and the nuclei of galaxies. These cosmic explosions are summarized in Table 23.3.

You have seen that the nuclei of galaxies are active. Such nuclei and quasars eject matter in the form of massive clouds, high-energy particles, and perhaps discrete objects (such as quasars themselves). Both active nuclei and quasars appear to have extremely small regions (less than a few light years in diameter) where the violent processes originate. In addition, the range of objects examined—from quasars to Seyferts to ordinary galaxies—may represent a sequence through which all galaxies evolve.

A key point to remember is that although we have many observations of quasars and

TABLE 23.3 The Range of Cosmic Violence

	Energy Output Rate (W)	Form	Estimated Lifetime (yr)	Total Energy Output (J)	Equivalent Mass Energy Loss (kg)
Nucleus of Milky Way Galaxy	10^{34}	Infrared	10^{8}	10^{51}	10^{34}
	10^{33}	Gamma ray			
Active Galaxy	10^{38}	Optical	10^{8}	10^{54}	10^{37}
	10^{39}	Infrared			
Quasar	10^{39}	Optical	10^{6}	10^{56}	10^{39}
	10^{42}	Infrared			
	10^{40}	Radio			

active galaxies, we still have *no* clear idea of their energy source—the demon in their cores.

This investigation has also revealed the pervasiveness of nonthermal sources of radiation. This type of emission requires the input of high-energy electrons in a small volume of space containing a magnetic field. The universe has two aspects: a slowly evolving one detectable by its thermal emission and an explosive one marked by the rapid release of a wide range of radiation.

Key Words

supermassive black hole
active galaxy
compact radio galaxy
extended radio galaxy
radio jet

head-tail radio galaxy
Seyfert galaxy
BL Lacertae (BL Lac) object
quasar

relativistic Doppler shift
superluminal motion
relativistic jet
gravitational lens
spinar

Review Questions

1. What observational evidence do we have that the synchrotron process produces some radiation from active galaxies and quasars? (Objective 1)
2. Contrast the evidence for violence in the Milky Way Galaxy with that for any active galaxy. (Objectives 1 and 7)
3. What evidence do we have that quasars are far away? (Objectives 2 and 4)
4. What observational evidence indicates that, for many quasars, their light must pass through clouds of thin, cool gases? (Objective 3)
5. What powers a quasar? (Objectives 3 and 5)
6. Evaluate the strengths and weaknesses of the model of quasars as objects shot from galaxies. (Objectives 4 and 5)

Problems

1. Calculate the Schwarzschild radius of a 10^8-solar-mass black hole. (Objective 5)
2. Quasar OQ 172 has a red shift of 3.53. What is its radial velocity? (Objective 6)
3. The double quasar has an emission red shift of 1.41. Calculate its radial velocity and its distance. (Objectives 6 and 8)
4. 3C 273 shows a blob expanding along its jet at about 0.8 milliarcsec/yr. Calculate its linear expansion rate, assuming that $H = 50$ km/sec/Mpc. Do the same for $H = 100$ km/sec/Mpc. (Objective 11)
5. From the information in problem 4, estimate the angle of the jet relative to our line of sight. Hint: Assume blobs expelled along the jet move at about c. (Objective 11)
6. The quasar PHL 938 has emission lines with a red shift 1.955 and one set of absorption lines at a red shift of 0.613. Assume the absorption lines come from an intervening cloud. Calculate the distance from us to that cloud compared to the distance to PHL 938. (Objectives 6 and 8)

BEYOND THIS BOOK . . .

Black Holes, Quasars, and the Universe, 2nd ed. (Houghton-Mifflin, Boston, 1980) by H. Shipman has an excellent presentation of the controversy over quasars.

* *Mercury*, November-December 1974, contains "The Quasar Controversy—An Interview with Caltech Astronomer Halton Arp." Also see *The Redshift Controversy* (Benjamin, Reading, MA, 1973), edited by G. Field, H. Arp, and J. Bahcall.
* "The Evolution of Quasars" by M. Schmidt and F. Bello in *Scientific American*, May 1971, argues for fast evolution of quasars in the early universe.
* Somewhat technical, but mostly descriptive articles in *Annual Reviews of Astronomy and Astrophysics* are: "The BL Lacertae Objects" by W. Stein, S. O'Dell, and P. Strittmatter, vol. 14 (1976), p. 173; "The Line Spectra of Quasi-Stellar Objects" by P. Strittmatter and R. Williams, vol. 14 (1976), p. 307; "Extended Extragalactic Radio Sources" by D. DeYoung, vol. 14 (1976), p. 447; and "Seyfert Galaxies" by D. Weedman, vol. 15 (1977), p. 69.
* R. Weymann discusses "Seyfert Galaxies" in *Scientific American*, January 1969.
* For a detailed look at the double quasar, read "The Discovery of a Gravitational Lens" by F. Chaffee in *Scientific American*, November 1980. Also useful are "Radio Astronomy by Very-Long-Baseline Interferometry" by A. C. S. Readhead, *Scientific American*, June 1982, and "Cosmic Jets" by R. D. Blandford, M. C. Begelman, and M. J. Rees, *Scientific American*, May 1982.

THE ORIGIN AND EVOLUTION OF THE UNIVERSE

24

LEARNING OBJECTIVES
After studying this chapter, you should be able to:

1. State the basic assumptions of cosmology.
2. Present, in a short paragraph, basic observations that have cosmological import.
3. Compare the Big Bang and Steady State models, especially the assumptions made for each and the ability of each to explain fundamental cosmological observations.
4. Describe briefly the observed properties of the cosmic background radiation.
5. Present at least one argument for ascribing a cosmic origin to the background radiation and explain how it is a natural consequence of a Big Bang model.
6. Discuss the importance and the impact of cosmic radiation for both Big Bang and Steady State models.
7. Outline the process of element formation in the standard Big Bang model, and cite at least one observation that supports the theoretical ideas.
8. Make use of energy arguments to explain particle production from photons in the young, hot universe.
9. Outline a history of matter and radiation from the time they both stopped interacting strongly to the present.
10. Explain the concept of Jeans length, and calculate it for a given temperature and density.
11. Describe a simple model of galaxy formation and pinpoint problems with this model.
12. Describe the importance of a possible mass for the neutrino for cosmology and galaxy formation.

CENTRAL QUESTION:
How have the physical properties of the universe changed since its origin in the Big Bang?

Some 15 billion years ago the cosmic bomb exploded. Perhaps you have seen movies of an H-bomb blast. Split seconds after detonation, an awesome fireball rips violently through the atmosphere. Our universe was born out of a similar fireball, but this was a cosmic fireball, the Big Bang, in whose violence all we now see was created; and it happened not just at one point, but everywhere.

That, in a nutshell, is a picture of our universe's creation accepted by many astronomers today. But before we discuss its history, let's ask how *the universe* can be defined.

If an astronomer considers all that can be seen with various telescopes, he or she is considering the *observable universe*. Yet this cannot be *all* of the universe; there are objects too faint and too far to be seen, and regions of the spectrum to which we and our instruments are so far blind. Other objects may be detected by their gravitational effects, such as a dark companion in a double-star system. So there is more than the observable universe: a *physical universe* that includes directly observable matter and those objects we detect by effects described by the laws of physics. The physical universe contains all matter (and energy) accessible to us, in principle, if not in practice, in any way. The reality of the physical universe rests on the assumption that local physical laws apply to the rest of the universe—a grand, but necessary, assumption.

Cosmology, the subject of this chapter, is the study of the nature and evolution of the physical universe. You can't talk about the evolution of the universe by simply describing what happens to each part; you must consider the universe as a unique whole. That's one of the problems of cosmology: We can tell a lot about stars simply because there are so many stars around, but we have only one cosmos to look at!

Cosmologists have been fed a meager diet of observational facts about the universe. Despite (or because of!) this lack they have been able to dream up many models of the universe. Some of the models have been quite bizarre, but only two have gathered a substantial following: the Steady State model and the Big Bang model. We pay little attention to the Steady State model in this chapter. We, and many other astronomers, believe that the present evidence indicates that the universe began in a Big Bang.

24.1 FUNDAMENTAL ASSUMPTIONS OF COSMOLOGY

One fact about the cosmos can be found in the simple observation that the sky is dark at night. This observation requires that the universe *evolve*, that it be dynamic, not static. Here's how this argument goes.

In 1826 the German astronomer Heinrich Olbers (1758–1840) asked why the sky is dark at night. Recall that Newton, in a dramatic split with ancient traditions, conceived that the universe was infinite in extent. Olbers took up Newton's infinite universe, filled it with stars, and concluded that the night sky should be as bright as the sun. But it is not. This observation lies at the heart of *Olbers' paradox*.

Olbers tackled a seemingly simple question: What is the total amount of starlight striking the earth? He assumed that (1) the universe is infinite, (2) it is uniformly populated by stars similar to the sun, (3) no interstellar material blocks out any of the stellar light, (4) the light diminishes as the inverse square of the distance, and (5) the universe (and the stars in it) does not evolve. Because this static universe is filled with stars of the same luminosity, an observer sees differences in the brightness because stars are at different distances. Observed brightness depends on the inverse square of the distance. But as you look at fainter stars, you see deeper into space, and the number of stars increases directly as the square of the distance (Fig. 24.1). The diminishing intensity (inverse square law) is just balanced by the increase in number of light sources (direct square law). Net result: The total light striking the earth is equivalent to that from one average star placed so close that it completely fills the sky.

Let's put this argument into a mathematical form. Suppose the universe is populated with stars of luminosity L distributed with a uniform density n (number per cubic light

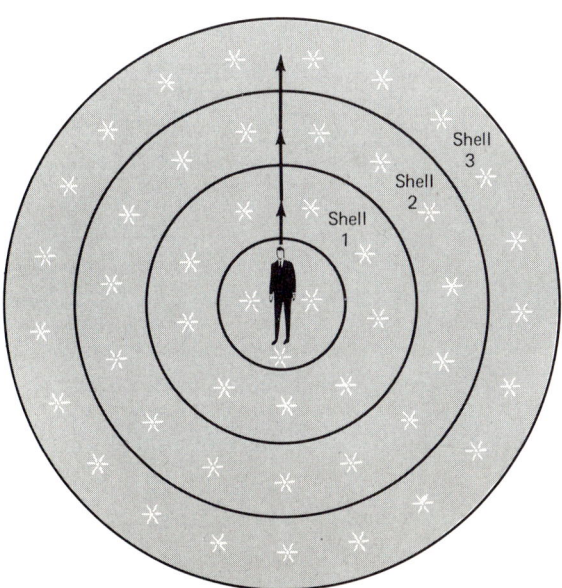

FIGURE 24.1 Geometry for Olbers' paradox. The stars are spread out uniformly, so each shell contains more stars at an increased rate of the square of the distance of the shell from the earth.

year). Divide the universe up into a series of concentric shells of thickness Δr. The flux of energy (ΔF) from the stars contained in the shell at distance r is the number of stars in the shell ($n 4\pi r^2 \Delta r$) times their luminosity (L) divided by the area over which the radiation is spread by the time it reaches the earth ($4\pi r^2$). So

$$\Delta F = n 4\pi r^2 \Delta r \frac{L}{4\pi r^2}$$

Note that the factors of $4\pi r^2$ cancel out. Now add up the flux from all the shells:

$$F = \sum \Delta F = \sum n \Delta r L = n L \sum \Delta r$$

(The symbol Σ means "summation of.") Since the factors n and L are constants (by assumption), we can take them out of the summation, and $\Sigma \Delta r$ is just the total distance to the last shell, which is infinity, since the shells are assumed to go on forever. So by this argument, the flux should be infinite. But that's not quite right, for we have treated the stars as if they were points, and have left out the fact that light from a single star directly behind another star can't reach us. Taking this blockage into account, the result is as stated above: the sky at any point should be the same brightness as the surface of a star.

The actual night sky, however, is not a luminous blend of stars but a dark expanse dusted by sparks of light. This gross discrepancy between model and observation suggests that one or more of the initial assumptions are false.

For aesthetic reasons most astronomers are reluctant to give up the assumption that the universe is uniform in its overall properties. Neither is there any good reason to doubt the inverse square law. The solution presented in Olbers' day was that dust blocks the distant starlight. But that doesn't work, for any energy absorbed by the dust would heat the dust, causing it to glow, returning the energy to space again. We are left with the last assumption. Olbers required that the universe be static. The light from distant stars can be reduced if the universe expands and the stars recede from us.

Consider that each star emits the same number of photons of the same energy each second. If the stars are receding, the number of photons reaching the earth each second is reduced, compared with the static case. Because the Doppler shift increases the wavelength, the energy of each photon is also reduced. So for both reasons the total energy arriving from the stars is diminished. If the expansion is uniform (so that the more distant stars have the greatest velocities), then the light from the farthest stars is diminished the most. (Today we would substitute "galaxies" for "stars" in explaining Olbers' paradox.) One possible escape from Olbers' dilemma is an expanding, rather than a static cosmos.

The idea that the universe evolves, that it is dynamic, appeared rather recently on the astronomical scene—essentially with Einstein (Chapter 5). To discuss this physical evolution in detail requires a few fundamental assumptions.

First, we assume the *universality of physical laws*. This assumption covers both local (the earth and solar system) and distant regions. It means that the physical laws we uncover here apply to all localities at all times and to the universe as a whole. A few key observations support this assumption. For example, the spectra of distant galaxies contain the same atomic spectral lines as those produced by elements found on the earth. So other galaxies are made of the same elements as here, put together in the same way. When

24.1 FUNDAMENTAL ASSUMPTIONS OF COSMOLOGY

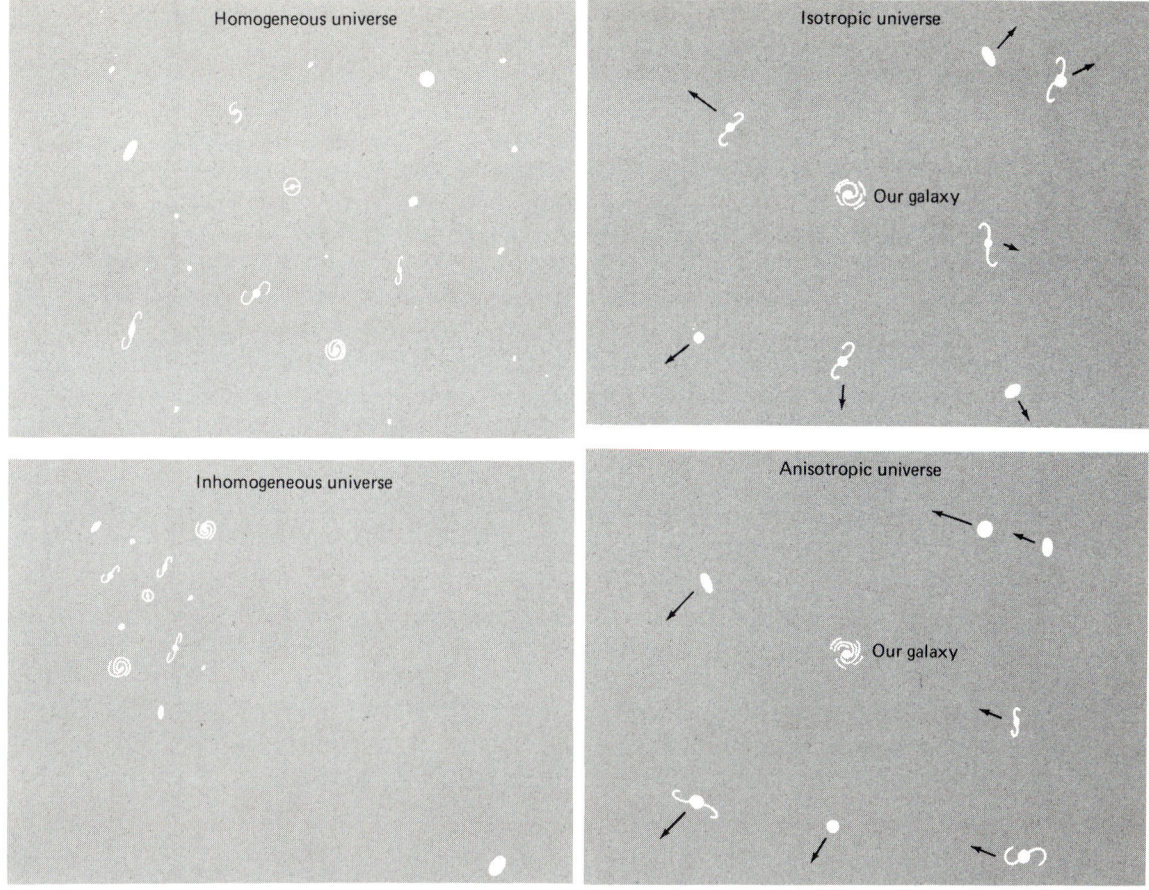

FIGURE 24.2 The homogeneous universe. Matter and radiation are spread out uniformly, if you look over large distances.

FIGURE 24.3 The isotropic universe. Space has the same properties in all directions. For example, the distribution of velocities of galaxies in the expanding universe is smooth, uniform, and the same in all directions.

applied to double stars, Newton's law of gravitation correctly describes their motion. And Kepler's third law applies to the rotational motion of galaxies.

Second, the universe is assumed to be *homogeneous* (Fig. 24.2). This means that matter and radiation are spread out uniformly, with no large gaps or bunches. This assumption is not strictly true, for clumps of matter, such as galaxies and stars, do exist. But the cosmologist assumes that the size of the clumping is much smaller than the size of the universe. It's like looking at the earth from space: Bumps, such as mountains, are too small to be seen, so the globe looks extremely smooth.

Third, the universe is assumed to be *isotropic* (Fig. 24.3). This idea is a bit more abstract; it relates to a quality of space itself, rather than to the matter in it. Here's one way to think of it. Space has the same properties in all directions. No direction or place in space can be distinguished from any other by any experiment or observation. The universe has no center in space, because there is no way to tell if you are there. No direction in space provides special rewards when taken. For example, your mass does not increase as you travel in one direction or decrease when you go another way. Another example is that the expansion of the universe is the same in all directions of space, and any observer sees the same Hubble law.

These assumptions can be summed up in one statement: The universe is *uniform*. All irregularities are ironed out. As the cosmologist Edward R. Harrison has quipped, the result is like that of the vanishing Cheshire cat in *Alice in Wonderland*: Everything is wiped out except the grin. Cosmological models that rest on the above assumptions ignore the structure and substance of planets, stars, and galaxies. Galaxies are considered, but only as tiny particles marking points in space, like a gas of atoms filling the universe. (Some would say the clusters of galaxies are the cosmic particles.) This gas is the cosmic grin. Such a mental simplification has real dangers, for like the Cheshire cat, what lies behind the grin may be crucial. The universe may not obey the laws we lay down for it or the models we develop about it. Observations must in the end validate our assumptions.

24.2 A BRIEF REVIEW OF COSMOLOGY

Chapter 5 outlined the rise of relativity and its impact on cosmological ideas. It also presented some fundamental cosmological observations. These observations did not fall into a grand scheme until explained by Einstein's general theory of relativity.

What *do* we know about the universe? First, the universe evolves; it is not static. Both the whole universe and its contents change with time.

Second, matter in the universe is *grouped*. Elementary particles (whatever they are) make up protons and electrons, which make up atoms. Atoms make up gases (molecular and atomic) and dust particles, which form stars, planets, and us. Stars come in clusters of stars, which are found in galaxies. And galaxies are grouped in clusters of galaxies, which in turn may congregate in superclusters of galaxies. And these, as far as we can tell, make up the universe itself.

Third, the universe is *expanding*. The observed rate of expansion now lies in the range 50 to 100 km/sec/Mpc. Einstein's theory of general relativity relates changes in the rate of expansion to the average density of matter (and energy) in the universe. This average density, in turn, determines the overall geometry of spacetime. Three possibilities exist for this geometry: (1) flat, an open universe with the geometry of Euclid, infinite in space and time; (2) spherical and closed, the cosmos being finite but unbounded in space and finite in time; (3) hyperbolic, again an open universe, infinite in space and time, but curved. Observations to date imply that the universe has an open geometry; but the question is still unsettled. (Einstein, as do many contemporary astronomers, had an aesthetic preference for a closed universe.)

Finally, we can estimate the ages of celestial objects. From the Hubble constant we find that the universe has been expanding for less than 20 billion years. Radioactive dating places the earth and the moon at an age of about 4.6 billion years. Models of stars evolving in globular clusters indicate the Galaxy's age is some 15 to 17 billion years. Many star clusters are younger. All these ages are less than the presumed age of the universe, and they fall in a natural evolutionary sequence, from the Galaxy to the earth and sun to the youngest star clusters. Here's a hint that the universe was born at a finite time in the past and evolved by known physical laws.

24.3 CONTEMPORARY COSMOLOGICAL MODELS

For almost 100 years ingenious theoreticians have devised a bewildering array of cosmological models. Almost all fall into one of two categories: (1) the *Big Bang model*, which is the standard model based on Einstein's general theory of relativity; and (2) the *Steady State model*, which requires some changes in standard relativistic ideas. (We consider the oscillating model, in which the universe expands, then contracts, only to expand again, as a Big Bang model. Think of it as a "bang-bang-bang . . ." model.)

The observational bases for the Big Bang and Steady State model are the same, but the interpretations of some observations have crucial differences that arise from the different philosophical and aesthetic grounds of each model. This section describes the two basic models of modern cosmological thought.

Steady State Model

Cosmological models rest on the assumptions that the universe is isotropic and homogeneous and described by the same physical laws everywhere. These assumptions taken together are sometimes called the *cosmological principle*; the universe appears uniform to all observers at every location. Now add another restriction: the universe appears the same not only at *all locations*, but also at *all times*. This statement is often called the *perfect cosmological principle*. It is the aesthetic basis for Steady State models of the universe. You can see how it gets its name. No matter where and when you look at the universe, its state is the same (Fig. 24.4). However, this principle does *not* mean that the universe does not change at all. Galaxies condense, stars ignite and flare out, but the cosmos retains the same general features, for example, the same numbers of galaxies of different types. The universe does evolve in the Steady State model, but in such a way that, overall, it has a similar appearance in all places and at all times.

Let's look at how strongly this philosophical premise affects the Steady State model. Because the universe is expanding, you might think that gaps of matter must appear as the galaxies spread apart (Fig. 24.5). So you could date different times in the universe's history by watching the average density of matter decrease. The Steady State model gets around this dilemma by allowing the continual creation of new matter to fill the gaps resulting from expansion. The new matter forms into new galaxies and new stars, so everything looks the same. Not much matter needs to be created to keep up with the expansion: only a single hydrogen atom per liter every billion years. This did not contradict any observation or experiment known at the time the model was proposed. As originally envisioned by Herman Bondi, Thomas Gold, and Fred Hoyle, the created matter springs forth everywhere, newborn from nothing. The continuous creation of matter marks a primary difference between the Steady State and the Big Bang model. If the universe is expanding, matter must be created to save the perfect cosmological principle. In other words, the adherence to a philosophical, aesthetic ideal demands an unusual physical consequence—the creation of matter from nothing.

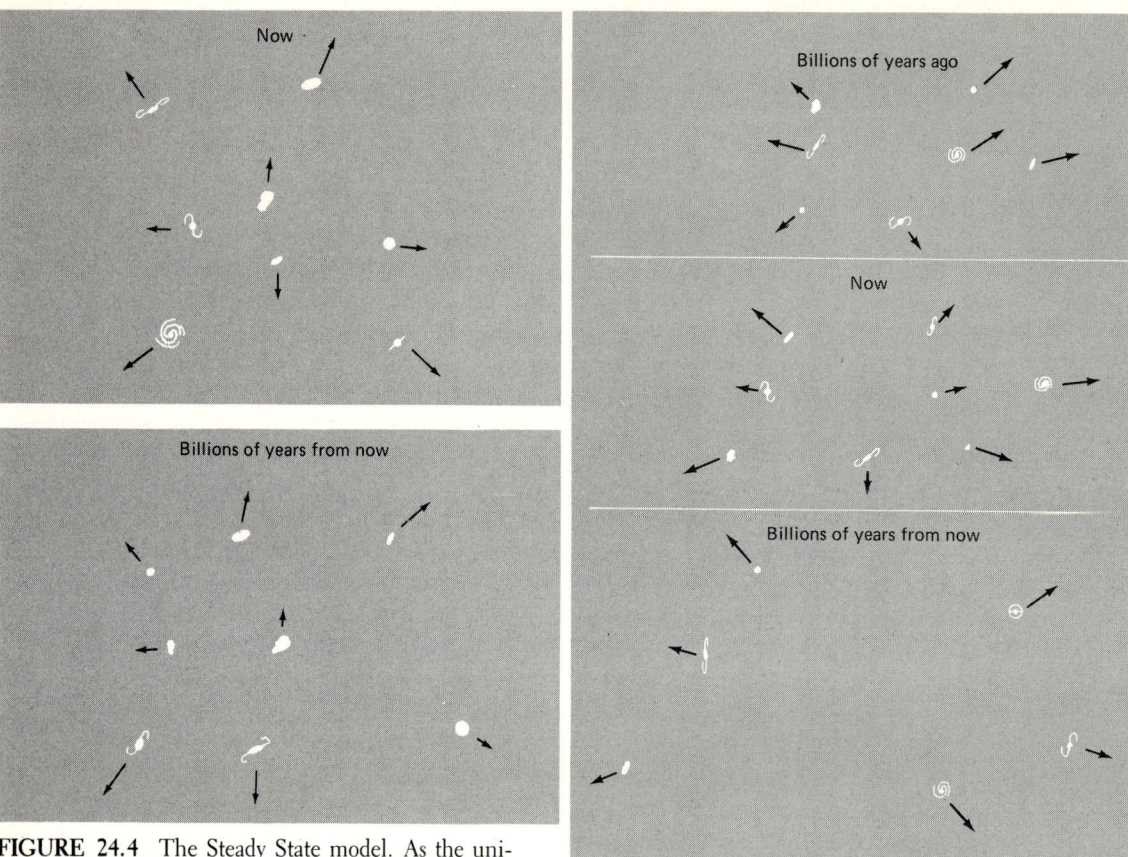

FIGURE 24.4 The Steady State model. As the universe expands, newly created matter forms into galaxies, so that the average density of matter remains the same.

FIGURE 24.5 The Big Bang model. As the universe expands, the average density of matter decreases.

Observational Problems with the Steady State Model

The Steady State model presents a pleasant picture of a fulfilled, eternal cosmos. Nasty questions, such as how the universe began, do not arise. But how well does this model stand up to observations?

First, consider the age of matter in the universe. Three possibilities exist: (1) It is infinitely old, (2) it has a finite age because it was created in a single event in the past, (3) different particles have different ages because they were created at different times. Which is correct? It is impossible to tell the age of any one particle, such as a proton. So we must look at matter in bulk. As gravity squeezes matter together, hydrogen transforms to heavier elements in a one-way process. If all matter were infinitely old, most of it would no longer be hydrogen. But observations show that the matter in the universe is mostly hydrogen. So the choice becomes between incremental creation (Steady State) and instantaneous creation (Big Bang).

The trouble (or virtue!) with instantaneous creation is that you can date it. Before 1952 the Big Bang model faced the following dilemma: The age of the universe estimated from the Hubble constant was only 2 billion years (H was thought to be about 500 km/sec/Mpc).

But the age of the earth was estimated at 4 billion years, that of the sun at 10 billion years, and the age of the oldest stars near 20 billion years. The universe appeared younger than the objects it contained! Steady State supporters took this discrepancy as evidence for their model. Since 1952 the Hubble constant has been redetermined to be smaller, and the age of the universe estimated from it has increased enough to include the oldest Population II stars, now believed to be 15 to 17 billion years old. So a major thorn has been taken out of the side of the Big Bang model, and the Steady State theorists have lost one of their main arguments.

The Steady State model also hits a snag because it predicts that as the galaxies spread apart, new galaxies are born. So young galaxies must be scattered among the old, and this age pattern for galaxies must be observable everywhere. But galaxies, those which are close enough to observe in detail, appear to have roughly the same age; they all have old stars, indicating that they formed at about the same time.

Second, the shape of the Hubble plot calculated for the Steady State model dips far below the Hubble plot for a flat universe, that is, the Steady State model requires an open universe. Though the evidence is that the universe presently is open, the observed Hubble plot for galaxies does not match that predicted by the Steady State theory.

Third, in the Steady State model everything, on a large scale, must remain the same, even the Hubble constant, $H = v/d$. The universe is expanding, however, so d for any galaxy must increase. To keep H the same, v must also increase. The Steady State model evokes a special force, not detectable in any other way, to accelerate galaxies to higher velocities as they move apart. Many astronomers find this objectionable, and it is another argument against the Steady State model.

Fourth, the distribution of quasars in space and time (Section 24.7) indicates that more quasars existed in the past than do now. This is contrary to the Steady State theory; it requires that the number of quasars at all times be the same.

Fifth, the original Steady State model predicted creation of both particles and antiparticles, such as electrons and positrons or protons and antiprotons. Occasionally, a particle should meet its antiparticle, annihilate, and produce gamma-ray photons (see Section 24.4). So there should be a certain amount of gamma rays coming from all directions in space, but this is not, in fact, observed.

The final blow, however, came from an accidental observation made in 1965. It dropped a bomb in the minds of cosmologists: the discovery of the cosmic background radiation, the still-resounding echo of the Big Bang.

The Big Bang Model Revisited

The universe is now expanding. Imagine it running backward. The galaxies and all matter within and without them would eventually come tightly together. The extreme compression would heat both matter and radiation to a very high temperature—high enough to break down all structure that was created previously. Atoms would break down into nuclei and electrons. Eventually nuclei would be broken apart into protons, neutrons, and electrons. In addition, the density of matter would be so great that photons could travel only a short distance before they were absorbed. As a result, the entire universe would be opaque to its own radiation.

Now, when matter is opaque to all radiation, it acts like a blackbody. The radiation from a blackbody exhibits a characteristic shape when you study its spectrum (Fig. 24.6).

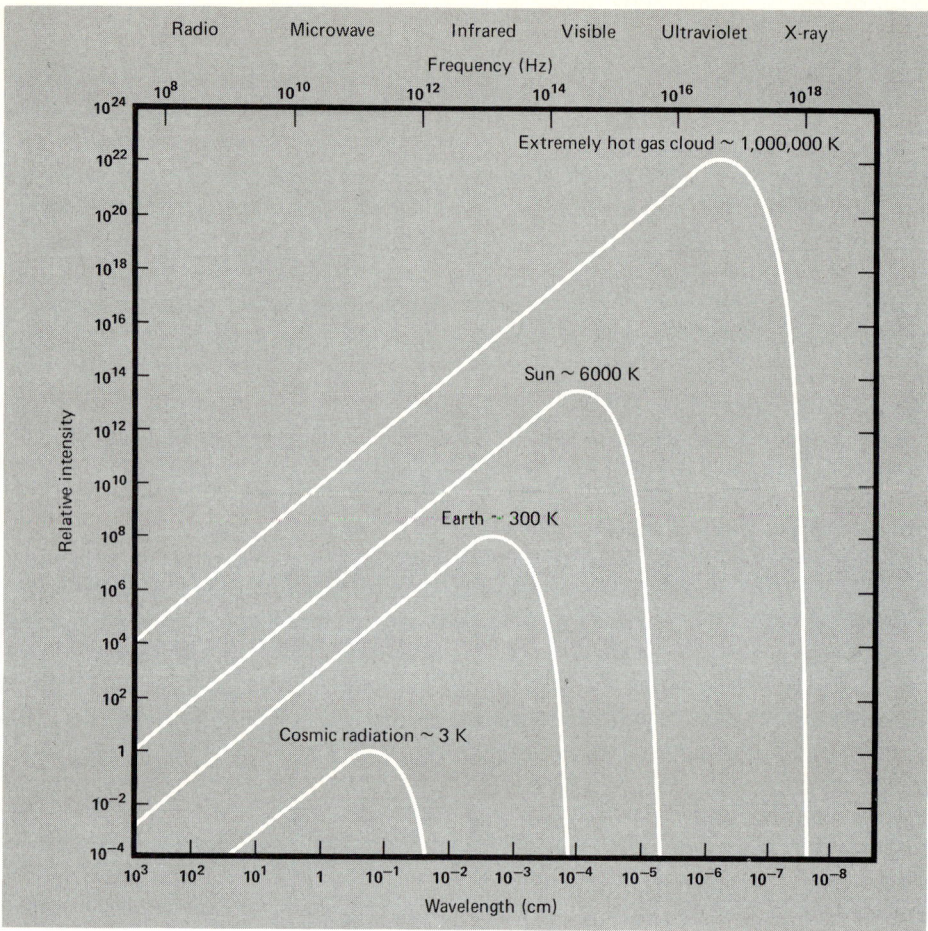

FIGURE 24.6 A comparison of the spectra of blackbodies at different temperatures. Note that cosmic radiation at about 3 K peaks in intensity at microwave wavelengths; hotter blackbodies peak at shorter wavelengths.

In an early, dense, hot state the universe acted like a blackbody radiator. If you could have been there, you would have seen a bright fog all around, like the sun's interior (but with the radiation mainly at even shorter wavelengths, because of the higher temperature).

Imagine the universe expanding from this infernal state. It is so hot that it expands in a violent rush like an explosion; hence the name *Big Bang* for this picture of the universe. As the universe expands, its overall density and temperature decrease. Eventually the temperature drops low enough so protons can capture electrons to form neutral hydrogen.

This event—the formation of neutral hydrogen from an ionized gas—marks a crucial stage in the evolution of the universe. No longer ionized, the universe becomes transparent to its own radiation; the light is freed of its close interaction with matter and freely speeds throughout space. The expansion dilutes it, and the radiation cools. It is red shifted, just like the light of distant galaxies, for indeed, being emitted long ago, it comes from far away, and Hubble's law predicts a large red shift. (Another way to see this: The

24.3 CONTEMPORARY COSMOLOGICAL MODELS

photons act like the particles of a gas, and when a gas expands, it cools, and the energy of its particles decreases. But the photons cannot slow down; for them to lose energy means that they shift to longer wavelengths.)

If the universe *did* begin in a hot Big Bang, debris (both matter and radiation) from the cosmic explosion of the Big Bang must now lie all around. The matter is obvious. But what about the radiation produced in the Big Bang? It's been red shifted to a fairly low temperature. Because of its low temperature, the radiation's wavelength will be long compared with that of light. But if the universe expanded uniformly, the graph of the radiation's spectrum should show the telltale blackbody shape. Have we seen such cosmic radiation? Yes! Its discovery supports the hot Big Bang model!

The Cosmic Blackbody Microwave Radiation

In 1964 Robert H. Dicke, P. James E. Peebles, Peter G. Roll, and David T. Wilkinson at Princeton University were pursuing the possible existence of leftover radiation from a Big Bang. As sometimes happens in science, their ideas had been foreshadowed by those of Herman Alpher, Hans Bethe, and George Gamow in an earlier version of the Big Bang model proposed in 1948. Gamow's original idea was to form all the elements in a young, hot universe. Calculations based on this idea predicted a temperature of 7 K for the remnant radiation today. The Princeton group attacked the problem from a different direction: What would happen if the universe went through a hot stage, so that a high temperature decomposed any heavy nuclei into elementary particles? If a primeval fireball occurred, cosmic blackbody radiation should survive today. Peebles calculated that this fossil radiation should have a blackbody temperature of roughly 10 K.

Just as Roll and Wilkinson were building apparatus to detect the radiation, Arno Penzias and Robert Wilson (Fig. 24.7), scientists with the Bell Telephone Laboratories in New Jersey, began a sensitive study of the radio emission from the Milky Way. They detected an annoying excess radiation in their special low-noise radio antenna. They knew the excess noise would limit their study of the Galaxy, so they set about to try to eliminate it.

They tuned their radio receiver to 7.35 cm (4080 MHz), where the radio noise from the Galaxy is very small. Still they picked up the static. Penzias and Wilson further

FIGURE 24.7 Arno Penzias (right) and Robert Wilson (left) standing near the horn antenna with which they discovered the cosmic background radiation. (*Courtesy Bell Laboratories.*)

discovered that the noise did not change in intensity with the direction in the sky, the time of day, or the season. Perplexed, they examined the antenna again and found a pair of pigeons roosting inside. These pigeons, oblivious to radio astronomy, had coated a part of the antenna with their droppings. Perhaps the coating could explain the excess noise. The birds were moved, their droppings cleaned out. But still the excess noise persisted, with an intensity equivalent to that of a blackbody at 3.5 K.

Penzias called Bernard Burke, a radio astronomer at MIT, to discuss matters other than the excess noise. Burke asked about the experiment, and Penzias explained the problem. Because Burke has heard about Peebles' work, he suggested that Penzias call the Princeton group. When he made contact, together they quickly concluded that the excess noise came from the cosmos—radiation left over from the Big Bang.

This intuitive, risky leap needed verification. After all, the Penzias and Wilson measurement was at a single wavelength. To establish the radiation as truly from a hot, dense stage in the universe's past, more observations were needed. Did the radiation have the characteristic blackbody shape in its spectrum and was it uniform in intensity around the sky? Soon Roll and Wilkinson added a point at another wavelength, with the intensity corresponding to a 2.8 K temperature, close to that observed by Penzias and Wilson. Other experimental groups contributed additional evidence (Fig. 24.8). The points crept up the long-wavelength side of the (hopefully) blackbody curve to the turnover point. At 3 K, the maximum of the blackbody curve occurs at about 1 mm wavelength. An observed point over the hump would clinch the argument, but that region lay at far-infrared wavelengths, where observations are notoriously difficult to make because of absorption by the earth's atmosphere.

Far-infrared observations made above most of the earth's atmosphere have confirmed that the spectrum of the cosmic radiation turns down at short wavelengths. Observations by a Berkeley group are plotted in Fig. 24.8. Other observations from a group at Queen Mary College, London, agree well. The blackbody temperature of the Queen Mary

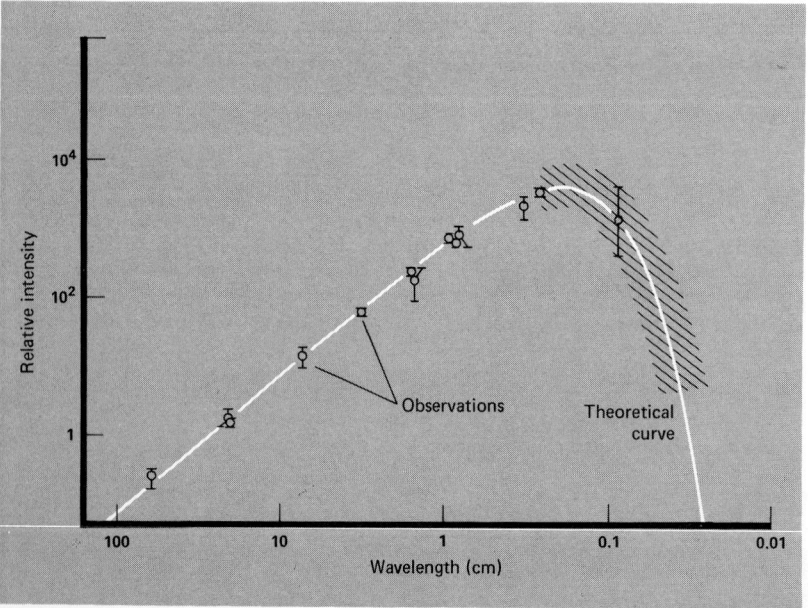

FIGURE 24.8 Observations of the cosmic microwave radiation. The points show measured intensities at radio and infrared wavelengths; the curve corresponds to a theoretical blackbody of 3 K. The shaded region corresponds to infrared observations by P. L. Richards and co-workers. (*Adapted from a diagram by P. J. E. Peebles.*)

24.3 CONTEMPORARY COSMOLOGICAL MODELS

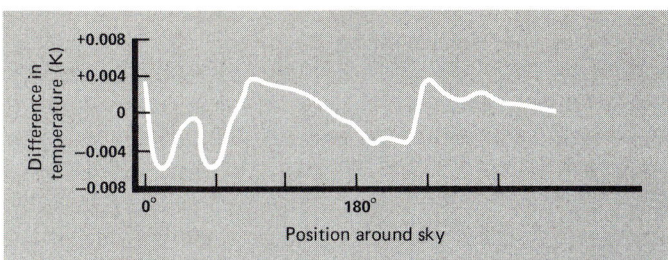

FIGURE 24.9 The measured uniformity in the intensity of the cosmic background radiation from different directions in the sky. Shown here is how little the measured temperature (an indication of intensity) varies around the sky. More than two years of observations went into this graph. (*Adapted from a diagram by R. B. Partridge.*)

College result is 2.94 K, that of the Berkeley group 2.99 K. All observations to date averaged together give a blackbody temperature of 2.89 K. For convenience, we'll take the radiation's blackbody temperature as simply 3 K.

The observations so far fit a blackbody spectrum pretty well, but not perfectly. Paul Richards and coworkers have reported a sharper peak in the observed spectrum than predicted by a theoretical blackbody spectrum. The discrepancy amounts to 10 percent and may be an experimental error, though the observers do not think this is likely. What will come of this discrepancy remains to be seen.

To further confirm its cosmic origin, the radiation should be isotropic, that is, have the same measured intensity from all directions in the sky. If the radiation is from a hot primeval state, the isotropy of the universe at that time would fix the isotropy of the radiation. (Recall that in Section 24.1 we *assumed* isotropy—that the universe looks the same in all directions—without experimental confirmation.) The Princeton group also made some of the first observations to determine the isotropy of the radiation. R. Bruce Partridge and David Wilkinson searched for a variation in the radiation's intensity, scanning around the sky (Fig. 24.9). After a year of collecting data, Partridge and Wilkinson concluded that the variation was no more than 0.5 percent.

This result suggested both that the radiation originated from the early universe and that, in fact, the newborn universe was very nearly isotropic. The lack of a regular variation ruled out a local solar system or galactic source of the radiation. The evidence—equal intensity from all directions in space—implies that the radiation is cosmic, existing everywhere in space, so that it arrives at the earth uniformly from every direction. It fills all space at all times. The isotropy observations clinch the interpretation of the radiation as cosmic.

This background radiation is usually given the long-winded name of *cosmic blackbody microwave radiation*. "Cosmic" because it comes from all directions in space; "blackbody" because of its spectral shape; and "microwave" because its spectrum peaks at millimeter wavelengths.

The cosmic radiation contains a wealth of information useful in constructing models of the universe. First, it rules out the simple Steady State model, for it contradicts the perfect cosmological principle. The argument goes like this. The radiation is now blackbody in character. That means it must have been emitted by an opaque, dense gas. The universe is not *now* opaque and dense. So it must have been different in the past, which is contrary to the assumption of the Steady State model.

Second, the radiation enables astronomers to glimpse the raw, young universe. We can conclude that the early universe was indeed quite homogeneous and isotropic.

Third, the present temperature, about 3 K, and the isotropy of the radiation set severe limits on the thermal history of the universe, that is, the change of the temperature of matter and radiation with time.

Fourth, the radiation's presence establishes an important marker for galaxy formation. Until the radiation and matter stopped their interaction, matter could not form any clumps of large size. Only after the ionized gas recombined could matter form clumps that eventually became stars and galaxies.

The discovery of the cosmic blackbody microwave radiation has a significance for cosmology as great as that of the discovery of the expansion of the universe (Section 5.4). Its present measured temperature (about 3 K), combined with Einstein's equations of general relativity, allows us to work out the evolution of the universe. The details of this evolution will be traced out in the next section: the standard Big Bang model that we believe has reasonable support in present observations.

Is the Cosmic Background Radiation Perfectly Isotropic?

Early observations indicated that, within experimental error, the cosmic background radiation was isotropic. This property implies that, at the time the universe became transparent, it was expanding isotropically and the matter was distributed homogeneously. If so, we can use the cosmic background radiation to define a reference frame against which we can measure our motion. If we take the cosmic radiation to be "at rest," then the earth's motion around the sun, plus the sun's motion around the Galaxy, plus the Galaxy's motion in the Local Group, plus the Local Group's motion in the local supercluster—all these motions together should show up as a Doppler shift. In the direction of the summed-up motions we'd see a blue shift, so the radiation would appear stronger and hotter. In the opposite direction we'd see a red shift, so the radiation would appear weaker and cooler. The radiation would appear slightly anisotropic.

Such an anisotropy has been observed. A research group at the University of California, Berkeley, has detected an anisotropy of a mere 3.5×10^{-3} K with the hottest spot in the direction of the constellation Leo, and the coolest 180° away in the direction of Aquarius. They interpret this anisotropy as due to a motion of the earth relative to the radiation of almost 400 km/sec. Other measurements have confirmed this result, with an average value of 360 km/sec.

What might account for this motion? Recall that Aaronson and coworkers (Section 22.5) found that the Local Group moves at 480 km/sec in the direction of the Virgo cluster. They note that the anisotropy measurement, after taking out the motion of the solar system and of the Galaxy, results in a Local Group velocity of 540 km/sec in a direction about 35° away from Virgo. The part of the motion toward Virgo is then 440 km/sec. The reasonable agreement of these results (which have experimental errors of roughly 50 to 75 km/sec) implies that the anisotropy arises largely from motions in the local supercluster.

24.4 THE PRIMEVAL FIREBALL

Since the discovery of the cosmic microwave background radiation, most astronomers accept a hot Big Bang model. With the addition of experimental and theoretical knowl-

24.4 THE PRIMEVAL FIREBALL

edge on how matter behaves under hot, dense conditions, theoreticians have been able to develop step-by-step details of the Big Bang.

Warning: Don't picture the Big Bang as happening in the "center" of the universe and expanding to fill it. The Big Bang involved the *entire* universe; every place in it was *at* creation, which was an event marking the beginning of time and space (Sections 5.2 and 5.4).

The Hot Start

Although the present temperature of the cosmic radiation is low, the amount of energy it contributes to the universe is large. Each cubic meter contains 4×10^{-14} J, equivalent ($E = mc^2$) to about 4×10^{-31} kg of matter. For comparison, if you take the material contained in the galaxies we can see and spread it uniformly around the universe, each cubic meter would contain 4×10^{-28} kg. So for each kilogram of galactic matter there are approximately 10^{+14} J of cosmic radiation. If the energy in the radiation could be used to heat up the matter, the temperature would be greater than 10^{12} K. Here is one clue about how hot the early universe might have been.

Another way to compare the cosmic radiation with matter is to compute the ratio of number of photons to number of nucleons (protons or neutrons). Blackbody radiation (Section 6.4) peaks at a wavelength

$$\lambda_{max} = \frac{2.9 \times 10^{-3}}{T} \, m$$

The energy of a photon of wavelength λ (Section 6.2) is

$$E_{ph} = \frac{hc}{\lambda}$$

So for the photons of wavelength $\lambda = \lambda_{max}$,

$$E_{ph} = \frac{hcT}{(2.9 \times 10^{-3})}$$

With $h = 6.6 \times 10^{-34}$, $c = 3 \times 10^8$, and $T = 2.7$ K, the energy of the photons at the peak of the blackbody curve is

$$E_{ph} = \frac{(6.6 \times 10^{-34})(3 \times 10^8)(2.7)}{(2.9 \times 10^{-3})}$$
$$= 2.0 \times 10^{-22} \, J/photon$$

The photons making up the blackbody radiation have a variety of energies, of course, but if we take this peak energy as typical, the number of photons needed to produce the 4×10^{-14} J of radiation contained in every cubic meter is

$$n_{photon} = \frac{4 \times 10^{-14} \text{ J/m}^3}{2.0 \times 10^{-22} \text{ J/photon}}$$

$$\approx 2 \times 10^8 \text{ photons/m}^3$$

Now consider the matter. The mass of one nucleon (a proton or neutron) is 1.7×10^{-27} kg. The 4×10^{-28} kg of matter contained in 1 m^3 is equivalent to

$$n_{nucleon} = \frac{4 \times 10^{-28} \text{ kg/m}^3}{1.7 \times 10^{-27} \text{ kg/nucleon}}$$

$$\approx 2 \times 10^{-1} \text{ nucleons/m}^3$$

The ratio of these two numbers is

$$\frac{n_{photon}}{n_{nucleon}} = \frac{2 \times 10^8}{2 \times 10^{-1}} = 10^9$$

So there are a billion times as many photons in the universe as nucleons.

One of the characteristics of the cosmic blackbody radiation is that the total number of photons remains the same, if account is taken of the expanding volume. Individual photons might be absorbed and reemitted by matter (in the early era when the universe was opaque), but the net change is zero. Similarly for the matter, protons might react with neutrons to make other elements, but the total number of nucleons remains the same. So the above ratio of photons to nucleons will stay constant as the universe expands (at least after the high temperature era when photon collisions created particles; see below).

The dominance of photons over matter requires a very hot origin, often called the *primeval fireball*. How hot was it? We don't really know, because we don't understand how matter behaves at temperatures greater than 10^{12} K. Models of the expansion constructed from Einstein's equations of general relativity indicate that a temperature of 10^{12} K corresponds to a time of about 10^{-4} sec after time zero. By "time zero" we mean the actual beginning of the present expansion—the Big Bang. At that time the universe had an average density of 5×10^{16} kg/m^3 (almost as dense as the atom's nucleus).

The primeval fireball produces such a rapid expansion that the temperature and density drop quickly. The contents of the universe (Table 24.1)—initially a flood of energetic light and heavy particles, such as protons, and their respective antiparticles—changes with the temperature and time. As mentioned earlier, the possibility that nucleosynthesis might occur on a cosmic scale motivated George Gamow (1904–1968) to investigate the situation for which he coined the term Big Bang. Only very limited, but very important nucleosynthesis occurred in the early universe: the formation of helium and deuterium and a little bit of other light elements.

In the young universe temperature plays a crucial controlling role. Each period of time since time zero can be matched with a corresponding temperature. Roughly, the universe's thermal history divides into four eras: (1) a *heavy-particle era* when massive parti-

24.4 THE PRIMEVAL FIREBALL

cles (protons and neutron) and their antiparticles dominate, (2) a *light-particle era* when electrons and positrons are continually created and destroyed, (3) a *radiation era* when most particles have vanished and radiation is the main form of energy, and (4) a *matter era*, in which we now live, when matter is the dominate form of energy.

Here's an outline of what follows: The universe in a few minutes after time zero expands and reaches temperatures and densities suitable for the formation of deuterium and helium. The Big Bang model predicts a helium abundance of 25 to 30 percent (by mass) and very little formation of heavier elements. (These are made later in stars.) If the Big Bang picture is correct, this abundance is the *minimum* amount of helium in the universe. About 700,000 years after the time of helium's formation, electrons and protons get together to form neutral atoms (almost all hydrogen and helium.) At this time matter can clump to form the first stars, galaxies, and quasars. The radiation no longer plays a dominant role in the universe's evolution.

Warning: Whenever we say the universe expands what we mean is that the distance between any pair of objects becomes larger. The universe itself may be infinite, and so have no real size. That overall geometry does not affect what we'll say below about the Big Bang.

TABLE 24.1 Particles That Play a Role in Cosmic Nucleosynthesis

Particle and Antiparticle	Symbol	Charge	Comments
Neutrino	ν	0	Massless(?) particles that travel at light speed; stable(?)
Antineutrino	$\bar{\nu}$	0	
Proton	p	+1	Nucleus of hydrogen; stable
Antiproton	$\bar{p}$	−1	
Electron	e^-	−1	Particles surrounding the nucleus of an atom; stable
Positron	e^+	+1	
Neutron	n	0	Decays to a proton and an electron in about 1000 sec
Antineutron	$\bar{n}$	0	
Photon	γ	0	Packet of radiation, electromagnetic energy
Deuteron	^{2}H	+1	Nucleus of deuterium, or "heavy hydrogen"; contains 1 proton, 1 neutron; stable
Helium 3	^{3}He	+2	Nucleus of an unusual type of helium; contains 2 protons, 1 neutron; stable
Helium 4	^{4}He	+2	Nucleus of ordinary helium; contains 2 protons, 2 neutrons; stable
Lithium 7	^{7}Li	+3	Nucleus of most abundant type of lithium; contains 3 protons, 4 neutrons; stable
Beryllium 7	^{7}Be	+4	Nucleus of most abundant type of beryllium; contains 4 protons, 3 neutrons; unstable

Creation of Matter from Photons

Before we unfold the story, we need to consider one key point: the creation of matter and antimatter from photons.

One consequence of the fact that there are about 10^9 photons per nucleon in the universe now is that some time in the past the energy density of radiation exceeded that in the form of matter. Though the *number* density of photons relative to that of nucleons has not changed, the *energy* density in radiation compared to that in matter was higher in the past. When the temperature of the radiation was higher, each photon had a higher energy. The energy contained in matter now is 4×10^{-11} J/m^3; that in radiation is 4×10^{-14} J/m^3. The ratio is about 1000:1 in favor of matter. But when the radiation temperature was about 1000 times higher (about 3000 K), the energy density in matter and radiation was about equal (the average energy of each photon is proportional to the temperature). Farther back in time, when the temperature was even higher, the radiation was dominant. At some point the temperature of the radiation was so high that photon collisions produced matter. This process can happen when the energy contained in two colliding photons equals or exceeds the rest-mass energy of the particles produced.

This production process is just the reverse of matter-antimatter annihilation, which completely converts rest mass into photons. So when photons make matter, they must do so in the form of a matter-antimatter pair, such as an electron and positron (Fig. 24.10).

Let's see what minimum temperature the radiation would need to have so that photons could produce an electron and a positron. We can call this the *particle threshold temperature*. The mass of an electron or positron is 9.1×10^{-31} kg, and the equivalent energy of the two together is

$$E = (m_{e^-} + m_{e^+})c^2$$
$$= (2 \times 9.1 \times 10^{-31})(3 \times 10^8)^2$$
$$= 1.6 \times 10^{-13} \text{ J}$$

Now for blackbody radiation the typical energy of a photon is roughly

$$E \approx kT$$

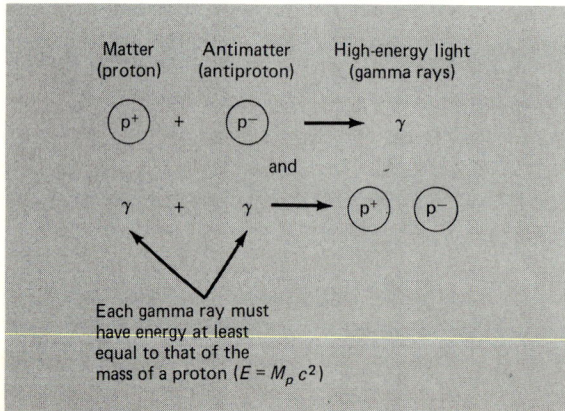

FIGURE 24.10 Matter and antimatter annihilation to make photons; particle and antiparticle production from photons.

24.4 THE PRIMEVAL FIREBALL

where k is Boltzmann's constant, 1.4×10^{-23} J/K. For a photon to have an energy of 1.6×10^{-13} J requires

$$T \approx \frac{E}{k}$$

$$\approx \frac{1.6 \times 10^{-13} \text{ J}}{1.4 \times 10^{-23} \text{ J/K}}$$

$$\approx 1.2 \times 10^{10} \text{ K}$$

You need temperatures of about 1.2×10^{10} K to produce electrons and positrons from photons.

For protons and antiprotons a greater temperature and energy are needed. Because a proton's mass is about 1800 times greater than that of an electron, the energy needed to form a proton-antiproton pair is $1800 \times (1.6 \times 10^{-13})$ J $= 3.0 \times 10^{-10}$ J, and the threshold temperature is therefore 2.2×10^{13} K. So to make a proton-antiproton pair requires two photons each having 1.5×10^{-10} J. That corresponds to a photon frequency of

$$\nu = \frac{E}{h}$$

$$= \frac{1.5 \times 10^{-10}}{6.6 \times 10^{-34}}$$

$$= 2.3 \times 10^{23} \text{ Hz}$$

and a wavelength of

$$\lambda = \frac{c}{\nu}$$

$$= \frac{3 \times 10^8}{2.3 \times 10^{23}}$$

$$= 1.3 \times 10^{-15} \text{ m}$$

$$= 1.3 \times 10^{-5} \text{ Å}$$

That wavelength is well into the gamma-ray range of the spectrum.

Other elementary particles have different threshold temperatures depending on their mass. Some are listed in Table 24.2. As time goes by and the universe expands, the

TABLE 24.2 Rest Energies of Some Elementary Particles

Particle	Symbol	Rest Energy (J)	Threshold Temperature (K)
Photon	γ	0	0
Electron	e^-	8.19×10^{-14}	5.9×10^9
Proton	p	1.50×10^{-10}	1.1×10^{13}
Neutron	n	1.51×10^{-10}	1.1×10^{13}

radiation's temperature goes down. As it falls below each threshold temperature, these particle pairs can no longer be produced. Once the radiation's temperature falls to about 10^{10} K, only electron-antielectron pairs can be produced by photons.

Particle production from photons plays a dominant role in the very young universe. Let's see what the model tells about those times.

Temperature Greater Than 10^{12} K, Time Less Than 0.0001 Sec after Time Zero

Photons produce pairs of particles and antiparticles. The photons have enough energy so that even protons, neutrons, and their antiparticles can be produced (Table 24.2). This is the *heavy-particle era*.

Annihilation also takes place, and the balance between annihilation and creation fixes the density of particles and antiparticles for the next stage. This balance between massive particle production and destruction marks the heavy-particle era. It does not last long, as the expansion is rapid and the temperature declines quickly.

A critical difficulty with the model here is that the average density is greater than 10^{17} kg/m^3—and we really do not know how matter behaves under such extreme densities (and we certainly do not know what happened at earlier times and higher densities). But whatever happens in detail, heavy particles are made at this stage.

So at the earliest times we can calculate, the universe is a smooth soup of high-energy light and massive particles (Fig. 24.11).

Temperature from 10^{12} to 5×10^9 K, Time from 0.0001 to 4 Sec

Annihilation of protons and neutrons (and other heavy particles) with their antiparticles continues. The remaining photons, however, lack the energy to create new heavy

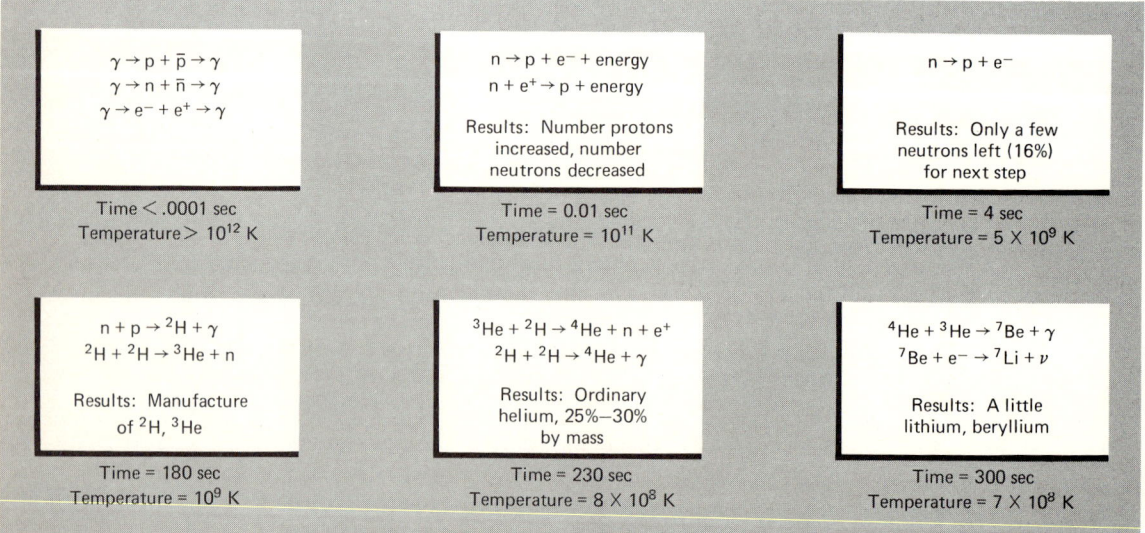

FIGURE 24.11 One possible sequence for nucleosynthesis in the Big Bang.

24.4 THE PRIMEVAL FIREBALL

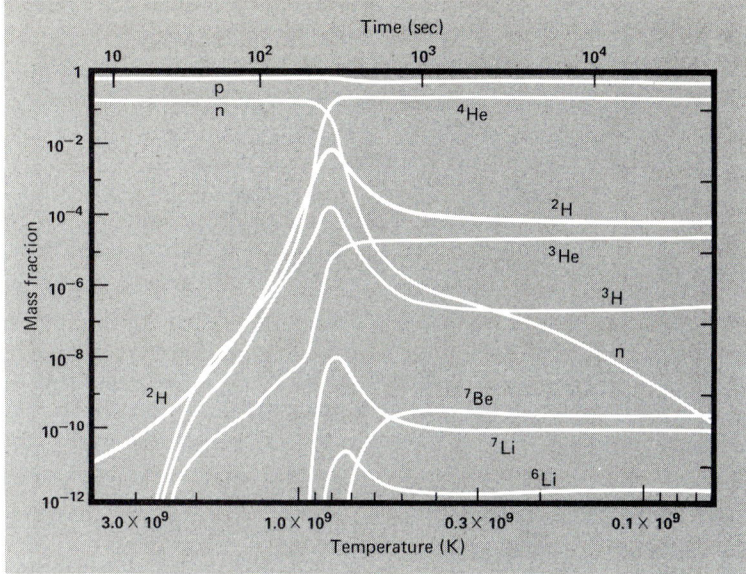

FIGURE 24.12 Details of the formation of low-mass elements in the Big Bang. The top axis gives the age of the universe from time zero; the bottom axis gives the temperature; and the vertical axis gives the abundance in terms of the fraction of the total mass. Note that all the nucleosynthesis takes place in a sharp blip between 100 and 1000 sec. (*Based on theoretical calculations by R. V. Wagoner.*)

particles. Only light particles—electrons—can be made, because these need less energy for creation. The universe enters the *light-particle era*.

But not all heavy particles are destroyed. Observations imply that the visible universe is now mostly matter. The present imbalance of matter over antimatter requires that a few extra protons and neutrons remain; that is, the early universe contained a slight excess of these nucleons over their antiparticles. (Why is still a baffling mystery.) When the temperature falls to 5×10^9 K, photons can no longer make electron-positron pairs. This temperature marks the end of the light-particle era and the beginning of the radiation era. The universe still expands, but less rapidly.

Early in this era the interactions of electrons, protons, neutrons, and neutrinos produce a balance between the number of protons and neutrons (Fig. 24.11). As the temperature decreases, the number of neutrons falls, while the number of protons rises. Neutrons, totaling 16 percent of the particles, can no longer be produced by interaction of protons with electrons or of antiprotons with positrons. The neutrons are now free to decay into protons and electrons.

Temperature 10^9 K, Time from 4 to 300 Sec

In this crucial period the remaining neutrons and protons react to form nuclei that can survive the still high temperatures. The most important reaction involves the combination of a neutron and a proton to form deuterium (^{2}H). All neutrons, except those that have decayed, end up in deuterium. Once the deuterium is produced, a proton-proton reaction (Fig. 24.11) creates normal helium (^{4}He) and also a little tritium (^{3}H). The net result is 25 to 30 percent of helium by mass. A little bit of beryllium (Be) and lithium (Li) is created by the combination of deuterium and tritium. Some deuterium is left over. Extremely little of the heavier elements is made (Fig. 24.12).

Warning: Do not confuse the term heavy element (which refers to elements like carbon, iron, uranium) with the term heavy particle (which refers to things like protons and neutrons).

The final helium and deuterium abundance depends on the rate of the universe's expansion. If the expansion rate had been any faster, less time would have been available for neutrons to decay, and more helium would have been produced. On the other hand, if the expansion rate had been very much slower, the neutrons that would have formed helium would have had a longer time to decay, so less helium would result. So the present helium abundance in the universe sets severe constraints on the expansion rate and density of the early universe. The helium abundance also serves as a test for the Big Bang cosmology. The model predicts that no celestial object can have a helium abundance of less than 25 to 30 percent. Because stars form helium, the abundance can be greater than this number, but not less.

Stop here for a moment and consider what you've just read: a blow-by-blow account of the first few minutes of creation based on the hot Big Bang model. Does it seem a little unreal? It's astounding that the model allows us to talk of such times with some confidence. Of course, it's always possible to make very accurate calculations. What is amazing is that we can fit our own universe to such a Big Bang model.

Temperature 3000 K, Time 21×10^{12} Sec

So far the temperature has been so high that all atoms in the universe have been ionized. At about 700,000 years after time zero the radiation's temperature has plunged to only 3000 K. The nuclei begin to capture electrons to form neutral atoms. This recombination process happens in a few thousand years. With few electrons, opacity processes such as free-free absorption and electron scattering are less effective. The matter becomes transparent to the radiation. Suddenly light breaks through. The matter and radiation are no longer locked together. We say they are *decoupled*. For the first time in the history of the cosmos, the sky turns dark.

Freed from this interaction, the radiation merrily expands with the universe. And because the matter is freed from the pressure of radiation, local gravity effects can take over. So the matter follows a different course from the light. Because of little local bumps in the generally smooth distribution of matter, a local region of slightly higher than average density does not expand as fast as the rest of the universe, owing to its self-gravity. Consequently, a local blob increases its density contrast, that is, the ratio of blob density to the density of the surrounding matter increases. This increases the self-gravity even more. Eventually the expansion of the blob stops and reverses. Clouds of matter condense out of the primeval fireball. The radiation era ends, and the matter era begins. The universe has a density of 10^{-24} kg/m^3. Radiation pressure prevented material condensation until the atoms recombined and the radiation and matter decoupled. This event flags the time when the galaxy formation could begin (see next section).

For a summary of these events, see Table 24.3.

The Evidence for the Big Bang

How do we know that the contemporary Big Bang cosmological picture is not mere speculation? The scenario for the hot early universe requires that we push our knowledge

24.4 THE PRIMEVAL FIREBALL

about matter and gravitation to shaky limits. What observational evidence supports such a stretching of present ideas?

First, and foremost, the model predicts that the universe expands. Second, the model predicts that blackbody radiation pervades the universe. Within the limits of observational errors, measurements confirm a blackbody spectrum with roughly a 3 K temperature. Measured in many directions of space, the radiation comes to the earth with a pretty uniform intensity. The background radiation has the attributes expected of a cosmic, hot origin.

Third, the model predicts that primeval helium was formed in the first few minutes of the universe's history and that the helium abundance should be 25 to 30 percent by mass. The Big Bang model sets this helium abundance as the basement level; any observation of a substantially lower amount would call the model into question. Because helium is formed in stars and some of it is ejected back into space, the present helium abundance can be larger than 25 to 30 percent.

Unfortunately, we can assess the present cosmic helium abundance only indirectly. Spacecraft have sampled the solar wind to find that 15 to 30 percent is helium nuclei by mass. As Chapter 6 pointed out, the solar photosphere is not hot enough to excite helium

TABLE 24.3 Sequence of Events in the Big Bang

Event	Time	Density (kg/m^3)	Temperature (K)	Comments
Creation	0	?	?	Not the province of present science; general relativity fails.
Heavy-particle era	10^{-44} sec	10^{97}	10^{33}	Photons make massive particles (such as protons) and antiparticles.
Light-particle era	10^{-4} sec	10^{17}	10^{12}	Photons have only enough energy to make light particles and antiparticles, such as electrons and positrons; protons and electrons combine to make neutrons.
Radiation era	10 sec	10^7	10^{10}	Few particles left in a sea of radiation; these partake in nucleosynthesis of deuterium, helium, lithium, and beryllium.
Matter era	10^6 yr	10^{-18}	3000	Ionized hydrogen recombines; cosmic radiation and matter decouple.
Now	10^{10} yr	10^{-31} (radiation)	3 (radiation)	Astronomers puzzle about creation.

lines for direct viewing with a spectroscope, so it's not possible to measure the helium abundance in the solar photosphere. The chromosphere's higher temperatures do excite helium atoms to emission, and the observed abundance there is estimated to be 38 percent. Theoretical models of stellar evolution place the helium abundance in the sun's interior at 17 to 28 percent. Similar calculations for Jupiter yield a helium abundance of roughly 20 to 30 percent. Voyager showed that the atmosphere of Jupiter contains 20 percent helium, that of Saturn 11 percent.

Population I O and B stars exhibit helium absorption lines that imply a helium abundance of approximately 30 to 34 percent. H II regions surrounding hot stars have helium emission lines at optical and radio wavelengths that give abundances from 26 to 29 percent. Planetary nebulas also have strong helium emission lines that imply a helium abundance of greater than 30 percent.

But many of the above objects are fairly young. They could have picked up helium made in stars. The best objects to search for primeval helium are the oldest stars now surviving, the Population II stars. Unfortunately, most Population II stars are much too cool to excite helium lines in their spectra. However, indirect evidence from star models indicates that these stars have helium abundances of about 30 percent.

Though the range of determinations is moderately large, it is remarkable that the helium abundance for a variety of celestial objects falls so close to the number predicted by the Big Bang model. The evidence for the compatibility of observations with the model is good, but not conclusive, and as is usual in astrophysics, no one observation clinches the affair. But in this case, the accumulation of evidence for the hot Big Bang is impressive.

24.5 THE END OF TIME

So much for the universe's past. What about its future? Recall (Chapter 5) that Einstein's general theory of relativity allows the cosmos to have one of three general geometries: hyperbolic, closed, or flat. If hyperbolic or flat, the universe will expand forever, and time will never end. If closed, however, the universe must eventually collapse, running backwards through the history outlined in the previous section.

Which fate will be ours? Chapter 5 mentioned two observational tests that indicate a hyperbolic universe. First, the measured value of the Hubble constant (H = 50 km/sec/Mpc) and Einstein's theory give a critical density for a closed universe; it's about 5×10^{-27} kg/m^3. The observed density seems to be much less, so, on the surface, this evidence implies that the universe is open. Second, the curvature of the Hubble plot for distant galaxies can indicate the universe's geometry. This Hubble test is tough to do, but so far, very weak evidence points to a hyperbolic geometry.

The standard hot Big Bang model provides another test, which is perhaps the strongest. The test rests on the observed present cosmic abundance of deuterium. How can the abundance of deuterium tell us whether the universe is open, closed, or flat? To do the Big Bang model calculations, you need to put in the *present* value of the Hubble constant, the *present* temperature of the cosmic radiation, and the *present* average density of the universe. The first two items are reasonably well known, but the third is not. The amount of helium which comes out of the Big Bang calculations does not depend very much on the value used for the present density of the universe. But the amount of deuterium that

24.5 THE END OF TIME

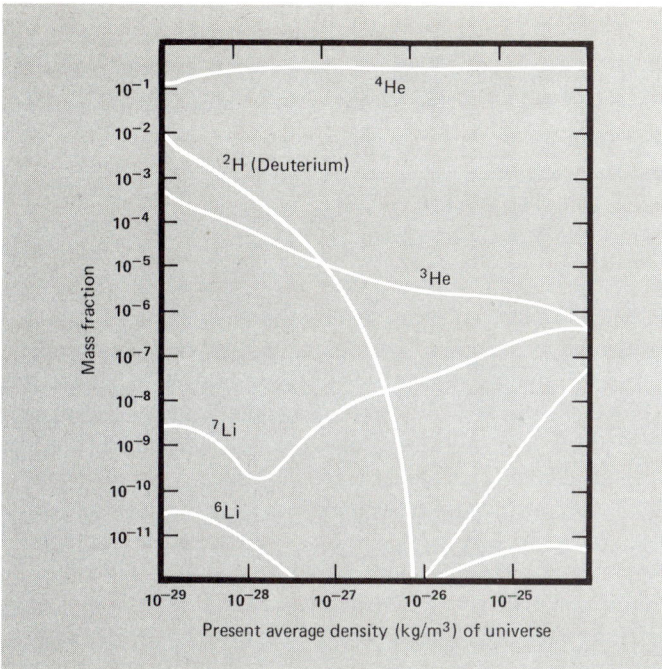

FIGURE 24.13 Theoretical calculations of the abundance of low-mass elements formed in the Big Bang as related to the average density of the universe now. Note how sharply the abundance of deuterium (^{2}H) varies with different present average densities. (*Based on theoretical calculations by R. V. Wagoner.*)

theoretical calculations predict does depend very sensitively on the value used for the present density (Fig. 24.13). Why? Basically because in a low-density universe the deuterium nuclei don't collide as often, and so few get converted to helium before the nuclear reactions cease. So we can use the model to turn the argument around. If we can measure the present cosmic abundance of deuterium, then we can check this number against that predicted by the Big Bang model (Fig. 24.13) for various densities. The model then gives the present average density of the universe, and we can compare that number with the critical density.

As usual, there's a hitch to this beautiful procedure. Unlike helium, deuterium easily burns up in fusion reactions in stars, even in cool main-sequence stars. So we can't look at stars to find the deuterium abundance. Even if we find deuterium in the interstellar medium, we have to estimate how much of that has been processed by stars.

The lines of atomic deuterium occur at about the same wavelengths as those of hydrogen, so they are difficult to detect. But molecules containing deuterium have lines shifted significantly in wavelength from those of the corresponding hydrogen-containing molecules. Unfortunately, the lines from the most abundant deuterium molecule, HD, lie in the far ultraviolet, and so are not observable from the surface of the earth. The Copernicus satellite carried an ultraviolet telescope that could detect spectral lines of the molecules H_2 and HD produced by interstellar clouds lying in front of bright O and B stars. Both molecules have been detected, and the value for the D to H ratio of the interstellar medium is about 2.0×10^{-5} by mass.

Beatrice Tinsley estimates about half the original deuterium has been burned up in stars, so the original deuterium abundance was about twice that observed by the Copernicus satellite, or 4×10^{-5} relative to hydrogen. Look at the Big Bang calculation (Fig.

24.13). A deuterium to hydrogen ratio of 4×10^{-5} implies a present cosmic density of 4×10^{-28} kg/m^3, considerably less than the critical density (5×10^{-27} kg/m^3). The conclusion from this test: The universe is open. It will expand forever.

Recent experiments—those which suggest that neutrinos have mass—may change this conclusion. We will discuss them in Section 24.8.

24.6 FROM BIG BANG TO GALAXIES

Up to 700,00 years after the Big Bang, gravity could not clump matter into stars and galaxies. The universe was opaque to radiation, so light and matter interacted strongly. Pressure from the radiation itself inhibited gravitational collapse. But once protons and electrons recombined, the universe became transparent to radiation, and no longer could radiation pressure stop gravity from doing its natural work. We refer to this time as the time of *decoupling*, when the radiation and matter were no longer coupled together.

Because we observe that the cosmic background radiation now arrives very uniformly from all directions in space, we know that at the time of decoupling the matter and radiation in the universe must have had a very uniform distribution.

But that's not the situation now. The universe contains matter clumped in planets, stars, clusters of stars, galaxies, and clusters of galaxies. Even the most spread-out of these systems of matter—clusters of galaxies—have average densities about 100 times greater than the average density of the universe.

So here's the crucial question: How did an originally very smooth universe become clumpy? This is a tough question to answer because systems with a wide range of masses developed from the Big Bang. The high end of this range includes galaxies, which have from 10^6 to 10^{12} solar masses, and the largest clusters of galaxies, which contain about 10^{15} solar masses. On the low end it may also include globular clusters, which have about 10^5 solar masses. A model of galaxy formation must account for the formation of masses from about 10^5 to 10^{15} solar masses.

Any model also faces another critical hurdle; it must operate effectively in at most a few billion years. When you look at very distant galaxies you peer back in time to when the universe was billions of years younger. To date, astronomers have seen objects (the quasars) as far as some 12 billion light years away. So the matter from the Big Bang must have formed into large clumps well before this time.

We'll describe the general models devised to explain the formation of galaxies, but we want to warn you that some of these ideas are mostly earnest speculations. In a few years our present concepts will probably change, but the central problem will very likely still be there.

Gravitational Instability

What makes matter form large clumps? The process involved is called *gravitational instability*. Imagine a very uniform, static gas. Picture a disturbance taking place so that a small, spherical region becomes slightly more dense than its surroundings. The disturbed patch attracts more matter to it and also condenses. These actions increase both the patch's density and gravity. The particles in the disturbed region (and those added to it) condense to form a gravitationally-bound object.

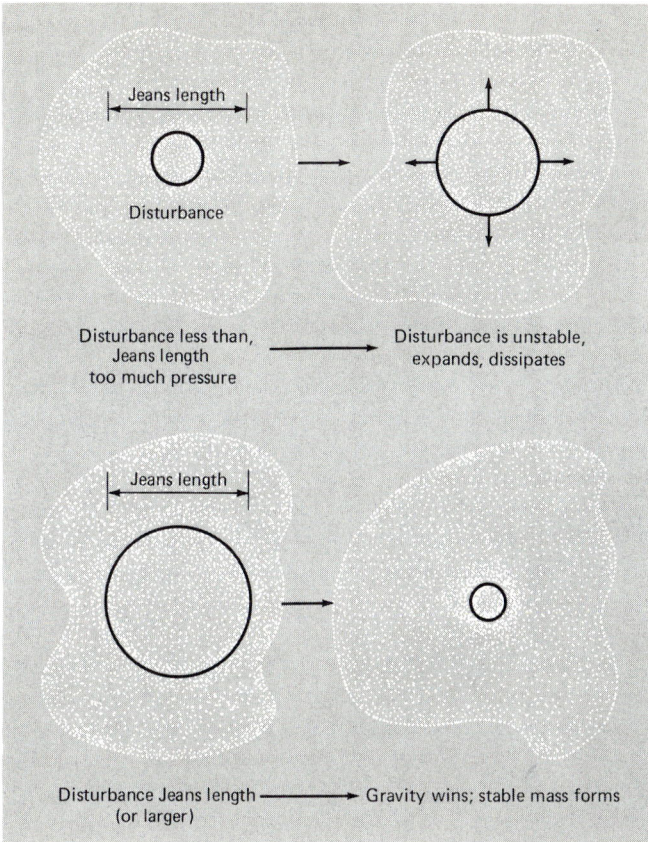

FIGURE 24.14 Gravitational instability and the Jeans length. The size of the Jeans length depends on the temperature and density of a gas. If a disturbance is smaller than the Jeans length, internal pressure dissipates it. If larger, the disturbed region has enough mass to overcome pressure and collapse gravitationally.

Early in this century Sir James Jeans showed that internal pressure plays a critical role in the development of gravitational instability. A disturbed patch of gas has a higher density than its surroundings, but if it has the same temperature, it also has a higher internal pressure. Like the gas inside a balloon, this pressure tends to push the condensed region apart.

Which wins out, the inward pull of gravity or the outward push of pressure? Jeans discovered that the outcome depends on the size of the disturbed region (Fig. 24.14). If the patch is too small, pressure overcomes gravity and makes the patch expand. If the region is large enough, gravity wins and the region condenses. The minimum size a disturbance must have to contract gravitationally depends on the pressure, and so on the temperature and density of the gas; it is called the *Jeans length*. If the disturbance is as large as or larger than the Jeans length, gravitational collapse must follow. If smaller, the disturbance simply results in the compression and expansion of the gas as in a sound wave. Such waves eventually dissipate, just as sound waves in the atmosphere die out if you turn off the source of the waves.

Jeans found that the temperature and density of the gas affected the Jeans length in the following way. The higher the temperature and the lower the density, the larger the Jeans length. This fact has important consequences for the formation of galaxies. As the universe expands, its density and temperature decrease. But the density decreases at a *faster rate* than the temperature does. So the Jeans length gets larger, which means that bigger and bigger disturbances are needed to ensure gravitational collapse. As the universe ages, it becomes harder to make disturbances that can result in the formation of galaxies.

A rough derivation of the Jeans length is not difficult. Suppose we have a cloud of gas with uniform density composed of particles with mass m at temperature T. Cut a sphere of radius R out of the cloud. Its mass is

$$M = \frac{4}{3}\pi R^3 \rho$$

The escape velocity (V_e) at the edge of the sphere (at R) is (Section 5.5)

$$V_e = \left(\frac{2GM}{R}\right)^{1/2}$$

So

$$V_e = \left(\frac{2G(4/3)\pi R^3 \rho}{R}\right)^{1/2}$$

$$= \left(\frac{8}{3}\pi GR^2\rho\right)^{1/2} R$$

Now we want the cloud's gravity to trap particles in it. They have a kinetic energy $\frac{1}{2}mV^2$ equal to $\frac{3}{2}kT$. Thus they move with a velocity on the order of

$$V = \left(\frac{3kT}{m}\right)^{1/2}$$

To hold the particles together, V_e must be at least equal to V, so

$$V_e = V$$

$$\left(\frac{8\pi}{3}GR^2\rho\right)^{1/2} = \left(\frac{3kT}{m}\right)^{1/2}$$

$$R = \left(\frac{9\pi}{8G}\frac{kT}{m\rho}\right)^{1/2}$$

Neglecting the factor $(9\pi/8)^{1/2}$, we obtain the Jeans length:

$$R_J \approx \left(\frac{kT}{Gm\rho}\right)^{1/2}$$

24.6 FROM BIG BANG TO GALAXIES

which is a good enough approximation for rough estimates. The mass contained in a sphere of this radius, the Jeans mass, is

$$M_J = \frac{4}{3}\pi \left(\frac{kT}{Gm\rho}\right)^{3/2} \rho$$

$$= \frac{4}{3}\pi \left(\frac{kT}{Gm}\right) \rho^{-1/2}$$

Let's apply this concept to the universe at the time of recombination and decoupling, when disturbances in the cosmic gas no longer have to fight radiation pressure to grow. At the time of recombination, $T \approx 3000$ K, $\rho \approx 10^{-24}$ kg/m^3, and if we consider hydrogen gas, $m = 1.7 \times 10^{-27}$ kg. Then

$$R_J \approx \left(\frac{(1.4 \times 10^{-23})(3000)}{(6.7 \times 10^{-11})(1.7 \times 10^{-27})(10^{-24})}\right)^{1/2}$$

$$\approx 6 \times 10^{20} \text{ m}$$

or about 64,000 ly. A sphere with that radius would contain

$$M = \frac{4}{3}\pi R^3 \rho$$

$$= \frac{4}{3}\pi (6 \times 10^{20})^3 (10^{-24})$$

$$= 9 \times 10^{38} \text{ kg}$$

$$= 5 \times 10^5 \text{ solar masses}$$

about the mass of a globular cluster or a dwarf galaxy.

This example shows that, at decoupling, disturbances need to be *at least* a few tens of thousands of light years in size in order to grow by gravitational instability. They could, of course, be much larger.

Relativity and Instabilities

In 1946 the Russian physicist Eugene Lifshiftz applied Jeans' analysis to disturbances in a model of the expanding universe that used Einstein's theory of general relativity. He discovered that Jeans' ideas applied virtually unchanged. Lifshiftz ran into a disturbing problem, however. The predicted rate of growth of gravitational instabilities was very slow, so slow that galaxies could just *barely form by now* from small disturbances in the young universe. Of course, if the disturbances were large enough in the beginning, they could grow into galaxies. But producing large initial disturbances is then the problem.

Lifshiftz made his analysis before the discovery of the cosmic radiation by Penzias and Wilson. The observation of this radiation forced astrophysicists to consider what happens

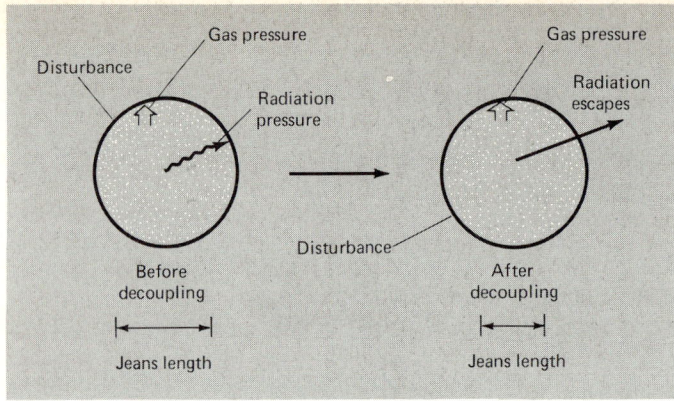

FIGURE 24.15 Decoupling and the growth of instabilities. Before decoupling radiation pressure adds to the internal pressure of a disturbed patch because the gas is opaque to the radiation and the radiation cannot escape. After decoupling the gas is transparent and radiation escapes, no longer contributing to the internal pressure. The Jeans length decreases, and the region can collapse.

to disturbances in a hot, dense universe filled with matter and radiation. They found that, before decoupling, the radiation played a powerful role in inhibiting the growth of disturbances. A dense patch in the early universe will have a high internal pressure because the radiation adds to the pressure force. Only *very* large disturbances would have any chance of growing so long as the matter and radiation interacted. The radiation pressure dissipates disturbances that have masses of 10^{12} solar masses (or less) up to the time of decoupling. So before the decoupling of matter and radiation, disturbances that contained roughly the mass of a galaxy could not grow.

Just after decoupling, it's a new show (Fig. 24.15). The radiation and the gas no longer interact, so radiation no longer helps out in the battle against gravity. Small disturbances, amounting to only 10^5 solar masses, can condense out of the gas, along with disturbances of greater mass. This result gives us some hope, for the large gravitationally bound masses we see now range from 10^5 to 10^{15} solar masses (clusters of galaxies). Disturbances of this size range can grow just after decoupling. So the time of decoupling, roughly 700,000 years after the Big Bang, marks the time when galaxy formation could take place in the young universe.

There's one real weakness in these ideas, and that is Lifshiftz's result. Even though disturbances can be unstable, they grow slowly, so slowly that the galaxies we see could hardly have formed by now, unless the disturbances were already fairly large in the beginning.

Turbulence and the Growth of Galaxies

But galaxies *have* formed and did so early in the history of the universe. If gravitational instability alone did not do the trick, what helped the process out? One contemporary answer, favored by a few astrophysicists, pictures *turbulence* as the critical aid to galaxy formation. This idea pictures the Big Bang as producing not only matter and radiation, but also turbulence, which speeded up the growth of galaxies.

Picture the young universe, before decoupling, filled with whirlpools of photons and gas particles. This turbulence was generated somehow very early in the Big Bang, perhaps in the first few minutes. The largest eddies contained 10^{12} solar masses and greater. Calculations show that these huge whirls experience very little friction and so persist for a long time—all the way up to decoupling. After decoupling, they end up quickly forming galaxies, spurred on by the turbulence (Fig. 24.16). This development may have been helped out by the generation of shock waves after decoupling. (Shock waves help to

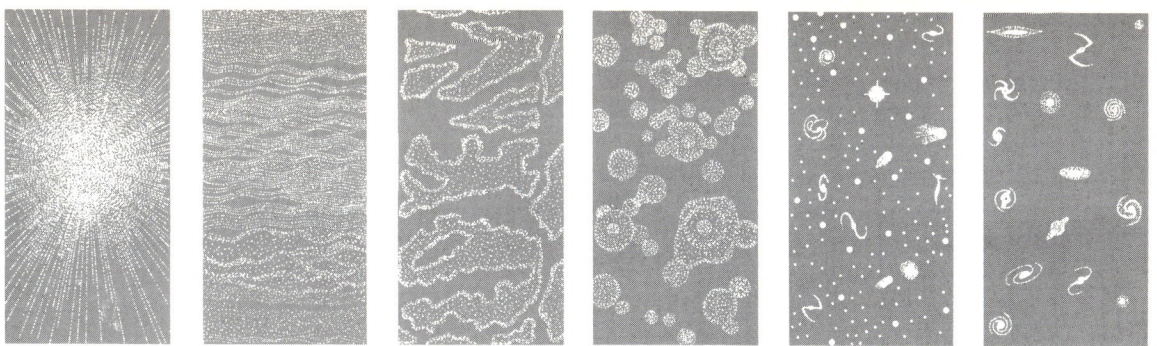

FIGURE 24.16 A complete time strip of the sequence from the Big Bang to now, highlighting galaxy formation. Aided by turbulence, irregularities in the Big Bang eventually condense by gravitational instability to form the variety of galaxies we see today.

compress material.) Also, large eddies may collide, cause shock waves, and break up. The smaller blobs of matter broken off in the crashes then form into galaxies.

You have noticed that we skipped over an important issue: How do eddies and disturbances with greater than 10^{12} solar masses end up as *clusters* of galaxies rather than as individual galaxies? There are some models, similar to that discussed above for individual galaxies, but we don't think anyone has really understood this yet.

The Formation of the Galaxy

Let us sketch a possible scenario for the formation of our own Galaxy. Assume that at decoupling a disturbance occurs, large enough to condense finally into the Galaxy. This disturbance would have to have a mass of at least 10^{11} solar masses. It forms a cloud that first expands with the universe. But as gravity pulls the material in the cloud together, the rate of expansion slows down. The cloud reaches a maximum size, about 150,000 ly (50,000 pc), and then rapidly falls together. This collapse takes place about 1 billion years after the Big Bang. Meanwhile, other clouds like the one that will form the Galaxy also reach their maximum size and then condense. All these clouds interact with each other gravitationally and set each other spinning slowly.

The spinning is important. As the Galaxy's cloud condenses, it spins faster and faster and flattens out into a pancake with a dense central condensation. During this collapse parts of the cloud become gravitationally unstable and fragment into stars. Whatever material did not get caught up in this first phase of star formation makes up the disk of the young Galaxy and then later becomes stars.

We can view the process of fragmentation in terms of the Jeans length. The temperatures and densities expected in the young Galaxy require large Jeans masses, at least a few hundred solar masses. (These masses are large because the temperature is high, approximately 1000 K.) The first generation of stars to form in the Galaxy could *not* have been ordinary stars, which typically do not exceed 100 solar masses. These superstars evolved quickly (in a million years), blew up as supernovas, and formed black holes or neutron stars. So we can't see them now. In their explosive deaths the superstars flung heavy elements into the material from which the next generation of stars formed. The addition of these heavy elements cooled off the remaining gas, and so the Jeans length became smaller. A disturbance of only a few solar masses could then grow. This stage of starbirth probably formed the oldest stars we see now, the Population II stars.

Whatever material did not make up the second generation of stars fell into the disk. Turbulence in the disk, shock waves generated by supernova explosions, or spiral density waves helped to form the third generation of stars, which we call Population I stars. One of these stars is now our sun, and the gravitational collapse of the cloud that formed the sun also resulted in the formation of the planets, including the earth.

The general processes of formation for other types of galaxies probably ran much the same as in the Milky Way Galaxy. The details are obscure, however, and most ideas about galaxy formation are speculative. Elliptical galaxies (Section 22.4) have little gas and dust and few young stars, in direct contrast to spirals. The lack of hot stars and of an interstellar medium falls into the scheme outlined above. Because of their short lifetimes, hot stars are present only if they are continually formed from the interstellar medium. A sparse amount of gas between the stars ensures the formation of few new stars.

Why do galaxies come in different types, and why did ellipticals change all their material into stars so early? These are some of the vexing questions about galaxy formation. Some theoreticians have suggested that the different forms and histories are related to the angular momentum of the original clump from which the galaxy forms. But the answers are not really known.

24.7 SPECULATIONS ON THE EVOLUTION OF QUASARS AND ACTIVE GALAXIES

How do quasars and active galaxies fit into this gravitational instability model for the formation of galaxies? Not very well, because we do not yet have clear ideas about the evolution of quasars and active galaxies, and what physical processes link them. However, if quasars' red shifts are cosmological, then quasars must be among the first-born, massive objects from the primeval fireball.

We can infer something general about the evolution of quasars by observing the most distant and so the youngest ones. The trouble here lies in the fact that the limited light-gathering power of telescopes permits a glimpse of only the brightest quasars of the ones that are farthest away. So we must extrapolate if we wish to count those of low brightness. Using studies of quasars at different distances to help make the extrapolation, Maarten Schmidt has concluded that quasars evolved rapidly when the universe was young. The evidence: the number of faint quasars increases more rapidly than we would expect if the distribution were uniform in space and time, that is, there are more quasars at great distances (large red shifts) than there are nearby. This observation tells us that the number of quasars was greater in the past than it is now. (Remember that as we look outward in space, we are also looking backward in time.) Schmidt concludes that about 8 to 9 billion years ago approximately 2000 times more quasars marked the heavens than exist now. And only 5 billion years ago, at about the time the earth was formed, the abundance was about 100 times greater than now.

Schmidt estimates that 15 million quasars, visible to the largest telescopes, dot the entire sky. The majority, however, are so far away that during the long time required for their photons to reach earth, the quasars themselves have expired. Like fireballs blasted from a display skyrocket, which are extinguished before they hit the ground, most quasars are now dark, massive cinders in space.

Now, if quasars are simply very young galaxies, then the above observations imply that the quasar stage of a galaxy does not last long—only a few hundred million years. Prior to the quasar stage the young galaxy develops a supermassive black hole in its nucleus—if, in fact, such demons power quasars and active galaxies.

This model, if correct, suggests that young galaxies formed with very dense cores of stars and gas. If the first stars were extremely massive (100 solar masses or more), they would evolve quickly and possibly form massive black holes. Nearby material could fall into them, and they might collide with each other to produce supermassive black holes. Rapid feeding of these black holes results in quasars. As the core material is swallowed up and thins out, the quasar becomes less active and develops into an active galaxy, such as a Seyfert.

From Schmidt's work it seems that the quasar stage of a galaxy's evolution began a few billion years after the Big Bang and lasted at most 10^8 years. A galaxy goes through an active galaxy phase for a few billion years. Then somehow the active phase ends, leaving a galaxy with a monstrous black hole in its nucleus! Unfortunately, we have only suggestive evidence that a black hole hides in the core of our Galaxy or others like it.

24.8 COSMIC NEUTRINOS WITH MASS: MASTERS OF THE UNIVERSE?

We have not yet paid our respects to the most populous particle in the universe—the ghostly neutrino. During the first second or so in the universe's history, the hot Big Bang model predicts that many neutrinos should be produced by particle and photon interactions. (Every time a neutron decays, for example, a neutrino is produced: $n \rightarrow p + e + \nu$.) Relic neutrinos should outnumber ordinary matter (protons, neutrons, and electrons) by about 10^9 to one. Every cubic meter of space should now contain some 10^8 relic neutrinos.

Now, if neutrinos lived up to their billing as massless particles that travel at the speed of light and interact feebly with ordinary matter, the story would end right here. All these cosmic neutrinos left over from the Big Bang would simply play a passive role in the universe now. But if these neutrinos have mass, even a little bit, they play a dominant role in the universe now and perhaps also in the past.

In 1980 two independent experiments produced indications that neutrinos may have a small mass, about 0.002 percent that of an electron. A mass of this order is predicted by some modern models of elementary particles (called grand unification theories, often referred to as GUTs). Even such a tiny mass for neutrinos makes all the difference in the universe.

First, massive neutrinos may have a combined mass great enough to close the universe.

Second, they can solve the riddle of galaxy formation. If, as the cosmic background radiation indicates, the universe was smooth (fluctuations no more than 1 part in 10^4), how did galaxies form so quickly? Neutrinos would have decoupled from matter much earlier than radiation, at a time of roughly 1 sec after the universe began. But, if neutrinos have mass, they would clump together by their own gravity in a few thousand years (well before regular matter could do so). After decoupling, the neutral matter could collapse and would quickly fall into the regions of neutrino clumps. This would trigger rapid galaxy formation. The size of the massive neutrino clumps has been calculated to be roughly 50 million ly, about the same scale as that for superclusters of galaxies.

Finally, massive neutrinos may also bind clusters of galaxies. Initially fast-moving neutrinos with mass would slow down as the universe expands to speeds of only a few kilometers per second. So they would be gravitationally captured by a cluster of galaxies (especially the larger ones) and could bind them against evaporation.

These ideas are still speculative because we don't have firm evidence that neutrinos do have mass. But if so, even the tiniest amount makes them masters of the universe. We would then have a cosmos dominated by matter unlike that which we can easily see.

SUMMARY

This chapter has emphasized theoretical ideas concerning the universe by spending more time with the cosmological models than with observations. The two main models presented were the Steady State model and the Big Bang model. We pointed out that present observations offer major stumbling blocks to the Steady State model.

The discovery of the cosmic blackbody microwave radiation boosted the Big Bang theory to its present status as the proper model of our universe. The isotropy of the radiation implies that our assumptions of isotropy and homogeneity are reasonable. The

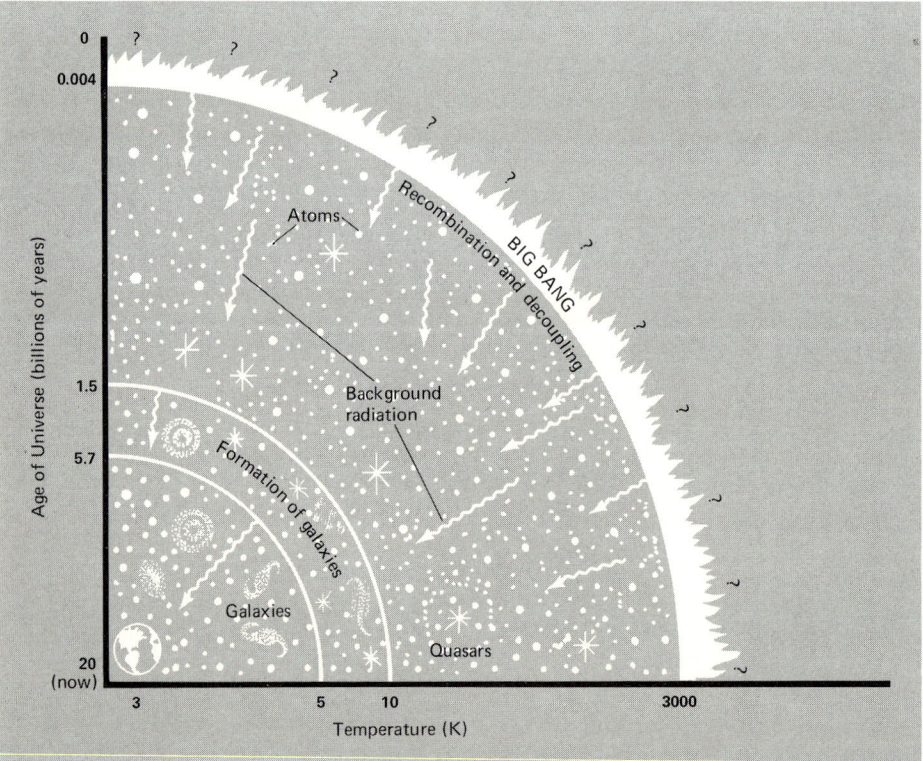

FIGURE 24.17 A schematic, visual history of the universe in the standard Big Bang model. We see this past in any direction we look into space.

present temperature of the radiation and the universe's expansion tell us that the universe was hot and dense in the past. The model also predicts the variation of the temperature with time. This permits an outline of the history of the universe that describes the interaction of radiation and matter.

The universe begins as a cosmic fireball (Fig. 24.17). In the first few seconds of the expansion high-energy radiation creates matter. In the first few minutes the matter interacts to form helium nuclei, about 25 to 30 percent by mass. The universe expands more slowly. Eventually nuclei capture electrons to form neutral atoms. At this point matter can be brought together by gravity to form galaxies and stars.

The broad outlines of this history do not depend strongly on the details of the Big Bang model. For example, whether the universe is hyperbolic, closed, or flat does not affect the general conclusions for the universe's past. In all these cases the universe had a hot Big Bang origin and still has about the same amount of helium as produced in early times.

Key Words

observable universe
physical universe
Olbers' paradox
homogeneous
isotropic
Steady State model
Big Bang model
cosmological principle
perfect cosmological principle

creation of matter
cosmic blackbody
   microwave radiation
primeval fireball
particle threshold temperature
heavy-particle era
light-particle era
radiation era

matter era
decoupling
recombination
deuterium abundance
gravitational instability
Jeans length
turbulence
cosmic neutrinos
neutrino mass

Review Questions

1. State in one sentence how the philosophical basis of the Steady State model differs from that of the Big Bang model. (Objectives 1 and 3)
2. Make a list of the fundamental cosmological observations. (Objective 2)
3. Interpret the observations in question 2 in the framework of the standard Big Bang model. (Objective 3)
4. Give one observational argument for asserting that the microwave background is cosmic in origin. (Objectives 4 and 5)
5. How does the discovery of the cosmic microwave background radiation "disprove" the Steady State model? (Objective 6)
6. List the elements that can be made in a hot Big Bang, and give one reason why no elements heavier than lithium and beryllium are manufactured. (Objective 9)
7. What observational evidence do we have that backs up the standard Big Bang model? (Objectives 3, 5, 6, 7, and 9)

Problems

1. If the cosmic background radiation has a blackbody temperature of 3 K, calculate the wavelength of its peak intensity. (Objective 4)
2. Calculate the *minimum* temperature needed in a photon gas in order to make proton-antiproton pairs. (Objective 8)

3. For a Hubble constant of 50 km/sec/Mpc and a cosmological density of 10^{-28} kg/m^3, calculate how many atoms would have to be created in a cubic meter each year to maintain a steady density despite the expansion. (Objective 3)
4. Within the Galaxy now the average density is about 10^{-21} kg/m^3 and the average temperature is about 100 K. Find the Jeans length and the corresponding Jeans mass. (Objective 10)
5. Calculate the wavelength of a typical photon at the time of decoupling. (Objective 9)

BEYOND THIS BOOK . . .

* J. Singh looks at the philosophical base of cosmology in *Great Ideas and Theories of Modern Cosmology* (Dover, New York, 1970).
* For a short introduction to relativistic ideas, try *Relativity and Cosmology*, 2nd ed. (Harper & Row, New York, 1977) by W. Kaufmann III.
* P. J. E. Peebles and D. Wilkinson discuss the discovery of the cosmic background radiation in "The Primeval Fireball," *Scientific American*, June 1967. Updates can be found in "The Cosmic Background Radiation" by A. Webster, *Scientific American*, August 1974, and "The Cosmic Background Radiation and the New Aether Drift" by R. Muller, *Scientific American*, May 1978.
* J. R. Gott, III, J. E. Gunn, D. N. Schramm, and B. Tinsley present an excellent discussion on "Will the Universe Expand Forever?" in *Scientific American*, March 1976. Their answer is yes.
* For a more comprehensive exposition of the origin of the universe in the standard Big Bang model, read *The First Three Minutes* (Basic Books, New York, 1977) by S. Weinberg.
* Robert Wilson presents a personal history of the discovery of the relic radiation in "The Cosmic Microwave Background Radiation," *Science*, vol. 205, August 1979, p. 866.
* A good recent exposition of cosmology and the origin of galaxies is contained in *The Big Bang* (Freeman, San Francisco, 1980) by Joseph Silk.
* For a discussion of the imbalance of matter over antimatter in the universe and the connection to particle physics, read "The Cosmic Asymmetry Between Matter and Antimatter" by F. Wilczek, *Scientific American*, December 1980.

PART FIVE
COSMIC SPECULATION

LIFE IN THE UNIVERSE

25

LEARNING OBJECTIVES

After studying this chapter, you should be able to:

1. Identify at least three characteristics of life as we know it.
2. State the central dogma of modern biology about life's origin.
3. Specify the chemical basis of life, and pinpoint those processes that resulted in the formation of the essential elements and the deposition of those materials in the earth.
4. Outline a model for prebiological chemical evolution on the young earth and describe what experiments support this model.
5. Outline the evidence for the early evolution of the simplest life on the earth.
6. Describe what effects the origin of life and its expansion had on the earth's environment.
7. Sketch a time line from the creation of the universe to the present, highlighting the crucial events of physical, chemical, and biological evolution that have led to the development of human beings on the earth.
8. Describe the main results of the Viking experiments.
9. Evaluate, in light of recent information from space probes and other observations, the possibility of extraterrestrial life in the solar system.
10. Assess the possibility of other planetary systems in our Galaxy, relying on both theoretical arguments and observational evidence.
11. Argue—with clearly stated assumptions, biases, and guesses—the possibility of life elsewhere in our Galaxy.

CENTRAL QUESTION:
What are the characteristics and possible origin of life as we know it, and what consequences follow from these for the possibility of life elsewhere in our Galaxy?

How did we get here? Is there anyone out there? To answer these questions, we must examine where we stand in the span of cosmic evolution. The origin of life on the earth (as understood by scientists today) involved a *natural* sequence of chemical and biological evolution. The basic material was there. It needed only to be put together in a special way. The special arrangement of molecules that make a living organism on the earth results naturally from the chemical properties of matter. The origin of life may be both natural and universal.

Are we alone? How life developed here gives us some grasp on whether or not life exists elsewhere. Can we estimate, on some reasonable basis, the chances that life exists elsewhere in our Galaxy or in other galaxies? To some degree we can, from basic physical, chemical, and astronomical knowledge. Admittedly the estimate is speculative and debatable, but it rests on a scientific basis.

This chapter will lead seemingly far astray from astronomy into the realm of molecular biology and biological evolution. These excursions are needed to investigate the essential features of life on the Spaceship Earth. They'll show how the probable origin of life followed this sequence: physical evolution, chemical evolution, and biological evolution. Each provides clues to whether life arose elsewhere in the universe. To make the jump from here to there relies on the assumption of the uniformity of physical laws. It rests on the hope that *we are typical*. This may not be true. *We may be unique*.

25.1 THE NATURE OF LIFE ON EARTH

How do you know if something is alive? The question must be dealt with before searching for life elsewhere in the universe. But it's a sticky one. Go out sometime and look for things that are alive, dead, or nonliving. Then consider how you came to these judgments. It's probable that you relied on some very subjective evaluations.

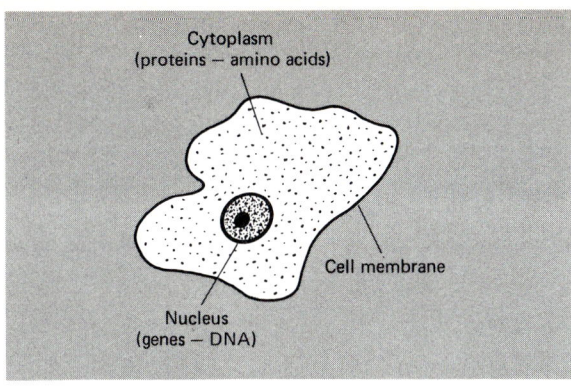

FIGURE 25.1 The key parts of a typical cell (simplified). The nucleus contains the genetic material, the protoplasm is the site of metabolic activity, and the cell membrane allows the passage of selected materials, such as food.

Rather than try for an exact definition of life, let's accept a useful rule of thumb: *Living things are those that reproduce, mutate, and reproduce the mutations*. What does this mean? First, living things have an *organization* that they pass on when they reproduce. So all living things are *organisms*. A simple example of an organism on the earth is a *cell* (Fig. 25.1). Second, reproduction does not involve simply the copying of an organism. Occasionally the offspring has a genetic difference from its parents—it exhibits a *mutation*. (The genes in a cell's nucleus carry the information code for how an organism is to be put together; if the genes change, the organism changes.)

Some mutations result in the quick death of the mutated organism. Others have no lethal effects, and the organism survives and transmits its new characteristic to its offspring. This is important, because mutation provides the possibility of change. Together with the process of *natural selection*, it leads to *evolution*, the development of organisms of greater complexity, better adapted to their environment.

Mutation and evolution are distinguishing features of life. The ability to mutate distinguishes living corals from nonliving rocks. The quartz crystals in rock are organized and duplicated in the growth of new rocks. But a flawed crystal *cannot* duplicate itself, and it is never better adapted to its environment. A mutant coral duplicates itself and not its nonmutant parent, and thus has a chance to adjust to the world around it.

All terrestrial organisms operate on the basis of interactions among relatively few types of molecules. These in turn consist of simple combinations of the most common chemical elements in the universe: hydrogen, carbon, nitrogen, oxygen, and a few others (Table 25.1). So all terrestrial organisms have a common chemical makeup, closely connected to the physical evolution of the universe, in that it involves the most common elements made in stars or in the Big Bang explosion creating the universe. (The exception is helium. Though the second most common element, helium is not found in living organisms, because it is chemically inactive.) This common chemistry suggests the central scientific idea for the origin of life. Life is a natural consequence of the evolution of the universe. In the view of modern evolutionary biology, the central dogma about life's origin is: *Life arose from nonlife*.

25.2 BIOCHEMISTRY SIMPLIFIED (A LOT!)

To understand life you must understand its chemistry, for chemical reactions provide the energy that keeps humans and the other creatures of the earth alive. The sun is the source of this life energy. Sunlight powers photosynthesis (Fig. 25.2) in plants, which take in

TABLE 25.1 Comparative Abundances (Percentage of Number of Atoms) of the Elements in the Earth's Crust, Solar System (Mostly the Sun), and Human Beings

Earth's Crust		Solar System		Human Beings	
Oxygen	55.1	Hydrogen	86.2	Hydrogen	63
Silicon	16.1	Helium	13.8	Oxygen	25
Hydrogen	15.5	Oxygen	7.7×10^{-4}	Carbon	10
Aluminum	4.9	Neon	4.3×10^{-4}	Nitrogen	1
Sodium	2.0	Carbon	3.5×10^{-4}	Calcium	0.5
Iron	1.6	Nitrogen	1.0×10^{-4}		
Magnesium	1.4	Silicon	3.0×10^{-4}		
Potassium	1.1				

Note: The earth's crust is dominated by the silicon and oxygen that are present as silicon compounds in rocks. The solar system's composition is dominated by the sun. The large amount of hydrogen in the human body is due to the fact that animal cells are about 85 percent water by mass. Note that if neon and helium (which are chemically inert substances) are scratched from the solar system list, the elements' rank is the same as that for us.

water and carbon dioxide to produce sugar (a food) and free oxygen. Sugars are carbohydrates, simple compounds of carbon, oxygen, and hydrogen. Animals eat the plants and breathe the oxygen, which burns the food slowly, to produce energy to keep them alive, growing, and reproducing. They produce water and carbon dioxide, which plants take in, and the cycle turns again. Almost all earth's organisms live on sunshine as the ultimate energy source.

What makes up these organisms? Chemically, you'd find that, aside from water, you are made largely of *proteins*—long molecule chains, the links of which are smaller subgroups called *amino acids* (Fig. 25.2). The basic units of chemical structure of an amino acid are the amino group, one nitrogen atom linked to two hydrogens (NH_2), and the carboxyl group, a carbon atom linked to two oxygens and a hydrogen (CO_2H). There are many different types of amino acids possible, but only 20 types make up all the proteins of the earth. The common elements of the universe—carbon, oxygen, nitrogen, and hydrogen—make up the basic units of life—proteins and amino acids.

Proteins work in two ways. They are the structural material of which cells are made, and they act as *enzymes*, which monitor and facilitate important chemical reactions in a cell. But how is the construction of cells and the functioning of the enzymes controlled?

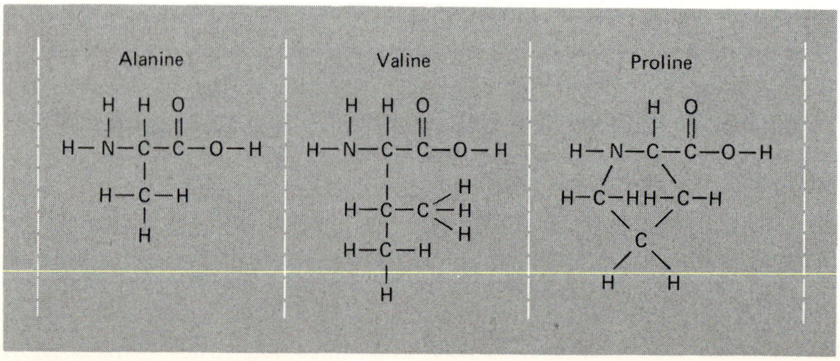

FIGURE 25.2 The chemical structure of three common amino acids found in life on the earth. Note that hydrogen, oxygen, carbon, and nitrogen—four of the most common elements in the cosmos—form the chemical basis.

That's the job of a different kind of long molecular chain found in the cell nucleus—*nucleic acid*. The most important nucleic acids are *deoxyribonucleic acid* (DNA) and *ribonucleic acid* (RNA). The nucleic acids are enormous molecules; a strand of human DNA contains *billions* of atoms and stretches *meters* in length. Chemically, nucleic acid consists of the bonding of four bases (carbon-nitrogen compounds called adenine, guanine, cytosine, and thymine in DNA or uracil in RNA), sugars (deoxyribose in DNA and ribose in RNA), and phosphates (compounds of phosphorus and oxygen) to form the chain (Fig. 25.3). So the most common atoms in nucleic acids are just those found in proteins—hydrogen, carbon, nitrogen, and oxygen—plus phosphorus.

DNA serves as the chemical blueprint that informs the protein in cells how to function. The nucleic acid molecule consists of two long molecular strands cemented together in a helical structure (Fig. 25.4). When a cell divides, the strands separate, and each half chemically constructs its complementary half, with the help of enzymes, to form a complete chain. So by unzipping and growing chemically, the nucleic acid duplicates itself and the chemical information on how to construct and run a cell. The offspring inherits this information. DNA forms the chemical basis of genetics, yet it comprises a small amount of an organism's total material. In your whole body you have only about one teaspoon of DNA!

DNA contains the instructions on how to make the proteins. But it needs to use proteins (enzymes) to reproduce itself. Here's the original chicken-and-egg question! Which came first, the self-replicating mechanism or the enzymes essential to make the mechanism work? Neither really—the complex interrelation of these molecules evolved from much simpler structures over a long period of time.

25.3 IDEAS ABOUT THE GENESIS OF LIFE ON THE EARTH

Early biologists, from evidence such as the oozing of maggots from decayed matter and the appearance of microorganisms in laboratory solutions, concluded that life can arise spontaneously from nonliving matter. The theory of *spontaneous generation of life* views life as born directly and rapidly from nonliving material.

In the nineteenth century Louis Pasteur (1822–1895) devised careful experiments that finally demolished spontaneous generation. By using scrupulously clean equipment and carefully controlled procedures, Pasteur in 1862 showed that the microorganisms appearing in laboratory soups actually came from the air. If the equipment was properly sterilized (pasteurized), the laboratory solutions would not grow microorganisms. Pasteur's experiments refuted the idea of spontaneous generation. Living matter could not suddenly spring from nonliving material.

Modern biology also sees life's origin as coming from nonliving material, but not spontaneously. Rather, it arises as a natural consequence of the slow processes of chemical and biological evolution.

Darwin's Long Thread to Humankind

In 1831 the British ship H.M.S. *Beagle* set sail for a five-year voyage around the world. Charles Darwin (1804–1882) was the ship's naturalist. During the voyage of the *Beagle*

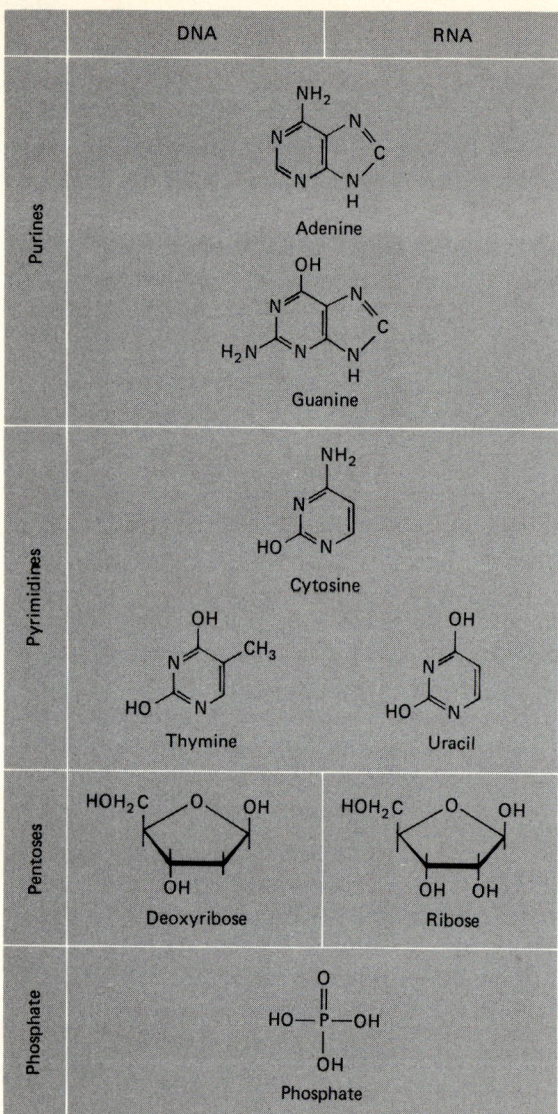

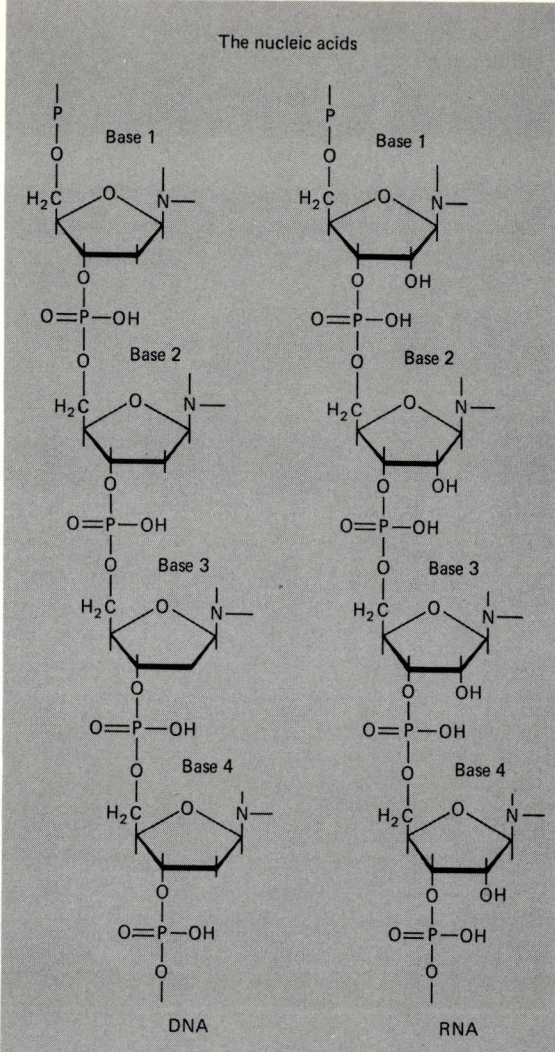

FIGURE 25.3 The chemical building blocks of DNA and RNA. Five chemical bases are involved: adenine, guanine, cytosine, thymine (in DNA) and uracil (in RNA). These bases are found linked in pairs (a purine with a pyrimidine) and connected by a backbone chain of phosphates and pentose sugars. The sugars are deoxyribose (in DNA) and ribose (in RNA). The phosphorus (in the phosphates) is the most complex atom in the chemical construction.

FIGURE 25.4 Schematic diagrams of one strand of the double-stranded DNA and RNA molecules. Each molecule consists of two strands in a helix, cross-linked by pairs of adenine, guanine, thymine (uracil), and cytosine. The bases cannot pair up to random; adenine pairs only with thymine (uracil) and guanine with cytosine. The nonrandom ordering provides the chemical basis of the DNA information coding.

Darwin made critical observations that led him to conclude that biological evolution had occurred and was now still occurring.

In *The Origin of Species* (published in 1859) Darwin declared that living species are adapted for survival in their environment. No single mutation can generate the evolution of a species. Some feedback mechanism is required to select mutations which make a species move with its environment. Although Darwin lacked modern knowledge of

25.3 IDEAS ABOUT THE GENESIS OF LIFE ON THE EARTH

genetics, he patiently unearthed the general feedback mechanism: *natural selection*. (Alfred Wallace also had this idea and, after an agreement with Darwin, both published their ideas at the same time.)

What is natural selection? Put simply, modern biologists view natural selection as a consequence of successful reproduction. Individuals well adapted to a local environment not only survive, but, more important, they usually succeed in producing offspring. Their reproductive success spreads their genetic material. Their children tend to survive and reproduce successfully. Eventually their genes dominate the local gene pool, and the survival and expansion of new genetic material results in evolution. The new genetic material comes from random mutations. Natural selection directs the random jumble of mutation into biological evolution.

The Darwin-Wallace theory implied a natural development from simple to complex organisms with the passage of time. This progression presupposes some primeval life form to start the process of biological evolution. At some point life arose and biological evolution began, but chemical evolution must have preceded it.

Darwin did have ideas about the stage before the organism. He wrote how "in some warm little pond, with all sorts of ammonia and phosphoric salts, light, heat, electricity, etc., present, . . . a protein compound was chemically formed ready to undergo still more complex changes." Here's the germ of the concept of chemical evolution. But under what conditions did the "little pond" suitable for forming life appear?

Clues from Geology

As geologists pare back rock layers, they reveal the geologic history of the earth. However, until this century the geologist had no clear-cut method of tagging the successive layers with a date. The layered rocks indicated a relative order only, for younger rocks are generally found on top of older ones. The advent of radioactive-decay dating techniques (Section 13.1) fixed the absolute dates of the geologic time sequence (and also determined the age of the earth as 4.6 billion years).

Until recently the fossil record faded at the hazy border (about 600 million years ago) between Cambrian and Precambrian times (Table 25.2). Extremely primitive life forms did not possess structures that could be fossilized easily. Careful microscopic inspection of ancient rock samples, however, reveals the remains of bacteria and algae from 1 to 3.5 billion years old. These provide important clues to life's evolution on earth.

One of the oldest set of rocks containing microfossils, called the *Fig Tree formation*, lies on the border between the Republic of South Africa and Swaziland. Radioactive dating places these rocks at about 3.2 billion years. The rocks contain evidence of two distinct life forms: rod-shaped structures resembling modern bacteria (Fig. 25.5) and round cells similar to modern blue-green algae.

Bacteria and blue-green algae both have a simple cell structure devoid of a cell nucleus and specialized parts. They reproduce by simply splitting in two. The lack of specialized cellular structure indicates that these fossils lie on the lowest rungs of the evolutionary ladder.

These self-sustaining organisms must have been able to manufacture food from inorganic substances. Photosynthesis provides the simplest chemical mechanism for freeing the organism from dependence on its local environment to produce food, and many modern bacteria and all modern blue-green algae are photosynthetic. So the Fig Tree

TABLE 25.2 Simplified Geologic Time Scale

Era	Age (millions of years)	Life
Cenezoic	Now	Homo sapiens dominates earth
	2–4	Homo sapiens appears
	58	Mammals appear
Mesozoic	63	End of dinosaurs
	135	Flying reptiles
	181	First bird
Paleozoic	239	Dinosaurs appear
	280	First reptiles and insects
	400	First amphibians
	410	First land plant fossils; first insect fossils
	460	First fish fossils
Cambrian	500	First plant fossils
Precambrian	600	
	3800	Oldest possible fossils; oldest rocks
	4500– 5000	Formation of earth

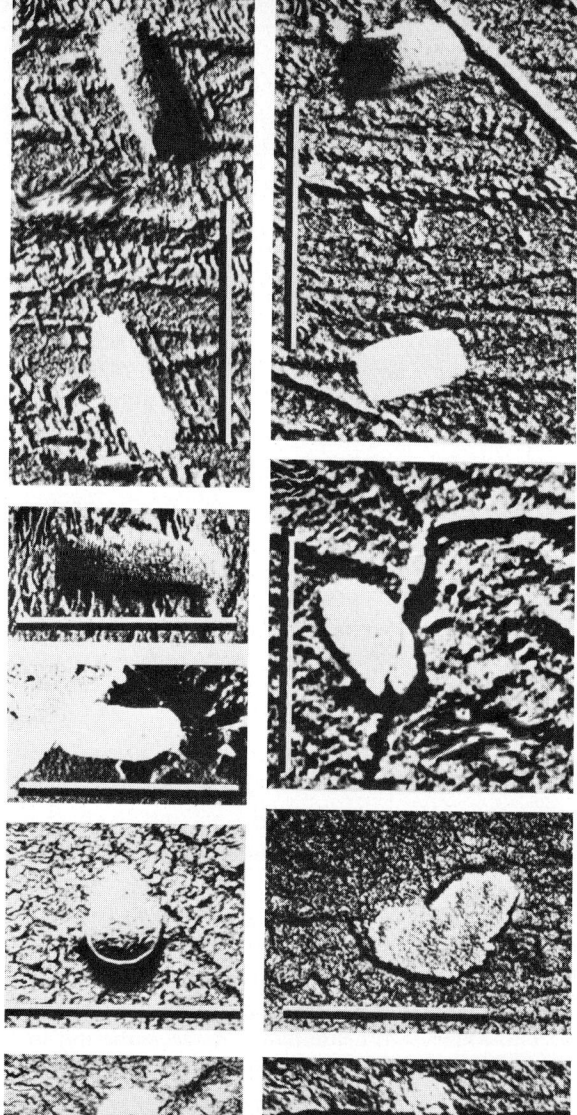

FIGURE 25.5 Some of the oldest fossils known. Found in the rocks of the Fig Tree formation in Africa, these relics have an age of about 3.2 billion years. Their shape is similar to that of modern bacteria. (*Courtesy E. Barghoorn.*)

fossils appear to represent a crucial stage in biological evolution, the development of photosynthesis at about 1.4 billion years after the formation of the earth.

Recently microfossils have been discovered at North Pole, Western Australia. These rocks are a bit older than the Fig Tree formation, about 3.5 billion years old. They contain spheres of carbon, which have shapes similar to the fossils in the Fig Tree rocks. The discoverers argue that the carbon spheres are biogenic in origin. At another location in Australia, a rock, also dated at 3.5 billion years, contains layered structures thought to be built by colonies of bacteria. These experiments suggest that early life may have been more widespread than previously thought.

Evidence for the next evolutionary step comes from the *Gunflint formation* along the

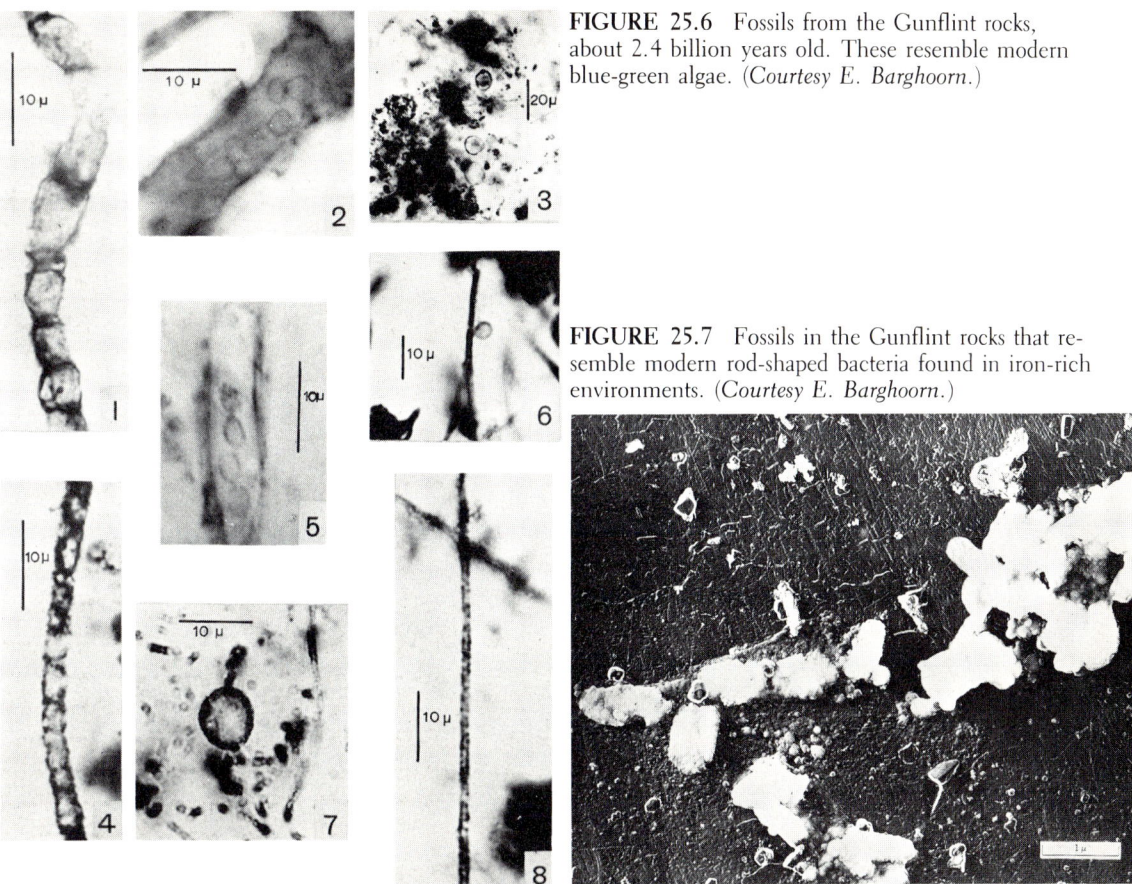

FIGURE 25.6 Fossils from the Gunflint rocks, about 2.4 billion years old. These resemble modern blue-green algae. (*Courtesy E. Barghoorn.*)

FIGURE 25.7 Fossils in the Gunflint rocks that resemble modern rod-shaped bacteria found in iron-rich environments. (*Courtesy E. Barghoorn.*)

shores of Lake Superior in western Ontario, Canada, for which radioactive dating sets a maximum age of about 2.4 billion years. Many fossils here show characteristic algal structures similar to modern photosynthetic blue-green algae (Fig. 25.6). Even more striking are structures suggestive of a cell nucleus—an indication that a larger proportion of a cell's material is DNA. With the advent of definitive genetic structures in the nucleus, a cell gains a greater potential for mutation and rapid evolution. The Gunflint fossils show that this level may have begun about 2 billion years ago.

The Gunflint rocks also contain fossils that look very much like rod-shaped bacteria (Fig. 25.7). A comparison with living bacteria indicate a resemblance to modern iron bacteria. The Gunflint rocks are rich in iron, which may have been deposited there by the biologic activity of the now-fossil bacteria.

The next document of the lower rungs of biological evolution comes from the *Bitter Springs formation* in the Northern Territory of Australia. The rocks found here are approximately 1 billion years old. Three of the fossils appear to resemble modern types of green algae. Unlike the simpler blue-green algae, the green algae contain complete cell nuclei. There is also evidence here for a new kind of reproduction. Rather than each cell simply duplicating its genetic material and splitting in two, now cells had developed the capacity to join together and mix the genetic material from different cells. The Bitter

Springs fossils display this capacity for increased genetic mixing; 2 billion years of evolution had passed before organisms attained this potential. Once it happened, biological evolution greatly accelerated, because the mixing of genetic material from different cells made it possible to form many new genetic combinations very quickly. The slow process of mutation in creating change was replaced by the rapid lottery of genetic mixing.

The fossil hunters have unearthed vital clues in the biological evolution of life. The fossil discoveries imply (1) that evolution takes a long time (*billions* of years) and (2) that chemical evolution must have been completed on the primeval earth no more than about 1 billion years after the earth formed in space, for DNA and proteins must exist before cells form. What must have been the conditions on the young earth for this chemical evolution to take place?

Clues from Astronomy

Let's step back a moment to reconsider the physical evolution of the cosmos. The universe began in a Big Bang some 15 to 20 billion years ago. The Big Bang produced hydrogen and helium (and traces of lighter elements). The other elements up to iron were produced in the cores of massive stars. When massive stars explode, heavier elements are made, and the star-manufactured materials are blown into space. So at least one generation of massive stars must have died before the elements necessary for life were available in interstellar clouds.

The sun is a later generation star, formed about 4.6 billion years ago. The birthdates of the earth and moon are essentially the same as that of the sun. This coincidence reflects the fact that the formation of the sun, earth, and moon occurred out of the same cloud of gas and dust (this is a good time to review the material in Chapters 13 and 19).

The sun reflects the average chemical composition of material in our Galaxy. The earth ended up with a composition quite different from that of the sun (Table 25.1). How did this happen? Recall (Chapter 19) that the nebular model for the origin of the solar system pictures the protoearth forming from the accretion of planetesimals. The chemical composition of the planetesimal material depends on the temperature in the solar nebula, as given by the condensation sequence (Table 19.5). Most of the gaseous and icy materials were left out in the agglomeration of the planetesimals to make the earth. At 1 AU it was too hot for them to condense.

Let's look a bit more closely at the earth's depletion of chemicals relative to the solar system in general (Table 25.3). Note that the most depleted elements are the inert noble gases (helium, neon, argon, krypton, and xenon), which condense only under very low temperatures and form few chemical compounds (and then only under special circumstances). Except for helium, atoms of the inert gases have large enough masses that little would be lost by atmospheric escape from the earth in 4.6 billion years. So the inert gases, since they are depleted now, must have been present initially only in small quantities.

Elemental depletion and the expectations from the condensation sequence give an estimate of the solar nebula's temperature at the earth's formation. For silicates and iron to condense, the temperature must have been less than 1500 K. Sulfur vapor reacts with iron at about 680 K. Sulfur in the earth is depleted (Table 25.3), but not by a large amount. So the sulfur-iron reaction began, but did not use up all the available sulfur. Conclusion: The earth formed at a temperature a little below 700 K, probably around 600 K or so.

25.3 IDEAS ABOUT THE GENESIS OF LIFE ON THE EARTH

But a 600 K condensation temperature presents a sticky problem with carbon. The condensation sequence predicts that carbon (in methane, CH_4) condenses at 120 K, far cooler than 600 K. So the planetesimals forming the earth contained little methane and therefore little carbon. But the carbon *is* here; otherwise, we wouldn't be!

As yet this carbon deposition problem is unsettled, but here's a possible way out. At high temperatures carbon combines with oxygen to form carbon monoxide (CO). Under the proper conditions carbon monoxide combines with hydrogen to form large hydrocarbons, such as those found in tar. If such reactions went on in the solar nebula, the tars would form on the grains that made up the protoearth. In addition, if ammonia (NH_3) and water are present, the chemical reactions result in the formation of organic molecules, such as amino acids. This process—formation of carbon molecules on grains—is one way carbon might have been incorporated into the earth.

Did this happen? We have an indirect clue. Some molecules found in interstellar clouds can be made by the same kinds of reactions. These include hydrogen cyanide (HCN), formaldehyde (H_2CO), formic acid (HCOOH), and cyanoacetylene (HC_3N). Perhaps these molecules are left over from the formation of a planetary system around a star. Admittedly, this is only circumstantial evidence.

We now think the earth had little primordial atmosphere. How did it gain a substantial one? If the earth's surface were hot at the time—perhaps even completely molten—the heat would gasify volatiles within the earth, which would rise upward to the surface. Volcanoes today outgas water, carbon dioxide, hydrogen sulfide, methane, and ammonia, for instance. We would expect pretty much the same materials to outgas from the primitive earth to form the *secondary atmosphere.*

How fast this outgassing took place depended on the crustal temperature. It's still a matter of debate whether or not the earth's crust was kept molten by impacts of planetesimals. That depends on the rate at which the planetesimals accumulated. If the rate was slow (the whole process taking more than 100 million years), the heat from the impacts would radiate into space fast enough to keep the temperature below the melting point of rock. If the rate was rapid (taking less than 100 million years), the infalling material would have melted as it accumulated. In either case, radioactive decay later heated the earth's interior.

TABLE 25.3 Depletion of Elements in the Earth

Element	Depletion (abundance in earth/abundance in solar system)
H	2.5×10^{-7}
He	10^{-14}
C	10^{-4}
N	1.3×10^{-6}
O	0.16
Ne	2.5×10^{-11}
S	0.32
Cl	0.20
Ar	5.0×10^{-7}
Kr	6.3×10^{-8}
Xe	3.2×10^{-7}

If the earth formed molten, then outgassing of its secondary atmosphere happened quickly, in 100 million years or less. Although a lot of gas was produced in this way, it still did not make the present atmosphere, for this secondary atmosphere lacked oxygen. In other words, the secondary atmosphere was probably nonoxidizing and hydrogen-rich.

The stage is set. What were the first steps in chemical evolution on the earth?

25.4 THE SPARK OF LIFE

In 1924 the Russian biochemist Aleksandr I. Oparin proposed that life on earth was the result of gradual chemical evolution. He also recognized that the atmosphere for such evolution must have contained hydrogen compounds. Oparin's work was not published in English until 1938. Before this happened, the English biologist J. B. S. Haldane had written in 1928 an article in which he proposed that

> . . . when ultraviolet light acts on a mixture of water, carbon dioxide, and ammonia, a variety of organic substances are made, including sugars, and apparently some of the materials from which proteins are built up. Before the origin of life they must have accumulated until the primitive oceans reached the consistency of a hot dilute soup.

How was the step taken from hydrogen compounds to organic compounds? Haldane had recognized that a nonoxidizing atmosphere required anaerobic rather than aerobic life forms. An *aerobe* uses the oxidation of sugar to maintain its metabolism and needs oxygen in the process. An *anaerobe* does not use oxidation to generate energy. An example of an anaerobe is the yeast used to ferment beer; the end product is alcohol instead of water. Compared with an aerobe, an anaerobe operates an inefficient energy-extraction machine. From the same amount of sugar, oxidation generates about 20 times as much energy as fermentation. If oxygen is available, aerobes have a survival advantage over anaerobes. The earth's early atmosphere may not have contained much oxygen, so it supported anaerobic organisms.

Energy Sources to Make Molecules

Before the creation of an organism, even more primitive organic substances must exist: DNA and proteins. To synthesize complex molecules from simpler ones requires free energy. Photosynthesis in plants now captures solar energy and stores it in the form of chemical bonds. On the young earth, sunlight was available, but plants were not.

Where did the energy needed for synthesis of complex molecules come from (Table 25.4)? One possibility is ultraviolet radiation from the sun. Theoretical studies of the sun indicate that the total solar energy striking the earth would have been about 25 percent less 4.6 billion years ago. But it is the ultraviolet energy, not the total solar energy, that matters for the origin of life. Solar ultraviolet at wavelengths less than 2200 Å plays a key role because it is absorbed by molecules and can cause the formation of more complex ones. (The ozone layer now filters out most of the ultraviolet radiation. Because of the lack of free oxygen, the secondary atmosphere, produced by outgassing, did not form an ozone layer.) As a pre-main-sequence star, the sun's surface temperature was less than

25.4 THE SPARK OF LIFE

now. So we might expect that in the past it emitted less ultraviolet, as a fraction of its total energy, than it does now. Estimates are 5 to 40 percent of the present values at 1500 to 2500 Å at a time 4.4 billion years ago. On the other hand, the sun may have had a more active chromosphere in the past, heated by sound or magnetic waves from a deep convection zone (Section 6.5). So despite the lower temperature and luminosity, the sun may have emitted more ultraviolet light (from its hotter chromosphere) in the past than it does now.

Radioactive decay can also release free energy. The rate now is about 1/100 the energy input from solar ultraviolet. But 4.6 billion years ago the fraction of radioactive isotopes in the earth's surface would have been roughly 3 times greater than now. Thus radioactive energy might have been comparable to solar energy on the early earth, unless the sun had an active chromosphere.

Heat from the earth's crust—lava from active volcanoes—also can drive chemical reactions. This energy source now generates little energy compared with radioactivity and sunlight. Perhaps 10 times more volcanic activity probably occurred just after the earth's crust solidified than now, but the total energy available from this source would still have been a fraction of that from solar energy and radioactive decay.

We should mention two other possible energy sources. The first is cosmic rays from both the sun and beyond the solar system (Section 9.1). The earth's magnetic field deflects the low-energy cosmic rays, so if the young earth had little or no magnetic field, the cosmic ray intensity would be greater than now, perhaps ten times more. If the sun were more active in the past, the cosmic ray intensity might have been higher still. However, the higher intensity still wouldn't amount to much compared with other energy sources.

The second other possible energy source is electric discharges, especially as lightning. On the earth now lightning accounts for almost as much free energy as short wavelength ultraviolet radiation. It is hard to estimate the lightning intensity on the primitive earth. Before the earth cooled enough for rain to fall, probably little lightning occurred near the earth's surface. Why? The regions of thunderclouds that generate the electric charges for lightning are in the ice zone, below 0°C. The turbulent circulation of the ice crystals somehow produces the huge accumulation of charge needed for a lightning bolt. When

TABLE 25.4 Present Energy Sources Averaged over the Earth

Source	Amount (W/m^2)
Total radiation from the sun	350
Ultraviolet light at wavelengths less than 2000 Å	0.12
Electric discharges	5×10^{-4}
Cosmic rays	2×10^{-6}
Radioactivity (to 35 km depth)	0.02
Volcanoes	1.8×10^{-4}

Source: Adapted from S. L. Miller and H. C. Urey, *Science*, vol. 130, p. 245, 1959 and S. W. Fox and K. Dose, *Molecular Evolution and the Origin of Life* (W. H. Freeman, San Francisco, 1972), chapter 3.

the earth was hotter, the ice zone was higher in the atmosphere, so any lightning flashed at much higher altitudes than now. When the earth had cooled enough for the first primeval rain, which filled the oceans, lightning storms may have ranged widely over the earth's surface. Then the energy from lightning probably had at least the relative importance it does now.

In summary, radioactivity from the crust, solar ultraviolet radiation, electric discharge, volcanic heat, and cosmic rays all provided sources of free energy available for molecular synthesis. Some of these energy sources came in spurts rather than at constant rates. But we must remember that all these energetic processes have the capacity to destroy as well as to help synthesize molecules. The balance between creation and destruction determines the number and kinds of molecules which could exist.

Synthesis of Simple Organic Molecules

The critical conditions for chemical synthesis are free energy and a nonoxidizing atmosphere—one that consists of hydrogen rather than oxygen compounds. Absence of oxygen is important, because free oxygen destroys organic compounds. Laboratory experiments validate this key point. When our *present* atmosphere (mostly nitrogen, oxygen, carbon dioxide, and water vapor) is subjected to electrical discharges and ultraviolet radiation, only very small amounts of organic compounds are produced. These experiments show that the primitive atmosphere in which chemical evolution took place was probably not like our atmosphere now.

In 1953 Stanley L. Miller and Harold C. Urey experimented with hydrogen-rich atmospheres. They sparked a gas mixture made up of various hydrogen compounds, such as ammonia, water, and methane (Fig. 25.8). An analysis of the products turned up a slew of amino acids, including four commonly found in terrestrial proteins. Ultraviolet light as the energy source for organic molecule production was found to have a much lower efficiency than spark discharges.

Here's a very simple outline of the chemical reactions involved. Adding energy to methane, ammonia, water, and carbon dioxide results in aldehydes (compounds with the same chemical ending, CHO, as formaldehyde, HCHO) and nitriles (large molecules containing hydrogen cyanide, HCN). Then aldehydes and nitriles combined with ammonia and water to produce amino acids. Formaldehyde can also produce sugar.

Many such experiments have been performed on *gaseous* mixtures (typically CH_4, NH_3, CO_2, H_2O, and H_2S) with energy provided from sources similar to those that might have been available on the primitive earth. One variation, heating of the above mixture in contact with silica (pure sand), results in a wider variety of amino acids (such as analine, valine, and proline; see Fig. 25.2). In another experiment a *solution* of ammonia and formaldehyde was heated to 185°C for 8 hours. Ten amino acids were found among the reaction products. These experiments, whether with gaseous mixtures or solutions, produce naturally *most* of the amino acids common in protein and *no* amino acids *not* found in modern protein.

We have concentrated so far on amino acids, because they form the building blocks of proteins. Other, somewhat more tentative experiments have tried for the synthesis of hydrocarbons, fatty acids, and sugars—all molecular pieces of living things. But what about the basics of DNA?

Experiments to simulate the primitive synthesis of the building blocks of DNA have

25.4 THE SPARK OF LIFE

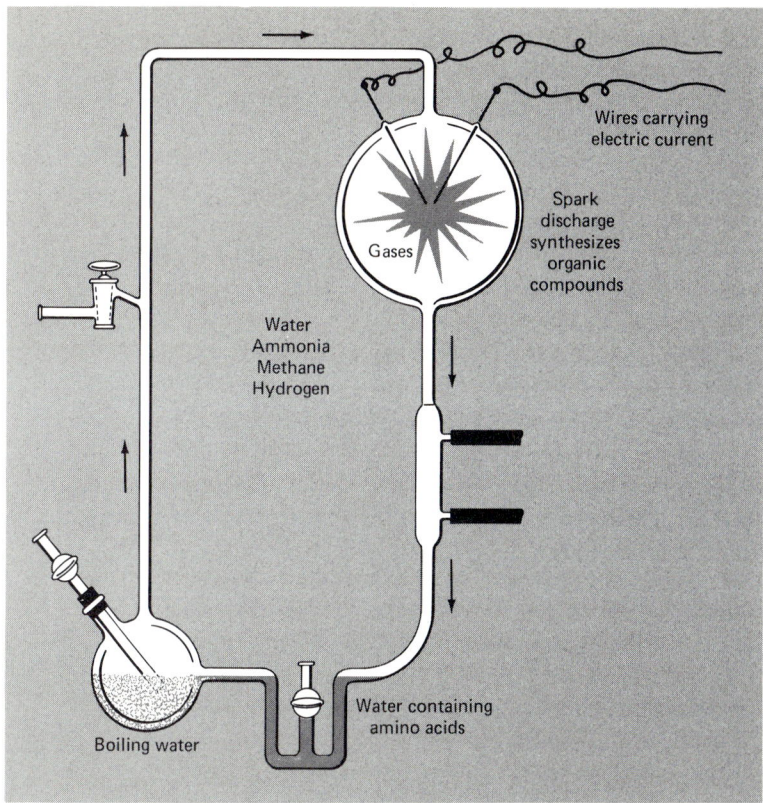

FIGURE 25.8 A schematic diagram of the Miller-Urey experiment. Electrical discharges were fired in a gas of water, ammonia, methane, and hydrogen. Output collected at the bottom included amino acids and fats.

been less extensive than those for amino acids. Recall that the DNA chain (Fig. 25.4) consists of bases (such as adenine and cytosine), a sugar (deoxyribose), and a phosphate. A key chemical precursor of DNA is hydrogen cyanide (HCN). For example, heating HCN in a water solution of ammonia produces adenine. Another critical compound is phosphate, whose presence is needed to synthesize some nucleotides and form a piece of the DNA backbone.

To sum up, the requirements for making the building blocks of life are a nonoxidizing atmosphere and free energy (ultraviolet light, lightning, heat, radioactivity). The processes appear to be natural chemical reactions in gases and liquids. Laboratory experiments show these results: H_2CO plus HCN gives amino acids; H_2CO ends up in sugars; and HCN comes out in nucleic acid bases. In addition, phosphates must also partake in the reaction. The simple organic molecules needed for life form naturally under plausible primitive earth conditions. The simple molecules can be cooked.

The Synthesis of Complex Molecules

What about the actual proteins and nucleic acids? Our knowledge of complex molecules is much shakier than of the simple molecules (and remember, we aren't yet anywhere near the complexity of living cells). Proteins and nucleic acids are not only huge molecules; they also have a special and precise architecture. How did the first of these macromolecules get together?

Let's first look at how simple molecules link together to form polymers (larger molecules containing repeating units of the smaller molecules). The process is called polymerization. To form proteins the amino group (H_2N) at one end of an amino acid joins to the acid group (COOH) at one end of another amino acid. The C and N link together to make a peptide bond. A water molecule (H plus OH) must be removed. Similarly, to put together a base, sugar, and phosphate to build a nucleic acid requires the removal of a water molecule at each bonding step. So for both proteins and nucleic acids, water must be removed to build them up. The simple molecules must be dried out to make polymers, which can be done with heat.

Sounds simple, but there's a hitch. In the scenario so far the simple molecules synthesized, perhaps in the atmosphere, fall into the primitive oceans, ponds, or lakes. Molecules on a dry, solid surface cannot participate in further synthesis, because they can't move around to meet other molecules. Furthermore, water serves to protect the new molecules from destruction by ultraviolet radiation. So whatever waters exist must catch the synthesized molecules and store them. But it's difficult to synthesize the more complex macromolecules in water, because water must be removed in the process. It's like trying to dry yourself with a towel while still taking a bath!

Difficult, but not impossible. J. D. Bernal in 1947 noted that tidal pools might just be a hospitable place for polymerization. When the tide rolls out, the water in small pockets evaporates in the sun and polymerization occurs. When the tide rolls in, the water dissolves the molecules, mixing them with other molecules. This scheme requires that the primitive earth had a moon to produce tides. This is an argument that the earth and moon formed together. (Does it mean that only planets with sizable moons develop life?)

In another possible scheme the polymerization goes on in water with enzymes or catalysts attaching themselves to water molecules in the macromolecules, thus drying them out, even though in water. Your cells do this type of water removal whenever they synthesize protein. But this scheme raises the question of where the first enzymes came from.

Experiments to make proteins and nucleic acids under prebiologic conditions have been somewhat successful. But a serious problem arises. At one end the chemist has the raw materials and at the other the nucleic acids and proteins to be produced. Energy sources drive this production. But the chemist cannot do this synthesis in one step. First, the chemist synthesizes small molecules. Then these small molecules must be put together to make large ones. To do this requires some selected mixture of simpler molecules. So as the experimenters move from a primitive soup to a more specialized one, they set conditions that may never have actually occurred on the young earth. Another problem is that the dehydration process does make very long, large molecules, but the larger the molecules, the less they have the orderly structure of modern protein molecules. The arrangements are not right.

To sum up, the precise complexity of proteins and nucleic acids makes their synthesis difficult. We do not yet understand the specific pathways to their original production.

25.5 AMINO ACIDS FROM SPACE

Meteorites provide some evidence to support the theories of natural synthesis of organic compounds. Of the three main classes of meteorites (Section 18.2), the minority are

25.5 AMINO ACIDS FROM SPACE

carbonaceous chondrites, which contain a relatively high percentage of carbon (2 percent). Only 36 carbonaceous chondrites are known to have fallen to earth. People have regularly speculated that some of the carbon contained in these meteorites might be organic in nature.

The best chance for the discovery of extraterrestrial organic materials occurs when the sample undergoes analysis in a scrupulously clean environment soon after its fall. That chance came in on September 28, 1969, when a meteorite fell in Murchison, Australia. This meteorite, a carbonaceous chondrite, was rushed to the Ames Research Laboratory of NASA and analyzed by a team of scientists headed by Cyril Ponnamperuma. The NASA research group discovered five amino acids common to living protein. The quantities were small, only a few micrograms of amino acids in each gram of the meteorite.

Some skeptics have scoffed, claiming that this experiment was mainly an example of the ease of contamination of extraterrestrial samples with earthly materials. However, the Ponnamperuma group has a clever rebuttal of this claim. Organic molecules exist in two distinct forms: right-handed ones and left-handed ones, depending on the direction of the twist of the linkage of the atoms. Almost all terrestrial organic molecules are left-handed, so earth-based contamination is expected to be left-handed. The Murchison meteorite contained just about equal quantities of right- and left-handed molecules, the left-handed forms predominating a little. This evidence strongly points away from terrestrial contamination and toward an extraterrestrial, nonbiological origin of the Murchison organic molecules. When nonbiological organic molecules are synthesized in a chemistry lab (rather than by an organism), they show an equal number of right-handed and left-handed forms.

Other samples of the Murchison meteorite have contained hydroxy acids in amounts comparable with the amounts of amino acids. (Hydroxy acids are formed when hydroxyl, OH, is added to a hydrocarbon in a particular way.) Amino acids can be formed by electric discharges in a mixture of ammonia, hydrogen cyanide, and aldehydes. Such discharges can also produce hydroxy acids from hydrogen cyanide and aldehydes. How much hydroxy acids are produced compared with amino acids depends upon the amount of ammonia; the more ammonia, the more amino acids and the fewer hydroxy acids. The detection of the hydroxy acids in the Murchison meteorite backs up the ideas of a nonbiological origin—electric discharges forming organic compounds.

In 1969 a Japanese scientific team discovered meteorites in the Antarctic, and since then more than 1000 samples have been collected. The Antarctic provides a clean, cold environment that is relatively unlikely to contaminate the meteorites with terrestrial materials. An analysis of one of those meteorites—a carbonaceous chondrite found near Allan Hills—shows it contains amino acids free of terrestrial contamination. It has only about 10 percent of the total amino acid content of the Murchison meteorite.

The glut of complex molecules discovered by radio astronomers (Section 9.1) lends further credibility to extraterrestrial, nonbiological formation of organic substances. Of particular importance to interstellar organic chemistry are the molecules formaldehyde, hydrogen cyanide, cyanoacetylene, formic acid, methyl alcohol, and methylacetylene. Each of these molecules has greater complexity and nonhydrogen composition than the previous one in the sequence. For example, cyanoacetylene (HC_3N) has a core of three carbon atoms that is the heart of many organic substances. The presence of cyanoacetylene strongly suggests the presence of acetylene (C_2H_2), which has now also been observed. Acetylene is an active compound with a tendency to form complex molecules, especially benzene (C_6H_6). The benzene ring provides the necessary links for many amino acids and for the nucleic building blocks of DNA. Also, as we mentioned in Section 25.4, formaldehyde and hydrogen cyanide can be chemically combined to make amino acids.

Radio astronomers have searched a few molecular clouds for amino acids. It's a difficult observation. The amino acid concentration in interstellar clouds is not expected to be large, and the radio frequencies of emission from amino acids are not well known. They have found none so far, but the detection efforts continue.

The important conclusion is this. The chemical evolution from simple compounds to complex organic substances occurs so naturally that it takes place even in the hostile environment of space, without biological aid.

25.6 FROM MOLECULE TO ORGANISM

Chemical evolution naturally, and perhaps inevitably, leads to the complex organic compounds that are the building blocks of proteins and nucleic acids. These are both needed to join together in a cell. How was that first cell made? Quite bluntly, no one knows. Fossils cannot give information about this crucial time. From them we do know what type of organisms populated the earth at about a billion years after our planet's formation. What happened before this time remains a topic for speculation.

Cells are not simple structures. A cell's activity depends on the fluid nature of its *protoplasm*. It consists of small particles and proteins suspended in water. The characteristics of the protoplasm provide some hints about the formation of the first cells.

Large protein molecules can surround themselves with a shell of water by attracting the water molecules with chemical bonds. By doing so the protein is concentrated in a smaller volume, but the water allows it to interact with other molecules. A number of protein molecules can unite, making a supermolecule encased in water. This resembles the protoplasm of present cells.

Another suggestion for grouping molecules comes from the ability of chains of amino acids to form little spheres when dissolved in water. These spheres might clump together to form larger organizations.

In either case the first cells might have formed at the surface of water (such as in a small pond) or at the interface of water and solid material (such as on rocks in a tidal pool). Although the details about the origin of the first cells are unknown, we do know that organisms appeared on the earth within a billion years after its formation.

These first organisms must have used food already present in their environment. They must also have been protected from incoming ultraviolet radiation. (Ultraviolet light destroys cells; it is sometimes used in hospitals for sterilization.) Remember, little oxygen existed in the atmosphere at this time, so there was no ozone layer then to cut out the ultraviolet radiation. Ultraviolet light penetrates only a few centimeters of water, however, so the early organisms could survive if below the surface of ponds, lakes, or oceans. They probably formed at the surface and then sank down.

Gradually ultraviolet light acted on water vapor to release some free oxygen to the atmosphere. The ozone layer began to build up. This process cut off more and more of the ultraviolet radiation and so cut down on the synthesis of organic compounds. The first organisms had their food supply cut off at this point. Millions probably died. But a few survived—those that were able to develop photosynthesis to make their food. More oxygen was added to the atmosphere, some by the organisms themselves, and finally the ozone layer was thick enough to shut out almost all the ultraviolet radiation.

Life multiplied rapidly after the ozone shield was up. Why? Partly it was because the ozone shield allowed plants to spread to land. But also the new availability of oxygen allowed more efficient metabolism of compounds.

Note the opposite roles played by ultraviolet radiation at different stages in the development of life. In the very early stages of molecular evolution, ultraviolet light works as a possible source of energy for the synthesis of more complex molecules from simple substances. But ultraviolet light is fatal to complex biological systems, for it can also break apart the larger and more fragile organic molecules and destroy cells. Organisms developing in water would have some protection, but before life could develop into complex cellular forms and spread over the earth, the ozone layer had to form to shield out the ultraviolet radiation. A controversy now rages in scientific and government circles on the possible long-range effects on the ozone layer of certain chemical pollutants being introduced into the atmosphere by people.

Let's recap the key events so far in the sequence of cosmic evolution. The physical evolution of the cosmos takes us up to the formation of the earth, along with the sun, from a cloud of interstellar gas and dust. The raw materials for life are available: the hydrogen from the Big Bang, the heavier elements from the generations of massive stars that preceded the birth of the sun. The process of the fusion of hydrogen in the sun's core guarantees that the earth will receive sunlight for a long time.

The scene shifts to chemical evolution on the earth. An atmosphere containing simple hydrogen compounds such as methane and ammonia is produced by outgassing from the interior. With the addition of free energy—mostly ultraviolet radiation from the sun—chemical evolution proceeds to make simple organic molecules. These eventually link up to make more complex ones, such as proteins and nucleic acids. Somehow these molecules get together to form the first organisms. Biological evolution begins. That evolution continues today.

The "somehow" that leads to biological evolution is not clearly known. But we do know that it *did* happen. With this scheme in mind, let's turn to the solar system and the Galaxy to investigate the possibility that we have cosmic neighbors.

25.7 THE SOLAR SYSTEM AS AN ABODE OF LIFE

So far we know of only one planet in the solar system that has life—the earth. What about the other eight members of the solar family? Keep in mind the conditions required for the existence of the simplest life forms. These are a temperature range from 0 to 100 °C (273 to 373 K), a solvent (commonly water), a carbon chemistry, free energy (from a star), protection from ultraviolet light, and a long time to evolve.

Terrestrial Planets

The terrestrial planets appear to be the best candidates for the existence of life. We'll consider Mercury, Venus, and Mars in their order from the sun, which is roughly the order of increasing probability as abodes of life. Of the four terrestrial planets, three are set in the *ecosphere* of the sun, the zone around the sun where the range of temperatures on a planet is suitable for life.

How thick is the ecosphere around a star? That depends on its luminosity. We can estimate its thickness roughly as follows. An airless planet with albedo A at distance R from a star with luminosity L has a temperature (Section 15.2)

$$T = \left(\frac{1-A}{D^2}\frac{L}{4\pi\sigma}\right)^{1/4}$$

If we assume the planet is perfectly absorbing (A = 0), use solar units for L and AUs for D, and plug in the constants, we obtain

$$T\text{(Kelvins)} = 394\left(\frac{L}{D^2}\right)^{1/4}$$

To use this equation to calculate D, we turn it around:

$$D\text{(AU)} = (394)^2 \frac{L^{1/2}}{T^2}$$

At the inner edge of the ecosphere $T = 373$ K; at the outer edge $T = 273$ K. So for the solar system,

$$D_{inner} = \frac{(394)^2(1)^{1/2}}{(373)^2} = 1.12 \text{ AU}$$

$$D_{outer} = \frac{(394)^2(1)^{1/2}}{(273)^2} = 2.08 \text{ AU}$$

and the thickness of the ecosphere is 0.96 AU. Note that this estimate ignores any greenhouse effect, which moves the edge of the ecosphere outward from the sun; reflection from ice coverings, which moves the edge inward; conduction of heat into the surface, which moves the edge inward; and other complicating effects. If it weren't for these effects, the earth, at 1 AU, would not be within the ecosphere of the sun!

For a star more luminous than the sun, the ecosphere is farther out and thicker. For Sirius A, for instance, with $L = 25\ L_{sun}$,

$$D_{outer} = \frac{(394)^2}{(273)^2}(25)^{1/2} = 10.41 \text{ AU}$$

$$D_{inner} = \frac{(394)^2}{(373)^2}(25)^{1/2} = 5.57 \text{ AU}$$

and its thickness is 4.8 AU. But Sirius has a shorter lifetime than the sun, which would be unfavorable to the development of life.

Mercury

The rotation and the revolution rates of Mercury (Section 15.6) result in a solar day 176 terrestrial days long. Every point on Mercury's surface receives an intense bath of

FIGURE 25.9 The bleak surface of Mercury. Every 176 earth days, a spot on the surface suffers the full blast of the sun. (*Courtesy NASA.*)

solar radiation each day, with noon temperatures topping 700 K. This scorching temperature is far above the maximum for life, unless the organisms carry a protective shield or live underground. The range of temperature, about 500 K from noon to midnight, would push life beyond the limits of survival. On the surface of Mercury it is a harsh place to live (Fig. 25.9).

Venus

Life also stands little chance on Venus. Day or night the surface temperature is about 700 K, and the surface pressure is about 100 times that on earth. That crushing atmosphere contains mostly carbon dioxide with sulfur compounds such as hydrogen sulfide, sulfur dioxide, and carbonyl sulfide (OCS). No water flows on the surface. The clouds contain little water; they may be droplets of highly concentrated sulfuric acid (Fig. 25.10). The surface looks mostly rocky with little soil. Venus is a burning desert, unsuitable for life.

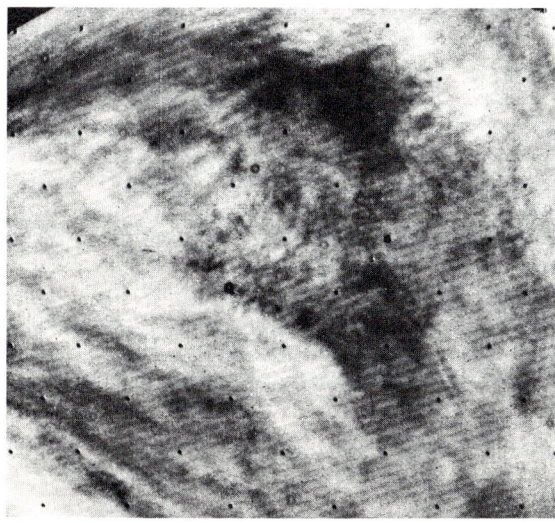

FIGURE 25.10 A close-up of the upper cloud layers of Venus in ultraviolet light. The clouds probably consist of water and sulfuric acid. (*Courtesy NASA.*)

FIGURE 25.11 A panoramic view of the Tharsis ridge region and Valles Marineris near the equatorial region of Mars. (*Courtesy NASA.*)

Mars

The fate of Martian life hinges on the abundance of surface water. The Viking missions found an atmospheric composition of about 95 percent carbon dioxide. The surface pressure is about 0.006 to 0.008 earth atmospheres, much less than that on the highest mountains on the earth (you have to go about 40 km up into the earth's atmosphere to find the pressure this low). At this low pressure liquid water cannot exist. As evidenced by the polar caps, both water and carbon dioxide on Mars are in solid ice form. Even in these regions, the abundant water needed for life probably does not exist. Mars is a dry planet; even in the "wet" polar regions, the water vapor in the atmosphere, if all condensed, would form a layer only 0.1 mm thick on the surface.

Perhaps water flowed on Mars in the past. East of the volcanic ridge dominated by Olympus Mons stretches the series of canyons (Fig. 25.11) which look like they have been eroded by liquid water. A possible source of water is a frost layer below the ground (called permafrost, since the temperature is always below freezing a short distance below the surface). Alternatively, some astronomers imagine that at an earlier epoch Mars had a denser atmosphere capable of holding water vapor sufficient to generate rainfall.

This speculation ties in with past volcanic activity. On the earth volcanoes spew out large volumes of carbon dioxide and water vapor. Possibly the violent geologic episode that spawned the volcanoes at the Tharsis Ridge also injected a significant amount of volcanic gases into the Martian atmosphere. An increased atmospheric pressure would have allowed water to flow on certain regions of the surface and to cut the meandering channels. That water may now be frozen below the surface, locked up in surface minerals, or at the polar caps.

Recurring deluges or meltings may explain the origin of the laminated terrain found in the polar regions (Fig. 25.12). There, stacks of thin plates of crustal material stand about 10 km tall and up to 200 km across. Because they exist only in the polar regions, where carbon dioxide and water ice form annually, the plates may have been built by the influx and outgo of these substances. A time of a denser atmosphere may have produced the laminated terrain along with the eroded channels.

FIGURE 25.12 Laminated terrain near the south polar cap of Mars. These formations of surface soil probably contain water ice. The photo covers an area 47 by 60 km; the pits are about 500 m deep. (*Courtesy NASA.*)

An eruption of interior gases may also have been combined with astronomical effects to change the Martian environment in the past. Owing to the gravitational attraction of the other planets, Mars's orbit varies in its average distance from the sun. So the amount of sunlight varies over 2-million-year periods. The variation in solar energy input affects the size of the polar caps. It is possible to imagine large polar caps accumulating during the colder periods and melting during the warmer ones, adding carbon dioxide and water to the atmosphere.

The direct test for Martian life came from the Viking landers' one spectroscopic and three biology experiments. Both landers contained an instrument called a mass spectrometer, designed to detect and measure the organic molecules that characterize life. The instruments had the sensitivity to detect organic compounds in a concentration of just a few parts in a billion. That's about 1 million bacteria (dead or alive) in a sample, far below the concentration found in desert soils on the earth. At both landing sites *no large organic molecules were found.*

In light of the lack of complex molecules in the soil, any apparently positive results of the landers' three biology experiments can all be explained by *chemical* reactions rather than *biological* ones. The results, taken as a whole, strongly point to the disappointing conclusion that life does not inhabit the top layers of the soil of Mars. Our extraterrestrial search for terrestrial life has failed so far.

Jovian Planets: Prebiological Worlds?

At first glance the giants of the solar system do not appear receptive to life. Their low temperatures and sunlike compositions seem to be inhospitable to life. Yet the low temperature and high gravity guarantees that even hydrogen has not been depleted from the atmospheres of the Jovian planets since their formation. The retention of atmospheric hydrogen implies the retention of all heavier elements, such as carbon, nitrogen, and oxygen. The natural expectation, supported in part by observations, is that the Jovian planets are engulfed by extensive atmospheres rich in hydrogen compounds such as water, methane, and ammonia. These compounds probably were the prebiological components of the earth's atmosphere.

We'll treat Jupiter in detail, partially because we know more about it than about the others. Jupiter's zones and belts, the result of atmospheric convection powered in part by the outflow of internal heat, are distinctly colored (Plate 10). Some observers have noted pastel shades of pink, red, and blue. Infrared and ultraviolet spectroscopy, both from the earth and from the Pioneer and Voyager spacecraft, have identified numerous kinds of molecules in Jupiter's atmosphere, including water, methane, ammonia, and molecular hydrogen, as well as acetylene, ethane, phosphine, and germane (Section 17.1). Carl Sagan and his coworkers have investigated theoretically the expected abundances of large carbon molecules constructed of the known compounds. Hydrocarbons with appropriate color appear possible under Jovian conditions. These organic molecules might be produced locally by spark discharges in the various cloud layers. Experiments find that spark discharges in a mixture of methane and ammonia at room temperature result in a carbon compound with a deep red color; and ultraviolet light shone on a mixture of methane, ammonia, water, and hydrogen sulfide at room temperature and pressure results in slush with a brownish-yellowish color. Other experiments have produced similar results: Energy added to a Jovian-like atmosphere promotes the manufacture of colored organic compounds.

The production of such compounds requires a fair amount of energy. Where can it come from? Jupiter's upper atmosphere receives a mere 0.1 W/m^2 from solar ultraviolet radiation (compared to 2.7 W/m^2 at the earth). At this energy level the efficiency of organic molecule production is low, so the total amount formed would not be high. Shock waves are a much more effective production mechanism; two possible sources on Jupiter are thunder and meteorite impacts. The Voyagers' confirmation of lightning in the Jovian atmosphere makes the scene look like that expected on the primitive earth. However, it is difficult to estimate the energy available from these sources. The flow of internal heat upward through the atmosphere might also provide some of the energy needed to synthesize large carbon molecules.

Although Jupiter's upper atmosphere is cold (150 K), the temperature just below the cloud tops is similar to the earth's (about 300 K); and high pressures do not significantly affect the genesis and survival of life. A layer of liquid ammonia that might exist about 50 km below the layer of frozen ammonia is a possible alternative solvent to water, particularly at temperatures below freezing. However, the Pioneer flybys found intense radiation close to Jupiter; this energetic radiation might inhibit the development of any life forms. Also, the continuous convection in the atmosphere probably carries any organic compounds down to lower, hotter regions where they would decompose.

Jupiter appears to be in a state of chemical evolution that is prebiological, but we have no idea if creatures of any kind now or will eventually develop.

25.8 THE GALAXY AS AN ABODE OF LIFE

Although the question of whether there is life in the solar system (outside the earth) has not yet been answered, some scientists have already turned to the Galaxy in search of

25.8 THE GALAXY AS AN ABODE OF LIFE

footprints in the stellar sand. The huge number of stars in the Galaxy implies the existence of some planets elsewhere, if the nebular model of planetary formation is correct (Chapter 19). Even if the probability of the genesis of life were slim, the number of possible habitats is so large that perhaps some extraterrestrial creature has viewed the dawn of its day. If life developed from the natural evolution of the inorganic, these processes must have also operated beyond the solar system. The elements of life are the most abundant in the cosmos, so there is no lack of proper ingredients. All that is required is the proper construction. This forming takes physics, chemistry, and—most important—time.

Cosmic Prospecting

To simply declare that extraterrestrial life exists is not sufficient to begin the search. You might argue in a similar fashion about the existence of gold in the grains of sand on all the beaches of the world: that the uncountable number of sand grains requires that some gold grains be mixed in. This reasoning is not much help to the gold prospector, who wants to know not only that gold exists, but also *how much* exists and *where* is the best place to hunt for it. Here we try to answer the same two questions concerning life in the Galaxy.

Civilizations of living creatures must evolve; that's part of cosmic evolution. So the numbers of intelligent civilizations in the Galaxy changes with time. At any one time the number of civilizations depends on the rate at which these civilizations are born and how long they last.

Here's an analogy. Suppose you are locked in a dark room filled with candles. A friend lights 1 candle every 15 min (4/h). Suppose each candle burns for 1 hr. How many candles are lit at any give time? During the first hour the number increases from 1 to 2 to 3 to 4. But just as the fifth one is lit, the first one goes out. As the sixth is lit, the second goes out. One goes out as each new one is lit, leaving 4 candles burning at any one time. Note that the number of observed candles (N_c) is equal to the rate of candle lighting R_c times the lifetime of one candle L_c, or

$$N_c = R_c L_c$$

Thus, if you know the average lifetime of a single candle and the rate at which they begin their life, you can anticipate the number lit at any time.

The same reasoning applies to the number of civilizations aflame in the Galaxy. If R_{ic} is the rate of formation of intelligent civilizations and L_{ic} is their lifetime, then

$$N_{ic} = R_{ic} L_{ic}$$

This equation was first put together, in somewhat different form, by radio astronomer Frank Drake, so it's called the *Drake equation*. It may be broken down into more specific factors, loosely independent of one another:

$$N_{ic} = R_* P_p P_e N_e P_l P_i L_{ic}$$

The meaning of each of these factors relates directly to important facets of cosmic evolution. R_* is the rate of star formation averaged over the age of the Galaxy. P_p is the probability that once a star has formed, it will possess planets. P_e is the probability that the star's ecosphere will exist long enough for life to form, and N_e is the number of planets in the ecosphere. P_l is the probability that a planet in a star's ecosphere will develop life, and P_i is the probability that biological evolution will lead finally to intelligent life. The final term, L_{ic}, is the lifetime of this intelligent civilization. Note that these factors fall into three categories: R_*, P_p, P_e, and N_e relate to astronomy and physical evolution; P_l and P_i relate to biology and to chemical and biological evolution; and L_{ic} derives from what we would call speculative sociology.

Astronomical Factors

The Galaxy contains a few times 10^{11} stars. These stars have formed over at least 15 billion years. So the average birthrate of stars from these figures is about 7/yr. However, the initial burst of star formation delivered a first generation composed mostly of massive stars that quickly spent their energy stores. Their violent ends ejected nearby elements into the currents of space. The next stellar generation acquired these elements and formed in a greater range of masses. Some of the second-generation material was also flung into space, but some of it remained, trapped by gravity to become white dwarfs, neutron stars, or black holes. A third generation of stars (our sun, for example) was born at a more leisurely rate than the first two generations. The slowdown of the birthrate pushes the initial estimate down to perhaps 1/yr. Recent work indicates that the birthrate in the past was probably no more than twice the current value. We adopt 1 for R_*.

What is the chance that one of these stars will develop a planetary system? Nebular models (Chapter 19) imply that many planets exist in the Galaxy. A collapsing gas and dust cloud must form either a star with a planetary system or a multiple-star system, perhaps also with planets. More than 50 percent of the stars in the Galaxy are in binary or other multiple-star systems. A planet in a multiple-star system may not have a stable orbit, so we exclude these from consideration. If we take planetary system versus a multiple-star system as an either-or proposition, P_p equals 0.5.

A star's ecosphere—the zone in which planets must lie to have conditions suitable for life—depends primarily on the luminosity of the star. The more luminous the star, the farther out the habitable zone starts (Fig. 25.13). The width of the ecosphere is also greater for luminous stars and thinner for less luminous ones. The ecosphere must persist long enough to foster the genesis and evolution of life.

Luminous O and B stars live out their normal main-sequence lives in about 100 million years, a time much shorter than the 4.6 billion years that elapsed while life evolved on the earth. By our standards these energy spendthrifts are improper parents. It seems unlikely that attendant planetary systems to such short-lived stars would have the time to develop life. Therefore, we consider only stars whose life spans are at least equal to the sun's—spectral class G or cooler, a choice that includes 98 percent of all the normal stars in the Galaxy. For stars cooler than spectral class K, the ecosphere is too small. If we throw out these cool stars, only about 8 percent of the total remains, so P_e equals 0.08. These stars are the good suns for life.

Michael Hart has emphasized that, because a star's luminosity changes with time, the ecosphere evolves too. His calculations indicate that the thickness of a *continuously* habitable ecosphere is very small. He defines the ecosphere as the zone in which an earthlike

25.8 THE GALAXY AS AN ABODE OF LIFE

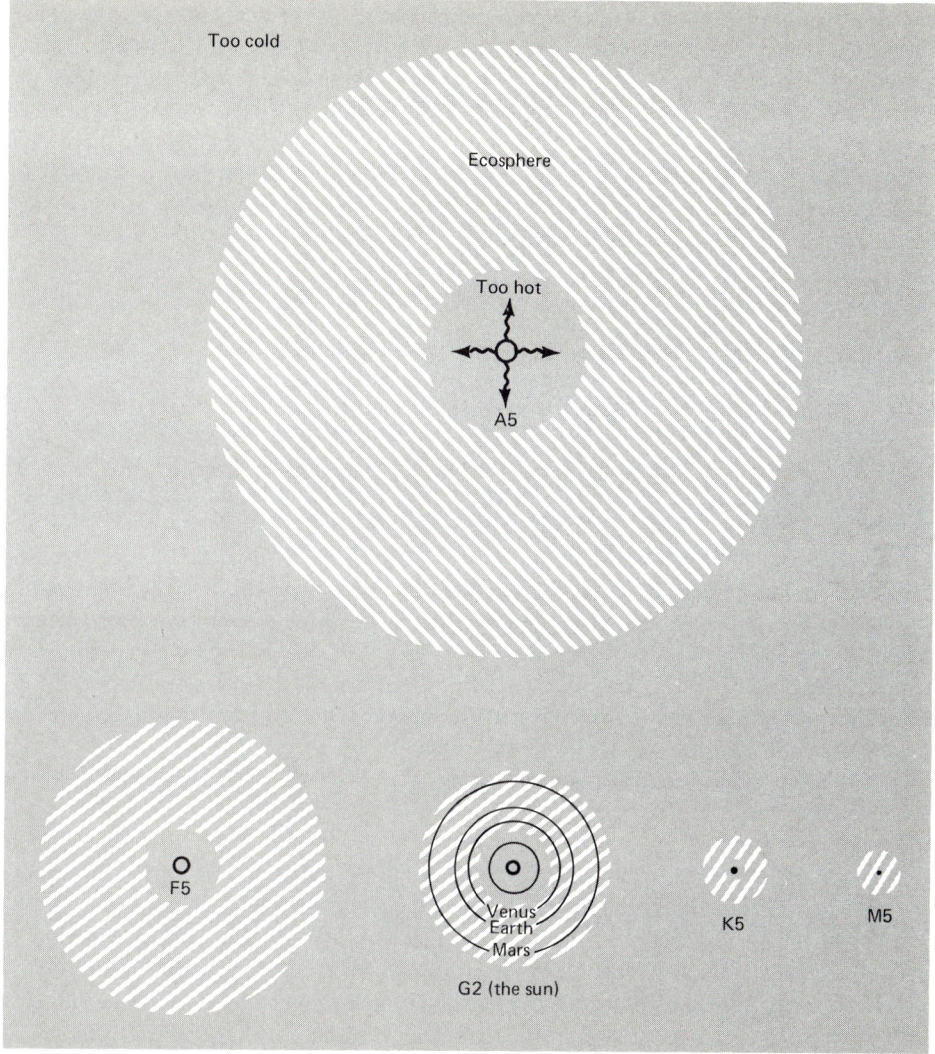

FIGURE 25.13 The sizes of stellar ecospheres for main-sequence stars of different spectral types. Note that the cooler the star, the thinner the ecosphere and the closer it lies to the star.

planet will *not* suffer from a runaway greenhouse effect early in its history nor runaway glaciation after it develops an oxygen-rich atmosphere. From this evolutionary viewpoint, Hart finds that for the sun the inner edge of the ecosphere lies at 0.958 AU, the outer edge at 1.004 AU, for a thickness of a meager 0.046 AU (20 times smaller than the rough calculation made in Section 25.7). Essentially no ecospheres exist for K and M stars; the maximum thickness for F9 stars (about 1.1 solar masses) is 0.069 AU. So P_e may be only 0.01 or less. These specific numbers probably shouldn't be taken too seriously, for many poorly known parameters go into Hart's models, but it is important to realize that the continuously habitable ecosphere is probably significantly smaller than might be thought at first.

How about the number of planets in the ecosphere N_e? We have only the example of our solar system in which three planets, Venus, Earth, and Mars, lie in that zone. (Hart would argue that only the earth lies in the ecosphere.) If the planetary formation processes in the nebular model are universal, we might expect other planetary systems to more or less resemble the solar system. Is this belief reasonable? Some years ago Stephen Dole attempted to model the results of a nebular-style formation that results in a Titius-Bode "law" for spacing planets. Richard Isaacman and Carl Sagan recomputed the model and found similar results. The planets consist of a Jovian-terrestrial mix. And a few planets orbit at the magic ecosphere distance of a solar-mass star. So N_e may range from 1 to 4 or so. We'll use 3.

Biological Factors

Here's another proposition. Is the existence of terrestrial life unique and the probability of life elsewhere zero? Or, is the earth typical, the normal result of cosmic evolution, so that the probability of any planet's developing life once the astronomical conditions are favorable is close to 1. Or, is some intermediate probability the correct one?

Appealing to the presumed uniformity of physical laws, assume that we are typical. Scientists strongly believe that the physical laws unraveled here locally apply to the rest of the universe. This *belief* is expressed in the cosmological principle (Section 24.1). Unfortunately, we have only ourselves as the proof of this supposition. However, laboratory experiments (Section 25.4) have shown the natural start of chemical evolution. With these as a guide, along with a feeling that the nature of the universe makes the start of chemical evolution inevitable, we choose P_l equal to 1.

But beware. It's easy enough to produce organic molecules, but much harder to put them together to form an organism. The step from the "warm pond" to the first cells is a big one!

It's extremely difficult to estimate the probability of the evolution of intelligent life once life has developed on a planet. Is intelligence inevitable? The development of multicellular organisms (which is a advantageous feature because it allows diversification of cells for greater efficiency) requires coordination between cells, that is, some kind of nervous system, for both sensory and motor coordination. Furthermore, the ability to learn appears at even simple levels and is probably advantageous for survival. For example, even protozoa can be trained to swim up a glass tube. Although instincts—tested patterns of behavior arrived at after much trial and error—are very efficient mechanisms for survival, a complex, changing environment can promote the success of a creature able to adapt to a variety of circumstances. The adaptive powers of a thinking organism appear so great to us that we think, if it is at all possible genetically, intelligence is very likely to be the ultimate result of natural selection. So we choose P_i equal to 1. This statement assumes that a comfortable environment persists for the billions of years needed for intelligence to develop.

Speculative Sociological Factors

How long can an advanced, technological civilization survive? Is intelligence flexible and complex enough to cope with adverse aspects of its technology? What happens when the fragile net of the environment is pushed beyond its breaking point? These are pressing questions now as a growing humanity demands energy and food at the edge of our present capabilities. Their answers may determine, for us, the lifetime of intelligent civilization.

Two tacks appear possible to navigate these winds of doom. The optimistic one views the heavy winds as short gusts that only momentarily hinder us on the course to the technological golden age. The pessimistic one peers at a gloomy, unrelenting storm ahead in which present civilization will soon founder. By our own example to date, the lifetime of an intelligent civilization may be only a few thousand years. But if it is possible that every advanced civilization steers clear of its problems, then it should survive as long as the parent star. For civilizations encircling a G star, their lifetimes may be about 10^{10} years. But their lifetimes may also be much, much shorter.

The Numbers Game

As we have progressed through the astronomical, biological, and sociological factors needed for a rough estimate of the number of extraterrestrial societies, the footing has become shakier. We have also ignored some important factors in the analysis, such as the possibility of stable planetary orbits in a binary star system. Our method was not intended to give precise results, but rough estimates, because exact answers are not yet possible. Personal biases also affect the discussion; we hope we made ours clear.

Now evaluating L_{ic}, we come up with

$$N_{ic} = 1 \times 0.5 \times 0.08 \times 3 \times 1 \times 1 \times L_{ic}$$
$$\approx 0.1 \, L_{ic}$$

if L_{ic} is expressed in years. The result depends critically on L_{ic}—how long our candle remains lit. (Note that the "years" in L_{ic} cancels those in R_*.) If we assume we are at the brink of destruction, then $L_{ic} \approx 10^3$ and $N_{ic} \approx 10^2$. Intelligent civilizations are few and far between. If we survive as long as the sun shines, then $L_{ic} \approx 10^{10}$ and $N_{ic} \approx 10^9$. In this case many stars in the Galaxy have fostered an intelligent civilization!

How seriously can you take these results? The weakness of this approach is that it is basically speculation. The results should be viewed very skeptically. We have delineated each of the values of the life factors in a reasonable and yet ultimately arbitrary manner. Our own example may be more special than we have been willing to admit. The genesis of terrestrial organisms moved in a crooked path through the junctions of geological time. At each crucial point one path was chosen out of many, one history written out of many potential ones. That life elsewhere should follow the same path is unlikely. The long flow of time to us can be retraced, but it cannot be repeated.

On the other hand, many other paths may also have led to a similar final result. For example, the eye has evolved independently at least three times, in insects, squids, and in our own ancestors. Intelligence, too, may be arrived at in different evolutionary ways. And just as insect and human eyes operate very differently in performing the same function, the mechanisms of intelligence may be very different in unrelated organisms.

25.9 NEIGHBORING SOLAR SYSTEMS?

What can we see of other planetary systems? Because a planet shines by reflected light from its parent star (and also because planets are small in size), the light from a perfectly reflecting planet near any star in the solar neighborhood would be too weak to detect from the earth. Also, a planet likely orbits very close to its local sun. The angular separation of the planet and the star, as seen from the earth, would be so small that the planetary gleam would be lost in the stellar glare. So we cannot *directly* observe other planets outside the solar system with earth-based telescopes.

Planetary Companions?

Instead of searching for the light from very large planets, we can hunt for the motion around the center of mass of the planet-star system. The visible star wobbles from side to side about the center of mass if a massive planet orbits it (Fig. 25.14). From the observed stellar wobble and an estimate of the stellar mass, we can estimate the mass of the invisible planetary companion by the same method used to measure binary star masses.

Barnard's star, the second nearest to the sun, appears to exhibit such a corkscrew motion in its proper motion across the sky. In 1963 Peter van de Kamp of the Sproul Observatory concluded that a planet about the mass of Jupiter could account for the size of the wobble. Later he proposed an alternative interpretation, which allows two planets, one about 80 percent the mass of Jupiter and the other 10 percent larger than Jupiter, to encircle the star. The lesser mass would be about 2.8 AU from the star, and the greater mass approximately 4.7 AU away. Persisting with his planet hunt, van de Kamp announced in 1974 that a possible jumbo planet, six times the mass of Jupiter, may orbit the star Epsilon Eridani. This star is not only one of the closest stars to the sun, but it is also quite similar to it in spectral class. Although the supposed dark companion is much larger than any planet in our solar system, it is certainly too low in mass to be a *bona fide* star.

Barnard's star is the one which best displays the telltale signal of invisible comrades. Some other nearby stars of solar mass show weaker evidence of motions that may result from dark companions (Table 25.5).

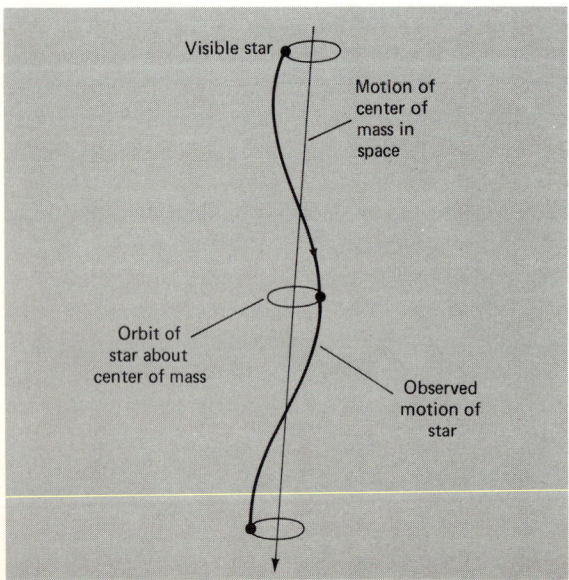

FIGURE 25.14 The path of a star in space with orbital motion around the center of mass of the star-planet system.

25.9 NEIGHBORING SOLAR SYSTEMS?

Warning: The observations required to detect planets around nearby stars are extremely difficult to make. The wiggles sought for are only about 0.01", or about 1/100 the size of a star's image on an astronomical photographic plate. Such minuscule changes could be caused by changes in the telescopes themselves, whether produced by self-aging or conscious effort (such as cleaning). George Gatewood of Allegheny Observatory has analyzed the errors in such observations and concluded that *no* good evidence supports the existence of Jovian-mass planets.

A curious fact emerges from such investigations. The sun is the only star which we know *for certain* does have a planetary system and does not have a companion star. This brings up a key question: What fraction of stars like our sun in the Galaxy do have stellar companions? We might hope to extrapolate from such data to predict the number which have planetary companions.

Helmut Abt and Saul Levy have searched 123 sunlike stars within 85 ly (26 pc) of the sun in an effort to find out whether or not they have companions. Their technique aimed at detecting regular Doppler shifts in the spectra of these stars. Over half (57 percent) of their stars had at least one stellar companion. What about the others? Most of the companions are of small mass, the number increasing as the mass gets smaller. By inference, Abt and Levy conclude that roughly $\frac{1}{6}$ to $\frac{1}{5}$ of sunlike stars could have planetary-size companions (but none were directly observed). This argument, if true, implies that the galaxy contains 15 to 20 billion stars with planets, and many of these stars resemble the sun.

Note that the above discussion refers in any case to Jovian-size planets. There is little hope of detecting terrestrial planets. Not only are the gravitational effects smaller, because of their smaller mass, but the effects would have to be disentangled from the effects of larger bodies in the same planetary system. Also there is a problem of distinguishing small effects, such as variations in radial velocity, from phenomena associated with the stars themselves. The sun, for example, is known to show periodic changes in apparent radial velocity due to pulsation-like phenomena, changes which are larger than the change produced by a terrestrial planet. So though detection of Jovian-size planets is conceivable, but difficult, it is unlikely we will ever be able to directly detect terrestrial planets around other stars.

TABLE 25.5 Selected Stars That May Have Invisible Companions with Less Than Stellar Mass

Name	Distance (ly)	Suspected Companion's Mass (Jupiter = 1.0)
Epsilon Eridani	10.7	6–50
61 Cygni	11.0	8
Barnard's star	5.9	1.1, 0.8
BD +43° 4305	16.9	10–30

Source: Peter van de Kamp, "Unseen Astrometric Companions of Stars," *Annual Review of Astronomy and Astrophysics*, vol. 13 (1975); and George Gatewood, "On the Astrometric Detection of Neighboring Planetary Systems," *Icarus*, vol. 27 (1976), p. 1.

Nearby Good Suns

Of the few hundred stars fairly close to the sun, three stand out, according to Stephen Dole, as good candidates for possessing planets suitable for life. The number of these nearby systems are consistent with the range of numbers developed in the preceding section for the probability of extraterrestrial life. The first is the star system closest to the earth, Alpha Centauri, a multiple system. If Alpha Centauri is encircled by the same number of planets our solar system possesses, one is likely to shelter life. Such a planet would be a true interstellar neighbor, because Alpha Centauri is only 4.3 ly (1.3 pc) from the sun.

Next in line, at a distance of 10.8 ly (3.3 pc), is the K star Epsilon Eridani. Its long main-sequence lifetime, greater than 10^{12} years, enhances the opportunity for life to be created. At almost the same distance from the solar system, 11.8 ly (3.6 pc), is the G dwarf star Tau Ceti, which has physical characteristics much like the sun's. These two stars are the most likely of those nearest to us to have planets with life.

SUMMARY

Assume that life as we know it on earth is typical in the sense that it has arisen naturally in the course of cosmic evolution. Then we can hope to find the trail of that evolution, which falls into three interconnected stages: physical, chemical, and biological evolution.

We can hunt down the traces of biological evolution on the earth. Fossil evidence implies that life has grown more complex on the earth and that the origin of life took a relatively long time—about a billion years to get to the first cell.

To search for evidence of the chemical evolution that preceded the biological evolution is hopeless. Here we rely on general theoretical ideas and some crucial laboratory experiments. These show that if the primitive atmosphere of the earth were rich in hydrogen compounds, then the addition of energy (solar ultraviolet, lightning, etc.) would naturally result in the formation of basic organic materials, such as amino acids. But ultraviolet radiation, perhaps necessary to start life, is harmful to it once it develops. Discovery of complex molecules in space and amino acids in meteorites support the idea that molecular formation is not a freak accident. Where in the solar system do conditions for prebiological evolution exist? Certainly not on the earth and other terrestrial planets. Evidence so far points to Jupiter as the most likely candidate.

Astronomical ideas underlie our understanding of physical evolution. For life to arise we need a planet (earth), a star (the sun), and the proper elements (hydrogen, carbon, nitrogen, oxygen, and some others). Where did these come from? From the dust and gas of the interstellar medium. Both dust and gas are the material lost by earlier stars, mostly in their violent ends. Nucleosynthesis in these explosions and normal fusion reactions in stars manufactured the chemical elements, except for hydrogen and helium. These were made in the first few minutes of the Big Bang.

This sequence in general seems appropriate for all the observable universe. So if life on earth is typical, then it may be common in our Galaxy and spread throughout the entire universe.

Key Words

physical evolution
chemical evolution
biological evolution
organism
mutation
cell
natural selection
protein
enzyme
amino acid
nucleic acid
DNA
spontaneous generation
Fig Tree fossils
Gunflint fossils
Bitter Springs fossils
aerobe
anaerobe
organic molecule
nonoxidizing atmosphere
polymerization
Murchison meteorite
protoplasm
ecosphere
Drake equation
Barnard's star
Alpha Centauri
Tau Ceti
Epsilon Eridani

Review Questions

1. Which of the following are necessary for the synthesis of molecules to occur on earth?
    a. Explosion of supernovas sometime in the past
    b. Availability of a large amount of energy
    c. Oxygen
    d. A well-equipped lab
    e. Water
    (Objectives 3, 4, and 7)

2. Which of the following are true?
    a. Astronomers have discovered living organisms in meteorites and interstellar clouds.
    b. Fossils of simple cells devoid of a cell nucleus or other complex structures have been discovered in rocks dated 3 billion years old.
    c. Before formation of primitive organic substances (such as DNA and proteins) could occur on earth, an ozone layer in the atmosphere had to develop to filter out the solar ultraviolet radiation.
    d. No planets have been detected associated with other stars, but they are presumed to exist because of our current thinking about the origin of the solar system as a noncatastrophic event.
    e. It is much easier to produce organic compounds, such as amino acids, in an atmosphere like that of Jupiter than in an atmosphere like that of the present-day earth.
    (Objectives 4, 5, 6, and 9)

3. What, more than anything else, is the factor that makes the number of planets with life probably very high (choose only one)?
    a. The huge number of suitable stars
    b. The high probability that organic molecules form from inorganic ones
    c. The very high probability that organisms will form from organic molecules
    d. The long lifetime of a technological civilization
    e. The fact that all stars have planetary systems
    (Objectives 10 and 11)

4. Life on Jupiter (mark all correct answers):
    a. As a possibility is suggested by the fact that the atmosphere of Jupiter has many of the compounds necessary for life.

b. Might not arise because the surface is so cold.
   c. If it existed, might use electrical discharges as an energy source.
   d. Would have an abundant source of energy available in the form of UV radiation from the sun.
   e. Could not exist because of all the poisonous compounds such as methane in the Jovian atmosphere.
   (Objective 9)
5. Life on earth revolves around carbon. Where did the carbon come from, and how did it get to the earth? (Objective 3)
6. What is the crucial role played by supernovas in the origin of life? (Objective 3)
7. Where in the solar system are the chances for life best? (Objective 9)
8. Suppose that radio astronomers tomorrow announced the discovery of amino acids in interstellar molecular clouds. How would that affect the discussion about the origin of life? (Objective 4)

Problems
1. Calculate the distance from an O star a blackbody would have to be placed to have a temperature equal to the freezing point of water. Then do it for the boiling point of water. Find the thickness of the ecosphere. (Objective 11)
2. Do the same calculation for an M star. (Objective 11)
3. Suppose you looked at the sun and Jupiter from a distance of 4 ly. What would their angular separation be? Is this visible with a ground-based telescope? With a space telescope with a 2-m diameter mirror? (Objective 10)
4. If you viewed the solar system edge on, what would be the maximum Doppler shift of sunlight reflected by Jupiter relative to that coming directly from the sun? (Objective 10)
5. What is *your* estimate of the number of intelligent civilizations in the Galaxy? (Objective 11)

BEYOND THIS BOOK . . .

* *The Cosmic Connection* (Dell, New York, 1973) by C. Sagan is an expansive view of life in the universe.
* The classic is *Intelligent Life in the Universe* (Holden-Day, San Francisco, 1966) by I. S. Shklovskii and C. Sagan.
* You can find a technical, comprehensive discussion of chemical evolution in *Molecular Evolution and the Origin of Life* (Freeman, San Francisco, 1972) by S. Fox and K. Dose.
* A dated, but classic book is *Origin of Life* (Dover, New York, 1953) by A. I. Oparin.
* For a detailed account of the Viking lander biology results, see "The Search for Life on Mars" by N. Horowitz, *Scientific American*, October 1977.
* C. Ponnamperuma has a nontechnical discussion about *The Origins of Life* (Dutton, New York, 1972).
* The September 1978 issue of *Scientific American* deals with biological evolution. Articles pertinent to this chapter are "Chemical Evolution and the Origin of Life" by R. Dickerson and "The Evolution of the Earliest Cells" by W. Schopf.
* Another recent book is *Life Beyond the Earth* (Morrow, New York, 1980) by G. Feinberg and R. Shapiro.
* A series of articles about life in the universe has been collected by D. Goldsmith in *The Quest for Extraterrestrial Life: A Book of Readings* (University Science Books, Mill Valley, CA, 1980).

FIRST SETI, THEN CETI

26

LEARNING OBJECTIVES
After studying this chapter, you should be able to:
1. Define SETI and CETI.
2. From the estimated lifetime of a technological civilization and the Drake equation, estimate the number of such civilizations in the Galaxy and the distance between them.
3. Estimate, on astronomical grounds, the probable range of values for the number of technological civilizations now in the Galaxy.
4. Estimate the time it would take to explore the Galaxy with von Neumann probes.
5. Argue that physical space travel is impractical for SETI.
6. Argue that radio is the best medium for SETI and CETI.
7. Argue that the microwave band from 1 to 10 GHz is the best radio band for SETI from the earth's surface.
8. Describe what is meant by the "cosmic water hole."
9. Outline a possible SETI strategy.
10. State the results of SETI searches to date.
11. Speculate about the consequences for humankind of contact with extraterrestrial intelligence.

CENTRAL QUESTION:
How can we search for evidence of
extraterrestrial intelligence?

What do the acronyms SETI and CETI mean? SETI stands for the Search for *Extra*Terrestrial Intelligence and CETI for *Com*munication with *Extra*Terrestrial Intelligence. The previous chapter concluded that life probably abounds in the Galaxy. If so, and if that life has evolved to the stage of a technological civilization, then two questions naturally provoke us: Where are they? How can we communicate with them?

This chapter deals with these two questions from an astronomical viewpoint. First, it investigates how close together (or far apart) technological civilizations might be in the Galaxy. Second, it looks at the modes of communication technically feasible today. The best chance is with radio, especially microwaves. Third, the chapter turns to possible search strategies through listening and to messages we have already sent and might want to send.

This chapter contains speculation almost entirely. We have no firm evidence for the existence of any civilization besides our own. But scientific speculation often bears fruit, for it frees our imagination for new ideas while tying it down to what we do know. We expect the general physical laws to hold true throughout the universe. These laws will affect whatever civilization arises whenever it does.

26.1 THE CENTRAL ISSUE: HOW MANY TECHNICAL CIVILIZATIONS?

So far we know of only one—us. Whether or not it makes sense to search for others depends on how many technologically advanced civilizations exist in the Galaxy *now*. (For this discussion, we take "technologically advanced" to mean creatures who can manipulate their environment at least to the extent that they can build communication equipment.) If the number is large, then on the average such civilizations will be closer together than if the number is small.

26.1 THE CENTRAL ISSUE: HOW MANY TECHNICAL CIVILIZATIONS?

Section 25.7 pointed out the key element in estimating the number of technological civilizations (N_{ic}) is their *lifetime* (L_{ic}). For example, if the lifetime is about 100 years (which is about how long we've had a technologically advanced human culture on earth), then the average distance between galactic civilizations is roughly 10,000 ly (3000 pc). That makes communication practically impossible. If we tried to signal by radio, for example, by sending out a message just at the moment our technology permitted, our civilization would have died while our words were still in transit. SETI and CETI are only possible if the number of civilizations is large and their lifetimes are long.

Note that N_{ic} is *not* a fixed number. It changes with time as the Galaxy and the objects within it evolve. For instance, for the first billion years of the Galaxy's existence, N_{ic} was probably zero, because life had not yet had time to evolve anywhere. So our estimates for now need not apply to the past or future.

Let's try to focus on a "best estimate" of N_{ic} for the Milky Way Galaxy. Broadly speaking, we can divide the options into four classes. N_{ic} may be (1) very large, (2) very small, (3) either very large or very small, or (4) neither very large nor very small. Let's look at each briefly.

N_{ic} Is Very Large

This conclusion requires that the lifetime of such civilizations is long. A simple argument can be made that the lifetimes *are* probably long. If a technical civilization can survive only a short time, then right now *we are it* in our Galaxy. We would have a unique status. But in the history of astronomy all claims for a special status for humankind have proved wrong. So the number of civilizations may be large and lifetimes long (if we assume we are typical, not unique), even if we don't know exactly how large or how long. By large, we mean, following the argument in Section 25.7, that $L_{ic} \approx 10^{10}$ yr, so $N_{ic} \approx 10^9$. Then many stars in this Galaxy, in this option, have fostered life.

N_{ic} Is Very Small

The conclusion that the number is very small—essentially only one, the earth—contradicts the idea that we are typical. But if you want to play it really conservatively and work from the evidence to date, it is easy, given the complexity of life on the earth, to conclude that it may be unique. As far as we know, we are a miracle. The discovery of life on Mars would have transformed life from a miracle to an ordinary matter. But the search for life on Mars has so far failed.

To argue that we are unique—flaunting one of the strongest lessons taught by the history of astronomical thought—requires that the Drake equation gives one. That does *not* mean that other civilizations never did or never will exist in our Galaxy. It *does* imply that no more than one civilization exists at a time. We are alone now and so was any advanced civilization in the past.

Michael Hart is one astronomer who holds to the "we are alone" view. His opinion is based partially on computer calculations of the evolution of the earth's atmosphere (Chapter 14). He finds a most delicate balance must be maintained to keep temperatures in a moderate range. Hart notes that if the earth had an orbit of radius 0.95 AU, rather than 1.00 AU, the greenhouse effect would run away and turn the earth into a Venus. On the other hand, if the earth orbited at 1.01 AU from the sun, glaciation would have iced

FOCUS 26.1 GREETINGS TO THE COSMOS

In March 1972 the Pioneer 10 space probe was launched to study Jupiter and then to fly out from the solar system forever. Astronomers Carl Sagan and Frank Drake persuaded NASA that Pioneer 10's odyssey was a rare opportunity to attempt interstellar communications, if the wandering spacecraft were ever intercepted by extraterrestrial intelligent beings. With the help of Linda Sagan, the two astronomers designed a plaque attached to Pioneer 10's antenna supports. Its purpose is to convey basic information about the solar system, its location in the Galaxy, and the inhabitants of the third planet.

The illustration presents two representative earthlings (A in Fig. F.24), the man's hand raised in what is hoped to be a universal gesture of peace. If not recognized as such, the

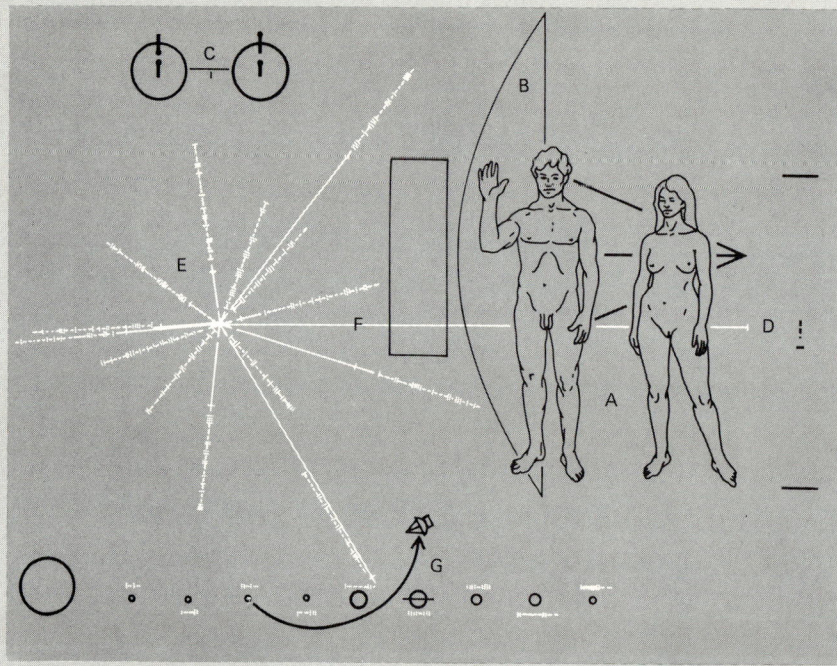

FIGURE F.24 The gold-etched plaque placed in Pioneer 10. (*Courtesy NASA.*)

up the earth 1.7 billion years ago. Our planet never would have gotten warm enough to foster the evolution of life.

Another argument, advanced by Frank Tipler, is that other civilizations, if they existed, could explore the Galaxy very quickly. We ourselves have already sent off spacecraft into interstellar space (Pioneer 10 and Voyager). If we had wished, we could have aimed these at some particular nearby star. We have the technology to launch interstellar probes and the technology for programming and controlling fairly sophisticated encounter missions. Tipler argues that in a relatively short time, we will have the technical capability to construct probes that can repair themselves. It is a small step from self-repair to self-reproduction. Such probes, called *von Neumann probes* after the mathematician who first

picture at least shows that humans have five digits on the hand, four fingers, and an opposing thumb. The size of the two people is scaled by an outline of the Pioneer 10 vehicle in the background (B). Another clue to the size is provided by C, which represents an atom of hydrogen, the most abundant element in the universe. On the right-hand side the atom is shown with the spins of the proton and the electron in the same direction; on the left-hand side the electron's spin is reversed. Recall (Section 9.1) that this change in the spin relationships is a change of energy state that results in the emission of 21-cm radiation. This wavelength sets a basic unit of length. For example, to the right of the woman is the binary symbol for 8 (D). If 21 cm is multiplied by the number, the result is 168 cm, the height of the woman.

The hydrogen atom also serves to establish a basic, precise unit of time for other information contained on the diagram. The average time required for the change of energy for the atom that results in the 21-cm radiation is well known. The starburst pattern (E) makes use of this information; 14 of the lines represent the directions and distances to specific pulsars, each of which has its own unique frequency indicated on the lines (again in binary form). The other line (F) gives the distance and direction of the center of the Galaxy from the sun. For specific details of the solar system, the hypothetical beings need only look at the bottom of the plaque (G). Here the illustration shows the spacecraft leaving the third planet as it veers off into space.

The two Voyager spacecraft, launched in 1977, also carry a message to the cosmos. This time the messages are the sounds of earth. A golden phonograph record is attached to each spacecraft, along with a stylus and pictorial instructions on how to use it to reproduce the sounds. Included are the word hello in a hundred languages, symbolic of an all-earth greeting. Other sounds include the noise of cities, bird calls, machinery, animal noises, the tolling of bells, all sorts of things deemed representative of life on earth. But will other species communicate by sound and so comprehend our message? Perhaps not sound exactly, but some kind of coded waveform is likely. The *meaning* of these sounds will not be easy for other species to determine—language is very specific to earth—but the idea that we communicate in this way may be apparent. Perhaps the record has more meaning for us, a way of saying to ourselves that we are reaching out into space in a human, not just technological, way.

discussed the theory of self-replicating machines, should be possible to construct within 100 years.

Suppose a von Neumann probe is launched to the nearest star. While it orbits that star making observations of its planets, it manufactures two copies of itself and sends them to other stars. These two reproduce themselves and send probes to four other stars, and so on. Each step in the process doubles the number of stars visited. After n steps, the number visited is 2^n. How many steps are needed to cover the entire Galaxy? To reach the 10^{11} stars in the Galaxy we need to have

$$2^n = 10^{11}$$

Taking the log of both sides of this equation, and solving for n,

$$n \log 2 = 11$$

$$n = \frac{11}{\log 2} = 36.5$$

So 37 steps would certainly cover the entire Galaxy.

How long would this take? The distance between stars is about 4 ly, or 4×10^{13} km. Suppose the probe travels at 30 km/sec (0.01 percent the speed of light). It will take 1.3×10^{12} seconds, or 40,000 years, to get there. If the probe can reproduce itself in a short time (even 100 years is short compared to 40,000 years), the 37 steps to reach all the stars will take only 1.6 million years. That's very short compared to the 15-billion-year lifetime of the Galaxy.

Tipler's argument, then, is that if *any* technological civilization had arisen in the past, von Neumann probes would have covered the Galaxy in a very short time. Such probes could be programmed not only to observe the planets in a star's system, but also to attempt communication with any life forms on them. Since there is no evidence for such probes in our solar system, and no attempts have been made to communicate with us, Tipler concludes that no other technical civilizations have arisen in the Galaxy. We are alone; $N_{ic} = 1$.

We don't hold to Hart's or Tipler's pessimistic view. But some astronomers do.

N_{ic} Is Either Very Large or Very Small

This compromise position admits to our ignorance—ignorance so immense that it allows the two extreme possibilities. Either we are alone or we have lots of company.

Basically you can adhere to both positions described above and remain open to the fact that just *one* piece of evidence for extraterrestrial life tips the case completely. Right now we only know that N_{ic} is at least one. Perhaps it is *only* one. But the existence of even one extraterrestrial life form, relatively nearby, would imply, from the argument of the uniformity of nature, that many exist.

N_{ic} Is Neither Very Large nor Very Small

This view accepts the lesson of astronomy and admits that, even though we do not yet have evidence to the contrary, it is unlikely that we are unique. At the same time it accepts the SETI results to date: no extraterrestrial signals have been discovered so far. If N_{ic} were very large, and if anybody cared to communicate, such signals should have been easily found.

The adherents to this position have used their own estimates of the astronomical, biological, and sociological factors to come up with a "best value" (guess!) for N_{ic} of about 10^6, with a range from 10^4 to 10^8. (JG favors this view, MZ the previous one.)

How Far to Our Galactic Neighbors?

Given a number for N_{ic}, how can we estimate the average distance to the nearest

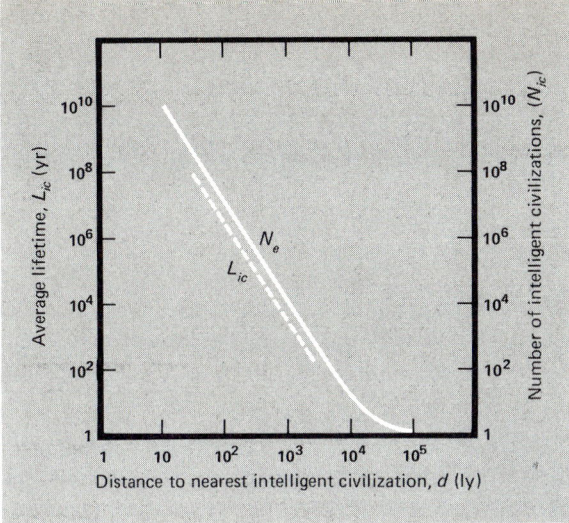

FIGURE 26.1 The relationship of the average lifetime of intelligent civilizations, L_{ic}, the total number of intelligent civilizations in the alaxy now, N_{ic}, and the distance to the nearest such civilization now, d. (*Adapted from a diagram by R. N. Bracewell.*)

civilization? Suppose we view the Galaxy as a cylinder of radius R and thickness h. Its volume is then $V = \pi R^2 h$. If $R = 60{,}000$ ly and $h = 2000$ ly, then

$$V = \pi (60{,}000 \text{ ly})^2 (2000 \text{ ly})$$
$$= 2 \times 10^{13} \text{ ly}^3$$

Suppose the civilizations are a distance d apart. The volume of a sphere containing just one civilization is $(4\pi/3)(d/2)^3$. The number of civilizations in the Galaxy is the total volume divided by the volume occupied by one:

$$N_{ic} = \frac{2 \times 10^{13} \text{ ly}^3}{(4\pi/3)(d/2)^3}$$
$$= \frac{4 \times 10^{13}}{d^3}$$

So

$$d = \frac{3.4 \times 10^4}{(N_{ic})^{1/3}} \text{ ly}$$

For example, if N_{ic} is very large, say 10^9, then $d = 34$ ly; our neighbors are only a few tens of light years away (Fig. 26.1). If N_{ic} is 10^6, the compromise guess, then $d = 340$ ly; our neighbors live a few hundred light years away. They are then just within reach. If N_{ic} is one (if we are it), we don't have any neighbors within 10^5 ly (30,000 pc), the size of the Galaxy.

Note that each of these choices implies a value for L_{ic} (Fig. 26.1). If N_{ic} is very small, L_{ic} is at most a few hundred years. We are then probably on the verge of extinction. If N_{ic} is very large, L_{ic} is 10^9 to 10^{10}. Civilizations then last as long as their suns. If N_{ic} is 10^6, then L_{ic} is roughly 10^5 to 10^6, and we still may have some time on the planet earth. (But recall, L_{ic} is an average, not necessarily the most probable lifetime. Maybe civilizations last *either* a very short time *or* an extremely long time, depending on whether or not they survive the crisis of atomic technology. In such an either-or situation, there might be none which actually last the average time.)

26.2 HOW TO SETI

This section assumes either the very large or the compromise estimate of N_{ic} and therefore that our neighbors are either very close or within reach. How can we reach or find civilizations tens to hundreds of light years away? There are three possible ways: space travel, unmanned probes, and radio communication.

Space Travel

The traditional, science fiction way to go—travel in a spacecraft—is impractical. The distances are too far, the energy is too much, and human lifetimes are too short (despite the popularity of *Star Trek*).

First, look at time and distance. Special relativity tells us that no spacecraft can travel faster than the speed of light. General relativity says that time passes more slowly for observers who are accelerated than for those who are not. Any real spacecraft must be accelerated and decelerated to reach its destination. So the clocks on the spacecraft will run slower than earth-based clocks if the acceleration exceeds 9.8 m/sec/sec. In addition, during the time the spacecraft is not accelerating, that is, coasting at a constant velocity, special relativity tells us that the ship's clocks run slower than earth's clocks. These relativistic effects make a long trip close to the speed of light livable for the crew, but not for the people left behind.

Here's an illustration. Suppose a spacecraft accelerates at 9.8 m/sec/sec for half a trip and then decelerates at the same rate for the second half. (Maintaining such an acceleration requires an increasing force as v gets higher, for the apparent mass increases as v approaches c.) Accelerating at 1 g for 1 year gets you very close to a cruising speed of c (9.8 m/sec/sec $\times$ 3 $\times$ 10^7 sec/yr = 3 $\times$ 10^8 m/sec = c). How far can the spacecraft travel within a crew's lifetime? Suppose we set 30 years, spacecraft time, as the maximum duration of the trip. Calculations with the formulas of relativity show that the spacecraft would reach 1565 ly (480 pc), the distance to the Orion Nebula, stop, and return in its 30 years. Back on earth, 3100 years would have gone by (Table 26.1).

Now, consider the energy requirements. Can we build such a rocket? No. Recall how a rocket works. It burns a propellant that pours out of its back end. The reaction force of the propellant on the rocket (Newton's third law) pushes the rocket forward. How much propellant is used and how fast it blows out of the rocket determines how fast the payload is going when the fuel runs out.

Our rockets now run on chemical fuels. These engines fall far short of the requirements for practical interstellar travel. Even if one can pull the same trick as Pioneer 10 and use Jupiter to assist in reaching the escape velocity of the solar system, about 30 km/sec (10^{-4} c), it would take roughly 40,000 years to reach Alpha Centauri.

Imagine that we could construct the most efficient rocket possible, one run by matter-antimatter annihilation (Fig. 26.2). Suppose the payload (computer, life-support systems,

TABLE 26.1 Relativistic Space Travel

Round Trip Duration for Crew (yr)	Elapsed Time on Earth (yr)	Distance Reached (ly)	Farthest Object Reached
1	1	0.059	Comets
10	24	9.8	Nearest stars
20	270	137	Hyades cluster
30	3100	1565	Orion Nebula
40	36×10^3	1.8×10^4	Globular cluster 47 Tuc
50	42×10^4	2.1×10^5	Magellanic Clouds
60	5×10^6	2.48×10^6	Galaxy Messier 33

Note: The spacecraft is assumed to accelerate at 9.8 m/sec/sec.
Source: Adapted from a table by S. V. Hoerner.

ten or so people, food, etc.) has a mass of 1000 metric tons. To accelerate it outward, decelerate at the destination, and return to the earth at speeds close to c requires 33,000 tons of fuel. At launch the rocket would have a mass of 34,000 tons! The total energy released by the fuel conversion would be 3×10^{24} J, as much energy as released by the sun in 0.01 sec. Put that way, it doesn't sound like much, but at $0.01 per kilowatt-hour (cheap by today's standards) it amounts to some 10^{16} dollars worth of energy. It is also enough energy to power the United States' present electrical consumption for 50,000 years.

If the cost isn't bad enough, the matter-antimatter rocket has another serious drawback. The annihilation emits copious gamma rays. So the rocket could not be launched from or near the earth. The intense gamma-ray exhaust would be extremely harmful to all living creatures.

These considerations all but rule out space travel for SETI. It is possible, but *very* expensive. Of course, we could cope with slower speeds and longer trip times if we could place the humans on board in suspended animation—a favorite trick of science fiction writers. But we don't have the technology for that yet.

Other hypothetical advanced civilizations in the Galaxy are bound by the same physical laws. So space travel probably won't work for them either, unless their lives are much longer than ours. Then they might take trips of a few tens of light years, if they can afford the cost and don't mind or can get around the boredom.

One way out of the time bind for space travel is space colonies—huge space ships that contain self-sustaining populations of people, plants, and animals. Traveling at 0.01 to 0.1 c, such colonies could reach the nearest stars in a few centuries. The initial voyagers

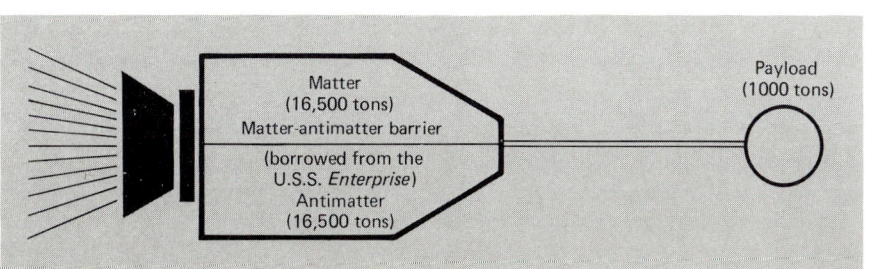

FIGURE 26.2 A hypothetical matter-antimatter starship. (Based on an idea by E. Purcell.)

would die before they reached their goal, but their offspring a few generations later would make the arrival. During the trip the people would carry out the same activities as they do on the spaceship earth: laughing, playing, loving, grieving. So the journey would be as interesting as our life now, but with a different and fascinating ultimate goal (more on space colonies in Chapter 27).

Interstellar Probes

If we can't go ourselves, why not send a robot probe? This probe could be fairly small, say a ton or so. It would be designed to search for radio/TV signals from an extraterrestrial civilization and flash a signal when it made a detection. Traveling at about $0.1\,c$, the probe could reach Alpha Centauri in less than 100 years. A detection signal would take only 4 years to return. Such a spacecraft plus fuel would have a mass of only about 2 tons.

To send a few such probes seems practical. However, if we wanted to bug all the nearby sunlike stars (after all, we don't know where ETI resides), it would be a vast, expensive project. About 10^3 sunlike stars lie within 100 ly (30 pc), about 10^6 within 1000 ly (300 pc). If we probed these by launching one interstellar craft per day, it would take 3000 years just to send the probes off. Not impossible, but surely impractical. However, once we are able to build self-replicating von Neumann probes, such a venture becomes practical.

Ronald Bracewell has suggested that advanced civilizations might use such probes to hunt for us and therefore that we should be on the lookout for them. That, of course, assumes that they are interested in searching for others.

Light

Searching physically for ETI is unrealistic, at least for now. We can't easily and cheaply make spacecraft travel at speeds very close to c. But we do have a messenger that does go precisely at c—light itself. Photons are cheap, fast, and available. Round-trip communication to the stars within a few hundred light years takes a few hundred years.

Light encompasses a broad energy spectrum, from radio to gamma-rays. What range is best for SETI? If we require that the ideal range must be the cheapest per photon (so we can send many at low cost), not be absorbed readily by the interstellar medium or planetary atmospheres, be easy to collect from a large area, be easy to transmit and detect, and be expected from advanced civilizations, then radio fits the bill the best.

Radio astronomers have constructed telescopes specially designed to detect very weak radio signals. These instruments can also be used to transmit, as is done in radar astronomy. So not only is radio the best range of wavelengths, but the basic technology is available now both to send and receive. With other civilizations, too, radio will probably be the first form of communication technology to be developed, because it's the easiest.

The Microwave Band

Radio itself covers a wide band of the electromagnetic spectrum. What range of frequencies is best? The choice of frequencies hinges on what part of the radio spectrum has the least background of natural noise, because we will be trying to detect weak signals. (Noise means, for example, the incessant jumble you hear when you tune your AM radio to a spot between stations.)

26.2 HOW TO SETI

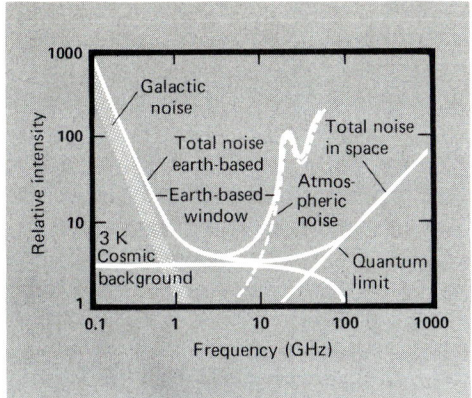

FIGURE 26.3 Natural background noise in the microwave region of the spectrum. At frequencies less than 1 GHz most of the noise comes from synchrotron emission in the Galaxy (labeled "galactic noise"). At frequencies greater than 100 GHz noise arises from the atomic nature of the materials that make up radio receivers ("quantum limit"). From 1 to 100 GHz some noise arrives from the 3 K cosmic radiation. The earth's atmosphere contributes at frequencies less than 1 and greater than 100 GHz. Adding these all together results in a low-noise dip at 1 to 10 GHz ("earth-based window"). (*Adapted from the Project Cyclops report.*)

Astronomy and physics naturally define a low-noise band (Fig. 26.3). At the low-frequency end of the radio spectrum (0.1 to 1 GHz) noise from the Galaxy (mostly synchrotron emission from high-speed electrons) dominates. At the high-frequency end (100 to 1000 GHz) noise in radio receivers, which comes from the quantum nature of matter and so cannot be eliminated, picks up. Between these two noise hills lies a valley of relative quiet (filled in a bit by the 3 K cosmic radiation, Section 24.3) from 1 to 100 GHz, part of the *microwave* region of the radio spectrum. Within this range the earth's atmosphere contributes a bit more noise at frequencies greater than 10 GHz. That leaves a low-noise window between 1 and 10 GHz for any receiver on a planet like the earth. A space receiver could have a wider window, from 1 to 100 GHz.

Note that this microwave window includes the radio lines from hydrogen (1420 MHz = 1.42 GHz) and the hydroxyl radical (1667 MHz = 1.667 GHz). Some people have argued that the range from 1.42 to 1.667 GHz is the best band to tune in or transmit for SETI. Why? Water dissociates into H and OH. As stated in the Project Cyclops report:

> Standing like the Om and the Um on either side of the gate, these two emissions of the dissociation products of water beacon all water-based life to search for its kind at the age-old meeting place of all species: the water hole.

The reckoning of physics and astronomy points to a narrow band in the radio spectrum. In it stand the signals of the medium of life. Is this a sign we should heed? Maybe, but perhaps this is being too romantic. Will interstellar species sense the need to meet at the interstellar water hole? We know of lines of other molecules that fall in the 1 to 10 GHz range, for example, formaldehyde (6.2 cm or 4.8 GHz), formic acid (18 cm or 1.2 GHz), and water itself (1.3 cm or 22 GHz). It is true that H and OH are two of the most common species in the interstellar medium; that may be a better reason for choosing the band bound by them rather than the fact that they are the dissociation products of water. But we may need to search the entire terrestrial microwave window, examine every common molecular emission line.

One last point. Any earth-based SETI system has to compete with sources of local radio interference. The 1 to 10 GHz band has been allocated for a number of uses, including aviation, meteorological aid, location finding, and space operations. Most of these transmit at low power. But satellite transmissions take up about half of the band. These are beamed down at the earth and so directly interfere with a search of the sky. At the World Administrative Radio Conference in Geneva in October 1979, a proposal to reserve the water-hole band for SETI was approved. This is where we should search first.

For a long-term SETI the interference problem might be considered serious enough to want to move the equipment off the earth. The far side of the moon is a good spot. Here we have available a wider microwave window and a complete shielding from earth's radio transmissions.

26.3 SETI STRATEGIES

How to begin SETI with the instruments at hand? First, we must decide if we are to spend our limited time listening or sending. Listening seems now to be the better choice. If technological civilizations live at least a thousand years, then we are mere infants. Most of them will be technologically superior to us, with better facilities for sending and receiving. Since sending is technically more difficult than receiving, we use our resources most efficiently if we listen. And if we listen, we might learn something immediately, whereas if we send, it may be a long time before a reply is received (and we would have to listen for that in any case). For SETI, then, it is better at first to receive and listen for evidence of others. Later, when technology improves, we can start sending.

Next, we must decide on the method of search. Do we make a target search of specific objects or scan the whole sky hunting for sources? Remember that a radio telescope sees only a small fraction of the sky at a time, roughly a circle a few arcminutes in size (called the telescope's beam). It takes roughly 10^8 circles an arcminute in size to cover the entire sky. Changing a telescope's sky position once a second, it would require only a few years to cover the sky completely, but that gives little time to listen at each position. On the other hand, a telescope's beam, although small in angular size, covers large regions of space. When a radio telescope points at some direction, it can pick up signals from a huge cone, extending far into space, with its tip at the telescope and its cross section larger at greater distances. The telescope can detect an object if it falls within this cone. However, the emission from a celestial object decreases as the inverse square of the distance. So although the expanding cone from the telescope picks up more objects at greater distances, their signals will be weaker, assuming they all have the same luminosity.

A target search mode could look at nearby sunlike stars. In our region of the Galaxy stars of types F, G, and K have an average density of $5.4 \times 10^{-4}/\mathrm{ly}^3$ (or $0.088/\mathrm{pc}^3$). The Galaxy's disk here is some 2000 ly (600 pc) thick. If we draw shells around the sun, we then get the number of sunlike stars in increasing volumes (Fig. 26.4). Roughly, we have 10^3 stars out to 100 ly (30 pc) and 10^6 to 1000 ly (300 pc). These are the stars we could search for radio emissions coming from technical civilizations. This search would be practical with present equipment. The Arecibo 300-m dish, for example, could detect its twin transmitting from a few thousand light years distance in a reasonable observing time. We could easily start with nearby stars and work out.

Unfortunately, we do not yet have a catalog of the positions of all sunlike stars to 1000 ly. A target survey out that far would require that optical astronomers put together such a catalog.

A whole-sky search would look for a different class of transmission. It may be that an

26.3 SETI STRATEGIES

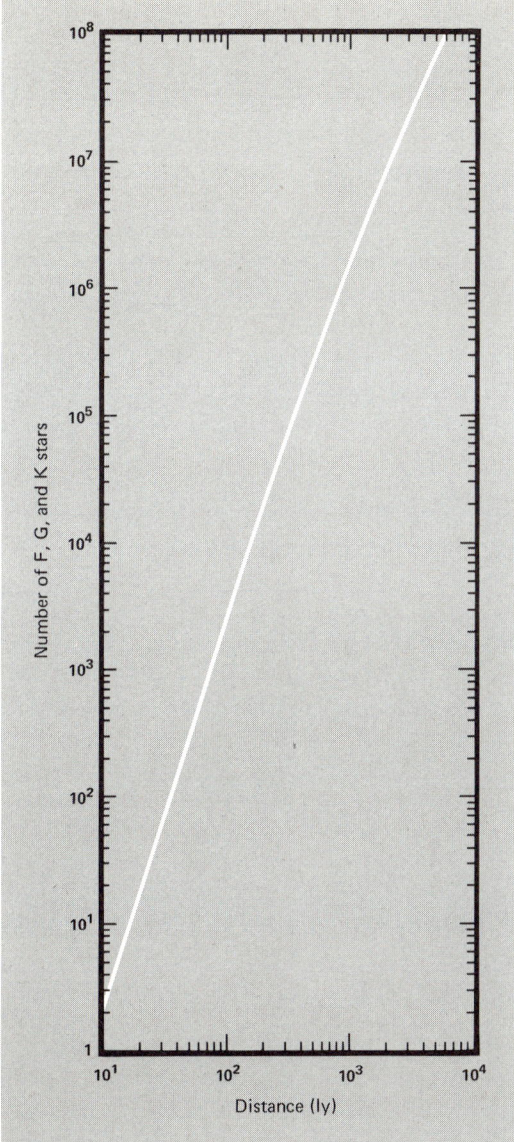

FIGURE 26.4 The number of F, G, and K stars as the distance from the sun increases.

advanced civilization far away sets up a powerful radio beacon just to signal their existence. This beacon might then be the brightest ETI radio source in the sky. That is, the luminosity function of ETI radio sources may resemble that for stars: many of low luminosity and a few very powerful ones. (The brightest stars you see in the sky are typically very luminous ones far away.) But we don't know, before we start, where to look for such luminous beacons. So we need to search the whole sky. (Actually, we'd probably do well enough simply to stay in the plane of the Milky Way.) To be thorough, the search must be sensitive enough to detect a radio beacon on the other side of the Galaxy. The weaker the signal (received flux) we wish to detect, the longer the search will take, for we would have to observe longer at each point to detect a weaker signal.

Note the basic operational difference between the two strategies. In a target search you spend a lot of time looking carefully at a few promising nearby stars. In a whole-sky search you spend a small amount of time on any one area and therefore can detect only strong sources. To start SETI, a mix of target and whole-sky search seems the best strategy for success.

How to effectively search the microwave band? Remember, we do not know the frequencies of our hypothetical transmitters. And even if they did send precisely at the hydrogen frequency, 1420 MHz, the signals would probably be Doppler shifted by rotation and revolution of their planet and of ours. So we need to search some interval in the radio band. With any radio receiver, when tuned to a particular frequency, we get not only that frequency, but also some small chunk of frequencies around it. The range in frequency around a tuned-in frequency detected by a radio receiver is called its *bandwidth*.

The problem here is that if we choose a very narrow bandwidth, we have many individual channels to tune to. For example, a 1 GHz bandwidth means ten channels from 1 to 10 GHz. A 1 MHz bandwidth means 10^4 channels, 1 KHz bandwidth 10^7 channels.

You might think the widest bandwidth possible helps the search; it means fewer channels in a given range. But wide bandwidths increase the noise picked up by the receiver, and so make it harder to detect a weak signal. Wide bandwidths also result in a loss of frequency resolution. If the transmitter produces a very sharp radio line, a wide-bandwidth receiver would smear it out into a small bump, making it harder to detect (Fig. 26.5).

Consider this analogy. The FM band in the United States runs from 98 to 108 MHz. Imagine your FM receiver had a bandwidth of 10 MHz and you tuned it to 103 MHz in the center of the band. You'd receive a station transmitting at that frequency, plus all the others in the range 98 to 108 MHz. You'd hear a cacophony rather than the symphony you wanted. Local FM stations are separated by 500 KHz (and FM receivers are built with smaller bandwidth) just to avoid such interference.

With a narrow bandwidth you must spend time, and receiving at each of many channels. You have a wide range to search and you cover a small fraction of it at each channel

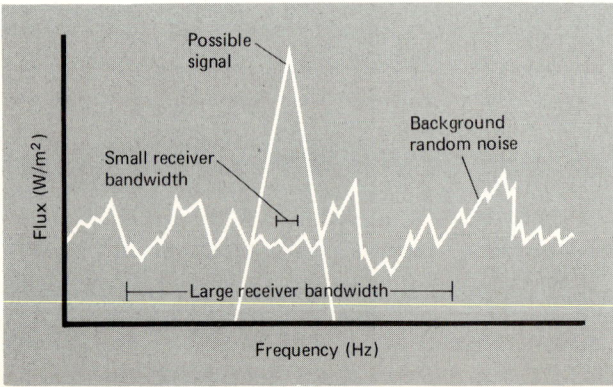

FIGURE 26.5 Bandwidth. Suppose a sharp signal exists on top of a random noise background. A receiver with a bandwidth much smaller than the width of the signal will reduce the noise—and also the signal. A receiver with a bandwidth much larger than the signal's width will also include so much noise that the signal will be reduced relative to the noise.

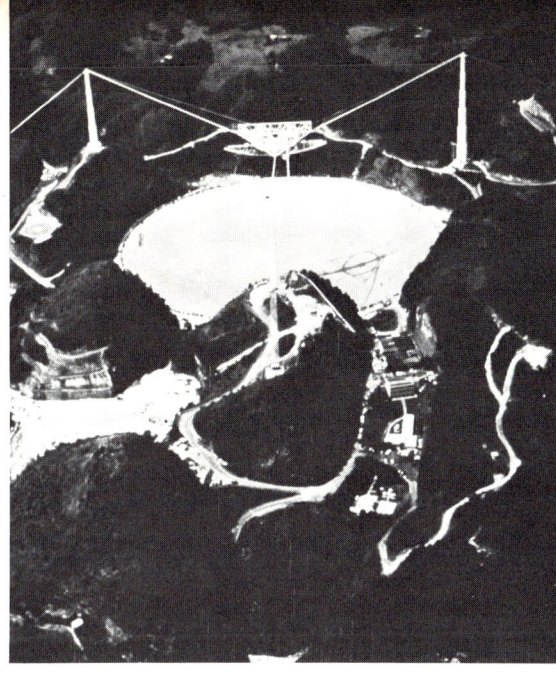

FIGURE 26.6 The 300-m Arecibo antenna in Puerto Rico. (*Courtesy F. Drake and Cornell University.*)

you tune to. Most radio telescopes can now detect several channels at a time, and a few can search as many as 1000 simultaneously. What is needed for SETI is a receiver that can detect a few *million* channels simultaneously. Recent advances in electronics make such a receiver seem possible, but one has not yet been constructed.

What would a SETI system look like? A radio/radar telescope like Arecibo fits the requirements fairly well (Fig. 26.6). It radiates about 7×10^{12} W at 2.38 GHz in the transmission (radar) mode. The Arecibo system could easily detect its duplicate placed at the distance of the star Alpha Ophiuchi, which is 60 ly (18 pc) from us (Fig. 26.7). With a little effort and a moderately long observing time, it could reach much farther, perhaps to a few thousand light years. (But we'd have to know where to look.) A signal emitted at a

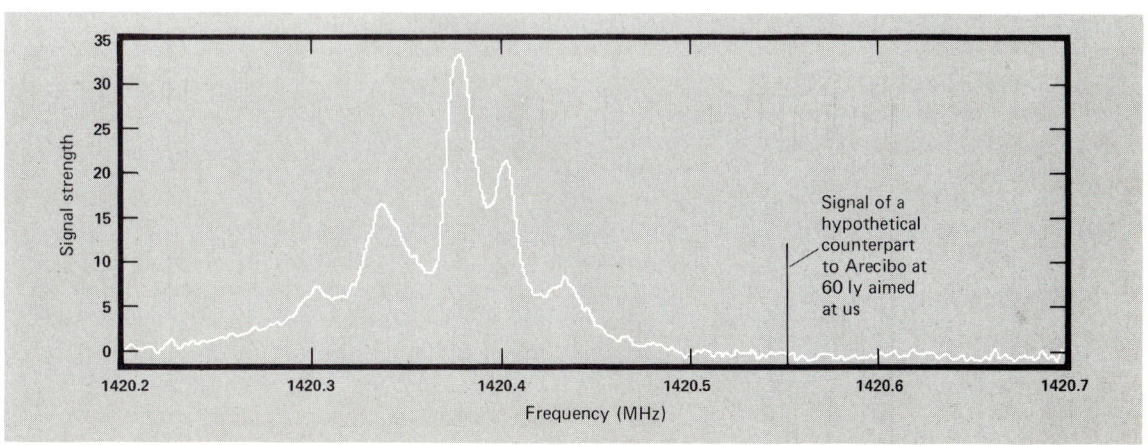

FIGURE 26.7 CETI by the Arceibo antenna. The present equipment is sensitive enough to detect easily a signal at 1420.55 MHz beamed at us from similar equipment 60 ly away. To the left of the artifical signal are lines from the 21-cm line of interstellar hydrogen. (*From the NASA SETI report.*)

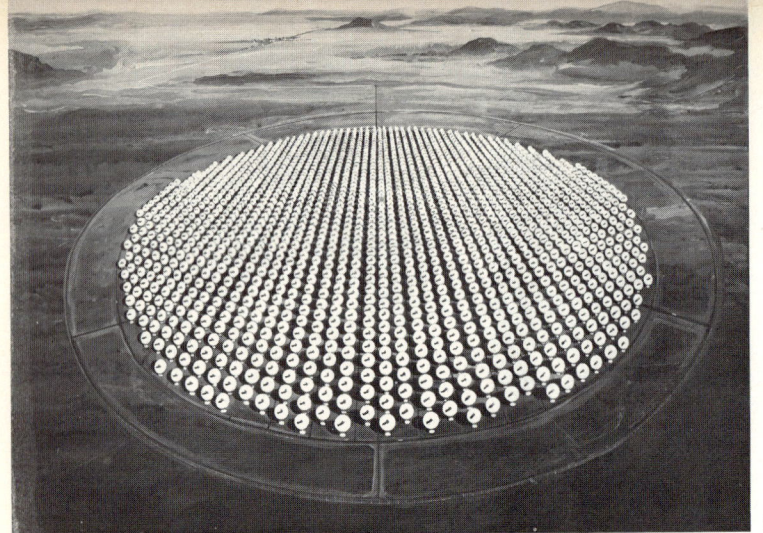

(a)

(b)

FIGURE 26.8 (a) An artist's conception of the Cyclops array, 16 km in diameter, viewed from the air, (*Courtesy NASA*). (b) The Cyclops array, at ground level, showing the individual 100-m antennas and the central control facility. (*Courtesy NASA.*)

frequency higher than 1420 MHz (the hydrogen line) by a twin Arecibo would easily stand out above galactic background noise when analyzed by the Arecibo receiver (which can handle 1024 channels at once).

Of course, if we have only one antenna operating for SETI, even full time, the search might take a long, long time. Recognizing this essential limitation, NASA helped to sponsor Project Cyclops, a study of the optimal, feasible design for a system to detect ETI, using state-of-the-art techniques. The result was truly a Cyclops: a plan for a circular array, 16 km in diameter, containing some thousands of 100-m radio antennas separated by 300 m (Fig. 26.8). The antennas would be coordinated by and feed information to a central computer. Its receiver could handle about a million channels at a time. The cost, dominated by the antenna structure, was estimated to be around $8 billion. To cover the entire sky, two Cyclops systems would have to be built, one for each hemisphere.

What are the capabilities of detecting ETI with such a monster system? Suppose we aimed it at the two nearest stars that might have a planetary system, Tau Ceti and Epsilon Eridani. Cyclops could detect a radio transmitter emitting only 500 W! (A typical radio station broadcasts more than 10,000 watts.) For a search over the water-hole frequency band of all sunlike stars within 100 ly (30 pc), Cyclops would take about a month; for those within 1000 ly (300 pc), about 50 years. To search the full sky in the water hole down to a flux level of 10^{-24} W/m^2 would take 12 days; to cover the 1 to 10 GHz band, 17 years. A modified, all-sky search strategy at flux levels between 5×10^{-25} W/m^2 at 1 GHz and 5×10^{-23} W/m^2 at 10 GHz would take only a year. Any Arecibo within a few hundred light years would be picked up in this way.

Suppose the Cyclops search fails to detect ETI. What would we have learned? First, we would have a better idea of the limits on N_{ic} and our place in cosmic evolution. Second, the search would generate a tremendous fallout of radio information of astronomical importance.

Cyclops is appealing to some even though it involves enormous costs. But, if you're willing to spend time rather than money, you don't need a Cyclops system. A modest 26-m antenna could cover the whole sky down to a flux level of roughly 10^{-23} W/m^2 in just 5 years with the development of very low noise receivers capable of handling 10^6 channels simultaneously. That's about as faint as Cyclops, but it does take longer.

Should we do it? Both of us agree that to *try* to answer the question "Are we alone?" is a worthy effort for humankind. But we differ as to timing. MZ says "now"; JG says "later" (at least for massive projects like Cyclops), when technology has improved a bit. You have to come to your own decision.

26.4 SETI RESULTS TO DATE

The first SETI, and perhaps the most famous, was the Project Ozma search done by Frank Drake. He used the National Radio Astronomy Observatory (NRAO) 26-m antenna at Green Bank, West Virginia. Drake monitored for four weeks the two nearby "best candidate" stars, Epsilon Eridani and Tau Ceti, at a frequency of 1420 MHz, the hydrogen line, which was suggested by Giuseppe Cocconi and Philip Morrison in 1959 as the natural wavelength for interstellar communication. Drake found no ETI signals. He estimated a sensitivity limit of 8×10^{-22} W/m^2, good enough to detect a transmitting Arecibo at 10 ly (3 pc) distance. It's interesting to note that Drake spent 2 weeks each on 2 stars to achieve a sensitivity only 1/100 as good as can be achieved now with a 25-m telescope that covers the full sky in 5 years. Technology has improved immensely in just a few years!

Eight years after Ozma, the Soviet astronomers V. S. Troitskii, A. M. Starodubstev, L. I. Gershtein, and V. L. Rakhin listened at 0.927 GHz to 11 stars, mostly G type, within 62 ly (19 pc) of the sun, and also to the Andromeda Galaxy (M 31). They found no ETI signals, down to a flux level of 2×10^{-21} W/m^2.

Paul Horowitz in 1978 reported on a very sensitive search using the Arecibo antenna. He examined 180 sunlike stars within 82 ly (25 pc) at 1420 MHz with a sensitivity of 4×10^{-27} W/m^2. That was good enough to detect an Arecibo 1210 ly (370 pc) away. No signals were detected.

The Ozma II project, carried out by Pat Palmer and Ben Zuckerman, searched some 600 stars and again resulted in no detection.

Other searches have been carried out sporadically by astronomers in the United States and the USSR since 1960. A few hundred stars have been observed, mostly at 1420 MHz. *No extraterrestrial signals have been detected to date.*

The lack of results is understandable. Less than 0.01 percent of the sky and the terrestrial microwave window has been searched. Yet the nondetection is annoying; it certainly doesn't generate optimism.

26.5 WHERE ARE THEY?

If life is common in the Galaxy, then the lifetimes of intelligent civilizations are long. All searches for extraterrestrial life have failed. All searches for signs of extraterrestrial intelligence have failed. So where are they?

At a session on extraterrestrial life at the 1979 Montreal meeting of the International Astronomical Union, the discussion revolved around the so-called four facts of ETI, which were used to support pessimistic views about the abundance of life in the Galaxy.

1. We are here, but we don't know for how long.
2. No life has been found on Mars.
3. No positive results have resulted from radio searches.
4. No aliens are here now.

One must certainly agree with the first three "facts." But can we be so sure about number 4? We have conducted fairly careful extraterrestrial searches for terrestrial life. But we've been very lax in mounting a terrestrial search for extraterrestrial life.

The speculation (and this is *really* speculation) goes like this. If other beings are advanced technologically, they could send observers here. The observers might be robots, which might be biologically engineered rather than mechanically constructed (the ultimate von Neumann probe!). They might be here to find out when we've reached the cultural maturity to take the shock of ETI contact. Of course, they work hard to remain undetected. That sounds like science fiction, but it is one way of answering the question Where are they. They *are* here, but *in disguise*!

This view takes off from the fact that on the earth contact between a technologically advanced culture and one less advanced usually results in disruption of the latter. That shock may be totally unintentional. For example, the Europeans who settled along the east coast of North America brought with them smallpox, to which the natives had no resistance. Many Indians died in smallpox epidemics. The Spanish brought horses with them in their exploration of the North American southwest. Some horses escaped or were taken by the Indians, and the introduction of the horse drastically transformed the Plains Indians' society. Contact with ETI may knock us for a loop. If they are wise enough—and they probably are if they exist—they might wish to stay out of touch until we are ready for them.

It is sometimes suggested that we ourselves are descendents of ETI colonists. But the fossil record securely indicates that we evolved from primitive life forms here on earth. And it's unlikely that life arose elsewhere and then was transported here as simple cells; the substances present in living organisms are quite similar to those in seawater, even down to the trace elements.

But many scientists do not think the "Where are they?" question is easily dismissed. As discussed in Section 26.1, if only one civilization had survived to a stage of space colonization and decided to spread out in the Galaxy, it could have spread everywhere by now. But we see no evidence of this, and therefore, these people say, we can conclude that ETI does not exist. The fact that there is no evidence as yet for ETI surely means something, but, in our opinion, what it means is not yet clear.

26.6 UFOS: EVIDENCE OF ETI?

The subject of UFOs is a sticky and emotional one, but we firmly believe it should be dealt with openly. UFO stands for *Unidentified Flying Object*. Actually, in many cases UFOs can be identified, are not flying, and are not physical objects.

Many people believe that UFOs are contacts with ETI, that is, that UFOs are spacecraft piloted by extraterrestrial creatures. For example, in a 1966 Gallup poll, 48 percent of those polled thought that "flying saucers" were real. In a survey done in 1968 for the University of Colorado Project (commonly known as the Condon Report) 24 percent of the adults polled believed that "some flying saucers have tried to communicate with us," 28 percent that "earth has been visited at least once in the past by beings from another world," 41 percent that "some UFOs have landed and left marks on the ground," and 40 percent that "people have seen space ships that do not come from this planet."

We would judge that a survey now would indicate people believe even more strongly in the ETI origin of UFOs. A survey similar to the 1966 Gallup poll was repeated in some astronomy classes at Harvard and at the University of New Mexico. Some 50 to 60 percent believed that "UFOs are real."

These attitudes are taken despite the fact that *no* evidence to date clearly demonstrates that UFOs are vehicles run by extraterrestrial beings.

People have been seeing strange phenomena in the sky for many years. UFOs perceived as flying saucers are actually a relatively new phenomenon. (The term "flying saucer" was coined after the famous Kenneth Arnold report in 1947.) The U.S. Air Force began to keep track of UFO sightings in 1948 and continued until 1969, usually under the name of Project Blue Book. During that time the Air Force received reports of 12,618 sightings of which 701 were judged to be "unidentified" (Fig. 26.9). (303 of these took place during the 1952 wave of sightings, which followed the release in 1951 of the movie *The Day the Earth Stood Still*. The increase in sightings may or may not have been related to the movie.)

What do astronomers and other scientists make of the "unidentified" reports? (Remember that only a small fraction of sightings were reported to the Air Force or anyone; 87 percent in the Colorado survey reported sightings only to family and friends.) Careful examinations of UFO reports and interviews with witnesses show that most UFO sightings can be easily explained by natural phenomena (many astronomical, such as the planets Venus and Jupiter) and that most UFO reports are *not* hoaxes. For example, of 887 reports to Blue Book in 1965, 246 were astronomical, 210 aircraft, 152 satellites, and only 34 hoaxes. In fact, most people who make UFO reports are genuinely convinced that they have seen something strange; they are not of a lunatic fringe. The remaining 245 were true UFOs in the sense that they were unidentified. (That doesn't necessarily make them the product of ETI, merely unidentifiable from the available evidence.)

← Motion

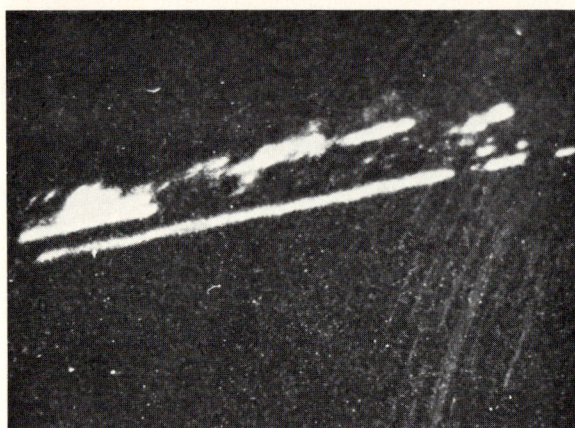

FIGURE 26.9 A "nocturnal light" UFO photographed in May 1973 with a Polaroid camera. This was a time exposure, so streaks result from the UFO's motion; the direction of motion is to the left. The UFO was described as "a silent ball of light that approached our position near Piedmont, Missouri, and turned." (*Courtesy H. Rutledge.*)

What of these reports that remain unidentified and unexplained after the analysis by trained investigators? They are the heart of the controversy. Some people, such as Donald Menzel, Edward Condon (both now deceased), and Philip Klass, believe that *all* UFO sightings can be explained by some natural phenomena. (Menzel even advised MZ in a letter not to deal with UFOs in an astronomy class.) Others—Allen Hynek and Peter Sturrock, for example—believe that something puzzling is contained in the residual unidentified reports. If it's not ETI, perhaps it is some new astronomical or meteorological phenomenon which we don't yet understand, because we don't have enough good data about it.

Hynek developed the classification scheme of UFO reports made famous in the movie *Close Encounters of the Third Kind*. He divides them generally into two major categories of those sighted at a distance and those sighted close enough for details to be seen. The distant category includes nocturnal lights (seen only at night), daylight disks (the object is seen as oval or disklike), and radar-visual (the object is both detected by radar and seen). The close category includes:

Close encounters of the first kind: A UFO is seen at close range, but does not interact with the environment.

Close encounters of the second kind: A UFO is seen along with physical effects (such as scorching and breaking) on both living and nonliving material.

Close encounters of the third kind: "Occupants" are reported in or about the UFO.

Close encounters of the fourth kind: Direct, intended contact and communication occur (as depicted at the end of the movie).

Hynek generally disregards contact cases because in almost all of them only one observer was involved. We agree that single-person UFO reports have low credibility because you have no way to cross-check witnesses. But Hynek does find that, in his judgment, the close encounter cases are the strangest and also tend to be from credible, reliable witnesses. Do these prove ETI? No, because, in our view, no hard physical evidence remained behind nor did any communication take place.

Most astronomers, ourselves included, will not be convinced that UFOs are evidence of ETI until beings from space show themselves publicly to many people, as in the movie *Close Encounters of the Third Kind* or the book *Childhood's End* by Arthur C. Clarke. Even then, the landing of a "flying saucer" with extraterrestrial beings in it does not necessarily explain all previous UFO sightings. (In fact, the creatures could well deny that they have anything to do with those!)

Frankly, it seems absurd that if UFOs are piloted by ETI, they just don't land to say "hello." If they have interstellar craft, they certainly have the technology and the smartness to keep out of our sight. Why should they tantalize us with random close encounters but not communication?

Of course, making such a statement involves applying human motives to nonhuman creatures. So let us offer a less anthropomorphic argument against UFOs as evidence of ETI. It's the Santa Claus argument, developed, as far as we know, by Carl Sagan. You know about Santa Claus. He visits some 10^8 houses in the United States within about 8 hours on December 24–25. Suppose he stays at each house only 1 sec (not even time for a good "Ho, ho, ho") and travels faster than light between them. It would take him some 10^8 sec—about 3 *years*—to make all the visits. Santa Claus is physically impossible.

Let's apply the same kind of argument to visits by ETI. Assume only one ETI visits a year as a UFO. The Galaxy contains about 10^{11} stars to visit. Assume that 10^6 advanced civilizations with interstellar flight capability exist in the Galaxy. Dividing up the places to visit equally, each sends spacecraft to 10^5 places. For one visit a year, then, even assuming instantaneous travel time (physically impossible, according to special relativity), each civilization would need to launch 10^5 vehicles a year (about 300 a day). That's unreasonable, especially given the practical difficulties of interstellar flight (Section 26.2). The above argument assumes one launch for each encounter, but there may be multiple encounters. And the existence of self-reproducing von Neumann probes would completely wipe out the Santa Claus argument. But even so, we remain skeptical.

Our skepticism about UFOs as evidence of ETI does not mean we're opposed to the study of UFOs. Contrary to the attitude of Condon, who saw no scientific value in studying UFOs, our attitude is to welcome serious scientific enquiry. We'd like to know what UFOs are. And many members of the American Astronomical Society agree. In a survey done by Peter Sturrock, 53 percent of the respondents replied that the UFO problem "certainly" or "probably" deserves scientific study. Older scientists were much more negative than younger ones. (We seem to qualify as "younger ones.")

26.7 CETI

So far we've dealt with the search. Suppose the discovery day arrives. What then? After picking up "here we are" signals, we'd probably want to enter into two-way communications.

That presumption may be wrong. Perhaps for political, social, religious, or economic reasons, we may decide to remain silent. Perhaps we may become afraid once we actually pick up ETI. But we believe a reply should be sent, because we would have a lot to learn—perhaps even how to avoid the potential planetary death of an all-out nuclear war.

Presumably they'll want to talk with us, out of curiosity, if for no other reason. That statement again has the shortcoming of ascribing human motives to other beings. But if they do have a beacon set up, it is reasonable that they desire communication.

How to carry on a conversation? Because of the vast distances involved, it would be very slow and one-sided. We might send a reply in the direction of their beacon. Then we wait—200 years if they are 100 ly away. Then we'd have to decode their answer and send out a reply. And wait again. Can humans cope with such a long-term project? Not humans as individuals, for of course they'd be dead before an answer arrived, but humans as political, social groups? Do we have sufficient vision and courage to sustain the project? For a fictional view of this problem, read *The Listeners* by James Gunn.

What might our messages look like? Imagine for a moment we pick up a beacon from within a few hundred light years. Our first response, probably, should be a beacon beamed at them at the same frequency at which we picked up their beacon. Our beacon should look obviously artificial, perhaps a rhythmic series of pulses and spaces broadcast for at least a month. If their beacon had a pulse pattern, we could just imitate it and send it back. Many years later we listen carefully to the reply. Suppose, as Bernard M. Oliver has speculated, it is a series of pulses and spaces repeated every 22 h 53 min. That time interval could be the length of their day. If we put in a zero for each space and a one for each pulse, we have a message in binary code, the same simple language used by computers (Fig. 26.10a). (Binary is the simplest code we know about, so we assume that a message we send or receive will be in binary.) The message seems a mess. It consists of 1271 ones and zeros. But wait—1271 is the product of two prime numbers, 31 and 41, so let's arrange the message in a picture with 31 characters down and 41 across. Use a blank for each zero and a dot for each pulse. We end up with a picture (Fig. 26.10b).

Now the really hard part: interpreting the message. Oliver contends that it says the creatures are erect bipeds, who reproduce sexually and may be mammals (region A). The circle and column of dots on the left (region B) represent their planetary system. The (male?) figure on the left points to the fourth planet, where they live. Down the extreme left-hand side, the planets are numbered, starting with a binary point to mark the beginning (C). The wavy line extending from the third planet (D) shows it has water with fishlike forms of marine life. Along the top (region E) are diagrams representing hydrogen, carbon, and oxygen atoms; so their life is based on a carbon chemistry. Above the hand of the right (female?) biped is the binary number for 6. This implies 6 digits on a hand and a base 12 counting system. Next to the creature (region F) is a size marker indicating the length of the creature. In its center lies the binary code for 11. The biped is 11 units of something tall. The universal unit of size, so some astronomers argue, is the 21-cm line of interstellar hydrogen. So the biped is 231 cm tall (7.7 feet).

Upon receipt and decoding of this interstellar radio postcard (it might be contained in the original beacon signal), we can send back our own in a similar format. In fact, we have already sent out such a message from Arecibo. The transmission was part of a dedication ceremony in 1974 and was done to demonstrate the capability of the instrument. It was beamed at the globular cluster M 13 in Hercules. It will arrive about 27,000 years from now, in roughly A.D. 29,000. Don't wait for the reply. (You wouldn't expect one from a cluster of Population II stars anyway. Think about that.) We have also sent more solid messages onboard the Pioneer 10 and Voyager spacecraft (Focus 26.1).

Suppose we do make contact with a civilization as close as Tau Ceti. And suppose we can convince each other that we are intelligent by adopting a logical code to transmit and decipher information. Each exchange of signals would take about 20 years; in a human lifetime only three round-trip communications could take place, and they would require

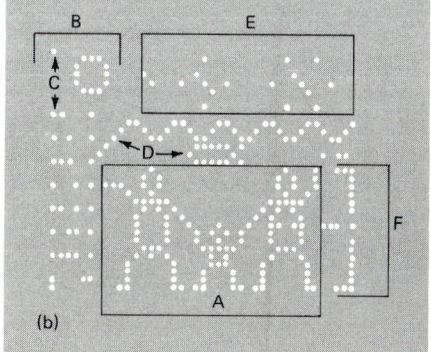

FIGURE 26.10 (a) A hypothetical binary message ("1" is on, "0" is off) that might be received from an extraterrestrial civilization. (b) A decoding of this signal into a decipherable picture. (*Adapted from a figure by B. M. Oliver.*)

great patience at both ends. If the nearest intelligent civilization is three times the distance to Tau Ceti, one greeting and response would span a lifetime.

Space and time effectively quarantine us as individuals from contact with our neighbors. Even if the Galaxy teems with life, each planetary system is a solitary unit fenced off by vast volume of space. We are very much alone in terms of continuous, direct contact.

SUMMARY

SETI can be done, but it requires patience. Physics and our present technology force us to use radio waves as the cheapest, fastest, most practical means to search. Astronomy and physics guide us to the radio band from 1 to 10 GHz, if the search is done from the earth, as the most effective wavelength range. Within that band lies the cosmic water hole—1420 to 1667 MHz—which may be the best place of all to search.

Once contact is made, we must then, as a planetary society, decide whether to enter into communication. Discovery of ETI will certainly be a shock. But how much we can learn! And perhaps the contact will be necessary to our survival, to our evolution as an intelligent species. Perhaps it is another phase in the process of cosmic evolution.

Key Words

SETI (Search for Extraterrestrial Intelligence)	von Neumann probe	Project Cyclops
CETI (Communication with Extraterrestrial Intelligence)	cosmic water hole	Project Ozma
	bandwidth	UFO (Unidentified Flying Object)

Review Questions

1. What is one argument that the number of technological civilizations is large? What is one argument that it is small, possibly only one? (Objective 2)
2. Why wouldn't we want to listen or transmit at *exactly* 21.11 cm? (Objectives 7 and 9)
3. Give one reason why all searches to date for ETI have failed. (Objective 10)
4. Criticize Arecibo's sending a message to the globular cluster M 13. *Hint:* What kinds of stars make up globular clusters? (Objective 10)

Problems

1. If there are 10^6 technological civilizations in the Galaxy, what is their average lifetime? (Objective 2)
2. Arecibo can transmit at about 10^{13} W. What is the flux of this signal at 10 ly? 100 ly? 1000 ly? 10,000 ly? (Objective 9)
3. If we transmitted at 1420 MHz, how much of a frequency shift would result from the earth's orbital velocity? What would the minimum bandwidth of our transmitter have to be so that there would always be some energy radiated at 1420 MHz? (Objective 9)
4. Assume a radio telescope has a beam projected on the sky of one square arcminute. Consider an all-sky search with each position observed for one hour. About how long would the search take? (Objective 9)
5. There are about 0.02 galaxies/Mpc^3 in the observable universe. How many steps would it take for von Neumann probes, if they duplicate at each galactic encounter, to reach every galaxy out to 3000 Mpc? (Objective 4)

BEYOND THIS BOOK . . .

* C. Sagan and F. Drake discuss SETI in "The Search for Extraterrestrial Intelligence," *Scientific American*, May 1975. More details can be found in NASA special report SP-419 (1977) with the same title, edited by P. Morrison, J. Billingham, and J. Wolfe.
* *Project Cyclops* (CR114445) is available from NASA Ames Research Center, Code LT, Moffett Field, Calif. 94035.
* *Interstellar Communication: Scientific Perspectives* (Houghton-Mifflin, Boston, 1974), edited by C. Ponnamperuma and A. Cameron, is an excellent source of a wide range of aspects of CETI. It contains a huge bibliography.
* Many UFO books are on the market. We suggest *UFOs: A Scientific Debate* (Norton, New York, 1972), edited by C. Sagan and T. Page, and the *Scientific Study of Unidentified Flying Objects* (Bantam Books, New York, 1969), otherwise known as the Condon Report.
* *The Search for Life in the Universe* (Benjamin/Cummings, Menlo Park, Calif. 1980) by D. Goldsmith and T. Owen is an up-to-date book on the subject.
* The magazine *Cosmic Search* (P.O. Box 293, Delaware, Ohio 43015) is devoted to the search for intelligent life beyond the earth.

COSMIC FUTURES

27

LEARNING OBJECTIVES
After studying this chapter, you should be able to:
1. Speculate, from an astronomical basis, on the future of the universe for open, closed, and flat models.
2. Speculate, from present astronomical models, on the future of the Galaxy.
3. Speculate, from ideas of stellar evolution, on the future of the sun, the earth, and other planets.
4. State what is meant by exponential growth and doubling time.
5. Present a specific example of exponential growth with a doubling time on the order of a human lifetime or less.
6. Describe what is meant by space colonization.
7. Argue that space colonization is physically possible.
8. Argue either for or against space colonization in human, economic, or personal terms.
9. Speculate on the future of the human race from an astronomical perspective.

CENTRAL QUESTION:
In what directions will cosmic evolution
unfold in the future?

What will the future bring for the earth, the sun, the Galaxy, the universe? We have touched upon these futures throughout the book. This chapter brings them together, lines them up, and considers them in the context of cosmic evolution.

Where do we stand in the course of cosmic evolution to date? Astronomy tells that we exist here 10 to 20 billion years after the Big Bang and 4.6 billion years after the birth of the sun and earth. Many stars have died, and had to die, so that life could arise on this planet. We, the erect biped called the human being, have been here only for 2 million years, the length of time it has taken for the light which we are seeing now from the Andromeda Galaxy to travel across the void to us. Many cosmic connections were made to get us here, now.

What of our future? We will speculate, from an astronomer's viewpoint, on the future of humankind, aided by what we understand so far of cosmic evolution. We encourage you to speculate, too. There is nothing wrong with this creative exercise of the mind. But keep your speculation disciplined! Fantasies about what *might* be must take into account what *is* and what we already know about the universe and how it works.

27.1 THE FUTURE OF THE UNIVERSE

What fate can we envision for the cosmos? Its future depends on whether the geometry of the universe is open or closed (Chapter 5). If we can find out the geometry of spacetime, Einstein's theory of general relativity predicts the future.

Open Models

Although geometrically different, both open models, flat and hyperbolic, have essentially the same futures. In a flat model the rate of expansion (the Hubble constant, H) gradually decreases until, at an infinite time in the future, it falls to zero. Then clusters of

27.1 THE FUTURE OF THE UNIVERSE

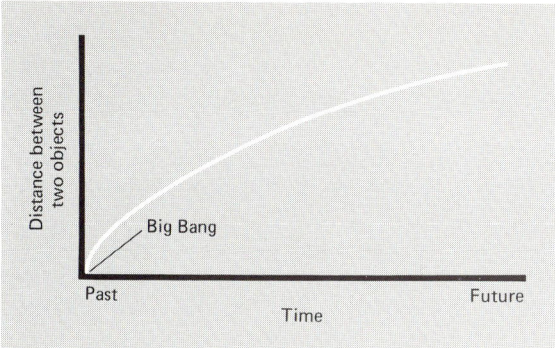

FIGURE 27.1 Expansion of a flat cosmos. This graph shows how the distance between two widely separated objects (such as galaxies) changes with time since the Big Bang.

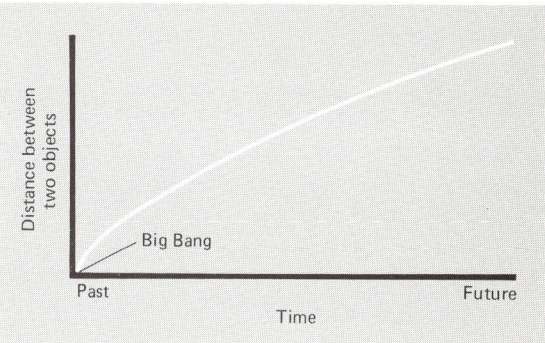

FIGURE 27.2 Expansion of a hyperbolic cosmos since the Big Bang.

galaxies get neither farther apart nor closer together, and the expansion grinds to a halt (Fig. 27.1). The flat case corresponds to a model in which the Big Bang provides exactly the escape velocity for all matter in the universe (Section 5.5).

In a hyperbolic model clusters of galaxies are still moving apart at an infinite time in the future, but more slowly than now (Fig. 27.2). H becomes very small, but never reaches zero. This case corresponds to a model in which the Big Bang imparts greater than escape velocity to all matter in the universe.

This book has argued that the universe is open according to present evidence (Chapters 5 and 24). Especially critical is the observation of the average density of matter (and energy) in the universe. If H equals 50 km/sec/Mpc, the critical cosmic density (Section 5.5) is 5×10^{-27} kg/m^3. If the actual density is less than this value, the universe is open; equal to it, flat; greater than it, closed.

Recall that it is exceedingly difficult to measure the universe's average density. Results to date imply a density of luminous matter of roughly a few times 10^{-28} kg/m^3, or about 1/10 the critical density. But the absence of evidence so far should not be taken necessarily as evidence for absence. The so-called missing matter, that needed to provide a closed universe, might exist in a dark form. Two possible sources are gas within clusters of galaxies and massive galactic halos of low-luminosity stars. Observations to date indicate that neither of these are sufficient. Other possibilities are black holes, either of stellar mass or supermassive ones, or neutrinos, if they have a mass.

Many people have an aesthetic preference for an open universe. They like its one-shot character. In contrast, Einstein preferred a closed universe, in part for its clear symmetry. Only further observations can tell which aesthetic preference is valid for the real universe.

Closed Models

Let's look at the universe's future in Einstein's view. If closed, the universe expands until clusters of galaxies reach a maximum separation. H falls to lower values and hits zero in a finite time. Then the universe contracts.

The value of H becomes increasingly negative and measures approach rather than recession. Eventually the universe ends in a crushing singularity (Fig. 27.3).

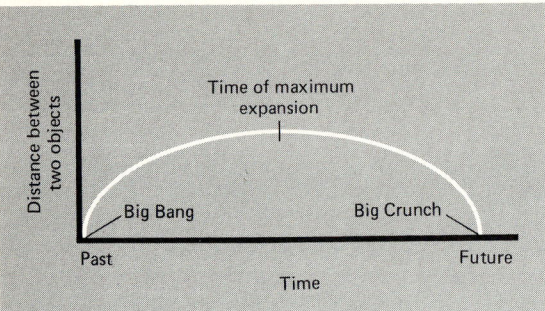 

FIGURE 27.3 Expansion and contraction of a closed cosmos. Note the symmetry of the curve on both sides of the time of maximum expansion, halfway through the cycle. A closed cosmos begins in a Big Bang and ends in a Big Crunch—the formation of the cosmic singularity.

FIGURE 27.4 An artist's idea of the microscopic appearance of spacetime. A slice of spacetime examined under very high magnification might show a foamlike structure, filled with holes, bumps, bridges, and tunnels.

Space will end, and so will time. This case corresponds to a model in which the Big Bang imparts less than escape velocity to matter.

Note the beautiful symmetry of a closed model. Also note the tremendous physical problems it creates. Time, space, and so physics end. In what sense can the very cosmos "end"? There is no way out, if you stick to the theory of relativity. Theoreticians have shown that a closed model, obeying relativity and not containing negative mass-energy, *must* develop a singularity.

The same problem crops up with black holes (Chapter 12). The gravitational collapse of a mass to a black hole can be looked at as a "laboratory model" of the contraction of a closed universe. The probable discovery of black holes validates this awesome prediction of general relativity. Black holes then raise the same sticky questions as a closed universe.

The difference lies not in the hard physical questions that result, but in what choices are open to an observer. For a black hole, an observer may choose to remain outside, and so find out nothing, or to dive in to find knowledge and a certain death.

In a closed universe observers have no choice: All are caught up in the cosmic black hole. And all black holes within the universe are rammed together in the contracting phase to join the ultimate singularity.

That prediction may signal the breakdown of Einstein's theory. Perhaps the paradox arises not because physics ends, but because our understanding of physics ends. A similar crisis hit physics at the beginning of this century concerning atoms and electromagnetic radiation. The development of a new physics, quantum theory, resolved that crisis. (Ironically, Einstein didn't believe in some aspects of quantum theory.) Some physicists, such as John Wheeler, believe that the prediction of singularities can be resolved by quantizing the general theory of relativity. In Einstein's view spacetime is perfectly smooth and continuous on all size scales. In a quantum view, on some very small size scale—physicists have argued for 10^{-35} m—spacetime loses its smooth appearance and breaks up into a violent, spongelike structure in which holes and bridges are created and destroyed in times on the order of 10^{-44} sec (Fig. 27.4). In this view dynamic spacetime collapses and restores itself naturally. Perhaps spacetime might undo itself and spring back from the collapse of a closed universe.

If the universe is closed, what part of the cycle are we in? In one model compatible

27.1 THE FUTURE OF THE UNIVERSE

TABLE 27.1 Properties of a Closed Model

Property	Value
Radius at maximum expansion	19×10^9 ly
Time from start to maximum	30×10^9 yr
Time from start to final crunch	60×10^9 yr
Radius today	13×10^9 yr
Age from Hubble constant	20×10^9 yr
Hubble constant today	49 km/sec/Mpc
Total amount of matter	3×10^{23} solar masses

Source: Adapted from a table by C. Misner, K. Thorne, J. Wheeler in *Gravitation*, W. H. Freeman and Co., San Francisco, 1973. This model is within the range of values known for H and q.

with observations (Table 27.1) the universe's cycle runs for 60 billion years. We are roughly 15 billion years into the cycle, or about 1/2 of the way to the time of maximum expansion.

Oscillating Models

Can the cosmic crunch be avoided? We don't know. But one set of cosmological models does presume that a closed universe survives the singularity. In these models the universe bounces outward at the end of its contraction, its contents somehow missing the singularity (Fig. 27.5). Then comes another phase of expansion, another contraction, another bounce, and so on. These are called the *oscillating models*. They not only have the appeal of symmetry, but also the beauty of no end or beginning for the cycles.

Each new cycle starts out fresh, since all matter and structure is destroyed in the contraction of the previous phase.

This model makes sense, in a strange way. If you accept the Big Bang, you've taken on the notion that the universe was *born out of a singularity*. Then its expansion is the mirror image of collapse. The evidence shows that the universe survives a singularity at birth, and this fact then implies that it can somehow survive a singularity at death. Expansion and contraction are just time-reversed sequences of the same physical process.

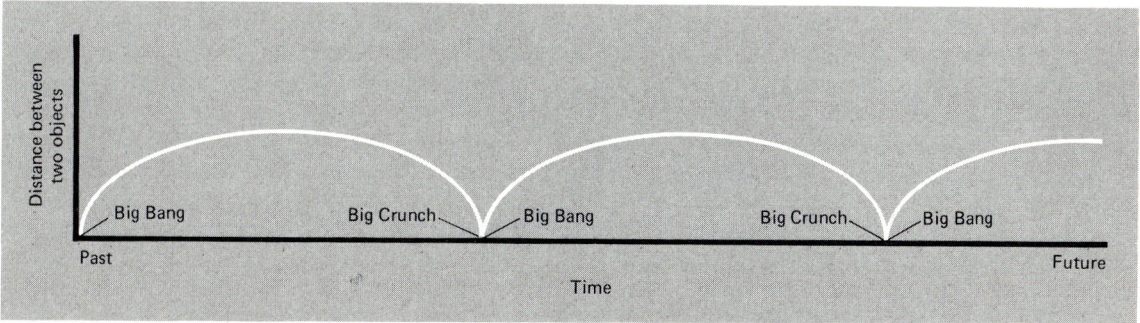

FIGURE 27.5 Expansion and contraction of an oscillating cosmos. Note the symmetry of the oscillation, with each Big Bang a mirror image of each Big Crunch.

Wheeler argues that the singularity marks the reprocessing of the universe. Matter is remade, time is reborn, physical "constants" are reconstructed, and even the physical laws are reshaped. We don't really know how to formulate these concepts in meaningful terms. But the universe may renew itself and continue the symmetry of its existence. It is also possible that alternating cycles are made of matter and antimatter; this would preserve another symmetry in the universe.

True? Perhaps. Calculations to date of the bounce using quantum ideas have come to ambiguous and contradictory conclusions. Certainly here is today's ultimate frontier of cosmology.

27.2 THE FUTURE OF THE MILKY WAY GALAXY

Let's step back from the cosmic vista to consider our own Milky Way Galaxy. If the universe is open, nothing exciting will happen to the Galaxy as a whole.

If the cosmos is closed, the Galaxy will be crushed with other matter some 45 billion years from now.

Aside from its cosmic context, the Galaxy's future (as has its past) largely comes about

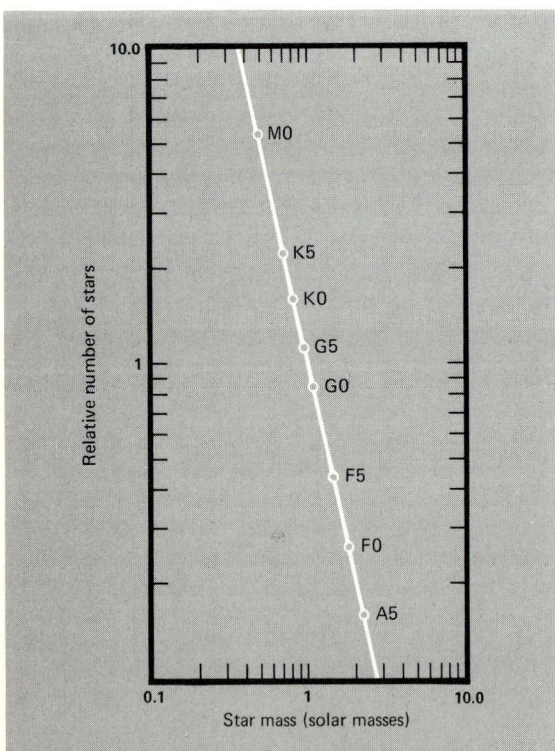

FIGURE 27.6 The numbers of stars of various masses in the Galaxy. The spectral types are indicated along the line. Note that the largest numbers of stars are the low-mass ones (spectral type M).

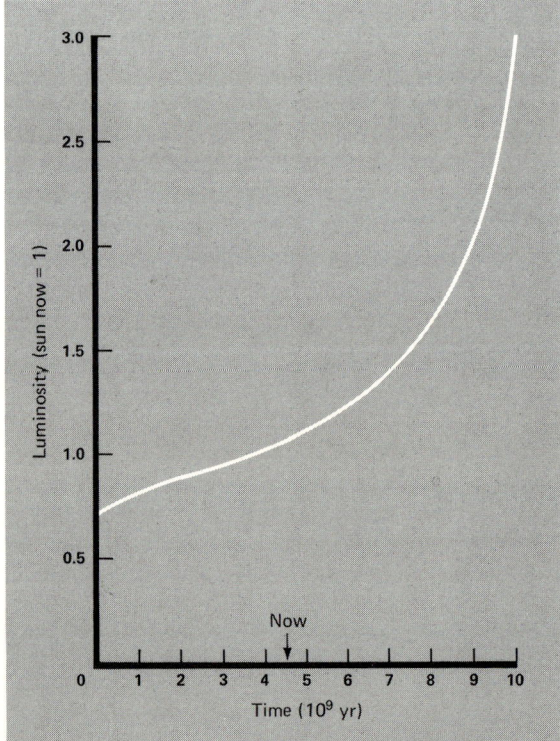

FIGURE 27.7 One theoretical calculation of the change in luminosity of our sun since its birth. Note that the luminosity is given in units of the sun's *present* luminosity. (*Based on calculations by I. Iben, Jr.*)

from the exchange of matter between stars and the interstellar medium. During the star phase matter is processed by thermonuclear fusion reactions (Section 10.10). In stardeath some of that material rejoins the interstellar medium (Section 11.6). In starbirth some of the matter is reprocessed by reincorporation into stars (Section 9.3).

As time goes by, the relative abundance of heavy elements in the interstellar medium will increase. But the total amount of gas and dust between the stars will go down as matter becomes locked in stellar corpses. Stars with less than 4 solar masses become white dwarfs, then black dwarfs. Those with greater mass become neutron stars or black holes.

Most stars in the Galaxy have low mass (Fig. 27.6). Main-sequence M stars, for instance, make up about 66 percent of the total number, K stars 13 percent, and G stars 8 percent. That means that almost 90 percent of the Galaxy's stars have masses less than 1 solar mass and life expectancies greater than the presumed age of the Galaxy. When these stars die (for M stars that will be about 100 billion years from now), they will shed only about 10 percent of their mass. Compared with all the stars, G through M stars contain about 86 percent of the Galaxy's stellar mass; so they will return only a small fraction of their material, relatively unprocessed, to the interstellar medium.

Massive stars do most of the recycling. Dying as supernovas, they blow back some 80 to 90 percent of their mass. But these stars make up only 4 percent of all stars and contain only 11 percent of the Galaxy's stellar mass. So although they recycle quickly because of their short lifetimes (less than a billion years), they don't recycle much of the overall mass.

The average recycling time for matter in our Galaxy is roughly 4×10^9 years. If 80 to 90 percent gets locked up in corpses each cycle, and if now approximately half the Galaxy's mass is in the interstellar medium, then our Galaxy will live some 20 billion years more before it has essentially stopped recycling; then only 0.03 percent of the present interstellar medium will remain.

27.3 THE FUTURE OF THE EARTH AND SUN

Models of stellar evolution and a comparison with H-R diagrams of stellar clusters indicate that our sun will become a red giant and then a white dwarf (Chapters 10 and 11). The future of the earth hangs on the fate of the sun.

Theoretical calculations indicate what changes we can expect for the sun (Fig. 27.7). About 4.5 billion years from now the sun will have 100 times its present luminosity; 5 billion years from now 500 to 1000 times its present luminosity and 70 times its present radius. A planet's surface temperature depends on the fourth root (Section 15.2) of the sun's luminosity. So 5 billion years from now the earth's surface temperature will be roughly 1400 K. Long before that time the oceans and the atmosphere will have boiled away. When the sun becomes a red giant, the earth will be seared of all life. From the earth the luminous red sun will cover some 35° in the sky!

With the ignition of helium core burning, the sun's luminosity will drop to some 100 present solar luminosities. Its surface temperature will rise to 5000 K. Still the earth will remain too hot for life, about 900 K. Then after some 50 million years, the sun will swell to a red giant again. This time it puffs off its outer layers, revealing the hot, dense core that will become a white dwarf. The atmosphere will rip past the earth, which will then be bathed in ultraviolet light, a final wash of sterilization. As the core cools, the sun's luminosity will drop. When it is 1/100 the sun's present luminosity, the earth's temperature will drop to a mere 100 K. The earth will then be a frozen corpse that will gradually become cooler and cooler as the sun's luminosity steadily declines.

A planet's surface temperature also depends on the inverse square root of its distance from the sun. Will moving to an outer planet or one of the moons of Jupiter or Saturn provide an escape from the solar blast? At Jupiter's distance the temperature will be a factor of 2.3 less than on the earth, at Saturn a factor of 3.1 less. Five billion years from now, when the sun is 500 times its present luminosity, the surface temperature of the Jovian satellites will be 600 K and that of Saturn's satellites 450 K, both too hot for us. Uranus would be 4.4 times cooler than the earth, a warm but tolerable 320 K. Neptune, 5.5 times cooler, would still be a frozen 255 K. So Uranus or one of its satellites might prove a temporary haven when the sun becomes a red giant.

But it won't be safe for long. When the sun begins helium burning and drops again in luminosity, the temperature at Uranus will fall to 200 K. Then we would have to move back to Saturn to find a comfortable temperature of about 300 K.

No planet will receive much heat from the dying sun. When it sinks to 0.01 its present luminosity, even Mercury will be a frigid 160 K.

27.4 THE FUTURE OF HUMANKIND

Before the sun dies, if we are still alive, we probably will have left our home planet—perhaps for other planets or perhaps not for planets at all. Predicting the future is always a dubious enterprise, but it seems certain that we will have to leave the earth. And we may have to leave long before the sun dies, because the Spaceship Earth is a finite resource. If we do not achieve a stable population for the human race we have no choice but to leave the earth.

Growth

The problem here is one of growth—*exponential growth*, the opposite process of exponential decay (Section 13.2). Here's an example to show how constant growth, which may appear small from year to year, rapidly gets out of hand. Consider bacteria that double every minute.

Suppose we start out with 1 bacterium. In 1 min it doubles into 2; 1 min later the 2 split into 4. With a constant doubling time of 1 min how long would it take for 1 bacterium to produce 4×10^9 bacteria, a number equal to the population of the world? Roughly 32 doubling times (since $2^{32} = 4 \times 10^9$), or 32 min, only half an hour!

Steady doubling characterizes exponential growth. It leads to rapid increases in numbers, even for small fractional growth rates. A rule of such growth is that the doubling time T (in years) for a growth rate P (percent in a year) is approximately $T = 70/P$. For example, if P is 10 percent a year, the doubling time is 7 years. In 1975 the world contained about 4 billion people, and the growth rate of the human population was 1.9 percent a year. So the doubling time is roughly 37 years. If the rate remained constant, the mass of people would equal the mass of earth in only 1600 years!

Constant doubling results in the rapid use of finite resources. Consider bacteria again. Suppose we place a bacterium in a bottle and note that the bottle is full at 12 noon. When was the bottle half full? One doubling time (1 min) earlier: 11:59. It was a quarter full two doubling times earlier at 11:58. Suppose at this time some far-sighted bacteria leaders got together and intensively searched for more living space. At 11:59 they find an empty bottle on the shelf, which doubles their total living room. When will that new bottle be filled? *12:01.*

When consumption grows exponentially, even enormous increases in resources are consumed in short times.

Consider oil consumption, which has increased at 7.04 percent per year from 1890 to 1970. If that rate were to hold constant and if all the earth were made of oil, it would be consumed in about 340 years, only 5 human lifetimes. Even if the oil consumption growth rate hit zero tomorrow, the earth's proven resources would last less than 200 years.

The earth is a small planet. Even if zero population growth were achieved, our resources would be consumed at present rates in only a few human lifetimes. We will need to leave our home planet for space, for energy, and for natural resources.

Space Colonization

Where to go when we leave the earth? The traditional science fiction view had us journeying to and colonizing other planets in our solar system. With the possible exception of Mars, we now realize that the other worlds in the solar system are not now habitable planets. Even if they were, don't forget the lesson of the bacteria in the bottle. Human population growth now doubles roughly every 40 years. At our present growth rate it would take only one doubling time, 40 years or so, to fill Mars. In only 1500 years (37 doubling times) we would have enough people to populate 10^{11} planets, one for every star in the Galaxy.

But there's no basic reason to restrict ourselves to living on planets at all. This fact has encouraged Gerard K. O'Neill to revive and develop an older idea (some aspects were foreseen by the Russian physicist Konstantin Tsiolkowsky almost 100 years ago): human habitations in space, often known as space colonies. O'Neill has aimed at making the dream of space colonization a reality with available technology.

We won't detail his plans here, but let's sketch the broad outlines. Stripped of luxuries, people need energy, air (oxygen), water, land, and (probably) gravity to live a comfortable life. With space colonies in orbit around the earth, somewhere between the earth and the moon, all these are available: energy from the sun, oxygen and raw materials from the moon, and water from the earth. (Water might also be collected from the asteroids or moons of Jupiter and Saturn.) Gravity can be simulated by rotating the space colony.

Key to this vision is harnessing solar power. A satellite solar power station has many advantages over a terrestrial one: no weather to worry about, no cycle of day and night, a zero-gravity environment in which it is easier to build large physical structures, less upkeep and so a longer life for the photovoltaic cells that convert sunlight into electrical energy. Overall, a space environment has about 4 times more (one factor of 2 for continuous daylight, another for absence of weather) solar energy falling on a square meter than does the earth, roughly 1 kW/m². So a collecting area of 10^6 m² (1 km by 1 km) can gather 10^6 kW. If the photovoltaic cells operate at 10 percent efficiency, the system develops 100 MW.

This power can be transmitted by microwaves. In a household microwave oven an electronic device converts electricity to microwaves; the same device can work in a solar satellite power station. At the receiving end, whether on the earth, moon, or in space, the

FIGURE 27.8 One possible model of a space colony, as suggested by G. K. O'Neill. The colony consists of twin cylinders, each 32 km long and 6.4 km in diameter and holding some 200,000 people. Windows and mirrors let in and control the sunlight to simulate night and day and the seasons. The cylinders rotate around their axes once every 2 min to create an earth-like gravity on the inside wall. (b) An interior landscape of a cylindrical space colony. (*Courtesy NASA.*)

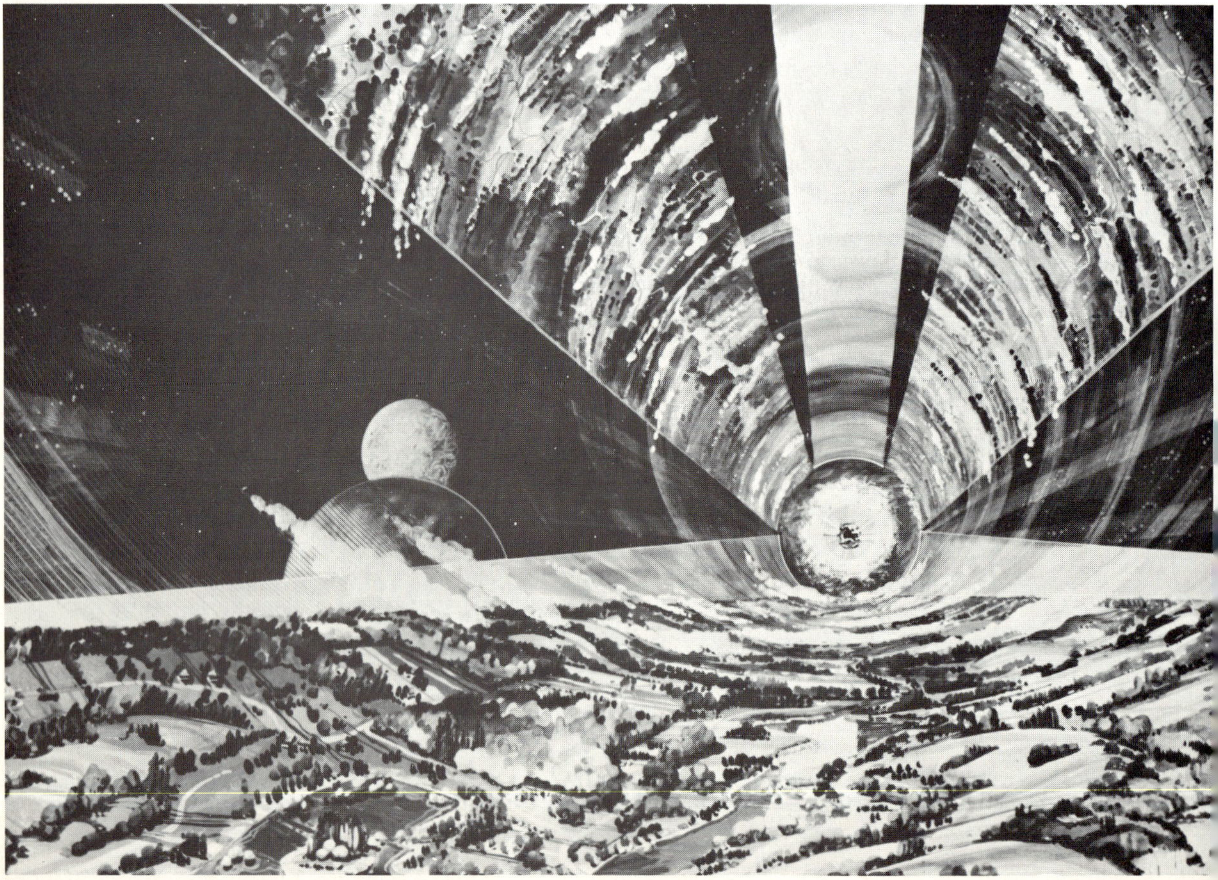

27.4 THE FUTURE OF HUMANKIND

microwaves would be converted back to electricity. Though reverse conversion devices aren't very efficient yet, they have been built.

The first space colonies and solar satellite stations would be built with resources from the earth. But earth's resources are limited, and it takes considerable energy to lift material into orbit. Expansion to many colonies requires breaking Mother Earth's umbilical cord. For raw resources for development, we can mine the moon. The moon's surface, as we know from the Apollo missions, contains abundant aluminum, titanium, oxygen, and silicon. It lacks water, so we would have to bring hydrogen from the earth to combine with lunar oxygen to make H_2O. A solar power station can support mining activities. Then a reliable, inexpensive way is needed to transport the lunar materials to the colonies' orbit. O'Neill and others envision the use of a magnetic linear accelerator that would boost payloads well above lunar escape velocity (about 2 km/sec) with accurate aim to the space colonies. Here they would be caught in a huge net and then used for manufacturing. It is critical to this concept that the space colonies become economically independent. They could bankroll themselves by selling power (beamed by microwaves) to the earth.

What might a space colony look like? The simplest design is a cylinder (Fig. 27.8). A cylinder some 3 km in diameter and 32 km long could support upward of 200,000 persons on its inner surface. Spinning once every 2 min to simulate gravity, the inside would alternate strips of land and windows. The windows would have shutters to simulate the seasons and day and night by controlling the influx of sunlight. The colony craft would be constructed of aluminum and titanium from the moon.

The space colonization ideas being batted around today make the colonies appear like potential utopias. And well they could be, compared with the earth. But only if we've learned the lessons of our interaction with the earth's environment. History teaches pessimism. For instance, consider the relatively uncontrolled, environmentally destructive growth of cities in the southwestern part of the United States. The lessons from the histories of east coast and west coast cities do not have much impact on these developing regions.

How long will this high frontier accommodate our population growth? At present growth rates the local space could handle 400 to 500 years of population doubling. That may sound long, but it's only about 7 human lifetimes. At present growth rates all the space between the sun and the earth would be saturated with people jammed together in about 3000 years. The moral of this extrapolation is that the human race will leave the solar system a few thousand years from now. Unless we change our ways.

Beyond the Solar System

Space colonies—"cities in space" as science fiction writer James Blish called them—could convey people between stars at speeds of $0.01 \, c$. Imagine that we send colonies out across the Galaxy. When each arrives at a suitable star, they could build a new colony ship from local resources and send it out to the next star. A few centuries might elapse from arrival to the next departure, and then a few centuries to glide to the next star. How long does it take to travel across the Galaxy at $0.01 \, c$? Take 120,000 ly (40,000 pc) for the diameter of the Galaxy. Then at $0.01 \, c$ the time in years is 100 times the distance in light years. This star-hopping process could carry people across the Galaxy and colonize it in roughly 12 million years.

Twelve million years is short compared with the age of the Galaxy and the time needed for life on the earth to evolve intelligence. So we could fill the Galaxy quickly, in cosmic terms. And so could any civilization that has achieved a technological level similar to ours.

This view brings up again the question of the number of technological civilizations in the Galaxy (Section 26.1). If colonization can happen so swiftly, and if L_{ic} is long, then it probably has happened, and "they" are everywhere. But if we take seriously the evidence of "their" absence, then L_{ic} must be very short, at least less than 10^7 years. But suppose we ignore that argument. If L_{ic} is 10^7 or so, some 10^3 waves of colonization have rippled through the Galaxy since its formation. Would we find artifacts from such former colonizations if we searched the right places? Michael Papagiannis has proposed that the place to look in our solar system is the asteroid belt, not the planets. He argues that the interstellar space travelers would ignore planets when they arrived at a new solar system. Instead, they would park at a suitable place and build another colony. Our asteroid belt contains raw materials for living (water) and construction (iron and nickel). The asteroids would be easier to mine than a planet because of the low gravity. Could we detect an abandoned colony? Probably not now. A colony would be about the size of a large asteroid and hard to distinguish from it with present instruments.

The Future of Technological Civilization

The Soviet radio astronomer N. S. Kardashev has classified technical civilizations by the amount of energy they control. Kardashev notes that in terrestrial history technological advances have been made possible by large changes in the energy budget per person. That may also apply to the evolution of other civilizations.

In Kardashev's classification, Type I civilizations have harnessed the energy resources of their planet. Our civilization now uses some 10^{13} W. The sun provides us with about 10^{17} W. When we have mastered that amount, we will become a Type I civilization. That day will come, perhaps, with the advent of space colonies.

Type II civilizations can utilize a substantial amount of the total energy from their parent sun, about 10^{26} W. It's likely that this stage must be achieved before the civilization can depart its home solar system.

Type III civilizations are interstellar communities with the capability of controlling the energy output of an entire galaxy, some 10^{37} W.

Within 100 years we may achieve Type I status. Within a few thousand years we could reach Type II. How? If we mine the moon and the asteroids, we could build a shell of space colonies an AU in radius to catch most of the sunlight. A small part of this energy, 1 percent or less, might be used for interstellar exploration. That could take the form of space colonies sent out on long interstellar trips. Not tied down to life on a planetary surface, they could orbit at a suitable distance from almost any good sun. The sun provides the energy, debris around it the material to produce more space colonies. Once these have achieved Type II status, they send off explorers to other stars. So we could slowly colonize the Galaxy and gain Type III status.

What next? By then a human lifetime may be much longer than now, perhaps 10 times longer. And perhaps by then we will have joined up with other galactic civilizations (if any) to make a galactic community with a common wisdom much, much greater than we command now. Our populations will grow. We might decide to move out to other galaxies. This step is a huge one. The nearest spiral galaxy, Andromeda, has a distance 20 times the diameter of the Milky Way Galaxy. At near light speeds intergalactic ships could shelter slowly aging beings and make intergalactic trips in millions of years. Once we arrive we might find others who, like us, have expanded to fill their galaxy. We would join up with them and others to create an intergalactic community.

This community would be a Type IV civilization, one that could manipulate the energy of clusters of galaxies, a sizable fraction of the total free radiative energy in the universe. It would take billions of years to develop.

Does the evolution of life, intelligence, and civilization end there? Cosmic evolution shows us continual change. A Type IV civilization is unlikely to remain static. You can imagine as well as we can what it might do. We will make only one prediction: The real future will likely be wilder than either you or we can imagine.

EPILOGUE

Why are we here? Quantum theory tells us that the isolated, detached observer does not exist. We are all participators, even the scientist making the most dispassionate of laboratory experiments. What greater participation could we have than to shape the evolution of the universe itself?

We are here because of our past. We are here to affect our future. If conditions in the Big Bang had not been just so, we would not be here. We are creatures that can look into space and within ourselves and join the inner and outer places in awareness. An awareness that revives our spirit and forms our future. Perhaps we are the means by which the universe becomes aware of itself—aware of the very large, the very small, and the very human.

Key Words
oscillating model
exponential growth
doubling time
space colony
satellite solar power station
Type I, II, and III civilizations

Review Questions
1. Why will the growth of heavy element abundance in the Galaxy eventually stop? (Objective 2)
2. What determines whether the universe will expand forever or eventually stop expanding and start collapsing? (Objective 1)
3. Why will the earth's temperature first increase, then decrease, and then increase again, before finally cooling off? (Objective 3)
4. What will space colonies use as a source of energy? Of material? (Objective 7)
5. Why is space colonization possible, but communication by space travel not possible (or at least impractical)? (Objective 7)

Problems
1. How luminous would the sun have to become to make Pluto a habitable temperature? (Objective 3)
2. Verify that it will take only about 300 years at present growth rates to fill the space between the earth and sun with people jammed together. (Objective 5)

3. Suppose we develop into a Type II civilization and completely surround the sun, absorbing all its energy. This energy, after utilization, must be reradiated into space. At what wavelength would you expect the maximum radiation, assuming that the sphere of space colonies surrounding the sun radiates like a blackbody? What kind of telescope might detect this sphere? (Objective 9)

BEYOND THIS BOOK . . .

* *The High Frontier* (Bantam Books, New York, 1977) by G. K. O'Neill and *Colonies in Space* (Warner Books, New York, 1977) by T. A. Heppenheimer discuss many aspects of space colonies.
* C. Sagan presents his speculations about the past and future of the human race in *The Garden of Eden* (Random House, New York, 1977).
* G. Verschuur discusses some astronomical futures in *Cosmic Catastrophies* (Addison-Wesley, Reading, Mass., 1978).
* *The End of the World* (Doubleday/Anchor, Garden City, N.Y., 1981) by Richard Morris treats several potential disasters of astronomical origin.

APPENDIX A UNITS

Powers of 10

Astronomers deal with quantities ranging from the truly microcosmic to the macrocosmic. To avoid having to write out and to read numbers such as 20,000,000,000 years (the age of the universe) or 149,600,000,000 meters (the distance to the sun), powers-of-10 notation is used. A positive exponent tells you how many times to multiply by 10, for example, $10^1 = 10$, $10^2 = 10 \times 10 = 100$. A negative exponent tells you how many times to divide by 10, for example, $10^{-2} = 1/100$. The only trick is to remember that $10^0 = 1$. Using powers-of-10 notation, the age of the universe is 2.0×10^{10} years, and the distance to the sun is 1.496×10^{11} meters.

The English and Metric Systems

You are familiar with the fundamental units of length, mass, and time in the English system: the yard, the pound, and the second. The other common units of the English system are often strange multiples of these fundamental units, such as the ton (2000 lb), the mile (1760 yd) the inch ($\frac{1}{36}$ yd), and the ounce ($\frac{1}{16}$ lb). Most of these units arose from accidental conventions and so have few logical relationships. Most of the world uses the much more rational metric system, with the following fundamental units:

Length: 1 meter (m)
Mass: 1 kilogram (kg)
Time: 1 second (sec or s)

This is the meter-kilogram-second, or mks, system. A slightly older system often used by astronomers is the centimeter-gram-second, or cgs, system, with the following fundamental units:

Length: 1 centimeter (cm)
Mass: 1 gram (g)
Time: 1 second (sec or s)

All of the unit relationships in the metric system are based on multiples of 10, so it is very easy to multiply, divide, and use powers-of-10 notation.

The contemporary standard for the meter uses the wavelength of orange light from krypton 86. The meter is defined as 1.65076363×10^6 times this standard wavelength. Any efficient laboratory can set up such a standard and use it accurately.

The multiples of the metric system and their associated prefixes are:

10^{-12} = pico-   (p)
10^{-9} = nano-   (n)
10^{-6} = micro- (μ)
10^{-3} = milli- (m)
10^{-2} = centi- (c)
10^{-1} = deci-   (d)
10^1 = deca-   (da)
10^2 = hecto- (h)
10^3 = kilo-   (k)
10^6 = mega-   (M)
10^9 = giga-   (G)
10^{12} = tetra- (T)

Some relationships between the metric and English system are:

Length:
1 kilometer (km) = 1000 m = 0.6214 mile
1 meter (m) = 1.094 yd = 39.37 inches
1 centimeter (cm) = 0.01 m = 0.3937 inch
1 millimeter (mm) = 0.001 m = 0.03937 inch
1 mile = 1.6093 km
1 inch = 2.5400 cm

Mass:
1 metric ton = 10^6 g = 1000 kg = 2.2046×10^3 lb
1 kilogram (kg) = 10^3 g = 2.2046 lb
1 gram (g) = 0.0353 oz = 0.0022046 lb
1 milligram (mg) = 0.001 g = 2.2046×10^{-6} lb
1 lb = 453.6 g
1 oz = 28.3495 g

Temperature Scales

Scales of temperature measurement are tagged by the freezing point and boiling point of water. In the United States the Fahrenheit (F) system is commonly used; water freezes at 32°F and boils at 212°F. In Europe the Celsius (formerly the Centigrade) system is the common temperature system; water freezes at 0°C and boils at 100°C. The Kelvin system is based on the idea of absolute zero, the temperature at which all random molecular motion ceases, which is −273.15°C. Since 0 K is at absolute zero, water freezes at 273 K and boils at 373 K. Note that the size of the degree is the same in both the Kelvin and Celsius systems (100 between the freezing and boiling points of water). To convert between the systems, recognize that 0 K = −273°C = −459°F (the Kelvin system never measures negative degrees) and that the Celsius and Kelvin degrees are larger than Fahrenheit degrees by the factor 180/100 = 9/5. Then the relationships between systems are:

$$K = °C + 273$$
$$°C = \frac{5}{9}(°F - 32)$$

A comparison of the three temperature scales is seen in Fig. A.1.

Astronomical Distances

Although astronomers do use the metric system, they encounter distances so large that other measures are often used. In the solar system the natural distance is the Astronomical Unit (AU), the average distance of the earth from the sun. The AU equals 1.496×10^8 km.

Beyond the solar system even the AU is too small to be convenient. So astronomers then use the light year or the parsec. The light year (ly) is the distance that light travels in one year. It equals 9.46×10^{12} km. A parsec (pc) equals 206,265 AU, or 3.09×10^{13} km, or 3.26 ly.

Beyond the Galaxy, astronomers often talk in multiples of parsecs. A thousand parsecs is called a kiloparsec (kpc), and a million parsecs is termed a megaparsec (Mpc).

Other Physical Units

Another important unit you will encounter is the speed of light (c), which equals 2.9979×10^5 km/sec.

The unit of energy in the cgs system is called the erg. A mass of 2 g traveling at 1 cm/sec has an energy of 1 erg. To illustrate how small an erg is, if you place one foot up one stair, you have expended about a billion ergs. In the mks system the energy unit is a joule (J), which equals 10^7 ergs. It takes about a joule of energy to lift an apple from the floor to a table.

Power is the amount of energy coming from an object per second, so it is measured in ergs per second. (Astronomers use the word *luminosity* to describe what most physicists would call power.) A convenient and familiar unit for power (or luminosity) is the watt (W), defined as 10^7 ergs/sec, or 1 J/sec.

This book uses the gauss (G) as the unit of magnetic field strength (rather than the mks unit of *telsa*, which equals 10^4 G). To give you a feel for a gauss, keep in mind that the earth's magnetic field is about 0.5 G.

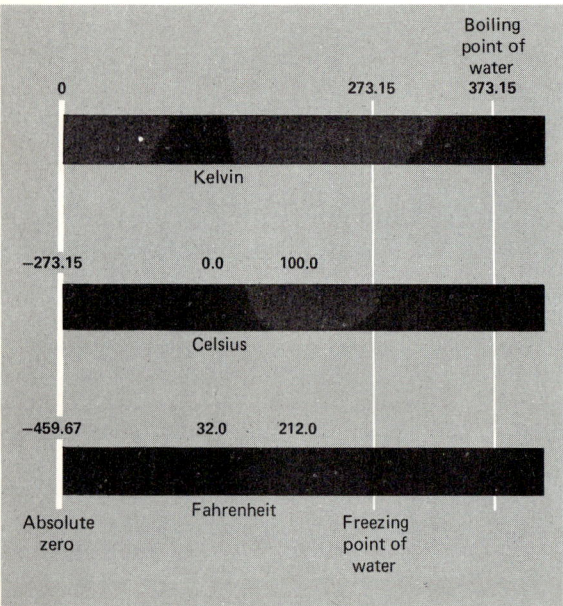

FIGURE A.1 A comparison of the Kelvin, Celsius, and Fahrenheit temperature scales.

APPENDIX B PLANETARY DATA

TABLE B.1 Planetary Rotation Rates and Inclinations of Rotation Axes

Planet	Rotation Period (equatorial)	Inclination of Equator to Orbital Plane	Method of Measurement
Mercury	58.65 days	Less than 28°	Radar Doppler shift
Venus	243.01 days (retrograde)	3°	Radar Doppler shift
Earth	23 h 56 min 4.1 sec	23° 27'	Star transits
Mars	24 h 37 min 22.6 sec	23° 59'	Optical features
Jupiter	9 h 50.5 min	3° 05'	Optical features
Saturn	10 h 14 min	26° 44'	Optical Doppler shift
Uranus	12–23 h	97° 55'	Optical Doppler shift
Neptune	18–22 h	28° 48'	Optical Doppler shift
Pluto	6.39 days	?	Optical light variations

TABLE B.2 Distances, Periods, and Orbital Velocities of the Planets

Planet	Semimajor Axis of Orbit* AU	Semimajor Axis of Orbit* 10^6 km	Sidereal Period Years†	Sidereal Period Days	Average Orbital Velocity (km/sec)
Mercury	0.387	57.9	0.240	87.97	47.9
Venus	0.723	108.2	0.615	224.7	35.1
Earth	1.000	149.6	1.000	365.26	29.8
Mars	1.523	227.9	1.881	687.0	24.1
Jupiter	5.203	778.3	11.86	4,333	13.1
Saturn	9.540	1427	29.46	10,759	9.7
Uranus	19.18	2869	84.10	30,685	6.8
Neptune	30.07	4498	164.8	60,188	5.4
Pluto	39.44	5900	248.4	90,700	4.7

*Same as average distance from sun.
†Tropical years, that is, the year of seasons.

TABLE B.3 Dimensions of the Planets

Planet	Nature of Measurement	Radius (km)	Radius (earth radii)	Method of Measurement
Mercury	Average radius	2,439	0.38	Radar and optical
Venus	Upper cloud layer	6,110	0.95	Optical
	Equatorial radius	6,050		Radar
Earth	Polar radius	6,356	1.00	Satellite
	Equatorial radius	6,378		
Mars	Polar radius	3,394	0.53	Optical
	Equatorial radius	3,407		
Jupiter	Polar radius	66,550	10.8	Optical
	Equatorial radius	70,850		
Saturn	Polar radius	53,450	8.9	Optical
	Equatorial radius	60,330		
Uranus	Polar radius	24,700	4.0	Optical
	Equatorial radius	25,400		
Neptune	Average radius	25,100	3.9	Optical
Pluto	Average radius	1,500	0.24	Speckle interferometry

TABLE B.4 Masses of the Planets

Planet	Mass (earth masses)	Mass (kg)	Density (kg/m^3)	Measured by Motions of
Mercury	0.05533	3.32×10^{23}	5400	Mariner 10
Venus	0.815	4.87×10^{24}	5200	Mariner 2 and 10
Earth	1.000	5.974×10^{24}	5500	Mariner 4
(Moon)	0.012	7.35×10^{22}	3300	Ranger series
Mars	0.1074	6.42×10^{23}	3900	Mariner 4, 6, and 7
Jupiter	317.9	1.899×10^{27}	1300	Satellites and asteroids
Saturn	95.1	5.69×10^{26}	680	Satellites
Uranus	14.56	8.69×10^{25}	1300	Satellites
Neptune	17.24	1.03×10^{26}	1600	Triton
Pluto	0.0018	1.1×10^{22}	~800	Charon

TABLE B.5 Atmospheric Gases

Planet	Gas
Mercury	Helium, hydrogen
Venus	Carbon dioxide, carbon monoxide, hydrogen chloride, hydrogen fluoride, water, argon, nitrogen, oxygen, hydrogen sulfide, sulfur dioxide, helium
Earth	Nitrogen, oxygen, water, argon, carbon dioxide, neon, helium, methane, krypton, nitrous oxide, ozone, xenon, hydrogen, radon
Mars	Carbon dioxide, carbon monoxide, water, oxygen, ozone, argon, nitrogen
Jupiter	Hydrogen, helium, methane, ammonia, water, carbon monoxide, acetylene, ethane, phosphine, germane
Saturn	Hydrogen, helium, methane, ammonia, acetylene, ethane, phosphine, propane
Titan	Nitrogen, methane, ethane, acetylene, ethylene, hydrogen cyanide
Uranus	Hydrogen, methane
Neptune	Hydrogen, methane, ethane
Pluto	Methane

Note: The results listed above are a combination of ground-based and satellite observations.

TABLE B.6 Satellites of Mars

Satellite	Distance from Center of Planet (10^3 km)	Sideral Period of Revolution (days)	Radius of Satellite (km)	Mass (planet = 1)	Bulk density (kg/m^3)
Phobos	9.37	0.3189	$14 \times 11 \times 9$	1.5×10^{-8}	1900
Deimos	23.52	1.262	$8 \times 6 \times 5$	3.1×10^{-9}	2100

TABLE B.7 Satellites of Jupiter

Satellite Name	Satellite Number	Distance from Jupiter 10^3 km	Distance from Jupiter Jupiter Radii	Orbital Period (days)	Radius of Satellite (km)	Mass (planet = 1)	Bulk Density (kg/m^3)
1979J3	J16	128	1.79	0.29	20	—	—
1979J1	J14	129	1.80	0.30	20	—	—
Almathea	J5	181	2.55	0.49	130 × 80	2×10^{-9}	3000
1979J2	J15	222	3.11	0.68	40	—	—
Io	J1	422	5.95	1.77	1820	4.7×10^{-5}	3500
Europa	J2	671	9.47	3.55	1500	2.6×10^{-5}	3000
Ganymede	J3	1,070	15.10	7.15	2640	7.8×10^{-5}	1900
Callisto	J4	1,880	26.60	16.70	2500	5.6×10^{-5}	1800
Leda	J13	11,110	156	240	—	5×10^{-13}	—
Himalia	J6	11,470	161	251	80	8.5×10^{-10}	1000
Lysithea	J10	11,710	164	260	—	1×10^{-12}	—
Elara	J7	11,740	165	260	—	4×10^{-11}	—
Ananke	J12	20,700	291	617	—	7×10^{-13}	—
Carme	J11	22,350	314	692	—	2×10^{-12}	—
Pasiphae	J8	23,300	327	735	—	8×10^{-12}	—
Sinope	J9	23,700	333	758	—	2×10^{-12}	—

TABLE B.8 Satellites of Saturn

Satellite	Distance from Saturn 10^3 km	Distance from Saturn Saturn Radii	Orbital Period (days)	Radius of Satellite (km)	Mass (planet = 1)	Bulk Density (kg/m^3)
1980S28	137.67	2.28	0.602	10 × 20	—	—
1980S27	139.35	2.31	0.613	70 × 50 × 40	—	—
1980S26	141.70	2.35	0.629	55 × 45 × 35	—	—
1980S3	151.42	2.51	0.694	70 × 60 × 50	—	—
1980S1	151.47	2.51	0.695	110 × 100 × 80	—	—
Mimas	185.54	3.08	0.942	195	6.6×10^{-8}	1200
Enceladus	238.04	3.95	1.370	250	1×10^{-7}	1200
Tethys	294.67	4.88	1.888	530	1.3×10^{-6}	1200
1980S13	294.87	4.88	1.888	17 × 14 × 13	—	—
1980S25	294.87	4.88	1.888	17 × 11 × 11	—	—
Dione	377.42	6.26	2.737	560	1.85×10^{-6}	1400
1980S6	377.42	6.26	2.737	18 × 16 × 15	—	—
Rhea	527.07	8.74	4.518	765	4.4×10^{-6}	1300
Titan	1,221.86	20.25	15.945	2560	2.36×10^{-4}	1880
Hyperion	1,481.00	24.55	21.277	205 × 130 × 110	—	—
Iapetus	3,560.80	59.02	79.33	730	3.3×10^{-6}	1200
Phoebe	12,954	214.7	550.45	~110	—	—

TABLE B.9 Satellites of Uranus

Satellite	Distance from Center of Planet (10^3 km)	Sidereal Period of Revolution (days)	Radius (km)
Ariel	191.8	2.52038	~700
Umbriel	267.3	4.14418	~600
Titania	438.7	8.70588	~800
Oberon	586.6	13.46326	~800
Miranda	130.1	1.414	~300

TABLE B.10 Satellites of Neptune

Satellite	Distance from Center of Planet (10^3 km)	Sidereal Period of Revolution (days)	Radius of Satellite (km)
Triton	653.6	5.87683	3600–5300
Nereid	5570	365	~300

APPENDIX C PHYSICAL CONSTANTS, ASTRONOMICAL DATA

Physical Constants

Gravitational constant
$G = 6.673 \times 10^{-11}$ newton·m²/kg²

Speed of light in a vacuum
$c = 2.9979 \times 10^8$ m/sec

Planck's constant
$h = 6.62618 \times 10^{-34}$ J·sec

Wien's constant
$\sigma_w = 0.0029$ m·K

Boltzmann's constant
$k = 1.3806 \times 10^{-23}$ J/K

Stefan-Boltzmann constant
$\sigma = 5.6697 \times 10^{-8}$ W/m²/K

Electron' mass
$m_e = 9.10956 \times 10^{-31}$ kg

Proton mass
$m_p = 1.6726 \times 10^{-27}$ kg $= 1836.1 m_e$

Neutron mass
$m_n = 1.6749 \times 10^{-27}$ kg

Mass of hydrogen atom
$m_H = 1.6735 \times 10^{-27}$ kg

Astronomical Data

Astronomical Unit
AU $= 1.4959789 \times 10^{11}$ m

Parsec
pc $= 206264.806$ AU
$= 3.2616$ ly
$= 3.0856 \times 10^{16}$ m

Light year
ly $= 9.46053 \times 10^{15}$ m
$= 6.324 \times 10^4$ AU

Sidereal year
y $= 3.155815 \times 10^7$ sec

Mass of sun
$M_\odot = 1.989 \times 10^{30}$ kg

Luminosity of sun
$L_\odot = 3.827 \times 10^{26}$ W

Solar constant
$S = 1370$ W/m²

Radius of sun
$R_\odot = 6.96 \times 10^5$ km

Mass of earth
$M_\oplus = 5.9742 \times 10^{24}$ kg

Equatorial radius of earth
$R_\oplus = 6.37814 \times 10^3$ km

Mass of moon
$M_M = 7.34 \times 10^{22}$ kg

Radius of moon
$R_M = 1.738 \times 10^3$ km

APPENDIX D STARS WITHIN 13 LIGHT YEARS

Name	Parallax (arcsec)	Distance (ly)	Spectral Type	Proper Motion (arcsec/y)	Apparent Visual Magnitude	Luminosity (sun = 1)
Sun			G2 V		−26.7	1.0
α Cen A	0.750	4.3	G2 V	3.68	−0.01	1.6
B			K0 V		+1.3	0.45
C	0.772	4.2	M5e		+11.0	0.00006
Barnard's star	0.552	5.9	M5 V	10.30	+9.5	0.00045
Wolf 359	0.431	7.6	M8e	4.84	+13.5	0.00002
Lalande 21185	0.402	8.1	M2 V	4.78	+7.5	0.0055
Luyten 726-8A	0.1387	8.4	M6e	3.35	+12.5	0.00006
B (UV Ceti)			M6e		+13.0	0.00004
Sirius A	0.377	8.6	A1 V	1.32	−1.5	23.5
B			wd		+8.7	0.003
Ross 154	0.345	9.4	M5e	0.74	+10.6	0.00048
Ross 248	0.314	10.3	M6e	1.82	+12.3	0.00011
ε Eri	0.303	10.7	K2 V	0.97	+3.7	0.30
Luyten 789-6	0.302	10.8	M7e	3.27	+12.2	0.00014
Ross 128	0.301	10.8	M5	1.40	+11.1	0.00036
61 Cyg A	0.292	11.2	K5 V	5.22	+5.2	0.083
B			K7 V		+6.0	0.040
ε Ind	0.291	11.2	K5 V	4.67	+4.7	0.13
Procyon A	0.287	11.4	F5 IV-V	1.25	+0.4	7.65
B			wd		+10.7	0.00055
Σ 2398 A	0.284	11.5	M3.5 V	2.29	+8.9	0.0028
B			M4 V		+9.7	0.0013
Groombridge 34 A	0.282	11.6	M1 V	2.91	+8.1	0.0058
B			M6 V		+11.0	0.00040
Lacaille 9352	0.279	11.7	M2 V	6.87	+7.4	0.013
τ Ceti	0.273	11.9	G8 V	1.92	+3.5	0.45
BD + 5° 1668	0.266	12.2	M5	3.73	+9.8	0.0015
L725-32 (YZ Ceti)	0.262	12.4	M5e	1.31	+11.6	0.0002
Lacaille 8760	0.260	12.5	M1	3.46	+6.7	0.028
Kapteyn's star	0.256	12.7	M0 V	8.79	+8.8	0.0040
Kruger 60 A	0.254	12.8	M4	0.87	+9.8	0.0017
B			M5e		+11.3	0.00044

Source: Based on a table by A. H. Batten in the *Observer's Handbook* (1980) of the Royal Astronomical Society of Canada.
Note: An e after the spectral type indicates emission lines in the spectrum.

APPENDIX E THE TWENTY BRIGHTEST STARS

Star	Name	Apparent Visual Magnitude	Spectral Type	Absolute Magnitude	Distance (ly)	Proper Motion (arcsec/y)
1. α CMa A	Sirius	−1.46	A1 V	+1.42	8.7	1.324
2. α Car	Canopus	−0.72	F0 I-II	−3.1	98	0.025
3. α Boo	Arcturus	−0.06	K2 III	−0.3	36	2.284
4. α Cen A	Rigil Kentaurus	+0.01	G2 V	+4.39	4.3	3.676
5. α Lyr	Vega	+0.04	A0 V	+0.5	26.5	0.345
6. α Aur	Capella	+0.05	G8 III (?)	−0.6	45	0.435
7. β Ori A	Rigel	+0.14	B8 Ia	−7.1	900	0.001
8. α CMi A	Procyon	+0.37	F5 IV-V	+2.7	11.3	1.250
9. α Ori	Betelgeuse	+0.41	M2 Iab	−5.6	520	0.028
10. α Eri	Achernar	+0.51	B3 V	−2.3	118	0.098
11. β Cen AB	Hadar	+0.63	B1 III	−5.2	490	0.035
12. α Aql	Altair	+0.77	A7 IV-V	+2.2	16.5	0.658
13. α Tau A	Aldebaran	+0.86	K5 III	−0.7	68	0.202
14. α Vir	Spica	+0.91	B1 V	−3.3	220	0.054
15. α Sco A	Antares	+0.92	M1 Ib	−5.1	520	0.029
16. α PsA	Fomalhaut	+1.15	A3 V	+2.0	22.6	0.367
17. β Gem	Pollux	+1.16	K0 III	+1.0	35	0.625
18. α Cyg	Deneb	+1.26	A2 Ia	−7.1	1600	0.003
19. β Cru	Beta Crucis	+1.28	B0.5 III	+4.6	490	0.049
20. α Leo A	Regulus	+1.36	B7 V	−0.7	87	0.248

Source: Based on a table compiled by Donald A. MacRae in the *Observer's Handbook* (1980) of the Royal Astronomical Society of Canada.

APPENDIX F CELESTIAL COORDINATE SYSTEMS

The study of the motions of celestial bodies necessitates a precise and standard system to designate positions on the sky. These *celestial coordinate systems* serve as fundamental guides in describing locations of astronomical objects, just as street addresses aid in finding a house in a city.

Celestial coordinate systems are based on the concept of the *celestial sphere* (Fig. F.1). Observations of the sky establish the illusion that the stars and other celestial bodies are attached to a large sphere. In the center of the celestial sphere is the earth. We can imagine the celestial sphere to be very much bigger than the earth, so any position on the earth is effectively at the center of the cosmic sphere. We can describe the location of a heavenly body by giving its location with respect to a coordinate system drawn on the celestial sphere.

Celestial coordinate systems are similar to the terrestrial coordinate system of *longitude* and *latitude* (Fig. F.2). The latitude of a place on the earth is its angular distance (measured in degrees, minutes, and seconds) north or south of the terrestrial equator (Fig. F.2a). Longitude is the angular distance (measured in the same units) east or west between the Greenwich meridian and the meridian of the desired location. (A meridian drawn on the earth's surface is any great circle that passes through both the north and south terrestrial poles; Fig. F.2b.) The location of any surface feature on the earth can be pinpointed by stating its longitude and latitude (Fig. F.2c). The basic characteristic of this spherical coordinate system is the measurement of two angles with respect to two fundamental reference circles, the equator and the Greenwich meridian. These reference

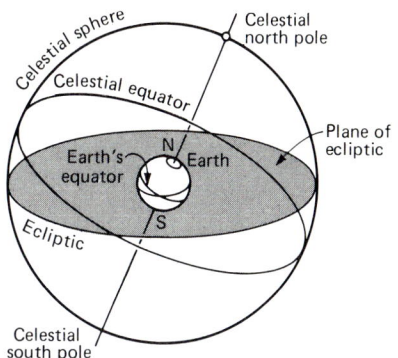

FIGURE F.1 A two-sphere model of the universe. The stars are attached to the celestial sphere, in whose center the sphere of the earth is located. The terrestrial north and south poles, defined by the earth's rotation, extend out to the celestial sphere to define the north and south celestial poles. Similarly, the celestial equator is defined by projecting the earth's equator onto the celestial sphere. [*Adapted from a figure by A. Strahler, The Earth Sciences, Harper & Row, New York, 1971.*]

circles are *great circles* on the sphere, that is, circles whose plane cuts through the center of the sphere.

On the celestial sphere the two systems of coordinates most commonly used by astronomers are called the *horizon system* and the *equatorial system*. Like the terrestrial system, both locate an object by giving two angles with respect to two reference circles.

The simpler system to describe is the horizon system (Fig. F.3). The fundamental reference circles of the horizon coordinate system are the horizon and the local celes-

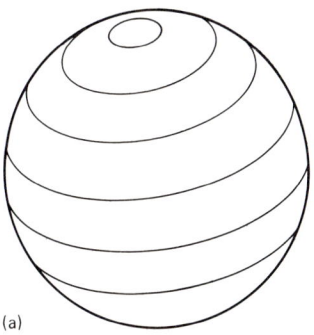
(a)

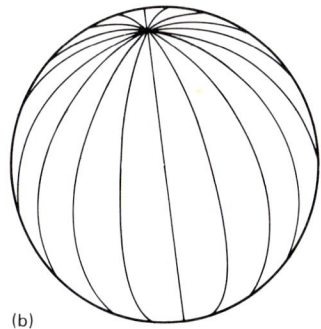
(b)

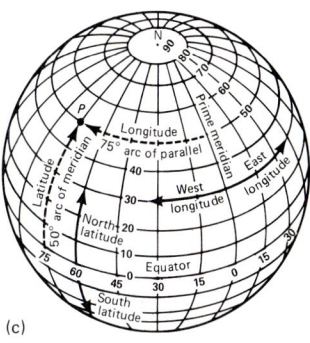
(c)

FIGURE F.2 The earth's coordinate grid of latitude and longitude. *(a)* The circles of latitude are parallel to the earth's equator. Latitude is the angular distance north or south of the equator. *(b)* The circles of longitude are great circles drawn through the north and south poles. Longitude is the angular distance east and west of the prime meridian (the meridian of Greenwich, England). *(c)* The two coordinates uniquely define any position on the earth's surface. [*Adapted from a figure by A. Strahler, The Earth Sciences, Harper & Row, New York, 1971.*]

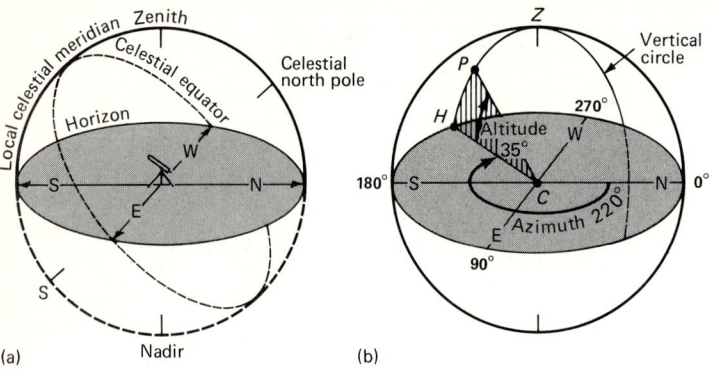

FIGURE F.3 The horizon coordinate system. (a) The two reference circles of the horizon system are the horizon and the celestial meridian. The position of the celestial equator and poles relative to the horizon (their altitude) depends on the latitude of the observer. (b) Altitude is the angular height of an object above the observer's horizon; it is analogous to terrestrial latitude. Azimuth is the angular distance along the horizon, measured counterclockwise from north. A vertical circle, a line drawn through the zenith (Z) and the object (P), intersects the horizon at H. The azimuth of P is measured from the north to H. [Adapted from a figure by A. Strahlar, The Earth Sciences, Harper & Row, New York, 1971.]

tial meridian. The *local celestial meridian* is the great circle passing through the north-south points on the horizon and through the point directly overhead, the *zenith*. It is the projection on the celestial sphere of the terrestrial meridian of the observer (Fig. F.3a). Because of the rotation of the earth, the celestial sphere appears to rotate, carrying the stars westward past the celestial meridian.

The *horizon* is the great circle on the celestial sphere 90° from the zenith. The apparent horizon as it is visible at sea most closely approximates this technical definition. Note that the term horizon does *not* have its common meaning as the line where the sky meets the landscape; that line is usually uneven and differs for different locations.

The two coordinates needed to specify the position of a heavenly object in the horizon system are called *altitude* and *azimuth* (Fig. F.3b). Both are measured relative to the fundamental reference circles, the horizon and the local celestial meridian. The altitude of an object is its distance in angular measure above the observer's horizon, measured along a great circle, called a *vertical circle*, through the zenith. Altitude ranges from 0° (the horizon) to 90° (the zenith). The azimuth of an object is the angle between the celestial meridian and a vertical circle through the object. It is measured along the horizon, clockwise from north through east, and ranges from 0° to 360°. So to determine azimuth, draw a great circle from the zenith through the object to the horizon. Then measure the angle clockwise along the horizon from the point due north to the point of intersection with this vertical circle. The azimuths of the cardinal compass points are north 0°, east 90°, south 180°, and west 270°.

The advantage of the horizon system is that it is simple to describe and easy to use. The reference circles are natural and specific for a given observer. It is easy to get an approximate notion of what is horizontal and where north is; a level and a compass can be used for greater accuracy. (This is the system used by surveyors.) So it is easy to tell someone "Look at azimuth 135°, altitude 45°, and you will find Jupiter." But this convenience and naturalness is also the horizon system's greatest drawback. At any given instant an object has a unique altitude and azimuth *for a particular observer*. But another observer some distance away would note a different meridian and horizon and correspondingly would assign a different altitude and azimuth to the same object. Because the altitudes and azimuths differ, considerable confusion would result if the first observer attempted to communicate the object's position to the second. In addition, the altitude and azimuth of an object change with time. So a large table of numbers would be required for each object to give its position at each instant of the day. Consequently, the horizon system is restricted in use to specific locations at specific times.

The *equatorial system* used by astronomers alleviates many of the deficiencies of the horizon system (Fig. F.4). The first fundamental reference circle in this system is the *celestial equator*, which is the great circle produced by the intersection with the celestial sphere of the plane through the earth's equator. The *north* and *south celestial poles* are 90° from the celestial equator and are directly above the north and south terrestrial poles (Fig. F.4a). The second reference circle in the equatorial system is the hour circle of the vernal equinox. An *hour circle* is a great circle through the celestial poles. The hour circle of the vernal equinox is the circle which crosses the equator at the location of the sun on the first day of spring. The apparent path of the sun relative to the stars is the great circle called the *ecliptic*, tilted $23\frac{1}{2}°$ with respect to the celestial equator. So the vernal equinox can also be described as one of the two points where the ecliptic and equator intersect (the other, the *autumnal equinox*, is the location of the sun on the first day of autumn).

The equatorial system also uses two coordinates to locate an object in the sky. The angular distance of an object north or south of the celestial equator, measured along an hour circle, is its *declination* (abbreviated Dec or represented by the symbol δ). Declination is analogous to terres-

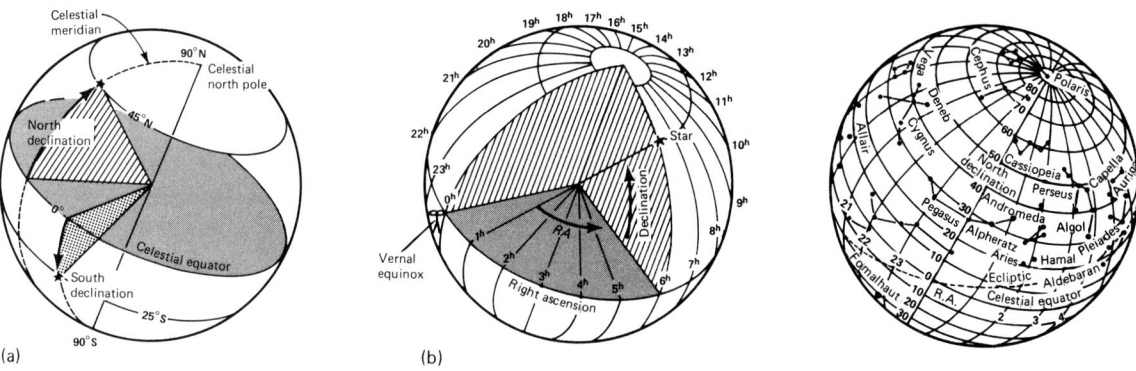

FIGURE F.4 The equatorial coordinate system. (a) Declination, analogous to latitude, is the angular distance of a celestial object above or below the celestial equator. (b) Right ascension, analogous to longitude, is the position along the celestial equator, usually measured in time units (Table A.11). (c) These two coordinates, attached to the celestial sphere, uniquely define the position of any celestial object. [Adapted from a figure by A. Strahler, *The Earth Sciences*, Harper & Row, New York, 1971.]

trial latitude (Fig. F.4a). It is measured in degrees, minutes, and seconds of arc. The second coordinate, called *right ascension* (abbreviated RA or represented by the symbol α), is the angular distance between the hour circle through an object and the hour circle of the vernal equinox, measured eastward (counterclockwise) along the celestial equator. Right ascension is analogous to terrestrial longitude (Fig. F.4b). It can be specified either in arc measure (degrees, minutes, and seconds of arc) or time units (hours, minutes, and seconds of time). Table F.1 gives the relationship between the two units of measure. To determine the right ascension of an object, draw a great circle (the hour circle) from the pole through the object to the celestial equator. Then measure the angle eastward (counterclockwise) from the vernal equinox to the point of intersection of the hour circle and the equator. For example, the sun's position on the day of the summer solstice is right ascension 6^h and declination $23\frac{1}{2}°$ north. Note that right ascension is measured in the same direction as the apparent motion of the sun with respect to the stars.

The advantage of the equatorial system is that it is a coordinate lattice fixed with the stars to the celestial sphere, and therefore the same at all local times and for all observers (Fig. F.4c). This system participates in the daily motion of the sky. Over short periods of time the right ascension and declination of the stars are constant, so they can be measured and written down in tables. Because the planets, sun, and moon move with respect to the stars, their celestial coordinates change slowly, but steadily. For example, the solar right ascension changes an average of about $1°$ (or 4^m) per day as the sun apparently travels eastward along the ecliptic. So specifying the position of these objects is not quite as simple as for the stars, but these motions can be calculated in advance. They are published in a yearly timetable called an *ephemeris* (*pl. ephemerides*) or *almanac*.

The stellar positions do change very slowly from year to year, partly because the stars are moving through space (Section 20.3). More importantly, the precession of the equinoxes (Focus 1.1), produced by the wobble of the earth's axis, reorients the position of the celestial equator and moves the vernal equinox reference point. As the celestial poles swing slowly around in circles, the equatorial coordinates of the stars change. The period of precession is about 26,000 years, so the annual variation is small, but steady. Consequently, any precision star catalog must be dated; this date is the *epoch* of the catalog. For example, if a catalog epoch is stated 1980.0 (or 1980), it means that the right ascensions and declinations refer to the positions of the stars with respect to the vernal equinox and celestial equator positions of January 1, 1980. Positions on other dates must be calculated taking into account the corrections due to precession and to the stars' proper motions.

TABLE F.1 The Relationship Between Time Angle Units and Arc Angle Units

Time Angle Units	Arc Angle Units
24^h	$360°$
1^h	$15°$
4^m	$1°$
1^m	$15'$
4^s	$1'$
1^s	$15''$

The key problem here is not to confuse the words referring to time units (minutes and seconds as fractions of an hour) with the same words when they are used to refer to angular units (minutes and seconds as fractions of a degree). Also, one must remember that, though time units are used, it is really an angle which is being measured, not a duration of time. The fundamental conversion from time to angular units arises from the fact that the earth turns through 360° in 24 hours.

APPENDIX G PERIODIC TABLE OF THE ELEMENTS

Group IA																	VIII
1 Hydrogen **H** 1	IIA											IIIA	IVA	VA	VIA	VIIA	2 Helium **He** 4
3 Lithium **Li** 7	4 Beryllium **Be** 9											5 Boron **B** 11	6 Carbon **C** 12	7 Nitrogen **N** 14	8 Oxygen **O** 16	9 Fluorine **F** 19	10 Neon **Ne** 20
11 Sodium **Na** 23	12 Magnesium **Mg** 24	IIIB	IVB	VB	VIB	VIIB	— VIII —			IB	IIB	13 Aluminum **Al** 27	14 Silicon **Si** 28	15 Phosphorus **P** 31	16 Sulfur **S** 32	17 Chlorine **Cl** 35	18 Argon **Ar** 40
19 Potassium **K** 39	20 Calcium **Ca** 40	21 Scandium **Sc** 45	22 Titanium **Ti** 48	23 Vanadium **V** 51	24 Chromium **Cr** 52	25 Manganese **Mn** 55	26 Iron **Fe** 56	27 Cobalt **Co** 59	28 Nickel **Ni** 59	29 Copper **Cu** 64	30 Zinc **Zn** 65	31 Gallium **Ga** 70	32 Germanium **Ge** 73	33 Arsenic **As** 75	34 Selenium **Se** 79	35 Bromine **Br** 80	36 Krypton **Kr** 84
37 Rubidium **Rb** 85	38 Strontium **Sr** 88	39 Yttrium **Y** 89	40 Zirconium **Zr** 91	41 Niobium **Nb** 93	42 Molybdenum **Mo** 96	43 Technetium **Tc** 99	44 Ruthenium **Ru** 101	45 Rhodium **Rh** 103	46 Palladium **Pd** 106	47 Silver **Ag** 108	48 Cadmium **Cd** 112	49 Indium **In** 115	50 Tin **Sn** 119	51 Antimony **Sb** 122	52 Tellurium **Te** 128	53 Iodine **I** 127	54 Xenon **Xe** 131
55 Cesium **Cs** 133	56 Barium **Ba** 137	57 *Lanthanum **La** 139	72 Hafnium **Hf** 178	73 Tantalum **Ta** 181	74 Wolfran (tungsten) **W** 184	75 Rhenium **Re** 186	76 Osmium **Os** 190	77 Iridium **Ir** 192	78 Platinum **Pt** 195	79 Gold **Au** 197	80 Mercury **Hg** 201	81 Thallium **Tl** 204	82 Lead **Pb** 207	83 Bismuth **Bi** 209	84 Polonium **Po** 210	85 Astatine **At** 210	86 Radon **Rn** 222
87 Francium **Fr** 223	88 Radium **Ra** 226	89 **Actinium **Ac** 227	Kurchatovium **Ku**	105 Hahnium **Ha**													

Atomic number — 11 Sodium ← Name
Symbol → **Na** 23 ← Approximate atomic weight (to nearest whole number)

Lanthanide series

6	58 Curium **Ce** 140	59 Praseodymium **Pr** 141	60 Neodymium **Nd** 144	61 Promethium **Pm** 147	62 Samarium **Sm** 150	63 Europium **Eu** 152	64 Gadolinium **Gd** 157	65 Terbium **Tb** 163	66 Dysprosium **Dy** 163	67 Holmium **Ho** 165	68 Erbium **Er** 167	69 Thulium **Tm** 169	70 Ytterbium **Yb** 173	71 Lutetium **Lu** 175

Actinide series

7	90 Thorium **Th** 232	91 Protactinium **Pa** 231	92 Uranium **U** 238	93 Neptunium **Np** 237	94 Plutonium **Pu** 242	95 Americium **Am** 243	96 Cerium **Cm** 247	97 Berkellium **Bk** 247	98 Californium **Cf** 251	99 Einsteinium **Es** 254	100 Fermium **Fm** 253	101 Mendelevium **Md** 256	102 Nobelium **No** 254	103 Lawrencium **Lr** 257

GLOSSARY

Å Abbreviation for angstrom.

absolute magnitude A measure of the brightness a star would have if it were to be placed at a standard distance of 10 pc (32.6 ly) from the earth; an astronomical measure of a star's luminosity.

absolute zero The temperature at which all molecular motion ceases; 0 Kelvins.

absorption (dark) line A color missing in a continuous spectrum because of the absorption of that color by atoms.

absorption-line spectrum Dark lines superimposed on a continuous spectrum.

acceleration The rate of change of velocity with time.

accretion The colliding and sticking together of small particles to make larger masses.

accretion atmosphere The original atmosphere of a planet, formed when it accreted from planetesimals.

accretion disk A disk-shaped collection of gas around an object (usually a neutron star or black hole) formed by transfer from a nearby companion star.

active galaxy A galaxy characterized by a nonthermal spectrum and a large energy output.

active region A region of the solar surface containing sunspots, flares, prominences, or other manifestations of solar activity.

active sun The state of the sun during times of high solar activity, as indicated by sunspots, flares, and active regions.

aerobe An organism that is dependent on free oxygen for its metabolism.

albedo A measure of an object's reflecting power; the ratio of reflected light to incoming light, where complete reflection gives an albedo of 1.

Allende meteorite A meteorite that fell in 1969 and was found to contain ^{26}Mg, evidence of a supernova event having occurred shortly before the formation of the solar system.

almanac See ephemeris.

Almathea One of the small, innermost satellites of Jupiter.

Alpha Centauri The star closest to the sun (4.3 ly), with almost the same luminosity and surface temperature.

alpha particle The nucleus of a helium atom.

altitude Angular distance above or below the horizon.

aluminum 26 (^{26}Al) A radioactive element of short half-life, the likely precursor of the excess ^{26}Mg found in some meteorites.

amino acids The building blocks of proteins, consisting mostly of carbon, nitrogen, oxygen, and hydrogen.

anaerobe An organism that does not depend on free oxygen for its metabolism.

angstrom (Å) A unit of length, equal to 10^{-10} m, commonly used to measure wavelengths of visible light.

angular momentum The tendency for bodies, because of their inertia, to keep spinning or orbiting; for a particle of mass m moving at speed v a distance r from a center of rotation, the angular momentum L is $L = mvr$.

angular separation The observed angular distance between two celestial objects, measured in degrees, minutes, and seconds of angular measure.

angular size The apparent size of an object in angular measure; the angular separation of two points on opposite sides of the object.

angular speed The rate of change of angular position of an object, measured in units of angular measure per unit of time, such as degrees/hour or arcseconds/year.

annular eclipse A solar eclipse where the moon's angular diameter is smaller than the sun's, so the sun's disk is not completely covered.

anorthosite A basaltic mineral composed of calcium and sodium with aluminum silicate; the predominant mineral of the lunar highlands.

antapex The point on the celestial sphere from which the sun appears to be receding, relative to nearby stars, exactly opposite the solar apex.

antimatter Matter made of antiparticles, those particles with opposite properties to ordinary particles; for example, an antiproton has the same mass as a proton but a negative charge.

apex The point on the celestial sphere toward which the sun appears to be moving, relative to the nearby stars, exactly opposite the solar antapex.

aphelion For a body orbiting the sun, the point on its orbit which is farthest from the sun.

apocenter The point on an elliptical orbit farthest from the focus.

apogee The point on its orbit where an earth satellite is farthest from the earth.

apparent magnitude A number indicating the brightness of a star (or any other celestial object) as seen from the earth; related to the logarithm of the flux.

arroyo Channel formed by fluid erosion, found on both the earth and Mars.

association A loose cluster of stars with a common origin (*see* OB association).

asteroid (minor planet) One of the several thousand very small objects of the solar system that revolve around the sun generally between the orbits of Mars and Jupiter.

asteroid belt The region lying between the orbits of Mars and Jupiter, containing the majority of asteroids.

asteroidal moon A small satellite of one of the outer planets with irregular shape and rocky composition, similar to an asteroid.

astrometric binary A binary system in which an invisible companion is detected by its effects on the orbital motion of the visible star.

Astronomical Unit (AU) The average distance between the earth and the sun; 149.6 million km or 8.3 light minutes.

atmosphere A gaseous envelope surrounding a planet, or the visible layers of a star; also a unit of pressure equal to the pressure of air at sea level on the earth's surface.

atmospheric escape The process by which gases may leave the atmosphere of a planet, made possible by the planet's low gravity, high temperature, or both.

atmospheric extinction The decrease in intensity of light caused by its passage through the atmosphere.

atom The smallest particle of an element that exhibits the chemical properties of the element.

AU Abbreviation for Astronomical Unit.

aurora A bright glow produced in the earth's atmosphere near the magnetic poles by charged particles from the sun striking and exciting atmospheric atoms.

autumnal equinox See equinox.

axis One of two or more reference lines in a coordinate system; the straight line through the poles about which a body rotates.

azimuth The angle along the horizon measured eastward from north.

Balmer series (lines) The set of transitions of electrons in a hydrogen atom between the second energy level and higher levels; also the set of absorption or emission lines corresponding to these transitions, which lies in the visible part of the spectrum.

bandwidth The range of frequencies or wavelengths which a radio receiver can detect at once.

Barnard's star A star 5.9 ly from the sun, thought to have a planetary companion.

barred spiral A spiral galaxy with a barlike structure across its center.

barycenter The center of mass of two revolving bodies.

basalt An igneous rock, composed of olivine and feldspar, that makes up much of the earth's lower crust.

baseline The side of a triangle (used in triangulation) whose length is known.

basin A large depression on the surface of a planet or planetary satellite (such as Mercury or the moon).

Becklin-Neugebauer object A strong infrared source near the Orion Nebula that is probably a protostar.

belt A dark band in the atmosphere of a Jovian planet marking descending gases of low pressure (*see* zone).

beta decay A process of radioactive decay in which a neutron disintegrates into a proton, an electron, and a neutrino.

beta particle An electron.

Beta Regio An upland plateau on the surface of Venus.

Betelgeuse A red supergiant star with a luminosity 8000 times that of the sun.

Big-Bang model A model for the evolution of the universe that postulates its origin from a hot, dense state that rapidly expanded to cooler, less dense states.

binary accretion model A model for the formation of the earth-moon system, in which both formed by accretion out of the same cloud of gas and dust.

binary galaxies Two galaxies bound together by gravity revolving around a common center of mass.

binary stars Two stars bound together by gravity revolving around a common center of mass.

binary X-ray source An X-ray source associated with a normal star, both orbiting a common center of mass; considered a possible candidate for a black hole.

biological evolution The process by which life develops from simple cells to complex organisms.

biosphere The regions of the earth where life exists.

Bitter Springs formation A rock formation in Australia, 1 billion years old, containing fossil traces of early algae with complete cell nuclei.

BL Lacertae (BL Lac) objects Objects, possibly related to active galaxies, characterized by rapid variability, no emission lines, nonthermal radiation, and strong and variable polarization.

black dwarf The cold remains of a white dwarf after all its thermal energy is exhausted.

black hole A mass that has collapsed to such a degree that the escape velocity from its surface is greater than the speed of light, so that light is trapped by the intense gravitational field.

blackbody A (hypothetical) perfect radiator of light that absorbs and reemits all radiation incident upon it; its light output depends only on its temperature.

blackbody spectrum The continuous spectrum emitted by a blackbody; the brightness at each wavelength is given by Planck's law.

blue shift Part of the Doppler effect; a decrease in the wavelength of the radiation emitted by an approaching celestial body; a shift toward the short-wavelength (blue) end of the spectrum.

Bode's law *See* Titius-Bode law.

Bohr model of the atom An energy-level scheme for explaining and predicting the wavelengths at which atoms emit and absorb radiation.

bolide An unusually bright meteor.

bolometer An instrument for measuring the total energy received from an object.

bolometric magnitude The magnitude of an object measured over all wavelengths.

Boltzmann's constant A number ($k = 1.3806 \times 10^{-8}$ J/K) relating the kinetic energy of the particles in a gas to its temperature ($K.E. = \frac{3}{2} kT$).

bound-bound transition A transition of an electron in an atom from one discrete, bound energy level to another.

bound-free transition A transition of an electron in an atom from a discrete, bound energy level to a free state, not bound to the atom.

Brackett-alpha line The emission or absorption line of the hydrogen atom involving a transition between the third and fourth energy levels; the first line in the Brackett series.

breccia Material composed of rock and mineral fragments cemented together; a common part of the lunar surface.

bright-line spectrum *See* emission-line spectrum.

bright nebula *See* diffuse nebula.

brightness The energy per unit area per second received from an object; the flux.

C-type asteroid An asteroid with a low albedo, suggestive of a high carbon content.

Caloris Basin A lowland area on the surface of Mercury, about 1300 km in diameter, resembling the maria basins on the moon.

canali (Italian "channels") Term used by Giovanni Schiaparelli to describe dark linear features seen on the surface of Mars.

capture model A model for the moon's origin, proposed about 1955, that pictures the moon as captured by the earth's gravity, after which it spiraled in toward the earth, reversed orbital direction, and spiraled outward.

carbon-nitrogen-oxygen (CNO) cycle A series of thermonuclear reactions taking place in a star's core, in which carbon, nitrogen, and oxygen aid the fusion of hydrogen into helium; a secondary energy production process in the sun, but the major process in high-mass main-sequence stars.

carbonaceous chondrites A type of meteorite containing chondrules embedded in material rich in carbon.

Cassegrain reflector A reflecting telescope that uses a curved secondary mirror to form an image through a hole in the primary mirror.

Cassini's division A gap about 2000 km wide in Saturn's rings, discovered in 1675 by Giovanni Cassini.

catastrophic model A model for the origin of the solar system in which an improbable event (usually collision with another star) led to the collection of gaseous materials that became the planets.

celestial equator The great circle on the celestial sphere halfway between the celestial poles.

celestial meridian *See* meridian.

celestial pole An imaginery projection of the earth's pole onto the celestial sphere; a point about which the apparent daily rotation of the stars takes place.

celestial sphere An imaginary sphere of very large radius centered on the earth on which celestial bodies appear fastened and against which their motions are charted.

cell The basic unit of living organisms, usually containing a nucleus surrounded by protoplasm both contained within the cell membrane.

Centaurus X-3 A pulsing binary X-ray source, thought to be a neutron star.

center of mass The balance point of a set of interacting or connected bodies.

central force A force directed along a line connecting the centers of two objects.

centripetal acceleration The acceleration toward the center of a curved path resulting from the change in direction of the velocity.

centripetal force A force required to divert a body from a straight path into a curved one and directed toward the center of the curve.

cepheid variable (cepheid) A star that varies in brightness as a result of a regular variation in size; a class of variable stars for which the star Delta Cephei is the prototype; their period-luminosity relation makes them useful distance indicators.

Ceres The first observed asteroid, discovered by Father Giuseppe Piazzi in 1801.

CETI An acronym for communication with extraterrestrial intelligence.

Chandrasekhar limit The maximum amount of mass for a white dwarf star, about 1.4 solar masses. This amount leads to the highest density and smallest radius for a star made of a degenerate electron gas; more than this and the star collapses.

Charon The satellite of Pluto.

chemical condensation sequence *See* condensation sequence.

chemical evolution The natural process operating on the primitive earth by which complex organic molecules are formed from simpler substances.

Chiron An asteroid located between the orbits of Saturn and Uranus.

chondrite A stony meteorite characterized by the presence of small, round silicate granules (chondrules).

chondrules Round silicate granules lacking volatile elements. Found in chondritic meteorites, or chondrites, they are believed to be primitive solar system materials.

chromosphere The part of the sun's atmosphere just above the photosphere. It is hotter and less dense than the photosphere. It creates the flash spectrum seen during eclipses.

circumpolar star For an observer north of the equator, a star that is continually above the northern horizon and never sets; for a southern observer, a star that never sets below the southern horizon.

circumstellar dust Dust surrounding a star.

closed geometry *See* spherical geometry.

closed model A model for the universe with a closed geometry.

cluster of galaxies A group of galaxies bound together by gravity.

CNO cycle *See* carbon-nitrogen-oxygen cycle.

co-orbital satellite A satellite in the same orbit as another satellite.

collisional excitation Excitation of an atom or ion to a higher energy level by collision with another particle.

color index The difference in the magnitudes of an object measured at two different wavelengths; a measure of the color and hence the temperature of a star.

color temperature The temperature assigned to an object by comparing its color with that of a blackbody.

coma The bright, visible head of a comet.

Coma cluster of galaxies A nearby cluster of galaxies in the direction of the constellation Coma Berenices.

comet A body of small mass, which revolves around the sun, usually in a highly elliptical orbit, and consisting, in the dirty-iceberg model, of small, solid particles (probably of rocky material) embedded in frozen gases.

compact radio galaxy A radio galaxy for which the radio emission comes from a region smaller than the optically observable galaxy.

compound A substance composed of two or more atoms bound together by chemical forces.

condensation The growth of small particles by the sticking together of atoms and molecules.

condensation sequence The order in which chemical compounds can form out of a gas cloud at specific densities and temperatures.

conduction The process of heat transfer by collisions of particles with their neighbors.

conjunction The time at which two celestial objects appear closest together in the sky.

conservation of energy A fundamental principle in physics that states that the total energy of an isolated system remains constant regardless of whatever internal energy changes may occur.

conservation of magnetic flux The physical principle stating that, for an ionized gas, the product of magnetic field strength times area is constant as the gas expands or contracts.

constellation An apparent arrangement of stars on the celestial sphere, usually named after ancient gods, heroes, animals, or mythological beings; now an agreed-upon region of the sky containing a group of stars.

continental drift Wegener's theory that the present continents were at one time a single landmass that fragmented and drifted apart.

continuous spectrum A spectrum showing emission at all wavelengths, unbroken by either absorption lines or emission lines.

contour map A diagram showing how the intensity of radiation varies over a region of the sky.

convection The transfer of energy by moving currents of a fluid.

convection zone A region in the outer part of the sun where energy is carried primarily by convection.

coplanar Lying in the same plane.

core (of the earth) The central region of the earth, which has a high density, is most likely liquid, and is believed to be composed of nickel and iron.

core (of the sun) The inner 25 percent of the sun's volume, where the temperature is great enough for thermonuclear reactions to take place.

core-mantle grains A model of interstellar dust particles consisting of a core of carbon or silicate material surrounded by a mantle of icy material.

corona The outermost region of the sun's atmosphere, consisting of thin, ionized gases at a temperature of about 1 million K.

coronal holes Regions of the solar corona of lower than average density and temperature.

coronal interstellar gas A component of the interstellar gas at high temperature and low density, conditions similar to those in the solar corona.

cosmic blackbody microwave radiation Radiation with a blackbody spectrum at a temperature of about 3 K permeating the universe; believed to be the remains of the primeval fireball in which the universe was created.

cosmic neutrinos Neutrinos that may fill the universe and hence play a major role in cosmology and galaxy formation.

cosmic ray Charged atomic particles moving in space with very high energies (the particles travel close to the speed of light); most originate beyond the solar system, but some are produced in solar flares.

cosmic water hole The region of the radio spectrum between the emission lines of H and OH, which may be the likely band for interstellar communication.

cosmological principle The statement that the universe, averaged over a large enough volume, appears the same from any location.

cosmology The study of the nature and evolution of the physical universe.

cosmos The universe considered as an orderly and harmonious system.

Crab Nebula A gaseous nebula in the constellation Taurus, a remnant of the supernova explosion of 1054, containing a pulsar at its center.

crater A circular depression of any size, usually caused by the impact by a solid body or by a surface eruption.

cratered terrain A region on a planet or satellite heavily covered with craters.

creation of matter An aspect of the Steady State model of the universe, in which matter is assumed to appear spontaneously to compensate for dilution caused by the expansion of the universe.

crescent moon The moon's phase when its elongation is less than 90°.

critical density The density needed to just close the universe and to slow the expansion velocity to zero after infinite time.

crust The thin, outermost surface layer of a planet. On the earth, it is composed of basaltic and granitic rocks.

Cygnus arm A segment of one of the spiral arms of the Galaxy, lying about 34,000 ly out from the center and containing a small branch, the Orion spur, in which the sun is located.

Cygnus X-1 A binary X-ray source in the constellation Cygnus thought to contain a black hole.

D line One of a pair of dark lines in the yellow region of the spectrum, produced by sodium.

dark line See absorption line.

dark-line spectrum See absorption-line spectrum.

dark nebula (dark cloud) A cloud of dust obscuring distant stars and hence appearing as a dark patch on the sky.

deceleration parameter A number indicating the rate at which the expansion velocity of the universe is decreasing.

declination The angular distance north or south of the celestial equator.

decoupling In the early universe, the cessation of interaction between radiation and matter due to the decrease of opacity caused by the recombination of hydrogen at about 3000 K.

deferent An ancient geometrical device used to account for the apparent eastward motion of the planets; a large circle, usually centered on the earth, that carries around a planet's epicycle.

degenerate electron gas Matter in which electrons are packed together as tightly as possible.

degenerate gas pressure A force exerted by highly dense, compacted matter that depends mostly on how dense the matter is and very little on its temperature.

degenerate neutron gas Matter made up of neutrons packed together as tightly as possible.

Deimos The smaller of the two moons of Mars.

density The amount of mass in a unit volume of an object.

density-wave model A model for the generation of spiral structure in galaxies which pictures density waves (similar to sound waves) plowing through the interstellar matter and sparking star formation.

deoxyribonucleic acid (DNA) The basic genetic material of life as we know it; a very large two-stranded, helical molecule consisting of subunits composed of bases, sugar, and phosphate.

detector A device that measures radiation, usually producing an electrical signal proportional to its intensity.

deuterium An isotope of hydrogen that has a nucleus consisting of one proton and one neutron.

differential rotation Rotation at different angular speeds at different distances from the center.

diffuse (bright) nebula A cloud of ionized gas, mostly hydrogen, with an emission-line spectrum.

dipole field A magnetic field with two poles, generated by a circulating loop of electric current.

dirty-iceberg cometary model A model for comets that pictures the nucleus as a compact, solid body of ices, embedded with rocky material, that turn into gases as a comet nears the sun, so creating the head and tail.

disk (of a galaxy) The flattened wheel of stars, gas, and dust outside the nucleus.

distance candle See standard candle.

distance pyramid The step-by-step process of obtaining distances throughout the universe starting with measurements in the solar system.

diurnal Daily.

DNA See deoxyribonucleic acid.

Doppler effect A change in the wavelength of waves from a source reaching an observer when the source and the observer are moving with respect to each other along the line of sight. The wavelength increases (red shift) with motion away from the observer and decreases (blue shift) with motion toward the observer.

doubling time The time it takes a reproducing population to double its size.

Drake equation A formula for estimating the number of intelligent civilizations in the Galaxy.

dust tail One of two tails of a comet, composed of dust expelled from the nucleus.

dynamo model A model for the formation of planetary magnetic fields, involving circulations in a conducting fluid in the planet's core.

dwarf A star of relatively low light output and relatively small size.

east The point on the horizon 90° (clockwise) from north.

eccentric An ancient geometrical device used to account for nonuniform planetary motion; a point offset from the center of circular motion.

eccentricity A measure of the deviation of an ellipse from a perfect circle.

eclipse The phenomenon of one body passing in front of another, cutting off its light (*see* solar eclipse, lunar eclipse).

eclipsing binary Two stars that revolve around a common center of mass, the orbits lying edge on to the line of sight, so that each star periodically passes in front of the other.

ecliptic The apparent yearly path on the celestial sphere of the sun with respect to the stars; also, the plane of the earth's orbit.

ecosphere The region around a star where an orbiting planet's surface temperature is within the range for life as we know it.

effective temperature The temperature assigned to a body equal to that of a blackbody that would emit the same total energy.

Einstein-Rosen bridge (wormhole) A connecting path between two separated regions of spacetime.

electromagnetic radiation Energy in the form of waves of varying electric and magnetic fields, including visible light, radio waves, X rays, infrared, and ultraviolet.

electromagnetic spectrum The range of all different wavelengths of electromagnetic radiation.

electron A negatively charged subatomic particle usually found in orbit around the nucleus of an atom.

electronic transition A change in the energy state of an atom as an electron moves from one energy level to another.

element A substance that is made of atoms with the same chemical properties and cannot be decomposed chemically into simpler substances.

ellipse A plane curve drawn such that the sum of the distances from a point on the curve to two fixed points is constant; an oval.

elliptical galaxy A gravitationally bound system of stars that has rotational symmetry but no spiral structure and that contains mainly old stars with little gas or dust.

elongation The angular separation of an object from the sun along the ecliptic.

emission (bright) line Light of a specific wavelength or color emitted by atoms; a sharp energy peak in a spectrum.

emission nebula *See* diffuse nebula.

empirical Derived from experiment or observation.

energy The ability to do work.

energy level One of the possible states of an atom, with a specific value of energy.

enzyme A protein that brings about or accelerates reactions at body temperatures without itelf undergoing destruction in the process.

ephemeris (*pl.* **ephemerides**) A table that gives the positions of celestial objects at various times.

epicycle An ancient geometrical device used by astronomers to account for the westward retrograde motion and other irregular motions of the planets; a small circle whose center moves along a larger one (the deferent).

epoch The date for which the positions in a star catalog are calculated.

Epsilon Eridini A nearby K-star (10.7 ly distant), a likely candidate for planetary companions.

equant An ancient geometrical device invented by Ptolemy to account for variations in retrograde motion; essentially an eccentric in which the center of the circle is not the center of uniform motion.

equation of state A relationship that describes the conditions in a physical system, such as an equation relating the pressure, temperature, and density of a gas.

equatorial bulge The excess diameter, about 43 km, of the earth through its equator compared with its poles.

equatorial system The system which specifies the position of an object on the celestial sphere by its right ascension and declination.

equilibrium A state of a physical system in which there is no overall change.

equinox Time of year of equal length of days and nights; the two times of the year when the sun crosses the celestial equator. The spring (vernal) equinox occurs about March 21 and the fall (autumnal) equinox about September 21. Also the points on the celestial sphere where the ecliptic and equator cross, the position of the sun on March 21 and September 21.

escape velocity The speed a body must achieve in order to break away from the gravity of another body and never return to it.

Euclidean geometry See flat geometry.

event A point in spacetime, the time *and* place at which something happens.

event horizon See Schwarzschild radius.

evolution Change with time.

evolutionary track The path on an H-R diagram indicating how the luminosity and effective temperature of a star change with time.

excitation The process of raising an atom to a higher energy level (*see* collisional excitation and photon excitation).

exosphere The topmost region of a planet's atmosphere, where particles in the atmosphere can escape into space.

expanding (3-kpc) arm A segment of spiral-arm structure encircling the center of our Galaxy at a distance of about 3 kpc (9800 ly). It appears to be moving toward us and away from the Galaxy's center.

exponential growth The increase in population by a certain percent each year, which eventually leads to very rapid growth, exceeding the resources that support the population.

extended radio galaxy A radio galaxy in which the source of radio radiation is larger than the optically observed galaxy.

extinction The dimming of light when it passes through some medium, such as the earth's atmosphere or interstellar material.

extragalactic Beyond the Galaxy.

eyepiece A magnifying lens used to view the image produced by the objective of a telescope.

fall See meteorite fall.

Fig Tree formation A rock formation about 3.2 billion years old containing fossil evidence of the earliest forms of life.

find See meteorite find.

first point of Aries The vernal equinox.

fission See nuclear fission.

fission model The earliest of the major models of the origin of the moon that suggests that a piece of the young, rapidly spinning, molten earth spiraled out into orbit and cooled down to form the moon.

flash spectrum The spectrum that appears immediately before the totality of a solar eclipse, as the normal absorption spectrum is replaced briefly by the chromosphere's own emission spectrum.

flat geometry The geometry of Euclid, in which parallel lines never meet, only one parallel line can be drawn through a point near another line, and the sum of the angles of a triangle is exactly equal to 180°.

fluorescence The process by which a high-energy photon is absorbed by an atom and reemitted as two or more photons of lower energy.

flux The amount of energy flowing through a given area in a given time (usually one second).

focal length The distance from a lens to the point at which it focuses parallel rays.

focus The point at which light is gathered in a telescope.

forbidden line A spectral line that results from a transition with a low probability of happening.

force A push or pull that changes an object's momentum.

forced motion One of Aristotle's two classes of motion, the other being *natural motion*.

frame of reference A set of axes with respect to which the position or motion of something can be described or physical laws are formulated.

Fraunhofer lines The name given to absorption lines in the spectrum of a star, especially the sun.

free-bound transition The change of an electron from a free state to a discrete, bound energy level of an atom.

free-fall time The time it takes a gas cloud to collapse to its center moving only in response to its own gravity, with no resisting internal pressure.

free-free transition A change of an electron, not bound to an atom, from one free energy state to another.

frequency The number of waves that pass a particular point in some time interval (usually a second).

full moon The moon's phase when it is in opposition to the sun, at an elongation of 180°.

fusion See nuclear fusion.

galactic cannibalism A process by which a galaxy at the center of a cluster of galaxies increases in mass by absorbing other galaxies through collisions and tidal disruption.

galactic cluster (open cluster) A small group, about ten to a few hundred, of gravitationally bound stars of Population I, found in or near the plane of the Galaxy.

galactic equator The great circle along the line of the Milky Way, marking the central plane of the Galaxy.

galactic latitude The angular distance north or south of the galactic equator.

galactic longitude The angular distance along the galactic equator from a zero point in the direction of the galactic center.

galactic rotation curve A graph showing how fast an object some distance from the center of a galaxy revolves around it.

galaxy A huge assembly (between a million and hundreds of millions) of stars, gas, and dust that is held together by gravity; spelled with a capital letter (the Galaxy), our own Milky Way Galaxy, containing the sun.

Galilean moons The four largest satellites of Jupiter (Io, Europa, Ganymede, Callisto), discovered by Galileo.

gamma ray A very high-energy photon, of wavelength shorter than that of X rays.

gas tail One of two tails of a comet, consisting of gas expelled from the nucleus.

gauss A measure of the strength of a magnetic field; the earth's field is about 1/2 gauss.

general theory of relativity The idea developed by Albert Einstein that mass and energy determine the geometry of spacetime and that any curvature of this spacetime shows itself by what we commonly call gravitational forces.

geocentric Centered on the earth.

geomagnetic axis The axis that connects the earth's magnetic poles. It is inclined about 20° from the geographic spin axis and does not pass through the earth's center.

giant A star of high luminosity and large size.

giant molecular cloud An interstellar cloud containing hydrogen and other molecules, with a mass of 10^5 solar masses and a diameter of a few tens of light years.

gibbous moon The moon's phase when its elongation is between 90° and 180°.

globular cluster A gravitationally bound group of about 10,000 to 100,000 metal-poor Population II stars, symmetrically shaped, found in the halo of the Galaxy.

gnomon An ancient instrument for measuring time; most simply, a stick stuck vertically into the ground whose shadow is used to indicate the sun's position with respect to the horizon.

granite A rock solidified from molten lava containing silicon, aluminum, oxygen, sodium, and potassium, making up much of the continents.

granule Brief-lived (3 to 10 min) bright spots that appear as a rough texture on the solar photosphere.

gravitation In Newtonian terms, a force between masses that is characterized by their acceleration toward each other. The size of the force depends directly on the product of the masses and inversely on the square of the distance between them.

gravitational collapse The unhindered contraction of any mass due to its own gravity.

gravitational energy Potential energy released by gravitational collapse.

gravitational field The property of space having the potential for producing gravitational force on objects within it, characterized by the acceleration of free masses.

gravitational instability The tendency for a disturbed region in a gas to undergo gravitational collapse.

gravitational lens A massive object such as a galaxy that bends light rays by its gravitational influence and so focuses them.

gravitational mass The amount of mass in an object determined by measuring the gravitational force it produces.

gravitational red shift The change to longer wavelengths that marks the loss of energy by a photon that leaves any mass.

great circle A circle on a sphere whose plane passes through the center of the sphere; it has the property of being the shortest distance between any two points on the sphere.

Great Red Spot A disturbance in the atmosphere of Jupiter, a region of high pressure and rising material.

Great Rift An apparent splitting of the Milky Way caused by a dust cloud absorbing light from distant stars.

greenhouse effect A warming of a planetary atmosphere caused by high opacity in the infrared which impedes escape of energy.

grooved terrain A region on a satellite consisting of ridges and grooves, probably caused by the tearing and shearing of the surface.

ground state The lowest energy level of an atom.

Gunflint formation A rock formation 2.4 billion years old containing fossil evidence of cells with definite genetic structure.

H I region A region of neutral hydrogen in interstellar space.

H II region A zone of ionized hydrogen in interstellar space, which usually forms a bright nebula around a hot, young star or cluster of hot stars.

H-alpha line The first line of the Balmer series, the set of transitions in a hydrogen atom between the second energy level and levels with higher energy. It lies in the red part of the visible spectrum.

H-R diagram *See* Hertzsprung-Russell diagram.

habitable zone Same as ecosphere.

half-life The time required for half of the radioactive atoms in a sample to disintegrate.

Halley's Comet A bright comet with a period of 76 years.

halo (of a galaxy) The spherical region around a galaxy, not including the disk or the nucleus. It contains globular clusters, some gas, and probably a few stray stars.

head-tail galaxy A radio galaxy that has a strong radio source around the visible galaxy (the head) and a narrow source extending from the galaxy (the tail).

heat The kinetic energy of random motion of particles in an object.

heavy elements In astronomy, the elements heavier than helium.

heavy-particle era The stage in the early history of the universe when the temperature was high enough (greater than 10^{12} K) so that heavy particles (protons and neutrons) could form from photons.

heliocentric Centered on the sun.

heliocentric parallax *See* trigonometric parallax.

GLOSSARY

helium flash The rapid burst of energy generation with which a star initiates helium burning by the triple-alpha process in the degenerate core of a low-mass red giant star.

Hertz A unit of frequency, equal to 1 cycle per second.

Hertzsprung-Russell (H-R) diagram A graphical representation of the classification of stars according to their spectral class (or color or surface temperature) and luminosity (or absolute magnitude). The physical properties of a star are correlated with its position on the diagram, so a star's evolution can be described by its change of position on the diagram with time.

heterosphere The region of the earth's atmosphere where the composition varies.

high-velocity star A star, generally of Population II, moving with a velocity greater than about 60 km/sec with respect to the sun and having rather eccentric and inclined orbit about the center of the Galaxy.

highlands The regions on the moon or another planet of relatively high elevation, usually more cratered than the lowlands and older.

Homo sapiens Humankind.

homogeneous Having a consistent and even distribution of matter; the same in all parts.

homosphere The region of the earth's atmosphere where the composition is homogeneous.

horizon The intersection with the sky of a plane tangent to the earth at the location of the observer.

horizon system The system which specifies the position of an object on the celestial sphere by its altitude and azimuth.

horizontal branch A portion of the Hertzsprung-Russell diagram reached by stars of low mass and of low metal abundance after the red giant stage and typically found in a globular cluster. It ranges from yellow to red stars all having roughly the same luminosity (about 100 times the sun's).

hour circle A great circle on the celestial sphere through the celestial poles.

Hubble classification scheme A method of assigning galaxy types on the basis of their appearance—presence or absence of spiral arms or a central bar and general shape.

Hubble constant The slope of the relation between velocity and distance for other galaxies; the constant of proportionality in the Hubble law.

Hubble law The relationship between recession velocity and distance for galaxies, expressed by the equation $v = Hd$, where H is the Hubble constant.

hydrogen corona A very thin outer atmosphere of the earth, extending to at least 60,000 km, containing largely hydrogen gas.

hydrostatic equilibrium An equilibrium characterized by the absence of mass motions, when pressure balances gravity.

hyperbolic geometry An alternative to Euclidean geometry constructed by N. I. Lobachevski on the premise that more than one parallel line can be drawn through a point near a straight line; the sum of the angles of a triangle drawn on a hyperbolic surface is always less than 180°.

icy materials Substances such as solid water, ammonia, and methane making up the outer planets.

igneous rock Rock formed by the cooling of molten lava.

image The apparent reproduction of an object produced by a lens which focuses light rays coming from the object.

impact theory The theory that craters are formed by the impact of solid objects on a surface.

inertia The resistance of an object to a force acting on it.

inertial mass Mass determined by subjecting an object to a known force and measuring the acceleration that results.

inferior conjunction The alignment of an inner plant (Mercury or Venus) with the sun when the planet is between the earth and sun.

infrared telescope A telescope specially designed for operation at infrared wavelengths. Although most optical telescopes can also be used in the infrared, better sensitivity is obtained if the telescope is constructed to avoid generation of infrared radiation which can fall on the detector.

instability strip A region in the H-R diagram where stars are unstable and undergo pulsations.

intercrater plains Flat areas on the surface of a planet devoid of craters.

interference The mixing together of waves.

interferometer *See* radio interferometer.

intergalactic medium The gas found between galaxies.

interplanetary medium The gas and dust found between the planets.

interstellar cloud A dense region of interstellar gas.

interstellar dust Solid particles found between the stars.

interstellar extinction curve A graph showing the amount of extinction at different wavelengths.

interstellar gas The gas between the stars.

interstellar medium All the gas, dust, and particles found between stars.

inverse beta decay The reverse process of neutron decay, in which electrons and protons are forced together to form neutrons and neutrinos.

inverse square law The statement that the flux of energy from a source decreases as the square of its distance.

ion An atom that has become electrically charged by the gain or loss of electrons.

ionization The process by which an atom loses or gains electrons.

ionized gas A gas that has been ionized, so that it contains free electrons and charged ions; a plasma.

ionosphere A layer of the earth's atmosphere, ranging from about 100 to 700 km above the surface, in which oxygen and nitrogen are ionized by sunlight, producing free electrons.

irons One of three main types of meteorites, typically made of about 90 percent iron and 9 percent nickel, with a trace of other elements.

irregular cluster of galaxies A cluster that has no symmetry and no concentrated, single center.

irregular galaxy A galaxy without spiral structure or rotational symmetry, containing mostly Population I stars and abundant gas and dust.

Ishtar Terra An upland plateau on the surface of Venus.

isotope One of two or more forms of the atom of an element, differing from one another in the number of neutrons, but all containing the same number of protons.

isotropic Having no preferred direction in space.

Jeans length The minimum size a disturbance in a gas must have to result in gravitational contraction. It depends on the pressure, temperature, and density of the medium.

Jovian planets Jupiter, Saturn, Uranus, and Neptune; planets of large size and mass, but low density.

Kelvin (absolute) temperature scale Temperature counted from absolute zero, the temperature at which all random molecular motion ceases. The freezing point of water is 273.15 K and the boiling point is 373.15 K.

Kepler's laws Three laws describing planetary motion.

kiloparsec (kpc) 1000 parsecs.

kinetic energy The ability to do work because of motion.

Kirchhoff's rules Descriptions of the types of spectra produced by solids, liquids, and gases.

kpc Abbreviation for kiloparsec.

KREEP A lunar material composed of potassium (K), rare-earth elements (REE), and phosphorus (P).

Large Magellanic Cloud (LMC) *See* Magellanic Clouds.

latitude Angular distance north or south of the equator.

LAWKI An acronym for life as we know it.

laws of motion The physical laws devised by Newton which describe the motions of objects in response to forces.

lens A curved piece of glass that brings light rays to a focus, forming an image of an object.

lenticular galaxy A galaxy similar to a spiral with a disk containing no gas and dust; also called S0 galaxy.

light curve A graph of a star's changing brightness with time.

light-gathering power The relative ability of a telescope to collect light; it depends on the area of the telescope objective.

light-particle era The stage in the early history of the universe when the temperature was high enough (greater than 5×10^9 K) so that light particles (electrons and positrons) could form from photons.

light ray The path of light through a medium.

light year The distance light travels in a year, about 6 trillion miles, or 9.46×10^{12} km.

lighthouse model A model for pulsars incorporating emission of radiation in a narrow cone which sweeps around as the pulsar rotates.

Local Group A gravitationally bound group of about 20 galaxies to which our Milky Way Galaxy belongs.

local supercluster The supercluster of galaxies in which the Local Group is located.

longitude Angular distance east or west of a meridian on a sphere; on the earth, the angular distance east or west of the meridian through Greenwich, England.

longitudinal wave A sound wave that moves in a push-pull motion. It can travel through solids, liquids, and gases with a velocity that depends on the density of the medium.

low-velocity star A star in our Galaxy which has a more or less circular orbit and generally shares the rotational motion of the sun about the galactic center and so has a low velocity relative to the sun (less than 60 km/sec).

lowlands Regions on the moon or planet of relatively low elevation; the maria.

luminosity The total rate at which radiative energy is given off by a celestial body, over all wavelengths. The sun's luminosity is about 4×10^{26} W.

luminosity class (galaxies) A scheme for putting galaxies in rough classes according to their luminosities.

luminosity class (stars) The categorization of stars that have the same surface temperatures but different sizes, resulting in different luminosities. It is based on the widths of dark lines in the star's spectrum, with giant stars having narrower lines than dwarf stars.

luminosity function For stars or galaxies, the relation describing the number of objects in each range of luminosity.

lunar eclipse An eclipse of the moon, caused by the moon passing through the shadow of the earth.

lunar occultation The passing of a star or other object behind the moon; an eclipse.

lunar soil A type of material on the lunar surface; fine particles created by the bombardment of the surface by meteorites.

Lyman series (lines) All transitions of electrons in a hydrogen atom to and from the lowest energy level, involving large energy changes; also the set of absorption or emission lines corresponding to these transitions, which lies in the ultraviolet part of the spectrum.

μ (μm) Abbreviation for micron.

M 17 A nearby H II region containing some very young stars and associated with a giant molecular cloud.

M 31 The nearest spiral galaxy to us; the Andromeda Galaxy.

M 33 A spiral galaxy, a member of the Local Group.

M-type asteroid An asteroid of low albedo indicative of a metallic composition.

Magellanic Clouds Two irregular neighboring galaxies, the Large Magellanic Cloud (LMC) and the Small Magellanic Cloud (SMC), visible from the Southern Hemisphere to the unaided eye; companions to our Galaxy.

magnesium 26 (^{26}Mg) An isotope of magnesium found in meteorites, produced by the radioactive decay of ^{26}Al.

magnetic field The property of space having the potential of exerting magnetic forces on bodies within it.

magnetic flux The product of magnetic field strength times area (*see* conservation of magnetic flux).

magnetic lines of force A graphic representation of a magnetic field showing its direction and, by the degree of packing of the lines, its intensity.

magnetosphere The region around a planet where particles from the solar wind are trapped by the planet's magnetic field.

magnifying power The ratio of apparent angular size of an object as seen through a telescope to its true angular size.

magnitude A measurement of an object's brightness; larger magnitudes represent fainter objects.

main sequence The principle series of stars in the Hertzsprung-Russell diagram. Such stars are converting hydrogen to helium in their cores by the proton-proton process or by the carbon-nitrogen-oxygen cycle; the longest stage of a star's active life.

main-sequence lifetime The time a star remains on the main sequence converting hydrogen into helium in its core.

major axis Maximum diameter across an ellipse.

mantle The major portion of the earth's interior below the crust, made of a plastic rock probably composed of olivine.

mare (*pl.* maria) Latin for "sea"; a lowland area on the moon that appears darker and smoother than the highland regions, probably formed by lava that solidified into basaltic rock about 3 to 3.5 billion years ago.

Mare Imbrium A round basin on the near side of the moon.

Mare Orientale A round basin on the far side of the moon, surrounded by concentric rings of mountains.

mascon An abnormal concentration of mass beneath the lunar maria. Mascons were detected by their effect on the orbits of moon-orbiting satellites.

maser *See* molecular maser.

mass The measurement of an object's resistance to change in its motion (inertial mass); a measure of the strength of gravitational force an object can produce (gravitational mass).

mass-luminosity ratio The ratio of mass to luminosity for a galaxy, abbreviated M/L, and usually expressed in solar units. For elliptical and spiral galaxies M/L is about 70 and 35, respectively.

mass-luminosity relation An empirical relation for main-sequence stars between a star's mass and its luminosity. Roughly, the luminosity is proportional to the third power of the mass.

matter era The stage in the history of the universe from the time of decoupling of matter and radiation (700,000 years after the beginning) until now, when matter is dominant over radiation, its self-gravity controlling the formation of galaxies and stars.

maximum elongation The greatest angular distance of an object from the sun along the ecliptic.

Maxwell Montes A mountain range on Venus, bordering the Ishtar Terra plateau.

mechanics A branch of physics that deals with forces and their effects on bodies.

megaparsec (Mpc) 1 million parsecs, or about 3.26 million light years.

megaton An explosive force equal to that of 1 million tons of TNT (about 4×10^{15} J).

meridian On the earth, any great circle through the poles; on the celestial sphere, the great circle through the zenith and the north and south points on the horizon.

mesosphere A region of the earth's atmosphere, extending from about 50 to 90 km, above the stratosphere and below the thermosphere.

metal-poor star A star with a fractional abundance of elements heavier than helium much less than that in the sun, belonging to Population II.

metal-rich star A star with a fractional abundance of elements heavier than helium comparable to or greater than that in the sun and other Population I stars.

metallic hydrogen The physical state of hydrogen under high pressure, in which the electrons are free to move around, as in a metal.

meteor The bright streak of light that occurs when a solid particle (a meteoroid) from space enters the earth's atmosphere and is heated by friction with atmospheric particles; sometimes called a falling star.

meteor shower The occurrence of a large number of meteors due to the encounter of the earth with a meteor stream or swarm.

meteor stream A collection of meteoroids spread out along the orbit of a former comet.

meteor swarm A concentration of meteoroids in a stream, which causes an unusually large shower when it encounters the earth.

meteorite A solid body from space that survives a passage through the earth's atmosphere and falls to the ground.

meteorite fall A meteorite seen to fall and immediately collected.

meteorite find A meteorite collected long after its fall. Falls are more likely to be metallic, as they are much easier to distinguish from ordinary rocks than are stony meteorites.

meteoroid A very small body, like a grain of sand or dust speck, moving through space.

micrometeorite A very small meteorite, slowed enough by its encounter with the atmosphere so that it does not burn up, but settles slowly to the ground.

micron (μ or μm) 10^{-6} meter.

microwaves A wave of electromagnetic radiation of wavelength about 1 mm to 10 cm, between the radio and infrared parts of the spectrum.

mid-oceanic ridge An almost continuous submarine mountain chain that extends some 64,000 km through the earth's ocean basins.

Milky Way The band of light that encircles the sky, caused by the blending of light from the many stars lying near the plane of the Galaxy; also sometimes used to refer to the Galaxy to which the sun belongs, the Milky Way Galaxy.

minute of arc 1/60 of a degree.

molecular maser A collection of molecules excited in such a way as to emit very strongly in the microwave part of the spectrum.

molecule A combination of two or more atoms bound together electrically; the smallest part of a compound that has the properties of that substance.

momentum The tendency of a body, because of its inertia, to keep moving in a straight line; quantitatively, the momentum of a mass m moving at velocity v is mv.

Mpc abbreviation for megaparsec.

Murchison meteorite A carbonaceous chondrite which fell to earth in 1969 and was found to contain amino acids.

mutation A basic change in the hereditary material, the genes, of an organism.

natal astrology The study of the supposed influence of the stars and planets upon human affairs based on their positions and relationships, especially at the time of an individual's birth.

natural motion One of Aristotle's two classes of motion, the other being forced motion; the motions of the four basic elements (earth, air, fire, and water), for example, the natural motion of earthlike material toward the center of the universe.

natural selection The process by which individuals with genes producing characteristics most well adapted to their environment have greater genetic representation in future generations.

nebula Latin for "cloud"; a cloud of interstellar gas and dust; formerly used to refer to any cloudlike object in the sky, such as a spiral nebula, which is really a galaxy.

nebular model A model for the origin of the solar system, in which an interstellar cloud of gas and dust collapsed gravitationally to form a flattened disk out of which the planets formed by accretion.

negative hydrogen ion A hydrogen atom with an electron added, making an ion with a net negative charge.

neutrino An elementary particle with little or no mass and no electric charge that travels at the speed of light and carries energy away during certain types of nuclear reactions.

neutron A subatomic particle about the mass of a proton with no electric charge; one of the main constituents of an atomic nucleus.

neutron star A star of extremely high density and small size (about 10 km in radius) that is composed mainly of very tightly packed neutrons.

new moon The moon's phase when it is in conjunction with the sun, at an elongation of 0°.

Newtonian reflector A reflecting telescope that uses a flat secondary mirror to place the image at the side of the telescope.

nonoxidizing atmosphere An atmosphere with no oxygen.

nonthermal radiation Emitted energy that is not characterized by a blackbody spectrum; usually used to refer to synchrotron radiation.

nonthermal spectrum A spectrum with a shape quite different from a blackbody spectrum, usually indicative of synchrotron radiation.

noon Midday; that time halfway between sunrise and sunset when the sun reaches its highest point in the sky with respect to the horizon.

Norma arm A segment of one of the spiral arms of the Galaxy that lies 11,000 to 13,000 ly from the sun toward the center of the Galaxy in the direction of the constellation Norma.

north magnetic pole One of the two points on a star or planet from which magnetic lines of force emanate and to which the north pole of a compass points.

nova Latin for "new"; a star that has a sudden outburst of energy, temporarily increasing its brightness hundreds to thousands of times. This term was also used in the past to refer to some stellar outbursts that modern astronomers now call supernovas.

nuclear bulge (of a galaxy) The central region of a spiral galaxy, spheroidal in shape, containing mainly old Population I stars.

nuclear fission A process in which a heavy nucleus is hit by a high-energy particle and splits into two or more lighter nuclei whose combined mass is less than the original, the missing mass being released as energy.

nuclear force A force effective between nucleons when they are very close together.

nuclear fusion A process in which nuclei of lighter elements are joined to make heavier ones whose combined mass is less than that of the constituents, the missing mass being released as energy.

nucleic aid A huge spiral-shaped molecule, commonly found in the nucleus of cells, that is the chemical foundation of genetic material.

nucleon A proton or neutron.

nucleosynthesis The chain of thermonuclear fusion processes by which hydrogen is converted to helium, helium to carbon, and so on through all the elements of the periodic table.

nucleus (of an atom) The massive central part of an atom, containing neutrons and protons, about which the electrons orbit.

nucleus (of a comet) The small, bright, starlike point in the head of a comet. It is believed to be a solid, small (a few tens of kilometers) mass of frozen gases with some rocky material embedded within it.

nucleus (of a galaxy) The very central portion of a galaxy, composed of old Population I stars, some gas and dust, and, for many galaxies, a concentrated source of nonthermal radiation.

number density The number of objects, such as atoms in a gas, in a unit volume.

OB association A group of O and B stars, formed together in space, but not currently held together by gravity.

OB subgroup A small cluster of stars within an OB association, closer in age than the association as a whole.

objective The main light-gathering lens or mirror of a telescope.

observable universe All of the objects in the universe that can be seen or detected with various types of telescopes.

occultation The eclipse of a star or planet by the moon or another planet.

Olbers' paradox The statement that if there were an infinite number of stars distributed uniformly in an infinite space, then the night sky would be as bright as the surface of a star, in obvious contrast to what is observed.

old Population I star A class of stars with high abundance of heavy elements, but fairly old, found in elliptical galaxies and the nuclear bulge of spiral galaxies.

Olympus Mons A very large shield volcano on the surface of Mars.

Oort's comet cloud The group of cometary bodies thought to be surrounding the solar system, left over from the time of its formation.

Oort's constant A number relating radial velocities of nearby stars, due to the differential rotation of the Galaxy, to their distances and directions.

opacity The property of a substance that hinders (by absorption or scattering) light passing through it; opposite of transparency; the amount of this absorption or scattering.

open cluster *See* galactic cluster.

open geometry *See* hyperbolic geometry.

open model A model for the universe with an open geometry.

opposition The configuration of a celestial body lying exactly opposite the sun in the sky as seen from the earth, at an elongation of 180°; the time at which this configuration occurs.

optics The branch of physics dealing with light and light-measuring instruments.

orbit The path of a body revolving around another.

orbital angular momentum The angular momentum of a revolving body; the product of a body's mass, orbital velocity, and the distance from the system's center of mass.

orbital inclination The angle between the orbital plane of a body and some reference plane. In the case of a planet in the solar system, the reference plane is that of the earth's orbit, the ecliptic; in the case of a satellite, it is usually the equatorial plane of the planet; for a double star, it is the plane of the sky, perpendicular to the line of sight.

organic Relating to that branch of chemistry concerned with the carbon compounds of living creatures.

organic molecules Molecules containing at least one carbon atom.

Orion Nebula A bright, nearby H II region in the constellation Orion, associated with a molecular cloud and a site of recent star formation.

Orion spur A small branch of the Cygnus arm in which the sun is located.

oscillating model A model for the evolution of the universe in which the universe alternately expands and contracts.

outgassing The extrusion of gases from the body of a planet after its formation.

ozone layer A layer of the earth's atmosphere about 40 to 60 km above the surface, characterized by a high content of ozone, O_3.

Pangaea Primeval continent from which the present ones separated.

parallax The change in an object's apparent position when viewed from two different locations; usually, a measure of a star's distance, half the angular change as seen from opposite sides of the earth's orbit.

parent meteor bodies Small, solid bodies, a few hundreds or thousands of kilometers in size, believed to be the source of nickel-iron meteorites. They formed early in the history of the solar system and then broke up through collisions.

parsec The distance an object would have to be from the earth so that its heliocentric parallax would be 1 second of arc; equal to 3.26 ly.

particle threshold temperature The temperature at which photons have energy enough to create a given type of particle.

Paschen series The set of spectral lines for the hydrogen atom involving energy transitions between level 4 and any higher level.

Pauli exclusion principle The statement that no two particles of certain types (such as electrons or neutrons) can be in the same place with the same energy states at the same time.

pc Abbreviation for parsec.

perfect cosmological principle The statement that the universe appears the same to an observer at all locations and at all times.

perfect gas law The relationship for an ideal gas of noninteracting particles connecting its pressure, density, and temperature.

pericenter The point in an elliptical orbit closest to the focus.

perigee The point in its orbit at which an earth satellite is closest to the earth.

perihelion The point at which a body orbiting the sun is nearest to it.

period The time interval for some regular event to take place, for example, the time required for one complete revolution of a body around another.

period-luminosity relationship For cepheid variables, a relation between the average luminosity and the time period over which the luminosity varies; the greater the luminosity, the longer the period.

periodic comets Comets that have relatively small elliptical orbits around the sun, with periods less than 200 years.

periodic variable A star whose light varies with time in a regular fashion.

Perseus arm A segment of one of the spiral arms of the Galaxy that lies about 10,000 ly (3 kpc) from the sun in the direction of the constellation Perseus.

phases of the moon The monthly cycle of the changes in the moon's appearance as seen from the earth.

Phobos The larger of the two moons of Mars.

photodissociation The breakup of a molecule by the absorption of light with enough energy to break the molecular bonds.

photometry Measurement of light intensity.

photon A discrete amount of light energy; the energy of a photon is related to the frequency of the light by the relation $E = h\nu$, where h is Planck's constant.

photon excitation Excitation of an atom or ion to a higher energy state by absorption of a photon.

photosphere The visible surface of the sun; the region of the solar atmosphere from which visible light escapes into space.

physical evolution The formation and development of the physical universe.

physical universe The directly observable matter and those objects we detect by effects described by the laws of physics.

pitch angle The angle between a spiral arm's direction and the direction of circular motion about the Galaxy.

Planck curve The spectrum of a blackbody radiator.

Planck's constant A number ($h = 6.62 \times 10^{-34}$ J·sec) relating the frequency of a photon to its energy.

planet From the Greek word for "wanderer"; any of the nine (so far known) large bodies that revolve around the sun; traditionally, any heavenly object that moved with respect to the stars (in this sense, the sun and the moon were also considered planets).

planetary nebula A thick shell of gas moving out from an extremely hot star. It is believed to be the outer layers of a red giant star thrown out into space, the core of which eventually becomes a white dwarf.

planetary ring A ring of small particles in orbit around one of the major planets; known to exist around Jupiter, Saturn, and Uranus (and perhaps Neptune).

planetesimal An asteroid-sized body that, in the formation of the solar system, combined with others to form the protoplanets.

plasma A hot gas consisting of ionized atoms and electrons.

plate tectonics The model for the evolution of a planet's crust involving motions and collision of large masses, the continental plates.

polar cap A white patch of water and carbon dioxide ice near the north or south poles of Mars which changes shape and size with the Martian seasons.

Polaris The present north pole star; the outermost star in the handle of the Little Dipper.

polarization A lining-up of the planes of vibration of light waves.

polymerization The stringing together of small molecular fragments into long molecular chains.

Population I star A class of stars found in the disk of a spiral galaxy, including the most luminous, hot, and young stars. They have a heavy element abundance similar to that of the sun (about 2 percent of the total). Old Population I stars are found in the nuclear bulge of spiral galaxies and in elliptical galaxies; young Population I stars are found especially in the spiral arms.

Population II star A class of stars found in globular clusters and the halos of galaxies. They are somewhat older than any Population I stars and contain a smaller abundance of heavy elements.

positron An antimatter electron; the same as an electron, but with a positive charge.

potential energy The ability to do work because of position. It is storable and can later be converted into other forms of energy.

PP chain See proton-proton chain.

precession of the equinoxes The slow westward motion of the equinox points on the sky relative to the stars of the zodiac because of the wobbling of the earth's spin axis.

pre-main-sequence evolutionary track The path of a point on the H-R diagram showing how the luminosity and surface temperature of a star change in the course of its pre-main-sequence evolution, from the time of formation to the initiation of nuclear fusion in its core.

pre-main-sequence star A star in the stage of evolution preceding nuclear burning in its core; its energy comes from gravitational contraction.

pressure The force per unit area exerted by a gas due to the random motion of its molecules.

primary The brighter of the two stars in a binary system.

primary atmosphere The original atmosphere of a planet collected at the time of its formation.

primeval fireball The hot, dense beginning of the universe in the Big-Bang model, when most of the energy was in the form of high-energy light.

principle of equivalence The fundamental idea in Einstein's general theory of relativity; the statement that one cannot distinguish between gravitational accelerations and other kinds of acceleration, or the equality of inertial mass and gravitational mass. A consequence is that gravitational forces can be made to vanish in a small region of spacetime by choosing an appropriate accelerated frame of reference.

prism A wedge-shaped glass used to disperse white light into a spectrum.

Project Cyclops A proposed array of radio telescopes designed to search for communications from extraterrestrial civilizations.

Project Ozma An early attempt to search for radio signals from extraterrestrial civilizations.

prominence Clouds of hydrogen gas lying above the photosphere of the sun near active regions.

proper motion The angular displacement of a star on the celestial sphere due to its motion through space (as opposed to the apparent motion of rising and setting).

protein A long chain of amino acids linked by hydrogen bonds.

protogalaxy A cloud with enough mass that it is destined to collapse gravitationally into a galaxy.

proton A massive, positively charged elementary particle; one of the main constituents of the nucleus of an atom.

proton-proton chain (PP chain) A series of thermonuclear reactions that occur in the interiors of stars, by which four hydrogen nuclei are fused into helium; believed to be the primary mode of energy production in the sun.

protoplanet A large mass formed by the accretion of planetesimals; the final stage in the formation of the planets out of the primeval nebula.

protoplasm The fluid within a cell.

protostar A collapsing mass of gas and dust out of which a star will be born (when thermonuclear reactions turn on) whose energy comes from gravitational contraction.

pulsar A radio source that emits signals in very short, regular bursts; believed to be a highly magnetic, rotating neutron star.

quantum (*pl.* quanta) A discrete packet of energy.

quarter moon The phase of the moon when it is at an elongation of 90° east (first quarter) or 90° west (last quarter).

quasar An intense, pointlike source of light (and sometimes radio waves or X-rays) that is characterized by a large red shift of the lines in its visible spectrum; thought to be related to active galaxies.

quasistellar source See quasar.

quiet sun The solar phenomena not associated with active regions.

quintessence The material of which the celestial spheres were supposed to be made according to Aristotle.

r-process See rapid process.

radar mapping The surveying of the geographical features of a planet's surface by the reflection of radio waves from the surface.

radial velocity The component of relative velocity that lies along the line of sight.

radiant The point on the celestial sphere from which a meteor stream appears to come.

radiation Usually refers to electromagnetic waves, such as light, radio, infrared, X ray, ultraviolet; also sometimes used to refer to atomic particles of high energy, such as electrons (beta radiation), helium nuclei (alpha radiation), etc.

radiation era The time in the early history of the universe, following the light-particle era, when radiation pressure prevents any gravitational condensations from forming.

radiative energy The capacity to do work that is carried by electromagnetic waves.

radio galaxy A galaxy that emits large amounts of radio energy, generally characterized by two giant lobes of emission situated on opposite ends of a line drawn through the nucleus.

radio interferometer An instrument that achieves high angular resolution by combining the signals from at least two radio telescopes.

radio jet A long, narrow region emitting radio energy that appears to be material ejected from a galaxy.

radio-line emission Sharp energy peaks at radio wavelengths, usually caused by low-energy transitions in atoms.

radio recombination line An emission line of an atom, such as hydrogen, in the radio part of the spectrum produced in a transition from one very high energy level to another nearby energy level following recombination of an electron with an ion.

radio telescope An instrument designed to collect and detect radio radiation from space.

radioactive dating A process that determines the age of an object from the amount of decay of radioactive elements within it.

radioactive decay The process by which an element fissions into lighter elements.

rapid process (r-process) The formation of very heavy elements by the rapid addition of neutrons to a nucleus followed by beta decay which takes place during a supernova explosion.

ray A bright streak emanating from a crater on the moon.

recombination The joining of an electron to an ion; the reverse of ionization.

red giant A large, cool star with a high luminosity and a low surface temperature of about 2000 to 3000 K.

red shift Part of the Doppler effect; an increase in the wavelength of the radiation received from a receding celestial body; a shift toward the long-wavelength (red) end of the spectrum.

red variables A class of cool stars variable in light output.

reddening The preferential scattering or absorption of blue light by small particles, allowing more red light to pass directly through.

reference frame A set of coordinates relative to which position and motion may be specified.

reflecting telescope A telescope that has a curved mirror as a primary light gatherer.

reflection The change in direction of a light ray hitting a surface but not penetrating it.

reflection nebula A bright cloud of gas and dust that is visible because of the reflection of starlight by the dust.

refracting telescope A telescope that uses glass lenses to gather light and form an image.

refraction The bending of a light ray when it goes from one medium to another, such as from air to glass.

regular cluster A cluster of galaxies that has a definite symmetry and well-defined center.

relativistic Doppler shift The wavelength shift due to the radial velocity of a source as calculated from the exact formula of special relativity; the result differs from that using the simple Doppler formula only when the velocity is close to that of light.

relativistic jet A beam of gas moving at speeds close to that of light.

relativity See special theory of relativity or general theory of relativity.

resolving power The ability of a telescope to separate close stars or to pick out fine details on celestial objects.

retrograde motion The apparent anomalous westward motion of a planet with respect to the stars, which occurs near the time of opposition (for an outer planet) or inferior conjunction (for an inner planet).

retrograde rotation Rotation from east to west, the opposite of the more common (direct) rotational direction from west to east.

revolution The motion of a body in orbit about another body.

rift valley A depression in the surface of a planet produced by the separation of land masses.

right ascension On the celestial sphere, the angular distance east of the vernal equinox.

ring See planetary ring.

Roche lobe The surface surrounding a double star separating the regions where the gravity of each star dominates.

rotation The turning of a body, such as a planet, on its axis.

rotation curve The relation between rotational velocity in a galaxy and distance from its center.

RR Lyrae star A class of giant, pulsating variable stars with periods of less than 1 day; they are Population II objects found in globular clusters and the galactic halo.

Rydberg constant A number relating the spacing of the energy levels of the hydrogen atom.

s-process *See* slow process.

S-type asteroid An asteroid with an albedo indicative of a silicate surface composition.

Sagittarius A East A small nonthermal radio source near the center of the Galaxy.

Sagittarius A West A thermal radio source (H II region) at the center of the Galaxy.

Sagittarius B A giant molecular cloud near the center of the Galaxy.

Sagittarius arm A segment of one of the spiral arms of the Galaxy that lies about 5000–6500 ly from the center of the Galaxy and appears in the direction of the constellation Sagittarius.

satellite solar power station A satellite designed to collect energy from the sun and transmit power to the earth.

scarp A long, vertical wall running across a flat plain.

scattering That part of extinction due to deflection of photons without absorption or change of wavelength.

Schwarzschild radius The critical size that a mass must reach to be dense enough to trap light by its gravity, that is, to become a black hole.

scientific model A mental picture based on physical principles and aesthetic ideas that attempts to explain what is observed in nature and predict what can be measured.

second of arc 1/3600 of a degree, or 1/60 of a minute of arc.

secondary The fainter of the two stars in a binary system.

secondary atmosphere An atmosphere of a planet formed by outgassing from the interior some time after the planet is formed.

seeing The unsteadiness of the earth's atmosphere that blurs telescopic images.

seismic wave A sound wave traveling through and across the earth that is produced by earthquakes.

seismometer An instrument used to detect earthquakes and moonquakes.

semimajor axis The distance from the center of an ellipse to its farthest point.

SETI An acronym for search for extraterrestrial intelligence.

sexagesimal system A counting system based on the number 60, such as 60 minutes in an hour, or 60 minutes of arc in 1°.

Seyfert galaxy A spiral galaxy with a bright nucleus showing broad emission lines in its spectrum; often a strong radio and infrared source as well.

shepherd satellites The two satellites of Saturn located just inside and outside the F-ring, which confine the F-ring particles to a narrow band.

shield volcano A type of volcano of large extent and shallow slope built up gradually by extrusion of lava.

shock wave A wave moving through a gas faster than the speed of sound, causing a sudden change in pressure and density as it passes by.

sidereal month The period of the moon's revolution around the earth with respect to a fixed direction in space; about 27.3 days.

sidereal period The time interval needed by a celestial body to complete one revolution around another with respect to a fixed direction in space.

signs of the zodiac The twelve equal angular divisions of 30° each into which the ecliptic is divided; each corresponds to a zodiacal constellation.

silicate A compound of silicon and oxygen with other elements, very common in rocks at the earth's surface.

singularity A theoretical point of zero volume and infinite density to which any mass that becomes a black hole must collapse, according to the general theory of relativity.

slow process (s-process) The formation of very heavy elements by the slow addition of neutrons to the nuclei of lighter elements followed by beta decay, a process that takes place in red giant stars.

Small Magellanic Cloud (SMC) *See* Magellanic Clouds.

sodium D line A strong absorption line in the spectrum of the sun and other stars produced by the element sodium.

solar day The time interval from noon to noon.

solar eclipse An eclipse of the sun by the moon, caused by the passage of the moon in front of the sun.

solar flare A sudden burst of energy (visible light, X rays, radio waves) and particles from the surface of the sun in an active region.

solar mass The mass of the sun, about 2×10^{30} kg.

solar material Material from the sun or having similar composition to material in the sun, mainly hydrogen and helium.

solar nebula Gas and dust in a disk around the young sun out of which the planets formed.

solar wind Charged particles, mostly protons and electrons, that escape into the sun's outer atmosphere at high speeds and stream out into the solar system.

solstice The time at which the day or the night is the longest. For the northern hemisphere, the summer solstice (about June 21) is the time of the longest day and the winter solstice (about December 21) is the time of the shortest day; the dates are opposite in the southern hemisphere. Also the positions on the ecliptic farthest north and south of the celestial equator; the location of the sun on June 21 and December 21.

south magnetic pole A point on a star or planet from which the magnetic lines of force emanate and to which the south pole of a compass points.

space A three-dimensional region in which objects exist and events occur and have relative direction and position.

space colony A large space station designed to be self-sufficient for a large group of people.

space velocity The total velocity of an object through space, combining the components of radial velocity and transverse velocity.

spacetime A continuous system of one time coordinate and three space coordinates by which events can be located and described.

spacetime diagram A graph showing the relation of events in space and time.

special theory of relativity Einstein's theory describing the relations between measurements of physical phenomena as viewed by observers who are in relative motion at constant velocities.

spectral class A designation of type of spectrum based on the strengths of various spectral lines; the common classes, in order of decreasing temperature, are O, B, A, F, G, K, and M.

spectral line A particular wavelength of light corresponding to some energy transition in an atom.

spectroscope An instrument for examining spectra.

spectroscopic binary Two stars revolving around a common center of mass that can be identified by periodic variations in the Doppler shift of the lines of their spectra.

spectroscopic parallax A technique for measuring distance by comparing the brightness of stars with their actual luminosities, as determined from their spectra.

spectroscopy The analysis of light by separating it by wavelength (color).

spectrum (*pl.* spectra) The array of colors or wavelengths obtained when light is dispersed, such as by a prism; the amount of energy given off by an object at every different wavelength.

speed The rate of motion; distance traveled per unit time.

spherical (closed) geometry An alternative to Euclidean geometry constructed by G. F. B. Riemann on the premise that no parallel lines can be drawn through a point near a straight line; the sum of the angles of a triangle drawn on a spherical surface is always greater than 180°.

spicule A jet of gas in the sun's chromosphere.

spin angular momentum The angular momentum of a rotating body; proportional to the body's mass, rotational velocity, and radius.

spinar A highly condensed, massive spinning object that may be the energy source for a quasar.

spiral arm A structure, part of a spiral pattern in a galaxy, composed of gas, dust, and young stars, that winds out from near the galaxy's center.

spiral galaxy A galaxy with spiral arms; the presumed shape of the Milky Way Galaxy.

spiral nebula An older term for a galaxy.

spiral tracer An object commonly found in spiral arms and so used to trace spiral structure, for example, Population I cepheids, H II regions, and OB stars.

spontaneous generation The natural origination of living things from lifeless matter.

sporadic meteor A meteor that occurs at random, not associated with a shower.

stadium An ancient Greek unit of length, probably about 0.2 km.

standard (distance) candle An astronomical object of a standard luminosity used to estimate distances.

star counting A technique used by Herschel to determine the extent of the Galaxy.

star model A table of values of the physical characteristics (such as temperature, density, and pressure) as a function of position within a star for a specified mass, chemical composition, and age, calculated from theoretical ideas of the basic physics of stars.

statistical parallax A parallax determined from the average proper motions of groups of stars.

Steady State model A model for evolution of the universe based on the perfect cosmological principle, which states that the universe looks basically the same to all observers at all times.

Stefan-Boltzmann law The relation for a blackbody between temperature (T) and the energy (E) emitted per unit area of its surface ($E = \sigma T^4$, where σ is the Stefan-Boltzmann constant).

stellar spectral sequence A classification scheme for stars based on the strength of various lines in their spectra; the sequence runs OBAFGKM, from hottest to coolest.

stellar wind The outflow of material from a star, similar to the solar wind.

stimulated emission Radiation produced by the effect of a photon stimulating an atom in an excited state to emit another photon of the same wavelength; both photons can stimulate other atoms to emit, and eventually a strong beam of radiation is built up. This is the process that produces a laser or maser.

stony meteorite A type of meteorite made of light silicate materials.

stony-iron meteorite A type of meteorite made of a blend of nickel-iron and silicate materials.

straight line The shortest distance between two points. On a sphere a great circle is a "straight line"; in curved spacetime some other curve may be, in this sense, a "straight line."

GLOSSARY

stratosphere A layer in the earth's atmosphere, above the troposphere and below the mesosphere, in which temperature changes with altitude are small and clouds are rare.

summer solstice See solstice.

sunspot A temporary cool region with a strong magnetic field in the sun's photosphere.

sunspot cycle The 11-year cycle during which the number of sunspots rises and falls.

supercluster A group of clusters of galaxies.

supergiant A massive star of large size and high luminosity.

supergiant elliptical (cD) galaxy The largest type of elliptical galaxy usually found in the center of a rich cluster of galaxies.

superior conjunction The configuration in which an inner planet (Mercury or Venus) lies in the same direction as the sun, but on the opposite side of the sun from the earth.

superluminal motion Motion apparently faster than the speed of light.

supermassive black holes A black hole of 10^6 solar masses or larger, which may exist at the center of some active galaxies or quasars.

supernova An explosion of a massive star, which increases its brightness hundreds of millions of times in a few days.

supernova remnant The expanding gas cloud left over after a supernova explosion, detectable at radio wavelengths long after it has faded from view at optical wavelengths.

supernova-triggered star formation A model in which stars are formed following compression of interstellar gas by the shock wave from a supernova.

surface temperature The temperature at the visible surface of a star.

synchrotron emission Radiation from accelerating charged particles (usually electrons) in a magnetic field. The wavelength of the emitted radiation depends on the strength of the magnetic field and the energy of the charged particles.

synodic month The time between similar configurations of the moon with respect to the sun, as seen from the earth, for example, the time from full moon to full moon.

synodic period The interval between successive similar lineups of a celestial body with the sun, for example, between oppositions.

tachyon Hypothetical particle that travels only faster than light speed.

Tau Ceti A nearby star (11.9 ly distant) similar to the sun, a likely candidate for planetary companions.

temperature A measure of the average speed of the particles in a substance.

temperature gradient The change in temperature in a unit change of distance.

terrestrial material Material having a composition similar to that of the earth, mainly rocky and metallic substances.

terrestrial planets The planets similar in composition to the earth: Mercury, Venus, Mars, the moon.

Tharsis ridge A region on Mars containing a cluster of volcanoes, including Olympus Mons.

thermal equilibrium Steady-state situation characterized by no large-scale temperature changes.

thermal radiation Electromagnetic radiation due to the fact that a body is hot; often characterized by a blackbody spectrum.

thermal velocity The average velocity of random motion of particles in a hot gas.

thermosphere A layer of the earth's atmosphere, above the mesosphere, heated by X rays and ultraviolet radiation from the sun.

threshold temperature See particle threshold temperature.

tidal friction The frictional force of the oceans on the earth as it turns beneath the tidal bulge, which causes the rotational speed to decrease and the day to lengthen.

tidal gravitational force The difference in gravitational force between different points in a body produced by another body, which results in a stretching force, causing the first body to be deformed.

tides The bulging of the oceans in the direction of the moon and the opposite direction due to the moon's tidal force, which causes an apparent rising and falling of the sea level twice each day as the earth rotates beneath the tidal bulge.

time A measure of the flow of events.

Titan Saturn's largest satellite; the first satellite found to have an atmosphere.

Titius-Bode "law" A nonphysical formula that gives an approximate distance of the planets from the sun. It works moderately well for the inner planets, even predicting a planet at the asteroid belt, but fails for the outer planets.

ton (metric) 1000 kilograms.

torque A twisting force, which acts to change a body's rotation.

transition (in an atom) A change in the electron arrangements in an atom, which involves a change in energy.

transverse fault A crack on the surface of a planet or satellite where the ground has moved sideways.

transverse velocity The component of velocity perpendicular to the line of sight.

transverse wave A wave in which the oscillatory motion is perpendicular to the direction of propagation. Transverse sound waves cannot travel through liquids.

Trapezium cluster The group of O and B stars at the center of the Orion Nebula.

trigonometric parallax A method of determining distances by measuring the angular position of an object as seen from the ends of a baseline having known length.

triple-alpha reaction A thermonuclear process in which three helium atoms (alpha particles) are fused into one carbon nucleus.

Triton Largest satellite of Neptune.

troposphere The lowest level of the earth's atmosphere, reaching 8 to 10 km from the surface, in which most of the weather takes place.

Tully-Fisher relation An observationally discovered relation between luminosity of a spiral galaxy and the width of its 21-cm hydrogen emission line.

turbulence Irregular and sometimes violent convective motion.

turnoff point That point on the H-R diagram of a cluster at which the main sequence appears to terminate at the high luminosity end.

T Tauri star A class of stars which are variable stars of spectral class G, K, and M still in their pre-main-sequence contraction stage.

21-cm line The emission line, at 21.15 cm wavelength, of neutral hydrogen gas produced by atoms changing direction of spin of their proton and electron from parallel to antiparallel; observations at 21 cm are used to map the spiral structure of the Galaxy.

two-sphere universe The basic premise of the celestial coordinate systems, that the universe is composed of two concentric spheres, the earth and the celestial sphere.

Type I, II, III civilizations Kardeshev's classification of extraterrestrial civilizations according to their ability to utilize the energy resources of their planet, star, and galaxy, respectively.

Type I, Type II supernovas A classification of supernovas by their spectral characteristics and light curves. Type II supernovas belong to Population I and Type I supernovas belong to Population II.

UFO An acronym for unidentified flying object.

uniformity of nature The assumption that objects in one part of the universe are pretty much the same as in another part, for example, that the cepheids in other galaxies are the same as those in the Milky Way Galaxy.

universal law of gravitation See gravitation.

universality of physical laws The assumption that the physical laws understood locally apply throughout the universe and perhaps to the universe as a whole.

universe The totality of all space and time; all that is, has been, and will be.

upland plateau A relatively uncratered region on the northern part of the surface of Venus.

Valles Marineris A large canyon on the surface of Mars.

Van Allen radiation belts Belts of charged particles around the earth trapped by its magnetic field.

vector A quantity that expresses magnitude and direction.

velocity The rate and direction at which distance is covered in some interval of time.

vernal equinox See equinox.

vertical circle On the celestial sphere, any great circle through the zenith.

Very Large Array (VLA) A large collection of radio telescopes in New Mexico operated together to obtain very high angular resolution on radio sources.

Virgo cluster of galaxies One of the nearest large clusters of galaxies.

visual binary Two stars that revolve around a common center of mass, both of which can be seen through a telescope so that their orbits can be plotted.

volatile A material, such as helium or methane, that vaporizes at low temperatures.

volcanic theory The theory of the formation of craters from cones left over from lava eruptions.

von Neuman probe A self-reproducing machine designed for exploration of the Galaxy.

watt A unit of power; 1 joule expended per second.

wavelength The distance between two successive peaks or troughs of a wave.

weight The total force on some mass produced by gravity.

white dwarf A small, dense star that has exhausted its nuclear fuel and shines from residual heat. Such stars have an upper mass limit of 1.4 solar masses and their interior is a degenerate electron gas.

white hole A singularity from which photons can emerge; a time-reversed black hole.

Widmanstätten figures Large crystal patterns that appear on the surface of iron meteorites when they are polished and etched.

Wien's law The relation between wavelength of maximum emission in a blackbody spectrum and temperature.

winter solstice See solstice.

wormhole See Einstein-Rosen bridge.

X rays High-energy electromagnetic radiation, of wavelength smaller than ultraviolet but larger than gamma-ray radiation.

X-ray burster An X-ray source which emits brief, but pow-

GLOSSARY

erful bursts of radiation; probably a neutron star in a close binary system.

young Population I stars A class of stars which are found in the disk of a spiral galaxy, especially in the spiral arms, have a heavy metal abundance similar to the suns, and are the youngest stars.

ZAMS An acronym for zero-age main sequence.

zenith The point on the celestial sphere that is located directly above the observer at 90° angular distance from the horizon.

zero-age main sequence (ZAMS) The position on the H-R diagram reached by a star once it derives most of its energy from thermonuclear reactions rather than from gravitational contraction.

zodiac The twelve constellations through which the sun travels in its yearly motion, as seen from the earth.

zodiacal light A glow along the ecliptic produced by reflection and scattering of sunlight by dust and gas particles in the solar system.

zone A light band in the atmosphere of 2 Jovian planets marking rising gases of higher pressure (*see* belt).

zone of avoidance A region near the plane of the Galaxy where very few other galaxies are visible because of obscuration by dust.

INDEX

Absolute magnitude, 260
   and cepheids, 715–716
   and distances to galaxies, 769, 781
   and distances to stars, 281
   and galactic clusters, 383–384
   and Hertzsprung-Russell diagram, 276–281
   and luminosity, 260
Absorption lines. *See* Spectra, dark line
Acceleration, 97
   centripetal, 108–110, 112–113
   and general relativity, 134–136
   gravity at earth's surface, 100, 111
Accretion, 682
Accretion atmosphere, 522
Accretion disk, 413, 837
Active regions of sun, 208, 217, 219
Adams, John C., 122
Aerobe, 894
Age
   of earth, 503, 506
   of lunar samples, 550
   of meteorites, 672–673
   of star clusters, 388
   of sun, 480
   of universe, 148
Aggregation of the planets, 689
Albedo, 515, 597, 656
Aldebaran, 268, 274
Alfonsine Tables, 74, 85
Algol, 289
Almagest, 47, 56
Almalthea, 621

Alpha Centauri, 276, 913
Altair, 277
Amino acids, 886, 898–899
Anaerobe, 896
Anasazi, 25–28
Andromeda Galaxy (Messier 31), 720, 769, 790–792, 793, 841
Angstrom (Å), 178
Angular measurement, 6
Angular momentum, 339
   and collapse of interstellar clouds, 339–341
   and distribution in solar system, 682, 685, 694
   and earth–moon system, 339–340
   and galaxies, 760–761, 875
Angular separation, 6
Angular speed, 9
Anorthrosites, 547
Antapex, 709
Antares, 266
Antimatter, 862–864
Apex, 709
Aphelion, 83, 558
Aphrodite Terra, 579–580
Apogee, 535
Apollo missions, 547
   and moon's age, 550
   and moon's history, 551–554
   and moon's interior, 550–551
   and moon's origin, 555
   and moon's surface, 547
Apparent magnitude, 255–257
Aquarius, 11

Aquila, 11, 700–701
Arcminutes, 6
Arcseconds, 6
Arecibo, 931
Argon, 481
Aries, 11, 15
Aristarchus, 36, 43
Aristarchus crater, 546
Aristotle, 39, 56
  de Caelo, 39
  and physics, 39–40
Asteroidal moons, 657
  of Jupiter, 621
Asteroid belt, 653–655
Asteroids, 651, 674
  formation of, 692
  orbits of, 653
  physical features of, 655
  sizes of, 655
Astrology, 36
Astronomia nova, 85
Astronomical distances, 956
Astronomical Unit (AU), 71, 165. *See also* Earth–sun distance
Astronomy, ancient, 5–6, 34–56
Atmosphere of earth, 508–514
Atmosphere of pressure, 509
Atmospheric escape, 512
Atmospheric extinction and reddening, 515
Atmospheric gases, 958
Atoms, 172
  and energy levels, 182
  excited, 183
  ionized, 183
  and matter, 172
  and spectra, 186
Auroras, 215, 219
  of Jupiter, 613
Autumnal equinox, 11

Baade, Walter, 721
Babylonian astronomy, 5, 34–36
Balmer, Johannes, 180
Balmer series, 180, 183, 269, 272, 279, 823
  and classification of stars, 272
B and V magnitudes, 262
Bandwidth, 930
Barnard's star, 912
Barred galaxies, 772
Barringer Crater, 670
Basalt, 503
Basins
  of earth, 517–519
  of moon, 544
Becklin-Neugebauer object, 325, 329, 345
Beta decay, 442. *See also* Inverse beta decay

Beta Region, 579–580
Betelgeuse, 249, 263–264, 274–275, 277
Big Bang, 859, 867, 914, 943
Big-bang model of the universe, 846, 853
  primeval fireball, 858, 860
Binary accretion model for the moon, 555
Binary galaxies, 785
Binary radio pulsars, 437
Binary stars, 122, 281–282, 284
  eclipsing, 289–290
  and mass-luminosity relation, 293–294
  and neutron stars, 439, 441, 464
  and novas, 412–413
  spectroscopic, 284, 287–289
  visual, 282, 284
Binary X-ray sources, 458, 461
  evolution of, 465–466
Biochemistry, 885
Biological evolution, 884, 914
Bitter Springs Formation, 891
Blackbody radiation, 189
  and big bang, 859
  and stars, 263–264, 266–267
  and sun, 189–191
  and thermal radiation, 424
Blackbody spectrum, 857
Black dwarf stars, 374
Black holes, 447, 451, 944
  and binary X-ray sources, 458, 470
  and Centaurus X-3, 463
  and Cygnus X-1, 461
  observing, 458
  and Schwarzschild radius, 449–450
  supermassive in Galaxy, 808, 837
  supermassive in quasars, 834
  theoretical properties, 455–458
BL Lacertae objects, 822, 835
Bohr, Neils, 182
Bohr model of atom, 182–186
Bolometer, 242
Bolometric magnitude, 261
Bound-bound transition, 186
Bound-free transition, 186
Brackett-Alpha line, 345
Brahe, Tycho, 74, 226
  and Kepler, 81
  model of the cosmos, 77
Breccias, 547
Bright nebulas, 301–305. *See also* H II regions
B-stars, 476, 758, 762

Callisto, 619
Caloris basin, 563–564
Canals of Mars, 589–590
Cancer, 11
Canis Major, 479

Cannon, Annie Jump, 270
Capricornus, 11
Capture model for the moon, 555
Carbonaceous chondrite, 671
Carbon dioxide, atmospheric, 523. *See also* Greenhouse effect
Carbon–nitrogen–oxygen cycle (CNO cycle), 203, 364, 375
Cassegrain telescope, 231
Cassini's division, 630
Catastrophic models for origin of solar system, 680, 682
Celestial coordinate systems, 963–965
Celestial pole, 7, 14
Celestial sphere, 7
Cell, 887
Centaurus A, 812, 817
　nucleus, 815
Centaurus X-3, 463
Center of mass, 114, 120, 283, 539
Central force, 107
Centripetal acceleration, 108–109, 112
Cepheids, 393, 715, 776
　and RR Lyrae stars, 716
　Type I and Type II, 716
Cepheid variables. *See* Cepheids
Ceres, 652
Chaco Canyon, 25–28, 425
Chamberlain, Thomas C., 682
Chandrasekhar limit, 403
Charon, 496–497, 643
Chemical condensation sequence, 686–689
Chemical evolution, 884, 914
Chemical properties of solar system, 678
Chiron, 656
Chondrites, 671
Condrules, 671, 693
Chromosphere of sun, 197
Circular orbit, 151
Circumpolar stars, 7
Closed geometry. *See* Spherical geometry
Closed universe, 153–154, 943
Coefficient of opacity, 361
Collisional excitation, 187
Color index, 262
Color temperature of a star, 264–265
Coma cluster of galaxies, 794, 796
Coma of comets, 661
Comet Halley. *See* Halley's comet
Comet Ikeya-Seki, 662
Comet Tempel, 668
Comet West, 662
Comets, 651, 658, 674
　cloud, 664
　dirty-iceberg model, 662, 669, 674
　orbits, 658
　physical characteristics, 660
Communication with extraterrestrial intelligence (CETI), 918, 937

Compact radio galaxies, 809, 811
Compounds, chemical, 172
Condensation, 682
Conduction, 207
Conjunction, 17–18, 61
Conservation of angular momentum, 340, 483, 536, 555
Conservation of energy, 168
Constellations, 5, 28
Continental drift, 518–520, 526
Continuous spectrum, 173
Contour maps, 238
Convection, 199, 207
Coorbital satellites, 624
Copernican model, 61, 123. *See also* Heliocentric model of the universe
　relative distances of planets, 70
　retrograde motion, 65–67
Copernicus, Nicolas, 60
　model of universe, 63, 88
Core
　of earth, 502
　of Jupiter, 611
　of Mars, 585
　of Mercury, 561
　of moon, 550
　of Saturn, 623
　of sun, 204, 218
　of Venus, 573
Core-mantle grain models, 329
Coronal holes, 216
Coronal interstellar gas, 308
Corona of sun, 199
Cosmic background radiation, 858
Cosmic blackbody microwave radiation, 855, 857, 878
Cosmic dust, 320, 330
Cosmic fireball, 858, 860. *See also* Big-bang model of the universe
Cosmic nucleosynthesis, 861
Cosmic rays, 319, 735, 895
Cosmological models, 55, 102. *See also* Big-bang model of the universe; Steady-state model
Cosmological principle, 851
Cosmology, 846–847
　Babylonian, 34
　of Copernicus, 61–63
　contemporary, 846, 851
　Greek, 36
　of Newton, 123
　of Ptolemy, 48–50
Cosmos, 38
　of Aristotle, 39
　Babylonian, 34
　of Eudoxus, 39
　expanding, 144
　heliocentric, 43–44, 61
　homogenous, 849

Cosmos (*Continued*)
  isotropic, 849
  of Ptolemy, 54, 56
Crab Nebula, 417, 425
  pulsar, 427, 432
  synchrotron process, 426
Cratered terrain
  of Callistro, 619
  of Ganymede, 617–618
  of Mars, 596
  of Mercury, 562–563
  of moon, 544, 547
  of Saturn's moons, 626–629
  of Venus, 577, 580
Cratering, 557
Craters, 544, 627
Critical density, 152–153
Crust of the earth, 502, 517
C-type asteroid, 656
Curtis, Heber D., 720
Cygnus, 600
Cygnus A, 812, 817
Cygnus arm, 742, 761
Cygnus X-1, 461, 470, 700

Daily motion
  of moon, 13
  of planets, 16
  of stars, 7
  of sun, 8–9
Dark clouds, 349
Dark nebulas, 321
Darwin, Charles, 887
DA white dwarfs, 404
Day, solar, 8
  on Mercury, 560
  on moon, 538
  on Venus, 572
De caelo. *See* Aristotle
Deceleration parameter, 155
Declination, 965
Decoupling, 866, 870
Deferents, 45
Deflection of starlight by the sun, 140
Degenerate gas, 371–372, 401–402
Degrees of angular measure, 6
Deimos, 597, 656
Delta Cephei, 714
De Magnete, 82
Density
  of earth, 501
  of moon, 540
  of planets, 486–487, 489
  of stars, 293
  of sun, 167–169
  of universe, 152–154
Density-wave model, 740, 755, 758

Deoxyribonucleic acid (DNA), 887
De revolutionibus, 60, 62, 78, 101
Detector for telescope, 233–234
Deuterium
  abundance of, 869
  in Big Bang, 866
Deuterium-hydrogen ratio, 683
Differential rotation, 745
Diffuse nebulas. *See* Bright nebulas; H II regions
Dione, 626–627
Dipole fields, 210
Disk (galactic), 722–723, 736
Distance pyramid, 775
D-lines of sodium, 174, 177
Doppler effect or shift, 285, 405, 462, 559, 571, 635, 752, 819, 824
Double quasar, 830
Doubling time, 948
Draco, 15
Drake Equation, 907, 911, 919
Dust. *See* Cosmic dust
Dust tails of comets, 661
Dwarf stars. *See* Main-sequence stars
Dynamic properties, 679
Dynamo model, 506

Earth
  age of, 503, 506
  atmosphere of, 486, 508, 514, 520
  crust of, 502, 517
  density, mass, size, 486
  energy sources in primitive, 894
  evolution of, 524
  future of, 947
  interior of, 502, 504–505
  magnetic field, 506
  mass and density, 500
  oceans of, 520
  revolution, 118
  rotation period, 117, 486
  secondary atmosphere, 893
  surface temperature of, 523
Earth–moon distance, 532
Earth–moon system, 539
Earthquakes, 517
Earth–sun distance, 46. *See also* Astronomical Unit
Eccentricity, 83
Eccentrics, 45, 50
Eclipses, 19–22, 48
  lunar, 20
  solar, 19
Eclipsing binary stars, 289–290
Eclipsing X-ray binary system, 459
Ecliptic, 9–10, 21, 28
Ecosphere, 908
  of sun, 902
Effective temperature, 267

Einstein, Albert, 128, 131
   gravitation, 133, 143, 156
   special theory of relativity, 131
Einstein-Rosen bridge, 458
Electromagnetic fields. *See* Fields, magnetic
Electromagnetic spectrum, 178
Electron, 172
   transitions of, 182
Elements, 172, 503
Ellipses, 82
Elliptical
   galaxies, 770
   orbit, 82, 151
Elongation, 17
Emission lines, 174. *See also* Spectra, bright line
Enceladus, 627
Energy, 168–169, 171
   absorption, 183
   emission, 183
   levels in atoms, 182
   sources in primitive earth, 894
   transport in solar interior, 207
Enuma Elish, 35, 56
Enzymes, 886
Epicycles, 45–48, 65, 72
Epsilon Eridani, 912, 914
Equant, 50, 61, 63
   point, 51
Equation of radiative transfer, 361
Equation of state, 358, 360
Equatorial coordinate system, A-10–A-11
Equinoxes, 11
   precession of, 14, 120
Equivalence, principle of, 134
Eratosthenes, 42, 46
Escape velocity, 150–152
Euclid, 130
Eudoxus, 39, 55
Europa, 493, 617
Evening "star," 17
Event, 133
Event horizon, 456
Evolution, 885, 914
   earth, 517–524
   Mars, 596
   Mercury, 565
   moon, 551, 557, 565
   stars, 354–393
   terrestrial planets, 557
Evolutionary track, 359
Exosphere, 512
Expanding arm (3-kpc arm), 743, 761
Exponential decay, 482
Exponential growth, 948
Extended radio galaxies, 809
Extinction,
   atmospheric, 515
   interstellar, 321, 323
Extraterrestrial life, 901–911, 918–923
Eyepiece of telescope, 229

Fajada butte, 26
Falling stars. *See* Meteors
Fields, magnetic, 209–212
   of earth, 506
   of moon, 551
   of sunspots, 209, 219
Fig Tree formation, 889
Fission model for the origin of the moon, 554
Flares, solar, 213, 219
Fluorescence, 302
Flux. *See also* Apparent magnitude
   of sun, 188
   of stars, 249–250, 255
Focal length, 228
Forced motion, 39, 99
   and Aristotle, 99
   and Einstein, 134
   and Galileo, 99
   and Kepler, 80–81
   and Newton, 107–115
Force Law. *See* Newton's Laws of Motion
Fornax cluster of galaxies, 794
Foucault, Jean, 117
Fraunhofer, Joseph von, 175
Fraunhofer lines, 175. *See also* Spectra, dark line
Free-bound transition, 303
Free-fall gravitational collapse, 332–333
Free-free absorption and emission, 187
Free-free process, 304
Frequency, 175
Fusion reactions, 203, 363–367

Galactic cannibalism, 799
Galactic clusters, 703. *See also* Open clusters
Galactic latitude, 751
Galactic longitude, 743
Galactic rotation curve, 748, 752
Galaxian luminosity function, 796
Galaxies
   active, 808, 822, 834, 876
   clusters, 790–798
   colors of, 788
   distances to, 775
   elliptical, 770
   and expansion of universe, 144
   formation of, 876
   gas and dust, 789
   head-tail, 815
   irregular, 772
   lenticular, 772
   luminosity classes of, 772
   mass of, 784
   radial velocity of, 778

Galaxies (*Continued*)
  radio, 809
  red shift of, 778
  ring, 774
  rotation curve of, 784, 786
  size of, 783
  spiral, 750, 771
  structure of, 750–751
  turbulence and growth of, 874
Galaxy. *See* Milky Way Galaxy
Galilei, Galileo, 92–103, 123
  description of terrestrial motion, 102
  Dialogue on the Two Chief World Systems, 101
  Letters on Sunspots, 95
  telescopic observations, 93–95, 230, 616, 702
Galle, Johann, 122
Gamma rays, 245, 807
Ganymede, 617
Gases
  degenerate, 371–372, 401–402
  number density, 170
  opacity of, 193
  ordinary, 170
  pressure, 170
  temperature, 170
Gas tails of comets, 661
Gemini, 11
General theory of relativity, 131, 133
Geocentric, 7, 19
Geocentric model of the universe, 48–52, 76
Geometry, 137–140
  Euclidian, 137
  flat, 130, 137, 868
  and Hubble's diagram, 154
  hyperbolic, 128, 137, 142, 145, 152, 157, 868
  of spacetime, 136
  and the universe, 143, 157, 868
Giant molecular clouds, 312
Gibbous phase, 13
Gigahertz (GHz), 178
Globular clusters, 382, 385, 469, 703, 734
  distribution in the sky, 717
Gnomon, 9
Gondwanaland, 518
Grains. *See* Cosmic dust
Granite, 503
Gravitation, 143, 156
Gravitational
  collapse, 332, 682
  instability, 870. *See also* Jeans length
  lens effect, 832
  mass, 134
  potential energy, 201
  red shift, 405, 408
Gravity, 107–109
Great circle, 130
Great nebula in Andromeda, 793. *See also* Andromeda galaxy

Great Red Spot, 494, 607
Great Rift, 700
Greek astronomy, 32, 36–39
Greenhouse effect, 516. *See also* Carbon dioxide, atmospheric
Grooved terrain, 617
Ground state of atoms, 183
Gum Nebula, 421
Gunflint fossils, 890

Hale, George, E., 209
Half-life, 480
Halley, Edmund, 104, 706
Halley's comet, 121, 658, 665
Halo, galactic, 722, 734, 736
H-alpha line, 303
Harmonic law. *See* Kepler's third law
Harmony of the spheres, 38, 78
Harvard spectral classification, 271
Heavy elements, 394–395
Heavy-particle era, 860, 864
Heliocentric, 60
Heliocentric model of the universe, 43, 63. *See also* Copernican model
Heliocentric stellar parallax, 72, 257, 259, 706
Helium, 196, 203
  in Big Bang, 866–867
  in stars, 274–275
  in sun, 196
  and triple-alpha reaction, 366
Helium flash, 372, 469
Herschel, Caroline, 122
Herschel, William, 122
  discovery of Uranus, 635
  surveying Milky Way, 704
Hertz (Hz), 178
Hertzsprung-Russell diagram, 275–277, 296
  and ages of clusters, 388
  and stellar evolution, 354, 380
  and time, 356
Heterosphere, 512
High-velocity stars, 749–750
Hipparchus, 44–45, 255
Homogenous universe, 849, 880
Homosphere, 512
Horizon coordinate system, 964
Horizontal branch, 380, 385. *See also* Hertzsprung-Russell diagram
H I region, 316
Horoscope, 36
Horsehead Nebula, 321
H II regions, 302–306, 316, 750. *See also* Bright Nebulas; Nebulas
Hubble, Edwin M., 144, 770
Hubble constant, 145, 148, 853
  Aaronson-Huchra-Mould procedure to, 782
  Sandage-Tammann procedure to, 781
Hubble diagram, 154

INDEX

Hubble law, 144–149, 157, 778, 780
Humankind, future of, 948
Huygens, Christian, 584
Hydrogen, 203, 357
   corona of earth, 512
Hydrostatic equilibrium, 336, 358, 360, 368
Hydroxyl maser, 314–315
Hyperbolic geometry, 128, 137, 142, 145, 152, 157, 868
Hyperion, 627

Iapetus, 626
Icarus, 655
Igneous rocks, 503
Images, 227
Impact theory for crater formation, 591
Inertia, 99
Inertial law. *See* Newton's laws of motion
Inertial mass, 134
Inferior conjunction, 572
Infrared astronomy, 242, 729, 806, 821
Infrared radiation, 243
Instability strip, 393
Intercloud gas, 308
Intercloud medium, 317
Intercrater plains, 562
Intergalactic matter, 800
Interplanetary debris, 651
Interplanetary dust, 674
Interplanetary gas, 651, 673–674
Interstellar atoms, 306
Interstellar clouds, 307–308
Interstellar dust, 320, 352
   and formation of molecules, 330
   nature of, 327
   observations of, 324
Interstellar extinction curve, 328
Interstellar gas, 300, 352
Interstellar hydrogen atoms, 307
Interstellar medium, 299, 317, 724, 736
Interstellar molecules, 310–311
Interstellar probes, 926
Interstellar sodium, 307
Inverse beta decay, 407. *See also* Beta decay
Inverse square law for gravity, 107
Inverse square law for light, 254
Io, 493, 613, 615
Ion, 173
Ionosphere, 511
Iron meteorites, 671–672
Irregular galaxies, 772
Ishtar Terra, 577, 579
Isotope, 172, 479
Isotropy, 857–858, 878

Jansky, Karl, 235
Jeans, James, 871
Jeans length, 871. *See also* Gravitational, instability

Jovian planets, 483, 485, 603, 905
Jupiter, 494, 605–614, 906
   atmosphere of, 487, 605, 609, 906
   auroras of, 613
   belts, 605
   composition of, 605
   density, mass, size, 487, 605
   diameter of, 605
   formation of, 688–690
   interior of, 608
   magnetosphere, 611
   physical characteristics, 605
   radio emission from, 611
   retrograde motion, 67
   rings of, 621
   rotation of, 606
   rotation period, 487
   satellites of, 94, 487, 613–621, 961
   zones, 605

Kant, Immanuel, 681
Kapteyn, Jacobus C., 713
Kapteyn universe, 713
Kepler, Johannes, 36, 63, 78, 87
Kepler's laws, 83–85, 165–166, 831
Kepler's model for solar system, 88
Kepler's second law, 83–84, 340
Kepler's third law, 84, 114, 116, 282, 292, 459, 488, 643, 745, 747
   and sun's mass, 118, 166
Kilohertz (kHz), 178
Kinetic energy, 168, 187
Kirchhoff's Rules, 175
Kleinmann-Low nebula, 325, 345
KREEP, 548, 550

Laplace, Marquis de, 681
Large Magellanic Cloud, 792
Latitude
   on earth, 963
   galactic, 751
Laurasia, 518
Laws of motion. *See* Newton's laws of motion
Leavitt, Henrietta, 270, 714
Lenses, 228
Lenticular galaxies, 772
Leo, 11
Libra, 11
Life, 884–885
Light, 926
Light curve, 714
   of a cepheid, 714–715
   of an eclipsing binary, 289–290
   of a nova, 409
   of Type I and Type II supernovas, 418
Light-gathering power, 231
Lighthouse model for pulsars, 435

Light-particle era, 860, 865
Light rays, 227
Light year, 144
Linblad, Bertil, 743
Local group of galaxies, 790–791, 793
Local supercluster, 797
Longitude
  on earth, 963
  galactic, 743
Longitudinal waves, 504
Long-period comets, 659
Loop Nebula, 421
Lowell, Percival, 589, 641
Low-energy cosmic rays, 320
Low-velocity stars, 749–750
Luminosity, 188, 252, 787. See also Absolute magnitude
  classes, 280, 772
  function, 723
Lunar day, 538
Lunar eclipse, 20
Lunar highlands, 544, 553
Lunar lowlands, 544, 553
Lunar occulations and angular diameters of stars, 267
Lunar samples, age of, 550
Lunar soil, 547
Lyman series, 183, 185, 270
Lyra, 15, 709

Maestlin, Michael, 81
Magellanic clouds, 713, 792
Magnetic fields and forces. See Fields, magnetic
Magnetic flux, 406
Magnetic neutron stars, 464
Magnetic poles, 506
Magnetosphere
  of earth, 508
  of Jupiter, 611
  of a pulsar, 436
  of Saturn, 634–635
Magnifying power, 232–233
Magnitudes, 255. See also Absolute magnitude; Apparent magnitude
Main-sequence stars, 276, 278, 296, 367–369
Mantle of the earth, 502
Mare basalts, 547
Mare Imbrium, 533, 544
Mare Orientale, 544, 552, 563
Maria, lunar, 543, 552
Mariner missions, 558, 561
Mars, 492, 904
  arroyos of, 593
  atmosphere of, 486, 586, 904
  canals, 589
  day, 584
  density, mass, size, 486, 584
  dust storms, 589
  evolutionary comparison with Venus and earth, 599
  evolution of, 596
  magnetic field of, 585
  orbit, 584
  physical characteristics, 583–586
  polar caps, 589
  retrograde motion, 14–16
  rotation period, 486
  sands of, 588
  satellites of, 486, 597, A-4
  seasons, 584
  southern hemisphere, 596
  surface of, 588, 591–592
  surface pressure of, 586
  surface temperature of, 586
  volcanoes, 594, 904
Mascon, 544
Maser, 315
  pumping, 315
  silicon monoxide, 316
  water, 316
Mass, 111
Massive protostellar collapse, 337
  signposts for birth of, 342
Mass-luminosity law for stars, 291–293
Mass-luminosity ratio of galaxies, 788
Mass-luminosity relation of cepheids, 713–716
Mathematical Composition of Claudius Ptolemy. See *Almagest*
Mathematical Principles of Natural Philosophy, 105
Matter, 850
Matter–antimatter annihilation, 862
Matter era, 861
Maximum elongation, 17
Maxwell Montes, 579
Mean parallaxes. See Statistical parallaxes
Mechanics, 97
Megahertz (MHz), 178
Megaparsec (Mpc), 145
Mercury, 490, 902
  atmosphere of, 486, 562, 566
  compared with moon, 563
  density, mass, size, 486, 561
  evolution of, 565
  interior of, 561
  magnetic field of, 566
  orbit of, 558
  physical characteristics, 557
  precession of orbit, 142
  retrograde motion, 17
  revolution, 486, 902
  rotation period, 486, 558–559, 902
  size, mass and density, 557
  solar day, 560
  surface, 561–562
  surface pressure, 562
  surface temperature, 561

Mesosphere, 509
Messier 13, 938
Messier 17, 345, 347
Messier 31. *See* Andromeda Galaxy
Messier 33, 793
Messier 87, 811
Metallic hydrogen, 611
Meteorites, 479, 665–666, 674, 692
   falls and finds, 669
   physical nature, 671
Meteoroids, 651, 666–667, 674
   physical characteristics, 667
Meteors, 665, 674
   streams and showers, 667–668
   swarm, 668
Meter–kilogram–second system (MKS), A-1
Micrometeorite, 666
Microwaves, 926
Microwave window, 927
Mid-oceanic ridge, 517
Milky Way Galaxy, 307, 669, 750
   black hole in core, 733
   CO distribution in, 725
   contents of, 703
   discovery of, 713
   disk of, 723, 740
   evolution of, 739, 759
   formation of, 760, 762, 875
   future of, 946
   halo of, 740
   and life, 906
   mass of, 747
   nucleus, 726, 740, 806–808
   radio emission from center, 726–728
   rotation of, 743
   structure of, 704, 740, 750, 761
   X rays from center, 731
Miller-Urey experiment, 896
Mimas, 627
Minkowski, Hermann, 133
Minutes of arc, 6
Miranda, 637
Models
   scientific, 32–33, 54–55, 86, 225
   of solar system formation, 680
   of universe, 942–945
Molecular clouds, 312, 317
   giant, 312
   and hydrogen gas, 724
Molecular masers, 314
Molecule, 172
Moon. *See also* Apollo missions
   age of, 550
   atmosphere of, 540, 566
   compared with Mercury, 565–566
   crescent, 13

   evolution of, 551, 557, 565
   gibbous, 13
   history of, 551–553
   interior of, 550
   magnetic field of, 551
   new, 13
   orbit and tides, 535
   orbit around earth, 532
   origin of, 554
   phases of, 13
   quarter, 13
   size and mass of, 538–539
   surface gravity, 540
   surface of, 543, 557
   surface temperature, 541
Moonquakes, 550
Morning "star," 17
Motions
   of moon, 12
   of planets, 13
   of stars, 7
   of sun, 8
Moulton, Forest, R., 682
Mountains of earth, 517
M-stars, 278
M-type asteroid, 656
Murchison meteorite, 899
Music of the spheres, 37–39
Mutations, 885
Mysterium cosmographicum, 81

Natal astrology. *See* Astrology
Natural motion, 39, 99, 129, 136
Natural selection, 885, 889
Nebular models for origin of solar system, 680–681, 684, 693
Nebulas, 322, 703. *See also* H II regions
   bright, 301–305
   dark, 321
   planetary, 394
Negative hydrogen ion, 193
Neighboring solar systems, 911–914
Neptune, 121, 496, 639
   atmosphere of, 487, 639
   density, mass, size, 487
   rotation period of, 487, 639
   satellites, 487, 640
Nereid, 640
Neutral hydrogen gas, 724
Neutrino, 204, 877
Neutron, 172
Neutron stars, 407–408, 428, 439, 464
Newton, Isaac, 92, 103, 123
   discovery of Neptune, 121
   and gravitation, 143, 156
   and moon-apple experiment, 103, 107, 109
Newtonian reflector, 231

Newton's law of gravitation, 108–113, 118–123, 166
Newton's laws of motion, 105–106
Nodes, 20
Noon, 8–9, 26
Norma arm, 743, 761
North Magnetic Pole of earth, 506
North Pole Star, 7. See also Polaris
Nova, 409–416
   distances to, 411
   light curve of, 409
   model of, 412
Nova Cygni 1975, 413
Nova Hercules, 408, 410
Nuclear bulge, galactic, 722, 736
Nuclear fission, 203
Nuclear fusion, 203
Nuclear jets of galaxies, 811–812, 818, 835
Nucleic acids, 887–888
Nucleosynthesis, 367, 441
   in Big Bang, 864–866
   in stars, 394
Nucleus
   atomic, 172
   of comet, 661
   of Milky Way, 726–734

OB associations, 348, 386, 703
Oberon, 637
Objective of telescope, 229
Observational astronomy, 226–245
Observations and Ptolemaic model, 52, 226
OB subgroups, 348
Oceans of the earth, 520
Olbers, Heinrich, 847
Olbers' paradox, 847
Old population I stars, 722, 726, 729
Olympus Mons, 594
On the Revolutions of the Heavenly Spheres. See *De revolutionibus*
Oort's cloud of comets, 664
Oort's constant, 746, 753
Opacity, 193–194, 359, 368
Open clusters, 382, 703
   distance to, 384
   and dust, 320
Open geometry. See Hyperbolic geometry
Open universe, 942
Ophiucus, 351
Opposition, 18
Optical astronomy, 229–233, 752
Optical telescopes, 226–231
Optics, 227
Organisms, 885
Origin of Species, 888
Orion, 5, 249
Orion Nebula, 301, 304, 312, 314, 325, 343, 345

Orion spur, 742
Oscillating model of universe, 945
O-stars, 476, 758, 762
Oxygen, 509
Ozone layer, 511
Ozonosphere, 511

Pangaea, 518
Parallax, 257
   heliocentric, 40, 257, 259
   spectroscopic, 279–280, 706
   statistical, 710, 713
   stellar, 40
   trigonometric, 257
Parent meteor bodies, 672
Parsec, 144, 259
Particle threshold temperature, 862
Paschen series, 186
Pauli exclusion principle, 372
Pendulum, 117
Penzias, Arno, 855
Perfect cosmological principal. See Steady state model
Perigee, 532
Perihelion, 83, 558
Periodic variable, 393, 714. See also Cepheids
Period-luminosity relationship, 715, 769, 776
Perseus arm, 742, 761
Phase of moon, 12–13
Phobos, 597, 656
Photodissociation, 599
Photons, 181, 219, 862. See also Light
   excitation, 187
Photosphere of sun, 194–196, 219
Photosynthesis, 885, 889
Physical evolution, 884, 914
Pioneer missions, 625, 633–634, 920
Pisces, 11
Planck, Max, 181
Planck curve, 263, 265
Planck's constant, 169, 181
Planetary
   atmospheres, 489
   density, 489
   distances and diameters, 488
   motion, 490
Planetary nebulas, 373
Planetesimals, 483–484, 526, 682, 685, 690, 692, 694
Planetoid, 655
Planets, 13
   aggregation of, 689
   data about, 486–487, 957–959
   formation of, 482
   mass of, 488, 958
   motion of, 13
   relative distance, 18
Plate tectonics, 519, 527

INDEX

Plato, 39, 55
Pleiades, 497
Pluto, 641
  atmosphere of, 487, 644
  density, mass, size, 487, 642–644
  rotation period, 487, 643
  satellite of (Charon), 487, 643
Polaris, 7, 14, 24, 41
Polarization, 323, 424
Polymerization, 898
Population I stars, 386, 722, 736, 759, 788
Population II stars, 386, 722, 736, 759, 788
Positron, 204
Potassium, 481
Potential energy, 168
Powers of ten, 955
Precession of equinoxes. See Equinoxes, precession of
Precession of Mercury's orbit, 142
Prehistoric astronomy, 25
Pre-main-sequence stars, 331, 334, 367, 375. See also Protostars
Pre-main-sequence evolutionary track, 331
Primary star, 284
Primeval fireball, 858, 860
Principia, 104
Principle of Equivalence, 134, 140
Prism, 228
Procyon B, 404
Project Cyclops, 932
Project Ozma, 933
Prominences, 213
Proper motions, 706, 708
Proteins, 886, 898
Proton, 172
Proton–proton chain, 203, 364
Protoplanets, 483–484, 685–686, 694
Protoplasm, 900
Protostars, 331, 335, 367, 683
Prutenic Tables, 74, 85
Ptolemy, Claudius, 47, 56
  model of the universe, 32, 50–52, 88, 225
  observations with model, 52–54
  size of the cosmos, 54
Pueblo Bonito, 25
Pulsars, 427–432. See also Neutron stars
  and intense magnetic fields, 435
  physical characteristics, 430
Pulsation, 430
Pythagoreans, 56

Quanta, 181. See also Photons; Light
Quasars, 806, 823–842
  and active galaxies compared, 834, 876
  cosmological distances of, 828
  emission lines, 826
  energy and size, 828
  energy sources, 833, 843
  light from, 826
  luminosity of, 827
  as nuclei of galaxies, 841
  observed properties, 825
  red shifts, 823, 825, 838–841
  spectrum of, 826–827
  supermassive black holes in, 834, 836
Quiet sun, 196–200
Quintessence, 39

Radar mapping of Venus, 577
Radial velocities, 145, 286, 707
Radian measure, 22
Radiant, 668
Radiation, 207
Radiation era, 860
Radiative energy, 168
Radioactive dating, 480, 503
Radioactive decay, 480, 550, 895
Radio astronomy, 236, 751, 806, 926
Radio galaxies, 809, 811–812
Radio interferometers, 241
Radio recombination lines, 303
Radio telescopes, 235, 237, 240
Rapid process, 443
Reaction law. See Newton's laws of motion
Reber, Grote, 236
Recombination, 302
Reddening, 321, 323, 515
Red giant stars, 371, 443, 947
Red shift, 778
Red variables, 393
Reflection, 227
Reflection nebulas, 321
Refraction, 227
Reinhold, Erasmus, 62, 73
Relativistic jets, 830
Relativity, 150. See also Einstein, Albert
Resolving power of telescopes, 231
Retrograde motion, 15–19, 28
  in Copernican model, 65
  in Ptolemaic model, 50
Rhea, 626–627
Rheticus, Georg F. B., 62
Ribonucleic acid (RNA), 887
Rich clusters of galaxies, 794
Rift valleys of Venus, 581
Rigel, 5, 263
Right ascension, 965
Roche lobe, 412
Rotation and pulsars, 430
RR Lyrae variables, 393, 716
Rubidium, 481
Rudolphine tables, 85
Russell, Henry N., 276

Rutherford, Ernest, 180
Rydberg constant, 184

Sagittarius A, 726, 807
Sagittarius A East, 726, 807
Sagittarius A West, 726, 733, 807
Sagittarius arm, 742, 761
Sagittarius B, 728
Satellites. *See satellites of individual planets*
Saturn, 494, 621–635
　atmosphere of, 487, 622, 626
　density, mass, size, 487, 621
　formation of, 691
　interior of, 623
　magnetosphere of, 634
　physical characteristics of, 621
　radiation belt, 634
　ring system of, 630–634
　rotation period, 487
　satellites of, 487, 623–630, A-5
Scarps, 562, 564
Schiaparelli, Giovanni, 589
Schwabe, Henrich, 209
Schwarzschild radius, 449–450, 454, 470. *See also* Black holes
Scientific models, 32–33, 54–55, 86, 225
Scorpius, 11, 700
Seasons, 10–11
Secondary star, 284
Seconds of arc, 6
Seeing, 514
Seismic waves, 504. *See also* Earthquakes; Moonquakes
Seismograph, 504
Semimajor axis, 83
Sequential model of massive star formation, 346
SETI, 918, 928, 939
Sexagesimal system, 6
Seyfert, Carl, 818
Seyfert galaxies, 818–821
　Classes I and II, 821, 835
Shapley, Harlow, 716
Shapley-Curtis debate, 721, 766
Shepherd satellites, 628
Shield volcanoes, 580
Shock wave, 335
Short period comets, 659
Sidereal month, 538
Sidereal periods in Copernican model, 68
Siderius nuncius, 95
Singularities, 449–450, 945. *See also* Black holes
Sirius, 291, 293, 295, 403
Skinny triangles, 22
Skylab, 199
Slipher, Vesto M., 144
Slow process, 443
Small Magellanic Cloud, 792

Sodium-D-lines. *See* Spectra, sodium-D-lines
Solar absorption-line spectrum, 194
Solar day, 8
　of Mercury, 560
　of moon, 538
　of Venus, 572
Solar eclipse, 19
Solar flares, 320
Solar-mass protostellar collapse, 335
Solar-mass stars, 369
　birth of, 347, 367
Solar neutrino experiment, 205
Solar system
　composition of, 679
　dynamical and chemical properties, 678
　elements of, 679
　origin of, 678
Solar wind, 200
Solstices, 11
South magnetic pole of earth, 506
Space astronomy, 243
Space colonies, 925, 949, 951
Spacetime, 132, 136, 458, 850
Spacetime diagram, 132, 152
　and black holes, 455
Space travel, 924
Spectra
　analysis of, 174
　of atoms, 186
　bright line, 174
　classification of stellar, 269–275
　continuous, 219
　dark line, 174, 219
　of galaxies, 778
　lines, 174
　sodium-D-lines, 174
　stellar, 273–275
Spectroscope, 174
Spectroscopic analysis, 175
Spectroscopic binary, 284
Spectroscopic distance method. *See* Spectroscopic parallaxes
Spectroscopic parallaxes, 279–280, 706
Spectroscopy, 171, 173
Spectrum, 173
Spectrum of gas tail of comet, 661
Spherical geometry, 138, 152, 157, 868
Spicules, 198
Spinar, 834
Spiral arms, 741, 755
Spiral galaxies, 771–772
Spiral nebulas, 719. *See also* Galaxies
Spiral tracers, 750
Spontaneous generation of life, 887
Sporadic meteors, 669
SS 433, 438–441
Standard candles, 776

INDEX

Starbirth, 299, 331, 352
  observational clues, 342
Stardeath, 400–444
Stars, 247, 353, A-8
  colors, temperatures, and sizes, 263–267, 281
  composition of, 357, 378
  energy generation, 363
  equations for structure, 360
  evolution of 5-solar-mass, 374
  evolution of one-solar-mass, 367–374
  evolution of very massive, 377
  formation of, 396
  lifetime, 295
  luminosity of, 252, 281
  models of, 359
  motion in space, 705–713
  motion in the sky, 7
  parallax, 257
  spectral classes, 269–275
  within 13 light years, A-7
Statistical parallaxes, 710, 713
Steady-state model, 846, 851
  problems with, 852
Stefan-Boltzmann law, 191–192
Stellar distances, 257
Stellar evolution, 354, 379–382, 396
Stellar parallax, 64
Stellar populations, 386, 721. *See also* Population I stars; Population II stars
Stellar winds, 331, 378
Stimulated emission, 315
Stony-iron meteorites, 671
Stratosphere, 509, 511
Strontium, 481
S-type asteroid, 656
Summer solstice, 11
Sun, 163, 279. *See also* Solar-mass stars
  angular size, 166
  birth of, 476
  core of, 204
  corona, 199
  density of, 169
  diameter, 166
  eclipse of, 19
  ecosphere of, 902
  fusion reactions in, 203–205
  future of, 947
  interior model, 362
  interior of, 200, 362
  lifetime of, 205
  luminosity, 188
  mass, 118, 166
  motion in sky, 8–12
  motion in space, 709
  opacity, 193
  physical characteristics, 165–169
  temperature, 189
Sunspots, 95, 208, 219
  cycle, 209
  maxima, 209
  minima, 209
  physical nature of, 213
Superclusters of galaxies, 796–798
Supergiant stars, 277, 296, 377
Superluminal motion, 830
Supernovas, 75, 376, 390, 409, 416–417, 432, 441
  frequency of, 420
  and interstellar medium, 317
  origin of, 420
  remnants of, 421, 425
  Type I and II, 418, 420, 442
Supersonic, 199
Synchrotron emission, 424
Synchrotron radiation, 811
Synodic month, 538
Synodic period, 68
Synthesis of organic molecules, 896

Tachyons, 132
Tangential velocity, 707
Target search mode, 928
Tau Ceti, 914, 938
Taurus, 11, 417–418
Telescopes, 93, 229
  functions, 231, 233
  infrared, 242
  magnifying power, 232
  radio, 235, 237, 240
  reflecting, 104, 229, 231
  refracting, 229, 231
  resolving power, 232
Temperature, 170–171
Temperature gradient, 359
Temperature scales, A-1
Terrestrial planets, 483, 485, 557, 605, 901
Tethys, 626–627
Tharsis ridge, 594
Thermal and nonthermal (synchrotron) emission, 422
Thermal emission, 424
Thermal equilibrium, 358, 360
Thermosphere, 510
3-kpc arm. *See* Expanding arm
Tidal friction, 536
Tidal gravitational forces, 452, 536
Tides, 535
Time, 8, 10, 13
Titan, 495, 625–626
Titania, 637
Titius-Bode law, 652
Tombaugh, Clyde, W., 641
Torque, 340
Trail of meteor, 666

Transverse faults, 618
Transverse velocity, 708
Transverse waves, 504
Trapezium, 325
Trigonometric parallax, 257. See also Heliocentric stellar parallax
Triple-alpha reaction, 366
Triton, 640
Troposphere, 509–511
T-Tauri stars, 349–351
Tully-Fisher relation, 782
Turbulence, 686
Turn-off point, 388
21-cm line, 751, 756
Tycho's supernova, 422

UFOs, 934
Ultraviolet radiation, 244
Umbriel, 637
Uniform acceleration, 97
Universe
 closed, 138
 expanding, 144, 147
 future of, 155, 868, 942
 geocentric, 48–52, 76
 heliocentric, 43, 63
 and Olbers' Paradox, 847
 open, 942
Upland plateaus of Venus, 577, 579
Uraniborg, 76
Uranus, 495, 635
 atmosphere, 487, 635
 density, mass, size, 487, 635
 rings of, 637
 rotation period, 487, 635
 satellites of, 487, 637, A-5
Urey, Harold, 686

Valles Marineris, 592, 597
Van Allen radiation belts, 506
Variable stars, 392. See also Cepheids
Vectors, 97
Vela X pulsar, 434
Velocity, 97, 130, 175
Venus, 491, 903
 atmosphere of, 486, 573, 575, 903
 clouds of, 574
 compared with earth, 582
 density, mass, size, 486, 573
 evolution of, 582
 magnetic field, 573
 phases of, 95
 physical characteristics, 570
 retrograde motion, 67
 revolution, 486, 571
 rift valleys, 581–582
 rotation period, 486, 571
 solar day, 572
 surface of, 575–576, 579–581
 surface pressure, 575
 surface temperature, 576
 volcanoes, 580
Vernal equinox, 11, 964
Very Large Array (VLA), 241, 726
Very Long Baseline Interferometry, 242
Viking missions, 583, 592
Virgo cluster of galaxies, 796
Visual binary stars, 284
Visual flux, 253
Visual luminosity, 253
Volatiles, 549, 678
Volcanoes, 580, 615, 895
von Neumann probes, 920
Voyager missions, 606, 614, 625, 629
Vulcan, 123

Wavelength, 177
Wegener, Alfred F., 518
Weight, 111
Weightless, 136
White dwarf stars, 277, 296, 400, 404, 947
 and general relativity, 404
 magnetic, 406
 mass–radius relation of, 401
 observations of, 403
White hole, 457
Widmanstätten figures, 672
Wien's law, 191
Wilson, Robert, 855
Winter solstice, 11, 25
Wolf-rayet stars, 378
Worldline, 133
Wormhole, 458

X-ray bursters, 466–470
X rays, 199, 233, 244, 466, 470, 731, 807, 811, 814, 838

Young population I stars, 722
Year, 9

Zero-age main sequence star, 368
Zodiac, 10–11, 28
Zodiacal light, 673
Zone of avoidance, 768

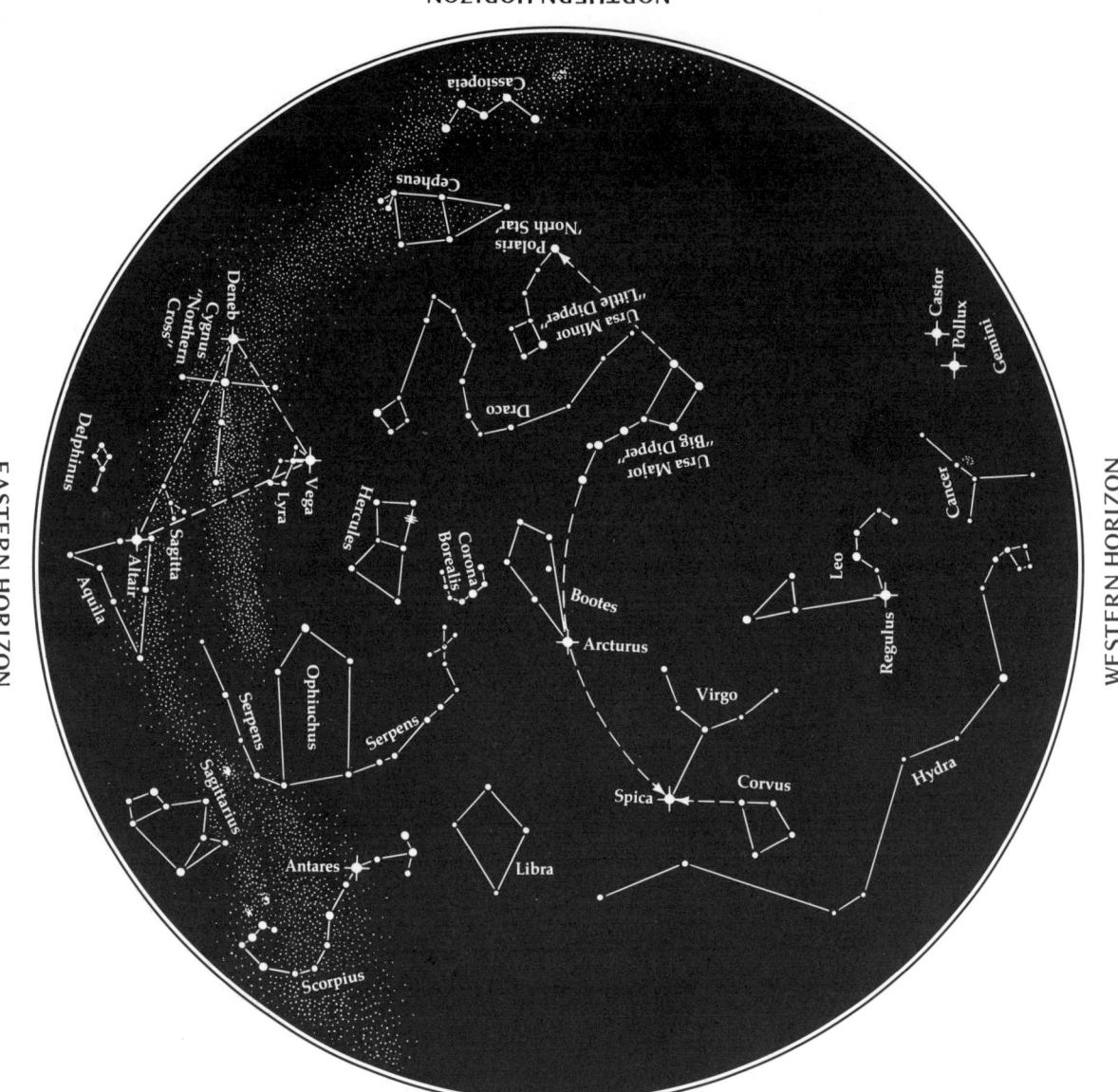

The Night Sky in JUNE